Evolution of African Mammals

EVOLUTION OF AFRICAN MAMMALS

Edited by
Vincent J. Maglio
and H. B. S. Cooke

Harvard University Press
Cambridge, Massachusetts
London, England
1978

Copyright © 1978 by the President and Fellows of Harvard
College
All rights reserved
Printed in the United States of America

Library of Congress Cataloging in Publication Data

Main entry under title:

Evolution of African mammals.

    Bibliography: p.
    Includes index.
    1. Mammals, Fossils.   2. Paleontology—Africa.
3. Mammals—Evolution.   I. Maglio, Vincent J.
II. Cooke, Herbert Basil Sutton.
QE881.E84   569′.096   77-19318
ISBN 0-674-27075-4

To the many who have contributed to the
study of paleontology and prehistory in
Africa

Geological research, though it has added numerous species to existing and extinct genera, and has made the intervals between some few groups less wide than they otherwise would have been, yet has done scarcely anything in breaking down the distinction between species, by connecting them together by numerous, fine, intermediate varieties; and this not having been effected, is probably the gravest and most obvious of all the many objections which may be urged against my views . . . For my part . . . I look at the natural geological record, as a history of the world imperfectly kept, and written in a changing dialect; of this history we possess the last volume only . . . Of this Volume, only here and there a short chapter has been preserved; and of each page, only here and there a few lines. Each word of the slowly-changing language, . . . being more or less different in the interrupted succession of chapters, may represent the apparently abruptly changed forms of life, entombed in our consecutive, but widely separated formations. On this view, the difficulties . . . [arising from the poor fossil record] are greatly diminished, or even disappear.

—Charles Darwin, 1859

Preface

One of the more remarkable effects of the postwar economic boom has been an explosion of intellectual investigation and a multifold increase in published data. This trend has manifested itself in nearly every area of human endeavor, including the quest for knowledge of the earth's historical past, both physical and biological. The study of man has figured prominently in this realm, along with the necessary collateral studies on the ecological setting within which our own species originated and evolved. Thus we have recently witnessed a burgeoning of investigative activities that, because of geographical history, have centered on the African continent. More than twice as many publications on African prehistory have appeared in the last two decades as in the previous two centuries.

Any worker who has attempted a synthesis of these voluminous data, whether he is just beginning the study of African prehistory or has been involved for years, has undoubtedly run headlong into a wall of intellectual resistance, a wall built up of countless details of morphology, distribution, and geological setting. It has, in fact, been extremely difficult to synthesize current knowledge of any particular vertebrate group without spending much time digesting the often unpalatable morsels dispersed over an ever-widening primary literature. Thus, few workers have been able to achieve an adequate understanding of the evolutionary events that shaped Africa and the modern world, even with respect to one, or at most a few, isolated groups.

For those of us who are currently pursuing such research, it has become increasingly clear that this situation cannot be allowed to persist. Limited access to scientific data perpetuates an undesirable narrowness and makes it ever more difficult to foster the infusion of ideas that can catalyze new insight and broader directions. No area of inquiry can remain closed and yet avoid stagnation. In African prehistory it has been only in the Hominidae, the one group that has received broad treatment in summary fashion, that widespread interest and diverse opinion have been brought to bear on complex evolutionary problems. Such interest is slight or nonexistent in the study of other groups, partially because of their inherently less anthropocentric value, but also because of the restricted and overtechnical presentation of available data. The description of a new species is, of course, a critical part of paleobiological studies, but once in a while we must step back to view our creations from a distance and to attempt broad correlations within the structure of related biological concepts. Such overviews are the only means of assessing the whole and of picking out our significant achievements and failures, as well as pinpointing the persistent problems.

Even more important is the dissemination of knowledge. No research is significant unto itself; it is significant only to the extent that it relates to man and the world at large. With the accelerating destruction of the natural world we have a responsibility to help preserve what is left. A firm understanding of the origin and evolution of life forms and of the establishment of the ecological balance of a major continental fauna, such as that of Africa, can be significant in achieving this end.

From a more practical point of view, numerous field workers are collecting fossil data in Africa, often without full knowledge of where the major problems lie, except, perhaps, for the one or two groups with which they have particular expertise. Data are lost each year simply because we do not know what kinds of evidence are most needed in each group. For example, elephant teeth are always collected, even if fragmentary, because we have learned from past studies that they hold the key to understanding this group. However, it is now abundantly clear that only cranial evidence will eventually resolve the many remaining problems in proboscidean evolution. Yet, unless exceptionally complete, such evidence is rarely gathered. Similarly with hippopotami, the mandibular symphysis has proved to be critical in the assessment of some major lineages and of evolutionary trends, but this part is not likely to be collected unless a tooth or root-bearing ramus is attached.

It was for these reasons that we set out to prepare a single volume that would summarize the current state of our knowledge on the origin and evolution of the class Mammalia in Africa. We decided to restrict our efforts to that class, primarily because of the scarcity of information on the lower vertebrate assemblages of the continent. Each of the 27 systematic chapters has been written by one or several of the major authorities who have contributed to the study of the group and who, because of their intimate knowledge of most of the relevant collections, can best evaluate our current state of understanding. Each contributor was requested to speculate where possible on phyletic relationships both among African forms and between these and their Eurasiatic ancestors or descendants. For each of the 15 mammalian orders treated, this volume tries to summarize the valid taxonomic groups as now envisaged, the origin of its various subgroups, its geographic distribution, major phyletic units, and specific evolutionary trends.

No attempt has been made to present each group according to a standard format. This would be not only extremely difficult but undesirable, given the nature of the various groups and the data that are currently available. Thus, for some small or poorly known groups, such as the Palaeomerycidae, Tubulidentata, and Embrithropoda, little more than an account of the fossil forms can be given. For other groups, where the data and stratigraphic distributions are more abundant, such as in the Bovidae, Proboscidea, and Suidae, a more detailed account of their origins and probable evolution is attempted. A number of orders have been subdivided into their component families for more detailed treatment.

In addition to the differences in content, the treatment in each chapter differs according to the philosophy of the authors. It was felt that if a rigid organization were attempted, the volume would lose more from the imposition of the editors' biased style and from their deficiencies of personal perception than it would gain from a more predictable, textbook-like structure. Several chapters, such as those on the Hyracoidea, Deinotherioidea, and Equidae, are presented in a rather formal species-by-species treatment with orderly diagnoses, discussions, and critical revisions, whereas others, such as the Bovidae and Insectivora, are presented less formally. Still others, such as the Hippopotamidae, are discussed almost in narrative fashion.

The information amassed here derives from two basic sources. Much of it stems from the published literature, where assessments must continually be made as new and more complete evidence is unearthed. But second, and more important, are the new materials and reinterpretations that are discussed here for the first time. These new data stem from the active research of the various authors and include information not otherwise available to the nonspecialist; accordingly, this should provide the reader with a more comprehensive analysis than would be possible elsewhere.

Chapter 1 is a general discussion of the modern African mammalian fauna and the varied ecological environments within which it is found. The author also presents some comments on interactions between man and various species of wildlife. Chapter 2 presents a concise review of the major geological deposits of the African Cenozoic, from which the rich faunal records of the continent have been derived, and provides a framework based on available evidence for absolute and relative dating. Each succeeding chapter deals with a particular mammalian group, or in some cases a few mammalian groups; those families with more complete fossil records are treated in separate chapters.

To round out the volume, the broader patterns of faunal evolution are synthesized in an attempt to cut across taxonomic boundaries and to visualize the interdependence of faunal events in the continent as a whole.

Knowledge of any historical phenomenon can never be conclusive, and the degree to which interpretations command our confidence is in direct proportion to the adequacy of the documented record. This is even more true in paleobiology than in other historical subjects, because of the very incomplete nature of the fossil record, even under the best of circumstances. The present volume is not intended as a definitive treatment of the subject. Rather, it is conceived as an interim report on investigations still in progress, investigations that will not be completed by any of the present contributors and, indeed, will never be truly completed. The scope of the volume is broad and its treatment necessarily limited. But it is hoped that the book will fulfill a great need and that it may be useful to the professional who cannot be an authority on every group, to the student who wishes to gain a wide understanding of a particular group as a basis for further research, to the evolutionary biologist and to the geologist who can deal with the larger implications of such data, and to all those interested in the processes and events that shaped the last great intact faunal province on the earth.

This book represents the combined efforts of very many individual investigators and of their colleagues, far too numerous to mention here. To all these people we extend our gratitude. In particular, we wish to thank the contributors for their efforts and for their cooperation in bringing this document to fruition. A special debt is owed to Rosanne Leidy for her invaluable help in preparing the manuscripts for publication. Typing of the final draft for considerable parts of the volume was done by Ms. Jean Olsen, Ms. Evelyn Wolff, and Mrs. L. O'Hearn.

Vincent J. Maglio   H. B. S. Cooke

Contributors

Andrews, Peter
Department of Palaeontology
British Museum (Natural History)

Barnes, Lawrence G.
Natural History Museum of Los Angeles County

Beden, Michel
Laboratoire de Paléontologie des Vertébrés
et Paléontologie humaine
Université de Poitiers

Bigalke, R. C.
Department of Nature Conservation
University of Stellenbosch

Black, Craig C.
Carnegie Museum of Natural History

Butler, Percy M.
Department of Zoology
Royal Holloway College
University of London

Churcher, C. S.
Department of Zoology
University of Toronto

Cooke, H. B. S.
Department of Geology
Dalhousie University

Coppens, Yves
Musée de l'Homme
Muséum National´ d'Histoire Naturelle

Coryndon, Shirley C. (deceased)
Department of Geology
Bristol University

Crompton, A. W.
Museum of Comparative Zoology
Harvard University

Delson, Eric
Department of Anthropology
Herbert H. Lehman College
City University of New York

Domning, Daryl P.
Department of Anatomy
College of Medicine
Howard University

Gentry, Alan W.
Department of Palaeontology
British Museum (Natural History)

Hamilton, W. Roger
Department of Paleontology
British Museum (Natural History)

Harris, John M.
Department of Palaeontology
Louis Leakey Memorial Institute
Nairobi, Kenya

Hooijer, Dirk A.
Rijksmuseum van Natuurlijke Historie
Leiden

Howell, F. Clark
Department of Anthropology
University of California
Berkeley

Jenkins, Farish A., Jr.
Museum of Comparative Zoology
Harvard University

Lavocat, René
Ecole Pratique des Hautes Etudes
Institut de Montpellier

Madden, Cary T.
University of Colorado Museum

Maglio, Vincent J.
437 Ashwood Lane
Kirkwood, Missouri

Meyer, Grant E.
Raymond M. Alf Museum
Webb School of California

Mitchell, Edward D.
Arctic Biological Station
Ste. Anne de Bellevue, Quebec

Patterson, Bryan
Museum of Comparative Zoology
Harvard University

Pilbeam, David R.
Department of Anthropology
Yale University

Richardson, M. L.
Department of Zoology
University of Toronto

Savage, Robert J. G.
Department of Geology
Bristol University

Simons, E. L.
Primate Center
Duke University

Tanner, Lloyd G.
Division of Vertebrate Paleontology
State Museum
University of Nebraska

Walker, Alan C.
Departments of Cell Biology and Anatomy
Johns Hopkins University

Wilkinson, Albert F.
Embajada Britanica
Quito, Equador

Contents

1

Present-Day Mammals of Africa

R. C. Bigalke

Faunistics

It has become almost a commonplace to write of Africa that it has a remarkably rich and diverse mammalian fauna. As it is mainly a tropical region, and tropical faunas in general are larger and more diverse than those of temperate zones, this is to be expected. Tropical Asia and Neotropica also have many mammals. There are approximately 740 species and 52 families in Africa (Bigalke 1972). For Neotropica, although it is somewhat smaller (7 million sq mi as against 11.5 million sq mi), Hershkovitz (1972) gives the surprisingly high figure of 810 species in 50 families. The considerably smaller Oriental Region has approximately 470 species and 40 families (Darlington 1957). By way of a general explanation, Darlington's observation (1957) that tropical faunal diversity results from the presence of both widely distributed and localized groups may be noted. Temperate zones, on the other hand, are populated mainly by widely distributed groups, and localized taxa are unimportant.

The question of most significance is to what extent Africa differs from other tropical areas. There are basic and well-known differences in faunal composition between Neotropica on the one hand and the Ethiopian and Oriental regions—the Old World tropics—on the other. These may be simply illustrated by table 1.1.

The proportions of bats, "gnawers," and other land mammals in Africa and the Oriental Region are similar, although the last group has rather more species in Africa. Bats make up much the same fraction of all three faunas. But in Neotropica almost half of the species are "gnawers" and other land mammals are poorly represented.

The close faunal relationship between Africa and the Oriental region is shown by the fact that 10 families occur only in these two zones. They are lorisids, cercopithecid monkeys, apes, pangolins, bamboo rats, Old World porcupines, viverrids, elephants, rhinoceroses, and chevrotians (porcupines and viverrids each have one species in southern Europe). In a few cases genera and even species are shared. Some of the families do, however, have fossil representatives in European deposits, indicating a wider original distribution.

For all that Africa is, in a sense, a typical tropical region and has strong links with tropical Asia, the mammal fauna is distinctive. Its variety is due to the presence of a great mixture of groups of varying zoogeographic affinities. In addition to the 10 families shared with the Oriental Region, 17 are shared in varying degrees with Eurasia, and in a few cases with other regions. Eight families—shrews, vesper-

Table 1.1 Proportions of land mammals in major zoogeographic regions.

	Africa (after Bigalke 1972)	Oriental region, excluding Philippines (after Darlington 1957)	Neotropica (after Hershkovitz 1972)
Species of land mammals (except gnawers and bats)	330 (44%)	180 (38%)	205 (25%)
Species of gnawers (Lagomorphs and rodents)	237 (32%)	135 (29%)	380 (47%)
Species of bats	174 (24%)	154 (33%)	222 (28%)

tilionid bats, lagomorphs, squirrels, cricetids, canids, mustelids, and cats—are widespread or worldwide. Distinction is enhanced by 15 families (almost a quarter of the total number) and two subfamilies that are endemic or virtually so. Fossil remains of five of these families, Thryonomyidae, Orycteropodidae, Procaviidae, Hippopotamidae, and Giraffidae, are, however, known from Eurasia, but thryonomyids and procaviids are believed to be of African origin (Cooke 1972).

The orders, families, and approximate numbers of species are listed in table 1.2. Figures in this table refer to the continent of Africa without the extreme northwest corner, north of the Atlas Mountains. Groups with the greatest number of genera and species are insectivores (Lipotyphla and Menotyphla), bats, primates, rodents, carnivores, and artiodactyls.

There are several notable features about the fauna (see also Keast 1972). The many unusual endemic insectivores include aquatic otter shrews, subterranean golden "moles," and the elephant shrews, many of them kangaroolike in form. Catarrhine monkeys are the dominant primates. The Cercopithecidae have no fewer than 47 species, most of them arboreal but some terrestrial. The endemic galagos are of particular interest, some of them only as big as mice. Chimpanzees and gorillas are distinctive great apes. Dominant rodents are Muridae, with about 79 species, and Cricetidae, with about 109 species. While by no means as important as in the Neotropical Region, the rodent group is extremely diverse in form. Ellerman (1940–41) considered that Africa must be considered the present headquarters of the order so far as variation in character goes. Anomaluridae ("scaly tails" or "flying squirrels") have converged with the "flying" sciurids (such as *Pteromys*) of the Northern Hemisphere. *Pedetes* is an archaic, monotypic, kangaroolike endemic. Cane rats (Thryonomyidae) resemble large South American caviomorphs. Bathyergids are numerous and successful subterranean vegetarians, whereas gundis (Ctenodactylidae) are morphologically and ecologically remarkably similar to hyraxes.

Among conventional canids and felids, *Lycaon*, occupying the wolf niche, and the long-legged, coursing cheetah stand out. *Proteles*, a small, insectivorous hyenalike animal, is endemic. The robust hyenas are survivors of a group of hunter-scavengers that was once much richer in species. Mustelids are poorly represented, but the Viverridae, with 37 species, have speciated actively and fill most of the small carnivore and omnivore niches.

The extremely specialized aardvark *Orycteropus* and the successful, terrestrial and arboreal, herbivorous dassies or hyraxes are endemics of great interest.

Finally, the spectrum of almost a hundred large ungulates, including elephants, rhinoceroses, giraffes, hippopotamuses, pigs, and more species of bovids (78) than are found anywhere else, give the African fauna much of its unique pre-Pleistocene character. As the land of big game par excellence, more than any other continent it has retained something of the richness and strangeness of the mammal fauna that dominated the world before the Pleistocene extinctions took their toll.

The complex geological and geomorphological changes, climatic fluctuations, evolutionary developments, and migrations of animal groups, of which the recent mammal fauna is the end product, are the subjects of succeeding chapters. Faunal evolution will not be dealt with here. But the role of topography, climate, and vegetation in preserving ancient, relict forms while also providing isolated habitats in which speciation was stimulated must be appreciated. It is therefore necessary to review the general ecological characteristics of the continent.

General Ecological Features

Topographically, Africa can be simply described as an eroded peneplain of largely uplifted, flat land (Keast 1972). Perhaps the most important physical features that affect, or have affected, mammal dis-

Table 1.2 Families and approximate numbers of species and superspecies of contemporary mammals of continental Ethiopian Africa. (After Bigalke 1972.)

Order and family	Species	Super-species
INSECTIVORA		
Potamogalidae (Otter Shrews)	3	3
Chrysochloridae (Golden Moles)	16	7
Erinaceidae (Hedgehogs)	6	3
Soricidae (Shrews)	56	41
Macroscelididae (Elephant Shrews)	13	7
CHIROPTERA		
Pteropodidae (Fruit Bats)	26	22
Rhinopomatidae (Mouse-tailed Bats)	2	2
Emballonuridae (Sheath-tailed Bats)	7	7
Nycteridae (Hollow-faced Bats)	11	7
Megadermatidae (Big-eared Bats)	2	2
Rhinolophidae (Horseshoe Bats)	17	16
Hipposideridae (Leaf-nosed Bats)	14	14
Vespertilionidae (Simple-nosed Bats)	64	?56
Molossidae (Mastiff Bats)	31	26
PRIMATES		
Lorisidae (Pottos)	2	2
Galagidae (Galagos)	6	4
Cercopithecidae (Monkeys)	47	20
Pongidae (Apes)	3	2
PHOLIDOTA		
Manidae (Scaly Anteaters)	4	3
LAGOMORPHA		
Leporidae (Hares)	10	8
RODENTIA		
Sciuridae (Squirrels)	31	17
Anomaluridae (Scaly Tails)	7	7
Pedetidae (Springhare)	1	1
Muridae		
Murinae	79	54
Cricetidae		
Dendromurinae	14	11
Otomyinae	12	11
Cricetinae	1	1
Gerbillinae	33	26
Lophiomyinae	1	1
Cricetomyinae	5	3
Petromyscinae	3	3
Microtinae	1	1

Table 1.2 *(continued)*

Order and family	Species	Super-species
Rhizomyidae (Bamboo rats)	2	2
Muscardinidae (Dormice)	7	5
Dipodidae (Jerboas)	3	3
Hystricidae (Porcupines)	5	3
Thryonomyidae (Cane Rats)	2	2
Petromyidae (Dassie Rats)	1	1
Bathyergidae (Mole Rats)	13	8
Ctenodactylidae (Gundis)	5	5
Spalacidae (Blind Mole Rats)	1	1
CARNIVORA		
Canidae (Jackals, etc.)	11	10
Mustelidae (Weasels, etc.)	7	6
Viverridae (Genets, etc.)	37	32
Hyaenidae (Hyenas)	3	2
Protelidae (Aardwolf)	1	1
Felidae (Cats)	10	10
TUBULIDENTATA		
Orycteropodidae (Aardvark)	1	1
PROBOSCIDEA		
Elephantidae (Elephants)	1	1
HYRACOIDEA		
Procaviidae (Dassies)	11	4
SIRENIA		
Trichechidae (Manatees)	1	1
Dugongidae (Dugongs)	1	1
PERISSODACTYLA		
Equidae (Zebras)	5	4
Rhinocerotidae (Rhinos)	2	2
ARTIODACTYLA		
Suidae (Pigs)	3	3
Hippopotamidae (Hippos)	2	2
Tragulidae (Chevrotain)	1	1
Giraffidae (Giraffe and Okapi)	2	2
Bovidae (Antelopes)	78	67

Note: Families in italics are endemic. The order Insectivora is retained here although Lipotyphla and Menotyphla (for Macroscelididae) are used in the Smithsonian identification manual, *The Mammals of Africa*. Similarly, the Potamogalidae are retained as a family in this text although Corbet (1971) places them in the Tenrecidae in the same manual.

tribution and speciation are the high, isolated mountains, most of them of volcanic origin, which were formed in East Africa and Ethiopia during the Tertiary, and the great valleys that developed as a result of rift faulting in East Africa, probably mainly toward the end of the Pliocene (Cooke 1972).

Being situated across the equator, Africa has a wide range of climates, with almost 80% of its surface area in the tropics and extending far enough beyond them to include temperate zones at both northern and southern ends. It is, however, predominantly a dry continent. Keast (1972) states that only 28% of the surface area has an annual rainfall in excess of 40 in, while 31% receives less than 5 in of rain per year. The corresponding figures for South America, for example, are 68% and 2%. Accordingly the vegetation of much of Africa falls within the broad categories of woodland, savanna, grassland, and low shrub steppe. Rough calculations by Keast (1972) indicate that tropical rain forest covers only 9% of the continent. Figures for the other major vegetation types are woodland and open forest, 31%; savanna, grassland, and steppe, 19%; and desertic areas, 30%. Oversimplifying somewhat and using a wide definition for the hard-worked term "savanna," it can be said that savanna covers about one-half of Africa. Most of the dominant groups of African mammals are savanna forms and more species occupy this biome than any other.

Biotic Zones and Mammal Communities

Although savanna areas are important by virtue of their size, forests are ancient environments of great significance, with a fauna of their own. The interspersion of forest, savanna, and arid zones, together with other more minor zones, is responsible for much of the faunal diversity. The principal biotic zones, delimited by Davis (1962) from the distribution of vegetation types and faunal regions, are shown in figure 1.1.

Forest

The forested regions may conveniently be divided into the extensive, more or less continuous blocks of tropical lowland forest and the small relic forest patches scattered over mountains and coastal lowlands as far as the southern extremity of the continent.

Lowland forest has long been considered the most distinct African biome. Its mammal fauna includes many specialized, ancient forms and must have evolved over a long period in a stable environment. Among the insectivores, the endemic otter shrews—

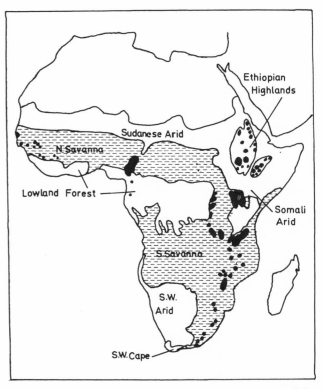

Figure 1.1 The main biotic zones of Africa south of the Sahara. Montane and coastal forests are in black. (After Davis 1962.)

traditionally considered a family, Potamogalidae, but classified as a subfamily of the Tenrecidae by Corbet (1971)—are an old forest group. They are highly adapted aquatic creatures that live in streams and feed on invertebrates and vertebrates. Soricids are well represented, partly because they are adapted to moist environments with dense vegetation. Their predominance in the tropics may also be due in part to their relatively recent immigration from Europe in the late Oligocene. Radiation further southward may have been restricted by competition from well-established endemic golden moles, Chrysochloridae, or, more likely, from elephant shrews, Macroscelididae (see Bigalke, forthcoming). The former is essentially a southern family not well represented in forest. The Macroscelididae are also mainly southern African but there are several forest forms such as the giant *Rhynchocyon* and a race of *Petrodromus tetradactylus* (Kingdon 1971).

As in other tropical regions, Megachiroptera are prominent members of the African forest fauna. As Kingdon (1971) points out, fruit bats are best represented in forests since they depend on a year-round supply of fruit from trees and shrubs. Some use is also made of nectar, pollen, and flowers and *Megaloglossus* is a specialized nectar feeder. *Myonycteris, Megaloglossus, Hypsignathus,* and *Epomops* are typical lowland forest genera. There are also a number of forest-dwelling Microchiroptera and Kingdon (1971) has drawn attention to the surprising correspondence between areas of distribution of these insectivorous bats—and of Megachiroptera—and those of other mammals, in spite of their mobility. One finds lowland forest species, relic forms confined to mountain refuges, and so on. Fleming (1975) considers insectivorous bats to be trophically much more important in the tropics than frugivorous and nectar-feeding species. However the vast numbers of colonial fruit bats, such as *Eidolon helvum,* which roost in some areas, must be of considerable local significance for seed dispersal and perhaps pollination.

The Lowland Forest Zone is the center of primate diversity. There are two families of prosimians. Pottos (Lorisidae) are specialized, somewhat slothlike, arboreal omnivores that are confined to this zone, although not only occupying high forest. The second family, Galagidae, is endemic to Africa. Four of the six species of bush babies are forest forms. In contrast to pottos they are agile and quick. Some, such as *Galagoides demidovii,* are little bigger than mice and their presence may have limited rodent radiation. The Cercopithecidae is the largest of the two families of higher primates in Africa and most species are confined to the forest biome. Colobus monkeys of the subfamily Colobinae are sluggish arboreal leaf eaters. Most Cercopithecinae are small to medium-sized, wholly or partly arboreal monkeys feeding on fruit or leaves. The genus *Papio* consists of large forms, baboons found both in forest and savanna, and the mandrill and drill, forest apes of the subgenus *Mandrillus.* Booth (1956) was perhaps the first to draw attention to the nature of niche differences in sympatric forest monkeys. Many of them use distinctly different levels in the canopy although there are also examples of considerable niche overlap. The great apes, Pongidae, like the lorisids and cercopithecids, are restricted to Africa and Asia and are one of the groups that provide evidence for past contact between forests of these regions. This is unlikely to have been later than the Oligocene or early Miocene (Misonne 1963). Since that time African and Oriental forms have diverged greatly. *Gorilla gorilla* is a herbivore strictly confined to forest while the more omnivorous chimpanzee *Pan troglodytes* is also found outside the Lowland Forest Zone in woodland and savanna. The pygmy form *P. paniscus* is a localized forest species.

The Pholidota are another of the old, specialized groups shared with the Oriental Region. While there are two partly arboreal forest pangolins, the order is not wholly confined to this biome. A large form occupies both forest and savanna and the remaining species inhabits savanna. All are ant and termite feeders.

The forest biome supports a distinctive rodent fauna, the most singular members of which are the Anomaluridae, scaly tails or "flying squirrels." This old, endemic family of arboreal, squirrellike gliding creatures is essentially confined to high forest. The animals feed on leaves, fruit, and in some cases perhaps insects. Sciuridae are particularly well represented by forms ranging from the mouselike pygmy squirrel *Myosciurus pumilio* to giant species of *Protoxerus,* 60 cm long and weighing a kilogram. Other forest genera of note are *Heliosciurus, Epixerus,* and *Funisciurus.* Dormice (Muscardinidae) of the genus *Graphiurus* are present and the tree porcupine *Atherurus* is an interesting, highly adapted forest-dwelling member of the Hystricidae. Murid rodents are, however, poorly represented and there are no really specialized forest species. This is probably because the group only entered Africa in the late Pliocene when many rodent niches were filled by small anomalurids, squirrels, galagos, and cricetids so that there was little stimulus to radiation (Misonne 1963; Kingdon 1971).[1] Forest-inhabiting genera include *Hybomys, Oenomys, Stochomys, Lophuromys,* and *Praomys morio. Cricetomys* is the most important representative of the Cricetidae but is not confined to forests.

Carnivora of the forests are limited in variety and there are few confined to this biome. The only Mustelids are the widespread ratel, *Mellivora,* and otters of the genera *Lutra* and *Aonyx,* the latter being represented in the Lowland Forest Zone by a distinct species, *A. congica.* Members of the Viverridae are the commonest carnivores. *Nandinia binotata,* the palm civet, is an ancient, mainly frugivorous forest species, the only member of the subfamily Nandiniinae. The Viverrinae includes several purely forest forms: *Poiana* (linsang), *Osbornictis* (aquatic civet), the giant genet *Genetta victoriae,* and other species of this genus. Mongooses of the genera *Bdeogale, Crossarchus,* and *Herpestes* are also confined to this biome. *Felis aurata* is the only true forest cat, but both the leopard *Panthera pardus* and *F. serval* are found in forested habitats.

One genus of the Hyracoidea, *Dendrohyrax,* is an arboreal forest form that also lives among rocks above the tree line on high mountains such as the Ruwenzori (Dorst and Dandelot 1970). The elephant

---

1 According to Lavocat (this volume) the murids are known from the middle Miocene. —Ed.

Loxodonta africana inhabits forests as well as most other habitats and a distinct small forest form, *cyclotis,* is recognized. Artiodactyls are represented by a fair range of species, but they are generally not plentiful in forested environments because of the limited food supply for ground dwellers. The giant forest hog *Hylochoerus* is essentially a forest species and the widespread bushpig *Potamochoerus* also occurs. The chevrotian *Hyemoschus* is an ancient specialized forest denizen of great interest, the only representative of the Tragulidae. *Choeropsis,* the pygmy hippopotamus, and *Okapia johnstoni* are two well-known forest specialists with restricted ranges. The duikers, Cephalophinae, are the dominant bovids. They are clearly of ancient lineage and have speciated to a remarkable extent. About 13 forest species are recognized, ranging from dwarf forms, e.g., *Cephalophus monticola,* to *C. sylvicultor,* which weighs about 75 kg. The bovid tribe Neotragini also contains tiny forest species, namely the royal and pygmy antelopes of the genus *Neotragus.* The bongo *Boocercus* is a true forest antelope and a small forest form of buffalo, *Syncerus,* also occurs. Species of wider distribution that tolerate forest conditions include bushbuck, *Tragelaphus scriptus,* and sitatunga, *T. spekei.* Terrestrial herbivores are largely dependent on fruits and leaves as grasses are very scarce on the poorly lit forest floor.

Since relatively little of the net production is available to forest herbivores, mammal biomass is characteristically low. For example, Collins (1959, quoted by Bourlière 1965) found the biomass of three ungulate species and seven species of primates in a forest in Ghana to be only 72.2 kg per sq km. For the ungulates alone, the figure was only 5 kg per sq km. Holloway (1962, quoted by Bourlière 1965) gives a figure of approximately 420 kg per sq km for a montane forest in Kenya.

The uniformity of the forest environment is such that the existence of well-marked regional differences in the fauna is rather surprising. Three centers of endemism are distinguished (Booth 1954; Misonne 1963). Liberia, or the Guinea forest block, is separated from the others by the Dahomey Gap. It has the most distinctive fauna. The Gabon (or Gabon-Cameroon) and Upper Congo (Ituri-Maniema forests) regions have sufficient faunal differences to suggest that they were separated by a nonforested corridor during dry periods in the late Pliocene or early Quaternary. Speciation in the three forest refuges during periods of isolation appears to have been a major factor in the enrichment of the African forest fauna.

Relic forests on the central and eastern African mountain massifs, in Zambia and Angola, as well as in coast lowlands and along escarpments down the east coast to South Africa, support mammal faunas that are, in general, impoverished versions of those of the lowland forest blocks. There are fewest species in forests most distant from the equator. This simplified statement requires some qualification. Kingdon (1971) shows that relic forests in northern East Africa have a homogeneous fauna very similar to that of the central African forests. Southern and coastal lowland forest relics on the other hand tend to retain older forms and are sufficiently different faunistically to suggest that they have been isolated for a much longer time. The presence of some endemic species on the high mountains of central and eastern Africa also indicates a considerable period of isolation. The faunal similarities between these widely separate mountains—Mount Cameroon, Ruwenzori, Elgon, Kenya, Kilimanjaro, and the Aberdares—show that they were once in contact through the intervening lowlands and the fauna is not a true montane one.

In summary, the forest biome is inhabited by ancient mammal groups that have survived in the refuge provided by this stable and favorable environment, and by more recent taxa. Together they make up a typical tropical fauna, rich in species but for the most part poor in numbers of individuals. Bats and arboreal animals—lorises, galagos, cercopithecid monkeys, anomalures, squirrels and other rodents, tree pangolins, tree hyrax, and viverrids—occupy the canopy and utilize its resources. Members of some groups that are mainly ground dwellers such as felids, small rodents, insectivores, and great apes also use the trees to some extent. Food supplies on the ground are limited and there are but small populations of terrestrial herbivores. These are specialized browsers and fruit eaters, solitary or with a simple social organization associated with the difficulties of communication in a closed environment. Most forest mammals are closely tied to this biome but the relic forests carry a surprisingly large array of these forms well beyond the tropics and deep into the Savanna Zone.

Savanna

The very distinct difference between African forest and savanna biomes was recognized by Sclater (1896) when he named the former the West African subregion and the latter the East and South African subregion. Within the extensive savanna biome there is a great diversity of vegetation types. From open temperate grasslands such as the South African high veld there are transitions through many plant associations of grass and woody shrubs and trees to dense woodlands. The boundaries between the savanna and arid zones are not clearcut and differences tend to be mainly a matter of degree of aridity. The vegetation of great areas of the arid zones is simply dry savanna.

The high constant humidity and even temperatures of the forest are in strong contrast to large daily and seasonal fluctuations of temperature and humidity in savanna. Droughts are common, the availability of water may play a critical role, and winds have a significance entirely lacking in forest. Structurally the environment is simpler than in forest and the fauna is dominated by mobile terrestrial animals. Grass is a major source of food for herbivores and seasonal changes in its availability and nutritional value affect them materially, in some cases inducing nomadic or migratory movement. Subterranean forms are common but arboreal and volant groups are unimportant. The openness of the vegetation facilitates visual communication and many mammals occur in groups with a complex social organization. Coursing predators, and not only those that stalk their prey, are found.

Approximately 40% of African mammal species occur in the savanna biome. They belong to a great variety of families, many of the largest of which are essentially savanna groups. This is true of chrysochlorids, which are most prominent in the south, their probable evolutionary center. They exploit invertebrates, mainly those living underground. Macroscelidids also have a preponderance of species living in the Savanna Zone. They shelter in dense vegetation and among rocks and feed on invertebrates and, in some cases, fruit. There are few African species of hedgehogs and most of them occupy the northern savanna and arid zones, as may be expected of an immigrant group from Europe. Shrews are not primarily a savanna family, but there are many species and individuals, especially in regions receiving more than 600 mm of rain a year (Meester 1962).

Most savanna bats are microchiropterans but a few fruit bats, for example species of *Epomophorus*, have a wide distribution in this biome. They tend to move a great deal in order to obtain the fruit on which they depend. There are few savanna primates but all tend to be numerous and successful. They include two species of *Galago*, baboons (*Papio*), the vervet monkey *Cercopithecus aethiops*, and the patas monkey *C.* (*Erythrocebus*) *patas*, all species that forage to a greater or lesser extent on the ground. Lagomorpha fill a small herbivore niche successfully in both savanna and arid zones.

The fossorial Bathyergidae are found almost entirely in the savanna and arid zones, where they exploit subterranean plant storage organs most successfully. No other rodent families are so exclusively associated with savanna, save perhaps the Pedetidae. The spring hare *Pedetes,* however, is also widespread in the Southwest Arid Zone. It is a burrowing, nocturnal grazing herbivore. Ground squirrels such as *Xerus* and partly arboreal genera like *Paraxerus* represent the Sciuridae. Many murids and most of the cricetids occupy the savanna biomes and form the backbone of the small omnivore and herbivore stratum in food chains. There are a great many burrowing species. Arboreal rodents include the tree rat *Thallomys, Praomys,* and *Thamnomys.* The Dendromurinae are a specialized subfamily of the Cricetidae, with semiprehensile tails and modified feet; they feed on grass seeds and insects and suspend their nests on plants. Rodents living on the ground surface may inhabit dense grass, e.g., Otomyinae, bush, e.g., *Aethomys,* or rocky situations, e.g., *Acomys.*

The savanna biome is the environment in which by far the most carnivores, perissodactyls, and artiodactyls occur. Small carnivores include most of the Viverridae, of which mongooses are especially typical savanna dwellers, preying on small vertebrates and invertebrates. There are also the few African mustelids—the skunklike *Ictonyx, Poecilogale* (weasels), and the rather wolverinelike ratel *Mellivora.* Felidae, from the small black-footed cat *Felis nigripes* to lion, leopard, and cheetah; small canids of the genera *Vulpes* (foxes), *Canis* (jackals), and the wolflike hunting dog *Lycaon pictus;* the termitophagous specialist *Proteles* (aardwolf); and three species of hyenas are all mainly inhabitants of savanna biomes. They hunt and scavenge in varying proportions, each species tending to take prey of a certain size range but accepting smaller or larger species in case of need and probably often competing with one another for food (Ewer 1973). Foxes and jackals may also take fruit when it is available.

Both the large grazing rhinoceros, *Ceratotherium,* and the smaller, mainly browsing *Diceros,* as well as the zebras, are savanna forms; in some cases they extend into the arid zone as well. Pigs are represented by the warthog, *Phacochoerus,* and the bushpig, the latter already mentioned as an inhabitant of forests. The giraffe utilizes both savanna and arid biomes and is most successful in the high browsing niche where the elephant is the only possible minor competitor. The elephant is or was widespread. Most species of Hyracoidea inhabit rocky habitats in this biome.

The large herbivore fauna of the savanna is dominated by Bovidae, of which approximately 80% of the species occur in this biome alone or both there and in arid zones. There is only one savanna duiker, *Sylvicapra*, but many small neotragine antelope occupy specialized microhabitats, for example rocks in the case of the klipspringer *Oreotragus* and thickets in the case of dik-diks (*Madoqua* spp.) and the suni *Nesotragus*. Among the Antilopini, gazelles and the springbok (*Antidorcas*) select open habitats and tend to be nomadic or migratory; the gerenuk *Litocranius* lives in bush. The impala (*Aepyceros*, Aepycerotinae) selects open savanna (Hirst 1975). Buffalo, *Syncerus* (Bovinae, tribe Bovini), are not habitat specialists but usually inhabit fairly well wooded country near water. Tragelaphine antelopes, which include bushbuck, kudu, nyala (*Tragelaphus*), and eland (*Taurotragus*), are in general characteristic of more or less closed savanna environments while reedbucks, waterbuck, and related forms (Reduncini) are associated with grassy habitats near water. Roan and sable antelopes (*Hippotragus*, Hippotragini) are found in various savanna vegetation types. The tribe Alcelaphini, consisting of blesbok, bontebok, and tsessebe (*Damaliscus*), the hartebeests, *Alcelaphus*, and the wildebeests, *Connochaetes*, are mainly medium-sized antelope of mesic grassland and savanna or woodland habitats. Many species tend to aggregate into large, irregularly nomadic or migratory groups and are thereby able to seek the best feeding grounds in areas of fluctuating environmental conditions (Estes 1974).

Within the savanna some common mammals are widely distributed, but there are sufficient faunal differences to warrant the separation of Northern and Southern Savanna zones. They meet at the "Sclater line" just north of the equator (Davis 1962). Expansion of forests during wet periods is likely to have separated northern from southern savannas and provided opportunities for faunal differentiation.

The northern savanna fauna is poorer than that of the south. Kingdon (1971) compares some mammals endemic to the two zones, and occupying broadly comparable habitats, in East Africa. He lists 10 northern as against 23 southern forms, also pointing out that most of the northern species are isolates of widely distributed genera and have related species in the south. These differences, Kingdon suggests, may be due to the relatively simple, homogeneous nature of the northern savanna vegetation belts, which are not divided into the complex mosaic pattern found in southern and eastern Africa. They may thus have provided fewer opportunities for spe-

ciation and may have been unattractive to potential colonizing species from the south.

Ecological Separation in Savanna Communities

Two aspects of the ecology of savanna communities deserve particular mention: the manner in which so many different large herbivore species are sustained, and the high biomasses supported. Keast (1972) has summarized the mechanisms that permit the continent to support so many herbivore species. In brief, and with some modifications to Keast's list, they are the following:

1. *Habitat selection.* Many species have well-developed preferences for particular vegetation formations. Lamprey (1963) has shown these quantitatively for 14 ungulate species in *Acacia* savanna in Tanganyika (Tanzania). Recently, Ferrar and Walker (1974) and Hirst (1975), using sophisticated mathematical techniques of analysis, have confirmed that most of the ungulates studied are more or less closely associated with certain vegetation types. They are therefore spatially separated from one another. The preferences of some species overlap, however, and fire and rainfall may modify intrinsic preferences. The proportion of total plant biomass contributed by woody plants, especially shrubs, was the most important single site characteristic responsible for habitat separation between the ungulates that Ferrar and Walker studied.

The role played by physiological characteristics of animals in influencing the choice of habitat is beginning to be appreciated. The water demands of the waterbuck, *Kobus ellypsiprymnus*, for example, are such that it cannot live far from sources of free water (Taylor, Spinage, and Lyman 1969). Oryx, *O. gazella*, and dorcas gazelle, *Gazella dorcas*, which inhabit extremely arid areas, have very low water turnovers (Macfarlane and Howard 1972).

Morphological (and probably also physiological) adaptations of lechwes, *Kobus leche* and *K. megaceros*, and the even more specialized sitatunga, *Tragelaphus spekei*, with elongated hooves, restrict them to floodplains and swamps. The klipspringer, *Oreotragus oreotragus*, is a specialized rock jumper.

Specializations also result in different microhabitats within a major vegetation formation being inhabited by different species. For example, dik-diks (*Madoqua* spp.) are confined to low, dense thickets in savannas.

2. *Feeding at different levels.* Giraffe, *Giraffa camelopardalis*, feed up to 18 ft above the ground, gerenuk, *Litocranius walleri*, at 4 to 8 ft, and kudu, *Tragelaphus strepsiceros*, at 3 to 6 ft.

3. *Food selection.* In place of the traditional categories of browsers and grazers, Hofmann and Stewart (1972) have classified 31 species of East African ruminants as bulk and roughage eaters (i.e., grazers—roughage, fresh grass, and dry region grazers); selectors of juicy, concentrated herbage (tree and shrub eaters, fruit and dicotyledon selectors), and intermediate feeders (some preferring grasses, others dicotyledons). Gwynne and Bell (1968), Field (1968), and Sinclair and Gwynne (1972) have demonstrated that grazing species may not only select different grass species but also varying proportions of plant parts, e.g., wildebeest take a high proportion of leaf, buffalo and topi less, and zebra select mainly stem material.

It is generally agreed, however, that there is often a considerable overlap in the diets of animals occupying the same area, and that in periods of food shortage direct competition is probably common.

4. *Seasonal patterns of environmental utilization.* Vesey-Fitzgerald (1960, 1965) has shown that the floodplains surrounding Lake Rukwa are used by 18 species of herbivorous ungulates—eight of them common—on a seasonally fluctuating basis. Different species use different communities at different times. The heaviest—elephant, hippopotamus, and buffalo—reduce the tall grassland of the seasonal swamps to a short-grass pasture that becomes available to the smaller antelope during the dry season.

In Serengeti National Park, an entirely different community, Gwynne and Bell (1968) describe another grazing succession. Zebra and buffalo enter long grass areas first during the dry season, use the stemmy material at the top of the herb layer and so prepare the way for topi and wildebeest, and finally for Thomson's gazelle; these follow successively and feed on the intermediate and lowest levels respectively, taking different components.

5. *Duplication of faunas in equivalent vegetation formations.* The ecologically similar but widely separated arid areas in southwestern, northeastern, and northern Africa have, to some extent, distinct mammal faunas. These species contribute to the large numbers inhabiting the continent.

Herbivore Biomasses

Records of high ungulate biomasses in savanna habitats have attracted much comment. Some early published figures were based on rather scanty data (see Talbot et al. 1965) but recent studies have a firmer foundation. Foster and Coe (1968) believe that the carrying capacity of the Nairobi Park is about 6,300 kg per sq km and state that this standing crop biomass is of the order one would expect in the acacia savanna habitat of East Africa. Watson, Graham, and Parker (1969) give somewhat lower figures of 4,027 kg per sq km and 3,907 kg per sq km respectively for the Serengeti Park ecosystem and the Loliondo Controlled Area in Tanzania. These data are for mixed populations of small- and medium-sized ungulates: zebra, giraffe, and various bovids such as wildebeest, hartebeest, eland, gazelle, and impala.

Field and Laws (1970) report a much higher figure from bushed grassland in the Queen Elizabeth National Park, Uganda. The mean year-round standing crop biomass over a four-year period was 29,490 kg per sq km. Of this 65% was contributed by hippo, 19% by buffalo, and 12% by elephant. A count by Watson and Turner (1965) in the Lake Manyara Park, Tanzania, where elephant and buffalo predominated, also revealed a high biomass of 21,870 kg per sq km. The presence of very heavy species apparently increases the total biomass considerably.

Comparison with domestic animal populations is instructive. Foster and Coe (1968) quote various authors. On tribal grazing land in East African savanna, domestic stock totaled 1,960 to 2,800 kg per sq km, and on managed European ranches, 3,728 to 5,600 kg per sq km. The average of all virgin ranges carrying domestic animals in the western United States is given as 3,448 kg per sq km while Petrides and Swank (1965) write that the best ranges there have a capacity of 4,300 kg per sq km.

In the Nairobi Park major predators have a biomass of only 1.4% of the total ungulate biomass. They are estimated to remove 15.5% of the total biomass, about 700 kg per sq km, per annum and this probably represents what man could take in the absence of predators (Foster and Coe 1968). At Loliondo 10% of the known populations could probably be removed without exceeding the sustained yield (Watson, Graham, and Parker 1969).

Arid Biomes

In addition to the Sahara Desert there are three distinct dry regions distinguished as the Southwest, Somali, and Sudanese Arid Zones. They are occupied by mammals adapted to heat and drought, some shared but many endemic to each region. This suggests a long history of aridity and periods of isolation during which local speciation led to the development of separate faunas.

Conditions are more stringent than in the savanna zones. Rainfall is low and unpredictable and as a result primary productivity is limited. Mammals are faced with serious problems of thermoregulation because of high ambient temperatures and a scarcity of water for evaporative cooling. Many small mammals escape by burrowing and emerge to feed during the cool night hours. There are only a few large species with the physiological adaptations necessary for survival and they must often move over great distances to find food. As in other very dry areas, many mammals are pale in color and reflect radiant energy efficiently.

The Southwest Arid Zone is the most distinct and endemic mammals are prominent in the fauna. They include five insectivores, two elephant shrews, *Elephantulus vandami* and *Macroscelides proboscideus,* and three golden moles, *Eremitalpa granti,* and two species of *Cryptochloris* (Meester 1965, 1971). Ten species of murids and cricetids, almost half of the total fauna of 24 species of small rodents, are endemic or almost endemic (Davis 1962). Gerbils are the most typical mice. The region is the habitat of the highly specialized crevice-dwelling rock rat *Petromus typicus* (Petromyidae), a hare (*Bunolagus*), and the gemsbok *Oryx gazella. Equus zebra* is found outside this zone only along the southern Cape mountain ranges.

Another group of species found only here and in the adjoining grassland subzone of the southern savanna are the hedgehog, *Erinaceus frontalis;* two ground squirrels, *Xerus;* two mongooses; *Felis nigripes; Vulpes chama* (silver fox); *Otocyon;* brown hyena; the extinct *Equus quagga;* springbok; red hartebeest; and black wildebeest (Meester 1965).

While the endemic fauna provides evidence for past isolation, some species and genera are shared with the Somali Arid Zone far to the northeast. Those discontinuously distributed in this way include the spring hare, *Pedetes;* caracal lynx, *Felis caracal;* aardwolf, *Proteles;* bat-eared fox, *Otocyon;* dik-dik, *Madoqua;* and *Oryx.* This interesting discontinuous distribution may be explained by assuming that a "drought corridor" linked southwestern with northeastern Africa during dry periods in the Pleistocene (Balinsky 1962). As Kingdon (1971) points out, it need only have consisted of areas of dry acacia bush and savanna rather than really arid communities.

The Somali Arid Zone in the horn of Africa has several interesting endemic mammals. They include a hedgehog, two elephant shrews, a ground squirrel and several gerbils, the peculiar maned rat *Lophiomys,* a hairless bathyergid, *Heterocephalus,* a ctenodactylid, a zebra, and several bovids, including the long-necked *Litocranius* and some gazelles. Some have spread into adjacent savanna areas. There is a fairly close relationship with the Sudanese Arid Zone but this has a much less well defined fauna than the others, being mainly a transitional area between the northern savanna and the Sahara.

As a harsh environment with a long history of extreme aridity (Cooke 1963) the Sahara supports a very small fauna. Typical mammals are a few species of hedgehogs and shrews, an elephant shrew, a dormouse, a few murines, about 15 gerbils, some dipodids, and ctenodactylids. Hares, hyraxes, two mustelids and two viverrids, several canids and felids, the wild ass, two large bovids, the addax, *Addax nasomaculatus,* the oryx, *O. dammah,* and about five species of gazelles complete the list. While some authors consider the mammals an impoverished Ethiopian fauna, there are regions in the Sahara in which the fauna has been described as transitional to that of the Near East (see Bigalke 1972).

There are few data on biomasses in arid areas. However, Bourlière (1965) gives estimates of 0.3 to 190 kg per sq km (0.003 to 1.9 kg per ha) for stony desert and 4 to 17 kg per sq km (0.04 to 0.17 kg per ha) for sandy desert in the Sahara.

Ethiopian Highlands

The large isolated block of mountainous country that dominates Ethiopia is an important center of endemism for mammals. The fauna includes unique forms such as the rodent *Muriculus;* the hamadryas and gelada baboons, *Papio hamadryas* and *P. gelada;* the Simenian fox, *Canis simensis;* a genet, *Genetta abyssinica;* the mountain nyala, *Tragelaphus buxtoni;* and the Ethiopian ibex, *Capra walie.*

Southwest Cape

This small zone at the southern end of the continent is famous for the distinctive Cape flora. A mediterranean climate with winter rains characterizes the western part, while to the east, summer rains become progressively more important. Mountains isolate the region except along the west coast, where it merges into the southwest arid biome. The Fynbos (macchia) vegetation is dominated by sclerophyllous shrubs, grasslike Restionaceae, and Cyperaceae and geophytes. Grass itself is unimportant. Soils tend to be poor and acid and there appears to be little food for mammals. The fauna is accordingly poor in species and in individuals and there is much sharing with adjacent parts of the Southwest Arid and Southern Savanna Zones. The region has, however, been poorly collected and much must still be learned.

There are not many insectivores—an elephant shrew, about five species of golden moles including one that is endemic, and three shrews. Bats comprise a fruit bat and about nine species of Microchiroptera, of which two vespertilionids are endemic. The small mammal fauna is dominated by Muridae and Cricetidae. Davis (1962) shows that 21 species are recorded. Three of them are endemic—*Praomys verreauxi, Acomys subspinosus,* and *Tatera afra*—and six are represented by distinct subspecies or isolated relic populations of savanna species. The remainder simply encroach into the southwest Cape. The porcupine, one or two dormice, and three species of Bathyergidae complete the list of rodents. The dune mole-rat *Bathyergus suillus* is endemic. The baboon is the only primate and *Orycteropus* occurs. Carnivores are surprisingly well represented by all families, the Hyaenidae only being extinct. Three canids, three Mustelidae, the aardwolf, two genets, two mongooses, leopard, caracal wild cat, and ? serval are recorded. Originally lion, elephant, black rhinoceros, mountain zebra, and hippopotamus were present. The dassie *Procavia capensis* is common. Artiodactyls consist mainly of small browsing species, namely Cape grysbok, *Raphicerus melanotis,* steenbok, grey duiker, klipspringer, and the vaal ribbok *Pelea.* The extinct bloubok *Hippotragus leucophaeus* and the rare bontebok *Damaliscus d. dorcas* are endemic to the region. Eland, red hartebeest, and buffalo were common in historic times.

Mammals of Madagascar

Madagascar is almost 250 mi from the African continent and is separated from it by an old, deep channel. Millot (1952) suggests that the island was colonized by a few ancestral African mammals early in its history when the channel was temporarily narrowed. From these a fascinating, typically insular fauna has evolved by adaptive radiation.

There are only 57 genera and about 94 species of mammals (including three introduced forms) belonging to the insectivores, primates, rodents, carnivores, and bats. The only ungulates are the bushpig *Potamochoerus,* believed to have been introduced from the mainland in historic times, and feral deer. A hippopotamus the size of *Choeropsis* is a common Pleistocene fossil.

Two species of *Suncus* are the only shrews. The Tenrecidae are the most important insectivores. There are two subfamilies, about 13 genera, and 29 species of these unique creatures. Many of them are hedgehoglike but there are fossorial species that resemble moles (*Oryzorictes*) and small shrewish forms.

Madagascar is famous for its lemurs, of which there are three endemic families. The Lemuridae are quite large and most are arboreal and mainly herbivorous. A subfamily contains the lesser lemurs, which resemble galagos and squirrels and take insects. Sluggish, leaf-eating, monkeylike species are placed in the family Indridae. The lemur-squirrel or aye-aye, *Daubentonia,* is a peculiar insectivorous creature classified in its own family.

The characteristic rodents are an endemic cricetid subfamily, the Nesomyinae. Like lemurs and tenrecs, they vary in form and size and have radiated to fill many niches. The only carnivores are a number of endemic and singular viverrids. Two genera are classified with the Asiatic palm civets (Hemigalinae). Bizarre and poorly known small mongooses of the endemic subfamily Galidiinae, with about seven species, correspond to the Herpestinae of the mainland while the large primitive catlike fossa *Cryptoprocta* (Cryptoproctinae) is a nocturnal forest dweller. Bats include both Micro- and Megachiroptera of six families. There are about 20 species but few are endemic.

Prehistoric Extinctions

The unique richness of the African mammal fauna is due to many factors. Most of the important ones have already been mentioned. They are the size of the continent, the large area within the tropics, the diversity of physiography and vegetation, the replication of arid zones, and the opportunities for faunal exchange with adjacent land masses and for speciation within the continent. How a fauna shaped by these influences has managed to survive virtually intact to modern times is a fascinating topic.

It is well known that many large mammals, particularly herbivores, rather mysteriously became extinct in Eurasia and North America toward the end of the Pleistocene, about 12,000 to 10,000 years ago, leaving these areas with greatly impoverished faunas. South America, too, suffered a drastic reduction, but extinction was only moderate in Australia, in both cases during the period 20,000 to 10,000 years B.C. (Keast 1972).

Africa is often said to have escaped these massive extinctions and to have a "pre-Pleistocene fauna." It did, however, suffer losses. Cooke (1972) points out that the Suidae and Bovidae in particular were greatly diminished. Martin (1966) writes, "It seems almost unbelievable, but the African plains game presently contain only about 60 per cent of the genera of large mammals encountered in the hand-axe faunas." Klein (1974) has recently drawn attention to significant megafaunal extinctions during the terminal Pleistocene in the southwest Cape. Simpson (1966, quoted in Keast 1972) does not, however, regard extinction at this time in Africa as significantly greater than what might be expected from a normal turnover rate. It certainly seems moderate by comparison with other regions.

How is the persistence of the African large mammal fauna to be explained? To summarize current views, Axelrod (1967) sees it as an expression of a continuing equable climate, stemming from the continent's transequatorial position. Martin (1966) has argued in favor of "overkill" by early man with newly developed weapons, the African fauna suffering less than that of other regions because it had evolved in the presence of man. Leakey (1966) advances cogent arguments against this view, asking among other things why more numerous hunters living after the extinction phase should have had such a small effect on the surviving fauna.

While the causes of extinction have yet to be determined and their small impact on the African fauna to be adequately explained, I agree with Klein (1974) that "the coincidence between extinctions and the relatively rapid and dramatic climatic change that characterized the end of the Pleistocene is too strong to ignore."

Recent Faunal History

From the end of the Pleistocene until comparatively recently the fauna appears to have suffered no major changes in composition, distribution, and abundance. Human influence first became marked in North Africa. The Egyptians probably captured and tamed elephants in early dynastic times (Carrington 1958) as well as domesticating geese, mongooses, the cat, and probably also gazelles, addax, oryx, and others (Zeuner 1963, Bigalke 1964). Their artists already used ivory in about 6000 B.C. (Sikes 1971). Reckless hunting by Roman colonists, capture for military use and for entertainment in Rome, and the gradual development of the Mediterranean coastline led to the extermination of elephants north of the Sahara. It seems likely that this process was completed in the first few centuries after the birth of Christ. Other large mammals were also gradually eliminated, the lion, for example, disappearing in the nineteenth century (Dorst 1970).

Commercial elephant hunting was a feature of the African scene for centuries before the rise of European influence. The slave and ivory trades were closely linked and Arab traders distributed muzzleloaders, thereby increasing the drain on elephant populations considerably. Large amounts of ivory were used in Europe for hundreds of years (Sikes 1971).

European settlers at the southern end of the continent decimated game herds during the eighteenth and nineteenth centuries and exterminated the bluebuck, *Hippotragus leucophaeus,* and also the quagga, *Equus quagga.* Commercial hunting and hunting for sport, both in southern Africa and farther north, became more efficient as firearms improved during the nineteenth century. Coupled with improved communications, the development of the continent, and increasing populations, hunting took a heavy toll of the teeming herds of wild animals so frequently described in the literature of the day. Elephants in particular had seriously declined virtually everywhere by the late nineteenth century when colonial administrations began to introduce firm control measures.

The rinderpest panzootic was another important influence. It killed millions of cattle and probably many more millions of wild ungulates throughout the continent at the turn of the century and is responsible for major current anomalies in the distribution of wildlife (De Vos and Lambrechts 1971).

At the beginning of the present century the ranges of many large mammals had been reduced and remaining populations had declined. Sidney (1965), Van der Merwe (1962), Dorst (1970), and others have written on this process. However, many parts of Africa south of the Sahara were still rich in wildlife until the Second World War and game can even now be found in large numbers outside reserves in countries such as Chad, Sudan, Congo Republic, Tanzania, Kenya, Uganda, Zambia, Angola, Rhodesia, and Botswana.

Probably the first game reserve to be proclaimed in Africa was the Pongola Reserve in the Transvaal, established in 1894 (Bigalke 1966). This was followed by Hluhluwe in Natal in 1897, the Sabi Game Reserve, Transvaal, in 1898 (subsequently, in 1926, it became the Kruger National Park), and, shortly after 1900, by various sanctuaries in East Africa. In 1925 the first African national park, the Albert, was created in the Congo and was the first step in a series that provided the eastern Congo with the best parks on the continent (Dorst 1970). These were managed as "total reserves." Visitors were allowed only in a limited area of the Albert Park and scientific research was actively pursued in all of them. Many other parks and reserves followed, most of the major new ones being created after the Second World War. Today there are few countries in Africa without nature conservation areas of some kind (see Guggisberg 1970 for list). Forest reserves have also played a most important role as sanctuaries for fauna, although they are often overlooked.

Utilization of Wild Mammals

Hunting Peoples

For all but a fraction of the million years and more during which man and his immediate predecessors have existed, hunting, fishing, and food gathering—the "'robber' economy of savagery" (Allan 1965)—have been his means of survival. In Africa there are still peoples who live largely or entirely by this means. The Bushmen of the Kalahari have remained surprisingly numerous, most of them living in Botswana and southwestern Africa. The Pygmies of the equatorial forests may be as common. In East Africa there are small numbers of Dorobo and other minor tribes who still live by hunting and honey gathering.

The economies of such peoples are finely balanced. They require big areas of land—perhaps 10 sq mi per person or more—but, given these, they can live on their production indefinitely, barring climatic catastrophe. There is evidence for deliberate limitation of their own populations in accordance with the resources available (Allan 1965). Allan is probably right in his conclusion that the hunters are likely to become extinct since their cultural patterns cannot be preserved without preserving also their isolation and their way of life by reserving vast areas in an increasingly land-hungry world, human "game reserves" that support tiny populations condemned by their technology to remain static or to starve. Perhaps such reserves will remain—there is every reason to urge that they should.

Cultivators

Cultivating peoples occupy all the sufficiently humid parts of Africa. Practically all of them supplement their diet by gathering wild plants, by fishing, or by hunting. Indeed, in many systems, a significant part of village subsistence is obtained by a full utilization of the surrounding bush, woodland or forest (Allan 1965). In West Africa much game is taken by "professional hunters" who live by selling meat on village and city markets. The smaller mammals of the forests are the most important sources of this "bush meat"; cane rats (*Thryonomys*), the giant rat, (*Cricetomys*), monkeys, bats, duikers, royal antelope (*Neotragus*), and others are commonly taken. Asibey (1971) estimates the annual yield from wild animals in Ghana at over 8,000 tons, valued at more than US$7,000,000. In Nigeria bush meat consumed was valued at £N10.2 million in 1965–66 (Charter 1971).

Virtually everywhere this resource is utilized without control and, in the face of habitat destruction accompanying increased cultivation and a ris-ing human population, it is unlikely to survive much longer. If, however, an elementary licensing system such as that recently introduced in Botswana (Child 1970) could be instituted and simple management applied while reliable methods of determining yields are worked out, a valuable source of the protein that most cultivators lack could be preserved (Bigalke 1975).

Pastoralists

People who live largely or entirely from their domestic animals occupy vast areas of Africa, including most of the north, parts of East Africa, and many other regions down to the southwestern extremity. Some rely also on agriculture to an extent. Many, for example the Masai, do not hunt to any significant degree for subsistence, and it is probably true that most African game survives in regions occupied by pastoralists.

While it is pleasing that this is so, the future of the remaining large assemblages of game outside parks and reserves is in jeopardy for this very reason. Schaefer-Kehnert and Brown (1975) have drawn attention to the implications for wildlife of the almost universal population increase among pastoral peoples and their flocks and herds, a well-known problem. Serious overgrazing has become the rule and governments are in general extremely loath to institute the controls needed to solve the problem. Direct competition between wild and domestic animals in overgrazed areas is likely to become more severe, to the detriment of game. Furthermore, undeveloped areas still rich in game are likely to be developed for pastoralists because this provides the easiest, if temporary, way out for governments unwilling or unable to be tough.

Game Farming

The large savanna ungulate biomasses, already mentioned, were first commented on in the 1950s by scientists such as Pearsall, Fraser Darling, Harrison Matthews, and Worthington, who were concerned about the future of African wildlife in states about to become politically independent (Bigalke 1966). The contrast between overgrazed tribal lands and game areas carrying higher biomasses, but in much better condition, led to support for the idea of utilizing wild animal populations for food.

No attempt will be made to review the considerable body of literature on the subject. Talbot et al. (1965) summarized available information almost a decade ago. Parker and Graham (1971) have put forward some trenchant criticisms of the popular view that traditional sources of protein are in short supply and that protein starvation could be alleviated in Africa by using the more efficient game populations instead of domestic animals. They believe the case that game is "better" than domestic stock in some areas to be unproven but concede that the diversity of African herbivores might be more productive of man's requirements than domestic species used exclusively. However, unless pastoral tribesmen change their attitudes under the pressure of necessity, they will continue to use domestic livestock even if the unowned and unownable wild animals are more efficient and productive. Tribal game farming is highly unlikely to take root.

Organized game cropping, mainly of the biggest species, has been a means of producing considerable quantities of edible, saleable meat in countries such as Uganda and Zambia and may still be so (there are no recent records). Game ranching on a limited scale has proved feasible and profitable in East Africa and Rhodesia and is now quite widespread in South Africa. In many cases, and especially in South Africa, only one or a few wild ungulate species are run in association with domestic animals.

Technological problems have been reviewed by Bigalke (1975). In spite of difficulties of grazing management, disturbance caused by sustained harvesting, disease conditions that preclude the marketing or export of fresh meat, and other disadvantages, some wild ungulates will probably be put to greater use in providing meat in the future, at least regionally and in association with domestic species. The greatest promise for productive game ranching per se seems to lie in large areas of savanna with mixed species populations (perhaps including domestic animals) utilized for sport hunting, trophies, and meat production. Whether it will become important probably depends largely on social and political factors rather than on ecological considerations. The ecological and economic prospects are bright enough to warrant the adoption of ranching with game as a respectable form of land use in undeveloped areas. If it is to succeed, more research is essential.

Tourism

Tourism based on national parks and reserves and on hunting has become an important industry in many parts of Africa. In 1968 Kenya was reckoned to have earned $30 million from overseas tourists (Thresher 1972) and the figure has since increased. Hunting safaris were worth £1.25 million and photographic tours £1.4 million in the three East African countries during 1966 (Clarke and Mitchell 1968). Child (1970) reports that tourism based on wildlife is worth R1 million per annum in Botswana, a poor country.

No doubt more complete and up-to-date statistics can be found. The point at issue is that wild mammals have developed an economic value of some magnitude that is now appreciated in most underdeveloped countries. This is likely to favor the retention of existing parks at least and may help to ensure that more are established. To maintain them, sophisticated scientific management will be needed.

Conclusion

Only in Africa can man still see something of the fantastic mammal fauna that populated the earth in the distant past. For centuries the fauna survived untouched and unaltered. In the past 200 years it has suffered decimation. Recently the rate of change has accelerated. Some species—fortunately only a few—have disappeared, while the ranges of most larger forms have been reduced by habitat loss, disease, or hunting. In order to conserve the remaining mammals for their cultural, aesthetic, and future survival value to man, social changes of great magnitude will be called for. They will entail effort in education and in increasing scientific knowledge upon which conservation and management can be based. The review of African mammal evolution in this volume will serve as a valuable reference work of basic importance to the task.

References

Allan, W. 1965. *The African husbandman.* Edinburgh: Oliver and Boyd.

Asibey, E. O. A. 1971. The present status of wildlife conservation in Ghana. *I.U.C.N. Publ.* n.s. 22:15–22.

Axelrod, D. I. 1967. Quaternary extinctions of large mammals. *Univ. Calif. Publ. Geol. Sci.* 74:1–42.

Balinsky, B. I. 1962. Patterns of animal distribution on the African continent. *Ann. Cape Prov. Mus.* 2:299–310.

Bigalke, R. 1966. South Africa's first game reserve. *Fauna and Flora* 17:13–18.

Bigalke, R. C. 1964. Can Africa produce new domestic animals? *New Scientist* 21(374):141–145.

———. 1966. Some thoughts on game farming. *Proc. Grassld. Soc. S. Afr.* 1:95–102.

———. 1972. The contemporary mammal fauna of Africa. In A. Keast, F. C. Erk, and B. Glass, eds. *Evolution, mammals and southern continents.* Albany: State University of New York Press.

———. 1975. Technological problems associated with the utilisation of terrestrial wild animals. *Proc. III World Conference on Animal Production.* Sydney: Sydney University Press.

———. Forthcoming. The biogeography and ecology of mammals in southern Africa. In M. J. A. Werger, ed. *Biogeography and ecology of southern Africa.* Amsterdam: Junk.

Booth, A. H. 1954. The Dahomey Gap and the mammalian fauna of the West African forests. *Rev. Zool. Bot. Afr.* 50 (3–4):305–314.

———. 1956. The distribution of primates in the Gold Coast. *J. W. Afr. Sci. Assoc.* 2:122–133.

Bourlière, F. 1965. Densities and biomasses of some ungulate populations in eastern Congo and Rwanda, with notes on population structure and lion/ungulate ratios. *Zool Afr.* 1(1):199–208.

Carrington, R. 1958. *Elephants.* London: Chatto and Windus.

Charter, J. R. 1971. Nigeria's wildlife: a forgotten national asset. *I.U.C.N. Publ.* n.s. 22:37.

Child, G. N. 1970. Wildlife utilization and management in Botswana. *Biol. Cons.* 3(1):18–22.

Clarke, R., and Mitchell, F. 1968. The economic value of hunting and outfitting in East Africa. *E. Afr. Agric. For. J.* 32(special issue):89–97.

Cooke, H. B. S. 1963. Pleistocene mammal faunas of Africa, with particular reference to southern Africa. In F. C. Howell, and F. Bourlière, eds. *African ecology and human evolution.* New York: Viking Fund Publications in Anthropology No. 36.

———. 1972. The fossil mammal fauna of Africa. In A. Keast, F. C. Erk, and B. Glass, eds. *Evolution, mammals, and southern continents.* Albany: State University of New York Press.

Corbet, G. B. 1971. *The mammals of Africa: an identification manual 1.2 sub family Potamogalinae.* Washington: Smithsonian Institution Press.

Darlington, P. G. 1957. *Zoogeography: the geographical distribution of animals.* New York: John Wiley.

Davis, D. H. S. 1962. Distribution patterns of South African Muridae, with notes on some of their fossil antecedents. *Ann. Cape Prov. Mus.* 2:56–76.

De Vos, V., and Lambrechts, M. C. 1971. Emerging aspects of wildlife diseases in southern Africa. *Proc. Symp. Nat. Cons. as a form of Land Use:* 97–109.

Dorst, J. 1970. *Before nature dies.* London: Collins.

Dorst, J., and Dandelot, P. 1970. *A field guide to the larger mammals of Africa.* London: Collins.

Ellerman, J. R. 1940–41. *The families and genera of living rodents,* 2 vols. London: British Museum (Natural History).

Estes, R. D. 1974. Social organisation of the African Bovidae. In V. Geist and F. Walther, eds. The behavior of ungulates and its relation to management. *U.U.C.N. Publ.* 24(1):166–205.

Ewer, R. F. 1973. *The carnivores.* London: Weidenfeld and Nicolson.

Ferrar, A. A., and Walker, B. H. 1974. An analysis of herbivore/habitat relationships in Kyle National Park, Rhodesia. *J. S. Afr. Wildl. Assoc.* 4:137–147.

Field, C. R. 1968. A comparative study of the food habits of some wild ungulates in the Queen Elizabeth National Park, Uganda. *Symp. Zool. Soc. Lond.* 21:135–151.

Field, C. R., and Laws, R. M. 1970. Studies on ungulate populations in the Queen Elizabeth National Park, Uganda. *J. Appl. Ecol.* 7:273–294.

Fleming, T. H. 1975. The role of small mammals in tropical ecosystems. In F. B. Golley, K. Petrusewicz, and L. Repzkawski, eds. *Small mammals; their productivity and population dynamics.* Cambridge: Cambridge University Press.

Foster, J. B., and Coe, M. J. 1968. The biomass of game animals in the Nairobi National Park, 1960–66. *J. Zool. Lond.* 155:413–425.

Gwynne, M. D., and Bell, R. H. V. 1968. Selection of vegetation components by grazing ungulates in the Serengeti National Park. *Nature* 220:390–393.

Guggisberg, C. A. W. 1970. *Man and wildlife.* London. Evans.

Hershkovitz, P. 1972. The recent mammals of the neotropical region. In A. Keast, F. C. Erk, and B. Glass, eds. *Evolution, mammals, and southern continents.* Albany: State University of New York Press.

Hirst, S. M. 1975. *Ungulate habitat relationships in a South African woodland/savanna ecosystem.* Wildlife Monographs 44. Washington, D.C.: The Wildlife Society.

Hofmann, R. R., and Stewart, D. R. M. 1972. Grazer or browser: a classification based on the stomach-structure and feeding habits of East African ruminants. *Mammalia* 36(2):226–240.

Keast, A. 1972. Comparison of contemporary mammal faunas of southern continents. In A. Keast, F. C. Erk, and B. Glass, eds. *Evolution, mammals, and southern continents.* Albany: State University of New York Press.

Kingdon, J. 1971. *East african mammals,* vol. 1. London; Academic Press.

Klein, R. G. 1974. A provisional statement on terminal Pleistocene megafaunal extinctions in the Cape Biotic Zone (Southern Cape Province, South Africa). *S. Afr. Archaeol. Soc.* Goodwin Series 2:39–45.

Lamprey, H. F. 1963. Ecological separation of the large mammal species in the Tarangire Game Reserve, Tanganyika. *E. Afr. Wildlife* 1:63–92.

Leakey, L. S. B. 1966. Africa and Pleistocene overkill? *Nature* 212:1615–1616.

MacFarlane, W. V., and Howard, B. 1972. Comparative water and energy economy of wild and domestic mammals. *Symp. Zool. Soc. Lond.* (1972) No. 31:261–296.

Martin, P. S. 1966. Africa and Pleistocene overkill. *Nature* 212:339–342.

Meester, J. 1962. The distribution of *Crocidura* Wagler in southern Africa. *Ann. Cape Prov. Mus.* 2:77–84.

———. 1965. The origins of the southern African mammal fauna. *Zool. Afr.* 1:87–95.

———. 1971. *The mammals of Africa: an identification manual. 1.3 Family Chrysochloridae.* Washington: Smithsonian Institution Press.

Millot, J. 1952. La faune malagache et le mythe gondwanien. *Mem. Inst. Sci. Madagascar.* sér. A, VII(1):1–36.

Misonne, X. 1963. Les Rongeurs du Ruwenzori et des regions voisines. *Exploration du Parc National Albert (Deuxième Série),* Fasc. 14. Bruxelles: Institut des Parcs Nationaux du Congo et du Rwanda.

Parker, I. S. C., and Graham, A. D. 1971. The ecological and economic basis for game ranching in Africa. In E. Duffey, and A. S. Watt, eds. *The scientific management of animal and plant communities for conservation.* 11th Symp. Brit. Ecol. Soc., Oxford: Blackwell.

Petrides, G. A., and Swank, W. G. 1965. Population densities and the range-carrying capacity for large mammals in Queen Elizabeth National Park, Uganda. *Zool. Afr.* 1(1):209–226.

Schaefer-Kehnert, W., and Brown, L. H. 1975. Economic and social aspects of animal production in relation to conservation and recreation. *Proc. III World Conference on Animal Production.* Sydney: Sydney University Press.

Sclater, W. L. 1896. The geography of mammals. IV. The Ethiopian region. *Geogr. J.* 7:282–296.

Sidney, J. 1965. The past and present distribution of some African ungulates. *Trans. Zool. Soc. Lond.* 30:1–396.

Sikes, S. K. 1971. *The natural history of the African elephant.* London: Weidenfeld and Nicolson.

Simpson, G. G. 1966. Mammalian evolution on the southern continents. *Neues. Jb. Geol. Palaeont. Abh.* B, 125:1–18.

Sinclair, A. R. E., and Gwynne, M. D. 1972. Food selection and competition in the East African buffalo (*Syncerus caffer* Sparrman). *E. Afr. Wildlife* 10:77–89.

Talbot, L. M.; Payne, W. J. A.; Ledger, H. P.; Verdcourt, L. D.; and Talbot, M. H. 1965. *The meat production potential of wild animals in Africa.* Farnham: Royal Commonwealth Agric. Bureau, Tech. Commun. No. 16.

Taylor, C. R.; Spinage, C. A.; and Lyman, C. P. 1969. Water relations of the waterbuck, an East African antelope. *Am. J. Physiol.* 217(2):630–634.

Thresher, P. 1972. African national parks and tourism—an interlinked future. *Biol. Cons.* 4(4):279–284.

Van der Merwe, N. J. 1962. The position of nature conservation in South Africa. *Koedoe* 5:1–122.

Vesey-Fitzgerald, D. F. 1960. Grazing succession among East African game animals. *J. Mamm.* 41:161–172.

———. 1965. The utilisation of natural pastures by wild animals in the Rukwa Valley, Tanganyika. *E. Afr. Wildlife* 3:38–48.

Watson, R. M.; Graham, A. D.; and Parker, I. S. C. 1969. A census of the large mammals of Loliondo controlled area, northern Tanzania. *E. Afr. Wildlife* 7:43–59.

Watson, R. M., and Turner, M. I. M. 1965. A count of the large mammals of the Lake Manyara National Park: results and discussion. *E. Afr. Wildlife* 3:95–98.

Zeuner, F. E. 1963. *History of domesticated animals.* London: Hutchinson.

2

Africa: The Physical Setting

H. B. S. Cooke

The Present Environment

The great continent of Africa, one-fifth of the land area of the earth, lies astride the equator and extends to very similar latitudes in the north (37°N) and the south (35°S), although the northern portion is twice as large as the southern part. It is a continent of contrasts, both in physical features and climate. Several mountain peaks in the eastern equatorial region rise to elevations greater than 5,000 m above sea level and a few carry permanent glaciers; in Egypt and in the Afar area of Ethiopia, on the other hand, there are depressions whose floors lie more than 100 m below sea level. The tropical forests of western equatorial Africa receive over 1,500 mm of rain a year, with a top figure of 10,000 mm in the mountains of Cameroon; but not far to the north the great Sahara Desert spans the continent from the Atlantic to the Red Sea and continues on into Arabia and the Middle East. Today this desert forms a significant barrier between the Mediterranean coastal region and the region to the south, conveniently dividing the continent into North Africa and sub-Saharan Africa.

From a physiographic viewpoint, Africa consists very largely of an elevated interior plateau, generally highest in the east and lowest in the west. The upland is disrupted by several important basins that have received large quantities of sediment derived from the uplifted rims. In sub-Saharan Africa, particularly in the south and on the east side, the elevated interior is commonly separated by a steep slope, or escarpment, from a relatively narrow coastal plain. Some of the desert bordering the Mediterranean is a broad, relatively low-lying area, reflecting its Mesozoic and Cenozoic marine history. Folded mountain chains, so usual in other parts of the world, are confined to the Atlas ranges in the northwest and the Cape ranges at the southern extremity of the continent. The Ethiopian massif and the East African plateau are surmounted by extensive lava flows and volcanic piles and these regions are also cut by the tectonic trenches of the Rift Valley System. The Red Sea Rift serves to separate Africa from Arabia, although the latter has been an integral part of the African block through most of its geological history. Related fracture systems and minor rifting exist in West Africa. Figure 2.1 shows the general relief and major morphological elements of the continent.

The varied climates of Africa are reflected in the vegetation patterns (figure 2.2). Environments range from hot and arid in the Sahara and Namib deserts to rainy tropical forests and, because of altitude, to cool mountain communities, including tun-

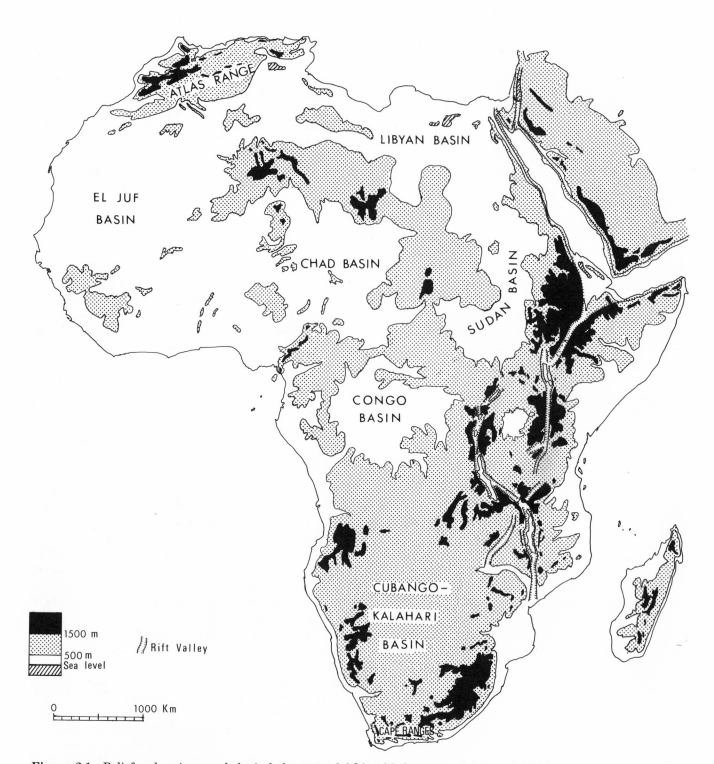

Figure 2.1 Relief and major morphological elements of Africa, Madagascar, and part of Arabia.

dra in some of the highest mountains. The rainfall is generally highest in summer but the southwest Cape and the northern fringe of the continent receive most of their rain in the winter. These two areas possess a Mediterranean type of vegetation, with low shrubs, evergreen bushes, and belts of forest. In North Africa olives are common but in the

Cape their place is taken by proteas. Both in the Atlas and in the Cape the high ranges prevent the rain from penetrating far into the continent, so that the Mediterranean macchia gives way to semidesert steppe, which borders the more arid desert areas. In the semidesert areas xerophytic shrubs are characteristic, together with tufts of *Aristida* grass form-

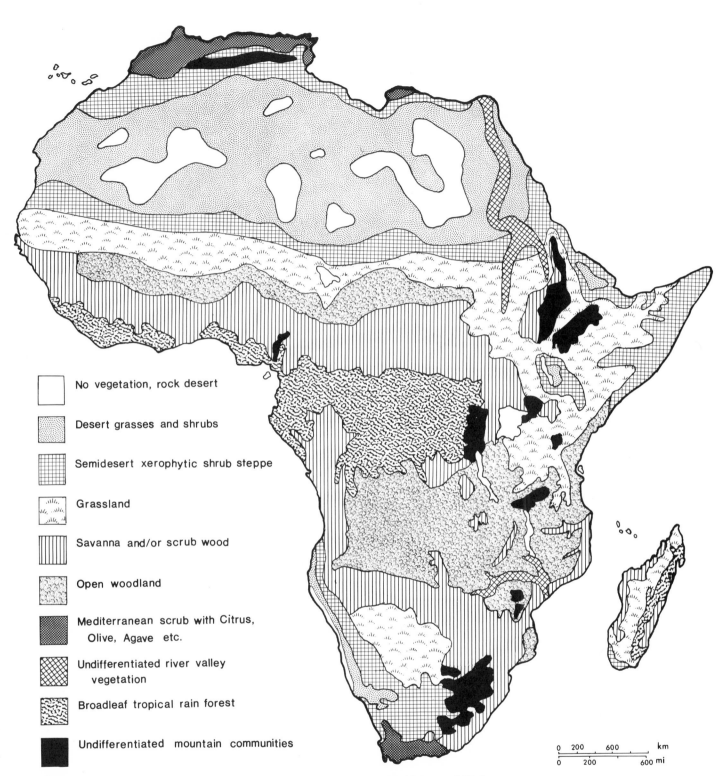

No vegetation, rock desert

Desert grasses and shrubs

Semidesert xerophytic shrub steppe

Grassland

Savanna and/or scrub wood

Open woodland

Mediterranean scrub with Citrus,
  Olive, Agave etc.

Undifferentiated river valley
  vegetation

Broadleaf tropical rain forest

Undifferentiated mountain communities

0   200   600   km
0   200   600 mi

Figure 2.2 General distribution of the main vegetation types in Africa and Madagascar.

ing an incomplete ground cover; there are some thorn trees, especially along the seasonal watercourses. This vegetation becomes scantier as the true desert belt is approached, being limited to occasional shrubs or thorn trees and tough desert grasses; in large areas of the desert, especially the rock desert and the "sand seas," vegetation of any kind is virtually lacking.

At the moister fringe of the semidesert belts the xerophytic shrubs disappear when the rainfall increases to about 250 mm a year, their decline being accompanied by an increase in the grass until it dominates the landscape as dry grassland or steppe, which may be treeless. However, the *Aristida* grass of the semidesert is replaced by other grasses, notably *Eragrostis* in the south and *Cenchrus* and *Chrysopogon* in the north. The cover remains incomplete and bare soil or sand is evident between the clumps of grass. In places, thorny *Acacia* and *Comiphora* form thickets or patches of woodland, or trees may be widely scattered over the landscape. As rainfall increases the grassland gives place to parkland with thorn trees, thorny bushes, and some deciduous trees scattered about or forming a light shady canopy above the grass.

Most characteristic of Africa is the extensive savanna and open woodland environment that surrounds the Tropical Forest Zone like a giant horseshoe. It is the typical "game reserve" landscape of Africa, usually a mixture of grass and trees but with a substantial range of variation according to changes in rainfall, seasonality, exposure, soil characteristics, and other factors, so that woodland and savanna, or even grass steppe, may occur within a small area. The grass is typically *Hyparrhenia* but may be *Themeda* or *Eragrostis*. Where the rainfall is below 750 mm, the trees or bushes are mainly thorn-bearing types (principally *Acacia*), but broad-leaved deciduous trees also occur and it is the latter that are most abundant where the rainfall approaches 1,000 mm. The moister woodland areas are dominated by a few genera of trees, notably *Isoberlinia*, *Brachystegia*, and *Julbernardia*. Many authorities believe that the climax vegetation of the savanna was originally richer in closed woodland, with relatively little grass, and that repeated fires have brought about the present situation.

Within the horseshoe of savanna and woodland lies the tropical forest, largely evergreen but with varying proportions of deciduous trees; the canopy is usually complete and layered, with the trees of the upper level as much as 50 to 60 m high. Toward the margins of the forest proper, the proportion of deciduous trees increases and forest-savanna mosaic forms a transitional zone. The typical tropical forest is essentially confined to low and moderate altitudes, but two other types of evergreen forest also occur. One of these is the rather limited evergreen forest of the Knysna area of the southern Cape, where the winter and summer rainfall regimes overlap. The other is the evergreen forest of the mountains, particularly in the East African and Ethiopian areas, where it normally lies higher than about 1,300 m above sea level. This montane forest differs from the lowland forest in the genera of trees represented and in the lower height of the tallest elements, but both the montane and Knysna forests contain a very similar spectrum of vegetation types. Forest of this kind, commonly termed "gallery forest," is found in many protected areas at lower altitudes and also extends along many watercourses in regions where the natural rainfall is inadequate to support it. These long ribbons of forest are an important feature of the drier landscapes.

Geological Development

With the exception of the Atlas and Cape ranges, Africa lacks the fold mountains of the other continents and has acquired its present form as a result of progressive uplift, gentle warping, volcanic activity, and faulting. Much of the exposed surface consists of Precambrian rocks or of Precambrian rocks hidden by a fairly thin veneer of Phanerozoic sediments (figure 2.3). The Precambrian "basement" consists of extensive areas of folded schists, gneisses, and granite rocks but also comprises thick sequences of sedimentary rocks that have suffered astonishingly little metamorphism and are in many cases gently folded or even still essentially horizontal. Nevertheless, the basement geology is complex and has involved several periods of strong deformation that are recognizable by belts within which radiometric ages tend to be fairly similar. Stabilization of the basement seems to have been progressive, with certain core areas having become rigid more than 2,500 m.y. ago; they are surrounded by younger orogenic belts. The last major metamorphism took place in the final Precambrian or earliest Cambrian (600 to 500 m.y. ago) and in parts of the continent there is no significant break between the Precambrian and the earliest Paleozoic sediments.

According to current views, Africa in the early Paleozoic was part of the southern supercontinent of Gondwanaland and both paleomagnetic evidence and the presence of glacial deposits in the western Sahara show that this was then the location of the south pole of the earth. By the Carboniferous the pole was close to the southern tip of Africa and extensive glaciation ushered in the lengthy cycle of deposition of the Karroo System. In the type area in South Africa this unit reaches a maximum thickness of 7,000 m and was deposited in a slowly subsiding basin. Overlying the glacial Dwyka Series are dark shales and grey sandstones, often rich in plant remains of the southern *Glossopteris* flora and also containing extensive coal deposits. By mid-Permian times slow equatorward movement of Africa had led to warming of the climate so that the Beaufort Series comprises colorful shales and yellowish sandstones indicative of deposition in an environment subject to periodic drying out and flooding. Fossil re-

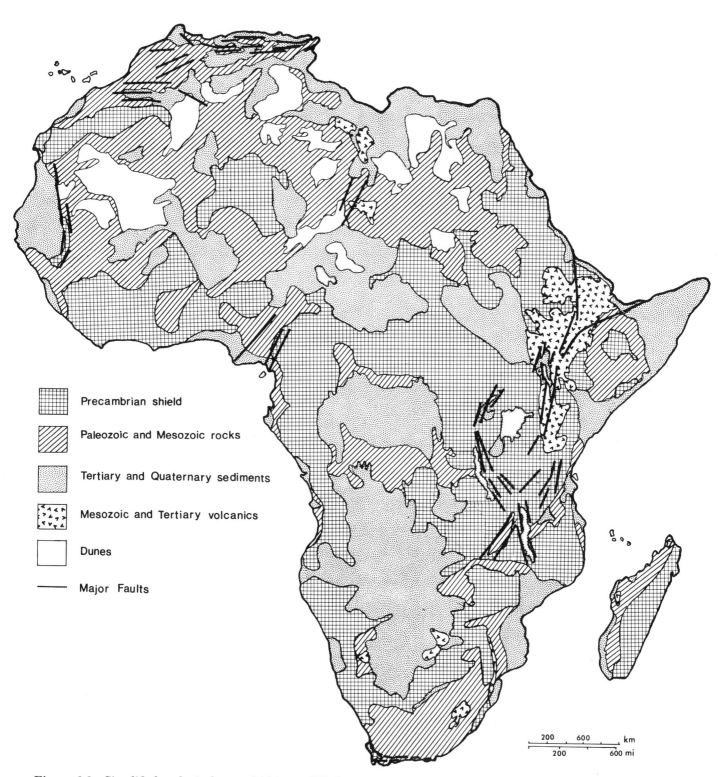

Figure 2.3 Simplified geological map of Africa and Madagascar.

mains are usually abundant in the Beaufort Series and the complex of reptiles shows increasingly mammallike features in the higher horizons, which are Triassic in age. The overlying Stormberg Series indicates progressive desiccation and the strata include much wind-blown sand of desertic character.

Two of the earliest true mammals so far known, *Erythrotherium parringtoni* and *Megazostrodon rudnerae,* come from a late Triassic unit of the Stormberg Series in Lesotho (Crompton 1964; Crompton and Jenkins 1968) and are assigned to the family Morganucodontidae, of which representatives are

Legend:
- Precambrian shield
- Paleozoic and Mesozoic rocks
- Tertiary and Quaternary sediments
- Mesozoic and Tertiary volcanics
- Dunes
- Major Faults

also known from Europe and China (Crompton 1974; see also this volume, Chapter 3). The Karroo cycle ended with the outpouring of up to 1,000 m of basaltic lavas in the early Jurassic, still preserved extensively in the highland of Lesotho. Although the complete succession is present only in the main basin of the Karroo, the system is widely represented by partial sequences elsewhere in southern and eastern Africa as far north as the equator, and there are equivalent beds in Madagascar as well as in other parts of Gondwanaland. In West Africa rocks of continental origin that were deposited at about the same time as the Karroo System are often included under the general term "Continental Intercalaire," used by French geologists, and in Egypt and the Sudan they have been included with the "Nubian Sandstone," although both these terms in the strict sense should apply only to rocks of later Mesozoic age.

In northern and western Africa the lower and middle Paleozoic are represented by extensive marine sequences related to fluctuations of the ancient Tethys Sea. At the end of the Carboniferous this region was largely emergent so that Permian and Triassic marine beds are restricted to a small part of Tunisia and Algeria. However, Permian marine deposits occur in western Madagascar, indicating the early stages of separation of that fragment of Gondwanaland, and Triassic marine rocks are also found there, as well as in the Sinai Peninsula and in eastern Arabia.

The Jurassic is marked by renewed encroachment of the northern Tethys Sea over Morocco, northern Algeria, and Tunisia, but this period is also notable for the development of an arm of the sea across Arabia and Ethiopia and along the eastern side of the present African continent as far south as central Mozambique, as well as western Madagascar. At Tendaguru in Tanzania marine and continental beds interfinger, and from a high Jurassic horizon an edentulous mammalian jaw was found and named *Brancatherulum tendagurense* by Dietrich (1927); it may be a paurodontid pantothere (Simpson 1928). Evaporites underlie mid-Jurassic marine beds in Tanzania and are presumably associated with the warping and rifting that led to separation of the "eastern" parts of Gondwanaland and early growth of the Indian Ocean. The rifting of the Kariba-Luangwa-Lower Zambezi troughs also probably dates back to the Jurassic, but the rifts did not spread, and they became the receptacles for Cretaceous sediments in which dinosaur remains occur. On the west side of Africa the Benue Rift was formed in the late Jurassic or early Cretaceous but also did not spread. At this time rifting began along the boundary between Africa and South America and lacustrine and saline deposits were formed in the early Cretaceous before the south Atlantic began to open up. Upper Cretaceous marine beds occur all the way down the Atlantic coast of Africa, but initially with a fauna different from that of the Tethys. North Africa was invaded extensively by the Tethys Sea, which in mid-Cretaceous times broke across the western part of the continent from Algeria to the Gulf of Guinea to link up with the young south Atlantic. Thus by mid-Cretaceous times Africa had for the first time achieved approximately its present borders, still with Arabia as an integral part of the continent, while both the Atlantic and the Indian oceans grew progressively wider. However, the distribution of late Cretaceous dinosaurs seems to demand maintenance of some kind of land link between Europe, South America, Africa, Madagascar, and India that is not easy to reconcile even with most recent interpretations of the palaeomagnetic evidence for the positions of the continents (e.g., Dietz and Holden 1970; Smith and Hallam 1970) but it is in fair agreement with the reconstruction by Smith, Briden, and Drewry (1973). These maps, however, reflect only plate distributions and are not paleogeographic maps in the strict sense.

Paleocene marine beds occur in North Africa, including reef formations that became important oil traps. It seems clear that by this time Africa (with Arabia still an integral part) was effectively isolated from the other continents. The borders of the "new" continent were invaded by the sea at many points, with the Tethys covering an extensive area in North Africa, while a large embayment of the Atlantic extended over Nigeria (figure 2.4). According to the paleomagnetic evidence, Africa was at this time still some 15° of latitude south of its present position so that the equator ran from Senegal to the northern tip of Ethiopia. This must have had an effect on the climate, but the implications have yet to be evaluated. The continent was effectively in its present position by the early Miocene. During the Eocene the Tethys withdrew progressively in the north so that by Oligocene times only two narrow belts of marine encroachment remained (figure 2.4).

Toward the end of the Oligocene, folding and elevation of the Atlas region led to the complete emergence of northwestern Africa and it is probably in the Oligocene that rifting of the Red Sea trough began and soon led to the almost complete isolation of Arabia by flooding of the trench in mid-Miocene times. In the Miocene the Tethys (Mediterranean) again encroached upon the continental margin, but the invasion was not nearly as extensive as in the

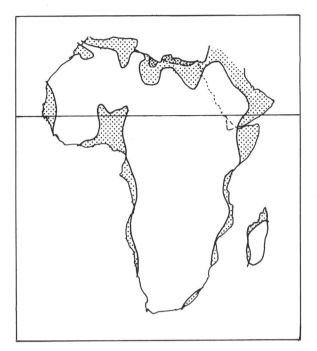

Figure 2.4 Epicontinental seas in Afro-Arabia during the Eocene (light stipple) and Oligocene (heavy stipple). The dashed line shows the incipient Red Sea Rift. The full horizontal line is the estimated Eocene equator.

Eocene. The principal mammal-bearing deposits in North Africa tend to lie fairly close to the fluctuating Tertiary shorelines. At the end of the Miocene the Mediterranean Basin was temporarily closed off from the Atlantic and evaporation led to substantial shrinking of the water body accompanied by the formation of extensive saline deposits. This phenomenon must have had significant effects on the climate and on animal life in the region.

Before the severance of Africa from its neighboring components of Gondwanaland, the continent appears to have been reduced to a landscape of fairly low relief, which L. C. King (1951, 1962) has named the Gondwana surface, still recognizable as patches on top of areas of greatest uplift. With the creation of a new coastline, rejuvenated rivers began the slow process of dissection of the uplifted surface and long stability led to the formation of the most extensive planation, variously called the African surface, mid-Tertiary surface, or Miocene peneplain. Substantial uplift in the Miocene has led to development of two levels of broad flat areas flanking the main river systems, and late Miocene uplift led to incision of the streams in the lower parts of their courses. The total elevation since the Cretaceous is difficult to estimate, but in sub-Saharan Africa post-Gondwana uplift of about 900 to 1,200 m seems to be demanded, with about half of it since the mid-Tertiary. The

major basins (figure 2.1) are not so much areas of sagging as areas that have lagged behind in the positive general uplift that has affected the continent since the disruption of Gondwanaland. The Congo Basin was successful in breaching its rising western limb, but the great Cubango-Kalahari Basin is largely one of internal drainage filled with debris that could not be carried away. The Chad Basin is also one of internal drainage containing much accumulated debris. Other small, closed basins have been created in the volcanic piles of the East African –Ethiopian area, but these are the product of more localized deformation associated with the volcanism and the fracture systems of the Great Rift Valley.

Major Fossil Mammal Localities

General

The principal localities that have yielded fossil mammals are shown in figure 2.5, although many Pleistocene sites have had to be omitted. There are no Tertiary localities and few Pleistocene sites in the heavily forested equatorial region, although Malembe near the mouth of the Congo borders on it. In general, the more important discoveries have been made in three areas: South Africa, East Africa, and the strip of land bordering the Mediterranean. The latter embodies two regions, the Atlas ranges and adjoining area of northern Algeria and Tunisia (known to the Arabs as the Maghreb) and the northern part of the Libyan and Egyptian deserts. In the Maghreb the fossil occurrences are generally isolated and restricted, especially in the zone of the Alpine folding, but in the desert area to the east are more extensive and are in some degree interrelated by the paleogeography.

Eocene—Oligocene

The oldest Cenozoic mammals so far described from Africa are of Eocene age and include some of the earliest marine mammals known. The first discovery was made by Schweinfurth in 1879, when he found remains of the primitive Cetacean *Dorudon* in the Fayum area of Egypt (Dames 1883, 1894). Schweinfurth's account of the geology was published in 1886, and the area was mapped by Beadnell (1901a, 1905), when he found terrestrial mammals, some of which he described himself (1901b, 1902) while others were published in a series of papers by Andrews, which culminated in his classic monograph (1906). Many expeditions have worked in the area and an excellent historical summary is given

Figure 2.5 Principal fossil mammal localities in Africa. The inset shows the East African area at an enlarged scale.

by Simons (1968), who led the most recent group from Yale University.

Eocene marine beds, consisting largely of massive or well-bedded limestones, sometimes laminated and carrying flint bands and concretions, are extensively developed in Egypt and Libya. Large foraminifera are commonly present. Marls and shales occur occasionally, becoming more frequent in the upper Eocene. Vertebrate remains are extremely rare except in the Fayum, where the Qasr el Sagha Formation consists largely of deltaic and interdeltaic deposits with occasional channel sands, and vertebrate fossils are not uncommon. There are interfingering marine horizons and the invertebrate fauna suggests a late Bartonian age. Several species of cetacean and the sirenian *Eotheroides* have been found, as well as fish, crocodiles, and the earliest known flightless bird. Plant debris and silicified logs are present and the terrestrial vertebrate remains tend to be associated with them. The land mammals include the primitive "proboscidians" *Moeritherium* and *Barytherium,* a hyaenodont and an anthracothere.

Overlying the Qasr el Sagha Formation is the Gebel Qatrani Formation, some 200 m thick, consisting of a wide variety of clastic sediments and rare lacustrine limestones. The lower 50 m are rich in fossil wood and most of the terrestrial vertebrates collected before 1961 came from the top of this zone. They are mainly large mammals, including the two Eocene "proboscidean" genera already mentioned as well as two further proboscideans, *Palaeomastodon* and *Phiomia.* Hyracoids are abundant and the earlier expeditions found a few rodents and very rare primates. Carnivores and anthracotheres are also present. A much thinner fossil wood zone occurring about 80 m higher in the sequence furnished substantial amounts of material to the Yale expeditions (Simons 1968), and important specimens are also derived from a thin horizon about midway between the Lower and Upper Fossil Wood Zones. The Yale expedition recovered many small mammals including rodents and important primates, largely from the higher horizons. Some 75 m above the Upper Fossil Wood Zone a basaltic lava caps some of the hills, clearly following an erosion interval of unknown duration, and the basalt has a K-Ar date of approximately 27 ± 3 m.y. The rich and varied fauna from the lower zone is generally considered to be early Oligocene in age but it is possible that the upper zone may be middle Oligocene. The sediments are interpreted as representing a channel floodplain complex laid down in an area of low relief under semiarid conditions, but with dense gallery forest along the stream channels and perhaps with some savanna existing between them. This interpretation is consistent with the general aspect of the fauna and would help to account for the presence of the important and varied primate fossils.

Eocene marine mammals have also been recorded from the Mokattam Hills near Cairo and from Buel Haderaut in Libya, but no land mammals are associated with them. Elsewhere in Africa sirenians have been reported from Somalia, both in Eocene and Oligocene beds (Savage 1969). A cetacean of mid-Eocene age is recorded from Ameki in Nigeria. Fragmentary remains of *Moeritherium* were found in a sequence of upper Eocene marine beds at M'Bodione Dadere in Senegal (Gorodiski and Lavocat 1953) and also in continental beds at Gao and at In Tafidet, both in Mali. However, terrestrial mammals of this age are not otherwise known from the sub-Saharan region.

The marine Eocene extends westward from Egypt into Libya and southern Tunisia, where it is well developed in the Sirte Basin. Occasional finds of sirenians have been made but the most important area is the Dor el Talha escarpment where terrestrial mammals were first recorded by Arambourg and Magnier (1961) and examined more extensively by Savage (1969, 1971). (Arambourg and Magnier called the locality Gebel Coquin, but the name Dor el Talha is preferred by Savage.) The lowest beds are calcilutites, with abundant oysters and other marine invertebrates. This is followed by 60 m of marls, the lower part of which has sandy channels, bituminous horizons, nodular layers, and gypsum sheets. Land mammals are rare but marine and aquatic mammals, fish, crocodiles, and turtles are abundant, as well as fossil wood and oyster bands. The upper part of the marls is more massive or poorly bedded, becoming thinly bedded at the top. Land mammals are abundant and are associated with fish and reptiles. Capping the marls is 30 m of unfossiliferous sandstone. The oysters in the calcilutites are correlated with those in the middle to late Eocene of Egypt. The lower marls contain sirenians and cetaceans, as well as *Moeritherium* and *Barytherium,* as in the Qasr el Sagha Formation of the Fayum and they are probably of equivalent age. Both these genera continue into the upper marls but without aquatic mammals, and this part of the sequence is most probably early Oligocene in age and equivalent to the fauna from the lower quarries in the Gebel Qatrani Formation of the Fayum.

In the Oligocene the Sirte Basin was still largely marine but there are two localities in the marginal area that have furnished terrestrial mammals of this period. The best fauna, although scrappy, is from Zella, 300 km northwest of Dor el Talha, where an estuarine conglomerate has yielded two of the Fayum proboscideans, a large hyracoid, a carnivore, and an anthracothere, as well as crocodile and turtle remains (Arambourg and Magnier 1961). Farther to the northwest, at Gebel Bou Gobrine in southern Tunisia, is another littoral zone sandstone from which scrappy remains were recovered, including the Fayum proboscidean *Phiomia* (Arambourg and Burollet 1962). (Arambourg and Magnier give it as "Bon" but the correct Arabic is "Bou.") The fossil-bearing horizon lies 100 m below a marine deposit with abundant oysters and pectens of early Miocene (Burdigalian) age and the mammals are certainly Oligocene.

Miocene of North Africa

In North Africa the Miocene sea advanced over a surface that had been folded during the late Oligocene, so there are several somewhat distinct basins of deposition. The sites with terrestrial mammals are related to the shoreline and tend to belong to fluviatile or estuarine environments. In Egypt fluviomarine and deltaic beds are exposed over a long distance and at three localities, in the Moghara Oasis and in Siwa and Wadi Faregh, they contain fossil mammals. Moghara has the richest fauna, with the proboscidean *Gomphotherium*, anthracotheres, perissodactyls, a felid, and a monkey, *Prohylobates*. The age is somewhat uncertain but is probably early Miocene.

Probably the richest and best site in North Africa is that of Gebel Zelten, 200 km northwest of Dor el Talha. It was first recorded by Arambourg and Magnier (1961) and has been worked more extensively by Savage (Savage and White 1965; Savage 1971). The regional setting and stratigraphy are well described by Selley (1968), who shows that the mammal-bearing sediments belonged to a coastal flood plain separating intertidal deposits to the north from the low-lying hinterland of the Sahara to the south. Mixed marine and continental deposits occur both below and above the vertebrate-bearing unit. The rich fauna is regarded by Savage as "Burdigalian"[1] and to indicate a savanna environment. It includes proboscideans, perissodactyls, giraffids, suids, ruminants, carnivores, and creodonts; but sirenians, fish, turtles, crocodiles, and birds have

also been found. The whole assemblage of terrestrial mammals is very similar to that from the early Miocene sites in East Africa but the smaller elements and micromammals are lacking at Gebel Zelten.

Some 200 km to the northeast of Gebel Zelten is the site of Qasr es Sahabi, where Miocene marine clays with invertebrates and fish occur in a sequence of estuarine, lacustrine, and fluviatile deposits containing well-preserved terrestrial and freshwater vertebrates (Petrocchi 1943). The collection includes a gomphotherelike elephant, *Stegotetrabelodon*, a mastodont, dinotheres, rhinoceros, an equid, anthracotheres, hippopotamus, suids, bovids, and carnivores. The collection has not been fully described but the age is probably very late Miocene. At El Haserat, 50 km north of Sahabi, an anthracothere has been found and may be of similar age, but this is speculative.

In Tunisia, north of Gebel Bou Gobrine, vertebrate fossils have been found at a number of localities, the most important of which is Bled ed Douarah some 40 km west of Gafsa (Robinson and Black 1969). Fossils occur at a number of horizons in the Beglia Formation, to which a "Vindobonian" age is assigned. The Bled ed Douarah fossils come mainly from a fluvio-deltaic complex, which passes northward around Sbeitla into dunes and then into a littoral sand facies at Djebel Cherichira (Biely et al. 1972). The fauna is rich and varied, including mastodonts, *Deinotherium*, suids, anthracotheres, birds, fish, and, in the upper levels, *Hipparion primigenium* (Forstén 1972). The birds indicate that both savanna and freshwater stream habitats are represented (Rich 1972). *Merycopotamus* is found both in the lower and in the upper faunal levels and changes indicate that a time lapse exists between the lower beds, without *Hipparion*, and the upper levels in which this equine occurs (Black 1972). The caprine bovid *Pachytragus* is present before *Hipparion* and is considered by Robinson (1972) to be of African origin. The freshwater fish are strongly African in affinities and contrast markedly with the Eurasian aspect of the modern freshwater fishes of Tunisia (Greenwood 1972). The Beglia Formation usually rests unconformably on earlier rocks, probably because of the Miocene tectonism associated with the Atlas region and its flanks. In the area around Djebel Mrhile and Henchir Beglia there are underlying "Burdigalian" deposits of the Mahmoud Formation that have yielded some fossil remains.

On the eastern flank of the Atlas, in the Oued (or Wadi) el Hammam Valley southeast of Oran, marine deposits with plentiful invertebrate fossils assigned to the lower Miocene are overlain uncon-

---

1 The use of European stage names like Burdigalian may be feasible in the marine associations, but their employment for terrestrial deposits in Africa is hazardous and attention is drawn here to this fact by inserting such terms in quotation marks in this text.

formably by some 400 m of alternating reddened sandstones, sands, shales, and conglomerates, with fossil mammal remains scattered through them; land snails also occur (Arambourg 1951). The continental beds belong to the Bou Hanifia Formation and are covered by marine sandstones and shales with oysters and gastropods. A tuff in the lower part of the Bou Hanifia Formation has been dated radiometrically at 12.18 ± 1.03 m.y. (Ameur, Jaeger, and Michaux 1976) and this provides an important reference point for correlation in the Maghreb. The fauna contains *Hipparion,* a rhinoceros, hyenas, *Orycteropus,* bovids, giraffids, rodents, ostrich, and turtle (Arambourg 1951, 1954a; Jaeger, Michaux, and David, 1973). It seems probable that these continental deposits belong to the same general environment as the Beglia Formation of Tunisia but if the absence of *Hipparion* in the lower part of the latter is real, the basal Beglia may be slightly older than the Bou Hanifia. Sporadic finds of *Hipparion* and *Mastodon* have been made elsewhere in Algeria and in Morocco at Gara Ziad, Melka el Ouidane (Camp-Berteaux), and Tadla Beni Amir, all probably later Miocene in age (Choubert and Ennouchi 1946; Jaeger, Michaux, and David 1973). A very similar fauna, richer than most of the others, occurs at Marceau in Algeria and is also later Miocene in age.

Miocene of Sub-Saharan Africa

In sub-Saharan Africa only two Miocene localities have been reported outside the East African region, Malembe and the Namib. A third, Bololo, lies just north of the Congo River 30 km from its mouth and has yielded scrappy mammal remains from a gravel bed of Quaternary age, but the recognizable fragments of molars of *Gomphotherium* suggest derivation from an unknown Miocene source. On the coast at Malembe, 80 km to the northwest, fossil mammals of mid-Miocene age occur in an estuarine phase of a marine sequence that includes fish, a cetacean, and a sirenian. The terrestrial mammals were presumably washed in and include *Gomphotherium,* chalicothere, carnivore, suid, and anthracothere remains (Hooijer 1963). Well to the south, in the Namib Desert area, fluvio-lacustrine and continental beds occur and have provided a limited but interesting fauna (Stromer 1926; Hopwood 1929). Rodents are the most important element and there are also insectivores and hyracoids. The larger elements include a small suid, a hyaenodont carnivore, and two bovids. The age is generally regarded as "Burdigalian."

The East African region possesses a substantial number of fossil mammal-bearing Miocene, Plio-cene, and Pleistocene deposits, the most important localities being shown in the inset to figure 2.5. The richness of the region in terrestrial fossils is largely a consequence of the volcanic and tectonic environment, which provided favorable basins as traps for sediment, rapid burial, and calcic or alkaline materials to prevent the more usual leaching and destruction of buried bones. The presence of volcanic materials in close stratigraphic association with fossil-bearing sediments has also provided unusually good radiometric controls for the evaluation of the succession.

As mentioned earlier, the African continent has experienced considerable uplift, especially in the zone through which runs the complex Rift Valley System. The rifts themselves follow old structural trends in the crust and some of the rifting dates back as far as the late Jurassic or early Cretaceous. By Oligocene times extensive planation had led to the exposure, in the East Africa region, of an erosion surface consisting dominantly of ancient granitic and metamorphic rocks. The continental divide probably lay in the region of the present Eastern Rift and the general altitude of the erosion surface may have been only in the order of 500 m above sea level, with residual masses rising above it (Cooke 1958). Saggerson and Baker (1965) showed the existence of a swell in the basement along the Eastern Rift and estimate that the residual mass of the Cherengani Hills has been uplifted as much as 1,700 m since the time of development of the subvolcanic surface; this would mean a former altitude of about 1,500 m above sea level for the divide at the time. Bishop and Trendall (1967) have examined the subvolcanic surface in Uganda and shown that a west-facing erosion scarp existed south of Moroto, rising some 600 m in 30 km and continuing beneath Mount Elgon. The drainage was westward from a watershed along the line of the residuals of the Chemorongi and Cherengani hills and there was modest relief in what they term the "Kyoga surface." This surface was tectonically modified by the growth of domes on the sites of some of the subsequent major volcanoes and by the development of local basinal structures as volcanism began in the early Miocene. The history of tectonic events in the Kenya Rift is well outlined by Baker and Wohlenburg (1971; Baker et al. 1971; Baker, Mohr, and Williams 1972) and the lava sequences have been described recently by King and Chapman (1972).

Miocene fossils were first found in 1909 at Koru, on the flanks of the old cone of Tinderet, and later at Karungu on the south side of Kisingere (Andrews 1911). The richest sites are on the northwest side of

Kisingere, on Rusinga Island and Mfwanganu Island, which lie near the entrance to the Kavirondo Gulf of Lake Victoria; on the north side of the gulf are other localities, Maboko and Ombo, that are less closely related to volcanic cones. Two other important localities occur on the flanks of Tinderet—Songhor to the north of Koru and the later Miocene site of Fort Ternan to the east. In Uganda there are two important sites some 250 to 300 km north of the Kavirondo Gulf, Napak, on the flanks of an old volcano, and Moroto, on the subvolcanic basement. Lesser sites in Kenya, such as Muruorot and Loperot, are not obviously related to former cones but are associated with extensive lava flows. The literature is abundant, but the initial history of discoveries is well set out by Le Gros Clark and Leakey (1951). The association of volcanic materials has made possible the determination of many radiometric ages (Bishop, Miller, and Fitch 1969; Van Couvering and Miller 1969), reflected in figure 2.6. The faunas have been listed by Leakey (1967) and Bishop (1967), with good bibliographies to that date.

Rusinga Island is by far the richest area, having yielded many thousands of fossils. The geology was first described in some detail by Kent (1944) and amplified by Shackleton (1951), Whitworth (1953), and McCall (1958). Van Couvering and Miller (1969) have made some amendments to the earlier stratigraphic interpretations and provided a convenient summary as well as radiometric dates. The island is underlain by the Kiahera Formation, consisting of brown clastic sediments passing into calcic volcanogenic sediments. Overlying this unit unconformably is the 40-m thick Rusinga Agglomerate, largely made up of blocks of nephelinite lava in a matrix of comminuted lava fragments, but with volcanic grits and silts interbedded locally; its radiometric age is 19.6 m.y. The Hiwegi Formation, 50 to 70 m thick, consists of a wide range of volcanic, clastic, and calcareous sediments, including silty beds that constitute the main fossil-bearing unit lying about 20 to 25 m above the base. The Hiwegi Formation is covered by the Kiangata Agglomerate, rich in nephelinite blocks, which in turn is capped by the Lunene Lava, a melanephelinite with a radiometric age of 16.5 m.y. The major fossil horizon in the Hiwegi is thus interpolated as close to 18.0 m.y., but fossils also occur in the upper Kiahera and in a unit known as the Kulu Beds, which lies in an erosion hollow cut into the Hiwegi Formation but whose exact age is not yet clear. The neighboring island of Mfwanganu has been described by Whitworth (1961) and the formations parallel those of Rusinga, although differing in detail.

Bishop (1963, 1968) has reviewed the distribution of the major fossil-bearing sites and shown that they reflect primarily conditions suitable for fossilization. He recognizes (1) fully lacustrine conditions predominating, with fish fossils present and a lithology mainly of fine-grained clastic sediments and water-deposited tuffs; (2) subaerial deposits with tuffs, fossil wood, and land gastropods, with interstratified water-borne sediments, resulting from temporary ponding and flash floods; (3) coarse clastics, including conglomerates, mixed with a variable proportion of fine volcanic ash as a matrix, usually resting on the basement and underlying the volcanic sequence proper.

The nonvolcanic sediments of type 3 accumulated in valleys and hollows on the irregular prevolcanic topography, some of which may have been the result of tectonic deformation preceding the volcanicity. The fossil assemblages are generally more or less rolled and fragmented, with small faunal elements lacking, but there are rare associated bones suggesting that carcases had been swept in occasionally. Some of the Napak sites, Moroto, Ombo, and Marewa, are regarded as of this type. Fully lacustrine environments, together with subaerial deposits, occur at Bukwa and at Karungu, Mfwanganu, and Rusinga. The remaining sites with well-preserved specimens owe the good preservation to fine-grained subaerial primary tuffs and there may be paleosols indicative of long periods of quiescence and weathering. These conditions prevailed at some of the Napak sites and at Songhor, Koru, and perhaps also at Maboko. The Moruorot deposits consist largely of subaerial tuffs sandwiched between lava flows. Loperot is a mixed environment with subaerial tuffs, fluviatile grits, and possibly lacustrine sediments beneath the Lower Turkana Basalt. Also somewhat different in character are the Miocene occurrences on the west and southwest sides of Lake Albert, where a thick sequence of fluvio-lacustrine sediments of Miocene to Pleistocene age filled the subsiding rift. The major occurrence of terrestrial mammals is at Karugamania, but some specimens are also found in the lower Nyamavi Beds (Lepersonne 1949; Hooijer 1963, 1970).

Although the majority of the East African faunas are traditionally regarded as "Burdigalian" because of the general resemblance to the Burdigalian faunas of Europe, Van Couvering (1972) argues convincingly that the radiometric dates show most of the sites to be, in fact, coeval with the Aquitanian of Europe. Thus the "Rusinga-like fauna" may have provided the ancestral stock for many of the characteristic Burdigalian mammals of Eurasia and North

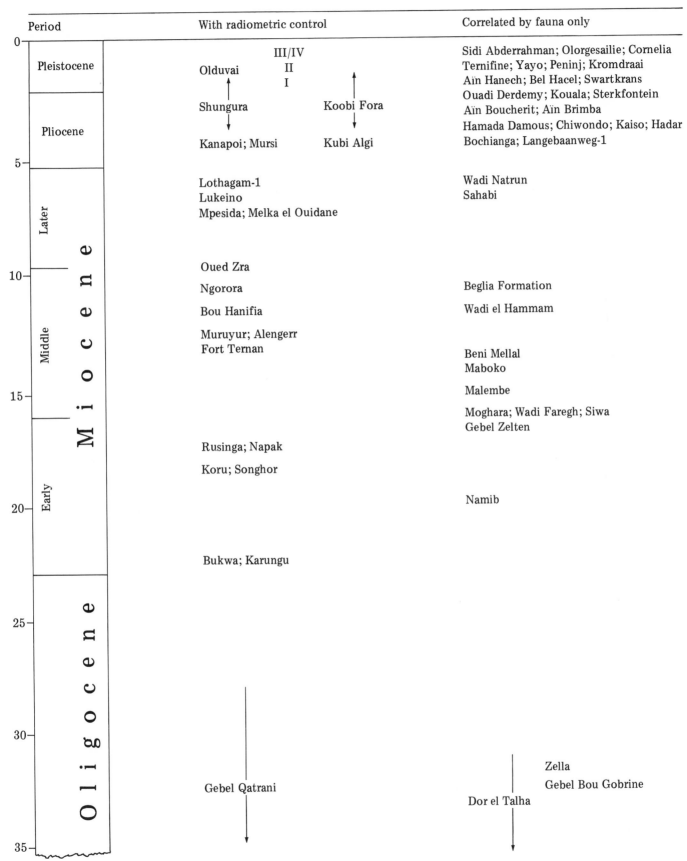

Figure 2.6 Age distribution of later Cenozoic mammal-bearing deposits in Africa.

America. On the other hand, the younger East African Miocene fauna at Fort Ternan (close to 14 m.y.) shows a substantial change from the Rusinga-like fauna and includes elements suggestive of Eurasian (especially Asian) immigrants (Gentry 1970).

Fort Ternan, formerly considered terminal Miocene, is now middle Miocene on the basis of the revised marine time scale, which places the Miocene-Pliocene boundary between 5.0 and 5.5 m.y. (Berggren 1971; Van Couvering 1972). Until less than a decade ago there was a substantial gap in the East African fossil record between Fort Ternan and the terminal Pliocene–early Pleistocene sites of Kenya, Tanzania, and Ethiopia. Recent mapping in the Kenya Rift north of Lake Hannington has disclosed a long sequence of volcanics, some 3,000 m thick, within which occur a number of sedimentary units of varying thickness (McCall, Baker, and Walsh 1967; Martyn 1967; Bishop and Chapman, 1970; Bishop et al. 1971). On the east side of the Tugen Hills successive faults step down the strata from the crest at about 2,800 m above sea level to the floor of Lake Baringo (1,000 m) while on the west side a dissected dip slope descends into the Kerio Valley. The succession rests upon the basement complex and the lower 1,800 m outcrops in the upper part of the main escarpment, capped by a useful marker, the Kabarnet Trachyte, with an age close to 7.0 m.y. The lower phonolites range back to more than 16 m.y. Twelve sedimentary formations of varying thickness occur at intervals through the lava sequence; five of them are older than the Pliocene boundary, namely the Alengerr, Muruyur (both older than 12 m.y.), Ngorora (>9.0 m.y. <12.0 m.y.), Mpesida (±7.0 m.y.), and Lukeino (±6.5 m.y.) (Bishop 1972; Bishop and Pickford 1975). A few mammals have come from the Alengerr Beds, including a rhinoceros that appears to belong to the Rusinga form *Dicerorhinus leakeyi* and not to the Fort Ternan *Paradiceros*. The Ngorora Formation is a few hundred meters thick and consists of five units of varying facies, ranging from coarse to fine (Bishop and Chapman 1970). It has a substantial fauna, including the earliest known hippopotamus, bovids somewhat similar to those of Fort Ternan, several different rhinoceroses, proboscideans, a large hyracoid, and a possible hominoid tooth; but much additional material awaits description by Pickford. The Mpesida Beds have a scanty fauna that includes *Hipparion* and also *Stegotetrabelodon orbus,* a proboscidean known otherwise only from Lothagam (Maglio 1970). The Lukeino Formation is quite extensive but has furnished only a limited mammalian fauna that, however, includes a hipparionid, gom-

phothere, bovids, hippopotamus, and a suid closely allied to the *Nyanzachoerus* from Lothagam. Although the lower part of the Lothagam sequence is most probably also late Miocene in age, this locality lies geographically west of Lake Turkana, well north of the sites near Lake Baringo, and is more conveniently considered below with the Pliocene occurrences (see figures 2.6 and 2.7).

Pliocene-Pleistocene of East Africa

In the Lake Baringo area, the later Miocene Lukeino Formation is overlain by the Kaparaina Basalts, some 450 m thick and having a radiometric age of approximately 5.4 m.y. (Pickford 1975). Younger than the basalts is an extensive complex of sedimentary beds, known as the Chemeron Formation, that occurs over a wide area west of Lake Baringo. There are three main areas, known as the Chemeron Basin, the Kipcherere Basin, and the northern extension, within which there are some differences in the details of the stratigraphy. The basal Chemeron includes bouldery torrent wash, but the higher units contain much tuffaceous material and consist of fluviatile and lacustrine clastic sediments that have been divided into five members (Martyn 1967). Petrified wood, reptiles, fish, and some mollusks have been recovered, with fish most notably present in Members 2 and 4. Mammalian fossils have been found in the basal member as well as in the two higher horizons, in Members 2 and 4, including a hominid temporal fragment from the latter (Tobias 1967). Although only a few elements of the fauna have been described, it has been fully listed (Bishop et al. 1971; Bishop 1972) and has distinct affinities with the faunas from Kanapoi and the Omo area. An upper limit of the age of the Chemeron Formation is set by an isotopic date of 2.0 m.y. from the overlying lavas.

Other sedimentary deposits also occur in the Baringo succession, of which two, the Aterir Beds and the Karmosit Beds, may be approximately coeval with part of the Chemeron Formation, although they can be dated independently (Bishop 1972). Substantially younger is the Kapthurin Formation, comprising 100 m of fluviatile and tuffaceous sediments within which mammalian fossils and Acheulian tools have been found (Fuchs 1950; Martyn 1969; M. Leakey et al. 1969). The age is younger than an underlying lava with a provisional age of 0.23 m.y. The fauna has resemblances to that of Olduvai Beds III to IV.

On the east side of Lake Baringo, in the Chemoigut Basin near Chesowanja, a hominid partial cranium was found in association with a limited

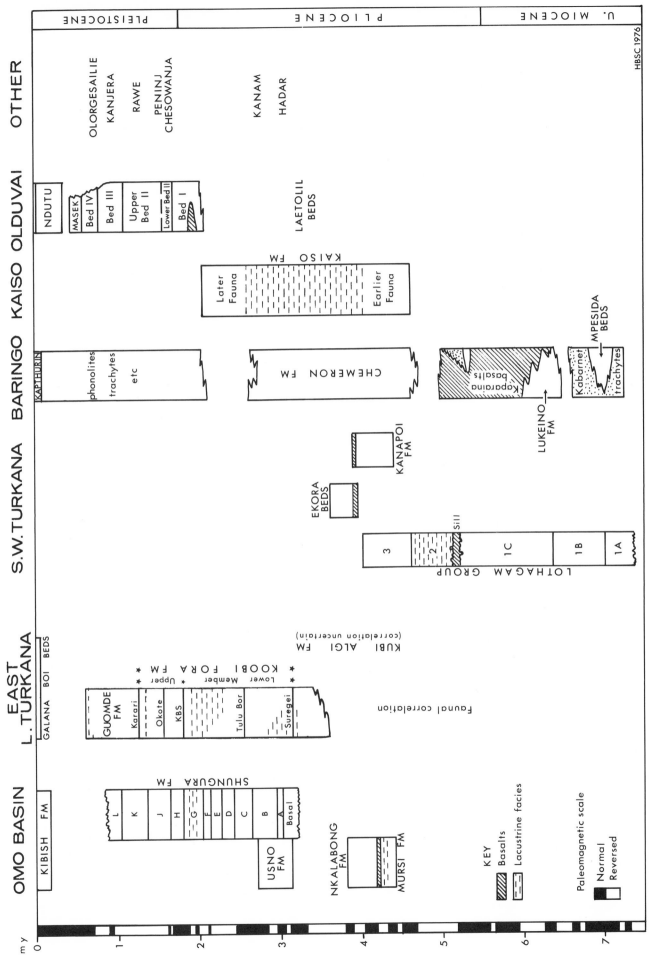

Figure 2.7 Provisional correlation of major later Miocene to Pleistocene deposits in East Africa. For the Koobi Fora Formation, the column represents a "best fit" correlation with Omo based on the faunas. The horizontal shading indicates dominantly lacustrine facies with few terrestrial fossils.

HBSC 1976

mammalian fauna (Carney et al. 1971). The fossil-bearing unit is 25 to 30 m thick and also contains stone tools recalling the Oldowan industry, while a younger unit has furnished Acheulian artifacts (Bishop, Pickford, and Hill 1975). The fauna suggests an age equivalent to some part of Bed II at Olduvai.

Some 200 km to the north of Lake Baringo in the drainage basin of the Kerio River there are three localities on the southwest side of Lake Turkana (formerly Lake Rudolf) that have yielded important collections of fossil mammals. The oldest fauna comes from Lothagam Hill, where Miocene volcanics, dated at 8.31 m.y., are overlain unconformably by a boulder conglomerate followed by a sedimentary sequence with a total thickness of about 600 m. The beds dip westward at moderately low angles and three major units are recognized (Patterson, Behrensmeyer, and Sill 1970). The lowest, Unit 1, consists of conglomerates, thinly bedded sandstones, silts, and occasional shales or tuffaceous shales, apparently representing fairly continuous and rapid deposition on an aggradational plain, perhaps deltaic; this is the major fossiliferous unit. Unit 2 comprises some 80 m of lacustrine clays and silts, almost devoid of fossil mammals. Unit 3 consists of at least 100 m of medium- to coarse-grained clastic sediments, probably of fluvial origin and containing a small number of mammalian fossils; the top is concealed. The contact between Units 1 and 2 is obscured by an intrusive basaltic sill, about 25 m thick, with an isotopic age of 3.71 m.y. As the sill postdates Unit 3, it provides a limiting age for the entire sequence, and it seems probable that the major fossiliferous unit, Lothagam-1, would be about 6.0 or 7.0 m.y. old and hence terminal Miocene (Maglio 1973). The fauna contains the most primitive known elephants (Maglio 1970, 1973), characteristic suids (Cooke and Ewer 1972), the last brachypotherine rhinoceroses (Hooijer and Patterson 1972), and primitive bovids with some resemblances to the Fort Ternan forms. The scanty fauna from Lothagam-3 resembles that of Kanapoi.

The Kanapoi locality lies 60 km south of Lothagam and has more than 70 m of gently dipping clastic sediments, the base of which is not exposed. Most of the fossil material comes from the lower part, which consists of clays, silts, and some sandstones, whereas the upper part is generally coarser. A basaltic lava caps the sediments and has a radiometric age close to 4.0 m.y. The fauna is decidedly different from that of Lothagam-1, with the appearance of many new genera and the disappearance of others (Patterson 1966; Maglio 1970; Cooke and Ewer 1972) but is essentially similar to that of Lothagam-3. At

Ekora, to the northeast of Kanapoi, another basalt is exposed and is overlain by a thin group of silty to gritty sediments, the top of which is concealed. These Ekora Beds have furnished relatively few fossils and the assemblage is like that of Kanapoi, although there are some differences in the proboscideans represented (Maglio 1970).

At the northern end of Lake Turkana there are two areas in which occur long sequences of strata covering a considerable time span, the Omo Basin to the north of the lake and the so-called "East Rudolf" or "East Turkana" region on the northeast side of the lake. In the southern Omo Basin of southwestern Ethiopia an isolated area at the foot of the Nkalabong highlands displays about 140 m of clays, silts, and sands, often mottled and stained by limonite ("Yellow Sands") together with occasional gypsiferous lenses, salt horizons, and concretionary shell beds. The sediments are capped by a basalt dated at 4.18 m.y. and the whole unit is known as the Mursi Formation (Butzer and Thurber 1969). A gravel horizon in the middle of the sediments has furnished a very limited fauna inseparable from that of Kanapoi. A number of isolated exposures of sediments and pyroclastics totaling 200 m in thickness, designated the Usno Formation (de Heinzelin and Brown 1969), are found 25 km to the east, near the confluence of the Usno stream with the Omo River. There is a thin basaltic lava intercalated near the base with a provisional age of 3.3 m.y. and a tripartite tuff 50 to 60 m below the top with a radiometric age of 2.64 m.y. Two localities in the Usno Formation, known respectively as "White Sands" and "Brown Sands," have furnished good fossil assemblages from the sandy beds and these can be correlated by magnetostratigraphy with the lower part of Member B of the Shungura Formation (Brown, pers. comm.).

The most extensive sequence outcrops on the west side of the Omo River for a distance of 60 km, although the major exposures are in the northern 20 km. These are the "Omo Beds," first worked extensively by Arambourg in 1932–33 (Arambourg 1947). They are now formally designated the Shungura Formation (Brown, de Heinzelin, and Howell 1970). The beds are tilted westward at about 10° to 25° and comprise more than 750 m of sands, silts, and clays together with tuffs that are often prominent and distinctive enough to be used as marker horizons. The marker tuffs have been assigned alphabetic designations and the whole sequence divided into members made up of the tuff and the overlying sediments; a Basal Member underlies Tuff A (figure 2.7). Radiometric ages from many of the tuffs have been combined with an excellent paleomagnetic record (Shuey, Brown, and Croes 1974) to provide a well-controlled time scale ranging from 3.1 m.y. for the Basal Member to 0.8 m.y. for the upper part of Member L (there is no member designated I, so J follows H). Within the members, cyclic lithologic sequences occur, interpreted as resulting from rhythmic fluviatile deposition and the complex elements of a river floodplain. Paleosols indicate temporary pauses in the general subsidence of the basin but the upper part of Member G is largely lacustrine. A rich and progressively changing fauna has been recovered and its stratigraphic distribution was reviewed recently by Coppens and Howell (1974; Howell and Coppens 1974). Stone tools occur in and above Member F. Lying unconformably on the earlier beds is a horizontal unit named the Kibish Formation, comprising about 100 m of mixed fluvial, deltaic, littoral, and lacustrine sediments ranging from the Holocene back to a tentative 130,000 years or more (Butzer, Brown, and Thurber 1969).

The East Turkana Plio-Pleistocene deposits extend from the Ethiopian border southward for at least 100 km on the east side of Lake Turkana and the belt varies from 10 to 40 km in width. A Miocene site exists in the region but has not yet been investigated in detail (Harris and Watkins 1974). The oldest Pliocene sediments lie in the southern part of the area and belong to the Kubi Algi Formation (Bowen and Vondra 1973). This consists of coarse clastic sediments lying unconformably on Mio-Pliocene volcanics, becoming finer upward through the 90-m thickness. Mammalian fossils are scarce but are sufficient to establish a general resemblance to that of Kanapoi and its correlatives (Maglio 1972). The upper limit of the Kubi Algi Formation is marked by laminated bentonitic tuffs and claystones, termed the Suregei Tuff Complex. This complex marks the base of the Koobi Fora Formation (Bowen and Vondra 1973), which is divided into lower and upper members at an apparent disconformity on top of a widespread tuffaceous unit known as the KBS Tuff. Because of partial isolation of the northern area around Ileret, the unit above the KBS Tuff in that area has been termed the Ileret member and is capped by the well-developed Chari Tuff. The total thickness of the Koobi Fora Formation is about 200 m and it consists of a complex of clastic sediments, ranging from claystones to conglomerates, together with molluskan limestones and other rare elements. The beds are considered to result from the interplay of fluvial, deltaic, and lacustrine environments. The formation seems in general to have been built by streams flowing from the east and northeast to form a prograding fluvio-deltaic complex, but episodes of lacustrine transgression and regression are ap-

parent. Although the tuffs are rarer than in the Omo Basin, they are important as marker horizons and some of them have been dated. The KBS Tuff was originally dated at $2.61 \pm .26$ m.y. by Fitch and Miller (1970) but the faunal evidence suggests correlation with Member F or G of the Shungura and hence a date closer to 1.8 to 1.9 m.y. (Cooke and Maglio 1972). Determinations by Curtis et al. (1975) suggested that there is more than one KBS Tuff and they report dates of $1.82 \pm .04$ and $1.60 \pm .05$ at two localities. Zircons have been dated by fission track methods at $2.44 \pm .08$ m.y. (Hurford, Gleadow, and Naeser 1976) and Fitch, Hooker, and Miller (1976) have made revised radiometric determinations giving an isochron age of 2.42 m.y. Preliminary paleomagnetic data can be interpreted as supporting this date, but the issue has not been resolved at the present time. The correlation of the faunas above the KBS Tuff are not in dispute and the East Turkana and Omo radiometric dates for this section of the record match very well. In figure 2.7 the radiometric dates have been used for the upper part but the positions of the KBS Tuff and the lower section are based on tentative faunal interpretations. Four faunal zones were originally established by Maglio (1972), but his list now needs some revision. The environment at East Turkana was conducive to excellent preservation and the fauna includes many important hominid fossils (Leakey 1970, 1975, 1976; Day 1975). Stone tools have been found in the KBS Tuff and in the upper part of the succession (Isaac, Harris, and Crader 1975). On the Ileret Ridge a 30-m thick, poorly fossiliferous unit, the Guomde Formation, overlies the Chari Tuff and consists dominantly of lacustrine molluscan limestone and dark laminated sandstones. Resting unconformably on the earlier formations in various parts of the area are thin diatomaceous siltstones termed the Galana Boi Beds, probably of late Pleistocene or Holocene age.

In the lower reaches of the Awash River in the central Afar region of Ethiopia a sequence of clays, sands, and gravels about 140 m thick constitutes the Hadar Formation (Taieb et al. 1972, 1976). These deposits have yielded substantial collections of fossils with strong resemblances to the fauna from the basal part of the Shungura Formation and have also proved to be rich in hominid remains (Johanson and Taieb 1976). A basalt flow in the lower half of the succession has been dated radiometrically as about 3.0 m.y., which accords well with the faunal evidence.

It is only during the past decade that most of the deposits described from the Baringo-Turkana Basin have become known. Until that time the only major sequence known was that of Olduvai Gorge, Tanzania, first described by Reck in 1914 and later by L. S. B. Leakey in 1951 after 20 years of work in the gorge. More intensive study since 1959 has led to new understanding of the fauna and archaeology (L. S. B. Leakey 1965; M. D. Leakey 1971), age of the deposits (Evernden and Curtis 1965; Curtis and Hay 1972), and a wealth of detail regarding the geology (Hay 1963, 1971, 1976). The deposits rest on granites and gneisses and the lowest unit recognized is a welded tuff, the Naabi Ignimbrite. Bed I overlies the ignimbrite and has a thickness in excess of 60 m. The lower part is pyroclastic and is largely covered by basalt, formerly thought to be the base of the sequence, and with a mean age of 1.96 m.y. The upper member of Bed I rests on the almost unweathered lava and comprises 10 to 40 m of tuffs and clays accumulated in and around a persistent saline lake with no outlet, so that the shoreline fluctuated greatly in response to precipitation and evaporation changes. The major fossil-bearing deposits were alternatively invaded by the lake, covered, and then exposed to dry out. The mean age of the tuffs from Bed I is 1.79 m.y. A conspicuous claystone and tuff, "Marker Bed B," terminates Bed I and is overlain by Bed II, the lower part of which is lithologically and faunistically similar to the upper member of Bed I. A zeolitized eolian tuff, designated the Lemuta Member of Bed II (Hay 1971) seems to mark a break in the sedimentation and above it sandstones are common and root casts occur in the claystones; lenticular conglomerates are widespread. Bed III usually consists of reddish-brown volcanic conglomerates, sandstones, and claystones cemented by zeolites and rather poor in fossil material. Bed IV may also be reddened but more commonly comprises grey clays, sandstones, and conglomerates, also poor in fossils. Beds III and IV together have a thickness of 20 to 50 m and represent dominantly a playa situation with stream channel deposits. Paleomagnetic studies show a change from reversed to normal polarity within Bed IV, probably representing the Brunhes-Matuyama transition and thus providing a date of close to 0.7 m.y. Bed IV is overlain by the Masek Beds (Hay 1971), consisting of 10 m of wind-worked nephelinite tuffs cemented by calcite and zeolites. As a result of major faulting after the deposition of the Masek Beds active erosion took place in the late Pleistocene and two formations, the Ndutu Beds and the Naisiusu Beds, were laid down on the plain adjacent to the gorge and remnants survive on the flanks and floor of the gorge itself (M. D. Leakey et al. 1972).

Immediately south of the Olduvai Gorge exposures along the Vogel River were named the Laetolil Beds by Kent (1941) and consist largely of fine tuffs from which a moderate fauna was described by Dietrich (1942, 1950). Recent work has yielded a number of important hominid fossils and has shown that the thickest section is as much as 130 m thick, while the fossiliferous zone has been bracketed by radiometric dates as between 3.59 and 3.77 m.y. old (M. D. Leakey et al. 1976).

To the north of Olduvai, on the western side of Lake Natron, Plio-Pleistocene lavas are overlain by a 100-m thick sedimentary sequence known as the Peninj group (Isaac 1965, 1967). The lower unit, the Humbu Formation, consists of sands, clays, and basaltic tuffs with mollusk, fish, and mammal fossils, including a hominid mandible, associated with a lower Acheulian industry. An interbedded flow of olivine basalt has normal polarity but the ages are ambiguous (Isaac and Curtis 1974). The faunal material resembles that of Bed II at Olduvai. The overlying Moinik Formation consists largely of lacustrine clays and sands grading both laterally and upward into tuffaceous shales and trachytic tuffs. Faulting terminated the depositional cycle.

In the western rift, where fossil mammal discoveries are rare, the most important finds have been in the Kaiso Formation on the east side of Lake Albert and continuing southward astride the Semliki River and the Kazinga Channel. The thick sequence in this trough includes unfossiliferous grits, conglomerates, sandstones, and occasional clays of the Kisegi Beds (Mohari Formation) passing upward through passage beds into the typical Kaiso Formation. The latter consists dominantly of clays, sandy clays, marls, micaceous silts, and fine sands within which there occur occasional ferruginous horizons from which, or just beneath which, virtually all the aquatic and vertebrate remains have come, as well as most of the Mollusca (Bishop 1965). The fossil mammals are derived primarily from two zones in the lower part of the Kaiso Formation and from a zone near the top of the formation, the faunas being termed the "earlier Kaiso" and the "later Kaiso" assemblages, respectively (Cooke and Coryndon 1970). The "earlier" assemblage suggests an age perhaps a little earlier than the typical Kanapoi fauna, whereas the "later" assemblage matches well with that of the Shungura Formation between Members C and G. The well-known *Elephas recki* occurs at Kaiso Village as part of the "later Kaiso" assemblage and belongs to the early part of Maglio's (1970) Stage 2. A gravel deposit at Kikagati on the Kagera River east of Lake Kivu (but not related to the Kaiso Formation) has yielded as its only fossil three molars of the earliest stage of *E. recki*.

Below Homa Mountain, on the south side of the

Kavirondo Gulf of Lake Victoria, Plio-Pleistocene sediments are exposed in a series of steep gullies. The oldest unit is the Kanam Beds (Kent 1941) consisting mainly of light brown lacustrine clays with interbedded fine tuffs and occasional gravels, grading laterally into shallow water and fluviatile sands and gravels. Fossil mammal remains have come from three areas, Kanam East, West, and Central, and also from a site called Kokkoth. Much of the fauna resembles that of the Kanapoi assemblage and suggests a similar age, but the collections also include more advanced elements that conflict with this view and indicate an age equivalent to the middle part of the Shungura Formation (ca 2.0 m.y.). It is possible that the collections sample different horizons and more work is required to resolve the problem. The Kanam Beds were faulted and eroded before the deposition of a younger unit of fluvio-lacustrine sediments known as the Rawe Beds (Kent 1941) comprising laminated shaly clays and fissile siltstones and sandstones, some of which have yielded fish fossils. Mammalian remains were found at several horizons and the fauna suggests broad equivalence to Bed II at Olduvai. To the northeast of Kanam another group of greenish tuffs and yellowish clays with limestone bands is known as the Kanjera Beds (Kent 1941). The fauna suggests that it is equivalent to the upper part of the Olduvai sequence, perhaps to Beds III and IV, and this is confirmed by the Acheulian hand axes from Kanjera. Two later Pleistocene lake terraces have been developed across the Kanjera Beds.

In the eastern (Kenya) rift there are a number of sites of middle to upper Pleistocene age that are important for their archaeological remains rather than their faunas. One of the most remarkable is at Olorgesailie, to the north of Lake Magadi, where hand axes litter the surface in profusion; it has been well described by Isaac (1966). The fauna is limited but resembles that of Kanjera and the top of the Olduvai sequence. Hand axes also occur in diatomites at Kariandusi in the Nakuru-Naivasha Basin and there is a complex of former lake and fluviatile sediments in this basin, originally described by Nilsson (1929) and by Leakey and Solomon (1931) but recently reconsidered by Isaac, Merrick, and Nelson (1972). Many other archaeological sites in East Africa are discussed by Cole (1963).

Pliocene-Pleistocene of North Africa

In Egypt the Pliocene is scantily represented and the only major fauna has come from the Wadi Natrun. Blanckenhorn (1901) described the section near Garet el Muluk, where some 25 to 30 m of flu-

vial and lacustrine sands and clays yielded fossils from three horizons, and assessed its age rather vaguely as Messinian to Pontian. The fauna, described mainly by Andrews (1902), Stromer (1913, 1914, 1920), and Tobien (1936), includes elements suggestive of an early Pliocene or even a terminal Miocene age, but others that seem more advanced; the site clearly needs reinvestigation. A few other sites in Egypt have furnished rare fossils, the most important of which is *Anancus osiris* from the Mena House, near Cairo (Arambourg 1946).

In the Maghreb region a number of late Pliocene or early Pleistocene sites occur but it is difficult to place them with certainty either in relation to the European or to the sub-Saharan sites. A key locality in this regard is Oued Fouarat (or Aïn el Arriss), 30 km northeast of Rabat, where marine sands with a Tortonian fauna are overlain by sands and gravels containing marine mollusks assigned to the "Moghrebian" stage, originally regarded by Choubert (1953) as equivalent to the Calabrian of the Mediterranean but now believed to be slightly earlier (Biberson 1964). Gravels in this deposit have yielded remains of the proboscideans *Anancus osiris* and *Mammuthus africanavus,* regarded by Arambourg (1970) as very characteristic of the North African "Villafranchian" fauna. At Oued Akrech, south of Rabat, a similar marine setting has furnished *M. africanavus,* together with *Hipparion libycum,* a large bovid, and a rhinoceros perhaps representing an early *Ceratotherium* (Arambourg and Choubert 1965). Substantially later is the middle Pleistocene complex at Sidi Abderrahman, near Casablanca, where marine gravels are covered by a calcareous dune that was attacked by an advancing sea, leaving against the eroded dune, and in karstic caves, a succession of both marine and terrestrial deposits. Acheulian tools are plentiful and correspond with those from the upper part of the Olduvai sequence. Human remains have been found but the associated fauna is generally poor; both *Loxodonta atlantica* and *Elephas iolensis* occur. Other archaeological sites and faunas are reviewed by Biberson (1961, 1967).

In Algeria the most important area lies 6 km northwest of St. Arnaud on the Constantine Plateau, where there is a series of marine beds of late Miocene age overlain by lacustrine and fluvio-lacustrine deposits (the St. Arnaud Lake Beds) in which fossil mammals occur at two horizons (Arambourg 1970). The lower horizon is now named Aïn Boucherit and has a fauna resembling that of Fouarat except that it includes *Equus numidicus* and *Hippopotamus,* so may be a little younger. The higher level, Aïn Hanech, is cut off from the main sequence by a fault and has a later "Villafranchian" fauna, including *Mammuthus meridionalis* and other elements that resemble the general faunas of the middle Shungura or lower Olduvai deposits. A third site in this area is at the cemetery of St. Arnaud, where sands overlying fluvio-lacustrine beds have yielded a limited fauna of Pliocene aspect, including a small equine, *Hipparion sitifense,* that also occurs at Mascara (southeast of Oran), Aïn el Hadj Baba (near Constantine), and in the earlier Kaiso assemblage in East Africa. Near Oran, in the Chelif Valley, marine beds of Astian and Calabrian type are overlain by dunes and middle Pleistocene terrestrial deposits. A few teeth of *Hipparion libycum* have been found in the supposed Calabrian. A little way to the east, at Djebel Bel Hacel, marine beds of Astian type are overlain by lacustrine marls and limestones and by fluvial conglomerates, all folded into an anticlinal structure. From these deposits some mammals have been found, including *Mammuthus meridionalis, Equus,* white rhinoceros, and several bovids, the whole fauna resembling that of Aïn Hanech and regarded as "upper Villafranchian" (Arambourg 1970). Important younger localities are Lake Karâr, to the southwest of Oran, and Palikao (or Ternifine), southwest of Mascara, both of which have yielded faunas resembling those of upper Bed II at Olduvai, although Ternifine is the earlier of the two North African sites (Arambourg 1954b, 1962).

In Tunisia the most important site is Garet (or Lac) Ichkeul, west of Bizerta, where Mio-Pliocene marine beds are overlain by a boulder bed and then by sands, grits, and shales (Arambourg 1970). The whole series is quite steeply tilted. The lower sands have furnished an excellent fauna resembling that of Fouarat and Aïn Boucherit. There is also a good pollen spectrum suggestive of some preglacial cooling (Arènes and Depape 1953). About 100 km to the southeast of Garet Ichkeul, or 45 km southeast of the city of Tunis, is a large valley known as Hamada Damous. In it are substantial exposures of estuarine marls and sands, with a dip of 7° to 9°, and vertebrate fossils occur at several levels; there are also interstratified oyster beds and correlation with Mediterranean stages may later become possible. The fauna (Coppens 1971a, b) includes *Anancus osiris* in the basal section and *Mammuthus africanavus* up to the middle of the sequence. The typical Kanapoi suid, *Nyanzachoerus jaegeri* (Coppens 1971 = *N. plicatus* Cooke and Ewer 1972) provides an interesting link with the East African assemblages. Another site in the same region, Djebel Mallah, midway between Garet Ichkeul and Hamada Damous, yields a similar fauna. There is also a third locality, Sidi Bou

Koufa, 25 km west of Hamada Damous. Coppens (1971b) correlates the lower part of the Hamada Damous section with Garet Ichkeul, the middle part with Aïn Brimba, and the upper part with Sidi Bou Koufa, still regarding the latter as upper "Villafranchian" despite the absence of *Mammuthus africanavus.* In southern Tunisia, some 5 km north of the oasis of Mannsoura, is the site of Aïn Brimba (Arambourg 1970). It lies at the foot of the tilted Cretaceous limestones that form the Gebel Tebaga and the fossils come from a complex of pink breccias, red shales, and greenish marls. The fauna is very similar to that of Garet Ichkeul.

Further to the south, in the Djourab depression northeast of Lake Chad, there are a number of localities that have yielded fossil mammals, apparently covering the whole range from the Pliocene to late Pleistocene; but the stratigraphic relations are not clear and the relative ages are assessed on the faunas (Coppens 1967). A broadly "Villafranchian" fauna was first reported from the area near Koro Toro (Abadie, Barbeau, and Coppens 1959; Coppens 1960). The oldest faunal assemblage, typified by Bochianga, Atoumanga, and Kolinga I, comes from a very fine white sand, often consolidated, underlain by clays, and overlain by diatomite. The fauna is poor but includes *Anancus osiris* and *Primelephas korotorensis,* suggesting an age roughly similar to Lothagam-1 in Kenya. A younger fauna, from Ouadi Derdemy and Koulá, comes from a greenish sandy clay or sometimes from an underlying clay. *Anancus* is present, together with *Primelephas* (at Koulá) and also both a progressive *Mammuthus africanavus* and *Elephas recki* (at Ouadi Derdemy). Although *Primelephas* is not otherwise known in this association, the occurrence of *Elephas recki* suggests an age equivalent to the lower part of the Shungura Formation in Ethiopia, and the progressive *Mammuthus africanavus* indicates an age younger than the Fouarat/Aïn Brimba assemblage but older than Aïn Hanech and Bel Hacel. This serves as a useful link between North Africa and sub-Saharan Africa. The large suid *Notochoerus* is also present, along with a hipparionid, a small hexaprotodont hippopotamus, and a sivathere. A younger fauna occurs at Yayo, typically with *Loxodonta atlantica* and a hippopotamus resembling the living form, but smaller, and it is roughly similar to the Ternifine assemblage; a problematical hominid, *Tchadanthropus uxoris,* also comes from Yayo (Coppens 1965). Younger levels are exposed around Ounianga Kebir 400 km northeast of Koro Toro, where the advanced *Elephas iolensis* has come from one horizon and *Loxodonta africana* from a higher level.

Later Pleistocene fossils occur at many other sites in the Chad Basin, indicating the existence of typical savanna elements and even forest in this area, at least intermittently, during most of the Pleistocene (Franz 1967). Many sites throughout the Sahara have faunas indicative of more humid conditions at times during the Pleistocene (Monod 1963).

Pliocene-Pleistocene of Southern Africa

In the Karonga district of northern Malawi, west of Lake Malawi, the Chiwondo Beds comprise a series of pale greenish to brown sandstones and marls with occasional calcareous horizons and vertebrate fossils that resemble elements of the East African Plio-Pleistocene assemblages (Clark, Stephens, and Coryndon 1966). The remains include crocodile and fish, cercopithecoids, a small giraffe, several bovids, hippopotamus, the typical Kanapoi suid *Nyanzachoerus jaegeri*, a *Notochoerus* (Mawby 1970), and three primitive proboscideans (Maglio 1970). The age may be a little younger than Kanapoi but is at least as old as the lower part of the Shungura Formation. Overlying the Chiwondo Beds are gravels and sands, termed the Chitimwe Beds, containing "Middle Stone Age" tools and also a fine elephant butchery site (Clark and Haynes 1970).

The oldest Pliocene fauna so far found in South Africa comes from phosphate quarries at Langebaanweg, 100 km north of Cape Town (Hendey 1970, 1973, 1976a, b). The major collections are from E Quarry, where the phosphate-bearing Varswater Formation overlies a clay of unknown age (Tankard 1975). The 20-m thick formation is divided into four units, the lowest of which (Bed 1) is a beach gravel containing phosphate rock pebbles and yielding marine invertebrates and vertebrates; Bed 2 is a medium-grade sand, mainly nonphosphatic, and contains numerous fossil vertebrates, most of them terrestrial. Bed 3a is a medium-grade phosphate sand of limited lateral extent with plentiful vertebrates, almost exclusively terrestrial. Bed 3b is a somewhat coarser phosphatic sand with few fossils. Overlying the Varswater Formation is from 2 to 40 m of unfossiliferous sand, largely eolian in character. The rich fauna, which has recently been listed by Hendey (1973), represents a remarkable cross-section of vertebrates, including sharks, skates, rays and fish, tortoises, penguins and many other birds, a monachine seal, and numerous terrestrial mammals. Carnivores are more common than usual and include the Eurasiatic *Percrocuta* and *Dinofelis* as well as the only remains of bear known south of the Sahara (*Agriotherium*). There are also several bovids of Eurasiatic aspect, like those of Lothagam

and Kanapoi, with which there are distinct faunal resemblances. Hendey (1976b) has also described the first fossil peccary to be found in Africa. The large suid *Nyanzachoerus* is represented by a variety close to the typical Kanapoi species *N. pattersoni*. The rhinoceros *Ceratotherium praecox* is present and is known from Lothagam-1, the Chemeron Formation, the Mpesida Beds, and the Mursi Formation in East Africa (Hooijer 1972, 1973). An *Hipparion* is found and has been referred to the Kaiso species *H. albertense;* it is not *"Stylohipparion."* A gomphothere is present and also *Mammuthus subplanifrons,* considered by Maglio and Hendey (1970) to be more primitive than the typical Vaal River material. The faunal resemblances thus suggest an age close to that of Kanapoi (i.e., 4.0 m.y.), but Langebaanweg might well be slightly older (Hendey 1970). Interest also attaches to the presence of remains of several small mammals. The overall environment involves interaction between a marine and fluviatile situation and the faunal assemblage implies grassland and riverine woodland. Curiously, both hippopotamus and crocodile are absent. It is possible, although speculative, that an assemblage of material from Klein Zee, near the coast a short distance south of the Orange River (Stromer 1932; Patterson 1965) may be of roughly comparable age.

Because of the fossil hominid remains that they contain, great interest attaches to the so-called "cave breccias" of the Transvaal and northern Cape Province. The first discovery was of *Australopithecus africanus* in the Buxton Lime Quarry at Taung, Cape Province, 130 km north of Kimberley (Dart 1925); but the specimen is that of a child and the first adult was found a decade later at Sterkfontein, 50 km west of Johannesburg, Transvaal (Broom 1936). Further discoveries were made at Kromdraai, 2 km east of Sterkfontein, in 1938, and at Swartkrans, 2 km west of Sterkfontein, in 1947. A complex of small pockets of varying ages occurs at Bolt's Farm, 1 km south of Swartkrans. Another important locality is Makapansgat, in the northern Transvaal, 250 km northeast of Johannesburg.

The geology of the Taung deposit is essentially different from that of the Transvaal cave breccias and has been described in some detail by Peabody (1954). The Harts River, a tributary of the Vaal River, flows on one side of a broad valley bounded on the west by a steep escarpment, 30 to 100 m high, formed of Precambrian dolomite limestones. Deltalike aprons have been built out as the result of evaporation of lime-charged water seeping from the cliff. Peabody recognized two major and two minor calc tufas at the Buxton Quarries, each partially eroded before the deposition of the next, partly overlapping, carapace of secondary limestone. Fissures and other openings developed in the tufas at various stages and were filled by sandy material, sometimes containing bone or stone implements, and then firmly cemented into a breccia. The *Australopithecus* skull and the rather scanty fossil material found close to it came from fissure fillings within the oldest (Thabaseek) tufa carapace and Peabody considered the fillings to predate the second (Norlim) carapace. After restudy of the site and the breccia, Butzer (1974b) suggests that the skull was penecontemporary with the second (Norlim) stage of tufa deposition.

The Transvaal cave breccias comprise cemented deposits laid down in subsurface caverns or enlarged fissures that were formed by solution in gently dipping Precambrian dolomitic limestones. The process of formation has been considered fully by Brain (1958). While a cavern is cut off from the surface but clear of the ground water table, deposition of flowstone and dipstone occurs, sometimes contaminated by insoluble products resulting from solution of the parent limestone; the result is a white, grey, or banded travertine. As soon as an opening to the surface develops, external debris enters at a slow rate and contaminates the secondary limestone deposit, producing a "Phase 1" breccia that may be rich in bones. With enlargement of the surface opening, external soil is introduced and the deposit soon changes to a clastic one with a calcareous cement, or "Phase 2" breccia. Although not as rich in bone as some of the Phase 1 breccias, it is the brownish or pinkish Phase 2 deposits that have been the main source of the tens of thousands of fossils recovered from these sites. The upper parts of the cave filling are often full of fragments or blocks of dolomitic limestone resulting from collapse of the edge of the opening or of the cave roof itself. Subsequent to their primary filling and cementation, the breccias themselves may be subject to erosion and cavities or fissures in them may be filled by later material. Undermining of the floor may result in collapse during filling, so the depositional history can be complex and difficult to interpret. The early breccia deposits have now lost most or all of their original roofs and the firmly cemented pinkish-brown breccia, originally formed at depth, is now exposed on the surface. The age of any particular deposit is dependent on the date of opening to the surface, which determines the start of its period of sampling of the environment. Correlation between sites and relative dating is dependent at present solely on the faunal material.

Approximately 170 mammalian species are repre-

sented in the cave breccias and there are a few reptiles and bird remains. Various analyses of the faunas have shown that the assemblages from Makapansgat and Sterkfontein have much in common and differ from the assemblages from Swartkrans and the Kromdraai "A" faunal site[2] (Ewer 1957; Wells 1962; Ewer and Cooke 1964). These are regarded as representing two faunal "stages" (Wells 1962), now termed the "Sterkfontein Faunal Span" and "Swartkrans Faunal Span" respectively (Cooke 1967). The fauna of the Sterkfontein Extension Site, where stone implements occur, is placed in the Swartkrans Faunal Span (Wells 1962; Cooke 1967); Vrba (1974) confirms the separation and also shows that a still younger breccia occurs at Sterkfontein. Wells (1969) has argued that the Kromdraai "A" fauna should be made an intermediate unit or be placed in the Vaal-Cornelia Faunal Span; this is accepted in the accompanying correlation chart (table 2.1). (The proposal by Hendey [1974] for establishing mammal "ages" for southern Africa is considered premature.) The position of Taung in relation to the Transvaal sites is difficult to assess both on account of the scantiness of the Taung fauna and the existence of some species at Taung that are not present at the other sites, probably because of ecological differences. Of the 20 species at Taung that do occur at other sites, more than half are present at Makapansgat and Sterkfontein and less than half at Swartkrans. Taung is thus tentatively placed in an intermediate position and it seems very improbable that it is younger than Swartkrans. Correlation between the South African cave breccias and the East African sequence is even more difficult, but at present it seems highly probable that both Sterkfontein and Makapansgat are older than Olduvai Bed I and Swartkrans is roughly coeval with Bed I and at least part of Bed II. Further data are required before correlation can be effective.

The Vaal River drains a large area in the southern Transvaal and Orange Free State, together with a small area in the northeastern Cape Province. Terrace gravels, at altitudes up to 90 m above the river and at distances up to 10 km from it, have been worked for diamonds for more than a century (du Toit 1907). The area was studied in some detail by Söhnge, Visser, and van Riet Lowe (1937), who dis-

2 The Kromdraai australopithecine skull came from a nearby site (Kromdraai "B") with a scanty fauna, not necessarily of the same age. At Swartkrans there are two breccias, an earlier "pink" breccia and a secondary "brown" breccia that may be considerably later (Brain, pers. comm.). It is possible that the analyses on available identifications include, in error, some of the later elements.

Table 2.1 Tentative grouping of main Plio-Pleistocene Deposits in Southern Africa.

Faunal Unit	Vaal River Basin	Cave deposits	Open sites	Central Africa
"Recent"	Minor erosion; soil formation Alluviation of flood terraces Dissection of earlier fill	Numerous caves	Surface	Nachikufu, etc.
Florisbad faunal span	Vertizol development Silt and loam deposition Alluviation of tributary sands and gravels Silts in Vaal floodplain Dissection of earlier fill	Wonderwerk, etc. Cave of Hearths	Vlakkraal Florisbad	Mumbwa cave Broken Hill cave Twin Rivers breccia Chelmer
Cornelia faunal span	Calcereous paleosol Alluvium in valleys Low-level terrace gravels "Younger Gravels" Bedrock dissection	Kromdraai	Hopefield Cornelia	Younger terrace gravels

Swartkrans
faunal span
	Local calcification
	Alluviation and reworking
	  of high-level "Older
	  Gravels"
		Swartkrans
		Sterkfontein
		Extension
				Older terrace gravels

Sterkfontein
faunal span
	Aggradation of original
	  high-level "Older
	  Gravels" in three stages
		Sterkfontein
		Makapansgat
				Oldest terrace gravels

Langebaanweg faunal span	Bedrock dissection of major valleys ?Upper Miocene/Pliocene		Langebaan phosphate sands	Chiwondo Beds

? Taung Bolt's Farm Complex (spanning the Cornelia through Sterkfontein faunal spans in the Cave deposits / Open sites region)

tinguished three sets of deposits, designated "older gravels," "younger gravels," and "youngest gravels," which they regarded as reflecting climatic changes. Their conclusions have been discussed or criticized by other workers but the basic stratigraphy is sound and the whole situation has been reconsidered recently by Butzer et al. (1973), whose interpretation is generally followed here (table 2.1). The older gravels lie essentially on three erosional platforms, the highest of which is approximately 90 m above present river level. The original gravels were rich in boulders of diabase but these were decomposed and the weathered products often removed by extensive eluviation that has resulted in considerable redistribution of the resistant elements as "potato gravels" with an enriched diamond content (Cooke 1947). Although only one fossil has ever been recorded as coming from these gravels, there are a number of archaic elements in the younger gravels that are believed to have been derived from them.

Following bedrock dissection of more than 10 m, the younger gravels were deposited on the flanks of the present river and three units can be recognized, styled from the base upward, as A, B, and C. Nearly all the fossil material comes from these deposits (Cooke 1949; Wells 1964), principally from units B and C, and the gravels also yield middle to late Acheulian tools in considerable abundance. The fill of younger gravels was then dissected and subsequently covered disconformably by "calcified sands," the age of which is not at present clear. Some of the fossils have come from the base of the sands. Tributary valleys are filled by or have flanking exposures of a series of alluvial silts, sands, and gravels to which the name Riverton Formation has been given (Butzer et al. 1973). Middle Stone Age tools occur in these beds and Later Stone Age tools on the surface. Fossil remains are rare.

Within the Vaal River Basin are a number of other important deposits of which the oldest are the

Cornelia Beds, first described by Van Hoepen in 1930 and recently considered in detail by Butzer (1974a). The deposits lie in the basin of the Skoonspruit, a tributary of the Vaal River and situated some 50 km southwest of Standerton. The total thickness of the type site is approximately 15 to 20 m and consists of clays, silts, and loam overlying a thin basal gravel or rubble layer from which have come weathered flakes and crude hand axes. The major fossiliferous unit, 1 to 8 m thick, lies upon this basal part of the sequence and is covered by an almost equal thickness of barren clays and silts, together forming Member 1; a second unit (Member 2) was deposited after an intervening erosion interval and is barren. The artifacts are generally crude and the assemblage is comparable with industries of upper Acheulian affinities from Olduvai Gorge, Olorgesailie, and other sites (Clark 1974). The fauna is equivalent to part of the younger gravels assemblage but also has remarkable links with that from the upper part (Beds III/IV) of the Olduvai sequence (Cooke 1974). This fauna also resembles that from Elandsfontein (Hopefield) in the southwestern Cape Province, where an old deposit of swampy character was exposed by the local removal of the sandy cover by wind action (Singer 1954; Mabbutt 1956; Butzer 1973). Acheulian tools occur with the fossils at Hopefield and include a hominid jaw and partial cranium (Singer and Wymer 1968).

Faunal assemblages clearly younger than that of Cornelia have been found at many sites in southern Africa and include only a few elements that are now extinct. Notable are the thermal springs at Florisbad and Vlakkraal in which Middle Stone Age artifacts occur together with fossil bones and teeth. The Florisbad site yielded a human skull and the lower peaty layer is older than 48,000 years B.P. The Vlakkraal fauna (Wells, Cooke, and Malan 1941) is very similar to that of Florisbad (see Cooke 1963) and forms the basis for a "Florisbad-Vlakkraal Faunal Span" following a "Vaal-Cornelia Faunal Span" (Cooke 1967). A deposit at Chelmer in southern Rhodesia is also placed here and it is probable that the Broken Hill site in Zambia is of roughly similar age. Numerous cave deposits containing some extinct elements are widespread in association with Middle Stone Age artifacts, whereas the fauna associated with Later Stone Age is typically one of living species. Klein (1974) has summarized the terminal Pleistocene extinctions in the southern Cape Province and this pattern may be characteristic of a wider area.

Intercorrelations between South Africa and East Africa and between the latter and North Africa are at present possible only in a broad sense and the principal inferences have been shown in figure 2.6. However, studies are advancing rapidly at the present time and more links between the major regions are coming to light, so it should not be many years before a much firmer framework of correlation is established. The overview of mammal taxa presented in this volume is an important step in this direction.

References

Abadie, J.; Barbeau, J.; and Coppens, Y. 1959. Une faune de vertébrés villafranchiens au Tchad. *C. R. Hebd. Séanc. Acad. Sci.* 248:3328–3330

Ameur, R. C.; Jaeger, J.-J.; and Michaux, J. 1976. Radiometric age of early Hipparion fauna in northwest Africa. *Nature* 261:38–39.

Andrews, C. W. 1902. Note on a Pliocene vertebrate fauna from the Wadi Natrun, Egypt. *Geol. Mag.* (4) 9:433–439.

_____. 1906. *A descriptive catalogue of the Tertiary Vertebrata of the Fayûm, Egypt.* London: British Museum.

_____. 1911. On a new species of Dinotherium (*Dinotherium hobleyi*) from British East Africa. *Proc. Zool. Soc. London* 1911:943–945.

Arambourg, C. 1946. *Anancus osiris,* un mastodonte nouveau du Pliocène inférieur d'Egypte. *Bull. Soc. Géol. Fr.* (5) 15:479–495.

_____. 1947. Contribution a l'étude géologique et paléontologique du bassin du Lac Rodolphe et de la Basse vallée de l'Omo. Deuxième partie, Paléontologie. *Mission Scient. Omo. Géol.-Anthrop.* 1:479–495.

_____. 1951. Observations sur les couches a *Hipparion* de la vallée de l'Oued el Hammam (Algérie) et sur l'époque d'apparition de la faune de vertébrés dite "Pontienne." *C. R. Hebd. Séanc. Acad. Sci.* 232 (26):2464–2466.

_____. 1952. Note preliminaire sur quelques éléphants fossiles de Berberie. *Bull. Mus. Hist. Nat., Paris,* Sér. 2, (24):407–418.

_____. 1954a. La faune a Hipparion de l'Oued el Hammam (Algérie). *Int. Géol. Congr.,* 19th Sess. 21:294–302.

_____. 1954b. L'hominien fossile de Ternifine (Algérie). *C. R. Hebd. Séanc. Acad. Sci.* 239 (15):893–895.

_____. 1962. Les faunes mammalogique du Pléistocène circummediterranean. *Quaternaria* 6:97–109.

_____. 1970. Les vertébrés du Pléistocène de l'Afrique du Nord. *Mém. Mus. Natl. Hist. Nat.,* Sér. 7, 10:1–126.

Arambourg, C., and Burollet, P. F. 1962. Restes de vertébrés oligocènes en Tunisie centrale. *C. R. Somm. Séanc. Géol. Fr.* 2:42–43.

Arambourg, C., and Choubert, G. 1965. Les faunes de mammifères de l'étage Moghrebien du Maroc occidental. *Notes Serv. Géol. Maroc* 185:29–33.

Arambourg, C., and Magnier, P. 1961. Gisements de vertébrés dans le bassin tertiare de Syrte (Libye). *C. R. Hebd. Séanc. Acad. Sci.* 252 (8):1181–1183.

Arènes, J., and Depape, G. 1953. Étude paléobotanique. In C. Arambourg. Contribution a l'étude des flores fossiles quaternaire de l'Afrique du Nord. *Arch. Mus. Natl. Hist. Nat. Paris,* Sér. 7, 2:7–85.

Baker, B. H.; Mohr, P. A.; and Williams, L. A. J. 1972. Geology of the eastern Rift System. *Geol. Soc. Amer., Spec. Pap.* 136:1–67.

Baker, B. H.; Williams, L. A. J.; Miller, J. A.; and Fitch, F. J. 1971. Sequence and geochronology of the Kenya Rift volcanics. *Tectonophysics* 11:191–215.

Baker, B. H., and Wohlenburg, J. 1971. Structure and evolution of the Kenya Rift Valley. *Nature* 229:538–542.

Beadnell, H. J. L. 1901a. Découvertes géologique récentes dans la vallée du Nil et le désert Libyen. *VIII Int. Geol. Congr., Paris* 2:839–866.

―――. 1901b. The Fayûm depression: a preliminary notice of the geology of a district in Egypt containing a new palaeogene vertebrate fauna. *Geol. Mag.* (4) 8:540–546.

―――. 1902. *A preliminary note on* Arsinoitherium zitteli *Beadn. from the Upper Eocene strata of Egypt.* Cairo: Survey Dept., Public Works Ministry.

―――. 1905. *The topography and geology of the Fayûm Province of Egypt.* Cairo: Survey Dept., Public Works Ministry.

Berggren, W. A. 1971. Tertiary boundaries and correlations. In B. M. Funnell, and W. R. Riedel, eds. *The micropalaeontology of oceans.* Cambridge: Cambridge University Press, pp. 693–809.

Biberson, P. 1961. Le cadre paléogéographique de la préhistoire du Maroc atlantique. *Publ. Serv. Antiq. Maroc* 16:1–235.

―――. 1964. La place des hommes du Paléolithique Marocain dans la chronologie du Pléistocéne atlantique. *Anthropologie* 68 (5–6):475–526.

―――. 1967. Some aspects of the lower Palaeolithic of northwest Africa. In W. W. Bishop, and J. D. Clark, eds. *Background to evolution in Africa.* Chicago: University of Chicago Press, pp. 447–475.

Biely, A.; Rakus, M.; Robinson, P.; and Salaj, J. 1972. Essai de correlation des formations Miocène au sud de la Dorsale Tunisienne. *Notes Serv. Géol. Tunis* 38 (7):73–92.

Bishop, W. W. 1963. The later Tertiary and Pleistocene in eastern equatorial Africa. In F. C. Howell, and F. Bourlière, eds. *African ecology and human evolution.* Chicago: University of Chicago Press, pp. 246–275.

―――. 1965. Quaternary geology and geomorphology in the Albertine Rift Valley, Uganda. *Geol. Soc. Am. Spec. Pap.* 84:293–321.

―――. 1967. The later Tertiary in East Africa—volcanics, sediments, and faunal inventory. In W. W. Bishop, and J. D. Clark, eds. *Background to evolution in Africa.* Chicago: University of Chicago Press, pp. 31–54.

―――. 1968. The evolution of fossil environments in East Africa. *Trans. Leic. Lit. Phil. Soc.* 62:22–44.

―――. 1972. Stratigraphic succession "versus" calibration in East Africa. In W. W. Bishop, and J. A. Miller, eds. *Calibration of hominoid evolution.* Edinburgh: Scottish Academic Press, pp. 219–246.

Bishop, W. W. and Chapman, G. R. 1970. Early Pliocene sediments and fossils from the northern Kenya Rift Valley. *Nature* 226:914–918.

Bishop, W. W.; Chapman, G. R.; Hill, A.; and Miller, J. A. 1971. A succession of Cainozoic vertebrate assemblages from the northern Kenya Rift Valley. *Nature* 233:389–394.

Bishop, W. W.; Miller, J. A.; and Fitch, F. J. 1969. New potassium-argon age determinations relevant to the Miocene fossil mammalian sequence in East Africa. *Am. J. Sci.* 267 (6):669–699.

Bishop, W. W., and Pickford, M. H. L. 1975. Geology, fauna and palaeoenvironments of the Ngorora Formation, Kenya Rift Valley. *Nature* 254:185–192.

Bishop, W. W.; Pickford, M. H. L.; and Hill, A. 1975. New evidence regarding the Quaternary geology, archaeology and hominids of Chesowanja, Kenya. *Nature* 258:204–208.

Bishop, W. W., and Trendall, A. F. 1967. Erosion surfaces, tectonics and volcanic activity in Uganda. *Quart. J. Geol. Soc. London* 122:385–420.

Black, C. C. 1972. A new species of *Merycopotamus* (Artiodactyla: Anthracotheriidae) from the late Miocene of Tunisia. *Notes Serv. Géol. Tunis* 37:5–39.

Blanckenhorn, M. 1901. Neues zur Geologie und Paläontologie Aegyptens—IV. Das Pliozän—und Quartärzeitalter in Aegypten ausschliesslich des Rothen Meergebietes. *Zeits. Deutsch. Geol. Ges.* 53:307–502.

Bowen, B. E., and Vondra, C. F. 1973. Stratigraphical relationships of the Plio-Pleistocene deposits, East Rudolf, Kenya. *Nature* 242:391–393.

Brain, C. K. 1958. The Transvaal ape-man-bearing cave deposits. *Transv. Mus. Mem.* 11:1–131.

Brock, A., and Isaac, G. Ll. 1974. Palaeomagnetic stratigraphy and chronology of hominid-bearing sediments east of Lake Rudolf, Kenya. *Nature* 247:334–348.

Broom, R. 1936. A new fossil anthropoid skull from South Africa. *Nature* 138:486–488.

Brown, F. H.; Heinzelin, J. de; and Howell, F. C. 1970. Pliocene/Pleistocene formations in the lower Omo Basin, southern Ethiopia. *Quaternaria* 13:247–268.

Butzer, K. W. 1973. Re-evaluation of the geology of the Elandsfontein (Hopefield) site, southwestern Cape, South Africa. *S. Afr. J. Sci.* 69:234–238.

―――. 1974a. Geology of the Cornelia Beds, northwestern Orange Free State. *Nas. Mus. Bloemfontein Mem.* 9:7–32.

―――. 1974b. Paleoecology of South African australopithecines: Taung revisited. *Curr. Anthrop.* 15:367–382; 413–416.

Butzer, K.; Brown, F. H.; and Thurber, D. L. 1969. Horizontal sediments of the lower Omo Valley: The Kibish Formation. *Quaternaria* 11:15–29.

Butzer, K.; Helgren, D. M.; Fock, G. J.; and Stuckenrath, R. 1973. Alluvial terraces of the lower Vaal River, South Africa: a reappraisal and reinvestigation. *J. Geol.* 81:341–362.

Butzer, K. W., and Thurber, D. L. 1969. Some late Cenozoic sedimentary formations of the lower Omo Basin. *Nature* 222:1132–1143.

Carney, J.; Hill, A.; Miller, J. A.; and Walker, A. 1971. Late australopithecine from Baringo District, Kenya. *Nature* 230:509–514.

Choubert, G. 1953. Les rapports entre les formations marines et continentale quaternaries. *Actes 4 Internat. Congr. INQUA* (Rome-Pisa):576–590.

Choubert, G., and Ennouchi, E. 1946. Premières preuves paléontologique de la presence du Pontien au Maroc. *C. R. Soc. Géol. Fr.* 1946:207–208.

Clark, J. D. 1974. The stone artefacts from Cornelia, O.F.S., South Africa. *Nas. Mus. Bloemfontein Mem.* 9: 33–61.

Clark, J. D., and Haynes, C. V. 1970. An elephant butchery site at Mwanganda's village, Karonga, Malawi, and its relevance for Palaeolithic archaeology. *World Archaeol.* I (3):390–411.

Clark, J. D.; Stephens, E. A.; and Coryndon, S. C. 1966. Pleistocene fossiliferous lake beds of the Malawi (Nyasa) Rift: a preliminary report. *Am. Anthrop.* 68 (2):46–87.

Cole, S. 1963. *The Prehistory of East Africa.* New York: Macmillan and Co.

Cooke, H. B. S. 1947. The development of the Vaal River and its deposits. *Trans. Geol. Soc. S. Afr.* 49:243–262.

———. 1949. Fossil mammals of the Vaal River deposits. *Mem. Geol. Surv. S. Afr.* 35 (3):1–117.

———. 1958. Observations relating to Quaternary environments in East and southern Africa. *Trans. Geol. Soc. S. Afr.* 60 (Annexure):1–73.

———. 1963. Pleistocene mammal faunas of Africa, with particular reference to southern Africa. In F. C. Howell, and F. Bourlière, eds. *African ecology and human evolution.* New York: Viking Fund Publications in Anthrop. No. 36:65–116.

———. 1967. The Pleistocene sequence in South Africa and problems of correlation. In W. W. Bishop, and J. D. Clark, eds. *Background to evolution in Africa.* Chicago: University of Chicago Press. Pp. 175–184.

———. 1974. The fossil mammals of Cornelia, O.F.S., South Africa. *Nas. Mus., Bloemfontein Mem.* 9:63–84.

Cooke, H. B. S., and Coryndon, S. C. 1970. Pleistocene mammals from the Kaiso Formation and other related deposits in Uganda. In L. S. B. Leakey and R. J. G. Savage, eds., *Fossil Vertebrates of Africa,* Edinburgh: Academic Press, vol. 2, pp. 107–224.

Cooke, H. B. S., and Ewer, R. F. 1972. Fossil Suidae from Kanapoi and Lothagam, northwestern Kenya. *Bull. Mus. Comp. Zool., Harvard* 143 (3):149–295.

Cooke, H. B. S., and Maglio, V. J. 1972. Plio-Pleistocene stratigraphy in East Africa in relation to proboscidean and suid evolution. In W. W. Bishop, and J. A. Miller, eds. *Calibration of hominoid evolution.* Edinburgh: Scottish Academic Press, pp. 303–329.

Coppens, Y. 1960. Le Quaternaire fossilifère de Koro-Toro (Tchad). Résultats d'une première mission. *C. R. Hebd. Seanc. Acad. Sci.* 251 (21):2385–2386.

———. 1965. L'hominien du Tchad. *C. R. Hebd. Séanc. Acad. Sci.* 260:2869–2871.

———. 1967. Les faunes de vertébrés Quaternaires du Tchad. In W. W. Bishop, and J. D. Clark, eds. *Background to evolution in Africa.* Chicago: University of Chicago Press, pp. 89–99.

———. 1971a. Une nouvelle éspece de suidé du Villafranchien de Tunisie, *Nyanzachoerus jaegeri* sp. nov. *C. R. Hebd. Séanc. Acad. Sci.*Sér. D, 272 (26):3264–3267.

———. 1971b. Les vertébrés Villafranchiens de Tunisie: gisements nouveaux, signification. *C. R. Hebd. Séanc. Acad. Sci.,* Sér. D, 273 (1):51–54.

Coppens, Y., and Howell, F. C. 1974. Les faunes mammifères fossiles des formations Plio-Pleistocene de l'Omo en Ethiopie (Proboscidea, Perissodactyla, Artiodactyla). *C. R. Hebd. Séanc. Acad. Sci.* Sér. D, 278:2275–2278.

Crompton, A. W. 1964. A preliminary description of a new mammal from the Upper Triassic of South Africa. *Proc. Zool. Soc. Lond.* 142:441–454.

———. 1974. The dentitions and relationships of the southern African Triassic mammals, *Erythrotherium parringtoni* and *Megazostrodon rudnerae. Bull. Brit. Mus. Nat. Hist. (Geol.)* 24 (7):400–437.

Crompton, A. W., and Jenkins, F. A. 1968. Molar occlusion in late Triassic mammals. *Biol. Rev.* 43:427–458.

Curtis, G. H.; Drake, T.; Cerling, T. E.; Cerling, B. L.; and Hampel, J. H. 1975. Age of KBS Tuff in Koobi Fora Formation, East Rudolf, Kenya. *Nature* 258:395–398.

Curtis, G. H., and Hay, R. L. 1972. Further geological studies and potassium-argon dating at Olduvai Gorge and Ngorongoro Crater. In W. W. Bishop, and J. A. Miller, eds. *Calibration of hominoid evolution.* Edinburgh: Scottish Academic Press, pp. 289–302.

Dames, W. 1883. Ueber eine Tertiäre Wirbeltier Fauna von der westlichen Insel des Birket el Qurun in Fayûm (Aegypten). *S. B. Preuss. Akad. Wiss.,* Berlin 1883:129–153.

———. 1894. Ueber Zeuglodonten aus Aegypten und die Beziehungen der Archaeoceten zu den übrigen Cetaceen. *Paläont. Abh., Jena (N.F.)* 1 (5):1–36.

Dart, R. A. 1925. *Australopithecus africanus:* the man-ape of South Africa. *Nature* 115:195–199.

Day, M. H. 1975. Hominid postcranial remains from the East Rudolf succession. In Y. Coppens, F. C. Howell, G. LL. Isaac, and R. E. F. Leakey, eds. *Earliest man and environments in the Lake Rudolf Basin.* Chicago: University of Chicago Press, pp. 507–521.

Dietrich, W. O. 1927. *Brancatherulum* n.g.—ein Proplacentalier aus dem obersten Jura des Tendaguru in Deutsch-Ostafrika. *Cbl. Min. Geol. Paläont.* 1927B:423–426.

———. 1942. Ältestquartäre Säugetiere aus der südlichen Serengeti, Deutsch-Ostafrika. *Paläontogr.* 94A:43–133.

———. 1950. Fossile Antilopen und Rinder Äquatorial afrikas (Material der Kohl-Larsen'schen Expeditionen). *Paläontogr.* 99A:1–62.

Dietz, R. S., and Holden, J. C. 1970. Reconstruction of Pangaea: breakup and dispersion of continents, Permian to Recent. *J. Geophys. Res.* 75:4939–4956.

du Toit, A. L. 1907. Geological survey of the eastern part of Griqualand West. *Geol. Comm. Cape of Good Hope: 11th Ann. Rpt.* (1906):87–176.

Evernden, J. F., and Curtis, G. H. 1965. The potassium-argon dating of late Cenozoic rocks in East Africa and Italy. *Curr. Anthrop.* 6 (4):343–364 (with comments pp. 364–385).

Ewer, R. F. 1957. Faunal evidence on the dating of the Australopithecinae. *Proc. 3rd Pan-Afr. Congr. Prehist.* (Livingstone, 1955):135–142.

Ewer, R. F., and Cooke, H. B. S. 1964. The Pleistocene mammals of southern Africa. In D. H. S. Davis, ed. *Ecological studies in southern Africa. Monographiae Biologicae* 14:35–48.

Fitch, F. J.; Hooker, P. J.; and Miller, J. A. 1976. $^{40}Ar/^{39}Ar$ dating of the KBS Tuff in Koobi Fora Formation, East Rudolf, Kenya. *Nature* 263:740–744.

Fitch, F. J., and Miller, J. A. 1970. Radiometric age determinations of Lake Rudolf artefact sites. *Nature* 226: 226–228.

Forstén, A. M. 1972. *Hipparion primigenium* from southern Tunisia. *Notes Serv. Géol. Maroc* 5 (1):7–28.

Franz, H. 1967. On the stratigraphy and evolution of climate in the Chad Basin during the Quaternary. In W. W. Bishop, and J. D. Clark, eds. *Background to evolution in Africa.* Chicago: University of Chicago Press, pp. 273–282.

Fuchs, V. E. 1950. Pleistocene events in the Baringo Basin. *Geol. Mag.* 87 (3):148–174.

Gentry, A. W. 1970. The Bovidae (Mammalia) of the Fort Ternan fossil fauna. *Fossil Vertebrates of Africa* 2: 243–323.

Gorodiski, A., and Lavocat, R. 1953. Première découverte de Mammifères dans le Tertiare (Lutétien) du Sénégal. *C. R. Somm. Géol. Fr.* 15:314–316.

Greenwood, P. H. 1972. Fish fossils from the late Miocene of Tunisia. *Notes Serv. Géol. Tunis* 37:41–72.

Harris, J. M., and Watkins, R. 1974. New early Miocene vertebrate locality near Lake Rudolf, Kenya. *Nature* 252:576–577.

Hay, R. L. 1963. Stratigraphy of Beds I through IV, Olduvai Gorge, Tanganyika. *Science* 139:829–833.

———. 1971. Geologic background of Beds I and II: stratigraphic summary. In M. D. Leakey, *Olduvai Gorge,* volume 3: *Excavation in Beds I and II, 1960–1963.* Cambridge: Cambridge University Press.

———. 1976. *Geology of the Olduvai Gorge.* Berkeley: University of California Press.

Heinzelin, J. de, and Brown, F. H. 1969. Some early Pleistocene deposits of the lower Omo Valley: the Usno Formation. *Quaternaria* 11:32–46.

Hendey, Q. B. 1970. The age of the fossiliferous deposits at Langebaanweg, Cape Province. *Ann. S. Afr. Mus.* 56 (3):119–131.

———. 1973. Fossil occurrences at Langebaanweg, Cape Province. *Nature* 244:13–14.

———. 1974. Faunal dating of the late Cenozoic of southern Africa, with special reference to the carnivores. *Quatern. Res.* 4:149–161.

———. 1976a. The Pliocene fossil occurrences at "E" Quarry, Langebaanweg, South Africa. *Ann. S. Afr. Mus.* 69 (9):215–247.

———. 1976b. Fossil peccary from the Pliocene of South Africa. *Science* 192:787–789.

Hoepen, E. C. N. van 1930. Fossiele perde van Cornelia, O.V.S. *Paleont. Navors. Nas. Mus. Bloemfontein* 2 (2):1–24.

Hooijer, D. A. 1963. Miocene Mammalia of Congo. *Ann. Mus. R. Congo Belge,* Ser. 8, 46:1–77.

———. 1970. Miocene Mammalia of Congo, a correction. *Ann. Mus. R. Congo Belge,* Ser. 8, 67:163–167.

———. 1972. A late Pliocene rhinoceros from Langebaanweg, Cape Province. *Ann. S. Afr. Mus.* 59 (9):151–191.

———. 1973. Additional Miocene to Pleistocene rhinoceroses of Africa. *Zool. Med. Leiden* 46:149–177.

Hooijer, D. A., and Patterson, B. 1972. Rhinoceroses from the Pliocene of northwestern Kenya. *Bull. Mus. Comp. Zool., Harvard* 144 (1):1–26.

Hopwood, A. T. 1929. New and little known mammals from the Miocene of Africa. *Am. Mus. Novit.* 344:1–9.

Howell, F. C., and Coppens, Y. 1974. Les faunes des mammifères fossiles des formations Plio-Pleistocène de l'Omo en Ethiopie (Tubulidentata, Hyracoidea, Lagomorpha, Rodentia, Chiroptera, Insectivora, Carnivora, Primates). *C. R. Hebd. Séanc. Acad. Sci.* Sér. D, 278:2421–2424.

Hurford, A. J.; Gleadow, A. J. W.; and Naeser, C. W. 1976. Fission-track dating of pumice from the KBS Tuff, East Rudolf, Kenya. *Nature* 263:738–740.

Isaac, G. LL. 1965. The stratigraphy of the Peninj Beds and the provenance of the Natron Australopithecine mandible. *Quaternaria* 7:101–130.

———. 1966. The geological history of the Olorgesailie area. *Actas 5 Congr. Panafr. Prehist. de Estudio del Cuaternario* 2:125–134.

———. 1967. The stratigraphy of the Peninj Group—early middle Pleistocene formations west of Lake Natron, Tanzania. In W. W. Bishop, and J. D. Clark, eds. *Background to evolution in Africa.* Chicago: University of Chicago Press, pp. 229–258.

Isaac, G. LL., and Curtis, G. H. 1974. Age of early Acheulian industries from the Peninj Group, Tanzania. *Nature* 249:624–627.

Isaac, G. LL.; Harris, J. W. K.; and Crader, D. 1975. Archaeological evidence from the Koobi Fora Formation. In Y. Coppens, F. C. Howell, G. LL. Isaac, and R. E. F. Leakey, eds. *Earliest man and environments in the Lake Rudolf Basin.* Chicago: University of Chicago Press, pp. 533–551.

Isaac, G. LL.; Merrick, H. V.; and Nelson, C. M. 1972. Stratigraphic and archaeological studies in the Lake Nakuru basin. In E. M. van Zinderen Bakker, ed. *Palaeoecology of Africa.* Cape Town: Balkema. vol. 6, pp. 225–232.

Jaeger, J.-J.; Michaux, J.; and David, B. 1973. Biochronologie du Miocène moyen et du supèrieur continental du Maghreb. *C. R. Hebd. Séanc. Acad. Sci.* Ser. D, 277: 2477–2480.

Johanson, D. C., and Taieb, M. 1976. Plio-Pleistocene hominid discoveries in Hadar, Ethiopia. *Nature* 260: 293–297.

Kent, P. E. 1941. The Pleistocene beds of Kanam and Kanjera, Kavirondo, Kenya. *Geol. Mag.* 79:117–132.

———. 1944. The Miocene beds of Kavirondo, Kenya. *Quart. J. Geol. Soc. Lond.* 100:85–116.

King, B. C., and Chapman, G. R. 1972. Volcanism of the Kenya Rift Valley. *Phil. Trans. Roy. Soc. London* (A) 271:185–208.

King, L. C. 1951. *South African Scenery*. Edinburgh: Oliver and Boyd.

———. 1962. *The morphology of the earth*. Edinburgh: Oliver and Boyd.

Klein, R. G. 1974. Provisional statement on terminal Pleistocene mammalian extinctions in the Cape biotic zone (Southern Cape Province, South Africa). *S. Afr. Archaeol. Soc.*, Goodwin Series 2:39–45.

Leakey, L. S. B. 1951. *Olduvai Gorge*. Cambridge: Cambridge University Press.

———. 1965. *Olduvai Gorge 1951–1961*. Volume 1: *Fauna and background*. Cambridge: Cambridge University Press.

———. 1967. Notes on the mammalian faunas from the Miocene and Pleistocene of East Africa. In W. W. Bishop, and J. D. Clark, eds. *Background to evolution in Africa*. Chicago: University of Chicago Press, pp. 7–29.

Leakey, L. S. B., and Solomon, J. D. 1931. East African archaeology. *Nature* 124:9.

Leakey, M.; Tobias, P. V.; Martyn, J. M.; and Leakey, R. E. F. 1969. An Acheulean industry with prepared core technique and the discovery of a contemporary hominid mandible at Lake Baringo, Kenya. *Proc. Prehist. Soc. Lond.* 3:48–76.

Leakey, M. D. 1971. *Olduvai Gorge,* Volume 3: *Excavations in Beds I and II, 1960–1963*. Cambridge: Cambridge University Press.

Leakey, M. D.; Hay, R. L.; Curtis, G. H.; Drake, R. E.; Jackes, M. K.; and White, T. D. 1976. Fossil hominids from the Laetolil Beds. *Nature* 262:460–466.

Leakey, M. D.; Hay, R. L.; Thurber, D. L.; Protsch, R.; and Berger, R. 1972. Stratigraphy, archaeology, and age of the Ndutu and Naisiusu Beds, Olduvai Gorge, Tanzania. *World Archaeol.* 3 (3):328–341.

Leakey, R. E. F. 1970. Fauna and artefacts from a new Plio-Pleistocene locality near Lake Rudolf in Kenya. *Nature* 226:223–224.

———. 1975. An overview of the East Rudolf Hominidae. In Y. Coppens, F. C. Howell, G. Ll. Isaac, and R. E. F. Leakey, eds. *Earliest man and environments in the Lake Rudolf Basin*. Chicago: University of Chicago Press, pp. 476–483.

———. 1976. New hominid fossils from the Koobi Fora Formation in northern Kenya. *Nature* 261:574–576.

Le Gros Clark, W. E., and Leakey, L. S. B. 1951. *The Miocene Hominoidea of East Africa*. Fossil Mammals of Africa, 1. British Museum (Nat. Hist.), London.

Lepersonne, J. 1949. Le fossé tectonique Lac Albert-Semliki-Lac Edouard. Résumé des observations géologique effectuées en 1938-1939-1940. *Ann. Soc. Géol. Belg.* 77:M 1–92.

Mabbutt, J. A. 1956. The physiographical surface geology of the Hopefield fossil site. *Tran. Roy. Soc. S. Afr.* 35:21–58.

Maglio, V. J. 1970. Early Elephantidae of Africa and a tentative correlation of African Plio-Pleistocene deposits. *Nature* 225:328–332.

———. 1972. Vertebrate faunas and chronology of hominid-bearing sediments east of Lake Rudolf, Kenya. *Nature* 239:379–385.

———. 1973. Origin and evolution of the Elephantidae. *Trans. Am. Phil. Soc.,* N.S. 63:1–149.

Maglio, V. J., and Hendey, Q. B. 1970. New evidence relating to the supposed stegolophodont ancestry of the Elephantidae. *S. Afr. Archaeol. Bull.* 25:85–87.

Martyn, J. M. 1967. Pleistocene deposits and new fossil localities in Kenya (with a note on the hominid remains by P. V. Tobias). *Nature* 215:476–479.

———. 1969. Notes on the geology of the Kapthurin Beds. In M. Leakey et al. 1969, *Proc. Prehist. Soc. Lond.* 3:48–76.

Mawby, J. E. 1970. Fossil vertebrates from northern Malawi: preliminary report. *Quaternaria* 13 (2):319–324.

McCall, G. J. H. 1958. Geology of the Gwasi area. *Geol. Surv. Kenya, Report* 45:1–88.

McCall, G. J. H.; Baker, B. H.; and Walsh, J. 1967. Late Tertiary and Quaternary sediments of the Kenya Rift Valley. In W. W. Bishop, and J. D. Clark, eds. *Background to evolution in Africa*. Chicago: University of Chicago Press, pp. 191–220.

Monod, T. 1963. The late Tertiary and Pleistocene in the Sahara and adjacent southerly regions. In F. C. Howell, and F. Bourlière, eds. *African ecology and human evolution*. New York: Viking Fund Publications in Anthropology. 36:117–229.

Nilsson, E. 1929. Preliminary report on the Quaternary geology of Mount Elgon and some parts of the Rift Valley. *Geologiska Fören. Forhand. Stockholm* 51:253–261.

Patterson, B. 1965. The fossil elephant shrews (family Macroscelididae) *Bull. Mus. Comp. Zool., Harvard* 133 (6):295–335.

———. 1966. A new locality for early Pleistocene fossils in northwestern Kenya. *Nature* 212:577–578.

Patterson, B.; Behrensmeyer, A. K.; and Sill, W. D. 1970. Geology and faunal correlations of a new Pliocene locality in northwestern Kenya. *Nature* 226:918–921.

Peabody, F. E. 1954. Travertines and cave deposits of the Kaap escarpment of South Africa and the type locality of *Australopithecus africanus* Dart. *Bull. Geol. Soc. Am.* 65:671–706.

Petrocchi, C. 1943. Sahabi, eine neue Seite in der Geschicht der Erde. *N. Jb. Min. Geol. Paläont.* 1943 B:1–9.

Pickford, M. 1975. Late Miocene sediments and fossils from the northern Kenya Rift Valley. *Nature* 256:279–284.

Rich, P. V. 1972. A fossil avifauna from the upper Miocene Beglia Formation of Tunisia. *Notes Serv. Géol. Tunis* 35:29–66.

Robinson, P. 1972. *Pachytragus solignaci,* a new species of Caprine Bovid from the late Miocene Beglia Formation of Tunisia. *Notes Serv. Géol. Tunis* 37:73–94.

Robinson, P., and Black, C. C. 1969. Note préliminaire sur les vertébrés fossiles du Vindobonien (formation Beglia) du Bled Douarah, Gouvernerat de Gafsa, Tunisie. *Notes Serv. Géol. Tunis* 31:67–70.

Saggerson, E. P., and Baker, B. H. 1965. Post-Jurassic erosion surfaces in eastern Kenya and their deformation in relation to rift structure. *Quart. J. Geol. Soc. Lond.* 121:51–68.

Savage, R. J. G. 1969. Early Tertiary mammal locality in southern Libya. *Proc. Geol. Soc. Lond.,* No. 1657:167–171.

———. 1971. Review of the fossil mammals of Libya. In *Symposium on the Geology of Libya,* University of Libya, pp. 215–225.

Savage, R. J. G., and White, M. E. 1965. Two mammal faunas from the early Tertiary of central Libya. *Proc. Geol. Soc. Lond.,* No. 1623:89–91.

Schweinfurth, G. 1886. Reise in das Depressionsgebiet im Umkreise des Fayûm. *Z. Ges. Erdk., Berlin* 21:96–149.

Selley, R. C. 1968. Near-shore marine and continental sediments of the Sirte basin, Libya. *Quart. J. Geol. Soc. Lond.* 124:419–460.

Shackleton, R. M. 1951. The Kavirondo Rift Valley. *Quart. J. Geol. Soc. Lond.* 106:345–392.

Shuey, R. T.; Brown, F. H.; and Croes, M. K. 1974. Magnetostratigraphy of the Shungura Formation, southwestern Ethiopia: fine structure of the lower Matuyama polarity epoch. *Earth Planet. Sci. Letters* 23:249–260.

Simons, E. L. 1968. Early Cenozoic mammalian faunas, Fayûm Province, Egypt. Part I—African Oligocene Mammals: Introduction, History of study, and faunal succession. *Peabody Mus. of Nat. Hist. Bull.* (Yale University) 28:1–21.

Simpson, G. G. 1928. Mesozoic Mammalia XI. *Brancatherulum tendagurense* Dietrich. *Am. J. Sci.,* Ser. 5 (15): 303–308.

Singer, R. 1954. The Saldanha skull from Hopefield, South Africa. *Am. J. Phys. Anthrop.* 12:345–362.

Singer, R., and Wymer, J. 1968. Archaeological investigations at the Saldanha skull site in South Africa. *S. Afr. Archaeol. Bull.* 23:63–74.

Smith, A. G.; Briden, J. C.; and Drewry, C. E. 1973. Phanerozoic world maps. In *Organisms and continents through time.* Palaeontological Soc., London, Special Papers in Palaeont., no. 12, pp. 1–42.

Smith, A. G., and Hallam, A. 1970. The fit of the southern continents. *Nature* 225:139–144.

Söhnge, P. G.; Visser, D. J. L.; and van Riet Lowe, C. 1937. The geology and archaeology of the Vaal River Basin. *Union S. Afr. Geol. Surv. Mem.* 39 (1 and 2):1–184.

Stromer, E. 1913. Mitteilungen über die Wirbeltierreste aus dem Mittelpliozän des Natrontales (Aegypten). *Zeits. Deutsch. Geol. Ges.* 65:350–372.

———. 1914. Mitteilungen über die Wirbeltierreste aus dem Mittelpliozän des Natrontales (Aegypten). 3. Artiodactyla:Bunodontia:Flusspferd. *Zeits. Deutsch. Geol. Ges.* 66:1–33.

———. 1920. Mitteilungen über die Wirbeltierreste aus dem Mittelpliozän des Natrontales (Aegypten). 5. Nachtrag zu 1. Affen. 6. Nachtrag zu 2. Raubtiere. *S. B. Bayer. Akad. Wiss.* 1920:345–370.

———. 1926. Reste land- und süsswasserbewohnender Wirbeltiere aus den Diamantenfeldern Deutsch-Südwestafrikas. In E. Kaiser. *Die Diamantenwüste Südwestafrikas.* Berlin: D. Reimer, pp. 107–153.

———. 1932. Reste süsswasser- und landbewohnender Wirbeltiere aus den Diamantfeldern Klein-Namaqualandes (Südwestafrika). *S. B. Bayer. Akad. Wiss.* 1931:17–47.

Taieb, M.; Coppens, Y.; Johanson, D. C.; and Kalb, J. 1972. Dépôts sedimentaire et faunes du Plio-Pléistocène de la basse vallée de l'Awash (Afar Central, Ethiopie). *C. R. Hebd. Séanc. Acad. Sci.,* Sér. D. 1275:819–822.

Taieb, M.; Johanson, D. C.; Coppens, Y.; and Aronson, J. L. 1976. Geological and Palaeontological background of Hadar hominid site, Ethiopia. *Nature* 260:289–293.

Tankard, A. J. 1975. Varswater Formation of the Langebaanweg-Saldanha area, Cape Province. *Trans. Geol. Soc. S. Afr.* 77:265–283.

Tobias, P. V. 1967. see Martyn, J. M. 1967.

Tobien, H. 1936. Mitteilungen über Wirbeltierreste aus dem Mittelpliozän des Natrontales (Aegypten). 7. Artiodactyla: Bunodontia: Suidae. *Zeits. Deutsch. Geol. Ges., Berlin* 88:42–53.

Van Couvering, J. A. 1972. Radiometric calibration of the European Neogene. In W. W. Bishop, and J. A. Miller, eds. *Calibration of hominoid evolution.* Edinburgh: Scottish Academic Press, pp. 247–271.

Van Couvering, J. A., and Miller, J. A. 1969. Miocene stratigraphy and age determinations, Rusinga Island, Kenya. *Nature* 221:628–632.

Vrba, E. 1974. Chronological and ecological implications of the fossil Bovidae of the Sterkfontein Australopithecine site. *Nature* 250:19–23.

Wells, L. H. 1962. Pleistocene faunas and the distribution of mammals in southern Africa. *Ann. Cape Prov. Mus. Nat. Hist.* 2:37–40.

———. 1964. The Vaal River "Younger Gravels" faunal assemblage: a revised list. *S. Afr. J. Sci.* 60:91–93.

———. 1969. Faunal subdivision of the Quaternary in southern Africa. *S. Afr. Archaeol. Bull.* 24:93–96.

Wells, L. H.; Cooke, H. B. S.; and Malan, B. D. 1941. The associated fauna and culture of the Vlakkraal thermal springs, O.F.S. *Trans. Roy. Soc. S. Afr.* 29:203–233.

Whitworth, T. 1953. A contribution to the geology of Rusinga Island, Kenya. *Quart. J. Geol. Soc. Lond.* 109:75–96.

———. 1961. The geology of Mfwanganu Island, western Kenya. *Overseas Geol. Mineral Resources* (Gt. Brit.) 8 (2):150–190.

Wood, A. E. 1968. Early Cenozoic mammalian faunas, Fayûm Province, Egypt. Part II—The African Oligocene Rodentia. *Peabody Mus. Nat. Hist. Bull.* (Yale University) 28:23–102.

Addendum

The earliest Tertiary land fauna so far known in Africa has recently been found in Morocco. It is clearly of Paleocene age and has yielded a small collection of microvertebrates, including two different Palaeoryctids, possible Miacid and Provivverine carnivores and other creodont/carnivore remains.

Capetta, H.; Jaeger, J.-J.; Sabatier, M.; Sige, B.; Sudre, J.; and Vianey-Liaud, M. 1978. Découverte dans le Paléocène du Maroc des plus anciens Mammifères eutherien d'Afrique. *Géobios* 11 (2): 257–263.

3

Mesozoic Mammals

A. W. Crompton and
F. A. Jenkins, Jr.

1 CM

Only three mammals have been positively identified from Mesozoic rocks in Africa. Two of these, *Megazostrodon rudnerae* (Crompton and Jenkins 1968) and *Erythrotherium parringtoni* (Crompton 1964), are from the late Triassic, and the remaining specimen, *Brancatherulum tendagurense* (Dietrich 1927), is from the late Jurassic. The Triassic specimens are important because they are among the oldest known mammals and because they are nearly complete skeletons. They provide critical information on the origin and development of several characteristic features of mammals such as a three-boned middle ear, diphyodonty, precise dental occlusion, regional differentiation in the vertebral column (notably in the anterior cervical and middorsal regions), erect or semierect limb posture, and specialization of the limb joints.

The Jurassic specimen consists of a poorly preserved, edentulous jaw. Its value is limited to indicating that mammals were present on the African continent during late Jurassic times.

Megazostrodon and Erythrotherium

Classification

These two genera are generally considered as nontherian mammals (figure 3.1), a group that has recently been reclassified by Hopson (1970). For convenience, part of the classification is repeated here.

Class: Mammalia
  Subclass: Prototheria
    Infraclass: Eotheria
      Order: Triconodonta
        Family: Morganucodontidae
    Genera included in this family are *Megazostrodon, Erythrotherium, Eozostrodon* (= *Morganucodon*)
      Order: Docodonta
    Infraclass: Ornithodelphia
      Order: Monotremata
    Infraclass: Allotheria
      Order: Multituberculata

Kermack, Mussett, and Rigney (1973) also have recently classified nontherian mammals, preferring the term Atheria to Prototheria. We are not in agreement with all aspects of their classification, i.e., the definition of the subclass Atheria, the referral of *Megazostrodon* to a family Sinoconodontidae and the inclusion of this family in a new suborder Morganucodonta, and the creation of a new suborder Eutriconodonta for both triconodontids and amphilestids. For this reason, we will use Hopson's classification until some of these issues can be discussed in greater detail.

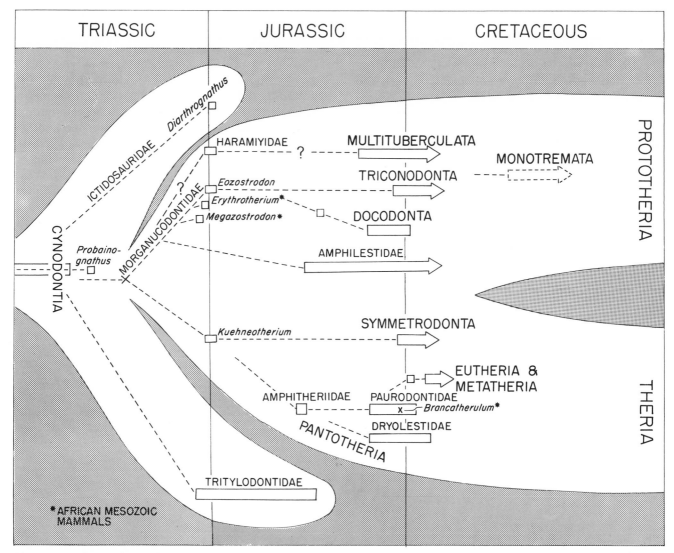

Figure 3.1 Interrelationships of Mesozoic mammals.

Locality Data

Erythrotherium parringtoni was discovered in 1962 in the upper Red Beds of the Stormberg Series (about 300 ft below the bottom of the Cave Sandstone) near the village of Tsekong, 4 mi southeast of Mafeteng in Lesotho (Crompton 1964).

Megazostrodon rudnerae was discovered in 1966 in the middle of the Red Bed sequence (400 ft below the Cave Sandstone), 3 km southeast of the trading store at Fort Hartley in the Quthing District of southern Lesotho (Crompton and Jenkins 1968). Although *Megazostrodon* appears to have come from a lower horizon than *Erythrotherium,* the relative age of the specimens is indeterminate because the Red Beds tend to increase in thickness in a southerly direction and thus a lower horizon in the south may correspond to a higher one in the north.

The Red Beds and overlying Cave Sandstone are generally considered to be Norian and/or Rhaetic (Cox 1973). The division between Cave Sandstone and Red Beds does not correspond to the division between the Norian and Rhaetic, and the age of the transition between these beds appears to differ from locality to locality.

If the middle Red Beds of southern Africa are identified as being of Norian age, then the two southern African Triassic mammals are older than the Rhaetic mammals of Europe (Hillaby 1967). However, because the European, Chinese, and southern African Triassic mammals are so similar and because the relative ages of the localities on the three continents cannot be accurately determined at present, it is perhaps safer simply to refer to the southern African mammals as being of late Triassic age.

Of the skulls of the *Erythrotherium* and *Megazos-*

trodon only the dentitions have been described (Crompton 1964, 1974; Crompton and Jenkins 1968; Hopson and Crompton 1969). Jenkins and Parrington (1976) surveyed the postcranial skeletons of *Megazostrodon* and *Erythrotherium,* together with the dissociated remains of the closely related *Eozostrodon* (= *Morganucodon*) from the Rhaetic fissure fillings of Britain. On the basis of these materials, which they concluded were structurally similar, they presented a skeletal reconstruction of a Triassic morganucodontid (figure 3.2).

Dentition

The skull of *Erythrotherium* is partially disarticulated and represents a juvenile. The dental formula (figure 3.3) appears to have been $\text{I}\frac{4}{3}\text{C}\frac{1}{1}\text{PM}\frac{4}{4}\text{M}\frac{3\,(+1?)}{4}$. Unerupted and partially formed second and fourth incisors are present and the last lower deciduous molar is still in place. The molars and premolars are of the typical morganucodontid pattern. The lower molar consists of four cusps in a row (from front to back, $\overline{\text{b}}\ \overline{\text{a}}\ \overline{\text{c}}\ \overline{\text{d}}$) and a well-developed cingulum supporting several cusps, including an enlarged cusp g (Kühneocone). The upper molars also have four cusps in a row (from

front to back, $\underline{\text{B}}\ \underline{\text{A}}\ \underline{\text{C}}\ \underline{\text{D}}$), and external and internal cingula with cuspules. The lower jaw lacks a pronounced angle and is more slender than that of *Eozostrodon*. A well-developed dentary condyle is present. A groove on the medial side of the dentary, containing the remnants of the postdentary bones, is evidence that both the reptilian and mammalian jaw articulations functioned alongside one another (as in other morganucodontids; Kermack, Mussett, and Rigney 1973). Consequently, these forms lacked a typically mammalian three-boned middle ear (Crompton and Parker 1978) which was isolated from the lower jaw.

In *Megazostrodon* (figure 3.4), incisors and canines are missing but the premolars and molars are exceptionally well preserved. The number and structure of the teeth, as well as the wear patterns on the molars, indicate that the animal was a young adult. The dental formula was $\text{I}\,?\,\text{C}\,?\,\text{PM}\frac{5}{5}\text{M}\frac{4}{4}$. The molars are, on the whole, similar to those of *Erythrotherium* and *Eozostrodon* except that the cingula and cingular cusps, especially those on the outer surface of the uppers and the inner surface of the lowers, are far larger than in the other morganucodontids. In the upper molars the three principal cusps, $\underline{\text{B}}$, $\underline{\text{A}}$, and $\underline{\text{C}}$, form a wide-angled triangle

Figure 3.2   Reconstruction of Triassic triconodont, based upon *Megazostrodon, Eozostrodon,* and *Erythrotherium.*

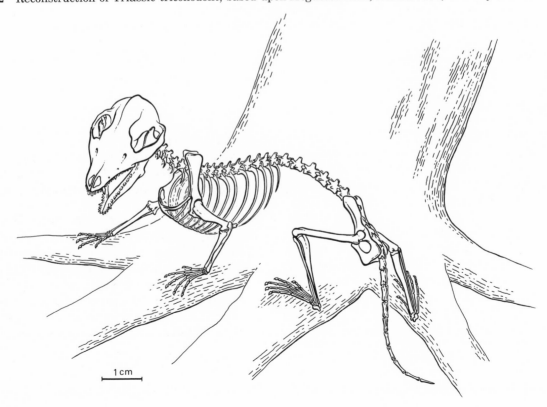

1 cm

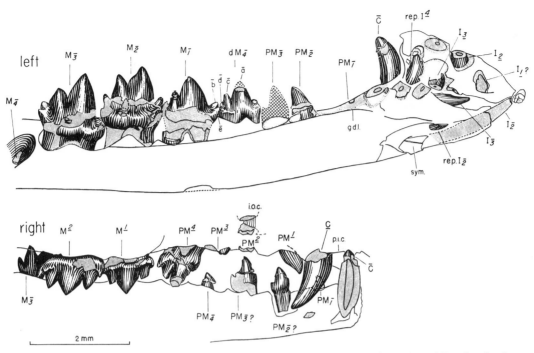

Figure 3.3   Internal view of the lower dentition and external view of the upper dentition of *Erythrotherium parringtoni.*

rather than being arranged in a straight line. The molar occlusal relationships of *Erythrotherium* and *Megazostrodon* are slightly different from that of *Eozostrodon.* In the African morganucodontids the main cusp of the lower (ā) occluded against and slightly in front of the internal surface of cusp B of the upper molar; the main upper molar cusp (A) occluded against the external surface of cusp d̄ of the lower. Moreover, the last upper premolar is relatively small. In the Chinese and European morganucodontids the last upper premolar is always larger than the first molar, and cusp (ā) of the lower molar occludes between cusp (B) and cusp (A) of the opposing uppers. Thus, the size of the last upper premolar and the pattern of molar occlusion differentiate the African from the Eurasian forms.

On the basis of skull size and tooth structure the morganucodontids appear to have been small insectivores with a premolar and molar dentition designed for puncturing and shearing. Diphyodonty and fixed relationships between teeth appear to have arisen in the advanced mammallike reptiles immediately ancestral to these late Triassic mammals. Hopson (1973) has argued that a diphyodont dentition may have been related to the origin of mammary glands and suckling behavior. This mode of feeding permits most skull growth to take place before the first set of milk teeth erupt. The limited amount of growth that takes place after this stage is only sufficient to accommodate a single replacement of milk teeth by permanent incisors, canines, and premolars as well as the addition of permanent molars.

Brain Size

Jerison (1973) has published a formula that expresses the relationship between brain and body size in vertebrates, viz., $E = KP^{2/3}$ (where E and P are brain and body weight in centimeter-gram units [grams or milliliters] and K, a proportionality constant). An average value for K for living mammals is 0.12 and for lower vertebrates, 0.007.

There are very few reliable data on brain/body ratios in mammallike reptiles. Part of the problem is the difficulty of determining how much of the braincase was occupied by the brain itself. One of the most mammallike of the cynodonts, *Probainognathus,* has a maximum brain volume of about 1.5 ml (= 1.5 g). The total body length of this form must have been about 30 to 35 cm, with a body weight between 1,000 and 1,500 g. If *Probainognathus* were a typical reptile, the expected brain size would have been between 0.7 and 0.9 ml, whereas if it were an average mammal it would have been between 12 and 16 ml. The brain size of *Probainognathus* falls well within the range of "average" reptiles.

Jerison (1973) has estimated that the brain volume of a late Jurassic triconodont was about one-half that of an "average" mammal. Recently dis-

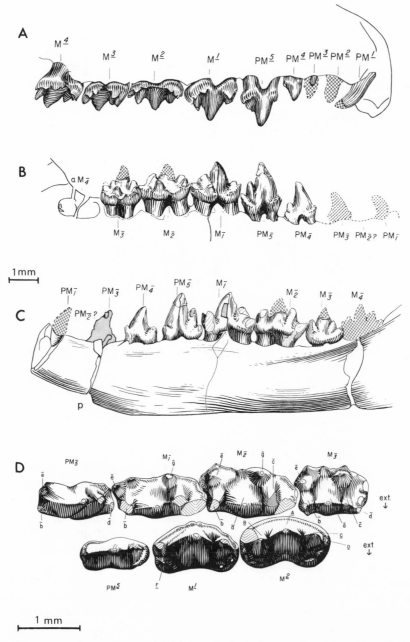

Figure 3.4 Dentition of *Megazostrodon rudnerae*. (*A*) External view of upper dentition, (*B* and *C*) internal and external views of lower dentition, (*D*) crown view lower PM5 through M3 and upper PM5 through M2.

covered early Cretaceous triconodonts appear to have had slightly larger brains than the Jurassic forms. As the brain size of an advanced mammallike reptile is less than one-eighth of that of an average mammal, there is a marked contrast in brain size between this form and the Jurassic and Cretaceous triconodonts.

Increase in brain size in early mammals is accompanied by a reorganization of the structure of the braincase. In advanced mammallike reptiles the cranial cavity anterior to the pituitary fossa nar-

rows rapidly in vertical dimension. The olfactory bulbs and tracts were contained in this narrow, cylindrical space formed between the frontals above and the orbitosphenoids below and laterally. A deep, vertical space, presumably filled by a cartilaginous interorbital septum, lies between the dorsally situated orbitosphenoids and the horizontal palatal bones. In mammals the expansion of the forebrain is accompanied by a ventral migration of the orbitosphenoid so that it becomes part of the basicranial axis in line with the basioccipital and basisphenoid;

concomitantly, the frontals are drawn down to form a major part of the lateral wall of the braincase. This has already taken place in Cretaceous triconodonts. The individual bones comprising the braincase in the Triassic triconodont *Megazostrodon* are dislocated, and it is therefore not possible to measure the intracranial volume accurately. However, the individual elements making up the braincase of morganucodontids appear to be almost identical to those of later triconodonts, an indication that the brain/body ratio was probably the same as, or close to, that of Jurassic triconodonts. In the transition from advanced mammallike reptiles to early mammals, a sudden and marked enlargement of the brain took place.

Postcranial Skeleton

Megazostrodon appears to have been 10 cm in head/body length (or 14 cm from snout to tip of tail) and weighed between 20 and 30 g (Jenkins and Parrington 1976). *Eozostrodon* and *Erythrotherium* are about 20% smaller.

The postcranial skeleton of early mammals, insofar as known, was structurally advanced over those of advanced cynodonts, i.e., those mammallike reptiles from which the first mammals appear to have arisen. Nevertheless, several primitive features are retained in the morganucodontid skeleton, such as a cynodontlike pectoral girdle that retains both coracoids and lacks a supraspinous fossa, the presence of a condyle on the humerus (rather than a trochlea) for articulation with the ulna, and the presence of an acetabulum composed of three more or less separate bony facets (iliac, ischial, and pubic).

Jenkins (1974) reviewed the longstanding question of terrestriality versus arboreality in the origin and evolution of mammals. Although various authors have argued that the earliest mammals were adapted for one or the other habitat, Jenkins pointed out that for many small mammals, at least, the locomotor repertoire required for both habitats is much the same and that from the perspective of a small mammal one set of substrates grades into the other. The primitive musculoskeletal adaptations of mammals are best interpreted in terms of providing a flexible range of postures and movements that are employed in active foraging over spatially complex substrates. Such a locomotor repertoire is required by both arboreal and terrestrial substrates. The earliest mammals, which were very small, might well be expected to have been behaviorally and structurally adapted for the kind of niche occupied today by some tree shrews or tenrecs.

Jenkins and Parrington (1976) interpreted vari-

ous skeletal features of morganucodontids either as adaptations for locomotion on spatially complex and even steep surfaces, such as might be encountered by mammals active in both terrestrial and arboreal environments, or as structural innovations that appear for the first time in mammalian ancestry (for which the adaptive significance is unclear). Among the former are the following features. (1) Small size; the incremental increase in energy consumption for running up steep inclines, as compared with that required for running on a level surface, is far less (relative to body size) in small than large animals (Taylor, Caldwell, and Rowntree 1972). (2) Limb posture; the femur and humerus appear to have been abducted from the sagittal plane, and the elbow and knee flexed, so that the center of gravity was held low with a relatively wide stance. (3) Well-developed ball and socket joints at the shoulder and hip, making possible limb movements over a wide range of excursions. (4) A pelvis of a generalized mammalian pattern (although it should be remembered that at least two lineages of mammallike reptiles, the tritylodontids and ictidosaurs, evolved basically the same structure). (5) A hallux with joint adaptations similar to those in *Tupaia* (Jenkins 1974), and therefore probably capable of independent, "opposable" movements to the same degree. (6) Claws laterally compressed and sharp, with prominent flexor tubercles, an adaptation common to climbing mammals (see Cartmill 1974). (7) Atlanto-occipital and atlanto-axial joints with all the major adaptations (with the exception of synostosis of the atlantal arches and intercentrum) for movement and stability characteristic of later mammals. (8) Enlargement of the midcervical vertebral canal. This is evidence of an increased size of the cervical intumescence of the spinal cord, correlated with the greater neuromuscular control of the forelimb characteristic of mammals. (9) Distinctive structural modifications in the mid-dorsal vertebrae, notably in zygapophyseal orientation ("diaphragmatic vertebra") and spinal process orientation ("anticlinal vertebra"), evidence of regional specialization of the epaxial musculature by means of which mammals typically localize flexion-extension movements in the mid-dorsal region. Extension and flexion are important elements in locomotion in small mammals (Jenkins 1974).

Biology

Following the general consensus that early mammals were probably capable of maintaining a constant body temperature, Jerison (1973) suggested that homeothermy enabled the first mammals to invade a nocturnal niche that in late Triassic times did not include reptiles. Crompton, Taylor, and Jagger (1978) surmised that early mammals were probably similar to modern-day hedgehogs and Madagascan tenrecs in maintaining a low body temperature (25 to 30°C) but possessing a reptilian metabolic rate three to five times lower than most living mammals. Not only is a reptilian metabolic rate economical and sufficient to sustain a constant low body temperature, but it is also suitable for a nocturnal existence. High ambient temperatures and solar radiation, coupled with diurnal activity and a low preferred body temperature, would have required inordinate amounts of evaporative heat loss to maintain a low body temperature. For this reason hedgehogs and tenrecs secrete themselves in burrows during the day. The high body temperature of typical mammals probably developed as an adaptation to a diurnal existence. Jerison (1973) pointed out that the invasion of a nocturnal niche by previously diurnal reptiles would have made extraordinary demands not only on their temperature regulation capacities but also on their sensory systems. Diurnal reptiles rely primarily on vision whereas most nocturnal mammals depend on developed auditory and olfactory senses. The dramatic enlargement of the olfactory bulb and forebrain in early mammals supports Jerison's view that these forms processed considerably more olfactory information than their mammallike reptile antecedents. Jerison (1973) suggested that the expansion of brains in early mammals was also, in part, related to greater neuronal complexity for auditory discrimination. Comparing mammals to reptiles, he pointed out, there is at least a tenfold increase in the neurons directly related to the auditory system at the thalamic level and up to a hundredfold increase in neurons at the cortical level.

No complete endocranial casts of Mesozoic mammals are currently available, and therefore it is not possible to document a marked increase in the brain areas processing auditory information in the earliest mammals as it is in the case of the enlarged olfactory bulbs. However, there is evidence of a marked and sudden enlargement of the cochlear region of the inner ear in early mammals. The ability of mammals to discriminate between frequencies covering a wide range depends in part on the length of the basilar membrane and associated organ of Corti. In mammals the cochlea is housed in the petrous portion of the petrosal bone. The ventral surface of the cochlea housing, or the promontory, is a marked feature on the ventral side of the skull in front of the fenestra ovalis in all early mammals.

This is in marked contrast to advanced mammallike reptiles where the position and size of the cochlea is not represented on the external surface of the bones surrounding the inner ear. Triassic mammals are much smaller than their known cynodont antecedents and it can perhaps be argued that because the cochlea is a sense organ it may not have undergone the same degree of reduction as the rest of the skull during the cynodont-mammal transition. However, a Jurassic survivor of the mammallike reptiles, *Oligokyphus* (which weighed about 1,000 g), is considerably smaller than its late Triassic antecedents (which weighed up to an estimated 40 kg) and shows no evidence of a relative increase in the size of the cochlear housing.

An important feature of the Triassic mammals is the acquisition of a new mammalian jaw joint between the squamosal and dentary. This joint exists alongside the old reptilian joint between the quadrate (= incus) and articular (= malleus). The dentary condyle is far larger than the old reptilian jaw joint and these bones could therefore be used solely to conduct vibrations to the inner ear that were received by the tympanic membrane. In the earliest mammals the enlarged cochlea and the auditory ossicles (freed from a jaw support function) are evidence of improved auditory acuity. This interpretation supports Jerison's (1973) hypothesis that an improvement in the sense of hearing would have been expected if the early mammals had invaded a nocturnal niche. The features of the postcranial skeleton are not inconsistent with the suggestion that early mammals were nocturnal insectivores.

The diversity within late Triassic morganucodontid mammals is limited, but they had a rather wide distribution—China, southern Africa, and Europe. They occur in great numbers in the British fissure fillings. Rapid dispersal into what was probably a worldwide distribution and limited diversity accords with the view that early mammals invaded a previously vacant nocturnal niche.

Relationships

It is generally accepted that the morganucodontids (figure 1.1) are the stock that gave rise to the Triconodonta of the late Jurassic and Cretaceous and to the Docodonta of the Jurassic (Hopson and Crompton 1969; Mills 1971; Parrington 1971; Crompton and Jenkins 1973). Little is known of the origin and relationship of the amphilestids from the late Jurassic; traditionally classified as a subfamily of triconodontids, they were given separate family status by Kühne (1958). It is even questionable whether they are triconodonts, for the occlusion of their postcanine teeth is different from that of most morganucodontids, typical triconodonts, and the therian mammals. The amphilestids probably arose from the same stock that gave rise to the morganucodontids and the enigmatic late Triassic haramiyids, the latter known only from isolated teeth. The haramiyids are possibly related to the largest group of prototherian mammals, the multituberculates (Hahn 1973). Kielan-Jaworowska (1970) has suggested that a relationship exists between monotremes and multituberculates. A third group of Triassic mammals, the Kuehneotheriidae, appear to have given rise to the therian mammals (Kermack, Kermack, and Mussett 1968) but are far less common in the British deposits and have not as yet been found outside Europe.

The interrelationship of the late Triassic mammals has not yet been clearly established. The morganucodontids and kuehneotheriids share a suite of osteological features: a diphyodont dentition with well-differentiated premolars and molars, a fixed relationship between upper and lower molars, a transverse component to jaw movement during dental occlusion, and a dentary-squamosal articulation. These features, which clearly differentiate these two groups of early mammals from all the mammallike reptiles, may also have been present in other Triassic mammals. Some workers (Hopson and Crompton 1969; Parrington 1971; Crompton and Jenkins 1973) have claimed that these features indicate close relationship between morganucodontids and kuehneotheriids. One or several of these osteological features may have evolved independently in other mammallike reptile lineages, e.g., the dentary-squamosal articulation in the ictidosaurs and tritylodontids (Crompton 1972), but the combination of these features probably arose only in the group that gave rise to both groups of Triassic mammals. When the kuehneotheriids and haramiyids are better known, it may be possible to increase this list to include features now only known in the morganucodontids, such as expansion of the forebrain and an enlarged cochlea.

Other workers (Kermack 1967; Mills 1971), however, have suggested that the differences in molar structure and occlusion not only between the morganucodontids and kuehneotheriids but also between the northern (Europe and China) and southern (Lesotho) morganucodontids are so great that they probably arose from different groups of mammallike reptiles. If this is correct, it implies that at least the two main groups of Triassic mammals have independent lineages extending well back into the Triassic and possibly even into the upper Permian. In this case many of the features considered diagnostic of mammals must have arisen independently at least twice, and even more if the haramiyids and amphilestids are not closely related to either the morganucodontids or kuehneotheriids. Only further study of available material and the discovery of new material will resolve this problem.

Associated Fauna

The upper Red Bed vertebrate fauna is dominated by small and large prosauropods (Charig, Attridge, and Crompton 1965). Less numerous elements are the advanced mammallike reptiles, tritylodonts (Ginsberg 1962), ictidosaurs (Crompton 1958), primitive crocodiles (Nash 1975), and a few thecodonts (Walker 1972). This fauna is typical of late Triassic terrestrial deposits of other continents. Morganucodontid mammals (Kermack, Mussett, and Rigney 1973), tritylodonts (Young 1947), prosauropod dinosaurs (Young 1951), and primitive ornithischian dinosaurs (Simmons 1965) are known from the Lufeng Beds of China. The Los Colorados Formation of Argentina (Bonaparte 1971) contains a similar array of tritylodontids, ictidosaurs, prosauropods, saurischian dinosaurs, primitive crocodiles, and thecodonts. Tritylodonts, saurischian and ornithischian dinosaurs, primitive crocodiles, and thecodonts are also known from the North American late Triassic deposits (Romer 1971). Prosauropod and other saurischian dinosaurs, morganucodontids, tritylodontids, and thecodonts are known in the late Triassic beds of Europe (Robinson 1957, 1971). In many cases the representatives of these various groups on different continents are almost identical. For example, among the mammals, *Erythrotherium* from southern Africa, *Eozostrodon* (= *Morganucodon*) *watsoni* from Wales, and *Eozostrodon* (= *Morganucodon*) *oehleri* from southern China are so similar that Kermack, Mussett, and Rigney (1973) have placed them all in the same genus. The bipedal prosauropods, such as *Thecodontosaurus* from Europe, *Massospondylus* from southern Africa, and *Yaleosaurus* from North America, are as difficult to tell apart as are the melanorosaurid prosauropods such as *Melanorosaurus* from southern Africa, *Riojasaurus* from South America, and *Sinosaurus* from China. The tritylodontids (*Tritylodon* from southern Africa, *Bienotherium* from China, and the undescribed tritylodontid from the Kayenta Formation of North America) are also very similar. The distribution of the late Triassic faunas and the close similarity between individual genera on what are now widely separated continents suggest that barriers that now exist between the continents were not present in late Triassic times.

Brancatherulum tendagurense

Classification

The classification of *Brancatherulum tendagurense* is difficult because the fossil consists only of a poorly preserved lower jaw.

Subclass: Theria
  Order: Pantotheria
    Family: Paurodontidae

The specimen is a damaged and edentulous right jaw and was first described by Branca in 1916. It was referred to the Pantotheria by Hennig (1919), fully described and named by Dietrich (1927), and further reviewed and discussed by Simpson (1928).

The only feature of diagnostic importance is a pantotherelike angle; the number of postcanine teeth (between six and eight) is relatively small. For these reasons it is usually referred to the Paurodontidae. (See Simpson 1928 for an account of the similarities between *Brancatherulum* and *Peramus* from the British Purbeckian deposits.) However, Kühne (1968) noted similarities in the back of the jaws of *Amphitherium, Brancatherulum, Peramus, Archaeotrigon,* and a specimen from the Kimmeridgian of Portugal that he refers to the genus *Peramus;* but he did not go so far as to suggest that these forms (which are usually placed in two separate families) should be placed in a single taxonomic unit. Clemens and Mills (1971) have recently published a review of the structure of the teeth and jaws of *Peramus* and they conclude that the phylogenetic relationships of *Brancatherulum* remain obscure.

Age and Associated Fauna

The Tendaguru deposits have also yielded a diverse and abundant dinosaurian fauna. These have been dated as being of late Kimmeridgian or early Tithonian age. The dinosaurs are similar to those of the Morrison Formation of North America, particularly the smaller saurischians, the heavy sauropods, and the stegosaurs (Colbert 1961), indicating that late Jurassic vertebrate faunas still had a worldwide distribution.

Our knowledge of Mesozoic mammals is based almost entirely upon remains from North America, Europe, and Asia and with the exception of the forms mentioned in this paper none are known from southern continents. It will not be possible to answer many fundamental questions about mammalian evolution, such as the origin and distribution of monotremes or the differentiation of marsupials and placentals from mammals of "therian grade," unless a concerted effort is made to search for Mesozoic mammals in Africa and other southern continents.

References

Bonaparte, J. F., 1971. Los tetrapodos del sector superior de la Formación Los Colorados, La Rioja, Argentina. (Triásico Superior) *Op. Lill. 22:*1–183.

Branca, W., 1916. Ein Säugetier?–Unterkiefer aus den Tendaguru–Schichten Wissens. Ergeb. Tendaguru–Expedition 1909–12. *Arch. Biontol. 4:*137–40.

Cartmill, M., 1974. Pads and claws in arboreal locomotion. In Jenkins, F. A., ed., *Primate locomotion,* New York: Academic Press, pp. 45–83.

Charig, A. J.; Attridge, J.; and Crompton, A. W., 1965. The origin of the sauropods and the classification of the Saurischia. *Proc. Linn. Soc. Lond. 176:*197–221.

Clemens, W. A., and Mills, J. R. E., 1971. Review of *Peramus tenuirostris* Owen (Eupantotheria, Mammalia). *Bull. Brit. Mus. Nat. Hist.* (Geol.) 20:89–113.

Colbert, E. H., 1961. *Dinosaurs, their discovery and their world.* New York: E. P. Dutton.

Cox, C. B., 1967. Changes in terrestrial vertebrate faunas during the Mesozoic. In W. B. Harland, et al., eds., *The Fossil Record.* London: Geological Society, pp. 77–89.

_____. 1973. Gondwanaland Triassic stratigraphy. *An. Acad. Brasil Ciênc. 45:*115–119.

Crompton, A. W., 1958. The cranial morphology of a new genus and species of ictidosaurian. *Proc. Zool. Soc. Lond.* 130:183–216.

_____. 1964. A preliminary description of a new mammal from the upper Triassic of South Africa. *Proc. Zool. Soc. Lond. 142:*441–452.

_____. 1968. In search of the insignificant. *Discovery 3:* 23–32.

_____. 1972. The evolution of the jaw articulation of cynodonts. In K. A. Joysey, and T. Kemp, eds., *Studies in vertebrate evolution.* Edinburgh: Oliver and Boyd pp. 231–251.

_____. 1974. The dentitions and relationships of the southern African Triassic mammals *Erythrotherium parringtoni* and *Megazostrodon rudnerae. Bull. Brit. Mus. Nat. Hist.* (Geol.) 24:397–437.

Crompton, A. W., and Jenkins, Jr., F. A., 1968. Molar occlusion in late Triassic mammals. *Biol. Rev. 43:*427–458.

_____. 1973. Mammals from reptiles; a review of mammalian origins. *Ann. Rev. of Earth and Planetary Sci. 1:*131–155.

Crompton, A. W., and Parker, P. 1978. *Evolution* of the mammalian masticatory system. *American Scientist 66:* 192–201.

Crompton, A. W.; Taylor, C. R.; and Jagger, J. A., 1978. Evolution of homeothermy in mammals. *Nature 272:* 333–336.

Dietrich, W. O., 1927. *Brancatherulum* n.g., ein proplacentalier aus dem obersten Jura des Tendaguru in Deutsch-Ostafrika: *Centralbl. Min. Geol. Pal. 1927(B):*423–426.

Ginsburg, L., 1962. *Likhoelia ellenbergeri,* Tritylodonte du Trias Supérieur du Basutoland (Afrique du Sud). *Ann. Paléont. 48:*179–194.

Hahn, G., 1973. Neue Zähne von Haramiyiden aus der

Deutschen Ober-Trias und ihre beziehungen zu den Multituberculaten. *Palaeontogr. 142*:1–15.

Hennig, E., 1919. Die entstehung des Säugerzahns und die Paläontologie. *Naturwiss. Wochenshrift,* N. F. *18* (No. 51):1–7.

Hillaby, J., 1967. Clues to evolution. *New Scientist 34*:163.

Hopson, J. A., 1970. The classification of nontherian mammals. *J. Mamm. 51*:1–9.

———. 1973. Endothermy, small size, and the origin of mammalian reproduction. *American Naturalist 107*: 446–452.

Hopson, J. A., and Crompton, A. W., 1969. Origin of mammals. In T. Dobzhansky; M. K. Hecht; and Steere, W. C., eds., *Evolutionary biology 3*:15–72.

Jenkins, F. A., Jr., 1974. Tree shrew locomotion and the origins of primate arborealism. In Jenkins, F. A. Jr., ed., *Primate locomotion.* New York: Academic Press, pp. 85–115.

Jenkins, F. A., Jr., and Parrington, F. R., 1976. The postcranial skeletons of the Triassic mammals—*Eozostrodon, Megazostrodon* and *Erythrotherium. Phil. Trans. Roy. Soc. Lond.* (B) *273*:387–431.

Jerison, H. J., 1973. *Evolution of the brain and intelligence.* New York: Academic Press.

Kermack, D. M.; Kermack, K. A.; and Mussett, F., 1968. The Welsh pantothere *Kuehneotherium praecursoris. J. Linn. Soc. Lond.* 47:407–423.

Kermack, K. A., 1967. The interrelation of early mammals. *J. Linn. Soc. Lond.* 47:241–249.

Kermack, K. A.; Mussett, F.; and Rigney, H. W., 1973. The lower jaw of *Morganucodon. J. Linn. Soc. Lond.* 53:87–175.

Kielan-Jaworowska, Z., 1970. Unknown structures in multituberculate skull. *Nature* 226:974–976.

Kühne, W. G., 1968. Kimmeridge mammals and their bearing on the phylogeny of the Mammalia. In E. T. Drake, ed., *Evolution and environment.* New Haven: Yale University Press, pp. 109–123.

Mills, J. R. E., 1971. The dentition of *Morganucodon.* In D. M. Kermack, and K. A. Kermack, eds., *Early Mammals.* London: Academic Press. pp. 29–62.

Nash, D. S., 1975. The morphology and relationships of a crocodilian, *Orthosuchus strombergi,* from the upper Triassic of Lesotho. *Ann. S. Afr. Mus. 67*:227–329.

Patterson, B., and Olson, E. C., 1961. A Tricondontid mammal from the Triassic of Yunnan. *Internat. Colloq. on the evolution of mammals.* Kon. Vlaamse Acad. Wetensch. Lett. Sch. Kunsten Belgié, part I: 129–191.

Parrington, F. R., 1971. On the upper Triassic Mammals. *Phil. Trans. Roy. Soc. Lond. Ser. B 261*:231–272.

Robinson, P., 1957. The Mesozoic fissures of the Bristol Channel area and their vertebrate faunas. *J. Linn. Soc. Lond.* 43:260–279.

———. 1971. A problem of faunal replacement on Permo-Triassic continents. *Palaeontology 14*:131–153.

Romer, A. S., 1969. Cynodont reptile with incipient mammalian jaw articulation. *Science 166*: 881–882.

———. 1971. Tetrapod vertebrates and Gondwanaland. *Proc. Int. Union Geol. Sci., Gondwana Symposium.* Pap. No. 2:111–124.

Simmons, D. J., 1965. The non-therapsid reptiles of the Lufeng Basin, Yunnan, China. *Fieldiana: Geology 15:* 1–93.

Simpson, G. G., 1928. Mesozoic Mammalia XI *Brancatherulum tendagurense.* Dietrich. *Am. J. Sci. 15*:303–308.

Taylor, C. R.; Caldwell, S. L.; and Rowntree, V. J., 1972. Running up and down hills: some consequences of size. *Science 178*:1096–1097.

Walker, A. D., 1972. New light on the origin of birds and crocodiles. *Nature 237*:257–263.

Young, C. C., 1947. Mammal-like reptiles from Lufeng, Yunnan, China. *Proc. Zool. Soc. Lond. 117*:537–97.

———. 1951. The Lufeng saurischian fauna in China. *Paleont. Sin. 134*:1–96.

4

Insectivora and Chiroptera

Percy M. Butler

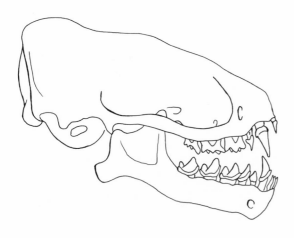

Several groups constitute the nonrodent micro-mammalian fauna. Insectivora is here used in a broad sense to denote an artificial assemblage of early offshoots from the eutherian stem (Butler 1972). The fossil history of the insectivores and bats is much less well known than that of most groups of larger mammals and their phylogeny in Africa has not yet passed the conjectural stage. At present they form a significant element of the fauna (21 genera of insectivores and 43 genera of bats), and this has probably been the case throughout the Tertiary. However, only a few localities have so far yielded remains of small mammals; the known distribution of the families of insectivores and bats is summarized in table 4.1.

Order Insectivora

Family Ptolemaiidae

RECORDED SPECIES:

Ptolemaia lyonsi Osborn 1908; Schlosser 1923; Van Valen 1966; Coombs 1971; Simons and Gingerich 1974. Early Oligocene, Fayum.

Qarunavus meyeri Simons and Gingerich 1974; Schlosser 1910, 1911; Matthew 1918; Butler 1969. Early Oligocene, Fayum.

Kelba quadreemae Savage 1975. Early Miocene, Rusinga.

The relationships of these very imperfectly known forms are still uncertain, even at the ordinal level. *Ptolemaia lyonsi* was based on a single mandible with worn teeth; a second specimen has recently been described by Simons and Gingerich (1974). A pair of mandibles with milk teeth and unworn M_1 and M_2 were referred to the same species by Schlosser (1910, 1911), but Matthew (1918) doubted whether they belonged to the same genus. They have now been named *Qaranavus meyeri* by Simons and Gingerich (1974). *Kelba quadreemae* was described from two isolated upper molars by Savage (1965), who regarded it as an arctocyonid. A complete upper dentition is now available and will soon be described by R. J. G. Savage.

Van Valen (1966) reviewed the various opinions on the relationships of *Ptolemaia,* which he placed near the Pantolestidae, a view first tentatively expressed by Schlosser (1923). In 1967 he included *Kelba* in the Ptolemaiidae without giving reasons, but presumably because of a general pantolestid similarity in the upper molar pattern (see also R. J. G. Savage, chapter 11, for further discussion); he placed the Ptolemaiidae next to the Pantolestidae in his superfamily Tupaioidea. Butler (1969) noted a

Table 4.1 Known distribution in time of the families of insectivores and bats on the African continent.

Family	Early Oligocene: Egypt	Early Miocene East Africa	Early Miocene S.W. Africa	Mio-Pliocene: Morocco	Pliocene East Africa	Pliocene S.W. Africa	Pliocene-Early Pleistocene: S. Africa	Early Pleistocene: Olduvai	Recent
Ptolemaiidae	X	X	—	—	—	—	—	—	—
Macroscelididae	X	X	X	—	—	X	X	X	X
Erinaceidae	—	X	—	X	—	—	X	X	X
Soricidae	—	?	—	X	X	—	X	X	X
Tenrecidae	—	X	—	—	—	—	—	—	X
Chrysochloridae	—	X	—	—	—	—	X	—	X
Pteropodidae	—	X	—	—	—	—	—	—	X
Vampyravus	X	—	—	—	—	—	—	—	—
Rhinopomatidae	—	—	—	—	—	—	—	—	X
Emballonuridae	—	X	—	—	—	—	—	—	X
Megadermatidae	—	X	—	X	—	—	—	X	X
Rhinolophidae	—	—	—	X	—	—	X	—	X
Hipposideridae	—	X	—	X	—	—	—	—	X
Nycteridae	—	—	—	—	X	—	—	—	X
Myzopodidae	—	—	—	—	—	—	—	X	—
Vespertilionidae	—	—	—	X	—	—	X	X	X
Molossidae	—	—	—	X	—	—	—	X	X

number of similarities in the teeth between the juvenile jaw of *Qarunavus* and the macroscelidid *Rhynchocyon*. Until more is known of the structure of these animals it seems hardly justifiable to classify them other than as Eutheria *insertae sedis*. Simons and Gingerich (1974) suggest a possible relationship of *Ptolemaia* to the Tubulidentata.

Family Macroscelididae

RECORDED SPECIES:

Metoldobotes stromeri Schlosser 1910, 1911; Matthew 1910, 1915; Patterson 1965. Early Oligocene, Fayum.

Rhynchocyon clarki Butler and Hopwood 1957; Butler 1969. Early Miocene, Rusinga, Songhor.

Rhynchocyon rusingae Butler 1969. Early Miocene, Rusinga, Songhor.

Macroscelididae indet. 1. Butler 1969. Early Miocene, Rusinga.

Macroscelididae indet. 2. Butler 1969. Early Miocene, Rusinga.

Myohyrax oswaldi Andrews 1914; Stromer 1926; Hopwood 1929; Whitworth 1954; Patterson 1965. Synonym: *M. doederleini* Stromer 1926. Early Miocene, Karungu, Koru, Rusinga, Namib.

Protypotheroides beetzi Stromer 1922, 1926; Hopwood 1929; Whitworth 1954; Patterson 1965. Synonym: *Myohyrax osborni* Hopwood 1929. Early Miocene, Namib.

Palaeothentoides africanus Stromer 1932; Butler and Hopwood 1957; Patterson 1965. Pliocene, Klein Zee.

Elephantulus fuscus leakeyi Butler and Greenwood 1976, 1965. Synonym: *Nasilio* sp. Late Pliocene to early Pleistocene, Olduvai, Makapansgat and probably other cave breccias in the Transvaal, where it has been recorded as *Elephantulus (Nasilio)* cf *brachyrhynchus*.

Elephantulus broomi Corbet and Hanks 1968; Broom 1937, 1938; Butler and Greenwood 1976. Synonym: *Elephantomys langi* Broom 1937. Late Pliocene to early Pleistocene, S. Africa (Schurveberg, Makapansgat), Olduvai. Most of the specimens referred to *Elephantulus langi* by De Graaff (1960) appear to be *E. antiquus*.

Elephantulus antiquus Broom 1948; Butler and Greenwood 1976. Late Pliocene to early Pleistocene, Bolt's Farm, Makapansgat, and other S. African cave breccias.

cf *Elephantulus antiquus* Butler and Greenwood 1976. Early Pleistocene, Olduvai.

Macroscelides proboscideus vagans Butler and Greenwood 1976. Late Pliocene, Makapansgat, Taung.

Mylomygale spiersi Broom 1948; De Graaff 1960; Patterson 1965. Late Pliocene, Taung, Sterkfontein.

This family, confined to Africa, occupies an isolated position among eutherian mammals, in recognition of which a separate order Macroscelidea has been proposed for it (Butler 1956b, Patterson 1965; ranked as a suborder of the Insectivora by Van Valen 1967). McKenna (1975) regards the Asiatic Paleocene-Oligocene family Anagalidae as macroscelidean, and believes that the Macroscelididae originated in Asia in or before the early Oligocene. He considers that the order Macroscelidea is related to the order Lagomorpha; Szalay (1977) goes further and includes the Macroscelidea in the Lagomorpha as a suborder.

The fossil history of the Macroscelididae in Africa, though fragmentary, is better known than that of other African "insectivore" families; it has been reviewed by Patterson (1965).

The oldest known species, *Metoldobotes stromeri* from the Fayum, was originally interpreted as a mixodectid but was recognized as macroscelidid by Patterson (1965). It is known only from a single specimen, a mandibular ramus of which the anterior tip is missing. The dental formula, as interpreted by Patterson, is $I_{3?}C_1P_4M_2$. The anterior part of the dentition, from P_3 forward, is abbreviated in comparison to the length of M_{1-2}, a resemblance to the living *Macroscelides* that is more likely to be due to parallel evolution than to direct relationship. There is an enlarged, procumbent incisor (I_3 or I_2), which was apparently not the most anterior tooth. The canine is a small tooth with a single blunt cusp. The cheek teeth are bunodont and low-crowned as in Miocene Rhynchocyoninae, and M_3 is absent as in most macroscelidids. Despite its early date, *Metoldobotes* is already too specialized to stand near the ancestry of any of the later members of the family. Patterson (1965) very tentatively referred it to the Macroscelidinae, a subfamily otherwise known only from the Pliocene and Pleistocene, but this carries little conviction. *Metoldobotes* seems best interpreted as an extinct offshoot from an Eocene macroscelidid stock, indicative of a radiation that had already taken place by early Oligocene time.

The fragment of a femur, tentatively referred to *Metoldobotes* by Schlosser (1911), seems to have little resemblance to living Macroscelididae, and the identification must be regarded as very improbable.

By the Miocene the Macroscelididae had diversified into at least two subfamilies, of which the Rhynchocyoninae still exist. Two Miocene species have been placed in the existing genus *Rhynchocyon* (Butler and Hopwood 1957, Butler 1969), but these show primitive characters that might justify at least a subgeneric separation. *R. clarki* is known from the anterior part of a skull, and a number of teeth and jaw fragments have been referred to this species mainly on the basis of size. The anterior dentition is unknown beyond P^1 and P_2. *R. rusingae* is known only from jaw fragments and an isolated M^1. The skull of *R. clarki* shows the characteristic rhynchocyonine flattening of the face, with a broad unperforated palate and long infraorbital canal, but it is primitive in some respects. The crowns of the teeth are less elevated than in living species, and the anterior hypoconid crest (cristid obliqua) meets the trigonid midway between the protoconid and the metaconid, instead of joining the metaconid; in these respects there is a resemblance to *Metoldobotes*. *R. rusingae* is larger than *R. clarki,* resembling the living species in size; it also agrees with living forms in the presence of a protostylid on the posterior ridge of P_2, P_3, and dP_3, and it may be near the direct ancestry of the two living species of *Rhynchocyon*.

The Myohyracinae appear to have specialized for a herbivorous diet, and were formerly regarded as hyracoid (e.g., Whitworth 1954); they were recognized as macroscelidid by Patterson (1965). *Myohyrax oswaldi* is known from the almost complete skull and mandible, some tarsals, vertebrae, and fragments of long bones and is thus the most completely known fossil species of the family. *Protypotheroides beetzi* is represented only by teeth and jaws. The skull of *Myohyrax* resembles those of Macroscelidinae rather than Rhynchocyoninae in the narrow facial region. The mandible is much deeper below the cheek teeth and more expanded in the region of the angle than in other members of the family, implying a greater development of the masseter musculature. The first two incisors in each jaw are enlarged and procumbent (perhaps a resemblance to *Metoldobotes*), and their enamel is confined to the labial surface. The cheek teeth are prismatic as in Macroscelidinae, but they have crown cement which does not occur in that subfamily. Wear of the upper cheek teeth produces fossettes, which are subdivided as in some species of *Elephantulus* (*E. antiquus, E. edwardi*). P^3 and P_3 are fully molariform, P^2 and P_2 partly so. M^2 is similar in size to M^1, and M_2 to M_1. Third molars are present as small teeth in both jaws. Thus the chewing area has been considerably expanded. *Protypotheroides* is larger and somewhat differently specialized. Its cheek teeth lack cement, but a fossette develops in the trigonids of P_4–M_2, due to a deepening of the valley between the paraconid and the metaconid. A second fossette in the talonid of these teeth separates the entoconid from the rather strongly developed hypoconulid.

Two problematic specimens (Butler 1969) may indicate other lines of evolution in the Miocene. One of them, from Rusinga, consists of two associated mandibles of a juvenile animal with milk molars. DP_3 has a metaconid, displaced lingually, and thus differs from *Rhynchocyon* which has a protostylid in a more buccal position on this tooth, but no metaconid. In Macroscelidinae the metaconid of dP_3 is indistinct or absent, but there is a metastylid situated more posteriorly and connected to the rudimentary hypoconid by the equivalent of the cristid obliqua. The unerupted M_1 of the Rusinga specimen can be seen in vertical cross-section; it is more high-crowned than in *Rhynchocyon*. Though tentatively referred to the Myohyracinae, this specimen may represent a hitherto unknown group.

The second problematic specimen, from Koru, consists of a fragment of mandible with unworn P_2 and P_3. P_2 has a protostylid as in some specimens of *Rhynchocyon,* but it also possesses a minute metaconid in a posterolingual position. On P_3 the protostylid is absent, but the metaconid is much better developed.

The Macroscelidinae form the largest subfamily at present, but they are unknown in the fossil record before the Pliocene (unless *Metoldobotes* is included). The taxonomy of the living species has been revised by Corbet and Hanks (1968), who recognize three genera, *Petrodromus* (one species), *Elephantulus* (including *Nasilio*) (nine species), and *Macroscelides* (one species). A tenth species of *Elephantulus, E. fuscus,* has since been distinguished by Corbet (1973). The Macroscelidinae cannot be derived from either of the Miocene subfamilies, though they perhaps show rather more resemblance to the Myohyracinae than to the Rhynchocyoninae; presumably their ancestry lies in a third group, at present unknown. By the time that the subfamily appears in the fossil record, the main radiation must have taken place, for *Macroscelides* and *Elephantulus* were well differentiated at Makapansgat, and a third, now extinct genus, *Palaeothentoides,* is known from Klein Zee. The most recent study of Pliocene-Pleistocene Macroscelidinae is by Butler and Greenwood (1976), who reported on specimens from Olduvai Bed I and from Makapansgat. The commonest species from Olduvai possesses a third molar in the lower jaw and is close to the living *Elephantulus (Nasilio) fuscus,* a species with a limited distribution around the southern end of Lake Nyasa. It occurs also at Makapansgat, and possibly at other Transvaal sites from which *"Nasilio"* has been reported but not examined in detail. However, the main species at Makapansgat, and probably in other Transvaal cave breccias, is *E. antiquus,* a distinctive species with some resemblance to *E. rufescens,* from East Africa, and to *E. edwardi,* from South Africa. Less common, but found both at Olduvai and a number of Transvaal localities, is *E. broomi* (= *E. langi*) a form which seems to resemble quite closely the living *E. intufi. Macroscelides* is represented at Makapansgat by a form only subspecifically different from the living *M. proboscideus;* the genus occurs at Taung, but apparently not at Olduvai. Vegetation, and therefore rainfall, seems to be a major factor limiting the distribution of the species of Macroscelidinae; at the extremes are *Petrodromus,* an inhabitant of forests and probably in a number of respects the most primitive surviving form, and *Macroscelides,* which is confined to areas with rainfall below

100 mm and specialized in many respects. The fossil history of *Petrodromus* is unknown.

M_3 is absent in most Macroscelidinae, but it occurs in the Pliocene *Palaeothentoides* and in three species of *Elephantulus* (*E. brachyrhynchus, E. fuscus, E. fuscipes*) that are placed by some authors (including Patterson 1965) in a separate genus, *Nasilio.* In most other characters the species of *Nasilio* are very close to *Elephantulus intufi, E. rupestris,* and *E. broomi;* indeed lower dentitions referred to *E. broomi* are indistinguishable from those of the early Pleistocene form of *E. fuscus* except for the absence of M_3. Outside the Macroscelidinae M_3 is present only in the Myohyracinae, where it is associated with M^3, a tooth that occurs in *Nasilio* only as a very rare individual abnormality. Whether M_3 is a primitive character in the family which has been lost in most lines, or whether it was lost early and then regained in a limited number of lines, is a question that cannot be answered on existing knowledge.

Another character of taxonomic interest is the form of P^2. In *E. intufi, E. rupestris, E. broomi,* and the species possessing M_3, as well as in *Macroscelides,* P^2 resembles P^3 in being a comparatively broad tooth with four roots and well-developed lingual cusps; in the remaining species of *Elephantulus,* including the late Pliocene-early Pleistocene *E. antiquus,* P^2 is narrower, with three or two roots, and it shows various degrees of reduction of the lingual cusps. Similar differences, though less marked, can be seen in P^3. It is not unlikely that the broad type of P^2 is the more primitive, and that the narrower type has come about by reduction; even within the species *E. (Nasilio) brachyrhynchus* there is much individual and regional variation in the size of the protocone of P^2, leading in extreme cases to its almost complete absence. Broom (1937) proposed the generic name *Elephantomys* for *E. langi* (i.e., *E. broomi*) and *E. intufi* because of their "molariform" P^2, but subsequently (1938) he realized that the type species of *Elephantulus, E. rupestris,* also had this character and therefore that *Elephantomys* was invalid. Although he did not create a name for the remaining species he continued to believe that the character has taxonomic value, and in this belief he was probably correct.

Undescribed material from Irhoud-Ocre, Morocco (about 2 m.y.), shows the presence there of a species of *Elephantulus,* perhaps ancestral to *E. rozeti* which now inhabits North Africa (J.-J. Jaeger, pers. commun.).

There remains *Mylomygale,* a problematic form from the late Pliocene of the Transvaal, known only by a lower jaw from Taung (Broom 1948) and an isolated P$_4$ from Sterkfontein (De Graaff 1960). It has high-crowned teeth like *Macroscelides,* but the dentition anterior to P$_3$ is even more abbreviated than in that genus. It agrees with *Protypotheroides* and differs from Macroscelidinae in the presence of a reentrant fold of enamel between the entoconid and the enlarged hypoconulid. *Myxomygale* needs restudy. Patterson (1965) proposed for it a new subfamily Myxomygalinae, but it might eventually prove to be a late survivor of the Myohyracinae.

From this very incomplete history the following tentative conclusions may be drawn. The Macroscelididae were a component of the Paleogene fauna of Africa, and there is an indication of a major radiation of the group that was already in progress in the Oligocene and continued into the Miocene. An unknown derivative of this radiation gave rise to the Macroscelidinae, which underwent a secondary radiation in the late Tertiary. *Rhynchocyon* is the only surviving relic of the earlier radiation.

Family Erinaceidae

RECORDED SPECIES:

Galerix africanus Butler 1956a. Early Miocene, Songhor, Rusinga.

Lanthanotherium sp., Butler 1969. Early Miocene, Songhor.

Amphechinus rusingensis Butler 1956a, 1969. Early Miocene, Rusinga, Songhor.

Gymnurechinus leakeyi Butler 1956a, 1969. Early Miocene, Rusinga.

Gymnurechinus camptolophus Butler 1956a. Early Miocene, Rusinga.

Gymnurechinus songhorensis Butler 1956a, 1969. Early Miocene, Songhor, ?Rusinga.

Protechinus salis Lavocat 1961. Mio-Pliocene, Beni Mellal.

Erinaceus broomi Butler and Greenwood 1973; Broom 1937, 1948. Synonym: *Atelerix major* Broom 1937. Early Pleistocene, Bolt's Farm, Olduvai.

Unlike the Macroscelididae, the Erinaceidae are widespread in the lower Tertiary of Europe, Asia, and North America, and they appear in Africa only as immigrants.

Whereas in the Miocene of Europe the Echinosoricinae were as numerous and varied as the Erinaceinae, in Africa they are represented by only half a dozen fragments of mandible from Songhor and Rusinga. The specimens have been provisionally placed in the European genera *Galerix* and *Lanthanotherium,* but the material is so incomplete that identification can be only tentative; it is not even certain that more than one species is represented. The evi-

dence for the existence of *Galerix* in the Mio-Pliocene of Morocco, consisting as it does of an isolated incisor tooth (Lavocat 1961), is very doubtful indeed.

Erinaceinae, on the other hand, are common in the early Miocene of East Africa. A species of *Amphechinus* (= *Palaeoerinaceus*) is represented by several specimens from Rusinga and Songhor, including the anterior part of a skull with associated lower jaw. Although comparatively late in date (the genus occurs in the Oligocene of Europe and China), *A. rusingensis* possesses some primitive characters unknown in the European forms—e.g., the relatively larger upper canines and the two-rooted P$_2$.

Most of the Miocene erinaceid material from East Africa has been placed in the genus *Gymnurechinus,* unknown outside Africa but the most common component of the insectivore fauna of Rusinga and Songhor. Several skulls have been found, and also part of a skeleton (Butler 1956a, 1969). Like *Amphechinus, Gymnurechinus* is at a level of evolution at which the dentition had already approached the modern erinaceine condition, but the skull and skeleton still retained many echinosoricine features. It has avoided a number of specializations that exclude *Amphechinus* from the direct ancestry of *Erinaceus* and thus cannot be derived from *Amphechinus;* it probably represents an offshoot from the unknown stock that gave rise to *Erinaceus.* Other derivatives of this stock make their appearance in Europe in the middle ("*Erinaceus*" *sansaniensis*) and late Miocene (*Mioechinus oeningensis*) (Butler 1948, 1956a), but the Oligocene ancestry is unknown.

Thus the African Miocene erinaceids probably did not come from Europe. A more likely source is Asia south of the Tethys Sea, but unfortunately no fossil insectivores have been obtained from that region.

The African Miocene erinaceids appear to have become extinct: although *Gymnurechinus* resembles *Erinaceus* in many ways it is too specialized in others to stand in the direct ancestry of the modern forms. *Protechinus salis* has been claimed as a possible derivative (Lavocat 1961); it has several primitive characters such as the transverse upper molars, but at the same time it has advanced to the modern erinaceine condition in the facial position of the lacrimal foramen and other features of the antorbital region. It is known only by a maxillary fragment, some incomplete mandibles, the lower dentition, and the upper cheek teeth. It could well be an early representative of the *Erinaceus* group that had invaded Africa from Europe or Asia in late Miocene time. *Postpalerinaceus vireti* Crusafont and Villalta (1948) from the early Pliocene of Spain shares some specializations with *Gymnurechinus* and could be a migrant from Africa.

At present most African Erinaceidae belong to the genus *Erinaceus*, of which they form a subgenus *Atelerix;* the only exceptions are single species of the predominantly Asiatic genera *Hemiechinus* and *Paraechinus* that are probably very late immigrants into northern Africa. *Atelerix* is represented in the fossil record by *E.* (*A.*) *broomi* (= *A. major* Broom), known from the anterior part of a skull from Bolt's Farm, Transvaal, and from numerous dental and skeletal remains from Bed I, Olduvai (early Pleistocene). An analysis of its characters (Butler and Greenwood 1973) shows that *E. broomi* shares many primitive characters with *E.* (*Atelerix*) *algirus,* a Mediterranean species, and with *E. europaeus;* at the same time it shares some more specialized characters with the existing *E.* (*A.*) *albiventris,* widespread in tropical Africa, and *E.* (*A.*) *sclateri* from Somalia. The evidence is consistent with an invasion from the north of a stock resembling *E. europaeus,* of which *E. algirus* is the least modified surviving descendant.

Thus there appear to have been at least two invasions of Erinaceidae into Africa, one at the beginning of the Miocene, the result of which subsequently died out, and the other perhaps at the beginning of the Pleistocene. The first was probably from southern Asia, the second from southwestern Asia or Europe. *Protechinus* may represent another invasion, perhaps late Miocene.

Family Soricidae

RECORDED SPECIES:

Crocidura sp. Butler and Hopwood 1957. Early Miocene, Rusinga (?).

"Sorex" dehmi africanus Lavocat 1961. A crocidurine (see Repenning 1967). Mio-Pliocene, Beni Mellal (Morocco).

Myosorex robinsoni Meester 1955; Butler and Greenwood 1965. Late Pliocene and early Pleistocene, Makapansgat, Sterkfontein, Bolt's Farm, Swartkrans, Kromdraai, Olduvai.

Sylvisorex cf *granti* Thomas. Butler and Greenwood (unpub.). Early Pleistocene, Olduvai.

Sylvisorex sp. Butler and Greenwood 1965 (as *Suncus* cf *lixus* (Thomas)). Early Pleistocene, Olduvai.

Suncus varilla (Thomas). Meester and Meyer 1972; Meester 1955 (as *Suncus* sp.). Late Pliocene and early Pleistocene, Makapansgat, Sterkfontein, Bolt's Farm, Sterkfontein Extension, Swartkrans, Kromdraai.

Suncus cf *varilla* Butler and Greenwood 1965 (as *S.* cf *orangiae* (Roberts)). Early Pleistocene, Olduvai.

Suncus infinitesimus (Heller). Meester and Meyer 1972.

Late Pliocene and early Pleistocene, Sterkfontein, Sterkfontein Extension, Gladysvale, Makapansgat.

Suncus cf *infinitesimus.* Butler and Greenwood 1965 (as *Suncus* sp. 1). Early Pleistocene, Olduvai.

Crocidura taungensis Broom 1948; Meester 1955; De Graaf 1960. Late Pliocene, Taung.

Crocidura cf *hindei* Thomas. Butler and Greenwood 1965. Early Pleistocene, Olduvai.

Crocidura cf *bicolor* Bocage. Davis and Meester (unpub.). Latest Pliocene, Bolt's Farm.

Diplomesodon fossorius Repenning 1965, 1967. Late Pliocene, Makapansgat.

Although the shrews are the most numerous insectivores in Africa at the present time, their fossil record in the Tertiary is extremely meager. All African shrews belong to the subfamily Crocidurinae, distinguished according to Repenning (1967) by their unpigmented teeth and by characters of the mandibular condyle and P_4. Heim de Balsac and Lamotte (1956) noted a number of resemblances of *Myosorex* to Soricinae and proposed to unite the two subfamilies Soricinae and Crocidurinae. Though *Myosorex* clearly stands apart from other African shrews, its resemblances to Soricinae may consist of primitive characters lost in other Crocidurinae.

Because the present distribution of Crocidurinae is mainly African and Oriental, Repenning (1967) believed that the group had originated south of the Tethys Sea, and entered Europe only in the Miocene. He identified the following European species as crocidurine: *"Sorex" pusilliformis* Doben-Florin (early Burdigalian), *"S." dehmi* Viret and Zapfe (late Burdigalian or Vindobonian), *Miosorex grivensis* (Deperet) (Vindobonian), and *Soricella discrepans* Doben-Florin (Burdigalian). The first of these has one lower tooth between I_1 and P_4, as in all living crocidurines except *Myosorex,* in which an additional minute tooth is normally present anterior to P_4. *Miosorex* and *Soricella* have two intermediate teeth and *"Sorex" dehmi* has three; reduction of teeth seems to take place in the region immediately anterior to P_4. The range of dental formulae shown by these presumably immigrant forms suggest a source area inhabited by a wide variety of crocidurine shrews. It is unlikely that this area was Africa, for despite the discovery in the East African early Miocene of a fair number of specimens of tenrecids and other small mammals of similar size to shrews, only one soricid specimen has been recorded. Even this specimen is doubtful; it is a surface find from Rusinga, a mandibular ramus containing only M_2 and a broken M_1, but showing the alveoli of the other teeth. Insofar as it can be compared, it closely resembles species of the *Crocidura flavescens–C. occidentalis* group, which are among the more advanced of living forms. The lack of primitive characters in such an early shrew is remarkable, and until more material is obtained the suspicion must remain that this specimen is of Recent origin.

If the Rusinga record is rejected, the oldest African shrew would be *"Sorex" dehmi africanus* from the Mio-Pliocene of Morocco, regarded by Lavocat (1961) as only subspecifically different from *"S." dehmi* of the Miocene of Europe. It is known from two mandibular rami. Although its dental formula is very primitive, there is as yet no evidence that this form is close to the ancestry of the numerous living African crocidurines. It does, however, show that the subfamily had reached Africa by the beginning of the Pliocene.

Several species are known from the Plio-Pleistocene of Olduvai and the Transvaal, but they are not sufficiently different from living forms to throw much light on the evolution of the group. For this it is necessary to fall back on a comparative study of the Recent species.

The genus *Myosorex* (in which *Surdisorex* is included as a subgenus) has a wide distribution in tropical Africa, where most of the species are confined to high altitudes (Heim de Balsac and Lamotte 1956, Heim de Balsac 1967, 1968). Two closely related species, however, extend to South Africa (Meester 1958): *M. cafer* and *M. varius,* which are believed to represent successive southward migrations during Pleistocene pluvials. Farther north, on the Uluguru Mountains of southern Tanzania, occurs *M. geatus* which seems to belong to the same group; Heim de Balsac (1967) has associated with it a population from the eastern border of Rhodesia, included by Meester (1958) in *M. cafer. M. robinsoni,* from the Plio-Pleistocene of the Transvaal, is an early member of this group of species. It has some characters that seem to exclude it from the direct ancestry of *M. cafer* and *M. varius* and may represent an earlier migration. A form similar to *M. robinsoni* is common in Bed I at Olduvai. According to Heim de Balsac (1967) the most primitive living species of the genus is *M. schalleri,* found in forests at low altitude south of Mount Ruwenzori, and an origin of the genus in equatorial Africa seems highly probable. *Myosorex* has fossorial adaptations and is confined to moist soil in forested areas or river banks.

Sylvisorex and *Suncus* differ from *Crocidura* in the possession of an additional small tooth in the maxilla anterior to P^4 (probably P^3), present also in most species of *Myosorex*. The species of *Sylvisorex* have been reviewed by Heim de Balsac and Lamotte (1957), who point to the frequency of archaic characters in the genus, and regard it as broadly ancestral to *Suncus* and at least in part to *Crocidura. Sylvisorex* is restricted to tropical Africa, where five of the eight species are confined to mountains, two live in low hygrophilous forest and only one (*S. megalura*) extends to savanna. Two species occur in Olduvai Bed I. One of these, known only from the lower jaw and dentition, except I_1, seems to be close to *S. granti* and *S. megalura.* The other is confined to the lower part of Bed I where it is the commonest shrew; it is known from numerous lower jaws and limb bones, but no upper teeth have been identified. It is a very distinct species, related apparently to *S. johnstoni* of West Africa, but much larger.

Whereas *Sylvisorex* is confined to Africa, *Suncus* has a wide distribution in Eurasia, and the greatest diversity of forms occurs in the Oriental Region where the genus probably originated. Some species there—e.g., *S. fellowes-gordoni* from Ceylon—have primitive characters like *Sylvisorex,* such as a narrow mandibular condyle and a basined M_3 talonid, and they cast some doubt on whether the two genera should be separated. Only six species of *Suncus* occur in Africa (Meester and Lambrechts 1971), including *S. murinus,* a commensal that has almost certainly been introduced in historical times. Fossils from Olduvai and the Transvaal cave breccias are close to the living *S. varilla* (including *S. orangiae*) and *S. infinitesimus* (including *S. chriseos*). Both these species are related to *S. etruscus,* an inhabitant of the Mediterranean region and southwestern Asia that has extended its range into West Africa south of the Sahara. It seems likely that this group of species entered Africa rather late.

Crocidura is by far the largest genus of African shrews, and indeed one of the largest genera of living mammals. The diversity of its species on the continent strongly suggests an African origin, though the genus is widely distributed in Asia and Europe. There is some indication that *Crocidura* is polyphyletic, having been derived from a number of members of the *Sylvisorex-Suncus* group by loss of P^3 (Heim de Balsac and Lamotte 1957). This tooth reappears as an atavism in several of the species (Meester 1953). Some of the most primitive species of *Crocidura,* such as *C. maurisca* and *C. bottegi,* are very close to *Sylvisorex,* while more advanced forms such as *C. bicolor* resemble *Suncus.* It is remarkable that in the early Pleistocene of Olduvai and the Transvaal *Crocidura* was less common than *Suncus* or *Myosorex.* The principal species at Olduvai, represented by about 20 mandibular fragments, seems to be a primitive relation of living species grouped around *C. hirta* and *C. hindei;* five additional specimens indicate the presence of three other species, probably *Crocidura* but too fragmentary to be identifiable. These figures may be contrasted with some 350 specimens from Olduvai referable to the genus *Suncus.* At Bolt's Farm, Transvaal, *Crocidura* is represented by a small species resembling *C. bicolor* (Davis and Meester, unpub.); a maxilla of another species from Taung was named *C. taungensis* by Broom (1948).

Paracrocidura and *Scutisorex* are specialized forms of uncertain affinities and with no fossil record. *Praesorex* is perhaps merely a large form of *Crocidura* (Heim de Balsac and Lamotte 1957).

The most remarkable fossil shrew so far reported from Africa is *Diplomesodon fossorius* Repenning (1965) from Makapansgat. It is known from more than 20 specimens, which show the mandible and the facial part of the skull. The type and only other species of the genus is *D. pulchellum* (Lichtenstein) from Central Asia, and the discovery of a South African species is of some zoogeographical interest. The two species agree in dental formula (they differ from *Crocidura* in having lost an antemolar from the maxilla), the short stout I_1, the short talonids of M_1 and M_2 and their general robustness, but they differ in some respects. In *D. fossorius* the ventral part of the mandibular condyle is much wider transversely, exceeding *Suncus* and the most advanced species of *Crocidura,* whereas in *D. pulchellum* it is narrower as in primitive species of *Crocidura;* the talonid of M_3 in *D. fossorius* is reduced to a cingulum, a greater degree of reduction than occurs in *Crocidura,* whereas in *D. pulchellum,* though simplified, it is no more reduced than in many species of *Crocidura* and *Suncus;* P_4 of *D. fossorius* has a well-developed metaconid, absent in *D. pulchellum,* and I_2 of *D. fossorius* is less reduced in comparison with P_4. In view of the prevalence of parallel evolution in the Soricidae, the generic reference of *D. fossorius* should be treated with great reserve; it could well be a derivative of the *Crocidura* group paralleling *Diplomesodon* in adaptations to dry environmental conditions.

Some soricid specimens have been obtained from Omo (H. Wesselman, pers. comm.), and also from Bulla Regia, Tunisia (ca 2.5 m.y.), and Irhoud-Ocre, Morocco (ca 2 m.y.) (J.-J. Jaeger, pers. comm.). None of this material has yet been described.

This survey underlines the lack of information about the small mammals of Africa in the later Tertiary. The existence of Soricidae in Africa in the early Miocene requires confirmation. Their high degree of specific differentiation, reaching in some cases to the generic level, implies that they have lived in Africa for a period much longer than the Pleistocene, and their presence in the Mio-Pliocene of Morocco confirms this. Some groups, however, might have entered relatively late, such as the *Suncus etruscus–S. varilla* group. There is an indication that the main source of immigrants was the Oriental Region.

Family Tenrecidae

RECORDED SPECIES:

Protenrec tricuspis Butler and Hopwood 1957; Butler 1969. Early Miocene, Songhor, Rusinga, Napak.

Erythrozootes chamerpes Butler and Hopwood 1957; Butler 1969. Early Miocene, Koru, Songhor, Napak.

Geogale aletris Butler and Hopwood 1957. Early Miocene, Rusinga.

The African Tenrecidae now consist of the two genera of the subfamily Potamogalinae: *Potamogale,* with a single species from the Congo forest region, and *Micropotamogale,* with two species isolated on widely separated mountains, Nimba and Ruwenzori. The remaining subfamilies, Tenrecinae, Oryzorictinae, and Geogalinae, are confined to Madagascar. The fossil record is restricted to the Miocene of East Africa.

Outside Africa, the Tenrecidae have usually been held to be related to the Solenodontidae of the West Indies and the Apternodontidae from the Oligocene of North America; Van Valen (1967) goes so far as to include the Apternodontidae within the Tenrecidae as a subfamily. *Butselia,* from the early Oligocene of Europe, was placed in the Tenrecoidea by Quinet and Misonne (1965), but additional material has shown that it falls better into the Plesiosoricidae (Butler 1972). Matthew (1913) traced the ancestry of the Tenrecidae to the North American Paleocene *Palaeoryctes;* on the other hand Butler (1956b) and McDowell (1958) pointed to their resemblance in many nonprimitive characters to other lipotyphlan insectivores, especially the Soricidae, and in a recent review Butler (1972) has argued that zalambdodont molar teeth do not form a satisfactory basis for insectivore classification.

Whatever their relationships may be, it is not disputed that the Tenrecidae are an ancient, isolated group. Though unknown before the Miocene, it is very likely that they formed part of the Paleogene African fauna. In the Miocene there were three very distinct genera, implying an earlier radiation (Butler and Hopwood 1957, Butler 1969). *Protenrec* is known from the antorbital part of a skull, the upper dentition as far forward as P³, and the mandible and lower dentition as far forward as the canine. Of *Erythrozootes* most of the skull is known, together with the entire upper dentition and the lower cheek teeth. *Geogale aletris* is represented by a single specimen consisting of the anterior part of a skull with worn teeth.

Protenrec and *Erythrozootes* agree in that the protocones of the upper cheek teeth are much better developed than in any living tenrecid except *Potamogale,* and the talonids of the lower teeth are correspondingly comparatively large. However, neither of these Miocene genera approaches the tribosphenic condition as closely as does *Potamogale,* and if zalambdodonty is taken to be the derivative state, *Potamogale* must be more primitive. *Micropotamogale,* on the other hand, is more completely zalambdodont than the fossil forms. The Miocene genera have well-developed lacrimal canals, lost in Potamogalinae perhaps in adaptation to their aquatic habits. *Protenrec* is uniquely primitive for a tenrecid in the intraorbital position of its lacrimal foramen; it also has a long infraorbital canal like that of erinaceids. *Erythrozootes* shows another resemblance to erinaceids in the ossification of the medial wall of its alisphenoid canal. Its tympanic cavity and basicranium closely resemble those of the Madagascan *Microgale* and *Geogale* and differ from Potamogalinae. *Erythrozootes* differs from *Protenrec* in dental formula, having only two upper premolars; it is a much larger animal, with blunter cheek teeth and a rugose cranial surface. The two genera are evidently neither closely related to the living subfamilies nor to each other. They indicate that a greater variety of tenrecids probably existed in Africa in the middle Tertiary, including the unknown ancestors of the living African and Madagascan forms.

In this connection *Geogale aletris,* the third Miocene form, is of interest. It shows several special resemblances to the living *G. aurita* from Madagascar: the enlarged first upper inciscors, widely separated from each other in the midline; reduction of the teeth anterior to P⁴; and the infraorbital canal, which is much longer than in other living Tenrecidae. Some differences may be significant, notably a backward prolongation of the palate in *G. aletris,* and parallelism cannot be ruled out. If the relationship is real, it would follow that *Geogale* reached Madagascar independently of the ancestor of the other Madagascan tenrecids, for it is too specialized to have given rise to them.

Family Chrysochloridae

RECORDED SPECIES:

Prochrysochloris miocaenicus Butler and Hopwood 1957; Butler 1969. Early Miocene, Songhor, Koru.

Proamblysomus antiquus Broom 1941. Plio-Pleistocene, Bolt's Farm.

Chlorotalpa spelea Broom 1941. Late Pliocene, Sterkfontein.

Amblysomus hamiltoni De Graaff 1958 (as *Chrysotricha hamiltoni*). Late Pliocene, Makapansgat.

Like the tenrecids, the golden moles are unknown before the Miocene, but presumably formed part of the early Tertiary African fauna. Their relationships are very obscure, mainly because of their extreme fossorial specialization. Their adaptations have been paralleled on other continents, notably in *Epoiocatherium,* a North American pholidote, *Necrolestes,* a South American marsupial, and *Notoryctes,* the Australian marsupial mole (Turnbull and Reed 1967). The Chrysochloridae are often grouped with the Tenrecidae, but the two families share very few characters that cannot be found in Soricidae and other Lipotyphla. Although zalambdodont cheek teeth occur in both families, the patterns differ in detail.

The Miocene *Prochrysochloris* is known from the facial part of the skull and the upper and lower dentition. The skull seems to be typically chrysochlorid. The teeth, however, have some features shared with Miocene tenrecids: V-shaped protocones on the upper molars, comparatively large talonids on the lower molars, and a maximum molar emphasis on M¹⁻². Such resemblances cannot be taken as implying anything more than an approach of both families towards the primitive lipotyphlan condition.

Classification of the living species has been discussed by several authors. Characters that have been used are the presence or absence of the posterior molars, the degree of reduction of the talonids of the lower cheek teeth, and proportions of the skull. Simonetta (1968) emphasized the size and shape of the head of the malleus (which is often greatly enlarged) and the morphology of the epitympanic recess in which it lies. He grouped the Chrysochloridae into three subfamilies with six genera and showed that loss of the lower molar talonids took place independently within each subfamily. Unfortunately Simonetta's classification cannot be applied to *Prochrysochloris,* the ear region of which is unknown. It resembles *Eremitalpa* in having a very broad, short face.

Three species have been described from the Plio-Pleistocene of the Transvaal, but their exact relationships to living species remain to be determined. Chrysochlorids have not been recorded from Olduvai.

The Pattern of Insectivore Evolution in Africa

The fossil record of African insectivores is so scanty that only a few broad and tentative conclusions can be drawn. In the Oligocene there appears to have been a fauna containing ptolemaiids and macroscelidids, and almost certainly also tenrecids and chrysochlorids, families that evolved on the continent during its period of isolation. These families persisted into the Miocene, by which time the Macroscelididae and Tenrecidae had undergone an adaptive radiation. The Miocene saw the appearance in Africa of the Erinaceidae, immigrants presumably from southern Asia. Whether the Soricidae entered from Asia at the same time requires confirmation.

A considerable faunistic change intervened between the early Miocene, as represented by the deposits in Kenya and Uganda, and the Plio-Pleistocene, as seen at Olduvai and the Transvaal. Ptolemaiidae disappear; Myohyracinae and Rhynchocyoninae are replaced by Macroscelidinae; Miocene erinaceids disappear and *Erinaceus* has invaded from the north; Soricidae become the dominant insectivores; Tenrecidae disappear; only the Chrysochloridae persist. Apart from Chrysochloridae, the only early Miocene insectivores that have survived into the modern fauna with little change are *Rhynchocyon* and the Potamogalinae, which are not represented in the Pliocene and Pleistocene record. A principal factor in such a change of fauna is most likely to have been the entry of mammals from the Oriental and Palearctic regions during the later Tertiary, but the lack of paleontological data on the small mammals of Africa and southern Asia of that time prohibits a more detailed analysis: the only direct evidence is the presence of Erinaceidae and Soricidae in the Mio-Pliocene of Morocco. Once the isolation of Africa had been ended, one can envisage a series of invasions. The early Pleistocene species of *Erinaceus* and *Suncus* are so closely related to extra-African forms that it is unlikely that they had been on the continent for very long, but the higher level of differentiation of most Soricidae suggests that their ancestors entered Africa much earlier, perhaps during the Miocene.

Order Chiroptera

RECORDED SPECIES:

Family Pteropodidae

Propotto leakeyi Simpson 1967; Walker 1969. Early Miocene, Songhor, probably Rusinga.

Family Microchiroptera *insertae sedis*

Vampyravus orientalis Schlosser 1910, 1911; Savage 1951; Russell and Sigé 1970. Synonym: *Provampyrus orientalis* Schlosser 1911. Early Oligocene, Fayum.
Indet. Lavocat 1961. Mio-Pliocene, Beni Mellal.

Family Emballonuridae

Taphozous incognita Butler and Hopwood 1957 (as *Saccolaimus incognita*). Early Miocene, Koru.
Indet. Butler 1969. Early Miocene, Rusinga.

Family Megadermatidae

Indet. Butler and Hopwood 1957. Early Miocene, Rusinga.
Afropterus gigas Lavocat 1961; Russell and Sigé 1970. Mio-Pliocene, Beni Mellal.
Cardioderma sp. Butler and Greenwood 1965. Early Pleistocene, Olduvai.

Family Rhinolophidae

Rhinolophus ferrumequinum millali Lavocat 1961. Mio-Pliocene, Beni Mellal.
Rhinolophus cf *capensis* Lichtenstein, De Graaf 1960. Late Pliocene, Makapansgat.

Family Hipposideridae

Hipposideros sp. Butler 1969. Early Miocene, Songhor.
Asellia (?) *vetus* Lavocat 1961. Mio-Pliocene, Beni Mellal.

Family Myzopodidae

Myzopoda sp. Butler and Greenwood (unpub.). Early Pleistocene, Olduvai.

Family Vespertilionidae

Indet. Lavocat 1961. Mio-Pliocene, Beni Mellal.
Myotis sp. Broom 1948. Plio-Pleistocene, Bolt's Farm.
Myotis sp. Butler and Greenwood (unpub.). Early Pleistocene, Olduvai.
Cf *Nycticeius* (*Scoteinus*) *schlieffeni* (Peters), Butler and Greenwood 1965. Early Pleistocene, Olduvai.
Cf *Pipistrellus* (*Scotozous*) *rueppelli* (Fischer), Butler and Greenwood 1965. Early Pleistocene, Olduvai.
Eptesicus cf *hottentotus* (Smith), Butler and Greenwood (unpub.). Early Pleistocene, Olduvai.
Miniopterus cf *schreibersi* (Kuhl) Butler and Greenwood (unpub.). Early Pleistocene, Olduvai.

Family Molossidae

Indet. Lavocat 1961. Mio-Pliocene, Beni Mellal.
Indet. (4 spp.). Butler and Greenwood (unpub.). Early
Pleistocene, Olduvai.

The paleontological record of the Chiroptera in
Africa, as in the world generally, is even more incomplete than that of the Insectivora. Hayman (1967)
distinguishes 187 species now living on the African
continent, of which 157 are endemic or extend only
to southern Arabia or to Madagascar and adjacent
islands. Those are placed in 43 genera representing
9 families, of which the Vespertilionidae is much the
largest, with 75 species in 14 genera. The next largest are the Molossidae with 31 species in 6 genera,
and the Pteropodidae with 27 species in 12 genera.

The oldest fossil bat described from Africa is *Vampyravus* (= *Provampyrus*) *orientalis,* a humerus from
the Oligocene of the Fayum. Though referred by
Schlosser (1911) to the American family Phyllostomatidae, its morphology falls outside the known
range of members of that family (Savage 1951) and
the relationships of *Vampyravus* are problematic;
Smith (1972) refers it tentatively to the Megadermatidae. Some chiropteran jaw material has recently been collected from the Fayum but not yet described (E. L. Simons, pers. comm.).

Four families have been recognized in the early
Miocene of East Africa. Of these, the Pteropodidae
are represented by *Propotto leakeyi,* originally described by Simpson (1967) as a lorisid, but recognized as pteropodid by Walker (1969). It is known
only from mandibular fragments. It shows some
primitive features: the cheek teeth are less spaced
than in most modern forms; the molars are very low
crowned and provided with small blunt cusps, and
the groove that runs along the length of the tooth in
modern forms is not developed. If *Archaeopteropus,*
from the Oligocene of Italy, is removed from the
Megachiroptera following Russell and Sigé (1970),
Propotto becomes the oldest known member of the
suborder. It does not seem likely, however, that the
Pteropodidae are of African origin, as the family
reaches its greatest diversity in the Oriental Region:
apart from the *Epomophorus* group, which is confined to Africa, the African Pteropodidae belong to
or are related to genera of mainly southern Asiatic
distribution.

The Emballonuridae are known from the early
Miocene by a skull fragment (Butler and Hopwood
1957) and a humerus (Butler 1969). This family has
a pantropical distribution and first appears in Europe in the early Oligocene. The Megadermatidae
appear in the late Eocene or early Oligocene of Europe and are now widely distributed in the Old
World tropics. A poorly preserved fragment of mandible from the early Miocene of Rusinga has been referred to the family, and Russell and Sigé (1970)
have suggested that *Afropterus gigas* Lavocat
(1961), based on some isolated molars from the Mio-
Pliocene of Morocco, is also a megadermatid. A maxilla and a mandible from Olduvai appear to represent an extinct species related to the existing
Cardioderma cor. The Hipposideridae are represented by a humerus from the Miocene of Songhor
(Butler 1969) and by a maxilla and a mandible from
the Mio-Pliocene of Morocco (Lavocat 1961). The
family goes back to the middle Eocene in Europe,
and at present is widespread in the Old World
tropics.

The Nycteridae, at present confined to tropical
Africa except for one Asiatic species, are known
from undescribed Pliocene material from Kanapoi
(B. Patterson, pers. comm.). The Rhinolophidae,
which date from the late Eocene or early Oligocene
of Europe, have been recorded in Africa only from
the Mio-Pliocene of Morocco (Lavocat 1961) and the
late Pliocene of Makapansgat (De Graaff 1960).

It is somewhat surprising that the two largest
families of living African bats, the Molossidae and
the Vespertilionidae, together comprising more
than half the existing species, have not been found
in the Miocene of East Africa, although they go back
to the beginning of the Oligocene in Europe. In view
of the small sample (four or five species) of bats so
far identified from the African Miocene, the absence
of these families may be a matter of chance. Certainly in the Mio-Pliocene of Morocco both families
were present, though known only from isolated
molar teeth (Lavocat 1961). In the early Pleistocene
of Olduvai, four species of Molossidae and five of
Vespertilionidae have been distinguished on the
basis of mandibles or humeri or both; thus the two
families had reached their present-day dominance.
A vespertilionid, *Myotis* sp. has been obtained from
Bolt's Farm, Transvaal (Broom 1948).

A humerus from Olduvai agrees very closely, except for its larger size, with that of *Myzopoda aurita,*
the only living species in the family Myzopodidae,
confined to Madagascar. This family, therefore, formerly existed on the African continent. *Myzopoda*
may be an offshoot from the Vespertilionidae.

The Rhinopomatidae, with one genus found in
arid regions of southern Asia and northern Africa,
are the only living family not represented in the fossil African record. They are probably late immigrants.

Because of the mobility conferred on them by their

powers of flight, the Chiroptera are less useful as paleogeographical indicators than most other orders of mammals. Six of the existing African families are known from the Paleogene of Europe: Emballonuridae, Megadermatidae, Rhinolophidae, Hipposideridae, Molossidae, and Vespertilionidae. Three of these were in Africa in the early Miocene (Emballonuridae, Megadermatidae, Hipposideridae), but whether they had crossed the Tethys Sea at an earlier date or whether they had entered only after a land connection with Eurasia had been formed is unknown. The Pteropodidae probably evolved south of Tethys, but their existing distribution suggests an Asiatic rather than an African source, and it may be supposed that they entered Africa with the Erinaceidae at the beginning of the Miocene. The Molossidae and the Vespertilionidae, which dominate at the present day, appear in the African record, together with the Rhinolophidae, only in the Mio-Pliocene; it is possible that like the Soricidae they entered at a relatively late date, though further discovery may well disprove this. The Nycteridae appear from their existing distribution to have originated in Africa. The peculiar Myzopodidae are probably also of African origin, but they survive only in Madagascar. The Rhinopomatidae may be late immigrants from southern Asia like *Paraechinus*.

Although the evidence is scanty, it does suggest a chiropteran history that parallels that of the Insectivora. *Vampyravus* is so far the only bat recorded from the Paleogene, when the continent was isolated and endemism must have predominated. Then at the beginning of the Miocene we can imagine a modernization brought about by the introduction of groups from Eurasia, followed by a second phase of modernization later in the Miocene. Faunistic interchange, particularly with the Oriental Region, must have played an important part in the later phases of bat evolution in Africa because many of the genera, and species groups within large genera, extend into southern Asia.

References

Andrews, C. W. 1914. On the lower Miocene vertebrates from British East Africa collected by Dr. Felix Oswald. *Q. J. Geol. Soc. Lond.* 70:163–186.

Broom, R. 1937. On some new Pleistocene mammals from limestone caves of the Transvaal. *S. Afr. J. Sci.* 33:750–768.

———. 1938. Note on the premolars of the elephant shrews. *Ann. Transv. Mus.* 19:251–252.

———. 1941. On two Pleistocene golden moles. *Ann. Transv. Mus.* 20:215–216.

———. 1948. Some South African Pliocene and Pleistocene mammals. *Ann. Transv. Mus.* 21:1–38.

Butler, P. M. 1948. On the evolution of the skull and teeth in the Erinaceidae, with special reference to fossil material in the British Museum. *Proc. Zool. Soc. Lond.* 118:446–500.

———. 1956a. Erinaceidae from the Miocene of East Africa. *Fossil Mammals of Africa,* 11. London: British Museum (Natural History).

———. 1956b. The skull of *Ictops* and the classification of the Insectivora. *Proc. Zool. Soc. Lond.* 126:453–481.

———. 1969. Insectivores and bats from the Miocene of East Africa: new material. In L. S. B. Leakey, ed., *Fossil vertebrates of Africa,* vol. 1. New York and London: Academic Press, pp. 1–38.

———. 1972. The problem of insectivore classification. In K. A. Joysey and T. S. Kemp, eds., *Studies in vertebrate evolution.* Edinburgh: Oliver & Boyd, pp. 253–265.

Butler, P. M., and M. Greenwood. 1965. Insectivora and Chiroptera. In L. S. B. Leakey, ed., *Olduvai Gorge 1951–1961.* Vol. 1: *Fauna and background.* Cambridge: Cambridge University Press, pp. 13–15.

———. 1973. The early Pleistocene hedgehog from Olduvai, Tanzania. In L. S. B. Leakey, R. J. G. Savage, and S. C. Coryndon, eds., *Fossil vertebrates of Africa,* vol. 3. London: Academic Press, pp. 7–42.

———. 1976. Lower Pleistocene elephant-shrews (Macroscelididae) from Olduvai and Makapansgat. In R. J. G. Savage, and S. C. Coryndon, eds., *Fossil Vertebrates of Africa,* vol. 4. London: Academic Press, pp. 1–56.

———. Early Pleistocene Soricidae and Chiroptera from Olduvai. In preparation.

Butler, P. M., and A. T. Hopwood. 1957. Insectivora and Chiroptera from the Miocene rocks of Kenya Colony. *Fossil Mammals of Africa* 13. London: British Museum (Natural History), 35 pp.

Coombs, N. C. 1971. Status of *Simidectes* (Insectivora, Pantolestoidea) of the late Eocene of North America. *Am. Mus. Novit.* 2455:1–41.

Corbet, G. B. 1973. Family Macroscelididae. In J. Meester, ed., *The mammals of Africa: an identification manual.* Washington: Smithsonian Institution, part 1.5.

Corbet, G. B., and J. Hanks. 1968. A revision of the elephant-shrews, family Macroscelididae. *Bull. Br. Mus. Nat. Hist.,* Zoology 16:47–111.

Crusafont Pairo, M., and J. F. de Villalta. 1948. Sur un nouveau *Palerinaceus* du Pontien d'Espagne. *Eclog. Geol. Helv.* 40:320–333.

Davis, H. D. S., and J. Meester. Report on the microfauna in the University of California collections from the South African cave breccias. In preparation.

De Graaff, G. 1958. A new chrysochlorid from Makapansgat. *Palaeont. Afr.* 5:21–27.

———. 1960. A preliminary investigation of the mammalian microfauna in Pleistocene deposits of caves in the Transvaal system. *Palaeont. Afr.* 7:59–118.

Hayman, R. W. 1967. Chiroptera. In J. Meester, ed., *Preliminary identification manual for African mammals.* Washington: Smithsonian Institution, part 11.

Heim de Balsac, H. 1967. Faits nouveaux concernant les *Myosorex* (Soricidae) de l'Afrique Orientale. *Mammalia* 31:610–628.

Heim de Balsac, H., and M. Lamotte. 1957. Évolution et phylogenie des Soricidés africains. II. La Lignée *Sylvisorex-Suncus-Crocidura. Mammalia* 21:15–49.

Hopwood, A. T. 1929. New and little-known mammals from the Miocene of Africa. *Am. Mus. Novit.* 344:1–9.

Lavocat, R. 1961. Le gisement de vertébrés miocènes de Beni Mellal (Maroc). Étude systematique de la faune de mammifères. *Notes Mém. Serv. Mines Carte géol. Maroc* 155:29–94.

McDowell, S. B. 1958. The Greater Antillean insectivores. *Bull. Am. Mus. Nat. Hist.* 115:115–214.

McKenna, M. C. 1975. Toward a phylogenetic classification of the Mammalia. In W. P. Luckett and F. S. Szalay, eds., *Phylogeny of the primates: a multidisciplinary approach.* New York and London: Plenum, pp. 21–46.

Matthew, W. D. 1910. Schlosser on Fayum mammals. A preliminary notice of Dr. Schlosser's studies upon the collection made in the Oligocene of Egypt for the Stuttgart Museum by Herr Markgraf. *Am. Nat.* 49:429–483.

———. 1913. A zalambdodont insectivore from the middle Eocene. *Bull. Am. Mus. Nat. Hist.* 32:307–314.

———. 1915. A revision of the Wasatch and Wind River faunas. Part IV—Entelonychia, Primates, Insectivora (part). *Bull. Am. Mus. Nat. Hist.* 34:429–483.

———. 1918. A revision of the Lower Eocene Wasatch and Wind River faunas. Part V.—Insectivora (continued), Glires, Edentata. *Bull. Am. Mus. Nat. Hist.* 38:565–657.

Meester, J. 1953. The genera of African shrews. *Ann. Transv. Mus.* 22:205–217.

———. 1955. Fossil shrews of South Africa. *Ann. Transv. Mus.* 22:271–278.

———. 1958. Variation in the shrew genus *Myosorex* in southern Africa. *J. Mammal.* 39:325–339.

Meester, J., and A. von W. Lambrechts. 1971. The southern African species of *Suncus* Ehrenberg (Mammalia: Soricidae.) *Ann. Transv. Mus.* 27: 1–14.

Meester, J., and I. J. Meyer. 1972. Fossil *Suncus* (Mammalia: Soricidae) from southern Africa. *Ann. Transv. Mus.* 27:269–277.

Osborn, H. F. 1908. New fossil mammals from the Fayum Oligocene, Egypt. *Bull. Am. Mus. Nat. Hist.* 24:265–272.

Patterson, B. 1965. The fossil elephant shrews (Family Macroscelididae). *Bull Mus. Comp. Zool. Harv.* 133:295–335.

Quinet, G. E., and X. Misonne. 1965. Les insectivores zalambdodontes de l'Oligocène inférieur belge. *Bull. Inst. Roy. Sci. Nat. Belg.* 41:1–15.

Repenning, C. A. 1965. An extinct shrew from the early Pleistocene of South Africa. *J. Mammal.* 46:189–196.

———. 1967. Subfamilies and genera of the Soricidae. *U. S. Geol. Surv.,* Professional Paper, 565. 74 pp.

Russell, D. E., and B. Sigé. 1970. Révision des chiroptères lutétiens de Messel (Hesse, Allemagne). *Palaeovert.* 3: 83–102.

Savage, D. E. 1951. A Miocene phyllostomatid bat from Colombia, South America. *Univ. Calif. Publs. Bull. Dep. Geol.* 28:357–366.

Savage, R. J. G. 1965. Fossil mammals of Africa:19. The Miocene Carnivora of East Africa. *Bull. Br. Mus. Nat. Hist.* 10:241–316.

Schlosser, M. 1910. Über einige fossile Saugetiere aus dem Oligocän von Ägypten. *Zool. Anz.* 53:500–508.

———. 1911. Beiträge zur Kentniss der Oligozänen Landsäugetiere aus dem Fayum: Aegypten. *Beitr. Paläont. Geol. Öst.-Ung.* 24:51–167.

———. 1923. Mammalia. In K. A. von Zittel, F. Broili, and M. Schlosser, eds., *Grundzuge der Paläontologie* (Paläozoologie), 4th ed. Munich: R. Oldenbough.

Simonetta, A. M. 1968. A new golden mole from Somalia with an appendix on the taxonomy of the family Chrysochloridae (Mammalia, Insectivora). *Monitore Zool. Ital.* (n.s.) 2 (supp.):27–55.

Simons, E. L., and P. D. Gingerich. 1974. New carnivorous mammals from the Oligocene of Egypt. *Ann. Geol. Surv. Egypt* 4:157–166.

Simpson, G. G. 1967. The Tertiary lorisiform primates of Africa. *Bull. Mus. Comp. Zool. Harvard* 136:39–62.

Smith, J. D. 1976. Chiropteran evolution. In *Biology of the Phyllostomatidae.* Special publications of the Texas Technical University 1:49–69.

Stromer, E. 1922. Erste Mitteilung über tertiäre Wirbeltier-Reste aus Deutsch-Südwestafrika. *Sber. Bayer. Akad. Wiss.* 1921:331–340.

———. 1926. Reste Land- und Süsswasser-bewohnender Wirbeltiere Deutsch-Südwestafrikas. In E. Kaiser, ed., *Die Diamantenwüste Südwestafrikas,* vol. 2. Berlin: Dietrich Riemer, pp. 107–153.

———. 1932. *Palaeothentoides africanus* nov. gen., nov. spec., ein erstes Beuteltier aus Afrika. *Sber. bayer. Akad. Wiss.* 1931:177–190.

Szalay, F. S. 1977. Phylogenetic relationships and a classification of the eutherian Mammalia. In *Major Patterns in Vertebrate Evolution,* M. K. Hecht, P. C. Goody, and B. M. Hecht, eds. *NATO Advanced Study Inst., Ser. A* 14:315–374.

Turnbull, W. D., and C. A. Reed. 1967. *Pseudochrysochloris,* a specialized burrowing mammal from the early Oligocene of Wyoming. *J. Paleont.* 41:623–631.

Van Valen, L. 1966. Deltatheridia, a new order of mammals. *Bull. Am. Mus. Nat. Hist.* 132:1–126.

———. 1967. New Paleocene insectivores and insectivore classification. *Bull. Am. Mus. Nat. Hist.* 135:217–284.

Walker, A. 1969. True affinities of *Propotto leakeyi* Simpson 1967. *Nature* 223: 647–648.

Whitworth, T. 1954. The Miocene hyracoids of East Africa. *Fossil Mammals of Africa, No. 7,* British Museum (Natural History). 58 pp.

5

Rodentia and Lagomorpha

René Lavocat

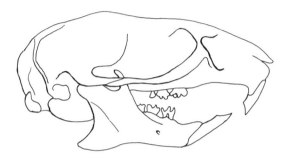

The Recent African rodent fauna is exceedingly complex. Some very important elements, for example, the Muroidea, Sciuroidea, and Gliroidea, have worldwide distribution; others, such as the Anomaluroidea, Pedetoidea, Ctenodactyloidea, and Bathyergoidea, are strictly limited to Africa. The Hystricidae and the genera *Petromus* and *Thryonomys* have long been considered to share very close morphological similarities with the South American rodents now placed in the Caviomorpha. Some authors even included *Thryonomys* and *Petromus* in subfamilies of the Caviomorpha. Transatlantic land bridges were usually assumed to have provided the opportunity for these New World forms to reach Africa in relatively recent times.

On the other hand, zoologists have long recognized the importance of land connections between Africa and Asia for understanding the origin of at least part of the African rodent fauna, but the extent of Asiatic influence is by no means perfectly clear.

Distribution of the Recent Fauna

The distribution of African rodent genera has been given by Ellerman (1940). Looking at the main pattern of this distribution we see a very sharp distinction between the Ethiopian and Palaearctic African regions. The Bathyergidae, Anomaluridae, Pedetidae, and Thryonomyidae are absent from the Palaearctic area, and the Sciuridae are represented there only by a single genus. Of the Muscardinidae, only two rather different genera occur here, one in each major biogeographic region. On the other hand, the Dipodidae and Ctenodactylidae are typically North African and can be found overlapping the Ethiopian zone only at a very narrow boundary. Only the Hystricidae, a few genera of Murinae, and three genera of Gerbillinae are really common to the two regions.

In the Ethiopian region the Murinae and Sciuridae are very abundant, usually with wide generic distribution. Among the Thryonomyidae, *Thryonomys* occurs through most of the zone, in contrast to *Petromus,* which is limited to a rather narrow region of Southwest Africa in what seems to be a relict distribution. The distribution given by Ellerman for the Bathyergidae is "From Sudan, Abyssinia and Somaliland, and from the Gold Coast to the Cape" (1940:79). For the Anomaluridae he gives "Africa, Western and Central: from Sierra Leone to Uganda, Tanganyika and Northern Rhodesia" (1940:536). The distribution of the Pedetidae is given as "Central and Southern Africa: from Kenya and Angola to

Cape Province" (p. 547). The Graphiurinae are present throughout Africa south of the Sahara.

Until recently a great many problems surrounded the African rodent fauna, such as origin, mode and time of immigration, and relationships. For example, which groups are autochthonous? When did the Cricetodontidae and the Muridae arrive in Africa? Must we recognize a relationship between *Thryonomys* and *Petromus* on the one hand and the Hystricidae on the other, and if so, what kind? With which of the Muroidea is the genus *Mystromys* related? What are the relationships between the Malagasy Nesomyidae and the African rodents? What roles have endemism and immigration played in the development of the African rodent fauna? What is the systematic position of the Bathyergidae? Is their infraorbital structure primitive or advanced? How does one explain the morphological similarities between *Thryonomys* and *Petromus* and the South American Caviomorpha?

It is not possible to answer these questions without recourse to the historical facts. But until recently the history of African rodents was very poorly known. Now, as a result of recent field studies and laboratory research, enough information is available to provide an understanding of at least the main stages of this history, even if we cannot yet perceive all the details.

Sites and Collections

Only those sites which have produced fossil rodents and those collections which have been studied are recorded here.

Oligocene. The oldest levels from which rodents are known in Africa are the early Oligocene deposits of the Fayum, United Arab Republic. First worked by Andrews and Beadnell in 1901, further collections were made by the American Museum of Natural History. This material was studied by Osborn (1908). Collections in Stuttgart and Munich were studied by Schlosser (1911). New excavations in 1961–67 by Simons produced material now housed in the Yale Peabody Museum and studied by A. E. Wood (1968).

Lower Miocene. Among the sites from this level are the Diamond Fields of Southwest Africa (Namib), studied by Stromer (1926). The material is now in the collections of the Munich Museum, although part of it was destroyed during World War II. Additional material in the American Museum was studied by Hopwood (1929).

From East Africa, sites from the Lake Victoria region include Karungu, collections from which are now in the British Museum, and were studied by Andrews (1914); Rusinga, Mfwanganu, Songhor, and Koru, collections of the National Museum of Kenya studied by MacInnes (1957) and by Lavocat (1967); Napak, Uganda, now in the Geology Department, Bedford College, London, and studied by Lavocat (1967); Bukwa, Uganda, of the Uganda Museum, studied by Lavocat (1967); Kirimun, northern Kenya, specimens in the University of California, Berkeley, and Loperot, collections of the National Museum of Kenya, both studied by Lavocat (1967).

Upper Miocene. In East Africa, materials are from Fort Ternan, collections of the National Museum of Kenya, and Nakali, in the University of Madrid, both under study by Lavocat.

In North Africa, the localities are Beni Mellal, now in Hautes Etudes, Paris, studied by Lavocat (1961), and by Jaeger; also from Oued Zra, in Morocco, under study by Jaeger.

Upper Pliocene. This includes material from Lake Ichkeul in Tunisia, collected and studied by Jaeger (1971a); Makapansgat and Sterkfontein Caves in South Africa, studied by Lavocat (1967b) and de Graaf (1960).

Pleistocene. In Morocco, specimens are from Jebel Irhoud, collected and studied by Jaeger (1971b). Olduvai Gorge in Tanzania has produced material of this age, studied briefly by Lavocat (1967b) and later by Jaeger. In South Africa, additional specimens from Swartkrans and Kromdraai have been studied by Lavocat (1967b) and de Graaf (1960). In Algeria, Ternifine is under study by Jaeger.

Definitions

The definition of the Hystricomorph, Sciuromorph, and Myomorph structure is found in every handbook related to the rodent's anatomy; it is not so easy to find the definition of sciurognathy and hystricognathy. In a few words, the difference is that in the sciurognath, the posterior half of the mandible, and especially its inferior border, is in the same longitudinal plane as the incisor, whereas in the perfectly hystricognath structure, this posterior half is clearly in a longitudinal plane external to the plane of the incisor. Sometimes, the interpretation can be somewhat dubious, at least at first sight. But on the whole, we have here an excellent anatomical characteristic.

Systematics of African Rodents

The revised diagnoses of important taxa established as a consequence of current research are given below. Table 5.1 lists all recognized genera and gives their distribution by geological age and locality, as well as the origin of holotype material.

Order Rodentia

Suborder Hystricognathi Tulberg 1899

DIAGNOSIS. Rodents with an infraorbital foramen generally of great size, of the hystricomorph type, but secondarily reduced in one fossorial family. Mandible always hystricognath. Pterygoid fossa open anteriorly, communicating with the orbitotemporal cavity, or secondarily with the endocranial cavity. Maxillary bone nearly excluding palatine from orbitotemporal floor. Middle ear primitively with a very prominent and ventrally free promontory (*Diamantomys, Lagostomus, Thryonomys*), evolving in some forms toward a condition in which tympanic sheet covers the medial half of the promontory. Teeth of a primarily pentalophodont type. Most members with P4 or dP4 plus three molars, but P[3] persists in several families; in Bathyergoidea from three to six teeth.

Infraorder Phiomorpha Lavocat 1962

DIAGNOSIS. Lower jaw always hystricognath. Infraorbital structure primarily hystricomorph, greatly modified and secondarily reduced in some Bathyergoidea. Molar teeth morphologically tetra- or pentalophodont; greatly simplified in the Bathyergidae. Generally four cheek teeth, rarely three, sometimes five.

DISTRIBUTION. Tropical to warm temperate of Old World.

Superfamily Thryonomyoidea Wood 1955

DIAGNOSIS. Superfamily of Phiomorpha with brachydont to hypsodont dentition in which crown is not subdivided into many cusps. Anterior palatine foramina generally well developed. Frontal sinus sometimes developed. Generally four cheek teeth, sometimes five (with persistence of P[3]) in the upper jaw. Upper incisors not extending far posteriorly.

DISTRIBUTION. Oligocene: Fayum; Miocene: East Africa, Namib, Morocco, Chios Island, Chinji of Indian Siwaliks; Recent: Ethiopian biogeographic region.

Family Phiomyidae Wood 1955

DIAGNOSIS. Lower molars with from three to five transverse and more or less complete crests; upper molars with four to five crests; cusps still well individualized. Milk teeth with delayed replacement in some forms, persistent in others. Molar form highly variable. P[3] or dP[3] present in some forms.

DISTRIBUTION. Oligocene: Fayum; Miocene: East Africa, Namib.

Family Thryonomyidae Pocock 1922

DIAGNOSIS. Thryonomyoidea with a muzzle of normal proportions; where known, masseter muscle insertion extending far in front of infraorbital foramen, and anterior palatine foramina well developed. Semihypsodont molars with well developed crests. The number of crests reduced in several forms.

DISTRIBUTION. Oligocene: Fayum; lower Miocene: East Africa, Namib; upper Miocene: Morocco, East Africa (Fort Ternan), Chinji, Indian Siwaliks; Recent: Ethiopian biogeographic region.

Family Diamantomyidae Schaub 1928

DIAGNOSIS. Thryonomyoidea with five upper cheek teeth, molariform, rather hypsodont and with secondary crests well developed.

DISTRIBUTION. Oligocene: Fayum; Miocene: East Africa, Namib.

Family Kenyamyidae Lavocat 1973

DIAGNOSIS. Thryonomyoidea with short masseteric insertion, four gliriform cheek teeth, with cusps weakly or not at all distinct, long and narrow crests, which may be low or high on a very brachydont crown.

DISTRIBUTION. Lower Miocene: East Africa.

Family Myophiomyidae Lavocat 1973

DIAGNOSIS. Small Thryonomyoidea. P[3] (or dP[3]) present at least in *Myophiomys*. Cheek teeth cricetodontoid with prominent cusps and lower crests.

DISTRIBUTION. Oligocene: Fayum; lower Miocene: East Africa, Namib.

Superfamily Bathyergoidea Osborn 1910

DIAGNOSIS. Fossorial Phiomorpha; infraorbital foramen primitively large; secondarily reduced in the Bathyergidae, very much so in the recent ones; incisors frequently very lengthened; number of cheek teeth varying, their structure simplified to extremely simplified.

DISTRIBUTION. Miocene: East Africa, Namib; Recent: Africa south and east of a line from Togo to Ethiopia.

Family Bathyergoididae Lavocat 1973

DIAGNOSIS. Bathyergoidea with rather conservative structure of cheek teeth. Incisors lengthened.

Table 5.1 Distribution of the species of rodents in African Tertiary and Pleistocene layers.

	Oligocene			Lower Miocene										Upper Miocene					Pliocene			Pleistocene				
	Low. Fayum A-F	Mid. Fayum G	Upp. Fayum I-R	Namib	Bukwa	Karungu	Mfwangano	Rusinga	Songhor	Koru	Napak	Lodwar	Kirimun	Beni Mellal	Bou Hanifia	Fort Ternan	Nakali	Oued Zra	Ichkeul	Taung	Sterkfontein	Makapan	Olduvai	Swartkrans	Kromdraai	Jebel Ichoud
Thryonomyoidea																										
PHYOMYIDAE																										
Andrewsimys parvus	–	–	–	–	–	–	–	X	–	–	–	–	–	–	–	–	–	–	–	–	–	–	–	–	–	–
Phiomys andrewsi	X	+	–	–	–	–	–	–	+	+	–	–	–	–	–	–	–	–	–	–	–	–	–	–	–	–
Phiomys paraphiomyoides	–	X	–	–	–	–	–	–	–	–	–	–	–	–	–	–	–	–	–	–	–	–	–	–	–	–
Phiomys aff. paraphiomyoides	–	–	+	–	–	–	–	–	–	–	–	–	–	–	–	–	–	–	–	–	–	–	–	–	–	–
Phiomys lavocati	X	–	–	–	–	–	–	–	–	–	–	–	–	–	–	–	–	–	–	–	–	–	–	–	–	–
THRYONOMYIDAE																										
Gaudeamus aegyptius	X	–	–	–	–	–	–	–	–	–	–	–	–	–	–	–	–	–	–	–	–	–	–	–	–	–
Paraphiomys simonsi	–	X	–	–	–	–	–	–	–	–	–	–	–	–	–	–	–	–	–	–	–	–	–	–	–	–
Paraphiomys pigotti	–	–	–	–	–	+	X	+	+	+	+	+	+	–	–	–	–	–	–	–	–	–	–	–	–	–
Paraphiomys stromeri stromeri	–	–	–	X	–	–	–	+	–	–	+	–	+	–	–	–	–	–	–	–	–	–	–	–	–	–
Paraphiomys stromeri hopwoodi	–	–	–	–	–	–	–	–	+	–	–	–	–	–	–	–	–	–	–	–	–	–	–	–	–	–
Paraphiomys occidentalis	–	–	–	–	–	–	–	–	–	–	–	–	–	–	–	X	–	–	–	–	–	–	–	–	–	–
Epiphyomy coryndoni	–	–	–	–	–	–	–	+	–	+	–	–	–	–	–	–	–	–	–	–	–	–	–	–	–	–
Petromus minor	–	–	–	–	–	–	–	–	–	–	–	–	–	–	–	–	–	–	–	X	–	–	–	–	–	–
DIAMANTOMYIDAE																										
Metaphiomys schaubi	X	–	–	–	–	–	–	–	–	–	–	–	–	–	–	–	–	–	–	–	–	–	–	–	–	–
Metaphiomys beadnelli	–	X	–	–	–	–	–	–	–	–	–	–	–	–	–	–	–	–	–	–	–	–	–	–	–	–
Diamantomys luederitzi	–	–	–	X	–	+	+	+	+	+	+	+	–	+	–	–	–	–	–	–	–	–	–	–	–	–
Pomonomys dubius	–	–	–	X	–	–	–	–	–	–	–	–	–	–	–	–	–	–	–	–	–	–	–	–	–	–
KENYAMYIDAE																										
Kenyamys mariae	–	–	–	–	–	–	+	–	X	–	–	–	–	–	–	–	–	–	–	–	–	–	–	–	–	–
Simonimys genovefae	–	–	–	–	–	–	+	+	X	+	+	–	–	–	–	–	–	–	–	–	–	–	–	–	–	–
MIOPHIOMYIDAE																										
Phiocricetomys minutus	–	X	–	–	–	–	–	–	–	–	–	–	–	–	–	–	–	–	–	–	–	–	–	–	–	–
Myophiomys arambourgi	–	–	–	–	–	–	–	+	+	X	–	+	–	–	–	–	–	–	–	–	–	–	–	–	–	–
Elmerimys woodi	–	–	–	–	–	–	–	–	X	–	–	–	–	–	–	–	–	–	–	–	–	–	–	–	–	–
Phiomyoides humilis	–	–	–	X	–	–	–	–	–	–	–	–	–	–	–	–	–	–	–	–	–	–	–	–	–	–
Bathyergoidea																										
BATHYERGOIDIDAE																										
Bathyergoides neotertiarius	–	–	–	X	–	–	–	+	+	+	–	–	–	–	–	–	–	–	–	–	–	–	–	–	–	–
BATHYERGIDAE																										
Proheliophobius leakeyi	–	–	–	–	–	–	+	X	?	–	–	–	–	–	–	–	–	–	–	–	–	–	–	–	–	–
Paracryptomys mackennae	–	–	–	X	–	–	–	–	–	–	–	–	–	–	–	–	–	–	–	–	–	–	–	–	–	–
Bathyergidae nov. sp.	–	–	–	–	–	–	–	–	–	–	–	–	–	–	–	–	–	X	–	–	–	–	–	–	–	–
Heterocephalus sp.	–	–	–	–	–	–	–	–	–	–	–	–	–	–	–	–	–	–	–	–	–	–	–	+	+	–
Cryptomys robertsi	–	–	–	–	–	–	–	–	–	–	–	–	–	–	–	–	–	–	–	+	+	+	–	+	+	–
Gypsorhychus darti	–	–	–	–	–	–	–	–	–	–	–	–	–	–	–	–	–	–	–	X	–	–	–	–	–	–
Gypsorhychus minor	–	–	–	–	–	–	–	–	–	–	–	–	–	–	–	–	–	–	–	X	–	–	–	–	–	–
Gypsorhychus makapani	–	–	–	–	–	–	–	–	–	–	–	–	–	–	–	–	–	–	–	–	–	X	–	–	–	–
Hystricoidea																										
HYSTRICIDAE																										
Hystrix	–	–	–	–	–	–	–	–	–	–	–	–	–	–	–	–	–	–	–	–	–	–	–	–	–	–
Hystrix cf. africae-australis	–	–	–	–	–	–	–	–	–	–	–	–	–	–	–	–	–	–	–	–	–	–	–	+	–	–
Hystrix major	–	–	–	–	–	–	–	–	–	–	–	–	–	–	–	–	–	–	–	–	–	–	–	+	–	–
Hystrix cristata	–	–	–	–	–	–	–	–	–	–	–	–	–	–	–	–	–	–	–	–	+	–	–	–	–	–
Hystrix sp.	–	–	–	–	–	–	–	–	–	–	–	–	–	–	–	–	–	–	–	–	–	–	–	–	+	–
Xenohystrix crassidens	–	–	–	–	–	–	–	–	–	–	–	–	–	–	–	–	–	–	–	–	–	+	–	–	–	–

Table 5.1   (*continued*)

	Oligocene					Lower Miocene								Upper Miocene					Pliocene				Pleistocene			
	Low. Fayum A-F	Mid. Fayum G	Upp. Fayum I-R	Namib	Bukwa	Karungu	Mfwangano	Rusinga	Songhor	Koru	Napak	Lodwar	Kirimun	Beni Mellal	Bou Hanifia	Fort Ternan	Nakali	Oued Zra	Ichkeul	Taung	Sterkfontein	Makapan	Olduvai	Swartkrans	Kromdraai	Jebel Ichoud
ANOMALURIDAE																										
Anomaluridae sp.	—	—	+	—	—	—	—	—	—	—	—	—	—	—	—	—	—	—	—	—	—	—	—	—	—	—
Paranomalurus bishopi	—	—	—	—	—	—	—	+	+	X	—	—	—	—	—	—	—	—	—	—	—	—	—	—	—	—
Paranomalurus soniae	—	—	—	—	—	—	+	+	X	+	+	—	—	—	—	—	—	—	—	—	—	—	—	—	—	—
Paranomalurus walkeri	—	—	—	—	—	—	+	—	X	—	+	—	—	—	—	—	—	—	—	—	—	—	—	—	—	—
Zenkerella wintoni	—	—	—	—	—	—	—	—	X	—	—	—	—	—	—	—	—	—	—	—	—	—	—	—	—	—
Pedetoidea																										
PEDITIDAE																										
Parapedetes namaquensis	—	—	—	X	—	—	—	—	—	—	—	—	—	—	—	—	—	—	—	—	—	—	—	—	—	—
Megapedetes pentadactylus	—	—	—	—	—	—	—	—	X	—	+	—	—	—	—	—	—	—	—	—	—	—	—	—	—	—
Megapedetes sp.	—	—	—	—	—	—	—	+	—	—	—	—	—	—	—	—	—	—	—	—	—	—	—	—	—	—
Pedetes gracile	—	—	—	—	—	—	—	—	—	—	—	—	—	—	—	—	—	—	—	X	—	—	—	—	—	—
Pedetes sp.	—	—	—	—	—	—	—	—	—	—	—	—	—	—	—	—	—	—	—	—	—	—	+	—	—	—
Muroidea																										
CRICETODONTIDAE																										
Afrocricetodon songhori songhori	—	—	—	—	—	—	+	—	X	—	+	—	—	—	—	—	—	—	—	—	—	—	—	—	—	—
Afrocricetodon songhori korui	—	—	—	—	—	—	—	—	X	—	—	—	—	—	—	—	—	—	—	—	—	—	—	—	—	—
Paratarsomys macinnesi	—	—	—	—	—	—	X	—	—	—	—	—	—	—	—	—	—	—	—	—	—	—	—	—	—	—
Notocricetodon petteri	—	—	—	—	—	—	+	X	+	—	—	—	—	—	—	—	—	—	—	—	—	—	—	—	—	—
CRICETIDAE																										
Kanisamys sp.	—	—	—	—	—	—	—	—	—	—	—	—	—	—	—	—	X	—	—	—	—	—	—	—	—	—
Ruscinomys sp.	—	—	—	—	—	—	—	—	—	—	—	—	—	—	—	—	—	—	X	—	—	—	—	—	—	—
Cricetodon atlasi	—	—	—	—	—	—	—	—	—	—	—	—	—	X	—	—	—	—	—	—	—	—	—	—	—	—
cf. Cricetodon	—	—	—	—	—	—	—	—	—	—	—	—	—	—	—	—	—	—	X	—	—	—	—	—	—	—
Myocricetodon cherifiensis	—	—	—	—	—	—	—	—	—	—	—	—	—	X	—	—	—	—	—	—	—	—	—	—	—	—
Myocricetodon sp.	—	—	—	—	—	—	—	—	—	—	—	—	—	—	—	—	—	—	X	—	—	—	—	—	—	—
Cricetidae indet. sp. indet.	—	—	—	—	—	—	—	—	—	—	—	—	—	—	—	—	—	—	X	—	—	—	—	—	—	—
Leakeymys ternani	—	—	—	—	—	—	—	—	—	—	—	—	—	—	—	X	—	—	—	—	—	—	—	—	—	—
Tatera sp.	—	—	—	—	—	—	—	—	—	—	—	—	—	—	—	—	—	—	—	+	—	—	—	—	—	—
Tatera sp.	—	—	—	—	—	—	—	—	—	—	—	—	—	—	—	—	—	—	—	—	+	—	—	—	—	—
Tatera cf. brantsi	—	—	—	—	—	—	—	—	—	—	—	—	—	—	—	—	—	—	—	—	+	+	—	—	+	—
Gerbillus sp.	—	—	—	—	—	—	—	—	—	—	—	—	—	—	—	—	—	—	—	—	+	—	—	—	—	—
Otomys kempi	—	—	—	—	—	—	—	—	—	—	—	—	—	—	—	—	—	—	—	—	+	—	—	—	—	—
Palaeotomys gracilis	—	—	—	—	—	—	—	—	—	—	—	—	—	—	—	—	—	—	—	+	+	+	—	—	+	—
Prototomys cambelli	—	—	—	—	—	—	—	—	—	—	—	—	—	—	—	—	—	—	—	X	—	—	—	—	—	—
Mystromys darti	—	—	—	—	—	—	—	—	—	—	—	—	—	—	—	—	—	—	—	—	—	X	—	—	—	—
Mystromys antiquus	—	—	—	—	—	—	—	—	—	—	—	—	—	—	—	—	—	—	—	X	—	—	—	—	—	—
Mystromys hausleitneri	—	—	—	—	—	—	—	—	—	—	—	—	—	—	—	—	—	—	—	—	+	+	—	—	+	—
MURIDAE																										
Dendromus antiquus	—	—	—	—	—	—	—	—	—	—	—	—	—	—	—	—	—	—	—	X	—	—	—	—	—	—
Dendromus cf. mesomelas	—	—	—	—	—	—	—	—	—	—	—	—	—	—	—	—	—	—	—	+	+	—	—	—	—	—
Dendromus sp.	—	—	—	—	—	—	—	—	—	—	—	—	—	—	—	—	—	—	—	—	—	—	+	—	—	—
Dendromus sp.	—	—	—	—	—	—	—	—	—	—	—	—	—	—	—	—	—	—	—	—	—	—	+	—	—	—
Malacothrix typica	—	—	—	—	—	—	—	—	—	—	—	—	—	—	—	—	—	—	—	—	+	—	—	—	—	—
Malacothrix makapani	—	—	—	—	—	—	—	—	—	—	—	—	—	—	—	—	—	—	—	—	—	X	—	—	—	—
Steatomys sp.	—	—	—	—	—	—	—	—	—	—	—	—	—	—	—	—	—	—	—	—	—	—	+	—	—	—
Steatomys cf. pratensis	—	—	—	—	—	—	—	—	—	—	—	—	—	—	—	—	—	—	—	—	—	—	+	—	—	—
Paraethomys anomalus	—	—	—	—	—	—	—	—	—	—	—	—	—	—	—	—	—	—	+	—	—	—	—	—	—	—
Muridae: sp. or aff. sp of recent genera	—	—	—	—	—	—	—	—	—	—	—	—	—	—	—	—	—	—	—	2	7	9	13	—	—	—
Progonomys cathalal	—	—	—	—	—	—	—	—	—	—	—	—	—	—	—	—	—	—	+	—	—	—	—	—	—	—

Table 5.1 (*continued*)

	Oligocene					Lower Miocene								Upper Miocene						Pliocene			Pleistocene			
	Low. Fayum A-F	Mid. Fayum G	Upp. Fayum I-R	Namib	Bukwa	Karungu	Mfwangano	Rusinga	Songhor	Koru	Napak	Lodwar	Kirimun	Beni Mellal	Bou Hanifia	Fort Ternan	Nakali	Oued Zra	Ichkeul	Taung	Sterkfontein	Makapan	Olduvai	Swartkrans	Kromdraai	Jebel Ichoud
Gliroidea																										
GLIRIDAE																										
Dryomys														X				X								
Gliridae sp.																		X								
Ctenodactyloidea																										
Africanomys pulcher														X												
Africanomys sp.																		X								
Metasayimys jebeli														X												
Irhoudia bohlini																										X
Sciuroidea																										
SCIURIDAE																										
Vulcanisciurus africanus						X	+	+	+																	
Sciuridae sp.						X																		+		
Heteroxerus														X												
Xerus sp.																								+		
Sciuridae sp.																								+		

Note: X indicates the localities from which the type specimens for each species were collected; + indicates other localities from which the species is recorded.

Infraorbital foramen widely open, not reduced; masseter inserted on the rim of the foramen, but not on the muzzle.

DISTRIBUTION. Miocene: East Africa, Namib.

Family Bathyergidae Waterhouse 1841

DIAGNOSIS. Bathyergoidea with structure of cheek teeth very or extremely simplified. Strong tendency to reduce infraorbital foramen, always very reduced in the Recent forms, and through which, part of the masseter only rarely passes.

DISTRIBUTION. Miocene: East Africa, Namib; Recent: as for the superfamily.

Superfamily Hystricoidea Gill 1872

DIAGNOSIS. Phiomorpha in which teeth have a strong tendency toward subdividing crests into many cusps. Replacement of dP^4 by P^4. Anterior palatine foramina generally very small. Tendency, very strongly realized in some genera, toward hypertrophy of nasal fossa and frontal or nasal sinuses.

DISTRIBUTION. Miocene: Siwaliks; upper Miocene: North Africa; Pontian-Pliocene: Eurasia; Recent: Africa, Mediterranean Europe, Asia mostly tropical.

Suborder indet.

Superfamily Anomaluroidea Gill 1872

Family Anomaluridae Gill 1872

DIAGNOSIS. Sciurognath rodents with greatly enlarged infraorbital foramen through which the masseter passes to be broadly inserted on the muzzle. Ascending ramus of orbital arch weak. Palatine bone noticeably contributing to the orbitotemporal floor. Pterygoid fossa not open anteriorly. Four cheek teeth with transverse crests of typical pattern. Middle ear typical with globular promontory.

DISTRIBUTION. Oligocene: Upper Fayum (one ulna); Miocene: East Africa; Recent: West and Central Africa, parts of East Africa.

Suborder indet.

Superfamily Pedetoidea Ellerman 1940

Family Pedetidae Owen 1847

DIAGNOSIS. Sciurognath rodents with greatly enlarged infraorbital foramen and strong masseteric insertion on muzzle. Strong jugal bone. Blind pterygoid fossa. Large bulla. Cheek teeth of two lobes connected by one longitudinal crest, lateral in position, either externally or internally.

DISTRIBUTION. Miocene: East Africa, Namib, Morocco; Recent: from Kenya and Angola to the Cape region.

Suborder Myomorpha Brandt 1855

Superfamily Muroidea Miller and Gidley 1918

DIAGNOSIS. Sciurognath rodents with two to three cheek teeth; infraorbital region of "myomorph" type (double masseteric insertion).

DISTRIBUTION. Oligocene to Recent: Holarctic; Miocene-Recent: Africa; Pleistocene to Recent: South America.

Family Cricetodontidae Schaub 1925

DIAGNOSIS. Infraorbital region primitive (low masseteric plate) to advanced (more vertical masseteric plate) in structure. Teeth of cricetodont type, with only two files of cusps, even in the upper teeth. Secondary cusps sometimes added to main ones.

DISTRIBUTION. Oligocene-Miocene: Eurasia; Miocene: Africa; Oligocene-Miocene: North America.

Subfamily Afrocricetodontinae Lavocat 1973

DIAGNOSIS. Miocene Cricetodontidae with primitive teeth and infraorbital structures.

DISTRIBUTION. Miocene: East Africa.

Family Nesomyidae Major 1897

African Muroidea, descendants of Afrocricetodontinae, with advanced skull or teeth.

Subfamily Nesomyinae Major 1897

DIAGNOSIS. Malagasy Nesomyidae, showing a wide spectrum of dental structures from bunodont to lamelliform.

DISTRIBUTION. Entirely restricted to Madagascar; not known before the Pliocene.

Subfamily Lophiomyinae Thomas 1897

DIAGNOSIS. Teeth of cricetid plan; armored skull. One genus only.

DISTRIBUTION. Ethiopia, Somalia, Kenya, and Sudan.

Subfamily Mystromyinae Lavocat 1973

DIAGNOSIS. Rather primitive cricetodontid plan; hypsodont tubercles; zygomasseteric structure advanced.

DISTRIBUTION. Very abundant in the oldest australopithecine caves of Southern Africa, Albany District, Natal, Transvaal; ? Tanzania.

Subfamily Tachyoryctinae Ellerman 1940

DIAGNOSIS. Fossorial Nesomyidae; cheek teeth lamelliform, infraorbital foramen high and rather small.

DISTRIBUTION. Pleistocene and Recent: Eastern Africa.

Subfamily Gerbillinae Alston 1876

DIAGNOSIS. Cheek teeth with cusps more or less connected giving transverse crests, separated by transverse valleys, more or less wide and open.

DISTRIBUTION. Pliocene and Pleistocene caves of South Africa; Pleistocene of Olduvai; Recent: Africa, Southwestern Asia.

Subfamily Otomyinae Thomas 1896

DIAGNOSIS. Cheek teeth composed of one file of narrow transverse lamellae.

DISTRIBUTION. Plio-Pleistocene: South African caves; Pleistocene: Olduvai. Recent: Africa, widely distributed south of Sahara.

The reference of the last six subfamilies to the Nesomyidae is only tentative because we cannot at the moment give evidence that they all are descendants of Afrocricetodontinae, rather than of later Cricetodontidae from Asia.

Family Muridae Gray 1827

DIAGNOSIS. Descendants of the Asiatic Cricetodontidae, originating in Asia; cusps of teeth primitively well distinct, little or not hypsodont.

DISTRIBUTION. Upper Miocene to Recent: Old World; Pleistocene to Recent: Australia.

Subfamily Murinae Murray 1866

DIAGNOSIS. Muridae with three files of cusps in the upper teeth.

DISTRIBUTION. Miocene: Asia, Europe, North Africa; Recent: Old World, some forms introduced elsewhere.

Subfamily Dendromurinae Allen 1939

DIAGNOSIS. Muridae with normally only one internal supplementary cusp in upper teeth; longitudinal crests of the Cricetodontidae lacking or strongly modified.

DISTRIBUTION. Upper Miocene: Ngorora; Recent: Africa south of the Sahara.

Suborder Myomorpha?

Superfamily Dipodoidea Weber 1904

DIAGNOSIS. Sciurognath rodents with cricetoid teeth, with or without P4; very enlarged infraorbital foramen, masseteric plate reduced and strictly ventral.

DISTRIBUTION. Oligocene and Pleistocene-Recent: Europe; Pliocene to Recent: Asia; Recent: Holarctic biogeographic region, including northern Africa.

Suborder indet.

Superfamily Gliroidea Simpson 1945

DIAGNOSIS. Sciurognath rodents of European origin, with infraorbital foramen frequently of rather great size; cheek teeth of very distinctive pattern, transverse crests usually low and frequently numerous. Generally four cheek teeth; P^3 present in *Gliravus*.

DISTRIBUTION. Glirinae: Eocene to Recent: Europe; Miocene: Beni Mellal, Morocco: Recent: North Africa, Asia. Graphiurinae: Recent: Africa, Ethiopian biogeographic region.

Suborder indet.

Superfamily Ctenodactyloidea Simpson 1945

DIAGNOSIS. Sciurognath rodents of Asiatic origin, with enlarged infraorbital foramen, masseteric place reduced and ventral, pterygoid fossa deep but blind, lacrimal bone of large size and of peculiar pattern; cheek dentition with P^4_4 and sometimes even P^3_3; tetralophodont structure of the teeth, very simplified in the Recent forms.

DISTRIBUTION. Oligocene to Miocene: Asia; Miocene to Recent: Africa, north of the Equator.

Suborder Sciuromorpha Brandt 1855

Superfamily Sciuroidea Gill 1872

DIAGNOSIS. Sciurognath rodents with "sciuromorph" infraorbital structure, brachydont cheek teeth $\left(\dfrac{5-4}{4} \right)$, retaining much of the primitive paramyid structure.

DISTRIBUTION. Oligocene to Recent: Europe, Asia; Miocene to Recent: North America; Miocene: East Africa, Beni Mellal; Recent: Africa.

Evolution of The African Rodents

The Ancient Stock

Oligocene. The most ancient rodent molars known in Africa are found in the Fayum, Egypt. At this site Eocene deposits occur, but until now only the Oligocene levels have furnished rodents. The same is true

for the Primates. R. J. G. Savage has collected rodent incisors from the late Eocene of Libya, and the rodents, as well as at least most of the other elements of the Oligocene fauna, are probably related to animals that reached Africa much earlier, perhaps very soon after the end of the Paleocene, and probably before the Lutetian. As for the rodents, it does seem probable that these early forms were some sort of Ischyromyidae, *sensu* G. G. Simpson.

The most recent excavations at the Fayum by Simons (1968) were extended to several levels in succession and appear to include levels from lower to upper Oligocene. But the upper Oligocene age of these sediments is inferred from the terminal Oligocene age of the basaltic covering; this basalt gives, of course, only an uppermost age limit. Thus it remains possible that the fossiliferous level could be older than is proposed by Simons.

The fossil rodent remains can be referred only to the superfamily Thryonomyoidea, and four of the five families included now in this superfamily are already present. Two genera, *Phiomys* and *Metaphiomys*, had been previously discovered by Osborn and referred by him (1908) to the European family Theridomyidae. Their pentalophodont molars have a structure which is very close to that of *Theridomys*, at least at first sight. Nevertheless, on dental characters, Wood (1968) rejects this identification. Moreover, the lower jaw of these forms is typically hystricognath, though that of the Theridomyidae is strictly sciurognath. Comparisons of the skulls, known from the Miocene deposits, strongly enhance these differences.

The genus *Phiomys*, the most primitive of the forms present in the Fayum, is known from several species, all with rather brachydont teeth, in which the tubercles are more fully developed than the crests. According to Wood (1968), the group shows a neat tendency to retain a permanent dP4; however, at that ancient time one can still observe the replacement of dP4 by P4. The skull of this Oligocene form is not known, but we believe one is known from the later Miocene deposits in Kenya. If we assume that the cranial characters did not evolve more than those of the teeth, then we can assume the following skull characters, already observed in the Miocene sample, to have been present in the Oligocene *Phiomys*: anterior palatine foramen of moderate length, infraorbital foramen already very enlarged, but the masseter impression limited to the dorsal half of the muzzle and less forward than, for example, in *Paraphiomys*. Wood believes that the genus *Phiomys* is the common ancestor of all the Thryonomyoidea, and that the ancestral form must be a still more primitive species of the genus, possibly in the African upper Eocene. I personally think that the ancestral form of the Thryonomyoidea, and moreover of the entire Phiomorpha complex, must be much older. Nevertheless, it must already have been very distinct from the Ischyromyidae, and already supplied with the fundamental characters of the group, at least since the Lutetian. There are several reasons for this belief: first, the structure of the skull of *Phiomys,* at least as we can observe it in the Miocene specimen, is clearly more advanced than that of the skull of the Miocene *Kenyamys,* of which the teeth are well advanced in a gliriform direction; second, the Kenyamyidae probably, and the Myophiomyidae certainly, must be referred to a dental stage very definitely more primitive than that of *Phiomys;* third, the teeth of the Oligocene genus *Gaudeamus* imply a more highly specialized level of evolution than the *Phiomys* structure already in that epoch; fourth, we have in the Lutetian of Pakistan a tooth which in my opinion (an opinion which is not shared indeed by other specialists who have seen the sample) is perfectly suitable as a very primitive Phiomorpha.

The Thryonomyidae are represented in the Oligocene by a single specimen of *Paraphiomys,* characterized by a predominance of the crest in semihypsodont molars. The genus *Gaudeamus* is also referred to this family. This last genus is characterized by a "pseudotaeniodont" dental structure with oblique transverse lamellae crossing the crown. This dental structure is derived from that of *Phiomys* by simplification and by translation of some crests which acquire new interconnections, with correlative building of valleys crossing the width of the tooth. The upper P^4 of this genus is very remarkable in that it shows a simplified structure closely similar to that formerly described by Wood (1949) and Wood and Patterson (1959) in some Oligocene South American rodents. At that time these authors thought that this was unique to this American group. Wood argues that this genus *Gaudeamus* was the ancestor of the present genus *Thryonomys.* But, in my opinion, it is already too advanced for that, with teeth even more simplified than they are in Thryonomys.

The genus *Metaphiomys,* of the family Diamantomyidae, is known by two Fayum species, one from the lower level, the other from the upper one, and this latter one seems very probably to have been derived from the former. Wood (1968) probably correctly relates this genus to the ancestry of the Miocene genus *Diamantomys.* The presence of an upper P^3 or dP^3 in both genera is a very interesting resemblance, but beyond that, all details of the peculiar structure of the teeth of *Diamantomys* suggest an origin in *Metaphiomys.* But, on the other hand, associated study of the lower jaw of *Metaphiomys* and of the small part known of its maxillary region, very clearly indicate that the skull still retains proportions which we could call normal, similar to those of *Phiomys,* and very different from the curiously lengthened proportions of the skull and mandible (mostly of the muzzle) of *Diamantomys.* The difference between the two genera is such that there must have been a rather long interval of time between the Fayum upper level and the *Diamantomys* levels from Kenya.

The genus *Phiocricetomys,* which represents the family Myophiomyidae, is remarkable in the strong convergence of its teeth with those of primitive Cricetodontidae. Both have the same number of teeth (three cheek teeth) and an acute character of the tubercles, but *Phiocricetomys* shows such teeth on a typically hystricognath jaw. Such as it is known, this Fayum fauna probably gives us only an incomplete picture of the rodent fauna of that time. The Miocene faunas of East and Southwest Africa that we will see below give us knowledge of many forms so clearly related to those of the Fayum that they must have a common origin with them, even though they are not yet known in the Fayum. Other forms, notwithstanding the fact that they are strictly African, cannot be related to the Phiomorpha, and their ancestors must have existed in Africa during the Oligocene. These ancestors must connect them with the primitive ischyromyid invasion. Of these ancestral forms we have only one indication in the Fayum, a portion of an ulna collected by Simons in the upper level and pertaining without possible doubt to an already flying anomalurid.

Miocene. The Miocene fossil sites of the Diamond Fields of Southwest Africa, and the probably contemporaneous deposits of Kenya and Uganda, have produced a considerable number and variety of specimens, affording an exceptional series of data fundamental to the understanding of the structure and evolution of the African rodent fauna. Except for the complete skeleton of *Parapedetes namaquensis,* a genus with affinities to the Recent *Pedetes,* the sites of Southwest Africa have furnished only teeth or very incomplete parts of jaws. The faunas had been published as early as 1922 by Stromer (1922, 1923, 1926). Part of the skull of a Bathyergoidea, unknown to Stromer and noticed although not described by Hopwood (1929), had been collected by Lang, and is now preserved in the American Museum in New York. It is a new genus of great interest, *Paracryptomys,* of which more will be said below dealing with the Bathyergoidea from Kenya. Several holotype specimens come from these sites: *Diamantomys, Pomonomys, Neosciuromys, Phiomyoides,* and *Bathyergoides* (the samples of *Phiomyoides* are lost; I consider *Neosciuromys* to be in fact a *Paraphiomys*). The new genera described by Hopwood (1929) fall into synonymy with *Paraphiomys.*

The first discovery of a rodent in the Miocene of East Africa occurred 60 years ago. The collector was unfortunately killed soon after by a crocodile. The sample was later described by Andrews (1914:163) under the name of *Paraphiomys pigotti,* after a fragmentary jaw with teeth. The faunas collected since that time under the direction of L. S. B. Leakey are prodigiously rich and varied; hundreds of specimens of the rodents alone, among which are the skull and skeleton of *Paraphiomys,* skulls of *Phiomys, Diamantomys,* and *Kenyamys* among the Thryonomyoidea; the skull and skeleton of *Bathyergoides;* and a skull of *Proheliophobius* for the Bathyergoidea. These two superfamilies represent all the Phiomorpha known at that time in Africa. It should be emphasized that the skulls are known by numerous samples. We must then add to this list the other African endemics, such as *Megapedetes,* related to the Recent *Pedetes,* known from skull and skeleton, and *Paranomalurus,* the first discovered fossil anomalurid, of which we have a complete skull, notwithstanding the fragility of the bones of this genus. And side by side with these antochthonous forms we see new immigrants that certainly took the opportunity of a recently established terrestrial connection with Asia to enter Africa. Those immigrants include the Afrocricetodontinae, African representatives of the worldwide family of Cricetodontidae, with three new genera, *Afrocricetodon, Notocricetodon* and *Paratarsomys.* These genera are known by a few skulls of which some are very complete. The other immigrants are the Sciuridae, known only by a small number of teeth and fragmentary jaws of two or three species referred to one or two new genera. Except for *Megapedetes,* already described by MacInnes (1957), the anatomical and systematic study of this important fauna has been published only recently (Lavocat 1973).

The study of *Paraphiomys* shows that its skull and skeleton are very close to the Recent *Thryonomys* and *Petromus* and clearly advanced by comparison with *Phiomys.* The masseter insertion is extended onto most of the lateral surface of the muzzle, and the anterior palatine foramina are extended back to the level of cheek teeth. Between the external and internal pterygoid there is a fossa opening anteriorly into the orbitotemporal cavity (a subordinal

character). In the molars, the crests dominate. In the upper teeth the metaloph is quite distinct and separated from the posteroloph by the fourth syncline in M^1; it is more posterior in M^2, and in M^3 it is fused with the posteroloph, while at the same time the mesoloph is lengthened up to the border of the tooth. This helps us to understand the structures of *Thryonomys* in which the vestigial metaloph is represented by a short crest on the anterior flank of the posteroloph. Also, the transition from a pentalophodont structure to a seemingly tetralophodont one in the South American Caviomorpha is more easily understood, and the new Deseadan genera *Branisamys, Sallamys,* and *Incamys* (Hoffstetter and Lavocat 1970) give evidence for a similar reduction process.

Besides *Paraphiomys,* we find the closely related new genus *Epiphiomys,* known by dental series, and characterized by the presence of a P^3 or dP^3. Remarkably enough, it has been discovered that, while the skull of *Phiomys* lacks a P^3, there are dentitions of a new genus, *Andrewsimys* (Lavocat 1973) with very similar molars, but retaining a P^3 or dP^3. Nevertheless, it does not seem that this situation is only the result of the loss of P^3 at old age in the forms where this tooth is absent. That assumption is certainly false in the case of *Paraphiomys* and *Epiphiomys* since we have very young, nearly newborn *Paraphiomys* skulls. One must more probably think of two evolving series, one without, the other retaining, P^3. D. S. Webb (pers. comm.) has pointed out that such a condition is true in North America among Sciuridae, two forms differing by the presence or absence of the P^3.

The Kenyamyidae, known by two genera of which one is known from a skull, do not show any indication of P^3. This is true even though the skull is more primitive than that of *Phiomys* in its infraorbital region, because no trace of insertion of the masseter on the muzzle is evident. Indeed, this muscle takes insertion, I believe, from all of the border of the large infraorbital foramen. It is true that the teeth, gliroid with rather high cutting crests over a low dental basis, indicate a rather high level of evolution compared to the *Phiomys* pattern. In such a species, in which the cranial characters are primitive but the molars advanced, the disappearance of P^3 is logical.

We do not know the upper dentition of all the Diamantomyidae and Myophiomyidae, but in the three genera where it is known, one Oligocene, *Metaphiomys,* the others Miocene, *Diamantomys* and *Myophiomys,* P^3 or dP^3 is present. This is perfectly logical for the Myophiomyidae in which the teeth, characterized by dominant and acute cusps, are of very cricetoid pattern, probably derived from a structure more primitive than that of *Phiomys,* which itself is already too lophodont. Also, the anterior palatine foramina observed in *Metaphiomys,* and deduced in *Myophiomys,* remain of the short type.

The Diamantomyinae, descendants of the Oligocene Metaphiomyinae, include two genera. *Diamantomys* is known from beautifully preserved skulls and some portions of the skeleton, *Pomonomys* by lower jaws only. This last genus is found, curiously enough, only in the Namib. What does appear immediately is the considerable lengthening of the muzzle and consequently of the masseter impression. Lengthening certainly also existed in *Pomonomys,* as is clear from the fact that the lower jaws of this genus are as much elongated as those of *Diamantomys.* The main peculiarity of *Pomonomys* is the presence of cement in the cheek teeth. The only Recent rodent in which I personally know such a lengthening of the muzzle and jaw in Africa is *Cricetomys,* for which the meaning of this lengthening is not known. Aside from that, the main structures of the Diamantomyinae are the same as those of *Paraphiomys.* The middle ear region, beautifully preserved, shows the greatest similarity not only with that of Recent *Thryonomys,* but also with those of the Caviomorpha, particularly *Lagostomus,* and this seems to me to be significant.

It seems that in East Africa *Paraphiomys* and *Diamantomys,* known from Loperot and Kirimun, persist without great change up to the level of Fort Ternan at least, and we find a rather advanced *Paraphiomys* species in the upper Miocene of Beni Mellal, in Morocco. But until now these are the only genera of Thryonomyoidea discovered in a level more recent than Burdigalian, the probable age of Songhor and Rusinga. There remains a great gap between the youngest level where they have been discovered and Recent time. So, while the genus *Paraphiomys* can reasonably be supposed the probable ancestor of the modern forms, we remain ignorant of the precise line of evolution from the Miocene Thryonomyoidea fauna up to the Recent fauna, as well as the time of extinction of the disappeared families. The presence of *Paraulacodus* genus very near to *Thryonomys* in the Chinji level of the Siwaliks induces us to guess the *Thryonomys* genus could begin at a similar epoch. The Recent fauna is very impoverished, consisting only of two genera, *Thryonomys* of great size, rather abundant and widespread, and *Petromus,* a relic form restricted to the rocky hills north and east of Namaqualand in South Africa.

In Miocene times we see this fauna at its best, be-fore it declines, probably under the pressure of the immigrants, particularly of the Afrocricetodontinae. The Myophiomyidae especially had a dental structure closely imitating those of these immigrants, and also by their small size were entirely comparable to the Afrocricetodontinae.

There does not exist in the lower Miocene of Africa any known form which could be said with certainty to be ancestral to the Hystricidae, a group now flourishing in Africa. The structures of this family are so entirely comparable to those of the other Phiomorpha that one must think of them as close relatives. With the current state of our knowledge one cannot indeed be certain that none of the known African fossil forms could be ancestral to them. The *Antherurus* group especially is really structurally very close to *Paraphiomys.* But, of course, the fact that as early as the Miocene one finds in the Indian Siwaliks good Hystricidae, leaves room for question. Perhaps the Hystricidae were the descendants of primitive Phiomorpha, already distinct and having migrated to Pakistan as early as the Eocene.

The Bathyergoidea are well represented by three genera; *Bathyergoides,* created by Stromer upon some tooth rows from Namib, and now well known from a skeleton and several skulls from Songhor, in Kenya, and Napak in Uganda; *Proheliophobius,* known by several skulls from East Africa, mostly from Rusinga; and *Paracryptomys* based on one rather incomplete portion of skull from Namib. These three perfectly fossorial genera are extremely important. *Proheliophobius* is allocated without problem to the family Bathyergidae, of which it is the first Tertiary specimen known. It is close to Recent genera of this family, such as *Heterocephalus* and *Heliophobius.* But a fundamental observation is that the infraorbital foramen is much less reduced than that of the Recent forms and was certainly functional; at the same time the teeth were nearly as strongly simplified as in the Recent genera.

Bathyergoides, of rather large size, is the type of a new family related to the Bathyergidae but clearly distinct, since its teeth retain vigorous structures. Being clearly comparable to those of the Phiomyidae and Thryonomyidae, these teeth, even if somewhat simplified, suggest the possibility of linking *Bathyergoides* with a Phiomorpha root in common with the Thryonomyoidea. But also, it permits us to understand the simplified structure of the Bathyergidae. Thus, these teeth give us the possibility of linking the Bathyergidae themselves to the same Phiomorpha root. The infraorbital foramen, large and certainly perfectly functional, itself refers to a very primitive stage of structure of this foramen, a stage of which fortunately the Thryonomyoidea *Kenyamys* gives us a good example. In *Kenyamys,* as well as in *Bathyergoides,* the masseter insertion does not extend forward on the muzzle. Structurally midway between Thryonomyoidea and Bathyergidae, *Bathyergoides* gives us the possibility to refer all the Bathyergoidea to the Phiomorpha and to consider the reduced size of the infraorbital foramen of the Bathyergidae as the result of a secondary reduction. This is also demonstrated by *Proheliophobius.* Here again, unfortunately, the history of the Bathyergoidea between the Miocene and Recent cannot be determined in detail, and little evidence indicates when and how *Bathyergoides* disappeared, for the only intermediate aged site is Fort Ternan. At this site is preserved the complete skull of a Bathyergidae, different from *Proheliophobius* without being its descendant, and in which the infraorbital foramen is already as much reduced as in present genera. As regards *Bathyergoides,* well adapted as it seems, the concurrence of the genus *Tachyoryctes,* a genus related to the Cricetodontidae, may have been the main reason of its extinction.

The fossils referred to the Pedetoidea, *Parapedetes* in the Namib, *Megapedetes* in East Africa (and also in the upper Miocene of Beni Mellal, Morocco), do not help in any way to break the systematic isolation of that superfamily. On the contrary, the analysis of a juvenile dentition of *Megapedetes* (Lavocat and Michaux 1966) induces us to acknowledge that starting from the primitive trigonodont structure, the original development and general disposition of the cusps in the cheek teeth have nothing to do, particularly, with that of the Phiomorpha. The fundamental structures of these teeth are already well established in the Miocene, but with a more brachydont pattern.

Three species of the new genus *Paranomalurus,* and probably also one of *Zonkerella,* represent the Anomaluroidea. A complete skull of *Paranomalurus,* the first skull known as a fossil in the family, demonstates that as early as the Miocene all of the essential anatomy of the Anomaluridae was established. The very small differences that do exist bear on the shape of the cranial roof, the space between the two rows of upper cheek teeth, and the still brachydont pattern of the cheek teeth where the ancestral trigonodonty is more easily recognized. An ulna demonstrates with clear evidence an adaptation to a gliding flight at least as good as that of the Recent species. If we look at the structures of the auditory region, they are extremely close to Recent forms, evidence that one cannot relate the Anomaluroidea to the Theridomyoidea, as I tentatively proposed 20 years ago (Lavocat 1951). So, as with the Pedetoidea, the Anomaluroidea are an independent superfamily of subordinal rank. Evidence suggests an origin in Africa where they are now living, even though their Oligocene record is extremely poor except for the Fayum ulna.

Miocene Immigrants

As I pointed out above, the Sciuridae and Cricetodontidae appear in the African Miocene as recent immigrants. The Sciuridae are rare, but even the Cricetontidae are not very abundant. These two families existed in Europe and Asia as early as the Oligocene, and the extraordinary convergent similarity of dental characters of the Myophiomyidae with those of the Cricetodontidae would be difficult to explain if a cricetodontid stock had also been present in the area where the evolution of the Phiomorpha had proceeded. The most logical solution, therefore, is to admit that these groups had taken advantage of the reestablishment of terrestrial connections to invade Africa. Of course, the hypothesis of an invasion by rafts before the establishment of these terrestrial connections cannot be totally excluded. But, in our actual state of knowledge, that hypothesis is unnecessary and strictly gratuitous.

There is little to say about the Sciuridae, except that they are present with at least two different forms. These are too different from the living Sciuridae in their dentitions to make possible the evaluation of their relations with them.

The Afrocricetodontinae are much better known, each of the three genera being known from skulls. These three genera, ranging in size from small to large (for the family), are characterized by a very primitive structure of the teeth and by the infraorbital foramen which corresponds to the structure of the Oligocene forms of Europe. The genus *Paratarsomys,* approximately of the size of recent *Mus musculus,* is so close to the Recent *Macrotarsomys* of Madagascar that one must evidently find in this *Paratarsomys,* or in very close relatives of this genus, the ancestors of the Malagasy rodents. On the contrary, we do not know any recent African form which could be directly related to the genera *Afrocricetodon* and *Notocricetodon.* Like *Paratarsomys,* these genera show an infraorbital foramen with a broad opening, and not a narrow vertical fissure, and a masseteric plate remaining nearer to the horizontal plane compared to the modernized form. As for the other groups, we lack forms of intermediate age. Nevertheless, the presence in the African Miocene of very primitive Cricetodontidae is enough to explain the presence and characters of recent

Mystromys, even if we do not actually know the direct ancestors or probable intermediates. There is no need to seek immediate relations with the recent South American or North European Cricetidae, such as was suggested by Ellerman (1940).

Coming to the Muridae *sensu stricto,* we do not find any indication of their presence in East Africa, either in the lower Miocene strata or, as yet, in the more recent middle Miocene of Fort Ternan.

I am now studying the Fort Ternan fauna, which is probably approximately of the same age as Beni Mellal (Morocco). Two Afrocricetodontinae, at least, are present: one seems very near *Afrocricetodon;* the other frankly hypsodont, could perhaps lead us toward the Nakali specimen discovered by E. Aguirre (pers. comm.), which I have compared with *Kanisamys* of the Siwaliks and Afghanistan. The Myocricetodontinae, so prominent in the Maghreb fauna, are not lacking at Fort Ternan: one small species is approximately the same size as *Myocricetodon parvus;* the other, comparable to *M. irhoudi,* seems to differ in cranial characters. *Leakeymys ternani* shows great similarities with the Gerbillinae, but a direct connection is not established. We will see later that, following J.-J. Jaeger (1975a), some Gerbillinae could result from the evolution of North African Myocricetodontinae. One species at least of Anomaluridae is present, very similar to the Burdigalian forms, as well as a small *Paraphiomys* species, The complete skull of a Bathyergidae recalls *Proheliophobius* of the Burdigalian, but the infraorbital foramen is narrow. *Megapedetes* is present. In the Miocene of Chios Island, H. Tobien (1968) has observed the presence of *Megapedetes,* and of a probably Phiomorpha, thus underscoring the extension of the African fauna, or at least a part of it, at that time.

The rodents of the late Miocene of East Africa are still very rare. At Nakali, admittedly "Vallesian," two specimens are provisionally referred to the Asiatic genus *Kanisamys.* At Ngorora (about 12 m.y.), a form said by J.-J. Jaeger to be the oldest Dendromurinae retains some characters of the Afrocricetodontinae, with which it should be directly connected. Following also J.-J. Jaeger, Bishop and Pickford (1975) point out two new Cricetid genera, small and of medium size, one new Phiomyid genus, a Sciurid and a Pedetid. Lastly, the faunal list of Lukeino (6.5 m.y.) (Pickford 1975) shows an Hystricidae indet., seemingly the oldest discovery at the moment in East Africa, and also another undetermined rodent. As other sites are now known, better information can be expected.

In the Pliocene of Lothagam 1, B. Patterson et al. (1970) record Anomaluridae and Thryonomyidae. Maglio (1974) notes *Hystrix* and *Tatera* from the Pliocene of Kanapoi. The Shungura formation of the Omo Group has furnished, mostly in Member B-10 (K/A 2.94 m.y.) but also in Member F (K/A 2.04 m.y.) a rather small number of samples, resulting nevertheless in a rather wide spectrum of genera:

All the genera which have been found are also present in eastern Africa to day. In contrast, in the northern African upper Pliocene localities, most of the fossil genera are now extinct, and very few are found in common with eastern African Pliocene/Pleistocene localities. All the fossil rodents are very similar to modern species, although there are some slightly archaic characters in the teeth, and this confirms what has been found in Bed I at Olduvai . . . the nonmurid fauna is more diversified in member F (than in member B). The presence of *Gerbillus, Jaculus, Heterocephalus,* and the leporid all indicate that this reflects an environmental change . . . Most of the microfauna from the L. 28 assemblage (of member F) argues for the spread of open arid conditions sometime before 2 m.y. . . . The B-10-3 assemblage is first of all characterized by an abundance of murids compared with other rodent taxa, but all suggest a wooded savanna (J.-J. Jaeger and H. B. Wesselman 1976).

The Olduvai site has recently furnished rodents in every bed of the formation. However, only those of Bed 1 (1.75 m.y.) have been described, although briefly, by me (Lavocat 1965, 1967). Recently J.-J. Jaeger (1976) has made a thorough study of the Muridae. Nine species, eight of which are new, are described. It is possible to distinguish them from the recent one by the size, the fossils being generally somewhat smaller, by more primitive characters of the stephanodonty and of the proportions between the teeth in some forms. All the genera still exist. On the whole, the fauna seems to indicate a more humid environment than the present one, such as lake shore with marshes, surrounded by arboreal savanna. With the murids are also present *Otomys,* the Gerbillids *Tatera* and *Gerbillus,* Dendromurinae, Sciuridae, *Pedetes, Hystrix, Heterocephalus,* a Rhizomyidae, Lagomorpha. According to J.-J. Jaeger (1976): "The Muridae are abundant in the median layers, where they are more numerous than the Gerbillinae. On the contrary, in the Upper Level FLK N 1 the proportions are reversed, except in the level 5. In the lower levels, *Tatera* is the only Gerbillid present. The modifications of the fauna point toward an aridification of the climatic conditions at the top of the Bed 1. Several genera of Muridae are common with those of South Africa, but probably specifically distinct." In Olduvai the Murinae are rather abundant; in South Africa they are present but the specimens seem much more rare, at least as judged by the collections available to me (Lavocat 1956, 1957, 1967b) or by those studied by de Graaff (1960). This rarity seems more evident in the oldest caves.

In South Africa *Mystromys,* a descendant of the Cricetodontidae, is very abundant in the supposedly older caves. It is unknown in Olduvai. It is probably in the early Pleistocene that we must locate the reversal in abundance of *Mystromys* and the Murinae in the southern region. Turning to other groups, the Gerbillinae, very abundant at Olduvai, are not rare at Taung and are present at Sterkfontein. The Dendromurinae are present in the two complexes. Also, *Pedetes, Hystrix,* and the Bathyergidae are present in the two regions. A remarkable fact that led me to place the genus *Otomys* among the relatives of the Afrocricetodontinae rather than among the Murinae is the great abundance of this genus everywhere, even in the levels with australopithecines that are poorest in Murinae.

Whatever the exact age relations between Olduvai and the australopithecine cave sites, it seems certain that the faunal differences are largely due to geographical isolation and to processes of local differentiation occurring over a rather long time. But we need far more evidence by which to follow the Pleistocene process of building the recent fauna. Moreover, modern distributions have certainly been modified by the intervention of man. Recent studies have shown significant transformations in relative species proportions as a consequence of the introduction of cultivation in several African regions.

North Africa (Palaearctic Region)

Compared with other parts of Africa, the Maghreb is a very peculiar region. The Saharan desert, even taking into account undoubted climatic variations, was probably established in ancient times and was not much easier to cross than a sea arm; according to Jaeger, the Maghreb showed strong characters of insularity as early as the lower Miocene. The oldest rodent deposit known in the Maghreb is Beni-Mellal, studied first by Lavocat (1961) and later by Jaeger (1971, 1976). Taking advantage of much better samples, Jaeger was able to improve upon the earlier results. The attribution of Beni Mellal to the Vallesian by Lavocat must be corrected following the research of Jaeger and Michaux (1976), who have given a clear demonstration that this layer is middle Miocene, probably about the same age as Fort Ternan in Kenya. Confirmation is given by study of a rhinoceros from Beni Mellal (Guerin, pers. comm.).

Two genera, *Megapedetes* and *Phiomys,* unknown later in the Maghreb, indicate some connection with other parts of Africa. Either some exchange was still possible at that time, or as Jaeger believes, these are the last relics of an earlier panafrican distribution. Also found is a Sciurid, described by Lavocat under the name of *Getuloxerus tadlae.* This generic name has been abandoned by H. de Bruijn in favour of *Atlantoxerus,* a recent Moroccan genus which, according to de Bruijn et al. (1970), has a perimediterranean Neogene distribution (two species in the Spanish Miocene, one in the Pliocene of Rhodes). Following Jaeger (1976) the Moroccan form *A. tadlae* should be the ancestor of *A. huvelini* from the Lower Pleistocene of Irhoud-Ocre, in Morocco, but not of the recent *A. getulus,* which is already known at Irhoud-Ocre and Sidi Abdallah. Anyway, the affinities of these Sciurids are much more with the other Mediterranean forms than with the recent or fossil forms of the African continent.

The Gliridae, unknown in the Miocene in the other parts of Africa, are represented at Beni Mellal by one species of *Microdyromys,* a genus known in the European Oligocene and Miocene. The species is rather distinct from the European ones, but may be the ancestor of a species from the lower or middle Miocene of Sardinia. It is interesting to remember that in this same island are also found Ctenodactylids bearing affinities with those of Beni Mellal. The genus disappears with the Miocene, and the Gliridae reappear only in Middle Pleistocene time, with *Eliomys* still present today.

The most widely represented groups are the Ctenodactylidae and the Cricetodontidae. The Ctenodactylidae were described by Lavocat provisionally under four genera. Taking advantage of better preserved material, Jaeger (1971b) has been able to regroup all within only two genera, *Africanomys* and *Metasayimys.* Black (1972) has even proposed retaining only one genus, *Africanomys,* with two species. It is well known that the only other place where fossils of this family are found is Asia, in Mongolia and the Siwaliks, at least as early as the Oligocene and up to the Miocene. It is certainly from Asia that the African population arrived, perhaps as early as the first opening of communications at the beginning of the Miocene. It is extremely important to note that the recent Ctenodactylids are strongly connected with arid biotopes. Therefore, the narrow belt limiting their migration zone in Miocene time seems to correspond with a well-defined climatic zone. Remarkably, the fossil Ctenodactylids are known neither in Europe nor in East Africa, notwithstanding the fact that this region has yielded Myocricetodontinae, those peculiar Cricetodonitidae with which the Ctenodactylids are associated in Beni Mellal and later horizons in the Maghreb, as well as in the Siwaliks.

The Cricetodontidae had been described by Lavocat under two distinct genera, *Cricetodon* and *Myocricetodon,* in two separate subfamilies. *Cricetodon* was supposed to contain the stock for *C. ibericus,* from Spain; *Myocricetodon* was an entirely new genus and type of a new subfamily, the Myocricetodontinae. Lavocat thought that it lay near the root of the Dendromurinae, a recent endemic African family. Jaeger (1971b) thinks, apparently correctly, that all the Cricetodontidae present in Beni Mellal are parts of the Myocricetodontinae. *C. atlasi* is given the new name of *Mellalomys,* and a new genus, *Dakkamys,* is described. (The fate of this group is discussed later in this chapter.) The layer of Testour, in Tunisia, worked by Robinson and Black (1973), has furnished *Africanomys,* plus one tooth given as the type of a new genus of Ctenodactylid, *Testouromys,* a Glirid, an Ochotonid, and *Mellalomys atlasi.* It is probably an age equivalent of Beni Mellal or Pataniak 6.

Seven layers recently discovered by Jaeger, Martin, Michaux, and Chabar Ameur in the Maghreb range from middle to upper Miocene in age and provide a partial understanding of the history of the population of the Maghreb. According to Jaeger (1971b, 1974, 1975a,b,c), who studied the problem intensively, *Metasayimys* disappears before the end of the Vallesian, while *Africanomys* is evolving toward the genus *Irhoudia,* characteristic of the Pliocene and lower Pleistocene of the Maghreb. Most of the characters of *Irhoudia* appear in *Africanomys* as early as Pataniak 6, somewhat younger than Beni Mellal. Nevertheless, the origin of the recent genera of Ctenodactylids is not clear and must probably be sought outside the Maghreb, in the zone of the recent Sahara. The recent distribution, still unexplained, certainly results from recent migrations, as *Irhoudia*—and this genus alone—is widely dispersed during lower Pleistocene time. After that, there is a gap.

Looking at the Gliridae, the *Microdyromys* lineage appears to be extinct in the upper Miocene, and the family reappears only in mid-Pleistocene times with *Eliomys* still extant.

The Dipodidae make a short appearance in Pataniak 6 as well as in the Beglia Formation of Tunisia, according to Robinson. Represented in Pataniak 6 by *Protalattaga* which bears strong affinities with *P. grabaui* from the Chinese Miocene, they call our attention to the Asiatic affinities of a part of the Maghreb fossil fauna. Later on, the group is known

from some Lower and mid-Pleistocene localities (Irhoud Ocre, Irhoud Derbala village, Sidi Abdallah). A genus very near *Mellalomys, Zramys* (also similar to *Cricetodon*), probably originates from the same ancestral group as *Mellalomys.* A Pliocene form similar to *Ruscinomys* is probably a result of *Zramys* evolution. The two *Myocricetodon* species of Beni Mellal, *M. parvus* and *M. cherifiensis* are, according to Jaeger, the beginnings of two groups differing in the size and structure of the molars. The group starting with *M. parvus* develops molars with one supernumerary cusp, extraordinarily similar to those of recent *Petromyscus;* but the structure of the palate, the zygomatic plate and the mandible of *Myocricetodon* are already more evolved than in *Petromyscus.* This observation definitely eliminates the idea of a possible relationship. Within the group evolving from *M. cherifiensis,* there is a rapid increase of size, and the crests of the pairs of cusps becomes transverse, resulting in a Gerbilline type of molar. Moreover, the Gerbillinae show most of the cranial characters of Myocricetodon, indeed with stronger specialization. Notable is the appearance in Amama 2 (Turolian) of the earliest known Gerbillinae, *Protatera.* Of course, it is not possible to show a direct connection between this genus and the described forms of Myocricetodontinae; perhaps we should search in the Saharan region to find the precise lineages. There is also a good probability that a part of the Gerbillinae—*Meriones,* for example, which is more Asiatic than African—may have originated in Asia where other Myocricetodontinae can be found. In favor of this view is the discovery in the Turolian of Austria of the primitive Gerbillinae *Epimeriones.* Thereafter, in the Pliocene and the Pleistocene, the Gerbillinae, essentially *Gerbillus* and *Meriones,* are mostly very abundant, even more so than the Muridae.

Very important is the fact that the earliest appearance of the Murinae in the Maghreb is by *Progonomys cathalai* itself, the very same species and genus as in Europe, and at the same time (Vallesian). This is probably the consequence of parallel migration along each border of the Mediterranean, starting from one common oriental source. Perhaps as a result of worse conditions, the evolutionary divergence seems less in the Maghreb. Nevertheless, the genus *Paraethomys,* known in North Africa mostly in the Plio-Pleistocene, already appears in the upper Miocene of Kendek el Ouaïch, perhaps as an offshoot of *Progonomys,* and is found also at Amama 2. The great interest of this genus is that it is known all around the Mediterranean. *P. miocaenicus,* from Kendek el Ouaïch, is also present at the same time at Librilla, in Spain. The Pliocene species *P. anomalus* described from Maritsa (Rhodes Island by de Bruijn et al. (1970) is also recorded in the Pliocene of Lake Ichkeul (Tunisia) (probable age about 35 m.y.). In the same Ichkeul locality are found *Mus musculus,* earliest appearance, a Cricetid similar to *Ruscinomys,* thought by Jaeger to be an offshoot of *Zramys,* and a Lagomorpha, *Prolagus sardus.*

The murids of the Plio-Pleistocene, specially studied by Jaeger (1975b), are always less numerous than the Gerbillinae, the comparative abundance of which increases with time. Their diversification is feeble, the limitation coming probably from climatic factors, particularly the absence of tropical humid climates. There is clear endemism. The Maghreb appears to be a marginal zone where representatives of European, Asiatic, or tropical African forms eventually entered, under conditions temporarily favoring admission, but thereafter mostly disappeared within a short time, doubtless under the pressure of severe climatic or biologic conditions. In terms of this sequence, Ternifine, of early Pleistocene age, appears as an important step, following a climatic crisis, and characterized by an important renewal of the fauna. The two lineages of *Paraethomys* and *Praomys* (Berberomys) (both probably direct offshoots from *Progonomys*) are more stable and push their lineages up to the middle and even upper Pleistocene. *Paraethomys* reaches the upper Pleistocene in the Irhoud Neanderthal, after dividing in two lines in the lower Pleistocene; *Praomys* (Berberomys), which ends with the middle Pleistocene, gives rise, through *P. pomeli,* to *Kritimys* of the mid-Pleistocene of Crete. Although dentally very similar to *Praomys* from South Africa, *Berberomys* does not seem to be directly connected with this form. *Pelomys,* known in the upper Pliocene of Rhodes, appears briefly in Amama 3 (Pliocene). Without doubt the common center of origin is the Nile Valley.

As has been said before, the lower Pleistocene of Ternifine marks a renewal after crisis and extinction. *Arvicanthis,* strictly an African genus, able to live in the arid zone as long as a place with water is available, is known in the middle and upper Pleistocene of Palestine and in the lower Pleistocene of Olduvai (Bed I). It appears in Ternifine, along with *Ellobius,* a remarkable fossorial Arvicolinae of the temperate warm steppic countries, now limited to Eastern Europe and parts of Asia, and also with *Meriones,* a Gerbillinae now essentially Asiatic and north African. Probably due to an adverse trend to aridity, *Arvicanthis* disappears before the Presoltanian, while *Ellobius* becomes extinct at the end of the same period. *Mus,* still extant, is known in the upper

Pliocene and Pleistocene by three species that are seemingly mutually exclusive. No Maghrebian origin for this genus can be supposed. *Lemniscomys,* now living from Tanzania to the Maghreb, appears only in the Presoltanian. *Apodemus,* mostly a European animal, appears in the Upper Paleolithic, probably via Palestine. Although a rather similar form is said to live in Orania in the upper Pliocene and lower Pleistocene, *Apodemus* certainly does not take origin from this form.

Hystrix, recorded as a fossil in Marceau (Vallesian), is found thereafter in the lower Pleistocene.

Lagomorpha

There is little to be said about the history of the Lagomorpha. As in other groups, this one took advantage of the opening of terrestrial connections to enter Africa in Miocene time. At that time, two Ochotonidae occur, *Austrolagomys* in the Namib and *Kenyalagomys* in East Africa. There also exists a lagomorph at Beni Mellal. The Leporidae, *Pronolagus* and *Lepus,* are represented since the Pleistocene. The Ochotonidae are now extinct. Some specialists believe that the species of *Lepus* from the Cape region is the same as that from Europe.

General Conclusions

At least three main stages can be distinguished in the African history of the rodents. The first stage, we assume, began very early in the Eocene, after an invasion of Africa by members of the Ischyromyidae (*sensu lato*). Such an invasion necessarily did occur but we remain ignorant of its exact mode. At least as early as the Lutetian this first invasion had given rise to the first Hystricognathi with all the essential characters of that group already developing. The proof of the possibility of such early differentiation is given by the fact that at the same time the essential characters of the Theridomorpha were already developed in Europe. A single molar from the Lutetian of Pakistan possibly represents an immigrant from this primitive group expanding out of Africa during a marine regression. This molar, still brachydont and primitive, already shows, I think, all the structural characters which would develop further in the more advanced Hystricognathi.

During or after the Lutetian, but certainly before the Oligocene, the main families of Hystricognathi became differentiated. The Oligocene fauna of the Fayum is apparently very incomplete, lacking some of the Phiomorpha, particularly the Bathyergoidea, and also lacking the Pedetoidea. In my opinion these groups are African endemics, direct offshoots of some of the Ischyromyidae which entered this continent during the Eocene. The hystricognath structure of the lower jaw in the suborder Hystricoganthi as well as the dental characters and the anatomy of the middle ear in the Hystricognathi, the Anomaluroidea, and in the Pedetoidea (also different of course between each group) are evolved characters and, as such, cannot preclude a common stem with the Theridomorpha; however, they afford enough evidence to separate these groups definitely from the Theridomorpha themselves. Previously there groups had been assumed to be very closely related. The early results of the researches of Parent (1971, 1976a, 1976b) on the middle ear region of rodents seem to confirm our early conclusions about the systematic value of this region. Thus, it is with some confidence that we make use of it in the problems of establishing systematic affinities (Lavocat and Parent 1971). I have suggested the possibility of a common ancestral (lower Eocene) group (Lavocat 1976).

With continental isolation, evolution of the Hystricognathi resulted in diversification with the Phiomorpha occupying various ecological niches and specializing particularly with respect to cranial and dental adaptations. The dentition tends toward hyposodonty in several lineages, as well as toward simplification of dental structures by diminution of the number of the crests or even by the nearly complete loss of most details of the crown. The result of this tendency in the extant Thryonomyidae is a notable example. As early as the Miocene, this tendency resulted in a very marked dental simplification in the Bathyergidae. At the same time, or perhaps earlier, in this family the infraorbital foramen, primitively probably of size comparable to that in *Kenyamys,* had already begun to regress. This results, at least in the living forms, in a complete regression of this foramen and of the adapted muscular structure. A similar condition is seen already in one genus from the late Miocene of Fort Ternan. In Miocene time this hystricognath autochthonous fauna was very well developed and flourishing. If geological events had not resulted in an early end to the isolation of Africa, no doubt this fauna would have given rise to a Recent diversity similar to that obtained in South America from the Santa Cruz fauna, also Miocene in age.

The second stage of this history involves a Miocene invasion from Asia. At this time we see an important difference between Mediterranean Africa (Palaearctic region) and the remaining and major part of the continent (Ethiopian region). In Ethiopian Africa, the invasion, known as early as the "Burdigalian," introduced several genera of Cricetodontidae and Sciuridae obviously originating in Asia. These cricetodontids are, in Miocene time, at an evolutionary level similar to that of Oligocene genera in Europe. Those of the European Miocene are much more evolved. So Africa provides information on the status of the Asiatic fauna from which it received its immigrants, that is, the southern part. We are thus induced to believe that in Miocene time it was possible to distinguish two faunal regions in Asia, a conservative southern one where anatomical characters were little evolved, and a northern one which was probably the reservoir of origin of Miocene migrations resulting in renewal of the cricetodontid fauna of Europe. The latter also probably faced intense selection pressures because of the more severe ecological conditions. This resulted in the anatomical modifications that we observe in the European species.

Probably at approximately the same time, Mediterranean Africa received from Asia its first Ctenodactyloidea and also without doubt the Myocricetodontinae. This area also received one glirid and one species of *Xerus* either from Europe or from Asia. The presence of *Paraphiomys* and of *Megapedetes* at Beni Mellal results either from the existence of limited faunal exchanges between the two African areas, or from the persistence of relics of an ancient common population.

It is probably also at about this time that the afrocridetodontine *Paratarsomys* or a species very close to it crossed the Mozambique Channel from East Africa to Madagascar and gave rise to the Nesomyinae. The latter subfamily presents such an harmonious adaptive range in the dentitions that I cannot personally admit that the different genera constituting this Malagasy group can be the result of several different invasions from Africa, as proposed by Petter (1962:301). It seems rather that only a single origin and subsequent diversifying evolution could have resulted in such a coherently varied population. Of course, only a very efficient filter such as a portion of sea could have prevented most of the contemporaneous mammals on continental Africa from entering Madagascar along with the Nesomyidae.

At the same time in continental Africa, the Afrocricetodontinae were evolving, giving rise as early as in the late Miocene to species with hypsodont tubercles. It may also have given rise to the Dendromurinae. Unfortunately we know too little of the late Miocene in East Africa to be able to follow the evolution of these forms. We can see that an invasion, beginning in the Miocene, explains the Recent cricetodontoid populations of Africa, but the details escape us. For the Gerbillinae, the problem remains very obscure as these animals live also in Asia. Which continent was the source of this group? Following Jaeger, perhaps both. It is not very clear and the hiatus between Fort Ternan and the Pleistocene remains too great for us to see any detailed phyletic relationships.

Similarly, in North Africa the problem of the detailed phylogeny of the Ctenodactyloidea is not entirely cleared up. There is also the problem of the descendants of the Myocricetodontinae, if there were any. Jaeger suggests that they probably gave rise to Gerbillinae.

Although we decidedly think that the Miocene invasion has been very efficient in competing against the autochthonous fauna of Africa and we can see in the present fauna results of the decline of the latter, there is at the moment no possibility to examine the stages of this decline because it happened precisely in a time period that is up to now unavailable in the fossil record of that continent.

The third important stage in the history of the African rodent fauna is that of the arrival of the Murinae. This event occurred here at the same time and with the same species as it did in Europe. But here also, except in Maghreb, we lack the necessary evidence prior to the late Pliocene relating to the details of subsequent development of this invasion.

As can be seen, we have learned much about Africa in recent years so that now we can outline the general lines and main stages in the establishment of the Rodentia on that continent. But important gaps remain to be filled before the details of this history can be written.

However, from the systematic point of view, the Recent and fossil anatomical documents now allow us to group the Thryonomyoidea, Hystricoidea, and Bathyergoidea under one infraorder, Phiomorpha. Regarding the Bathyergoidea, the paleontological evidence permits us to solve the important problems on the origin and meaning of their dental and infraorbital structures, and thus to pull them out of their systematic isolation. For the Muroidea as well, their African history sheds a new light over their general systematics, not only in Africa but also in the entire world. This evidence is important particularly by showing more clearly the relationships between the Cricetodontidae and the Recent Muroidea, especially the Murinae. Paleontology confirms the antiquity of the Malagasy population of Rodentia. It confirms also the systematic isolation of the Anomaluroidea and Pedetoidea. With such evidence we can follow in the Ctenodactyloidea and Myocrice-todontinae the special relationships between southern Asia and Mediterranean Africa.

From the biogeographic and systematic points of view, the most important consequence of a much better knowledge of the African rodents is probably the confirmation given to the opinions of the early anatomists and zoologists about the close relations between the African (and Asiatic) Phiomorpha and the South American Caviomorpha. Regardless of the conclusions that one wishes to draw, the anatomical similarities uniting these two infraorders are a fact. I should repeat here what I have already written in the conclusions of a more detailed study of the East African Miocene Rodentia (Lavocat 1973): We have with the greatest tenacity searched for anatomical structures that would result in definite diagnostic differences between these two infraorders, but without success. Structures and associations of structures are the same in both, even when comparisons with other groups demonstrate that these structures are effectively not functionally related. In my opinion, such an identity cannot be the effect of mere chance, because the probability of identical association of structures that are not necessarily linked is much too small. We consider as an unavoidable consequence of this statement the necessity of admitting that one of the two faunas is a phyletic derivative of the ancestors of the other. In such a situation, the study of the elements involved shows that the place of origin must be in Africa, and the derived fauna in South America. Because the rodents are known in South America as early as the early Oligocene, the migration is necessarily earlier, and one is led to place it in the Eocene. At that time the crossing of the Atlantic Ocean by rafts was easier owing to the shorter distance between the two opposite shores.

We must record here that results of comparative study of the muscular system, and also of the parasitology, confirms those of the osteological studies. Wood (1955) has sustained, and is still sustaining, a quite different point of view, according to which the South American rodents should have come not from Africa, but from North America, also by rafting. His arguments, bearing essentially on the impossibility of a migration through the Bering Straits, remain perfectly conclusive, but they are not conclusive against the hypothesis of a transatlantic migration. Even recently, mariners are said to have found living rodents on trees rafted as far as 2,000 km from the shores (Chaline, pers. comm.). On the other hand, I think that the comparative analysis of the dental structures in the Oligocene rodents of the two continents strongly supports my opinion that their primitive and ancestral structures were identical. Wood (1972) recently made known the discovery in North America of an Eocene rodent with an hystricognath lower jaw. This discovery is evidently of major significance. Nevertheless, I do not believe that it can demonstrate in any decisive manner the North American origin of the Caviomorpha. One isolated anatomical character cannot be given the same weight as the association of many anatomical features. In fact, it is only the discovery of this total association that was convincing enough to induce me to abandon the thesis supported by Wood, a thesis that I strongly supported until then. In any event, it is evident that additional information about the rodents of the two sides of the Atlantic is of prime importance and that it is critical that we discover in Africa rodents from Eocene time. Even in view of all the other blanks in our knowledge of the history of the rodents in Africa, I believe that the most needed discovery is that of this Eocene fauna.

I would like to extend my gratitude to D. S. Webb and to V. J. Maglio for comments on the manuscript and for discussions on the English translation.

References

Andrews, C. W. 1914. On lower Miocene vertebrates from British East Africa. *Quart. J. Geol. Soc.* 70:163–186.

Bishop, W. W., and Pickford, M. 1975. Geology, fauna and palaeoenvironments of the Ngorora Formation, Kenya Rift Valley. *Nature* 254:185–192.

Black, Craig C. 1972. Review of fossil rodents from the Neogene Siwalik Beds of India and Pakistan. *Paleont.* 15(2):238–266.

Bruijn, H. de; Dawson, Mary R.; and Mein, P. 1970. Upper Pliocene Rodentia, Lagomorpha and Insectivora (Mammalia) from the isle of Rhodes (Greece). I, II, and III. *Proc. Koninklijke Ned. Akad. Wetens.*, ser. B., 73(5):535–584.

Bugge, J. 1971. The cephalic arterial system in New World hystricomorphs, and in bathyergoids, with special reference to the systematic classification of the rodents. *Acta Anat.* 80:516–536.

Chabbar Ameur, R.; Jaeger, J.-J.; and Michaux, J. 1976. Radiometric age of early *Hipparion* fauna in Northwest Africa. *Nature* 261:38–39.

Colbert, E. H. 1933. Two new rodents from the Lower Siwalik beds of India. *Am. Mus. Novit.* 633:1–6.

Durette-Desset, M. C. 1971. Essai de classification des Nématodes Héligmosomes. Corrélations avec la Paléobiogéographie des hôtes. *Mém. Mus. Natl. Hist. Nat.*, n.s., sér. A, 69:1–126.

Ellerman, J. R. 1940. *The families and genera of living rodents*. London: British Museum (Nat. Hist.), Vol. 1:1–689; Vol. 11:1–690.

de Graaff, G. 1960. A preliminary investigation of the mammalian microfauna in Pleistocene deposits of caves in the Transvaal System. *Palaeont. Afr.* 7:59–118.

Hinton, M. A. C. 1933. Diagnoses of new genera and species of rodents from the Indian Tertiary deposits. *Ann. Mag. Nat. Hist.* 10(12):620–622.

Hoffstetter, R., and Lavocat, R. 1970. Découverte dans le Déséadien de Bolivie de genres pentalophodontes appuyant les affinités africaines des rongeurs caviomorphes. *C. R. Hebd. Acad. Sci.* (Paris) 271(D): 172–175.

Hopwood, A. T. 1929. New and little known mammals from the Miocene of Africa. *Am. Mus. Novit.* 344:1–9.

Jaeger, J.-J. 1971a. Les micromammifères du "Villafranchien" inférieur du lac Ichkeul (Tunisie): données stratigraphiques et biogéographiques nouvelles. *C. R. Hebd. Acad. Sci.* (Paris) 273(D):562–565.

———. 1971b. Un Cténodactylidé (Mammalia, Rodentia) nouveau, *Irhoudia bohlini* nov. g; n. sp. du Pléistocène inférieur du Maroc. Rapports avec les formes actuelles et fossiles. *Notes Serv. Géol. Maroc* 31(237):113–140.

———. 1974. Nouvelles faunes de Rongeurs du Miocène supérieur d'Afrique Nord Occidentale. *Ann. Geol. Surv. Egypt* 4:263–268.

———. 1975a. Les Rongeurs du Miocène moyen et supérieur du Maghreb. Thèse Doctorat Sc. Nat. Languedoc.

———. 1975b. Les Muridae du Pliocène et du Pléistocène du Maghreb. Thèse Doctorat Sc. Nat. Languedoc.

———. 1975c. Origine et évolution du genre *Ellobius* au Maghreb. Thèse Doctorat Sc. Nat. Languedoc.

———. 1976a. Les Rongeurs (Mammalia, Rodentia) du Pléistocène inférieur d'Olduvai Bed I (Tanzanie). Ire Partie: Les Muridés. In R. J. G. Savage and S. C. Coryndon, eds., *Fossil Vertebrates of Africa*, vol. 4. New York and London: Academic Press, pp. 58–120.

———. 1976b. Les Rongeurs du Miocène de Beni Mellal. *Palaeovert.* 7(4):91–100.

Jaeger, J.-J., and Martin, J. 1971. Découverte au Maroc des premiers micromammifères du Pontien d'Afrique. *C. R. Hebd. Acad. Sci.* (Paris) 272(D):2155–2158.

Jaeger, J.-J.; Michaux, J.; and David, B. 1973. Biochronologie du Miocène moyen et supérieur continental du Maghreb. *C. R. Hebd. Acad. Sci.* (Paris) 277(D):2477–2481.

Jaeger, J.-J.; Michaux, J.; and Thaler, L. 1975. Présence d'un rongeur muridé nouveau, *Paraethomys miocaenicus* nov. sp., dans le Turolien supérieur du Maroc et d'Espagne. Implications paléogéographiques. *C. R. Hebd. Acad. Sci.* (Paris) 280(D):1673–1675.

Jaeger, J.-J., and Wesselman, H. B. 1976. Fossil remains of Micromammals from the Omo Group Deposits. In Y. Coppens, F. Clark Howell, G. Ll. Isaac, and R. E. F. Leakey, eds., *Earliest man and environments of the Lake Rudolf Basin*. Chicago and London: University of Chicago Press, pp. 351–360.

Landry, S. O. 1957. The interrelationships of the New and Old World Hystricomorph rodents. *Univ. Calif. Pub. Zool.* 56(1):1–118.

Lavocat, R. 1951. *Revision de la faune des mammiferes Oligocenes d'Auvergne et du Velay.* Paris: Editions "Sciences et Avenir."

———. 1956. La faune de rongeurs des grottes à Australopithèques. *Palaeont. Afr.* 4:69–75.

———. 1957. Sur l'age des faunes de rongeurs des grottes à Australopithèques. *Proc. Third Pan. Afr. Congr. Prehist.,* Livingstone, 1955:133–134.

———. 1959. Origine et affinités des Rongeurs de la sousfamille des Dendromurinae. *C. R. Hebd. Acad. Sci.* (Paris) 248:13.

———. 1961. Le gisement de vertébrés fossiles de Beni Mellal. *Notes Mém. Serv. Géol. Maroc* 155:1–144.

———. 1962. Réflexions sur l'origine et la structure du groupe des Rongeurs. *Colloques Int. C.N.R.S.* 163:491–501.

———. 1967a. Les microfaunes du Néogène d'Afrique orientale et leurs rapport avec celles de la région paléarctique. In W. W. Bishop and J. D. Clark, eds., *Background to evolution in Africa.* Chicago: University of Chicago Press, pp. 57–66.

———. 1967b. Les microfaunes du quaternaire ancien d'Afrique orientale et australe. In W. W. Bishop and J. D. Clark, eds., *Background to evolution in Africa.* Chicago: University of Chicago Press, pp. 67–72.

———. 1969. La systématique des Rongeurs hystricomorphes et la dérive des continents. *C. R. Hebd. Acad. Sci.* (Paris) 269:1496–97.

———. 1971. Affinités systématiques des Caviomorphes et des Phiomorphes et origine africaine des Caviomorphes. *Ann. Acad. Brasil. Cienc.* Supp. 43: 515–522.

———. 1973. Les Rongeurs du Miocene d'Afrique Orientale. 1. Miocène inférieur. *Mém. Trav. E.P.H.E.,* Institut de Montpellier, 1:1–284.

———. 1974. What is an Hystricomorph? In I. W. Rowlands and Barbara J. Weir, eds., *The biology of Hystricomorph rodents.* London, New York, and San Francisco: Academic Press, pp. 7–20, 55–60.

———. 1976. Rongeurs caviomorphes de l'Oligocene de Bolivie: II. Rongeurs du Bassin Déséadien de Salla-Luribay. *Palaeovert.* 7(3):1–90.

Lavocat, R., and Michaux, J. 1966. Interprétation de la structure dentaire des Rongeurs africains de la famille des Pédetidés. *C. R. Hebd. Acad. Sci.* (Paris) 262:1677–79.

Lavocat, R., and Parent, J. P. 1971. Valeur systématique de la région de l'oreille moyenne chez les Rongeurs. *C. R. Hebd. Acad. Sci.* (Paris) 273(D):1478–1480.

Leakey, L. S. B. 1965. *Olduvai Gorge 1951–1961,* vol. 1. Cambridge University Press, pp. 1–188.

MacInnes, D. G. 1953. The Miocene and Pliocene Lagomorpha of East Africa. In *Fossil Mammals of Africa.* London: Brit. Mus. (Nat. Hist.), 6:1–30.

———. 1957. A new Miocene rodent from East Africa. In *Fossil Mammals of Africa.* London: Brit. Mus. (Nat. Hist.) 12:1–35.

Maglio, V. J. 1974. Late Tertiary fossil vertebrate succes-sions in the Northern Gregory rift, East Africa. *Ann. Geol. Surv. Egypt* 4:269–286.

Mossman, H. W., and Luckett, W. P. 1968. Phylogenetic relationships of the African mole-rat, *Bathyergus janetta,* as indicated by the fetal membranes. *Am. Zool.* 8:806.

Osborn, H. F. 1908. New fossil mammals from the Fayum Oligocene. *Bull. Am. Mus. Nat. Hist.* 24:265–272.

Parent, J. P. 1976a. La région auditive des Rongeurs sciurognathes. Caractères anatomiques fondamentaus. *C. R. Hebd. Acad. Sci.* (Paris) 282(D):2183–85.

———. 1976b. Disposition fondamentale et variabilité de la région auditive des Rongeurs hystricognathes. *C. R. Hebd. Acad. Sci.* (Paris). 283(D):243–245.

Petter, F. 1962. Monophylétisme ou polyphylétisme des Rongeurs malgaches. *Colloques Int. C.N.R.S.* 104:301–310.

———. 1964. Affinités du genre *Cricetomys.* Une nouvelle famille de Rongeurs Cricetidés, les Cricetomyinae. *C. R. Hebd. Acad. Sci.* (Paris) 258:6515–18.

———. 1968. Un muridé quaternaire nouveau d'Algérie, *Paraethomys filfilae.* Ses rapports avec les Muridés actuels. *Mammalia* (Paris) 32(1):54–59.

Pickford, M. J. L. 1975. Late Miocene sediments and fossils from the Northern Kenya Rift Valley. *Nature* 256:279–284.

Robinson, P., and Black, C. C. 1973. A small Miocene faunule from near Testour, Beja Gouvernorat, Tunisia. Livre Jubilaire M. Solignac, *Ann. Mines Géol.* (Tunis). 26: 445–449.

———. 1974. Vertebrate faunas from the Neogene of Tunisia. *Ann. Geol. Surv. Egypt* 4:319–332.

Schaub, S. 1953. Remarks on the distribution and classification of the "Hystricomorpha". *Verh. Naturf. Ges. Basel* 64:389–400.

Schlosser, M. 1911. Beiträge zur Kentinss der Oligozänes Landsäugetiere aus dem Fayum: Aegypten. *Beitr. Paläont. Geol. Öst.-Ung.* 24:51–167.

Simons, E. L. 1968. Early Cenozoic mammalian faunas Fayum Province, Egypt. Part I: African Oligocene Mammals: introduction, history of study and faunal succession. *Bull. Yale Peabody Mus. Nat. Hist.* 28:1–23.

Stromer, E. 1922. Erste Mitteilung über Tertiäre Wirbeltier-Reste aus Deutsch-Südwestafrikas. *S. B. Bayer Akad. Wiss.* (München 1921):331–340.

———. 1923. Bemerkungen über ersten Landwirbeltier-Reste aus dem Tertiäre Deutsch-Südwestafrikas. *Palont. Zeitschr.* 5:226–228.

———. 1926. Reste Land-und Süsswasser-Bewohnender Wirbeltiere aus den Diamantfeldern Deutsch-Südwest Afrikas. In E. Kaiser, ed., *Die Diamantenwuste Südwestafrikas,* vol. 2. Berlin: D. Reimer, pp. 107–153.

Tobien, H. 1968. Paläontologische Ausgrabungen nach jungtertiären Wirbeltiere auf des Insel Chios (Griechenland) und bei Maragheb (NW Iran). *Sond. Iahr. Verein. Fr. Univ. Mainz*:51–58.

Walker, A. 1968. The Lower Miocene Fossil site of Bukwa Sebei. *Uganda J.* 32(2):149–156.

Wood, A. E. 1949. A new Oligocene rodent genus from Patagonia. *Am. Mus. Novit.* 1435:1–54.

_____. 1950. Porcupines, paleogeography and parallelism. *Evolution* 4(1):87–98.

_____. 1955. A revised classification of the Rodents. *J. Mammal.* 36:167–187.

_____. 1962. The juvenile tooth patterns of certain African rodents. *J. Mammal.* 43:310–331.

_____. 1968. Early Cenozoic mammalian faunas Fayum Providence, Egypt. Part II: The African Oligocene Rodentia. *Bull. Yale Peabody Mus. Nat. Hist.* 28:29–105.

_____. 1972. An Eocene hystricognathous rodent from Texas. It's significance in interpretation of continental drift. *Science* 175:1250–1251.

_____. 1973. Eocene Rodents, Pruett formation, Southwest Texas; their pertinence to the origin of the South American Caviomorpha. *The Pearce-Sellards Series* (Austin: Texas Memorial Museum) 20:1–40.

_____. 1974. The evolution of the Old World and New World Hystricomorphs. In I. W. Rowlands and Barbara J. Weir, eds., *The biology of Hystricomorph rodents.* London, New York, and San Francisco: Academic Press, pp. 21–54, 55–60.

Wood, A. E., and Patterson, B. 1959. The rodents of the Deseadan Oligocene of Patagonia and the beginnings of South American rodent evolution. *Bull. Mus. Comp. Zool.* 120:281–428.

_____. 1970. Relationships among hystricognathous and hystricomorphous Rodents. *Mammalia* 34(4):628–639.

6

Prosimian Primates

Alan C. Walker

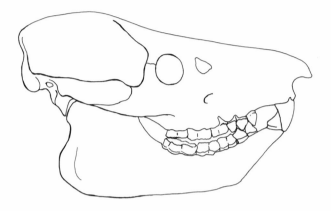

The living African prosimians, although members of the single family Lorisidae, fall into two clearly defined subfamilies, the Galaginae (bush babies) and the Lorisinae (pottos). Lorisines also live today in Asia, but the bush babies are restricted to Africa. The lemurs of Madagascar are fairly closely related to the lorisids, but without fossil evidence of lemurs in Africa and with none of the subfossil Madagascan lemurs more than 3,000 years old, speculation on the relationships between these two groups would be out of place in this volume.

Among the living prosimian primates, the Madagascan lemurs show the greatest evolutionary diversity. However, the very recently extinct subfossil lemurs testify to an even greater range of adaptations in this group. The living species range from small, bush-babylike, insectivorous or gum-eating species to larger and often diurnal, folivorous, or frugivorous species. Reviews of the general biology and behavior are given by Doyle (1974) and Martin (1972).

The extinct species were all larger than their living relatives. Tattersall (1975) and Tattersall and Schwartz (1974) have reviewed the cranial and dental evidence and Walker (1974a) has reviewed the postcranial evidence. These extinct forms included three species of large folivore that were both cranially and postcranially very similar to living koalas. Another genus had species that evolved extremely elongated forelimbs and is presumed to have lived a rather slothlike existence, being a folivore that suspended itself from hooklike hands and feet. Yet another group developed characteristics of the cercopithecoid monkeys, two species being somewhat baboonlike and one more geladalike. All the extinct species must be considered, in evolutionary terms, as biologically contemporaneous with the living Madagascan lemurs. Their extraordinary radiation must be largely caused by their long isolation on Madagascar and their consequent lack of competition with other primates until the recent arrival of human beings.

By contrast, the lorisids have produced only two basic types, galagos and lorises. Both are relatively successful; the galagos are among the most common small mammals in some parts of Africa. The galagos are active, leaping, mainly insectivorous or gum-eating forms, whereas the lorises are slow-moving, cautious stalkers of, usually, active prey.

The fossil record of prosimian evolution in Africa is restricted in the time ranges sampled, geographic coverage, and the numbers of actual specimens. All three factors add difficulties to the usual strictures affecting paleontological interpretations. Most of the material is from the early Miocene; one specimen is from the middle Miocene and a few from the Plio-Pleistocene. Geographic restrictions are even greater, with all specimens coming from East Africa. The specimens themselves are mostly teeth and associated mandibular or maxillary parts, but some more complete cranial remains are known. Isolated postcranial bones are known of some forms and, more importantly, associated dental and postcranial bones have been found for a few species.

All the fossils can be allocated to the family Lorisidae, but most are galagines. Lorisine remains are quite scarce. Accounts dealing with the original material are few and a complete bibliography is given here. There are no fossil members of the family known outside Africa. *Indraloris* from the Indian Siwaliks (Lewis 1933, Tattersall 1970) is most probably not a lorisine (Tattersall 1969), but confirmation of this, despite very complete new material, unfortunately is not forthcoming.

The Collections

The fossil lorisid material from Kenya is all housed in the National Museum of Kenya, Nairobi, with the exception of two small pieces collected in 1931 that are in the British Museum (Natural History), London. The Uganda fossils are kept at the Uganda Museum, Kampala. The Olduvai Gorge fossils from Tanzania are presently housed in the National Museum of Kenya under an arrangement with the Tanzanian government. The main Kenya collections were made by Dr. A. T. Hopwood, Dr. D. G. MacInnes, Dr. L. S. B. and Dr. M. D. Leakey, Dr. P. J. Andrews, and Dr. M. Pickford. The Olduvai fossils were collected by L. S. B. and M. D. Leakey. The Uganda material was collected by Dr. W. W. Bishop and the Ethiopian fossils by Dr. H. B. Wesselman.

History of Nomenclature

The fossils from the Plio-Pleistocene are referred to extant species or are indeterminate. The Miocene lorisids are all of extinct taxa and the following brief historial review gives an outline of the manner in which taxonomic ideas have developed to date.

The first specimens were collected by Hopwood at Koru in 1931 but were not described until 1967. The first described specimen was the type mandible of *Progalago dorae* (MacInnes 1943) and confusion began at the outset with MacInnes referring the specimen to the Galaginae in the text and to the Anaptomorphidae in the taxonomic introduction.

It is clear, however, that MacInnes thought that the specimen was probably in the direct ancestral line of modern bush babies. This specimen was for nearly ten years the only recognized lorisid fossil. The British-Kenya Miocene Expedition collected more specimens from sites in the Kavirondo Gulf that were documented by Clark and Thomas (1952). More specimens were added to the hypodigm of *P. dorae* and two new species of the genus were established, *P. robustus* and *P. minor,* all being assigned to the Galaginae. In 1956 Clark described a nearly complete cranium found by M. D. Leakey on Rusinga Island. He recognized that it was from a taxon at least specifically and probably generically distinct from that of which a partial cranium existed, but it was assigned to *Progalago* for convenience (Clark 1956). Clark pointed out its clear resemblances to the crania of living lorisines, but he never pursued this observation to even tentative conclusions.

Dr. W. W. Bishop discovered a facial fragment of a prosimian cranium at Napak in Uganda and Leakey (1962) placed this specimen in the Galaginae, making it the type of *Mioeuoticus bishopi.* Leakey claimed that the fossil showed special resemblances with *Euoticus,* a subgenus of *Galago* that includes the needle-clawed bush babies, but this appears to have been a misconception that was partly based on a misunderstanding of the way in which the palate of the specimen was distorted. Simpson (1967), in the most comprehensive review as yet attempted, set up *Progalago songhorensis* for some material previously placed in *P. dorae* and erected a new genus, *Komba,* for the two smallest species. He referred the nearly complete cranium (KNM-RU 2052) to *P. dorae* and suggested that *Mioeuoticus bishopi* was an extremely closely related form and further that the latter genus was "dubious." Three mandibular specimens that comprised the hypodigm of a new genus and species, *Propotto leakeyi,* were later removed from the order (Walker 1969a; see chapter 3, this volume.). Simpson pointed out the difficulties of assigning material to either subfamily without postcranial evidence.

With nonassociated postcranial material covering the total size range expected from the dental material (Walker 1970), I suggested that even postcranial evidence might not be sufficient to place fossils in either subfamily because characters that are of value when dealing with modern taxa may not help when dealing with earlier and especially ancestral species. At that time I also followed Simpson's suggestion and relegated *Mioeuoticus* to a synonym of *Progalago.* With new fossils available, especially new maxillary specimens and associated cranial and postcranial material, I have more recently suggested (Walker 1974b) that the postcranial bones of early galagines only have been found and further that cranium KNM-RU 2052 was not of *P. dorae.* This, the most complete specimen so far, together with the Napak facial fragment, I now believe to be lorisine. The name *Mioeuoticus* will have to be resurrected, therefore, and the cranium will become the type of a second species of that genus. The new specimens also demonstrate that while my methods of assigning postcranial bones to taxa based on dental material might be successful, the names given in the first attempt were not all correct because the taxonomy based primarily on dental evidence was faulty. This attempt will now have to be repeated. New fossils have been recovered since my last review of the group. The most important is a maxillary fragment from Fort Ternan that, while showing resemblances to *Mioeuoticus,* is clearly not from any species previously sampled. This specimen is called here "lorisine ?*gen. et sp. nov.*"

Prosimian Fossil Sites

The Miocene fossil sites (table 6.1) are all associated with central volcanoes to the west of the East-

ern Rift Valley. Details of the ages of these sites as determined stratigraphically or radioisotopically are given in Cooke (chapter 2, this volume), and Bishop (1971, 1972), and Bishop, Miller, and Fitch (1969). There has been a tendency in the past for all the early Miocene sites to be considered together to gain an overall view of the fauna of those times. Differences in the faunas between sites can be accounted for by many factors including ecological, sedimentary, taphonomical and temporal ones, but given the short times represented by the thicknesses of strata at any one site it is almost inevitable that there will be differences because of the age of the deposits. Andrews and Van Couvering (1975), while noting that other factors might be influencing the situation, give some evidence that a geographic barrier might have existed between the northern sites of Koru, Songhor, and Napak and the southern ones of Rusinga Island and Mwafanganu in the form of a longstanding river along the line of the Kavirondo Rift. This might have been the case, but I feel that temporal differences of the order of a million years that are indicated by the radioisotopic age determinations would have more effect. The evidence from the prosimian fossils helps a little here. The most common fossils of prosimians are mandibles and

Table 6.1   Fossil lorisid sites.

Site	Approximate age in million years B.P.	Species
Songhor, Kenya	20	*Komba minor* *Komba robustus* *Progalago songhorensis* *Progalago dorae* *Mioeuoticus* sp. nov.
Koru, Kenya	19.5	*Komba robustus* *Progalago dorae*
Rusinga Island, Kenya	18	*Komba minor* *Komba robustus* *Progalago songhorensis* *Mioeuoticus* sp. nov.
Napak, Uganda	18	*Komba minor* *Komba robustus* *Progalago dorae* *Mioeuoticus bishopi*
Moroti I, Uganda	probably about 18	*Komba minor*
Mwafanganu, Kenya	18	*Komba robustus* *Progalago dorae*
Fort Ternan, Kenya	14	lorisine ?gen. et sp. nov.
Kapchebrit, Kenya	probably about 4	galagine indet.
Omo (Shungura Formation) (Jaeger & Wesselman, 1976	3	*Galago* cf. *senegalensis* lorisid indet.
Olduvai Bed I	2	*Galago senegalensis*

Table 6.2 Mandibular depths at various tooth levels of *Komba robustus* (in mm).

Sites	At P$_4$	At M$_1$	At M$_2$	At M$_3$
Earlier: Songhor, Koru	mean 4.2 N = 4 range 3.8–4.5	mean 4.3 N = 6 range 3.8–5.0	mean 4.3 N = 6 range 4.0–4.9	mean 4.4 N = 5 range 4.0–4.6
Later: Napak, Rusinga Island Mwafanganu	mean 5.3 N = 1	mean 5.1 N = 3 range 5.0–5.1	mean 4.9 N = 5 range 4.4–5.3	mean 5.0 N = 5 range 4.4–5.5

teeth of *Komba robustus;* if those from Koru and Songhor are taken, as shown by the potassium/argon dates, as an older sample and those from Napak, Rusinga, and Mwafanganu as a somewhat younger one, then definite differences are seen between the two samples. Table 6.2 gives measurements of the depth of the mandibles at different tooth levels in the two samples. Although the samples are not large, the mandibles from the later sites are, on the whole, a little deeper than those from the earlier ones. Tooth measurements give the same picture with the hint that the third molars have not increased in size with time as much as the other teeth. The molar teeth in the later sample also show less pointed cusps and have relatively wider trigonids and talonids than in the earlier one.

Samples from the later sites are poor, but it is hoped that with continuing field work the samples will increase and the gaps in the temporal record will be filled. Fort Ternan offers the best possibility at present for the finding of prosimian fossils that will show whether the early Miocene forms contributed to later lorisid evolution. The single, undescribed specimen from that site can be regarded as intermediate between the earlier *Mioeuoticus* species and living African lorisines, especially *Arctocebus.* Later lorisine and galagine fossils should be recovered with careful excavation and sieving.

Taxonomy

The taxonomy given here for the Miocene lorisids is based primarily on Simpson (1967), but discoveries of more complete specimens, and especially associated upper and lower dentitions, have necessitated some changes in his scheme.

Family Lorisidae Gregory 1915
  Subfamily Galaginae Mivart 1864
    Progalago MacInnes 1943

      P. dorae MacInnes 1943
      P. songhorensis Simpson 1967
    Komba Simpson 1967
      K. robustus (Clark and Thomas 1952)
      K. minor (Clark and Thomas 1952)
  Subfamily Lorisinae Flower and Lydekker 1891
    Mioeuoticus Leakey 1962
      M. bishopi Leakey 1962
      M. sp. nov.
    ?gen. et sp. nov.

Among the species that I consider to be galagine, the size range is about that met with in living species of the genus *Galago.* *K. minor* is the size of *G. demidovii,* *K. robustus* about the size of *G. alleni, P. songhorensis* a little smaller than *G. crassicaudatus* and *P. dorae* a little larger than *G. crassicaudatus.* Among the species considered as lorisines, *M. bishopi* is about the size of *Arctocebus calabarensis,* the *Mioeuoticus* species represented by cranium KNM-RU 2052 is a little larger, about the size of *Perodicticus potto,* and the Fort Ternan lorisine is intermediate in size. The overlap in size of teeth of some species has caused some confusion in the past and may well cause more in the future. Also, the size of some, if not most, of the species increases with time so that, for instance, the molars of *P. songhorensis* from Songhor can be the size of those of *K. robustus* from Mwafanganu. The morphological characters of the teeth tend to be related with size, as well, which means that differences tend to become indistinct unless carefully distinguished. Mandibular shape, however, seems to be a good indicator of generic affinity within the galagine fossils.

One small point of taxonomic note is the hint given in Simpson (1967) that there might be a species larger than *P. dorae* that is represented by one tooth (P$_2$). This specimen, first described by Clark and Thomas (1952), is a deciduous canine of a hominoid, probably a left lower deciduous canine of *Dendropithecus macinnesi.*

Features of the Miocene Fossils

Over thirty mandibular specimens are known that give a fair indication of the form of the mandible for all galagine species. Only one specimen appears to be from *Mioeuoticus,* and this is only a fragmentary mandible. All specimens that preserve those parts show that the symphysis was unfused as in modern species and that the anterior six teeth were placed in the procumbent position necessary for the support of the grooming comb. Clark and Thomas (1952) thought that this procumbency was not to the degree found in modern lorisines, but Simpson (1967) and Walker (1969b) found that there were no features that differed from the modern condition. New specimens now show that the roots of the anterior grooming comb teeth are compressed mesiodistally and closely packed, leaving little doubt that the grooming apparatus was fully developed in the Miocene species.

One feature that distinguishes the mandibles of *Komba* and *Progalago* species and that might be of considerable importance is the mandibular depth. In species of *Komba* the mandibular body is of roughly equal depth along its length, while in species of *Progalago* the mandibular body deepens posteriorly so that the maximum depth is close to the angle, and the impression is gained that the body becomes thin and platelike posteriorly. This latter condition is only found in *Perodicticus* and *Nycticebus* in Recent species; the other two lorisine genera have mandibular bodies that are even in depth as in *Galago* and

Komba. This is evidently not a phenomenon related to size because *Galago crassicaudatus,* the largest bush baby, also has a body of even depth along its length. One fragment of mandible (KNM-RU 1885) that was previously considered to be of *P. dorae* is too large for that species and I consider it to belong to a species of *Mioeuoticus.* It increases in depth posteriorly only very slightly. An attempt is made in diagrammatic form in figure 6.1 to show this feature in the Miocene species and to compare it with the two basic mandibular types found in living lorisids.

Teeth that can be assigned with some degree of certainty are listed in table 6.3. The fact that no crowns of the grooming comb teeth are known indicates that they were fragile and perhaps can be taken as more evidence that the comb was as fully developed as in modern species. Alveoli for the upper anterior teeth are only known for the lorisine species, suggesting that the supporting bone in the more commonly preserved galagine specimens was not very strong. A single M^2 together with part of the maxilla is all we have of the upper dentition and facial skeleton of *K. minor;* this may be a collecting artifact but is more likely a reflection of the small size of that species. Although lorisine specimens are few, those which we do have are the best preserved in the whole Miocene lorisid collection.

In general the lower teeth of the Miocene species closely resemble those of their modern counterparts. Differences between *Komba* and *Progalago* species are mostly related to the more cuspidate nature of

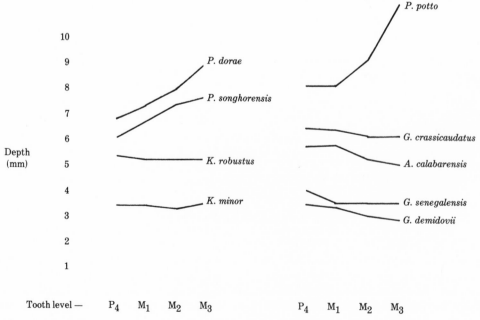

Figure 6.1 Depths of mandibles at various tooth levels in Miocene lorisids, with comparable values for some living species.

Table 6.3   Known parts of the upper and lower dentitions of the Miocene lorisids.

Species	P2	P3	P4	M1	M2	M3	
Komba minor	—	—	—	—	X	—	(upper)
	—	—	X	X	X	X	(lower)
Komba robustus	—	—	X	X	X	X	(upper)
	—	X	X	X	X	X	(lower)
Progalago dorae	—	X	X	X	X	X	(upper)
	—	X	X	X	X	—	(lower)
Progalago songhorensis	X	—	X	X	X	X	(upper)
	—	X	X	X	X	X	(lower)
Mioeuoticus bishopi	X	—	X	X	X	X	(upper)
Mioeuoticus sp. nov.	—	—	X	X	X	X	(upper)
	—	—	—	—	X	X	(lower)
lorisine ?gen. et sp. nov.	—	X	X	X	X	X	(upper)

the teeth of the former and the more bunodont character of the teeth of the latter. Similar, but slightly less marked differences are found between the teeth of species of living galagos, and the differences between the two currently known Miocene genera may also be those of size. The later specimens of *Komba robustus,* which are nearly as large as the earlier specimens of *Progalago songhorensis,* have more bunodont molars than the earlier ones and even the proportions of the trigonids to talonids approximate those of the *P. songhorensis* specimens of similar size. This might be taken as evidence that the generic differences are not well founded, but the shape of the mandible is a clear indicator of the genus, whatever the age of the specimen. The lower teeth referred to *Mioeuoticus* have very expanded trigonids and talonids that would imply upper teeth of quadrate shape.

The increased number of specimens of the upper teeth in the past few years has greatly helped to clarify some of the difficulties in earlier attempts at a comprehensive taxonomy. Associated upper and lower dentitions of *Progalago songhorensis* and *Komba robustus* have helped enormously and have practically removed the objections of Simpson (1967) that the attributions of the upper dentitions are uncertain and that they make for difficulties in a taxonomy based essentially on lower dentitions. The single specimen of an upper tooth of *K. minor* is so small that it can only belong to that species. The upper teeth now referred to *Progalago dorae* are similar to, but larger than, those known to belong to *P. songhorensis* and are of the size predicted from the lower teeth. In general the upper teeth of *Komba* and *Progalago* species are very similar to those of

modern *Galago* species, but the P^4 is not molariform and the hypocone, though distinct, is more closely applied to the protocone. The upper molars of *Komba robustus* are more cuspidate and have a more distinct hypocone than those of the two *Progalago* species. M^3 reduction as found in modern *Galago* is seen in the fossil galagines, but not to a similar degree. Upper molars of *Mioeuoticus* are very different from those of the galagine species, with M^1 and M^2 square in outline and bearing well-developed buccal and distobuccal cingula. M^3 in species of this genus has a reduced hypocone and its outline is thus more triangular, but the distobuccal cingulum is still present. The new specimen from Fort Ternan, while resembling *Mioeuoticus*, shows some greater similarities with modern lorisines, particularly African ones. The resemblances of the upper dentition of *Mioeuoticus* to *Perodicticus* that Clark (1956) pointed out are not striking and to my mind the similarities are with *Arctocebus*. Figure 6.2 illustrates some of the known specimens.

The facial and palatal parts of *Progalago* and *Komba* species known at present are, for the most part, only fragments that surround the tooth roots. One fragmentary specimen of *K. robustus* from Kathwanga, Rusinga Island, has part of the orbital margin preserved; the orbits were as enlarged in the Miocene species as in the living ones. A partial cranium with natural endocast is referred to *K. robustus* on the basis of size. A regression line was calculated between lower molar tooth row length and the width between the temporomandibular joints for many individuals of several different species of lorisines. On the basis of this regression, the expected widths between the temporomandibular joints of the

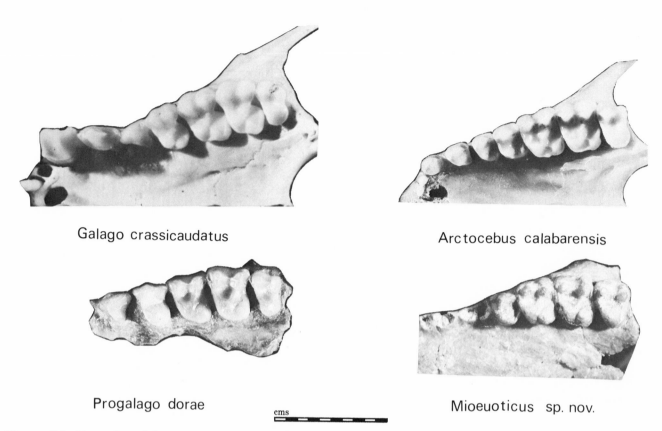

Galago crassicaudatus

Arctocebus calabarensis

Progalago dorae

cms

Mioeuoticus sp. nov.

Figure 6.2 Examples of the upper dentition of Miocene lorisids compared with similar-sized modern representatives. (Scale in mm.)

Miocene species are: *K. minor,* 12.5 mm; *K. robustus,* between 16.5 and 17.5 mm; *Progalago songhorensis,* 20 mm; and *P. dorae,* 22.5 mm. The value measured on the cranial specimen (KNM-RU 1940) is 16.0 mm, making the specimen likely to be *Komba robustus.* The cranium shows clear affinities with *Galago* species in its light construction, lack of strong temporal ridges, strong inflation of the bulla, and moderate basicranial flexion.

Both the type of *Mioeuoticus bishopi* and the nearly complete cranium of *Mioeuoticus* are well preserved. Among the features that indicate lorisine affinities are the strongly constructed cranium with very well-marked temporal ridges, the upwardly directed orbits, the weakly inflated bulla and mastoid, the wide internal nares, and the slight basicranial flexion.

In both specimens that have the auditory bulla preserved, the details are clearly lorisid and not lemurlike. The tympanic cavity is divided by a bony plate into an outer tympanic and an inner hypotympanic. All the Madagascan prosimians lack plates and hence have a single cavity. The ectotympanic ring is applied to the bulla wall in both specimens, again as in the lorisids and unlike the Madagascan

prosimian condition. In detail the structure and position of the ring in the specimen of *Mioeuoticus* is most closely matched by that in modern *Galago crassicaudatus* (Clark 1956), while that seen in the *Komba robustus* specimen is closest to *Galago senegalensis.*

The first limb bones of the Miocene lorisids were described by Walker in 1970, but not one bone was associated with any dental material and reference to species was undertaken on the basis of size. Since then new associated material has been found. Many parts of the postcranial skeleton are known for several of the species, but to date no limb bones of the lorisines are known. In most respects the limb bones can be matched closely by those of modern species of *Galago.* The only major difference in the postcranial skeleton appears to be in the tarsus. The great elongation of the calcaneum and navicular seen in modern *Galago* species is not seen in the Miocene fossils. Three complete calcanea are known that show that this element was elongated in the Miocene species, but not to the degree found in *Galago.* The simplest conclusion to be drawn from this evidence is that the locomotion of the Miocene species was broadly similar to that of modern galagos, with leaping the major form of locomotion, but that tarsal elongation had not progressed to the degree found today. The exact locomotor function of the elongated tarsus remains a mystery to me, but the obvious explanation—that it is related to hopping—may be correct.

Evolutionary Relationships

Living lorisid primates can be separated clearly into two groups; in Africa these are represented by the galagines (species of the genus *Galago*) and the lorisines (*Arctocebus calabarensis* and *Perodicticus potto*). Both basic types are nocturnal, arboreal, and basically predatory. The galagines are adapted to catch their prey (mainly insects) by leaping and the lorisines by stealth (Walker 1969c). The basic differences in the locomotor patterns are reflected in the postcranial skeletons. The two species of living African lorisines are restricted to tropical evergreen forest while species of *Galago* are found in a variety of habitats, the only basic requirement being trees of some sort. Many workers, including me, prefer to place all the living bush babies in the genus *Galago*, but some prefer separate generic status for some

species. Among the species for which separate genera are sometimes called are *G. demidovii* (*Hemigalago* or *Galagoides*) and *G. elegantulus* (*Euoticus*). It may be that species presently placed in *Galago* are not very closely related and that they share a large number of ancestral characters, but in the absence of fossil evidence this cannot be proved and the simplest solution is to regard them as a reasonably closely related species group.

The fossil evidence indicates an African origin for the family and that the characters that distinguish the family from any family of Madagascan prosimians were already present in the early Miocene. On the other hand, the differences between the lorisids and the Madagascan lemurs are not so extreme as to call for placing them in two different infraorders (Martin, 1972). The East African fossil record also shows that the two subfamilies, Galaginae and Lorisinae, were established by the early Miocene. The few lorisine fossils seem to be closer to African lorisines than Asian ones, but it could be that living African species retain more ancestral characters than the living Asian ones, or that the Miocene fossil lorisines are close to the ancestral

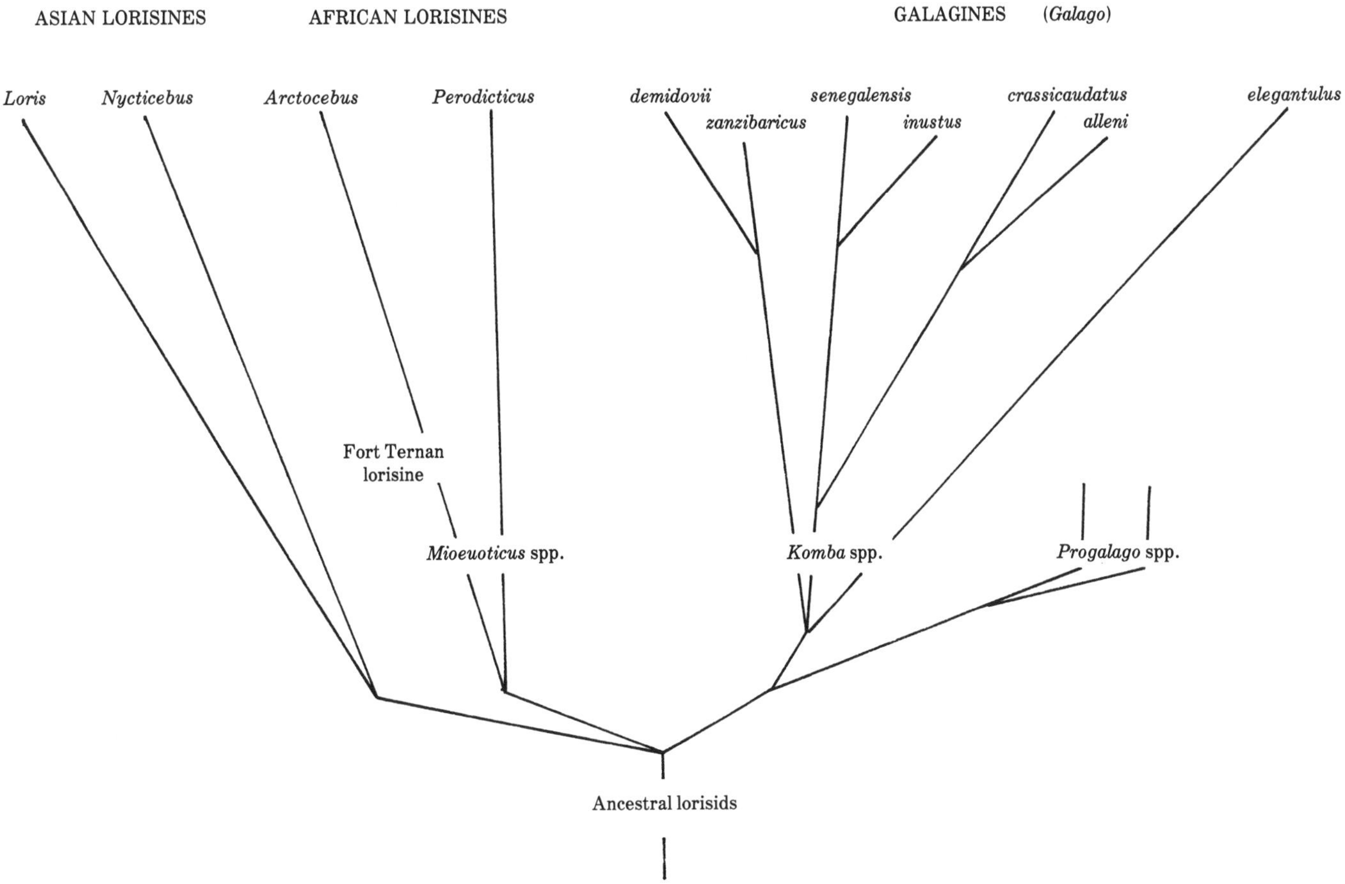

Figure 6.3 Tentative phylogenetic scheme of lorisid evolution.

stock of all lorisines. Because the early Miocene gal-agine fossils cover a size range that is practically identical to that covered by living species of the genus *Galago,* it is tempting to suggest that the fossils document the beginnings of the radiation of the galagos. However, it is difficult to prove that any one of the Miocene species is directly ancestral to any one of the living forms, although the shape of the mandible might be taken to show that species of *Komba* are in or near the ancestry of living *Galago* species while species of *Progalago* are not. Goodman and his colleagues (1974) have presented some immunological evidence of the relationships within the Lorisidae that shows that three species of *Galago, G. demidovii, G. senegalensis,* and *G. crassicaudatus,* are all reasonably closely related (although *G. demidovii* might be slightly more divergent from the other two species). They further indicate that the two Asian lorisines and the two African lorisines form two separate stocks that are immunologically as different from each other as each is from the *Galago* stock. They point out that the splitting between *G. senegalensis* and *G. crassicaudatus* is more recent than that between these two species and *G. demidovii* and also that this split is more recent than either the *Nycticebus-Loris* split or the *Perodicticus-Arctocebus* split.

Within the genus *Galago,* resemblances are between *G. demidovii* and *G. zanzibaricus, G. inustus* and *G. senegalensis,* and, perhaps, *G. crassicaudatus* and *G. alleni. G. elegantulus* appears to differ somewhat from the other species in such areas as dentition, palate shape, nail conformation, and pelage. The later East African fossils, presently assigned to modern species, only establish that *G. senegalensis* existed in at least part of its present range over 2 million years ago.

A speculative phylogeny is outlined in Figure 6.3. As I have indicated, the fossil evidence on which this is based is, to say the least, meager, but further collections from middle to late Miocene sites and further comparative anatomical and biochemical studies on living species of lorisid primates will show whether the broad outlines given here are correct.

I would like to thank P. J. Andrews, the late W. W. Bishop and M. Pickford for permission to examine fossils collected by them. H. B. Wesselman kindly showed me the specimens he collected in Ethiopia. I thank the director and trustees of the National Museum of Kenya for permission to study fossils in their collections.

References

Andrews, P. J., and J. A. H. Van Couvering. 1975. Palaeoenvironments in the East African Miocene. In F. Szalay, ed., Approaches to primate paleobiology. *Contrib. Primatol.* 5:62–103.

Bishop, W. W. 1971. The late Cenozoic history of East Africa in relation to hominoid evolution. In K. K. Turekian, ed., *The late Cenozoic glacial ages.* New Haven: Yale University Press, pp. 494–527.

———. 1972. Stratigraphic succession versus calibration in East Africa. In W. W. Bishop and J. A. Miller, eds., *Calibration of hominoid evolution.* Edinburgh: Scottish Academic Press, pp. 219–246.

Bishop, W. W., J. A. Miller, and F. J. Fitch. 1969. New potassium-argon age determinations relevant to the Miocene fossil mammal sequence in East Africa. *Am. J. Sci.* 267:669–699.

Clark, W. E. Le Gros. 1956. A Miocene lemuroid skull from East Africa. *Fossil Mammals of Africa* (British Museum, London) 9:1–9.

Clark, W. E. Le Gros, and D. P. Thomas. 1952. The Miocene lemuroids of East Africa. Fossil mammals of Africa (British Museum, London) 5:1–20.

Doyle, G. A. 1974. Behavior of prosimians. In A. M. Schrier and F. Stollnitz, eds., *Behavior of nonhuman primates,* vol. 5. New York: Academic Press, pp. 155–353.

Goodman, M., W. Farris, W. P. Moore, W. Pruchodko, W. Poulik, and M. W. Sorenson. 1974. Immunodiffusion systematics of the Primates II: findings on *Tarsius,* Lorisidae and Tupaiidae. In R. D. Martin, G. A. Doyle and A. C. Walker, eds., *Prosimian biology.* London: Duckworth, pp. 881–890.

Jaeger, J.-J., and H. B. Wesselman. 1976. Fossil remains of micromammals from the Omo group deposits. In Y. Coppens, F. C. Howell, G. Ll. Isaac, and R. E. F. Leakey, eds., *Earliest man and environments in the Lake Rudolf Basin.* Chicago: Chicago University Press, pp. 351–360.

Leakey, L. S. B. 1962. *Primates.* In W. W. Bishop: The mammalian fauna and geomorphological relations of the Napak volcanics, Karamoja. *Rec. Geol. Surv. Uganda* 6–9.

Lewis, G. E. 1933. Preliminary notice of a new genus of lemuroid from the Siwaliks. *Am. J. Sci.* 26:134–138.

MacInnes, D. G. 1943. Notes on the East African Miocene primates. *J. East Afr. Uganda Nat. Hist. Soc.* 17:141–181.

Martin, R. D. 1972. Adaptive radiation and behaviour of the Malagasy lemurs. *Phil. Trans. Roy. Soc.* (B) 264:295–352.

Simpson, G. G. 1967. The Tertiary lorisiform primates of Africa. *Bull. Mus. Comp. Zool. Harvard* 136:39–61.

Tattersall, I. 1969. More on the ecology of *Ramapithecus.* Nature 224:821–822.

———. 1970. A mandible of *Indraloris* (Primates, Lorisidae) from the Miocene of India. *Postilla,* Yale University 123:1–10.

———. 1975. Notes on the cranial anatomy of the subfossil Malagasy lemurs. In I. Tattersall and R. W. Sussman, eds., *Lemur biology.* New York: Plenum Press, pp. 111–124.

Tattersall, I., and J. M. Schwartz. 1974. Craniodental morphology and the systematics of the Malagasy lemurs

(Primates, Prosimii). *Anthrop. Papers, Am. Mus. Nat. Hist.* 52(3):137–192.

Walker, A. 1969a. True affinities of *Propotto leakeyi,* Simpson 1967. *Nature* 223:647–648.

———. 1969b. New evidence from Uganda regarding the dentition of Miocene Lorisidae. *Uganda J.* 33:90–91.

———. 1969c. The locomotion of the lorises, with special reference to the potto. *E. Afr. Wildlife J.* 7:1–5.

———. 1970. Post-cranial remains of the Miocene Lorisidae of East Africa. *Am. J. Phys. Anthropol.* 33:249–261.

———. 1974a. Locomotor adaptations in past and present prosimian primates. In F. A. Jenkins, Jr., ed., *Primate locomotion.* New York: Academic Press, pp. 349–381.

———. 1974b. A review of the Miocene Lorisidae of East Africa. In R. D. Martin, G. A. Doyle, and A. C. Walker eds., *Prosimian biology.* London: Duckworth, pp. 435–447.

7

Cercopithecidae and Parapithecidae

E. L. Simons and E. Delson

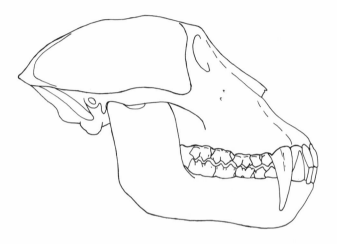

The higher primates of the suborder Anthropoidea represent the more progressive division of the primate order related to and including man himself, as distinguished from the lower primates or suborder Prosimii. The New World monkeys or Ceboidea (Platyrrhini) today occur in South and Central America and apparently represent a radiation of the neotropical region that may well have been distinct from Old World forms (Catarrhini) at least since early Eocene times. The Old World Anthropoidea, which includes monkeys, apes, and man, occurs as a range of both living and fossil species throughout Africa and Eurasia.

Both groups of Anthropoidea share a number of readily identified features. The brain is large and convoluted, eye sockets are fully formed, the rostrum is typically foreshortened, and the jaw rami and frontal bones fuse together in embryonic or early juvenile stages. The front lower premolar (P_2 in New World monkeys and P_3 in Old World Anthropoidea) is often but not always anterolaterally extended, bears comparatively thick enamel, and may serve as a hone that sharpens the posterior blade of the upper canine. In man and certain ceboids this function is reduced.

Following Simpson (1945), almost all authorities divide the catarrhine infraorder into two living superfamilies, Hominoidea, including three families—Hominidae, Pongidae, and Hylobatidae—and Cercopithecoidea, with a single family—Cercopithecidae. The extinct African Oligocene monkeys of the family Parapithecidae are here considered as possibly referable to Cercopithecoidea, Simons considering them closer and Delson further from living Old World monkeys. Delson and Andrews (1975) have revised the foregoing classification somewhat, but for this review we will remain taxonomically conservative. As a group the Old World higher primates show little or no expansion of the auditory bulla, but among all extant species the tympanic bone is drawn out into an elongated external auditory meatus, part of the roof of which is formed by the squamosal bone. The postorbital plate is composed of elements of the jugal, frontal, and alisphenoid bones, with maxillaries contributing a small portion to the medial orbital floor. The dental formula is typically $\frac{2.1.2.3}{2.1.2.3} \times 2$, but as would be expected for early forms, in *Apidium* and *Parapithecus* it is $\frac{2.1.3.3}{2.1.3.3} \times 2$. Postcranial adaptations vary with modes of locomotion (Simons 1972:86–92).

The history of monkeys in Africa is little understood but it is generally supposed that the earliest phases of the development of cercopithecoids took place there (von Koenigswald 1969, Simons 1970, 1972, Delson 1975a). Nevertheless, these animals are not abundant in Miocene and earlier Pliocene deposits known from that continent. In addition, most of the African Plio-Pleistocene species described so far differ markedly from present-day African monkeys. Some North African fossil monkeys of the later Tertiary may well have had strong affinities with circum-Mediterranean faunas, while others are linked with those of sub-Saharan Africa (Delson 1973, 1975a, b).

Moreover, the only Oligocene monkeys known from Africa, species of *Parapithecus* and *Apidium,* are too disjunct in time from the much later occurring series of African fossil monkeys for their exact phyletic position, in relation to the latter, to be known. That is, one cannot be certain (without intermediate connecting links) whether the Egyptian Oligocene Parapithecidae gave rise directly to modern Cercopithecidae or are relatively distantly related to their ancestors. We can be definite, however, that the Parapithecidae had reached the grade of organization of higher primates and are not Prosimii (Simons 1974; Conroy, Simons, and Schwartz 1975; Delson 1975c; Conroy 1976).

Being approximately 28 to 30 m.y. old, *Parapithecus* and *Apidium* species naturally show a blend of "primitive" and "advanced" features that cause them to resemble in some respects the ceboid monkeys (Conroy 1976). Nevertheless, our view is that the ceboid-parapithecid similarities—most are no more than retention of prosimian or ancestral anthropoidean features—reflect only their general level of organization. In terms of *grade,* the ceboids, cercopithecids, and parapithecids are monkeys, rather than prosimians or apes, but this might reflect nothing more than parallel advance from a common genotype last shared some time between early Eocene to late Eocene. Pending more complete analysis of not only parapithecids (under way by Simons) but also of ceboids by Rosenberger (e.g. 1977), relationships among the three monkey groups will continue to be uncertain.

The derivation of South American monkeys from an early Tertiary African stock is a hypothesis that has gained a certain fashion recently. The transit postulated to have occurred first for rodents and then for monkeys by rafting across the South Atlantic in Eocene times (Lavocat 1969, Hoffstetter 1972) is wholly untenable. The South Atlantic was then perhaps two-thirds as broad as it is now (see Phillips and Forsythe 1972). Any small Eocene primates in West Africa postulated to have been washed out to sea clinging on masses of vegetation would surely have died of thirst and exposure long before the hypothetical slow transit to South America by ocean currents was completed: approximate crossing time today based on drift card transport in the South Equatorial Current is an absolute minimum of 60 days (Sheltima 1971). To this must be added an unknown time of random wandering between formation of rafts in rivers and their possible insertion into ocean currents. Even allowing mid-Tertiary South Atlantic widths of two-thirds to one-half the present and omitting time lost in random wandering, the corresponding 30 to 40 days minimum for current transport across would prevent small primate survival (see Simons 1977).

Several studies covering the fossil cercopithecid monkeys of Africa have recently been completed. These include Delson (1973) on the circum-Mediterranean species with comments on the East African Miocene forms; Jolly (1970, 1972) on the genus *Theropithecus* (=*Simopithecus*) and on the adaptive significance of the *Theropithecus* functional complex; the various reviews of Freedman (1957–1976) and of Maier (1970–1972) on the South African fossil monkeys; studies by R. E. F. Leakey (1969), von Koenigswald (1969), M. G. Leakey (1976), M. G. and R. E. F. Leakey (1973a, b, 1976) and Eck (1976, 1977) on East African monkeys; and finally general reviews of all fossil monkeys by Simons (1970, 1972) and Delson (1975a).

Family Cercopithecidae—Miocene to Pleistocene Monkeys

The essential taxonomic characters of the cercopithecids have been summarized most recently by Delson (1975a), while more detailed anatomical treatments are presented by Hill (1966, 1970, 1974; see also Kingdon, 1977). In brief, the two subfamilies differ in such cranial features as relative interorbital width and facial length (nasal bone length, lacrimal relationships), the colobines having shorter and broader faces than cercopithecines. Dentally, the colobines are characterized by high relief on the molars, short trigonids on lowers, relatively straight-sided molars and narrow incisors, as well as several minor features; African forms often also show a reduction of the P^3 protocone and/or the P_4 metaconid and may have the distal lophid on M_3 wider than the mesial. The cercopithecines fall into three groups dentally: most distinct are the African guenons and relatives, tribe Cercopithecini, which have lost the hypoconulid on M_3 as well as on dP_4– M_2, but share with other cercopithecines enlarged incisors (related to a frugivorous diet), longer trigonids, and low molar relief; they also preserve enamel on the lingual surfaces of the lower incisors and share relatively straight-sided molars with colobines (except in *Allenopithecus*). The tribe Papionini includes both African and Eurasian forms that in a majority of cases appear to be dentally the most conservative among cercopithecids. Baboons and mangabeys, as well as macaques, typically have large M_3 hypoconulids, low relief, and moderate to long trigonids, as well as large incisors and more "flaring" or sloping-sided molars, but they also share a reduction (or complete lack) of enamel on the lingual face of the lower incisors, probably an adaptation to a gliriform "self-sharpening" incisor wear pattern. The genus *Theropithecus* has further converged on colobines from a "typical papionin" dental pattern. Cheek-tooth relief is increased and the trigonid is somewhat shortened. Postcranial differences among cercopithecids appear to reflect habitus more clearly than heritage. For example, moderate to extreme terrestriality has arisen several times in parallel. Brief comments on each genus are presented here: see Szalay and Delson (in press) for greater detail and documentation.

Subfamily Cercopithecinae Gray 1825

Tribe Cercopithecini

Genus *Cercopithecus* Linnaeus 1758

Although the Early Miocene African fauna mainly represents a forest biotype, this environment is poorly sampled in the later Neogene. Modern *Cercopithecus* is represented by numerous species, of which most are restricted to forest; it is likely that members of the genus would rarely be found as fossils. Only one set of *Cercopithecus* fossils has been described in any detail, by Eck and Howell (1972). They recovered a partial mandible and isolated teeth from several levels in the Omo Group ranging in age from ca 3.0–1.5 m.y., but were unable to associate these remains clearly with any modern species. Additional fossils referable to *Cercopithecus* have been recovered from the East Rudolf area (see M. G. Leakey 1976). The group of sites to the east of Lake Turkana (formerly Lake Rudolf) will continue to be denoted here by their original East Rudolf locality designations now embedded in the paleontological literature and should not, in our opinion, be subject to fluctuating name changes. *Cercopithecus* also occurs at Kanam East (Delson, unpublished). No fossils referable to either *Allenopithecus* (which would be recognizable from its distinctive dental morphol-

ogy) or *Erythrocebus* (also with distinctive large teeth and postcrania) have been reported.

Tribe Papionini Burnett 1828

Genus *Macaca* Lacépède 1799

Fossil finds indicate that the genus *Macaca* was European or perhaps circum-Mediterranean in origin. Today it has achieved what appears to be the most widespread distribution of any Old World monkey, ranging discontinuously from Gibraltar and North Africa across India and Southeast Asia to the northern island of Japan. The Japanese macaque, *Macaca fuscata,* is adapted to the most rigorous environment tolerated in the wild by any living nonhuman primate.

Although *Macaca* has an extensive fossil record in Europe, it is scarce in Asia and only poorly known in North Africa. Two important collections from the latter region which are provisionally referred to the genus (about as much on geography as on morphology) represent its earliest occurrences. Stromer (1920) described a group of fragmentary gnathic specimens from Wadi Natrun, Egypt (6? m.y.) which he termed *Aulaxinuus libycus,* employing as the generic name a term previously applied to some European fossils (Cocchi 1872) but which had been shown to be synonymous with *Macaca* (Ristori 1890). This locality has also yielded the colobine *Libypithecus* and a small assemblage of other mammals, including *Lutra, Sivachoerus* (*Nyanzachoerus*), a distinctive rodent and other taxa, discussed in more detail under *Libypithecus. Macaca* is essentially a highly conservative papionin, so that its identification in the fossil record is fraught with probable error, but no characters in the known sample (about five partial mandibles, two maxillae, and a dozen isolated teeth) preclude assignment to this genus, as a probably distinct species *M. libyca.*

Perhaps slightly earlier than the Natrun local fauna (7? m.y.) is one from Marceau, Algeria, which is strangely dominated by cercopithecids. Arambourg (1959) reported these fossils, but he both confused the age of Marceau with that of the much older Oued el Hammam and named only a single cercopithecid species, *Macaca flandrini.* Delson (1973, 1975a) showed that the type and several other teeth of this form were colobine, possibly sharing certain dental features with other African forms, and he referred the species to "?*Colobus,*" employing the modern name as a "form-genus" pending further data. The remainder of the sample, about forty mainly isolated teeth, are clearly papionin but are otherwise indeterminate. Pending their more detailed study by Delson (in progress), at least one species of cf. *Macaca* may be recognized here. The single M₃ from Ongoliba, Zaïre, described by Hooijer (1963, 1970) as "cf. *Mesopithecus* c.q. *Macaca*" is very similar in size and morphology to those from Marceau and might represent the same taxon, being probably of similar age as well. Delson (1975b) has suggested that the Marceau and Ongoliba populations might be part of one or more widespread species of early papionin which ranged across the Sahara region before it suffered Final Miocene dessication and the macaques differentiated from sub-Saharan papionins.

Genus *Parapapio* Jones 1937

Of the six papionin genera represented in the Plio-Pleistocene of sub-Saharan Africa, *Parapapio* is certainly the most "primitive," providing something of a model for the origin of the group. It is difficult to differentiate it from *Macaca* morphologically, as each has retained many of the characters that probably were present in Late Miocene ancestral papionins. The bulk of known *Parapapio* specimens is from southern Africa (South Africa, also Angola), but recently some fragmentary remains have been recognized in eastern Africa as well.

Among the major distinguishing characters of *Parapapio* are rather straight profile of the muzzle dorsum in both sexes from nasion to rhinion (or to nasospinale), rather than the steeper anteorbital drop seen in *Papio* and *Cercocebus;* glabella and supraorbital tori not projecting; weak development of temporal lines and slender cheek bones, suggesting gracile masticatory musculature; lack of strongly developed maxillary or mandibular fossae (in most cases); and lack of sexual dimorphism in teeth (other

than the canine-P₃ complex) or in skull shape. Many of the features that distinguish *Parapapio* and *Papio* cranially can be seen in figure 7.1. Muzzle shape may vary individually, but there does appear to be a fairly uniform pattern of distinction from *Papio* and *Cercocebus* overall, although both these genera might conceivably have evolved from early species of *Parapapio*-like animals.

Some three hundred to five hundred specimens of *Parapapio* are now known from South African sites, which include Sterkfontein, Makapansgat, Bolt's Farm, and Taung. Specimens of the genus have also been reported at Swartkrans, Kromdraai, and at Leba, Angola, but the amount of material from these latter sites is very small. Studies so far have divided *Parapapio* into four species—*P. broomi, P. jonesi, P. whitei,* and *P. antiquus*. However, it is not always entirely possible to decide to which of these species an individual specimen belongs (compare, e.g., Maier 1971a and Freedman 1976). Much of the definition of species at Sterkfontein (type site for all species but *P. antiquus*) and Makapan has rested on absolute tooth size, but recent studies emphasizing cranial (and dental) proportions arrive at differing conclusions. Although the known variation appears to require several species, their precise differentiation is still unclear. *Parapapio antiquus* is known only from Taung and appears cranially somewhat more distinctive.

In eastern Africa, at least one (perhaps two) species of small to medium-sized papionin is known from several sites in the 3–4 m.y. range: Kanapoi, Lothagam-3, Laetolil, Hadar, and perhaps Kubi Algi. Later, generally even more fragmentary material has been reported from Omo, the Koobi Fora

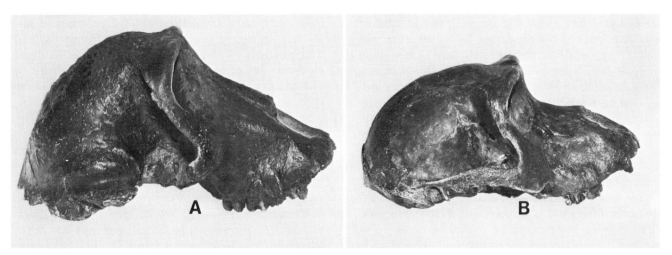

Figure 7.1   Comparison of casts of (*A*) male *Parapapio* cf. *broomi* and (*B*) female *Papio wellsi* from South African Pleistocene deposits. Both genera often occur in association with *Australopithecus*. The profile of the nasal bones in *Parapapio* is much less concave than in *Papio*. (Scale ×½ approximately.) (Photo by A. H. Coleman.)

Formation at East Rudolf, and perhaps Olduvai. Patterson (1968) originally referred the Kanapoi jaw to *P. jonesi,* but M. G. and R. Leakey (1976) thought the specimen indeterminate, while describing others from the noted sites as *Cercocebus* sp. (see also M. G. Leakey 1976, Eck 1976). It would seem that most of the earlier fossils, at least, are best referred to *Parapapio* at present, but detailed analysis is badly needed. The genus *Parapapio* should have great potential as a transcontinental indicator of stratigraphic age in sub-Saharan Africa. Unfortunately, its paleobiology is less certain, as no postcranial elements have yet been unquestionably allocated, but there is some suggestion of greater arboreality than among *Papio* species.

Genus *Cercocebus* E. Geoffroy 1812

The living mangabeys are divided into two main groups that appear to differ minimally in characters of dental wear and development of facial fossae. They are basically arboreal, although some forms come to the ground regularly. Species of this genus are, as expected from their arboreality, rare in the fossil record. Specimens from Makapan previously identified as *Parapapio jonesi* have recently been suggested to be *Cercocebus* sp. (Eisenhart 1975), while other small papionins from Omo and Koobi Fora may also represent this genus, as do probably some isolated teeth from Kanam.

Genus *Papio* Müller 1773

Typical baboons, usually considered to represent the genus *Papio,* are the dominant open-country cercopithecid in modern Africa. Their close relatives in West African forests, the mandrills and drills, are often termed *Mandrillus,* but Delson and Napier (1976) have shown that the term *Papio* was applied to a mandrill before being used for a "savanna" baboon. They requested the International Commission on Zoological Nomenclature to rule on this problem. Pending such a ruling, the generic term *Papio* will be used here for both groups of baboons, which may well be congeneric (Delson 1975a). Studies of behavior, genetics, and morphology show that all the "savannah" baboons are quite closely related, but it appears that two species may be recognized today: *P. cynocephalus* in most of sub-Saharan Africa and *P. hamadryas* in eastern Ethiopia and Arabia, probably in the final phases of speciating from the former.

Papio is rather rarer than might be expected in the fossil record, especially in eastern Africa. This is probably due to the past widespread presence of *Theropithecus,* which Jolly (1970, 1972) suggested may have inhabited wetter grasslands. In the latter environment it may only recently have been replaced by *Papio,* species of which may now, in turn, be adapting to drier savanna from a previous forest-fringe habitat. Fossils apparently referable to the modern species have been reported from Olduvai (horizon uncertain) and occur throughout the Omo sequence (Eck 1976, 1977), while *Papio robinsoni* from numerous South African sites is probably best considered at most a temporal subspecies of *P. cynocephalus.* The characters utilized by Freedman (1957) in diagnosing *P. robinsoni* are minimal compared to the range of variation seen in the modern polytypic biospecies. On the other hand, *Papio baringensis* R. Leakey (1969), may well represent a distinctive species whose most diagnostic characters are small incisors, marked postorbital constriction, and some differences in facial proportions (see M. G. and R. Leakey 1976).

All the fossil remains discussed above have about the cranial size found in most living baboon populations, but a group of smaller specimens is also known from several South African sites. These fossils show the diagnostic facial features of *Papio* (such as deep facial fossae, steep anteorbital drop, and relatively projecting brows), but are smaller than or comparable to individuals of the smallest modern subspecies *P. cynocephalus* "*kindae.*" *P. izodi* and *P. wellsi* have been reported from Taung, while *P. angusticeps* is known from several of the younger sites (Swartkrans, Kromdraai, and others), and one juvenile skull was reported from Sterkfontein (Eisenhart 1975). The relationships among these nominal species are still uncertain.

Modern *Papio* species, excluding drills and mandrills (of whom there is no definite fossil record), are highly terrestrial animals, presumably like their fossil congeners. Unfortunately, the presence of numerous species of similar size in the South African cave deposits (where direct associations of skeletal elements are almost unknown) makes it difficult to allocate the few limb bones reported there to specific species or even to particular genera.

Genus *Dinopithecus* Broom 1937

This monotypic genus is known essentially from only Swartkrans, South Africa. It was among the largest of known cercopithecids; only the biggest *Theropithecus* species of the past and perhaps some modern *Papio* (mandrills) are larger. The type specimen of *Dinopithecus ingens,* and perhaps one other fossil, is from Schurweberg, but the several dozen other specimens are from Swartkrans (see Freedman 1957); additional undescribed remains may be present at Leba, Angola. Arambourg (1947) named some teeth from Omo *D. brumpti,* but most of these are now known to be referable to *Theropithecus.* The morphology of the skull and teeth of *Dinopithecus ingens* is most like that of *Papio,* from which it is hard to separate. Like *Papio,* but unlike *Parapapio,* the skull was very large and rugged, and the female at least had a rather long muzzle somewhat reminiscent of large "savanna" baboons; maxillary fossae are lacking. In females, the upper tooth row is arranged in a U-shaped outline narrowing somewhat to the rear. The males may have been similar but with less posterior narrowing. The tooth morphology of *D. ingens* most closely resembles that of *Gorgopithecus major,* but the latter species is smaller and apparently shows little or no sexual dimorphism in tooth size. Cresting on the skull of *Dinopithecus* is pronounced: both sexes show a large nuchal crest, and the temporal lines are strong. The males have well-developed sagittal crests and large postglenoid processes. The large molars often show many accessory cuspules, and the genus is characterized in general by a tendency to show large, broad cheek teeth coupled with a relatively short lower front premolar possessing a particularly large anterior fovea (Freedman 1957).

Genus *Gorgopithecus* Broom and Robinson 1949

The one species now placed in the genus *Gorgopithecus* was originally described by Broom in 1940 as *Parapapio major;* later Broom and Robinson (1949) elevated it to generic level. The type consists of only two teeth, an upper second and third molar that are considerably worn and of unidentifiable sex. However, there seems to be no other South African fossil monkey species that falls into the same size range as *Gorgopithecus.* Skulls of this genus are about the size of those of male *Papio cynocephalus ursinus.* Freedman (1957) pointed out that the type of *Gorgopithecus* is a particularly unfortunate specimen because of its incompleteness, but concluded that two other Kromdraai specimens, on which most of the diagnosis of this animal is based, presumably belonged to the same species as the type. *Gorgopithecus* appears to be restricted to the Kromdraai "faunal" site, or Kromdraai A; the only cranium of this species is illustrated in figure 7.2.

Compared with modern *Papio,* the muzzle of *Gorgopithecus* is short and the brain case is rather longer. Another significant difference is that in *Gorgopithecus* the anterior insertion of the zygomatic processes has an almost vertical face, whereas that of *Papio cynocephalus ursinus* slopes backward, giving the impression of a shorter snout in the latter than would otherwise be apparent. Even so, with the

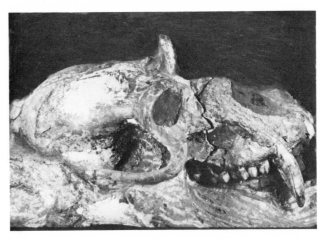

Figure 7.2 Oblique view of the cranium of *Gorgopithecus major,* a large *Papio*-like baboon from Kromdraai (A), South Africa. The middle part of the face in this unique skull has been largely reconstructed in plaster. (Scale ×⅓ approximately.) (Photo by R. Ciochon.)

muzzle starting more or less directly under the orbit in *Gorgopithecus,* the snout appears short; this feature is combined with a vertically high and transversely narrow rostrum. Other important characteristics of *G. major* include an apparent lack of sexual dimorphism in either skull or tooth size (but only a few specimens are known and not all can be sexed by the dimorphic canine-P_3 complex) and the presence of strong supraorbital ridges and deep maxillary fossae, without mandibular fossae. Only one skull is known, and although this is nearly complete, the facial region has been badly "reconstructed" in plaster (Delson 1975a), a point not mentioned by either of the major describers which has confused interpretations by later workers. Delson (1975a) suggested placing *Gorgopithecus* as a subgenus of *Dinopithecus,* mainly on grounds of size, but this view cannot yet be substantiated by shared derived morphological features—that is, by "characters of common descent." *Gorgopithecus* is retained here as a full genus, pending revision. Both *Gorgopithecus* and *Dinopithecus* are probably most closely related to *Papio,* and Maier (1971b; also Eisenhart, pers. comm.) suggested that they may be synonyms of the modern genus, but until this view can be documented, it need not be accepted. Freedman (1957, 1976), on the other hand, has suggested that both large extinct forms have some dental features specially in common with *Theropithecus,* but Delson (1973, 1975a; also Jolly 1972) has refuted this opinion. Allocation of postcranial elements to these two taxa would go far toward clarifying their mode of life as well, perhaps, as their phyletic affinities.

Genus *Theropithecus* I. Geoffroy 1843

Theropithecus today is represented by a single species, *T. gelada,* occupying a relict distribution in the Ethiopian highlands where it feeds extensively on dry grass blades, seeds, and rhizomes. Relatives of the gelada have long been known from the fossil record of the African Plio-Pleistocene, under the name *Simopithecus* Andrews 1916. Jolly (1972, 1970) has recently revised these forms. He has suggested that they should best be ranked as a subgenus of *Theropithecus,* a view we and most, but not all, primatologists now accept (compare, for instance, M. G. and R. Leakey 1973a, Freedman 1976).

In addition to dental morphology, the major special features that link the modern and fossil forms include reduction of the anterior dentition (which grows proportionally smaller in larger populations); upright mandibular ramus, higher in larger forms and, in turn, linked to a deep and short face; sagittal crest placed relatively anteriorly if present; elongate forelimb, especially humerus; elbow complex with expanded, back-tilted ulnar olecranon and small, retroflected medial epicondyle of the humerus; and short, stout phalanges. These characters reflect the dietary and postural-locomotor adaptations of *Theropithecus:* a concentration on tough foods, requiring little incisal preparation, but prolonged trituration by complex cheek teeth (see figure 7.3A) in turn necessitating a musculature that brings the greatest pressure to bear on the distal teeth. These monkeys present an extreme terrestriality—living geladas hardly ever enter trees but instead walk long distances in search of food and then eat it in a sit-

ting posture, passing small food items to the mouth with the prehensile hands. The presence of most of the same morphological features in the larger fossil races or species suggests that they, too, were strongly terrestrial graminivores. Jolly (1972) was further able to demonstrate that most *Theropithecus* (*Simopithecus*) remains occur in waterside habitats, where seasonal rainfall fluctuations would have produced extensive open grasslands.

A number of fossil populations have been recovered and numerous taxa named, mostly reviewed in the cited papers. The oldest specimens are isolated lower molars: one from Lothagam-3, Kenya (Patterson et al. 1970), about 4 m.y. old, has a morphology rather typical of early *Theropithecus.* Another find is from Aïn Jourdel, Algeria, probably slightly younger and morphologically more distinctive than at Lothagam (this was described as *Cynocephalus atlanticus* by Thomas 1884; see also Delson 1975a, and in preparation). Two species of *Theropithecus* (*Simopithecus*) can be recognized from the later Pliocene through the Pleistocene. *T.* (*S.*) *darti* occurs at Makapan South Africa, and perhaps also at Hadar and East Rudolf in eastern Africa. It is somewhat larger than the modern species (in turn smaller than most *Papio* species) and shows only slight incisor reduction. Of the foregoing numerous dental-gnathic elements are known, but no postcranials have been reported. Relatively recently, juvenile and adult female skulls from Makapan were described by Maier (1972) and Freedman (1976), respectively (see figure 7.3B). *T.* (*S.*) *oswaldi* was described first from Kanjera, Kenya, but is now known as several sub-

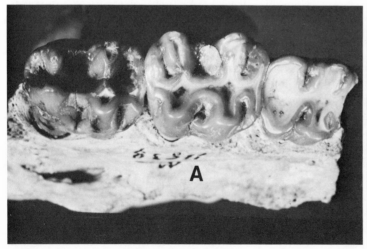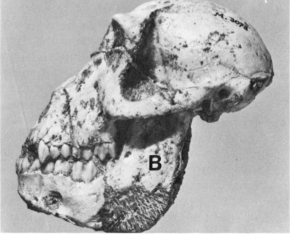

Figure 7.3 (A) *Theropithecus oswaldi,* female, upper right molars from Kanjera, Kenya, East Africa, with thick, infolded enamel and differential wear on successive molars (scale ×2). (B) *Theropithecus darti,* lateral view of the skull of a female juvenile from Makpansgat, South Africa, showing the flattened face and elongated ascending mandibular ramus characteristic of members of this genus (scale ×⁴/₉ approximately). (Photos by E. Delson and W. Eisenhart.)

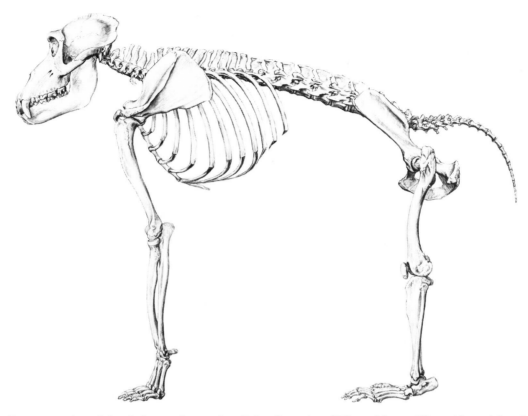

Figure 7.4   Reconstruction of the skeleton of an extinct "giant" species of *Theropithecus* (*Simopithecus*) found at Olduvai Gorge, Tanzania. At least in part for allometric reasons the forelimbs are distinctly longer than the hind-limbs. The skull has been restored with too much resemblance to *Papio;* the canines should be smaller, the face shorter and deeper, and the ascending ramus of the jaw should be at more of a right angle to the horizontal. (Scale ×⅛.) (Drawing by Jay H. Matternes, © National Geographic Society.)

species from numerous East African localities, as well as in at least one site each in northern and southern Africa. The several populations vary greatly in size, and a number of allometric (and possibly temporal) trends can be discerned, mostly from teeth, but also from the cranial and postcranial material known from Kanjera, Olorgesailie, and Olduvai, especially. In the largest forms, the cheek teeth increase in size in proportion to the mandible and cranium, but the anterior dentition is hardly if at all larger than in much smaller individuals of the modern species.

To illustrate this pattern, the modern *T. gelada* has molars (and canines) about equal in size to those of neighboring *Papio hamadryas,* but rather smaller incisors. In *T. darti* from Makapan, the molars are larger, while the incisors are about the same size, but the male canines are still tall and compressed. In the smaller *T.* (*S.*) *oswaldi,* the incisors are smaller still, while in the larger forms, the cheek teeth are very long, the incisors even smaller and the canines robust but short, with reduced P_3 mesiobuccal flange for honing the upper canine. The most

extreme case is the mandible from the lower Ndutu Beds at Olduvai Gorge (above Bed IV, perhaps 300,000 years old; Hay 1976). It was termed *Simopithecus jonathoni* by Leakey and Whitworth (1958). Although previously considered a female, M. G. and R. Leakey (1973a) suggested that it may have been a male with canines even more reduced and turned somewhat laterally, away from occlusion with P_3; this specimen is nearly the size of a female gorilla mandible and may represent the maximum size of any known cercopithecid. Eck (1977) has also discussed a cranially distinctive species from the earlier Omo deposits as *Theropithecus brumpti* (Arambourg 1947).

Jolly (1972) has described a number of complete and partial postcranial elements that underline the affinities and adaptations of *T.* (*Simopithecus*). A nearly complete male skeleton has more recently been recovered in Bed II at Olduvai, (M. and R. Leakey 1973a), but never fully described. A reconstruction of it has been attempted by J. Matternes (figure 7.4), but the canine and skull are probably too *Papio*-like. The great elongation of the forelimb was used by Jolly as evidence of extreme terrestrial adaptation, like that seen in *Gorilla,* in turn suggesting postural ability comparable to *Theropithecus gelada.* Moreover, the phalanges are about the length of those in much smaller male *Papio,* but considerably more thick or robust, and the elbow and shoulder joints are constructed as in geladas and other terrestrial primates.

The extinction of such a successful and widespread taxon is always a problem of some interest, and in this case, several suggestions have been put forward. In a number of localities, especially Olorgesailie, there is evidence for *Theropithecus* having been extensively hunted by humans with an Acheulean culture. In addition, changing climatic conditions may have at once reduced the extent of wet grassland and extended drier savanna, to which previously forest-fringe *Papio* species became better adapted. Eventually, only the relict and probably ancient population of *T. gelada* survived in upland Ethiopia. Interestingly, *Theropithecus* individuals appear behaviorally subordinate to *Papio* in their infrequent interactions.

Subfamily Colobinae Blyth 1875

Genus *Colobus* Illiger 1811

Colobus is the only living African colobine, including four main species in three groups, often ranked as subgenera. Fossil remains are rare and most fragmentary. A skull of uncertain age of the modern *C. guereza* was reported by Simons (1967) from Sudan, and partial dentitions are known from the Omo, East Rudolf, and Kanam deposits. As indicated above, Delson (1975a, 1973) showed that the type and several other isolated teeth of *Macaca flandrini* Arambourg, 1959, from Marceau, Algeria, were colobine in morphology. These specimens are similar in size and shape to *Cercopithecoides williamsi,* but are also morphologically close to certain living colobine species. To emphasize their uncertain allocation, they may be termed ?*Colobus flandrini.* Except for unpublished remains from the Ngorora Formation, central Kenya, these constitute the oldest known colobine in Africa.

Genus *Libypithecus* Stromer 1913

The most completely preserved skull of a North African late Tertiary monkey is that of *Libypithecus markgrafi* Stromer (1913) from a probably late Turolian faunal equivalent (age estimated at 6 m.y. by Cooke and Maglio 1972) at Gar Maluk, Wadi Natrun, northern Egypt (see figure 7.5). With it at the same site have been found a lower M_1 probably of

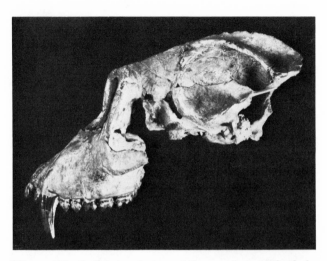

Figure 7.5   Left lateral view of the cranium of the holotype of *Libypithecus markgrafi,* a colobine with possibly semiterrestrial habits from the Latest Miocene Wadi Natrun deposits of northern Egypt. (Scale ×0.6.) (Photo by E. Delson.)

the same species, as well as mandibular and maxillary fragments of a species of *Macaca.* Because of its completeness and its unique combination of anatomical features the relationships of this species have been much discussed (see Stromer 1913, Edinger 1938, Jolly 1967, Hill 1970, and Simons 1970, 1972). Delson (1973, 1975a) has reviewed the history of study and the anatomy of this single skull in detail and has added some interesting clarification of the animal.

The male skull is nearly complete, although because of missing parts the mid and lower face is insecurely joined to the neurocranium. Moreover, the basicranium is extensively damaged, and there was antemortem trauma to the right canine and premaxilla. The dentition is anatomically colobine, the incisors small, but the molars increase in size to the rear. The face is projecting and somewhat narrow for a colobine, the supraorbital torus weak, and the sagittal and nuchal crests strong, meeting at the raised inion. The size and shape of the skull is quite similar to that of *Colobus badius* (as suggested by Jolly 1967), but also close in form to the larger genus *Cercopithecoides* and in dental size to those of *Mesopithecus.* The similarity to *Dolichopithecus ruscinensis* suggested by Jolly (1967) was refuted by Delson, as were past considerations by various authors of relationships to *Papio* (partly based on incorrect assessment of its size) and to *Theropithecus* (Hill 1970). The endocranial cast was studied by Edinger (1938) and Radinsky (1974), who found basic resemblances to other colobines.

There are no postcranial elements known, so that

locomotor adaptation is uncertain, and the dental evidence is inconsistent. The rearward placement of temporal musculature suggests, following Jolly (1970), an emphasis on incisal food preparation, as in mandrills; but the incisors are small, as is typical for colobines, while the largest molar is the third, suggesting cheek tooth emphasis on crushing or slicing. Thus, the diet was probably different from that of modern colobines. Delson (1975a, 1973) was unable to determine the direction of that difference, although in some ways it parallels the dental adaptation of *Theropithecus.* The paleoecology of Wadi Natrun indicates that the region (although much better watered 6 m.y. ago than at present) was surrounded by open country inhabited by ostrich, mastodon, and *Hipparion,* as well as macaque. The aridity that produced the aeolian excavation of the Egyptian Western Depressions (Qattara, Natrun, Fayum) was probably underway by this time, as a continuation of the longer term Saharan aridity. Thus, although the shores of Natrun Lake were probably wooded, the open country might have been suited to some sort of terrestrial feeding, possibly on small objects by *Libypithecus.* On the other hand, despite the generally circum-Mediterranean nature of the associated mammals, the affinities of this species are clearly with the African colobines *Colobus, Cercopithecoides,* and *Paracolobus,* none of which was terrestrial. Therefore a strong terrestrial adaptation appears to be unlikely.

Genus *Cercopithecoides* Mollett 1947

Cercopithecoides is one of the most widespread African monkeys, occurring in almost all the southern African localities from Makapan (the type site) through Kromdraai to Swartkrans "b," as well as at East Rudolf (Freedman 1957, 1976; M. G. and R. Leakey 1973b). The South African species, *C. williamsi* (figure 7.6), is larger than any modern colobine, with a slightly long and narrow face, short premaxilla, deep ophryonic groove, and pronounced sexual dimorphism in cranial shape (including the face). The temporal musculature is weak, with no evidence of sagittal cresting, and the nuchal muscles are also poorly developed. The mandibular ramus is slightly back tilted, with no enlargement of the gonial region. No postcranial elements have yet been allocated to *C. williamsi,* as is also the case with most other South African cercopithecid species. Specimens previously termed *C. molletti,* as well as the neurocranium made the holotype of ?*Parapapio coronatus,* are now referred to *Cercopithecoides williamsi,* which may be clearly placed in Colobinae on

Figure 7.6 Dorsal view of the most completely preserved skull of *Cercopithecoides williamsi,* a male specimen from Makapansgat, South Africa. (Photo by W. Eisenhart.)

both cranial and dental morphology, despite previous uncertainty.

The affinities of *Cercopithecoides* within Colobinae are not completely certain, however. There appears to be little evolutionary change in the South African species over a long time range, despite some variability at Makapansgat. M. G. and R. Leakey (1973b) referred to the genus a skull and mandible from East Rudolf that are certainly distinct specifically but could well be close to *C. williamsi* phyletically. Overall dental and facial morphology appears to relate it most closely to *Paracolobus* (Delson 1973), but detailed comparative studies remain to be undertaken.

Genus *Paracolobus* R. Leakey 1969

The type species of the genus *Paracolobus, P. chemeroni,* was described by Richard Leakey (1969) from a find of a remarkably well-preserved skeleton in the Chemeron Beds west of Lake Baringo, Kenya. The partial skull of this specimen is illustrated in figure 7.7. Geological evidence from this succession indicates that the skeleton is about 4 m.y. old, while the associated fauna is comparable to 2-m.y.-old levels at Omo, suggesting a channel filling.

Paracolobus is a comparatively large form showing definite ties with the colobine monkeys. An estimate of skull length is that it probably exceeded 16 cm, and length of the hindlimb from the head of the femur to the distal extremities of the toes would have been approximately 0.75 m. These measurements exceed those of any other known colobine and

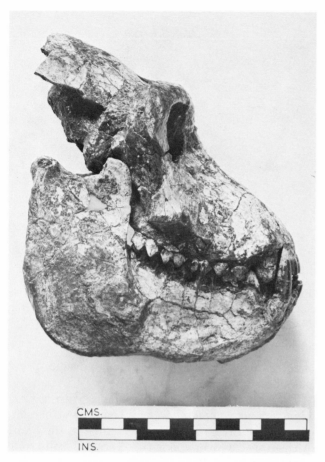

Figure 7.7 Lateral view of the cranium of *Paracolobus,* a large extinct colobine monkey from late Pliocene deposits in the Baringo Basin, Kenya. (Courtesy Kenya Nat. Mus.)

are comparable with the general gigantism of many East African Pleistocene and Pliocene mammals.

Considering dental function, mandibular anatomy of *Paracolobus* corresponds well with what would be expected of a large folivorous monkey. In *Paracolobus,* the face is deep, the horizontal line of the tooth row is dropped well below the basicranium, and the axis of the high mandibular ramus is at right angles to the corpus.

In these features, among several others, the probable male of *Paracolobus* resembles males of the genus *Colobus.* It differs from the latter cranially in having a longer snout and very broad face, comparable to *Pygathrix* species; laterally, the face has a smoothly rounded outline that is little like that of *Papio* (figure 7.7). Another difference from most modern *Colobus* is that the nasal aperture is transversely quite narrow and vertically unusually high. R. E. F. Leakey (1969) appears to believe that, in general, South African *Cercopithecoides* is more similar to some modern *Colobus* than is *Paracolobus chemeroni,* but the two are probably closely related.

The postcranial skeleton of this unusual find still awaits comprehensive analysis, but a few notes can be added. The olecranon process of the ulna is curved anteriorly to a line drawn through the long axis of the remaining shaft, which, together with a humero-femoral index below 100, could suggest that this was an arboreally adapted species; even so, a similar low index can be found in the patas monkey, a terrestrial form. Despite the retroflected humeral medial epicondyle, the long phalanges seem to confirm probable arboreality.

Additional colobine material has been recovered from several localities in both eastern and southern Africa which may be referable to *Paracolobus* or a similar genus. Fragmentary remains from Makapan are larger than *Cercopithecoides* from that site, while several mandibles and teeth from Laetolil are quite similar to *P. chemeroni.* A cranium and mandible from Omo was described by M. G. and R. Leakey (1973b) as a new, unnamed genus, but it and more recently recovered specimens from Omo (Eck 1977), East Rudolf, and Hadar indicate the presence of two or three colobine species of differing size during the Plio-Pleistocene interval, whose full analysis is awaited.

Subfamily *incertae sedis*

Genus *Victoriapithecus* von Koenigswald, 1969

Victoriapithecus is now known by a sample of over one-hundred mostly isolated teeth, but including several partial mandibles and about a dozen postcranial elements, almost all from a Middle Miocene East African locality on Maboko Island, Lake Victoria, Kenya. Five of these mandibular fragments are illustrated in figure 7.8. Although von Koenigswald (1969) stated that the holotype of *V. macinnesi* was from Rusinga, its attached matrix and the lack of additional specimens from anywhere on that island make this improbable; instead, this mandible may have also come from Maboko. Macinnes (1943) described and figured it, mentioning also some other teeth and indicating that at least one specimen was from Rusinga, without being more specific; he tentatively referred the jaw to *Mesopithecus* sp., then the oldest known cercopithecid. Unfortunately, several recent authors have continued to employ uncritically this mistaken generic allocation. As Delson (1975a) indicated, von Koenigswald only described part of the original collection available to him, and numerous additional specimens have been recovered more recently on Maboko Island by Pilbeam, Andrews, and others.

In his report, von Koenigswald (1969) established

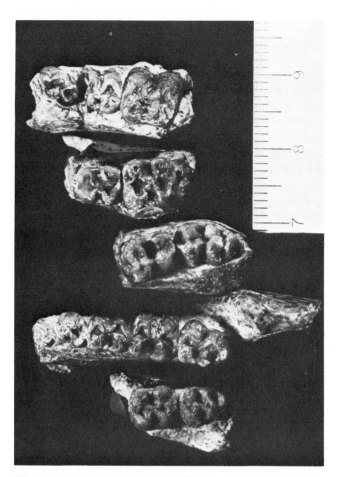

Figure 7.8   Occlusal views of *Victoriapithecus* mandibles from the Miocene of Maboko Island, Lake Victoria, Kenya. Center specimen is *"V." leakeyi*. (Photo by E. Delson.)

two species of *Victoriapithecus, V. macinnesi,* based on the mandible mentioned above, and *V. leakeyi,* based on an isolated upper molar. Delson (1974, 1975a) reviewed the finds, suggesting that von Koenigswald's allocations were confusing, but that two different morphologies could be recognized on grounds other than size. *V. macinnesi* includes most of the dental remains, with both large and small individuals characterized by squarish molars with short trigonids (like colobines) but shallow lingual notches (like cercopithecines) and also with P_4 turned slightly oblique to the molar row. One lower jaw and perhaps some additional isolated teeth present narrower, more elongate molar crowns, also with shallow notches but longer trigonids, more like the macaque type of tooth thought to be ancestral for Cercopithecidae. The upper molar holotype of *"V." leakeyi* (which, along with one other tooth, preserves the crista obliqua) appears to occlude with some of these, rather than with *V. macinnesi* of any size. It may be suggested (following Delson) that the latter

species had begun to develop a colobine tooth pattern, and it is possible that the two "morphs" represent different genera, if not subfamilies (in a vertical classification), but further analysis of sample variability in size and wear pattern is required. Both morphologies (species, genera?) are nearly fully bilophodont and are probably near the ancestry of later cercopithecids.

The several postcranial elements mostly derive from one species showing phalangeal robusticity and a macaquelike adaptation of its elbow joint. One humerus and perhaps one phalanx appear more "arboreal" in function, and are somewhat similar to those of modern *Colobus*. Delson has attributed the former, larger elements to *"V." leakeyi* and the latter to *V. macinnesi,* but without better association, such allocation is uncertain.

Additional specimens possibly referable to *Victoriapithecus* are known from Ombo, Loperot, and perhaps Napak, Uganda. From the latter site, Pilbeam and Walker (1968) described one upper molar of cercopithecine aspect and a frontal bone of uncertain allocation.

Genus *Prohylobates* Fourtau 1918

The species *Prohylobates tandyi* was described in 1918 by the French geologist Réné Fourtau. It was one of a series of species of Miocene vertebrate fossils collected from the ridges north of Hatayet al Moghara, on the northwest side of Wadi Moghara, Egypt, U.A.R. In addition to the type specimen of *P. tandyi,* Fourtau described two other mandibular fragments under the name *?Dryopithecus mogharensis.* In recent years Remane (1965) and Le Gros Clark and Leakey (1951) have recognized that the molars of the three Wadi Moghara primates resemble those of cercopithecoid monkeys but, because they had not seen the originals, their remarks were cautious. Simons (1969) recently restudied these three finds and showed that they are not apes, but all have the mandibular anatomy of cercopithecoid monkeys. It seems reasonable to conclude that they all belong to the same species because insofar as they can be compared there seems to be no anatomical difference between the three, the size disparity is not great, and they come from the same general locality and age. *Prohylobates tandyi* should thus be sustained as a taxonomically valid name for what seems to be the oldest monkey described to date.

Initially, Remane (1924), who considered the significance of Fourtau's Miocene Egyptian primates at some length, believed that *Prohylobates* was rather closely related to *Propliopithecus.* Nevertheless Remane was the first to point out that the entire set of characteristics by which Fourtau attempted to relate *Prohylobates* to *Hylobates* was not taxonomically important and that they do not indicate in any way that the two held an ancestor-descendant relationship. On the other hand, Remane himself, working from Fourtau's poor photographic plates, did not realize that the external border of the mandible had been broken away anteriorly so that the mesial alveolus and root of P_3 had been lost. Concluding instead that P_3 of *Prohylobates* was single-rooted, he considered the form quite distinctive, because this feature is found commonly elsewhere among higher primates only in *Homo*. By 1965, however, he had reversed his position that *Prohylobates* was related to the ape *Propliopithecus,* concluding instead that its affinities might lie with the colobines.

The molar teeth of *Prohylobates tandyi* are not as completely bilophodont as those of *Victoriapithecus* or later cercopithecids, but the basic pattern is present. It may be best seen on M_2 of the holotype, despite its wear: the mesial lophid is worn away (if ever present), the distal one is poorly developed, and a low crest links the metaconid and hypoconid—the ancient cristid obliqua. In addition, there may have been a small median hypoconulid on M_2, but this cuspule could also have been a variable minor accessory feature, as sometimes occurs in *Theropithecus* species. The M_3 in *Prohylobates tandyi* is shorter than M_2 but damaged distally, and the P_4 is somewhat oblique to the molar row, as in *Victoriapithecus macinnesi* and also some modern colobines. The corpus is somewhat deep compared to crown height, suggesting possible colobine affinities to Simons (1969). Recently, a new specimen from Jebel Zelten, Libya, has been identified as a possible *Prohylobates,* differing from *P. tandyi* in larger size and elongate M_3 (Delson, 1977).

Oligocene Monkeys

Superfamily ?Cercopithecoidea or Parapithecoidea

Family Parapithecidae Schlosser 1911

Catarrhine primates from the Oligocene and Early Miocene have been found only in Africa. Many sites for fossil mammals in Europe and Asia date to this period (from about 16 to about 37 m.y. ago) but none have yielded any cercopithecoid monkey fossils. This suggests that before the later Miocene about 12 m.y. ago no monkeys had reached Eurasia (von Koenigswald 1969; Simons 1970, 1972; Delson 1975b). Monkeys are not well known either from most

Miocene localities in Africa, while in these numerous sites fossil apes are quite common. Only at Maboko Island in Lake Victoria do apes and monkeys occur together abundantly. Today in most parts of forested Africa monkeys are quite common while apes are absent or rare. The lack of monkeys in the earlier Miocene of Africa may have been caused by the fact that, like bovids, their adaptive radiation had not yet come. Evidently monkeys were not adapted to the habitats represented in the earlier Miocene of Kenya and Uganda.

Going farther back to the Oligocene sites of the Old World the whole of our knowledge of Anthropoidea comes from the Fayum of Egypt. No primates are known definitely from the Oligocene of Europe and the sole variety reported from this epoch in Asia, "Kansupithecus" is now dated much younger by associated fauna. In the later Eocene deposits of Burma, however, fossils of two genera, *Pondaungia* and *Amphipithecus,* show various resemblances to Egyptian Fayum primates but are so incompletely known that the exact degree of this relationship is uncertain. Insofar as the affinity is real, their similarities do suggest some sort of Eocene faunal interconnections between Africa and Asia.

The origin of the Cercopithecidae has been considered in several recent papers (e.g., Simons 1969, 1972, 1974; Delson 1975c). Although we differ on their interpretation, the Fayum parapithecids clearly are relevant to solution of this problem. Before indicating the possible alternative viewpoints, it is useful to review parapithecid morphology and adaptations.

The Parapithecidae as currently understood includes two genera, *Parapithecus* and *Apidium,* each with two known species, one from the middle and one from the upper Fayum fossiliferous horizons. As a result of the six Yale expeditions to Egypt directed by Simons from 1961 to 1967, a large series of new specimens has been found. These make it clear that parapithecids are animals of primitive anthropoidean grade, thus monkeys in adaptive terms, whether or not they are so in a phyletic sense. Among the shared features that define the family are dental formula of 2.1.3.3, as in most cebids; early fusion of mandibular symphysis (and metophic suture); P_2 semicaniniform, smaller and simpler than P_{3-4}; upper molars with well-developed hypocone and large bulbous conules, not linked by crests to the main cusps; upper premolars probably with central conule; lower molars waisted, with reduced buccal cingulum, hypoconulid on midline; P_{3-4} with small metaconid; mandible shallowing slightly anteriorly. This combination of features demonstrates both the monkey grade of parapithecids and their taxonomic distinction from cercopithecid or ceboid monkeys.

One of the more important new facts about species of these two genera is that *Apidium* and *Parapithecus* are shown to be the most common African Oligocene primates. Including isolated teeth, there are about two hundred finds of *Apidium* and about a fifth as many of *Parapithecus*. All the Fayum ape specimens put together do not equal as many specimens as those of *Parapithecus* alone. Parapithecids are, in fact, the commonest Fayum mammals. They differ in a number of their features from the modern African monkeys, but on the other hand they do not have many specializations that would distinguish them from what would be hypothetically expected stages in the ancestry of Old World monkeys. In this regard they are unlike such abundant earlier primates as *Plesiadapis,* where extreme crossing specializations such as incisor, canine, and premolar loss—coupled with hypertrophy of remaining incisors—renders ancestral status to later primates impossible. The parapithecids also have several distinctive, but presumably "primitive," features that resemble tooth morphology in a broad range of Paleocene and Eocene primates. *Parapithecus* especially bears many resemblances to Miocene-Recent Cercopithecoidea. Unlike all the other Fayum primates, *Parapithecus* has the principal cusps of the upper and lower molars arranged in a quadrate pattern that could be a foreshadowing of the quadrate teeth of Miocene-Recent cercopithecoids, but its one or two specialized or derived characters might eliminate it as an immediate ancestor. Of course, no Oligocene mammal species can be proved to be directly ancestral to a living mammal species. What is of interest regarding *Parapithecus* is whether it is possible to take it as broadly representing a stage in the ancestry of the Old World monkeys. Recovery of forms intermediate in age between *Parapithecus* at about 30 m.y. and *Victoriapithecus* at 15 m.y. might aid in resolving this question. Undue stressing of the parapithecids as direct Old World monkey ancestors is as unwise as similar emphasis on the view that they cannot be so related. Because we have approximated these extremes in the past and because this is a review article, we will summarize a middle view.

Genus *Apidium* Osborn 1908

Apidium was the first Oligocene "monkey" to be discovered. When it was described Osborn was not certain even of its ordinal affinities. The name, meaning "Little Apis" (the Ptolemaic sacred bull) suggests that at least when the name was coined, he considered it an artiodactyl. The type specimen of *Apidium phiomense* was found by Richard Markgraf in the upper fossil wood zone of the Jebel el Qatrani Formation, but the exact locality was not recorded except as northeast of Quarry A. Extensive prospecting by Yale University expeditions in this zone has demonstrated that fossils in the upper Fayum levels are so rare, except in Quarries I and M—the only known places where *A. phiomense* occurs—that the type specimen probably came from one or the other of these quarry areas, or from their same level "to the N. E. of Quarry A."

Simons (1962) described a smaller and older *Apidium* species, *A. moustafai,* known from a hypodigm of a few dozen jaws found in Quarry G. This site is located stratigraphically between the two fossil wood zones. During the several Yale expeditions, *A. phiomense* proved to be the most common mammal found at Quarry I. The material now available includes many limb bones, scores of teeth and mandibles, three partial frontals, petrosals, and several maxillae. The lower molars increase in length from M1 to M3, and all show the centroconid (mesoconid) on the cristid obliqua between hypoconid and metaconid. The upper molars are also polycuspidate, with strong conules, often a pericone and a "protoconule" (called cusp *a* by Kay 1977) mesial to the paraconule; all cusps are nearly the same size, although the protocone is generally larger. The eruption pattern has been described by Conroy, Simons, and Schwartz (1975).

A partial cranial reconstruction of *A. phiomense* made possible by the discoveries of Simons' expeditions, is illustrated in figure 7.9.

Preliminary cranial and postcranial studies suggest that *A. phiomense* had a short, rather marmosetlike face with small canines relative to those of *Parapithecus* and similar to those of *Callithrix* (see figure 7.9). Like *Aegyptopithecus* ear structure was at approximately the ceboid grade of organization (see Gingerich 1973). Postcranials indicate that the hind feet were adapted for springing. The olecranon process is not retroflexed but rather the ulna is bowed forward as in highly arboreal primates. *Apidium phiomense* was apparently about the size of the present-day owl monkey, *Aotus trivirgatus.* Conroy (1976) implied that the general locomotor adaptation may have been similar to *Samiri* or *Cebus.*

The affinities of *Apidium* were much discussed when there was but one specimen. It seemed condylarthlike to some (e.g., Hürzeler 1968), but condylarths have not otherwise been found in Africa. Simons (1960) stressed that in its polycuspidation

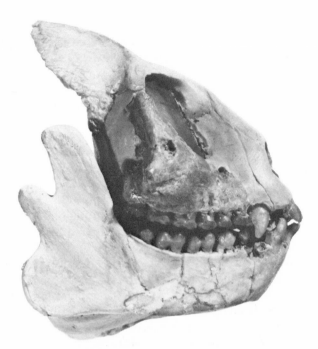

Figure 7.9 Photo reconstruction of the lateral view of the face of *Apidium phiomense*, primitive catarrhine from the Oligocene deposits of Egypt. The view is based on the American Museum frontal, Cairo Museum maxilla, and a mandible at Yale (rostrum and posterior mandible restored from other specimens at Yale). (Scale ×2.5.) (Photo by A. H. Coleman.)

and possession of a lower molar centroconid *Apidium* resembles *Oreopithecus* from the late Miocene of Tuscany (but see below). Hürzeler (1968) suggested that *Apidium* is not a primate, giving as reasons the facts that the molar tooth structure resembles that of nonprimates and that the mandibular ramus is low or shallow, a nonprimate characteristic. Nevertheless, he has admitted a striking resemblance between *Apidium* and *Oreopithecus*. Inspection of the type of *Apidium phiomense* shows that it is a juvenile; it has M_3 unerupted and the characteristic juvenile striated bone—indicating that the animal's immaturity was the reason for the shallowness of the horizontal ramus of this mandible. It would have deepened with increasing age, as demonstrated now by numerous adult specimens of both species.

Scores of jaws of *Apidium* and *Parapithecus* have been found in the Fayum by recent Yale expeditions. These new materials resolve without doubt the affinities of *Apidium*. In January 1967, at Fayum Quarry I, a possibly associated group of skull fragments of *A. phiomense* was found by Simons, together with palatal fragments and upper teeth. One of the fragments of this find is a partial frontal bone

preserving the interorbital pillar. The frontals are completely fused, showing no trace of the metopic suture between bilateral frontals commonly present in prosimians and nonprimates. This frontal fragment is exactly like that from the Fayum described by Simons (1959), and consequently the latter more complete frontal, which gives evidence of the presence of postorbital closure can be assigned to *A. phiomense*. The upper teeth of *Apidium* are anteroposteriorly short and have well-delineated separate cusps with distinct paraconule and metaconule. These teeth admit the possibility, just as the lower teeth do, of a topological conversion into those of *Oreopithecus*, particularly because those of *Oreopithecus* (unlike any other later member of the Catarrhini) retain well-defined paraconule and metaconule cusps. Nevertheless, about 20 m.y. separate *Apidium* and *Oreopithecus*. The analogies in the upper and lower tooth structure may thus be parallelisms, primitive retentions, or indications of real phyletic affinity.

Genus *Parapithecus* Schlosser 1910

The genera *Parapithecus* and *Propliopithecus* (and *Moeripithecus*) were described briefly by Schlosser in 1910 and then analysed in detail by him in 1911. *Parapithecus fraasi* Schlosser has the most complete type specimen of all the African Oligocene primates described in the early part of this century. It consists of a mandible that preserved both bodies and part of the rami at the back. In the type dentition were preserved a pair of central incisors and posterior to these seven teeth on the left, and six on the right (one anterior premolar had fallen out). As long as the type was the sole specimen of *Parapithecus*, it remained problematical because the dental formula was subject to various interpretations, while the morphology of the cheek teeth was somewhat unlike anything else among primates. Apart from a vague and distinct resemblance to the contemporary Fayum ape *Propliopithecus*, little tallying of affinities was possible, other than that this was definitely a primate dentition. Nevertheless, the specimen was endlessly discussed and associations ranging from ties with the modern prosimian *Tarsius* to an ancestral relationship for present-day man were proposed. At the anterior midline of this mandible, still the only specimen of this species, two small incisors are preserved. Their very small size showed finally that they are the central pair, but this was also accepted from the start. Lateral to these were two much larger teeth which were variously interpreted as large lateral incisors or as canines (see figure 7.10). Few if any students who wrote about it had access to the original specimen, depending instead on crown-view photographs that did not show the extreme size disparity between the two seemingly most anterior pairs of lower teeth. Had this discrepancy been understood at the start, the problem of what were the canines of this animal would have been solved before new finds were made. In writing the original description of this mandible, Schlosser wavered somewhat as to whether the lower dental formula was 1.1.3.3 as in the living tarsier or 2.1.2.3 as in the present-day catarrhines. This set the stage for the two principal interpretations of *Parapithecus:* (1) that it was a tarsioid prosimian, or (2) that it was a higher primate. Gregory (1920) even went so far as to suggest it for hominid ancestry.

In fact, neither of these interpretations can be supported because neither of the dental formulas was correct. The new finds of *Parapithecus* and *Apidium* show them to be closely related forms, although generically distinct. Moreover, all the jaws that preserve part of the symphyseal region show sockets for *two* pairs of lower incisors. In Schlosser's type specimen the delicate alveolar border in the symphyseal region must have been broken away at or before the time of collection, and the two horizontal rami as well as the two central incisors were glued together in such a way as to suggest that the specimen did not have lateral incisors. This in turn led to the errors in interpreting the dental formula that continued to be perpetrated up to and including Kälin's monograph (1961) on the Fayum primates. Also, it was believed that the symphysis was unfused—because the damaged area had destroyed evidence of the incisor alveoli, the two central incisors were simply glued to the front of the mandible. Because the mandibular bodies were glued together with a central fragment missing, they diverged

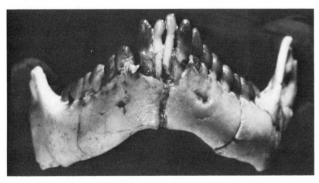

Figure 7.10 Anterior face of the symphysis of the type mandible of *Parapithecus fraasi*, primitive catarrhine from the Oligocene Fayum of Egypt. There are no contacts between the two sides of the lower jaw or with the anterior teeth. (Scale ×2.5.) (Photo by E. Delson.)

markedly to the rear. This pronounced V-shaped arrangement of the tooth rows was taken to be a tarsioid resemblance. Actually, the new jaws, with their fused symphyses, are slightly more U-shaped.

Most of the new *Parapithecus* finds are 15 to 20% larger than the type specimen of *Parapithecus fraasi,* which was evidently found at a lower level than Quarry I, where the new species occurs. This new parapithecid has been named *P. grangeri* (Simons 1974). In addition to size, it appears to differ from *P. fraasi* in having slightly smaller M_3, larger metaconids on P_{3-4}, cusps slightly less bulbous on lower molars, and mandibular corpus relatively deeper distally.

Parapithecus shares with *Apidium* the basic dental features listed above as characterizing the family. It differs from *Apidium* in the short M_3, relatively larger canine and premolars (compared to molars), presence of paraconid variable, and lack of the centroconid. A few isolated upper teeth appear to be referable to *P. grangeri,* as they occlude with the lowers and differ from those of *Apidium* in such features as lack of pericone and "protoconule," smaller paraconule and stronger buccal cingulum on the molars, and smaller conule and relatively taller paracone on the premolars. The presence of strong, relatively bulbous, isolated conules on these upper cheek teeth does suggest enough similarity with *Apidium* for them to be referred to *Parapithecus* pending final revision of the family. In dental and mandibular size, *P. grangeri* is comparable to the smallest living cercopithecid, *Cercopithecus (Miopithecus) talapoin,* with which it shares some dental similarities (relative M_3 size, essentially four-cusped lower molars). The known postcranial elements from Yale Fayum Quarry I are homogeneous (except for a few large bones referred to *Aegyptopithecus*), and all have been referred to *Apidium,* as the frequency of isolated teeth and jaws shows that this is the most common small mammal at Quarry I (Conroy 1976). In consequence, no limb elements of *Parapithecus grangeri* have yet been definitely separated out from those presumed to be of *Apidium.*

Relationships of the Parapithecids

Most previous authors have each interpreted differently the phyletic position of *Parapithecus* and *Apidium* and thus we have refrained from commenting on this matter above. In addition to the shared monkey grade features of the two genera, the several detailed morphological resemblances outlined previously convince us that they are each other's closest relatives (or sister taxa), requiring that they be classified in a single higher taxon, to which family rank is usually assigned (see Kälin, 1961, Simons 1974). This view contrasts with that of Kay (1977) who has tentatively suggested that *Apidium* may be more closely related to the *Aegyptopithecus* group of Fayum primates.

Two major phyletic linkages have been suggested for the parapithecids: *Apidium* with *Oreopithecus* and *Parapithecus* with Cercopithecidae. As noted briefly above, *Apidium* does share with *Oreopithecus* the presence of a centroconid on the lower molars, which moreover are waisted somewhat in both genera; in addition, conules are present on upper molars in both (Gregory 1922; Simons 1960, 1972; Delson and Andrews 1975). On the other hand, the placement and relative size of conules are different in both, and the presence of premolar conules in *Apidium*, which tends to link it strongly with *Parapithecus*, further separates it from the enigmatic Tuscan primate. At present Delson considers it unlikely that there was a common ancestor shared by *Apidium* and *Oreopithecus* more recently than the ancestor of all catarrhines, while Simons is prepared to stand by his 1960 assessment that a structural similarity exists but that in view of the great time separation and the lack of any intervening relevant fossils, that a relationship is hinted at but by no means confirmed. Simons was mainly concerned with refuting doubts, when there was only one very incomplete specimen, that *Apidium* was a primate, by showing structural analogies with undoubted primates such as *Oreopithecus* (Piveteau 1957, Hürzeler 1958).

Turning to the question of cercopithecid origins introduced at the beginning of this section, both authors have presented their views recently (Simons 1974, Delson 1975c). Simons has suggested that the observed similarities between *Parapithecus grangeri* and *C. talapoin* are too strong to be merely convergence, while Delson argued that the latter was indeed the case, especially as the talapoin is a probably derived member of a dentally clearly derived group of monkeys. The presence of three premolars in parapithecids and of a tympanic without ossified external meatus in *Aegyptopithecus* would imply convergent development of both tubular tympanic and C'/P_3 honing in living cercopithecids and hominoids. Simons (1974) has accepted these implications, while Delson has argued that there is no evidence for either case of nonhomology of these complexes in modern forms. However, the primate fossil record documents six cases of independent P_2^2 loss and at least three of separate development of a tubular auditory meatus (although none are as similar as are those of modern hominoids and cercopithecoids). More recently, Gingerich (1975) reported a structural "sequence" of gradual reduction of anterior premolars and distal shifting of upper canine honing in the adapid genera *Adapis* (with P_1 of moderate size, smaller than P_2, and honing upper C) *Leptadapis* (P_1 quite small, the upper *C* honing on lower C, P_1, and the front of P_2) and *Cercamonius* (known only from P_4 to M_2, but with P_2 single-rooted and P_3 probably enlarged and obliquely emplanted, suggesting upper C honing). This might offer an analog to the way some parapithecid might have reduced P_2, even if that tooth were involved in canine honing already, as tentatively suggested by Delson (1975c). A possible counter to this argument might rest on the fact that although smaller and simpler than P_{3-4}, P_2 in parapithecids has not yet been reduced in height to the extent of indicating its probable subsequent loss. Moreover, the presence of large conules on the upper molars and especially premolars allocated to *Parapithecus* could represent an autapomorphic condition or "crossing specialization" not now seen in cercopithecoids were it not for the fact that we do not know the starting point for these tooth structures in basal Old World monkey ancestors. That is, like many Paleocene and Eocene forms that show acessory conules, these structures in *Parapithecus* and *Apidium* could have been derived from their ancestors and later lost in their descendants. Just because modern cercopithecoids have such extremely simplified tooth crowns, one cannot safely rule out that their ancestors had more complex tooth-crown anatomy.

The interpretation of the possibility of relationship between parapithecids and cercopithecids thus rests on variant interpretations of characters and character complexes. Clearly, no known parapithecid can be proven to be directly ancestral to Neogene Old World monkeys, and we know so little of African Oligocene mammals generally that it seems improbable that exact ancestors for later forms have been found. It is still possible, but not proven that parapithecids more or less broadly represent a stage in the ancestry of such monkeys, or of catarrhines in general. For resolution of the question of cercopithecid origins we are concerned not only with the nature of the early parapithecids but also require knowledge of temporally intermediate (or Oligocene alternate) possible ancestors. If cercopithecids shared a common ancestor with the parapithecids more recently than they did with the forerunners of living apes, either before some of the distinctive parapithecid morphology appeared or after it was lost, then a number of characters developed in parallel. On the other hand, if parapithecids are a side branch of catarrhine evolution, with no living close relatives, the similarities between the teeth of *Parapithecus* and *Cercopithecus* may best be viewed as adaptations to similar ecologies. As yet, this dilemma can not be fully resolved.[1]

E. Delson was partially supported and assisted in the preparation of this manuscript of NSF Grant BNS74-13258 AO1 and CUNY Faculty Research Award Program Grants 11152, 11483 and 11785; previous grants from Columbia University, the Wenner-Gren Foundation, and the National Geographic Society permitted him to undertake research on which this summary is based. E. L. Simons carried out research while an Alexander von Humboldt prize recipient at the University of Kassel, awarded by the German Federal Republic. Results reported here by Simons were also supported in part by NSF Grants G-18102, GP-433, GP-3547, GA-723, GA-11145, and Smithsonian Foreign Currency Program Grants No. 5, 23, and 1841. Field research was also supported through the contribution of personnel and equipment from the Cairo Geological Museum and the Geological Survey of Egypt. Alfred Rosenberger provided helpful comments on the manuscript.

References

Andrews, C. W. 1916. Note on a new baboon (*Simopithecus oswaldi*, gen. et sp. n.) from the (?)Pliocene of British East Africa. *Ann. Mag. Nat. Hist.* 18:410–419.

Arambourg, C. 1947. Contribution à l'étude géologique et paléontologique du bassin du Lac Rodolphe et de la basse vallée de l'Omo: Deuxième partie, paléontologie. *Mission Sci. Omo 1932–1933*, I. Géol. Anth. 1(3):231–562.

_____. 1959. Vertébrés continentaux du Miocène supérieur de l'Afrique du Nord. *Mém. Serv. Carte Géol. Algérie n.s., Pal.* 4:1–161.

Broom, R. 1937. On some new Pleistocene mammals from limestone caves of the Transvaal. *S. Afr. J. Sci.* 33:750–768.

Broom, R., and J. T. Robinson. 1949. A new type of fossil baboon, *Gorgopithecus major*. *Proc. Zool. Soc. London* 119:379–386

Clark, W. E. Le Gros, and L. S. B. Leakey. 1951. The Miocene Hominoidea of East Africa. *Fossil Mammals of Africa* 1, Brit. Mus. (Nat. Hist.), pp. 1–117.

Cocchi, I. 1872. Su di due Scimmie fossile italiane. *Bolletino Comitato Geol. d'Italia.* 3(3,4):59–71.

[1] We have not discussed the African Oreopithecoidea mentioned as occurring at Fort Ternan, Kenya, by L. S. B. Leakey (1968b) and at Maboko Island, Lake Victoria, Kenya, by von Koenigswald (1969), who gave the latter find the generic name *Mabokopithecus,* because Simons, on a recent visit to the Kenya National Museum, Nairobi, determined that these specimens do not belong to primates. Therefore, no evidence remains that the *Oreopithecus* group was ever distributed in Africa.

Conroy, G. C. 1976. Primate postcranial remains from the Oligocene of Egypt. *Contrib. Primatol.* 8:1–134.

Delson, E. 1973. Fossil Colobine monkeys of the Circum-Mediterranean Region and the evolutionary history of the Cercopithecidae (Primates, Mammalia). Thesis, Columbia University.

_____. 1974. The oldest known fossil Cercopithecidae. *Am. J. Phys. Anthropol.* (abst.) 41:474–475.

_____. 1975a. Evolutionary history of the Cercopithecidae. *Contrib. Primatol.* 5:167–217.

_____. 1975b. Paleoecology and zoogeography of the Old World monkeys. In R. Tuttle, ed., *Primate Functional Morphology and Evolution.* The Hague, Mouton, pp. 37–64.

_____. 1975c. Toward the origin of Old World monkeys. *Actes C.N.R.S. Coll. Instl.* No. 218, *Evol. Verts.,* pp. 839–850.

_____. 1977. A new species of ?*Prohylobates* from the Early Miocene of Libya. *Am. J. Phys. Anthropol.* (abst.). 47:126.

Delson, E. and P. Andrews. 1975. Evolution and interrelationships of the catarrhine primates. In W. P. Luckett and F. S. Szalay, eds., *Phylogeny of the primates.* New York: Plenum, pp. 405–446.

Delson, E., and P. H. Napier. 1976. Request for the determination of the generic names of the baboon and the mandrill (Mammalia:Primates, Cercopithecidae). *Bull. Zool. Nomencl.* 33:46–60.

Eck, Gerald G. 1976. Cercopithecidae from Omo Group deposits. In Y. Coppens, F. C. Howell, G. L. Issac and R. E. F. Leakey, eds., *Earliest man and environments in the Lake Rudolf Basin.* Chicago: University of Chicago Press, pp. 332–334.

Eck, G. G. 1977. Diversity and frequency distribution of Omo Group Cercopithecoidea. *J. Hum. Evol.* 6:55–63.

Eck, Gerald G., and F. C. Howell. 1973. New fossil *Cercopithecus* material from the lower Omo basin, Ethiopia. *Folia Primatol.* 18:325–355.

Edinger, T. 1938. Mitteilungen über Wirbeltierreste aus dem Mittel-Pliozän des Natrontales (Ägypten). 9. Das Gehirn des *Libypithecus. Zbl. Min. Geol. Pal. Stuttgart* 8:122–128.

Eisenhart, William. 1975. A review of the fossil cercopithecoids from Makapansgat and Sterkfontein, South Africa. *Am. J. Phys. Anthropol.* (abst.) 42:299.

Fourtau, R. 1918. *Contribution à l'étude des Vertebres Miocène de l'Égypte.* Cairo: Egypt Surv. Dept.

Freedman, L. 1957. The fossil Cercopithecoidea of South Africa. *Ann. Trans. Mus.* 23:121–262.

_____. 1961. New Cercopithecoid fossils including a new species, from Taung, Cape Province, South Africa. *Ann. S. Afr. Mus.* 46:1–14.

_____. 1971. A new checklist of fossil Cercopithecoidea of South Africa. *Palaeont. Afr.* 13:109–110.

_____. 1976. South African fossil Cercopithcoidea: a reassessment including a description of new material from Makapansgat, Sterkfontein, and Taung. *J. Hum. Evol.* 5:297–315.

Gingerich, P. D. 1973. Anatomy of the temporal bone in the Oligocene anthropoid *Apidium* and the origin of the Anthropoidea. *Folia primat.* 19:329–337.

———. 1974. Stratigraphic record of early Eocene *Hyopsodus* and geometry of mammalian phylogeny. *Nature* 248:107–109.

———. 1975. *Cercamonius:* A new genus of Adapidae (Mammalia, Primates) from the Late Eocene of southern France, and its significance for the origin of higher primates. *Contrib. Mus. Paleont., Univ. Michigan* 24:163–170.

Gregory, W. K. 1920. On the structure and relations of *Notharctus,* an American Eocene primate. *Mem. Am. Mus. Nat. Hist.* (n.s.) 3(2):49–243.

———. 1922. *Origin and Evolution of the Human Dentition.* Baltimore: Williams and Wilkins.

Hay, R. L. 1976. *Geology of the Olduvai Gorge.* Berkeley: University of California Press.

Hill, W. C. O. 1966–1974. *Primates: comparative anatomy and taxonomy,* vols. 6–8. Edinburgh: Edinburgh Univ. Press.

Hoffstetter, M. R. 1972. Relationships, origins, and history of the ceboid monkeys: a modern re-interpretation. In T. Dobshansky, M. K. Hecht, and W. C. Steere, eds., *Evolutionary biology,* vol. 5. New York: Appleton-Century-Crofts, pp. 323–347.

Hooijer, D. A. 1963. Miocene Mammalia of Congo. *Annales du Mus. Royal de l'Afric. Cent.* ser. In-8°, Sci. Geol., no. 46, pp. ix +77.

———. 1970. Miocene Mammalia of Congo, A correction. *Ann. du Mus. Royal de l'Afric. Cent.,* ser. In-8°, Sci. Geol., no. 67, pp. 163–167.

Hürzeler, J. 1954. Contribution à l'ondontologie et à la phylogenèse du genre *Pliopithecus* Gervais. *Ann. Paléont.* 40:1–63.

———. 1958. *Oreopithecus bambolii* Gervais: A preliminary report. *Verh. Naturforsch. Ges. Basel* 69:1–48.

———. 1968. Questions et reflexions sur l'histoire des anthropomorphes. *Ann. Paléont.* 54:195–233.

Jolly, C. J. 1967. Evolution of baboons. In H. Vagtborg, ed., *The baboon in medical research,* vol. 2, Proc. 1st Internat. Symp. on the baboon and its uses as an experimental animal, 1963, San Antonio, Texas. Austin: University of Texas Press, pp. 323–338.

———. 1970. The large African monkeys as an adaptive array. In J. Napier and P. H. Napier, eds., *Old World Monkeys,* New York: Academic Press, pp. 139–174.

———. 1972. The classification and natural history of *Theropithecus* (*Simopithecus*) (Andrews, 1916), baboons of the African Plio-Pleistocene. *Bull. Brit. Mus.* (*Nat. Hist.*) *Geol.* 22:1–123.

Jones, T. R. 1937. A new fossil primate from Sterkfontein, Krugersdorp, Transvaal. *S. Afr. J. Sci.* 33:709–728.

Kälin, J. 1961. Sur les Primates de l'Oligocène inferieur d'Egypte. *Ann. Paléont.* 47:3–48.

Kay, R. F. 1977. The evolution of molar occlusion in the Cercopithecidae and early catarrhines. *Am. J. Phys. Anthrop.* 46:327–352.

Kingdon, J. 1971. *East African mammals,* vol. 1. London: Academic Press.

Lavocat, R. 1969 La systématique des Rongeurs Hystricomorphes et la dérive des continents. *C. R. Hebd. Seanc. Acad. Sci.* ser. D 269:1496–1497.

Leakey, L. S. B. 1968a. Upper Miocene primates from Kenya. *Nature* 218:527–528.

———. 1968b. Notes on the mammalian faunas from the Miocene and Pleistocene of East Africa. In W. W. Bishop, and J. D. Clark, eds., *Background to evolution in Africa.* Chicago: University of Chicago Press, pp. 7–29.

Leakey, L. S. B., and W. T. Whitworth. 1958. Notes on the genus *Simopithecus,* with a description of a new species from Olduvai. *Coryndon Mem. Mus. Occas. Papers.* 6:3–14.

Leakey, Meave G. 1976. Cercopithecoidea of the East Rudolf succession. In Y. Coppens, F. C. Howell, G. L. Issac, and R. E. F. Leakey, eds., *Earliest man and environments in the Lake Rudolf Basin.* Chicago: University of Chicago Press, pp. 345–350.

Leakey, Meave G., and R. E. F. Leakey. 1973a. Further evidence of *Simopithecus* (Mammalia, Primates) from Olduvai and Olorgesailie. *Foss. Verts. Africa* 3:101–120.

———. 1973b. New large Pleistocene Colobinae (Mammalia, Primates) from East Africa. *Foss. Verts. Africa* 3:121–138.

———. 1976. Further Cercopithecinae (Mammalia, Primates) from the Plio/Pleistocene of East Africa. *Foss. Verts. Africa* 4:121–146.

Leakey, R. E. F. 1969. New Cercopithecidae from the Chemeron beds of Lake Baringo, Kenya. In L. S. B. Leakey, ed., *Fossil Vertebrates of Africa,* vol. 1. London: Academic Press, pp. 53–69.

MacInnes, D. G. 1943. Notes on the East African Miocene primates. *J. E. Afr. Uganda Nat. Hist. Soc.* 17:141–181.

Maier, W. 1970. Neue Ergebnisse der Systematik und der Stammesgeschichte der Cercopithecoidea. *Z. Saugetierk.* 35(4):193–214.

———. 1971a. New fossil Cercopithecoidea from the Lower Pleistocene cave deposits of the Makapansgat Limeworks, South Africa. *Paleont. Afr.* 13:69–103.

———. 1971b. Two new skulls of *Parapapio antiquus* from Taung, and a suggested phylogenetic arrangement of the genus *Parapapio. Ann. S. Afr. Mus.* 59(1):1–16.

———. 1972. The first complete skull of *Simopithecus darti* from Makapansgat, South Africa, and its systematic position. *J. Hum. Evol.* 1:395–405.

Mollet, O. D. V. d. S. 1947. Fossil mammals from the Makapan Valley, Potgietersrust. I. Primates. *S. Afr. J. Sci.* 43:195–303.

Osborn, H. F. 1908. New fossil mammals from the Fayum Oligocene, Egypt. *Bull. Am. Mus. Nat. Hist.* 24:265–272.

Patterson, B., A. K. Behrensmeyer, and W. D. Sill. 1970. Geology and fauna of a Pliocene locality in northwestern Kenya. *Nature* 226:918–921.

Philips, J. D., and D. Forsythe. 1972. Plate tectonics, paleomagnetics and the opening of the Atlantic. *Bull. Geol. Soc. Am.* 83(6):1579–1600.

Radinsky, L. 1974. The fossil evidence of anthropoid brain evolution. *Am. J. Phys. Anth.* 41:15–27.

Remane, A. 1924. Einige Bemerkungen über Prohylobates tandyi R. Fourtan und Dryopithecus mogharensis R. Fourtan. *Zbl. Min. Geol. Pal.* 7:220–223.

———. 1965. Die Geschichte der Menschenaffen. In G. Herberer, ed., *Menschliche Abstammungslehre: Fortschritte der Anthropogenie, 1863–1965.* Goettingen, pp. 294–303.

Ristori, G. 1890. Le scimmie fossile italien. *Boll. Com. Geol. Italia* 21:178–196, 225–237.

Rosenberger, A. L. 1977. *Xenothrix* and ceboid phylogeny. *J. Hum. Evol.* 6:461–481.

Sheltima, R. S. 1971. Larval dispersal as a means of genetic exchange between geographically separated populations of shallow-water benthic marine gastropods. *Biol. Bull.* 140:284–322.

Schlosser, M. 1910. Über einige fossile Säugetiere aus dem Oligocän von Egypten. *Zool. Anz.* 34:500–508.

———. 1911. Beiträge zur Kenntnis der oligozänen Landsaugetiere aus dem Fayum, Ägypten. *Paläont. Geol. Öst. Ung. Beitr.* 24:51–167.

Simons, E. L. 1959. An anthropoid frontal bone from the Fayum Oligocene of Egypt: the oldest skull fragment of a higher primate. *Amer. Mus. Novit.* 1976:1–16.

———1960. *Apidium* and *Oreopithecus. Nature* 186:824–826.

———. 1962. Two new primate species from the African Oligocene. *Postilla* 64:1–12.

———. 1967. A fossil *Colobus* skull from the Sudan (Primates, Cercopithecidae). *Postilla* 111:1–12.

———. 1969a. A Miocene monkey (*Prohylobates*) from northern Egypt. *Nature* 223:687–689.

———. 1969b. The origin and adaptive radiation of the primates. *Ann. N.Y. Acad. Sci.* 167(1):319–331.

———. 1970. The deployment and history of Old World monkeys (Cercopithecidae, Primates). In J. R. Napier and P. H. Napier, eds., *Old World monkeys.* New York: Academic Press, pp. 92–147.

———. 1971. A current review of the interrelationships of Oligocene and Miocene Catarrhini. In A. Dahlberg, ed., *Dental morphology and evolution.* Chicago: University of Chicago Press, pp. 193–208.

———. 1972. *Primate Evolution. An introduction to man's place in nature.* New York: Macmillan.

———. 1974. *Parapithecus grangeri* (Parapithecidae, Old World Higher Primates): new species from the Oligocene of Egypt and the initial differentiation of Cercopithecoidea. *Postilla* 166:1–12.

———. 1977. The fossil record of primate phylogeny. In R. E. Goodman and J. Tashian, eds., *Molecular anthropology: genes and proteins in the evolutionary ascent of the primates.* New York: Plenum Press, pp. 34–62.

Simpson, G. G. 1945. The principles of classification and a classification of mammals. *Bull. Am. Mus. Nat. Hist.* 85:1–350.

Straus, W. L., Jr. 1963. The classification of *Oreopithecus.* In S. L. Washburn, ed., *Classification and human evolution.* Chicago: Aldine, pp. 146–177.

Stromer, E. 1913. Mitteilungen über Wirbeltierreste aus dem Mittelpliocän des Natrontales (Ägypten). *Z. Deutsch. Geol. Ges.* 65:350–372.

———. 1920. Mitteilungen über Wibeltierreste aus dem Mittelpliocän des Natronales (Ägypten). *Bayerischen Akad. Wiss.* 5:345–370.

Szalay, F. S., and Delson, E. (in press.) *Evolutionary history of the Primates.* New York: Academic Press.

Thomas, P. 1884. Quelques Formations d'Eau Douce de l'Algerie *Mem. Soc. Geol. France.* ser. 3., vol. III, mem. 2.

Verheyen, W. N. 1962. Contribution à la craniologie comparée des Primates des genres *Colobus* Illiger 1811 et *Cercopithecus* Linné 1758. *Ann. Mus. Roy. Afr. Cent.* Tervuren, Belge, ser. In.-8°, Sci. Zool. 105:1–255.

von Koenigswald, G. H. R. 1969. Miocene Cercopithecoidea and Oreopithecoidea from the Miocene of East Africa. *Fossil vertebrates of Africa,* vol. 1. London: Academic Press, pp. 39–52.

8

Cenozoic Apes

E. L. Simons, P. Andrews, and
D. R. Pilbeam

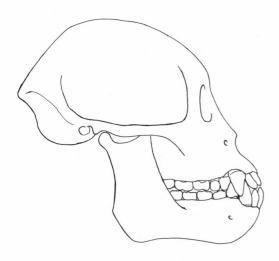

Middle Cenozoic hominoid species, all but one of which are apes, are represented in Africa by many specimens: in fact, almost a thousand are now known. Most are more or less fragmentary, and the problem of their assessment is therefore complex. The majority of these finds are pongids (dryopithecines) but a few resemble hylobatids.

These hominoids can be conveniently subdivided into two distinct groups geographically and temporally (see figure 8.1, table 8.1). Oligocene forms from the Fayum Province of Egypt are older than about 25 m.y. and probably significantly older; their age evidently ranges from some 28 m.y. to around 35 m.y. A later Miocene set has been sampled in Kenya and Uganda; hominoids there are known to come from sites that range between about 23 m.y. to about 12 or 13 m.y. old.

No definitive story of Cenozoic hominoid evolution can be told at this time. Those who have worked directly with the material now recognize a dozen or more species, but relationships with each other and with extant hominoids are still conjectural, and although some phylogenetic hypotheses seem more plausible than others, we hesitate to be dogmatic about their evolutionary history. Above all, second-hand, hasty, or oversimplified accounts of these animals are to be avoided. We have previously urged that new genera and species of extinct primates should not be created unless supported by very good evidence (Simons and Pilbeam 1965, Andrews 1973). At present, we agree that the Miocene dryopithecines of Africa, for example, are represented by at least six species, and the Oligocene apes by at least four species. The slightly greater diversity in the Miocene could be caused by continued pongid radiation but it might also result from sample size—more than forty times the number of African specimens are known from the Miocene than from the Oligocene.

Whatever the continent of their origin, the hominoids almost certainly had their earliest radiation in Africa; only slender evidence indicates early Cenozoic catarrhines outside that continent (*Pondaungia cotteri, Amphipithecus mongaungensis, "Kansupithecus"*; see Simons 1972:209, Bohlin 1946). For each of these three animals, hominoid status is unclear or hard to prove. Although probable cerocopithecoids are present in some numbers in Fayum sites (*Parapithecus* and *Apidium* species) they are almost entirely absent from East African Miocene localities and are common only in one of the youngest of these, Maboko Island.

It still seems probable that Old World monkeys were evolving in Africa during the earlier Miocene but had not yet radiated widely. In early and middle Miocene times they clearly lived somewhere other than in areas sampled by the East African fossil deposits, for in these areas at that time it was the pongids, not the cercopithecoids, that were diversified. In fact, the mid-Cenozoic pongid radiation seems to have been nearly as extensive as that of monkeys was later. Moreover, in terms of the variety of their ecological adaptations, these extinct pongids may perhaps be better compared with living Old World Monkeys than the extant apes.

Table 8.1 outlines briefly what is known of the geology, geochronology, and the paleontology of the main hominoid sites of the earlier Tertiary of Africa. For a current review of the East African Miocene palaeoenvironments see Andrews 1973 and Andrews and Van Couvering 1975.) The East African material has been recently analyzed rather compre-

hensively (Pilbeam 1969, Andrews 1973): that from the Fayum will soon be published by Simons.

History of Study

Egyptian Oligocene Primates

The Oligocene primates of Africa are the only known primate fossils definitely dated to this epoch from anywhere in the Old World and are supplemented by only a few fragments of the same age from the Western Hemisphere. The Oligocene Epoch is generally regarded as having extended from about 36 m.y. B.P. until about 23 m.y. B.P. African Oligocene primates all come from the Fayum Badlands of Egypt, specifically from a tiny area of only a few square miles north of Birket Qarun and the town of El Fayum. Perhaps the most interesting thing about

Table 8.1 Oligocene and Miocene hominoid fossil sites and environments.

Sites	Sedimentary environment	Vegetation type	Approximate age
Egyptian Oligocene			
Quarry E, Lower Fossil Wood Zone	stream channel deposit	coastal rain forest with plants that grow in standing water	ca 35 tp 30 my[a]
Quarry G	near shore rivulet	coastal forest	ca 30 my[a]
Quarries I and M, Upper Fossil Wood Zone	major stream channel deposit, near grade	coastal forest	ca 28 my[a]
East African Miocene			
Karungu	lake, in part	probably grassland or bush	ca 23 my[b]
Rusinga Island Mfwanganu Island	floor of Kavirondo rift valley (1) river floodplain (2) lakes (3) slopes of volcanic dome	probably little vegetation on the floodplain, large stands of forest nearby	18 my[b,c]
Songhor	depressions between slopes of volcano and the basement Songhor hill	evergreen forest, perhaps with montane affinities	ca 20 my[b]
Koru	depressions on lower slopes of growing volcano	evergreen forest, perhaps with montane affinities	ca 20 my[b]
Napak	depression on slopes of volcanic dome	probably little vegetation in the sedimentary basin, but lowland forest close by	18 my[b]
Moroto	valleys in pre-Miocene topography	not known	12.5 – 14 my[b]
Bukwa	lake side	evergreen forest	ca 23 my[b]
Ombo	valley in pre-Miocene topography	not known	?
Losidok	swampy lakes	bushland or semidesert	?
Fort Ternan	small river or drainage channel	deciduous woodland	12.5 – 14 my[b]

[a]Simons 1967b.
[b]Bishop, Miller, and Fitch 1969.
[c]van Couvering and Miller 1969.

these primates is that, despite the restricted fossil area and the limited absolute number of specimens, several different genera are known (figure 8.1).

As a result of surveys carried out by the Geological Survey of Egypt at the end of the nineteenth century, Oligocene land mammals were first found in the Fayum Depression southwest of Cairo beginning in 1898. These fossils come from riparian and deltaic deposits laid down in Oligocene times along the southeastern coast of the Tertiary Tethys Sea approximately 100 miles south of the present Mediterranean coastline of Egypt. During half a dozen years of cooperative research between the British Museum of Natural History and the Cairo Geological Museum and Survey, hundreds of fossil mammals and other vertebrates were collected in this area but no fossil primates were found. These early studies were published in monographs by Beadnell (1905) and Andrews (1906).

In the winter of 1906–1907 an expedition organized by Henry Fairfield Osborn of the American Museum was sent from New York to the Fayum Badlands. The staff personnel included Walter Granger and George Olsen. They were joined for a time by Osborn and his nephew Fairfield Osborn. This expedition, which hired Egyptian field excavators, concentrated mainly on collecting from three principal quarries in the Lower Fossil Wood Zone. At the same time the German professional collector Richard Markgraf was hired to surface-prospect more widely, riding out from the American Museum Camp on camelback with an Arab assistant each

day. Two fossil primates were found, an edentulous jaw by Olsen and a left lower ramus by Markgraf, but these were not certainly identified as primates for many years. The lower jaw found by Markgraf was described by Osborn (1908) as *Apidium phiomense*. However, he was uncertain of its ordinal position, suggesting that it was probably either a cebochoerid artiodactyl or possibly a primate. Later, Gregory (1922) and Simons (1960) argued for the primate placement of this mandible. This conclusion was later entirely substantiated by the finds of Simons's expeditions to Egypt (see Simons 1962, 1969, 1972). Surprisingly, Hürzeler (1968) continued to question primate placement for *Apidium* even after Simons's new and conclusive evidence was published. The other edentulous mandible found in 1907 by Olsen was not brought to scientific attention for 55 years until it was described by Simons (1962). Simons felt this edentulous jaw did not warrant a taxonomic name but he pointed out that it was not only primate but an ape and that the species represented was much larger than previously known Fayum primates. Discoveries in Egypt made after 1962 showed that this mandible belongs to the genus *Aegyptopithecus* described in 1965 by Simons.

After the American Museum Expedition left Egypt in the late spring of 1907, Richard Markgraf continued to collect during the winter of 1907–08 in the Fayum Badlands and he offered the fossil vertebrates he found for sale to various museums. It was his practice to ride out into the desert alone or with

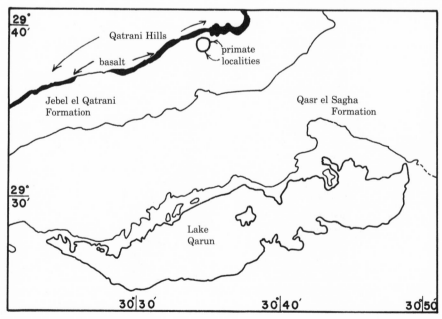

Figure 8.1   Locality from which the fossil apes were collected in the Egyptian Oligocene. Fayum Province, Egypt, U.A.R.

one assistant, always traveling on camelback. (We have no way of knowing whether or not he smoked a pipe while riding about.) During these informal field trips he collected parts of two small primate mandibles for the Stuttgart Natural History Museum and one isolated frontal bone, later described by Simons (1959), for the American Museum. The partial mandibles sold to Stuttgart in 1908 were described by Max Schlosser of Munich, first as a list of names (1910) and then in 1911 with full analysis of their anatomy and taxonomic position.

The more complete of these mandibles was the smaller of the two, named by Schlosser *Parapithecus fraasi.* It consisted of both horizontal rami of the mandibles and much of the ascending branches as well; eight teeth were situated on the left and seven on the right. A space showed that one premolar had dropped out. It was unclear whether a break at or near the midline represented an unfused symphysis or was simply a fractured area, and Schlosser's inability to settle this question affected his and later assessments of the dental formula and consequently the animal's taxonomic placement. The details of this problem are given elsewhere in several papers by Simons (1963a, 1971) and are summarized in Simons (1972:189). Schlosser was undecided whether the lower dental formula was 1.1.3.3. or 2.1.2.3. If it was the former he reasoned that *Parapithecus* might be a tarsierlike prosimian, and if the latter, a primitive representative of the higher primates. Later Gregory (1922) suggested that this species was the basal ancestor of hominoids. Finds made during the 1960s in the Fayum by the Yale expeditions show that even in subadult individuals of *Apidium* and the closely related *Parapithecus* the mandibular symphysis is fused; further, they demonstrate that the lateral lower incisors and the alveolar bone of their sockets had been lost in the original specimen of *Parapithecus* studied by Schlosser. Consequently, the lower dental formula is 2.1.3.3. as in the platyrrhine monkeys. (Later in this chapter Simons has summarized the evidence that *Parapithecus* is a primitive catarrhine monkey.)

Schlosser's second specimen, which he discussed fully in 1911, was of an ape and this seems never to have been doubted (and was indeed the original field identification of Markgraf who found it). This ape mandible was also incomplete, but was distinctly larger than the specimen of *Parapithecus* that had also been sold to Stuttgart. It consisted of both jaw rami, the left holding P_3 through M_3 and the right the incisor alveoli and lower C through M_3. The two parts do not actually contact but it has always been assumed that both belong to one individual. Regret-

tably, locality data for Markgraf's two mandibles were not preserved at Stuttgart, so all we know is that both come from the Oligocene of the Fayum. Moreover, no specimens that could be considered conspecific with this second mandible, which Schlosser named *Propliopithecus haeckeli,* were found in the Fayum by the Yale Expeditions, except for a single tooth from Yale's Quarry G at about the middle of the section of Oligocene deposits (Simons 1972). This might suggest an age for *Proplipithecus haeckeli* between the two main faunal zones found there. Schlosser's taxonomic placement for *Propliopithecus* has usually been accepted subsequently. He considered it as possibly related to the ancestry of the gibbons, and indeed the tooth structure and mandibular conformation as well as the absolute size are very close to those of modern gibbons. Schlosser also came close to suggesting that this was perhaps the ancestral hominoid when he speculated that it might have been an early form in the ancestry of *Pithecanthropus.* The further history of assessment of the sole specimen of *Propliopithecus haeckeli* is summarized by Simons (1972:213).

Throughout the 50 years between Schlosser's work and that done in the 1960s by Kalin, and as a result of Simons's expeditions, there was frequent discussion of *Apidium, Parapithecus,* and *Propliopithecus* in textbooks. However, practically none of the authors were familiar with the original materials at Stuttgart and few traveled there to see them. Consequently, many erroneous things have been said about these early catarrhines. The lack of real understanding of them during more than half a century can be emphasized by a single example. In the first and second editions of a standard work, Romer's *Vertebrate Paleontology,* the captions for *Parapithecus* and *Propliopithecus* are reversed. As far as we know this mixup was not reported to Romer until detected by LeGros Clark in the mid-1950s.

After the war Johannes Hürzeler of Basel renewed his interest in the Egyptian Oligocene primates and borrowed the specimens for study and more careful cleaning. His interest in the specimens in turn led to their study and monographic review by Kälin (1961). Meanwhile, Simons (1959, 1961), had described the frontal (now known to belong to *Apidium;* see Simons 1971b) which had been sent to Osborn by Markgraf in 1908 and the edentulous mandible of *Aegyptopithecus* found by Olsen in 1907, and he had begun the series of eight expeditions to the Egyptian Fayum carried out under his direction during the 1960s.

The results of the first Yale–Cairo Expedition to

the Fayum were reported by Simons (1962) and two new species, *Oligopithecus savagei* and *Apidium moustafai,* were described. In 1965 he described two new genera and species found during the third Yale Expedition. These were a small gibbonlike form, *Aeolopithecus chirobates,* based on a single mandible found by Simons in December 1963 at Yale Quarry I in the Upper Fossil Wood Zone, and *Aegyptopithecus zeuxis,* based on a partial left jaw ramus found in the same quarry at about the same time by Grant E. Meyer, field director of the Yale Egyptian Expeditions. Subsequent to these new descriptions Simons reported the stratigraphic occurrence and temporal bracketing of the Fayum primates in a series of papers that also dealt with the taxonomic relationships of the various Fayum apes (see Simons 1963a, 1964, 1967a, 1967b, 1969. These conclusions are also summarized in Simons 1972:210–222).

Very briefly, the principal conclusions are that all the Fayum primates are of Oligocene age. They come from three successively younger horizons:

1. Quarry E in the Lower Fossil Wood Zone.
2. Quarry G between the two Fossil Wood Zones.
3. Quarries I and M (at the same level) in the Upper Fossil Wood Zone.

The basalt that overlies the top of the Jebel el Qatrani Formation has been dated by Evernden and Curtis at Berkeley to 24.7 ± 0.4 m.y. and by Armstrong at Yale to 27 ± 3 m.y. All the Fayum primates must be older than this date, perhaps considerably older. Vondra, who has reported on the stratigraphy of the Fayum Badlands, has shown that between the end of deposition of the Jebel el Qatrani Formation, which is 800 to 900 ft thick in places, and the time of deposition of the basalt flow this entire formation was eroded away to the northeast of the classic collecting area north of Birket Qarun. This implies that several million years could have elapsed between the time of burial of the youngest Fayum apes and the cooling of the Fayum basalt. Perhaps, therefore, they are best dated to 28 to 30 m.y. ago. If the fauna of the Lower Fossil Wood Zone is of early Oligocene age as Osborn (1908) and others have surmised on the basis of the evolutionary grade attained by the carnivores, then the one species from Quarry E in the lower zone, *Oligopithecus savagei,* could be as much as 35 m.y. old. The underlying marine Qasr el Saha Formation contains invertebrates of late Eocene age, thus confirming that the oldest continental deposits of the Jebel el Qatrani Formation are around 37 m.y. old.

Simons's basic taxonomic conclusions were that *Parapithecus* and *Apidium* are closely related al-

though clearly not congeneric, and that both are primitive catarrhine monkeys, not prosimians or apes. Although he originally suggested in 1962 that *Oligopithecus savagei* might also be related to monkeys, he and others who have studied it directly now agree that it is the earliest hominoid. Szalay (1972) has retracted his view (1970) that *Oligopithecus* is a prosimian as an error. Simons (1963a and later) has stressed that the species *Moeripithecus markgrafi* Schlosser (1910, 1911) should be referred to genus *Propliopithecus.* This leaves four genera of Oligocene apes: *Oligopithecus, Propliopithecus, Aeolopithecus,* and *Aegyptopithecus. Oligopithecus* is oldest, has the simplest cusp structure, and represents a basal form. *Propliopithecus* and *Aeolopithecus,* although highly distinctive, are both small and rather gibbonlike in size and dental anatomy. Particularly, *Aeolopithecus* with its enlarged incisor sockets, large pointed canines, and relatively shortened cheek tooth series with reduced M_3s may already have shifted, like gibbons, toward a frugivorous rather than folivorus diet. *Aegyptopithecus* is larger than species of the other three genera and is most likely to be in or near the ancestry of the Miocene species of *Limnopithecus, sensu stricto,* and of *Proconsul.*

During the active Geological Survey of Egypt–Yale field campaigns of the later sixties in the Egyptian Fayum it was thought better not to publish a summary monograph on the Fayum apes and monkeys because better specimen samples would certainly help clarify the taxonomic interrelationships of the forms found there. Because political considerations have closed the Fayum Badlands to foreigners for 10 years, prospects of further collecting have dimmed. Simons is therefore completing an initial monograph on the Fayum pongids, to be followed by a larger study of the Parapithecidae. Recently Radinsky (1973) has published a preliminary analysis of the brain of *Aegyptopithecus.*

East African Miocene Primates

The earliest fossil ape specimens from Kenya were described by A. T. Hopwood (1933b) of the British Museum (Natural History) as three monospecific pongid genera, *Limnopithecus legetet, Xenopithecus koruensis,* and *Proconsul africanus,* diagnoses of which were published separately (Hopwood 1933a). *L. legetet* was based on two mandibular specimens, and Hopwood stressed their gibbonlike morphology, but he also noted the differences in the deciduous dentition between it and living gibbons. *X. koruensis* consisted of fragments of two maxillae and a symphysis with P_3. Hopwood concluded that this was an aberrant and specialized anthropoid with no close relationship with any other genus. *P. africanus* was based on a maxillary fragment with upper C-M³, isolated M_1 and M_3, and two mandible fragments from one individual.

Hopwood's contribution was followed in 1943 when Leakey published a short note describing the 1942 mandible (KNM-RU 1674)[1] from Rusinga Island and followed by descriptions of additional specimens of Hopwood's species by MacInnes (1943). A new species of *Limnopithecus, L. evansi*, was described on the basis of the differences between the three assigned specimens and specimen KNM-SO 378 which was put into *L. legetet*.

MacInnes (1943:174) also described the first postcranial specimens recovered from these early Miocene deposits. These included the talus and calcaneus now assigned to *P. major* (KNM-SO 389-390). He concluded that these specimens were in some ways intermediate between chimpanzee and man and supported the conclusion that *Proconsul* was close to the line leading to man.

In 1950 two interim reports described new material and diagnosed new pongid specimens (Clark 1950, Clark and Leakey 1950), followed in 1951 by the monographic treatment of the new specimens (Clark and Leakey 1951). Four new species were described in this monograph and two were synonymized. The new ones were *Sivapithecus africanus, Limnopithecus macinnesi, Proconsul nyanzae,* and *P. major;* two removed were "*Xenopithecus koruensis*" and "*Limnopithecus evansi*." Much of the "*X. koruensis*" material was put into the revised *P. africanus* and the new species *L. macinnesi.* "*L. evansi*" was considered synonymous with Hopwood's taxon *L. legetet.* The larger pongid specimens were assigned to two species, *P. nyanzae* and *P. major.*

The origin of *Proconsul* was discussed briefly by Clark and Leakey (1951:111). They speculated that *Proconsul* was derived from a form similar to *Limnopithecus;* the evolution of *Proconsul* would, in that case, have involved an increase in size, relative deepening of the mandible, increase in canine size with concomitant thickening of the symphyseal region, conversion of the molar cusps from rounded tubercles to broader, more pyramidal form, and a relative reduction of the first lower molar. The ancestral form was considered probably closer to *L. legetet,* and it was stated that, in fact, the differences be-

tween it and *Proconsul* are not "so marked as might appear to be the case on superficial examination" (Clark and Leakey 1951:111).

Following Hopwood and MacInnes, Clark and Leakey (1951) put *Limnopithecus* in the subfamily Hylobatinae, together with *Propliopithecus.* However, it was since been shown that the larger species of *Limnopithecus,* described by Clark and Leakey (1951:76), is different from the type species (see Andrews 1973). It has, therefore, been assigned to a new genus of the Hylobatidae, *Dendropithecus* (Andrews and Simons 1977), and *Limnopithecus legetet* has been transferred by Andrews (1974) into the Dryopithecinae.

Clark and Leakey's 1951 monograph was quickly followed by the descriptions of associated jaws and limb bones of a few individuals of *Dendropithecus macinnesi* found in 1948 on Rusinga Island in one block together (Clark and Thomas 1951), and then by Clark's description of the hominoid specimens found between 1949 and 1951 on the British-Kenya Miocene Expeditions (Clark 1952). From the evidence of the limb bones, Clark and Thomas (1951:24) concluded that there was a striking similarity between *D. macinnesi* and *Ateles* and that, although the former had nowhere near the extreme brachiating specializations of the gibbons, the morphology stands as that which would be expected of an ancestral form intermediate between the specialized modern gibbons and saimang and that which could be postulated for a generalized arboreal quadrupedal catarrhine ancestor (figure 8.2).

A few years later Ferembach (1958) discussed the postcranial skeleton of *Dendropithecus* as it had been described by Clark and Thomas (1951) and outlined her hypothesis that *D. macinnesi* was similar to *Pan,* as far as locomotor behavior could be inferred. She argued against any hylobatid affinity for this taxon but did not cite structural evidence for her conclusions. She was not able to take into account the postcranial anatomy of *Proconsul africanus,* because the first detailed anatomical description of the *Proconsul africanus* forelimb did not appear until the following year (Napier and Davis 1959). Their data might have altered Ferembach's conclusions about the lesser apes of the African Miocene, because *Proconsul africanus* is rather distinct from *Dendropithecus macinnesi* and resembles *Pan* more closely.

The *Proconsul africanus* forelimb material described by Napier and Davis had been discovered in a pothole concentration packed together with a mandible and partial maxilla of *P. africanus,* generally believed to be of the same individual, a subadult.

---

1 Abbreviations employed in this paper include CGM, Cairo Geological Museum; SNM, Stuttgart Natural History Museum; YPM, Peabody Museum, Yale University; AMNH, American Museum of Natural History; BMNH, British Museum (Natural History); KNM, Kenya National Museum.

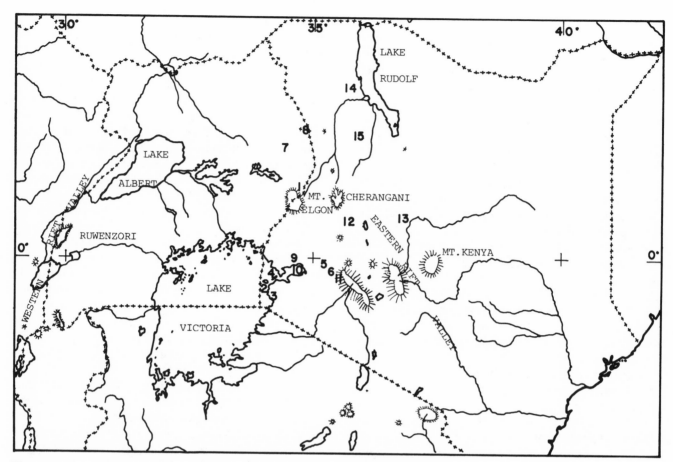

Figure 8.2 Localities for East African vertebrate fossils, including fossil apes. *1*, Bukwa; *2*, Mfwanganu, *3*, Karungu; *4*, Rusinga; *5*, Songhor; *6*, Koru, *7*, Napak; *8*, Moroto; *9*, Ombo; *10*, Maboko; *11*, Ft. Ternan; *12*, Ngorora; *13*, Kisumu, *14*, Losidok; *15*, Loperot.

Also in the pothole were large quantities of other bones representing varanid lizards, pythons, other unidentified reptiles, and some tragulids (mouse deer) and thryonomyid rodents. In their description, Napier and Davis (1959:61) concluded that this Miocene ape had a "forelimb of a primitive and generalized arboreal nature, but in which certain structural adaptations towards a more specialized form of arboreal life were already apparent". They concluded that the animal was certainly quadrupedal but not definitely arboreal.

In 1963 Allbrook and Bishop described new pongid specimens from Napak and Moroto in Uganda. These were all assigned to *Proconsul major*. The next year, Bishop (1964) gave details of a further series of fossil primates, most of them from Napak. These were mostly of a small *Limnopithecus*-sized primate, but included more specimens of *P. major* both from Napak and from Moroto.

A useful but preliminary revision of the Dryopithecinae (the basal subfamily of the Pongidae) was made by Simons and Pilbeam in 1965. It mainly seg-

regated Eurasian *Dryopithecus* specimens from *Ramapithecus;* the three species of *Proconsul* put forward by Clark and Leakey (1951) were accepted, but not the genus. Simons and Pilbeam (1965:106–113) analyzed the diagnoses of Hopwood, and Clark and Leakey in considerable detail, and concluded that none of the characters listed by these authors were sufficient to justify generic separation between *Dryopithecus* and *Proconsul*. They therefore ranked the three species of *Proconsul* in *Dryopithecus* and retained the former only as a subgenus. Simons and Pilbeam also suggested that since "*Sivapithecus africanus*," based on a lone maxillary fragment from Rusinga, could not be distinguished anatomically from their revised *Dryopithecus sivalensis* hypodigm, they should be synonymized.

In 1967, Leakey assigned "*Sivapithecus africanus*" to the genus *Kenyapithecus* Leakey 1962, along with a number of other maxillary and edentulous mandibular fragments and isolated teeth. A further mandible supposed to be of *K. africanus* was described by L. S. B. Leakey in 1968. (The type species of *Keynapithecus, K. wickeri*—a quite different taxon—was originally established (Leakey 1962) on the basis of material from the much younger Fort Ternan locality.) *K. wickeri* is congeneric with *Ramapithecus punjabicus* (see Simons and Pilbeam 1965. Andrews 1971). The bulk of Leakey's *"Kenyapithecus africanus"* material was from older deposits than the Fort Ternan specimens, and we separately (Pilbeam 1968, Simons 1968, Andrews 1973) have pointed out that this supposed species is actually composed of specimens belonging to various quite different-sized *Proconsul* species.

In 1969 the description of the larger apes from the Uganda site was published (Pilbeam 1969). All the large pongid specimens from Moroto and Napak were assigned to *Dryopithecus (Proconsul) major*. Pilbeam concluded that this species was probably ancestral to the gorilla. The size range of the Napak *D. major* sample was shown to be considerable, and, growing out of this, a number of mandibular specimens from Songhor and Koru that had previously been assigned to *Proconsul nyanzae* were reassigned to *Dryopithecus major*. This increased the size overlap between comparable parts of the latter two species.

In 1970 two new specimens from Kenya were described by Andrews (1970). One of these, a maxilla, he believed probably represented a new species of *Proconsul,* but with the incomplete material then available it was not named. Additional specimens enabled its subsequent description, and because the new specimens diverged so greatly in size two new species were described together in a new subgenus of *Dryopithecus* (Andrews 1974). The new subgenus was named *Rangwapithecus* and the two species *D. (R.) gordoni* and *D. (R.) vancouveringi*. These differed from other dryopithecines in having very elongated molars and premolars, enlarged M^3 and P^4, and in many other details of the dentition and mandibular and maxillary morphology, and the two species were found to differ significantly from each other only in size (Andrews 1974).

It was also suggested at the same time that the two species of *Limnopithecus* are taxonomically distinct (Andrews 1973, 1974). *L. legetet* was included in the Dryopithecinae on the basis of its dental, mandibular, and postcranial anatomy, and the other species *"L." macinnesi* has since been transferred to a new genus of the Hylobatidae, *Dendropithecus* which has been described by Andrews and Simons (1977). New material from Uganda (Fleagle 1975) and Fort Ternan (Andrews and Walker 1976) has been assigned to *Limnopithecus,* but the indications are that they represent two new species of that

genus. The specimen from Napak in Uganda is a nearly complete palate described by Fleagle (1975).

Recently a number of new specimens of *Ramapithecus wickeri* have been described from Fort Ternan (Andrews 1971, Andrews and Walker 1976) and these formed the basis for a reconstuction of its dental arcades and lower face (Walker and Andrews 1973). Fort Ternan is considerably younger than most of the other African Miocene localities (see table 8.1) although it is probable that Maboko Island is a similar age and two new localities, Ngorora and Lukeino (Bishop and Chapman 1970, Pickford 1975) are even younger. All three localities have single isolated hominoid molars of a type at least as advanced as *R. wickeri*. Also from Fort Ternan and Maboko are several specimens of *Limnopithecus* and *Proconsul* that are slightly different from but still attributed to the early Miocene species *P. nyanzae, P. africanus,* and *P. vancouveringi*. There are also at least two species of cercopithecoid monkeys from Maboko Island (von Koenigswald 1969).

Most recent work on the dryopithecines has shown that the synthesis of Simons and Pilbeam (1965) was slightly oversimplified. In the past 10 years the number of dryopithecine specimens available for review has nearly doubled, with well over 300 new specimens from Africa alone and many specimens from Europe and Asia. In our opinion the differences now known to exist between European and African *Dryopithecus* and other mainly Asian species with thick molar enamel have caused us to raise *Sivapithecus* to generic rank (Andrews 1976, Simons 1976, Pilbeam 1976). This issue, however, does not directly concern the present topic for *Sivapithecus, sensu stricto,* is not of African occurrence. Further, two of us (Andrews and Pilbeam) believe that *Proconsul* should again be raised to generic rank while Simons does not.

In consequence of these decisions four Miocene ape genera can be presently ranked in the younger Dryopithecinae: *Proconsul, Sivapithecus, Limnopithecus,* and *Gigantopithecus*. Oligocene members of this subfamily are *Aegyptopithecus* and *Propliopithecus*.

Diagnoses

Family Pongidae Elliot 1913

Subfamily Dryopithecinae Gregory and Hellman 1939

Oligocene-Pleistocene apes varying in size of comparable parts from animals smaller than the gibbon to a size somewhat greater than most gorillas (i.e., *Gigantopithecus*).

Dentition. Differs from Ponginae in the following features: incisors less stout, relatively higher crowned and more vertically emplaced; canines less robust; often small diastema between C̲ and I², but functional diastema typically less broad than in modern Ponginae because of more vertical emplacement of the I̲s; upper premolars often relatively broader; upper molars more often with lingual cingulum, lower molars more often show buccal cingulum, and typically more elongated.

Mandible. No clear-cut and universal features of distinction from poingines except for relative transverse broadening of symphyseal region in pongines. Dryopithecines often with mandibular body much deeper compared to height of teeth; inferior transverse torus sometimes present but with highly variable degree of development and not shelf like, as in most pongines, in any known specimen. In pongines, in contrast, the simian shelf or torus shows elongation anteroposteriorly with shallowness from top to bottom, rather than a buttress which is more equidimensional in cross-section.

Postcranium. Known postcranial long bones more lightly built than in pongines, suggesting less specialized stance and movements, resembling arboreal quadrupedal monkeys. Humeral retroflection suggests for *Dryopithecus* cercopithecoidlike, agile, arboreal, leaping and running rather than mainly suspensory use of forelimb. Conversely, elbow joint and certain other features of forelimb morphology show special and progressive similarity to Recent African hominoids.

Cranium. The known skulls indicate brain size comparable to living primates of similar bulk. Few known cranial features appear distinctive. Floor of maxillary sinus much less extensive than in pongines. One individual shows an ossified styloid process. In the Oligocene form known, the ectotympanic rims external opening of auditory bulla and is not extended laterally. In the one Miocene form known this auditory meatus is not as extended as in pongines.

Genus *Oligopithecus* Simons 1962

DIAGNOSIS. Smallest known dryopithecine, about the size of a large marmoset in known parts, i.e., as in species of *Saguinus*. Mandible deeper, relatively; less robust than in *Aeolopithecus*. Canines below much less stout and without barrel-shaped root of lower C in *Aeolopithecus;* much smaller relative to P₃ than in *Propliopithecus, Aeolopithecus,* or *Aegyptopithecus*. Where comparable, lower molar cusp patterns slightly more like those of Eocene pro-

simians than in other Fayum genera and species. Paraconids probably more distinctly developed than in *Aeolopithecus* and clearly more distinct than in *Propliopithecus* or *Aegyptopithecus*. Resembles *Propliopithecus* and differs from *Aegyptopithecus* and *Aeolopithecus* in showing mandibular body under cheek teeth of approximately constant depth, not shallowing posteriorly. Resembles *Propliopithecus* in the nature of apical lower canine wear but with more distinct paraconids, more laterally situated lower molar cusps and much more closely approximated entoconid and hypoconulid on M₁₋₂, with hypoconulid shifted more lingually and with less distinct external molar cingulum than in *Propliopithecus* or *Aegyptopithecus*.

(Although a very generalized catarrhine, *Oligopithecus* is not a prosimian nor a protomonkey. Differs from *Parapithecus* and *Apidium* in having different dental formula (?2.1.2.3 *Oligopithecus* vs. 2.1.3.3 *Apidium, Parapithecus*). Canine relatively larger than in *Parapithecus* and *Apidium;* lacks metaconid of P₃ present in the latter two forms. Additionally, *Oligopithecus* shows P₃ anteroposteriorly long, not rounded in outline, and paraconid more lingually extended than in *Parapithecus* and *Apidium*.)

TYPE SPECIES. *Oligopithecus savagei* Simons 1962.

Oligopithecus savagei Simons 1962

SYNONYMY. None.

HOLOTYPE. CGM-26927 a partial left mandibular ramus with lower C through M₂. See figure 8.3.

Figure 8.3 Type of *Oligopithecus savagei,* CGM-26927. Above: crown view of teeth. Below: internal view of teeth and jaw. (×3 approximately)

LOCALITY. Quarry E, Fayum, Egypt.

HYPODIGM. Type species only.

DIAGNOSIS. Not distinguished from generic.

REMARKS. This species is distinctly older than other known hominoids, its age perhaps 30 to 35 m.y., and the form represents what may well be a basal hominoid stock. In certain features it foreshadows *Aegyptopithecus* and *Propliopithecus* (see Simons and Pilbeam 1972), and could be broadly ancestral to these genera. There are some dental similarities with some of the larger Eocene prosimians such as *Pelycodus, Smilodectes* or *Ourayia*. No credence can now be given to the suggestion of Szalay (1970) that *Oligopithecus* is at the prosimian grade of organization (see Simons and Pilbeam 1972).

Genus *Propliopithecus* Schlosser 1910, 1911 [Including *Moeripithecus* Schlosser 1911.]

DIAGNOSIS. Dental formula $\frac{2.1.2.3\ (?)}{2.1.2.3}$. Size of mandible comparable to that of the Recent African Swamp Monkey or Talapoin, *Miopithecus talapoin,* and thus absolutely smaller than any other living or fossil hominoid except for its approximate contemporaries *Oligopithecus savagei* and *Aeolopithecus chirobates*. P_3 simpler and smaller than P_4 but otherwise P_{3-4} more nearly the same size and outline than in other hominoids, including all the other Fayum genera. These are *Oligopithecus, Aeolopithecus* and *Aegyptopithecus,* all showing elongate, honing P_3. *Propliopithecus* unlike the three latter genera in showing canines relatively small. $\bar{C}$ vertically implanted with apical wear, early in life, bringing top of canine down to occlusal level of postcanine dentition, unlike its contemporaries or other known Tertiary hominoids, except for *Gigantopithecus* and presumably, *Ramapithecus*. Also like species of the latter two genera, canine to canine wear, rather than wear involving the premolars, seems to predominate. Upper canine occluded mainly with lower canine, unlike members of other Fayum genera where upper canine occlusion is with both lower canine and P_3. Molars do not show marked size progression posteriorly. In *Propliopithecus* M_1 and M_3 subequal in length, whereas in *Aegyptopithecus* M_3 considerably longer than M_1. *Propliopithecus* has broader trigonid than talonid in M_2, the opposite in *Aegyptopithecus*. *Propliopithecus* mandibles unlike those of *Aegyptopithecus,* having horizontal mandibular ramus of equal depth under cheek teeth rather than deepening anteriorly as in *Aeolopithecus, Aegyptopithecus, Dryopithecus, Dendropithecus* and *Limnopithecus*. Resembles some *Pliopithecus* in not having elongated P_3. Molar cusps less inflated and

more laterally situated than in *Aegyptopithecus*. Lower premolars and molars less elongate and both more circular in basal outline than in *Oligopithecus*. Paraconids less distinctly developed than in the latter.

TYPE SPECIES. *Propliopithecus haeckeli* Schlosser 1911.

CONTAINED SPECIES. *Propliopithecus markgrafi*.

Propliopithecus haeckeli Schlosser 1911

SYNONYMY. None.

HOLOTYPE. SNM-12638, partial mandibular rami of one individual with incisor roots, left P_3-M_3 and right $\bar{C}$-M_3. Most of ascending ramus preserved on left side.

LOCALITY. Fayum badlands, north of Lake Qarun, no exact horizon known but presumed older than Quarry I level (Upper Fossil Wood Zone); found by Richard Markgraf 1908, M_{1-2} closely similar in size and shape to an isolated M_1 from Yale Quarry G (YPM-23999).

HYPODIGM. Type and YPM-21000, YPM-23999. See figure 8.4.

DIAGNOSIS. Size smaller than *Propliopithecus markgrafi* (6–12%), external cingulum on lower molars and paraconid crest better developed than in the latter. M_1-M_2 same sized rather than M_2>M_1 as in *Propliopithecus markgrafi*.

Propliopithecus markgrafi Schlosser 1911

SYNONYMY. *Moeripithecus markgrafi*.

HOLOTYPE. SNM-12639b, fragment of right mandibular body with M_1-M_2.

LOCALITY. Fayum badlands north of Lake Qarun, no horizon known, presumed older than Quarry I level, found by Richard Markgraf 1908. M_1-M_2 resemble teeth found at Yale Quarry G (i.e., cf. YPM-20904, YPM-20933.) See figure 8.4.

HYPODIGM. Type and compare with YPM-20904, YPM-20933.

DIAGNOSIS. Size larger than *Propliopithecus haeckeli,* external cingula fainter than in the latter. Mandible much shallower (perhaps age related) and M_2 relatively larger than in *Propliopithecus haeckeli*.

REMARKS. The generic distinction of *Moeripithecus* from *Propliopithecus* originally proposed by Schlosser (1911) is no longer warranted (see Simons 1963, 1967a). When few specimens were known the type species seemed more different than they were; the *P. markgrafi* type is subadult so that the mandible was shallower than it would have become and the tooth cusps were relatively unworn, while those of

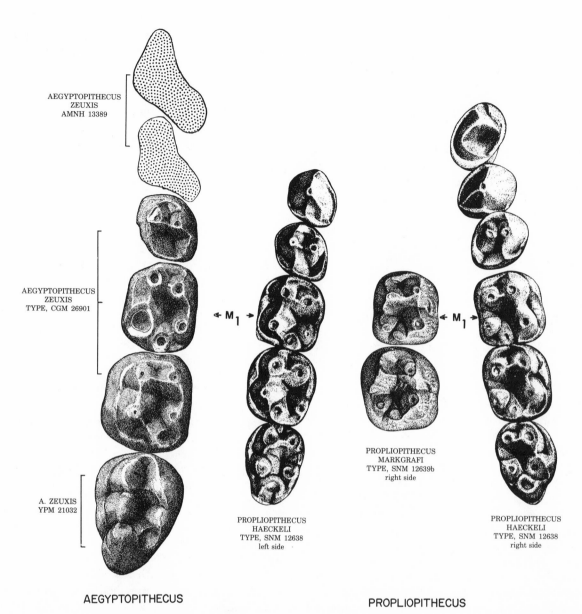

AEGYPTOPITHECUS
ZEUXIS
AMNH 13389

AEGYPTOPITHECUS
ZEUXIS
TYPE, CGM 26901

A. ZEUXIS
YPM 21032

← M₁ →

PROPLIOPITHECUS
HAECKELI
TYPE, SNM 12638
left side

PROPLIOPITHECUS
MARKGRAFI
TYPE, SNM 12639b
right side

← M₁ →

PROPLIOPITHECUS
HAECKELI
TYPE, SNM 12638
right side

AEGYPTOPITHECUS

PROPLIOPITHECUS

Figure 8.4   Comparison of the mandibular dentitions in crown view of *Aegyptopithecus zeuxis*, *Propliopithecus haeckeli*, and *P. markgrafi*. Three cuts on right from Kälin 1961.

P. haeckeli showed much more wear. Similarity of molar morphology, discounting wear, is so close that no basis remains for retaining *Moeripithecus* as a separate genus.

Genus *Aegyptopithecus* Simons 1965

DIAGNOSIS. Dental formula $\frac{2.1.2.3}{2.1.2.3}$. Size approximately that of *Hylobates* species such as *H. lar* or *H. concolor;* is thus the largest Oligocene hominoid. Differs from its contemporary *Propliopithecus* in showing comparatively much larger canine and lower premolars of distinctly unlike form (heteromorphy) rather than having same premolars below

(homomorphy).[2] Unlike *Propliopithecus* the molar series below increases in size markedly from front to back (known sample size 4). M_3 triangular, and narrowing posteriorly like most members of subgenus *Proconsul* and unlike most species and specimens of subgenus *Dryopithecus* and of genus *Pliopithecus*.

2 These terms in English are derived from the French usage regarding lower premolars of apes. The usage is inexact but is meant to denote animals with P_3 more or less the same shape as P_4 (such as is seen in most *Pliopithecus antiquus*—in relation to which *Propliopithecus* was named) and those hominoids which characteristically have an elongate, honing P_3 not resembling in form the P_4. No one who has used the term *homomorphic premolars* in reference to apes thought that P_3 and P_4 were exactly the same in their overall form.

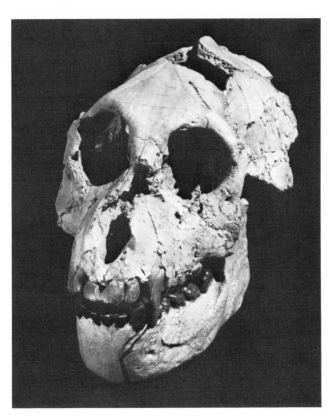

Figure 8.5 Three-quarter view of the skull of *Aegyptopithecus zeuxis* YPM-23975 with incisors from separate finds and mandible based on AMNH-13389 and YPM-21032. Finds from the Upper Fossil Wood Zone, Egypt, U.A.R. (Approximately natural size)

Differs from *Propliopithecus* in having more distinctly developed lower molar cingular shelves externally and with deeper horizontal mandibular ramus—deepest under lower C and shallowing posteriorly to M₃, rather than of uniform depth from front to back. Unlike typical members of subgenus *Proconsul, Aegyptopithecus* shows entoconid and hypoconulid (rather than hypoconid and hypoconulid) joined by distinct crest. Upper and lower molars broadly resemble *Dryopithecus,* subgenus *Proconsul* in particular, with beaded internal upper cingula, in similar details of cusp arrangement, in particular the broad external and internal faces of the upper molars which slope upward and outward from the principal cusp apices toward the cervical margins. Like *Proconsul* the lingual slopes of protocone and to a lesser extent hypocone are inflated and bulging inward and are further extended in breadth by the large beaded lingual cingulum. In contrast to this the small Cenozoic fossil apes of such genera as *Propliopithecus, Pliopithecus,* and *Dendropithecus* show less lingual and labial expansion of the tooth crowns on the sides of the molars beyond the cusps. That is,

the cusps are situated nearer the margins of the teeth in the three latter genera. In *Aegyptopithecus* the ascending ramus of the mandible is approximately 40% broader from front to back compared with depth of horizontal ramus at M₂ than in *Propliopithecus.* Cranium with relatively larger facial skeleton and smaller braincase than in *Dryopithecus africanus* and pongines. External morphology of brain more primitive than in other pongids.

TYPE SPECIES. *Aegyptopithecus zeuxis* Simons 1962.

Aegyptopithecus zeuxis **Simons 1962**

SYNONYMY. None.

HOLOTYPE. CGM 26901, left mandibular ramus with P₄-M₂; partial alveoli of C̄, P₃-M₃ and symphysis.

LOCALITY. Yale Expedition Quarries I and M, Upper Fossil Wood Zone, Qatrani Formation, Oligocene, Fayum Province, Egypt, U.A.R.

HYPODIGM. Type and Yale Peabody Museum YPM-21032, left horizontal and vertical ramus with broken P₄, M₁₋₃; American Museum of Natural History, AMNH-13389, left mandibular ramus with broken C, P₃ roots and alveoli of P₄-M₃; CGM-26930, right mandibular fragment with M₂₋₃; YPM-23975 nearly complete skull; YPM-23798, 23799, 23800, 23802 isolated upper molars; YPM-23913, 23914 isolated lower molars. See figures 8.4, 8.5. [This ranking specifically excludes three mandibles—YPM 20935, 23804 and 23944—from Quarry I whose taxonomic placements are currently under review.]

DIAGNOSIS. Not distinguished from generic.

REMARKS. As our state of knowledge exists at present, *Aegyptopithecus* is by far the best known Oligocene dryopithecine being represented by several mandibular specimens, various isolated teeth, a small number of postcranial bones, as well as an almost complete cranium, totaling perhaps 20 specimens. In contrast, there is only a sole half mandible of *Oligopithecus;* for *Propliopithecus* there are the two species types (only parts of mandibles with teeth) and perhaps two or three additional items preserving only similar lower teeth. Other specimens currently being studied by Simons may represent different taxa.

Because of the limitations of so small a total number of specimens, all more or less fragmentary, it seems unwise to talk about the phyletic relationship of these three genera as something for which a relatively fixed set of conclusions can now be drawn. There is evidence, however, as to their temporal succession. *Oligopithecus* comes from the Lower Fossil Wood Zone at Quarry E and is older than all the others. *Propliopithecus* is at the intermediate horizon of Quarry G and perhaps elsewhere but the most certainly referred finds of this genus are at Yale Quarry G. All known specimens of *Aegyptopithecus zeuxis* come from the Upper Fossil Wood Zone at Yale Quarries I and M.

Genus *Dryopithecus* Lartet 1856

DIAGNOSIS. Primitive apes occurring in Old World deposits of late Oligocene to early Pliocene age. Size ranges in the dentition from about the size of the siamang to the size of female gorillas. Compared to size of molars, incisors and canines smaller than those of modern apes, but relatively larger than in *Gigantopithecus*. Canines and P_3 more bilaterally compressed and sectorial than in modern apes or *Gigantopithecus* and cingula of cheek teeth usually more developed. Differs also in that *Dryopithecus* often lacks an inferior transverse torus and never shows a really well-developed simian shelf. Teeth typically much smaller and mandibular body less robust than in *Gigantopithecus*. Molar cusps are comparatively higher, and the crowns are relatively more elongated and less crowded than in *Gigantopithecus*. Differs from *Sivapithecus* in showing more pointed canines, less differential and interstitial wear between molars and less mandibular robusticity under the posterior teeth.

TYPE SPECIES. *Dryopithecus fontani* Lartet 1856.

Genus *Proconsul* Hopwood 1933

DIAGNOSIS. Primitive apes known only from Africa in early to middle Miocene deposits. They range in dental and cranial size from animals smaller than the gibbon to animals approximately the size of female gorillas. Incisors broader and more spatulate and canines less bilaterally compressed than in *Dryopithecus*. The buccal cusp of P^3 is relatively projecting. Upper molars have well marked occlusal ridges, usually with at least slight development of the protoconule, and upper molar lingual cingula are proment, unlike *Dryopithecus* and *Sivapithecus*. Cusp projection and crenulations of occlusal surfaces are greater than in *Dryopithecus* and *Sivapithecus*. Lower molars always with distinct buccal cingula and greater cusp projection than in *Dryopithecus* and *Sivapithecus*. M_3 elongated, with massive development of the hypoconulid. Genial pit of mandible directed more inferiorly than in *Dryopithecus* and *Sivapithecus*, and there is a large symphyseal superior transverse torus and absence of inferior torus.

TYPE SPECIES. *Proconsul africanus* Hopwood 1933.

Subgenus *Proconsul* Hopwood 1933

DIAGNOSIS. A group of species covering the full size range of the genus. Incisors are broader and lower crowned than those of (*Rangwapithecus*), the premolars smaller with more projecting cusps, the upper molars relatively broad, and the lower molars less elongated. The M^3 is often reduced and the M_3 often narrows distally so that it is triangular in outline. Cingula are absent or small on the premolars. The body of the mandible is slightly more robust than in (*Rangwapithecus*), the alveolar processes of the maxilla are considerably more robust, and the floor of the maxillary sinus is reduced in extent.

TYPE SPECIES. *Proconsul africanus* Hopwood 1933.

Subgenus *Rangwapithecus* Andrews 1974

DIAGNOSIS. A group of species approximating in size to the gibbon and siamang. Incisors high crowned and relatively very narrow compared with (*Proconsul*). Upper molars and premolars elongated, the molars usually longer (mesiodistally) than broad, low cusped, and the occlusal surface often has more secondary wrinkling than in (*Proconsul*). Upper molars increase in size from M^1 to M^3 and premolars from P^3 to P^4 unlike (*Proconsul*). No reduction of M^3. Lower molars and premolars also elongated. The molars have a marked wear gradient, such that M1 may have dentine exposed on the occlusal surface when M3 is only just coming into wear, unlike the condition in (*Proconsul*). Strong lingual cingula are developed on all the upper molars and premolars, and the premolars also have a prominent distal cingulum. Zygomatic process is set very low over $M^{1,2}$. The floor of the maxillary sinus is greatly extended. Mandibular body and symphysis relatively deep and robust.

TYPE SPECIES. *Proconsul (Rangwapithecus) gordoni* (Andrews) 1974.

Proconsul africanus Hopwood 1933b

SYNONYMY. *Xenopithecus koruensis* Hopwood 1933a, *Proconsul africanus* Clark and Leakey 1950, *Dryopithecus (Proconsul) africanus* Simons and Pilbeam 1965.

HOLOTYPE. BMNH M-14084, left maxilla with $\underline{C}$-M^3.

LOCALITY. Koru, Kenya.

DISTRIBUTION. Early Miocene. Mainly distributed in the sites centered on Rangwa: Rusinga and

Mfwanganu Islands. It is also present, but less common, at Koru and Songhor.

HYPODIGM. 118 specimens which cover the complete dentition and mandible, and the maxillary, frontal, and temporal regions of the skull. A nearly complete forelimb and many other postcranial fragments are also known. See figures 8.6, 8.7.

DIAGNOSIS. A species of *Proconsul* intermediate in dental size between the siamang and pygmy chimpanzee. Cingula well developed in maxillary cheek-teeth, particularly mesially and lingually; buccal cusp of P³ strongly projecting; occlusal ridges and protoconule well developed on the upper molars; M³ typically much reduced particularly metacone and hypocone. Total length of upper premolar-molar series less than 40 mm, lower less than 45 mm. Skull lightly built, relatively orthognathous, lacking brow ridges of *Pan* species. Subarcuate fossa for the petrosal lobule of cerebellum present. Postcran-

ial skeleton with some features characteristic of *Pan,* particularly the development of the deltoid crest and the medial epicondyle of the humerus and the conformation of the distal articular surface of the humerus.

REMARKS. The original material assigned to this species (Hopwood 1933b, MacInnes 1943) has since been divided into three species (Clark and Leakey 1951, Pilbeam 1969). At present the type specimen (BMNH M-14081) of the prior named "*Xenopithecus koruensis*" is included in *Proconsul africanus* (see, Clark and Leakey 1951 and Andrews 1973), but while such an allocation of this specimen is somewhat uncertain its isolated position and possibly aberrant nature provides little basis for questioning Clark and Leakey's suppression of the term.

Recent unpublished research by A. C. Walker on the forelimb bones of *P. africanus* has emphasized its similarity with the chimpanzee, as have studies

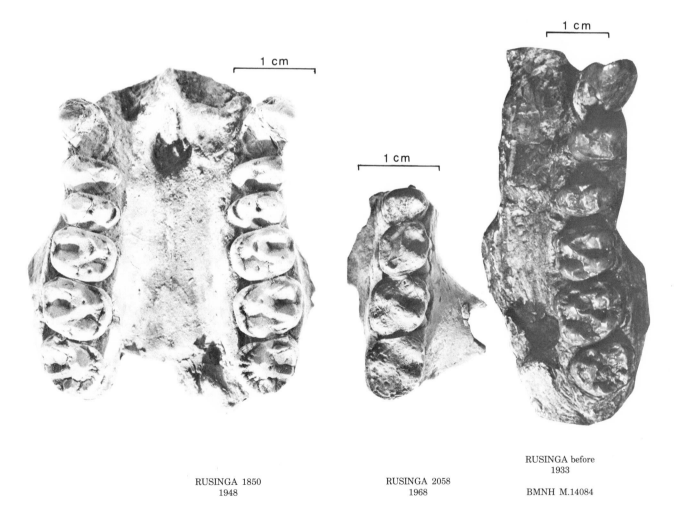

RUSINGA 1850
1948

RUSINGA 2058
1968

RUSINGA before
1933

BMNH M.14084

Figure 8.6 Comparison of the upper dentition of the types of *Proconsul africanus* and *P. vancouveringi* with the most complete dentition of *Dendropithecus macinnesi*. All from Rusinga Island, Kenya, East Africa.

Figure 8.7   Facial view of the skull of *Proconsul africanus* discovered in 1948 on Rusinga Island, Lake Victoria, Kenya, by Mary D. Leakey. Slightly less than natural size.

by Conroy and Fleagle (1972) and Zwell and Conroy (1973). Similarly the conformation of the distal articular surface of the ulna has been suggested by Lewis (1972) to be similar to that of modern pongids and more advanced toward them than in modern hylobatids. In view of the hylobatid affinities of *Dendropitheus macinnesi* (to be discussed below), this evidence points strongly to a dichotomy between the two families having already been established at this stage. Pilbeam (1969) went further than this and postulated that *P. africanus* might have been ancestral to one later pongid species, the chimpanzee. Such a choice, however, was made before the evidence for a greater variety of Miocene pongids species in East Africa was presented in detail (Andrews 1973, 1974).

A total of 118 numbered specimens representing at least 54 individuals are known for *P. africanus*. Most of these specimens come from Rusinga Island, where 89 specimens are so far identified. These can be assigned to a total of at least 41 individuals on

the basis of minimum numbers. This sample is fairly homogeneous, both in size and morphology. An additional 23 specimens from Mfwanganu and Songhor (representing 10 individuals) are closely similar to the Rusinga specimens. The type specimen (BMNH M-14084) and additional specimens from Koru are again very similar to the Rusinga sample, but the original type specimen of *"Xenopithecus koruensis"* (BMNH M-14081) can only provisionally be considered to be *P. africanus* on the basis of size.

Proconsul nyanzae Le Gros Clark and Leakey 1950

SYNONYMY. *Proconsul africanus* MacInnes 1943, *Proconsul nyanzae* Clark and Leakey 1950, *Sivapithecus africanus* Clark and Leakey 1950, *Kenyapithecus africanus* Leakey 1967.

HOLOTYPE. BMNH M-16647, distorted maxilla with upper dentition lacking the incisors

LOCALITY. Rusinga Island, Kenya

DISTRIBUTION. Early Miocene. Entirely distributed in the Rangwa sites on Rusinga and Mfwanganu Islands and Karungu.

HYPODIGM. 103 specimens which cover the complete dentition, mandible, and maxilla, and parts of the face. Only a few postcranial fragments of limb bone can be attributed to this species with any certainty. See figure 8.11.

DIAGNOSIS. A species of *Proconsul* approximating in dental size to the chimpanzee. Canines well developed and often about as large relative to palate and mandible as in the chimpanzee. Strongly sexually dimorphic, particularly in the canines. Lingual cingulum of upper molars beaded, posterior cingulum well developed. M^1 very small relative to M^2; M^3 slightly smaller than M^2 but not reduced morphologically and much bigger than M^1. P^3–M^3 length between 40 and 50 mm; P_3–M_3 length 45–55 mm. Maximum depth of body greater than that of *D. africanus* (25 mm), but variable. Symphysis and body relatively gracile.

Proconsul major Le Gros Clark and Leakey 1950

SYNONYMY. *Proconsul major* Clark and Leakey 1950.

HOLOTYPE. BMNH M-16648, right mandible with P_4-M_3.

LOCALITY. Songhor, Kenya.

DISTRIBUTION. Early Miocene. Never abundant anywhere, it is most common at the sites around Tinderet: Songhor and Koru. It is also common at Napak and occurs at Moroto in Uganda. Some specimens from the northern Kenya sites on Losidok Hill are provisionally assigned to this species.

HYPODIGM. 53 specimens from Kenya and a further 22 specimens from Uganda are known which cover the dentition except for the lower incisors, the mandible except for the ascending ramus, and the maxillary and nasal regions of the skull. Few postcranials are known, but there are several vertebrae assigned to this species. See figure 8.8.

DIAGNOSIS. A species of *Proconsul* with dentition approximately the size of that of the female gorilla. Largest African species of *Proconsul*; length of P_3-M_3 may exceed 65 mm. Compared to tooth size, mandibular body much more massive than in *P. nyanzae* and symphysis more massive than in *S. indicus* which otherwise approaches *P. major* in size. M_3 typically larger and longer compared to M_2; much larger than M_1; only slightly larger than M^1, the M^1 significantly larger than M^1 of *P. nyanzae* at 0.02 probability; cingula usually more distinct than in gorilla and simian shelf not developed; crowns of cheek teeth perceptibly more wrinkled than in *S. indicus*.

REMARKS. Early workers included specimens from both these species in the hypodigm of *Proconsul africanus*. Hopwood (1933) included the earliest found mandible, BMNH M-14086, in *P. africanus*. Clark and Leakey (1951) reassigned it to *P. nyanzae*, and Pilbeam (1969) reassigned it to *D. major*. Clark and Leakey's type and other specimens of *P. nyanzae* were initially described by MacInnes as *P. africanus*. The size divisions between the three species of *Proconsul* were made by Clark and Leakey (1951) on the basis of much enlarged samples, and they have stood the test of time. They were accepted by Simons and Pilbeam (1965), Pilbeam (1969), and Andrews (1973) and we have retained them here.

Of the two populations of large pongid in the Kenya Miocene, one, *P. nyanzae*, was centered on the southern sites around the Rangwa volcano (Rusinga and Mfwanganu Islands and Karungu) and the other, *P. major*, on the northern sites around the Tinderet volcano (Songhor and Koru) and Napak and Moroto. It is not possible to determine with certainty on the available dental evidence whether these were subspecies belonging to one highly variable species, or separate but obviously closely related species. As separate samples they are moderately variable, but put together as a combined sample they are considerably more variable than the gorilla. Significance tests show a probability of only 2% that the northern and southern samples represent the same statistical population and justifies maintaining their division into two species on the basis of size. When more postcranial bones are known a definite answer may be forthcoming. In the meantime, on the basis of the differences between them, we have decided to retain them as two closely related species.

Based on the arbitrary assumption that *P. nyanzae* occurs at Rusinga and its neighboring sites (Mfwanganu and Karungu), and *P. major* occurs at Songhor, Koru, and Napak, the numbers of individuals are as follows:

P. nyanzae: 103 specimens representing at least 51 individuals from Rusinga; it also occurs at Karungu (2) and Mfwanganu (3).
P. major: 53 specimens representing no less than 15 individuals, from Songhor; the species also occurs at Koru (1), Losidok (3), Rusinga (1), and Napak.

Three large primate specimens from Losidok and Moruarot in northern Kenya are provisionally assigned to *P. major* because of their size. There are a number of distinctive morphological features in these specimens, notably the robustness of the canine, the mesiodistal compression of the upper molars, and the low, rounded conformation of the molar cusps. More complete material, if found, should throw some light on the distinctiveness of these specimens, but they are too scanty themselves to justify specific separation from *P. major* at this stage.

Proconsul (Rangwapithecus) gordoni (Andrews) 1974

SYNONYMY. *Dryopithecus (Rangwapithecus) gordoni* Andrews 1974.

HOLOTYPE KNM-SO 700, maxilla with complete upper dentition except for the incisors.

LOCALITY. Songhor, Kenya.

DISTRIBUTION. Early Miocene. Most common at

Figure 8.8 Palate and upper dentition of *Proconsul major* from Moroto, Uganda.

the Tinderet site of Songhor but a few specimens are known from the Rangwa volcano sites of Rusinga and Mfwanganu Islands.

HYPODIGM. 79 specimens. These cover the complete dentition, the mandible except for the ascending ramus, the maxilla and palate. A few postcranial fragments may belong to this species. See figures 8.9, 8.10, 8.11.

DIAGNOSIS. A species of *Proconsul* intermediate in dental size between the siamang and pygmy chimpanzee; subequal in size to *P. africanus*. Upper teeth similar to those of *P. vancouveringi* (as described in the subgeneric diagnosis) differing only in larger size. In the lower dentition the incisors are very high crowned and narrow; the canine is high crowned and bilaterally compressed; the P_3 is very elongated; the M_1–M_3 are also elongated, the cusps are low, the buccal cusps are divided by deep buccal sulci, as in the gorilla, the occlusal ridges and buccal cingulum are poorly defined, and secondary wrinkling is often present. P_3–M_3 lengths are, upper 40 mm, and lower 44 mm for single specimens. M_1–M_3 length is 29 mm (see figure 8.10).

Proconsal (Rangwapithecus) vancouveringi (Andrews) 1974

SYNONYMY. *Dryopithecus (Rangwapitheus) vancouveringi* Andrews 1974.

HOLOTYPE. KNM-RU 2058, left maxilla with crowns of P_4-M_3.

LOCALITY. Rusinga Island, Kenya.

DISTRIBUTION. Early to ?middle Miocene at the sites centered ?n Rangwa, Rusinga, and Mfwanganu islands, and the Tinderet site of Songhor. The species also occurs at Maboko Island.

HYPODIGM. Seven specimens. These cover only the upper postcanine dentition and parts of the maxilla. Also referred provisionally to this species are three specimens from Maboko Island. See figure 8.6.

DIAGNOSIS. A small species of *Proconsul* approximately the dental size of the siamang. It is like *P. gordoni* in morphology as defined in the subgeneric diagnosis, but differs from it in size. The M^1 is significantly different in size from the M^1 of *P. gordoni* at greater than the 0.001 level of probability. The upper tooth row lengths are M^1–M^3, 22 mm, and P^3–M^3, estimated at 31 mm for a single specimen.

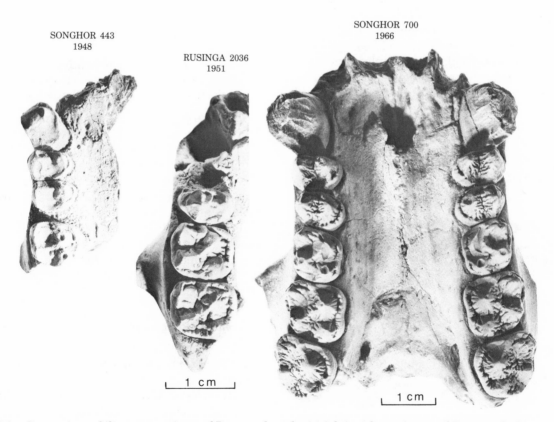

SONGHOR 443
1948

RUSINGA 2036
1951

SONGHOR 700
1966

|_ 1 cm _|

|_ 1 cm _|

Figure 8.9 Comparison of the type specimen of *Proconsul gordoni* (right) with specimens of *Proconsul africanus* (center) and *Limnopithecus legetet* (left). All from Kenya.

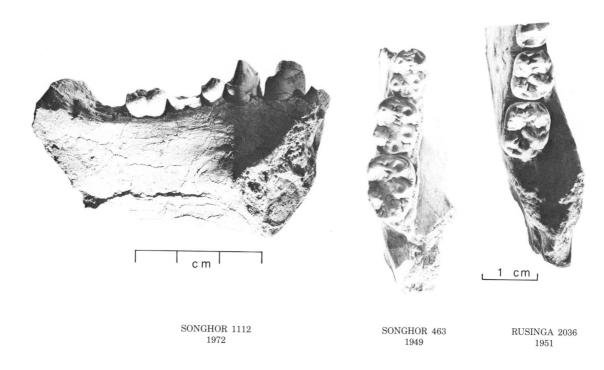

SONGHOR 1112
1972

SONGHOR 463
1949

RUSINGA 2036
1951

Figure 8.10   A new mandible of *Proconsul gordoni* (left) from Songhor compared with mandibular molars in *Proconsul gordoni* (center) and *P. africanus* (right).

REMARKS. A few specimens referred here to *Proconsul gordoni* and *P. vancouveringi* have previously been assigned to *"Proconsul" africanus* and *Limnopithecus" macinnesi* (Clark and Leakey 1951). At that stage there were too few specimens for anyone to have recognized the differences between them and the species that had then been diagnosed, although in the description of one specimen (KNM-SO 401) its aberrant features were noted and the comment added that it might turn out to be a new species (Clark and Leakey 1951:33). Much of the new additional material came from Leakey's 1966 Songhor expedition and from Van Couvering's 1967 collection at Rusinga Island.

The morphology of these two species is almost identical and evidently they were closely related. They differ considerably in size, and significance tests on the M^1 dimensions show it to be unlikely (probability 0.001) that the two samples could be from one population (Andrews 1974). The molars are elongated teeth, longer than broad, and increase in size greatly from M^1 to M^3. Both upper premolars have strong lingual and distal cingula and are also relatively elongated (Andrews 1970). The lower dentition of *P. (Rangwapithecus) gordoni* has strikingly high crowned incisors, and this is probably correlated with the relatively deep symphysis. Posteriorly the mandibular ramus becomes much shallower. The lower canines are also highly crowned.

The P_3 is anteroposteriorly elongated and is unusual in morphology, having the lingual edge nearly straight and parallel to the buccal edge. This gives it a characteristic daggerlike shape probably correlated with the massive distal flange of the upper canine already commented on (Andrews 1970). The lower molars tend to be elongated and their occlusal surface is long and narrow because of the strongly indented buccal edge of the crown. These features show up in highly exaggerated form in the dP_4.

Proconsul (Rangwapithecus) gordoni is represented by 79 specimens, mostly from Songhor but also from Rusinga and Mfwangano Islands. The numbers of individuals represented are at least 17 from Songhor, and one each from Rusinga and Mfwangano. *P. (R.) vancouveringi* is represented by only seven specimens from Rusinga, Mfwangano, and Songhor. In addition, three specimens from Maboko Island almost certainly belong to this species.

Genus *Limnopithecus* Hopwood 1933b

DIAGNOSIS. Teeth slightly smaller than in modern gibbons. Central incisors relatively large and broad. Canines well developed, the lower with short mesial ridge and very asymmetrical. P^3–P^4 low crowned, cusps of approximately equal size. P_3 single cusped but low crowned, P_4 elongated. Molar cusps low and rounded. Distinct buccal cingulum on the lower molars, occlusal ridges poorly developed, and distinct size increase from M_1–M_3. Distinct lingual cingulum on the upper molars, cingulum not crenulated, slight protoconule developed, occlusal ridges moderately well defined, and increase of molar size in sequence M^1–M^3–M^2. The M^3 is slightly reduced in size. P_3–M_3 length averages 26–27mm. The mandible is relatively robust.

TYPE SPECIES. *Limnopithecus legetet* Hopwood 1933b.

Limnopithecus legetet Hopwood 1933b.

SYNONYMY. *Limnopithecus evansi* MacInnes 1943, *Pliopithecus* (*Limnopithecus*) *legetet* Simons 1965.

HOLOTYPE. BMNH M-14079, distorted mandible with M_1–M_2.

LOCALITY. Koru, Kenya.

DISTRIBUTION. East African deposits dated between 23 and 17 million years old: Bukwa, Napak, Songhor, Koru, Rusinga Island.

HYPODIGM. 116 specimens which cover the complete dentition, the mandible except for the ascending ramus, and the maxilla excluding the palate. No cranial fragments are known. See figures 8.9, 8.11.

DIAGNOSIS. Not distinguished from generic.

REMARKS. The taxon *Limnopithecus legetet* was first described for very inadequate material from Koru (Hopwood 1933a). It was said at that time that it was a gibbonlike primate, in contrast to the chimpanzee-like morphology of *Proconsul africanus*. Little evidence was put forward to substantiate this view, but the contrary evidence of the differences between the milk dentitions of *Limnopithecus legetet* and *Hylobates* was pointed out (Hopwood 1933b). *Limnopithecus legetet* resembles pongids rather than hylobatids in the milk dentition.

A second species of *Limnopithecus* was later described by MacInnes (1943). Unfortunately, because the type specimens were so poor MacInnes designated as a neotype of *L. legetet* a specimen that, with the better material available today, can now be assigned as *Dendropithecus macinnesi*, and the specimens that he described as "*L. evansi*" are in fact typical *L. legetet*. The latter was recognized when "*L. evansi*" was sunk into *L. legetet* (Clark and Leakey 1951).

Further discoveries of *Limnopithecus* confused the situation still more. The description of a third species, *L. macinnesi* (Clark and Leakey 1951), which also was said to have many hylobatine characters, seemed to confirm the family classification of *Limno-*

pithecus. However, in their discussion on the origin of *Proconsul*, Clark and Leakey (1951:111) concluded that an ancestor for *Proconsul* might have looked something like *Limnopithecus* and that the two genera were really not so different. They considered that *L. macinnesi* was divergently specialized in its C-P3 morphology, but that as these specializations were absent in *L. legetet*, the species could serve as a good model for the ancestral conditions of *Proconsul*. This difficulty in segregating *Limnopithecus* raises the question of whether it is of a valid genus. Placement of the species in Pongidae is warranted based on the dental similarities between *L. legetet* and *Proconsul africanus* that Clark and Leakey noted, and the differences in C-P3 and mandibular regions between *Limnopithecus legetet* and *L. macinnesi*. Moreover, the similarities of the available postcranial bones between *L. legetet* and *Proconsul africanus* (both different from those of *Limnopithecus macinnesi*), further reinforce the view that *L. legetet* should be classified with the Pongidae rather that the Hylobatidae. It is so dealt with here as a genus in the Dryopithecinae (Pongidae).

A total of 36 individuals based on 116 specimens are now recognized for *L. legetet*. Most of these specimens come from Songhor, where a total of 94 specimens can be assigned to at least 19 individuals on the basis of minimum numbers. There are additional samples from other sites, all closely resembling the Songhor specimens both in size and morphology; these are from Koru (6 individuals), Rusinga (8), Maboko (1), Ombo (1), and Williams Flat (1), based on a total of 22 specimens. This species is also abundant at Napak, in Uganda, where it is slightly smaller in average size than at Songhor. Only one of the specimens, a palate from Napak IV, has yet been described (Fleagle 1975). It is considerably smaller than *L. legetet* from Kenya and may belong to a new species of *Limnopithecus*. Fleagle commented on the broad, arcuate palate, the broad nose, the great orbital breadth, and the shallow and short face, demonstrating that the Napak palate resembles modern gibbons in these features.

Finally, 14 specimens at present are attributed to *L. legetet* from Fort Ternan, Kenya (Andrews and Walker 1976). These are relatively large specimens, fitting into the top half of the *L. legetet* range. The molars are more elongate, the cingula are reduced in size and M_3 is reduced compared with *L. legetet*, but they share most of the derived characters that link *L. legetet* with the dryopithecines rather than with "*L.*" *macinnesi* and the hylobatids. It is possible that when more material is available the material from Fort Ternan will be assigned to a new species of

SONGHOR 1112
1972

(Formerly ascribed to
"Kenyapithecus africanus")
RUSINGA 2087
1967

RUSINGA 2036
1951

RUSINGA 2015
1951

KORU 8
1950

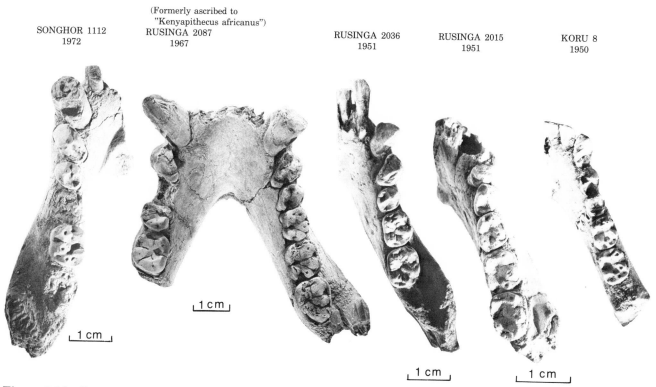

Figure 8.11   Comparison of *Limnopithecus* mandibular dentition (extreme right) with, from right to left, *Dendropithecus, Proconsul africanus, P. nyanzae,* and *P. gordoni.*

Limnopithecus, but such a course is not justified at present so it is retained here in *L. legetet.*

Family Hylobatidae Gill 1872

Genus *Aeolopithecus* Simons 1965

DIAGNOSIS. Lower dental formula 2.1.2.3. Second smallest extinct hominoid slightly smaller than *Propliopithecus.* Mandibular measurements about one half those of a modern gibbon. Differs from contemporary *Propliopithecus* in showing marked premolar heteromorphy (honing P_3), from *Oligopithecus* and *Propliopithecus* in much larger canines, and from the latter in having much larger and more procumbent incisors. Differs from both *Aegyptopithecus* and all other known hominoid genera in having tooth rows more divergent posteriorly. Compared to other early apes shows relatively higher and deeper genial fossa. (This means that the superior and inferior transverse tori are more shelflike and extended slightly further back.) Horizontal ramus of mandible shallows rapidly to the rear as in *Aegyptopithecus* instead of remaining about same depth as in *Propliopithecus* and *Oligopithecus.* Differs from *Pliopithecus* in having relatively larger canines, greater degree of premolar heteromorphy, and probably

comparatively larger incisors. Resembles some *Hylobates* and differs from *Aegyptopithecus,* European Miocene *Pliopithecus,* and East African *Dendropithecus* (see below) in having reduced M_3 relative to M_1 and M_2.

TYPE SPECIES. *Aeolopithecus chirobates* Simons 1965.

***Aeolopithecus chirobates* Simons 1965**

SYNONYMY. None.

HOLOTYPE. CGM-26923, complete horizontal rami of mandible fused at symphysis, with incisor alveoli and left and right lower C through M_3.

LOCALITY. Yale Expedition Quarry I, Upper Fossil Wood Zone, Qatrani Formation, Oligocene, Fayum Province, Egypt, U.A.R.

HYPODIGM. Type specimen. See figure 8.12.

DIAGNOSIS. Not distinguished from generic.

REMARKS. This species is tentatively assigned to Hylobatidae, although we realize that this may need to be changed with the acquisition of more material. We are concious of the fact that the genus could equally well be placed as *incertae sedis* among the Hominoidea. *Aeolopithecus* is the only early hominoid for which a possibley frugivorous dietary mode can be inferred, based on the relatively very large

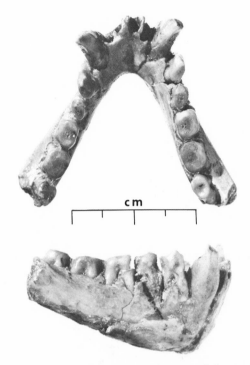

Figure 8.12 Occlusal and lateral views of the teeth and mandible of the type and only specimen of *Aeolopithecus chirobates* from Yale Quarry I in the Fayum badlands of Egypt, U.A.R.

anterior teeth and reduced cheek teeth, especially M₃.

Genus *Dendropithecus* Andrews and Simons 1977

DIAGNOSIS. small anthropoid apes approximating in dental size to the siamang. Incisors high crowned and strongly mesiodistally compressed. Canines bladelike in males with double mesial grooves, showing striking degree of sexual dimorphism. P₃ sectorial *sensu strictu* as in gibbons. P³ has strongly projecting buccal cusp. Lower molars have the cusps arranged around the periphery of the crown, not situated a little more inward as in *Limnopithecus*, cusps connected by well-defined ridges and enclosing large trigonid and talonid basins. Buccal cingulum of lower molars slightly developed, not like *Limnopithecus* where the cingulua are well developed. Upper molars of simple construction. M₃ slightly larger than M₂. M³ usually reduced. Palate long and narrow and maxillary sinus well developed. Body and symphysis of mandible robust, well-developed superior transverse torus and usually also an inferior torus projecting posteriorly at least to the extent of the superior torus and often further, unlike *Limnopithecus* where inferior transverse torus not developed. Dentally very similar to *Pliopithecus,* but

having higher crowned incisors, more strongly bilaterally compressed canines, more sectorial P₃, more projecting buccal cusp on P³, and lower molars with stronger ridge formation and deliniation of the talonid basin. Differs from *Dryopithecus* and *Pliopithecus* postcranially, particularly in the length and slenderness of the long bones. All the postcranial bones lack conspicuous muscular markings. Humerus straight-shafted, not retroflexed as in *Dryopithecus* and lacks the entepicondylar foramen and broad distal condyles of *Propliopithecus*.

TYPE SPECIES. *Dendropithecus macinnesi* (Clark and Leakey) 1950.

Dendropithecus macinnesi (Clark and Leakey) 1950

SYNONYMY. *Limnopithecus macinnesi* (Clark and Leakey) 1950, *Pliopithecus* (*Limnopithecus*) *macinnesi* Simons 1965.

HOLOTYPE. BMNH-M 16650, the greater part of a mandible containing the premolar-molar series except for the first molar.

LOCALITY. Wakondu, Rusinga Island, Kenya.

DISTRIBUTION. Late Oligocene to early Miocene. The species is most common in the southern sites centered around Rangwa: Rusinga and Mfwangano Island and Karungu, whereas occurrence is much less common at the sites centered on Tinderet: Songhor, and Koru.

HYPODIGM. 160 specimens which demonstrate the complete dentition, the mandible lacking the ascending ramus, and the maxilla and palate. Postcranial long bones are known in association with dentition.

DIAGNOSIS. As for the genus.

REMARKS. As discussed above, the type species of *Limnopithecus, L. legetet,* is more nearly like species of *Dryopithecus* (subgenus *Proconsul*) in dental and possibly postcranial features than heretofore recognized. A second species, *L. macinnesi,* originally diagnosed and referred to the genus by Clark and Leakey (1950), resembles European Miocene *Pliopithecus* in dental morphology. However, it differs markedly from the latter in postcranial structure, particularly in humeral anatomy in which the two are distinctly different, preventing their reference to the same genus. Consequently a new genus, *Dendropithecus*[3] was proposed for this East African species by Andrews and Simons (1977). The anatomical distinctiveness of this genus is important because the species *D. macinnesi* is the only Miocene primate

_____

3 From the Greek *dent-* ("tree, branch") and *pithecos* ("trickster, ape"), in analogy with *Dryopithecus, Oreopithecus,* etc.

whose limb structure indicates that the forelimb was used primarily in suspensory activities rather than having had the generalized quadrupedal use indicated by the humeri of other Miocene apes. Moreover, the humeral dissimilarities between *Pliopithecus* and *Dendropithecus* show that dentally very similar apes were already differentiated widely in locomotor adaptations by middle Miocene times. Such evidence points up the probability that other dentally similar early ape species may eventually prove adaptively varied. This, in turn, weighs against overly simplistic theories requiring derivation of present-day hominoid locomotor types from a single, generalized pattern of limb structure throught to characterise Miocene apes. The anatomical differences in the humerus that separate *Pliopithecus* and *Dendropithecus* are considerably greater than those which exist between humeri in *Homo* and *Pongo* or *Alouatta* and *Ateles,* for instance, while dentally the former pair are much more alike than either of the latter two sets. The comparatively elongated and straight-shafted humerus of *Dendropithecus* suggests that *D. macinnesi* may lie closer to the ancestral line of the modern hylobatids than does any other fossil.

Dendropithecus macinnesi was first described in 1950, as a new species of hylobatid by Clark and Leakey. Prior to its description, a few specimens belonging to this species had been referred by Hopwood (1933) to "*Xenopithecus koruensis*," but most of Clark and Leakey's hypodigm consisted of newly discovered specimens. The species was included in the genus *Limnopithecus* partly on the basis of size and partly because of some relatively superficial dental resemblances between it and the type species of the genus, *L. legetet.* An extended analysis of *Dendropithecus macinnesi* by Clark and Leakey (1951) was published a year after their description of it. In 1948, L. S. B. Leakey found a series of limb bones in a block of red limestone on Rusinga Island, Kenya, in direct association with the jaws and teeth of at least four individuals of *D. macinnesi.* These materials formed the basis for a study also published in 1951 by LeGros Clark and Thomas (1951). These authors determined that *D. macinnesi* strikingly resembled modern South American *Ateles;* they remarked further: "it is well recognized that the spider monkey of the New World shows certain structural modifications which parallel the line of development followed by the gibbons of the Old World. Thus if, as now appears, *Limnopithecus* [here read *Dendropithecus*] is to be regarded as a 'gibbon in the making', it is not surprising that it should display these resemblances to the spider monkey."

Collections made in the period since 1951 have greatly expanded the number of known jaws and teeth of this species. At present it is the best known fossil ape from the Kenya Miocene. Dentally this ape is quite distinct from other African Miocene species, particularly in the anterior dentition. The incisors are narrow bladed and high crowned. The canine is strongly bilaterally compressed, and, while its index of sexual dimorphism (see Garn et al. 1967) is greater than that of the gorilla, it is a projecting bladelike tooth in both sexes. Similarly the lower third premolar is also bilaterally compressed, and in its occlusion against the upper canine acts as a honing mechanism for sectorial wear as defined in 1970 by Every (1970); it is similar in general morphology to the sectorial premolar of monkeys and gibbons. The premolars and molars of *Dendropithecus macinnesi* have the cusps set at the edge of the occlusal surface and connected by relatively well defined ridges. Consequently, the trigon and talonid basins are large and well defined.

Dendropithecus macinnesi is best represented at Rusinga Island. In addition there are several specimens of isolated teeth from Songhor that can be attributed with reasonable certainty to *D. macinnesi.* These include, particularly, upper canines and molars, along with other specimens of more uncertain status. In molar size they resemble *Limnopithecus legetet,* but in the molar morphology and $\overline{C}$-P_3 development they resemble *Dendropithecus macinnesi.* These almost certainly belong to the latter species and probably differ from the Rusinga population at no more than the subspecies level at most.

A total of 78 individuals of *D. macinnesi* are recognized of which the majority (58) are from Rusinga Island. The 58 Rusinga individuals (minimum number) are based on 123 registered specimens. In addition, there are five individuals from Mfwanganu Island, two from the Koru site, one from Karungu and twelve from Songhor, (minimum number estimate) from a total of 37 specimens. Many of the specimen associations are good, for instance those for sites R 3A and R 113, but for the most part these estimates are based on the minimum number needed to account for the specimens within the limits imposed by individual size and degree of tooth wear.

Pongid Distributions

The Fayum Oligocene primates are from a relatively restricted geographical region, although they sample a considerable time span. The paleoecology of the sites has been reviewed by Simons (in Simons and Wood 1968) and by Vondra and Bowen (1974).

The sum of the evidence points to mixed habitats; the mammals, reptiles and fish faunas suggest forested conditions, a conclusion supported by the few identifiable plant remains. Sedimentological evidence suggests, however, that open country was nearby. A coastal, well-watered environment is indicated, with lakes and meandering streams; subtropical to tropical gallery forests were probably widespread throughout the area, interspersed with savannas in the interfluve regions. (Unfortunately crucial pollen evidence has been lost due to long-term leaching before the fossiliferous beds were exposed.) The abundance of parapithecine monkeys relative to hominoid apes suggests that the former were more abundant and perhaps more "social" than the apes at that time.

The East African species distributions divide into two or possibly three groupings. The most characteristic of these is the group found at the Kenya sites centered around the Miocene volcano of Kisingiri, now known as Rangwa. These sites are Rusinga Island to the north, Mfwangano Island to the west, and Karungu to the south. The pongid assemblage consists of *Proconsul africanus, P. nyanzae,* and *Dendropithecus macinnesi* as the three most common species. In addition, *Limnopithecus legetet* is represented by a mandible and some isolated teeth, *Proconsul vancouveringi* by two maxillary fragments, and *P. gordoni* by a partial upper jaw and possibly some isolated teeth. There are thus six species of higher primate known from these sites, three common ones and three that are much less common.

This pattern is repeated, but with different proportions of species, at the other main group of sites in Kenya, such as those around the Miocene volcano of Tinderet. These are at Songhor to the west and Koru to the south, and the characteristic pongid assemblage consists of *Limnopithecus legetet, Proconsul gordoni,* and *P. major.* Also present, but far less numerous, are *P. africanus, P. vancouveringi,* and *Dendropithecus macinnesi.* The early Miocene site of Napak in Uganda has *Limnopithecus legetet* and *Proconsul major,* as do Songhor and Koru; and of the two other Uganda sites, Moroto has *P. major* only and Bukwa has *Limnopithecus legetet.*

It is not possible at this stage to say exactly what factors are operating to produce these contrasts. It is unlikely that time differences are important. The Rangwa sites from which primates are known span a period from earlier than 22.5 ± 0.4 m.y. at Karungu and earlier than 19.6 ± 0.2 m.y. at Mfwangano to approximately 18 m.y. at Rusinga (Bishop, Miller, and Fitch, 1969; Van Couvering and Miller 1969). The Tinderet sites are all around 19.5 to 20

m.y., but the Uganda sites with the same primates range from greater than 22 ± 0.2 m.y. at Bukwa to older than 17.8 ± 0.4 m.y. at Napak. The site groups overlap entirely, and there is no division by species into older and younger sites.

Geographical differences or ecological barriers between the Miocene sites are also difficult to substantiate, although there is a possibility that there could have been a barrier between the Rangwa sites and the rest. The former sites are on the lower slopes of a volcano that stood on the southeastern shoulder of the developing Kavirondo Rift Valley. There is every likelihood that there was a moderately large river flowing westward through the rift valley from the Kenya dome which at that time was probably part of the east-west continental divide (see Andrews and Van Couvering 1975). The Tinderet volcanos are banked up against the Nyando escarpment that forms the north side of the Kavirondo Rift Valley, so it is very likely that the main tributary of this postulated river flowed south of Tinderet. This would place the Tinderet sites to the north of such a river, with the Rangwa sites to the south, and if the headwaters rose to any considerable altitude, then the river could be expected to constitute a considerable barrier.

The faunal differences between Rusinga and Songhor, the two most representative sites of each group, are very extensive. Large mammals like elephants, rhinoceroses, and anthracotheres are represented at Songhor by only one or two specimens each, whereas they are common on Rusinga. Smaller mammals like the ochotonid lagomorph *Kenyalagomys* and the hyracoid *Megalohyrax* are known from Rusinga but not at all from Songhor (although they do occur in the Uganda sites). Other genera like the insectivores *Prochrysochloris* and *Protenrec* are known from Songhor and not from Rusinga. Many genera have one species at Rusinga and a different species at Songhor, as is the case, for instance, in *Dorcatherium* and *Gymnurechinus.* Finally, many taxa differ subspecifically. These faunal contrasts must indicate some degree of ecological difference between the sites, although probably not a major difference in habitat. There are enough taxa in common to suggest that both sites were similar at least in part, and the nature of the fauna and flora strongly suggests that there were extensive forests in the vicinity of both. In the absence of temporal difference, and with this apparently minor ecological distinction, it seems an almost inevitable conclusion that there must have been some kind of geographical barrier between these sites (see Andrews and Van Couvering 1975).

The East African fossil localities probably are only sampling one small corner of the total African range of Miocene primate species. The populations in Miocene East Africa may well have been peripheral and hence particularly susceptible to speciation (Mayr 1963) and, if their range were itself divided up by geographic barriers as suggested above, speciation would occur still more readily. Evidence of contemporary speciation in the higher primates may be seen in the species pairs *Proconsul nyanzae/Proconsul major,* and *Proconsul gordoni/Proconsul vancouveringi.* The members of each pair are very similar to each other, and differ principally in size and geographical distribution. The species of the former pair have similarities with the third species of the subgenus, *Proconsul africanus,* reflecting an older level of speciation.

Discussion and Conclusions

The paleoenvironments inferred for Middle Cenozoic hominoids seem mainly to have been forested, at least during the Oligocene and the early Miocene. On the other hand, the middle Miocene locality at Fort Ternan, Kenya, preserves faunas that could imply more open habitats (open woodland or woodland savanna) see Andrews and Walker (1976). Thus the hominoids are most likely to have evolved originally as forest species and were presumably arboreal; such postcranial evidence as is known supports this conclusion. Phylogenetic relationships among fossil species, and between them and living forms, remain obscure. This is for two principal reasons: the large gaps in the fossil record, particularly during late Miocene and Pliocene times; and the patchiness in terms of body parts sampled, a majority of known specimens representing jaws and teeth.

Oligopithecus savagei is the oldest known catarrhine primate, as well as the most primitive. Some features of occlusal morphology hint at ties with prosimians, although *Oligopithecus* had clearly evolved to the anthropoid level. It may be regarded as ancestral to, or representative of, the ancestors of all later hominoids. The exact stratigraphic positions of the type specimens of the two *Propliopithecus* species are unknown. However, circumstantial evidence suggests a position at the middle of the Jebel el Qatrani Formation and they could have therefore been sampled from species that are descendants, at least broadly, of *Oligopithecus savagei.*

There are a number of differences between *Propliopithecus* species and the probably younger species *Aegyptopithecus zeuxis,* but it is unclear

whether these rule out an ancestor-descendant relationship. At present, the simplest hypothesis is that *Oligopithecus, Propliopithecus,* and *Aegyptopithecus* species are broadly part of an evolving continuum characterized by increased body and canine size. Changes in occlusal morphology also occur through time. Of course any scheme deduced may well be oversimplified, for it is certainly true that finds in the Fayum region to date could not possibly reflect the full diversity of African Oligocene Primates.

The nearly complete cranium of *Aegyptopithecus* has provided an endocranial mould which shows distinct advance toward the condition seen in later catarrhines (see Radinsky 1973). This skull is a blend of primitive and advanced features and it makes an almost perfect morphological intermediate between Eocene prosimian skulls such as those of *Notharctus* and the cranium of *Proconsul africanus* found at Gumba on Rusinga in 1948. Frontal and symphyseal fusion as well as postorbital closure qualify *Aegyptopithecus* as a higher primate. Interestingly this cranium does not show a single feature allying it with the so-called tarsioids of the Holarctic Eocene.

Dental crown patterns cannot fully be made out in *Aeolopithecus* since there is but one specimen, in which solution or erosion of the enamel has obscured important features of occlusal morphology. Nevertheless the proportions of the relatively complete mandible and the relative sizes and shapes of the teeth, as well as the presence of a shelflike inferior transverse torus, are resemblances to the hylobatids. Particularly, the very large incisor sockets and large barrel-shaped canines, seen together with relatively reduced molars, the whole set in mandibular bodies which diverge posteriorly at a higher angle than in any known contemporary or later hominoid, point to "front of the mouth" emphasis during food maceration. This all suggests possible frugivory and a short, small, gibbonlike face and snout. Perhaps of greatest interest is that *Aeolopithecus* is very differently sized and proportioned from *Aegyptopithecus* and the relative sizes and shapes of its teeth are quite different from *Propliopithecus* as well. Such differences emphasize that a radiation of the apes had been under way for some time when their fossil record is first revealed in the Fayum badlands.

Aegyptopithecus zeuxis can be viewed as in or near the ancestry of at least some of the early Miocene pongids of East Africa. These themselves exhibit considerable diversity and, despite the large number of specimens, little can yet be said about their phylogenetic relationships. However, *Proconsul major* and *P. nyanzae* could be interpreted as having relatively recently descended from a common ancestor, as may well be the case also for *P. vancouveringi* and *P. gordoni.* Some of the early Miocene species are represented also in the middle Miocene: *P. africanus, P. nyanzae, P. vancouveringi,* and *Limnopithecus legetet* are found at Fort Ternan and Maboko. Also appearing in middle Miocene times is *Ramapithecus wickeri,* a hominoid that exhibits many features suggesting that it is a basal hominid, as it has been classified by most students who have studied it directly.

Along with advances in cranial and dental features, the Miocene species are probably more specialized postcranially than those of the Oligocene. Thus certain features found in *Proconsul* species suggest affinities with living pongids, particularly the African ones, although it is also clear that the Miocene forms were more lightly built and agile than their putative descendants.

Some of the African dryopithecines, or species very like the known ones, are probably ancestral to middle Miocene pongids of Eurasia. Thus, *Proconsul nyanzae* is very similar to *D. fontani* of Europe and *D. laietanus* from Spain; *Proconsul major* makes an excellent ancestor for *Sivapithecus indicus,* another Asian pongid. Recent finds by Tobien and co-workers at a new Miocene locality in Turkey (Andrews and Tobien 1977) provides evidence for two forms intermediate in morphology between the African early Miocene and Eurasian middle and late Miocene forms.

Aeolopithecus from the Fayum and the presumed hylobatid from the East African Miocene, *Dendropithecus macinnesi,* are not particularly similar except for small size. However, *Dendropithecus* is distinct from its contemporaries among *Proconsul* and *Limnopithecus* species in cranial, dental, and postcranial morphology and the differences point to its closer association with the living hylobatids. It is also related in some way to the middle Miocene *Pliopithecus* species of Europe, although perhaps not as an ancestor or descendant, but more as is a cousin. In the near absence of possible hylobatids from the later Tertiary of Asia, little more can be said about relationships other than the fact that hylobatids and pongids seem to have been clearly distinct by early Miocene times, and may even have been so in the Oligocene.

Ties between known Miocene pongids and later species are highly likely to have existed though precise relationships are unclear. Thus, *Gigantopithecus bilaspurensis* and *G. blacki* are probably descended from *Sivapithecus indicus;* and the orangutan, *Pongo pygmaeus,* most likely has evolved from another Asian dryopithecine. Now that dryopithecine diversity in East Africa is known to be greater than previously thought, precise relationships with the modern African pongids become even more uncertain. Probably though, the ancestors of the living forms are to be found among these early Miocene materials.

The authors would like to acknowledge the financial assistance provided (in connection with the preparation of this and prior studies on fossil apes) by the following grants. Simons received N.S.F. Grants in Earth Sciences, G-19012, GP-433, GP-3547, GA-11145, GA-723; Smithsonian Foreign Currency Awards numbers 5, 23, and 1841; and Wenner-Gren Foundation numbers 1572 and 1744. Pilbeam received N.S.F. GS-3240 and Wenner-Gren numbers 1693, 2159, 2301; Andrews Wenner-Gren number 2604. Simons should like to acknowledge grants from the Boise Fund, Oxford University, and an Alexander von Humboldt Senior Scientist's Prize from the German Federal Republic that allowed him to complete the second revision of this manuscript in 1976, while on leave from Yale University.

We also thank Richard Leakey, National Museum, Nairobi; J.-P. Lehman, National Museum of Natural History, Paris, and Darwish Alfar, the director of the Cairo Museum of Geology, for access to specimens in their respective collections which were used in connection with this review.

References

Allbrook, D., and W. W. Bishop. 1963. New fossil hominoid from Uganda. *Nature* 197:1187–90.

Andrews, C. W. 1906. *A descriptive catalogue of the Tertiary Vertebrata of the Fayûm Egypt.* Brit. Mus. (Nat. Hist.) Monog., 324 pp.

Andrews, P. 1970. Two new fossil primates from the lower Miocene of Kenya. *Nature* 228:537–540.

———. 1971. *Ramapithecus wickeri* mandible from Fort Ternan, Kenya. *Nature* 231:192–194.

———. 1973. Miocene primates (Pongidae, Hylobatidae) of East Africa. Ph.D. thesis, Cambridge University. 511 pp.

———. 1974. New species of *Dryopithecus* from Kenya. *Nature* 249:188–190, 680.

———. 1976. Taxonomy and relationships of fossil apes. *Int. Cong. Primat. Soc.* 5.

Andrews, P., and E. L. Simons. 1977. A new African Miocene gibbon-like genus, *Dendropithecus* (Hominoidea, Primates) with distinctive postcranial adaptations: Its significance to origin of Hylobatidae. *Folia Primat.* 28:161–168.

Andrews, P. and Tobien, H. 1977. New Miocene locality in Turkey with evidence on the origin of *Ramapithecus* and *Sivapithecus. Nature* 268: 699–701.

Andrews, P., and J. Van Couvering. 1975. Paleoenvironments in the East African Miocene. In F. S. Szalay, ed., *Approaches to primate paleobiology,* Basel: Karger Press, pp. 62–103.

Andrews, and A. Walker. 1976. The primate and other fauna from Fort Ternan, Kenya. In G. LL. Isaac, and E. R. McCown, eds., *Human origins.* Menlo Park, Calif.: Benjamin, pp. 274–304.

Bishop, W. W. 1964. More fossil primates and other Miocene mammals from north-east Uganda. *Nature* 203:1327–1331.

Bishop, W. W., J. A. Miller, and F. J. Fitch. 1969. New potassium/argon age determinations relevant to the Miocene faunal and volcanic sequence in East Africa. *Am. J. Sci.* 267:269–299.

Bishop, W. W., and G. R. Chapman. 1970. Early Pliocene sediments and fossils from the Northern Kenya Rift Valley. *Nature* 226:914–918.

Beadnell, H. J. L. 1905. *The topography and geology of the Fayum province of Egypt.* Cairo: Egypt Surv. Dept., Pub. Works Min. Publ.

Benda, L., and J. E. Meulenkamp. 1972. Discussion on biostratigraphic correlations in the eastern Mediterranean Neogene. *Z. Deut. Geol. Gesell.* 123:559–564.

Bohlin, B. 1946. The fossil mammals from the Tertiary deposits of Taben-buluk, western Kansu. *Palaeont. Sin.* n.s. C(8b):1–259.

Clark, W. E. LeGros. 1950. New palaeontological evidence bearing on the Hominoidea. *Quart. J. Geol. Soc. Lond.* 105:225–264.

———. 1952. Report on fossil hominoid material collected by the British-Kenya Miocene Expedition 1949–1951. *Proc. Zool. Soc. Lond.* 136:359–373.

Clark, W. E. LeGros, and L. S. B. Leakey. 1950. Diagnoses of East African Miocene Hominoidea. *Quart. J. Geol. Soc. Lond.* 105:260–263.

———. 1951. The Miocene Hominoidea of East Africa. *Fossil Mammals of Africa,* no. 1, Brit. Mus. (Nat. Hist.).

Clark, W. E. LeGros, and D. P. Thomas. 1951. Associated jaws and limb bones of *Limnopithecus macinnesi. Fossil Mammals of Africa,* no. 3, Brit. Mus. (Nat. Hist.).

Conroy, G. C., and J. Fleagle. 1972. Locomotor behaviour in living and fossil pongids. *Nature* 237:103–104.

Delson, E., and P. Andrews. 1976. Evolution and interrelationships of the catarrhine primates. In W. P. Luckett and F. S. Szalay, eds., *Phylogeny of the primates,* New York: Plenum, pp. 405–446.

Every, R. G. 1970. Sharpness of teeth in man and other primates. *Postilla* 143:1–30.

Ferembach, D. 1958. Les limnopitheques du Kenya. *Ann. Paléont.* 44:240–249.

Fleagle, J. 1975. A small gibbon-like hominoid from the Miocene of Uganda. *Folia Primat.* 24:1–15.

Garn, S. M., A. B. Lewis, D. R. Swindler, and R. S. Kerewsky. 1967. Genetic control of sexual dimorphism in tooth size. *J. Dent. Res.* 46(5):963–972.

Gill, T. 1872. Arrangements of the families of mammals with analytical tables. *Smithsonian Misc. Coll.* 11(1):1–98.

Gregory, W. K. 1922. *The origin and evolution of the human dentition.* Baltimore: Williams and Wilkins.

Gregory, W. K., M. Hellman, and G. E. Lewis. 1938. Fossil anthropoids of the Yale-Cambridge India expedition of 1935. *Carnegie Inst.* (Washington) 495:1–27.

Hopwood, A. T. 1933a. Miocene primates from British East Africa. *Ann. Mag. Nat. Hist.,* ser. 10, 11:96–98.

———. 1933b. Miocene primates from Kenya. *J. Linn. Soc. Zool.* (London) 38:437–464.

Hürtzeler, J. 1968. Questions et réflexions sur l'histoire des Anthropomorphes. *Ann. Paléont.* 54(2):11–41.

Kälin, J. 1961. Sur les primates de l'Oligocene inférieur d'Egypte. *Ann. Paléont.* 74:1–48.

Koenigswald, G. H. R. von. 1969. Miocene Cercopithecoidea and Oreopithecoidea from the Miocene of East Africa. In L. S. B. Leakey, ed., *Fossil vertebrates of Africa,* vol. 1, London and New York: Academic Press, 102 pp.

Lartet, E. 1856. Note sur un grand singe fossile qui se rattache au groupe des singes supérieurs. *C. R. Acad. Sci.* (Paris) 43:219–223.

Leakey, L. S. B. 1943. A Miocene anthropoid mandible from Rusinga, Kenya. *Nature* 152:319–320.

———. 1962. A new lower Pliocene fossil primate from Kenya. *Ann. Mag. Nat. Hist.* 13 (4):689–696.

———. 1967. An early Miocene member of Hominidae. *Nature* 213:155–163.

———. 1968. Lower dentition of *Kenyapithecus africanus. Nature* 217:827–830.

Lewis, O. J. 1972. Evolution of the hominoid wrist. In R. Tuttle, ed., *The Functional and Evolutionary Biology of Primates,* Chicago: Aldine pp. 207–222.

MacInnes, D. G. 1943. Notes on the East African Miocene primates. *J. East Afr. Uganda Nat. Hist. Soc.* 17:141–181.

Mayr, E. 1963. *Animal species and evolution.* Cambridge, Mass.: Belknap Press.

Napier, J. R., and P. R. Davis. 1959. The fore-limb skeleton and associated remains of *Proconsul africanus. Fossil Mammals of Africa,* no. 16, Brit. Mus. (Nat. Hist.), pp. 1–69.

Osborn, H. F. 1908. New fossil mammals from the Fayum Oligocene of Egypt. *Bull. Am. Mus. Nat. Hist.* 24:265–272.

Pickford, M. 1975. Late Miocene sediments and fossils from the northern Kenya Rift Valley. *Nature* 256:279–284.

Pilbeam, D. 1968. The earliest hominids. *Nature* 219:1335–38.

———. 1969. Tertiary Pongidae of East Africa: evolutionary relationships and taxonomy. *Bull. Peabody Mus.* (Yale) 31:1–185.

———. 1976. Neogene hominoids of South Asia and the origins of Hominidae. In Les plus anciens hominides, Colloque VI, IX Congres, *Union Int. Sci. Prehist. et Protohist.,* pp. 39–59, Nice (1976).

Radinsky, L. 1973. *Aegyptopithecus* endocasts: oldest record of a pongid brain. *Am. J. Phys. Anthrop.* 39(2):239–247.

Schlosser, M. 1910. Über einige fossile Säugetiere aus dem Oligocän von Ägypten. *Zool. Anz.* 35:500–508.

———. 1911. Beiträge zur Kenntnis der Oligozänen Landsäugetiere aus dem Fayum (Ägypten). *Paläont. Geol. Öst.-Ung. Orients,* Beitr. 2, 24:51–167.

Simons, E. L. 1959. An anthropoid frontal bone from the Fayum Oligocene of Egypt. *Am. Mus. Novit.* 1967:1–16.

_____. 1960. *Apidium* and *Oreopithecus*. *Nature* 186(4727):824–826.

_____. 1961. An anthropoid mandible from the Oligocene Fayum beds of Egypt. *Am. Mus. Novit.* 2051:1–5.

_____. 1962. Two new primate species from the African Oligocene. *Postilla* 64:1–12.

_____. 1963a. A critical reappraisal of Tertiary primates. In J. Buettner-Janusch, ed., *Genetic and evolutionary biology of the primates.* London and New York: Academic Press, pp. 65–129.

_____. 1963b. Some fallacies in the study of hominid phylogeny. *Science* 141:879–889.

_____. 1964. The early relatives of man. *Sci. Amer.* 211(1):50–62.

_____. 1965. New fossil apes from Egypt and the initial differentiation of Hominoidea. *Nature* 205 (4967):135–139.

_____. 1967a. Review of the phyletic interrelationships of Oligocene and Miocene Old World Anthropoidea. *Problems actuels de palaeontologie (Evolution des vertebres). Coll. Internat. Cent. Nat. Recherche Sci.* 163:597–602.

_____. 1967b. New evidence on the anatomy of the earliest catarrhine primates. In D. Starck, R. Schneider, and H. J. Kuhn, eds., *Neue Ergebnisse der Primatologie.* Stuttgart: Fischer Verlag, pp. 15–18.

_____. 1968a. On the mandible of *Ramapithecus.* In G. G. Kuhn, ed., *Evolution and hominization,* 2d ed. Stuttgart: Fischer Verlag, pp. 139–149.

_____. 1968b. African Oligocene mammals: introduction, history of study, and faunal succession. In E. L. Simons, and A. E. Wood, Early Cenozoic Mammalian Faunas: Fayum Province, Egypt. Part I. *Bull. Peabody Mus.* (Yale) 28:1–21.

_____. 1969. The origin and radiation of the primates. *Ann. N.Y. Acad. Sci.* 167:319–331.

_____. 1971a. Relationships of *Amphiphithecus* and *Oligopithecus. Nature* 232:489–491.

_____. 1971b. A current review of the interrelationships of Oligocene and Miocene Catarrhini. In A. A. Dahlberg, ed., *Dental morphology and evolution.* Chicago: University of Chicago Press, pp. 193–208.

_____. 1972. *Primate evolution. An introduction to man's place in nature.* New York: Macmillan, 306 pp.

_____. 1974. *Parapithecus grangeri* (Parapithecidae, Old World Higher Primates): new species from the Oligocene of Egypt and the initial differentiation of Cercopithecoidea. *Postilla* 166:1–12.

_____. 1976. Relationships between *Dryopithecus, Sivapithecus* and *Ramapithecus* and their bearing on hominid origins. In *Les plus anciens hominidés,* Colloque VI, IX Congrès, *Union Int. Sci. Prehist. et Protohist.,* pp. 60–67.

Simons, E. L., and D. Pilbeam. 1965. Preliminary revision of Dryopithecinae (Pongidae, Anthropoidea). *Folia Primatol.* 3:81–152.

_____. 1972. Hominoid paleoprimatology. In R. Tuttle, ed., *The functional and evolutionary biology of the primates.* Chicago: Aldine, pp. 36–62.

Szalay, F. S. 1970. Late Eocene *Amphipithecus* and origins of catarrhine primates. *Nature* 227:355–357.

Van Couvering, J. A., and J. A. Miller. 1969. Miocene stratigraphy and age determination, Rusinga Island, Kenya. *Nature* 221:628–632.

Vondra, C. F., and B. B. Bowen. 1974. Paleoenvironmental interpretations of the Oligocene Jebel el Qatrani Formation, Fayum Depression, Egypt (UAR). In Essays on African Paleontology, *Ann. Geol. Surv. Egypt,* 2(3).

Zwell, M., and G. C. Conroy. 1973. Multivariate analysis of the *Dryopithecus africanus* forelimb. *Nature* 244:373–375.

Zapfe, H. 1960. Die Primatenfunde aus der miozänen Spaltenfüllung von Neudorf an der March (Děvinská Nova Ves), Tschechoslowakei. *Schweiz. Pal. Abh.* 78:4–293.

9

Ramapithecus (Hominidae, Hominoidea)

E. L. Simons and D. R. Pilbeam

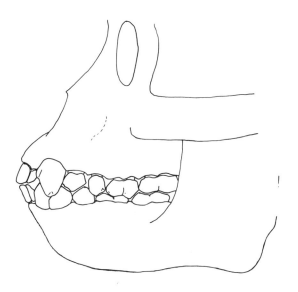

The only known definite *Ramapithecus* specimens recovered in Africa come from two sites in Kenya: Fort Ternan and Maboko Island. The maxillae of one individual from Fort Ternan are similar enough to *Ramapithecus* specimens discovered earlier in India and Pakistan to be considered congeneric; further isolated teeth and mandibular fragments from the African sites have been included in the hypodigm, but with less certainty. The find at Fort Ternan was originally named *Kenyapithecus wickeri* by Leakey (1962) but essentially all authorities now agree, following Simons (1963), that *Kenyapithecus* is a junior synonym of *Ramapithecus*.

Ramapithecus, a form that may well be the earliest hominid (Simons 1961, 1972), was first recognized in India. A right maxilla from Haritalyangar containing crowns, alveoli, or roots of six teeth was described by Lewis in 1934 as *Ramapithecus brevirostris*. Lewis suggested that *Ramapithecus* might be a hominid, having a relatively reduced premaxillary region and anterior dentition, hominidlike canine roots and cheek tooth crowns, and a posteriorly divergent dental arcade. Lewis (1937a) stated that *Ramapithecus* was, in a number of important features, intermediate between Miocene pongids and Plio-Pleistocene *Australopithecus* and *Homo*. Although, regrettably, his final conclusions were never published, Lewis in his Yale Ph.D. dissertation (1937c), clearly stated his belief that *Ramapithecus* was ancestral to *Australopithecus* and *Homo*. In 1938, Gregory, Hellman, and Lewis wrote: "while the Siwalik genus *Ramapithecus* and the South African *Australopithecus* were still simians by definition, they were almost at the human threshold, at least in respect to their known anatomical characteristics."

Simons (1963) suggested that a second maxilla from Haritalyangar in India, previously classified as *Dryopithecus punjabicus,* was in fact a *Ramapithecus*. In 1964, Simons transferred a number of mandibles from other taxa to *Ramapithecus,* and the material as a whole was reviewed by Simons and Pilbeam (1965). The paradigm, as it was then, was discussed by Simons in 1968; other specimens have since been added (Pilbeam 1969, Simons and Pilbeam 1972). For reasons outlined by Simons (1964) the correct species nomen is *R. punjabicus*.

Ramapithecus specimens described to date from the Indian subcontinent come mainly from Haritalyangar in India and the Potwar Plateau region of Pakistan (Colbert 1935). They are found in Siwalik Group rocks of the Chinji and Nagri formations (Lewis 1937b). (Undescribed material at Frankfurt is said to come from the younger Dhok Pathan Formation, Koenigswald 1973). No absolute ages are available at present in Asia; faunal comparisons suggest ages of between ∼ 12 and 14 m.y. for Chinji Formation faunas, between ∼ 9 and ∼ 10 m.y. for definite Nagri Formation faunas, and ∼ 7 to ∼ 9 m.y. for Dhok Pathan Formation faunas (Simons, Pilbeam, and Boyer 1971, Hussain 1973, Delson 1973). Where *Ramapithecus* specimens are of known stratigraphic provenance (Simons 1968), age limits of around 8 to 14 m.y. can be inferred.[1]

A specimen of *Ramapithecus* from a site near Athens, Greece, that is called Pyrgos Wassilissis or Tour la Reine, has a probable age of around 9 m.y. It has been discussed in detail by Koenigswald (1972) and Simons (1977a) and adds important new information to our understanding of *Ramapithecus*.

Specimens of *Ramapithecus* (three isolated teeth) were collected from Maboko Island prior to the 1950s, but were unrecognized as such until 1973. One (an M[1]) was described as *Sivapithecus africanus* by Clark and Leakey (1951).

In 1961, Heslon Mukiri, working with a group led by L. S. B. Leakey, found several specimens of a new hominoid at Fort Ternan in Kenya (right and left maxillae, left upper canine, right lower second molar). Leakey believed they came from one individual, which he described as *Kenyapithecus wickeri* (Leakey 1962). In 1962, more material was recovered. All these specimens are mentioned by Andrews (1971) and Walker and Andrews (1973), and discussed in detail in Andrews and Walker (1976).

In 1967, Leakey expanded his diagnosis of *Kenyapithecus wickeri* to include an upper central incisor found in 1962. In the following year, Leakey (1968) described a symphyseal fragment, with left P_3 and P_4 crowns, canine, and incisor alveoli, also found in 1962, as a *Dryopithecus,* mainly because of the presence of a marked inferior transverse torus and mesiodistally elongated P_3.

Beginning in 1963, one of us (Simons) and then other workers concluded that *K. wickeri* is an African *Ramapithecus* (most widely used binomen *R. wickeri*–Simons 1964, Pilbeam 1966, Andrews 1971). The two species have a number of morphologically very similar features, particularly cheek tooth occlusal morphology, tooth root and sinus relations,

and tooth proportions. Therefore, the relatively more complete Eurasian material can, when utilized with care, be helpful in analyzing the Kenya *Ramapithecus*.

Recently, more information has become available on the Fort Ternan specimens (Andrews 1971, Walker and Andrews 1973, Andrews and Walker 1976). The following material is now known: left and right maxillae (KNM-FT 46 and 47) of one individual; a symphyseal fragment, probably from the same individual (FT 45); an isolated lower molar—possibly M_2, probably of a second individual (FT 48); an isolated lower right canine (FT 3318); a right mandibular fragment (FT 7). This material is described and figured in a number of publications (Leakey 1962, 1967, 1968, Andrews 1971, Walker and Andrews 1973, Andrews and Walker 1976). One other undescribed M[1] specimen from Fort Ternan (Clark and Leakey 1951) and three upper and lower canines from Maboko Island are also probably *R. wickeri*.

Of some importance is the fact that the central incisor (FT 49) associated by Leakey (1967) with the maxillae (FT 46 and 47) can now be shown clearly to be a *Dryopithecus* and not a *Ramapithecus*. Discussions of dental proportions in *Ramapithecus* based on FT 49 are therefore meaningless.

The age of the Fort Ternan deposits has been determined within rather close limits (Bishop, Miller, and Fitch 1969). They are underlain by a biotite-rich crystal-lithic tuff potassium-argon dated at 14.7 ± 0.4 m.y. and 14.2 ± 0.2 m.y., and capped by a porphyritic phonolite yielding $^{40}Ar/^{39}Ar$ ages of 12.6 ± 0.7 m.y. and 12.5 ± 0.4 m.y. There are certain similarities between the faunas of the Kenya site and those from the Chinji Formation of India and Pakistan (Simons 1969, Gentry 1970), although the Fort Ternan fauna may be older than the bulk of the material from India and Pakistan (for example, see Aguirre 1972:517). Recent work by one of us (Pilbeam), as yet unpublished, at Maboko Island suggests an age similar to or slightly older than that of Fort Ternan.

Even the Fort Ternan fauna has been only partially described (Leakey 1968, Gentry 1970, Churcher 1970, Lavocat 1964, Hooijer 1968, Andrews and Walker 1976) but it is thought to contain both forest and open country elements. The latter are more plentiful than in early Miocene East African faunas (Andrews and Walker 1976), suggesting that woodland and grassland were more abundant in the middle Miocene of East Africa, at least at Fort Ternan.

---

1 Koenigswald (1973) has recently stated that the original *R. brevirostris* maxilla from Haritalyangar is geologically younger than Nagri, as originally believed by Lewis (1934). Lewis himself (1937a) clearly indicated that the specimen was associated rather with an older Nagri age fauna, and this point was merely reiterated by Simons in 1964 and 1968.

Description

A recent composite reconstruction of *R. wickeri* has been attempted by Walker and Andrews (1973; for further discussion see Andrews 1971 and Andrews and Walker 1976). Based on the actual root sockets the reconstructed specimen has very small incisors and small canines, relative to cheek tooth size. Morphologically, the canine crowns are relatively pongidlike rather than resembling those of hominids; however they are relatively lower crowned than in pongids. Cheek teeth are low-cusped with thick enamel, wearing flat without early enamel perforation; occlusal contact facets are oriented predominantly in a transverse direction and develop early in life. The face is relatively flat, with an abbreviated snout unlike that known for any ape. The tooth rows are reconstructed by them with negative posterior divergence in the region of the premolars; the mandibular body (based on FT 45) is shallow and robust; the symphysis is long, inclined markedly to the occlusal plane, and exhibits a pronounced inferior transverse torus. Unfortunately, FT 45 is crushed dorsoventrally, which has probably affected symphyseal orientation, depth, and perhaps other features too.

Four Eurasian specimens, YPM-13799 (the original *R. brevirostris* maxilla from Haritalyangar, India, on which the first discussions of *Ramapithecus* were based), the Pyrgos mandible from near Athens, Greece, the Çandir mandible from Turkey (Tekkaya 1975), and the Gandakas mandible from Pakistan (Pilbeam et al. 1977) can in each case yield important and generally better information for such reconstructions but, of course, only two of these had been published on by 1972. The Greek mandible (Pyrgos) has a complete and relatively undistorted ventral border, showing corpora gently diverging from a very narrow incisal region. The maxilla, YPM-13799, although not reaching across the midline, is also useful. As one of us (Simons 1961) has shown, total incisor breadth can be estimated within reasonable limits because the right lateral incisor root and part of the central incisor alveolus are preserved; it is unlikely that the bi-incisor breadth exceeded 20 mm.[2] The palate in YPM-13799 does not reach the midline, no trace of either interpalatine suture nor palatine foramen being preserved. This fact, together with knowledge of probable incisor

breadth, means that tooth rows must have diverged posteriorly to a distinct degree as originally deduced by Simons (1961) and could not have been parallel (as indicated by the drawing in Genet-Varcin 1969:96, fig. 47c). The divergence posteriorly of the tooth rows cannot be decreased by inserting large incisors, for the root of I² and socket of I¹ in YPM-13799 prove that these teeth were small, as noted by Simons (1961) and as originally pointed out by Lewis as long ago as 1934. Simons' figure 2 (1961) shows that even if the palatal fragment and its mirror image are contacted at the midline the angle of posterior divergence would not narrow significantly. In fact, posterior divergence is increased above that of his reconstruction if the bi-incisal breadth is as low as 20 mm. Comparatively small incisor size is probable in *Ramapithecus* as already indicated in YPM 13799. Small incisors or incisor roots or sockets are also seen in a range of newer *Ramapithecus* finds. Such evidences include the demonstration by Walker and Andrews (1973) that Fort Ternan, Kenya, *Ramapithecus* had remarkably small lower incisors.

The strictures of Vogel (1975) that Simons (1961) made a reconstruction of *Ramapithecus* that has a "falsifying and misleading effect" are themselves misleading because of Vogel's own misapprehensions. Most of his errors arise from his having published research on figures and not on the original fossil, whose correct orientation can be easily determined by visual inspection. Vogel states that Simons' figure 1 is a correct orientation—not the reconstruction of his figure 2. To the contrary Simons' figure 1 (1961) is a photograph that is taken out of

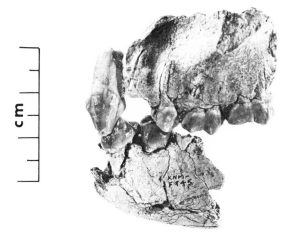

Figure 9.1 The *Ramapithecus* lower face and teeth from Fort Ternan, Kenya. (Scale in cm.) (Photograph courtesy of Peter Andrews.)

2 The reconstruction by Genet-Varcin (1969:96, fig. 47c) is clearly in error, with the hypothetically restored incisors being drawn in far too large and broad. Her drawing of the lateral incisor, for instance, has a reconstructed root twice as broad as the broken-off root actually sketched in this same drawing.

line with the correct orientation; the palatal floor can be seen to arch up out of sight, the incisor root-long axes diverge sharply out away from the mid-sagittal plane, and part of the rostral wall of the face above the molars appears exposed in this photograph. Thus, this figure is not taken in the correct orientation for determination of palatal arcade arrangement by mirror-image reconstruction, but was published to show the clearest details of tooth crown structure (see also Simons 1977b).

Two palates that are of *Ramapithecus* (or a closely related genus, as believed by their describer) have recently been reported from Rudabánya, northeast Hungary, by Kretzoi (1975) and have been seen and studied independently by both of us. The best of the Hungarian finds is Rud-12, an upper left jaw along which the midline suture can be traced for more than 3 cm. This allows for an upper tooth row reconstruction by mirror image photography. Rud-12 preserves central incisor through M^1, lacking only the lateral incisor (whose socket is clearly indicated). This specimen also shows that the palatal floor is flat, not arched up and away from the teeth. The angle of posterior divergence of the upper tooth rows in Rud-12 is 19° while that measured from Simons' reconstruction (1961) is 21°. Consequently both palates agree in their degreee of divergence allowing for the possibility of 2° to 3° range of error in measurement, coupled with even greater presumed individual variation. A hypothetical reconstruction of the Fort Ternan *Ramapithecus* published by Walker and Andrews (1973), however, has cheek tooth rows that diverge posteriorly at an angle of only 10°. This has to be considered out of line with the angles listed above. Consequently, the validity of their reconstruction has to be examined. Because the *Ramapithecus* symphyseal fragment from which they deduced tooth row divergence ends at P_4, the angle of posterior divergence has to be read by lines drawn through the four premolar protocones. These lines diverge at an angle of −8°: that is, they are posteriorly convergent. However, the published illustration of these authors shows widely divergent M_{1-3} of both sides attached to the symphyseal fragment and its mirror image, so that taken together the reconstituted arcades do diverge posteriorly at an angle of 14°. Nevertheless, there is no contact between the premolar bearing and molar bearing fragments from Fort Ternan that could confirm such a curved alignment of teeth. Further, there is no other independent evidence that any Tertiary hominoids have existed that had anterior mandible and teeth that converged posteriorly and then diverged. Because upper and lower arcades must open out at

about the same angle in order to occlude, any hypothetical reconstruction of the Fort Ternan palate and teeth, to be in harmony with the degree of divergence posteriorly of lower teeth, should give approximately the same angle above and below—in this case 14°, not 10°. As we shall show below this angle is only two-thirds of that in other *Ramapithecus*.

Since the early 1960s it has become apparent that the angles which describe the posteriorly divergent parabolas outlining dental arcades in extinct hominoids vary considerably but that only in modern apes are tooth rows essentially parallel. Simons and Chopra (1969:139) remark:

The degree of posterior divergence of hominoid tooth-rows can be expressed as the percentage ratio of the breadth between the insides of the canines to that between the second molars. In members of *Homo* and *Australopithecus* this percentage is typically below 50%. In modern pongids the approximate range is between 70-110%, but primitive apes . . . had posteriorly divergent rami with the percentage running between 55% to 65%. The percentages for *Gigantopithecus* are between those for primitive hominoids and those for *Australopithecus,* being as follows in the four known mandibles: 54.8%, 55.5%, 51.1%, 44.6%.

The most recently reported finds of *Ramapithecus* mandibles fortunately can settle the issue of mandibular arcade arrangement. In the Pyrgos, Çandir, and Gandakas mandibles both horizontal rami are preserved around the symphysis and distortion of their degrees of lateral spacing and of tooth row divergence posteriorly seems to be minimal. Moreover, because in all three canine roots or root sockets are preserved and M_2 is in place on one or both sides, it is possible to measure degree of angular divergence posteriorly and to calculate the ratio of the intercanine breadth divided into the inter-M_2 breadth. These measures, which are practically the same in all three, corroborate one another (Table 9.1). They also tend to corroborate the accuracy of the angles of posterior tooth row divergence illustrated through mirror imaging for the Yale palate in 1961 by Simons (21°) and the Rudabánya, Hungary, palate (19°).

Thus, in five *Ramapithecus* finds where the posterior divergence can be deduced, the measure ranges from 19° to 22° with a mean at about 20°. The divergence hypothesized for the Fort Ternan palatal arcade of 10° is in error as is the divergence of 14° determined for the Fort Ternan mandible. The posterior convergence of −8° in the alignment of the premolar protoconids in the Fort Ternan symphyseal fragment shows that these teeth are not correctly aligned in the 1973 reconstruction. For instance in the Çandir, Turkey, mandible the angle of

Table 9.1 Comparison of *Ramapithecus* mandible.

Site of *Ramapithecus* mandible	Çandir, Turkey	Gandakas, Pakistan	Purgos, Greece
Percentage ratio: breadth between the insides of canines and insides of M₂s	50%	40%	50%
Angle of posterior divergence of the parabola lying above the center of tooth rows and horizontal rami	20°	22°	21°

posterior divergence calculated through the premolar protoconids alone is 14° divergence. This makes a total of 22° discrepancy of alignment between the reconstructed premolar placement of Walker and Andrews (1973) and that seen in the Çandir mandible, where the premolars concerned are fixed in place in a complete and undistorted jaw.

All the available evidence about the various larger living and extinct hominoid genera leads to one main conclusion: the angle of posterior divergence and amount of lateral spacing of tooth rows that describes the dental arcades and that accounts for their parabolic outline, when determined for the various hominoid genera, puts them into an orderly range of form with modern *Homo* and modern pongids at opposite ends. Both *Ramapithecus* (nearer to *Australopithecus*) and *Gigantopithecus* (farther from *Australopithecus*) lie between *Australopithecus* and *Dryopithecus*. Simons (1972) shows how Miocene apes, with tooth rows that were only slightly divergent posteriorly, could serve as the starting point for two different trends: (1) the trend toward the condition of modern pongids has been to widen out the anterior part of the dentition, so that larger or more widely spaced incisors broaden the anterior end and thus make it more U-shaped; and (2)conversely, the hominid trend, through time, has been to widen the tooth rows posteriorly. With a posterior divergence angle of around 20° and an intercanine–inter-M₂ ratio of about 47%, *Ramapithecus* has the appropriate arcade arrangement to be expected for a basal hominid.

Discussion

Ramapithecus wickeri presents a unique constellation of hominid-pongid features. Some are resemblances to *Dryopithecus* and extant pongids (especially *Pan paniscus*), strongly suggesting derivation of the hominid from a Miocene dryopithecine. For example, the canine-premolar complex differs from that found in *Australopithecus* and *Homo*. Although small, the canines are not incisiform, and the distal

border of upper C occludes with a mesially elongated P₃. P₃ has two cusps, although the lingual cusp is small—it resembles P₃ in some *Pan* and *Gorilla* individuals in being partially molarized—though still honing with upper C.

Several other *Ramapithecus wickeri* characters differ somewhat from living and fossil Pongidae and from the Plio-Pleistocene Hominidae and appear intermediate between them; for example, arcade shape, symphyseal morphology, and such dental characters as relative size of P₃ metaconid (see Andrews 1971, Walker and Andrews 1973, Simons 1972, Pilbeam 1972, Conroy and Pilbeam 1974, Andrews and Walker 1976).

Compared to earlier and penecontemporaneous dryopithecines, known now from many hundreds of specimens, *Ramapithecus*, including *R. wickeri*, differs in a whole complex of features from these extinct apes. The differences appear to relate mainly or even exclusively to the presence in *Ramapithecus* of a masticatory apparatus different from that seen in any dryopithecine or pongine, save for *Gigantopithecus*, one adapted to powerful chewing and heavy wear; it seems likely that powerful transverse movements were most important in mastication.

Dental proportions, with small—even tiny—anterior teeth and relatively large cheek teeth point to a complex like that seen in *Theropithecus* (Jolly 1972), *Hadropithecus* (Jolly 1970, Tattersall 1973), *Australopithecus* (Robinson 1956, Pilbeam and Gould 1974, Wolpoff 1973), *Gigantopithecus* (Simons and Ettel 1970, Pilbeam 1970), and some nonprimates (e.g., the giant panda and the sea otter). Body size is unknown in *Ramapithecus*, but in overall size and proportions of maxilla and mandible, the fossil is closest in size to *Pan paniscus;* however, *Ramapithecus* has very much larger cheek teeth than the pygmy chimpanzee (see Simons 1969, 1972).

The mandibular body in *R. wickeri* is shallow and robust, and markedly more massively built than in *Pan paniscus;* the symphysis is long and well buttressed. Malar processes flare laterally from a marked canine fossa. As Walker and Andrews (1973) have demonstrated, *Ramapithecus wickeri* was a narrow-snouted species with a broad, flat face.

The cheek tooth roots are firmly set in dorsoventrally thickened alveolar bone, the maxillary sinuses remaining wholly above the root tips. The enamel is thicker than in pongid teeth; cusps are low, and wear produces a flat occlusal surface before the enamel perforates.

All these features suggest that *R. wickeri* (as well as Eurasian *Ramapithecus*) was adapted for powerful mastication in which transverse movements were particularly important (Pilbeam 1969, Andrews 1971, Walker and Andrews 1973, Pilbeam 1972, Simons 1972, Conroy and Pilbeam 1975); dental proportions indicate that the incisors were relatively unimportant in food preparation and that cheek teeth acted principally like a dental mill. It can be inferred very strongly that food eaten was highly resistant to mastication, although it is not possible to be more precise as to what it was.

Evolutionary Relationships and Classification

Ramapithecus clearly exhibits ties with the Dryopithecinae, and also interesting resemblances to *Pan paniscus,* the living pongid most similar to *Ramapithecus* and many *Dryopithecus* in body size. This is hardly surprising. In many features, though, *Ramapithecus* is uniquely different morphologically and presumably adaptively from these apes. The masticatory complex of *Ramapithecus* is most similar to that of *Australopithecus,* as are many features of occlusal morphology. It differs in canine morphology and arcade shape, but not, in our opinion, sufficiently to rule out an ancestral-descendant relationship between the two genera.

The considerable time gap between the youngest *Ramapithecus* (around 8 m.y.) and the bulk of the later African hominid material (equal to or less than ~4 m.y.) makes definite phylogenetic conclusions impossible to prove. However, we do consider it highly probable (say, a 75% likelihood) that *Ramapithecus* is ancestral to later Cenozoic hominids.

Whether or not *Ramapithecus* is classified as a hominid is partly a matter of taste, because the line between Pongidae and Hominidae is arbitrary. Moreover, *Ramapithecus* need not be an ancestor of *Australopithecus* or *Homo* to be a hominid. However, it is general paleontological practice to include as first members of families the earliest forms to provide structural evidence for the adaptive shift(s) that later came to characterize the family. In our view, *Ramapithecus* foreshadows the essential features of the dental mechanism of later hominids and is thus likely to be at least broadly ancestral to undoubted late Cenozoic hominids. Therefore we believe that, as has usually been done, it should continue to be recognized as the earliest detected and detectable member of Hominidae.

References

Aguirre, E. 1972. Les rapports phylétiques de *Ramapithecus* et de *Kenyapithecus* et l'origine des Hominidés. *L'Anthrop.* 76:501–523.

Andrews, P. 1971. *Ramapithecus wickeri* mandible from Fort Ternan, Kenya. *Nature* 230:192–194.

———. 1973. Miocene primates (Pongidae, Hylobatidae) of East Africa. Ph.D. thesis, Cambridge University.

Andrews, P., and A. C. Walker. 1976. The primate and other fauna from Fort Ternan, Kenya. In G. LL. Issac and E. R. McCown, eds., *Human Origins.* Menlo Park, Calif: Benjamin, pp. 279–304.

Bishop, W. W., J. A. Miller, and F. J. Fitch. 1969. New potassium-argon age determinations relevant to the Miocene fossil mammal sequence in East Africa. *Am. J. Sci.* 267:669–699.

Churcher, C. S. 1970. Two new Upper Miocene Giraffids from Fort Ternan, Kenya, East Africa. In L. S. B. Leakey and R. J. G. Savage, eds., *Fossil vertebrates of Africa,* vol. 2, London: Academic Press, pp. 1–105.

Clark, W. E. LeGros, and L. S. B. Leakey. 1951. The Miocene Hominoidea of East Africa. *Fossil Mammals of Africa* 1, Brit. Mus. (Nat. Hist.).

Colbert, E. H. 1935. Siwalik mammals in the American Museum of Natural History. *Trans. Am. Phil. Soc.* 26:1–401.

Conroy, G. C., and D. R. Pilbeam. 1975. *Ramapithecus:* a review of its hominid status. In R. H. Tuttle, ed., *World anthropology. Paleoanthropology, morphology and paleoecology,* pp. 59–86, The Hague: Mouton, 1975.

Delson, E. 1973. Fossil colobine monkeys of the circum-Mediterranean region and the evolutionary history of the Cercopithecidae (Primates, Mammalia), Ph.D. thesis, Columbia University.

Genet-Varcin, E. 1969. À la recherche du primate ancêtre de l'homme. Paris: Boubée et Cie.

Gentry, A. 1970. The Bovidae (mammals) of the Fort Ternan fossil fauna. In L. S. B. Leakey and R. J. G. Savage, eds., *Fossil Vertebrates of Africa,* vol. 2. London: Academic Press, pp. 243–323.

Gregory, W. K., M. Hellman, and G. E. Lewis. 1938. Fossil anthropoids of the Yale-Cambridge Indian expedition of 1935. *Carnegie Inst. Publ.* 495:1–27.

Hooijer, D. 1968. A rhinoceros from the Late Miocene of Fort Ternan, Kenya. *Zool. Meded.* 43:77–82.

Hussain, S. T. 1973. Appearance of *Hipparion* in the Tertiary of the Siwalik Hills of North India, Kashmir and Pakistan. *Nature* 246:531.

Jolly, C. J. 1970. *Hadropithecus:* a lemuroid small-object feeder. *Man* 5:619–626.

———. 1972. The classification and natural history of *Theropithecus (Simopithecus)* (Andrews, 1916), baboons of the African Plio-Pleistocene. *Bull. Brit. Mus. (Nat. Hist.)* 22.

Koenigswald, G. H. R. von. 1972. Ein Unterkiefer eines fossilen Hominoiden aus dem unterpliozän Greichenlands. *Kon. Ned. Akad. Wetensch.* 75:385–394.

———. 1973. *Australopithecus, Meganthropus* and *Ramapithecus*. *J. Hum. Evol.* 2:487–491.

Kretzoi, M. 1975. New ramapithecines and *Pliopithecus* from the Lower Pliocene of Rudabánya in north-eastern Hungary. *Nature* 257:578.

Lavocat, R. 1964. Fossil rodents from Fort Ternan, Kenya. *Nature* 202:1131.

Leakey, L. S. B. 1962. A new Lower Pliocene fossil primate from Kenya. *Ann. Mag. Nat. Hist.* 4:686–696.

———. 1967. An early Miocene member of Hominidae. *Nature* 213:155–163.

———. 1968. Upper Miocene primates from Kenya. *Nature* 218:527–528.

Lewis, G. E. 1934. Preliminary notice of manlike apes from India. *Am. J. Sci.* 27:161–181.

———. 1937a. Taxonomic syllabus of Siwalik fossil anthropoids. *Am. J. Sci.* 34:139–147.

———1937b. A new Siwalik correlation. *Am. J. Sci.* 33:191–204.

———. 1937c. Siwalik fossil anthropoids. Ph.D. thesis, Yale University.

Pilbeam, D. R. 1966. Notes on *Ramapithecus,* the earliest known hominid, and *Drypithecus*. *Am. J. Phys. Anthropol.* 25:1–5.

———. 1969. A newly recognised mandible of *Ramapithecus*. *Nature* 222:1093–1094.

———. 1970. *Gigantopithecus* and the origin of Hominidae. *Nature* 225:516–519.

———. 1972. *The ascent of man*. New York: Macmillan.

Pilbeam, D. R., and Gould, S. J. 1975. Size and scaling in hominoid evolution. *Science* 186:892–901.

———. 1975. Allometry and early hominids. (Reply to discussion) *Science* 189:64.

Robinson, J. T. 1956. The dentition of the Australopithecinae. *Transv. Mus. Mem.* 9:1–178.

Simons, E. L. 1961. The phyletic position of *Ramapithecus*. *Postilla* (Yale) 57:1–20.

———. 1963. Some fallacies in the study of hominid phylogeny. *Science* 141:879–889.

———. 1964. On the mandible of *Ramapithecus*. *Proc. Nat. Acad. Sci.* 51:528–535.

———. 1968. A source for dental comparison of *Ramapithecus* with *Australopithecus* and *Homo*. *S. Afr. J. Sci.* 64:92–112.

———. 1969. The late Miocene hominid from Fort Ternan, Kenya. *Nature* 221:448–451.

———. 1972. *Primate evolution*. New York: Macmillan.

———. 1977a. Europe's earliest hominids. *Nature* (in press).

———. 1977b. Reconstructing *Ramapithecus*. *Folia Primatol*. (in press).

Simons, E. L., and Chopra, S. R. K. 1969. A preliminary announcement of a new *Gigantopithecus* species from India. *Proc. 2nd. Int. Congr. Primat.,* Basel and New York: Karger, vol. 2, pp. 135–142.

Simons, E. L., and Ettel, P. 1970. *Gigantopithecus*. *Sci. Amer.* 222:76–85.

Simons, E. L., and Pilbeam, D. R. 1965. Preliminary revision of the Dryopithecinae (Pongidae, Anthropoidea). *Folia Primatol.* 3:81–152.

———. 1972. Hominoid paleoprimatology. In R. H. Tuttle, ed., *The functional and evolutionary biology of primates*. Chicago: Aldine.

Simons, E. L., Pilbeam, D. R. and Boyer, S. J. 1971. Appearance of *Hipparion* in the Teritary of the Siwalik Hills of North India, Kashmir and West Pakistan. *Nature* 229:408–409.

Tattersall, I. 1973. Cranial anatomy of the Archaeolemurinae (Lemuroidea, Primates). *Anthrop. Pap. Am. Mus. Nat. Hist.* 52(1).

Tekkaya, I. 1975. A new species of Tortonian anthropoid (Primates: Mammalia) from Anatolia. *Maden Tetkik Arama Enstitutu, Bull.* 148.

Vogel, C. 1975. Remarks on the reconstruction of the dental arcade of *Ramapithecus*. In R. H. Tuttle, ed., *World anthropology. Paleoanthropology, morphology and paleoecology*. The Hague: Mouton, pp. 87–98.

Walker, A. C., and Andrews, P. 1973. Reconstruction of the dental arcades of *Ramapithecus wickeri*. *Nature* 224:313–314.

Wolpoff, M. H. 1973. Posterior tooth size, body size, and diet in South African gracile australopithecines. *Am. J. Phys. Anthropol.* 39:375–393.

10

Hominidae

F. C. Howell

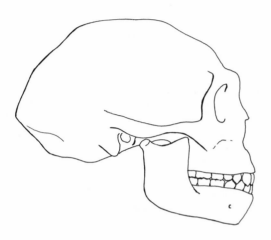

Just over 50 years ago the African continent yielded its first fossil evidence for a distinctive ancient human population, when a largely complete cranium was fortuitously recovered from a commercial lead mine at Broken Bill, Zambia (then Northern Rhodesia). It was the first clear demonstration—although artifacts of Paleolithic type had been known since the previous century—that the continent might preserve a Quaternary fossil record of ancient humanity. In 1925 R. A. Dart's recognition of a primitive hominid skull, recovered, along with other primate fossils, during commercial lime-quarrying at Buxton, near Tau'ng (northern Cape Province), was exposed to an unreceptive and skeptical world of science.

Over the past half century, parts of the African continent have increasingly become primary foci for investigations into the origin and evolution of Hominidae and the development of hominid cultural capabilities. Skeletal parts of Hominidae are now known from nearly every range of late Cenozoic time on the continent. The ever-accelerating pace of intensified field researches suggests that, unless thwarted by nationalistic emotionalism, the African continent will afford the fullest documentation of hominid biological and cultural evolution in the world of the late Cenozoic.

Stratigraphic and Geographic Setting

Vast areas of the African continent (over 30 million km²) still remain inadequately explored; their paleoanthropological significance is consequently still largely or even wholly unknown. Most of the fossil record of Cenozoic Hominidae derives from restricted portions of the Maghreb and from eastern and southern Africa (figure 10.1). However, abundant archeological evidence attests to extensive distributions of human populations, especially during the middle and late Quaternary, over much of the continent (Clark 1967). Hence, the fossil record and its geographic representation reflect for the most part those portions of the continent most intensively colonized by Europeans and where conditions have been suitable for the preservation and subsequent exposure of organic remains. Such areas are also of fundamental importance to the development of lithostratigraphic and biostratigraphic successions and, in eastern Africa particularly, they have provided the most adequate bases for relative and absolute age calibration.

The oldest occurrences of Hominidae are restricted to sub-Saharan Africa. In eastern Africa all known occurrences are related to the Rift Valley

Figure 10.1   Principal African localities of late Cenozoic age yielding skeletal remains of Hominidae.

System in Ethiopia, Kenya, and Tanzania. In southern Africa all known occurrences are related to karstic fissure and cavern infillings (Brain 1958). The oldest substantiated occurrence is late Miocene. The final Pliocene is well documented. In both eastern and southern Africa this fossil record continues into

the earlier Quaternary.[1] The latter record is most well documented in eastern Africa and only imper-

1 The Tertiary-Quaternary boundary is set at ~ 1.8 m.y. following most current biostratigraphic conclusions and correlations from marine micropaleontology and magnetostratigraphy (see Berggren 1973).

fectly in the Maghreb as it passes up into the mid-Quaternary.[2] Unfortunately, the late Quaternary is sparsely documented everywhere in the continent (probably largely because of the relative paucity of occupied caves and rock shelters). Overall, the most extensive documentation, and one still unique to Africa, is for the Pliocene-Pleistocene time range between ~ 3 and 1 m.y. ago.

Useful recent general discussions of the age, associations, and contexts of Pliocene-Pleistocene Hominidae, with particular reference to eastern Africa, are provided by Maglio (1970), Howell (1972), Cooke and Maglio (1972), and by Bishop (1971, 1973). Very useful summaries of the Neogene in eastern Africa are provided by Bishop (1972; see also 1967, 1971).

For the later Neogene and earlier Quaternary time range, two eastern African localities particularly provide critical evidence for the absolute and relative ages, contexts, and vertebrate associations of early Hominidae. Olduvai Gorge in the Serengeti (northern Tanzania) and the lower Omo basin (southern Ethiopia) expose substantial sedimentary accumulations (100 m and 1.100 m, respectively) in association with volcanic rocks and tephra. Conventional potassium-argon radiometry (total degassing method) and intensive paleomagnetic sampling afford a consistent and concordant geochronology in association with a rich diversity of biostratigraphic data. These two broadly continuous successions provide, at least for the moment, the essential basis for correlating other relevant occurrences in eastern Africa.

The temporal relationships of the best known early hominid successions of eastern Africa are set out in figure 10.2. These successions are internally and externally consistent on the basis of conventional K-Ar determinations, magnetostratigraphy, and biostratigraphy.[3]

Leaving aside later mid-Miocene occurrences of *Ramapithecus*[4] or related taxa, the hominoid nature and distinctiveness of which are fully apparent although the hominid affinities are still controversial (Simons and Pilbeam chapter 9, this volume), there is only the scantiest of evidence for Hominidae prior to the Pliocene. However, three occurrences in the central Kenya Rift Valley afford suggestive evidence of Hominidae in the upper Miocene.

The Ngorora Formation, 215–400 m thick, outcrops among the Tugen Hills east of the Elgeyo Escarpment and west of Lake Baringo (Bishop and Chapman 1970, Bishop et al. 1971). The formation occurs in a structural basin floored by phonolites which are ~ 12 m.y. old; it comprises five fossiliferous members that represent largely fluviatile but also freshwater and alkaline lacustrine sedimentation. A diverse and substantial vertebrate (and invertebrate) faunal assemblage, the mammals of which are of 'Vallesian' age, is represented particularly in the middle three members (B, C, D) (Bishop and Pickford 1975). A hominoid upper molar, considered to have hominid affinities, occurred in the middle member (C), which represented a period of oscillating freshwater lacustrine sedimentation. The minimum age of this occurrence is set by the age (9.6–9.8 m.y.) of overlying tuffaceous sediments and the later Ewalel Phonolite (~ 8.5 m.y.).

The Lukeino Formation lies east of the Tugen Hills and below the fault bounding the east side of the Ngorora Formation (Pickford 1975). It overlies the Kabarnet Trachyte Formation (7 m.y. old) and is overlain by the Kaparaina Basalt Formation (~ 5.4 m.y.). The formation, comprising four members, is some 130 m thick and represents deposition under fluviatile and fresh to weakly saline and lacustrine conditions. Invertebrates and plant fossils are common in several members. The richly fossiliferous lower members yield one of the two most important

---

2 The boundary between the lower and middle Quaternary is here set at ~ 700,000 yr, at the Brunhes-Matuyama polarity epoch boundary, following most recent suggestions and the worldwide applicability of this criterion (see Butzer 1974a).

3 A persistent, confounding exception has been the relative and absolute age(s) of the several formations/members of the East Turkana succession, northern Kenya. In particular the substantial age of 2.6–3.0 m.y. attributed to the Lower Member, Koobi Fora Formation (Brock and Isaac 1974; Findlater et al. 1974) is not congruent with comparative biostratigraphic evidence (Brown, Howell, and Eck 1978; White and Harris 1977); and the nonconventional K-Ar age determinations (employing total 40Ar/39Ar degassing and age spectrum analyses) have yielded equivocal results, in part as a consequence of some unwarranted and untested assumptions. Subsequently the KBS Tuff, taken as the upper boundary of the Lower Member, has been shown to have an age of ~ 1.8 m.y. (Curtis et al. 1975) and further paleomagnetic studies (Hillhouse et al. 1977) now indicate that this part of the succession falls within the Olduvai

Normal Event. Further radiometric dating, coupled with analysis of glass shards and trace elements in tuffs (by T. E. and B. W. Cerling, G. H. Curtis, R. E. Drake and F. H. Brown, pers. comm.), have demonstrated that: the Chari Tuff (the upper boundary of the Upper Member) = Omo Tuff L with an age of ~ 1.35 m.y.; the Koobi Fora Tuff Complex (~ 1.36 m.y.), Okote Tuff (~ 1.45 m.y.) and Ileret Tuff Complex are all distinct, though perhaps of broadly comparable age; several distinct tuffs have been labelled 'KBS' of which the type and several other occurrences = Omo Tuff H2 (~ 1.8 m.y.); and five distinct tuffs have been labelled Tulu Bor of which one, at least, probably = Omo Tuff B-10 (2.9 m.y.). Evidence is now sufficient to demonstrate that at least some of the lower Omo and East Turkana tephra derive from a common source area.

4 The hominid status of *Ramapithecus* (ex-*Kenyapithecus*) *wickeri,* from the mid-Miocene of Fort Ternan (Kenya) is still disputed, and it is accordingly not considered here (see Simons and Pilbeam, chapter 9).

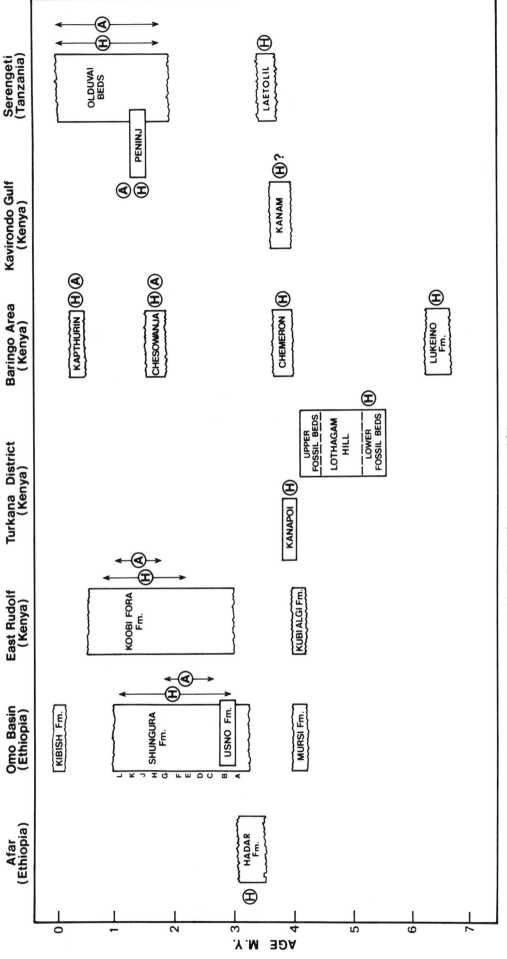

Figure 10.2 Relative and absolute ages of successions yielding Hominidae in eastern Africa.

Turolian age vertebrate assemblages (~ 6.5 m.y.)[5] yet recovered in eastern Africa. A hominoid lower molar that has distinct hominid resemblances was found in the lowest member (A).

The Baringo area has also yielded isolated occurrences of australopithecines of Plio-Pleistocene age from the Chemeron and Chesowanja localities and of *Homo* of mid-Pleistocene age from the Kapthurin Formation (Bishop et al. 1971).

Fragmentary remains (a partial mandible) attributed to Hominidae are known from uppermost Miocene sediments exposed at Lothagam Hill, an isolated fault block to the southwest of Lake Turkana (hitherto Lake Rudolf), Kenya (Patterson, Behrensmeyer, and Sill 1970, Behrensmeyer 1976b). The sediments comprising the Lothagam Group overlie Miocene volcanics; the youngest are ~ 8.3 m.y. old and the middle to upper formational units are intruded by a sill that is ~ 3.7 m.y. old. A "best fit" estimated age for the oldest vertebrate faunal assemblage from the lowest or Lothagam 1 formation is about 6 m.y. That formation, nearly 500 m thick, represents a long succession of fluvial and fluviodeltaic sedimentation. Its rich and diverse vertebrate assemblage, another of the best 'Turolian' age faunas ever recovered in eastern Africa, derives from the upper units (B and C) of the formation, and the hominid specimen derives from the upper part of the highest unit C (Behrensmeyer 1976b).

The Kanapoi Formation, some 75 km south of Lothagam, outcrops to the west of the Kerio River. It represents some 70 m of fluvial and lacustrine sedimentation overlain by a basalt that is ~ 4 m.y. old (Behrensmeyer 1976b). This locality has afforded a very extensive vertebrate fauna of lower Pliocene age and a single fragment (distal humerus) of a hominid (Patterson and Howells 1966).

The Omo succession, extending from >4 m.y. to ~ 1 m.y. ago, yields hominids, usually in fragmentary condition, largely from the 3–1 m.y. time range. The East Turkana succession yields hominids, in quantity and often in an excellent state of preservation, for the most part from the ~ 2 m.y.$-<1$ m.y. time range. The Olduvai succession yields hominids from the ~ 1.8–1.7 m.y. range, as well as from the late Quaternary (M. D. Leakey 1978). The Peninj occurrence is most probably comparable in age, on the basis of correlative radiometric, magnetostratigraphic, and biostratigraphic grounds, with the pre-

Lemuta Member portion of the Olduvai succession, broadly correlative with the Olduvai Normal Event (Isaac and Curtis 1974).

The Hadar Formation, in the west-central Afar Basin, is exposed extensively below the east Ethiopian plateau adjacent to the Awash River. The formation has an aggregate thickness of some 140 m and comprises fluviodeltaic, lake margin, and lacustrine sediments with volcanic tuffs and at least one lava flow (Taieb 1974b, Taieb et al. 1972, 1976). Four members are recognized in the formation (Taieb et al. 1976), and rich and diverse assemblages of fossil vertebrates, including Hominidae (Taieb et al. 1974, 1975a, 1975b, Johanson and Taieb 1976, Johanson, Taieb and Gray 1978) occur in the three upper members. Initial K/Ar age determinations on whole rock (basalt) and several tuffs suggest an age for the bulk of the formation of ~ 3 m.y. or slightly less; this assessment is supported by preliminary paleomagnetic determinations (Aronson et al. 1977).

The Laetolil Beds, exposed in and adjacent to the Garusi River, outcrop on the eastern Serengeti plateau above the Eyasi Rift escarpment. They have long been known to yield a diverse vertebrate fauna with elements more primitive than the oldest vertebrates from adjacent Olduvai Gorge (Dietrich 1942, 1945, 1950). Remains of Hominidae (a maxilla fragment) were first recovered from this area by L. Kohl-Larsen in 1939 (Kohl-Larsen 1943, A. Remane, in Weinert 1950). The initial geological researches by Kent (1942) provided a stratigraphic basis recently amplified and extended by R. L. Hay in conjunction with paleoanthropological and related field studies by M. D. Leakey and associates (1976). The Laetolil Beds disconformably underlie middle and upper Quaternary sediments (Ngaloba Beds), with Acheulian and later archaeological materials, patchy occurrences of Beds I and II of the Olduvai Gorge succession, lavas from local vents that are 2.4 m.y. old, and a distinctive massive marker tuff. Overlying basement complex metamorphics, the Laetolil Beds are aeolian tuffs cemented by zeolites representing a single depositional facies and derived from the sodic alkaline volcanic source of Sadiman. The paleoenvironment was upland, semiarid, and probably sparsely vegetated, as indicated by the wind-worked sediments and the dry savanna faunal assemblages. The upper fifth of this succession is fossiliferous, including a number of jaws and teeth of Hominidae, bracketed by K/Ar ages of 3.35 and 3.75 m.y. These hominids are among the oldest and best dated yet known in Africa.

The several hominid partial crania recovered by L. Kohl-Larsen in 1935 from sediments between the

---

5 The Miocene-Pliocene boundary is set at 5 m.y., following most current biostratigraphic conclusions and correlations from marine micropaleontology (see Berggren and Van Couvering 1974; Berggren 1973, also 1969, 1971, 1972; cf. Funnell 1964).

Mumba Hills and Lake Eyasi, in the floor of the Eyasi trough (Tanzania), have been customarily considered to be of late Quaternary age (Reck and Kohl-Larsen 1936, L. S. B. Leakey 1936b, L. S. B. Leakey and Reeve 1946). Amino acid racemization analysis of hominid bone suggests an age of ~ 34,000 yr (Bada and Protsch 1973). Such an approximation is quite in keeping with the essentially modern aspect of the associated mammals (Dietrich 1939) and is not controverted by the artifactual associations.

The type specimens of most Pliocene-Pleistocene Hominidae derive from cemented infillings of fissures, sinkholes, and caves of the South African Highveld and the Transvaal Plateau Basin. Initial guesses and later more reasonably founded estimates of their relative and correlative ages were made by Ewer (1956, 1957) and Cooke (1963, 1967). However, opinions have been widely divergent over the years. Direct radiometric age calibration, including the use of fission track (MacDougall and Price 1974), now appears to be inapplicable. Geomorphic estimates employing nick-point recession criteria (Partridge 1973) are generally unreliable if not ill-founded. Thus, only biostratigraphic methods now appear to produce consistent results. Such approaches have defined the successive temporal relationships of vertebrate assemblages associated with Hominidae at most of these localities (Ewer 1956, Cooke 1963, 1967, Hendey 1974). All the occurrences fall within a protracted faunal span, sometimes termed Makapanian. Limits to age spans can be estimated on the basis of the temporal duration (and association) of some mammal taxa (particularly proboscideans, suids, bovids, carnivores, and cercopithecoids, as well as microvertebrates) in eastern African successions for which there is direct radiometric and magnetostratigraphic control. Employing this approach and the rate of faunal change in the protracted Omo succession drawn from measures of faunal resemblance (Shuey et al. 1978), it is possible to arrive at a set of projected, "best-fit" age assessments (see figure 10.3). Preliminary paleomagnetic results from the Makapan Limeworks, where a reversed to normal change is recorded in Member II (below the main fossiliferous member III) (Brock et al. 1977), suggests an age between 2.8 and 3.3 m.y.

Along the Atlantic littoral of Morocco a protracted succession of marine and interrelated continental sediments apparently encompasses much of the late Cenozoic (Biberson 1961, 1963, 1971). The middle and younger units of this important sequence sometimes preserve vertebrate fossils, very often substantial artifact assemblages, and, thus far infrequently,

remains of Hominidae (figure 10.4). Hominids occur in time-stratigraphic units successively designated (older to younger) Amirian, Tensiftian, Presoltanian, and Soltanian (Biberson 1964, 1970, Jaeger 1975b). Whereas the stratigraphic relationships of these formations are quite well defined, there is little or no direct evidence to establish their absolute ages. U-series measurements (cf. Stearns and Thurber 1965) give, at best, limiting ages (Kaufman et al. 1971). Relevant radiocarbon control on the late Quaternary time range is still scant, except for the long and consistent series of determinations for the Haua Fteah Cave, Cyrenaican Libya (McBurney 1967; see also 1961, 1962). The very important hominid specimens from Djebel Irhoud (Morocco) are, unfortunately, still of unknown age, although the associated macro- and microvertebrate fauna is of late Quaternary aspect.

Several southern African localities have yielded largely incomplete cranial or jaw parts of Hominidae. On the basis of the associated vertebrate assemblages these can be attributed to older (Cornelian) or younger (Florisian) faunal spans (Cooke 1963, 1967, Wells 1962, Hendey 1974), which are usually considered as middle to late Quaternary. Unfortunately, the actual ages and durations of these "faunal spans" are largely unknown, although some radiocarbon determinations (Beaumont and Vogel 1972, Vogel and Beaumont 1972) and amino acid racemization analyses (Bada et al. 1974) afford some limiting values. However, the measure of uncertainty is clearly apparent from figure 10.5. The oldest occurrences are undoubtedly those from Saldanha (Butzer 1973), as well as, perhaps, Broken Hill (Klein 1973) and the Cave of Hearths. On faunal grounds, all antedate the Florisian faunal span.

The lack of consistent and reliable methods of age assessment for the middle and late Quaternary time ranges severely inhibits temporal placement of such hominid samples and thus inferences drawn from comparative morphological studies and phylogenetic interpretation.

Description

Hominidae gen. et sp. indet. (A)

Hominoid fossils of Vallesian age (Ngorora) and of Turolian age (Lukeino, Lothagam 1-C) have been considered by several workers to be attributable to Hominidae. These several upper Miocene specimens deviate distinctively from known penecontemporaneous or antecedent pongids. They are here referred to Hominidae gen. et. sp. indet. (A). Their affinities

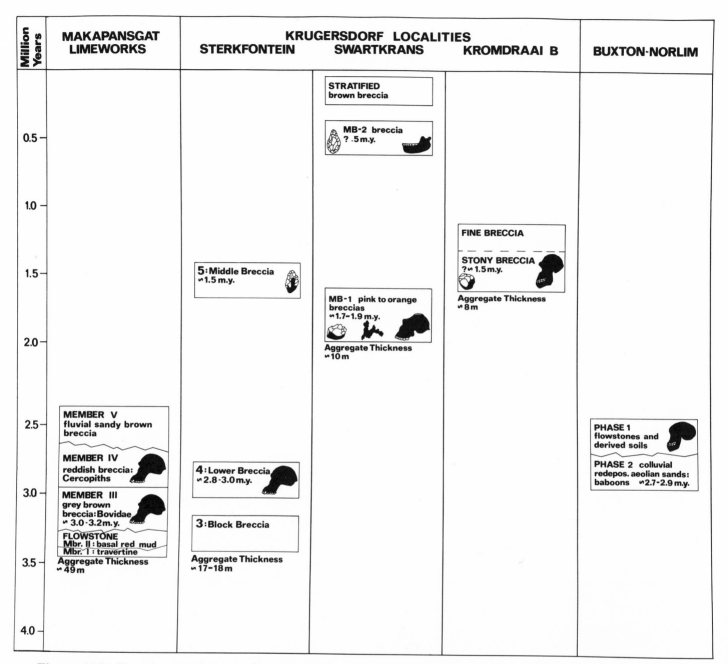

Figure 10.3 Tentative correlation and suggested age relationships of South African localities yielding remains of *Australopithecus* and *Homo*.

will remain uncertain until more completely preserved specimens are recovered from sedimentary formations of that time range.

The Ngorora specimen is a left upper molar, probably M^2 because it exhibits metacone reduction and preserves both mesial and distal interproximal facets. It is rolled and the roots are broken away. It is low-crowned with strong lingual flare, and has four rather low and quite widely separated cusps. The protocone and hypocone are large and of similar size. There is a low oblique crest and a shallow talon basin. The mesiolingual angle of the protocone bears a doubled Carabelli's pit. The anterior fovea is very small, centrally situated, seemingly with ill-expressed buccal limb that passes onto the mesial slope of the paracone. The posterior fovea is small, shifted distolingually, and more or less pitlike; the weak lingual limb extends into the distolingual slope of hypocone. There is a short, open buccal groove. The lingual groove is short, has a mesially open V shape, and lacks any associated structural features.

This specimen diverges from dryopithecine homologues

in having similar-sized lingual cusps, strong lingual flare, and reduced lingual cingular remnants, and low crown with blunt cusps. In these respects it approximates more closely hominid structure.

The Lukeino specimen (KNM-LU-335) is a left lower molar, either M₁ or, probably M₂ (P. Andrews in Pickford 1975). It is a crown lacking root development. The crown is short and broad and generally smaller than *Australopithecus* homologues. The cusps are low and rounded, with very thick enamel. The buccal margin of the crown has pronounced lateral flare and deep expanded mesial and distal grooves with pitlike inferior terminations overlain

by fine enamel extensions. The lingual margin is quite vertical, without a lingual groove. The anterior fovea (trigonid basin) is small, quite deep, and situated well distal of the low but substantial mesial marginal ridge; it is bounded distally by enamel extensions between protoconid and metaconid; an accessory lingual limb cuts into the mesial slope of the metaconid. The talonid basin is also small, narrow, and impinged on by the talonid cusps. The posterior fovea is a deep pit, situated distolingually to the enamel extensions of the posterior talonid cusps, and transected by a deep grove passing back from the talonid basin. There is no true distal marginal ridge. The primary

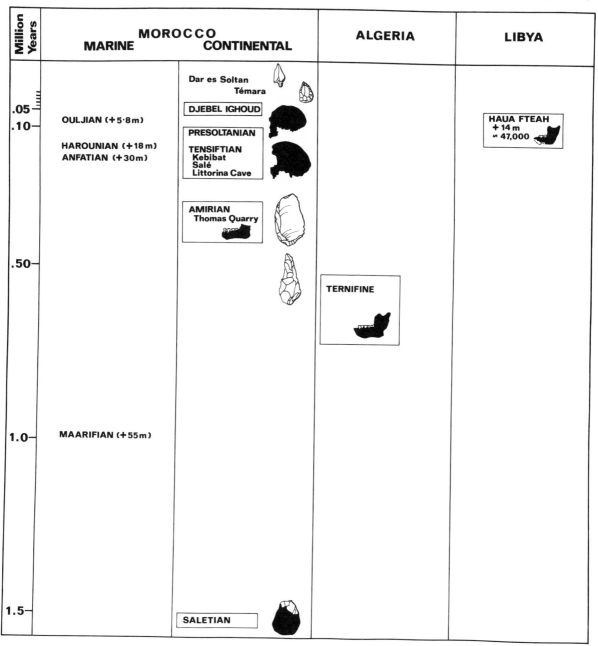

Figure 10.4 Tentative correlation and suggested age relationships of North African localities yielding remains of Hominidae. (For Djebel Ighoud read Djebel Irhoud).

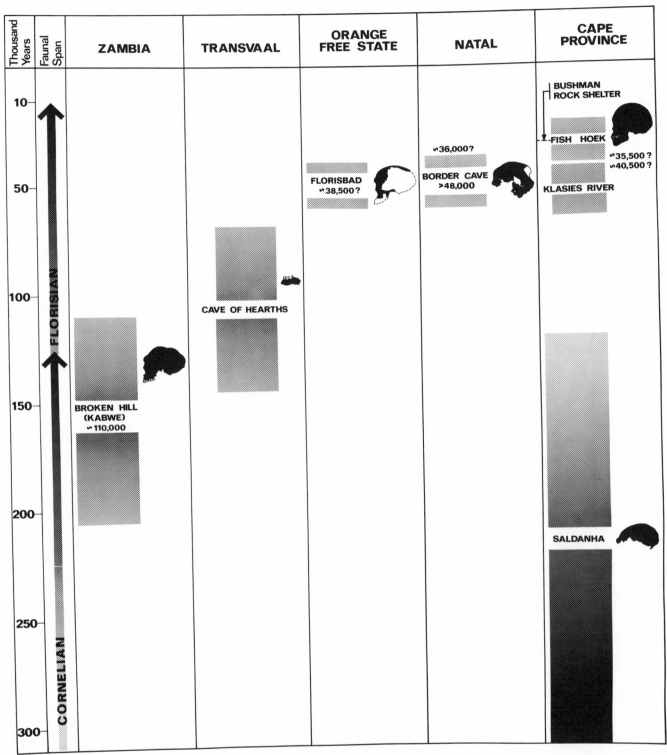

Figure 10.5 Tentative correlation and suggested age relationships of localities of later Quaternary age in southern Africa yielding remains of *Homo sapiens*.

fissure pattern is Y-5, but nearly approximates a +, with metaconid-hypoconid contact nearly transversely, rather than longitudinally. The largest cusp is the metaconid, followed by hypoconid and protoconid. The hypoconulid is substantial and situated slightly distobuccally. There is a broad distobuccal oblique contact between the entoconid

and hypoconid. The metaconid is almost subdivided by an enamel wrinkle on its distal slope.

This specimen diverges from dryopithecine homologues in having different proportions of the trigonid and talonid basin (trigonid basins are short and talonid basin long in dryopithecines), in its strong buccal margin flare and pitlike cingular remnants related to the buccal grooves, and in the marked enamel thickness. Although smaller than *Australopithecus* homologues, there are many structural resemblances to the robust australopithecine condition.

The Lothagam specimen (KNM-LT-329) has still to be described, but has been attributed by Patterson (in Patterson, Behrensmeyer, and Sill 1970) to *Australopithecus* cf *africanus*. It represents a part of the right mandibular body, broken anterior to M_1, preserving the heavily worn M_1 and the roots of M_2, and the area immediately posterior to, but with no trace of, the sockets of M_3. The root of the anterior portion of the ascending ramus rises just behind M_2, and the extramolar sulcus appears to have been relatively narrow. The specimen is not robust and lacks the lateral toral and internal basilar thickening characteristic even of small australopithecines. M_1 is as small as, or smaller than its homologue in the latter specimens.

Hominidae gen. et sp. indet. (B)

Hominid remains, the affinities of which have still to be determined, are known from the Laetolil Beds (Tanzania) and from several members of the Hadar Formation, Afar (Ethiopia). The specimens date from the middle Pliocene. Pending their full comparative study and evaluation these specimens are all referred to as Hominidae gen. et sp. indet. (B).

Upper and lower jaw parts as well as associated and isolated teeth of 20 individuals have been found in the Laetolil Beds. The first specimen, a palate fragment with upper C alveolus and premolar crowns, was designated *Meganthropus africanus* (Remane 1951, Şenyürek 1955). (The generic designation is preempted by a still ill-known Javanese hominid—the type of which included only lower dentition—and the specific designation by a species of *Australopithecus*.) Robinson (1953b) showed that this specimen had some close resemblances to the South African sample of *A. africanus*. Remane's study showed, however, the distinctive features of the upper premolars, including the three-rooted P³. Recently remains of some 14 individuals from 8 localities have been reported by M. D. Leakey et al. (1976) and White (1977). The total hominid sample from the Laetolil Beds now comprises the palate fragment, 4 partial mandibles, 2 mandible fragments, and over 40 associated or isolated teeth, including deciduous and permanent elements. Other cranial parts and postcrania are also known. Some salient morphological features are set out briefly below.

The juvenile mandible (Hom. 2, that of an individual with erupting M1's and full deciduous dentition) has a deep symphysis and posteriorly shallowing mandibular body with robust lateral torus and rounded lateral alveolar prominence; the alveolar planum is long and concave, and there are incipient superior transverse and small inferior transverse tori. An adult mandible, Hom. 4 (damaged anteriorly) shows a deepened symphysis and posteriorly shallowing body; the body lacks strong lateral torus development, with only a weak lateral prominence and oblique line, weak intertoral sulcus and marginal torus passing into the anterior marginal tubercle, and sharp, elongate basal tuberosities. The ramus arises at a high level opposite M2. The mental foramen is set below the anterior root of P_4, more than halfway below the alveolar margin. The internal body also lacks strong tori or robustness. There are weak anterior and posterior subalveolar fossae. Behind the level of M_1 the base of the body flares outward. The anterior symphysis has a swollen median eminence below which it slopes steeply backward; there are deep submental incisures. The alveolar planum is narrow, and strongly concave; the superior transverse torus is rounded, well developed, and extends back to the level of M_1; the inferior transverse torus is also well developed and passes in back of P_4 to merge with the alveolar prominence. Paired mental spines and an associated median intergenioglossal crest-groove occur above the torus. The tooth rows are anteriorly convergent, and the anterior dental arcade has a convex intercanine contour.

Most elements of the deciduous and permanent dentition are represented in the sample. Of the deciduous dentition the upper and lower molars are most like those of *A. africanus* or *Homo* in size and proportions. The dm² shows a weak lingual groove, moderate buccal groove with terminal pit, weak parastylar groove and ridge, weak Carabelli grooves on the lingual face of protoconid, strong oblique crest, deep anterior fovea with strong buccal limb, and centrally disposed, triradiate posterior fovea.

The upper dc is asymmetric, with higher mesial than distal margins, stronger mesial than distal lingual crests, lingual ridge set mesial of the crown axis, and weak buccal and lingual distal grooves. The dm¹ shows weak buccal grooves on the lingual face of protocone, strong oblique crest linking protocone and metacone, superior enamel extension on the buccal face of paracone, and paracone with strong paramolar tubercle.

There is no post lower dc diastema. The lower dc is asymmetrical, high mesially, with salient tip, and low, distal-marginal cuspid. The dm_1 is of nonrobust australopith type, with mesiobuccally set protoconid with sloping buccal margin, trace of a mesial accessory cuspule, an open mesiolingually directed anterior fovea, large metaconid not paired with protoconid, large hypoconid and small entoconid, deep but short buccal groove, and short but large and pitlike lingual groove. The dm_2 of nonrobust australopith type, with Y-5 pattern and substantial metaconid-hypoconid contact; crown constricted across trigonid cusps, and broadened distally with marked buccal swelling of distobuccal root and outer slope of large hypoconid; small hypoconulid and entoconid and no accessory cusps; anterior fovea simple, short, symmetrical, and transversely disposed; posterior fovea with limbs impinging on distal slopes of hypoconulid and entoconid; distal marginal ridge low, not robust; short, well-defined buccal grooves lacking pits; protoconid without protostylid or related structures.

The original Garusi maxilla fragment reveals strong subnasal prognathism, marked guttering of the floor of the nasal aperture, small-rooted incisiors, and very large alveolus for the upper C. Permanent dentition with large-crowned upper incisors and large, robust rooted and large-crowned upper and lower canines, small-crowned upper premolars and molars, P_3 long and narrow, P_4 small and scarcely molarized, small lower molars. Upper I's have strong buccal surface curvature, and prolongation of enamel rootward; lingual surface with strong relief including basal tubercle(s), with or without enamel extensions, marginal ridges and associated grooves. Upper C is moderately asymmetrical, with a nearly centrally situated cusp, buccal surface with slight mesial ridge and adjacent groove, lingual surface with moderately to strongly expressed mesial, central and distal ridges, and adjacent grooves or pits. Upper premolars of simple structure, with $P^3 > P^4$. P^3 is three-rooted, with mesial and distal buccal grooves, and prolongation of enamel rootward; strong development of mesiobuccal angle of crown, buccal swelling, well-developed anterior and posterior marginal ridges; large anterior and posterior foveae. P^4 is two-rooted, without enlargement of talon, but with expanded distolingual angle of crown. P_3 is elongate, with a strong mesiolingually set anterior fovea; bicuspid, with cusps approximated, and smaller, more distally disposed metaconid; large talonid has developed distolingual angle, lingually set posterior fovea, and strong distal marginal ridge; incurved buccal surface with strong mesial and distal grooves and attendant ridges. P_4 without marked molarization; short, strong buccal grooves; strong development of disto-lingual angle of crown; large talonid basin. Lower molars show a Y-5 pattern, protoconid swollen and/or with protostylid groove-ridge structure and lacks extra cusps or split cusps. Upper molars have Carabelli pits on large protocone, short V-shaped lingual groove, buccal groove weak or absent, small anterior fovea.

Hominid remains occur in association with rich fossil vertebrate assemblages in the three upper members of the Hadar Formation, Afar (Ethiopia). In the base of the uppermost, or Kada Hadar Member, a partial skeleton of *Australopithecus* aff. *africanus* has been recovered at one locality. In the middle Denen Dora Member, two principal fossil horizons have yielded hominid remains from four localities, here tentatively attributed to Hominidae gen. et sp. indet (B). In the next to lowest, or Sidi Hakoma Member, hominid remains occur in the two lowest and the uppermost sedimentary units, in the former instance at seven localities, and in the latter instance at five localities; these derive from five or six stratigraphically distinct horizons. At at least six of these localities the hominid remains are attributable to *A.* aff. *africanus*. The hominid remains from seven or eight other localities are here tentatively attributed to Hominidae gen. et sp. indet. (B). If these distinctions are meaningful it is perhaps of interest that there are no definite occurrences of more than one taxon in the upper Kada Hadar Member and the next lower Denen Dora Member; whereas the still lower Sidi Hakoma Member affords distinctive hominid taxa at more than a single stratigraphic horizon, though perhaps not at the same locality.

All hominid specimens tentatively attributed here to Hominidae gen. et sp. indet. (B) from the next lowest Sidi Hakoma Member comprise maxillary (AL-199, AL-200) or incomplete mandibular (AL-266, AL-277) parts (Johanson and Taieb 1976).

All specimens are small or at best of moderate size. The maxillary dental arcade is relatively long, the tooth rows subparallel, and the anterior arch broad. One specimen (AL-200-1) has strong I^2 to C diastemata. The upper incisors are broad, the canines relatively and absolutely large, and the premolars and molars very small to small, usually below *A. africanus* homologues. $P^3 > P^4$, and P^4 is small compared to M^1 and to the *A. africanus* condition. P_3 is moderately small compared to M_1. The subnasal region is long and prognathous, with subnasal guttering. The palate shallow, and the malar process of the maxilla arises above the anterior molar; thick enamel on all teeth. The upper I's have slight lingual relief; upper C is strongly asymmetrical, and distinctive lingual surface morphology; upper P's with simple morphology; $M^1 < M^2 < M^3$; Upper M's with simple crown structure, no cingular remnants on protocone. Mandible small, not particularly robust, but thickened basal portion, with moderate alveolar planum, moderate superior and inferior transverse tori. The tooth rows are slightly divergent posteriorly, and the anterior dental arch slightly curved. Small lower I's; robust, long-rooted lower C with distolingual structures. Lower P's with buccal groove(s); P_3 has large anterior fovea, strong protoconid, with or without approximated metaconid; P_4 only slightly molarized. Lower M's with Y-5 or +5 pattern, slightly to moderately swollen protoconid, absent to weak buccal cingulum remnants; M_3 may have split metaconid and/or protoconid, and even doubled sixth cusp.

Hominid specimens from the upper part of the middle and the lower part of the uppermost Denen Dora Member include adult and juvenile jaw parts, several partial crania, deciduous and permanent teeth, and many upper and lower limb elements. Probably over a dozen juveniles and adults are represented at one locality (AL-333 and AL-333 W) alone. There are a number of resemblances with specimens here referred to Hominidae gen. et sp. indet. (B) from the underlying (Sidi Hakoma) member of the Hadar Formation, including several mandibles with partial dentition, as well as isolated teeth. Elements of the permanent as well as the deciduous dentition, in the latter case including associated maxilla and mandible and much of an infant cranium, bear strong resemblances to Laetolil specimens. Some proportions and morphology of certain postcranial elements, both of the upper and lower limb, appear to differ from the *Australopithecus* morphological pattern. There is no substantive evidence to demonstrate conclusively the presence of a robust australopithecine (*A. boisei*) within the hominid sample from this member.

It is premature at this juncture to insist on the distinctiveness of the Laetolil and Hadar hominid specimens from the genus *Australopithecus sensu lato.* However, suggestions of differences in both dental morphology and the morphology of some postcranial parts from that taxon, including *A. africanus* are demonstrably represented in the Hadar succession. Most recently the hominid sample here referred to Hominidae gen. et sp. indet. (B) has been assigned to a new, primitive species of *Australopithecus, A. afarensis* (see Johanson, White, Leakey, and Coppens in *Kirklandia* 28 [1978]), an assignment with which I concur. Resolving this problem is important because it bears directly not only on the identification and antiquity of the genus *Homo,* but also on the phylogenetic affinities of *Homo* with *Australopithecus.*

Australopithecus R. A. Dart 1925

The extinct genus *Australopithecus* was first announced and described briefly by Dart (1925). It was not then given formal diagnosis. The provenience and associations of the specimen are summarized by Peabody (1954) and Butzer (1974b). The type specimen is a partial skull, comprising the frontofacial region, right basii cranii, and mandible, and a natural endocranial cast (but lacking bones of the cranial vault). It is a juvenile with full deciduous dentition and almost fully erupted unworn first permanent molars. Although the literature on this specimen is extensive, it has never received the thorough morphological description and comparative study it merits. The salient morphological features have been set out principally by Dart (1925, 1929, 1930–31, 1934, 1940b, 1948b), Broom, (1925a, 1925b, 1929a, Broom and Schepers 1946), Hrdlicka (1925), Keith (1931), and Abel (1931); Robinson (1956) has published details of the dentition and Holloway (1970a) an accurate assessment of endocranial volume. Much of the initial and protracted controversy over the resemblances and affinities of this specimen largely resulted from inadequate and inappropriate comparative study, its incompleteness, and its juvenile age.

Keith (1931:53) concluded that "in all its essential features *Australopithecus* is an anthropoid ape,"

and "the features wherein *Australopithecus* departs from living African anthropoids and makes an approach towards man cannot be permitted to outweigh the predominance of its anthropoid affinities."

It is a sad reflection of the then immature data, method, and theory of paleoanthropology and comparative primatology that such dispute and controversy could persist when, within 10 to 15 years after the discovery, the fundamentally hominid features of the dentition had been established by Dart, Broom, Gregory (1930; see also Gregory and Hellman 1939a,b,c), Adloff (1932; see also 1931), Bennejeant (1936; see also 1953), and Le Gros Clark (1940, 1947, and especially 1952). Moreover, Sollas (1926) clearly established the nonpongid features of the craniofacial morphology. However, the significance of this distinctive total morphological pattern was largely overlooked by both Keith (1931) and Abel (1931), both of whom employed substantial samples of juvenile pongids in their comparative studies. Obviously, some influential and respected members of the scientific community were frankly unprepared and unwilling to accept such an unexpected discovery and its attendant phylogenetic implications. Only the subsequent recovery of additional cranial, jaw, and postcranial parts of adult individuals through the pioneering efforts of Robert Broom, with the subsequent cooperation of J. T. Robinson as well as R. A. Dart, eventually convinced all but the most dedicated skeptics that *Australopithecus* was a primitive representative of the family Hominidae.

Salient nonpongid features of the type of *Australopithecus africanus* are enumerated below.

The cranial vault is overall long and narrow; the parietals are elongated, as is the frontal; the frontal is high, protrusive anteriorly, rounded and arched, but narrow; the occipital is shortened, and its nuchal plane shifted inferiorly, with a correspondingly low situation of the parieto-occipital junction (lambda); the squamous portion of the temporal is arched and elevated anteriorly, and there is a broad sphenoparietal sutural junction (at pterion); the anterior (orbital) portion of the cranial base is deflected inferiorly, the spheno-occipital area substantially angulated, and the foramen magnum set relatively forward.

The common trunk of the middle meningeal artery is short and bifurcates (into anterior and posterior branches) on the lower, incurved surface of the temporal squama, and hence on the inferior surface of the temporal lobe. The transverse and sigmoid sinuses are well developed. The endocranial capacity (natural endocranial cast) is ~ 404 cc (R. L. Holloway 1970b) and, assuming an additional 8% subsequent growth, would have had an adult capacity of ~ 440 cc (estimates by Dart 1925, 1948b; Zuckerman 1928; and Broom and Schepers 1946 were all substantially higher; the estimates by Keith 1931[6] and especially by Abel 1931 were much nearer that of Holloway); this value is small compared to modern man, but relatively large compared to modern apes (except male gorillas). The endocranial cast configuration and proportions are distinctive in respect to: overall increased height and decreased breadth, the base being especially reduced; the expanded parietal cortical area (especially posteriorly) and the laterally expanded temporoparietal regions; the posterior displacement of the occipital lobe and visual cortex, reflected in the disposition of the lunate sulcus, and protrusion posteriorly of the occipital pole over the cerebellum, which is relatively low and narrow; and in the prominent frontal rostrum, enlarged frontal, and especially prefrontal, area, including expanded and prominent frontal convolutions, and the inferior portion of the precentral cortical motor areas.

The facial skeleton is set beneath the vault, is shortened, and lacks strong prognathism; the components of upper facial triangle are correspondingly shortened; however, the face is relatively wide with a robust zygomatic arch (and probably, well-developed masseter muscles); the frontomalar process is shifted anteriorly, set obliquely with reference to the zygomatic process, and the malomaxillary suture has an anterosuperior disposition; the glabellar region is well developed, but there is no frontal torus expressed, a reflection of delayed maturation; the nasal root (nasion) is set low, on the slope below glabella, posteriorly near the frontal pole of the cerebrum and is narrow, producing a reduced interorbital space; the nasal bones are short and wide, the lower half of the internasal suture still unfused, and are arched superiorly, do not extend superiorly into the frontal, do not broaden inferiorly and are not prolonged below the level of the orbits; the orbits are small, nearly circular, and are contracted beneath the frontal; the ethmoids are not expanded laterally; the lacrimal fossa faces posteriorly; the nasal aperture is relatively small, elevated, lacks a nasal spine, and its floor passes gradually into the surface of the maxilla; there is partial persistence on the face and surface of the palate of the premaxillary suture. The occlusal plane is set high.

In general the deciduous dental arches are notably short and broad, and the incisor-canine dentition small. The maxillary dental arcade is parabolic, relatively long, wide across the first permanent molars, shortened and slightly arcuate anteriorly. The mandible body is short, robust, widened anteriorly, with fairly low and moderately vertical symphysis, moderate alveolar planum, well-developed superior and weak inferior transverse tori, and fairly narrow high ramus. The deciduous dentition, especially incisors and canines, is very substantially worn, but the newly erupted first permanent molars are essentially unworn. Thus, the deciduous dentition was retained for a protracted period prior to permanent molar eruption (and M2 was probably substantially delayed after M1, to judge from the position of the latter near the ramus); all reflect prolonged infancy and maturation.[7] The small vertically set incisors and small canines, and the lack of maxillary (precanine) and mandibular (postcanine) diastemata are of note. Overall the dentition is especially indicative of hominid affinity; thus, Gregory (1930) pointed out that 20 of 26 features of the deciduous dentition were most manlike and only 3 were uniquely shared with African apes. The incisor proportions are $di^1 > di^2$, and $di_1 < di_2$. The pattern of dc wear is distinctively nonpongid, and these teeth apparently did not interlock. The upper dc is spatulate, with obtuse tip, subconcave lingual surface, weak basal eminence, low lingual ridge, and adjacent depressions. The lower dc (absolutely and compared to dm_1) is low and short, blunt and spatulate, laterally convex, with lingual ridge, distal cusplet, and weak adjacent grooves. The crowns of the upper molars are short and wide. The dm^1 is molariform, with shorter lingual than buccal face, four-cusped, with enlarged mesiobuccal angle and a groove separating a parastyle, a very weak Carabelli's pit, and lacks a posterior fovea. The dm^2, which closely resembles M^1, is four-cusped with small paracone and metacone, reduced hypocone, and lacks both Carabelli's pit and posterior fovea. The dm_1 is a nonsectorial, molarized tooth, with distinctive wear, and higher and wider trigonid than talonid. It is basically five-cusped with a mesial accessory cuspule (of cingular origin, and *not* a paraconid), with lingually shifted protoconid having a flattened buccal slope and low enamel line, oblique protostylid pit, asymmetrically situated, deep and mesially open, lingually situated anterior fovea, high protoconid-metaconid crest, and well-developed mesiobuccal groove. The dm_2, much larger than dm_1, is five-cusped, with an asymmetrical Y-5 pattern with metaconid-hypoconid contact, an accessory sixth cusp distally, strongly convex buccal face with protuberant hypoconid, small main and accessory anterior foveae, well-marked protostylid pit and attendant ridge, and distal buccal groove and associated pit. M^1 is four-cusped, with paramolar cusp on the buccal face of the paracone, Carabelli groove and mesiolingual pit complex, and an anterior fovea. The M_1 is five-cusped, with modified Y pattern and alignment of the mesial limbs of the transverse grooves, an accessory sixth cusp, medial lingual accessory cuspule, large anterior fovea, deep buccal grooves, and associated protostylid pit and ridge.

Over a decade elapsed before other skeletal remains of adult hominids comparable to *Australopithecus* were recovered at Sterkfontein, Transvaal, in South Africa. However, initially only a single

---

6 Keith (1931: 67) considered, at one time, that "brain volume . . . is our best guide to mental ability" and stressed that "in the matter of brain endowment *Australopithecus* stands far above the chimpanzee; it was the equal, if not the superior of the gorilla, but falls far short of any standard which can be regarded as human or even prehuman." Of course *Australopithecus* body size, juvenile or adult, was then utterly unsuspected and the relevance of brain size compared to extant pongines of known body size and weight could not be explored.

7 On this basis a more human (rather than pongid) growth pattern is presumed, and therefore an age of 6.5 ± 1 yr has been inferred for the specimen (Mann 1968, 1975).

element (M_1) of the permanent dentition could be compared in these two samples. It required another decade before the same locality yielded a deciduous dentition that allowed detailed comparisons (Broom 1947, Broom, Robinson, and Schepers 1950). Within that interval, adult (1938) and then juvenile (1941) hominid remains, including most of the lower deciduous dentition of a related though somewhat morphologically distinctive hominid were recovered at the nearby Kromdraai B locality (Broom 1941, Broom and Schepers 1946). Later additional evidence of species diversity among such early Hominidae was obtained elsewhere in South Africa (Swartkrans and Makapansgat Limeworks, Transvaal) and from various localities in eastern Africa.

The Genus *Australopithecus*

A genus (extinct) of Hominidae distinguished by relatively small cranial capacity, averaging less than 500 cc, and enhanced encephalization (E.Q. = 3.4–3.7; C.E. = 40.0–46.0)[8] compared with Pongidae; species means average from about 440 cc to 520 cc; endocranial cast is distinguished by expanded height, consequent on parietal and temporal lobe enlargement; posteriorly displaced lunate sulcus; temporal lobes expanded, with protuberant tips; frontal lobes, especially orbital surfaces, enlarged; enlarged, more convoluted third inferior frontal gyrus; middle meningeal artery with low bifurcation of common trunk.

Cranial vault relatively thin-walled, in larger-vaulted and massive-toothed species with strong ectocranial superstructures and often marked pneumatization, supraorbital tori present, continuous, and substantially developed; orbits moderately to fairly high, and lower mean height than Pongidae; massive-toothed species often with fronto-parietal development of low sagittal crest (in some species slightly continuous with nuchal crest or occipital torus) in presumed males; foramen magnum, in juveniles and adults, with anterior situation on cranium, and occipital condyles set well behind sagittal midpoint of cranial length and in the transverse plane of biporial axis; shortened cranial base; no styloid process; nuchal planum of occipital extended only short distance above Frankfurt Horizontal, more inferiorly than posteriorly facing, and with inion displaced inferiorly and approximated to Frankfurt Horizontal; slight occipital torus or, in other species, low independent nuchal crest; pyramidal mastoid process, even in immature individuals; shallow, wide mandibular fossa of hominid (not pongid) conformation, with salient articular tubercle, entoglenoid process, and sometimes a substantial postglenoid process; elevated position of temporal bone, including external auditory meatus,

with reference to basis cranii; relatively to absolutely massive, robust mandible, mental eminence normally absent, and variably strong development of alveolar planum and superior and inferior transverse tori; internal mandibular arch variably V- to blunt U-shaped, and dental arcade parabolic, without diastemata; reduced incisors, canines moderate- to small-crowned, long-rooted, spatulate, noninterlocking, exhibiting tip downward wear; large premolar and molar teeth, with notable buccolingual expansion; P_3 bicuspid, with notable subequal cusps; molar series often progressively increased in size from M1 → M3, though often $M^3 < M^2$; nonsectorial, variably molarized dm_1 with well-developed talonid.

Limb proportions, stature, and body weight are very imperfectly known and variably estimated; proportionately more massive upper than lower extremities; forearm and thigh elongate; intermembral and femorohumeral indices probably in or at top of *Homo* range; brachial index probably at top or above *Homo* range; stature generally short, variably estimated from 130–160 cm among smaller and larger species; body weight variably estimated from 22–40 kg to 45–50 or even 75–100 kg, according to species.

Clavicle still unknown; scapula imperfectly known, with relatively narrow, symmetrically oval glenoid; robust, posteriorly curved coracoid with marked biceps origin, scapular spine set low on a perhaps relatively narrow blade. Humerus moderately to extraordinarily robust and length varying from ~265–335 mm, depending on species; robust, well-rounded head, well-defined anatomical neck; large well-delineated bicipital groove; well-defined to strong greater and lesser tuberosities and attendant crests; well-developed insertions for short scapular mm insertions; pectoralis major crest prominent and distally continuous with deltoid impression; moderate to strong lateral epicondyle and supraepicondylar ridge, and brachioradialis crest; medial epicondyle prominent and twisted; separation of trochlea and capitulum by variably keellike crest, proximally prolonged between depressions above these structures; shallow, broad olecranon fossa; transverse axis of trochlea distally set relative to axis of inferior humeral shaft; well-rounded capitulum. Ulna long, attenuated, dorsoventrally curved; lacking mediolateral sinuosity; weak interosseous border; weak concavity for radial tuberosity; flat, unbuttressed ulnar tuberosity; weak supinator crest; lacking posterior projection of proximal posterior border of shaft; expanded distal end and head, latter with only weakly convex crescentic articular surface.

Radius quite inadequately known, fragments at best, and undescribed. Manus very imperfectly known (represented by a single capitate; a thumb metacarpal; a middle phalanx; and associated incomplete metacarpal 2 and two proximal phalanges, the latter undescribed). Capitate (small species) small, short, broad with expanded head, metacarpal 2–3 articulation (pongid), and small navicular and trapezoid articular facets; metacarpal 1 (large species) short, stout, longitudinally curved, with beaklike laterally oblique head, and shallow sesamoid grooves, and base with extended medial articular margin; middle phalanx (? species) short, stout, longitudinally curved, waisted, with

---

8 Like other such measures, the Encephalization Quotient (E.Q.) of Jerison (1973) and the Constant of Encephalization (C.E.) of Hemmer (1971) are ratios of actual to expected brain size that take into account estimates of body size.

strong flexor insertion crests; metacarpal 2 long, slender, head continues axis of shaft, double sesamoid grooves, no distal articular dorsal shelf; proximal phalanges 2 and 5 relatively short and long, respectively.

Vertebral column with most elements (except cervical) represented, and by parts of several individuals, including large and small species. Skeletal trunk length[9] (projected values) markedly to substantially less than *Pan.* Axis (large species) robust, rounded neural arch, body thickened inferiorly, marked anterior sagittal keel, flat and quite concave superior articular facet, inferior articular facet not offset laterally. Thoracolumbar series similar in known structure between species, except larger (transversely, not superoinferiorly) and more robust, with longer processes and proportionately larger neural canal in the larger species; thoracics with successively smaller and more heart-shaped bodies superiorly; small demifacets for rib capitulum, and tuberculum facet on single transverse process; T-12 may show thoracolumbar morphology (in transverse processes, superior and inferior articular processes, and capitulum facet); the neural spines variably inferiorly slanted or long and straight, and relatively wide inferior notch, well-developed mammillary process, small transverse process. As many as six lumbar vertebrae (if so, with L-1 of thoracolumber type, with lumbar-type superior articular facets), centra well developed and progressively widened transversely and shallowed anteroposteriorly from L-1 to L-5/6, promontory and lumbar lordosis indicated by wedge-shaped (posterior body height less than anterior body height) lower lumbars; hominid type articular facets (superior facing medioposteriorly, inferior facing lateroanteriorly); slender neural spines; slender, relatively high pedicles, relatively large intervertebral notches and vertebral foramina; transverse processes very long and slender, laterally and superiorly directed in the midlumbar series.

Pelvis, known from three adolescents and four adults of several species from four localities; usually incomplete and/or distorted innominate portions, only one specimen largely complete (with much of associated vertebral column), though distorted. Sacrum (only partially known, small species), proportionately wide, with small vertebral bodies and large, long lateral masses; small lumbosacral articular surface; anterior part of sacral pelvic surface horizontally disposed, and sacrum set angulated to long axis of vertebral column; auricular surface relatively small—disposed over most of first and only portion of second lateral mass—in proportion to pelvis size; perhaps delayed fusion of vertebral bodies in comparison to fusion of lateral masses. Innominate distinguished by broadened, shortened, and twisted ilium; modified sigmoid curvature of iliac crest, convex anteriorly and concave posteriorly, with very substantial lateral flare; relatively small ratio of minimum to maximum iliac breadth; anterior superior

9 This value reflects postcervical trunk length (Biegert and Mauer 1972), overall proportional values of which are relatively accurately inferred from measures projected from only several "representative" thoracolumbar vertebrae.

iliac spine protuberant, massive, with slight mesiad deviation; prominent anterior inferior iliac spine and shallow iliopsoas groove; relatively large iliac fossa; posteriorly situated gluteal surface, its posterior half markedly concave; well-developed acetabulospinous buttress; thickened rather diffuse cortical bone to produce acetabulocristal buttress, anteriorly displaced (hence a pseudo-iliac pillar), at least sometimes with small, distinct iliac cristal tubercle, set distal of the latter buttress, splitline pattern of hominid type, except for acetabulospinous and acetabulocristal distinctions; acetabulum relatively and absolutely small, relatively caudally and ventrally situated, and approximated to auricular area, and variably shallow to moderately deep, usually with prominent anterior wall; auricular area (sacroiliac joint) posteriorly and inferiorly situated, relatively small, acutely angulated superior and inferior surfaces, with substantial relief and exhibiting substantial iliac tubercle development, posterior margin of ilium short with approximated posterior superior and inferior iliac spines; prominent sciatic notch showing hominid-type sexual dimorphism; well-developed ischial spine; rather elongate pubis, superior ramus and body slender, and shortened pubic symphysis; ischium relatively and absolutely short, with tuberosity not set perpendicular to axis of ischium, approximated to or more distal of acetabulum, and variably low and elongate or robust, and protuberant (according to species).

Femur varies perhaps from 300–350 mm and 370–410 mm long according to species; robust; increased femoral obliquity (femorocondylar angle), with adducted valgus stance; small femoral head; femoral neck long, broad-based, inferiorly buttressed, nearly horizontal superior border, and anteroposteriorly compressed; reduced angulation between neck and shaft; obturator externus tendon groove (indicative of femoral hyperextension); intertrochanteric line absent or usually weakly expressed and femoral tubercle small (suggestive of ill-differentiated iliofemoral portion of capsular ligament); reduced lateral flare of low, moderate-sized greater trochanter; deep trochanteric fossa and low gluteal tuberosity; posteromedial disposition of lesser trochanter; weak to moderately expressed spiral line; linea aspera variably developed, usually without marked lipping, and related to pilaster or strong femoral crest; distal shaft robust; lateral condyle larger than medial, with flattened arc of curvature; patellar surface concave with accentuated lateral lip (inhibiting lateral patellar displacement in quadriceps femoris action); variably narrow to broadened, but high intercondylar notch, forwardly prolonged, with distinct cruciate ligament impressions at superior end.

Tibia undescribed or affinity of referred specimens indeterminate. Fibula as yet unknown or undescribed.

Pes scarcely known, except for an incomplete talus, a right metatarsal, and (perhaps) a proximal and distal phalanx (never fully described). Talus overall small, with relatively broad, transversely symmetrically curved superior (distal tibial) articular surface: lateral (fibular) articular facet slightly curved, moderate in size; medial (malleolar) articular facet high, flattened and essentially vertical; short relatively wide, unconstricted neck, markedly angulated relative to talar body; head with low torsion angle, and broad, flattened gentle curvature medially and laterally of navicular articular surface.

Species of *Australopithecus*

Despite the recovery of hundreds of specimens of Pliocene-Pleistocene Hominidae at an ever-accelerating rate, substantial disagreement persists about their taxonomic status. But the problem is becoming more clearly defined through more adequate delineation of temporal relationships of hominid samples, quantitative assessments of their ranges of variation, and recognition of lineages (chronospecies) with attendant reduction of synonymy.

By 1950 as many as three genera and five species of early Hominidae were recognized by Broom (1950). Subsequently, largely through the efforts of Robinson (1954, 1956), the (essentially) South African samples were synonomized into two genera, two species and four subspecies. Two lineages, one (*Australopithecus*) with affinities to *Homo,* the other (*Paranthropus*) once considered ancestral but ultimately aberrant and extinct without issue, were recognized by Robinson (1962, 1963a). The former genus was duly sunk and assigned to *Homo,* as *Homo africanus,* an ancestral species of the *Homo* lineage (Robinson 1967). These lineages were presumed to have their bases on inferred ecological differences and dietary adjustments (the so-called dietary hypothesis).

The bases for these alleged differences have not gone unchallenged. Le Gros Clark (1964:167), who was personally familiar with or aware of most of the evidence then available, considered "that such anatomical differences as there may be between the fossil remains . . . , though they may possibly justify subspecific, or even specific, distinctions, are not sufficient to justify generic distinctions"; later he noted that "without doubt a great deal of unnecessary confusion has been introduced into the story of the australopithecines by the failure to recognize their high variability, thereby arbitrarily creating new species and genera that cannot be properly validated" (1967:48).

Through thorough comparative studies, Tobias (1967a) sought to demonstrate the fundamentally similar structure of australopithecines—samples referable to *Australopithecus*—in contrast to the structure exemplified by those samples referable to the genus *Homo*. The differences among australopithecine samples were considered to be at most specific, and three species were recognized and defined; a small gracile form, *Australopithecus africanus;* a

succedent, more robust species, *A. robustus;* and an extremely robust form (exclusively East African), considered to be penecontemporaneous, *A. boisei.*

More recently, and following in part different procedures, Campbell (1972) has also sought to recognize the basic generic pattern—biologically and behaviorally or culturally—distinctive of *Australopithecus.* He has suggested two divergent lineages or paleospecies within the genus—one simple and restricted to eastern Africa, *A. boisei;* the other, *A. africanus,* complex with four subspecies, represented by two subspecies in southern Africa, one in eastern Africa, and one in insular southeastern Asia.

Brace (1967, 1973) and Wolpoff (1971 a,b, 1974a), among others, have suggested that differences between samples may largely reflect sexual dimorphism or at best a subspecies distinction. This attractive though facile explanation fails, however, to consider the morphological differences in cranium and dentition (see below), which are not merely of size but also of proportions and structure. Thus, Robinson and Steudel (1973: 520; also Robinson 1969) have shown that analyses of proportions along the tooth row, which reflect in particular proportional differences between postcanine teeth and anterior teeth, "which in all cases discriminate well between the living hominoid samples used, in every case also discriminate well between *Paranthropus* (= *A. robustus*) and *H. africanus* (= *A. africanus*)." Similarly, Greene (1975: 86) has shown that the dental dimorphism in samples attributed to those two taxa "as measured by mean differences is greater than expected in either chimpanzee or modern man in both maxillary and mandibular comparisons and it exceeds that of the gorilla in the maxillary. The magnitude of the differences between the two taxa cannot be explained by sexual dimorphism of the sort found in either modern man or chimpanzee" and "it is also improbable that the differences can be explained in terms of something comparable to gorilla sexual dimorphism." Bilsborough (1972:482) has reached essentially the same conclusion and considers, as do some other authors (especially Robinson), that there are "marked differences between gracile and robust australopithecines, with the separation between these two groups often greater than that between *A. africanus* and all members of the genus *Homo.*"

So far as known all representatives of the genus *Australopithecus* are basically similar in postcranial morphology. There are, however, usually consistent differences from the total morphological pattern characteristic of the genus *Homo*—in the lower limb, in the axial skeleton, and in the forelimb. These differences between lesser taxa of the genus largely appear to reflect body size and robusticity. The principal differences between the lesser taxa are in cranial and dental morphology.

The principal features characteristic of each of the several species of *Australopithecus* are set out below, followed by a discussion of their geographic and temporal distributions.

Australopithecus africanus Dart 1925

A species of the (extinct) genus *Australopithecus* distinguished by small cranial capacity, between 425 and 485 cc (mean 440 cc); E.Q. about 3.5 and C.E. about 40. Cranium generally lightly constructed, long and ovoid, with reduced postorbital constriction; parieto-occipital region evenly curved, with substantial anteroposterior curvature of parietals, and steep parietotemporal walls; sagittal cresting absent or exceptionally rare, and then of limited extent; occipital with lower (cerebellar) scale less curved and larger than upper (cerebral) scale, former with variably salient transverse occipital torus, broader laterally than medially, and situated above the level of the internal occipital protuberance; low nuchal muscle insertion area on occipital, with base of external occipital protuberance approximately aligned with Frankfurt Horizontal; cerebellar fossae quite deep and rounded, internal occipital crest well-defined, and internal occipital protuberance approximated to opisthion; relatively backward situation of occipital condyles, and foramen magnum facing inferoposteriorly; condyles small and generally untapered anteriorly; laterally disposed, large pyramidal mastoid process, expanded supramastoid crest, strong occipitomastoid crest; moderate pneumatization; cranial base expanded with relatively low angle of the petrous axis to the median sagittal plane; spheno-occipital region only moderately shortened; sphenosquamosal suture transects entoglenoid process, dividing it into alisphenoid and squamosal portions; internal venous drainage via lateral sinus septum (hence transverse → sigmoid sinus → jugular bulb); substantial undercutting of lateral part of posterior wall of pyramid in posterior cranial fossae; deep, large middle cranial fossae, with undercut, robust lesser wings of sphenoid, steep undercut anteromedial portion of pyramid behind foramen ovale, and lacking Sylvian crest formation; anterior cranial fossa with high orbital roof, flattened anteriorly, quite deeply set cribriform plate of ethmoid, low crista galli, slender internal frontal crest.

Calvarium elevated above upper margin of orbit due to high hafting of calvaria to facial skeleton; continuous supraorbital torus, with raised but inferiorly disposed glabellar area, and supraglabellar depression; relatively narrow interorbital space; orbits semirounded; broad ethmolacrimal articulation; rather low situation of nasion; nasal margin elevated, quite sharply defined, bordered inferolaterally by moderate continuation of canine pillars that pass superiorly into flattened maxillary planum; flattened intercanine subnasal area; usually slight to well-marked nasal sill, some subnasal guttering, probably no anterior nasal spine; face moderately to substantially prognathous due to projection of subnasal maxilla with anterior dentition forward of frontal plane; persistence of premaxillary suture; robust, laterally flared malozygomatic region, with straight (unnotched) inferior border; malomaxillary area flat, inclined inferiorly; zygomatic pillar situated above M^1 or even approximated to P^4; palate with steep shelving far anteriorly, lacking a median palatine torus; molar portion of occlusal plane parallel to the Frankfurt Horizontal, the anterior portion slightly superiorly angled; anterior teeth set in a somewhat curved line. Mandible with ramus of moderate height, somewhat posteriorly inclined with strong endocoronoid and moderate endocondyloid crests; body generally robust with substantial lateral and marginal tori; symphysis with flattened anterior surface, posteriorly inclined with broad and moderate to substantial alveolar planum, moderate to marked superior and inferior transverse tori; internal mandibular arch contour may approximate a V shape.

Deciduous dentition with small spatulate d^c having thickened basal lingual ridge with attendant depressions; d_c with small (or no) mesial and larger distal cusplets, and both buccal and stronger lingual grooves; dm^1 longer than broad, strong mesiobuccal angle of crown with mesial accessory cusp, probably with Carabelli's cusp, strong trigon (oblique) crest, and strong buccal grooves; dm_1 incompletely molarized with five-cusped, strong protoconid with expanded, sloping buccal face, and inferiorly expanded enamel line, mesial accessory cusplet, deep, lingually situated anterior fovea with ill-developed anterior wall, large centrally situated metaconid, talonid relatively short compared to large trigonid and with less developed cusps (hypoconulid, entoconid); dm^2 four-cusped with relatively small hypocone, salient buccal cusps, some even substantial expression of Carabelli complex, ill-developed distal wall of posterior fovea; dm_2 with symmetrical Y-5 pattern, sixth cusp present or not, weak, oblique accessory groove distal of anterior fovea, trace of buccal cingulum, little secondary fissuration.

Permanent dentition with harmoniously proportioned anterior and postcanine teeth; anterior teeth (crowns) relatively larger, and posterior teeth (roots) relatively less robust than in other species; premolars and molars of moderate size, and not buccolingually broad as in other species; upper incisors moderately shovel-shaped, I^1 with buccal grooves and flattened buccal face; lower incisors with substantial lingual hollowing, no (I_1) or weak (I_2) lingual grooves, and without lingual tubercle; upper C asymmetrical with pointed, projecting apex, strong vertical curvature, weak lingual grooves and adjacent ridges, except prominent vertical lingual ridge, sometimes with small lingual tubercle; lower C large relative to other species, strongly asymmetrical, strong distal buccal groove, marked lingual (especially distal) grooves and prominent lingual ridge, wide flat gingival eminence, large distal lingual cusplet; P^3 sometimes $> P^4$, usually two- but occasionally three-rooted, variably smooth to slightly crenulated enamel, P^3 with well-defined buccal grooves, crown usually wider at occlusal margin than cervical line, P⁴ with ill-developed talon, equally expressed buccal grooves, and lacking distal lingual groove; P₃ < P₄; P₃ of fundamental hominid structure, rather markedly asymmetrical, lingual cusp well-defined, set rather mesial, buccal and lingual triangular ridges well expressed, buccal grooves well defined; P₄ with well-defined cusps; only moderate molarization, may be cuspule formation distal to posterior fovea and mesial to anterior fovea, foveae well developed, both buccal grooves well defined.

Permanent molars overall smaller in length and breadth dimensions than other species, M³ customarily < M², usually fully expressed Carabelli's complex, some parastyle development, buccal groove not strongly expressed, M³ frequently with typical upper molar structure, but enhanced tendency for central cuspule(s) expression, and variable development of anterior fovea. Lower molars, usually with substantial to moderate expression of buccal cingulum (protostylid complex); M₁ lacking typical Y-5 pattern, with small metaconid-hypoconid contact, and variably with sixth cusp, and only moderate secondary fissuration; M₂ with nearly +5 fissure pattern, often with sixth cusp, sometimes an accessory anterior transverse groove; M₃ always with near + fissure pattern, lingually

displaced longitudinal fissure, sometimes an accessory anterior transverse groove, frequent median accessory cusps, and rare secondary fissuration.

Limb proportions, stature and body weight imperfectly known and variably estimated; a small, lightly built species, stature perhaps 130–145 cm and body weight averaging about 35 kg and ranging from 20–45 kg; skeletal trunk length 38–40 cm. Postcranial morphology, so far as known, essentially same as for the genus except for its relatively gracile condition.

The small species (*africanus*) of *Australopithecus* occurs in deposits of Pliocene age in southern and in eastern Africa. It has never been found outside the Afro-Arabian continent and, at least for the moment, its distribution is distinctively sub-Saharan.

In southern Africa it is documented from the cemented infillings (breccias) of two caves in the Transvaal (Sterkfontein, Makapan Limeworks) and the original site (Tau'ng) in the northern Cape Province. The latter site afforded the holotype of this species. None of those situations provides convincingly sound evidence either for the contemporaneous pres-

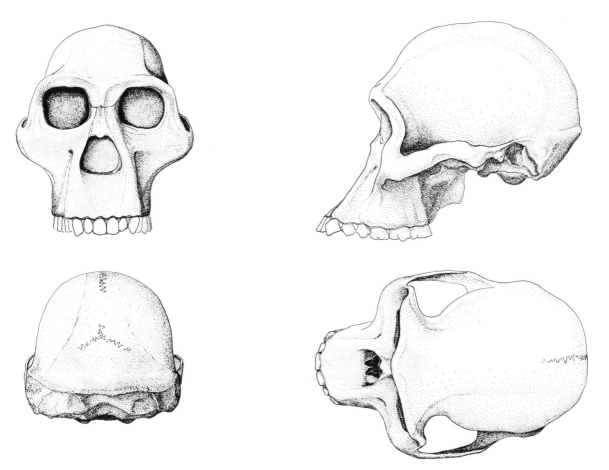

Figure 10.6 *Australopithecus africanus*. Facial, lateral, posterior, and superior views of restored cranium (based on specimens from Sterkfontein). Two-fifths natural size.

ence of another hominid species or for direct association of lithic or other materials indicative of recognizably culturally patterned behavior.[10]

At Sterkfontein all skeletal remains certainly attributable to *A. africanus* occur in the Lower Breccia (Unit B of Butzer 1978). The overall representation of a diversity of vertebrate species and their body parts, as well as the preponderance of certain medium- to large-sized bovids, strongly suggests that the accumulation was the consequence of carnivore (perhaps sabertooth?) predatory activities (Vrba 1975, 1976).

The distinctive composition and body part representation of various mammalian taxa, especially Bovidae, from the hominid-bearing infilling at Makapan Limeworks was first noted and stressed by Dart (1957a). The species spectrum and represented body parts indeed suggests, as in the case of Sterkfontein (Lower Breccia), that large carnivores were probably largely responsible for this richly fossiliferous accumulation (Vrba 1975).

In eastern Africa this species is documented in the Turkana (formerly Rudolf) basin, probably in the Baringo area and perhaps at Olduvai Gorge. Isolated specimens from Kanapoi (a distal humerus) and from Lothagam (a mandible fragment), Turkana district, northern Kenya, have also been tentatively attributed to the species (Patterson and Howells 1967, Patterson, Behrensmeyer, and Sill 1970). However, the exact affinities of the former must remain rather uncertain as there are scant bases for comparison.[11] The Lothagam specimen may indeed represent *Australopithecus*—and if so is the oldest known representative of the genus—but its specific affinities are still undetermined and it stands quite alone in respect to its very substantial antiquity (probably final Miocene).

The isolated temporal bone from the Chemeron

Formation, Baringo area, Kenya, is apparently of very substantial antiquity, perhaps comparable to the Kanapoi occurrence. The specimen is *Australopithecus*-like in overall morphology (Martyn and Tobias 1967), but diverges from the morphology found in *A. boisei*. It most closely resembles *A. africanus* and is here very tentatively referred to that species.

The skeletal remains attributed to this species from various localities are set out in table 10.1. Those specimens only tentatively attributed to the taxon are underlined.

The single most important sample is from Sterkfontein; it represents the remains of probably over 40 individuals. The Makapan Limeworks sample is much more fragmentary and represents only some 7 to 12 individuals (Mann 1975). The type specimen from Tau'ng, possessing a full deciduous dentition and erupting first permanent molars, is the best preserved juvenile known of this species. It affords an adequate basis for comparison with subadults from Sterkfontein (notably STS-2, STS-18, and STS-24), but unfortunately, only a single specimen (MLD-5) from Makapan Limeworks. Each larger sample shows a comparable mean age at death of approximately 22 years, and in each there are few or no infants or "middle-aged" adults over 40 (Mann 1975, 1976).

These samples provide a substantial basis for evaluating basic craniofacial morphology, dental structure, size, and proportions—including variability and (limited) extent of sexual dimorphism—and overall morphology of the axial skeleton and pelvis. However, they afford only partial and very limited insights into the proportions and morphology of the upper and lower limbs. This deficit severely hinders comparisons with those body parts known in other Plio-Pleistocene Hominidae from elsewhere in Africa, East Rudolf and Olduvai Gorge.

The Omo succession, southern Ethiopia, yields largely fragmentary jaw parts, associated or isolated teeth, a few postcranial parts which have been referred by Howell and Coppens (1976a) to *A. africanus*. These derive from Members B, C, D, E, and, perhaps, F and (lower) G of the Shungura Formation. The last three members yield evidence for the presence of another species of the genus. Associated or isolated teeth from the adjacent Usno Formation (Howell 1969a, 1969b), somewhat older than the oldest hominids from the Shungura Formation, have also been tentatively attributed to *A. africanus*. They may well ultimately prove to represent a distinct though related taxon, like *A. afarensis*.

10 Substantial numbers of stone artifacts of Developed Oldowan or even earliest Acheulian type do occur in the Middle Breccia (Unit C of Butzer 1978) in the Sterkfontein Extension (west pit) exposures (Robinson 1961, Robinson and Mason 1962) which are demonstrably *younger* than the australopithecine-bearing Lower Breccia (Unit B of Butzer.). Unit C has yielded undisputable skeletal remains of *Homo*.

11 Several distal humeri of early Hominidae are in fact known (see tables 10.1 to 10.3). However, only the specimen from Kromdraai, associated with the type of *A. robustus*, can be precisely identified. McHenry (1973), and more recently McHenry and Corruccini (1975a), have clearly shown substantial differences between three of the known early hominid humeri; it is particularly interesting that the oldest, that from Kanapoi, is the most modernly humanlike of the lot. Moreover two specimens, either known (Kromdraai 1517) or considered (ER-739) to represent an *Australopithecus,* are the most pongidlike.

No hominid remains from the Koobi Fora Formation, East Rudolf, Kenya, have as yet been formally attributed to *A. africanus*. However, I consider that upper and lower postcranial parts (ER-1500) from different localities in the Lower Member, and probably others (ER-1503/1504, ER-1476) from localities in the earlier part of the Upper Member, are perhaps assignable to *A. africanus*. They are tentatively so assigned here, with the caveat that these body parts are either not adequately known or appreciated among associated paratypes of this species.

A number of hominid postcranial parts from the oldest fossiliferous sediments of Olduvai Gorge have been attributed to the taxon *Homo habilis* (L. S. B. Leakey, Tobias, and Napier 1964). However, no clear evidence suggests that only a single taxon is represented by these specimens. In fact, several studies (Day and Wood 1968, Oxnard 1972, 1973, Lisowski, Albrecht, and Oxnard 1974, Wood 1974b) suggest that some or all of these specimens reveal either australopithecine or pongid features, hence phenetic affinities. These arguments and my own observations on the relevant specimens are sufficiently convincing that the latter (clavicle, Hom. 48; radius, Hom. 49; tibia and fibula, Hom. 35; pes, Hom. 8; and metatarsals, Hom. 43) are here tentatively attributed to *A. africanus*. However, these body parts are either ill- or unknown in paratype samples of this taxon; therefore their small size, proportions, and un-*Homo* features suggest this attribution more than any appropriately detailed comparative analytical study of comparable body parts.

In 1973 and 1974 very important finds of *Australopithecus* were made in the Hadar Formation, central Afar, eastern Ethiopia (Taieb 1974a, Taieb et al. 1975b). The specimens derive from the upper (Kada Hadar Member) unit of a 140-m long sequence of lacustrine and perilacustrine sediments. They occur above a basalt having an age of 3.01 m.y. The upper unit has afforded 52 parts of a very small female skeleton (AL 288-1) including cranial fragments, mandible with dentition, much of the upper limb (except the manus), partial ribs, and a number of vertebrae, sacrum, and innominate, and much of a lower limb (only talus of the pes). This extraordinary discovery will provide much needed data on body proportions, stature, and weight in this species, as well as abundant new information on joint morphology and hence new insight into locomotor adaptations (Taieb et al. 1976).

The partial skeleton (AL 288-1), that of a female individual, is the most complete specimen of *A. africanus*. The cranial bones are thin and lack any trace of superstructures, though the M_3s are erupted and in wear. The mandible is robust, though small, markedly V-shaped, with narrow incisal region and moderate alveolar planum. P_3 has a strongly inclined buccal surface and is nearly sectorial, with scant metaconid expression. The postcranial skeleton is *Australopithecus*-like overall. The ilium is somewhat higher, and the anterior border relatively more straight than in Sts. 14, and the acetabulum is quite shallow. The femora (like those of AL 128, 129) are small, with flattened, elongated neck, slight trochanteric lateral flare, intertrochanteric line, quadrate tubercle, pronounced spiral line, and lesser trochanter quite to moderately medially situated; distally there is a high bicondylar angle, flattened and elongated lateral condyle, and the lateral lip of the patellar groove is elevated. The tibiae show a pronounced tuberosity, well-developed intercondyloid fossae and strong intercondyloid eminence, slightly retroflexed head with scant tibial torsion, and distinct expression of soleal line and interosseous membrane attachment. A humerofemoral index of about 84 is estimated from the AL 288 specimen.

This is the youngest hominid occurrence in the Hadar Formation, falling within the normal interval between the Mammoth and Kaena reversed events, with an age of ~2.95 m.y. In many respects these remains are intermediate between *A. afarensis* and *A. africanus*.

Australopithecus boisei (Leakey) 1959

A species of the (extinct) genus *Australopithecus*, apparently restricted in distribution to eastern Africa, and distinguished by small cranial capacity, between 500 and 530 cc; endocranial cast showing marked parietal expansion, both transversely and vertically.

Cranium overall robustly constructed, though relatively thin-walled, with substantial to extensive pneumatization; vault long, spheroidal in shape, with strong postorbital constriction; variable development of ectocranial superstructures according to sex; males with sagittal crest and anteriorly and posteriorly divergent temporal crests, the latter producing a "bare area" bounded by temporal and nuchal crests; males with substantial nuchal crest, simple medially and laterally, but compound (temporal-nuchal crest) intermediately, and with strong posteroinferiorly disposed external occipital protuberance and nuchal musculature and related occipital impressions; occipital bone with larger and flatter upper than lower scale, superoposteriorly shifted inion (but below the Frankfurt Horizontal); vault with steep ascent of parieto-occipital plane from the external occipital protuberance; parietal with reduced coronal margin curvature; foramen magnum shortened, set well forward, and oriented nearly directly downward; small occipital condyles, small condylar groove; temporal with large inwardly directed mastoid process, extensively pneumatized, strong supramastoid

Table 10.1 Remains attributed to *A. africanus*.

| Elements | Tau'ng | Sterkfontein | Makapan Limeworks | Omo Basin, Ethiopia | | East Turkana: Koobi Fora Formation[a] | Chemeron | Olduvai Gorge: Bed I |
				Shungura Formation	Usno. Formation			
Crania	1 (+ mand.)	1511; Sts, 5, 17, 19, 20, 25, 26, 58, 60, 67, 71, 1006; StW-13	MLD. 1, 3, 10, 37/38				1 (f)	
Maxillae		1512, 1514; Sts. 2, 8 10, 12, 13, 27, 29, 32, 35, 42, 52, 53, 57, 61, 63, 64, 66, 69, 70; StW-18	MLD. 6, 9, 11 MLC. 1	Omo 75 (G-13)				
Mandibles		1515, 1516 1522; Sts. 7, 18, 24, 36, 38, 41, 52, 62; StW-14	MLD. 2, 4/ 18/24, 19/27, 22/34, 29, 40	Omo 18 (C), Omo 860 (F), Omo 222 (G-5) L.427 (G-4) Omo 75 (G-13)				
Dentition		69	4	G 1-13 = 10 locs. (24); F = 9 locs. (36); E = 7 locs. (10); D = 8 locs. (10); C = 16 locs. (38); B = 5 locs. (12)	2 locs. (21)			

[a]For East Turkana hominids, different type styles indicate different stratigraphic horizons, as follows: roman = Sub-Chari, Upper Member, Koobi Fora Formation; italic = Lower Member, Koobi Fora Formation; boldface = Sub-L/M Tuff Complex, sub-Okote, and sub-Koobi Fora Tuff, Upper Member, Koobi Fora Formation.

crest confluent with expanded zygomatic root; large robust entoglenoid process formed wholly by squamous temporal (perforated by foramen spinosum), small and sometimes doubled postglenoid process closely approximated to tympanic plate; cranial base substantially shortened, with marked angulation of petrous axis to median plane, extensively pneumatized sphenoid body, and massive, pneumatized pterygoid process; parietal with broad, low Sylvian crest; petrous pyramids well angulated, with steep nearly vertical posterior surface, hollowed laterally with sharp superior margin; internal occipital squama with low cruciate eminence, large though flattened cerebellar fossae, and upwardly shifted internal occipital protuberance; internal venous drainage (sometimes) lacking lateral sinus system, without transverse sinus contribution to sigmoid sinus; middle meningeal artery (sometimes) with reduced anterior and expanded posterior trunks.

Facial skeleton hafted to calvaria at low level (reflected in low supraorbital height index); frontal markedly flattened, with broad, hollow frontal trigone behind the large to massive supraorbital torus; torus with extensive frontal sinus development, and distinction in orientation of glabellar and superciliary parts, the latter inferolaterally inclined; frontal process of zygomatic anteriorly disposed, hollowed superoinferiorly, and with marked inferior and anterior slope to zygomatic body; interorbital region broad, inflated with pneumatized frontoethmoidal air cells; midfacial region hollowed and pyriform aperture and lower portions of nasal bones in depressed disposition; nasion nearly coincident with glabella; nasal bones high, prominent superiorly, and inferiorly narrowed, constricted and recurved; pyriform aperture long, with high posteriorly set anterior nasal spine; maxillaries large, both vertically and transversely, with extensive maxillary sinuses, in accord with substantial facial length and breadth, and inferior position of intraorbital foramen;

Table 10.1 (*continued*)

| Elements | Tau'ng | Sterkfontein | Makapan Limeworks | Omo Basin, Ethiopia | | East Turkana: Koobi Fora Formation[a] | Chemeron | Olduvai Gorge: Bed I |
				Shungura Formation	Usno. Formation			
Postcranials								
Clavicle								H. 48
Scapula		Sts. 7		Omo 119 (D)		ER-1500 o		
Humerus		Sts. 7				ER-1500 l	;1504/03	
Ulna						ER-1500 f, i		
Radius		Sts. 68				ER-1500 e, k		H. 49
Manus								
Carpals		Sts. 1526						
Meta-carpals								
Phalanges				Omo 18 (C)				
Vertebral Column		Sts. 14 (15 + 3 sac.); StW-8 (4) Sts. 73 (1)						
Pelvis		Sts. 14; 65	MLD 7/8; 25					
Femur		Sts. 14, 34, 1513; StW-25		L.754 (G-4)		ER-1500 b, d; 1503/04;		
Tibia						ER-1500 a, c, h, j, r; 1476 b, c		H.35
Fibula						ER-1500 g		
Pes								H.8
Talus						ER-1476a		
Calcaneum								
Tarsals								
Meta-tarsals						ER-1500 m		? H.43 (2)
Phalanges								

upper and lower facial prognathism reduced, with nearly vertical disposition of inferior part of nasoalveolar clivus, and deep subnasal maxilla and associated, elongate tooth roots; maxillary with conjoined alveolar juga of upper C/P^3 and root of zygomatic process which is anteriorly disposed between the upper premolar roots; lack of facial hollowing or sulcus above upper P; zygomatic process with incurved (incisive) lower border, hence inframalar notch, and substantial to massive masseteric musculature origins, even with encroachment anteriorly onto maxilla; palate generally very deep with steep shelving downward anterior to incisive foramen, and sometimes with abbreviated palatine torus; adult individuals with posterosuperior slope of occlusal plane; superior dental arcade parabolic with moderate curvature in disposition of ante-premolar dentition.

Mandible with ramus set just posterior to perpendicular to body axis; and with anterior border originating anterior to M_2 or to M_3, with swelling passing posteriorly from M_1; endocoronoid, and especially endocondyloid buttresses generally markedly developed; thickened mandibular angle with generally well-marked internal pterygoid and masseter insertion areas; deep body, and especially anteriorly, markedly to extremely robust, with expanded lateral and marginal tori; symphysis posteriorly inclined with nearly convex labial surface, marked to pronounced alveolar planum, superior and/or inferior transverse tori prominent to massive, and former prolonged distally to premolar region or beyond, usually with deep median fossa, mental foramen usually anteriorly situated, between P_3/P_4 or at P_4 mesial root, and relatively inferiorly situated due to elongate premolar and molar roots; mandibular arch moderately to markedly constricted anteriorly with variably posteriorly divergent bodies; anterior dentition vertically set and exhibiting mesial drift and/or crowding; premolar-molar dentition with progressive posterior lingual situation of back teeth.

Deciduous dentition, with upper dentition almost unknown, except for very large dm^2 with greatly complicated fissure and cusp pattern; lower anterior dentition poorly known, but d_c small-crowned, with moderate marginal ridges and lingual basal swelling; dm_1 and dm_2 large to exceptionally large; molarized dm_1, five-cusped, talonid with mesial cusps approximated, elongate protoconid with sloping buccal face, mesial accessory cusplet, mesial margin obliquely set, and obliquely disposed anterior fovea, hypoconid with strongly swollen buccal face, small hypoconulid aligned longitudinally with protoconid, and small, oblique posterior fovea; dm_2, five cusps, approximating a +5 pattern, with broad hy^d to en^d contact, small distolingual sixth cusp, narrow anterior fovea, and lacking noteworthy cingular derivatives.

Permanent dentition with markedly disharmoniously proportioned anterior and postcanine teeth; anterior teeth (roots) broadly comparable in size to, and posterior teeth (crowns) extremely robust compared to other species; upper C absolutely and relatively small, hence exceptionally low upper C/P^3 and upper C/P^4 ratios; premolars and molars of large to extremely large size, and buccolingually as well as in the case of posterior molars mesiodistally greatly enlarged; P3 larger than P4; M^3 variably (barely) smaller than, equal to, or larger than M^2; M_3 usually larger than M_2. Upper dentition with I's generally lacking notable surface relief; upper C asymmetric in form, with distal cusplet, strong lingual tubercle, well-separated parallel lingual grooves with intervening basal swellings, buccal face with swollen sub-cervical enamel; P^3 three-rooted, with large buccal cusp, weak buccal grooves, very high shape index; P^4 three-rooted, comparably sized buccal and lingual cusps, with expanded crown, especially distobuccally, and well-defined distobuccal groove, marked transverse diameter at occlusal line compared to that at occlusal surface; upper M's with weak buccal cingular traces (buccostyle), but may be some weak expression lingually of Carabelli complex, including extension onto hypocone. Lower dentition with long, robust-rooted anterior teeth, and greatly enlarged premolar-molar teeth; lower I's small-crowned, with simple lingual surface, ill-expressed lingual grooves, and weak to moderate labial grooves; lower C of moderate size, with asymmetrical crown and distolingual cusplet, poorly developed buccal grooves, lingual grooves variably slight to moderately deep, moderate median lingual ridge, no real lingual basal tubercle, and enamel bulge labially above cervical line; premolars, especially P_4 and all lower M's substantially enlarged, both lengthwise and especially buccolingually, and often with partially or extensively fused and platelike roots; P_3 large, but essentially as in other species; P_4 *strongly* molarized, with expansion of buccal and lingual angles of talonid, and accessory cuspules on distal margin, posterior fovea large, buccal grooves weak to moderate, except distal normally well expressed; lower molar dentition large to exceptionally large compared to other species, and especially the distal molars; enamel thick as on last premolars, and generally low, massive cusps, with primary fissuration deep and well delineated; often M_2 and M_3 with accessory cuspule formation, including divided metaconid, also hypoconid, and expression of sixth and seventh cusps, and buccal cingular remnants customarily no more developed than pits.

Stature, body weight and proportions still incompletely known or estimated; stature estimates, on basis of referred, partial upper (humerus) and lower (femur, tibia) postcranials have suggested values from 148–168 cm, and generally larger than other species; body weight estimates, on basis of incomplete postcranials from 50–65 kg to even as much as nearly 100 kg; forelimb relatively long relative to lower limb, arm long relative to thigh; overall absolutely massive body build; except for ultrarobustness, postcranial morphology essentially same as for the genus.

This exceptionally robust species (*boisei*) of *Australopithecus* occurs in final Pliocene and earlier Pleistocene deposits in eastern Africa (table 10.2). For the moment its distribution is only documented in this portion of the continent and only in sedimentary situations adjacent to the Eastern Rift Valley. In most instances the species is more or less directly associated with another hominid species as well as with lithic materials and sometimes (at Olduvai Gorge and Koobi Fora) even occupational situations reflecting culturally patterned behavior (whose is still a subject for debate).

The holotype of this species of *Australopithecus* is an almost complete cranium with full maxillary dentition of a young adult individual (L. S. B. Leakey 1959, Tobias 1967a). The specimen was recovered at Olduvai Gorge (Tanzania) from site FLK in sediments of Bed I from a land surface (weak paleosol) formed toward the southeastern margin of the lake. This surface preserved an ancient hominid occupational situation, including patterned distributions of remains of a diversity of vertebrate species and associated artifacts of an Oldowan stone industry (M. D. Leakey 1971). The general and specific stratigraphic situation (Hay in M. D. Leakey 1971) of the occurrence is exceptionally well documented, and its radiometric age (Curtis and Hay 1972) and placement in the geomagnetic polarity scale (Grommé and Hay 1971, Brock and Hay, 1976) are extremely well defined.

Additional hominid remains have since been recovered on or in a surficial relation to this same occupational situation. They include Hom. 6 (cranial vault fragments, and three upper teeth), Hom. 44 (an M^1), and Hom. 35 (tibia and fibula, largely complete) (cf. L. S. B. Leakey 1960). These remains have at one time or another been considered to represent a hominid other than *A. boisei* and are so considered here (see below). The remains of Hom. 35 (once attributed to Hom. 6) have been briefly described

Table 10.2   Remains attributed to *A. boisei*.

Elements	Olduvai Gorge	Peninj	Chesowanja	East Turkana[a]	Omo Basin
Crania	H.5		CH-1	ER-732, 733, **406, 407,** *417* **814, 1478, 1170**	Omo 323 (G-6/7) L.338y(E)
Maxillae				ER-405, **1804**	
Mandibles		X		ER-404, **725, 726, 728, 729, 801, 805, 818, 819, 727, 1468, 1816, 812, 403, 810, 1477,** *1469, 1482, 1803,* **1806, 3230**	Omo 44, 57 (E); L.7,74A (G-5)
Dentition	H.3; 26; 30 (?); 38; 15			ER-802, 1467, **1171, 816, 998, 1479, 1509**	E=1 loc. (7); F=3 locs. (4); G1–13=3 locs. (11)
Postcranials Clavicle Scapula Humerus Ulna Radius Manus Carpals Metacarpals Phalanges Vertebral column Pelvis	Postcranials Clavicle Scapula Humerus Ulna Radius			ER-739, 740 L. 40 (E); Omo 141 (G-3)	
Femur	H.20			**738, 815, 993,** 1463, 1465, 1503, **1505, 1592,** *3728*	
Tibia Fibula Pes Talus Calcaneum Tarsals Metatarsals Phalanges				741	

[a]For East Turkana hominids, different type styles indicate different stratigraphic horizons, as follows: roman = Sub-Chari, Upper Member, Koobi Fora Formation; italic = Lower Member, Koobi Fora Formation; boldface = Sub-L/M Tuff Complex, sub-Okote, and sub-Koobi Fora Tuff, Upper Member, Koobi Fora Formation.

(Davis 1964), but have never been assigned to a particular hominid taxon. They too are considered here to represent a hominid different from *A. boisei* (see above; cf. Wood 1974b). If this interpretation is correct, this site provides evidence for the most direct and immediate temporal sympatric association of two taxa of early Hominidae.

Hominid remains from some five other localities at Olduvai Gorge very probably represent *A. boisei*

(M. D. Leakey 1978). They include two upper deciduous teeth[12] (Hom. 3) from the upper part of the infilling of a large stream channel, rich in vertebrate fossils and yielding a Developed Oldowan industry

12 These specimens represent an upper dc and dm² (cf Dahlberg 1960); dm² has been variously identified as another deciduous or permanent tooth by L. S. B. Leakey (1958), Robinson (1960), and von Koenigswald (1960).

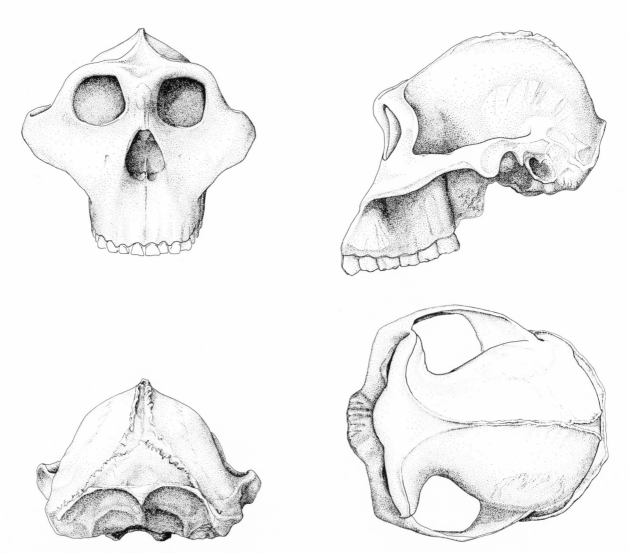

Figure 10.7 *Australopithecus boisei.* Facial, lateral, posterior, and superior views of restored cranium (based on specimens from Olduvai Gorge and Koobi Fora). Two-fifths natural size.

(M. D. Leakey 1971) at site BK (uppermost Bed II); an M_2 and two incisors (Hom. 38) at site SC, yielding a Developed Oldowan industry (related to Tuff IID, upper Bed II); two upper molars and a canine (Hom. 15) associated with a few artifacts at site MNK from a reworked tuff in a former lake margin context (lower part of upper Bed II); an M_3 (Hom. 26) at site FLK-West (probably derived from lower Bed II); and a proximal femur fragment (Hom. 20) at site HWK (Castle) (probably derived from lower Bed II). The latter specimen has been preliminarily described by Day (1969a) and attributed to this species of *Australopithecus.*

All these specimens seem to document the presence of *A. boisei* from near the base of Bed I to close to the top of Bed II, a time span of close to 0.75 m.y.

A partial hominid cranium, comprising most of the facial skeleton, anterior part of the cranial base,

and much of the right upper dentition is known from the Chemoigut Beds, Chesowanja (Kenya) (Carney et al. 1971). Its age has been a subject of concern, but is now considered to be (broadly) earlier Pleistocene, and not particularly late within that time range (Bishop, Hill, and Pickford 1975). It has generally been considered attributable to *Australopithecus,* although conflicting views have been expressed by Szalay (1971) and Walker (1972a) as to its probable extent of distortion and its presumed endocranial capacity. I consider it to represent *A. boisei.*

Olduvai Gorge has not yet yielded lower jaw parts of this species. The well-preserved adult mandible with complete dentition recovered from the Humbu Formation, at Peninj, Natron basin Tanzania, (Isaac 1965, 1967, was the first such specimen to be attributed to the species (L. S. B. Leakey and M. D. Leakey 1964).

The several members of the Koobi Fora Formation, East Turkana (Kenya) have yielded numerous and often remarkably well-preserved skeletal remains attributable to *A. boisei* (R. E. F. Leakey 1970, 1971, 1972, 1973, 1974) (table 10.2). They include complete and partial crania (8 individuals), maxillae (2 individuals), mandibles (20 individuals), isolated and associated teeth (7 individuals), partial humeri (3 individuals), femora (8 individuals) and tibiae (1 or more individuals) (see R. E. F. Leakey, Mungai, and Walker 1971, 1972, R. E. F. Leakey and Walker 1973, Walker 1972b) as well as very probably other specimens (especially postcranial) still to be taxonomically assigned.[13] Of these body parts of nearly 50 individuals, 3 derive from the sediments of the Lower Member (below the "KBS" tuff), 30 derive from sediments of the lower part of the Upper Member (below the Koobi Fora, Okote, or Lower/Middle Tuff Complexes), and 15 from sediments of the upper part of the Upper Member in or above these tuffs and below the Chari tuff.

In the Lower Member *A. boisei* is indeed rare, about as infrequent as *A. africanus* appears to have been. A species attributed to *Homo,* on the other hand, is more than five times as common. In the lower part of the Upper Member *A. boisei* is half again as common as a species attributed to *Homo,* whereas in the upper part of that member each taxon is about equally common.

This extremely valuable, indeed unique sample spans about a million years, from ~ 1 m.y. to perhaps ~ 2 m.y. (depending on the age of the "KBS Tuff" or tuffs; see footnote 3, this chapter; Fitch and Miller 1970, Findlater et al. 1974, Curtis et al. 1975).

Assigning postcranial parts to this taxon is not always straight-forward because several other hominid taxa are represented in the Koobi Fora succession. Moreover, the postcranial skeleton of the genus *Australopithecus* is still incompletely known from the South African paratype samples, so comparable body parts are often unavailable for comparative study. However, such attributions can be attempted by using first the morphological divergence from the structure characteristic of *Homo* and second the occurrence of several specimens found in close proximity and derived from the same fossiliferous horizon. Such hominid "clusters" are known in several instances (Behrensmeyer 1976a), for example, in

two floodplain paleoenvironments above the Middle Tuff in the Ileret area (Area 1) in which postcrania and mandibles occurred, as well as in another distributary paleoenvironment, just below the Lower Tuff (Area 6A) in which cranial and jaw parts occurred. Cranial and jaw parts and lower limb bones also occurred in the Koobi Fora area (Area 105) in a channel above the KBS Tuff.

Skeletal remains attributable to *A. boisei* are not common in the Omo Succession, southern Ethiopia (Howell and Coppens 1976a). The first firm documentation of this taxon is in Member E of the Shungura Formation, and comprises only a partial mandible (Omo Locality 57), a partial juvenile cranium (Locality 338y) (Rak and Howell 1978), a series of associated permanent teeth (Locality 338x), as well as a complete ulna (Locality 40). Complete ulnae are otherwise unknown in *Australopithecus,* but the distinctive morphology, size, and proportions of this specimen diverge from the *Homo* condition (Howell and Wood 1974, McHenry, Corruccini, and Howell 1976), and are also in accord with other aspects of upper limb morphology known for *A. boisei.* Only several isolated teeth from three localities represent this taxon in Member F of the Shungura Formation. In the lower fluviatile sediments (units G-3 and G-5) of Member G, isolated or associated teeth from three localities and mandibles from two other localities (7A and 74A), as well as a partial cranium, represent this taxon. It has not been surely documented in younger horizons in the Omo Succession.

Australopithecus crassidens (Broom) 1949

A species of the (extinct) genus *Australopithecus,* known only from southern Africa, distinguished by small cranial capacity, documented at 530 cc, but range of variation unknown; endocranial cast showing expanded parietal and temporal cortex and relatively decreased striate and parastriate (visual) cortex, enlarged and rounded inferior frontal convolution, cerebellar lobe well developed, situated well under cerebrum, and triangular with anteriorly and transversely expanded lateral lobe.

Overall cranium is robust but less than *A. boisei* and more than *A. africanus,* with ectocranial superstructures and substantial pneumatization; vault thin-walled, spheroidal, with strong postorbital constriction; males and probably females as well with simple sagittal crest in parieto-occipital region, with moderately keeled crests for anterior portions of temporal musculature, extending about halfway across width of orbit before diverging posteriorly; inferior temporal line most commonly separated from superior nuchal line, and with confluent occipitomastoid crest; supramastoid crest not greatly enlarged, and not confluent with zygoma root; mastoid process extensively pneumatized; base of external occipital protuberance essentially coterminous with Frankfurt Horizontal;

---

13 These remains are attributed to *A. boisei* on the basis of preliminary identifications by R. E. F. Leakey, A. C. Walker, B. A. Wood, and M. H. Day, and my own observations on the original specimens.

foramen magnum moderately long; digastric and occipital arterial sulci separated often by a substantial eminence; cranial base shortened, particularly spheno-occipital portion; moderately elevated angulation of petrous temporal to median sagittal plane; low axis of tympanic plate and petrous temporal; entoglenoid process large, robust, and formed wholly by squamous temporal; postglenoid process moderately large, but posterior articular surface of mandibular fossa formed by medial portion of tympanic plate; internal venous drainage lacking lateral sinus system, as in *A. boisei,* without transverse sinus contribution to sigmoid sinus; middle meningeal artery with early separation of anterior and posterior rami, large anterior division of posterior ramus, and anterior ramus divided into anterior and posterior branches in Sylvian region.

Facial skeleton hafted to calvaria at a relatively low level (and as reflected in low supraorbital height index); vault rises only slightly above upper margin of orbits, with low flat frontal and broad, hollowed frontal trigone behind prominent glabella and strong horizontally disposed supraorbital torus; torus thicker vertically than in *A. boisei,* and its lateral portion presents both upper and anterior surfaces, unlike *A. boisei;* broad, not markedly elongated face; pyriform aperture and nasal bones set in a midfacial hollow due to the forward projection of the malozygomatic elements; high, anterior position of nasion close below glabella; nasal bones with central constriction, substantial superior and more marked inferior broadening, with latter and inturned, nasal bridge rather elevated with nasals meeting at a quite wide angle; pyriform aperture not markedly elongated, with lower border set well back of the plane of the lateral margin; weak, posteriorly set anterior nasal spine formed from incisor crest; broad, moderately high maxillae; infraorbital foramen set relatively low below inferior orbital margin; infraorbital area of maxilla with anterior (canine jugum) and posterior (supra-P^4 zygomatic root) crests bounding narrow vertical maxillary sulcus; zygomatic process passes inferiorly in essentially a straight line into alveolar process; large, vertically deep zygomatic with small anterior and larger lateral facing surfaces; prognathic subnasal maxilla; moderately deep palate, shallowing markedly forward of the molar dentition, and with small median palatine torus; occlusal plane aligned parallel to Frankfurt Horizontal above molar dentition, but well upcurved anterior to P^4; premolar-molar tooth rows quite strongly divergent; sockets of upper anterior teeth transversely aligned or in slight curve, but tooth crowns disposed in a curve.

Mandible with vertically set, wide, high ramus; coronoid process prominent, posteriorly curved; sigmoid notch deep, constricted; salient, thickened crest along anterior margin of ramus for temporalis m. ligament; endocondyloid-pharyngeal crest(s) strong and endocoronoid crest generally massive, passing anteroinferiorly of coronoid process, and inferiorly confluent as triangular torus; with deep intercristal triangular planum; well-developed external condylar crest; broad, extensive retromolar space; lateral mandibular tori generally prominently developed; mental foramen set relatively inferiorly on body, in rela-

tion to elongated tooth roots; relatively to markedly V-shaped mandibular arch contour, in relation to enlarged premolar-molar dentition, and relatively small anterior dentition; outer symphysis flattened superiorly, with inferior third curved posteroinferiorly, and lacking all, or most principal elements of bony chin formation; symphysis overall robust, generally with massive inferior transverse torus and prominent to very robust superior transverse torus; alveolar planum usually substantial, posteroinferiorly inclined and variably concave; genioglossal fossae often deep, and with or without mental spines on inferior transverse torus; lingual mandibular tori substantial, especially that anteriorly confluent with inferior transverse torus.

Deciduous dentition, with anterior upper teeth poorly known; dm^1 of essentially modern four-cusped hominid type, with protuberant buccal face of paracone, deep anterior and posterior foveae and central fossa, without accessory mesiobuccal cusp; dm^2 much like M^1, four-cusped, but with relatively larger hypocone, well-defined primary fissure pattern, and without secondary fissuration, anterior fovea with buccal limb, median buccal groove with terminal pit. di_2 larger than di_1, latter with slight shovel shape; d_c with asymmetrical crown, cingulum higher mesially and lower distally, forming distinct cuspule, maximum length more than half of crown height, buccal face with strong distal groove, lingual face with weaker mesial and well-defined distal groove; dm_1 with talonid cusps well differentiated, fully molariform five-cusped tooth, accessory cuspule mesial to pr^d, pr^d relatively small and approximated to me^d with high intercuspal crest, primary fissure system of dryopithecine type, but tendency for alignment of lingual and buccal mesial limbs, straight main fissure set parallel to short axis of crown, anterior fovea deep and with high anterior face, and smaller than in *A. africanus,* posterior fovea ill-developed or absent; dm_2 generally larger than in *A. africanus,* and similar in size to, or smaller than *A. boisei,* fully molariform, symmetrical Y pattern, with tuberculum sextum and posterior fovea, anterior fovea doubled, mesial buccal groove with terminal pit.

Permanent dentition with disharmoniously proportioned anterior and postcanine teeth; anterior teeth (roots) comparable in size to, and premolar-molar crowns robust compared to *A. africanus;* small canines with correspondingly low lower C/P_3 and C/P_4 ratios; premolars and molars large, with notable buccolingual expansion; M_3 usually larger than M_2. Upper dentition with upper I's moderately shovel-shaped, gingival eminence and substantial lingual grooves; upper C without asymmetric crown lacking vertical curvature, buccal face nearly featureless, lingual face with double grooves, basal eminence well developed and marginal ridges well defined; premolars with P^4 considerably larger than P^3, often with double buccal roots, not markedly crenulated, fissure pattern usually deeply incised; P^3 with weak buccal grooves; crown relatively wider at cervical line; P^4 with strong talon, distal buccal groove, and well-defined distal buccal cusp, commonly with distal lingual groove; upper M's usually with weak Carabelli complex, including small though deep mesiolingual pit and associated shallow grooves; M^1 with strong buccal groove that is weaker on M^2 and M^3, no parastyle development, anterior fovea represented by wide shallow depression and confluent with central fossa; M^3 may be much modified from upper molar pattern, especially distal cusps (particularly metacone reduction), secondary crenulation, and development of accessory cuspules.

Lower dentition with long, robust rooted anterior teeth, small canines, enlarged premolar-molar teeth; lower I's small-crowned, only slightly shovel-shaped with shallow lingual fossa, weak marginal ridges, slight lingual grooves, no or only slight lingual tubercle; lower C small crowned, robust root, crown asymmetrical with high mesiosuperior margin and final slight upward slope to apex, small to diminutive distal lingual cusplet, buccal face nearly featureless though distal buccal groove distinct, lingual face with stronger distal than very weak mesial groove, with slight vertical ridge distal to midline, and without tubercle enlargement of basal eminence, enamel markedly increases in thickness above cervical line; P_3 substantially smaller than P_4; P_3 essentially as in *A. africanus* though less asymmetrical, weaker lingual cusp, buccal face with less relief and lingual face featureless, moderate triangular ridges between pr^d and me^d, small anterior and larger posterior foveae, latter with doubled cusplets or thickened distal wall; P_4 with strong distal buccal groove, weak or absent mesial buccal groove, low cusps, no real triangular ridges between foveae, small anterior and large posterior foveae, latter with well-developed distal wall and usually with some cuspule development; lower molar dentition large, and increasing in size posteriorly, with absent to weak buccal cingulum remnants; M_1 with symmetrical *Dryopithecus* pattern, sometimes with reduced me^d to hy^d contact and near alignment of lingual and mesial-buccal limbs, and without extra anterior transverse fissure, usually with sixth distal cusp; M_2 with more asymmetrical Y-fissure pattern, with alignment of lingual and mesial-buccal limbs, distortion of distal limb of Y in central fossa, without marked secondary fissuration, commonly with small median lingual accessory cuspule partially linked to me^d, usually with sixth distal cusp; M_3 with fissure pattern as in anterior molars, or further simplified to rather sinuous longitudinal groove, sometimes with more complicated pattern with accessory fissures, and with subdivisions of hy^d and en^d, and even of hy^{ld}.

Stature, body weight, and proportions very incompletely known and estimated from very incomplete postcrania; a robust species, larger than *A. africanus* but less massively built than *A. boisei;* stature perhaps 154–157 cm and body weight probably some 45–50 kg; skeletal trunk length 42–46 cm. Postcranial morphology less well known than in other species, and so far as known essentially the same as for the genus.

This robust species (*crassidens*) of *Australopithecus* is thus far known only from the Swartkrans lo-

cality, Krugersdorp area, Transvaal. This site has a complex history of infilling and related movements of portions of the dolomite in which the cave systems were developed. Recent investigations at this site have not only confirmed a complex history, but have also revealed a substantially more prolonged period of infilling than previously envisioned (Brain 1976b and pers. comm., Vrba 1975, 1976, Butzer 1978). These findings have important implications for the provenience, associations, and relative ages of the numerous hominid remains recovered there.

The Swartkrans cave system comprises an outer cave filled with (a) pink breccia, an inner cave filled with (c) brown (stratified) breccia, and (b) an intermediate orange breccia, at the base of the inner cave's infilling. The inner and outer caves are separated by a fallen roof block forming a floor barrier between them, which blocks an underlying lower cave.

Australopithecine remains have derived from the infilling (a), or from the intermediate unit (b) below the infilling (c). Recent work at the site has shown that the outer cave infilling (a) is itself complex, as those hominids and certain vertebrate taxa only occur in a localized area (Brain 1976b, Vrba 1975, 1976). Between a portion of that infilling (now termed Member A) and a younger infilling (Member B) there occurred a prolonged period of erosion and of shaft opening into the site (Brain 1976b, Butzer 1976b).

The robust australopithecine *A. crassidens* and a distinctive, more archaic vertebrate fauna is restricted to the initial infilling (Member A). Three hominid specimens referred to the genus *Homo* (Clarke, Howell, and Brain 1970, Clarke 1977), as well as artifacts of (Developed) Oldowan type, derive from the same infilling. The australopithecine sample comprises over 200 body parts including cranial parts, mandibles, numerous isolated teeth, and postcranials (table 10.3). These represent the remains of over 75 individuals, over 40% of which are immature (Mann 1975). The remainder are mature, with none older than their mid-30s. Of the total hominid sample isolated or associated teeth are most common, whereas cranial and jaw parts are about half as common. The postcranial skeleton is represented by only a dozen pieces including fragmentary parts of the axial skeleton and upper and lower limbs.

Among the total fossil assemblage (excluding microvertebrates) in this infilling, hominids (14%) and cercopithecoids (13%), represented by 4 taxa, are about equally common (Brain 1976b). There are 16 taxa of carnivores, which comprise only 10% of the total assemblage. Several other large mammals—

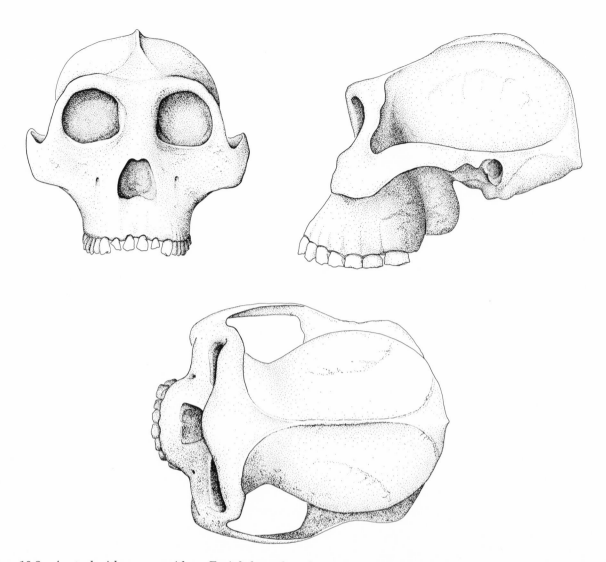

Figure 10.8 *Australopithecus crassidens*. Facial, lateral, and superior views of restored cranium (based on specimens from Swartkrans). Two-fifths natural size.

including two species of pigs, three species of equid, a sivathere—and a few porcupines and many specimens of two species of hyrax altogether compose 20% of the assemblage. Nearly half (43%) of the assemblage is represented by 16 taxa of bovids (Vrba 1975, 1976), which are largely of medium body size with a high percentage of juveniles. This evidence, along with the condition of bone and body part representation in the assemblage, strongly supports the conclusion that predators, particularly leopard (*Panthera pardus*) but perhaps also sabertooths, were largely responsible for this fossil accumulation (Brain 1970, 1973).

The younger infilling (Member B) of the Pink Breccia has yielded more advanced Paleolithic artifacts of an Acheulian industry as well as a single hominid specimen referred to the genus *Homo*. This specimen, the SK.15 mandible, is the type of *Tel-*

anthropus capensis (Broom 1949, Broom and Robinson 1950a) and was subsequently transferred by Robinson (1961) to the genus *Homo* and referred by him to *H. erectus* (see below).

The representation of postcranial body parts makes comparison of *A. crassidens* with other *Australopithecus* samples difficult. There are fundamental similarities in proportions and overall structure in the innominates and the femora of the several samples, here regarded as species, of *A. crassidens, A. boisei,* and *A. africanus*. Zihlman (1971), Zihlman and Hunter (1972), Lovejoy (1974, 1975, 1976b, 1978), and McHenry (1975b,d) have all argued, with good reason, that the postcranial skeleton, except perhaps for certain size- and sample-related differences, does not adequately distinguish these several early hominids. Robinson (1972) disagrees, particularly about samples he considers to represent *A. africanus*

(Sterkfontein, Makapan Limeworks) and *A. (Paranthropus) robustus* (Kromdraai, Swartkrans). Pilbeam and Gould (1974) consider that "the three australopithecines are, in a number of features, scaled variants of the 'same' animal." The various other body parts (hand bones, vertebrae) are too frequently incomparable or at least as yet unstudied in an analytical comparative manner. The principal morphological differences between these several samples, which are here considered to be of specific rank, are reflected in certain discernible and often mensurable characters of the craniofacial skeleton and the deciduous and permanent dentition.

Australopithecus robustus (Broom) 1938

A species of the (extinct) genus *Australopithecus,* restricted in distribution to southern Africa and known to date only from the Kromdraai B site (Transvaal), distinguished by small cranial capacity, very incompletely known but perhaps comparable to that of other "robust" species; endocranial form and proportions hardly known, but temporal lobe shows usual australopithecine expansion and conformation.

Cranium imperfectly known from a young adult and maxillary-mandibular parts of several infants or juveniles; cranial vault structure, form, and proportions still unknown; relatively low position of lateral portion of nuchal crest; supramastoid crest well-developed, and forms

Table 10.3   Remains attributed to *A. crassidens* and *A. robustus.*

Elements	Swartkrans	Kromdraai-B
Crania	SK. 46, 47, 48, 49, 52, 54, 79, 83, 848, 869, 14003; SKW. 11, 29	1517
Endocranial cast	SK. 1585	
Maxillae (with dentition)	SK. 11, 12, 13, 21, 55a, 57, 65, 66, 826, 831a, 838a, 839, 845, 877, 881, 1512, 1590, 1592, 14080, 14129; SKW. 12	1602
Mandibles (with dentition)	SK. 6, 10, 12, 23, 25, 34, 55b, 61, 62, 63, 64, 74a, 81, 841a, 843, 844, 852, 858, 861, 862, 869, 876, 1648, 1586, 1514, 1587, 1588, 3978; SKW.5	1517, 1536
Dentition (isolated or associated)	111	1603, 1604, 1600, 1601 (10 teeth)
Postcranials		
Clavicle		
Scapula		
Humerus		1517 }
Ulna		1517 }
Radius		
Manus		
Carpals		
Metacarpals	SK. 84 (1), 14147	
Phalanges		
Vertebral column	SK. 853 (L), 854 (Ax), 3981 (T,L), 14002 (T,L)	
Pelvis	SK. 50,3155b	1605
Femur	SK. 82, 97	
Tibia		
Fibula		
Pes		
Talus		1517
Calcaneum		
Tarsals		
Metatarsals		
Phalanges		1517 (2)

anteriorly strong posterior root of zygoma; mastoid crest robust, salient; markedly thickened mastoid angle of parietal; mastoid portion of temporal extensively pneumatized; foramen magnum very broad; entoglenoid process formed by squamous temporal, foramen spinosum situated on alisphenoid, and relatively approximated to foramen ovale; mandibular fossa very long, quite broad, and deep, with prominent and robust articular eminence; medial wall formed by extension of temporal; moderately developed postglenoid process; tympanic plate well-developed, but mostly unthickened, with marked inferior prolongation to form posterior wall of articular fossa; probably (?) an occipital-marginal venous sinus drainage pattern; middle meningeal artery with prolonged common stem, bifurcation over lateral portion of temporal lobe, and probably substantial extension of posterior trunk into the occipital area; petrous pyramid with gradual, not steep or concave slope to middle cranial fossa, and flattish shallow area anterior to its medial tip posterior to foramen ovale; posterior aspect of pyramid with some undercutting, and probably some overhang of petrous crest laterally of the posterior cranial fossa; moderate Sylvian crest.

Facial skeleton incompletely preserved; midface broad and flat, with infraorbital region depressed medially and anterolaterally protrusive; infraorbital foramen moderately inferiorly situated; straight inferior margin of inframalar area; malomaxillary with thickened, robust root, originating above P^3/P^4 and extending back to M^1; malar and zygomatic process laterally flared but relatively slender, midface distinguished by separation into anterior and lateral facies, separated by a canine jugum, which passes into maxillary infraorbital facies; infraorbital foramen groove, bounded posteriorly by flattened supra-P^4 ridge; lateral nasal aperture rounded; no sharply defined, but rather guttered nasal sill; anterior nasal spine represented by small bony crest; slight subnasal prognathism; shallow palate, with median palatine torus expressed anteriorly. Mandible robust, more so than in *A. africanus* and broadly comparable to that of *A. crassidens*.

Deciduous dentition, with maxillary teeth still unknown; di_2 lacking buccal or lingual features, including lingual tubercle and marginal ridges; d_c with symmetrical crown, cingulum relatively high mesially and distally, strong horizontal buccal convexity, weak buccal grooves, well-defined lingual grooves, small distal cusplet, moderate lingual tubercle; dm_1 relatively large, in lower part of *A. crassidens* range, smaller than *A. boisei*, and larger, particularly longer, than in *A. africanus;* with broadened trigonid, lacking mesial accessory cuspule, and without mesial accessory buccal groove on protoconid, hypoconid lacking protuberant buccal face, large symmetrical and transversely disposed anterior fovea, no posterior fovea, approximated apices of protoconid and metaconid, asymmetrical Y-5 pattern with small me^d–hy^d contact; large buccally disposed hypoconulid, and lacking cingular derivatives; dm_2 narrower, and shorter or as short as *A. crassidens,* longer than *A. africanus,* typical Y-5 cusp pattern, tuberculum sextum minute or absent, sometimes with median lingual accessory cusp, well-developed symmetrical anterior fovea, poor or absent accessory (distal) anterior

fovea, lacking metaconid deflecting wrinkle, but may have hypoconid deflecting wrinkle.

Permanent dentition with incisor/canine elements largely unknown; P^3 large, three-rooted, smaller than *A. boisei* and within *A. crassidens* range, and of similar morphology to the latter except buccal grooves more pronounced; P^4 large, three-rooted, smaller than *A. boisei,* but in lower *A. crassidens* range, though broad, with mesially displaced protocone, talon moderately developed with buccal cuspule and slight lingual groove, secondary fissuration rather ill-expressed. Upper molars with $M^1 < M^2 < M^3$, overall smaller than *A. boisei,* below or within *A. crassidens* range, lacking Carabelli complex; M^3 with simple, relatively unreduced crown morphology. Lower premolars broad, essentially within *A. crassidens* size range, and of similar morphology, smaller than *A. boisei,* and larger than *A. africanus,* P_3 with two incompletely fused roots, P_4 two-rooted, very substantially molarized with expanded talonid. Lower molars with $M_1 < M_2 < M_3$, usually in lower range of *A. crassidens,* and smaller than *A. boisei;* M_1 with typical Y fissure pattern, variably with sixth cusp, median lingual accessory cusp present, and mesial buccal groove with deep pit; M_2 with atypical Y pattern, distal accessory cuspule, large transverse anterior fovea, mesial buccal groove with pit; M_3 with strongly rectangular shape, vertical sides, well-defined simple cusps, nearly + fissure pattern, sixth cusp and split hypoconulid, mesial buccal groove with elongate pit.

Postcranial skeleton scarcely known from fragmentary parts of upper limb (distal humerus, proximal ulna, metacarpal and two manus-phalanges), innominate fragment, and lower limb (talus, two pes-phalanges). Apparently less robust limb skeleton than in *A. crassidens,* and perhaps more comparable to some *A. africanus;* body weight unknown, stature estimated as probably not less than 150 cm; known postcranial morphology essentially the same as for the genus.

The fortunate recovery (Broom 1938a, 1936b, Broom and Schepers 1946) of hominid remains in 1938 from Kromdraai (Site B), one of several fossiliferous infillings at that locality (Brain 1958, 1975), afforded the first association of *Australopithecus* cranial and postcranial parts. At the time of the discovery adult cranial and dental parts of the previously known species, *A. africanus,* were just becoming known at nearby Sterkfontein, but were still quite fragmentary. However, morphological differences in comparable body parts—especially the facial skeleton and the morphology and size of permanent teeth—strongly suggested that the Kromdraai specimen represented a different hominid taxon. That interpretation was later confirmed by the recovery at the same locality of a juvenile hominid mandible with a deciduous dentition morphologically distinct from that of the type of *A. africanus* (Broom 1941). It was not until 1947 that upper deciduous teeth of *A. africanus* became known from the Sterkfontein site (Broom 1947), and still later that lower deciduous teeth were described from the same locality (Broom, Robinson, and Schepers 1950). Only over a decade after the initial find at Kromdraai, was the polyspecific character of these early Hominidae clearly demonstrated.

A series of features of the adult cranial base and facial skeleton, as well as certain features and sizes and proportions of the permanent premolar-molar dentition, as well as the inferred size of the incisor-canine dentition (known from alveoli and roots only) of *A. robustus,* diverge from the *A. africanus* condition. This difference is further confirmed by the deciduous dentition, particularly the lower milk molar morphology (cf. Broom and Schepers 1946, Robinson 1954, 1956).

Unfortunately most of the several postcrania associated with the Kromdraai type specimen (TM-1517) lack comparable counterparts in the other South African australopithecine samples. An exception is a partial, though well-preserved probably female ilium (TM-1605); it seems to be of comparable morphology to, but somewhat larger than the STS-14 female specimen of *A. africanus* and is overall apparently quite similar to *A. crassidens* (Robinson 1972), which is probably male (cf. McHenry 1975d). Unfortunately, there are no counterparts for either the upper limb elements, metacarpals (2) of *A. crassidens* representing different digits than those known from *A. robustus,* or for several portions of the foot skeleton of *A. robustus.*

However, both upper limb and lower limb parts of *A. robustus* have some counterparts among some eastern African samples of early Hominidae, including specimens considered to represent *A. boisei* and others attributed to the genus *Homo.* McHenry (1973, 1975b, McHenry and Corruccini 1975a), has clearly demonstrated the distinction in (distal) humeral morphology between *A. robustus* and (presumed) *A. boisei.* Comparisons are still lacking for the metacarpal and for the phalanges of the pes. However, the talus is generally broadly similar to a specimen from Olduvai (Hom. 8) and diverges from the structure considered characteristic of *Homo* (Day and Wood 1968).

Three infillings occur at the Kromdraai site (Brain 1975). The main faunal site (A), lacking hominid remains, is now known to be of a different age from the hominid-bearing infilling of Kromdraai B. At least two infants, two older children, and two adolescents or young adults are represented among the Site B sample (Mann 1975). All were recovered from the upper part of the infilling. Hominids comprise 7% of the larger mammalian fossils (excluding micromammals), cercopithecoids (3 species) 43%, carnivores (8 species) 17%, bovids (3 or more species)

28%, and three other mammals 5%. Site B was an elongate solution gallery and the infilling accumulated along its slope. The extremely fragmented bone accumulation and the high juvenile bovid percentages with predominantly low weight species represented suggest that hominid hunting practices were at least in part responsible for the accumulation, although carnivores also occupied the site as their coprolites show (Brain 1975, Vrba 1975). However, fewer than half a dozen possible stone artifacts have been recovered at Kromdraai B.

Homo Linnaeus 1758

Few problems in paleoanthropology have been more persistently troublesome than recognizing and defining the genus *Homo* because of its fragmentary hominid fossil record, the tendency to oversplit hominid taxa and emphasize differences rather than similarities, and the lack of suitable methods to resolve the relative and absolute ages of fossil samples. However, the relative ages of relevant fossil hominids are now more reasonably and consistently estimated through more precise identification and comparison of associated vertebrate fossils. A variety of chemical tests on specimens of uncertain provenience or with unclear associations frequently enables their relative ages to be more precisely determined. The refinement and more extensive application of radiometric dating techniques, particularly radiocarbon (and amino acid racemization) for later phases of the Quaternary, and potassium-argon as well as paleomagnetics for earlier phases of the Quaternary and the Pliocene, allow a generally reliable absolute time scale for late Cenozoic events in temperate and tropical continental latitudes and the sediment-covered floors of major ocean basins.

Abandoning outmoded typological concepts and applying principles of the modern synthetic theory of evolution have led to substantial revisions of hominid taxonomy. Such revisions have not only affected the genus *Australopithecus,* but particularly extensive synonomy of specimens now assigned to *Homo erectus* and *Homo sapiens.*

Intensive field investigations in the past 15 years, particularly in eastern Africa, have resulted in the recovery of samples of hominid fossils from situations with well-defined contexts, associations, and relative and absolute ages. Several of these finds are of late Lower Pleistocene age[14] and fall within the known range of variation of *H. erectus,* including as-

---

14 That is antedating the Brunhes Normal Epoch and falling within the later part of the Matuyama Reversed Epoch. The base of the Middle Pleistocene is here considered to coincide broadly with the Matuyama-Brunhes boundary (cf. Butzer 1974a).

sociated skeletal parts that substantially add to understanding its morphology.

Other occurrences are of earliest Pleistocene age and demonstrably antedate *H. erectus,* heretofore the oldest well-defined species of the genus *Homo.* These specimens have an important bearing on the intriguing question of the time and place of the origin of genus *Homo* and, consequently, on the definition of the genus.

Although recent authors have sought to understand its emergence, the literature provides few inclusive definitions of *Homo.* Le Gros Clark (1964) was among the first to offer a very general definition. L. S. B. Leakey, Tobias and Napier (1964) and Tobias (1973a) modified and substantially extended that definition to encompass hominid fossils from Olduvai attributed by them to a new, early species, *Homo habilis.* Robinson (1968) offered another definition, although his hypodigm was different; Campbell (1974a) has also produced another with a somewhat different hypodigm.

Each characterization (not diagnosis) of *Homo* is somewhat similar. They stress various features of endocranial volume, craniofacial morphology, and usually postcranial structure indicative of posture and gait. However, none adequately characterizes the fundamental overall structural and adaptive pattern of *Homo,* which is only now becoming apparent in the light of the many important new discoveries in Africa. In fact, as most of this fossil material has still to be fully described and compared with other samples, it is premature to specify all its features, although at least the following principal features can be included:

Stature and body weight greater, often substantially more than *Australopithecus;* lower limb more elongated and upper limb probably relatively shorter than in *Australopithecus;* limb skeleton adapted for fully upright stance and habitual bipedal gait; hand capable of full precision grip, with well-developed opposable pollex; endocranial volume substantially enlarged above *Australopithecus* (though brain to body size ratios probably similar) and over 600 cc, and ranging two to three times those values in successive species; substantial changes in cerebral proportions, particularly in expansion of parietal and superior temporal regions, as well as parts of the frontal region; cranial vault expanded as a consequence of increased brain size, with enlarged parietals and temporal squama and progressively updomed frontal; superior temporal lines rarely approach the midsagittal plane and never form sagittal crest; reduced postorbital constriction; nuchal area smaller than in *Australopithecus* and progressively reduced along with occipital curvature changes and torus reductions in successive species; occipital condyles set relatively forward on cranial base; supraorbital torus

variably developed but, unlike the thin prominent structure in *Australopithecus,* often prominent, thick, and continuous, with supratoral sulcus in early species, variably reduced and with separate elements in successive species; facial skeleton reduced, only moderately prognathic to orthognathic, and smaller than *Australopithecus* in proportion to neurocranium; zygomatic arch moderately to lightly built, and temporal fossa of variable size but not as large as in *Australopithecus;* distinct margin between subnasal maxillary surface and floor of piriform aperture; maxillary anterior teeth arranged in parabolic curve, usually without pronounced diastema; mandible generally with U-shaped internal contour, body less deep and robust, due to smaller tooth roots, than in *Australopithecus,* reduced inner symphysial buttressing and alveolar planum, with progressive tendency in successive species to develop external incurvature and mental trigone; anterior and posterior dentition harmoniously proportioned; canine teeth moderate- to small-crowned, wear down from tip, and usually do not overlap after initiation of wear; P_3 bicuspid with tendency to reduce lingual cusp and to formation of single root; P_4 not molarized as in *Australopithecus* species; molar size variable, with tendency to reduce M3 as well as some cusps and cingular remnants and to simplify primary groove patterns in successive species; deciduous dentition in which dm_1 incompletely molarized, lingually displaced and frequently open anterior fovea, tuberculum molare, and (in primitive species) paraconid sometimes preserved.

Most (though not all) workers would agree that adaptation for, and dependence upon culturally patterned behavior is a diagnostic characteristic of the genus.

Homo habilis Leakey, Tobias, and Napier 1964

Early Hominidae with trends in craniofacial, dental, and postcranial morphology that approximate the genus *Homo* but that antedate and differ morphologically from *Homo erectus* are now known from several sub-Saharan African localities: Swartkrans (South Africa), Olduvai Gorge (Tanzania), Koobi Fora (Kenya), and probably the Shungura Formation, lower Omo Basin (Ethiopia).

Because most of these specimens are still being studied, I will not discuss all morphological features of this hominid taxon. However, I describe some salient characteristics briefly, using the hypodigm inventoried in table 10.4.

Stature and body weight unknown or as yet unestimated, but greater than *A. africanus* and probably within the range of *H. erectus;* extent of sexual dimorphism unknown, but may have been considerable; upper limb poorly known, except for possibly the proximal humerus (undescribed), and portions of the manus,[15] with pollex ca-

---

15 These are juvenile specimens of Olduvai Hom. 7, and include nine terminal, middle, and proximal phalanges, three carpals, and one metacarpal 2 (Day 1976b).

Table 10.4 Remains attributed to *Homo habilis*.

Elements	East Turkana[a]	Omo Basin	Olduvai Gorge	Swartkrans	Kanam
Crania	ER-*1470; 1590;* *3732; 1813.*	L.894-1 (cf.)	H.7; 13; 16; ?24	SK. 847 SK. 27	
Maxillae					
Mandibles	*1483* (?); *1802*		cf. H.7, 4; 37	SK. 45	? mandible frag.
Dentition	*1590*		H.16; ? 21, 27, 39, 41, 44, 45	SK. 2365	
Postcranials					
Clavicle					
Scapula					
Humerus	*1473* (?)				
Ulna					
Radius					
Manus			H.7		
Carpals					
Metacarpals					
Phalanges					
Vertebral Column					
Pelvis	*3228*				
Femur	*1481; 1475; 1472*				
Tibia	*1481; 1471*				
Fibula	*1481*				
Pes					
Talus					
Calcaneum					
Tarsals					
Metatarsals					
Phalanges					

[a]For East Turkana hominids, different type styles indicate different stratigraphic horizons, as follows: roman = Sub-Chari, Upper Member, Koobi Fora Formation; italic = Lower Member, Koobi Fora Formation; boldface = Sub-L/M Tuff Complex, sub-Okote, and sub-Koobi Fora Tuff, Upper Member, Koobi Fora Formation.

pable of rotary movements and adapted to power and precision grip postures, strong, flat-nailed fingers, with broad spatulate terminal phalanges, markedly constricted, wide-based middle phalanges with strong flexor mm insertions and broad, stout proximal phalanges with well-defined trochlear articular surfaces; most of lower limb, except pes, known; innominate broadly similar to that of *H. erectus;* femur relatively slender, elongate (395–400), and straight (frontal view) with distal taper of shaft, and enlarged epiphyses; enlarged head about two-thirds of a sphere; neck set at obtuse angle to shaft (as in other *Homo* species), shortened (compared to *Australopithecus*), anteverted, and of oval cross-section with vertical height greater than anteroposterior thickness; greater trochanter without strong lateral flare, with buttress passing onto neck, anterior aspect with femoral tubercle, and well-defined hip muscle insertions essentially as in other *Homo* species; lesser trochanter directed posteromedially, may or may not be joined by low intertrochanteric crest to greater trochanter; neck with or without notable obturator externus groove; large trochanteric fossa (for obturator externus); proximal shaft platymeric with convex medial border (as in *H. erectus*); linea aspera present, and variably developed and compound, with lateral contribution

superiorly from hypotrochanteric crest; femorocondylar angle as in other *Homo* species; popliteal surface with weak supracondylar ridges; elongated lateral condyle, with more steeply sloping lateral patellar articular surface (compared to surface on medial condyle), its tibial articular surface flattened and closely convexly curved posteriorly, and with well-defined attachments for (fibular) collateral ligament and (lateral) gastrocnemius head; medial condyle with smaller patellar articular surface, tibial articular surface more evenly curved overall, less closely curved posteriorly, and grooved antero-posteriorly; deep, wide intercondylar fossa, partial tibia, and fibula (fragment) suggest morphology much like other *Homo* species.

Mean cranial capacity (about 680 cc[16]) larger than any known species of *Australopithecus* and considerably less than the mean capacity for *Homo erectus;* notable brain expansion in parietal, temporal, and (some) frontal areas, with attendant changes in endocranial proportions.

Cranial morphology still incompletely known from several partially preserved adult specimens. Vault long ovoid shape, with moderate postorbital constriction; parietals

16 Based on four specimens from Olduvai Gorge, ER-1470 from Koobi Fora, and Sangiran 4 from Java.

nearly evenly domed with eminences expressed, and probably without sagittal keeling; bimastoid breadth greater than biparietal breadth; superior temporal lines well-separated from sagittal suture and no sagittal crest formation; occipital with rounded contour inferior of lambda, only moderate flexion between squamal and nuchal portions; variability of occipital superstructures ill-known, sometimes weakly expressed occipital torus bilaterally thickened (aligned with supramastoid crest), with only slight supra- and infratoral depressions, but may be very substantial continuous torus and attendant depressions; temporal morphology ill-known, mastoid probably large and extensively pneumatized, suprameatal crest may be pronounced, zygomatic root with substantial anterolateral curvature, postglenoid tubercle prominent, mandibular fossa wide and moderately long antero-posteriorly. Cranial base very incompletely known, generally narrower than *Australopithecus* species; middle cranial fossa only moderately excavated, but temporal surface of greater wing of sphenoid strongly concave vertically, and orbital segment set at nearly right angle to squamous segment; sphenosquamosal suture may separate foramen spinosum perforating temporal from foramen ovale perforating alisphenoid (unlike *A. africanus* or other *Homo* species); middle meningeal system with long common trunk, strong parietal ramification of posterior branch, and parietal ramification of posterior portion of anterior branch.

Frontal region with supraorbital torus present, variably developed with anteroinferiorly expanded, and slightly depressed glabellar region passing smoothly (sometimes without supratoral groove) upward into frontal squama, lateral (supraorbital segment) of torus may thin laterally from protuberant supraciliary segment or maintain substantial robusticity as it slopes inferolaterally and is demarcated from the squama by a supratoral sulcus; no sharply defined posterior margin of zygomatic process of frontal, but ill-defined superior temporal line arises laterally and passes in smooth anteriorly then posteriorly directed curve toward stephanion; frontal sinus substantial and prolonged in superior, posterior, and lateral dimensions.

Facial skeleton only imperfectly known from several adult specimens. Reduced facial height and breadths compared with *Australopithecus* species; middle and lower face flat (lateral view); piriform aperture margins everted, well-defined superiorly but blending into nasoalveolar clivus inferiorly; sill-like to rounded inferior piriform margin, sometimes with distinct anterior nasal spine, with marked angulation between nasal floor and broad subnasal maxillary planum that lacks guttering; moderate to slight subnasal prognathism, nasal bones concave (lateral profile), variably somewhat flattened or prominent and rounded, projecting forward and raised above the frontal process of the maxilla; frontonasal suture variably situated, but may be situated superiorly toward glabella; inflexure in lateral orbital margin, so not evenly concave; lateral half of inferior orbital margin projected anteriorly; zygomatic process of maxilla robust, arising at level of M^1 but extending to M^2, rather anteriorly inclined and curves

evenly or more sharply laterally from the alveolar process to a nearly horizontal position, with variable development of masseteric tubercle; canine juga not pronounced; palate relatively deep anteriorly and broadened across incisor region, with alveoli for anterior dentition (I's and C's) set in gentle curve, with incisors (particularly the medial) set anteriorly of the canine alveoli.

Upper dentition, including several deciduous teeth, associated directly with crania referred to this taxon in as many as four instances, as well as with associated deciduous (5) and permanent (12) teeth of a single individual and several other isolated upper molars of other specimens, all of which are largely undescribed: d_c asymmetrical, marked lingual surface relief; dm^1 smaller than *Australopithecus* homologues, with strong paramolar tubercle; dm^2 smaller than to within size range of *Australopithecus* homologues, with well-defined trigon basin and associated crests, small linear anterior fovea, protocone with Carabelli furrows, large hypocone, large centrally set posterior fovea. I^1 very large crown, weak labial surface relief, slight lingual surface relief, and without basal tubercle formation; I^2 with rather stronger lingual relief, especially distal marginal ridge, but also without basal tubercle formation; upper C crown large to very large, with marked lingual relief including mesial, distal, and central ridges and adjacent furrows, usually without basal tubercle formation; premolars within or below *A. africanus* size range; P^3 may have partial third root; both with simple occlusal morphology, with anterior and posterior foveae well expressed, salient to strong (especially mesial) marginal ridges, buccal grooves usually present though not so pronounced as in *A. africanus*. Molars within or below *A. africanus* size range and within lower limits of *A. boisei* and *robustus* and generally narrower; M^3 may be reduced, in both length and breadth, in comparison with M^2; molars with no or weak Carabelli structures, anterior fovea variably developed, posterior fovea present with limb(s) displaced onto adjacent posterior cusp(s), distal marginal ridge usually well-defined, robust, cusps may have (even substantial) secondary wrinkling.

Mandibular parts of nearly 10 specimens, none complete, but several with much of the lower dentition, have been considered to represent such a taxon in Africa. Another largely complete lower permanent dentition, associated with much of an upper dentition and some portions of a vault, of another individual is also known. Two isolated permanent lower molars from two sites may also represent this taxon. The lower deciduous dentition is still wholly unknown. Mandible with moderately retreating symphysis, slight or no mental trigone or mandibular incurvature, reduced symphysial height, and reduced superior and (especially) inferior transverse tori, but substantial alveolar planum, rather shortened tooth roots and attendant higher situation of mental foramen; reduced height and robusticity of mandibular body, especially the supporting alveolar bone; broadened, U-shaped lower dental arcade, shortened premolar-molar rows, and altered prelacteon: postlacteon proportions by comparison with *Australopithecus* species. Lower dentition with incisors relatively enlarged, moderately long-rooted, without marked lingual surface relief. Canines quite large absolutely (and relative to lower premolars), crown asymmetrical with steeper distal than mesial margin, labial surface with basal ridge prolonged as mesial and distal crests, lingual surface with substantial relief, including mesial and distal marginal ridges, and attendant grooves, arising from a basal swelling or even a tubercle. Premolars generally narrower than or within the range of *A. africanus*. P_3 bicuspid with strong salient buccal and usually weaker and approximated lingual cusp; it is linked by a transverse crest with an anterior buccal ridge, anterior and (especially) posterior foveae well-developed, the latter distolingually oriented, thickened distal marginal ridge, buccal cingular remnants generally absent. P_4 bicuspid, with paired cusps, small anterior fovea, with variably developed mesial marginal ridge, talonid elongated and substantial especially distolingually (and hence crown truncated distobuccally), with thickened and even tuberculated distal marginal ridge, substantial pitlike posterior fovea, mesio-buccal surface of crown may be swollen and/or preserve groovelike cingular remnants.

Molar size intermediate between lower range in *A. africanus* and upper range in *H. erectus*, with some tendency toward mesiodistal elongation and in general toward buccolingual narrowing of the crowns; M_3 longer than M_2. M_1 with Y-5 pattern, with usually substantial me^d to hy^d contact, a tendency to align stem and mesiobuccal limbs, and strong divergence of distobuccal limb, hy^{1d} large and centrally disposed, anterior fovea and mesial marginal ridge of small to moderate size only, posterior fovea at best pitlike, may be weak protostylar pit. M_2 rather similar to M_1, but may have broadened trigonid, but with reduced me^d to hy^d contact, hence closer approach to +5 pattern, small en^d, may be some development of a sixth cusp (or of a divided en^d to hy^{1d}) and of a (partially) split me^d, anterior fovea present and variable in size, as also the mesial marginal ridge, posterior fovea (usually?) absent, pr^d swollen and may preserve protostylar ridge. M_3 not unlike M_2 in groove pattern, but may have smaller cusps, especially hy^{1d}, with sixth cusp and split me^d, and lacking posterior fovea, anterior fovea may be substantial, and protostylid sometimes quite strongly developed.

The first, assumedly ancient hominid fossil attributed to the genus *Homo* was the symphysial portion of a mandible recovered from Kanam West, near Homa Mountain, Kenya. L. S. B. Leakey (1935, 1936a) assigned the specimen to a new species, *Homo kanamensis*. Boswell (1935) cast doubt on the provenience and hence the faunal associations and antiquity of the specimen. However, this problem has never been finally resolved despite efforts (Oakley 1960) to apply radiometric assays to the specimen and to the Kanam vertebrate fauna with which it was thought to be associated. On morphological grounds Tobias (1960, 1962) concluded that the pathological specimen had morphological affinities

with an advanced species of the genus *Homo* and was indeed far from really primitive.

The second, and again presumably ancient, hominid fossil ultimately attributed to the genus *Homo* comprised several specimens recovered in association with *A. crassidens* at the Swartkrans locality. The initial and type specimen, which Broom and Robinson (1949, also 1950a, 1952) designated *Telanthropus capensis*, was a largely complete mandible with five molar teeth (SK. 15). Altogether, some six specimens have been considered to represent this second hominid at this locality (Robinson 1953a): a mandible fragment with two worn molars (SK. 45), an isolated left P_3 (SK. 18a), a proximal radius (SK. 18b), the buccal half of a P_4 (SK.43), a maxilla with worn and incomplete I^2, P^3, and P^4 (SK. 80), which has now been shown (Clarke, Howell, and Brain 1970, Clarke and Howell 1972) to be part of a craniofacial fragment (SK. 847) previously considered as a robust australopithecine.

Another crushed juvenile cranium with partial dentition (SK. 27) and two worn upper premolars (SK. 2635) are also now correctly assigned to *Homo* sp. (Clarke 1977); three individuals of this taxon are now represented in the Swartkrans Member A sample.

The evidence for (Robinson 1953a, Gutgesell 1970) and against (Wolpoff 1968, 1971a,b) multiple hominid taxa at Swartkrans has been a subject of continuing debate. The new composite cranium (SK. 847, including SK. 80) now convincingly demonstrates (contra Wolpoff 1971b) a total morphological pattern that approximates that of a species of genus *Homo* and diverges significantly from *A. crassidens*, the predominant hominid represented at Swartkrans.

The type specimen of the taxon *Telanthropus capensis*, the mandible SK. 15, is now known to derive from a younger infilling (Member B) of the Swartkrans Cave than Member A, which yields abundant australopithecine remains (Brain 1976b). Although the age of this later infilling is still unknown, because an extended episode of erosion preceded it, it yields Acheulian stone artifacts, and its vertebrate fauna is suggestive of the Cornelian Faunal Span (Vrba 1975), it is sometime in the Middle Pleistocene. Robinson (1961) had earlier transferred the remains attributed to *T. capensis* to *H. erectus*, the most appropriate attribution for the type mandible, SK. 15, as well as the P_3 (SK. 18a) and the closely associated proximal radius (SK. 18b).

The Extension Site, or West Pit, of the Sterkfontein locality represents a breccia infilling (Member 5) that is demonstrably younger than the subjacent

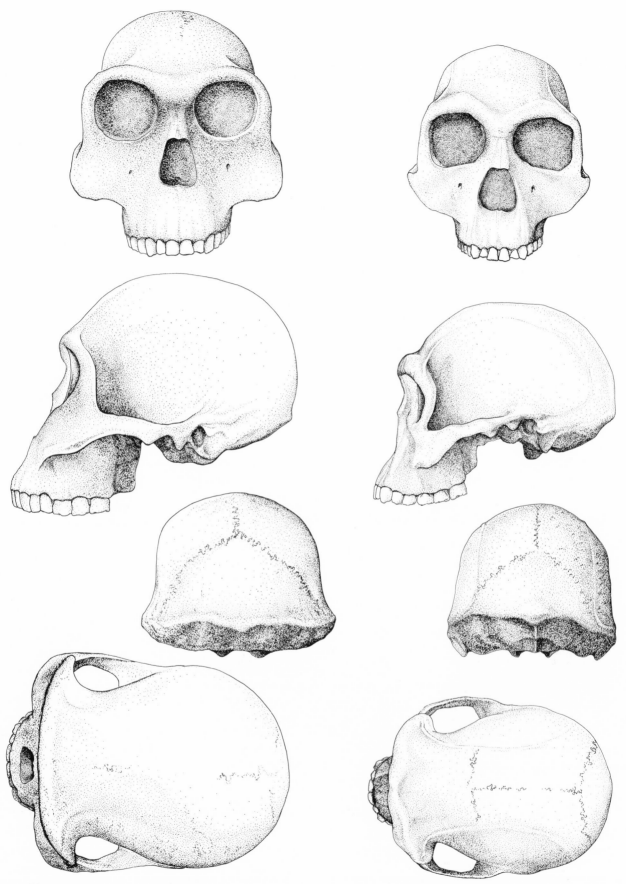

Figure 10.9 *Homo habilis*. Facial, lateral, posterior, and superior views of restored crania of presumed male (left) and female (right) (based on specimens ER-1470 and ER-1813 respectively). Two-fifths natural size.

australopithecine-rich infilling (Member 4). Its stratigraphic relationships and age have been questioned. Some half dozen specimens of hominid teeth and a fragmentary juvenile maxilla, as well as a number of stone artifacts, were recovered from this horizon in 1957–58 (Robinson and Mason 1957, 1962).

It was originally concluded that this red-brown "Middle Breccia" overlaid the lower, australopithecine-breccia unconformably (Robinson and Mason 1962); this assumption has since been confirmed by work at the Sterkfontein locality (Tobias and Hughes 1969, Butzer 1978)—most probably, a substantial lapse of time intervened between the two accumulations. The faunal assemblages from these successive infillings have not yet been fully compared; however, Vrba's analysis (1974, 1976) of the Bovidae has clearly demonstrated marked differences in species composition indicative of markedly different ages. The evidence is still inadequate to provide an accurate estimate of age for the younger assemblage which might be as little as 0.5 m.y., or as much as 1.0 m.y. or so.

Stone artifacts occur in substantial abundance only in this part of the Sterkfontein site and many have been recovered *in situ* from this breccia. None are known from the underlying Member 4, which yields abundant *A. africanus* remains. A comparison of the artifact sample with other early lithic assemblages, particularly those from Olduvai Gorge, reveals broad similarity in tool types, although proportions vary, with the Developed Oldowan (M. D. Leakey 1970, cf. also 1971). In the faunal assemblage the bovids are represented by very fragmented remains of a diversity of species of varied weight classes, among which juveniles form only a small percentage. Vrba (1975) suggests that this occurrence probably represents a hominid occupation site and that such remains may have been scavenged from carnivore kills.

Robinson (in Robinson and Mason 1957, 1962, Robinson 1958) attributed the first extension site hominid remains to *A. africanus,* largely on the basis of their size, as well as the morphology of the lower canine. These remains included a juvenile maxilla with M^1, dm^2, and dm^1 (SE. 225), and M^2 (SE. 1508), another probable M^2 (SE. 1579), an incomplete P^3 (SE. 2396), and a $\overline{C}$ (SE. 1937). Tobias (1965) questioned this attribution and suggested that some, even all, might be attributed to genus *Homo.* The recent discovery of most of a hominid cranium (STW-53) from Member 5 throws important new light on this problem (Hughes and Tobias 1977). Its craniofacial morphology and upper dentition diverge significantly from any *Australopithecus*

species and instead closely resemble that of various specimens, including SK. 847, here attributed to *Homo habilis.*

Just over a decade ago L. S. B. Leakey, Tobias, and Napier (1964) defined a new species, *Homo habilis,* and in so doing also redefined the genus *Homo.* The hypodigm of the new taxon was considered to comprise remains of as many as eight individuals, all but one recovered from the pre-Lemuta Member sediments of Beds I and lower II, Olduvai Gorge. Radiometric dating (Curtis and Hay 1972) and paleomagnetic studies (Grommé and Hay 1971, Brock and Hay 1976) now date these specimens at 1.6 to 1.8 m.y.

The type specimen of *Homo habilis* comprises cranial fragments (both partial parietals, occipital and other vault fragments, and petrous fragments), mandible with dentition through M_2, and several hand bones of a juvenile (Hom. 7)[17] recovered from or eroded from an occupation site (FLK. N.N.I) in Bed I. The other individuals considered as paratypes were Hom. 4, Hom. 6, Hom. 8, and possibly Hom. 35[18] (all from Bed I); and Hom. 13 from Bed II, found slightly above the aeolian tuff of the Lemuta Member. Two other specimens were referred to the taxon, Hom. 16, from site FLK. II, Maiko Gully, at the base of Bed II, and Hom. 14, cranial fragments of a juvenile individual probably derived from the same horizon as Hom. 13. More recently parts of eight other individuals, three from lower Bed II[19] and five from Bed I,[20] have been recovered and are attributed to this taxon (cf. M. D. Leakey 1976a).

Not all these specimens are unambiguously attributable to an ancient species of the genus *Homo.* Thus, Hom. 24, a crushed and only partially restored cranium considered by M. D. Leakey, Clarke, and Leakey (1971) to represent *H. habilis,* has some definite dental resemblances to such an early species of *Homo,* but seems to have a small cranial capacity, perhaps under 600 cc, and some features of the facial skeleton that seem to resemble *A. africanus.* However, the condition of the specimen severely limits its accurate reconstruction. The attribution of other hominid postcranial specimens from Bed I to such a taxon has also been questioned by Wood (1974b) and by Day (1976a, 1976b). These include Hom. 8, 43, 48,

17 An upper molar (left M^1 germ) was originally considered to belong to this individual, but is manifestly another, still younger individual, now designated Hom. 45.
18 Hom. 35 comprises largely complete adult left tibia and fibula, once attributed to Hom. 6, but now considered another individual, and by the writer, another taxon; they were recovered on the same occupation surface at site FLK. 1 as the type skull of *A. boisei.*
19 Hom. 27, 37, and 41.
20 Hom. 21, 24, 39, 44, and 46.

and 49, from FLK. N.N.I, Hom. 35 from FLK. I, and Hom. 10 from FLK. N.I (level 5). All these specimens are distinguished by their small size and by some morphological details that often approximate the condition found in *Australopithecus,* especially *A. africanus.* They might well represent that taxon, but as they are still under study and their affinities still undetermined none have been included in the foregoing discussion of the morphology of *Homo habilis.*

Soon after the introduction of *H. habilis* Tobias and von Koenigswald (1964) made some interesting and important comparisons between specimens of that hypodigm and some early hominid specimens from Java (Indonesia). The type mandible (Hom. 7) is extremely similar in mandibular morphology and in the size and morphology of its dentition to hominid specimens from the Pucangan and the basal Kabuh formations, Sangiran, central Java, referred to the taxon *Meganthropus palaeojavanicus* (von Koenigswald 1950).[21] They concluded that these hominids represented the same hominid evolutionary grade, and one advanced beyond that represented by *Australopithecus africanus.* Robinson (1953b, 1955) had earlier argued that these Javan specimens were most similar to a robust species of *Australopithecus.* Von Koenigswald (1973) has since admitted the "australopithecoid" nature of the Javan form, the morphology of which is nonetheless advanced over that of any African species of *Australopithecus.*

A paratype specimen (Hom. 13) attributed to *Homo habilis* was shown by the same authors to have strong resemblances in jaw size and overall morphology and in its dentition to another set of Javanese fossil hominids. Von Koenigswald (1950) attributed this series of specimens to *Homo (Pithecanthropus) modjokertensis,* the type specimen of which is a juvenile calvaria recovered at Perning, Modjokerto, eastern Java (von Koenigswald 1936, see also 1940). Altogether six specimens[22] have been attributed to this taxon, although there is scarcely any morphological basis for comparing the five adult specimens with the juvenile remains representing the type. All these specimens are considered to have derived from the Pucangan Formation, although the exact provenience of some is frankly quite uncertain. This hominid shows some distinct differences in mandibular morphology, size of tooth crowns and roots, and morphology of the permanent dentition from the type (Hom. 7) of *H. habilis* and specimens attributed to *Meganthropus palaeojavanicus.* Tobias and von Koenigswald (1964) considered that those hominids therefore represented another, more advanced evolutionary grade that more closely approached *Homo erectus.*[23] They concluded that "we think it wisest to keep the Bed II hominines separate from the Bed I habilines." Robinson (1965, cf. Tobias and Robinson 1966) similarly concluded that "In terms of the available evidence it would seem that there is more reason for associating the Bed I group of specimens with *Australopithecus* (*africanus*) and the Bed II group with *Homo erectus* than there is for associating the Bed I and II groups with each other." Nonetheless, the ages of these African hominid specimens are closely similar, and the geologically youngest specimens attributed to *H. habilis* may be substantially older than the oldest remains (at Olduvai Gorge) attributed to *H. erectus.* In Java, although the contextual evidence leaves much to be desired, it appears that the two grades—*palaeojavanicus* and *modjokertensis*—are at least in part penecontemporaneous. If the preliminary results from radiometric (K/Ar) dating on these formations in Java (Jacob and Curtis 1971, Jacob 1972, G. H. Curtis, pers. comm.) are correct, those hominids *may be* 0.5 to 1.0 m.y. older than *H. erectus.* Although initial paleomagnetic determinations seem to add some support to this interpretation (F. H. Brown, pers. comm.), because of the lack of accurate provenience data the precise temporal relationships and ages of these important hominid specimens are still largely unknown.

An extremely informative hominid fossil sample relevant to the elucidation of the origin of the genus *Homo* derives from the sub-KBS unit (= Lower

---

21 The hypodigm of this taxon comprises three principal specimens, all mandible fragments, listed as Sangiran 6 (two specimens, of the right and left sides, same individual, "*Meganthropus* A") and Sangiran 8 ("*Meganthropus* B," Marks 1952) in the inventory by Jacob (1973, also 1975a, 1975b). Other still unpublished isolated teeth (Sangiran 7a), permanent as well as deciduous, probably also represent this taxon (von Koenigswald 1950).

22 According to the inventory in Jacob (1973, also 1975a, 1975b; cf. von Koenigswald 1950) including Perning 1 (type); Sangiran 1b (= "Pithecanthropus" B), a hemimandible; Sangiran 4 (= "Pithecanthropus" 4), partial calvaria and palate; Sangiran 5 (= "P. dubius"), a mandible fragment; Sangiran 9 (= "P. C"), a

right mandible including the symphysis; and Sangiran 22, another mandible fragment (Sartono 1975). Von Koenigswald (1969) considers that Sangiran 5 and Sangiran 9, along with an isolated upper molar, should be attributed to a distinct species, *Homo (Pithecanthropus) dubius,* largely on the basis of lower premolar root structure.

23 However, the calvarial morphology of Sangiran 4 (Weidenreich 1945) certainly appears to diverge significantly from the fragmentarily preserved calvaria of Olduvai Hom. 13. It is still unclear to what extent some of these differences may reflect sexual dimorphism, but some surely do not.

Member) of the Koobi Fora Formation, East Turkana (Kenya). This sedimentary unit probably covers a substantial time span according to conventional potassium-argon age determination on tuffs capping it (Curtis et al. 1975). One specimen (ER-1590) from Area 12 derives from some 15 m below the tuff designated "KBS" there; that tuff in nearby Area 10 has an age of 1.54 to 1.6 m.y. Assuming the tuff has a mean age of ~ 1.57 m.y. and reasonable rates of sedimentation, the specimen might be about 1.8 to 1.9 m.y. old. The other specimens, as many as six cranial and jaw parts and postcranial parts including several lower limb bones of a single individual, derive from Area 131 below the Karari escarpment. Here the overlying "KBS Tuff" is 1.82 ± 0.06 m.y. old. Using that age and reasonable sedimentation rates, the oldest specimen (ER-1470, a fairly complete cranium), should be between 1.9 and 2.0 m.y. old. However, the various specimens here attributed to this hypodigm occur at four principal horizons throughout some 40 m of section so the total time span is probably on the order of 1.9 to 2.1 m.y. Some specimens appear to occur on or in some relation to an erosional or diastemic interface and might be somewhat or even considerably older.

This valuable hominid sample throws important new light on the structure and organization of an ancient species attributable to genus *Homo*. It includes overall cranial morphology and endocranial capacity (ER-1470, 1590, 1813, 3732), facial structure (ER-1470), upper (deciduous and permanent) dentition (ER-1590), mandible structure (ER-817, 1483, 1801, 1802), and permanent lower dentition (ER-1801, 1802, 1462) and the size and structure, if not the exact proportions of the lower limb (ER-1472, 1475, 1481, 1471). All the specimens depart significantly from the characteristic *Australopithecus* structure; on the contrary, there are important approximations to a condition resembling genus *Homo*.

Other hominid remains are also documented in this section. Several fragmentary specimens from low (ER-1474) and higher (ER-1803) in the section are of indeterminate affinities. Two mandible specimens (ER-1482, 1469), from low and higher in the section, represent *Australopithecus boisei*. Associated upper and lower postcranial parts of a single individual (ER-1500) from near the top of the section resemble *Australopithecus*, and I have tentatively referred them to *A. africanus* (see preceding description).

The upper members of the Shungura Formation, lower Omo Basin (Ethiopia) are generally less richly fossiliferous than the underlying members. Hominid fossils are quite rare. The most important such specimen is a fragmentary hominid cranium with premolar-molar dentition (specimen Local. 894-1) that derives from the uppermost part of Member G (G-28). The occurrence underlies a tuff dated ~1.84 m.y. (Brown 1972, Brown and Nash 1976), and falls in the lower portion of the Olduvai Event within the Matuyama Reversed Epoch (Shuey, Brown, and Croes 1974, Brown and Shuey 1976).

The size, proportions, and preserved morphology of the premolar-molar dentition resembles that of other specimens attributed to this taxon. Enough of the vault bones and parts of the base are preserved to suggest an endocranial capacity substantially larger than in species of *Australopithecus*. The cranial morphology, particularly that of the occipital and inferior temporal areas, also diverge from that of *Australopithecus* and thus approach the structure considered to characterize an early species of *Homo* (Boaz and Howell 1977).

Many hominid skeletal remains attributed to such an ancient species of the genus *Homo* fall within the time range of 2.0 and 1.5 m.y. Some, particularly from the Koobi Fora Formation, may well be still older. Campbell (1973) believes that "the boundaries of sequent taxa . . . should be conventionally agreed upon time-lines, rather than diagnostic morphological features." He does "not believe that the morphological distance between the mean of *A. africanus* and *H. erectus* is sufficient to justify creation of another species between them in the lineage" (Campbell 1972b, also 1963). Thus, he considers that such hominids between 2.0 and 1.3 m.y. should be assigned to *A. africanus,* as distinctive chrono- and geographic subspecies, *habilis* (in Africa) and *modjokertensis* (in southeastern Asia). Campbell (1973) arbitrarily defines "*Homo* as those hominids leading to modern man and less than 1.3 million years of age," which had such elaborated cultural developments as language and technology that enabled effective occupation and exploitation of Eurasian temperate latitude habitats (1972).

I consider, however, that these early hominids, which are not yet *H. erectus* and which significantly diverge in total morphological pattern of craniofacial, dental, and postcranial morphology from *A. africanus,* represent a distinct taxon. It is a valid species in its own right, although the question of synonymy has not yet been resolved. Moreover, there appears to have been considerable phyletic evolution, especially in craniofacial and dental morphology, within this chronospecies. This is evident both in Indonesia and at Olduvai Gorge and may well also prove to be the case in the Koobi Fora succession.

Despite recent advances toward documenting the emergence of the genus *Homo* the gaps in the fossil record leave both the locale and the time in question. Eastern Africa now affords the best evidence for the appearance of the genus *Homo* about 2 m.y. ago. However, there is suggestive evidence of an almost simultaneous distribution/dispersal into the Asian subtropics.

Homo erectus (Dubois) 1894

A species (extinct) of the genus *Homo,* known from Eurasia and Africa, distinguished by very substantial enlargement of brain size, mean (1,020 cc)[24] nearly twice that of large *Australopithecus* species and substantially above that of antecedent species of genus *Homo;* endocranial cast with essentially (modern) human fissuration pattern; generally low, flattened frontal region and prominent frontal keel; expanded (unilaterally) inferior frontal region, with wide separation from anterior temporal lobe; expanded precentral cortical area; exposed anterior insular area, in whole or part related to substantial development of Sylvian crest; lack of approximation between temporal lobe and cerebellum; absolutely and relatively narrow temporal lobe, tapering anteromedially, with poorly expanded inferior temporal area, salient superior temporal area, and posteriorly expanded posterior part of middle temporal convolution; expanded (unilaterally) inferior parietal region (supramarginal area); cerebellum with ipsilateral asymmetry, and cerebrum largely symmetrical, but with contralateral asymmetry expressed particularly in parieto-occipital and inferior frontal regions.

Cranial length greater than *Australopithecus;* vault bones of substantial to massive thickness, as expressed in outer and inner tables, cranial suture closure apparently earlier than in *H. sapiens,* and with coronal preceding sagittal suture closure; maximum breadth of vault at or toward the cranial base, usually coincident with biauricular breadth, with lesser bitemporal and biparietal dimensions; substantial postorbital constriction; low to more moderately arched vault, with low receding frontal (with or without notable frontal tuberosity), longitudinally flattened parietal, and occipital with marked to substantial angulation between upper (squama) and lower (nuchal) scales; usually distinct sagittal thickening, especially in bregmatic area, often associated with marked parasagittal depression; parietal smaller, more rectangular and transversely more curved than in *H. sapiens,* usually with sub-

stantial Sylvian crest along its endocranial sphenoidal angle; middle meningeal vessels characterized by low division from main trunk, large and prolonged superior temporal ramus, absent to small and short inferior temporal ramus, and small, poorly ramified frontoparietal ramus; occipital with upper shorter than lower scale, squama more curved than in *H. sapiens,* occipital torus strongly to massively developed, usually thickest centrally and where strong triangular prominence may be developed at external inion and be more or less confluent with external occipital crest, usually with supratoral sulcus (above) and faint to well-delineated superior nuchal line (below), and passing to mastoid region and supramastoid crest (often) via an angular torus; cerebellar fossae substantially smaller than cerebral fossae, hence cruciate eminence inferiorly situated and substantially below level of inion, which is customarily coincident with opisthocranion; foramen magnum angulated between its nuchal and basisphenoidal planes; temporal squama with long, flattish parietal margin, low in proportion to length, often too deep, acutely angulated parietal notch; cerebral surface of squama low, long with very broad squamous suture; robust zygomatic root with broad, shallow sulcus, robust supramastoid crest; mandibular fossa deep, short, set relatively laterally, with preglenoid planum or sometimes articular tubercle, lacking postglenoid process, and without entoglenoid process (medial wall formed by squama); auditory meatus on or just above nasion-opisthion line, relatively wide and variable in shape; tympanic plate set medially, overlaid by suprameatal tegmen formed by zygomatic root, and horizontally positioned parallel to basal surface of skull; axis of plate nearly at right angle to midsagittal plane; suprameatal spine variable; mastoid process variable in size, may be large with projecting anterior portion and the posterior portion forming a lateral bulge of the cranial wall; supramastoid and mastoid crests well developed, with intervening sulcus; petrous pyramid with lateral part of base distinctly overlapped by tympanic plate, with latter forming obtuse angle with pyramid; internal pyramid low, flat without marked relief of anterior and posterior surfaces; foramen lacerum and petro-occipital fissure may be absent; carotid canal generally smaller, jugular fossa flatter, narrower, and shorter, and sigmoid sulcus shallower and narrower (and fails to encroach on mastoid area and adjacent occipital (jugular process), without strong infratemporal crest or marked angulation of temporal and infratemporal facies, with low narrow cerebral facies, sometimes with accessory foramen ovale; pneumatization moderate to substantial, especially in mastoid area and adjacent occipital (jugular process), sphenoid and pterygoid process, ethmoid, and maxilla, frontal sinus variable, if present usually simple and not prolonged superiorly or laterally.

Facial skeleton with heavy, projecting suprafacial torus, supraorbitals continuous with glabellar torus, usually with well-defined surfaces, with or without supratoral sulcus; supraorbital margin thick, rounded with distinct supraorbital tubercle; infraorbital margin rounded, and at

---

24 Altogether some 21 substantially preserved calvaria have been referred to this hominid species. The Choukoutien sample ($N = 6$) has a mean value of 1,060 cc. The Java sample ($N = 4$) of "typical" *H. erectus* has a mean value of 806 cc whereas the Java sample ($N = 8$) of *H. erectus soloensis* has a mean value of 1,092 cc. The mean value in the text is based on these samples as well as the Lantian (China) specimen and two from Africa—the calvaria of Olduvai Hom. 9 and Salé (Morocco). The known range is 780–1,225 cc.

same level as orbital floor; lacrimal groove variably present or absent; face relatively small though broad, moderately to slightly prognathous, bones robust; nasal bones wide, with scant difference between upper and least width, with broad, relatively high transverse saddle arc; maxilla with strong anterior facies and alveolar process, long canine jugum, narrow infraorbital sulcus, malar pillar broad, rounded, with submalar notch; high zygomatic with arch below level of Frankfurt Horizontal, forward-facing malar facies, sometimes with zygomaxillary tuberosity.

Mandible overall more robust than in *H. sapiens;* basal robusticity reduced but alveolar portion thick, and higher alveolar and lower basilar portion than in *H. sapiens;* strong lateral tori; digastric fossa long, narrow and set on inferior border of body; strong marginal torus and anterior marginal torus; often with multiple mental foramina (more than two, and up to five); sometimes with (substantial) mandibular torus; symphysis with low inclination angle, reduced inferior and (especially) superior transverse tori, and alveolar planum; incipient mental trigone and faint symphyseal tuber, but lacking anterior mandibular incurvature and mentum osseum; often with marked submental incisure, separated by inferiorly projecting crestlike triangular symphysis; ramus steep, broad, everted posteriorly, often with marked masseteric fossa, blunted and everted gonial angle, strong endo- and ecto-condyloid crests, broad and thick coronoid process, semilunar notch of moderate to substantial depth; alveolar arch rather horseshoe-shaped, long, and relatively narrowly curved with rounded, projecting anterior dental segment.

Upper deciduous dentition largely unknown; lower deciduous dentition with stout d_is, having low marginal ridges, raised gingival eminences, robust and long roots; d_c with mesially set apex, lingual and buccal grooves, cingulum well developed and ascends mesially, distal cusplet and small mesial cusplet present; dm_1 elongate, narrow, with talonid longer than trigonid, trigonid three-cusped including mesial accessory cuspule, pr^d with flattened buccal face prolonged inferiorly down mesial root, anterior fovea lingually disposed with largely open mesial wall, talonid three-cusped; dm_2 with buccal grooves weak, (median) lingual groove, five-cusped with Y-5 pattern largely as in M_1 hy^d with expanded buccal face, may be well-developed pr^d to me^d crest closing anterior fovea distally, and may be accessory me^d cusplet.

Permanent dentition with moderately large crowns and roots (particularly incisors and canines), canine size generally similar to those of smaller species of *Australopithecus* and larger than in *Homo sapiens,* and premolars and molars significantly smaller than in *Australopithecus,* but rather larger than in *Homo sapiens;* upper I's with strong marginal ridges, swollen gingival eminence or even a basal tubercle present, lingual fossa substantial; upper C with large projecting crown, symmetrical shape with well-defined apex, lingual face with substantial relief including paired, convergent lingual grooves, (mesial) lingual ridge, slight gingival eminence and distinct cingulum on mesial and distal faces; upper premolars with weak buccal

grooves; P^3 incompletely three-rooted, asymmetrical crown shape with expanded mesiobuccal angle, large buccal cusp; P^4 smaller than P^3, more symmetrical crown shape, equal-sized cusps with lingual cusp mesially displaced, without talon enlargement; upper molars with $M^3 < M^1 < M^2$; anterior molars four-cusped, strong paracone, traces of buccal cingulum derivatives, Carabelli complex practically absent, triangular-shaped, weak anterior fovea, oblique crest poorly expressed, posterior fovea present, posterior molars with tendency (sometimes) toward root fusion; M^3 reduced in size, with both metacone and hypocone small, and crown form approaching a triangular shape.

Lower I's with shallow lingual fossae, weak gingival eminence; lower C not large, asymmetrical, with pointed apex or even approaching incisiform shape, cingulum preserved; P_3 with asymmetrical, oblong shape due to strong distolingual angle and but slight mesiobuccal angle, moderate buccal grooves, salient pr^d and smaller, buccally disposed, low me^d with adjacent thickened lingual margin, well-defined anterior and posterior foveae, one-rooted or with mesiolingual cleft (Tome's root form); P_4, with moderately expressed subequal main cusps, me^d shifted mesially, talonid moderately developed without cuspule formation, foveae wide and shallow with posterior larger than anterior, buccal groove moderately developed, some tendency to root division; lower molars with $M_2 \gtreqless M_1 > M_3$ usually, with well-defined buccal grooves, moderate to absent lingual groove, strong slope to buccal face, variable traces of protostylid development, usually Y-5 pattern with some tendency toward me^d reduction and approach to +5 pattern, infrequently with sixth cusp on posterior molars, shallow, distally open anterior fovea, posterior fovea variable in expression or absent, M_3 sometimes with substantial reduction of hy^{ld} and en^d, and variability in crown morphology, tendency toward root fusion; some populations with marked tendency toward pulp cavity expansion ("taurodontism"), and with substantial secondary enamel crenulation.

Stature, limb proportions, and body weight still very imperfectly known or estimated; stature imperfectly estimated (femora) as 144 to 156 cm; estimated humerofemoral index ~80; body weight estimated around 53 + kg.

Postcranial skeleton only partially known, and in only two important instances (ER-803 from Koobi Fora and Hom. 28 from Olduvai Gorge) are directly associated parts of the same individual. Clavicle seemingly not notably distinctive from *H. sapiens;* scapula unknown; humerus apparently distinguished by thick walls, narrow medullary cavity, slender distal portion of shaft, strongly developed deltoid tuberosity; ulna largely undescribed, but differing from *H. sapiens* in several distinctive ways; radius largely unknown; manus practically unknown, except for broad, undistinctive lunate; proximal phalanges long and straight.

Vertebral column practically unknown (except for two associated, incomplete elements of the cervical and thoracic series); both small and similar to *H. sapiens* counterparts.

Lower limb most adequately known part of postcranial skeleton, although tarsus and metatarsus ill-represented. Innominate (based on preserved female specimen) stoutly built, muscular and ligamentous markings prominent, and distinguished from *H. sapiens* in iliac and ischial structure; ilium large and broad with sinuous iliac blade set at wide angle to acetabulum, large acetabulum, exceptionally robust, vertically directed iliac pillar, heavily buttressed acetabuloiliac crest, prominent anterior inferior iliac spine (and origins of iliofemoral ligament and straight head of rectus femoris), probably prominent, anteriorly disposed anterior superior iliac spine, iliacus groove shallow and wide, iliopectineal line short, auricular surface small, obtuse, set low and obliquely, with large iliac tuberosity; ischium medially rotated, and small relative to acetabulum size, with attendant deep, narrow groove, prominent hamstring muscle origins.

Femur known from various individuals and localities, none preserving superior and inferior diaphyses; differing from *H. sapiens* in a number of features of the shaft and proximal portion; relatively short, nearly straight with only slight anterior convexity below middle of shaft; flattened in subtrochanteric and popliteal areas; substantial convexity of medial border of upper half of shaft with crestlike inner and outer borders; markedly distal situation of least shaft breadth; variably sharp linea aspera with only low pilaster; moderate to narrow medullary canal, sometimes with thick shaft walls; disposition and form of lesser trochanter unknown; intertrochanteric and pectineal lines absent or weak; persistence of hypotrochanteric crest to greater trochanter base with formation (sometimes) of hypotrochanteric tubercle; torsion (anteversion) of femoral neck; femoral neck thickly buttressed inferiorly, with (sometimes) coarse, more diffusely organized trabeculae than in *H. sapiens;* most of neck and head unknown, but large femoral head inferred (from acetabulum).

Knee joint structure unknown because distal femur lacking and proximal tibia unknown or undescribed; tibia with straight, slender shaft, anterior border rounded, and medially inclined, and without salient edge, strong soleal insertion markings, overall morphology similar to *H. sapiens;* fibula only fragmentarily known; pes imperfectly known, but with marked resemblances to *H. sapiens* in known morphology; constituents of ankle joint known, but neither described nor functionally evaluated. Talus small, with short neck set at small angle to axis of body, high torsion angle of head, head convex in both long and short axes, margins of trochlear surface symmetrically rounded, medial malleolar facet not prolonged onto talar neck; calcaneum and other tarsals unknown; known metatarsals (parts of 4, 5, and most of 3) comparable to *H. sapiens* counterparts, with metatarsal 3 having medial torsion, dorsoventral buttressing, expanded base; phalanges incompletely known but apparently like *H. sapiens* counterparts.

The extinct hominid species, *Homo* (ex-*Pithecanthropus*) *erectus,* was initially established on the basis of a calotte[25] recovered in 1891 from fluviatile deposits of the Kabuh Formation (Trinil Beds), exposed along the Solo River in central Java (Dubois 1894b). The species became only adequately known and its distinctiveness thoroughly established nearly half a century later with the recovery of additional, better preserved calvaria and other skeletal parts from the Sangiran Dome (von Koenigswald 1940) and the Ngandong locality, along the Solo River (Oppenoorth 1932, Weidenreich 1951) in Java and from 1923 onward from excavations (Black et al. 1933, Black 1934) of the extensive cavern infilling (Locality 1) at Choukoutien (Hopei), North China, which afforded an unprecedented hominid sample—including calvaria, mandibles, teeth, and postcrania from nearly 50 adults and juveniles—meticulously described by Black (1931) and Weidenreich (1936, 1937, 1941b, 1943). The recovery of additional remains from several localities in Java (Jacob 1973, 1975a,b, Sartono 1961, 1971, 1975) and from renewed excavations at the Choukoutien locality (Woo and Chao 1954, Woo and Chao 1959, Chiu et al. 1973) as well as the Lantian locality, Shensi (Woo 1964a, 1964b, 1964c, 1966) have further increased the Asian sample of this hominid species and our understanding of its morphology and variability.

Diverse skeletal parts attributable to *H. erectus* are known from some dozen localities in northern, eastern, and southern Africa (table 10.5). North African localities have yielded calvaria, jaw parts, and teeth, whereas only a mandible and isolated teeth are certainly known from South Africa. In East Africa, Olduvai Gorge (Tanzania) and Koobi Fora (Kenya) have afforded a rich sample of skull and jaw parts, teeth, and postcrania, including some associated parts of individual skeletons.

Many nomina (table 10.6) have been applied to specimens now reasonably assigned to this species. By a decade ago the assignment of the species *erectus* to the genus *Homo* was quite commonly accepted because the morphology did not justify distinction at the generic level (see Howell 1960, Campbell 1963, 1965).

The first hominid remains in Africa to show *H. erectus* affinities were recovered in 1933 from a quarry in the consolidated dune of Kébibat, south of Rabat, Morocco (Marçais 1934). The relative age of

25 However, a fragmentary juvenile mandible attributed to the species, and the first hominid remains found, was recovered from similar deposits south of Trinil at the Kedung Brubus locality in 1890 (Dubois 1891, Tobias 1966). I think that the several femora attributed to this taxon by Dubois (1926a, 1926b, 1932, 1934) cannot on comparative morphological grounds represent this extinct species (cf. Day and Molleson 1973, Day 1973b).

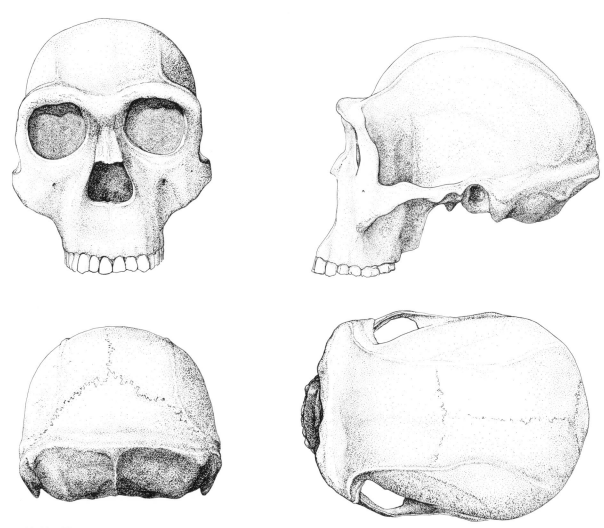

Figure 10.10 *Homo erectus*. Facial, lateral, posterior, and superior views of restored cranium (based on specimen ER-3733 from Koobi Fora). Two-fifths natural size.

these specimens has been discussed on various occasions by Neuville and Ruhlmann (1942), Choubert and Marçais (1947), Lecointre (1960), and Biberson (1963, 1964). It is now at least clear that this formation represents a continental littoral accumulation during the (middle) Tensiftian stage, or at latest to the subsequent Presoltanian stage, which are generally considered to correspond to the Penultimate Glacial stage of Europe (cf. Biberson 1961, 1964). Thorium/uranium measurements on mollusca from an overlying horizon suggest an age > 200,000 years (Stearns and Thurber 1965). Study of these remains by Vallois (1960, also 1945) revealed their similar mandibular morphology and proportions and the resemblance of the size and morphology of the upper and lower dentition to *H. erectus* (cf. also Arambourg 1963). Another partial hominid mandible of (earlier) Tensiftian age (Biberson 1956) with *H. erectus* features (Arambourg and Biberson 1955, 1956) was re-

covered in 1955 from Littorina Cave, in the Schneider Pit, at the huge Sidi Abderrahman quarry system southwest of Casablanca (Biberson 1956, 1971). Also of younger Tensiftian stage age is a hominid calvaria and partial maxilla with $I^2 \rightarrow M^2$ recovered from a quarry just north of Salé, Atlantic Morocco; its cranial morphology is distinctively *H. erectus* (Jaeger 1973, 1975a).

A left mandible, preserving $P_4 \rightarrow M_3$ (Ennouchi 1969a, 1970), with *H. erectus* morphology particularly in the dentition, has also been recently recovered from still older continental deposits probably of the final Amirian stage or later (Saussé 1975) in the Thomas Quarry 1, southeast of Sidi Abderrahman.[26]

26 Some cranial parts and 11 upper permanent teeth reported by Ennouchi (1972) from a nearby locality (Thomas Quarry 2) were recovered from a solution cavity in the Amirian age dunes and hence must be of more recent geologic age (cf. Jaeger 1975a, 1975b), perhaps Tensiftian.

Table 10.5   Remains attributed to *Homo erectus*.

Elements	North Africa			Eastern Africa		South Africa
	Morocco	Algeria	Ethiopia	East Turkana[a]	Olduvai	
Crania	Salé, Sidi Abderrahman–Thomas Quarry	Ternifine (par.), Yayo (Chad)?	L.996 (ff)–Omo, Gombore II, Melka Kontouré (ff)	1466 (f), 1805, 1821 (f), 3733	H.2, 9, 12, ?25 Ndutu	
Maxillae						
Mandibles	Sidi Abderrahman–Thomas Quarry,–Littorina Cave	Ternifine (3)		807, 1814 731, 820, 992 1811, 1805, **730,** **1501, 1502 +** **1812, 1507**	H.11 H.22, 23, 51	SK.15; (+ 18, 43 teeth)
Dentition		Ternifine (7)		732(6), 808(8), 809(3), 803(2)		STE.(7)
Postcranials						
Clavicle						
Scapula						
Humerus			Gombore (?)	1591?		
Ulna					H.36	
Radius						? SK.18
Manus						
Carpals						
Metacarpals						
Phalanges						
Vertebral column						
Pelvis					H.28	
Associated postcranials				803 (upper and lower), 164 (manus and phalanges)		
Femur				737, 1809	H.28; ?34	
Tibia				?1810; 813B	?H.34	
Fibula						
Pes						
Talus				813A		
Calcaneum						
Tarsals						
Metatarsals						
Phalanges						

[a]For East Turkana hominids, different type styles indicate different stratigraphic horizons, as follows: roman = Sub-Chari, Upper Member, Koobi Fora Formation; italic = Lower Member, Koobi Fora Formation; boldface = Sub-L/M Tuff Complex, sub-Okote, and sub-Koobi Fora Tuff, Upper Member, Koobi Fora Formation.

These remains, along with the specimens from artesian lake sediments at Ternifine, near Palikao, Algeria, are the oldest known hominids, and the earliest documented occurrence of *H. erectus* in northern Africa. The Ternifine specimens include three mandibles (with dentition), a parietal, and two mandibular and seven maxillary teeth; three of the latter are deciduous molars (Arambourg 1963). The richly fossiliferous deposits yielding these speci-

mens have been considered to be correlative with the Amirian stage of Atlantic Morocco (cf. Biberson 1963). However, there is scarcely any appropriate biostratigraphic basis for precisely assessing the relative and absolute age of this locality and its important fossils. Most Ternifine macro- and micromammals are more primitive than those of other "middle" Pleistocene fossiliferous localities in the Maghreb and still have certain important resemblances to faunal assemblages attributed to the provincial "upper Villafranchian" (Arambourg 1962, Jaeger 1969, 1975a, 1975b). However, Jaeger, using micromammal evidence, considers that a substantial time gap, perhaps as much as 0.3 m.y. might exist between the latter and the Ternifine locality. Some macro- and microfaunal evidence, admittedly indirect and inferential, suggests that Ternifine may be younger than the Ubeidiya locality (Israel), the fossiliferous horizons of which probably antedate the Brunhes Normal Epoch on the basis of K/Ar and paleomagnetic measurements (Horowitz et al. 1973, Siedner and Horowitz 1974). Unfortunately, no direct evidence permits the Ternifine locality to be placed either before or after the Brunhes-Matuyama epoch boundary, at just under 0.7 m.y. All the occurrences of *H. erectus* in the Maghreb are associated with an Acheulian stone industry.

The hominid frontofacial fragment from Yayo, in northern Chad, was initially considered to be an australopithecine (Coppens 1961, 1962, 1965, 1966). However, the specimen is poorly preserved and its facial morphology diverges in important aspects from *Australopithecus;* it approaches more closely the structure characteristic of *Homo.* The relative age of the occurrence is based solely on a few associated and often temporally undiagnostic mammal species, including a loxodont elephant (Coppens 1965) which is inseparable from *L. africana* (Maglio 1973), a species largely of middle to late Pleistocene and, of course, more recent age.

A substantial sample of *Homo erectus* remains is now known from eastern Africa. The largest collections are from the post-Lemuta portion of the fluviolacustrine deposits at Olduvai Gorge (Tanzania) (Hay in M. D. Leakey 1971), and from the upper part (sub-Chari unit) of the Koobi Fora Formation of East Turkana (Kenya).

Remains of 10 individuals attributable to *H. erectus* have been recovered from Olduvai, 2 of which derive from Bed II and 7 from Bed IV.[27] Apparently all or most of these hominid remains occur within the later part of the Matuyama Reversed Epoch and are older than ~ 700,000 yr (Hay 1976, A. Cox, pers. comm.). The remains are thus probably very broadly contemporaneous with typical *H. erectus* from Java (Jacob 1973, G. H. Curtis, pers. comm.). Most if not all of the *H. erectus* specimens from Olduvai appear to be associated with an Acheulian stone industry (cf. M. D. Leakey 1978).

This species is very well documented in the upper Koobi Fora Formation, East Turkana. The sample comprises body parts of over 20 individuals, including cranial parts of 5 individuals, mandibles of 9 individuals, associated teeth of 4 individuals, forelimb parts of 3 individuals, and lower limb parts of 6 individuals (cf. inventory in R. E. F. Leakey 1976a). These specimens are about equally divided in prove-

[27] Bed IV was formerly subdivided into units IVa and IVb, but the latter (upper part) is now given separate formational status and termed the Masek Beds (Hay in M. D. Leakey 1971). Hom. 23 derives from these beds. Hom. 34, the affinities of which are unclear because the remains are pathological, derives from Bed III or the base of Bed IV.

Table 10.6 Synonomy of major nomina ascribed to remains now attributed to *Homo erectus.*

	Source	Locality
AFRICA		
Homo leakeyi	Heberer 1963	Olduvai Hom. 9
Telanthropus capensis	Broom and Robinson 1949	SK-15 (Swartkrans)
Atlanthropus mauritanicus	Arambourg 1954	Ternifine, Sidi Abderrahman-Littorina Cave
Tchadanthropus uxoris	Coppens 1965	Yayo, Chad
ASIA		
Sinanthropus pekinensis	Black 1927	Choukoutien, Locality 1
Sinanthropus lantianensis	Woo 1964a, b, c	Chenchiawo, Lantian Shensi; Gongwanyling, Lantian Shensi
Sinanthropus officinalis	von Koenigswald 1953	Hong Kong "Drug Stores" (1935)
Homo (Javanthropus) soloensis	Oppenoorth 1932	Ngandong, Java
Anthropopithecus erectus	Dubois 1892	Trinil, Java
Pithecanthropus erectus	Dubois 1894b	Trinil, Java

nience between the lower or Koobi Fora Unit, and the upper or Chari Unit of the upper part of the formation. They range in age from < 1.6 m.y. (the age of the type KBS tuff in Area 105) to >1.3 m.y. (the age of the Chari tuff from Area 1) (Curtis et al. 1975).[28] The sample broadly spans the same time range as Olduvai Gorge, except that some still younger occurrences are present at Olduvai.

The Olduvai and Koobi Fora samples contribute a wealth of new data on African populations of *H. erectus,* as well as important new evidence on the structure and variation of previously unknown body parts.

Extremely fragmentary remains attributable to *H. erectus* are known from two localities in Ethiopia. Member K of the Shungura Formation, lower Omo Basin, has yielded parietal and temporal fragments (L. 996-17) with the characteristic morphology of this species (Howell, Boaz, and Coppens, in preparation). The remains date from toward the close of the Matuyama Reversed Epoch, in that portion preceding the Jaramillo Normal Event, and hence are between ~ 1.0 and 1.2 m.y. old (Shuey, Brown, and Croes 1974). One of the Acheulian occupation places (Gomboré II) at Melka Kontouré, in the Ethiopian highlands south of Addis Ababa, has also yielded a partial parietal bone; its morphology suggests this species (Chavaillon, Brahani, and Coppens 1974). The age of this occurrence is unknown but is generally considered to fall within the (later) middle Pleistocene (Chavaillon 1973).

A single locality in South Africa has afforded hominid remains attributable to *H. erectus.* A mandible (SK. 15), two lower premolars (SK. 18, 43), an associated proximal radius (SK. 18), all considered to represent the same individual, derive from a post-australopithecine infilling (Member B) at the Swartkrans locality (Brain 1976b, Butzer 1976b). The associated fauna (Vrba 1975, 1976) suggests a substantially more recent age, presumably later middle Pleistocene, than the initial, australopithecine-rich infilling.

The faunal assemblage of this infilling is distinctive in the predominance of smaller, low-weight bovid species and the high percentage of juveniles. Vrba (1975) considers this pattern to indicate predation, with a specialization on smaller prey, and to result from hominid hunting activities. This infilling has yielded a number of stone artifacts that resemble the Developed Oldowan industry of eastern Africa (M. D. Leakey 1970).

28 The best age estimate for the intermediate Koobi Fora Tuff, Okote Tuff and Lower Middle Tuff Complex (Ileret) is ~ 1.5 m.y. (Fitch and Miller 1975).

Descriptive, comparative, and analytical studies are still insufficient to permit any full explication of the structural transformations that occurred between the earliest species of *Homo* and *Homo erectus.* However, the total morphological pattern and the adaptive grade of the latter can only be evaluated by referring to the structure characteristic of the former species. The fundamental distinctions between *H. erectus* and *Australopithecus* have long been recognized and are now well established. Useful summaries are found in Le Gros Clark (1964, 1967), Howells (1966), Brace (1967), and the recent overview by Tobias (1974). Other useful, general discussions are those by Weiner (1958), Coon (1962), Kurth (1965), Ferembach (1965), Howells (1973), Gieseler (1974), and Pilbeam (1975). Changes in the dentition have been discussed by Robinson (1956), Wolpoff (1971c), Pilbeam (1972a), and Petit-Maire and Charon (1972); changes in cranial morphology by Weidenreich (1941a, 1947), Heintz (1967), and Bilsborough (1973); changes in the middle meningeal vascularization pattern by Saban (1977); changes in brain size and certain proportions, largely on the basis of endocranial casts, by Tobias (1971a), Holloway (1972a, 1974, 1975), and Lestrel and Read (1973); and changes in the postcranial skeleton by McHenry (1975b,c). Some of the cultural aspects of the problem are discussed by Campbell (1972), Howell (1972), Isaac (1972a,b), and Freeman (1975a).

The locomotor skeleton of *H. erectus* has been sometimes considered to be morphologically and functionally indistinguishable from that of *H. sapiens.* Certain evidence suggests that conclusion is faulty; substantially expanded knowledge of the postcranial skeleton of this species now confirms a number of distinctive morphological features, particularly of the lower limb. Nonetheless, the functional locomotor capabilities appear to have been generally similar to adaptations in *H. sapiens* for habitual erect posture and efficient striding bipedal gait. These adaptations include an alternating pelvic tilt mechanism, a powerful hip extensor mechanism for erecting or raising the trunk, a pelvic rest mechanism, posterior displacement of the center of gravity behind the hip joints, transfer of body weight to the pelvis through the sacral suspensory mechanism, and powerful hip flexion and knee extension (Day 1973b). Other portions of the lower limb are less well known, particularly the foot skeleton. Nonetheless, the tibia has strong soleus muscle attachments; the talus shows the form and joint morphology of a propulsive, arched foot; and the form, buttressing, and torsion of the third metatarsal is also in keeping with the full development of a transverse arch and adaptations to withstand propulsive forces (Day 1973b).

The axial skeleton and the forelimb are still imperfectly known in this species, although the latter (except the hand) is now more fully documented in the fossil record. Although much of the relevant skeletal material is not yet fully published or even analyzed, the forelimb morphology also seems to have diverged in distinctive ways from the *H. sapiens* pattern. However, the functional significance of these divergences is still unclear. The overall close approximation to the *H. sapiens* condition is noteworthy by comparison with the many divergences from the structure characteristic of *Australopithecus*.

The most obtrusive distinctions between *H. erectus* and *H. sapiens* are craniofacial and dental morphology. The differences are many and complex, and their interrelationships are poorly understood. It is generally considered that transformations in the form and proportions of the cranial vault are, in complex ways, related to the growth, proportional alteration, and probable reorganization of the cerebral cortex and underlying structures. However, the remarkable reduction in thickness of vault bones and the modification of cranial superstructures in size, extent, and form are also not well understood as yet. The reduction of the facial skeleton and mandible is, in part, associated with changes in the absolute and relative size of teeth, particularly roots and their supporting bone; and associated areas of stress absorption and transmission. However, other changes in facial structure and, particularly, in dental morphology are largely uninvestigated.

Homo sapiens Linnaeus 1758

The origin of *Homo sapiens* is still largely unknown. The source, the time, and the area of the differentiation of the species, as well as the circumstances under which that transformation occurred, all need further investigation. Nevertheless, the fossil record in Africa does provide examples of a number of Pleistocene hominids that are relevant to the understanding of the earlier history of *H. sapiens* and deserve to be attributed to that species.

Surprisingly there appears to be no inclusive definition of our own polytypic species. The extensive relevant literature reveals an unexpected lack of concern with the biological distinctiveness of a now-dominant mammalian species. Most characterizations, including those of philosophers and theologians and even a number of biologists, are based on single or very few traits. Philosophers'

characterizations (see Adler 1967) employ behavioral attributes to establish a "difference in kind" between human and nonhuman animals that may be relevant at the familial and generic level. Probably most such attributes are impossible to infer from the prehistoric record, although some might be relevant to the delineation of the species *sapiens*.

Other definitions are largely or wholly biological and suffer from various deficiencies. First is the emphasis on single traits—particularly brain size and cranial capacity—rather than a total morphological pattern. Le Gros Clark (1964) supplied such a definition, portions of which remain useful. However, he excluded from the hypodigm those antecedent human populations of middle and earlier upper Pleistocene age which nonetheless have some features suggestive of the species. The temporal duration of the species, and data relevant to its origins, are thus avoided by this restricted definition.

No attempt is made here to seek to provide an inclusive definition of the species *sapiens*. However, certain evolutionary trends and derived features—behavioral as well as morphological—are apparent when comparisons are made between late Pleistocene/Recent populations of *Homo sapiens sapiens* and the antecedent human species, *Homo erectus*. Ultimately a more complete knowledge of these trends and of the more important derived features will afford a basis for assessing the temporal duration of the species *sapiens*.

General prolongation of principal life spans; delayed maturation, with protracted infancy and childhood; pronounced adolescent growth spurt; less advanced skeletal maturation at end of fetal life (in respect to appearance of ossification centers, and stage of ossification). Probably decreased intrapopulational variability in sexual dimorphism. Tendency toward regionally vestigial hair, hairless rosy lips, and elaboration of certain mimetic musculature.

Appendicular skeleton adapted for habitual, fully orthograde posture and striding bipedal gait. Postcrania with reduced thickness of cortical bone and some enlargement of medullary cavities compared to *H. erectus*. Strong, elongated lower limbs with capabilities for full extension at knee and a distinctive plantigrade foot. Center of gravity at sacral 2 vertebra with line of weight (at ease position) passing just anterior to atlanto-occipital joint, just behind the hip joint, slightly in front of the center of knee joint, and about midway between the heel and metatarsal heads in front of the ankle joint. Vertebral column with marked secondary, convex-forwards-facing curvatures, and extreme development of sacral promontory; sacrum with large, pitted articular surface, enlarged erector spinae, and antirotator ligament attachments. Pelvis of distinctive form with sacrum and pubic symphysis approximated and at nearly right angles to the trunk; outer surface of ilium strongly concave-convex, with marked gluteal lines and well-developed iliac pillar and tuberosity of the iliac crest; pelvic tilt mechanism associated with distinctive disposition and proportions of lesser gluteal muscles; approximated anterior spinous processes; enhanced concavity and inward direction of iliac fossa; large, deep acetabulum and associated expanded spherical head of femur with expanded glenoid lip.

Femur elongated, inclined medially in valgus position; enlarged head, moderately long neck, and distinctive trabecular system; greater trochanter laterally expanded and heightened; strong, well-defined linea aspera; flattened, anteroposteriorly elongated lateral condyle; accentuated lateral ridge of patellar groove. Knee joint enlarged in relation to weight-bearing; distinctive "screw-home" and "locking" mechanism of the medial condyle through internal rotation, and unlocking in stance position through lateral rotation, facilitated through enlarged popliteus muscle.

Tibia elongated, with sharp anterior crest; distal articular surface with expanded posterior border. Fibula with inferiorly expanded maleolus. Foot with inner, longitudinal and transverse metatarsal arches, integrity of which is maintained by interosseous and plantar ligaments, plantaraponeurosis, and spring ligament; heel-toe stride with weight bearing in erect posture and bipedal gait via heel (the calcaneum having broadened inferior surface of tuberosity and a developed external tubercle), outer margin and ball (metatarsal heads) of foot; heel-raising in "push off" facilitated by enlarged soleus muscle; ankle with substantial movements of inversion (facilitated by plantar flexors, especially tibialis posterior) and of eversion (via peroneal muscles); obtuse plantar angle formed by transverse axis of metatarsal 1 to axes of metatarsals 2–5; shortened digits 2–5; hallux relatively and absolutely large, especially its phalanx 2, and inferiorly rotated; distal ends of metatarsals 2–5 narrowed and elevated; plantar surfaces of phalanges slightly flattened.

Hand with power and enhanced precision grips; thumb opposition with saddle-shaped base and relative elongation.

Brain enlarged by comparison with *H. erectus,* with a mean value about 1,350 cc. Cerebral asymmetry, with language centers, predominantly in dominant left hemisphere. Substantially expanded parietotemporal association areas, and enlarged third inferior frontal convolution. Cerebral cortex with enhanced secondary fissuration and greater opercularization.

Cranial bones substantially reduced in thickness in comparison with *H. erectus;* overall retardation of cranial suture closure, and sagittal tending to close prior to coronal suture (the reverse of *H. erectus*). Cranial vault enlarged, elevated especially in the frontal and parietal regions with attendant altered curvatures of those bones; parietal with development of tubera, and biparietal breadth greater than or equal to biauricular breadth; occipital region rounded, with reduced angulation, and increasing expansion between its upper (squama) and lower (nuchal) segments; expansion of cerebellar fossae and attendant reduction of cerebral fossae of occipital, with corresponding

upward shift of cruciate eminence and its approximation to the position of external inion; attendant progressive shift of inion toward opisthion, and of opisthocranion toward lambda; reduction of nuchal musculature, and attachment areas situated increasingly below occiput; reduction to loss of sagittal keeling, particularly in the frontal and progressively the parietal, and attendant disappearance of parasagittal flattening. Overall reduction and shape transformations of cranial superstructures; occipital torus reduced in lateral extent, and progressive loss of angular torus, with decrease of salience of central portion, and of related supra- and infratoral sulci; development of posteroinferiorly directed inion and of the several nuchal lines; differentiation of continuous, linear supraorbital torus into glabellar versus bilateral elements, and of supraciliary versus supraorbital elements, with progressive reduction of supraciliaries, and reduction to flattening of supraorbitals, and attendant reduction and slenderizing of malar process of frontal.

Shortened cranial base, with increased flexion (kyphosis) of basicranial axis in respect to its prechordal and chordal segments; foramen magnum and occipital condyles set far forward on cranial base and face inferoanteriorly. Expanded superior extent and curvature of temporal squama; mastoid process present but variable in size, with protrusive anterior portion, and flattened posterior portion, the whole structure tending to be elevated above the level of the cranial wall; deep digastric groove; usually deep, narrow mastoid notch; reduction of supramastoid and mastoid crests, and their associated sulci; mandibular fossa increasingly underlies lateral part of middle cranial fossa; "entoglenoid" process tends to be formed by inferior (spinous) expansion of sphenoid to form inner wall of mandibular fossa; transverse axes of tympanic plate and pyramid aligned, with straight to slightly curved conjoint axis; tympanic plate more frontally (coronally) oriented, its axis forming a more or less acute angle with the midsagittal plane, and the plate thinned, especially its medial portion; temporal usually with styloid and (especially) with vaginal processes developed; reduction of robusticity and expansion of greater wing of sphenoid, and sphenoid with increasing flexion between temporal and infratemporal facies, often with development of infratemporal crest; cerebral surface of sphenoid expanded and increasingly broadly curved; shortening and elevation of lateral part of middle cranial fossa, and that fossa substantially undercuts sphenoid wing; internal petrous pyramid salient and narrow with well-defined anterior and posterior surfaces; middle meningeal vessels with reduction of posterior ramus (often merely to a minor branch), and expansion of anterior segment as the principal ramus having abundant ramification; tendency toward overall enlargement of carotid canal, jugular fossa, and sigmoid sulcus in relation to intracranial circulation.

Facial skeleton progressively set well under prechordal segment of cranial base, with orbits substantially overlapped by frontal lobes of cerebrum; progressive facial shortening and reduction in height; progressive development of rather vaulted orbital roof, and of crestlike superior orbital margin; zygomatic process elevated and parallels rather than descends below the Frankfurt Horizontal; inferior nasal aperture usually with crestlike sill, often associated with well-developed nasal spine; early disappearance of intermaxillary sutures; reduction of alveolar processes of maxilla and of mandible (in association with reduction in tooth crown and root size), with attendant effects on lower facial prognathism and submalar structure; canine jugum narrowed and shortened, and development of canine fossa with attendant changes of root of zygomatic process, which is progressively flattened and posteriorly inclined.

Mandible with reduction of alveolar process, both in the symphysial area and the main body in association with decreased length and robusticity of tooth roots; body robusticity reduced particularly in regard to lateral and alveolar tori, as well as basal reduction, and lingual shift of digastric fossae; symphysis with reduction of alveolar planum, and superior (especially) and inferior transverse tori; labial surface with expansion of mental trigone, bounded by expanded mental fossae, and enhanced development of anterior mental incurvature concomitant with development of bony chin; ascending ramus overall thinner, with notable reduction of endocoronoid and endocondyloid crests.

Dentition with accelerated replacement of deciduous by permanent teeth, permanent canine commonly erupts prior to second molar, and protracted eruption time of third molars; molars show tendency toward root reduction and fusion; reduction in length and robustness of incisor roots; reduction and sometimes even loss (agenesis) of I^2 and $M3$'s; permanent canines more slender, with upper and lower canines more similar, upper C with high narrow crown, and lower C with pointed tip; lower premolars more symmetrical with shortened talonid, and P_3 usually with shortened, peglike root; premolars and molars without substantial wrinkling, and often with reduction of cingular remnants and of marginal groove system; molars with simplification of crown morphology, with variable cusp reduction (metaconids and hypoconulids in lower molars, hypocones and metacones in upper molars), and with variable transformation of primary groove system (to $+5$, $+4$, and other patterns).

Scattered, diverse, and still quite scant evidence now tends to suggest that the origins of *H. sapiens* is to be sought among human populations of middle Pleistocene age. These populations, by definition, would thus have already been in existence during some part of the Brunhes Normal Epoch, less than 700,000 years ago. Unfortunately the temporal relationships between late *H. erectus* populations and earliest representatives of *H. sapiens* are still very ill-defined. Nonetheless most, if not all, students consider it likely that *H. erectus* gave rise to *H. sapiens*.

The most useful relevant fossil record is found by hominid occurrences in western Europe and in northwestern Africa (Morocco). In western Europe emergent *H. sapiens* (subsp. indet.) fossils are documented from fluviatile sites attributed to the late Holsteinian interglacial stage (Swanscombe, Steinheim) and from various cave infillings generally attributed to several successive phases of the Penultimate Glacial Complex (Montmaurin, Orgnac, La Chaise, Lazaret). Not only do these hominid remains reveal features that diverge from the distinctive *H. erectus* morphology, but certainly (or possibly) older specimens, associated with Biharian-type mammal faunas and dating from the earlier part of the Brunhes Normal Epoch, show some similarly divergent features. These specimens are from Mauer (Germany), Vértesszöllös (Hungary), Petralona (Macedonia), and perhaps even Vergranne (Doubs, France). Consequently the *erectus-sapiens* transition may ultimately prove to fall within the earlier part of the middle Pleistocene.

In the Maghreb a number of isolated hominid finds from the Moroccan littoral derive from sedimentary formations considered to represent the upper part of the middle Pleistocene. (There is, however, no paleomagnetic control against which their age may be judged.) These are specimens attributed to the marine Anfatian stage (Salé), the lower and middle continental Tensiftian stage (Littorina Cave, Sidi Abderrahman; Kébibat, and Thomas ["2"] Quarry, Casablanca, respectively); the Thomas ("1")[29] Quarry (Casablanca) mandible is apparently of unknown, but perhaps mid-Tensiftian age. The time span encompassed by these finds is still uncertain; but, on the basis of Th/U measurements it may be estimated to extend from > 0.3 to > 0.1 m.y., if not older.

The incomplete Salé cranium features are overall broadly within the known range of *H. erectus* morphology. However, the occipital (increased curvature between upper and lower scales, reduced occipital torus and associated supratoral depression, inferior displacement of inion relative to opisthocranium) and parietals (expanded parietal bosses, lack of a Sylvian crest, and reduced temporal portion of the middle meningeal vascular system) reveal features divergent toward the *sapiens* condition. The partial mandibles from Littorina Cave (Sidi Abderrahman) (Arambourg and Biberson 1954) and from Thomas ("1") Quarry (Casablanca) (Sausse 1975), so far as they are preserved, show both dental and

29 Both individuals "1" and "2" from the Thomas Quarry (Casablanca) derive from a solution cavity in Amirian-age eolianite, the infilling of which is surely of Tensiftian age, and perhaps the middle to later part of that stage (Jaeger 1975).

gnathic features that fall broadly within the range of *H. erectus*.

The undescribed second hominid (2) recovered from the Thomas Quarry near Casablanca, is also subadult. It preserves much of the frontal, including the orbital and nasal area, much of the upper dentition, and part of the parietal (Ennouchi 1972). It is the first such specimen to preserve the upper facial region and the dentition and is particularly useful in revealing both craniofacial and dental morphology, which provide a basis for comparison with the Kébibat specimen (see below). The supraorbital torus is strongly developed, and most dental elements reveal a number of still nonmodern *sapiens* features, although morphological details are not yet available.

The midadolescent hominid specimen from Kébibat (Mifsud-Giudice Quarry), near Rabat, comprises portions of the fragmented vault, left maxilla, and much of the mandible (Saban 1975; Vallois 1945, 1960). It is a mosaic of more primitive, *erectus*-like, and more advanced *sapiens*-like features.

Unfortunately the cranial vault was fragmented during discovery and its overall morphology and endocranial size is unknown. The occipital is well curved sagittally without marked flexion of upper and lower scales; the occipital torus is practically absent, with only a slightly swollen superior nuchal line, a broad, very low, upwardly convexly curved V-shaped inion with suprainial fossa; the endinion is set 1 cm below the external inion, and the cerebrocerebellar limit (evidenced by the lateral sinus) is low (below the level of the superior nuchal line); the occipital lobe fossae are broad and deep, and the retrocalcarine crest separates their superior and inferior eminences, and cerebral asymmetry is evidenced through reduction of the right hemisphere. The middle meningeal arterial system shows a strong, relatively well-arborized inferior temporal branch, but still relatively weakly arborized and not extensive frontoparietal branch.

The preserved portion of lower face reveals marked subnasal height, weak but broad subnasal depression bounded by the medial incisor and canine crests, and a depression between upper C and P^4 corresponding to the inferior extent of the maxillary sulcus; the inferior nasal region is evenly concave, with laterally situated prenasal fossae, and weak anterior nasal spine; the upper dental arcade approaches a parabolic curvature, with deep palate and reduced palatal area.

The mandible shows anterior shortening and posterior broadening; the external surface of the body shows slight relief of toral and intertoral sulcal features and reduced basilare, dentale and premolar-molar related trajectories; a (doubled) mental foramen set back of P_4, narrow extramolar sulcus with anterior margin of ramus arising mesial to M_3; the (lingual) alveolar prominence is well developed, broad, and well separated by an abbreviated internal oblique line from concave submandibular fossa below, and

the reduced inferior surface of the body is flattened with weak digastric mm insertion area in the lower $C-P_3$ region; the symphysis is robust, its external surface vertically oriented, with a small symphyseal tuber flanked bilaterally by weak anterior mandibular incurvatures, but without true bony chin; the alveolar planum is long, gently sloping, well below which is a low superior transverse torus and a weakly thickened basal area, forming the reduced inferior transverse torus, between which are the narrow fossa and related genial tubercle complex.

The dentition is large overall; the lower incisors are set nearly transversely with the canines; the upper I's have concave lingual surfaces, moderate basal tubercles, and slight marginal ridges; the lower I's are strong, thick, with concave lingual surfaces but poorly developed lingual tubercles and no marked marginal ridges; the upper C is rather large-crowned and has a swollen lingual base with mesial, (strong) distal, and median crests; the lower C is rather incisiform; the upper Ps are single-rooted, P^4 with rather complex occlusal morphology. The P_3 crown is asymmetrical with prominent me^d, the double-rooted P_4 has a well-developed talonid; the molars are quite taurodont, with M3 > M1 > M2; the upper M's lack hypocone reduction, have swollen crown bases, and accessory crests masking the distinctness of the cusps; the lower M's have a well-developed Y-5 pattern, except M_3 a with +6 pattern.

The Kapthurin Beds, Baringo (Kenya), may provide evidence of such an early representative of *H. sapiens* in eastern Africa. The remains derive from the middle (Middle Torrent Wash) of five members of this 100-m thick fluviatile and tuffaceous formation. The remains occurred in association with a limited mammal fauna and an Acheulian industry distinguished by the use of raw material (lava) worked with a proto-Levallois (Victoria West) technique of flake production. The age of this occurrence is still in doubt. An underlying lava has yielded a seemingly reliable age of ~230,000 yr (W. W. Bishop, pers. comm.); however, overlying tuffs are preliminarily dated at more than twice that age. At any rate, an age within the upper part of the middle Pleistocene is probably a reasonable estimate of its relative antiquity.

The human remains, a nearly complete mandible of a young adult, and several postcrania have been briefly described by Tobias (in M. D. Leakey et al. 1969).

The mandible body is robust and low with parallel superior and inferior borders. The single mental foramen occurs below P_4 or between P_3 and P_4. The ascending ramus is broad, shortened, and set nearly perpendicular to the body; the extramolar sulcus is moderately broad. The symphysis is set at an acute angle of obliquity, its anterior surface sloping posteriorly. The mental trigone is only faintly indicated, as is an anterior mandibular incurvature, to produce a slight trace of bony chin. Lingually the alveolar planum is well-defined, and a superior and inferior transverse torus are moderately developed. A genial fossa is indicated, but the genial processes are ill-developed. The dentition is undescribed, but the M₂ is larger than M₁.

This specimen appears to have a less evolved morphology than the Kébibat specimen and hence perhaps approaches rather more closely that of the specimen from Littorina Cave, Sidi Abderrahman. In any case, it reveals some important divergences from *H. erectus.*

It is apparent from the upper and lower facial skeleton and the size, robustness, and proportions as well as morphological details of the dentition that these hominid remains diverge in significant ways from the usual *H. erectus* morphological pattern. Unfortunately the cranial vault is almost unknown, except for the fragmented Kébibat individual, and postcranial parts are equally ill-represented.[30]

A partial hominid cranium recovered from deposits exposed about seasonal Lake Ndutu, Serengeti Plains (Tanzania), is also relevant to the problem of the emergence of *H. sapiens.* It occurred in the upper part of the Masek Beds, beneath the Norkilili Member Tuff, in sediments rich in bone and lithic artifacts associated with at least two occupation surfaces (Mturi 1976). The industrial affinities are uncertain, but could represent a facies of Acheulian. These beds, and the Masek Beds at Olduvai, have afforded amino acid racemization ages of ∼ 0.5 m.y.

The hominid cranium, which comprises much of the vault and middle portions of the facial skeleton (Clarke 1976), is broadly *erectus*-like in overall morphology—particularly its relatively small size, thickness of vault elements, outline in superior views, curvature of the frontal bone, and probable well-developed supraorbital torus, angulated form of occipital and strong occipital torus, and form of the mastoid region. It diverges from the *erectus* condition and approximates *sapiens* morphology in the verticality of the lateral vault, pronounced parietal bosses, absence of sagittal torus development, reduction in size and extent of supramastoid crest, and perhaps also in basicranial morphology and upper facial form.

Stringer (1974) has recently projected the following set of cranial features for an early form of *H. sapiens.* His inferences of morphotype structure are largely derived from multivariate morphometric analyses of still primitive (that is, nonmodern) *sapiens* crania from geological horizons mostly younger than those in question here.

30 However, a nearly complete right ulna, a metatarsal, and several possibly hominid (manus) phalanges associated with the Kapthurin mandible, Baringo (Kenya), are as yet undescribed.

Long low skulls; prominent glabellar, supraorbital regions; frontal low, narrow and high nasion subtense fraction, strong and wide supraorbital torus, uniformly thick, and with robust circumorbital bone architecture; flattened, short parietal region, with small bregma-asterion chord; short occipital chord; biasteriomic breadth and lambda-opisthion subtense large; probably reduced, but persistent occipital torus; broad skull base, with large biauricular breadth (and bicondylar breadth of mandible); facial region broad, massive around orbits, and cheek high and palate broad; nasal breadth not great; face long, but nasal height not a high proportion of it; probably flat midfacial region, prognathous below subspinale, and lacking anterior projection of nasion and nasal area.

Evidently this is a reasonable, though still indirect characterization of the principal cranial features to be expected in an early *H. sapiens* population, and some features of jaws and dentition (briefly set out previously) might be added. However, without well-preserved specimens including associated skull parts and the almost total lack of postcrania from well-defined stratigraphic contexts, delineating the origin of *H. sapiens* must remain one of the major problems in human paleontology.

Human populations demonstrably of early upper Pleistocene age, or even late middle Pleistocene age, are still very poorly known throughout Africa. Moreover, relative ages are quite uncertain, as are often stratigraphic contexts and cultural associations. Absolute ages of various specimens continue to be the subject of intensified research efforts. No well-defined or acceptable paleoclimatic scheme exists for the later Pleistocene in these subtropical and tropical latitudes. A vertebrate biostratigraphy is still at best approximate, "gross" rather than "fine" in resolution; in southern Africa it comprises ill-defined "faunal spans" or "mammal ages" of still undetermined duration.

On morphological grounds at least two extinct, seemingly allopatric subspecies of *H. sapiens* are now usually recognized. *H. sapiens rhodesiensis* was apparently restricted to southern Africa, and the type skull was the first nonmodern human fossil recovered on the African continent. *H. sapiens neanderthalensis* is still known in Africa only from the south Mediterranean and the Atlantic littoral of Morocco, although there is no reason to exclude the distribution of such populations substantially southward, particularly when the Saharan area represented much less of a barrier. Some recent evidence now suggests the presence of another, distinctive subspecies in the eastern African area perhaps about the time of or subsequent to the transition from middle to upper Pleistocene. The known specimens are inventoried in table 10.7.

Table 10.7 Remains attributed to *Homo sapiens neanderthalensis* and *rhodesiensis*.

Elements	Northern Africa	Eastern Africa	Southern Africa
Crania	Djebel Irhoud 1+2, Mor.; Témara, Mor.; Taforalt, Mor. (ff)?	Eyasi 1+2, Tanzania	Hopefield, Cape (1); Broken Hill (Kabwe) 1+3, Zambia
Maxillae	M-el'Aliya, Mor. (1)		Broken Hill 2, Zambia
Mandibles	Haua Fteah, Libya (2); Djebel Irhoud 3, Mor. (1); Témara	Kapthurin, Baringo	Cave of Hearths Hopefield (ff)
Dentition	Témara; M-el'Aliya, Mor. (1)		
Postcranials		E R = 999 (femur), Kapthurin (ulna)	Broken Hill, Zambia-10 postcranials of three individuals

Homo sapiens rhodesiensis Woodward 1921

The age and temporal duration of this extinct subspecies are still poorly known. At the moment the temporal position and relationships of specimens comprising the hypodigm of the subspecies are based almost wholly on comparative vertebrate biostratigraphy. Conventional radiometric dating procedures have proved ineffective or inappropriate. Recent promising developments in the amino acid racemization method afford at least age approximations for some representatives of the hypodigm.

Only three, or possibly four, hominid finds from sub-Saharan Africa are customarily attributed to this subspecies. These specimens are from Kabwe (formerly Broken Hill), Zambia, representing the type; from Elandsfontein, Saldanha Bay, and the Cave of Hearths, Transvaal (South Africa); and from Lake Eyasi (Tanzania).

Because the Kabwe (Broken Hill) human remains (representing at least three and probably four individuals) were recovered in 1921 from a shallow mine shaft in a limestone hill during opencast mining, their geologic age has been a continuing concern of paleoanthropology. Over the past quarter century some progress has been made toward resolution of this problem, although considerable uncertainty still remains and perhaps always will because of the type of occurrence and the circumstances of the find. The ore body was composed of lead (below) and zinc (above), but as the cranium (four times as rich in zinc as lead) has been documented as having been recovered in the *lowest* level under pure lead, it must have been incorporated within a zinc "pocket" within the lead carbonate (Clark et al. 1950). The association of the various other skeletal parts, including another maxilla and postcranials, with the type cranium has been disputed (cf. Hrdlicka

1930a,b); but, for the most part these either have high lead contents, or about equal contents of lead and zinc, and hence must derive from the same stratigraphic source. This appears also to be true of the nonhuman vertebrates and the lithic and bone artifact assemblage (Oakley 1957, Clark 1960). The identity of the cultural materials is still in dispute, although it was at one time considered to be of "middle Stone Age" character, which suggested an upper Pleistocene age—perhaps some 40,000 to 50,000 years ago. The vertebrate assemblage is usually attributed to the Florisbad-Vlakkraal Faunal Span (cf. Cooke 1967). Nearly 30 vertebrate taxa have been identified but the assemblage still requires restudy. However, of 24 larger species a minimum of 6 (25%) are surely extinct. At any rate that fact tends to suggest that the occurrence may well antedate the upper Pleistocene Gamblian (Klein 1973). Recently Bada et al. (1974) has suggested a preliminary upper Pleistocene age of ∼110,000 yr on the basis of amino acid racemization of hominid bones from the site.

The other southern African locality yielding *H. sapiens rhodesiensis* cranial remains is Elandsfontein (or Hopefield), near Saldanha Bay (Republic of South Africa). The stratigraphic occurrence of the hominid remains, artifact assemblage, and rich vertebrate fauna has posed a problem since the discovery of the locality. However, controlled excavations (Singer and Wymer 1968) and critical geological studies, first by Mabbutt (1956) and more recently by Butzer (1973b), appear to have resolved most of the problems. A local stratigraphy is definable, but any correlation with external events is unfortunately minimal. The vertebrate fossil and associated artifactual occurrences accumulated on a calcareous lower Duricrust horizon, subsequently enriched in iron (Butzer 1973b), which was part of shallow pans and associated streams behind an extensive cemented coastal dune forming a barrier to the sea. It apparently represented "a fairly smooth sand slope traversed by braiding channels connecting shallow pans which might dry out seasonally" (Mabbutt 1956). The environment appears to have been mesic overall, predominantly fluvial, with little or no eolian activity; palynological analyses suggest an open, bushveld-type vegetation, with little macchia, but with aquatic plants. A subsequent history of sand accumulations, ferruginization, land surface formation, and deflation is recognizable. Although wetter and drier climatic oscillations have been recognized their significance and external correlation are still unknown.

Because radiocarbon determinations (on bone) only afford a minimal age ($> 35,000$ yr) for the fossil assemblage, any age assessment can only be made on the basis of comparative biostratigraphy. The vertebrate assemblage is diverse: 48 mammal species have been recognized. There are 36 species of larger mammals of which 20 (55%) are extinct. This number is very substantially higher than either the Kabwe assemblage or any faunal assemblage of demonstrably upper Pleistocene age in southern Africa.

The Elandsfontein fauna is frequently compared with faunal assemblages recovered from the Younger Gravels Complex, lower Vaal River Basin (Wells 1964) and that from the Cornelia area of the upper Vaal drainage (Butzer, Clark, and Cooke 1974). The Younger Gravels assemblage, excluding obviously older Plio-Pleistocene elements, comprises 30 species of large mammals of which at least 14 (47%) are extinct. The Cornelia assemblage comprises 24 species of large mammals of which 16 (66%) are extinct. These occurrences share 15 species of which 11 are extinct. Although it is impossible to correlate the fossiliferous alluvia of the upper and lower Vaal valley (Butzer et al. 1973), in each instance there is sufficient geomorphical and stratigraphic evidence to indicate a pre-upper Pleistocene age. However, which portion(s) and how much of the middle Pleistocene record is represented is difficult to ascertain. Elandsfontein shares 16 species overall, of which 12 are extinct, with the Vaal Younger Gravels assemblage. It shares 11 species overall, of which 8 are extinct, with the Cornelia assemblage. The number of extinct species, of all species shared by each of these pairs of assemblages, is almost identical (72% to 75%). In relation to the size of the smaller of any two assemblages compared, the order of sharing (Simpson's coefficient of similarity) is Younger

Gravels: Cornelia = 62.5%, Elandsfontein: Younger Gravels = 53%, Elandsfontein: Cornelia = 46%, indicating a rather closer (and expected) resemblance between the former pair, and rather similar degrees of resemblance between the latter two pairs. Altogether there are 22 extinct species in the combined assemblages, of which about 65% are known both north and south of the Sahara in middle Pleistocene contexts. Thus almost overwhelming paleobiological evidence indicates a substantial and middle Pleistocene age for these occurrences. The species resemblance (45%) of the Kabwe fauna to that of Elandsfontein is about the same as that of Elandsfontein to Cornelia. This would tend to support an older, rather than younger (upper Pleistocene) age for the former Kabwe occurrence, although presumably the resemblance is such that an age younger than Elandsfontein is to be expected. Thus, the duration of this hominid subspecies might well span several hundred thousand years, from the (later) middle to (earlier) upper Pleistocene.

The absolute and relative age of the partial hominid mandible, associated with an evolved Acheulian industry at the Cave of Hearths (Transvaal), are also unknown. The associated fauna comprises only 18 species, of which only 5 are definitely extinct. All the species represented are found in one or another of the larger mammal assemblages already discussed. The available evidence, and the fragmentary nature of the fossils which severely limits accurate species identification, makes it impossible to assess the relative age of the occurrence biostratigraphically.

The several partial hominid cranial parts recovered from sediments exposed along the shores of Lake Eyasi (Tanzania) are frequently considered to represent the same or a related human subspecies. Protsch (1975) has recently suggested an age as young as $\sim 34,000$ B.P. for these human remains on the basis of a C^{14} determination. If confirmed, and if these fragmented remains indeed represent the *rhodesiensis* subspecies, its duration would be extended well into the later part of the upper Pleistocene. At least 35 mammal taxa are recorded from this occurrence (Dietrich 1939, Cooke 1963), of which 29 are larger species. Only 2 extinct species (*Theropithecus* cf *oswaldi* and *Pelorovis* cf *antiquus* are recorded, and the fauna is essentially modern in aspect, in keeping with a late upper Pleistocene age.

The principal skeletal features of *H. sapiens rhodesiensis* are summarized below.

Postcranial skeleton known from incomplete elements of upper (humerus) and lower (sacrum, innominates, femora, tibia) limbs. Overall morphology within modern human range of variation. Stature (estimated from lower limb parts) in some, presumably male individuals 170–178 cm. Distal humerus without noteworthy features. Lower limb robust with features characteristic of adaptation to modern human bipedal gait pattern. Sacrum distinguished by rather small neural canal, relatively long and narrow, and only slightly curved; ventral surface markedly broadened inferiorly; auricular surface with large, massive superior and inferior limbs, and expansion onto dorsal surface; ventral spines form strong dorsal crest. Innominate with great iliac height, strongly concave gluteal facies, and shallow, flattened iliac facies; broad, medially oriented anterior (precristal) facies of ilium with massive anterior superior iliac spine; iliac pillar well developed and surmounted by liplike iliac tubercle; acetabulum large, strongly concave, with prominent marginal lip; broad ischial ramus and prominent ischial spine. Femur robust, and probably elongate; large spherical head, and deep fovea; neck short, and neck-diaphysis angle as in modern *sapiens;* greater trochanter with well-defined and modernly disposed hip muscle insertion areas, enlarged trochanteric fossa, strong gluteal tubercle, well-defined intertrochanteric crest; lesser trochanter large and distally situated; shaft straight, nonplatymeric, with well-marked linea aspera; distal end massive, with modern human disposition, size, and proportions of condyles. Tibia long, slender, and almost straight; internal condyle shallow, nearly circular, and backwardly inclined; deep, well-defined posterior, and wide anterior intercondylar notches; soleus and gastronemius muscle origins prominently defined; marked bilateral concavity of shaft below condyles; fibular facet elevated inferiorly.

Endocranial capacity over 1,250 cc. Endocranial cast with expansion of lower parietal region, posterior temporal area, and fronto-orbital area; prefrontal, upper parietal, and inferior temporal areas not expanded as in modern human condition; middle portion of temporal lobe narrow vertically and anterior part of lobe thickened rather than tapering.

Cranial vault very long, ovoid, relatively low, with maximum height behind bregma; vault bones reduced in thickness compared to *H. erectus* condition. Frontal low, slightly curved transversely, sometimes with weak metopic ridge; strong frontal constriction; superior temporal lines may be strongly defined and pass into strong to massive external angular processes; supraorbital torus well developed and may be massive, comprising swollen, inferiorly depressed glabellar area, passing up into frontal by a shallow postglabellar depression, and strong biarched lateral segments formed by fused supraciliary-supraorbital elements with an upwardly everted disposition and supratoral depression. Sagittal curvature low, sometimes with notable pre- and postbregmatic and prelambdoid depressions; parietals relatively short, not strongly arched, with median sagittal ridge and parasagittal depression, rounded lateral facies, diffuse and centrally situated eminences, and prominent mastoid angle (hence great biasterionic breadth). Occipital moderately high, curved transversely with wide squamous portion, expanded angle between upper and lower scales, the latter quite flat and inferiorly disposed; continuous occipital torus, with narrowed lateral segment, the superior border of which is delimited by a supratoral sulcus, and the inferior border curved, with well-defined superior nuchal lines bounding concave nuchal plane; angular torus present, as long, narrow elevation passing into superior temporal line and associated with strong occipitomastoid crest; foramen magnum long relative to breadth, occipital condyles long, narrow, outwardly displaced, and deep condyloid fossa.

Temporal with superiorly expanded squamous portion; marked infratemporal crest; deep temporal fossae, with broad based root of zygoma, and straight (unbowed) zygomatic arch; mandibular fossa wide and strongly concave, set perpendicular to midsagittal plane, with broad, low articular eminence, substantial postglenoid tubercle, and long tympanic plate; tympanic and petrous aligned in same axis; mastoid process large, elongate, downwardly curved, with blunt tip, mastoid crest forming continuation of angular torus, and pronounced, short supramastoid crest separated from latter by deep, narrow supramastoid sulcus, narrow, deep mastoid notch, and digastric fossae. Cranial base with large basal angle, steep clivus, relatively short, flat basioccipital; petrous forwardly set, with reduction of lacerate foramen, narrow foramen ovale in base of pterygoid process, and large carotid canal-foramen set mediodistal of foramen spinosum; deep wide pterygoid fossae. Anterior cranial fossa large, horizontally disposed with well-developed internal frontal crest; middle cranial fossa relatively shallow, broad, with only slight posterior projection of orbitosphenoid wing; posterior cranial fossa very shallow, low, and elongate. Middle meningeal arterial system preserving substantial posterior ramus, emerging at low level and supplying obelion and lambda areas, and posterior inferior temporal ramus with fine ramifications; frontoparietal ramus with extensive ramification over entire bregmatic-obelionic area, with well-developed obelionic side branch.

Facial skeleton long, absolutely and relative to base and face length, particularly as a consequence of high, broad maxillary alveolar processes and attendant facial, especially alveolar, prognathism. Orbits large, deep (because of supraorbital development), approximately quadrangular, with stout, slightly arcuate superior margin, and weak supraorbital notch; lacrimal quadrate, with wide, concave sulcus. Wide interorbital distance with strong expanded naso-frontal processes of maxilla; nasal bones moderately broad, quite long, transversely truncated at nasofrontal suture, slightly arched transversely, and slightly concave in profile to produce nasal bridge of moderate height; nasal aperture broad and rounded, inferior sill with guttering and adjacent ridges, and bifid anterior nasal spine. Malars massive and low set, with prominent tuberosity and without anteroinferior deflection, inferior border recedes smoothly posteriorly to form orbital floor; broad flat ascending process with backwardly directed anterior orbital border, slightly convex posterior border. Maxilla with long wide ascending processes, suborbital surface inflated, and only faint to weak expression of sub-

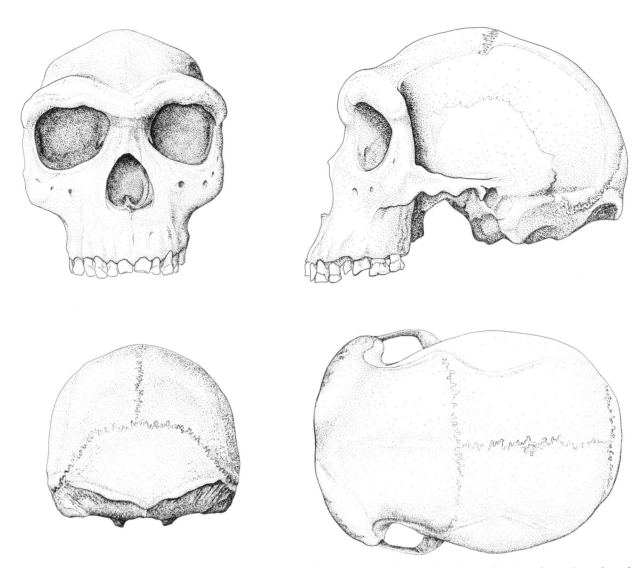

Figure 10.11 *Homo sapiens rhodesiensis.* Facial, lateral, posterior, and superior views of restored cranium (based on a specimen from Kabwe, Zambia). Two-fifths natural size.

orbital (canine) depression; low disposition of infraorbital foramen. Palate U-shaped, broad posteriorly, and very deep, with high, robust alveolar processes. Upper I sockets vertically set and incisor roots robust; macrodont premolar-molar series without taurodontism; M³ reduction.

The type (BM-E-686) represents the most completely preserved cranial remains known of this subspecies. The sample also has various postcranial parts, including upper limb portions (partial humerus, sacrum, ilium, femora, tibia = E-898, E-699, E-689, E-720, and E-691, respectively) of the same individual. Some postcrania and a maxilla (cf. Wells in Clark et al. 1950) exhibit the same fundamental morphological pattern, but are less robust and surely (innominate, F. 719) or probably represent female individuals. Thus, a partial maxilla (E. 687), though large, has a shorter subnasal region, shal-

lower palate, rather more transverse orientation of zygomatic process, and inferiorly inclined and slightly hollowed infraorbital surface. Unfortunately the extreme wear and extensive caries have obliterated the morphological details and even the individual sizes of the upper teeth.

The recovery of a hominid cranium at Elandsfontein (Saldanha Bay) clearly demonstrated that the Kabwe specimen was not unique. It evidences the variability within the subspecies. There is a "striking resemblance" (Singer 1954) between these specimens in overall size, contours, and most features of morphology. The differences are quite trivial and include in the Elandsfontein specimen a distinct ophyryonic groove; less broad supraorbital torus, with its anterior edge evenly curved outward; prominent, more angular shape of median frontal ridge;

smaller superior scale of occipital; and reduced angular torus.

A juvenile right hemimandible from the Cave of Hearths (Bed 3) has been tentatively referred by Tobias (1971) to this subspecies, and this is a reasonable attribution. (A proximal radius may or may not derive from the same horizon.) It represents a preteenage individual, with only slight P_3 and M_2 wear, and preserves only P_3, M_1, and M_2, and the broken roots of the anterior teeth.

The body is shallow, but robust (in part because of its juvenile age), with parallel upper and lower borders; the superior lateral torus, extending back to the lateral prominence, is well developed; the marginal torus is weakly developed and separated from the lateral torus by a faint intertoral sulcus; the single mental foramen is set below P_4. On the internal surface there is a strong alveolar prominence, thickened lingual basal margin, and weak mylohyoid line. There is only slight alveolar prognathism. The symphysis is low, with an acute inclination angle, and having its anterior surface inclined at a comparable angle to the basal margin as in the Kébibat and Témara specimens; a mental trigone is moderately developed, particularly its lower portion, and a slight anterior mandibular incurvature is present. The lingual face of the symphysis exhibits an alveolar planum (comparable to that in Kébibat) and moderate superior transverse torus; the genial processes are ill-defined. The anterior tooth roots are large, and the molar series (M_3 is congenetically absent) large, relatively narrow, elongate as in the aforesaid specimens from the Maghreb; the molars have a + 5 groove pattern. P_3 is bicuspid, with weak me[d], small anterior and (deeper) posterior foveae, slightly expanded talonid area, and slight buccal grooves.

Wells (1957) and Tobias (1968) among others have noted some resemblances of the several Eyasi human cranial remains, originally attributed by Weinert (1939) to an *erectus*-like form (*Africanthropus njarascensis*, a junior homonymn already assigned to the Florisbad cranium), to the *rhodesiensis* subspecies. (Even Weinert recognized some similarities to Broken Hill.) The very badly broken and heavily mineralized Eyasi specimens represent parts of at least two and probably three individuals. Parts of one cranium (No. 1)[31] have been reconstructed, and include a fair amount of the vault, some of the base, and several upper teeth; it has been considered to be a *rhodesiensis* female. Other specimens are largely fragments, including half an occipital (No. 2).

The cranial reconstruction shows a low, flat frontal with postorbital constriction and sharp-margined supraorbital

31 An age of 34,000 yr (isoleucine racemization determination) has recently been suggested for these remains by Protsch (1975).

torus and expanded glabella; lack of marked occipital angulation and some occipital torus development; relatively small mastoid process; reduced angulation of the petrous axis; high maxilla, with subnasal guttering and probably high palate; upper C is large-crowned, with strong mesiobuccal curvature, strong mesiolingual swelling, and lingual grooves. The partial occipital, more robust than the other specimen, is also more openly curved, preserves the supramastoid crest, and an occipital torus passing toward it by an expanded angular region, and an external inionlike development of the central segment of the torus.

Though fragmentary these specimens do exhibit some *rhodesiensis* characteristics. They also merit further description and comparative study.

Homo sapiens neanderthalensis King 1864

Skeletal remains attributed to Neanderthal peoples are uncommon in Mediterranean Africa and, like the Mousterian industry with which they are associated, are unknown into or south of the Sahara (Balout 1965a, 1965b). Three localities, all cave infillings, have thus far yielded skull parts attributed to this subspecies—Haua Fteah, in Cyrenaican Libya; the Mugharet el'Aliya, near Tangier, and the Djebel Irhoud, southeast of Safi, in northern and southern Morocco, respectively. Recently human remains have been recovered from another littoral cave complex, Dar-es-Soltan, close to Rabat, but their affinities have still to be specified.

The specimens from the Haua Fteah, two left mandible fragments of a young adult and adolescent, were recovered from an interface between levels 32–33 in association with a Mousterian industry of Levallois facies (McBurney 1967). Radiocarbon determinations suggest an age of ~ 47,000 yr B.P.

The overall morphology and special features of the specimens suggest Neanderthal affinities (McBurney, Trevor, and Wells 1953, Tobias, in McBurney 1967) and include reduced height and accentuated breadth of the ramus, shallow sigmoid notch, substantial angulation between intraramal crests related to enlarged ramal breadth, enlarged ramal recess; and some characteristics of the dentition, especially + 5 molar pattern, enlarged anterior fovea, and enhanced trigonid-talonid breadth ratio.

Although the specimens are very fragmentary there appear to be similarities with southwest Asian Neanderthals from the Tabūn, Amud (Israel), and Shanidar (Iraq) sites.

Two calvaria and a juvenile mandible representative of *neanderthalensis* have been recovered from the lower infilling of a karstic cave in a Mesozoic limestone hill, Djebel Irhoud (Morocco). Because they were found while mining for barite, precise contextual and associational details are largely lacking (Ennouchi 1966, also 1962a, 1963). However, a substantial and diverse vertebrate assemblage of upper Pleistocene (pre-Soltanian or Soltanian) aspect, with several now-extinct species, was also present, along with traces of fire (burnt bones and artifacts), and a substantial, finely fashioned Mousterian industry of Levallois facies (Balout 1970). The principal morphological features of the three specimens have been summarized by Ennouchi (1962b, 1968, 1969b) and the endocranial casts examined by Anthony (1966).

The cranial vault is long and low with moderately thick vault bones. The endocranial volume is 1,450–1,500 cc with a large, low endocast, maximum breadth falling in the posterior temporal area, narrow frontal lobes with the anterior pole prolonged ventrally, and strong overlap of the cerebellum by the cerebral occipital lobes. The parietals are long and low, and laterally broadened with swollen bosses at midheight. The frontal is broad, moderately to more steeply curved sagittally (no. 2 subadult), with some frontal constriction. The occipital is extended posteriorly with a long upper scale and is somewhat constricted transversely to produce a "bun shape"; the limbs of the lambdoid suture have a strongly obtuse angulation. The occipital torus is well expressed in its central portion, with stronger infratoral than supratoral sulcus, and thins and fades out laterally on the occipital. The temporal squama is low with a gentle curvature; the zygomatic roots are robust and prominent; the mastoid process is large, swollen, inferiorly directed and pointed at its tip, with a strong mastoid crest; the digastric groove is deep and opens laterally; the mandibular fossa is broad and deep, and the tympanic plate robust. The supraorbital torus is well developed, with a robust glabella and biarched lateral elements; the supraciliary is stronger than the supraorbital portion and set off by a supratoral sulcus, and the supraorbital more flattened, less robust and (in No. 2) partially separate from the supraciliary. The facial skeleton is only preserved in No. 1. Nasion is depressed below glabella, with elevated nasofrontal suture; the interorbital space is broad, and the orbits low, broad, and deep. The nasal aperture is long, very broad inferiorly, and bounded by strong nasal processes of the maxilla. The anterior face of the maxilla forms an angle with the malars, without marked suborbital depression, and the malozygomatic has a strong inferior inflexion. There is alveolar prognathism, and the palate is large, especially broad and deep, but not long; only upper tooth roots are preserved and these are large and robust. A juvenile mandible (No. 3) preserves M_1 and dm_2, the roots of other deciduous teeth, and the unformed or unerupted lower C, premolars, and incisor, suggesting an age between 7 and 8 years. The dental arch is parabolic, but transversely rather flattened anteriorly. The body is high anteriorly and lower below M_1 (a juvenile feature), with indications of lateral and marginal tori; the alveolar torus is indicated

lingually, as is the submaxillary depression. The ascending ramus is relatively broad, vertically disposed, with high coronoid process and slightly obtuse gonial angle. The symphysis is nearly vertical, with traces of mental tubercles, but without a true mentum osseum; the lingual surface preserves the digastric fossae and basal triangle inferiorly, and the genial processes are also indicated. The preserved teeth (or roots) are large. Lower C appears to have been incisiform. The dm_2 has a Y-5 pattern and a sixth cusp, and substantial wrinkling, as does M_1; there is no cingulum on these teeth, but the buccal surface is swollen.

These remains differ from most European and some southwest Asian Neanderthal populations and, like those from es-Skhūl and from Djebel Qafzeh (Israel), exhibit anatomically modern features particularly in respect to the facial skeleton. Ferembach (1972) has even suggested, rightly I believe, that the total morphological pattern suggests affinity with later human populations associated with the Ibero-Maurusian industry of the Maghreb.

The skeletal morphology of the peoples responsible for the Aterian industry, widespread across northern Africa and the Sahara in later Pleistocene times, has remained almost unknown. Exceptions have been a small, and uninformative parietal fragment from Pigeon Cave (level D), Taforalt, Beni-Snassen Mountains (Roche 1953), and a facial fragment and teeth of two individuals from Mugharet el-'Aliya (High Cave, layer 5), near Cape Spartel, Tangier, Morocco (Şenyürek 1940, Howe 1967), in each instance with final Aterian associations. The latter at least reveals some facial and dental features that are anatomically nonmodern and most closely resemble Neanderthal morphology.

The el-'Aliya remains (Şenyürek 1940, Briggs 1955, Howe 1967) are so fragmentary as to provide very limited morphological details. Nevertheless, a child's maxilla fragment reveals a horizontal nasal floor with ill-defined inferior margin, blunt nasal spine, and subnasal guttering; the zygomatic process slopes posterolaterally, and the suborbital surface is bulging rather than concave. The palate is curved, with thick alveolar processes and large tooth sockets. An upper C shows buccal asymmetry, reduced mesial and distal marginal triangular prominences, lingual surface with basal tubercle and partial median ridge, and a high crown index; a P^3 is bicuspid, with marginal buccal and lingual swellings, and mesial and distal marginal triangular prominences. A worn M^2 of an older adult is large, robust, and long-rooted.

Most recently human remains, in some quantity, have been recovered from the infilling of one of the karstic caves in the dunal sandstones of the Dar-es-Soltan, near Rabat, an area previously known from

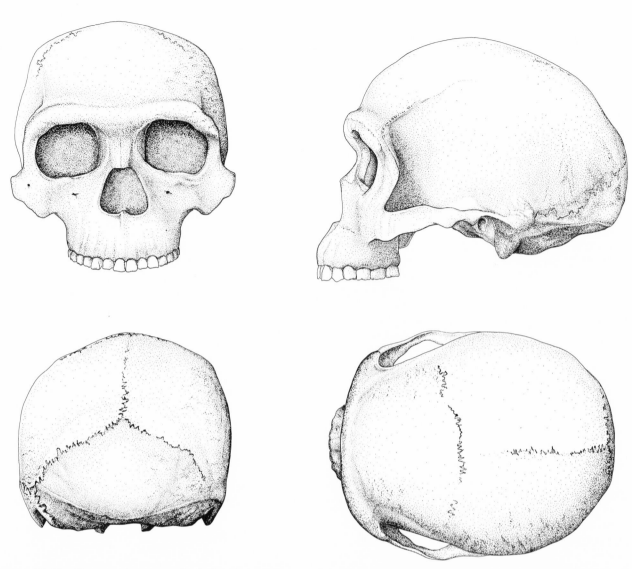

Figure 10.12 *Homo sapiens neanderthalaenis.* Facial, lateral, posterior, and superior views of restored cranium (based on a specimen from Djebel Irhoud, Morocco). Two-fifths natural size.

Ruhlman (1951) to preserve an Aterian to Neolithic industry of Capsian tradition cultural succession. The later phase of the Aterian industry was previously shown by C[14] determination to be more than 30,000 years old.[32] The remains, including maxillary, mandible, and vault parts of at least two adolescents and adults, derive from the lower levels of the infilling associated with an Aterian assemblage (Débenath 1972, 1975, Ferembach 1976b). Although morphological details are not published, these human remains seem to have Neanderthal affinities.

A frontal shows a low sagittal curvature, without median ridge, prominent glabella, strong supraorbitals, and broad interorbital region. Mandibles are reported to be ro-

bust, with strong alveolar processes. The dentition appears to be strongly taurodont, but no details on the dental or maxillary morphology are available.

The Témara mandible (Vallois and Roche 1958) was recovered from Contrebandiers Cave, southwest of Rabat. It was considered to derive from remnants of an older, brecciated infilling of a cave cut in what was thought to be a cliff of the Harounian transgressive stage (Biberson 1964). This older infilling was presumed to date from the preceding continental Presoltanian stage, of upper middle Pleistocene age.

The specimen is almost complete, with full permanent dentition (except an M_3) and parts of both ascending rami. The body is robust and overall similar to the Kébibat specimen, though rather larger, except for its stronger external oblique line and rather greater molar region mas-

32 Camps (1974, also 1968) has discussed the available radiocarbon determinations of the age of the Aterian industry, the "middle" phase of which is probably older than 40,000 years.

siveness. The basilar portion is reduced in robusticity and the digastric fossae have the same disposition as in Kébibat. The symphysis is quite vertically disposed and distinguished from the Kébibat specimen by its reduced alveolar planum and markedly reduced superior and inferior transverse tori; the internal surface lacks any trace of a fossa or processes related to the genial musculature. The external surface shows an incipient mental trigon and slight anterior mandibular incurvature. The ascending ramus was broad, quite low, arising from the body anterior to M_3, with shallow sigmoid notch and well-rounded gonial angle.

The dentition is large overall, the length of the premolar-molar series comparable to the Kébibat specimen. However, there are no remnants of cingulum in the Témara specimen, and the molars are somewhat larger and decrease in size posteriorly. All (except right M_1) have a Y-5 pattern; the left M_3 also has a sixth cusp. P_4 is strongly molariform, with mesially shifted me^d, reduced trigonid crest, a large talonid with expanded central fossa, and some indications of cusp formation distally. P_3 is less asymmetrical than in Kébibat, with the anterior transverse crest shifted mesially, and anterior foveal area expanded. The canine is large, with asymmetrical occlusal surface, and strong lingual crests. The incisors have large crowns, their lingual surfaces slightly concave with swollen base, but otherwise scant surface relief.

Further researches at the Contrebandiers site (Roche and Texier 1976) have revealed its successive infillings and human occupations. Sixteen horizons are represented that postdate the Ouljian transgression of earlier late Quaternary age. The initial human mandible, as well as other newly found fragmentary human remains (Ferembach 1976a), including the parieto-occipital portion of a cranium, all derive from an upper Aterian occupation in level 9.

Homo sapiens afer Linnaeus 1758

The late Pleistocene evolution of Hominidae in sub-Saharan Africa is still largely a matter of speculation. Although a fair number of human remains attributed on one basis or another to that time range have been recovered over the past half century, their primary contexts and varied associations are often poorly, if at all, established. The absence of any well-established stratigraphic framework for the upper Pleistocene in sub-Saharan Africa has been a major obstacle. The literature is large, frequently typological, and unduly speculative considering the limited facts available. It has been most recently and critically reviewed by Rightmire (1974, 1975). Earlier considerations of the available evidence and its varied interpretation are found in Galloway (1937c), Dart (1940a), Wells (1952, 1959), Tobias (1961), and Brothwell (1963).

The concept of a Boskop race, *H. sapiens capensis*, has a long and complex history. Galloway (1937b,c) first sought to define the morphological parameters of such a prehistoric population. The problem was evaluated by Singer (1958), who was among the first to appreciate the very weak basis on which the concept had been established. In recent years some progress has been made toward the establishment of a stratigraphic succession for the Upper Pleistocene of southern Africa, most notably in the southern Cape and Orange Vaal drainages (Butzer and Helgren 1972, Butzer 1973a, Butzer et al. 1973), to elucidate paleoenvironments and paleoclimates (van Zinderen Bakker and Butzer 1973), to clarify the faunal and cultural associations of human fossils, and to assess their "absolute ages" by radiometric and other methods (Vogel and Beaumont 1972, Beaumont and Vogel 1972, Protsch 1975, Bada and Deems 1975). Despite some progress many uncertainties and unresolved problems remain.

Wells (1969, 1972) has suggested that all (or most?) southern African human remains of later Pleistocene age should be attributed to the subspecies *H. sapiens afer* (Linnaeus 1758), a taxon that includes the ancestral stock of which African Bushmen and Negroes are "divergent specializations." Protsch (1975) has most recently followed essentially the same procedure. Unfortunately, however, he has chosen to retain *H. sapiens capensis* as a category that includes an arbitrary aggregate of supposedly older but anatomically modern human remains considered to be "Boskop" or "Boskopoid." Not only does he consider such a group as ancestral to subsequent African populations (cf *H. s. afer*), but also proposes "the world-wide evolution of all earliest anatomically modern fossil hominids from *Homo sapiens capensis* of Africa." As there is scarcely any morphological justification for this procedure or conclusion, it is not followed here.

The antiquity of this subspecies is still in doubt. Human remains of latest Pleistocene age attributed to this population include specimens from Bushman Rock Shelter, Transvaal (~29,000 B.P.), Mumbwa, Zambia (~20,000 B.P.), Lukenya Hill, Kenya (~18,000 B.P.), and Matjes River, South Africa (~10,000 B.P.) (Protsch 1975). Older representatives are considered to include specimens from Border Cave, Natal[33] (~45,000 B.P., and perhaps ~60,000 B.P.), Fish Hoek, Cape, South Africa (~35,000 B.P.), Florisbad, Orange Free State (~39,000 B.P. or

---

33 The partial skeleton from Tuinplaas, Springbok Flats (Transvaal) (Broom 1929b, Schepers 1941, Toerien and Hughes 1955) is sometimes considered as similar to the Border Cave adult. Unfortunately scant evidence of any sort exists that permits an accurate assessment of either its relative or absolute age (Wells 1959, Protsch 1975).

Table 10.8 Remains attributed to *Homo sapiens afer/sapiens*.

Elements	Northeast Africa		Eastern Africa		Zambia	South Africa
	Egypt	Sudan	Ethiopia	Kenya		
Crania	Kom Ombo-2 (ff)	Singa	Kibish Fm., Mb. 1 3+ crania	Kanjera; 40 parts of 4 crania Lukenya Hill	Mumbwa 1, 2, 3, 4 - all but 2 with associated postcranials	Border Cave; Boskop; some pc's; Fish Hoek, some pc's; Klasies River (ff); Florisbad
Maxillae Mandibles		Wadi Haifa - 1	Dire Dawa			Klasies River (2)
Dentition Postcranials			Kibish 1– partial vert. col., upper and lower limbs			
Upper limb Axial skeleton Lower limb				Kanjera; pc fragments; innominate		

> 44,000 B.P.), Klasies River Mouth Cave (1), Cape, South Africa (estimated ~ 70,000–80,000 B.P. by aspartic acid racemization). The mammal assemblages from the latter two sites correspond to the Florisbad–Vlakkraal Faunal Span (Cooke 1963) in which as many as perhaps 9 (Florisbad) to 4 (Klasies 1) (Klein 1975) large mammals represent extinct species. I also assign the human specimens from the (lower) Kibish Formation, lower Omo Basin, southern Ethiopia, to this same subspecies (see below).

All human remains attributed to this subspecies show a total morphological pattern that is overall of anatomically modern aspect. In most instances, as Singer (1958), Wells (1969), and Rightmire (1975, 1976) have noted, there is substantial overlap with the range of variability found among late prehistoric or extant indigenous human populations of sub-Saharan Africa. The skeletal remains attributed to this species are listed in table 10.8.

Very probably the demonstrably oldest (cf. Bada and Deems 1975) documented human specimens referrable to this subspecies are those fragmentary remains which derive from several (earlier) upper Pleistocene occupation horizons overlying a 6 to 8 m raised beach, at the Klasies River Mouth 1 cave. The specimens have been only briefly discussed by Singer and Smith (1969) and referred to by Wells (1972). Singer and Smith suggest that two distinct populations may be represented in the sample.

One mandible is small, though robust with body of constant height, and thick lower border, lacking a well-defined bony chin, and having large teeth with Y-5 patterned molars. Another mandible is gracile, with small taurodont teeth, and molars with + 5 pattern. A frontal is mentioned by Wells (1972) to have supraorbital structure similar to an adult individual from Border Cave.

Radiocarbon age determinations indicating minimum age and amino acid racemization studies clearly show that the Border Cave human remains have a substantial Upper Pleistocene antiquity (Beaumont and Bossier 1972, Beaumont 1973, Protsch 1975). The several specimens are associated either with a "Final Middle Stone Age" (Pietersburg) assemblage (infant) or a pre-"Early Late Stone Age" assemblage (adults), not recovered unfortunately in the most optimum circumstances to guarantee contextual and associational details (Cooke, Malan, and Wells 1945). Cooke, Malan, and Wells (1945) and de Villiers (1973) have described the salient features of these several individuals.

An adult calvarium preserves portions of the fronto-parietal-temporal regions; the vault bones are only moderately thick, the vault long, ovoid, broad, and moderately high; endocranial capacity in excess of 1,400 cc. The frontal is broad, of moderate height, curving evenly upward, with moderate frontal eminences, and faint median frontal ridge; glabella is somewhat swollen, and the substantial supraciliary eminences are separated from the thickened supraorbital margins; frontal sinuses are small. The parietals are well curved, and show moderate bossing. Temporal with broad mastoid of moderate length; shallow digastric fossa; mastoid and supramastoid crests well developed and shallow supramastoid groove; tympanic plate thickened. Interorbital region wide, and nasion depressed. Zygoma robust, convex, but not large.

Mandible, lacking dentition, of another individual, not robust, with shallow symphysis, lacking alveolar prognathism, and with a well-developed mental protuberance, internal symphysis with faint superior transverse torus, and shallow digastric fossae. An infant skeleton—preserving skull fragments, mandible and some postcrania— shows features of mandibular morphology especially which are essentially similar to modern Negro infants, as well as further mandible and postcranial resemblances to Bushmen.

This site has more recently yielded additional human remains from the Middle Stone Age occupations in the Third White Ash. Using radiocarbon determinations from higher horizons and inferences from rates of sedimentation, it is projected that this earlier MSA occupation may be approximately 80,000–90,000 years old (Butzer et al. 1978). The specimen, much of a mandible with some anterior and cheek teeth, resembles *H. sapiens afer* in all metrical and nonmetrical features (de Villiers 1976).

The provenience and associations of the Fish Hoek skull (Cape) are still in doubt, although Protsch (1974a) has recently suggested it has a very substantial antiquity ($\sim$ 35,000 B.P.), probably was associated with a fully developed "Middle Stone Age" (Stillbay) industry, and should be considered a representative of the same human group as the Border Cave adult individual. Keith (1931) fully described the specimen.

The partial cranium from Florisbad (Orange Free State, South Africa) has been the subject of continued interest and discussion since its discovery in 1932. Its context, in relation to the eye of a thermal spring, and the question of its association with other vertebrate fossils and distinctive "Middle Stone Age" industry (Hagenstad variant), have been largely resolved through chemical assays (Oakley 1954, 1957) and radiometric age determination (Protsch 1974b, Bada, Protsch, and Schroeder 1973). All appear to have been associated and seem to be of similar age—40,000 to even 50,000 B.P. However, the affinities of the hominid cranium are still in question; some authors consider the specimen to have resemblances to *H. sapiens rhodesiensis,* while others consider it to be more anatomically modern and hence another and more recent subspecies. The latter interpretation appears more reasonable, al-

though admittedly the specimen requires further reconstruction and restudy. Its salient features are set out briefly below (cf. Dreyer 1935, 1936, 1938, 1947, Drennan 1937, Galloway 1937, Singer 1956).

Cranial vault relatively low, even sagittal curvature, thick roofing bones, Frontal long, flatly curved, and very broad with low median frontal ridge; moderate frontal constriction; no ophryonic groove; faint frontal bosses. Parietals probably without marked bosses, but with interparietal groove; moderate auriculobregmatic height with vertex at bregma. Glabella prominent, inferiorly disposed, and confluent with thinner, bipartite wide supraorbital torus, moderately strong and convex supraciliary segment, and robust, convex, and rather everted supraorbital element, separated by well-defined supraorbital sulcus. Nasion inset below glabella, nasal bridge broad and probably flattened. Low, rectangular orbits, with orbital floor depressed below inferior orbital margin. Facial skeleton quite prognathous, malar large, set nearly perpendicular to maxillary body, and laterally directed, inframalar process strongly posterolaterally directed; substantial suborbital (canine) fossa. Endocranial cast with expansion of frontal and fronto-orbital areas, but apparently less-developed upper prefrontal region.

The Bushman Rock Shelter (Transvaal) affords a long "Middle Stone Age" (Pietersburg) succession (layers 43–48) overlain by sparse "Later Stone Age" occupations (layers 27–30) (Protsch and de Villiers 1974). An infant partial mandible was recovered in association with early LSA (between layers 14–18). Its overall contour, form of the mental protuberance, and the morphology of the genial area differs from the Border Cave infant, and the specimen appears to show some specific Negro affinities.

The specimen is larger than a Bush infant of comparable age. The contour of the body is angulated (between the intercanine distance and the rest of the body) and its inferior border concave; the ramus is relatively broad, with deepened mandibular notch and slightly everted angle. The symphysial profile is concave, the mental process well-defined, as are the mental fossae, and a crest along the inferior symphysial surface; there is a slight alveolar planum and faint superior transverse torus.

In eastern and northeastern Africa the hominid fossil record of the later Pleistocene is extremely scanty. The partial cranium from Singa, lower Blue Nile, and a badly preserved mandible, lacking tooth crowns, from Wadi Halfa (Sudan), are generally considered to be of late Pleistocene age, $\sim$ 17,000 and $\sim$ 15,000 B.P., respectively. Both are associated with industries considered to be of "Middle Stone Age" affinity. Of broadly comparable age, $\sim$ 17,600 years, are fragments of a cranial vault from the Lukenya Hill rock shelter (south-central Kenya) associated with a "Later Stone Age" industry. The partial mandible from "Porcupine Cave," near Dire Dawa (Ethiopia), is presumably also of later Pleistocene age, but without any firm dating. The several human remains from the Kibish Formation, lower Omo Basin (Ethiopia), are considered to be substantially older than any of these specimens.

Human cranial remains, and some associated postcrania, were recovered (in 1967) from a situation considered to represent the lower part (Member 1) of the Kibish Formation, lower Omo Basin, southwestern Ethiopia (R. E. F. Leakey, Butzer, and Day 1969). This unit of the Kibish Formation was accumulated when Lake Turkana stood some 60 m higher than its present (+ 375 m) level, and flooded the whole of the lower Omo Valley (Butzer et al. 1972). At least two (of three) specimens are thought to derive from different localities related to the upper sedimentary units (5 and/or 6, or 7 units) of that member. As they have comparable N and U values they are considered to be broadly contemporaneous. An age as old as ~ 130,000 years has been suggested for the specimens on the basis of Th/U measurements (Butzer and Thurber 1969). However, the reliability of this method has still to be adequately demonstrated; even if this determination appears reasonable, it is nonetheless unconfirmed. The radiocarbon determinations (on shell) from overlying members, with a minimum age > 37,000 years, have also been questioned. The mammal fauna associated with one (No. 1) of the specimens, with few species represented and none of them extinct, is frankly unhelpful and also unconvincing of any very remote antiquity.

The principal features of the more complete calvarium (lacking its face and much of the base), an adult of advanced age (Omo 2 from the PAS site), are briefly set out here (following Day 1969b, 1971).

Vault bones are thick, and the vault is long and wide with maximum breadth bimastoid. Cranial capacity is greater than 1,400 cc. Middle meningeal system with strongly ramified frontoparietal ramus and reduced inferior temporal ramus. The frontal is moderately elevated, though sloping posterosuperiorly, with vertex at bregma; and with slight midfrontal ridge. There is no evidence, although the lower frontal region is incomplete, of a supraorbital torus, the supraciliary was high and at least somewhat swollen, and the supraorbital area planar, bounded laterosuperiorly by very thickened inferior prolongation of the superior temporal line; postorbital constriction is slight. Sagittal profile of parietal elevated, with midsagittal depression (anteriorly) and midsagittal elevation and depression successively (posteriorly); parietal eminences well developed and superiorly situated. Oc-

cipital with moderate sagittal curvature, high superior scale, and flattened, anteroinferiorly inclined lower scale with deep nuchal mm impressions; robust, high occipital torus restricted to cental third of occipital, most protuberant medially (where surmounted by shallow supratoral depression), with rounded inferior margin, reducing in height laterally to pass into the plane of the occipital and decreasingly into the supramastoid crest. Internal occipital protuberance low and well separated from external inion (opisthocranion). Temporal with low, gently curved, elongate squama, and wide zygomatic root; low swollen mastoid process only slightly inturned inferiorly at broad, truncated tip; swollen, well-defined supramastoid crest; deep, narrow, and elongate digastric groove, and well-defined occipitomastoid crest; elliptical external auditory meatus, with anteriorly inclined axis; tympanic very robust, obliquely anterosuperiorly to posteroinferiorly oriented, very thickened, rounded medially forming posterior wall of broad, deep mandibular fossa; postglenoid process well developed, triangular; prominent, very robust articular eminence.

The Kibish 1 specimen (from the KHS site) includes a partial cranial vault, hemimandible, and some postcranial elements. The latter fall wholly within the modern *sapiens* range of variation.

The Kibish 1 calvarium differs from that of Kibish 2 in its less robust, higher vault, absence of midsagittal ridges and attendant depressions; transversely and superiorly expanded parietals, with maximum breadth situated low on the parietals; broad open occipital curvature, and marked reduction of occipital torus, and low set internal inion corresponding in level with the external occipital protuberance; inferiorly prolonged and tapered mastoid process. The mandible fragment has a well-developed bony chin and digastric fossa set posteriorly behind the symphysis. The cranial fragments of Kibish 3, another adult individual, are said to resemble this specimen.

Day (1971: 34) considered the Kibish Formation human remains are closely contemporaneous, and as "representatives of the African segment of evolving Upper Middle Pleistocene *Homo sapiens* that show a diversity of skull form at least as wide as that known for Upper Pleistocene sapiens from other parts of the world." Later, as a result of multivariate analyses Day (1973a) concluded that specimens 1 and 2 were different from one another, as well as distinct from several modern *Homo sapiens* populations. Although he suggested an "intermediate group" to include these specimens and various other middle Pleistocene hominids of Asia (Ngandong) and sub-Saharan Africa (Kabwe, Elandsfontein) little evidence supports such a conclusion. Almost every comparable morphological feature diverges not only from the evolved *H. erectus soloensis* condition, but also from that presumably character-istic of *H. sapiens rhodesiensis*. Stringer's (1974) analysis employing D² distances has subsequently demonstrated that specimens 1 and 2 are not only morphologically dissimilar, and hence may well *not* represent the same population, but also that their closest morphological resemblances are with *sapiens* cranial remains of late Pleistocene age. I completely agree with that conclusion.

These specimens from the Kibish Formation un-questionably represent a subspecies of *Homo sap-iens*, but a subspecies morphologically different from and very probably more recent than *H. sapiens rhodesiensis*. Their temporal position is still inse-curely established and merits further investigation. The lack of this important datum severely inhibits assessment of their affinities with African human populations of later Pleistocene age.

The Lukenya Hill human remains comprise only a partial frontal and parietal of an adult individual. Gramly and Rightmire (1973) have noted its resem-blances to African Negro specimens.

The frontal is quite low, receding, with weak curvature; the supraciliary ridges are well marked, merging with a prominent glabella, and passing laterally into a thickened zygomatic process of the frontal; the frontal sinus is large; the superior orbital margin is thick, rounded, and there is a shallow supraorbital notch. The superior temporal line is faintly expressed. The parietal is thick and shows slight bossing; several wormian bones are present. The nasal root is flattened, the nasion is only slightly depressed, and the frontal process of the maxilla is broad and transversely oriented.

The partial (right) mandible from "Porcupine Cave," Dire Dawa, is unfortunately poorly preserved as are its five teeth, which lack crowns and any enamel (Vallois 1951).

The specimen is characterized by a robust body; the infe-rior border of the body is substantially thickened, and the digastric fossae are large, transversely elongated, and situated inferiorly. The external symphysis is vertical, without bony chin development, and lacks alveolar progna-thism; there is a slight alveolar planum, an inferior trans-verse torus, and submandibular incisure. The premolar-molar series was apparently large, with M₃ not reduced, and without taurodontism.

The mandible from Wadi Halfa (site 6B28) is a badly eroded specimen without tooth crowns and without useful morphological details (Armelagos 1964).

The Singa cranium is the most complete hominid specimen from the later Pleistocene of northeastern Africa. At least one extinct species, *Pelorovis anti-quus,* may be associated with the level from which the artifacts and human cranium derived. Wood-ward (1938) and Wells (in Arkell et al. 1951) have equally directed attention to its overall Bushman-like character. The specimen is a nearly complete brain case, lacking the facial skeleton and the cra-nial base.

The vault is moderately long, but very broad, with strong parietal eminences; the transverse contour between the eminences is flattened, and there is a broad interparie-tal depression. The cranial capacity perhaps exceeds 1,500 cc. The vault is high, with maximum height at bregma, and the back steeply curved inferiorly to short, rather flat-tened occipital. The frontal is only moderately elevated, with median frontal ridge, indistinct frontal bosses, and some constriction behind the enlarged angular processes; glabella is only slightly swollen, and the arched supracil-iary ridges merge with it medially and laterally pass into the thinner supraorbital elements to produce a continuous, but rather strongly biarched supraorbital torus. On the oc-cipital the external inion and nuchal lines are only faintly delineated. The temporal is low, with swelling along the sphenotemporal suture; the mastoid process is narrow, ta-pering inferiorly, and short, and the digastric fossa is large and broadly open; the mastoid and supramastoid crests are prominent, separated by a narrow sulcus; the mandibular fossa is deep, with convex articular eminence, and thick-ened tympanic plate.

The lateral orbital margin is set well back of the interor-bital region, with the zygomatic process of the frontal slop-ing inferoposteriorly; the interorbital distance is wide, with nasion slightly inset below the glabella; the frontal processes of the maxilla face more anteriorly than later-ally; the nasal bones are narrow and nearly flat, rather than elevated. The orbits are wide and relatively low.

Discussion

Adaptive Grade and Structural-Functional Zones

Simpson (1963) diagrammed a representation of the adaptive grade and structural-functional zones of *ex-tant* Hominoidea. His representation combined a dendogram on an adaptive grid; his "hominid zone" referred only to the genus *Homo*. The fossil record of African Hominidae is examined here in the same way, particularly with regard to shared primitive (symplesiomorphic) and shared derived (synapomor-phic) characters, as well as nonshared (autopomor-phic) characters (Tattersall and Eldredge 1977).

Ramapithecus

Those hominoid primates which are the most likely ancestral source for Hominidae are customarily attributed to ramapithecines (Simons 1977, Conroy and Pilbeam 1975), probably even a single genus, *Ramapithecus*.[34] Such hominoids, of still indeterminate familial affinity, are represented in both eastern Africa and southern Eurasia, in pre-Vallesian (upper "Vindobonian") time, and persist into Vallesian time in southern Asia. These hominoids thus spanned some 5 or 6 m.y., between ~ 13 to 8 m.y. ago.

For the moment *Ramapithecus* is known largely from gnathic and dental elements.[35] There is thus no direct evidence of body proportions, overall body size (although it is considered, on other grounds, to be approximately that of a pygmy chimp), or of locomotor skeleton structure. Cranial morphology is unknown and there is no evidence of brain size, form, or proportions. Nonetheless the overall morphological pattern of the jaws and dentition diverges from that of penecontemporaneous pongids—that is, species of *Dryopithecus* and *Sivapithecus*.

Ramapithecus is distinguished by deep, robust body of maxilla; zygomatic arch laterally flared (suggesting large temporal fossae and temporalis musculature) and set anteriorly at M^1 level (suggesting wide mandibular ramus, expanded area of masticatory musculature attachment, and enhanced power arms in mastication), with well-defined fossa above upper P roots; abbreviated premaxilla; mandible body shallow and shallows posteriorly, robust, shortened anteriorly (anterolateral angle at P_3 root) with variably shallow and elongate, or more erect, shortened, heavily buttressed symphysis with extensive alveolar planum, substantial superior transverse torus, and large to very robust inferior transverse torus extending near or to M_1, and associated deep intertoral fossa; ascending ramus with markedly anterior origin behind M_1, probably vertically set and broad; premolar-molar rows aligned with only slight posterior divergence; lower incisors small, slightly to hardly procumbent, with the alveoli lingual to lower C; lower C alveolus set lingual to anterior root of P_3 and upper C not incorporated in cheek tooth series. Upper and lower cheek teeth with low crown relief, thick enamel, much interproximal wear, steep wear gradient, and transverse abrasion indicative of strong lateral component in mastication. I's unknown; upper C with compressed root, relatively small, conical crown projecting beyond plane of cheek teeth, with lingual cingulum and mesial tubercle, substantial lingual relief, with wear against both mesial and distal margins; P^3 three-rooted; upper molars steep-

sided, with expanded occlusal surfaces and low subconical to blunt cusps set at margins of crown, buccal grooves variably expressed, no or only slightest cingulum remnants on angle of protocone.

I's unknown; lower C with long, straight compressed root, conical crown with distal cingulum, shallow mesiolingual groove, short mesial ridge, wear along whole of distal margin and apex; P_3 elongate, oval-shaped low crown, set markedly obliquely to tooth row, sectorial or incipiently bicuspid (me^d), with small mesial (honing) and distal wear facets; P_4 bicuspid, and well-developed talonid basin; lower M's with Y-5 pattern, no or only buccal cingular remnants.

This interesting hominoid primate has been found to occur in relation to a diversity of paleoenvironments—floodplain and other streamside or lake margin situations. The vertebrate associations often seem to indicate relatively open country habitats, perhaps wooded to shrub savanna, but woodland and even forested situations are also suggested. It has been reasonably inferred that the primary feeding adaptation may have been foraging in forest floor or more open forest-edge, even woodland situations (Pilbeam 1976, Simons 1976, 1977). The extent to which animal protein supplemented a basically vegetal diet is unknown and in dispute.

There is an almost complete dearth of evidence on the adaptive grade of protohominids from the Turolian and Ruscinian time ranges—that is, between ~ 8 and nearly 4 m.y. ago. However, the scant evidence of jaw structure and dentition suggest affinities with *Ramapithecus* as well as with the subsequent australopithecines; however, a fragment of an upper limb (distal humerus from Kanapoi) diverges from the known australopithecine condition (see below). Its significance is not yet apparent—that is, whether the australopithecine condition had not yet been attained and this condition is therefore primitive, or whether it merely indicates one of several different configurations of locomotor diversification at that distant time.

Australopithecus

The hominid status of the extinct genus *Australopithecus* is now firmly established. The morphological distinctiveness of most known aspects of its skeletal structure from *Homo* is also now well documented. However, the overall adaptive grade and structural-functional zone of this interesting hominid taxon is still disputed. In my view this results from undue and often unfounded speculation without an adequate empirical data base and a persistently pervasive anthropocentric concern with the source for the ultimate origins of our own species, *Homo sapiens*.

34 Possibly some three species may be represented, including *wickeri* (East Africa, Anatolia), *punjabicus* (Siwaliks), and *hungaricus* (Pannonian Basin, and possibly Catalonian prelittoral depressions). *R. punjabicus* appears to have been a succedent species to *R. wickeri*.

35 As yet unreported postcranials, attributable to *Ramapithecus,* are apparently now known from middle Siwaliks formation in Pakistan (D. Pilbeam, pers. comm.).

The differences in stature and in body size between small, gracile and the larger, more massive robust *Australopithecus* species were substantial—from 100–145 cm and 20–30 kg, to 145–165 cm and 45–90 kg. There was some overlap in stature, if not perhaps in body weight. Sexual dimorphism is documented as substantial in all species, perhaps particularly so in the more robust taxa. Body proportions were distinctive in respect to upper limb elongation, including both upper arm and forearm elements.

Although clearly hominid and not pongid in total body structure, the genus represents a distinct adaptive grade. Elements of the forelimb were distinctive not only in their elongation, proportions, and general robusticity, but also in joint morphology and presumably also in associated flexor and extensor musculature. The conclusion can only be that the forelimb was employed in unfamiliar non-*Homo* behaviors. It has been suggested that these may have included supportive functions in respect to postural or active locomotor activities (though not necessarily suspensory postures, and knuckle-walking is apparently excluded by some aspects of elbow joint and digital morphology).

The genus is characterized by unmistakable adaptation of the musculoskeletal system—of the axial skeleton, pelvis and several elements and joints of the lower limb—for at least facultative, and more likely habitual erect posture and bipedal gait. However, a mosaic of morphologies is represented in *Australopithecus* structure (Howell, Washburn and Ciochon 1978) (table 10.9); the knee and ankle joints exhibit structural features most similar to those of *Homo,* as does perhaps also the transversely and longitudinally arched foot. The lower lumbar region, sacroiliac joint, hip joint, and pelvis form and proportions are all unique and exhibit a total morphological pattern of hominid rather than pongid type, but it is a type distinct from that of *Homo.* This finding suggests differences in gait—its phases and cadence—and the associated mechanisms of muscle support, balance, and limb movement.

Australopithecus dentition and its supporting gnathic elements—the permanent incisor-canine-P$_3$ complex, and their deciduous precursors—are non-pongid. Overall these are synapomorphic characters shared with *Ramapithecus*. However, the morphological pattern of all deciduous and permanent teeth, the relative proportions of anterior teeth to cheek (premolar-molar) teeth, and the size of the molars relative to body size as well as absolutely, distinguish this genus from *Homo.* Such features are fundamentally similar in all *Australopithecus* species and only morphological details and relative body size

should be noted. However, between antecedent and succedent taxa a complex of changes reflects an adaptation toward an enhanced crushing and grinding masticatory system (including flattened maxillae, reduced lower facial prognathism, (probably) earlier disappearance of the premaxillary suture, altered disposition of masticatory muscular reenforcement systems of the mandibular ramus, more vertically implanted upper incisors, flatter incisor wear bevels, earlier edge-to-edge bite, accelerated and enhanced edge-to-edge canine occlusion and reduction of canine crown height, enlarged, robust tooth roots with enlarged interroot plates, lower and more rounded cheek teeth cusps, and a lower grind angle between upper and lower cheek teeth (Wallace 1978). This trend is also reflected in the development of cranial vault superstructures and their expression, in several aspects of facial structure and proportions, and in the hafting of the facial skeleton to the cranial base and the configuration of the cranial vault and its several components. These represent a set of autopomorphic features distinctive of later species of the genus. Unfortunately there is as yet relatively little comparative data on growth and maturation in *Australopithecus* which might reveal more fully the ontogenetic background of adult morphology. Maturation appears to have been delayed and the pattern of (permanent) dental eruption, and perhaps also its timing, closely approximates that of *Homo* (although the eruption of P$_4$ is probably delayed in comparison with M$_2$).

Australopithecus species appear to have been important elements in their contemporary biota. In the Koobi Fora succession, where presumably natural or accidental death populations are largely sampled, an extinct *Theropithecus* species is as common as larger carnivores, other cercopithecids, and hominids combined. However, hominid fossils are as common as either of the other two groups; among hominids, except within the sub-KBS stratigraphic unit, the robust *A. boisei* is more frequent than *Homo* species by about 2:1. This same pattern of *A. boisei* predominance is found in the middle members (E, F, and lower G) of the Shungura Formation.

At Swartkrans, where carnivore predation or scavenging was doubtless responsible mostly, if not wholly, for the vertebrate accumulations, hominid and cercopithecoid primates (four species) occur in equal abundance, about 26% each of the total species (excluding micromammals) represented. *Australopithecus* (*crassidens*) is nearly 50 times as frequent as *Homo* in the hominid sample; over 40% are immature individuals. Only just over 10% of the sample represents older adult individuals (age 31–40) and

Table 10.9 Principal morphometric affinities of australopithecine postcranial elements.

Elements	Pongidae		Hominidae	
	Pongo	Pan/Gorilla	Homo	Unique
Shoulder Girdle & Forelimb				
Clavicle (Old Hom. 48)	Oxnard 1975 a,b		Napier 1965; Day 1976, 1978; Vallois 1976	
Scapula (STS 7)	Oxnard 1975 a,b; and Broom, Robinson, and Schepers 1950			Robinson 1972
Humerus (STS 7; TM 1517 (Krom); ER 739, 1504, 740)		Robinson 1972; Straus 1948; McHenry and Corruccini 1975 a	Robinson 1972; Le Gros Clark 1947	Day 1976; McHenry and Corruccini 1975a; McHenry 1973; Day, Leakey, Walker and Wood 1976
Gleno-humeral Joint (STS 7)		Ciochon and Corruccini 1976		
Ulna (Omo L 40)		McHenry, Corruccini, and Howell 1976		
Elbow Joint (TM 1517 (Krom); ER 739, 1504, 740; Omo L 40)			Day 1976; Le Gros Clark 1947; Broom and Schepers 1946	McHenry, Corruccini, and Howell 1976
Manus (Old Hom. 7 [?Aust.])		Napier 1962	Day 1976, 1978	
Carpals (TM 1526 [Sterk.])		Lewis 1973	Robinson 1972; Le Gros Clark 1947	
Metacarpals (SK 84, SKW 14147)		Robinson 1972; Napier 1959; Rightmire 1972	Day and Scheuer 1973	
Axial Skeleton				
Vertebral Column (STS 14; SK 853, 854, 3981 a,b [? Aust.])			Robinson 1972; Robinson 1970	Robinson 1972
Lower Limb				
Pelvis (STS 14; STS 65, MLD 7/8, 25 SK 50, 3155 b, TM 1605 [Krom.]; AL 288)		Robinson 1972	Robinson 1972; Dart 1949, 1958; Le Gros Clark 1955; Brain, Vrba, and Robinson 1974; McHenry 1975 d; Johanson and Taieb 1976; McHenry and Corruccini 1975b	

mean age at death was about 17. Evidently *Australopithecus* was a common species and a species subject to strong predation. At Kromdraai (site B) six *Australopithecus* (*robustus*) individuals are represented, four of which are immature. Three cercopithecoid primate species are represented in the same part of the infilling by 22 individuals, nearly 4 times as numerous as *Australopithecus*. This vertebrate accumulation differs from that of Swartkrans in significant ways (Brain 1978, Vrba 1976) and suggests a different origin, probably in some substantial part resulting from hominid hunting activities. Whatever

Table 10.9 (*continued*)

Elements	Pongidae		Hominidae	
	Pongo	Pan/Gorilla	Homo	Unique
Sacroiliac Joint (STS 14; AL 288)			Robinson 1972; Johanson and Taieb 1976	
Femur (AL 288; SK 82, 97, Old Hom. 20; STS 34, TM 1513 [Sterk.]; ER 738, 815, 993, 1463, 1465, 1503, 1505; ER 1500; AL 129)			Le Gros Clark 1947; Broom and Schepers 1946; Lovejoy and Heiple 1970, 1971; Heiple and Lovejoy 1971; Wolpoff 1976 d	Day 1973 b, 1976, 1969; Robinson 1972; Day, Leakey, Walker, and Wood 1976; Walker 1973 b; Johanson and Coppens 1976; McHenry and Corruccini 1976
Hip Joint (STS 14; SK 50, 3155b, TM 1605 [Krom.]; SK 82, 97, Old Hom. 20)		Robinson 1972	Robinson 1972; Lovejoy and Heiple 1970; Zihlman and Hunter 1972	
Tibia & Fibula (ER 1500; AL 129; Old Hom. 35 [?Aust.]; ER 1476			Day 1976, 1978; Day, Leakey, Walker, and Wood 1976; Johanson and Taieb 1976; Johanson and Coppens 1976	Davis 1964
Knee Joint (ER 1500; AL 129)			Day, Leakey, Walker, and Wood 1976; Johanson and Taieb 1976; Johanson and Coppens 1976	
Pes (Old Hom. 8)	Oxnard 1975 a,b	Lewis 1972	Day 1976; Day and Napier 1964; Archibald, Lovejoy, and Heiple 1972; Corruccini 1975 a	Day and Wood 1968
Tarsals (TM 1517 [Krom.]; ER 1476)	Oxnard 1975 a	Robinson 1972; Le Gros Clark 1947	Corruccini 1975 a	Day 1976, 1978; Day and Wood 1968
Phalanges (TM 1517 [Krom.]; Old Hom. 10)			Day 1976, 1967, 1974 b; Day and Napier 1966	

its cause, this occurrence reflects substantial predation, depending on the time represented by the upper part of this site's infilling.

The picture is less clear in regard to *A. africanus*. Presumably all three known occurrences in southern Africa constitute vertebrate accumulations largely and perhaps wholly the consequence of carnivore predation and scavenging.

At the holotype locality of Ta'ung the vertebrate associations are unfortunately unclear and no infer-

ences are possible. At both Sterkfontein and Makapan Limeworks *Australopithecus africanus* occurs in association with a diversity of other vertebrates. *Australopithecus* is a relatively small percentage of each total vertebrate accumulation. Of the 25 to 40 *Australopithecus* individuals at Sterkfontein, about 20% are immature. By contrast, over 600 specimens of cercopithecoid primates are known, comprising 7 species, predominantly of the extinct genus *Parapapio*. At Makapan Limeworks the faunal spectrum is quite different from that of Sterkfontein, and both *Australopithecus* and cercopithecoid primates are substantially less common, both relatively and absolutely. Only some 8 to 12 *Australopithecus africanus* individuals are represented; about 20% are immature. There are less than 300 specimens of 7 species of cercopithecoids, also mostly *Parapapio*. At both sites nearly 29% of the *Australopithecus* sample represents older adult individuals (age 31–40) and mean age at death is 22, in each instance significantly higher than that recorded at Swartkrans.

Thus, there are marked differences in each of these occurrences by comparison with the situation recorded at Swartkrans. In all these situations *Australopithecus* species are prey and not predators, the hunted and not the hunters.

Australopithecus, at least in part, has been assumed to have practiced hunting and to have been capable of using and making tools. Empirical evidence supporting this interpretation for *A. africanus* is entirely lacking and, in the case of succedent robust species, is ambiguous at best as the simultaneous, sympatric presence of *Homo*, often in direct association, is now well documented. Thus, the genus is probably best regarded as having occupied a collecting-foraging niche, *perhaps* scavenging as well as opportunistically taking vertebrate prey. Food was probably prepared intraorally without the benefit of tools.

Homo

Conceptions of the structural-functional zone and adaptive grade of genus *Homo* derive from knowledge of *Homo sapiens,* particularly as exemplified by recent subspecies. However, the species has had a substantial evolutionary history, and its antecedent species must also be considered.

Increasing and convincing evidence now suggests that the origin of *Homo* is to be sought within the earlier Quaternary if not the later Pliocene. By that time the pelvis and lower limb skeleton are known to deviate from the structure characteristic of *Australopithecus* and to approximate closely that of recognized species attributed to *Homo*. Modifications in cranial form and proportion, facial structure, and dentition size and morphology are also documented. The mandible is reduced in robusticity, teeth are smaller, some cusps are reduced, crowns have thin occlusal enamel, a helicoidal occlusal wear plane is present, and there is an increased rate of molar approximal attrition (Wallace 1978). Some cerebral enlargement and attendant changes in endocranial form and proportions are found. Sexual dimorphism was still substantial. The total morphological pattern indicates a new adaptive grade, divergent from that of *Australopithecus,* and appropriately recognized as that of *Homo*.

The functional behavioral aspects of this adaptive transformation are still ill-defined. Transformations in pelvic size and proportions seem to indicate larger-brained newborns. Pelvic and lower limb morphology indicate efficient, habitual bipedal walking and a striding gait. Modifications of the masticatory system and dentition have been interpreted to reflect extraoral preparation of food through tool use (Wallace 1978).

Artifacts and associated residues in contexts of known age increasingly attest to culturally patterned behaviors at least by 2.1 to 2.0 m.y. ago. Stone-tool manufacture, activity-specific locales including "home bases," and a meat-eating diet are demonstrated. Cooperative exploitation of vertebrate food resources through regular, organized hunting, food sharing, and attendant sexual division of labor and differentiated social roles might be inferred. Shortly thereafter the paleobehavioral record is elaborated through the extraordinary occupation sites in the lower horizons of Olduvai Gorge, a unique documentation. There are indications of extra-African dispersal of hominids, attributable to *Homo*, into subtropical Asia about that time. *Australopithecus* species persisted, seemingly only in Africa, for nearly a million years as the phyletic evolution of *Homo* progressed.

Coexistence of Hominid Taxa

Fossil evidence of the coexistence of several hominid taxa has been treated by some authors as inconsequential if not utterly dismissed. The competitive exclusion principle has been invoked as a basis for the impossibility of multiple coexistent hominid taxa, and the single-species hypothesis, which maintains that "because of cultural adaptation, all hominid species occupy the same, extremely broad, adaptive niche" (Wolpoff 1971a), proposed to account for the existence and "continued survival of only one hominid lineage." However, as a consequence of continued recovery of hominid skeletal parts in Africa, coexistence of several hominid taxa is now well documented at least for the Plio-Pleistocene time range. In most instances species of *Australopithecus* coexist with one or another pre-*sapiens* species of genus *Homo*.

At Swartkrans (Transvaal), in the Member A infilling, three individuals attributed to *Homo* sp. occur in association with some 90 individuals of a robust australopithecine, *A. crassidens* (Clarke and Howell 1972; Clarke 1977).

At Olduvai Gorge (Tanzania) several hominid taxa, including *Homo* sp. ("*habilis*") and *Homo erectus* (which appears first in the uppermost reaches of Bed II) and *Australopithecus boisei*, occur within Beds I and II (M. D. Leakey 1978). At several well-defined geological horizons at least two hominid taxa are recorded in circumstances that suggest both contemporaneity and sympatry. In relation to Tuff I^B at least four *Homo* sp. individuals occur with at least three *A. boisei* individuals; below Tuff I^C at least two *Homo* sp. individuals occur with an *A. boisei* individual and (perhaps) with an individual representing *A. africanus*. In lower Bed II, above Tuff I^F, three *A. boisei* occur in a horizon yielding a *Homo* sp.; between Tuff II^A and II^B remains of an *A. boisei* individual occur in sediments yielding at least five *Homo* sp. *A. boisei* last occurs in uppermost Bed II, above Tuff II^D, in sediments yielding two *Homo erectus* individuals.

At East Turkana (Kenya), the lower (sub-KBS) and upper (sub-Koobi Fora and sub-Chari) members of the Koobi Fora Formation provide sufficient evidence of the association of the genus *Homo* with at least one species of *Australopithecus*, *A. boisei*, and (doubtfully) *A. africanus* as well. In the Lower Member (sub-KBS unit) *Homo* sp. is some four times as common as *A. boisei* and at several localities these taxa have been found in comparable geological horizons. In the sub-Koobi Fora unit of the Upper Member *Homo* sp. and *A. boisei* occur initially in about equal frequency and these taxa are found in comparable geological horizons. In the upper part of the sub-Koobi Fora unit *H. erectus* occurs in about the same frequency as *A. boisei;* in at least two localities (areas 105 and 130) these taxa have been found in immediate proximity. *H. erectus* is demonstrably penecontemporaneous with *A. boisei* in this sedimentary formation (R. E. F. Leakey and Walker 1976). In the upper sub-Chari unit of the Upper Member *A. boisei* is more than twice as common as *H. erectus;* these taxa have been found in the same general collecting localities in comparable geological horizons.

In the Shungura Formation, lower Omo Basin (Ethiopia), *A. africanus* appears to overlap *A. boisei* (Members E–lower G), and it is clear that a *Homo* sp. coexisted with *A. boisei* in the uppermost part (Members upper G–H) of that succession (Howell and Coppens 1976a).

The geological occurrence of early hominid fossils in Africa and their morphological features now leaves no doubt that at least two hominid taxa coexisted temporally, even sympatrically, and very probably through a substantial span of late Pliocene and earlier Quaternary time. Evidently Hominidae experienced a substantial adaptive radiation at least as early as Pliocene times, if not previously.

Extinctions

At various times extinctions have been proposed or inferred to account for the evolutionary history of Hominidae. However, with the previously very incomplete nature of the fossil record and its probable important gaps, such inferences have been necessarily largely speculative.

Within the species *Homo sapiens* there is scarcely any direct evidence of the extinction of particular subspecies or populations within Africa. However, various workers have questioned the temporal and phylogenetic affinities of specimens which would now commonly be attributed to several populations, of middle to late Quaternary age, of *H. sapiens*. The extinction of particular populations has been at least implied, if not explicitly suggested. Thus, most recently, Protsch (1975) has favored the origin of the subspecies *rhodesiensis* from an antecedent subspecies *capensis,* and the ultimate extinction of *rhodesiensis* during *H. sapiens afer* times. However, the temporal relationships of particular fossil specimens attributed to these several species are still far from resolved. It is at least if not more than likely that the *rhodesiensis* subspecies was ancestral rather than descendant.

Similarly, *H. erectus* has been sometimes presumed to have been a descendant, ultimately extinct species derived from a more *sapiens*-like ancestral stock. However, scant evidence supports that interpretation, and the known temporal span of *H. erectus* largely and probably wholly antedates that of *H. sapiens*.

The most convincing evidence for extinction within Hominidae is of one or more species of *Australopithecus*. The disappearance from the fossil record of *A. boisei* and of *A. robustus* and *crassidens*, at a time when *Homo erectus* was already present and well-established, is now recognized at several localities in sub-Saharan Africa. Various speculations have been offered about the basis for the disappearance of robust *Australopithecus* species, largely framed in terms of diet, behavior, and competition. Their disappearance is also part of a larger pattern of (mammalian) faunal turnover and replacement within the continent around a million years ago.

Cultural Associations

Cultural associations and occupational residues in their natural contexts yield the only direct evidences of hominid behavior and adaptation in the later Cenozoic. Inferences and speculations on cultural capabilities, object use and tool-making, predatory behavior and subsistence activities, and habitat utilization ultimately derive only from such hominid occurrences.

The behavioral capabilities and adaptive adjustments of emerging Hominidae have frequently been speculated about. However, for the moment three localities in eastern Africa—Ngorora, Lukeino, and

Lothagam—considered to yield Hominidae (gen. et sp. indet. A.) of later Miocene age provide no evidence of any sort that indicates distinctively hominid behavior. However, the paleoenvironmental settings were unlike those of recent pongids.

Several localities of middle Pliocene age—including the Hadar, Usno and basal Shungura formations of the Omo, Kanapoi, and Laetolil—from which hominid remains have been recovered yield no traces of culturally patterned behavior. These include occurrences of *Australopithecus afarensis* as well as some of *Australopithecus africanus*.

Dart (1956, 1957a,b,c; cf. Tobias 1967b, Wolberg 1970) considered that the extraordinary bone accumulations (in the lower gray breccia, now termed Member 3) at Makapansgat resulted primarily from the predatory activities and selective behavior of *A. africanus*. However, the occurrence itself, and the disproportions in body parts and in species represented, may equally well represent predominately carnivore predation and scavenging (Brain 1976b). For the moment there is no conclusive evidence of hominid culturally patterned behavior, either in respect to the bone accumulations proper or to the presence of bone or stone utilization or transformation into tools or weapons.[36] Similarly, Vrba (1975) has suggested that the bone accumulation at Sterkfontein (main site) probably results from "predators that were specialized on large prey," perhaps sabertooth cats. This australopithecine occurrence also fails to yield any evidence of artifacts in bone or stone.

The first substantial, widespread, and incontrovertible evidence for the utilization of stone as a medium to reproduce tools and for the accumulation of occupational residues at circumscribed locations dates to about 2 m.y. B.P. In almost all known instances there is direct or indirect evidence of at least two (and possibly three) hominid species more or less closely related with these occurrences. These occurrences are not all of the same sort, and it is most probable that their differences overall reflect at least some earlier hominid culturally patterned behaviors and adaptations.

Numerous artifact occurrences are recorded from the lowest units of Mb.F of the Shungura Formation, Omo Basin, in or above Tuff F, which are 2.04 m.y. old (Merrick and Merrick 1976, Merrick 1976, Chavaillon 1976). These occur largely in relation to stream channels, including a

braided river system, and under substantially drier conditions than recorded earlier in the Omo succession. Both cf *Homo* sp. and *A. boisei* are represented in this sedimentary unit, but no hominid remains are immediately associated with any of the artifact localities. The artifact occurrences are often of very low density, but spatially widely distributed; several high density occurrences (comparable to those at Olduvai Gorge) are also known, however, and one or two appear to be *in situ* occupations. Artifacts are almost always small, in a limited number of raw materials, and both large and small tools are either infrequent or wholly absent. Other artifact occurrences are also known, always in derived condition, in channel gravels in the lower units of Member G that are ~1.9 m.y. old.

Some half dozen artifactual occurrences are documented in the Lower Member of the Koobi Fora Formation, East Turkana, in channel situations within the KBS Tuff, or in one instance below that tuff (Isaac, Harris, and Crader 1976, Isaac 1976b). These occurrences range in age from ~1.8 to 1.6 m.y. (Curtis et al. 1975). Although hominids (both *A. boisei* and *Homo* sp. and possibly, but rarely, *A. africanus*) occur in the same horizons, and broadly the same general areas as the artifact concentrations, none have been found in direct association. Several distinct forms of occurrence, apparently always proximal to stream channels, include several, usually low-density artifact scatters of limited extent; an artifact and fragmented bone concentration (including parts of nearly 10 ungulate species) of substantial size that suggest an occupation site or camp; and a smaller occurrence of much of a single hippo with a few tools and other artifacts that suggest a butchery site. All the occurrences are distinguished by the generally low density of artifacts, high frequency of flakes and flaking debris, rarity of larger shaped (core) tools, the absence or near absence of small (or larger) flake tools, and limited variety of raw material that must have been brought from at least a short distance away as it was not locally available. The lithic industry represented is overall quite comparable to the Oldowan of Olduvai Gorge. Overall, as Isaac (1971, 1976a,b, 1978) has stressed, the available evidence suggests the utilization of localized streamside habitats, the exploitation and transport of raw materials, the existence of home bases or "camps," tool manufacture and use, and meat-eating with transport of food sources; these features also suggest food sharing and imply some levels of cooperative behavior.

The occupational sites in Beds I and II, Olduvai Gorge, yield the most numerous and contextually rich artifactual and cultural residues associated with early Hominidae (M. D. Leakey 1971). Eighteen hominid occupation situations are documented within Bed I sediments, and 63 in Bed II sediments; both encompass a total time span of approximately 1 m.y. (Hay 1976, M. D. Leakey 1978). From this restricted and persistent minisedimentary basin derives the most complete history yet found of an earlier Pleistocene record of hominid adaptation and cultural capabilities.

In the earlier portion of this succession at least two and perhaps three hominid taxa are represented and are spe-

---

36 However, some undoubted quartzite flakes and pieces showing signs of utilization from the upper pink breccia (Member 4) at this locality do suggest some stone fracture and usage, even if definitive tool forms appear to be absent. However, the context and associations are still unclear.

cifically associated with occupational residues (M. D. Leakey 1978). In direct association with two Oldowan industry sites is the robust australopithecine, *A. boisei;* in one instance *A. boisei* is associated with postcranial parts considered perhaps to be *A. africanus. A. boisei* is also associated in at least one occurrence with a Developed Oldowan industry in the highest reaches of Bed II, some 0.75 m.y. later. An early species of the genus *Homo, Homo habilis,* occurs at 11 occupational situations from lower-most Bed I (6 occurrences), through mid-Bed II (2), lower Bed II (1), and middle Bed II (2)—a time span of ~ 0.2 m.y. In at least two occurrences in Beds I and mid-II this taxon is directly associated with *A. boisei,* and in other situations temporal contemporaneity is demonstrable on stratigraphic grounds.

The hominid occupation occurrences and their associated residues uniquely testify to the cultural capabilities, habitat adjustments, and overall adaptation of early hominids within this time span. Unfortunately these occurrences *do not* afford a basis for directly assessing such capabilities and differential adjustments between the several hominid taxa represented through this time range; their differential behaviors are not readily discernible from these varied occupational situations. Within the Oldowan industry occurrences with which all *Homo habilis* or almost all *A. boisei* individuals are associated, there are substantial differences, at least in degree, in the spatial distribution, form, and extent, of occurrences, which suggests spatial segregation of activities; the density of artifactual and/or food debris residues, including transported (distal) or more immediate (proximal) elements; composition, frequency, and respective body part representation of vertebrate taxa; and the overall composition, diversity, and character of the artifacts—shaped tools (whether core tools or flake tools obtained from cores), utilized pieces, or waste products, and manuports which by their very presence, and sometimes character, demonstrate the artificial activity of hominid presence. Limited sorts of lithic raw materials were selected, sometimes transported, and used to fashion particular implement categories and forms. Very probably home bases (of several activity types), workshop sites, butchery sites, and even other occurrences of undecipherable activities are represented within this unique spectrum (cf. Isaac 1971, 1976a,b, 1978). It has been suggested that probably the bulk of these and later occurrences (in Bed II) at Olduvai represent dry season occupation sites (Speth and Davis 1976).

The Developed Oldowan industry succeeds the Oldowan in lower Bed II and occurs thereafter through even Beds III and IV. From middle Bed II (above the Lemuta Member) onward it occurs in parallel with an Acheulian industry. A single hominid occurrence, *A. boisei,* is directly associated with a Developed Oldowan assemblage. That industry shows some shifts in the exploitation and transport of raw material, an increase in tool types, the appearance of new sorts of shaped tools, overall greater artifact diversification perhaps related to functional specialization, and greater variation in assemblage composition from one occurrence to another. Occupation sites show an increase in the proportion of artifacts to bone refuse, increased bone breakage and fracture, and differences in prey species represented.

In the Upper Member of the Koobi Fora Formation, East Turkana, there are several artifact occurrences (Karari industry) below the Okote Tuff Complex (1.45 m.y.) and many occurrences between the Okote Tuff Complex and the overlying Karari-Chari Tuffs (1.3–1.2 m.y.) (Harris and Isaac 1976). Hominid skeletal parts are abundant within this portion of the succession, but direct associations of hominid remains with artifact occurrences are infrequent. In one instance there is a direct, proximal occurrence of *A. boisei* and *Homo erectus* remains with the distinctive Karari industry; in another instance *A. boisei* occurs with a scatter of such artifacts above a more substantial occupation occurrence. The Karari industry, which shows some similarities to, but other features distinctive from the Developed Oldowan, shows very substantial assemblage diversity and variety in tool types. It is known from at least 50 sites, a number of which preserve a variety of bone debris and evidently represent either home bases or butchery sites.

The upper levels of the Chemoigut Formation, Baringo area, is a marginal saline lake setting associated with inflow channels; robust australopithecine cranial parts occur in two of some five horizons yielding an Oldowan industry (Bishop, Hill, and Pickford 1975, Harris and Bishop 1976).

Several artifacts of a Developed Oldowan type (M. D. Leakey 1970) also occur in the Member A infilling of the Swartkrans locality. The associated hominids there are robust australopithecines, in abundance, and three individuals of *Homo cf habilis.* The context of the artifact-hominid association is still unclear, but various evidence strongly suggests that carnivore activity, particularly leopard predation, may have been largely responsible for the vertebrate accumulation (Brain 1968, 1970, 1976b, 1978, Vrba 1975).

A few artifacts of undetermined industrial affinity are known from the Kromdraai B locality, the type occurrence of *A. robustus.* Both Brain (1975) and Vrba (1975) have suggested that the vertebrate accumulation and its condition indicated possible predation and believe that hominid hunters were at least in part responsible.

It is too often assumed that *Homo erectus* is predominantly associated with occurrences of the Acheulian industry, at least in Africa (and Europe). However, the relationships are not yet that firmly and widely established. And the cultural associations are surely different in eastern Asia.

It is now known that the earliest manifestation of the Acheulian, at least in eastern Africa, is of earlier Quaternary age. At Olduvai (M. D. Leakey 1975) and in the adjacent westerly reaches of the Natron Basin (Isaac and Curtis 1974), the industry appears only slightly less than 1.5 m.y. ago, about 1.2 to 1.3 m.y. ago. In the Koobi Fora area of East Turkana *H. erectus* skeletal parts (R. E. F. Leakey 1976, R. E. F.

Leakey and Walker 1976) are known from this and the subsequent time range, up to about 1.2 m.y. or so, but there is apparently no evidence of Acheulian associations or even the presence of the industry. At Olduvai an Acheulian industry is present throughout the upper reaches of Bed II and is represented in Beds III, IV, and the Masek Beds. The distinctive Developed Oldowan industry is, however, also present throughout this same time range (M. D. Leakey 1975). Direct associations of *H. erectus* are unknown in Bed II—although the species is demonstrably present by Tuff II^D times—and only two occurrences in lower and upper Bed IV and probably one occurrence in the Masek Beds afford such associations. A recently reported hominid cranium with *H. erectus* features occurs in the Lake Ndutu area, probably in the upper (Norkilili Member) of the Masek Beds, in association with a presumed Acheulian industry (Mturi 1976, Clarke 1976). The only other direct association now known in eastern Africa is from the Gomboré II locality, at Melka Kunturé (upper Awash Valley, Ethiopia) (Chavaillon, Brahimi, and Coppens 1974).

In southern Africa direct associations of *H. erectus* and the Acheulian are still unknown, unless the *Homo* aff. *erectus* mandible (SK 15) from the Member B infilling at Swartkrans should prove to occur in an artifactual context that demonstrably represents that industry (C. K. Brain, pers. comm.).

Although diverse Acheulian occurrences of varying ages are recorded in the Maghreb (cf. Freeman 1975) only three confirm the presence of *H. erectus*. Thus, "older" Acheulian bifaces and spheroids, as well as some half dozen mammal species, were recovered with the Thomas "2" Quarry hominid cranium (Ennouchi 1972). However, the nature of this association remains unknown. The fragmentary hominid mandible from Littorina Cave, Sidi Abderrahman, occurred in a lenslike filling, along with a small variety of mammal remains, mostly ungulates, all suggesting hyena accumulation (Biberson 1956, 1961, 1964). A very substantial evolved Acheulian industry (stage VI of Biberson) occurred in a partially contiguous infilling and was essentially contemporaneous though presumably not directly associated with a part of the bone accumulation. The artifact assemblage is only about 250 pieces, but is quite diversified in implement types and often sophisticated in the refinement of some tool classes and the elaborateness of retouch.

Ternifine is usually considered to afford the most direct evidence for a *H. erectus*–Acheulian industry association in the Maghreb. The site was a small,

springfed lake; the substantial vertebrate fauna, hominid cranial remains, and Acheulian artifacts were all recovered from basal clays and overlying sands. However, there are no available details on these occurrences and either their spatial or their vertical relationships and associations with each other. The industry (Balout, Biberson, and Tixier 1967) includes a substantial diversity of large shaped tools and heavy-duty tools, as well as some small flakes and flake tools. The industry is most like an early phase (stage III of Biberson) of the Moroccan Acheulian; comparisons with assemblages from eastern Africa, particularly Olduvai, have still to be made.

The distribution, principal occurrences, relative and absolute ages, contexts and associations, principal characteristics, and distinctive features of the Acheulian industry have been discussed extensively by Howell and Clark (1963), M. D. Leakey (1971), Isaac (1975), Clark (1975), and Freeman (1975a). Isaac (1975) is a very useful, critical comparative evaluation of the Acheulian industry and its occurrences with other earlier Paleolithic industries in eastern Africa. There is, as has been often noted, an apparent "quantum change" (Isaac 1969, 1972a) with the full-blown appearance of its distinctive large cutting-edge implements (hand axes and cleavers). The association, or lack of association, of these distinctive tool forms with butchery practices and meat consumption remains one of the more perplexing problems in prehistoric archaeology. Clark (1975) has recognized four major "culture areas" of the Acheulian "technocomplex," two in continental Africa (open savanna versus closed savanna forest zones) and two in the Mediterranean basin (Maghreb versus Levant); presumably, western Europe, north of the Pyrenees, is a fifth area. He has included, for important reasons, the Developed Oldowan (but see M. D. Leakey 1971 for a contrary view) within the Acheulian, as one of its four principal variants or industrial facies. As Clark notes, each variant "is characterized by quantitative and qualitative differences in the composition of the lithic assemblages," and each "exhibits regional differences in the composition of individual aggregates as well as in technique, style, and mean measurements though categories and classes of artifacts remain the same." He considers, as I do, that differences between tool kits reflect largely (but not entirely) different sets of human activities, though they are still very difficult to ascertain. Clearly the central problem now is to seek to recognize and to decipher more completely the similarities and the differences between Oldowan industry settings, contexts, and associations and those of the Acheulian technocomplex.

For the moment, no direct cultural associations are known anywhere on the African continent with those hominids here considered to represent early representatives of *Homo sapiens*.

Paleoenvironmental Settings

The paleoenvironmental settings and adjustments of Cenozoic Hominidae have been the subject of much speculation and inference, especially in secondary literature. In particular, the open savanna world of emerging Hominidae has been frequently contrasted with the assumed evergreen forested habitat of Tertiary (and recent) African Pongidae. And, in Africa, the vicissitudes of the Quaternary world of Hominidae were once extrapolated from a theoretical paleoclimatic framework that has not been substantiated by subsequent, intensified field investigations. Only in recent years have paleoenvironmental settings and adjustments of Cenozoic Hominidae been sought to be established empirically. The evidence is still relatively scant and incomplete, but at least a beginning has been made.

Miocene Hominidae. Both Baringo area localities, Ngorora and Lukeino, that have yielded presumed Miocene Hominidae, are situated in fault basins related to a developing rift valley system. Ngorora (Bishop and Pickford 1975) was one of a series of three adjacent lakes in (presumably) closed basins. These were disjunctive fresh to alkaline water bodies, separated by emergent alluvial plains of low relief subject to weathering processes and erosion, which oscillated in depth and expanse, seasonally or under longer term climatic fluctuations, with evidence of rhythmic deposition and seasonal dessication. The vertebrates from the hominid-bearing member (C) include aquatic birds, reptiles, and mammals indicating an open or lightly vegetated habitat.

Lukeino (Pickford, 1975) was a fresh to weakly saline lake, with the usual aquatic life (molluscs, diatoms, algae, fish, crocodilians, and hippo), and lake margin and associated fossiliferous inflow channel deposits. The mammal fauna of the hominid-bearing member (A) is very diverse; it includes six carnivores (1 aquatic), 3 proboscidians, 2 hipparionids, and 2 rhinos, a chalicothere, a suid, hippo (abundant), 2 aardvarks, several rodents, and as many as 5 bovid tribes, the latter with both water-loving and open-country species. The composition of the fauna, as well as the diversity of preserved plant remains,

indicates substantial variety of habitats from lake margin, to wooded or bush fringed watercourses, and more open savanna grassland.

These temporally distinct, but geographically proximal situations in a similar rift-associated setting evidently provide only limited insight into habitat adjustments of protohominids. They do indicate the importance of permanent to semipermanent water bodies and a diversity of proximal natural plant and animal communities in relation to streamside and open country environments. The regional setting was one of emergent rift escarpments, horst and graben structure, and adjacent periodic volcanism.

In contrast, the Lothagam occurrence (upper Member C) reflects fluvial to fluviodeltaic deposition in a riverine setting with channel beds and associated overbank and flood basin sediments (Behrensmeyer 1976b). The paleoenvironmental setting was evidently substantially different from this area of south Turkana today; it is now subdesertic, and its major rivers (Turkwel and Kerio) drain into the southern Rudolf Basin only seasonally. This difference is also reflected in the diverse fauna of aquatic reptiles and mammals, which include 2 cercopithecoids, 4 carnivores, 4 proboscideans, 2 or 3 hipparionids, 2 rhinos, 2 suids, 2 hippos, a rare giraffid, an aardvark, an (nonvolant) anomalurid rodent, and over a dozen bovids representing 7 tribes all of which indicate substantial habitat diversity.

Australopithecus. Conceptions of the paleoenvironmental settings occupied by *Australopithecus* have naturally enough derived almost wholly from the inland plateau sinkhole and cavern infillings of South Africa in which remains of those hominids may occur so abundantly. However, the nature of those infillings and their former settings have only recently been elucidated through renewed study. The recovery of species of *Australopithecus* elsewhere in eastern Africa has also substantially expanded a previously very imperfect picture.

The regional geomorphological settings of the three principal situations sampled—the Makapan Valley in the northern Transvaal, the Blaubank Valley in the central Transvaal, and the Gaap escarpment in the northern Cape—are markedly different. The australopithecine-bearing infillings patently span a very substantial range of time—certainly over 1 m.y. on biostratigraphic grounds—and it is now apparent that such situations were not only locally but temporally dissimilar.

Butzer's insightful study (1974b, 1978) has shown that the Tau'ng setting for the type specimen of

A. africanus was subhumid to humid, with mean precipitation values estimated from 25% to 50% greater than present. The microvertebrates from the fissure filling, which may well derive from an earlier dry phase of colluviation, definitely reflect more xeric conditions; that assemblage is totally devoid of any large- to medium-sized species.

The Makapan Limeworks infilling, which is extraordinarily thick, yields an uncommonly diversified (over 70 species) and abundant mammal fauna (with *A. africanus*) in the lower gray breccia (Member 3) that overlies a basal sandy infilling (Member 2) of fluviatile origin. The total infilling must encompass a very long time span, perhaps several hundred thousand years. The origin of the rich bone beds, which differ from the overlying pink and brown colluvial sediments (Members 4 and 5) in faunal representation (though *A. africanus* is still present) as well as sedimentological characteristics, is unresolved; a combination of colluvial and fluvial agencies, and the accidental falling or washing in of animal carcasses or body parts is reasonably postulated (Butzer 1978, pers. comm.). The paleoenvironment, to judge from the vertebrate assemblage, was at least as much and probably even more mesic than at present, with diverse wooded, bush, and more open habitats. The lower, or Unit B, infilling with *A. africanus* at Sterkfontein has been considered to have accumulated under essentially dry, "interpluvial" conditions. However, the sediments reflect colluviation and fairly constant washing effects, and among some 50 mammal species represented, the bovids (Vrba 1975) include a number of bush-loving and water-dependent species, in addition to some 50% antilopine and alcelaphine species that prefer open country. Thus, there is evidence there too of this hominid species in an open-country situation more mesic than often considered heretofore.

In the Shungura and Hadar formations (Ethiopia), *A. africanus/afarensis* occurs in other paleoenvironmental settings. The Hadar Formation is a part of the west central Afar Sedimentary Basin and represents lacustrine and lake margin depositional environments and associated inflowing stream channels (Taieb et al. 1976). Vertebrates, including *A. afarensis* (in lower and middle members), occur largely in lake margin sands and in channel fills. The very well-preserved and diverse vertebrate assemblage strongly suggests a variety and complexity of habitats.

In the lower Omo Basin *A. africanus* occurs in relation to riverine fluviatile, flood plain, and braided streams in the middle through (lower) upper members of the Shungura Formation (Howell and Coppens 1974, 1976a,b). Faunal diversity is great throughout these horizons and indicates a substantial variety of habitats. Vegetation in the earlier time range (2.5 m.y.), on pollen evidence (Bonnefille 1976), "included closed and/or open woodland, tree/shrub grassland, grassland, and some shrub thicket and shrub steppe" (Carr 1976). By contrast, 0.5 m.y. later, toward a time of establishment of a braided stream regime, total diversity of plant taxa is markedly reduced, species of mesic habitat affiliation decrease markedly, as does the percentage of pollen from woody plants, and the proportion of pollen with plains habitat affiliation increases markedly. These changes suggest "more xerophytic plant community types of grassland, tree/shrub grassland, shrub thicket, and shrub steppe" (Carr 1976). This rather profound environmental change is wholly corroborated by the microvertebrate fossil record (Jaeger and Wesselman 1976). A robust australopithecine (*A. boisei*) appears for the first time in the Omo succession at about the time of this environmental change, as does also an early species of genus *Homo*.

In the Koobi Fora area (East Turkana) only a few specimens, probably no more than three or four, which have been found both just above and below the "KBS" tuff, can be very tentatively attributed to *A. africanus*. These would appear to derive from both fluviatile channel and lake margin situations.

Robust australopithecines (*A. boisei*) are common in the Koobi Fora Formation. Behrensmeyer (1976a) has suggested that *A. boisei* was more frequently a denizen of fluviatile habitats, including fringing forest and adjacent bush, whereas *A. boisei* and one or more species of *Homo* occurred about equally frequently in lake margin settings. However, the frequency of occurrence may have been more or as much a reflection of hominid density within depositional members, rather than as a direct reflection of favored local paleoenvironments (table 10.10).

In *sub-KBS Tuff* time, *A. boisei* is about one-fourth as frequent as an early species of *Homo*, the paleoenvironmental setting having been largely, if not wholly lake margin.

In *sub-Koobi Fora Tuff* time, initially, hominids are apparently recovered only in lake margin situations. *A. boisei* is half again as frequent as is a species of *Homo*, but the former species is overall nearly twice as common as the latter. Later, the total hominid sample is more than twice as great and both lake margin and fluviatile depositional environments are sampled. Here *A. boisei* is about six times as common as *Homo* (*erectus*) in the fluviatile environments, whereas there is nearly equal representation

Table 10.10 Overall frequency of hominid taxa and their inferred paleo-environmental relationships in the Koobi Fora succession.

Horizon	Australopithecus boisei	Homo spp.
Sub-Chari		
Identifiable (1974)	15 (9 + 6)[a]	6 (4 + 2)
Fluviatile[b]	12	4
Lake margin[b]	2	1
Sub-Koobi Fora		
Upper		
Identifiable (1974)	21 (16 + 5)	11 (7 + 4)
Fluviatile[b]	12	2
Lake margin[b]	5	4
Lower		
Identifiable (1974)	9 (4 + 5)	5 (3 + 2)
Fluviatile[b]	—	—
Lake margin[b]	6	4
Sub-KBS		
Identifiable (1974)	4	15
Fluviatile[b]	—	—
Lake margin[b]	2	9
Total identifiable	49	37

a. Cranial and postcranial parts in parentheses.
b. Fluviatile and lake margin associations after A. K. Behrensmeyer.

of each taxon in lake margin situations. At about that time a pollen spectrum (Bonnefille 1976) reveals a pronounced representation of allochthonous pollen (over 50% not represented in the East Turkana area at present), a high percentage of woody species, and a very high proportion of mesic compared to plains pollen. The spectrum suggests a more humid or cooler (or both) climate with substantial density of woodland and bush proximal to the lake; this climate is far different from the present subdesertic steppe of the area.

In *sub-Chari Tuff* time, hominids are well represented, though perhaps less common or less well-preserved than previously. *A. boisei* is more than twice as common as *Homo (erectus)*, and is found three times more often in fluviatile as well as lake margin situations; *H. erectus*, as expected, is most frequently found in fluviatile situations.

In East Turkana *A. boisei* was initially rare (during sub-KBS Tuff time); later (throughout sub-Koobi Fora and even sub-Chari time), *A. boisei* seems to predominate over *Homo* spp. Although paleoenvironments remain poorly known the increasing dominance of *A. boisei* appears to have occurred in a markedly mesic situation. Whether in spite of or because of a shift to increasingly fluviatile depositional regimes, this situation contrasts interestingly with Olduvai Gorge. At Olduvai (see below) *A. boisei* occurs throughout the earlier part of the succession,

persisting toward the summit of Bed II; however *A. boisei* is always infrequent and is about a third as common as *Homo* spp., presumably regardless of depositional situation. The survival and adjustments of this species, which evidently became extinct rather more than 1 m.y. ago in the Olduvai Basin, are still ill-appreciated, as is also true in the East Turkana instance, in respect to apparent environmental vicissitudes during the earlier Quaternary in East Africa.

In southern Africa a robust australopithecine, *A. robustus/crassidens,* persists after *A. africanus,* with which it has never been found there in direct association. It has been suggested that the robust species was once associated with substantially more mesic habitats than its antecedent species, *A. africanus.* This assumption does not, however, appear to be correct. The Swartkrans infilling appears to have been rapid and subsequently extensive, with periodic flooding and ponding. The vertebrate fauna is not only substantially different in species composition from the demonstrably older australopithecine-bearing breccias (at Makapan Limeworks and nearby Sterkfontein), but yields a much higher percentage of bovid species favoring more open grassland habitats (Vrba 1975). At this locality a robust australopithecine is almost 50 times as frequent as a species of *Homo*. The depositional circumstances of the type *A. robustus* situation at Kromdraai (B) are still to be elucidated; however, among the few reported vertebrate taxa (20) there is again a very high percentage of open-country bovids (Vrba 1975, 1976).

Homo habilis. A species of *Homo* was first found directly associated and hence contemporaneous with *Australopithecus* at the Swartkrans locality, which was probably less mesic overall than older cavern infillings yielding *A. africanus.*

Olduvai Gorge was the first locality in Africa to yield really substantial evidence of a species of *Homo,* designated *Homo habilis* by L. S. B. Leakey, Tobias, and Napier (1964), antedating *H. erectus. H. habilis* was penecontemporaneous and evidently sympatric with the robust australopithecine, *A. boisei,* but the latter species is always much more rare in the fossil record. During this period Olduvai was a closed lake basin at the western foot of active volcanic highlands (Hay 1976). The lake was shallow and saline to alkaline with sodium carbonate–bicarbonate-rich waters. All hominids and 18 known hominid occupation sites in Bed I are clustered about the eastern lake margin, frequently in swamps or at marsh edges, and probably next to the inlets of freshwater streams inflowing from the uplands. The Serengeti is now a mosaic of grassland species with *Commiphora* scrub and scattered *Acacia*. The climate was then also semiarid, but the avifauna and microvertebrate fauna, particularly rodents, indicate more mesic conditions than those of the present. However, there were fluctuations within this pattern (Cerling, Hay and O'Neil 1977), with moister conditions documented (by the rodent fauna and palynology, R. Bonnefille, pers. comm.) in the intervals below Tuffs I^B and I^D, with more woody species (including montane *Podocarpus*) present; drier open steppic conditions subsequently prevailed below Tuff I^F. The significance of these shifts in local paleoenvironments—from drier to moister and back to drier with attendant expansion-contraction-expansion of the Olduvai lake—for hominids, occupation sites, and site residues have still to be elucidated. Arid to semiarid conditions persisted into Bed II times; the lake remaining without an outlet, highly saline with trona formation, zeolitization, strong evapotranspiration with caliche and dolomite formation, and evidence of extensive aeolian activity in sedimentation suggestive of a reduced vegetation mat. *Homo habilis* is present to below Tuff II^B and associated with Oldowan occupation sites mostly near lake margins. Dry conditions with substantial aeolian activity is evidenced toward the end of this time, about 1.65 m.y. ago, in the Lemuta Member which divides lower from middle-upper Bed II (Hay 1976).

At East Turkana this *Homo* species occurs, apparently almost exclusively, in lake margin situations above and below the KBS Tuff(s)—that is, in the Lower Member and the inferior portion of the Upper Member, Koobi Fora Formation. Artifact occurrences, essentially of Oldowan type, without fossil hominids directly in association, occur in and below the KBS Tuff usually in tuff-choked distributary channels proximal to a lake margin setting (Isaac, Harris, and Crader 1976, Isaac 1976b).

In the Shungura Formation an occurrence of this species is documented in uppermost Member G (G-28) at a time broadly equivalent to its earliest known presence at Olduvai. The depositional environment was the poorly vegetated shore of a shallow saline lake (Boaz and Howell 1977).

Homo erectus. The most direct evidence of paleoenvironmental settings of *Homo erectus* derive from occurrences of this species at Olduvai Gorge and East Turkana.

H. erectus first occurs at Olduvai in the highest sediments of Bed II, around the level of Tuff II^D (Hay 1976, M. D. Leakey 1978). The Olduvai lake was then markedly smaller, about a third of its former

extent, and its position shifted eastward. Approximately 1.2 m.y. ago it disappeared altogether as a consequence of faulting and was replaced by an alluvial plain with only local and ephemeral ponds and areas of marshlands. Developed Oldowan (B) and Acheulian industries coexisted at this time, the former for the most part in fluviolacustrine situations, and the latter almost exclusively in more distal, inland settings. *H. erectus* remains, not really common, appear to be more closely associated with stream channels and their infillings, but direct association with artifact occurrences are scant. This species persists through Beds III and IV, up to about 0.7 m.y. ago. The depositional environment was that of alluvial plains of relatively gentle slope with drainage predominantly from the western fluviatile sources in stream channels, and this is almost wholly the case for archeological occurrences as well, whether Acheulian or Developed Oldowan (C), the bulk of which are related to the main drainageway. As in the case of uppermost Bed II the paleoclimate appears to have been semiarid and overall relatively dry. *H. erectus,* or a derivative, may also be represented in the younger Masek Beds, which were accumulated on a gentle alluvial plain in association with streams draining into a small though perennial saline lake or pond, under drier conditions, with enhanced aeolian activity, probably like the present. Rare hominid fossils and rare Acheulian occurrences are generally associated with the main drainageway.

In East Turkana *H. erectus* occurs in the Upper Member of the Koobi Fora Formation, and is known from the upper part of the sub-Koobi Fora Tuff unit apparently through the sub-Chari unit. Nearly all these occurrences are related to fluviatile depositional environments. Hominid occupation sites, represented by the Karari industry, occur in the middle part of this succession, for the most part in relation to areally localized tuff complexes, that of the Okote Tuff, Karari area, and the Lower-Middle Tuff, Ileret area. In the eastern Karari area, the occurrences are associated in and near stream channels and floodplains; in the northern Ileret area, occurrences are associated with low-energy floodplains and delta floodplains. At one locality there is direct evidence of the close association of both *Homo* cf *erectus* and *Australopithecus boisei* in a horizon with a nearby occupation site that yields artifacts of Oldowan affinity (Harris and Herbich 1978, Harris and Isaac 1976).

In southern Africa *Homo* aff. *erectus* occurs in the middle member (B) of the Swartkrans infilling. No detailed paleoclimatic evidence is forthcoming, but

Vrba's bovid researches (1975) do not show any marked differences from the lower member infilling, and rather more than 70% of that element of the fauna comprises antilopines and alcelaphines. From the perhaps penecontemporaneous or slightly older middle breccia (Extension site) of Sterkfontein those open-country species comprise over 80% of the bovid fauna. Preliminary results of pollen analysis suggest "that the Middle Breccia was formed when the surrounding area was a vast, open grassland, and probably when the climate was drier than at present" (Horowitz 1975).

Elsewhere in Africa, notably the Maghreb, the paleoenvironmental settings of *H. erectus* remain almost unknown, although geological contexts and vertebrate associations are somewhat understood. Very broad and only limited inferences may be afforded by the megafaunal associations and the still very scant microfaunal assemblages. No locality has afforded palynological spectra as yet. Moist, cool paleoclimates and still colder conditions have been suggested for the Amirian and Tensiftian stages in Morocco.

Homo sapiens. There is scarcely any direct paleoclimatic evidence available from African localities yielding earlier representatives of *Homo sapiens.* At the Hopefield locality Butzer (1973b) inferred predominantly fluvial depositional processes, and he cites palynological studies that indicate an open bushveld vegetation including aquatic plants, but with scant macchia. In the geologically younger Florisbad situation there is evidence of dry and warm conditions initially, with Karroo-like vegetation at the time of the presence of *H. sapiens* before progressively wetter and cooler climates in the late Quaternary (van Zinderen Bakker 1957).

Predictably, the most substantial and sound body of African paleoclimatic evidence is that for the end-Quaternary (and, of course, the Holocene). The former is relevant to the paleoenvironmental settings of *H. sapiens afer* populations, although paleoclimatic evidence is general and site-specific data is almost unknown for particular occurrences of either that subspecies or its antecedent subspecies. The overall evidence and its implications have been discussed recently by Street and Grove (1976), Livingstone (1975), and Butzer et al. (1972); Bakker and Butzer (1973) and Bakker (1976) have reviewed the evidence for southern Africa. Burke, Durotyne, and Whiteman (1971) and Williams (1975) have emphasized the occurrence of one or more intervals of substantial and prolonged aridity; and Hamilton (1976) has discussed the implications of distributions of forest plants and animals. The African Quaternary

still lacks a paleoclimatic framework, but the evidence now available for the late and post-Quaternary indicates a pattern that could have recurred cyclically over the past several million years.

During late Last Glacial times (~21,000–12,500 B.P.) intertropical Africa experienced widespread aridity, with attendant expansion of desertic and semidesertic conditions, far south of present limits, marked reduction in distributions and continuity of equatorial lowland forest, and minimal lake levels in closed basins. Mean annual precipitation may have been reduced from 50% to 90% (East Africa) to 15% to 20% (Sahel) of present values, and mean annual temperature decreased by at least 3°C, and perhaps more.

In contrast, northern and southern Africa experienced more humid and cooler climatic conditions, with expanded lake levels, mean annual precipitation perhaps over 120% of present values, and mean annual temperature reduced by 6°C.

The paleoclimatic implications, according to Street and Grove (1976) are that the meridional temperature gradient was increased (particularly in the northern hemisphere); the northeast trades were more active; surface temperatures of upwelling areas of the Indian and Atlantic oceans were 6–8°C cooler; and the Atlantic oceanic polar front was substantially displaced southward (and the Mediterranean winter rainfall belt displaced over northern Africa). By implication, most African temperatures were 4–8°C cooler, global moisture transport was substantially decreased, climatic belts were contracted toward the meteorological equator, and the Intertropical Convergence Zone shifted south of the geographic equator.

During late Glacial times (~12,500–10,000 B.P.), with warming ocean temperatures, equatorial forest vegetation was reestablished extensively across intertropical Africa, and extended desertic southern Saharan areas were invaded by grassland and various scrub or thornbush savanna plant communities. These conditions were accentuated within the next several thousand years with markedly expanded lake levels, further restriction of desertic areas, and precipitation enhanced by 150–165% (East Africa) to as much as 200–400% (Saharan Zone). These conditions are also known to have prevailed in intertropical Africa previously during Late Quaternary times before the late Last Glacial. In the inland plateau of the southern subcontinent, open shrub (Karroo) type vegetation is documented for warmer, drier intervals and closed grass veld vegetation for cooler, moister intervals within this general time range (cf. van Zinderen Bakker 1976).

Along the southern coast, Last Glacial times record decreased temperatures and variably increased precipitation with evidence in places of substantial forest extension, particularly in regard to notably lowered sea levels, but elsewhere extensive open grassland paleoenvironments (well-documented by mammalian species and associations; R. G. Klein, pers. comm.); open grassland preceded the establishment of evergreen forest in late glacial and postglacial times. Klein's work (pers. comm.) demonstrates, in the southern Cape area at least, changes from open → closed → open vegetational communities, on the basis of bovid frequencies, from the onset through Last Interglacial times into early Last Glacial times—presumably when *Homo sapiens rhodesiensis* populations were present in the southern subcontinent.

The author is most appreciative of the hospitality and assistance afforded by many colleagues, over some two decades, at the following institutions in which specimens of African Cenozoic hominid remains are housed: Department of Anatomy, University of Cape Town; National Museum, Bloemfontein; Transvaal Museum, Pretoria; Department of Anatomy, University of the Witwatersrand, Johannesburg; National Museum of Kenya, Nairobi; National Museum, Addis Ababa; Muséum National d'Histoire Naturelle, Paris; Musée de l'Homme, Paris; Institut de Paléontologie Humaine, Paris; British Museum (Natural History), London; Cleveland Museum of Natural History, Cleveland.

The support of the Wenner-Gren Foundation for Anthropological Research and the National Science Foundation is gratefully acknowledged. All illustrations were prepared by Judith Ogden, to whom the author is particularly grateful.

References

Abel, W. 1931. Kritische Untersuchungen über *Australopithecus africanus* Dart. *Morph. Jhbk.* 65:539–640.

Adler, M. 1967. *The Difference of Man and the Difference it Makes*. 395 pages. New York: Holt, Rinehart and Winston.

Adloff, P. 1931. Über die Ursprung des Menschen im Lichte der Gebissforschung. *Schr. d. Königsburg. Gelehr Ges.*, N. Ki, 8:299–312.

———. 1932. Das Gebiss von *Australopithecus africanus* Dart. Einige ergänzende Bemerkungen zum Eckzahn problem. *Zts. f. Anat. u. Entwlgsgesch.* 97:145–156.

Aguirre, E. 1970. Identificacion de "*Paranthropus*" en Makapansgat. *Cronica XI Congreso Nacional de Arquelogia, Madrid,* pp. 98–124.

Andrews, P., and I. Tekkaya. 1976. *Ramapithecus* in Kenya and Turkey. In IXe *Congrès, Union Internationale des Sciences Préhistoriques et Protohistoriques, Nice, September 1976.*

Andrews, P., and A. Walker. 1976. The primate and other fauna from Fort Ternan, Kenya. In *Human origins: Louis Leakey and the East African evidence*. Menlo Park, California: W. A. Benjamin, pp. 278–304.

Anthony, J. 1966. Premières observations sur le moulage endocranien des hommes fossiles du Jebel Irhoud (Maroc). *C. R. Acad. Sci.* (Paris) 262-D:556–558.

Armelagos, G. J. 1964. A fossilized mandible from near Wadi Halfa, Sudan. *Man,* 64:12–13.

Arambourg, C. 1954. L'hominien fossile de Ternifine (Algérie). *C. R. Acad. Sci.,*(Paris) 239:893–894.

———. 1962. Les faunes mammalogiques du Pléistocène circum mediterranéen. *Quaternaria* 6:97–110.

———. 1963. Le gisement de Ternifine: *l'Atlanthropus* de Ternifine. *Arch. l'Inst. de Paléont. humaine,* Paris, 32:37–190.

———, and P. Biberson. 1955. Découverte de vestiges humains dans la carrière de Sidi Abderrahman, près de Casablanca. *C. R. Acad. Sci.* (Paris) 240:1661–1663.

———, and P. Biberson. 1956. The fossil human remains from the Paleolithic site of Sidi Abderrahman (Morocco). *Am. J. Phys. Anthrop.* 14:467–490.

———, and Y. Coppens. 1967. Sur la découverte dans le Pléistocène inférieur de la vallée de l'Omo (Ethiopie) d'une mandibule d'australopithécien. *C. R. Acad. Sci.* (Paris) 265-D:589–590.

———, and Y. Coppens. 1968. Découverte d'un australopithécien nouveau dans les gisements de l'Omo (Ethiopie). *So. Afr. J. Sci.,* 64:58–59.

Archibald, J. D., C. O. Lovejoy, and K. G. Heiple. 1972. Implications of relative robusticity in the Olduvai metatarsus. *Am. J. Phys. Anthrop.* 37:93–96.

Arkell, A. J., D. M. A. Bate, L. H. Wells, and A. D. Lacaille. 1951. The Pleistocene fauna of two Blue Nile sites. *Fossil Mammals of Africa,* 2. London: British Museum (Natural History).

Aronson, J. L., T. J. Schmitt, R. C. Walter, M. Taieb, J. J. Tiercelin, D. C. Johanson, C. W. Naeser and A. E. M. Nairn. 1977. New geochronologic and paleomagnetic data for the hominid-bearing Hadar Formation of Ethiopia. *Nature,* 267:323–327.

Bada, J. L., and L. Deems. 1975. Accuracy of dates beyond the ^{14}C dating limit using the aspartic acid racemisation reaction. *Nature* 255:218–219.

Bada, J. L., and R. Protsch. 1973. Racemisation reaction of aspartic acid and its use in dating fossil bones. *Proc. Nat. Acad. Sci.* 70:1331–1334.

Bada, J. L., R. Protsch, and R. A. Schroeder. The racemisation reaction of isoleucine used as a palaeotemperature indicator. *Nature* 241:394–395.

Bada, J. L., R. A. Schroeder, R. Protsch, and R. Berger. 1974. Concordance of collagen based radiocarbon and aspartic-acid racemization ages. *Proc. Nat. Acad. Sci.* 71:914–917.

Bakker, E. M. van Zinderen. 1957. A pollen analytical investigation of the Florisbad deposits (South Africa). *Proc. Pan-Afr. Congr. Prehist., 3rd, Livingstone, 1955,* ed. J. D. Clark. London: Chatto and Windus, pp. 56–67.

———. 1976. The evolution of Late Quaternary Palaeoclimates of southern Africa. *Palaeoecology of Africa* 9. 160–202.

———, and K. W. Butzer. 1973. Quaternary environmental changes in southern Africa. *Soil Science* 116:236–248.

Balout, L. 1965a. Données nouvelles sur le probleme du Moustérien en Afrique du Nord. In *Actas del V Congreso Panafricano de Prehistoria y de Estudio del Cuaternario,* vol. 1. Santa Cruz de Tenerife: Museo Arqueologico, pp. 137–143.

———. 1965b. Le Moustérien du Maghreb. *Quaternaria* 7:43–58.

———. 1970. L'industrie neandertalienne du Djebel Irhoud (Maroc). *Fundamenta,* Reiha A, Band 2 (A. Rust Festschrift):57–60.

———, P. Biberson, and J. Tixier. 1967. L'Acheuléen

de Ternifine (Algérie). Gisement de l'Atlanthrope. *L'Anthropologie* 71:217–237.

Beaumont, P. B. 1973. Border Cave—a progress report. *So. Afr. J. Sci.* 69:41–46.

———, and A. K. Boshier. 1972. Some comments on recent findings at Border Cave, northern Natal. *So. Afr. J. Sci.* 68:22–24.

———, and J. C. Vogel. 1977. On a new radiocarbon chronology for Africa south of the equator. *African Studies,* 31:65–89, 155–182.

Behrensmeyer, A. K. 1975. Taphonomy and paleoecology in the hominid fossil record. *Yrbk. Phys. Anthrop.* 19:36–49.

———. 1976a. Fossil assemblages in relation to sedimentary environments in the East Rudolf succession. In *Earliest man and environments in the Lake Rudolf basin,* ed. Y. Coppens, F. C. Howell, G. Ll. Isaac, and R. E. F. Leakey. Chicago: University of Chicago Press, pp. 383–401.

———. 1976b. Lothagam Hill, Kanapoi, and Ekora: a general summary of stratigraphy and faunas. In *Earliest man and environments in the Lake Rudolf basin.* ed. Y. Coppens, F. C. Howell, G. Ll. Isaac, and R. E. F. Leakey. Chicago: University of Chicago Press, pp. 163–170.

Bennejeant, C. 1936. La dentition de *l'Australopithecus africanus* (Dart). *Mammalia* 1:8–14.

———. 1953. Les dentures temporaires des Primates. *Bull. Mém. Soc. d'Anthrop.* (Paris), sér. 10, 4:11–44.

Berggren, W. A. 1969. Cenozoic chronostratigraphy, planktonic foramineral zonation and the radiometric time-scale. *Nature* 224:1072–1975.

———. 1971. Tertiary boundaries. In *Micropaleontology of Oceans,* ed. B. F. Funnell and W. R. Riedel. Cambridge: Cambridge University Press, pp. 693–809.

———. 1972. A Cenozoic time-scale—some implications for regional geology and paleobiogeography. *Leithaia* 5:195–215.

———. 1973. The Pliocene time scale: calibration of planktonic foraminiferal and calcareous nannoplankton zones. *Nature* 243:391–397.

———, and J. A. Van Couvering. 1974. The late Neogene. Biostratigraphy, geochronology and paleoclimatology of the last 15 million years in marine and continental sequences. *Palaeogeog., Palaeoclim., Palaeoecol.* 16:1–216.

Biberson, P. 1956. Le gisement de l'Atlanthrope de Sidi Abderrahman (Casablanca). *Bull. d'Archéol. Marocaine* 1:39–91.

———. 1961. Le cadre paléogeographique de la préhistoire du Maroc atlantique. *Publ. Serv. Antiq. Maroc,* Rabat, 16:1–235.

———. 1963. Quelques précisions sur les classifications du Quaternaire marocain. *Bull. Soc. Géol. France,* sér. 7, 5:607–616.

———. 1964. La place des hommes du Paléolithique marocain dans la chronologie du Pléistocène atlantique. *L'Anthropologie* 68:475–526.

———. 1971. Essai de rédefinition des cycles climati-

ques du Quaternaire continental du Maroc. *Bull. Assoc. Franc. l'étude Quatern.,* (Paris) 1(26):3–13.

Biegert, J., and R. Mauer. 1972. Rumpfskelettlänge, Allometrien und Körperproportionen bei catarrhinen Primaten. *Folia Primat.* 17:142–156.

Bilsborough, A. 1972. Anagenesis in hominid evolution. *Man* 7:481–483.

———. 1973. A multivariate study of evolutionary change in the hominid cranial vault and some evolution rates. *J. Hum. Evol.* 2:387–403.

Bishop, W. W. 1967. The later Tertiary in East Africa: volcanics, sediments and faunal inventory. In *Background to evolution in Africa,* ed. W. W. Bishop and J. D. Clark. Chicago: University of Chicago Press, pp. 31–56.

———. 1971. The late Cenozoic history of East Africa in relation to hominoid evolution. In *Late Cenozoic Glacial Ages,* ed. K. K. Turekian. New Haven: Yale University Press, pp. 493–527.

———. 1972. Stratigraphic succession "versus" calibration in East Africa. In *Calibration of hominoid evolution,* ed. W. W. Bishop and J. A. Miller. Edinburgh: Scottish Academic Press, pp. 219–246.

———. 1973. The tempo of human evolution. *Nature* 244:405–409.

———, and G. R. Chapman. 1970. Early Pliocene sediments and fossils from the northern Kenya Rift Valley. *Nature* 226:914–918.

———, G. R. Chapman, A. Hill, and J. A. Miller. 1971. Succession of Cainozoic vertebrate assemblages from the northern Kenya Rift Valley. *Nature* 233:389–394.

———, M. Pickford, and A. Hill. 1975. New evidence regarding the Quaternary geology, archeology and hominids of Chesowanja, Kenya. *Nature* 258:204–208.

———, and M. H. Pickford. 1975. Geology, fauna and palaeoenvironments of the Ngorora Formation, Kenya Rift Valley. *Nature* 254:185–192.

Black, D. 1927. On a lower molar hominid tooth from the Chou Kou Tien deposits. *Palaeont. Sinica* 7D:1–28.

———. 1931. On an adolescent skull of *Sinanthropus pekinensis* in comparison with an adult skull of the same species and with other hominid skulls, recent and fossil. *Palaeont. Sin.,* ser. D, vol. 7, fasc. 2:1–144.

———. 1933a. On an endocranial cast of the adolescent *Sinanthropus* skull. *Proc. Roy. Soc. London,* Ser. B, 112:263:276.

———. 1933b. The brain cast of *Sinanthropus*—a review. *J. Comp. Neurol.* 57:361–368.

———. 1934. On the discovery, morphology and environment of *Sinanthropus pekinensis.* *Phil. Trans. Roy. Soc. London,* 223-B:57–120.

———, T. de Chardin, C. C. Young, and W. C. Pei. 1933. Fossil man in China. The Choukoutien cave deposits with a synopsis of our present knowledge of the late Cenozoic of China. *Mem. Geol. Surv. China,* ser. A, 11:1–158.

Boaz, N. T., and F. C. Howell. 1977. A gracile hominid cranium from upper Member G of the Shungura Formation, Ethiopia. *Am. J. Phys. Anthrop.* 46:93–108.

Bonnefille, R. 1976. Implications of pollen assemblage

from the Koobi Fora formation, East Rudolf, Kenya. *Nature* 264:403–407.

————, F. H. Brown, J. Chavaillon, Y. Coppens, P. Haesaerts, J. de Heinzelin, and F. C. Howell. 1973a. Situation stratigraphique des localités à Hominidés des gisements Plio-Pléistocènes de l'Omo en Ethiopie (membres de base A, B, C, D, et J). *C. R. Acad. Sci.* (Paris) 276:2781–2784.

————, F. H. Brown, J. Chavaillon, Y. Coppens, P. Haesaerts, J. de Heinzelin, and F. C. Howell. 1973b. Situation stratigraphique des localités à Hominidés des gisements Plio-Pléistocènes de l'Omo en Ethiopie (membres E, F, G, et H). *C. R. Acad. Sci.* (Paris) 276:2879–2882.

Boswell, P. G. H. 1935. Human remains from Kanam and Kanjera, Kenya Colony. *Nature* 135:371.

Brace, C. L. 1967. *The stages of human evolution.* Englewood Cliffs, N.J.: Prentice-Hall.

————. 1973. Sexual dimorphism in human evolution. *Yrbk. Phys. Anthrop.* (1972) 16:50–68.

Brain, C. K. 1958. The Transvaal ape-man-bearing cave deposits. *Transv. Mus.* (Pretoria) *Mem.* 11:1–131.

————. 1968. Who killed the Swartkrans ape-men? *So. Afr. Mus. Assoc. Bull.* 9:127–139.

————. 1970. New finds at the Swartkrans australopithecine site. *Nature* 225:1112–19.

————. 1972. An attempt to reconstruct the behaviour of australopithecines: the evidence for interpersonal violence. *Zool. Afr.* 7:379–401.

————. 1973. The significance of Swartkrans. *J. So. Afr. Biol. Soc.* 13:7–23.

————. 1975. An interpretation of the bone assemblage from the Kromdraai australopithecine site, South Africa. In *Paleoanthropology: morphology and paleoecology,* ed. R. H. Tuttle. The Hague: Mouton, pp. 225–243.

————. 1976a. Some principles in the interpretation of bone accumulations associated with man. In *Human origins: Louis Leakey and the East African evidence,* ed. G. Ll. Issac and E. R. McCown. Menlo Park, California: W. A. Benjamin, pp. 97–116.

————. 1976b. A re-interpretation of the Swartkrans site and its remains. *So. Afr. J. Sci.* 72:141–146.

————. 1978. Some aspects of the South African australopithecine sites and their bone accumulations. In *Early African Hominids,* ed. C. J. Jolly. London: Duckworth (in press).

————, E. S. Vrba, and J. T. Robinson. 1974. A new hominid innominate bone from Swartkrans. *Ann. Transv. Mus.* 29(5):55–63.

Briggs, L. C. 1955. The Stone Age races of northwest Africa. *Bull. Am. Sch. Prehist. Res.* (Peabody Museum, Harvard) 18.

Brock, A. and R. L. Hay. 1976. The Olduvai event at Olduvai Gorge. *Earth and Planentary Science Letters,* 29:126–130.

Brock, A. and G. Ll. Isaac. 1974. Palaeomagnetic stratigraphy and chronology of hominid-bearing sediments east of Lake Rudolf, Kenya. *Nature,* 247:344–348.

Brock, A., P. L. McFadden and T. C. Partridge. 1977. Preliminary palaeomagnetic results from Makapansgat

and Swartkrans. *Nature,* 266:249–250.

Broom, R. 1925a. Some notes on the Taung skull. *Nature* 115:569–571.

————. 1925b. On the newly discovered South African man-ape. *Nat. Hist.* 25:409–418.

————. 1929a. Note on the milk dentition of *Australopithecus. Proc. Zool. Soc. London* 1928:85–88.

————. 1929b. The Transvaal fossil human skeleton. *Nature* 123:415–416.

————. 1938a. The Pleistocene anthropoid apes of South Africa. *Nature* 142:377–379.

————. 1938b. Further evidence on the structure of the South African Pleistocene anthropoids. *Nature* 142:897–899.

————. 1939. The dentition of the Transvaal Pleistocene anthropoids, *Plesianthropus* and *Paranthropus. Ann. Transv. Mus.* 19:303–314.

————. 1941. The milk molars of man and the anthropoids. *So. Afr. Dent. J.* 15:314–316.

————. 1947. The upper milk molars of the ape-man, *Plesianthropus. Nature* 159:602.

————. 1949. Thumb of the Swartkrans ape-man. *Nature* 164:841–842.

————. 1950. The genera and species of the South African fossil ape men. *Am. J. Phys. Anthrop.* 8:1–13.

————, and J. T. Robinson. 1949. A new type of fossil man. *Nature* 164:322–323.

————, and J. T. Robinson. 1950a. Man contemporaneous with the Swartkrans ape-man. *Am. J. Phys. Anthrop.* 8:151–156.

————, and J. T. Robinson. 1950b. Notes on the pelves of the fossil ape-men. *Am. J. Phys. Anthrop.* 8:489–494.

————, and J. T. Robinson, 1950c. See Broom, Robinson and Schepers, 1950.

————, and J. T. Robinson. 1952. Swartkrans ape-man, *Paranthropus crassidens. Transv. Mus.,* Pretoria, *Mem. 6.*

————, J. T. Robinson, and G. W. H. Schepers. 1950. Sterkfontein ape-man *Plesianthropus. Transv. Mus.* (Pretoria) *Mem.* 4, pp. 1–117.

————, and G. W. H. Schepers. 1946. The South African fossil ape-men, the Australopithecinae. *Transv. Mus.* (Pretoria) *Mem.* 2, pp. 1–272.

Brothwell, D. R. 1963. Evidence of early population change in central and southern Africa: doubts and problems. *Man* 63:101–104.

Brown, F. H. 1972. Radiometric dating of sedimentary formations in the lower Omo valley, Ethiopia. In *Calibration of hominoid evolution,* ed. W. W. Bishop and J. A. Miller. Edinburgh: Scottish Academic Press, pp. 272–287.

————, and W. P. Nash. 1976. Radiometric dating and tuff mineralogy of Omo Group deposits. In *Earliest man and environments in the Lake Rudolf Basin,* ed. Y. Coppens, F. C. Howell, G. Ll. Isaac, and R. E. F. Leakey. Chicago: University of Chicago Press, pp. 50–63.

————, and R. T. Shuey. 1976. Magnetostratigraphy of the Shungura and Usno Formations, lower Omo valley, Ethiopia. In *Earliest man and environments in the Lake Rudolf Basin,* ed. Y. Coppens, F. C. Howell, G. Ll. Isaac, and R. E. F. Leakey. Chicago: University of Chicago Press, pp. 64–78.

———, F. C. Howell, and G. G. Eck. 1978. Observations on problems of correlation of late Cenozoic hominid-bearing formations in the north Rudolf basin. In *Geological background to fossil man,* ed. W. W. Bishop. London: Geological Society of London. (in press).

Burke, K., A. B. Durotyne, and A. J. Whiteman. 1971. A dry phase south of the Sahara 20,000 years ago. *W. Afr. J. Archaeol.* 1:1–8.

Burns, P. E. 1971. New determination of australopithecine height. *Nature* 232:350.

Butzer, K. W. 1973a. Geology of Nelson Bay Cave, Robberg, South Africa. *So. Afr. Archaeol. Bull.* 28:97–110.

———. 1973b. Re-evaluation of the geology of the Elandsfontein (Hopefield) site, south-western Cape, South Africa. *So. Afr. J. Sci.* 69:234–238.

———. 1974a. Geological and ecological perspectives on the Middle Pleistocene. *Quat. Res.* 4:136–148.

———. 1974b. Paleoecology of South African australopithecines: Taung revisited. *Current Anthrop.* 15:367–388.

———. 1976a. The Mursi, Nkalabong and Kibish Formations, lower Omo basin, Ethiopia. In *Earliest man and environments in the Lake Rudolf Basin,* ed. Y. Coppens, F. C. Howell, G. Ll. Isaac, and R. E. F. Leakey. Chicago: University of Chicago Press, pp. 12–23.

———. 1976b. Lithostratigraphy of the Swartkrans Formation. *So. Afr. J. Sci.* 72:136–141.

———. 1978. Geo-ecological perspectives on early hominid evolution. In *Early African Hominids,* ed. C. J. Jolly. London: Duckworth (in press).

———, P. B. Beaumont, and J. C. Vogel. 1978. Lithostratigraphy of Border Cave, KwaZulu, South Africa: a Middle Stone Age sequence beginning ca. 195,000 B. P. *J. Archaeol. Sci.* 5 (in press).

———, and D. M. Helgren. 1972. Late Cenozoic evolution of the Cape coast between Knysna and Cape St. Francis, South Africa. *Quat. Res.* 2:143–169.

———, D. M. Helgren, G. J. Fock, and R. Stuckenrath. 1973. Alluvial terraces of the lower Vaal river, South Africa: a reappraisal and reinvestigation. *J. Geol.* 81:341–362. (and 82:663–667. 1974).

———, J. D. Clark, and H. B. S. Cooke. 1974. The geology, archeology and fossil mammals of the Cornelia Beds, O.F.S. *Nat. Mus.* (Bloemfontein) *Mem.* 9.

———, G. Ll. Isaac, J. L. Richardson, and C. Washbourn-Kamau. 1972. Radiocarbon dating of East African lake levels. *Science* 175:1069–1076.

———, and D. L. Thurber. 1969. Some late Cenozoic sedimentary formations of the lower Omo basin. *Nature* 222:1132–1137.

Campbell, B. G. 1962. The systematics of man. *Nature* 194:225–232.

———. 1963. Quantitative taxonomy and human evolution. In *Classification and Human Evolution,* ed. S. L. Washburn. Chicago: Aldine, pp. 50–74.

———. 1965. The nomenclature of the Hominidae, including a definitive list of hominid taxa. *Roy. Anthrop. Inst. G. Brit. Ireland,* occasional paper 22.

———. 1972. Man for all seasons. In *Sexual selection and the descent of man, 1871–1971,* ed. B. G. Campbell. Chicago: Aldine, pp. 40–58.

———. 1973. New concepts in physical anthropology: fossil man *Ann. Rev. Anthrop.* 1:27–54.

———. 1974a. A new taxonomy of fossil man. *Yearbook of Physical Anthropology* (1973) 17:195–201.

———. 1974b. *Human Evolution. An Introduction to Man's Adaptations,* 2nd ed. Chicago: Aldine.

———. 1978. Some problems in hominid classification and nomenclature. In *Early African Hominids,* ed. C. J. Jolly. London: Duckworth (in press).

Camps, G. 1968. Tableau chronologique de la préhistoire recente du Nord de l'Afrique. Première synthèse des datations absolues obtenués par le carbone 14. *Bull. Soc. Préhist. Franç.* 75:609–622.

———. 1974. Nouvelles remarques sur l'âge de l'Atérian. *Bull. Soc. Préhist. Franç. C. R. Séances Ménsuelles* 71(6):163–164.

Carney, J., A. Hill, J. A. Miller, and A. Walker. 1971. Late australopithecine from Baringo district, Kenya. *Nature* 230:509–514.

Carr, C. J. 1976. Plant ecological variation and pattern in the lower Omo basin. In *Earliest Man and Environments in the Lake Rudolf Basin,* ed. Y. Coppens, F. C. Howell, G. Ll. Isaac and R. E. F. Leakey. Chicago: University of Chicago Press, pp. 432–467.

Cerling, T. E., R. L. Hay and J. R. O'Neil. 1977. Isotopic evidence for dramatic climatic changes in East Africa during the Pleistocene. *Nature* 267:137–138.

Chavaillon, J. 1971. Etat actuel de la préhistoire ancienne dans la vallée de l'Omo (Ethiopie). *Archeologia* 38:33–43.

———. 1973. Chronologie des niveaux paléolithiques de Melka-Kontouré. *C. R. Acad. Sci.,* Paris 276-D:1533–1536.

———. 1975. Le site paléolithique ancien l'Omo 84 (Ethiopie). *Document pour servir à l'histoire des civilisations Ethiopiennes* 6:9–18. Paris: C.N.R.S.

———. 1976. Evidence for the technical practices of early Pleistocene hominids, Shungura Formation, lower Omo valley, Ethiopia. In *Earliest man and environments in the Lake Rudolf Basin,* ed. Y. Coppens, F. C. Howell, G. Ll. Issac, and R. E. F. Leakey. Chicago: University of Chicago Press, pp. 565–573.

———, C. Brahimi, and Y. Coppens. 1974. Première découverte d'hominidé dans l'un des sites Acheuléens de Melka-Kontouré (Ethiopie). *C. R. Acad. Sci.* (Paris) 278-D:3299–3302.

Chiu, C., Y. Gu, Y. Zhang, and S. Chang. 1973. Newly discovered *Sinanthropus* remains and stone artifacts at Choukoutien. *Vertr. Palasiatica* 11:109–131 (in Chinese).

Choubert, G., and J. Marçais. 1947. Le Quaternaire des environs de Rabat et l'âge de l'homme de Rabat. *C. R. Acad. Sci.* (Paris) 224:1645–1647.

Ciochon, R. L., and R. S. Corruccini. 1976. Shoulder joint of Sterkfontein *Australopithecus. So. Afr. J. Sci.* 72:80–82.

Clark, J. D. 1960. Further excavations at Broken Hill,

Northern Rhodesia. *J. Roy. Anth. Inst.* 89:201–232 (appendix on fauna by L. S. B. Leakey).

_____. 1967. *Atlas of African prehistory.* Chicago: University of Chicago Press.

_____. 1972. Palaeolithic butchery practices. In *Man, settlement, and urbanism,* ed. P. J. Ucko, R. Tringham, and G. W. Dimbleby. London: Duckworth, pp. 149–156.

_____. 1975. A comparison of the late Acheulian industries of Africa and the Middle East. In *After the Australopithecines,* ed. K. W. Butzer and G. Ll. Isaac. The Hague: Mouton, pp. 605–660.

_____, D. R. Brothwell, R. Powers, and K. P. Oakley. 1968. Rhodesian man: notes on a new femur fragment. *Man* 3:105–111.

_____, K. P. Oakley, L. H. Wells, and J. A. C. McClelland. 1950. New studies on Rhodesian man. *J. Roy. Anthrop. Inst.* 77:7–32.

Clarke, R. J. 1976. New cranium of *Homo erectus* from Late Ndutu, Tanzania. *Nature* 262:485–487.

_____, and F. C. Howell. 1972. Affinities of the Swartkrans 847 hominid cranium. *Am. J. Phys. Anthrop.* 37:319–336.

_____, F. C. Howell, and C. K. Brain. 1970. More evidence of an advanced hominid at Swartkrans. *Nature* 225:1219–1222.

_____. 1977. A juvenile cranium and some adult teeth of early *Homo* from Swartkrans, Transvaal. *So. Afr. J. Sci.* 73:46–49.

Conroy, G. C. and D. Pilbeam. 1975. *Ramapithecus:* a review of its hominid status. In *Paleoanthropology: Morphology and Paleoecology,* ed. R. H. Tuttle. The Hague: Mouton, pp. 59–86.

Cooke, H. B. S. 1963. Pleistocene mammal faunas of Africa, with particular reference to southern Africa. In *African ecology and human evolution,* ed. F. C. Howell and F. Bourliere. Chicago: Aldine, pp. 65–116.

_____. 1967. The Pleistocene sequence in South Africa and problems of correlation. In *Background to Evolution in Africa,* ed. W. W. Bishop and J. D. Clark. Chicago: University of Chicago Press, pp. 175–184.

_____, and V. J. Maglio. 1972. Plio-Pleistocene stratigraphy in East Africa in relation to proboscidean and suid evolution. In *Calibration of Hominoid Evolution,* ed. W. W. Bishop and J. A. Miller. Edinburgh: Scottish Academic Press, pp. 303–330.

_____, B. D. Malan, and L. H. Wells. 1945. Fossil man in the Lebombo mountains, South Africa: the "Border Cave," Ingwavuma district, Zululand. *Man* 3:6–13.

Coon, C. S. 1962. *The origin of races.* New York: Knopf.

Coppens, Y. 1961. Découverte d'un australopithecine dans le Villafranchian du Tchad. *C. R. Acad. Sci.* (Paris) 252:3851–52.

_____. 1962. Découverte d'un australopithecine dans le Villafranchian du Tchad. *Colloques Int. Cent. Natu. Rech. Scient.* (Paris) 104:455–459.

_____. 1965. L'hominien du Tchad. *C. R. Acad. Sci.* (Paris) 260:2869–71.

_____. 1966. Le Tchadanthropus. *L'Anthropologie* 70:5–16.

_____.1970a. Localisations dans le temps et dans l'espace des restes d'Hominides des formations Plio-Pleistocene de l'Omo (Ethiopie). *C. R. Acad. Sci.* (Paris) 271-D:1968–71.

_____. 1970b. Les restes d'Hominides des séries inférieurs et moyennes des formations Plio-Villafranchiennes de l'Omo en Ethiopie. *C. R. Acad. Sci.* (Paris) 271-D:2286–89.

_____. 1971. Les restes d'Hominides des series supérieurs des formations Plio-Villafranchiennes de l'Omo en Ethiopie. *C. R. Acad. Sci.* (Paris) 272-D:36–39.

_____. 1973a. Les restes d'Hominides des séries inférieures et moyennes des formations Plio-Villafranchiennes de l'Omo en Ethiopie (recoltes 1970, 1971 et 1972). *C. R. Acad. Sci.* (Paris) 276:1823–26.

_____. 1973b. Les restes d'Hominides des séries supérieures des formations Plio-Villafranchiennes de l'Omo en Ethiopie (recoltes 1970, 1971 et 1972). *C. R. Acad. Sci.* (Paris) 276:1981–84.

Corruccini, R. S. 1975. Morphometric assessment of australopithecine post-cranial affinities. *System. Zool.* 24:226–233.

_____. 1976. Multivariate allometry and australopithecine variation. *Evolution.* 30:558–563.

Curtis, G. H., R. Drake, T. E. Cerling, and J. Hampel. 1975. Age of the KBS Tuff in the Koobi Fora Formation, East Rudolf, Kenya. *Nature* 258:395–398.

Curtis, G. H., and R. L. Hay. 1972. Further geological studies and potassium-argon dating at Olduvai Gorge and Ngorongoro Crater. In *Calibration of hominoid evolution,* eds. W. W. Bishop and J. A. Miller. Edinburgh: Scottish Academic Press, pp. 289–301.

Dahlberg, A. A. 1960. The Olduvai giant hominid tooth. *Nature* 188:962.

Dart, R. A. 1925. *Australopithecus africanus:* the man-ape of South Africa. *Nature* 115:195–199.

_____. 1929. A note on the Taungs skull. *So. Afr. J. Sci.* 26:648–658.

_____. 1930–31. I caratteri dell' *Australopithecus africanus. Arch. p. Antropologia & Etnologia* 60-1:287–295.

_____. 1934. The dentition of *Australopithecus africanus. Folia Anat. Jap.* 12:207–221.

_____. 1940a. Recent discoveries bearing on human history in southern Africa. *J. Roy. Anthrop. Inst.* 70:13–27.

_____. 1940b. The status of *Australopithecus. Am. J. Phys. Anthrop.* 36:167–186.

_____. 1948a. The first human mandible from the Cave of Hearths, Makapansgat. *So. Afr. Archaeol. Bull.* 3:96–98.

_____. 1948b. The infancy of *Australopithecus.* In *Robert Broom Commemorative Volume,* ed. A. L. du Toit, Special Publication of the Royal Society of South Africa, Cape Town, pp. 143–152.

_____. 1949. The first pelvic bones of *Australopithecus prometheus:* preliminary note. *Am. J. Phys. Anthrop.,* 7:255–258.

_____. 1956. Cultural status of the South African man-apes. *Smithsonian Rep.* 1956:317–338.

_____. 1957a. The osteodontokeratic culture of *Aus-*

tralopithecus prometheus. Transv. Mus. (Pretoria) *Mem.* 10:1–105.

———1957b. The Makapansgat australopithecine osteodontokeratic culture. *Proc., Third Pan-African Congress on Prehistory*, Livingstone, 1955:161–171.

———. 1958. A further adolescent australopithecine ilium from Makapansgat. *Am. J. Phys. Anthrop.*, 15: 473–480.

Davis, P. R. 1964. Hominid fossils from Bed I, Olduvai Gorge, Tanganyika: A tibia and fibula. *Nature* 201:967–968.

Day, M. H. 1967. Olduvai hominid 10: a multivariate analysis. *Nature* 215:323–324.

———. 1969a. Femoral fragment of a robust australopithecine from Olduvai Gorge, Tanzania. *Nature* 221:230–233.

———. 1969b. Omo human skeletal remains. *Nature* 222:1140–1143.

———. 1971. The Omo human skeletal remains. In *The Origin of Homo sapiens,* ed. F. Bordes. Paris: UNESCO, pp. 31–35.

———. 1973a. The development of *Homo sapiens. Accad. naz. dei Lincei,* Rome, 370 (Quaderno N. 182 Problemi Attuali di Scienza e di Cultura, L'Origine dell'Uomo: 87–95.

———. 1973b. Locomotor features of the lower limb in hominids. *Symp. Zool. Soc. London,* 33:29–51.

———. 1974a. *Homo sapiens. Encyclopaedia Britannica* 3(8):1043–1052.

———. 1974b. The interpolation of isolated fossil foot bones into a discriminant analysis—a reply. *Am. J. Phys. Anthrop.* 41:233–236.

———. 1976a. Hominid postcranial remains from the East Rudolf succession—a review. In *Earliest man and environments in the Lake Rudolf Basin,* ed. Y. Coppens, F. C. Howell, G. Ll. Isaac, and R. E. F. Leakey. Chicago: University of Chicago Press, pp. 507–521.

———. 1976b. Hominid postcranial material from Bed I, Olduvai Gorge. In *Human Origins: L. S. B. Leakey and the East African evidence,* ed. G. Ll. Isaac and E. R. McCown. Menlo Park, California: W. A. Benjamin, pp. 362–374.

———. 1978. Functional interpretation of the morphology of postcranial remains of early African hominids. In *Early African Hominids,* ed. C. J. Jolly. London: Duckworth (in press).

———, R. E. F. Leakey, A. C. Walker and B. A. Wood. 1976. New hominids from East Turkana, Kenya. *Am. J. Phys. Anthrop.* 45:369–436.

———, and T. I. Molleson. 1973. The Trinil femora. In *Symposia of the Society for the study of human biology,* vol. 11 (*Human Evolution*). London: Taylor and Francis, pp. 127–154.

———, and J. R. Napier. 1964. Hominid fossils from Bed I, Olduvai Gorge, Tanganyika. Fossil foot bones. *Nature* 201:968–970.

———, and J. R. Napier. 1966. A hominid toe bone from Bed I, Olduvai Gorge, Tanzania. *Nature* 211:929–930.

———, and J. L. Scheuer. 1973. SKW 14147: a new homi-

nid metacarpal from Swartkrans. *J. Hum. Evol.* 2: 429–438.

———, and B. A. Wood. 1968. Functional affinities of the Olduvai hominid 8 talus. *Man* 3:440–445.

Débenath, A. 1972. Nouvelles fouilles à Dar es Soltane (Champ de tir d'El Menzeh) près de Rabat (Maroc). Note préliminaire. *Bull. Soc. Préhist. Franç.* 69:178–179.

———. 1975. Découverte de restes humains probablement atériens à Dar es Soltane (Maroc.) *C. R. Acad. Sci.* (Paris) 281:875–876.

Delson, E. 1978. Models of early hominid phylogeny. In *Early African Hominids.* ed. C. J. Jolly London: Duckworth (in press).

———, N. Eldredge, and I. Tattersall. 1977. Reconstruction of hominid phylogeny: a testable framework based on cladistic analysis. *J. Hum. Evol.* (G. H. R. von Koenigswald Festschrift) 6:263–278.

de Villiers, H. 1973. Human skeletal remains from Border Cave, Ingwavuma district, KwaZulu, South Africa. *Ann. Transv. Mus.* (Pretoria) 28:229–256.

———. 1976. A second adult human mandible from the Border Cave, IngwaVuma district, KwaZulu, South Africa. *So. Afr. J. Sci.* 72:212–215.

Dietrich, W. O. 1939. Zur Stratigraphie der *Africanthropus* fauna. *Zentralblatt f. Miner., Geol. u. Paläont.* Abt. B, 1939:1–9.

———. 1942. Ältestquartäre Säugetiere aus der südlichen Serengeti, Deutsch-Ostafrika. *Palaeont.* 94(A):43–133.

———. 1945. Nashornreste aus dem Quartär Deutsch-Ostafrika. *Palaeont.* 96(A):46–90.

———. 1950. Fossile Antilopen und Rinder Aquatorialafrikas. *Palaeont.* 99(A):1–62.

Drennan, M. R. 1937. The Florisbad skull and brain cast. *Trans. Roy. Soc. So. Afr.* 25:103–114.

———. 1953. A preliminary note on the Saldanha skull. *So. Afr. J. Sci.* 50:7–11.

———, and R. Singer. 1955. A mandibular fragment, probably of the Saldanha skull. *Nature* 175:364–365.

Dreyer, T. F. 1935. A human skull from Florisbad, Orange Free State, with a note on the endocranial cast by C. U. Ariens Kappers. *Proc. K. Ned. Akad. Wet. Amsterdam* 38:119–128.

———. 1936. The endocranial cast of the Florisbad skull —a correction. *Sool. Navors. Nas. Mus.* (Bloemfontein) 1(5):21–23.

———. 1938. The fissuration of the frontal endocranial cast of the Florisbad skull compared with that of the Rhodesian skull. *Zts. f. Rassenkunde* 8:193–198.

———. 1947. Further observations on the Florisbad skull. *Sool. Navors. Nas. Mus.* (Bloemfontein) 1(15):183–190.

Dubois, E. 1891. Palaeontologische onderzoekingen op Java. *Versl. Mijnw.* (Batavia) 3:12–14; 4:12–15.

———. 1892. Voorloopig bericht omtrent het onderzoek naar de pleistocene en tertiaire vertebraten-fauna van Sumatra en Java, gedurende het jaar 1890. *Natuurk. Tijdschr. Nederl.-Indië,* 51:93–100.

———. 1894a. Palaeontologische onderzoekingen op Java. *Versl. Mijnw.* (Batavia) 4:14–18.

———. 1894b. *Pithecanthropus erectus.* Eine menschenähnliche Übergangsform aus Java. *Batavia* (and in

Jaarboek v. h. Mijnwezen in Nederl. Oost Indie, 24:5–77, 1895, Amsterdam).

———. 1926a. On the principal characters of the femur of *Pithecanthropus erectus. Proc. K. Ned. Akad. Wet. Amst.* 29:730–743.

———. 1926b. Figures of the femur of *Pithecanthropus erectus. Proc. K. Ned. Akad. Wet. Amst.* 29:1275–1277.

———. 1932. The distinct organization of *Pithecanthropus* of which the femur bears evidence, now confirmed from other individuals of the described species. *Proc. K. Ned. Akad. Wet. Amst.* 35:716–722.

———. 1934. New evidence of the distinct organization of *Pithecanthropus. Proc. K. Ned. Akad. Wet. Amst.* 37: 139–145.

Eldredge, N., and I. Tattersall. 1975. Evolutionary models, phylogenetic reconstruction, and another look at hominid phylogeny. In *Approaches to primate paleobiology,* ed. F. S. Szalay. *Contributions to primatology,* vol. 5. Basel: Karger, pp. 218–242.

Ennouchi, E. 1962a. Un crâne d'homme ancien au Jebel Irhoud (Maroc). *C. R. Acad. Sci.* (Paris) 254-D:4330–32.

———. 1962b. Un néandertalien: l'homme du Jebel Irhoud (Maroc). *L'Anthropologie* 66:279–299.

———. 1963. Les néanderthaliennes du Jebel Irhoud (Maroc). *C. R. Acad. Sci.* (Paris) 256-D:2459–60.

———. 1966. Le site du Jebel Irhoud (Maroc). In *Actas del V Congreso Panafricano de Prehistoria y de Estudio del Cuaternario,* vol. 2. Santa Cruz de Tenerife: Museo Arqueologico, pp. 53–60.

———. 1968. Le deuxième crâne de l'homme d'Irhoud. *Ann. de Paléont.* (Vert.) 54:117–128.

———. 1969a. Découverte d'un Pithécanthropien au Maroc. *C. R. Acad. Sci.* (Paris) 269-D:753–765.

———. 1969b. Présence d'un enfant néanderthalien au Jebel Irhoud (Maroc). *Ann. de Paléont.* (Vert.) 55:251–265.

———. 1970. Un nouvel archanthropien au Maroc. *Ann. de Paléont.* (Vert.) 56:95–107.

———. 1972. Nouvelle découverte d'un archantropien au Maroc. *C. R. Acad. Sci.* (Paris) 274-D:3088–90.

Ewer, R. F. 1956. The dating of the Australopithecinae: faunal evidence. *So. Afr. Archaeol. Bull.* 11:41–45.

———. 1957. Faunal evidence on the dating of the Australopithecinae. *Proc. Third Pan-Afr. Congress on Prehistory,* Livingstone, 1955, ed. J. D. Clark. London: Chatto and Winds, pp. 135–142.

Ferembach, D. 1965. *Homo erectus. Bull. Societe d'études et de Récherches Préhistoriques Institut Pratique de Préhistoire* 14:1–15.

———. 1972. L'ancêtre de l'homme du Paléolithique supérieur était-il néandertalien? In *The origin of* Homo sapiens, ed. F. Bordes. Paris: UNESCO, pp. 73–80.

Ferembach, D. 1976a. Les restes humains Atériens de Témara (campagne 1975). *Bull. et Mém. Soc. d'Anthrop. Paris,* sér. 13, 3:175–180.

———. 1976b. Les restes humains de la grotte de Dar-es-Soltane 2 (Maroc) campagne 1975. *Bull. et Mém. Soc. d'Anthrop. Paris,* sér. 13, 3:183–193.

Findlater, I. C., F. J. Fitch, J. A. Miller, and R. T. Watkins. 1974. Dating of the rock succession containing fossil hominids at East Rudolf, Kenya. *Nature* 251:213–215.

Fitch, F. J., and J. A. Miller. 1970. Radioisotopic age determinations of Lake Rudolf artefact site. *Nature* 226:226–228.

Fitch, F. J. and J. A. Miller. 1976. Conventional potassium-argon and argon-40/argon-39 dating of volcanic rocks from East Rudolf. In *Earliest man and environments in the Lake Rudolf Basin,* ed. Y. Coppens, F. C. Howell, G. Ll. Isaac, and R. E. F. Leakey. Chicago: University of Chicago Press, pp. 123–147.

Fitch, F. J., P. J. Hooker, and J. A. Miller. 1976. ^{40}Ar/^{39}Ar dating of the KBS Tuff in Koobi Fora Formation, East Rudolf, Kenya. *Nature* 263:740–744.

Frazer, J. F. D. 1973. Gestation period for *Australopithecus. Nature* 242:347.

Freeman, L. G. 1975a. By their works you shall know them: cultural developments in the Paleolithic. In *Hominisation and behavior,* ed. G. Kurth and I. Eibl-Eibesfeldt. Stuttgart: G. Fischer, pp. 234–261.

———. 1975b. Acheulian sites and stratigraphy in Iberia and the Maghreb. In *After the Australopithecines,* ed. K. W. Butzer and G. Ll. Isaac. The Hague: Mouton, pp. 661–743.

Funnell, B. M. 1964. The Tertiary Period. In The Phanerozoic time scale: a symposium. *Quart. J. Geol. Soc. London* 120:179–191.

Galloway, A. 1937a. The nature and status of the Florisbad skull as revealed by its nonmetrical features. *Am. J. Phys. Anthrop.* 23:1–17.

———. 1937b. The characteristics of the skull of the Boskop physical type. *Am. J. Phys. Anthrop.* 23:31–47.

———. 1937c. Man in Africa in the light of recent discoveries. *So. Afr. J. Sci.* 34:89–120.

Genet-Varcin, E. 1966. Conjectures sur l'allure générale des Australopitheques. *Soc. Préhist. Franç., C. R. Séances Mens.* 63:cvi–cvii.

———. 1969. Structure et comportement des australopitheques d'après certains os post-craniens. *Ann. de Paléont.* (Vert.) 55:139–148.

Gieseler, W. 1974. Die Fossilgeschichte des Menschen. In *Die Evolution der Organism.* 3e Auflage. Band III. Stuttgart: G. Fischer, pp. 171–517.

Gillman, J. 1929. A review of some Bush and Bantu sacra, with special reference to the sacra of Rhodesian man and Boskop man. *So. Afr. J. Sci.* 26:602–622.

Gramly, R. M., and G. P. Rightmire. 1973. A fragmentary cranium and dated Later Stone Age assemblage from Lukenya Hill, Kenya. *Man,* ser. 2, 8:571–579.

Greene, D. L. 1975. Gorilla dental sexual dimorphism and early hominid taxonomy. *Symposium Fourth Intern. Congress of Primatology* 3:82–100.

Gregory, W. K. 1930. The origin of man from a brachiating anthropoid stock. *Science* 71:645–650.

———. 1939. The bearing of Dr. Broom's and Dr. Dart's discoveries on the origin of man. *Annual Proceedings, Associated Scientific and Technical Societies of South Africa, Johannesburg* 1938–1939:25–57.

———, and M. Hellman. 1939a. Evidence of the australopithecine man-apes on the origin of man. *Science* 88: 615–616.

———, and M. Hellman. 1939b. The dentition of the extinct South African man-ape *Australopithecus (Plesianthropus) transvaalensis* Broom. A comparative and phylogenetic study. *Ann. Transv. Mus.* 19:339–373.

———, and M. Hellman. 1939c. The South African fossil man-apes and the origin of the human dentition. *J. Am. Dent. Assoc.* 26:558–564.

Grommé, C. S., and R. L. Hay. 1971. Geomagnetic polarity epochs: age and duration of the Olduvai normal polarity event. *Earth and Planetary Science Letters* 10:179–185.

Groves, C. R., and V. Mazák. 1975. An approach to the taxonomy of the Hominidae: gracile Villafranchian hominids of Africa. *Časopis pro Mineralogü a Geologü*, roč. 20, c. 3:225–247.

Gutgesell, V. J. 1970. "*Telanthropus*" and the single species hypothesis: A reexamination. *Am. Anthrop.* 72:565–576.

Hamilton, A. 1976. The significance of patterns of distribution shown by forest plants and animals in tropical Africa for the reconstruction of Upper Pleistocene palaeoenvironments. A review. *Palaeoecology of Africa*, 9:63–97.

Harris, J. W. K., and W. W. Bishop. 1976. Sites and assemblages from the early Pleistocene beds of Karari and Chesowanja. *IXᵉ Congres Internationale des Sciences Préhistorique et Protohistorique, Nice 1976*.

Harris, J. W. K., and I. Herbich. 1978. Aspects of early Pleistocene hominid behaviour at East Lake Rudolf, Kenya. In *Geological background to fossil man,* ed. W. W. Bishop. London: Geological Society of London (in press).

Harris, J. W. K. and G. Ll. Isaac. 1976. The Karari industry: early Pleistocene archaeological materials from the terrain east of Lake Rudolf, Kenya. *Nature* 262:102–107.

Hay, R. L. 1971. Geologic background of Beds I and II. Stratigraphic summary. In *Olduvai Gorge. 3. Excavations in Beds I and II, 1960–1963,* by M. D. Leakey. Cambridge: Cambridge University Press, pp. 9–18.

———. 1976. *Geology of the Olduvai Gorge. A study of sedimentation in a semiarid basin.* Berkeley: University of California Press.

Heberer, G. 1963. Die Ur- und Frühmenschen funde in Afrika. 2. Die Olduvai-Funde. *Kosmos, Stuttgart* 58:84–88 (1962).

de Heinzelin, J., F. H. Brown, and F. C. Howell. 1970. Pliocene/Pleistocene formations in the lower Omo basin, southern Ethiopia. *Quaternaria* 13:247–268.

de Heinzelin, J., P. Haesaerts, and F. C. Howell. 1976. Plio-Pleistocene formations of the lower Omo basin, with particular reference to the Shungura Formation. In *Earliest Man and Environments in the Lake Rudolf Basin,* ed. Y. Coppens, F. C. Howell, G. Ll. Issac, and R. E. F. Leakey. Chicago: University of Chicago Press, pp. 24–49.

Heintz, N. 1967. Evolution de la hauteur maximale du frontal, du pariétal de l'occipital chez les hominidés. *Ann. Paléont.* (*Vert.*) 53:51–75.

Heiple, K. G., and C. O. Lovejoy. 1971. The distal femoral anatomy of *Australopithecus. Am. J. Phys. Anthrop.* 35:75–84.

Helmuth, H. 1968. Körperhöhe und Gliedmassenproportionen der *Australopithecinen. Zts. Morph. v. Anthrop.* 60:147–155.

Hemmer, H. 1971. Beitrag zur Erfassung der progressiven Cephalisation bei Primaten. *Proc. Third Intern. Congr. Primatology, Zürich 1970.* Basel: Karger, pp. 90–107.

Hendey, Q. B. 1974. Faunal dating of the late Cenozoic of southern Africa, with special reference to the Carnivora. *Quat. Res.* 4:149–161.

Hillhouse, J. W., J. W. M. Ndombi, A. Cox and A. Brock. 1977. Additional results on palaeomagnetic stratigraphy of the Koobi Fora Formation, east of Lake Turkana (Lake Rudolf), Kenya. *Nature,* 265:411–415.

Holloway, R. 1970a. Australopithecine endocast (Taung specimen, 1924): a new volume determination. *Science* 168:966–968.

———. 1970b. New endocranial values for the australopithecines. *Nature* 227:199–200.

———. 1972a. Australopithecine endocasts, brain evolution in the Hominoidea, and a model of hominid evolution. In *The Functional and Evolutionary Biology of Primates,* ed. R. Tuttle. Chicago and New York: Aldine-Atherton, pp. 185–203.

———. 1972b. New australopithecine endocast, SK 1585, from Swartkrans, South Africa. *Am. J. Phys. Anthrop.* 37:173–186.

———. 1973a. Endocranial volumes of early African hominids, and the role of the brain in human mosaic evolution. *J. Hum. Evol.* 2:448–459.

———. 1973b. New endocranial values for the East African early hominids. *Nature* 243:97–99.

———. 1974. The casts of fossil hominid brains. *Sci. Am.* 231:106–115.

———. 1975. Early hominid endocasts: volumes, morphology and significance for hominid evolution. In *Primate functional morphology and evolution,* ed. R. H. Tuttle. The Hague: Mouton, pp. 393–416.

———. 1976. Some problems of hominid brain endocast reconstruction, allometry and neural reorganization. In *IXᵉ Congrès, Union Internationale des Sciences Préhistoriques et Protohistoriques, Nice 1976*.

———. 1978. Problems of brain endocast interpretation and African hominid evolution. In *Early African Hominids,* ed. C. J. Jolly. London: Duckworth (in press).

———, and P. V. Tobias. 1966. Cranial capacity of the Olduvai Bed I hominine. *Nature* 210:1108–10.

Horowitz, A. 1975. Preliminary paleoenvironmental implications of pollen analysis of Middle Breccia from Sterkfontein. *Nature* 228:417–418.

———, G. Siedner, and O. Bar-Yosef. 1973. Radiometric dating of the Ubeidiya formation, Jordan valley, Israel. *Nature* 242:186–187.

Howe, B. 1967. The Palaeolithic of Tangier, Morocco. Excavations at Cape Ashakar, 1939–1947. *Bull. Am. School of Prehist. Res.* (Peabody Museum, Harvard) 22.

Howell, F. C. 1960. European and northwest African Middle Pleistocene hominids. *Current Anthrop.* 1:195–232.

———. 1969a. Remains of Hominidae from Pliocene/Pleistocene formations in the lower Omo basin, Ethiopia. *Nature* 223:1234–39.

———. 1969b. Hominid teeth from White Sands and Brown Sands localities, lower Omo basin (Ethiopia). *Quaternaria* 11:47–54.

———. 1972. Pliocene/Pleistocene Hominidae in Eastern Africa: absolute and relative ages. In *Calibration of hominoid evolution,* ed. W. W. Bishop and J. A. Miller. Edinburgh: Scottish Academic Press, pp. 331–368.

———. 1976. Overview of the Pliocene and earlier Pleistocene of the lower Omo basin, southern Ethiopia. In *Human origins: L. S. B. Leakey and the East African evidence,* ed. G. Ll. Isaac and E. R. McCown. Menlo Park, Ca.: W. A. Benjamin, pp. 226–268.

———. 1978. Overview of the Pliocene and earlier Pleistocene of the lower Omo basin, southern Ethiopia. In *Early African Hominids,* ed. C. J. Jolly. London: Duckworth (in press).

———, and J. D. Clark. 1963. Acheulian hunter-gatherers of sub-Saharan Africa. In *African ecology and human evolution,* ed. F. C. Howell and F. Bourlière. Chicago: Aldine, pp. 458–533.

———, and Y. Coppens. 1973. Deciduous teeth of Hominidae from the Pliocene/Pleistocene of the lower Omo basin, Ethiopia. *J. Hum. Evol.* (R. A. Dart Memorial Issue) 2:461–472.

———, and Y. Coppens (with J. de Heinzelin). 1974. Inventory of remains of Hominidae from Pliocene/Pleistocene formations of the lower Omo basin. Ethiopia (1967–1972). *Amer. J. Phys. Anthrop.* 40:1–16.

———, and Y. Coppens. 1976a. An overview of Hominidae from the Omo succession, Ethiopia. In *Earliest man and environments in the Lake Rudolf Basin,* ed. Y. Coppens, F. C. Howell, G. Ll. Isaac, and R. E. F. Leakey. Chicago: University of Chicago Press, pp. 522–532.

———, and Y. Coppens. 1976b. Les Hominidés de l'Omo. In *IXᵉ Congrès, Union Internationale des Sciences Préhistoriques et Protohistoriques, Nice, September 1976.*

———, and B. A. Wood. 1974. Early hominid ulna from the Omo basin, Ethiopia. *Nature* 249:174–176.

———, S. L. Washburn, and R. L. Ciochon. 1978. Relationship of *Australopithecus* and *Homo. J. Hum. Evol.* (in press).

Howells, W. W. 1966. *Homo erectus. Scientific American* 215(5):46–53.

———. 1973. Evolution of the Genus *Homo.* Reading, Mass.: Addison-Wesley.

Hrdlička, A. 1925. The Taungs ape. *Am. J. Phys. Anthrop.* 8:379–392.

———. 1930a. The Rhodesian man. *Am. J. Phys. Anthrop.* 9:173–204.

———. 1930b. The skeletal remains of early man. *Smithsonian Misc. Coll.* 83:1–379.

Hughes, A. R., and P. V. Tobias. 1977. A fossil skull probably of the genus *Homo* from Sterkfontein, Transvaal. *Nature* 265:310–312.

Hurford, A. J., A. J. W. Gleadow, and C. W. Naeser. 1976. Fission-track dating of pumice from the KBS Tuff, East Rudolf, Kenya. *Nature* 263:738–740.

Isaac, G. Ll. 1965. The stratigraphy of the Peninj beds and the provenance of the Natron australopithecine mandible. *Quaternaria* 7:101–130.

———. 1967. The stratigraphy of the Peninj Group—early Middle Pleistocene formations west of Lake Natron, Tanzania. In *Background to evolution in Africa,* ed. W. W. Bishop and J. D. Clark. Chicago: University of Chicago Press, pp. 229–254.

———. 1969. Studies of early culture in East Africa. *World Archaeol.* 1:1–28.

———. 1971. The diet of early man: aspects of archaeological evidence from lower and middle Pleistocene sites in Africa. *World Archaeol.* 2:278–299.

———. 1972a. Chronology and tempo of cultural change during the Pleistocene. In *Calibration of hominoid evolution,* ed. W. W. Bishop and J. A. Miller. Edinburgh: Scottish Academic Press, pp. 381–430.

———. 1972b. Early phases of human behaviour: models in Lower Palaeolithic archaeology. In *Models in archaeology,* ed. D. L. Clarke. London: Methuen, pp. 167–199.

———. 1975. Stratigraphy and cultural patterns in East Africa during the middle ranges of Pleistocene time. In *After the Australopithecines,* ed. K. W. Butzer and G. Ll. Isaac. The Hague: Mouton, pp. 495–542.

———. 1976a. The activities of early African hominids: a review of archaeological evidence from the time span two and a half million years ago. In *Human origins: L. S. B. Leakey and the East African evidence,* ed. G. Ll. Isaac and E. R. McCown. Menlo Park, California: W. A. Benjamin, pp. 483–520.

———. 1976b. Plio-Pleistocene artifact assemblages from East Rudolf, Kenya. In *Earliest man and environments in the Lake Rudolf Basin,* ed. Y. Coppens, F. C. Howell, G. Ll. Isaac, and R. E. F. Leakey. Chicago: University of Chicago Press, pp. 552–564.

———. 1978. The archaeological evidence for the activities of early African hominids. In *Early African Hominids.* ed. C. J. Jolly. London: Duckworth. (in press).

———, and G. H. Curtis. 1974. Age of early Acheulian industries from the Peninj Group, Tanzania. *Nature* 249:624–627.

———, J. W. K. Harris, and D. Crader. 1976. Archaeological evidence from the Koobi Fora Formation. In *Earliest man and environments in the Lake Rudolf Basin,* ed. Y. Coppens, F. C. Howell, G. Ll. Isaac, and R. E. F. Leakey. Chicago: University of Chicago Press, pp. 533–551.

Jacob, T. 1972. The absolute date of the Djetis beds at Modjokerto. *Antiquity* 47:148.

———. 1973. Palaeoanthropological discoveries in Indonesia with special reference to the finds of the last two decades. *J. Hum. Evol.* 2:473–486.

———. 1975a. Morphology and paleoecology of early man in Java. In *Paleoanthropology: morphology and paleoecology,* ed. R. H. Tuttle. The Hague: Mouton, pp. 311–326.

———. 1975b. The pithecanthropines of Indonesia. *Bull. Mém. Soc. d'Anthrop. Paris,* sér. 13, 2:243–256.

———, and G. H. Curtis. 1971. Preliminary potassium-argon dating of early man in Java. *Cont. Univ. Calif. Archaeol. Res. Fac.* 12:50.

Jaeger, J.-J. 1969. Les rongeurs du Pléistocène moyen de Ternifine (Algerie). *C. R. Acad. Sci.* (Paris) 269:1492–1495.

––––––. 1973. Un pithecanthrope évolue. *La Recherche* 39:1006–07.

––––––. 1975a. Découverte d'un crane d'hominide dans le Pléistocène moyen du Maroc. In *Problèmes actuels de Paléontologie-Evolution des Vértébres.* Colloque international C.N.R.S., no. 218. Paris: C.N.R.S., pp. 897–902.

––––––. 1975b. The mammalian faunas and hominid fossils of the Middle Pleistocene of the Maghreb. In *After the Australopithecines. Stratigraphy, ecology and culture change in the middle Pleistocene,* ed. K. W. Butzer and G. Ll. Isaac. The Hague: Mouton, pp. 399–418.

Jaeger, J.-J. and H. B. Wesselman. 1976. Fossil remains of micromammals from the Omo Group deposits. In *Earliest Man and Environments in the Lake Rudolf Basin,* ed. Y. Coppens, F. C. Howell, G. Ll. Isaac and R. E. F. Leakey, Chicago: University of Chicago Press, pp. 351–360.

Jerison, H. J. 1973. *Evolution of the brain and intelligence.* New York: Academic Press.

––––––. 1975. Fossil evidence of the evolution of the human brain. *Ann. Rev. Anthrop.* 4:27–58.

Johanson, D. C. and Y. Coppens. 1976. A preliminary anatomical diagnosis of the first Plio/Pleistocene hominid discoveries in the central Afar, Ethiopia. *Am. J. Phys. Anthrop.* 45:217–234.

––––––, Y. Coppens, and M. Taieb. 1976. Pliocene hominid remains from Hadar, central Afar, Ethiopia. In *IXe Congrès, Union Internationale des Sciences Préhistoriques et Protohistoriques, Nice, September 1976.*

––––––, and M. Taieb. 1976. Plio-Pleistocene hominid discoveries in Hadar, Ethiopia. *Nature* 260:293–297.

––––––, M. Taieb and B. T. Gray. 1978. Geological framework of the Pliocene Hadar Formation (Afar, Ethiopia) with notes on paleontology including hominids. In *Geological Background to Fossil Man in Africa,* ed. W. W. Bishop, London: Geological Society of London (in press).

Kalin, J. 1969. Über *Paranthropus robustus* Broom. *Arch. d. Julius-Klaus-Stifting* 24:162–187.

Kaufman, A., W. S. Broecker, T.-L. Ku, and D. L. Thurber. 1971. The status of U-series methods of dating mollusks. *Geochim. Cosmochim. Acta* 35:1155–84.

Kay, R. F., and H. M. McHenry. 1973. Humerus of robust *Australopithecus. Science* 182:396.

Keith, A. 1931. *New discoveries relating to the antiquity of man.* London: Williams and Norgate.

Kent, P. E. 1942. The recent history and the Pleistocene deposits of the plateau north of Lake Eyasi, Tanganyika. *Geol. Mag.* 78:173–184.

Klein, R. G. 1973. Geological antiquity of Rhodesian man. *Nature* 244:311–312.

––––––. 1975. Middle Stone Age man-animal relationships in southern Africa: evidence from Die Kelders and Klasies River Mouth. *Science* 190:265–267.

von Koenigswald, G. H. R. 1936. Ein fossiler Hominide aus dem Altpleistocän Ostjavas. *De Ingenieur in Nederl.-Indie* 8:149–157.

––––––. 1940. Neue *Pithecanthropus*-Funde 1936–1938.

Ein Beitrag zur Kenntnis der Praehominiden. *Weten Meded., Dienst van den Mijnbouw in Nederlandsch-Indie* (Batavia) 28:1–232.

––––––. 1950. Fossil hominids from the lower Pleistocene of Java. *Proc. Intern. Geol. Congress, Great Britain 1949* 9:59–66.

––––––. 1953. *Gigantopithecus blacki* von Koenigswald, a giant fossil hominid from the Pleistocene of southern China. *Anthrop. Pap. Amer. Mus.* 43:291–325 (1952).

––––––. 1960. Remarks on a fossil human molar from Olduvai, East Africa. *Proc. K. Ned. Akad. Wet. Amst.,* B, 63:20–25.

––––––. 1969. Java: Prae-Trinil man. *Proc. VIII Intern. Congr. Anthrop. Ethnol. Sciences,* Tokyo-Kyoto, 1968, 1:104–105.

––––––. 1973. *Australopithecus, Meganthropus* and *Ramapithecus. J. Hum. Evol.* 2:487–491.

Kohl-Larsen, L. 1940. Die Fundstätte des *Africanthropus. Natur und Volk* 70:487–499, 550–556.

––––––. 1943. *Auf den Spuren des Vormenschen. Forschungen, Fahrten und Ergebnisse in Deutsch-Ostafrika,* 2 vol. (esp. pp. 378–386).

Kramp, P. 1936. Die topographischen Verhältnisse der menschlichen Schädelbäsis (lage der condyli occipitales, des foramen magnum, der pori acustici externi und der processus mastoides) mit besonderer Berücksichtigung der Fossilfunde von La Chapell-aux-Saints, Steinheim a.d. Murr und Broken Hill (*Homo rhodesiensis*). *Anthrop. Anz.* 13:112–130.

Kurth, G. 1965. Die (Eu) Homininen. In *Menschliche Abstammungslehre,* ed. G. Heberer, Stuttgart: G. Fischer. pp. 357–425.

Leakey, L. S. B. 1935. *The Stone Age races of Kenya.* Oxford:Oxford University Press.

––––––. 1936a. Fossil human remains from Kanam and Kanjera, Kenya Colony. *Nature* 138:643.

––––––. 1936b. A new fossil skull from Eyasi, East Africa. *Nature* 138:1082–83.

––––––. 1958. Recent discoveries at Olduvai Gorge, Tanganyika. *Nature* 181:1099–1103.

––––––. 1959. A new fossil skull from Olduvai. *Nature* 184:491–493.

––––––. 1960. Recent discoveries at Olduvai Gorge. *Nature* 188:1050–52.

––––––, and M. D. Leakey. 1964. Recent discoveries of fossil hominids in Tanganyika: at Olduvai and near Lake Natron. *Nature* 202:5–7.

––––––, and W. H. Reeve. 1946. I. Report on a visit to the site of the Eyasi skull found by Dr. Kohl-Larsen. II. Geological report on the site of Dr. Kohl-Larsen's discovery of a fossil human skull, Lake Eyasi, Tanganyika Territory. *J. East. Afr. Nat. Hist. Soc.* 19:40–50.

––––––, P. V. Tobias, and J. R. Napier. 1964. A new species of the genus *Homo* from Olduvai Gorge. *Nature* 202:7–9.

Leakey, M. D. 1967. Preliminary survey of the cultural material from Beds I and II, Olduvai Gorge, Tanzania. In *Background to evolution in Africa,* ed. W. W. Bishop and J. D. Clark. Chicago: University of Chicago Press, pp. 417–446.

———. 1970. Stone artefacts from Swartkrans. *Nature* 225:1221–1225.

———. 1971. *Olduvai Gorge. 3. Excavations in Beds I and II, 1960–1963*. Cambridge: Cambridge University Press.

———. 1975. Cultural patterns in the Olduvai sequence. In *After the Australopithecines*, ed. K. W. Butzer and G. Ll. Isaac. The Hague: Mouton, pp. 477–494.

———. 1976. The early hominids of Olduvai Gorge and the Laetolil Beds. In *IXᵉ Congrès, Union Internationale des Sciences Préhistoriques et Protohistoriques, Nice, September 1976*.

———. 1978. Olduvai fossil hominids—their stratigraphic positions and associations. In *Early African Hominids*, ed. C. J. Jolly. London: Duckworth (in press).

———, Clarke, R. J., and L. S. B. Leakey. 1971. New hominid skull from Bed I, Olduvai Gorge, Tanzania. *Nature* 232:308–312.

———, R. L. Hay, G. H. Curtis, R. E. Drake, M. K. Jackes, and T. D. White. 1976. Fossil hominids from the Laetolil Beds. *Nature* 262:460–466.

———, P. V. Tobias, J. E. Martyn, and R. E. F. Leakey. 1969. An Acheulean industry with prepared core technique and the discovery of a contemporary hominid at Lake Baringo, Kenya. *Proc. Prehist. Soc.* 35:48–76.

Leakey, R. E. F. 1970. New hominid remains and early artefacts from northern Kenya. *Nature* 226:223–224.

———. 1971. Further evidence of lower Pleistocene hominids from East Rudolf, north Kenya, 1970. *Nature* 231:241–245.

———. 1972. Further evidence of lower Pleistocene hominids from East Rudolf, north Kenya, 1971. *Nature* 237:264–269.

———. 1973. Further evidence of lower Pleistocene hominids from East Rudolf, north Kenya, 1972. *Nature* 242:170–173.

———. 1974. Further evidence of lower Pleistocene hominids from East Rudolf, north Kenya, 1973. *Nature* 248:653–656.

———. 1976a. An overview of the East Rudolf Hominidae. In *Earliest man and environments in the Lake Rudolf Basin*, ed. Y. Coppens, F. C. Howell, G. Ll. Isaac and R. E. F. Leakey. Chicago: University of Chicago Press, pp. 476–483.

———. 1976b. New hominid fossils from the Koobi Fora Formation in northern Kenya. *Nature* 261:574–576.

———, K. W. Butzer, and M. H. Day. 1969. Early *Homo sapiens* remains from the Omo river region of southwest Ethiopia. *Nature* 222:1132–38.

———, J. M. Mungai, and A. C. Walker. 1971. New australopithecines from East Rudolf, Kenya. *Am. J. Phys. Anthrop.* 35:175–186.

———, J. M. Mungai, and A. C. Walker. 1972. New australopithecines from East Rudolf, Kenya. II. *Am. J. Phys. Anthrop.* 36:235–251.

———, and A. C. Walker. 1973. New australopithecines from East Rudolf, Kenya. III. *Am. J. Phys. Anthrop.* 39:205–222.

———, and A. C. Walker. 1976. *Australopithecus, Homo erectus*, and the single species hypothesis. *Nature* 261:572–574.

———, and B. A. Wood. 1974. A hominid mandible from East Rudolf, Kenya. *Am. J. Phys. Anthrop.* 41:245–250.

Lecointre, G. 1960. Le gisement de l'homme de Rabat. *Bull. d'Archéol. Marocaine* 3:55–85.

Le Gros Clark, W. E. 1928. Rhodesian man. *Man* 28:206–207.

———. 1940. Palaeontological evidence bearing on human evolution. *Biological Reviews* 15:202–230.

———. 1947. Observations on the anatomy of the fossil Australopithecinae. *J. Anat.* 81:300–333.

———. 1952. Hominid characters of the australopithecine dentition. *J. Roy. Anthrop. Inst.* 80:37–54.

———. 1955. The os innominatum of the recent Ponginae with special reference to that of the Australopithecinae. *Am. J. Phys. Anthrop.* 13:19–27.

———. 1964. *The fossil evidence for human evolution*, 2nd ed. Chicago: University of Chicago Press.

———. 1967. *Man-apes or Ape-men?* New York: Holt, Rinehart and Winston.

Lestrel, P. E., and D. W. Read. 1973. Hominid cranial capacity versus time: a regression approach. *J. Hum. Evol.* 2:405–411.

Leutenegger, W. 1972. Newborn size and pelvic dimensions of *Australopithecus*. *Nature* 240:568–569.

———. 1973a. Encephalization in australopithecines: a new estimate. *Folia primat.* 19:9–17.

———. 1973b. Gestation period and birth weight of *Australopithecus*. *Nature* 243:548.

———. 1974. Functional aspects of pelvic morphology of simian primates. *J. Hum. Evol.* 3:207–222.

Lewis, O. J. 1972. The evolution of the hallucial tarsometatarsal joints in the Anthropoidea. *Am. J. Phys. Anthrop.* 37:13–34.

———. 1973. The hominoid os capitatum, with special reference to the fossil bones from Sterkfontein and Olduvai Gorge. *J. Hum. Evol.* 2:1–11.

Lisowski, F. P., G. H. Albrecht, and C. E. Oxnard. 1974. The form of the talus in some higher primates: a multivariate study. *Am. J. Phys. Anthrop.* 41:191–215.

Livingstone, D. A. 1975. Late Quaternary climatic change in Africa. *Ann. Rev. Ecol. Syst.* 6:249–280.

Lovejoy, C. O. 1974. The gait of australopithecines. *Yrbk. Phys. Anthropol.* (1973) 17:147–161.

———. 1975. Biomechanical perspectives on the lower limb of early hominids. In *Primate functional morphology and evolution*, ed. R. H. Tuttle. The Hague: Mouton, pp. 291–326.

———. 1976. The locomotor skeleton of basal Pleistocene hominids. In *IXᵉ Congès, Union Internationale des Sciences Préhistoriques et Protohistoriques, Nice, September 1976*.

———. 1978. A biomechanical review of the locomotor diversity of early hominids. In *Early African Hominids*, ed. C. J. Jolly. London: Duckworth (in press).

———, and K. G. Heiple. 1970. A reconstruction of the femur of *Australopithecus africanus*. *Am. J. Phys. Anthrop.* 32:33–40.

———, and K. G. Heiple. 1972. Proximal femoral anatomy of *Australopithecus*. *Nature* 235:175–176.

———, K. G. Heiple, and A. H. Burstein. 1973. The gait of *Australopithecus*. *Am. J. Phys. Anthrop.* 38:757–780.

Mabbutt, J. A. 1956. The physiography and surface geology of the Hopefield fossil site. *Trans. Roy. Soc. So. Afr.* 35:21–58.

———. 1957. The physical background to the Hopefield discoveries. In *Proceedings Third Pan-African Congress on Prehistory, Livingstone, 1955*, ed. J. D. Clark. London: Chatto and Windus, pp. 68–75.

McBurney, C. B. M. 1961. Absolute age of Pleistocene and Holocene deposits in the Haua Fteah. *Nature* 192:685–686.

———. 1962. Absolute chronology of the Palaeolithic in eastern Libya and the problem of Upper Palaeolithic origins. *Adv. Sci.* 18:494–497.

———. 1967. *The Haua Fteah (Cyrenaica), and the Stone Age of the south-east Mediterranean*. Cambridge: Cambridge University Press.

———, J. C. Trevor, and L. H. Wells. 1953. The Haua Fteah fossil jaw. *J. Roy. Anthrop. Inst.* 83:71–85.

MacDougall, C., and P. B. Price. 1974. Attempt to date early South African hominids by using fission tracks in calcite. *Science* 185:943–944.

McHenry, H. M. 1973. Early hominid humerus from East Rudolf, Kenya. *Science* 180:739–741.

———. 1974. How large were the australopithecines? *Am. J. Phys. Anthrop.* 40:329–340.

———. 1975a. Fossil hominid body weight and brain size. *Nature* 254:686–688.

———. 1975b. Fossils and the mosaic nature of human evolution. *Science* 190:425–431.

———. 1975c. The ischium and hip extensor mechanism in human evolution. *Am. J. Phys. Anthrop.* 43:39–46.

———. 1975d. A new pelvic fragment from Swartkrans and the relationship between the robust and gracile australopithecines. *Am. J. Phys. Anthrop.* 43:245–262.

———, and R. S. Corruccini. 1975a. Distal humerus in hominoid evolution. *Folia primat.* 23:227–244.

———, and R. S. Corruccini. 1975b. Multivariate analysis of early hominid pelvic bones. *Am. J. Phys. Anthrop.* 43:263–370.

———, and R. S. Corruccini. 1976. Fossil hominid femora and the evolution of walking. *Nature* 259:567–568.

———, R. S. Corruccini, and F. C. Howell. 1976. Analysis of an early hominid ulna from the Omo basin. Ethiopia. *Am. J. Phys. Anthrop.* 44:295–304.

McKinley, K. R. 1971. Suvivorship in gracile and robust australopithecines: a demographic comparison and a proposed birth model. *Am. J. Phys. Anthrop.* 34:417–426.

Maglio, V. J. 1970. Early Elephantidae of Africa and a tentative correlation of African Plio-Pleistocene deposits. *Nature* 225:328–332.

———. 1973. Origin and evolution of the Elephantidae. *Trans. Am. Philos. Soc.* n.s. 63(3):1–149.

Mann, A. E. 1968. The paleodemography of *Australopith-ecus*. (Ph.D. Thesis. University of California, Berkeley.) Ann Arbor: University Microfilms.

———. 1975. Some paleodemographic aspects of the South African australopithecines. *Univ. Pennsylvania Publ. Anthrop.* 1:1–171.

———. 1976. Australopithecine demography. In *IXe Congrès, Union Internationale des Sciences Préhistoriques et Protohistoriques, Nice, September 1976*.

———. 1978. Australopithecine demographic patterns. In *Early African Hominids*, ed. C. J. Jolly. London: Duckworth (in press).

Marçais, J. 1934. Découverte de restes humains fossiles dans les gres quaternaires de Rabat (Maroc). *L'Anthropologie* 44:579–583.

Marks, P. 1952. Preliminary note on the discovery of a new jaw of *Meganthropus* von Koenigswald in the lower Middle Pleistocene of Sangiran, central Java. *Indonesian J. of Natural Science,* 109:26–33.

Martyn, J. E. and P. V. Tobias. 1967. Pleistocene deposits and new fossil localities in Kenya. *Nature,* 215:476–480.

Merrick, H. V. 1976. Recent archaeological research in the Plio-Pleistocene deposits of the lower Omo valley, southwestern Ethiopia. In *Human origins. Louis Leakey and the East African Evidence*, ed. G. Ll. Isaac and E. R. McCown. Menlo Park, California: W. A. Benjamin, pp. 461–482.

———, and J. P. S. Merrick. 1976. Archaeological occurrences of earlier Pleistocene age from the Shungura Formation. In *Earliest man and environments in the Lake Rudolf Basin*, ed. Y. Coppens, F. C. Howell, G. Ll. Isaac, and R. E. F. Leakey. Chicago: University of Chicago Press, pp. 574–584.

Mollison, T. 1937. Die Verletzungen am Schädel und den Gliedmassenknochen des Rhodesiafundes. *Anthrop. Anz.* 14:229–234.

Morant, G. M. 1928. Studies of Palaeolithic man. III. The Rhodesian skull and its relations to Neanderthaloid and modern types. *Ann. Eugenics* 3:337–360.

Mturi, A. A. 1976. New hominid from Lake Ndutu, Tanzania. *Nature* 262:484–485.

Napier, J. R. 1959. Fossil metacarpals from Swartkrans. *Fossil Mammals of Africa* 17:1–18. London: British Museum (Nat. Hist.).

Napier, J. 1962. Fossil handbones from Olduvai Gorge. *Nature* 196:409–411.

Napier, J. R. 1965. Comment. *Current Anthrop.* 5: 402–403.

———. 1963. The locomotor functions of hominids. In *Classification and human evolution*, ed. S. L. Washburn, Chicago: Aldine, pp. 178–189.

Neuville, R., and A. Ruhlmann. 1942. L'âge de l'homme fossile de Rabat. *Bull. Soc. d'Anthrop. de Paris*, sér. 9, 3:74–88.

Oakley, K. P. 1954. Study tour of early hominid sites in southern Africa, 1953. *So. Afr. Archaeol. Bull.* 9:75–87.

———. 1957. The dating of the Broken Hill, Florisbad and Saldanha skulls. In *Proceedings Third Pan-African Congress on Prehistory, Livingstone, 1955*, ed. J. D. Clark. London: Chatto and Windus, pp. 76–79.

_____. 1958. The dating of Broken Hill (Rhodesian Man). In *Hundert Jahre Neanderthaler. Neanderthal Centenary, 1856–1956,* ed. G. H. R. von Koenigswald. Utrecht: Kemink en Zoon, pp. 265–266.

_____. 1960. The Kanam jaw. *Nature* 185:945.

Oppennorth, W. F. F. 1932. *Homo (Javanthropus)* soloensis, een Pleistoceene mensch van Java. *Weten. Medel. Dienst van den Mijnbouw in Nederlandischindie, Batavia* 20:49–63.

Oxnard, C. E. 1968a. A note on the Olduvai clavicular fragment. *Am. J. Phys. Anthrop.* 29:429–432.

_____. 1968b. A note on the fragmentary Sterkfontein scapula. *Am. J. Phys. Anthrop.* 28:213–218.

_____. 1972. Some African fossil foot bones: a note on the interpolation of fossils into a matrix of extant species. *Am. J. Phys. Anthrop.* 37:3–12.

_____. 1973. Functional inferences from morphometrics: problems posed by diversity and uniqueness among the primates. *Syst. Zool.* 22:409–424.

_____. 1975a. The place of the australopithecines in human evolution: grounds for doubt? *Nature* 258:389–395.

_____. 1975b. *Uniqueness and Diversity in Human Evolution. Morphometric Studies of Australopithecines.* Chicago: University of Chicago Press.

Partridge, T. C. 1973. Geomorphological dating of cave opening at Makapansgat, Sterkfontein, Swartkrans and Taung. *Nature* 240:75–79.

Patterson, B., A. K. Behrensmeyer, and W. D. Sill. 1970. Geology and fauna of a new Pliocene locality in northwestern Kenya. *Nature* 226:918–921.

Patterson, B., and W. W. Howells. 1967. Hominid humeral fragment from early Pleistocene of northwestern Kenya. *Science* 156:64–66.

Peabody, F. E. 1954. Travertines and cave deposits of the Kaap escarpment of South Africa, and the type locality of *Australopithecus africanus. Bull. Geol. Soc. Amer.* 65:671–706.

Petit-Maire, N., and M. Charon. 1972. Tendances evolutives de la denture inferieure permanente des hominides du Quaternaire. *C. R. Acad. Sci.* (Paris) 274-D:365–368.

Pickford, M. 1975. Late Miocene sediments and fossils from the northern Kenya Rift Valley. *Nature* 256:279–284.

Pilbeam, D. 1969. Early Hominidae and cranial capacity. *Nature* 224:386.

_____. 1972a. Adaptive response of hominids to their environment as ascertained by fossil evidence. *Soc. Biol.* 19:15–127.

_____. 1972b. *The ascent of man. An introduction to human evolution.* New York: Macmillan.

_____. 1975. Middle Pleistocene Hominids. In *After the Australopithecines.* ed. K. W. Butzer and G. Ll. Issac. The Hague: Mouton, pp. 809–856.

_____. 1976. Neogene hominoids of South Asia and the origins of Hominidae. In *IXᵉ Congrès, Union Internationale des Sciences Préhistoriques et Protohistoriques, Nice, September 1976.*

_____. 1978. Recognizing specific diversity in heterogeneous fossil samples. In *Early African Hominids,* ed. C. J. Jolly. London: Duckworth (in press).

_____, and S. J. Gould. 1974. Size and scaling in human evolution. *Science* 186:892–901.

_____, and J. R. Vaisnys. 1975. Hypothesis testing in paleoanthropology. In *Paleoanthropology: morphology and paleoecology,* ed. R. H. Tuttle. The Hague: Mouton, pp. 3–13, 15–18.

_____, and M. Zwell. 1973. The single species hypothesis, sexual dimporphism, and variability in early hominids. *Yrbk. Phys. Anthrop. (1972)* 16:69–79.

Price, J. L., and T. I. Molleson. 1974. A radiographic examination of the left temporal bone of Kabwe man. Broken Hill mine, Zambia. *J. Archaeol. Sci.* 1:285–289.

Preuschoft, H. 1971. Body posture and mode of locomotion in early Pleistocene hominids. *Folia Primat.* 14:209–240.

Protsch, R. 1974a. The Fish Hoek hominid: another member of the basic *Homo sapiens afer. Anthrop. Ans.* 34:241–249.

_____. 1974b. Florisbad: its paleoanthropology, chronology and archaeology. *Homo* 25:68–78.

_____. 1975. The absolute dating of Upper Pleistocene sub-Saharan fossil hominids and their place in human evolution. *J. Hum. Evol.* 4:297–322.

_____. 1976. The position of the Eyasi and Garusi hominids in East Africa. In *IXᵉ Congrès, Union Internationale des Sciences Préhistoriques et Protohistoriques, Nice, September 1976.*

_____, and H. de Villiers. 1974. Bushman Rock Shelter, Origstad, Eastern Transvaal, South Africa. *J. Hum. Evol.* 3:387–396.

Pycraft, W. P. 1928a. Some suggestions for the analysis of the os coxa in man. *Man* 28:201–205.

_____. 1928b. *Rhodesian man and associated remains.* London: British Museum (National History).

Rak, Y. and F. C. Howell. 1978. Cranium of a juvenile *Australopithecus boisei* from the lower Omo basin, Ethiopia. *Am. J. Phys. Anthrop.* 48:345–366.

Read, D. W. 1975. Hominid teeth and their relationship to hominid phylogeny. *Am. J. Phys. Anthrop.* 42:105–126.

Reck, H., and L. Kohl-Larsen. 1936. Erster Überblick über die jungdiluvialen Tier und Menschenfunde Dr. Kohl-Larsen's im nordöstlichen Teil des Njarasa-Grabens (Ostafrika) und die geologischen Verhältnisse des Fundgebietes. *Geol. Rundschau* 27:401–441.

Remane, A. 1951. Die Zähne des *Meganthropus africanus. Zts. f. Morph. u. Anthrop.* 42:311–329.

_____. 1954. Structure and relationships of *Meganthropus africanus. Am. J. Phys. Anthrop.* 12:123–126.

_____. 1959. Die primitivesten Menschenformen (Australopithecinae) und des Problem des tertiären Menschen. *Schr. d. Naturw. f. Schleswig-Holstein* 29:3–10.

Rightmire, G. P. 1972. Multivariate analysis of an early hominid metacarpal from Swartkrans. *Science* 176:159–161.

_____. 1974. *The later Pleistocene and recent evolution of man in Africa.* New York: MSS Modular Publications, Module 27.

_____. 1975. Problems in the study of Later Pleistocene man in Africa. *Am. Anthrop.* 77:28–52.

_____. 1976. Relationships of Middle and Upper Pleistocene hominids from sub-Saharan Africa. *Nature* 260:238–240.

Robinson, J. T. 1953a. *Telanthropus* and its phylogenetic significance. *Am. J. Phys. Anthrop.* 11:445–502.

_____. 1953b. *Meganthropus,* australopithecines and hominids. *Am. J. Phys. Anthrop.* 11:1–38.

_____. 1954. The genera and species of the Australopithecinae. *Am. J. Phys. Anthrop.* 12:181–200.

_____. 1955. Further remarks on the relationship between "*Meganthropus*" and australopithecines. *Am. J. Phys. Anthrop.* 13:429–446.

_____. 1956. The dentition of the Australopithecinae. *Transvaal Mus.* (Pretoria) *Mem.* 9:1–179.

_____. 1958. The Sterkfontein tool-maker. *The Leech* (Johannesburg) 28:94–100.

_____. 1960. An alternative interpretation of the supposed giant deciduous hominid tooth from Olduvai. *Nature* 185:407–408.

_____. 1961. The australopithecines and their bearing on the origin of man and of stone tool-making. *So. Afr. J. Sci.* 57:3–13.

_____. 1962. The origin and adaptive radiation of the australopithecines. In *Evolution and hominisation,* ed. G. Kurth. Stuttgart: G. Fischer, pp. 120–140.

_____. 1963a. Adaptive radiation in the australopithecines and the origin of man. In *African ecology and human evolution,* ed. F. C. Howell and F. Bourlière. Chicago: Aldine, pp. 385–416.

_____. 1963b. Australopithecines, culture and phylogeny. *Am. J. Phys. Anthrop.* 21:595–605.

_____. 1965. *Homo habilis* and the australopithecines. *Nature* 205:121–124.

_____. 1967. Variation and the taxonomy of early hominids. In *Evolutionary biology,* vol. 1. New York: Appleton-Century-Crofts, pp. 69–100.

_____. 1968. The origin and adaptive radiation of the australopithecines. In *Evolution and hominisation,* 2nd ed., ed. G. Kurth. Stuttgart: G. Fischer, pp. 150–175.

_____. 1969. Dentition and adaptation in early hominids. *Intern. Congr. Anthop. Ethnol. Sciences, 8th,* Kyoto, Tokyo, 1968, pp. 302–305.

_____. 1970. Two early hominid vertebrae from Swartkrans. *Nature* 225:1217–1219.

_____. 1972. *Early hominid posture and locomotion.* Chicago: University of Chicago Press.

_____. 1978. Evidence for locomotor difference between gracile and robust early hominids from South Africa. In *Early African Hominids,* ed. C. J. Jolly. London: Duckworth (in press).

_____, and R. J. Mason. 1957. Occurrence of stone artefacts with *Australopithecus* at Sterkfontein. *Nature* 180:521–524.

_____, and R. J. Mason. 1962. Australopithecines and artefacts at Sterkfontein. *So. Afr. Archaeol. Bull.* 17:87–125.

_____, and K. Steudel. 1973. Multivariate discriminant analysis of dental data bearing on early hominid affinities. *J. Hum. Evol.* 2:509–527.

Roche, J. 1953. La grotte de Taforalt. *L'Anthropologie* 57:375–380.

_____, and J.-P. Texier. 1976. Découverte de restes humaines dans un niveau atérien supérieur de la grotte des Contrebandiers- a Temara (Moroc). *C. R. Acad. Sci.* (Paris) 282-D(1):45–47.

Romer, A. S. 1925. *Australopithecus* not a chimpanzee. *Science* 71:482–483.

Ruhlmann, A. 1951. La grotte préhistorique de Dar es Soltane. *Collection Heperis, Institut des Haute Etudes Morocaines,* 111–210.

Saban, R. 1972. Les hommes fossiles du Maghreb. *L'Ouest Med.* 25:2443–2458.

_____. 1975. Les restes humains de Rabat (Kébibat). *Ann. de Paléont.* (*Vért.*) 61:151–207.

_____. 1976. A propos des traces vasculaires endocraniennes chez l'homme de Rabat. In *IXe Congrès, Union Internationale des Sciences Préhistoriques et Protohistoriques, Nice, September 1976.*

_____. 1977. Les impressions vasculaires pariétales endocrâniennes dans la lignée des Hominidés. *C. R. Acad. Sci.* (Paris), 284:803–806.

Sakka, M. 1976. Exocrane des Australopithèques (valuer semeiologique). In *IXe Congrès, Union Internationale des Sciences Préhistoriques et Protohistoriques, Nice, September 1976.*

Sartono, S. 1961. Notes on a new find of a *Pithecanthropus* mandible. *Publikasi Teknik Seri Paleontologi, Departemen Perindustrian Dasar-Pertambangan, Djawatan Geologi, Republik Indonesia, Bandung,* no. 2.

_____. 1971. Observations on a new skull of *Pithecanthropus erectus* (*Pithecanthropus* VIII) from Sangiran, central Java. *Proc. K. Ned. Adad. Wet. Amst.,* B, 74:185–194.

_____. 1975. Implications arising from *Pithecanthropus* VIII. In *Paleoanthropology: morphology and paleoecology,* ed. R. H. Tuttle. *World Anthropology Series.* The Hague: Mouton, pp. 327–360.

Saussé, F. 1975. La mandibule atlanthropienne de la carrière Thomas I (Casablanca). *L'Anthropologie* 79:81–112.

_____. 1976. Les pithécanthropiens Marocains. In *IXe Congès, Union Internationale des Sciences Préhistoriqes et Protohistoriques, Nice, September 1976.*

Schepers, G. W. H. 1941. The mandible of the Transvaal fossil human skeleton from Springbok Flats. *Ann. Transv. Mus.* 20:253–271.

Şenyürek, M. S. 1940. Fossil man in Tangier. *Papers, Peabody Mus. Am. Archaeol. Ethnol.* (Harvard University) 16(3):1–27.

_____. 1941. The dentition of *Plesianthropus* and *Paranthropus.* Ann. Transv. Mus. 20:293–302.

———. 1955. A note on the teeth of *Meganthropus africanus* Weinert from Tanganyika Territory. *Belletën* 19:1–55.

Shuey, R. T., Brown, F. H., and M. K. Croes. 1974. Magnetostratigraphy of the Shungura Formation, southwestern Ethiopia: fine structure of the lower Matuyama polarity epoch. *Earth and Planetary Science Letters* 23: 249–260.

———, F. H. Brown, G. G. Eck and F. C. Howell. 1978. A statistical approach to temporal biostratigraphy. In *Geological Background to Fossil Man in Africa,* ed. W. W. Bishop. London: Geological Society of London. (in press).

Siedner, G., and A. Horowitz. 1974. Radiometric ages of late Cainozoic basalts from northern Israel: chronostratigraphic implications. *Nature* 250:23–26.

Simons, E. L. 1976. Relationships between *Dryopithecus, Sivapithecus* and *Ramapithecus*. In *IXᵉ Congrès, Union Internationale des Sciences Préhistoriques et Protohistoriques, Nice, September 1976.*

———. 1977. *Ramapithecus*. *Sci. Amer.,* 236:28–35.

Simpson, G. G. 1963. The meaning of taxonomic statements. In *Classification and Human Evolution,* ed. S. L. Washburn. Chicago: Aldine, pp. 1–31.

Singer, R. 1954. The Saldanha skull from Hopefield, South Africa. *Am. J. Phys. Anthrop.* 12:345–362.

———. 1956. Man and mammals in South Africa (with special reference to Saldanha man). *J. Paleont. Soc. India* 1:120–130.

———. 1958. The Rhodesian, Florisbad and Saldanha skulls. In *Hundert Jahre Neanderthaler. Neanderthal Centenary, 1856–1956,* ed. G. H. R. von Koenigswald. Utrecht: Kemink en Zoon, pp. 52–62.

———, and P. Smith. 1969. Some human remains associated with the Middle Stone Age at Klassies River South Africa. *Am. J. Phys. Anthrop.* 31:256 (abstract).

———, and J. Wymer. 1968. Archaeological investigations at the Saldanha skull site in South Africa. *So. Afr. Archaeol. Bull* 25:63–74.

Sollas, W. J. 1926. A sagittal section of the skull of *Australopithecus africanus*. *Quart. J. Geol. Soc. London* 82:1–11.

Speth, J. D., and D. D. Davis. 1976. Seasonal variability in early hominid predation. *Science* 192:441–445.

Straus, W. L. 1948. The humerus of *Paranthropus robustus*. *Am. J. Phys. Anthrop.* 6:285–312.

Stearns, C. E., and D. L. Thurber. 1965. Th²³⁰/U²³⁴ dates of late Pleistocene marine fossils from the Mediterranean and Moroccan littorals. *Quaternaria* 7:29–42.

Street, F. A., and A. T. Grove. 1976. Environmental and climatic implications of late Quaternary lake-level fluctuations in Africa. *Nature* 261:385–390.

Stringer, C. B. 1974. Population relationships of later Pleistocene hominids: a multivariate study of available crania. *J. Archaeol. Sci.* 1:317–342.

Szalay, F. S. 1971. Biological level of organization of the Chesowanja robust australopithecine. *Nature* 234:229–230.

Taieb, M. 1974a. Découverte d'hominidés en Afar central. *Le Courrier du C.N.R.S.* (Paris) 11:3–8.

———. 1974b. *Evolution Quaternaire du Bassin de l'Awash (Rift Ethiopien et Afar)*. 2 volumes. Université de Paris VI. Thèse, Docteur es Sciences Naturelles.

———, Y. Coppens, D. C. Johanson and R. Bonnefille. 1975. Hominidés de l'Afar central, Ethiopie (site d'Hadar, campagne 1973). *Bull. et Mém. Soc. d'Anthrop. Paris,* sér. 13, 2:117–124.

———, Y. Coppens, D. C. Johanson, and J. Kalb. 1972. Dépôts sédimentaires et faunes du Plio-Pléistocène de basse vallée de l'Awash (Afár central, Ethiopie). *C. R. Acad. Sci.* (Paris) 275:819–822.

———, D. C. Johanson and Y. Coppens. 1975. Expedition internationale de l'Afar, Ethiopie (3ᵉ campagne 1974): découverte d'hominidés Plio-Pléistocènes à Hadar. *C. R. Acad. Sci.* (Paris) 281:1297–1300.

———, D. C. Johanson, Y. Coppens, and J. L. Aronson. 1976. Geological and paleontological background of Hadar hominid site, Afar, Ethiopia. *Nature* 260:289–293.

———, D. C. Johanson, Y. Coppens, R. Bonnefille, and J. Kalb. 1974. Découverte d'hominidés dans les séries Plio-Pléistocènes d'Hadar (bassin de l'Awash; Afar, Ethiopie). *C. R. Acad. Sci.* (Paris) 279:735–738.

Tattersall, I. and N. Eldredge. 1977. Fact, theory and fantasy in human paleontology. *Amer. Scientist,* 65:204–211.

Tobias, P. V. 1960. The Kanam jaw. *Nature* 185:946–947.

———. 1961. New evidence and new views on the evolution of man in Africa. *So. Afr. J. Sci.* 57:25–38.

———. 1962. A reexamination of the Kanam mandible. *Actes du IVᵉ Congrès Panafricain de Préhistoire et de l'Étude du Quaternaire* 1:341–360.

———. 1964. The Olduvai Bed I hominine with special reference to its cranial capacity. *Nature* 202:3–4.

———. 1965. *Homo habilis*. Encyclopedia Britannica Book of the Year 1965:252–255.

———. 1966. A re-examination of the Kedung Brubus mandible. *Zool. Meded.* (Leiden) 41(22):307–320.

———. 1967a. *Olduvai Gorge, vol. 2. The cranium and maxillary dentition* of Australopithecus (Zinjanthropus) boisei. Cambridge: Cambridge University Press.

———. 1967b. Cultural hominisation among the earliest African Pleistocene hominids. *Proc. Prehist. Soc.* 13: 367–376.

———. 1968. The pattern of venous sinus grooves in the robust australopithecines and other fossil and modern hominoids. In *Anthropology and Human Genetics*. ed. E. Schwidetsky. Stuttgart: G. Fischer, pp. 1–10.

———. 1971a. *The brain in hominid evolution*. New York: Columbia University Press.

———. 1971b. Does the form of the inner contour of the mandible distinguish between *Australopithecus* and *Homo*? In *Perspectives in palaeoanthropology: D. Sen Festschrift Volume,* ed. A. K. Ghosh. Calcutta: K. L. Mukhopadhyoy, pp. 9–17.

———. 1971c. Human skeletal remains from the Cave of

Hearths, Makapansgat, Transvaal. *Am. J. Phys. Anthrop.* 34:335–368.

————. 1972. "Dished faces," brain size and early hominids. *Nature* 239:468–469.

————. 1973a. Darwin's prediction and the African emergence of the genus *Homo. Atti del Colloquio Internationale sul Tema: "L'Origine dell "Uomo," Accadmia Nazionale dei Lincei,* Roma, 370, Quaterno N. 182:63–85.

————. 1973b. Implications of the new age estimates of the early South African hominids. *Nature* 246:79–83.

————. 1974. *Homo erectus. Encyclopedia Britannica* 8:1030–1036.

————. 1975. Brain evolution in the Hominoidea. In *Primate Functional Morphology and Evolution,* ed. R. H. Tuttle. The Hague: Mouton, pp. 353–392.

————, and A. R. Hughes. 1969. The new Witwatersrand University excavation at Sterkfontein. *So. Afr. Archaeol. Bull.* 24:158–169.

————, and G. H. R. von Koenigswald. 1964. A comparison between the Olduvai hominines and those of Java and some implications for hominid phylogeny. *Nature* 204:515–518.

————, and J. T. Robinson. 1966. The distinctiveness of *Homo habilis. Nature* 209:953–960.

Toerien, M. J., and A. R. Hughes. 1955. The limb bones of Springbok Flats man. *So. Afr. J. Sci.* 52:125–128.

Vallois, H. V. 1945. L'homme fossile de Rabat. *C. R. Acad. Sci.* (Paris) 221:669–671.

————. 1951. La mandibule humaine fossile de la grotte du Porc-Epic près Dire-Daoua (Abyssinie). *L'Anthropologie* 55:231–238.

————. 1954. La capacité cranienne chez le primates supérieurs et le "rubicon cerebrale." *C. R. Acad. Sci.* (Paris) 238:1349–1351.

————. 1960. L'homme de Rabat. *Bull. d'Archéol. Marocaine* 3:87–91.

————. 1976. L'interprétation de l'omoplate du *Plesianthropus transvaalensis. L'Anthropologie,* 80:229–242.

————, and J. Roche. 1958. La mandibule acheuléene de Témara, Maroc. *C. R. Acad. Sci.* (Paris), 246:3113–3116.

Vogel, J. C., and P. B. Beaumont. 1972. Revised radiocarbon chronology for the Stone Age in South Africa. *Nature* 237:50–51.

Vrba, E. S. 1974. Chronological and ecological implications of the fossil Bovidae at the Sterkfontein australopithecine site. *Nature* 250:19–23.

————. 1975. Some evidence of chronology and palaeoecology of Sterkfontein, Swartkrans and Kromdraai from the fossil Bovidae. *Nature* 254:301–304.

————. 1976. The fossil Bovidae of Sterkfontein, Swartkrans and Kromdraai. *Transv. Mus. (Pretorial) Mem. 21.*

Walker, A. C. 1972a. Chesowanja australopithecine. *Nature* 238:108–109.

————. 1972b. New *Australopithecus* femora from East Rudolf, Kenya. *J. Hum. Evol.* 2:545–555.

Wallace, J. A. 1973. Tooth chipping in the australopithecines. *Nature* 244:117–118.

————. 1975. Dietary adaptations of *Australopithecus* and early *Homo.* In *Paleoanthropology: morphology and paleoecology,* ed. R. H. Tuttle. The Hague: Mouton, pp. 203–223.

————. 1978. Evolutionary trends in the early hominid dentition: a study in paleoanatomy. In *Early African Hominids,* ed. C. J. Jolly. London: Duckworth (in press).

Weidenreich, F. 1936. The mandible of *Sinanthropus pekinensis:* a comparative study. *Palaeont. Sinica,* ser. D, 7(3):1–162.

————. 1937. The dentition of *Sinanthropus pekinensis:* a comparative odontography of hominids. Atlas and Text, 2 vols. *Palaeont. Sinica,* n. S. D, no. 1 (whole series no. 101).

————. 1940. The torus occipitalis and related structures and their transformation in the course of human evolution. *Bull. Geol. Soc. China* 19:479–559.

————. 1941a. The brain and its role in the phylogenetic transformation of the human skull. *Trans. Am. Philos. Soc.* 31:321–442.

————. 1941b. The extremity bones of *Sinanthropus pekinensis. Palaeont. Sinica,* n. S. D, no. 5 (whole series no. 115):1–150.

————. 1943. The skull of *Sinanthropus pekinensis:* a comparative study on a primitive hominid skull. *Palaeont. Sinica,* n. s. D, no. 10 (whole series no. 127):1–484.

————. 1945. Giant early man from Java and south China. *Anthrop. Papers Am. Mus. Nat. Hist.* 40(1): 134 pages.

————. 1947. The trend of human evolution. *Evolution* 1:221–236.

————. 1951. Morphology of Solo man. *Anthrop. Papers Am. Mus. Nat. Hist.* 43(3):201–290.

Weiner, J. S. 1958. The pattern of evolutionary development of the genus *Homo. So. Afr. J. Sci.,* 23:111–120.

Weinert, H. 1950. Uber die neuen Vor- und Frühmenschenfunde aus Afrika, Java, China and Frankreich. *Zts. f. Morph. u. Anthrop.,* 42:113–148.

Weinert, H., H. Bauermeister and A. Remane. 1939. *Africanthropus njarasensis.* Beschreibung und phyletische *Einordnung* des ersten Affenmenschen aus Ostafrika. *Zts. f. Morph. v. Anthrop.* 38:252–308.

Wells, L. H. 1950. The Border Cave skull, Ingwavuma district, Zululand. *Am. J. Phys. Anthrop.* 8:241–243 (photos and note only).

————. 1952. Fossil man in southern Africa. *Man* 52:36–37.

————. 1957. The place of the Broken Hill skull among human types. In *Proceedings of the Third Pan-African Congress on Prehistory, Livingstone, 1955,* ed. J. D. Clark. London: Chatto and Windos, pp. 172–174.

————. 1958. On fitting a mandible to the Broken Hill skull. *So. Afr. J. Med. Sci.* 23:125–134.

————. 1959. The problem of Middle Stone Age man in southern Africa. *Man* 59:158–160.

————. 1962. Pleistocene faunas and the distribution of mammals in southern Africa. *Ann. Cape Prov. Mus.* 2:37–40.

_____. 1964. The Vaal river "Younger Gravels" faunal assemblage: a revised list. *So. Afr. J. Sci.* 60:91–93.

_____. 1969. *Homo sapiens afer* Linn.—content and earliest representatives. *So. Afr. Archaeol. Bull.* 24: 172–173.

_____. 1972. Late Stone Age and Middle Stone Age toolmakers. *So. Afr. Archaeol. Bull.* 27:5–9.

White, T. D. 1977. New fossil hominids from Laetolil, Tanzania. *Am. J. Phys. Anthrop.* 46:197–229.

_____, and J. M. Harris. 1977. Suid evolution and correlation of African hominid localities. *Science* 198: 13–21.

Williams, M. A. J. 1975. Late Pleistocene tropical aridity synchronous in both hemispheres? *Nature* 253:617–618.

Wolberg, D. L. 1970. The hypothesized osteodontokeratic culture of the Australopithecinae: a look at the evidence and the opinions. *Current Anthrop.* 11:23–37.

Wolpoff, M. H. 1968. *"Telanthropus"* and the single species hypothesis. *Am. Anthrop.* 70:447–493.

_____. 1971a. Competitive exclusion among lower Pleistocene hominids: the single species hypothesis. *Man* 6:601–614.

_____. 1971b. The evidence for multiple hominid taxa at Swartkrans. *Am. Anthrop.* 72:576–607.

_____. 1971c. Metric trends in hominid dental evolution. *Case Western Reserve University Studies in Anthropology, 2.*

_____. 1974a. The evidence for two australopithecine lineages in South Africa. *Yrbk. Phys. Anthrop. (1973)* 17:113–139.

_____. 1974b. Sagittal cresting in the South African australopithecines. *Am. J. Phys. Anthrop.* 40:397–408.

_____. 1975. Sexual dimorphism in the australopithecines. In *Paleoanthropology: morphology and paleoecology,* ed. R. H. Tuttle. The Hague: Mouton, pp. 245–284.

_____. 1976a. Some aspects of the evolution of early hominid sexual dimorphism. *Current Anthrop.* 17:579–606.

_____. 1976b. Evolutionary aspects of hominid tooth size reduction and early hominid dental variation. In *IXᵉ Congres, Union Internationale des Sciences Préhistoriques et Protohistoriques, Nice, September 1976.*

_____. 1976c. Fossil hominid femora. *Nature,* 264:812–813.

_____. 1978. Analogies and interpretation in paleoanthropology. In *Early African Hominidae,* ed. C. J. Jolly. London: Duckworth. (in press).

_____, and C. Lovejoy. 1975. A rediagnosis of the genus *Australopithecus. J. Hum. Evol.* 4:275–276.

Woo, J. K. 1964a. Mandible of the *Sinanthropus*-type discovered at Lantian, Shansi—*Sinanthropus lantianensis. Vertebr. Palasiat.* 8:1–17 (in Chinese).

_____. 1964b. Mandible of the *Sinanthropus*-type discovered at Lantian, Shensi—*Sinanthropus lantianensis. Sci. Sinica* 13:801–811.

_____. 1964c. Mandible of *Sinanthropus lantianensis. Current Anthrop.* 5:98–101.

_____. 1966. The hominid skull of Lantian, Shansi. *Vertebr. Palasiat.* 10:1–22.

_____, and L. P. Chao. 1954. New discoveries about *Sinanthropus pekinensis* in Choukoutien. *Acta Scie. Sinica* 3:335–351. (Also in *Acta Palaeont. Sinica* 2:267–288.)

_____, and L. P. Chao. 1959. New discovery of *Sinanthropus* mandible from Choukoutien. *Vert. Palasiat.* 3:169–172.

Wood, B. A. 1974a. Evidence on the locomotor pattern of *Homo* from the early Pleistocene of Kenya. *Nature* 251:135–136.

_____. 1974b. Olduvai Bed I post-cranial fossils: a reassessment. *J. Hum. Evol.* 3:373–378.

Woodward, A. S. 1938. A fossil skull of an ancestral Bushman from the Anglo-Egyptian Sudan. *Antiquity* 12:193–195.

Zihlman, A. L. 1971. The question of locomotor differences in *Australopithecus.* In *Proc. 3rd Int. Congr. Primat., Zurich, 1970.* Basel: Karger, 1:54–66.

_____. 1976. Sexual dimorphism and its behavioral implications in early hominids. In *Xᵉ Congrès, Union Internationale des Sciences Préhistoriques et Protohistoriques, Nice, September 1976.*

_____. 1978. Interpretations of early hominid locomotion. In *Early African Hominids.* ed. C. J. Jolly. London: Duckworth (in press).

_____, and W. S. Hunter. 1972. A biomechanical interpretation of the pelvis of *Australopithecus. Folia Primatol.* 18:1–19.

Zuckerman, S. 1928. Age-changes in the chimpanzee with special reference to growth of brain, eruption of teeth and estimation of age, with a note on the Taungs ape. *Proc. Zool. Soc. London* 1928:1–42.

11

Carnivora

R. J. G. Savage

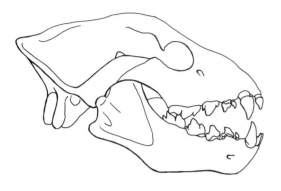

All fossil carnivorous mammals—that is, members of the creodont family Hyaenodontidae and seven of the eleven families of true carnivora—found in Africa are reviewed in this chapter. The carnivores (even without the creodonts) are the second largest order of mammals. Creodonts and carnivores together have 441 known genera (table 11.10), of which some three-quarters are extinct. However, the total number of African genera is only 69 (15% of the world total), and of these 29 are extinct (8% of the world total). Carnivores are always rare fossils compared with herbivores.

The bulk of our knowledge of fossil mammals comes from North, East, and South Africa. The South African sites are almost all Plio-Pleistocene, with one in the Miocene. The East African sites range through from early Miocene to late Pleistocene, almost continuously. The North African sites are Eocene-Oligocene and Miocene in Egypt, Libya, and Tunisia, with mainly Pleistocene sites in Morocco and Algeria.

There is much undescribed Miocene and Plio-Pleistocene material from East Africa; a preliminary survey will extend our knowledge of the range of known stocks rather than significantly add major new taxa. Many North African sites have not been described in detail or analyzed faunistically, and much reliance has had to be placed on review papers that do not detail occurrences at individual sites. "Barbary" is used to cover these ambiguities in the tables; in tables 11.10 and 11.11 the Pleistocene occurrences read slightly low because of the lumping of sites. The lower faunal successions of Omo, East Turkana, Olduvai, and Transvaal are here taken as early Pleistocene; all appear to extend beyond the 2.7 m.y. level usually taken as the base of the Pleistocene. However, the faunas with carnivores are probably almost all at levels above this datum. The order of sites within each time boundary in the tables is of necessity arbitrary.

During research for this chapter I have been struck by the uncertainty of many of the identifications and the amount of confusion in the literature regarding the occurrence of individual taxa at each site. This is especially true of the Pleistocene data; one writer may mention one species at a site, and another writer may mention another species at the same site, and it is often impossible to tell whether they might, in fact, be referring to the same specimen. The confusion is further confounded by the plethora of preliminary reports that today vitiate the literature. When little material is at hand, or when material is plentiful, taxonomic decisions are usually easy; one can either say nothing or give a firm answer. The intermediate condition is worst, when several solutions present themselves and none is provable. The result is a profusion of sp., aff., cf, and ? species, particularly for carnivores.

Unlike some contributors to this volume I have not been able to travel and see all the material reviewed, nor indeed would it have been feasible had monies been available, because over 140 species are involved. Hence my judgments have often had to be made from intuition rather than from experience. At some time or another I have seen almost all the Oligocene and Miocene specimens of African carnivores; my personal knowledge of Pliocene and Pleistocene material is more limited.

No attempt is made to give a full bibliography of references to the carnivore literature. The reader should refer to Hopwood and Hollyfield (1952) for all papers up to 1950 and to Cross and Maglio (1975) for contributions from 1950 to 1972. Welbourne (1969) has provided a supplement for Quaternary faunas.

The major studies dealing with carnivore faunas are found in Cooke (1964), Ewer (1955, 1956a, 1956b, 1956c, 1967), Hendey (1974), Leakey (1967), Petter (1973), and Savage (1965).

The Fossil Record

Suborder Creodonta Cope 1875

The major papers dealing with African creodonts are Andrews (1906), Osborn (1909), Savage (1965, 1973), Schlosser (1911) and Van Valen (1967). All known African creodonts belong to the family Hyaenodontidae, a family characterized by having molars $\frac{1 + 2}{2 + 3}$ specialized as carnassial shearing teeth (table 11.1). Savage (1965) recorded *Kelba* as a Miocene arctocyonid on the basis of an isolated molar; a complete upper dentition is now known and the affinities are still obscure. Van Valen (1967,

Table 11.1   Distribution of African Hyaenodontidae.

Time	Site	Metasinopa ethiopica	M. fraasi	M. napaki	Anasinopa leakeyi	Masrasector aegypticum	Dissopsalis pyroclasticus	Teratodon spekei	T. enigmae	Apterodon	A. altidens	A. macrognathus	A. minutus	Leakitherium hiwegi	Pterodon africanus	P. leptognathus	P. phiomensis	P. kaiseri	P. zadoki	Hyaenodon andrewsi	H. matthewi	H. pilgrimi	Hyainailouros nyanzae	H. fourtaui	H. sp.	Megistotherium osteothlastes	Hyaenodont indet.
Middle Miocene	Kaboor	—	—	—	—	—	+	—	—	—	—	—	—	—	—	—	—	—	—	—	—	—	—	—	—	—	—
	Fort Ternan	—	—	—	?	—	—	—	—	—	—	—	—	—	—	—	—	—	—	—	—	—	—	—	+	—	—
	Gafsa (L)	—	—	—	—	—	—	—	—	—	—	—	—	—	—	—	—	—	—	—	—	—	—	—	—	—	+
Early Miocene	Moghara	—	—	—	—	—	—	—	—	—	—	—	—	—	—	—	—	—	—	—	—	—	—	+	—	—	—
	Gebel Zelten	—	—	—	?	—	—	—	—	—	—	—	—	—	—	—	—	—	—	—	—	—	—	—	—	+	—
	Maboko	—	—	—	+	—	—	—	—	—	—	—	—	—	—	—	—	—	—	—	—	—	—	—	—	—	—
	Muruarot 2	—	—	—	+	—	—	—	—	—	—	—	—	—	—	—	—	—	—	+	—	—	—	—	—	—	—
	Ombo	—	—	—	—	—	—	—	—	—	—	—	—	—	—	—	—	—	—	+	—	—	+	—	—	—	—
	Mfwangano	—	—	—	+	—	—	—	—	—	—	—	—	—	—	—	—	—	—	—	—	—	—	—	—	—	—
	Rusinga 2	—	—	—	—	—	—	—	—	—	—	—	—	—	—	—	—	—	—	—	—	—	—	+	—	—	—
	R. 106, 31	—	—	—	+	—	—	—	—	—	—	—	—	—	—	—	—	—	—	—	—	—	—	—	—	—	—
	Rusinga	—	—	—	—	—	—	—	—	—	+	+	—	—	+	+	—	+	+	+	—	+	+	—	—	—	—
	R. 1, 1a, 3, 12, 18	—	—	—	—	—	—	—	—	—	—	+	—	—	—	—	—	+	—	+	—	—	+	—	—	—	—
	Napak I, IV, V	—	—	+	—	—	—	—	—	—	—	—	—	—	—	+	—	—	—	—	—	—	—	—	—	—	—
	Napak IIa	—	—	—	—	—	—	—	—	—	—	—	—	—	—	—	—	—	—	—	—	—	—	+	—	—	—
	Koru	—	—	—	—	—	—	+	—	—	—	—	—	—	—	—	—	—	—	—	—	—	—	—	—	—	—
	Songhor	—	—	—	—	—	—	+	+	—	—	—	—	—	—	—	—	—	—	—	—	+	+	+	—	—	—
	Karungu	—	—	+	—	—	—	—	—	—	—	—	—	—	—	—	—	—	—	—	+	—	—	—	—	—	—
	Lüderitz	—	—	—	—	—	—	—	—	—	—	—	—	—	—	—	—	—	—	—	+	—	—	—	—	—	—
Oligocene	Gebel Qatrani	+	+	—	—	+	—	—	—	—	+	+	+	—	+	+	+	—	—	+	—	—	—	—	—	—	—
	Dor el Talha	—	—	—	—	—	—	—	—	—	—	—	—	—	—	—	—	—	—	—	—	—	—	—	—	—	+
Eocene	Qasr es Sagha	—	—	—	—	—	—	—	—	+	—	—	—	—	—	—	—	—	—	—	—	—	—	—	—	—	—

Table 11.2 Distribution of Proviverrinae.

Time	Europe	Africa	Asia	North America
Miocene		*Dissopsalis* *Anasinopa* *Metasinopa* *Teratodon*	*Dissopsalis*	
Oligocene	*Cynohyaenodon*	*Metasinopa*		*Propterodon*
Eocene	*Cynohyaenodon* *Paracynohyaenodon* *Proviverra* *Quercytherium*		*Propterodon*	
	Cynohyaenodon *Prodissopsalis* *Prototomus* *Proviverra*			*Proviverra* *Tritemnodon*
	Prototomus			*Arfia* *Prototomus* *Tritemnodon*

p. 228) suggested that *Kelba* is as close to *Pantolestes* as to oxyclaenines and favored putting it in the Pantolestidae; he would also derive *Ptolemaia* from Pantolestidae. In the next sentence he placed *Kelba* in the Ptolemaiidae; this mood persists for the classification (p. 260). P. M. Butler (pers. comm.) also favors association of *Kelba* with *Ptolemaia*. I would not place much weight on Van Valen's view because he inclines to throw out innumerable suggestions on affinities without clear supporting evidence—no doubt in the long run one of them might turn out to be right. The most puzzling feature of the *Kelba* dentition is the fully molariform P⁴. Simpson (1937) recognized two subfamilies of the Pantolestidae, differentiated on P⁴; the Pantolestinae have a P⁴ which is little enlarged and lacks a metacone; the Pentacodontinae have a much enlarged P⁴ with distinct metacone, large protocone, expanded cingulum, and styles are small or absent. By definition the Pantolestinae cannot include the enigmatic genus *Kelba,* and in only two genera of the Pentacodontinae is P⁴ known. These are *Coriphagus* and *Aphronorus,* and in both the paracone is much larger than the metacone, while in *Kelba* the cones are equal. This and other differences make it unlikely that *Kelba* can be included in the Pentacodontinae (= Pentacondontidae of Van Valen 1967). No upper dentition of *Ptolemaia* is described, and hence I would not rule out the possibility of association with the taxon. Alternatively *Kelba* may turn out to be an aberrant condylarth as I originally suggested (1965). Savage also in 1965 described a new genus *Teratodon* from the East African Miocene; he placed this with the French

Quercytherium in a new family Teratodontidae. Van Valen (1967) agreed that this status was possible, but preferred to include these two genera along with the proviverrines in his subfamily Hyaenodontinae in which he recognized no tribal subdivisions. In this chapter I include *Teratodon* with the Proviverrinae as a subfamily of the Hyaenodontidae. *Apterodon* is another controversial African creodont. Van Valen (1966) placed it in the Mesonychidae and thus in the order Condylarthra. Szalay (1967) would not accept it as a mesonychid and gave it tribal status within the Hyaenodontidae, suggesting an origin within the proviverrines. Van Valen (1967) accepted Szalay's conclusions. Here it is accepted as an hyaenodontid and is discussed for convenience amongst the hyaenodontines.

Thus the African creodonts can be seen to belong to one of two subfamilies of Hyaenodontidae, the Proviverrinae or the Hyaenodontinae. Two other subfamilies are recognized on other continents, the Limnocyoninae from the Eocene of North America and Europe, a short-faced stock derived from proviverrines, and the Machaeroidinae from the North American Eocene, a felidlike stock possibly derived from limnocyonines.

Family Hyaenodontidae Leidy 1869

Subfamily Proviverrinae Matthew 1909

The distribution in time and space of all proviverrine genera is shown on table 11.2. The detailed distribution of the African species is given in table 11.1; seven species in four genera are recorded. The subfamily is distinguished by having upper molars in which the paracone and metacone are well separated and lower molars that possess a metaconid.

Metasinopa is reported with two species in the early Oligocene of the Fayum, and a third species from the early Miocene of Kenya. Van Valen (1967) considered that *M. fraasi* could be derived from *M. ethiopica* and thought the genus not clearly distinct from the European Eocene genera *Paracynohyaenodon* and *Prodissopsalis*. My preference is to allow the genus to stand and accept that it is probably descended from the European Eocene stock.

Simons and Gingerich (1974) described a new proviverrine hyaenodont from the Jebel el Qatrani Formation of the Fayum; they named it *Masrasector aegypticum* and pointed out the similarity which it bears to *Anasinopa* from Kenya.

Anasinopa is the most common creodont found in the East African Miocene faunas. The genus is known only from one species, and Van Valen (1967) synonymized it with *Paracynohyaenodon,* another genus based solely on the genotype, and which occurs in the late Eocene of France. Only the lower molars of *Paracynohyaenodon* are known; these have a number of characters in common with *Anasinopa* (see Savage 1965, Van Valen 1967). Probably the most significant difference between the two genera is the angle of carnassial cusps. In *Paracyno-*

hyaenodon the paraconid-protoconid-metaconid angle on M_2 and M_3 is 70° to 71°, while in *Anasinopa* it ranges from 55° to 60°.

Dissopsalis is represented in Africa by only one occurrence and at a level that is probably comparable with the level of the genotype in the Chinji Stage (middle Miocene) of the Siwaliks.

Teratodon must belong to a very aberrant sideline, and it is likely that *Quercytherium* from France has the same ancestry.

Subfamily Hyaenodontinae Trouessart 1885

The distribution in time and space of all hyaenodontine genera is shown in table 11.3. The detailed distribution of the African species is given in table 11.1; 15 species in 6 genera are recorded. The subfamily is distinguished by having upper molars in which the paracone and metacone are more or less connate, and lower molars which lack a metaconid and in which the talonid is either reduced or absent.

Apterodon has the distinction of being the earliest recorded creodont in Africa; Simons and Gingerich (1976) record a new species *A. saghensis* from the late Eocene Qasr el Sagha Formation of the Fayum.

The genus is better known from three Fayum species in the early Oligocene. The genus is so aberrant that, as reported above, its inclusion within the Creodonta has been doubted; European species are

Table 11.3   Distribution of Hyaenodontinae.

	Europe	Africa	Asia	North America
Miocene	*Hyainailouros*			
	Hyainailouros			
	Hyainailouros	*Hyaenodon* *Hyainailouros* *Leakitherium* *Megistotherium* *Pterodon*[a]	*Megistotherium*	
Oligocene	*Hyaenodon*			
	Apterodon *Hyaenodon*		*Hyaenodon*	*Hyaenodon*
	Apterodon *Hyaenodon* *Pterodon*	*Apterodon* *Hyaenodon* *Pterodon*	*Megalopterodon* *Hyaenodon*	*Hemipsalodon* *Hyaenodon* *Pterodon*
Eocene	*Apterodon* *Hyaenodon* *Pterodon*	*Apterodon*	*Hyaenodon* *Pterodon*	*Hyaenodon* *Pterodon*
	Apterodon *Hyaenodon* *Pterodon*			

[a]*Pterodon* includes *Metapterodon*.

known from middle Eocene through to middle Oligocene. The remaining five genera, *Pterodon, Leakitherium, Hyaenodon, Hyainailouros,* and *Megistotherium,* comprise the bulk of the African creodonts ranging from early Oligocene to middle Miocene. All five genera are closely related, and unfortunately the unevenness of the preservation makes comparisons very difficult. *Megistotherium* is best known from a skull, *Hyaenodon* is known only from lower dentitions, and *Leakitherium* from upper dentitions. Van Valen (1967) regarded *Metapterodon* as synonymous with *Pterodon*. In 1965 Savage placed *Pterodon biincisivus* in the genus *Metapterodon*. Clearly the genotypes of the two genera (*Pterodon dasyuroides* and *Metapterodon kaiseri*) are very close and on this basis I concur with Van Valen in now synonymizing them under *Pterodon*. The Fayum *Pterodon* and *Hyaenodon* species could have been derived from earlier European species. *Leakitherium* is distinguished from *Pterodon* in having connate paracone and metacone on M¹ and M², a feature seen in the deciduous dentition of *Pterodon dasyuroides;* thus *Leakitherium* can be seen as a late survivor from a pre-*Pterodon* stock.

Hyainailouros was for a long time a genus of uncertain affinity because of the incompleteness of the remains. It is now clearly accepted (see Beaumont 1970) as a large hyaenodontine, based on the European Miocene *H. sulzeri.* To this genus may also be referred a tooth from Moghara as *H. fourtaui* and some teeth from Kenya as *H. nyanzae.* Savage (1973) described some canines from Rusinga and Fort Ternan as hyaenodontine, and because of their size they may be referred to as *Hyainailouros* sp.

Megistotherium is the largest of all creodonts and is based on a virtually complete skull from Gebel Zelten, Libya. Though comparison with *Hyainailouros* is not easy, sufficient differences in the dentition justify generic separation. Also in the genus *Megistotherium* is placed a specimen from Bugti, Pakistan.

Among the undescribed creodont material is a hyaenodontid from the early Oligocene of Dor el Talha, Libya, a proviverrine (possibly *Anasinopa*) from the early Miocene of Gebel Zelten, and other proviverrines (close to *Anasinopa* and *Dissopsalis*) from the middle Miocene of Fort Ternan. Robinson and Black (1969) report a hyaenodontid from the lower levels (middle Miocene) of the Beglia Formation of Gafsa, southern Tunisia.

Hyaenodontids are well represented in Eocene faunas of Europe, especially Quercy, Egerkingen, and Geiseltal. The earliest known African creodonts of Fayum could readily be derived from these faunas, though the record is not good enough to establish clear lines of descent. The African Miocene creodonts suggest evolution from the earlier Oligocene faunas, with a reinvasion of Eurasia by advanced giant forms, *Hyainailouros* and *Megistotherium*. Creodonts remained the dominant carnivores much longer in Africa than in Eurasia, probably because of the late arrival of fissipeds.

Suborder Fissipeda Blumenbach 1791

Family Canidae Gray 1821

The present-day canid fauna of Africa comprise 5 genera and 13 species; 8 of these species have a fossil record (indicated by an asterisk).

Canis	**adustus, *aureus, gallaensis, lamperti, *mesomelas, semensis*
Vulpes	**chama, pallida, ruppellii, *vulpes*
Fennecus	**zerda*
Lycaon	**pictus*
Otocyon	**megalotis*

Unfortunately, canids are not common in the African fossil record, especially in pre-Pleistocene formations. Their distribution is shown in tables 11.4 and 11.5. The earliest are amphicyonines from the early Miocene; they are represented by two species of *Hecubides* from East Africa (Savage 1965) and by *Afrocyon* from Gebel Zelten (Arambourg 1961). From a similar level at Malembe in the Congo, Hooijer (1963) reported a canine tooth that Savage referred to an amphicyonine. *Hecubides* is close to

Table 11.4 Distribution of African Miocene and Pliocene Canidae.

Time	Site	Canid	*Hecubides euryodon*	*H. macrodon*	Amphicyonine	*Afrocyon burolleti*
Pliocene	Langebaanweg	+				
Late Miocene	Gafsa (U)	+				
Middle Miocene	Ngorora	+				
Early Miocene	Gebel Zelten					+
	Malembe				+	
	Mfwangano		+			
	R 31			+		
	Rusinga		+			
	Napak I, IV		+			
	Koru		+			
	Songhor		+			

Table 11.5 Distribution of African Pleistocene Canidae.

Time	Site	Canis †africanus	C. †atrox	C. †brevirostris	C. †terblanchei	C. mesomelas	C. adustus	C. aureus	C. sp.	Vulpes †pattisoni	V. †pulcher	V. chama	V. vulpes	V. sp.	Fennecus zerda	Lycaon pictus	Otocyon †recki	O. megalotis	Canid indet
Late Pleistocene	Broken Hill	—	—	—	—	—	+	—	—	—	—	—	—	—	—	—	—	—	—
	Mumbwa	—	—	—	—	—	+	—	—	—	—	—	—	—	—	—	—	—	—
	Barbary	—	—	—	—	—	—	+	—	—	—	+	—	—	+	+	—	—	—
	Swartklip	—	—	—	—	+	—	—	—	—	—	+	—	—	—	+	—	—	—
	Melkbos	—	—	—	—	—	+	—	—	—	—	—	—	—	—	—	—	—	—
	Cave of Hearths	—	—	—	—	—	+	—	—	—	—	+	—	—	—	+	—	+	—
	Vlakkraal	—	—	—	—	—	—	—	—	—	—	+	—	—	—	+	—	+	—
	Gamble's	—	—	—	—	+	—	—	—	—	—	—	—	—	—	+	—	—	—
Middle Pleistocene	Olduvai IV	?	—	—	—	?	—	—	—	—	—	—	—	—	—	—	—	—	—
	Hopefield	—	—	—	+	+	?	—	—	—	—	+	—	—	—	+	—	+	—
Early Pleistocene	Barbary	—	+	—	—	—	—	—	—	—	—	—	—	—	—	—	—	—	—
	Bolt's Farm	—	—	—	—	?	—	—	—	—	—	—	—	—	—	—	—	—	+
	East Turkana	—	—	—	+	+	—	—	—	—	—	+	—	—	—	—	—	—	+
	Kromdraai	—	+	—	—	+	—	—	—	—	—	+	—	—	—	—	—	—	—
	Coopers	—	—	+	—	—	—	—	—	—	—	—	—	—	—	—	—	—	—
	Swartkrans	—	—	—	—	+	—	—	—	—	—	?	—	—	—	—	—	—	—
	Olduvai II	+	—	—	—	—	—	—	—	—	—	—	—	—	—	—	—	—	—
	Olduvai I	+	—	—	+	—	—	—	—	—	—	—	—	—	—	—	+	—	—
	Ngarusi	?	—	—	+	—	—	—	—	—	—	—	—	—	—	—	—	—	—
	Makapansgat	—	—	—	—	—	?	—	—	—	—	—	—	—	—	+	—	—	—
	Taung	—	—	—	—	—	?	—	—	—	+	—	—	—	—	—	—	—	—
	Sterkfontein	—	—	+	—	+	—	—	—	—	—	—	—	—	—	—	—	—	—
	Baard's Quarry	—	—	—	—	—	—	+	—	—	—	—	—	—	—	—	—	—	—

†Extinct species.

the European genus *Amphicyon* from which it possibly originated. *Afrocyon* is known only from a mandibular fragment. Robinson and Black (1969) reported a large canid from the upper level (late Miocene) of the Beglia Formation of Gafsa, southern Tunisia. M. Pickford (pers. comm.) reports on the occurrence of a large amphicyoninelike canid from the Ngorora Formation of Kenya. The absence of recognizable canids at Langebaanweg is surprising.

The distribution of the Pleistocene canids is shown in table 11.5; seven species are now extinct. Numbers are too small to suggest any major extinction during the period, and the table rather gives the impression of gradual replacement of species.

Ewer (1956b) has admirably reviewed our knowledge of South African canids. The jackals are the most abundant canids in the fossil record; *C. mesomelas* (black-backed jackal) has the longest record, going back to the earliest Pleistocene. The four extinct species of *Canis* tabulated in table 11.5 are distinct species. *C. antiquus* is omitted; the type is aberrant according to Ewer and other material attributed to it is probably *C. mesomelas. C. adustus*

is found only in late Pleistocene of East Africa and Rhodesia, with a possible record in South Africa at Hopefield (middle Pleistocene). *C. aureus* (Asiatic jackal) is understandably limited to North Africa. *Canis africanus* was originally described in 1928 from a skull in the Olduvai fauna and recently a second similar skull has been recovered from Bed I; the species shows affinities with the wolf *C. lupus*. There are two early Pleistocene *Vulpes* species and the records of living species are limited to the late Pleistocene. The fennec is restricted today to the Sahara and its fringes; it has no fossil record south of the Sahara. *Lycaon* (hunting dog) has a distribution across Africa from Algeria to the Cape, from middle Pleistocene times to the present. The absence of canids from Omo and their rarity in the East Turkana succession is notable.

In this paper I follow Ewer (1965) in not separating *Otocyon recki* generically from the living *O. megalotis*. The differences between these two species (see Petter 1973) are no greater than those distinguishing species of the genus *Canis*.

There is no clue on the links between the Miocene

and the Pleistocene canids. Of the five Pleistocene genera, three are exclusively African (*Fennecus, Lycaon,* and *Otocyon*) with records going well back into the Pleistocene, which would suggest a radiation in Pliocene times. The Miocene taxa probably all belong to the Amphicyoninae; some students would include this taxon within the Ursidae (Ginsburg 1961) and others would raise it to family rank (Hunt 1972). Until the detailed studies of Hunt are published I have left it as a canid subfamily. The amphicyonines range in Eurasia from late Oligocene through to late Miocene and are then replaced by the Caninae. Although the ancestry of the *Lycaon* and *Otocyon* are unknown, it is unlikely that any of the living African canids could be derived from amphicyonines, and thus they represent stocks that reached Africa in late Miocene or early Pliocene times. *"Agriotherium"* from Langebaanweg (see the next section) could be the last survivor of the amphicyonines.

Family Ursidae Gray 1825

Bears are an omnivorous stock that branched off the canids and have radiated as a small family over Eurasia and the Americas. The African record is limited to the Pleistocene and Recent of the Barbary Coast (table 11.6). Arambourg (1933) has reviewed the occurrences in Morocco, Algeria, and Tunisia; all belong to *Ursus arctos* and Arambourg was able to recognize subspecific evolution from the middle into the late Pleistocene. There are numerous historical records of bears in the Atlas Mountains and the last report is of an animal killed in 1841 near Tetuan in Morocco. *Ursus* must have made the crossing from Spain during a low sea level in middle Pleistocene times.

The only other ursid record from Africa is the astonishing report of *Agriotherium* from Langebaanweg in Cape Province (Hendey 1972). The identification was originally based on some tooth fragments and a partial ulna. Now relatively complete skull material is available which confirms the identification with *Agriotherium*. However, I doubt the assignment of *Agriotherium* to the Ursidae; I believe it could turn out to be an amphicyonid. Until more evidence is available, I propose regarding the occurrence as Carnivore *incertae sedis*.

Family Mustelidae Swainson 1835

Mustelids are the largest family of carnivores with over 100 recorded genera, yet they are repre-

sented in Africa today by only 8 genera and only 6 genera are known in the fossil record. There are only 7 genera from South America—the family distribution is essentially holarctic and their place in the Old World tropics is mostly taken by viverrids.

No evidence of an early Miocene mustelid is found in any part of Africa, and the first appearances are a "mustelid" reported by Robinson and Black (1969) from the upper level of the Beglia Formation (late Miocene) of Gafsa, southern Tunisia, as well as from the Beni Mellal site in Morocco, that may be of similar age (Lavocat 1961).

Three subfamilies of mustelid are recorded from Africa: mustelines, mellivorines, and lutrines. The mustelines are represented in the Pleistocene by an African polecat at Hopefield, referred to the living species *Ictonyx striatus* (Hendey 1974), and by a weasel *Mustela nivalis* from a Palaeolithic site in Algeria (Romer 1928). The mellivorines are a small subfamily erected to accommodate the honey badger, living today only in Africa but with reports of fossil taxa from the late Miocene and Pliocene of North America and Asia. Identification of mellivorines as distinct from some mustelines (especially genus *Gulo*) is very difficult without complete cranial and dental material, and indeed the validity of the subfamily status could be argued. M. Pickford (pers. comm.) reports a new species of *Mellivora* from the Ngorora Formation of Kenya. The living species *Mellivora capensis* is recorded from late Pleistocene sites in South Africa (Hendey 1969), and Mellivorinae indet. occur in the Omo and East Turkana sequences.

All other African fossil mustelids are of Plio-Pleistocene age and belong to the otter subfamily Lutrinae, the most widespread of the mustelids and a stock without analogues among the viverrids. The aquatic habit of the otters presumably greatly aids their widespread distribution and probably also their preservation as fossils. There are a few records of living species of *Lutra* in the late Pleistocene, while two extinct species are recorded from Wadi Natrun in Egypt, though their validity is ambiguous. *Lutra* has recently been recorded from Olduvai, East Turkana and Omo. *Aonyx*, the clawless otter, inhabits rivers and lakes in Africa; its fossils have been found in North, South, and East Africa in sites always remote from the coasts. Most interesting, however, is the extinct genus *Enhydriodon*, which is recorded from Plio-Pleistocene sites of Southwest Africa, Baringo, Kanapoi, and Omo. *Enhydriodon* is a large otter with massive cheek teeth. Its closest affinities are probably with *Aonyx* (see Pilgrim 1932) and possibly like *Aonyx* it was a crabeater. The pres-

Table 11.6 Distribution of African fossil Mustelidae and Ursidae.

Time	Site	Mustelid	Mustela nivalis	Ictonyx striatus	Mellivora capensis	M. aff. †punjabiensis	M. sp. nov.	Lutra †libyca	L. hessica	L. lutra	L. maculicollis	L. sp.	Aonyx capensis	A. sp.	†Enhydriodon africanus	E. pattersoni	E. sp.	†Agriotherium africanum	Ursus arctos
Late Pleistocene	Barbary	—	+	—	—	—	—	—	—	+	—	—	—	—	—	—	—	—	+
	Swartklip	—	—	—	+	—	—	—	—	—	—	—	+	—	—	—	—	—	—
	Wonderwerk	—	—	—	—	—	—	—	—	—	—	—	+	—	—	—	—	—	—
	Florisbad	—	—	—	—	—	—	—	—	—	—	—	+	—	—	—	—	—	—
	Gamble's	—	—	—	—	—	—	—	—	—	+	—	+	—	—	—	—	—	—
Middle Pleistocene	Barbary	—	—	—	—	—	—	—	—	—	—	—	—	—	—	—	—	—	+
	Hopefield	—	—	+	+	—	—	—	—	—	—	—	—	—	—	—	—	—	—
Early Pleistocene	Baard's Quarry	—	—	—	?	—	—	—	—	—	—	—	—	—	—	—	—	—	—
	Bolt's Farm	—	—	—	—	—	—	—	—	—	—	—	+	—	—	—	—	—	—
	Olduvai II	—	—	—	—	—	—	—	—	—	+	—	—	—	—	—	—	—	—
	Olduvai I	—	—	—	—	—	—	—	—	—	+	—	—	—	—	—	—	—	—
	E. Turkana	—	—	—	—	—	—	+	—	—	+	—	—	—	—	+	—	—	—
	Omo	—	—	—	—	—	—	+	—	—	+	—	—	—	—	+	—	—	—
	Hadar	—	—	—	—	—	—	—	—	—	—	—	—	—	—	+	—	—	—
Pliocene	Kanapoi	—	—	—	—	—	—	—	—	—	—	—	—	—	+	—	—	—	—
	Langebaanweg	—	—	—	+	—	—	—	—	—	—	—	—	—	+	—	—	?	—
	Klein Zee	—	—	—	—	—	—	—	—	—	—	—	—	—	+	—	—	—	—
	Wadi Natrun	—	—	—	—	—	+	+	—	—	—	+	—	—	—	—	—	—	—
	Lukeino	—	—	—	—	—	—	—	—	—	—	—	—	—	—	+	—	—	—
Late Miocene	Ngorora	—	—	—	—	—	+	—	—	—	—	—	—	—	—	—	—	—	—
	Gafsa (U)	+	—	—	—	—	—	—	—	—	—	—	—	—	—	—	—	—	—
	Beni Mellal	+	—	—	—	—	—	—	—	—	—	—	—	—	—	—	—	—	—

†Extinct genus or species.

ence of species of *Enhydriodon,* from widely spaced Plio-Pleistocene sites in Southwest Africa and in East Africa, with others in Italy (Monte Bamboli) and the Siwaliks points to a former widespread distribution of the genus in space and time (late Miocene to Pleistocene).

The present African species of mustelids are listed here; those with a fossil record are marked with an asterisk.

Mustelinae
  Ictonyx   kalaharicus, orangiae, striatus
  Mustela   *nivalis, numidica, putorius
  Poecilictis libyca
  Poecilogale albinucha
Lutrinae
  Aonyz   *capensis
  Lutra   *lutra, *maculicollis
  Paraonyx   congica, microdon, philippsi
Mellivorinae
  Mellivora   *capensis

Family Viverridae Gray 1821

The Viverridae are today by far the largest family of carnivores in Africa with 16 genera, excluding 7 endemic to Madagascar. The list of living species (excluding Madagascar) follows; those with a fossil record bear an asterisk.

Viverrinae
  Poiana   richardsonii
  Genetta   *genetta, abyssinica, maculata, servalina, tigrina, victoriae
  Osbornictis   piscivora
  Civettictis   civetta
Paradoxurinae
  Nandinia   binotata
Herpestinae
  Herpestes   *ichneumon, *cauui, *sanguineus, dentifer, granti, ignitus, nigratus, ochracea, pulverulentus, ratlamuchi, ruddi
  Helogale   brunnula, dybowskii, hirtula, ivori, macmillani, mimetra, parvula, percivali, un-

	dulata, varia, vetula, victorina
Atilax	*paludinosus
Mungos	*mungo*
Crossarchus	*alexandri, ansorgei, gambianus, obscurus*
Ichneumia	*albicauda
Bdeogale	*crassicauda, jacksoni, nigripes, puisa, tenuis*
Rhynchogale	*caniceps, melleri*
Cynictis	*penicillata, *selousi, *sengaani*
Xenogale	*microdon*
Suricata	*suricatta

The family is ancient, originating from the Miacidae probably in late Eocene times in Eurasia. On dental characters alone, it is impossible to distinguish late viverravine miacids from early viverrids, and the diagnostic skull characters are often lacking. There are no known miacids in Africa and the first migrants, probably in earliest Miocene, were already viverrids (table 11.7).

Only two subfamilies have a fossil record in Africa, the Viverrinae (genets and civets) and Herpestinae (mongooses). These are readily distinguished on soft anatomy, but not easily distinguished in fossil material. The viverrines have long faces, no postorbital bar, short external auditory meatus, and trenchant molars; they lack a protocone on P³. The herpestines have short faces, a postorbital bar, long external auditory meatus, less trenchant molars, and a protocone on P³. Genera are as difficult as subfamilies to diagnose on fossil material, and the profusion of subgenera in the literature does not help the problem. Recent discoveries in the early Miocene of the Kavirondo and in the Plio-Pleistocene of Southwest Africa, Omo, and East Rudolf add considerably to our knowledge of the family. Herpestines have been more abundant and diverse than the viverrines throughout the stratigraphic record, with the exception of Pliocene times, where the record is poor.

From the early Miocene of Kenya the genus *Kichechia* is well known. This taxon is classified by Savage (1965) as a herpestine. The only other early Miocene African viverrid is a small species from Napak that may be a stenoplesictine, a stock which some students regard as viverrid and others as felid. A few more viverrids from middle and late Miocene of Africa are as yet undescribed. The Fort Ternan fossils include a *Genetta*-like viverrine and a viverrid is reported from Gafsa that Robinson and Black (1974) assign to the Asiatic genus *Tungurictis*. The Pliocene of Southwest Africa has recently yielded three viverrids (Hendey 1974) and one is recorded from Lothagam (Patterson et al. 1970). The diversity of taxa increases in the early Pleistocene, with numerous species, especially of herpestines, in both South and East Africa. The best records are in the early levels at Olduvai (Petter 1973) and Omo in East Africa, with good representation in the upper Pleistocene deposits of South Africa (Cooke 1964).

The Pleistocene record contains 4 viverrine and 8 herpestine genera; the present-day fauna has 4 viverrines and 11 herpestines. The Herpestinae comprise only 11 extant genera; of these 10 are exclusively African, with only *Herpestes* extending into Asia where it has radiated into 10 species ranging from Persia through India to China, Nepal, and Malayasia. Two extinct herpestine genera are known: *Kichechia* from the Kenyan early Miocene (Burdigalian) and *Herpestides* from the European earliest Miocene (Aquitanian) (Beaumont 1967). The picture for the Herpestinae looks like a migration from Europe to Africa in earliest Miocene times and a radiation within Africa during Miocene, Pliocene, and Pleistocene, and a recent (probably late Pleistocene) migration of *Herpestes* from Africa into Asia. The picture for the Viverrinae is less clear. There is some evidence that they existed in Africa by middle Miocene times, about the same time as their appearance in Eurasia, but they have not radiated in Africa to the same extent as in Asia.

The only carnivores on Madagascar are viverrids, and they have no true fossil record on the island. They comprise three subfamilies that are undoubtedly of Tertiary origin, probably colonizing the island sporadically from Miocene times onward. The Hemigalinae probably have a viverrine origin and are also known from Asia; the Galidiinae and Cryptoproctinae are known only from Madagascar and both have more in common with the herpestines than with the viverrines. Lamberton (1939) reported a subfossil fossa from Madagascar that he named *Cryptoprocta spelea;* it is probably only subspecifically different at most from the living *C. ferox.*

Family Hyaenidae Gray 1869

I use Ewer's admirable reappraisal of the fossil hyaenids of Africa (1967) as the basis of this account. At the time Ewer wrote, no hyaenids earlier than Pliocene were known from Africa and although the Miocene additions are not yet described, they are important in viewing the overall evolution of the stock. In addition there are now the hyaenids from Langebaanweg (Hendey 1974) and unpublished discoveries from East Rudolf and Omo.

The Hyaenidae are one of the smallest carnivore families; in Africa there are today only three genera

Table 11.7   Distribution of African fossil Viverridae.

Time	Site	Viverrid	Genetta genetta	G. sp.	Viverra †leakeyi	Civettictus civetta	C. sp.	†Pseudocivetta ingens	†Kichechia zamanae	Suricata suricatta	S. †major	Herpestes †primitivus	H. †delibis	H. †mesotes	H. †palaeoserengetensis	H. †palaeogracilis	H. sanguineus	H. ichneumon	H. cauui	H. sp.	Helogale sp.	Atilax paludinosus	A. sp.	Mungos †dietrichi	M. †minutus	M. sp.	Crossarchus transvaalensis	C. sp.	Ichneumia albicauda	Cynictis selousi	C. penicellata
Late Pleistocene	Barbary	—	+	—	—	—	—	—	—	—	—	—	—	—	—	—	—	+	—	—	—	—	—	—	—	—	—	—	—	—	—
	Swartklip	—	—	—	—	—	—	—	—	—	—	—	—	—	—	—	—	+	—	—	—	—	—	—	—	—	—	—	—	—	—
	Broken Hill	—	—	—	—	—	—	—	—	—	—	—	—	—	—	—	—	—	—	—	—	—	?	—	—	—	—	—	—	?	—
	Bulawayo W. W.	—	—	—	—	—	—	—	—	—	—	—	—	—	—	—	—	—	+	—	+	—	—	—	—	—	—	—	—	—	—
	Florisbad	—	—	—	—	—	—	—	—	—	—	—	—	—	—	—	—	+	—	—	—	—	—	—	—	—	—	—	—	—	—
	Cave of Hearths	—	—	—	—	—	—	—	—	+	—	—	—	—	—	—	—	—	+	—	—	—	—	—	—	—	—	—	—	—	—
	Eyasi	—	—	—	—	—	—	—	—	—	—	—	—	—	—	—	—	—	—	—	—	—	+	—	—	—	—	—	—	—	—
	Gamble's	—	—	—	—	—	—	—	—	—	—	—	—	—	—	—	—	—	—	—	—	—	—	—	—	—	—	—	—	—	—
Middle Pleistocene	Hopefield	—	—	—	—	+	—	—	—	+	—	—	—	—	—	—	—	+	—	—	—	—	—	—	—	—	—	—	—	—	—
Early Pleistocene	Bolt's Farm	—	—	—	—	—	—	—	?	—	—	—	—	—	—	—	—	—	—	—	—	—	—	—	—	—	+	—	—	—	—
	Kromdraai	—	—	—	—	—	+	—	—	—	—	—	—	+	—	—	—	—	—	—	—	—	+	—	—	—	—	?	—	—	+
	Swartkrans	—	—	—	—	—	—	—	—	—	—	—	—	—	—	—	—	—	—	—	—	—	+	—	+	—	—	—	—	—	—
	Olduvai II	—	—	—	—	—	—	+	—	—	—	—	—	—	—	—	—	—	—	—	—	—	—	—	—	—	—	—	—	—	+
	Makapansgat	—	—	—	—	—	—	—	—	—	—	—	—	—	—	—	—	—	—	—	—	—	—	—	—	—	—	—	—	—	+
	Olduvai I	—	—	+	—	—	—	+	—	—	—	+	+	—	?	?	—	—	+	—	—	—	+	+	+	—	+	?	—	—	—
	Ngarusi	—	—	—	—	—	—	—	—	—	—	—	—	+	+	—	—	—	—	—	—	—	—	+	—	—	—	—	—	—	—
	Laetolil	—	—	—	+	—	—	—	—	—	—	—	—	—	—	—	—	—	—	—	—	—	—	+	—	—	—	—	—	—	—
	East Turkana	—	+	—	—	—	—	+	—	—	—	—	—	—	—	—	—	—	—	—	—	—	—	—	—	—	—	—	—	—	—
	Omo	—	+	+	+	—	+	—	—	—	—	—	—	—	—	—	—	—	—	—	—	+	—	—	—	—	+	—	—	—	—
Pliocene	Langebaanweg	—	—	+	+	—	—	—	—	—	—	—	—	—	—	—	—	—	—	—	+	—	—	—	—	—	—	—	—	—	—
	Lothagam	—	—	—	—	+	—	—	—	—	—	—	—	—	—	—	—	—	—	—	—	—	—	—	—	—	—	—	—	—	—
Late Miocene	Ngorora	?	—	—	—	—	—	—	—	—	—	—	—	—	—	—	—	—	—	—	—	—	—	—	—	—	—	—	—	—	—
Middle Miocene	Gafsa (L)	+	—	—	—	—	—	—	—	—	—	—	—	—	—	—	—	—	—	—	—	—	—	—	—	—	—	—	—	—	—
	Fort Ternan	+	—	—	—	—	—	—	—	—	—	—	—	—	—	—	—	—	—	—	—	—	—	—	—	—	—	—	—	—	—
Early Miocene	Mfwangano	—	—	—	—	—	—	—	+	—	—	—	—	—	—	—	—	—	—	—	—	—	—	—	—	—	—	—	—	—	—
	Napak IV, V	+	—	—	—	—	—	—	+	—	—	—	—	—	—	—	—	—	—	—	—	—	—	—	—	—	—	—	—	—	—
	Muruarot	—	—	—	—	—	—	—	+	—	—	—	—	—	—	—	—	—	—	—	—	—	—	—	—	—	—	—	—	—	—
	Rusinga	—	—	—	—	—	—	—	+	—	—	—	—	—	—	—	—	—	—	—	—	—	—	—	—	—	—	—	—	—	—
	Songhor	—	—	—	—	—	—	—	+	—	—	—	—	—	—	—	—	—	—	—	—	—	—	—	—	—	—	—	—	—	—

†Extinct genus or species.

and four species. These are *Crocuta crocuta,* the spotted hyaena, *Hyaena hyaena,* the striped hyaena, and *H. brunnea,* the brown hyaena. The fourth is *Proteles cristatus,* the aardwolf. Authorities have differing views on the affinity of this strange beast; some place it in the Viverridae, some in the Hyaenidae, and yet others in a family of its own. As it has no fossil record the solution is immaterial to our discussion and I include it here for convenience rather than from conviction.

The five records to date of Miocene hyaenids are given in table 11.8. The earliest is an immature mandible from Fort Ternan; at the moment it can only be said that this specimen is more likely to be a hyaenid than a felid. Robinson and Black (1969) report that the lower level of the Beglia Formation at Gafsa in Tunisia has yielded two species of *Hyaena*. I have not seen the specimens but doubt that they can belong to the genus *Hyaena;* if so, then it becomes one of the longest living carnivore genera. From Wadi el Hammam (Bou Hanifia) Arambourg (1959) has reported *H. algeriensis;* Ewer (1967) comments that it is very different from all other species of *Hyaena* and also different from *Crocuta.* Indeed she makes a case for a separate genus but does not take the step. Kurtén (1957) established a case for the recognition of a subgenus of *Crocuta* to accommodate a number of species from upper Miocene and Pliocene sites, the name *Percrocuta* being chosen from a plethora listed by Kretzoi. Thenius (1966) elevated *Percrocuta* to generic rank and included within the taxon *P. algeriensis*. Crusafont and Aguirre (1971) described *P. tobieni* from Ngorora and Hendey (1974) described *P. australis* from Langebaanweg, while Howell and Petter report a fur-

ther species from Omo. Hendey discusses the evolutionary and distributional features of the genus, now becoming better known. The origins are still obscure; the case for an African origin would be strengthened if the Fort Ternan species turns out to be a *Percrocuta* sp. The suggested lineage *Percrocuta* to *Crocuta* is not supported by the Langebaanweg evidence and so the origin of *Crocuta* is still open to debate.

Table 11.8 Distribution of African fossil Hyaenidae.

Time	Site	Hyaenid	†*Ictitherium* sp.	†*Percrocuta algeriensis*	*P. tobieni*	*P. australis*	*P.* sp.	†*Hyaenictis preforfex*	†*Leecyaena forfex*	†*Euryboas silberbergi*	*E. namaquensis*	*Hyaena* †*abronia*	*H.* †*bellax*	*H. hyaena*	*H. brunnea*	*H.* sp.	*Crocuta crocuta*	*C.* sp.
Late Pleistocene	Barbary	—	—	—	—	—	—	—	—	—	—	—	—	+	—	—	+	—
	Swartklip	—	—	—	—	—	—	—	—	—	—	—	—	—	+	—	—	—
	Melkbos	—	—	—	—	—	—	—	—	—	—	—	—	—	+	—	—	—
	Mumbwa	—	—	—	—	—	—	—	—	—	—	—	—	—	+	—	—	—
	Broken Hill	—	—	—	—	—	—	—	—	—	—	—	—	—	—	—	+	—
	Bulawayo W.W.	—	—	—	—	—	—	—	—	—	—	—	—	—	—	—	+	—
	Vlakkraal	—	—	—	—	—	—	—	—	—	—	—	—	—	—	—	+	—
	Florisbad	—	—	—	—	—	—	—	—	—	—	—	—	—	+	—	—	—
	Eyasi	—	—	—	—	—	—	—	—	—	—	—	—	—	—	—	+	—
Middle Pleistocene	Hopefield	—	—	—	—	—	—	—	—	—	—	—	—	+	—	—	+	—
	Barbary	—	—	—	—	—	—	—	—	—	—	—	—	+	—	—	—	—
	Olduvai IV	—	—	—	—	—	—	—	—	—	—	—	—	?	—	—	—	—
Early Pleistocene	Gladysvale	—	—	—	—	—	—	—	—	—	—	—	+	—	—	—	—	—
	Bolt's Farm	—	—	—	—	—	—	—	—	—	—	—	+	—	—	—	—	—
	Kromdraai	—	—	—	—	—	—	—	—	—	—	—	+	—	—	—	—	—
	Swartkrans	—	—	—	—	—	—	+	+	—	—	—	—	—	—	—	+	—
	Olduvai II	—	—	—	—	—	—	—	—	—	—	—	—	+	—	—	+	—
	Makapansgat	—	—	—	—	—	—	—	—	—	—	—	+	?	—	—	—	—
	Sterkfontein	—	—	—	—	—	—	—	—	+	—	—	—	—	?	—	—	—
	Olduvai I	—	—	—	—	—	—	—	—	—	—	—	+	—	—	—	+	+
	Laetolil	—	—	—	—	—	—	—	—	—	—	—	?	—	—	—	?	—
	Kanam	—	—	—	—	—	—	—	—	—	—	—	—	—	—	?	—	—
	Barbary	—	—	—	—	—	—	—	—	—	—	—	—	—	—	?	—	—
	Chad	—	—	—	—	—	—	—	—	—	—	—	?	—	—	—	—	—
	Baard's Quarry	—	—	—	—	—	—	—	—	—	—	—	+	—	+	—	+	—
	Hadar	—	—	—	—	—	—	—	—	—	—	—	—	—	+	—	—	—
	East Turkana	—	—	—	—	—	—	—	—	—	—	—	—	—	+	—	+	—
	Omo	—	—	—	—	—	+	—	—	—	—	—	—	—	+	—	+	—
Pliocene	Lothagam	+	—	—	—	—	—	—	—	—	—	—	—	—	—	—	—	—
	Kanapoi	—	—	—	—	—	—	—	—	—	—	—	—	—	+	—	—	—
	Klein Zee	—	—	—	—	—	—	—	—	+	—	—	—	—	—	—	—	—
	Langebaanweg	—	—	—	—	+	—	+	—	—	+	—	—	—	?	—	+	—
Late Miocene	Ngorora	—	—	—	+	—	—	—	—	—	—	—	—	—	—	—	—	—
	Wadi el Hamman	—	—	+	—	—	—	—	—	—	—	—	—	—	—	—	—	—
	Beni Mellal	—	+	—	—	—	—	—	—	—	—	—	—	—	—	—	—	—
Middle Miocene	Gafsa (L)	—	—	—	—	—	—	—	—	—	—	—	—	—	—	—	?	—
	Fort Ternan	+	—	—	—	—	—	—	—	—	—	—	—	—	—	—	—	—

†Extinct genus or species.

Also from Langebaanweg Hendey (1974) reports *Hyaenictis preforfex,* and he further includes *Leecyaena forfex* from Swartkrans in *Hyaenictis.* Though I agree that both species may belong to the same genus, the case for the existence of *Hyaenictis* in South Africa is not strong and on balance I favor Ewer's choice of *Leecyaena,* though more evidence is needed. From Kanapoi and Langebaanweg *Hyaena* is recorded for the first time in Africa, possibly as an immigrant from north of the Tethys, with *Ictitherium* ancestry. An *Ictitherium* is reported from Beni Mellal (Lavocat 1961); at the time the Beni Mellal fauna was described it was thought to be Pontian but Robinson and Black (1973) have shown that it is likely to be much earlier.

This leaves the Pleistocene record. The maximum diversity is seen in the early Pleistocene with nine species, and three each in the middle and late Pleistocene. *Hyaena* is represented in the early Pleistocene by both living species and a third species *H. bellax* which may be attributable to the genus. A *Hyaena* species has recently been reported from Afar (Taieb et al. 1972). *Crocuta crocuta* also appears at this level, earlier apparently than in Eurasia. As well as these contemporary genera, there are the two extinct genera. Ewer (1967) placed Stromer's *H. namaquensis* tentatively in the genus *Lycyaena,* a genus which is also represented by *L. silberbergi* from Sterkfontein and Swartkrans. Ewer apparently had not seen Kurtén (1960) when she wrote her paper; Kurtén makes a case for the Sterkfontein and Swartkrans species being *Euryboas,* a European Villafranchian genus. His case seems to me convincing and more likely than the Pontian genus *Lycyaena.* For convenience, we can also include the poorly known Klein Zee species as *E. namaquensis* and probably also assign the Gafsa hyaenid to *Euryboas.*

The hyaenids are unique in their development of bone-crushing dental specializations, albeit in varying degrees. The family has its origins in the Viverridae and *Progenetta* from the European Burdigalian may be the earliest member of the Hyaenidae (Beaumont 1967). *Crocuta* is reported from Chinji in India, *Hyaena* from the Pliocene of southern France, and *Leecyaena* from the Pliocene of China. *Crocuta* became widespread only as a single species in Africa, though in Eurasia it diversified into many species. In the absence of better Miocene records it is not yet possible to establish evolutionary lines and migratory routes. Clearly an interchange occurred between the three continents during the Pleistocene, Pliocene, and at least part of the Miocene.

Family Felidae Gray 1821

Although the felids are not a large family, with only four living genera, they have a good fossil record (table 11.9). Because they are the most abundantly found carnivores in Africa, they are important stratigraphically.

The living felids of Africa comprise 3 genera and 11 species. The status of genus and subgenus is often questioned in the felines; here I follow Simpson (1945), and those species with an African fossil record bear an asterisk.

Felis (Felis)	**lybica, chaus, margarita*
Felis (Leptailurus)	**serval, brachyura*
Felis (Microfelis)	*nigripes*
Felis (Profelis)	*aurata*
Felis (Lynx)	**caracal*
Panthera (Panthera)	**leo, *pardus*
Acinonyx	**jubatus*

The grouping of genera within the Felidae is much disputed. Simpson's 1945 grouping is not entirely satisfactory in light of present knowledge. Beaumont's 1964 grouping is interesting and has much to recommend it morphologically but is not as satisfactory stratigraphically. This is not the place to reclassify the felids; only 5 out of 40 fossil felid genera are represented in Africa. Three of these are indubitably machairodontines (*Megantereon, Machairodus,* and *Homotherium*); only the status of *Metailurus* and of *Therailurus* need be discussed in detail.

Metailurus is recorded from sites in the early Miocene of Kenya (Savage 1965); there is a metapodial from Sinda in the Congo which Savage (in Hooijer 1963) reported was felid and could belong to *Pseudaelurus;* it could equally well and is more likely to belong to *Metailurus.* The Kenya beds with *Metailurus* are dated at around 18 to 22 m.y., which would make *M. africanus* the earliest *Metailurus* occurrence; others are in the middle Miocene of Tung Gur, Mongolia, and in the late Miocene of China and Pikermi. Beaumont (1964) and others have proposed that *Metailurus* is descended from *Pseudaelurus;* on morphological grounds this looks feasible but there are stratigraphic difficulties. Either *Metailurus* has an Oligocene origin in a primitive felid separate from the *Pseudaelurus* lineage, or *Metailurus africanus* is not related to the Eurasian later Miocene taxa. On balance I feel the latter alternative is the more likely, in which case Kretzoi's name *Afrosmilus* is available. The name-changing, however, does nothing to solve the problem of its origin. The earliest *Pseudaelurus* species in the Burdigalian of Europe is

Table 11.9 Distribution of African fossil Felidae.

Time	Site	Felid	*†Afrosmilus africanus*	*†Dinofelis barlowi*	*D. piveteaui*	*D. diastemata*	*D. sp.*	*†Megantereon whitei*	*M. gracile*	*M. eurynodon*	*M. sp.*	*†Machairodus transvaalensis*	*M. aphanistus*	*M. sp.*	*†Homotherium ethiopicum*	*H. problematicus*	*H. sp.*	*Acinonyx jubatus*	*A. sp.*	*Felis (Sivafelis) †obscura*	*F. (Felis) lybica*	*F. (Lynx) †issiodorensis*	*F. caracal*	*F. lynx*	*F. (Leptailurus) serval*	*Panthera (Leo) leo*	*P. (Panthera) pardus*	*P. (P.) crassidens*	*P. sp.*
Late Pleistocene	Barbary	—	—	—	—	—	—	—	—	—	—	—	—	—	—	—	—	+	—	—	+	—	+	+	+	+	+	—	—
	Swartklip	—	—	—	—	—	—	—	—	—	—	—	—	—	—	—	—	?	—	—	—	?	—	+	—	—	—	—	—
	Melkbos	—	—	—	—	—	—	—	—	—	—	—	—	—	—	—	—	—	—	—	—	—	—	—	—	—	+	—	—
	Olduvai V	—	—	—	—	—	—	—	—	—	—	—	—	—	—	—	—	—	—	—	—	—	—	—	—	—	—	+	—
	Broken Hill	—	—	—	—	—	—	—	—	—	—	?	—	—	—	—	—	—	—	—	—	—	—	—	+	+	+	—	—
	Bulawayo W.W.	—	—	—	—	—	—	—	—	—	—	—	—	—	—	—	—	—	—	—	—	—	—	—	—	+	—	—	—
	Cave of Hearths	—	—	—	—	—	—	—	—	—	—	—	—	—	—	—	—	—	—	—	—	—	—	—	+	—	+	—	—
	Gamble's	—	—	—	—	—	—	—	—	—	—	—	—	—	—	—	—	—	—	—	—	—	—	—	—	—	+	—	—
	Eyasi	—	—	—	—	—	—	—	—	—	—	—	—	—	—	—	—	—	—	—	—	—	—	—	+	—	+	—	—
Middle Pleistocene	Barbary	—	—	—	—	—	—	+	—	—	—	—	—	—	—	—	—	—	—	—	—	—	—	—	—	—	+	—	—
	Hopefield	—	—	—	—	—	?	—	—	—	—	—	—	—	—	—	—	—	—	—	—	—	—	—	+	—	+	—	—
	Olduvai IV	—	—	—	—	—	—	—	—	—	—	—	—	—	—	—	—	—	—	?	—	—	—	—	—	?	—	+	—
Early Pleistocene	Gladysvale	—	—	+	—	—	—	—	—	—	—	—	—	—	—	—	—	—	—	—	—	—	—	—	—	?	—	—	—
	Bolt's Farm	—	—	+	—	—	—	—	—	—	—	—	—	—	—	—	—	—	—	—	—	—	—	—	—	+	+	—	—
	Kromdraai	+	—	—	+	—	?	—	+	—	—	—	—	—	—	—	—	—	—	—	—	—	—	—	—	?	—	+	—
	Swartkrans	—	—	—	—	+	?	—	+	—	—	—	—	—	—	—	—	—	—	—	—	—	—	—	—	?	+	—	—
	Olduvai II	+	—	—	—	—	—	—	—	—	—	?	—	—	—	—	—	+	—	—	—	—	—	—	+	+	+	?	+
	Makapansgat	—	—	+	—	—	—	—	—	—	—	—	—	+	—	—	—	—	—	—	—	—	—	—	—	—	—	—	—
	Schurveberg	—	—	—	—	+	—	—	—	—	—	—	—	—	—	—	—	—	—	—	—	—	—	—	—	—	—	—	—
	Sterkfontein	—	—	+	—	—	—	+	—	—	+	—	—	—	—	—	—	—	—	—	—	—	—	—	—	—	+	—	—
	Olduvai I	+	—	—	—	—	+	—	—	+	—	—	—	—	—	—	—	?	—	—	—	—	—	—	—	+	+	?	—
	Laetolil	—	—	—	—	—	—	—	—	—	—	—	—	—	—	—	—	—	—	—	—	?	—	—	?	+	—	—	—
	Omo	+	—	—	—	—	+	—	—	+	—	—	—	—	+	+	—	+	—	—	—	—	+	—	—	—	+	+	—
	Oran	+	—	—	—	—	—	—	—	—	—	—	—	—	—	—	—	—	—	—	—	—	—	—	—	—	—	—	—
	East Turkana	—	—	+	?	—	+	—	—	+	—	—	—	—	—	—	—	+	—	—	—	—	—	—	—	+	+	+	—
Pliocene	Langebaanweg	—	—	—	+	—	—	—	—	—	—	+	—	—	—	—	—	+	—	—	+	—	+	—	—	—	—	—	—
	Wadi Natrun	—	—	—	—	—	—	—	—	—	+	—	—	—	—	—	—	—	—	—	—	—	—	—	—	—	—	—	—
	Lothagam	+	—	—	—	—	—	—	—	—	—	—	—	—	—	—	—	—	—	—	—	—	—	—	—	—	—	—	—
	Kanapoi	+	—	—	—	—	—	—	—	—	—	—	—	—	—	—	—	—	—	—	—	—	—	—	—	—	—	—	—
Late Miocene	Lukeino	+	—	—	—	—	—	—	—	—	—	—	—	—	—	—	—	—	—	—	—	—	—	—	—	—	—	—	—
	Gafsa (U)	—	—	—	—	—	—	—	—	—	—	—	+	—	—	—	—	—	—	—	—	—	—	—	—	—	—	—	—
Early Miocene	Sinda	+	—	—	—	—	—	—	—	—	—	—	—	—	—	—	—	—	—	—	—	—	—	—	—	—	—	—	—
	Rusinga	—	+	—	—	—	—	—	—	—	—	—	—	—	—	—	—	—	—	—	—	—	—	—	—	—	—	—	—
	Songhor	—	+	—	—	—	—	—	—	—	—	—	—	—	—	—	—	—	—	—	—	—	—	—	—	—	—	—	—
	Karungu	—	+	—	—	—	—	—	—	—	—	—	—	—	—	—	—	—	—	—	—	—	—	—	—	—	—	—	—

† Extinct genus or species.

much more primitive than the African *Afrosmilus*. The origin must be sought in the Oligocene, and here we have no evidence from Africa; and of the Eurasian taxa *Nimravus* comes closest. Species of *Nimravus* in the Phosphorites du Quercy (Piveteau 1931) possess no characters that would exclude them from *Afrosmilus* ancestry.

The early Pleistocene contains four genera of sabertooths, of which *Therailurus* is certainly the most primitive. The genus is based on *T. diastemata* from Roussillon, an Astian site in France. Ewer (1955) has analyzed the characters of the African and French species, which are all of approximately similar age. On balance *T. diastemata* is the most primitive, but its slightly broader upper canine probably excludes it from ancestry of the African species. Hemmer (1965) has argued that *Therailurus* is a junior synonym of *Dinofelis;* he united the European and African species of *Therailurus* with the Chinese form *Dinofelis abeli,* which is much more advanced than the known *Therailurus* species and is usually regarded as coming from the Pontian stage. Zdansky (1924), in the preface to the original paper naming the species, wrote that though the majority of fossils described originated from the *Hipparion* clays, he was including some whose age was unknown and those comprised all specimens from Honan. *D. abeli* is recorded as coming from Honan and it seems probable that it is Pleistocene in age. If this is accepted, then the evolution of the *Therailurus-Dinofelis* series makes stratigraphic as well as morphologic sense, and on grounds of priority the name *Dinofelis* should be adopted. It is tempting to look to *Afrosmilus* for ancestry of *Dinofelis* but impossible to prove.

The true sabertooths or Machairodontinae are confined to the Pliocene and early Pleistocene in Africa, with a few doubtful records from late Miocene and middle Pleistocene. The subfamily is characterized by a dentition in which the upper canines progressively enlarge and the lower canines are progressively reduced; P_3^3 reduced or absent; P^4 with vestigial protocone; M_1 with shallow notch between paraconid and protoconid; crenulation is common on teeth. Details of these characters, especially the shape and size of the upper canine, distinguish the genera. *Megantereon* is a true sabertooth that makes its first appearance in earliest Pleistocene. *"Megantereon" praecox* from the Nagri stage in the Siwaliks is poorly known and more likely to be allied to *Sansanosmilus* of the Eurasian Miocene than to *Megantereon.* The abundance of species found in the Asiatic Pleistocene compared with one from Europe inclined Ewer to favor an Asiatic origin for *Megantereon.* From Africa, *M. eurynodon* is well known but the other species are all based on poor material.

Machairodus is one of the best known genera of sabertooths; its record extends back to the late Miocene in Europe and Asia, but it is doubtful if the genus survived beyond the early Pleistocene. Almost all the African records are suspect largely because of fragmentary evidence. Robinson and Black (1969) report a *Machairodus* in the late Miocene of the Beglia Formation near Gafsa. From the Pliocene of Wadi Natrun Stromer (1913) reported *M. aphanistus,* a species characteristic of Pontian sites of Europe and Iran; the identification is based on a toothless fragment. Hendey (1974) recorded a good specimen from Langebaanweg and left it as *Mach-*

airodus sp; *M. transvaalensis* is based on an upper canine and may be the same species. *Machairodus* species are also reported from Olduvai II, Hopefield, and Broken Hill, all very dubious. Arambourg (1962) reported *Machairodus* from early and middle Pleistocene sites on the Barbary Coast.

Homotherium ethiopicum from the Omo rests on a mandibular fragment with two broken teeth (Arambourg 1947), but recent work by Howell and colleagues has revealed further specimens of the genus. Further south at East Turkana, Meave Leakey reports the presence of *Homotherium.* Hendey (1974) has undoubted *Homotherium* from Langebaanweg and the specimen described by Colling (1972) as *Megantereon problematicus* from Makapansgat may be conspecific. Ewer (1955) mentioned two very worn lower carnassial teeth from Kromdraai that she considered could belong to *Epimachairodus* (here regarded as a junior synonym of *Homotherium*); I prefer to regard these teeth as Machairodontinae *incertae sedis. Homotherium* is thought by many authors to be a descendant of *Machairodus;* it is equally possible if less likely to be a descendant of *Dinofelis.*

Two feline genera are well represented, *Panthera* and *Felis* (see Ewer 1956a). *Panthera* has a good early Pleistocene record at numerous sites. Here *P. leo* is taken to include subspecies *P. l. spelaeus* and *P. l. shawi. P. pardus* is reported from Laetolil, Olduvai Bed I, Omo, and many later sites. *P. crassidens* has a more restricted distribution. *Felis* is by contrast very rare in the early Pleistocene sites, though there is a plentiful record of the genus in the middle and late Pleistocene. Hendey (1974) reports two *Felis* species from the Pliocene of Langebaanweg, implying that the genus must have reached Africa before the Pleistocene. The major problem in establishing the origin of *Panthera* and *Felis* is distinguishing the genera from each other as well as from numerous others. *Felis* may go back as far as the late Miocene in Europe and Asia. *Panthera* has a reliable Villafranchian record in Europe but not earlier; it could be of African origin, but without a Pliocene record this is pure speculation.

Acinonyx appears in the early Pleistocene of Omo and there are later Pleistocene records from East, South, and North Africa; the genus is recognized in the Villafranchian of both Europe and Asia.

Family Phocidae Gray 1825

Only two living phocids frequent African waters. The Mediterranean seal *Monachus monachus* is known from the Mediterranean and northwest shores of Africa, while the southern elephant seal *Mirounga leonina* is recorded from the coasts of South Africa. The southern fur seal *Arctocephalus pusillus* is known around the South African shores and is the only African otarid.

Fossil otarids are unknown in Africa and only two records of phocid exist. One is from the Pliocene of Wadi Natrun in Egypt where Stromer (1913) recorded the monk seal *Pristiphoca occitana.* The second is of *Prionodelphis capensis,* reported by Hendey and Repenning (1972) from similar levels at Langebaanweg in South Africa. The genus *Prionodelphis* was erected by Frenguelli for some Pliocene teeth from Argentina, then thought to be cetacean. Hendey considered his fossils to have their closest affinity to the poorly known South American species *P. rovereti.* Among recent genera he emphasized comparison with the monk seal *Monachus* and with the leopard seal *Hydrurga;* the latter inhabits southern oceans today reaching the shores of South America, New Zealand, and southern Australia. *Monachus* and its closely allied genera appear to have been always limited to tropical and subtropical waters.

Conclusions

The distribution of creodont and carnivore families is set out in table 11.10. The mustelids and canids lead the numbers of such genera in the world, both extinct and in total, though first place among living genera goes to the viverrids. In Africa viverrids have a clear lead for living genera and overall totals, but the hyaenodonts, hyaenids, and felids outnumber them in extinct genera. The last two columns record the number of African fossil species (*sensu lato*) in each family. Hyaenodonts, canids, viverrids, and felids all have between 22 and 30 taxa, leaving 16 mustelids and 17 hyaenids. This accounts for all save 2 phocids and 1 ursid. But the occurrence of fossils of these families is quite different. Felids are the most abundant with over 100 occurrences, then come viverrids, hyaenids, canids, hyaenodontids, and mustelids in decreasing order.

The family distributions are analyzed stratigraphically in table 11.11. The problem in drawing conclusions is the high proportion of low figures. The geological record is too uneven to show trends; the only levels with adequate records are early Miocene and early and late Pleistocene. However some broad features are apparent, such as the replacement of creodonts by carnivores. In the Eocene and Oligocene only creodonts are represented. In early Miocene creodonts outnumber carnivores; by middle Miocene carnivores are about as abundant as creodonts and by late Miocene the creodonts are extinct. This spread of true carnivores occurred later in Africa than elsewhere. The Miocene of Africa has nine genera of creodonts, while Asia has two and Europe only one (and the Eurasian records are probably caused by reinvasion from Africa).

Table 11.10 Distribution of families of Carnivores.

Family	Genera			African Genera				African Total	
	Extinct	Living	Total	Extinct	Fossil[a]	Living	Total	Fossil Species[b]	Occurrences[c]
Oxyaenidae	11	—	11	—	—	—	—	—	—
Hyaenodontidae	36	—	36	11	11	—	11	26	49
Miacidae	16	—	16	—	—	—	—	—	—
Canidae	76	11	87	2	7	5	7	23	59
Otaridae	10	5	15	—	—	1	1	—	—
Odobenidae	5	1	6	—	—	—	—	—	—
Ursidae	11	6	17	1	1	1	2	2	2
Procyonidae	14	8	22	—	—	—	—	—	—
Mustelidae	78	27	105	1	6	8	9	16	33
Phocidae	14	10	24	2	2	2	4	2	2
Viverridae	7	37	44	2	13	16	19	30	64
Hyaenidae	10	3	13	5	7	3	8	17	56
Felidae	41	4	45	5	8	3	8	28	105
Totals	329	112	441	29	55	39	69	144	370

a. Genera with a fossil record, whether living or extinct.
b. Total species in fossil record, including indeterminate taxa.
c. Total sites at which species of the family are recorded.

Within the Pleistocene Ewer (1967) has pointed to the replacement of an early Pleistocene radiation of hyaenids and sabertooths by felines in mid and late Pleistocene. This is seen as an adaptational change-over; the scavenging, bone-crushing hyaenids and the meat-slicing of the predatory sabertooths are combined to a degree in the felines *Felis* and *Panthera*.

The origins of the African carnivore faunas are marked by big gaps in knowledge, especially of Asia.

Paleogeographically it is probable that most of the migrants came to Africa via Asia (see Coryndon and Savage 1973), but most of the paleontological evidence comes from Europe. The first migrants were the hyaenodont creodonts probably in late Eocene, from either Europe or Asia, providing the ancestors of the Fayum faunas. During the Oligocene the hyaenodonts evolved and spread across much of the African continent. In late Oligocene or early Miocene times the second wave of migrants arrived,

Table 11.11 Stratigraphic distribution of African carnivore families.

Time	Hyaenodontidae	Canidae	Ursidae	Mustelidae	Phocidae	Viverridae	Hyaenidae	Felidae	Total (Genera / Species / Occurrences)
Late Pleistocene	— / — / —	5 / 8 / 18	1 / 1 / 1	4 / 5 / 8	— / — / —	5 / 7 / 12	2 / 3 / 10	4 / 9 / 23	21 / 33 / 72
Middle Pleistocene	— / — / —	3 / 6 / 7	1 / 1 / 1	2 / 2 / 2	— / — / —	3 / 3 / 3	2 / 3 / 4	4 / 7 / 9	15 / 22 / 26
Early Pleistocene	— / — / —	3 / 11 / 23	— / — / —	4 / 5 / 11	— / — / —	12 / 25 / 36	5 / 9 / 29	8 / 21 / 59	32 / 71 / 158
Pliocene	— / — / —	1 / 1 / 1	— / ? / —	4 / 6 / 9	2 / 2 / 2	4 / 4 / 4	5 / 6 / 8	5 / 7 / 8	21 / 26 / 32
Late Miocene	— / — / —	1 / 1 / 1	— / — / —	2 / 2 / 3	— / — / —	1 / 1 / 1	2 / 3 / 3	2 / 2 / 2	8 / 9 / 10
Middle Miocene	3 / 4 / 4	1 / 1 / 1	— / — / —	— / — / —	— / — / —	1 / 2 / 2	2 / 2 / 2	— / — / —	7 / 9 / 9
Early Miocene	8 / 15 / 32	2 / 4 / 8	— / — / —	— / — / —	— / — / —	2 / 2 / 6	— / — / —	2 / 2 / 4	14 / 23 / 50
Oligocene	5 / 10 / 11	— / — / —	— / — / —	— / — / —	— / — / —	— / — / —	— / — / —	— / — / —	4 / 10 / 11
Eocene	1 / 1 / 1	— / — / —	— / — / —	— / — / —	— / — / —	— / — / —	— / — / —	— / — / —	1 / 1 / 1
Total occurrences	48	59	2	33	2	64	56	105	369

Note: Figures in top left of each compartment represent genera totals. Figures in middle of each compartment represent species totals. Figures in lower right of each compartment represent number of occurrences. Species include indeterminate taxa: occurrences used as in table 11.10.

with amphicyonines, herpestines, hyaenids, and sabertooth felids; their route is unknown but Coryndon and Savage (1973) suggested southwest Asia on the basis of generic similarities revealed by frequency distributions. It was also in early Miocene times that creodonts left Africa via Arabia for India (Bugti) and Europe. In middle Miocene times it is likely that mellivorines and viverrines arrived from India via Arabia, just before the Red Sea rift opened, and in the reverse direction the creodont *Dissopsalis* migrated to India. By late Miocene the creodonts had been replaced by the spread of true carnivores in Africa, and sea otters migrated around the coasts of Africa. In Pliocene times the carnivores that had been evolving in Africa were joined by new migrants —canids, sabertooths, and possibly the ancestors of the felines. During the Pliocene there appears to have been an increased interchange of faunas between Europe, Asia, and Africa; the African mammal faunas have their lowest endemism (50%) in the Pliocene, as shown by Coryndon and Savage (1973). In Pleistocene times the endemism rises abruptly as the Sahara begins to exert influence. The faunas south of the Sahara evolve in isolation from Eurasia, and those along the northern fringes receive European elements.

This account of origins and migrations is unhappily more conjecture than fact. Only many more facts will substantially improve it. Pliny wrote nearly two thousand years ago *"Ex Africa semper aliquid novi"*; it is still true today.

I owe much to the kindness of Meave Leakey in showing me and discussing with me so many of the new Pleistocene discoveries from East Turkana and for allowing me to record these in the lists. Thanks for similar facilities are due to Clark Howell for the Omo material and to Martin Pickford for the Baringo material. I am most grateful to Meave Leakey for her kindness in reading the manuscript, for her critical comments, and for information to update the Pleistocene carnivore data. Thanks are also due to the Natural Environment Research Council for continued support of field work in Africa and to the many colleagues with whom I have had the opportunity of discussing the faunas. Mrs. Joyce Rowland typed the manuscript and Mrs. Mary Hayes assisted with the tables and bibliography.

References

Andrews, C. W. 1906. *A descriptive catalogue of the Tertiary Vertebrata of the Fayum, Egypt*. London: British Museum (National History), pp. 1–324.

Arambourg, C. 1933. Revision des ours fossiles de l'Afrique du Nord. *Ann. Mus. Hist. Nat. Marseille* 25:249–299.

———. 1947. Contributions à l'étude géologique et paléontologique du bassin du Lac Rudolphe et de la basse vallée de l'Omo. Deuxieme Partie. Paléontologie. *Mission Scien. d'Omo*, 1(3):231–562.

———. 1959. Vertébrés continentaux du Miocène supérieur de l'Afrique du Nord. *Publ. Serv. Carte Géol. Alger., Paléontologie*. Mém. 4:1–159.

———. 1961. Note préliminaire sur quelques Vertébrés nouveaux du Burdigalien de Libye. *C. R. Soc. Géol. Fr.*, Paris, 1961:107–109.

———. 1962. Les faunes mammalogiques du Pléistocéne circumméditerranéen. *Quarternaria* 6:97–109.

Beaumont, G. de. 1964. Remarques sur la classification des Felidae. *Eclog. Geol. Helv.* 57(2):837–845.

———. 1967. Observations sur les herpestinae (Viverridae, Carnivora) de l'Oligocène supérieure avec quelques remarques sur des Hyaenidae du Neogène. *Arch. Sci. Genève* 20(1):79–108.

———. 1970. Le problème de la position taxonomique de *Hyaenailouros* Biedermann (Mammalia). *Bull. Soc. Vaud. Sci. Nat.* 70(8):357–363.

Broom, R. 1939. A preliminary account of the Pleistocene Carnivores of the Transvaal caves. *Ann. Transv. Mus.* 19:331–338.

Cooke, H. B. S. 1964. Pleistocene mammal faunas of Africa, with particular reference to southern Africa. In F. C. Howell and F. Bourlière, eds. *African ecology and human evolution* Chicago: Aldine, pp. 65–116.

Coryndon, S. C., and R. J. G. Savage. 1973. Origins and affinities of African mammal faunas. In *Organisms and continents through time*. Special Papers in Palaeontology 12, Systematics Assoc. Publication 9:121–135.

Cross, M. W., and V. J. Maglio. 1975. *A bibliography of the fossil mammals of Africa 1950–1972*. Princeton: Dept. of Geology, Princeton University, pp. 1–291.

Crusafont, M., and E. Aguirre. 1971. A new species of Percrocuta from the middle Miocene of Kenya. *Abh. hess. L-Amt. Bodenforsch.* 69:51–58.

Deitrich, W. O. 1942. Ältestquartäre Säugetiere aus der südlichen Serengeti, Deutsch-Ostafrika. *Paläont.* 94:43–133.

Ewer, R. F. 1955. The fossil carnivores of the Transvaal caves: Machairodontinae. *Proc. Zool. Soc. Lond.* 125:587–615.

———. 1956a. The fossil carnivores of the Transvaal Caves: Felinae. *Proc. Zool. Soc. Lond.* 126:83–95.

———. 1956b. The fossil carnivores of the Transvaal Caves: Canidae. *Proc. Zool. Soc. Lond.* 126:97–119.

———. 1956c. The fossil carnivores of the Transvaal Caves: two new Viverrids, together with some general considerations. *Proc. Zool. Soc. Lond.* 126:259–274.

———. 1965. Carnivora. In L. S. B. Leakey et al., eds., *Olduvai Gorge 1951–61*, vol. 1. Cambridge: Cambridge University Press, pp. 19–22.

———. 1967. The fossil Hyaenids of Africa—a reappraisal. In W. W. Bishop and J. D. Clark, eds., *Background to evolution in Africa*. Chicago: University of Chicago Press, pp. 109–123.

Frenguelli, J. 1922. *Prionodelphis rovereti*, un representante de la familia "Squalodontidae" en el paranense superior de Entre Rios. *Bol. Ac. Cordoba*,

Argentina 25:491–500.

Ginsburg, L. 1961. La fauna des carnivores miocènes de Sansan (Gers). *Mém. Mus. Natl. Hist. Nat.* C9:1–190.

Hemmer, H. 1965. Zur Nomenklatur und Verbreitung des Genus *Dinofelis* Zdansky, 1924 (*Therailurus* Piveteau, 1948). *Palaeont. Afr.* 9:75–89.

Hendey, Q. B. 1969. Quaternary vertebrate fossil sites in the South-western Cape Province. *S. Afr. Archaeol. Bull.* 24:(3,4):96–105.

———. 1970a. A review of the geology and palaeontology of the Plio/Pleistocene deposits at Langebaanweg, Cape Province. *Ann. S. Afr. Mus.* 56:75–117.

———. 1970b. The age of the fossiliferous deposits at Langebaanweg, Cape Province. *Ann. S. Afr. Mus.* 56:119–131.

———. 1972. A Pliocene Ursid from South Africa. *Ann. S. Afr. Mus.* 59:115–132.

———. 1974. The late Cenozoic Carnivora of the Southwestern Cape Province. *Ann. S. Afr. Mus.* 63:1–369.

———, and C. A. Repenning. 1972. A Pliocene phocid from South Africa. *Ann. S. Afr. Mus.* 59(4):71–98.

Hooijer, D. A. 1963. Miocene Mammalia of Congo. *Ann. Mus. Roy. Afr. Cent.* Serie in 8°, Sci. Geol. 46:1–77.

Hopwood, A. T., and J. P. Hollyfield. 1952. An annotated bibliography of the fossil mammals of Africa (1742–1950). *Fossil Mammals of Africa* 8:1–194, British Museum (Natural History) London.

Howell, C. 1968. Omo research expedition. *Nature* 219:567–572.

Hunt, R. M. 1972. Miocene Amphicyonids from the Agate Spring Quarries, Sioux County, Nebraska. *Amer. Mus. Novit.* 2506:1–39.

Kurtén, B. 1957. *Percrocuta* Kretzoi (Mammalia, Carnivora), a group of Neogene hyenas. *Act. Zoo. Cracov.* 2(16):375–404.

———. 1960. The age of the Australopithecinae. *Stockh. Contr. Geol.* 6:9–22.

Kurtén, B., and E. Anderson. 1974. Association of *Ursus arctos* and *Arctodus simus* (Mammalia: Ursidae) in the late Pleistocene of Wyoming. *Breviora* 426:1–6.

Lamberton, C. 1939. Contribution a l'étude de la fauna subfossile de Madagascar. *Mém. Acad. Malgache* 27(4, 5):156–193.

Lavocat, R. 1961. Le Gisement de Vertébrés Miocènes de Beni Mellal (Maroc). Étude Systématique de la Faune de Mammifères et Conclusions Générales. *Notes Mém. Serv. Mines Carte Geol. Maroc* 155:1–44.

Leakey, L. S. B. 1967. Notes on the mammalian faunas from the Miocene & Pleistocene of East Africa. In W. W. Bishop and J. D. Clark, eds., *Background to evolution in Africa.* Chicago: University of Chicago Press, pp. 7–29.

Leakey, M. D. 1971. *Olduvai Gorge,* vol. 3. Cambridge: Cambridge University Press, pp. 1–306.

Maglio, V. J. 1972. Vertebrate faunas and chronology of hominid-bearing sediments east of Lake Rudolf, Kenya. *Nature* 239:379–385.

Osborn, H. F. 1909. New carnivorous mammals from the Fayûm Oligocene, Egypt. *Bull. Am. Mus. Nat. Hist.* 26:415–424.

Patterson, B., A. Behrensmeyer, and W. D. Sill. 1970. Geology and fauna of a new Pliocene locality in northwestern Kenya. *Nature* 226:918–921.

Petter, G. 1973. Carnivores Pleistocènes du Ravin d'Olduvai (Tanzanie). In L. S. B. Leakey, R. J. G. Savage, and S. C. Coryndon, eds., *Fossil vertebrates of Africa,* vol. 3. London: Academic Press, pp. 43–100.

Pilgrim, G. E. 1932. The fossil carnivora of India. *Mem. Geol. Surv. Ind: Palaeo. Ind.* n.s. 18:1–232.

Piveteau, J. 1931. Les chats des Phosphorites du Quercy. *Ann. Paléont.* 20:107–163.

Robinson, P., and C. C. Black. 1969. Note préliminaire sur les vertébrés fossiles du Vindobonien (formation Béglia), du Bled Douarah, Gouvernorat de Gafsa, Tunisie. *Notes Serv. Géol. Tunis* 31: 67–70.

———. 1973. A small Miocene faunule from near Testour, Beja Gouvernorat, Tunisia. Livre Jubilaire M. Solignac, *Ann. Mines Géol. Tunis* 26:445–449.

———. 1974. Vertebrate faunas from the Neogene of Tunisia. *Ann. Geol. Surv. Egypt* 4:319–332.

Romer, A. S. 1928. Pleistocene mammals of Algeria. Fauna of the paleolithic station of Mechta-el-Arbi. *Bull. Logan. Mus.* 1:80–163.

Savage, R. J. G. 1965. Fossil mammals of Africa: 19. The Miocene Carnivora of East Africa. *Bull. Br. Mus. Nat. Hist.* 10:242–316.

———. 1973. *Megistotherium,* gigantic Hyaenodont from Miocene of Gebel Zelten, Libya. *Bull. Br. Mus. Nat. Hist.* 22:485–511.

Schlosser, M. 1911. Die Affen, Lemuren, Chiropteren, Insectivoren, Marsupialier, Creodonten und Carnivoren des europaischen Tertiars, 1. *Beitr. Paläont. Geol. Öst.-Ung.* 6:1–227.

Simons, E. L. 1968. Early Cenozoic mammalian faunas Fayum Province, Egypt. *Bull. Peabody Mus. Nat. Hist.* 28:1–105.

Simons, E. L., and P. D. Gingerich. 1974. New carnivorous mammals from the Oligocene of Egypt. *Ann. Geol. Surv. Egypt.* 4:157–166.

———. 1976. A new species of *Apterodon* (Mammalia, Creodonta) from the Upper Eocene Qasr el Sagha Formation of Egypt. *Postilla* 168:1–9.

Simpson, G. G. 1937. The Fort Union of the Crazy Mountain Field, Montana and its mammalian faunas. *Bull. U.S. Natl. Mus.* 169:1–287.

———. 1945. The principles of classification and a classification of mammals. *Bull. Am. Mus. Nat. Hist.* 85:1–350.

Stromer, E. 1913. Mitteilungen über die Wirbeltiere aus dem Mittelpliozän des Natrontales (Aegypten). *Zeit. Deut. Geol. Ges.* 65:350–372.

Szalay, F. S. 1967. The affinities of *Apterodon* (Mammalia, Deltatheridia, Hyaenodontidae). *Amer. Mus. Novit.* 2293:2–17.

Taieb, M., Y. Coppens, D. C. Johanson, and J. Kalb. 1972. Dépôts sédimentaires et faunes du Plio-pléistocène de la basse vallée de l'Awash (Afar) central, Ethiopie. *C. R. Hebd. Acad. Sci.* (Paris) 275D:819–822.

Thenius, E. 1966. Zur stammesgeschichte der Hyanen (Carnivora, Mammalia). *Z. Saugetierk.* 31(4):293–300.

Van Valen, L. 1966. Deltatheridia, a new order of mammals. *Bull. Am. Mus. Nat. Hist.* 132:1–126.

————. 1967. New Paleocene insectivores and insectivore classification. *Bull. Am. Mus. Nat. Hist.* 135:217–284.

Welbourne, R. G. 1969. Bibliography of Quaternary African palaeontology. *Palaeont. Afr.* 12:151–202.

Zdansky, O. 1924. Jungtertiäre Carnivoren Chinas, *Palaeont. Sin.* C2(1):1–149.

12

Pholidota and Tubulidentata

Bryan Patterson

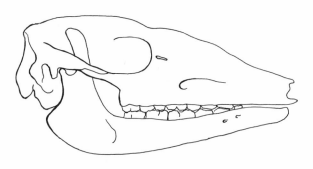

Order Pholidota

Until very recently no fragments representing pangolins had been reported from any fossil deposit in Africa. However, a very late Pleistocene specimen, questionably referred to the living species *temmincki,* has been recorded by Klein (1972:137) as occurring in the Nelson Bay Cave (unit YGL, dated at ca 18,000 B.P.), and Hendey (1974) has listed specifically unidentified postcranial bones from the Pliocene at Langabaanweg, in Bed 2 of the Varswater Formation (E Quarry), and a braincase from the middle Pleistocene at Elandsfontien. Even without these welcome indications of its former presence, however, the group would have deserved mention here, because, as has long been recognized, it is evident from the morphology of the Recent species that the African representatives differ in a number of ways from the Asian, which surely bespeaks a modest evolutionary radiation of some African antiquity.

If extinct pangolins were broadly similar in habits to the living—and there is no reason to suppose the contrary—their scarcity as fossils is not surprising. Most of the Recent species are forest-dwelling, and all except the strictly arboreal forms are partially subterranean. Forest and arboreal mammals are in general less well represented in the fossil record than inhabitants of open country. The bones of animals dying in burrows are particularly liable to destruction by the action of plant roots and percolating rain water. A further factor may militate against their entry into the record. Pangolins lack teeth—and probably have for a long time—which may well affect both their collection and recognition. It is quite possible that fragments of them may lie undetected among miscellaneous lots of material that await identification.

The seven living pangolin species have had a checkered taxonomic history, ranging from inclusion of them all within a single genus, *Manis* (e.g., Simpson 1945), to the recognition of three subfamilies and six genera (Pocock 1924) for their reception. As Emry has remarked (1970:461), the real situation appears to lie between these extremes. The African and Asian species contrast as shown in Table 12.1 (Jentink 1882, Pocock 1924, Weber 1928, Grassé 1955, Emry 1970, Segall 1973). As Pocock noted, the African group appears to be more advanced than the Asian.

Some of these distinctions, although constant, are relatively trivial, but others, such as the xiphisternal and ossicular differences, are of generic if not higher significance. Within the two divisions of the family thus delimited it is possible, in the case of one

Table 12.1   Contrasting Features of Asian and African Pangolins

Asian	African
Xiphisternum with either a short shovel-shaped cartilaginous extension or a short oval cartilaginous extension divided for most of its length (*aurita*).	Xiphisternum enormously prolonged by two cartilaginous extensions running back to the end of the rib cage or beyond before uniting, and then giving rise to two other extensions that double back anteriorly to unite at the level of the kidneys.
Malleus with small lamina; incus with short, well separated crura; stapes with comparatively long columella, head extending well over it.	Malleus with larger lamina and different orientation of head; incus with longer, more approximated crura; stapes with very short columella, head scarcely extending over it.
Pinna of ear represented by a flap or by a thickened rim of skin.	Pinna absent.
Hairs at bases of scales persistent in adults.	Hairs at bases of scales lacking in adults.
Median row of scales continuing to end of tail, scale arrangement on tail nearly always symmetrical.	Median row of scales not continuing to end of tail, dividing into two asymmetrical rows toward the distal extremity.

Table 12.2   Contrasting Features of Two Subgroups of African Pangolins

temmincki-giganteus	*tricuspis-tetradactyla*
Body heavily built; tail short, without terminal pad.	Body lightly built; tail very long, with terminal pad.
Lachrymal absent.	Lachrymal present.
Eyes small.	Eyes larger.
Manus with upper surface scaled, first digit not reduced, fifth digit not enlarged.	Manus with upper surface hairy, first digit greatly reduced and essentially functionless, fifth digit large.
Pes with first digit not reduced, fifth not enlarged.	Pes with first digit greatly reduced, fifth large.

of them, to subdivide further. The Asian species are each clearly distinct, but Pocock's generic separation of them all (1924:723) rests, it seems to me, on slender evidence. In the African species two distinct subgroups contrast in the manner shown in Table 12.2 (Jentink, Pocock, Emry).

The arrangement outlined above dates from Jentink, whose conclusions have been amply buttressed by subsequent authors employing additional characters. The question of how the diversity is to be treated taxonomically remains. Emry, although thoroughly dissatisfied with the assignation of all living species to *Manis*, refrained from subdivision until more thorough study could be undertaken. I believe, however, that the evidence now on hand is

sufficient for the purpose. Furthermore, with three subfamilies on the books and a genus available for all species except *giganteus,* it is scarcely possible to further burden the nomenclature; one has only to select among published names. As Emry has expressed the situation: "The three Asiatic species seem to be different enough from those of Africa to warrant generic distinction of these two groups, and the four African species could possibly be further subdivided at the subgeneric (if not the generic) level." My own view of the degree of diversity coincides very closely. Given the facts, it would be possible to recognize within the Pholidota either two subfamilies, with the African containing two genera, or two genera, with the African comprising two subgenera. I prefer the latter alternative for several reasons: classification should for so long as possible be kept as simple as the evidence permits; pangolins are, after all, a closely knit group; and the few imperfectly known extinct genera could not be fitted into subfamilial divisions. Accordingly, the arrangement suggested for the living species is:

Manis Linnaeus 1758 (including *M. pentadactyla, M. aurita, M. javanica*), Asia.
  Phataginus Rafinesque 1821, Africa
    P. (Phataginus) Rafinesque 1821 (including *P. (P.) tricuspis, P. (P.) longicaudata*).
    P. (Smutsia) Gray 1865 (including *P. (S.) temmincki, P. (S.) giganteus*).

The large Pleistocene Indian species *lydekkeri* Trouessart probably and the very large Pleistocene *palaeojavanica* Dubois of Indonesia almost certainly are referable to *Manis*. The forms represented by the Eocene-Oligocene and Miocene European fragments, both named—*Necromanis, Leptomanis* (of somewhat dubious reference), *Teutomanis*—and unnamed, cannot at present be brought into relationship with any of the surviving representatives of the order, which is also true of the North American early Oligocene *Patriomanis*.

These northern records provide good evidence that the order was not an "old African" group in the sense that proboscideans, hyracoids, and others were. Pangolins probably reached Africa from Eurasia, together with such orders as the Carnivora and Perissodactyla, late in Oligocene time. Such a date of entry probably would have afforded sufficient time for the degree of diversification attained by the living African species and for their divergence as a group from their Asian relatives.

There are several views about the origin of pangolins. Matthew (1918) saw in the extinct North American Palaeanodonta a group representing the ancestry of both the Pholidota and the Xenarthra. Simpson (1931), emphasizing the xenarthran features, excluded them from the pholidotan ancestry. Emry, emphasizing the pholidotan features, placed the palaeanodont families Epoicotheriidae and Metacheiromyidae in the Pholidota, referred the newly discovered North American *Patriomanis* to the Manidae, and rejected xenarthran affinities. His stimulating arguments are rather persuasive, but I am not entirely convinced. *Patriomanis* is a manid (new and as yet undescribed material kindly shown me by Dr. Emry decisively demonstrate this, contrary to my earlier suspicion; Patterson 1975:220), but it may have been an immigrant from Eurasia rather than a descendant of some as yet unknown North American palaeanodont stock. Epoicotheriids and metacheiromyids do show resemblances to the Xenarthra as well as to the Pholidota and also have characters of their own. My position, at the moment of writing at least, is close to Matthew's. There seem to be three groups, which may well have had a common ancestry in the later Cretaceous (there was plenty of time for such an ancestral stock to have come into existence; Patterson 1956:33–34), that radiated during the Cenozoic with varying degrees of success in South America, North America, and the Old World. Whether one regards these groups as distinct orders or as suborders of an order Edentata is of minor moment. Further pursuit of this question here, however, would take us far from Africa and out of the framework of this volume.

Order Tubulidentata

The living aardvark, *Orycteropus afer* (Pallas), the sole surviving representative of an isolated order of mammals, is sometimes cited as an example of a major taxon consisting of a single evolutionary lineage. It is now becoming apparent that the actual history of the group was rather more complex. First appearing in the early Miocene of Kenya, members of the order spread to Europe and southern Asia during the Miocene and early Pliocene but failed to maintain themselves there. One aberrant form occurs in Holocene swamp deposits in Madagascar. Fossils from the early Eocene of both North and South America and from the French Phosphorites have been referred to the order, but the evidence for these assignments is not convincing. The only really well known extinct species is *Orycteropus gaudryi*, represented by a score or more of specimens, including most of the skeleton, nearly all from the "classical Pontain" (Turolian) of Samos. Some 50 isolated bones of *Plesiorycteropus madagascariensis* have been recorded, but important parts of the skeleton remain to be discovered. *Leptorycteropus guilielmi* and *Myorycteropus africanus* are each known from parts of the skeleton of one individual. The other extinct taxa are known from one or a few specimens, or from various isolated fragments that have collectively been assigned to a species (e.g., Pickford 1975). In all, about 150 individuals have been reported. The spotty occurrence of aardvarks through time, as in the case of the pangolins, to some degree may be caused by partially subterranean habits, again supposing that extinct forms resembled their surviving relative to an extent.

This article has largely been condensed and the illustrations taken from my review of the order, which may be consulted for further data and additional references (Patterson 1975).

The Fossil Record

Four genera of aardvarks are at present known:

Leptorycteropus Patterson (1975). Represented by an incomplete skeleton from Lothagam-1, Member B, Kenya, this is the most generalized member of the order discovered to date. The maxillaries do not extend forward into an elongate snout, the palate is not grooved, and the unfused but rather stout mandibular symphysis extends back as far as the anterior end of P_2. The canine is comparatively large and is essentially in series with a normal eutherian number of cheek teeth. The molars resemble those of the living species, but the teeth anterior to them have their tubular central portions surrounded by a ring of cement. The limb bones (figures 12.1, 12.2, 12.3) are more slenderly constructed than those of *Orycteropus;* the humerus lacks a large V-shaped deltopectoral area; and the tibia is straighter and its cnemial crest less developed.

Myorycteropus MacInnes (1956). Known from parts of a skeleton from the early Miocene of Rusinga Island and from fragments found there and in deposits of comparable age on Mfwangano Island, Kenya, and at Napak in Uganda. The dentition is similar to that of *Orycteropus*. The skeleton in numerous respects differs decidedly from that of the surviving genus and even more from that of *Lep-*

torycteropus (figures 12.1 to 12.4). The ascending ramus of the mandible is inclined more posteriorly. In the anterior extremity the scapular spine bears an expansion on the prespinous side opposite the metacromion, the clavicle is relatively larger, the humerus is more laterally bowed, with a relatively larger deltopectoral area and a much wider distal end, the radius and ulna are likewise wider distally, and the manus is relatively larger. In the posterior extremity the femur has a more slender shaft, especially proximally, a more distinct neck and a much larger third trochanter, the shaft of the tibia is more compressed transversely and the cnemial crest terminates in a blunt tubercle, and the pes, like the manus, was relatively larger.

Pickford (1975) has, without discussion, placed *Myorycteropus* in the synonymy of *Orycteropus*. The two are indistinguishable in the dentition and in the structure of the various fragments referred by him to *M. africanus,* but the differences both in structure and proportions revealed by the partial skeleton of the type specimen indicate an animal rather differently adapted than *Orycteropus* and certainly not ancestral to it. The few adequately known extinct species of *Orycteropus* are much closer to *O. afer* than they are to *M. africanus.* MacInnes' genus is, I believe, valid.

Orycteropus Cuvier. A number of fossil forms, named and unnamed as to species, have been referred to the genus. Those references based solely on

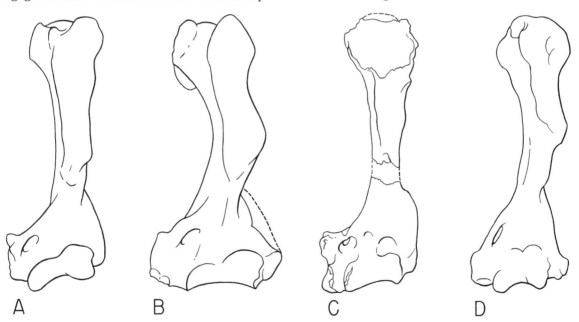

Figure 12.1 Anterior views of left humeri of *Orycteropus* (A), *Myorycteropus* (B), *Leptorycteropus* (C), and *Plesiorycteropus* (D). This and the three following figures are not to scale: *O. afer* is approximately twice the size of the others. *Orycteropus* and *Myorycteropus* are after MacInnes 1956 (fig. 4B modified); *Plesiorycteropus* is from Lamberton 1946 (figs. 1D, 3D, and 4D reversed).

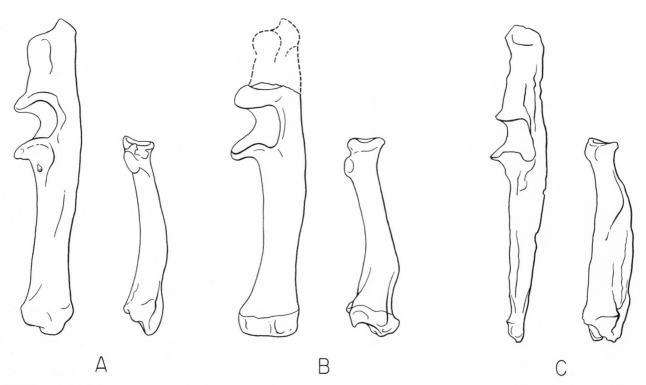

Figure 12.2 Lateral views of left ulnae and medial views of left radii of *Orycteropus* (A), *Myorycteropus* (B), and *Plesiorycteropus* (C). Not to scale.

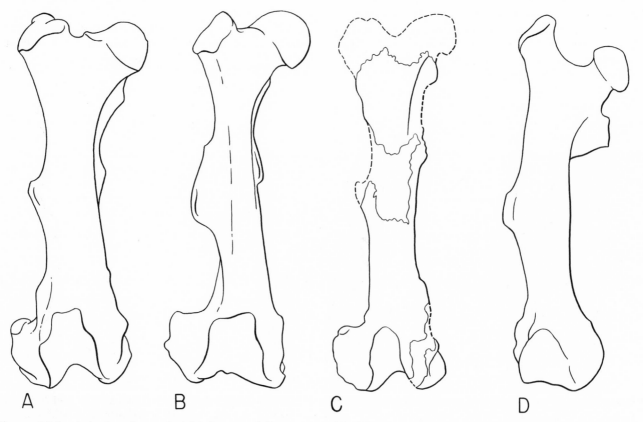

Figure 12.3 Anterior views of right femora of *Orycteropus* (A), *Myorycteropus* (B), *Leptorycteropus* (C), and *Plesiorycteropus* (D). Not to scale.

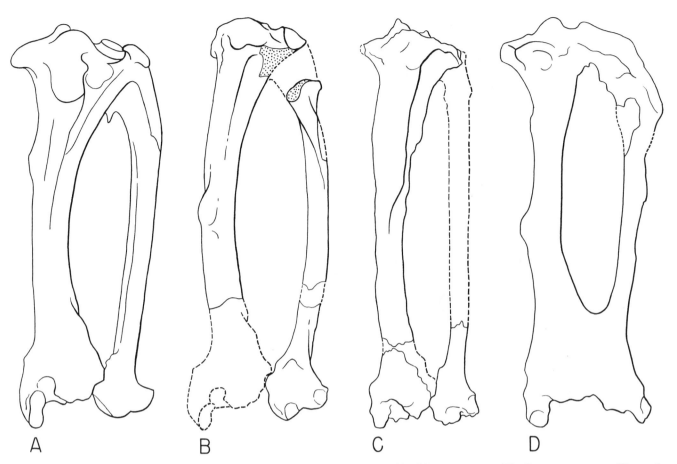

Figure 12.4 Anterior views of left tibiae-fibulae of *Orycteropus* (A), *Myorycteropus* (B), *Leptorycteropus* (C), and *Plesiorycteropus* (D). Not to scale.

isolated teeth, maxillary, mandibular, and postcranial fragments must now, unfortunately, be regarded with reserve. *Leptorycteropus,* for the molars, and *Myorycteropus,* for the cheek teeth as a whole, reveal that orycteropodines differing from *Orycteropus* in cranial and postcranial characters may have teeth so similar as to make generic discrimination on such evidence impossible.

The following may be placed in the genus with confidence: *Orycteropus mauritanicus* Arambourg, middle Miocene, Algeria; *O. gaudryi* Major, late Miocene, Samos, Iran, USSR; *O. depereti* Helbing, early Pliocene, France; *O. pottieri* Ozansoy (possibly the same as *depereti*—no comparable parts are known), early Pliocene, Turkey; *O.* sp. Dietrich, Laetolil, Tanzania; *O. crassidens* MacInnes (regarded as a subspecies of *afer* by Pickford), Pleistocene, Rusinga Island, Kanjera, and East Rudolf, Kenya.

The remainder should, I believe, be queried until more diagnostic, associated materials become available: *O.?* sp. MacInnes, early Miocene, Koru, Kenya; *O.? minutus* Pickford, early Miocene, Songhor, Rusinga and Mfwangano Islands, Kenya; *O.?* sp. Ga-

bunia, middle Miocene, Caucasus, USSR; *O.? chemeldoi* Pickford, middle Miocene, Tugen Hills, and Fort Ternan, Kenya; *O.?* sp. Pickford, middle Miocene, Siwalik Hills, Pakistan; *O.? browni* Colbert and *O.?pilgrimi* Colbert, late Miocene, Siwalik Hills, Pakistan; *O.?* sp. (large) and *O.* sp.? (very small) Pickford, late Miocene, vicinity of the Tugen Hills, Kenya; *O.?* sp. Kitching, Makapansgat, Union of South Africa; and the other African Pliocene and Pleistocene fragments that have been mentioned but not described.

The uncertain generic status of these queried forms is regrettable, but I believe that it can only be resolved by discoveries of relatively complete individuals. It is obvious that the late Miocene *Leptorycteropus* had an ancestry—as well as possible collateral relatives—going back at least as far as, and probably farther than, the ancestry of *Orycteropus,* and as likely as not that the *Myorycteropus* lineage persisted beyond the early Miocene. If either of these forms had been known only from molar teeth in maxillary and mandibular fragments and nondiagnostic, unassociated postcranial odds and ends their distinctiveness would certainly not have been recognized. In the East African early Miocene *O.? minutus* and late Miocene *O.? chemeldoi* we have species notable for the narrowness of the lower molars. These may or may not represent genera at present unrecognizable. The isolated molar recorded by MacInnes from Koru indicates the presence in the early Miocene of a form much larger than *O? minutus.*

Pickford (1975:82) places all named Eurasian species in an *O. gaudryi* group, implying an extra-African radiation stemming from that species. This may indeed have been the case, although I do not think we can at present be sure. Pickford's own discovery (recorded in a postscript to his paper) of orycteropodine phalanges in the Chinji demonstrates the presence of members of the subfamily in Asia well before the deposition of the Samos sediments. In fact, as matters now stand, we cannot even exclude the possibility that an aardvark other than *Orycteropus* succeeded in emigrating to Eurasia.

Plesiorycteropus Filhol (Lamberton 1946). The most isolated tubulidentate, morphologically as well as geographically, is *P. madagascariensis,* a recently extinct species whose remains occur in several of the superficial deposits of the island. For one of these a C^{14} date of 1035 ± 50 B.P. is available. *Plesiorycteropus* differs from all African and Eurasian forms, insofar as these are known, in numerous characters. For example, the skull (figure 12.5) was evidently short, the facial region being smaller than the cranial; a postorbital process is lacking, and the glenoid articulation is low, situated on a downward-projecting process of the squamosal (figure 12.6). It differs radically from *Orycteropus,* in which the articulation is high on the skull and the ascending ramus well developed and rather camellike in structure. The jaw is unknown; to judge from the position of the articulation, it was probably much reduced, as in myrmecophagous mammals generally. No teeth

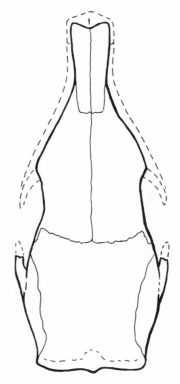

Figure 12.5 *Plesiorycteropus madagascariensis.* Tentative reconstruction of the skull in dorsal view. Based on the type and an unnumbered specimen in the Muséum Nationale d'Histoire Naturelle and on Lamberton 1946: pl. 1, fig. c. (Approx. ×1/1.)

of *Plesiorycteropus* are known; they were probably greatly reduced or absent. The rather striking differences in limb bone structure are evident in figures 12.1 to 12.4. Despite certain resemblances to myrmecophagids, dasypodids, and manids, the basic structure is orycteropodid, and there can, I think, be no doubt as to the ordinal affinities.

Discussion

Each of these four genera represents a distinct lineage (figure 12.7). *Leptorycteropus* is a persistently

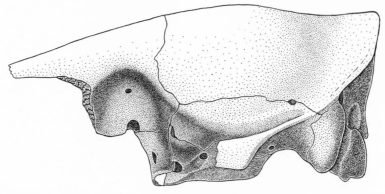

Figure 12.6 *Plesiorycteropus madagascariensis.* Left lateral view of cranium. Unnumbered specimen in Muséum Nationale d'Historie Naturelle. (×3/2.)

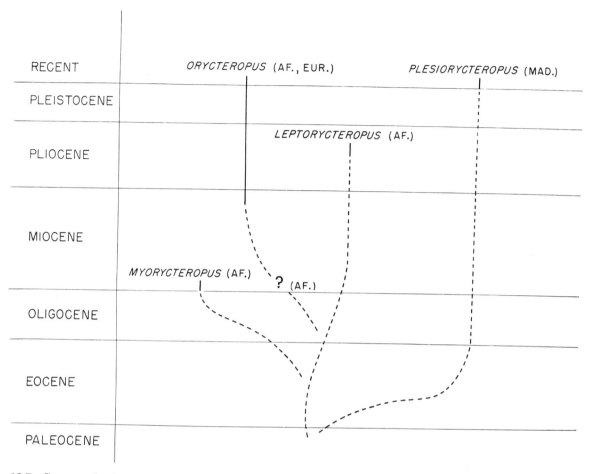

Figure 12.7 Suggested relationships of the known orycteropodid genera.

primitive form. *Myorycteropus* is precociously specialized and could not have been involved in the ancestry of the surviving genus. Like that of *Orycteropus,* its lineage evolved toward acquisition of a fossorial habitus, but in a rather different way at an earlier date. Within the *Orycteropus* lineage there is evidence of some increase in size and in digging capability. *O. afer* is about twice as large as the known species of the extinct genera and even the relatively small *O. gaudryi* is decidedly larger. *Plesiorycteropus* is sufficiently far apart from the African and Eurasian forms to indicate that its ancestry had a long independent history, probably reaching Madagascar during the Eocene, at about the same time as the ancestral lemuroids. Evolution within the lineage has resulted in a terminal form so distinct as to warrant recognition of two subfamilies, Orycteropodinae and Plesiorycteropodinae. The extinct tubulidentates rather clearly reveal a modest radiation within the order. They also suggest differences in adaptation and behavior. *Orycteropus after,* an inhabitant primarily of savanna grasslands although extending into forested and arid areas, is an accomplished digger that constructs burrows and tears

open the nests of mound-building termites. The animal digs with great rapidity, excavating with the forefeet and shifting the earth back by powerful movements of the hind feet and tail. Nocturnal in habits, smell and hearing are the predominant senses. Escape from predators is accomplished mainly by digging, the extremely thick and tough skin serving as a protection while the animal is burying itself. Aardvarks, according to reports, are capable of traveling between 10 and 30 km per night. All accounts agree that the primary food consists of termites, although, in common with other myrmecophagous mammals, the diet is varied to a minor degree. On the evidence there can be no doubt that *O. afer* is a well-adapted, fully committed myrmecophage. The several extinct species doubtless resembled the living one in most behavioral respects. The presence of aardvarks congeneric with the surviving species in deposits of early Ruscinian age in southern France and in Anatolia is a clear indication that the climate in those areas was then sufficiently warm to permit the existence of large termite populations. To what extent did the extinct genera resemble the living one? Before taking up the question some consideration of the myrmecophagous adaption is necessary.

The evolutionary history of termites and ants approximately coincides with that of eutherian and metatherian mammals. The colonies of numerous termite and some ant species are enormous, numbering in the hundreds of thousands and even millions of individuals. Numerous omnivorous, insectivorous, and carnivorous forms of vertebrates take advantage of this resource, the majority of them opportunely by taking foragers or alates during nuptial flights. To penetrate to the interiors of the colonies, to tap the riches at their source, and to depend upon them as the major article of diet—in other words to become a fully committed myremecophage—requires the evolution of special equipment. This equipment (Griffiths 1968) consists of a very extensible, vermiform or ribbonlike tongue lubricated by sticky secretions from greatly enlarged salivary glands, a highly developed olfactory sense, a stomach with a very muscular pyloric area that acts as a "gizzard" for grinding the food (in echidnas the stomach is simple and the food is ground by the spiny base of the tongue acting against spines on the palate), and fossorial limbs for breaching the walls of nests. The adaptive complex has been attained in somewhat different ways in the several myrmecophagous groups, but its function is similar in all. It is a device for the rapid transfer of small prey to the stomach in large quantities.

Teeth are superfluous to the adaptation. *Orycteropus* apart, fully committed myrmecophagous mammals have either lost their teeth entirely or have reduced them greatly. The masticatory musculature is accordingly much reduced, with accompanying reduction of the temporal fossa, postorbital process, and, in some cases, the zygomatic arch. The ramus of the mandible becomes diminished to a rodlike structure with very little in the way of an ascending ramus, and the condyle and glenoid articulation are brought down in one way or another nearly or quite to the level of the palate. This combination of osteological characters, together with a median groove in the bony palate (not present in myrmecophagids), permits ready recognition of any extinct myrmecophage possessing it. However, using these criteria one could not be sure that *Orycteropus* was a committed myrmecophage were it known only in the fossil state. This curious anomaly is discussed below.

Turning to what can reasonably be reconstructed regarding the adaptations of the extinct aardvarks, it can be said at once that *Leptorycteropus* was not specialized for myrmecophagy. What is known of the structure of the skull reveals nothing of the osteological combination associated with that method of feeding. The temporal fossa, postorbital, and jugal processes are relatively well developed, the palate is not grooved, the mandible has an extensive, firm symphysis, and a rather large canine is present, together with a full complement of cheek teeth. The limb bones indicate an animal capable of digging but not highly specialized for it; *Leptorycteropus* may have dug its own burrows but was certainly not as proficient as *Orycteropus*. Like *Dasypus*, it may, in addition to going to ground, have escaped its enemies by quick dashes into dense thickets. A thick and tough hide, supposing this to have been common to the order, would, like the armadillo carapace, have protected it from plant spines and thorns. The impression conveyed is of an omnivorous form that was a faster runner than the living species. Like other omnivores it no doubt ate termites and ants, but was not dependent on them to a major degree.

Because *Leptorycteropus* had the characteristic tubulidentate tooth structure I would judge that acquisition of this had nothing whatever to do with myrmecophagy. Like the dental structure of the Xenarthra it may have evolved early in the Tertiary.

Myorycteropus, in respect of fossorial adaptation, was as far ahead of *O. afer* as *Leptorycteropus* was behind it, at least as regards the forelimb. The pelvis and hind leg bones are not as robustly constructed as those of *Orycteropus*, which suggests that the hind legs and tail were not as much employed in digging. In all probability it constructed burrows. It was beyond any doubt perfectly capable of attacking termite and ant nests, but its degree of commitment to myrmecophagy is at present impossible to assess. Nothing is known of the skull other than a maxillary fragment, and this is not complete enough to reveal whether or not a median groove was present on the palate. The ascending ramus of the mandible slopes posteriorly considerably more than does that of *Orycteropus* and the condyle is lower. On this very scanty evidence one might suspect that *Myorycteropus* was some way along the road, but this is not certain. The jaw, dentition aside, is not as a whole very different from that of *Dasypus*, and there is nothing to indicate that the diet of *Myorycteropus* could not have been comparably diverse—and 488 species have been recorded in the *Dasypus* diet. Only if remains of possible descendants are found in later Miocene or Pliocene deposits will it be possible to determine whether or not a trend toward marked reduction of the jaw and elimination of the dentition was under way in the lineage.

The diet of *Plesiorycteropus* is much less uncertain. Alone among tubulidentates it displays cranial characters associated with a definite commitment to myrmecophagy, of which the most significant is the carrying down, somewhat in the pangolin manner, of the glenoid articulation on a descending zygomatic process of the squamosal. This implies that the posterior portion of the ramus was much reduced and the teeth diminished or perhaps lost entirely. Whatever the factor or factors that operate to maintain the size of the mandible and the persistence of fully functional teeth in *Orycteropus* may be, they were not involved in the evolution of the Malagasy form. The leg bones suggest an animal capable of digging and of jumping and do not bar climbing ability. At least one of the localities at which its remains are found was densely forested when the sediments there were laid down. The termite fauna of Madagascar includes both tree- and ground-dwelling species. *Plesiorycteropus* could have taken advantage of both; it may well have been the most versatile aardvark.

It begins to appear that tubulidentates played roles in Africa and Madagascar analogous to those of the armadillos and anteaters in South America. During the early Tertiary they may have been more numerous and diversified than during the later. Earlier they would have encountered little competition in the omnivore zone, and their only mammalian predators would have been hyaenodontid creodonts. With the arrival of the Carnivora they would have been faced not only with new predators but also with direct competition—a number of the smaller African carnivores are insectivorous-omnivorous and some are fossorial as well.

The intriguing question about *Orycteropus* is why, alone among committed myrmecophagous mammals, it possesses fully functional cheek teeth and a high, camellike ascending ramus of the mandible. Retention of these features is surely related to the diet, but to what item or items of it? Other than termites and ants, very little in the way of animal food has been recorded: locusts, "other insects," adult beetles, scarabeid larvae and pupae, and a mouse. The teeth are no doubt employed in the comminution of some of these items, but, judging from other myrmecophagous mammals, they were not essential; the well-developed pyloric "gizzard" could perform the task without them. Pangolins, for example, eat adult beetles, and captive specimens of *Manis pentadactyla* have taken mice and young rats, which were of course swallowed whole and dealt with in the interior. Unless the animal diet of *O. afer* is far more varied than all reports indicate the explanation does not lie here.

A possible clue is provided by a vegetable item in the diet. The living aardvark is known to eat the fruit of a species of Cucurbitaceae, *Cucumis humifructus*. The association between the two is sufficiently close to be described as symbiotic (Meeuse 1963). The plant coincides in range with the animal and is almost exclusively found near old aardvark holes; being geocarpic, it requires loose soil for the penetration of its long peduncles, at the ends of which the globose, tough-skinned fruits develop. Some of the numerous seeds pass undamaged through the animal's alimentary tract. Aardvarks bury their feces, frequently in the earth they excavate. The plant clearly benefits from and indeed may depend on the association. Some evidence suggests that the chief advantage to the animal is that the fruits provide it with an important source of water. The gape and the nature of the jaw articulation would permit taking the fruit into the mouth, where it could be crushed by the teeth. I have no other suggestion to offer, and the matter obviously requires further investigation. Apparently no one has presented a captive aardvark deprived of water with the fruit and recorded how the animal dealt with it.

Relationships

It has for some time been held that the Tubulidentata are ungulate in affinities, their ancestry going back to the late Cretaceous and early Tertiary order Condylarthra. The newer data available for both condylarthrans and tubulidentates in no way contradict this view and in various respects strengthen it. Resemblances displayed by *Plesiorycteropus* to pangolins and xenarthrans seem clearly to be convergent. Whatever the relationships of these two groups and of the extinct Palaeanodonta may eventually prove to be (see Pholidota, above), it is virtually certain that they are not with the aardvarks. The Tubulidentata, on available evidence, appear to be an "old African" group. As condylarth derivatives the aardvarks presumably came into existence in Paleocene time, together with other indigenous African ungulate orders, during the early Tertiary isolation of the continent.

References

Emry, R. J. 1970. A North American Oligocene pangolin and other additions to the Pholidota. *Bull. Am. Mus. Nat. Hist.* 42:455–510.

Hendey, Q. B. 1973. Fossil occurrences at Langebaanweg, Cape Province. *Nature* 244:13–14.

Grassé, P-P. 1955. Ordre des pholidotes. In P-P. Grassé, ed., *Traité de Zoologie* 17(2):1267–1284. Paris: Masson et Cie.

Griffiths, M. 1968. *Echidnas.* Oxford: Pergamon Press, pp. i–ix, 1–282.

Jentink, F. A. 1882. Revision of the Manidae in the Leyden Museum. *Notes Leyden Mus.* 4:193–209.

Klein, R. G. 1972. The late Quaternary mammalian fauna of Nelson Bay Cave (Cape Province, South Africa): its importance for megafaunal extinctions and environmental and cultural changes. *Quat. Res.* 2:134–142.

Lamberton, C. 1946. Contribution à la connaissance de la faune subfossile de Madagascar. Note XV. Le *Plesiorycteropus madagascariensis* Filhol. *Bull. Acad. Malgache* 25:25–53.

MacInnes, D. G. 1956. Fossil Tubulidentata from East Africa. *Fossil Mammals of Africa.* Brit. Mus. (Nat. Hist.) 10:1–38.

Matthew, W. D. 1918. Insectivora (continued), Glires, Endentata. In W. D. Matthew and W. Granger, eds. A revision of the Lower Eocene Wasatch and Wind River faunas. *Bull. Am. Mus. Nat. Hist.* 38:565–657.

Meeuse, A. D. J. 1963. A possible case of interdependence between a mammal and a higher plant. *Arch. Neerland. Zool.* 13 (suppl. 1):314–318.

Patterson, B. 1956. Early Cretaceous mammals and the evolution of mammalian molar teeth. *Fieldiana:* Geol. 13:1–105.

———. 1975. The fossil aardvarks (Mammalia: Tubulidentata). *Bull. Mus. Comp. Zool.* 147:185–237.

Pickford, M. 1975. New fossil Orycteropodidae (Mammalia: Tubulidentata) from East Africa. *Orycteropus minutus* sp. nov. and *Orycteropus chemeldoi* sp. nov. *Netherl. J. Zool.* 25:57–88.

Pocock, R. I. 1924. The external characters of the pangolins (Manidae). *Proc. Zool. Soc. London* 1924:707–723.

Segall, W. 1973. Characteristics of the ear, especially of the middle ear, in fossorial mammals, compared with those in the Manidae. *Acta Anat.* 86:96–110.

Simpson, G. G. 1931. *Metacheiromys* and the Edentata. *Bull. Am. Mus. Nat. Hist.* 59:295–381.

———. 1945. The principles of classification and a classification of mammals. *Bull. Am. Mus. Nat. Hist.* 85:i–xvi, 1–350.

Weber, M. 1928. Die Säugetiere. *Einführung in die Anatomie und Systematik der recenten und fossilen Mammalia,* vol. 2. Jena: Gustav Fischer, pp. 173–187.

13
Embrithropoda

Lloyd G. Tanner

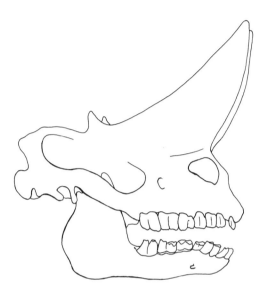

The arsinoitheres are probably the most mysterious and exotic of the fossil mammals found in the Fayum Province, Egypt. So unique are these beasts that they have been placed in a distinct order, Embrithropoda, by Andrews (1906). The name is a combination of *Arsinoe,* after Queen Arsinoe, who had a palace on Lake Moeris near the place where the first remains were found, and *therium,* the Greek word meaning beast. Their remains are recorded only from a single geographic locality, the Fayum area of the western Sahara, and a single geologic unit, the Jebel el Qatrani Formation. No candidates for likely ancestry have yet been discovered and they disappeared, prior to the Miocene epoch, as curiously as they arose, leaving no known descendants. There is always the hope that some of the continental deposits deeper into the western desert may hold the secret for the beginnings of the animal.

The most significant collections of arsinoithere remains can be found in the Geological Museum in Cairo, Egypt, the British Museum (Natural History), the Museum of Natural History in Stuttgart, Germany, the American Museum of Natural History, and more recently at the Yale Peabody Museum. The first and largest collection, now at the Geological Museum in Cairo, was made for the most part by H. J. L. Beadnell, under the supervision of Captain H. G. Lyons, director general of the (British) Egyptian Surveys. Beadnell mapped and explored along the eastern and northern borders of the Fayum Depression in 1898. Dr. C. W. Andrews, from the British Museum (Natural History), joined Beadnell in 1901 to continue exploration. From 1901 to 1904 Beadnell's collecting parties concentrated on the Upper and Lower Fossil Wood Zones and along with many other exciting finds, discovered the type specimen of *Arsinoitherium zitteli* Beadnell. Beadnell (1902) was quick to describe and illustrate such a huge and strange beast as the arsinoithere, and cooperative research continued on these incredible creatures then being discovered in the desert. In 1906, Andrews prepared his monograph, *Descriptive Catalogue of Tertiary Vertebrata of the Fayum, Egypt,* a good portion of which deals with arsinoithere specimens in the Cairo Museum and the British Museum.

Prospects for finding additional new and bizarre fossils quickly drew other expeditions to Egypt. A private collector, Richard Markgraf, working for Dr. Eberhard Fraas of the Stuttgart Museum, was responsible for a considerable collection of arsinoithere material now in Germany. Markgraf joined Henry Fairfield Osborn's American Museum of Natural History expedition in 1906. After some exploration in the area, Osborn departed, leaving Walter Granger and George Olson to continue more extensive excavations. Their quarrying resulted in a substantial collection of arsinoithere and other large animal remains now in the American Museum of Natural History. Markgraf was an exceptional collector and his techniques for collecting smaller fossils have been used by later workers with remarkable success (Simons and Wood 1968).

In more recent years, the Yale Museum expeditions to the Fayum collected many additional fossils and geologic data, concentrating mainly on localities yielding fossil primates. Although arsinoitheres are rather rare and extremely bulky to collect, several specimens were collected during the past few years. One skull, collected by the Yale parties from the Garet el Gindi area, is the best and most complete specimen recovered since the expeditions of Beadnell and Andrews during the 1900s.

Occurrence

Andrews (1906, viii, xi) and other earlier scientists regarded the "Fluvio-Marine Series" to be of the late Eocene; therefore arsinoitheres were thought to be of Eocene age also. However, in conjunction with the Yale expeditions, Dr. Carl Vondra and Bruce Bowen have been instrumental in providing a better understanding of the biostratigraphic occurrence of all the Fayum vertebrates, including the arsinoitheres. They considered the Jebel el Qatrani Formation to be of Oligocene age. Arsinoitheres are found throughout the formation, but seem to be more abundant in the lower portions. However, they do occur well above Quarry G and at the Quarry M level (Vondra, pers. comm.). Historical explanation of letter designations of the Yale Peabody Museum sites is given by Simons (1968, pt. 1, pp. 19–20).

Description

Two species are known of this unique order Embrithropoda: *Arsinoitherium zitteli* Beadnell (1902), and *A. andrewsi* Lankester (1903). With the exception of size, both species share the following distinctive characters: the very prominent pair of horns, supported primarily by the nasals alone, are the most immediately striking feature of the skull, a much smaller pair of horn cores occurs over the orbits. The occipital crest is slightly elevated and the occipital condyles, which are very prominent, are placed well behind the crest. The anterior bosses are relatively more roughened ventrally. There is a slight protuberance located medially, at the anterior

nasal opening of the type specimen. However, another, more complete specimen (CGM-8463) has a prenasal bar. At the base of each boss are a depression for labial muscles and grooves for nutritive arteries. The openings for the eye and ear are nearly horizontal; however, the narial opening is located slightly above this plane.

The upper permanent dentition is very much different from other ungulates (figure 13.1). Arsinoitheres had a full set of teeth, I_3^3, C_1^1, P_4^4, M_3^3. The

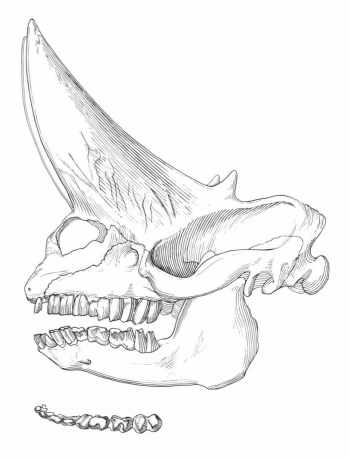

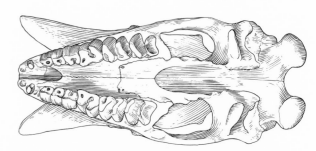

Figure 13.1 *Arsinoitherium zitteli* Beadnell, BMNH M-8463, skull and mandible found articulated, lateral and palatal views (tips of horns restored). (1/10th natural size. After Andrews 1906.)

Table 13.1   Comparative measurements of *Arsinoitherium* skulls (in mm).

Dimension	A. zitteli Type CGM=8130	A. zitteli BMNH M-8463	A. cf. zitteli YPM-30546	A. andrewsi Type BMNH M-8461
Length from condyle to tip of snout	740	770	—	—
Length from condyle to tip of horn	784	1000	—	—
Length from condyle to tip of small horn	389	380	—	—
Width at zygomatic arch	316	335	—	—
Greatest width of occipital surface	246	(260)	—	—
Greatest width between tips of small horns	232	245	—	—
Greatest width, combined bases of large horns	246	270	—	—
Greatest width of each horn at base (anteroposterior)	152	220	—	—
Greatest width of each horn at base	115	145	—	—
Greatest width above auditory opening	310	330	—	—
Length from basion to posterior border of palate	340	360	—	—
Length from basion to tip of snout	685	700	—	—
Length from border of palate to tip of snout	353	350	—	—
Height from tip of snout to end of large horns	590	720	—	—
Height of angle of pterygoids to roof of skull between small horns	(327)	340	—	—
Width between outer angles of condyles	(210)	240	—	—
Width of foramen magnum	80	100	—	—
Height of foramen magnum	55	55	—	—
Distance between tips of large horns	250	(370)	—	—
Length from fork to tip of large horns	210	(405)	—	—
Length from upper edge of nasal opening to tips of large horns	357	53	—	—
Transverse length of dental series				
P^1	—	—	—	25
P^2	26	26	27.2	34
P^3	—	28	30.2	36
P^4	—	32	41.1	41
M^1	52	53	43.8	69
M^2	(40+)	58	56.0	78
M^3	—	44	48.5	82
Length of molar series	167	165	—	—

Source: after Andrews (1906).
Notes: ( ) = approximate; BMNH M: British Museum Natural History catalog; CGM: Geological Museum, Cairo, catalog.

incisors and canines were not enlarged and, except for a short diastema between the upper median incisors, the tooth series was closed and continuous. The upper dentition is remarkably hypsodont, with well-developed metalophs and protolophs, and a less well-developed ectoloph. The lower dentition had a double-V pattern and was, like the uppers, quite hypsodont. Such developed hypsodonty is distinctly different from other large animals living during early Oligocene times.

Considerable size variation does exist among specimens of *A. zitteli* (tables 13.1, 13.2). Because the type specimen was based on a young animal, mature skull measurements are presented. Previously published measurements are included as well as illustrations of the types, because original data are not readily available.

Size is of primary concern for the validity of the species *A. andrewsi* Lankester. Because both *A. zitteli* and *A. andrewsi* were contemporaneous and are not significantly different, first impressions are that arsinoithere remains must be of a single species, *A. zitteli*. In spite of this, it is difficult to consider specimens such as the type mandible (BMNH M-8461) of *A. andrewsi*, which is nearly half again as large as a mature specimen of *A. zitteli*, to be of the same species (Andrews 1906:86; see also table 13.2). The huge pelvis (CGM-8413) (Andrews 1906:77) on display in the Cairo Museum is such a spectacular specimen when compared to a pelvis referred to *A. zitteli* (BMNH M-8812) that it must also be considered to be of a larger, separate species (figures 13.2, 13.3). The greatest width of the pelvis of CGM-8413, is 1,277 mm, while the greatest width of the referred

Table 13.2 Comparative measurements of mandibles of *Arsinoitherium* (in mm).

Dimension	*A. zitteli* BMNH M-8463	*A. andrewsi* Type BMNH M-8461
Length	—	730
Height at coronoid	315	435
Width of condyles	102	121
Width of ascending ramus	170	225
Depth of ramus at M_3	82	112
Width between posterior angles	295	—
Length of symphysis	180	215
Transverse length of dental series		
P_1	—	—
P_2	26	35
P_3	28	37
P_4	30	41
M_1	50	72
M_2	62	(70+)
M_3	52	(70+)

Source: After Andrews (1906).
Note: ( ) = approximate.

A. zitteli pelvis is 1,110 mm. The American Museum collections contain an unusually large upper first molar (no number) from the north end of Quarry B, collected by Granger and Olson, that can also be referred to *A. andrewsi*. The transverse diameter of this tooth is 64.5 mm and its anteroposterior diameter is 73.4 mm. A mature left upper first molar from a skull found near Garet el Gindi (YPM-30456) has the following dimensions: transverse diameter, 48.3 mm, anteroposterior diameter, 48.4 mm. Because no distinctions can be made at this point other than size, the evidence indicates a much larger arsinoithere existed and further study and comparison of individual variation are needed before *A. andrewsi* can be considered a discarded synonym.

The evolutionary story of the *Arsinoitherium* can not be concluded at this point. We do not have enough evidence to draw positive conclusions as to the origin of their ancestors and subsequent cause of their extinctions.

I appreciate the opportunity given to me as a member of the Yale Peabody Museum Expedition to the Fayum, where I was able to visit many of the localities that have yielded the remains of arsinoitheres. I am also grateful to our friends at the Geological Museum, Cairo, especially B. el Khashab. They were especially helpful in locating specimens in the collection. Through the courtesy extended me by the staff at the American Museum of Natural History, especially Bobb Schaeffer, I was able to examine the arsinoithere remains in that collection. Elwyn Simons and Grant Meyer of the Yale Peabody Museum aided me in many ways, and I am especially grateful to Carl Vondra and Bruce Bowen for stratigraphic information. Funds for travel were furnished in part by the University of Nebraska Research Council. Mary Tanner prepared the illustrations.

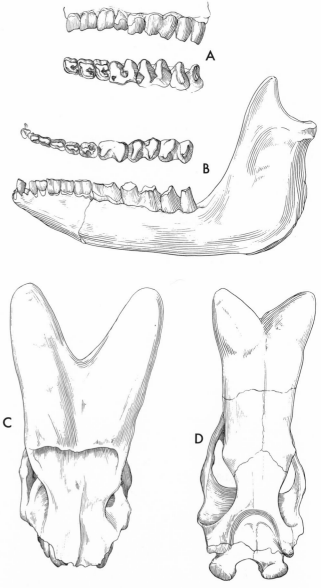

Figure 13.2 (*A*) *Arsinoitherium andrewsi* Lankester, BMNH M-8461, left maxilla, lateral and palatal view (1/8th natural size); (*B*) BMNH M-8461, mandible, palatal and lateral views (1/8th natural size); (*C*) *Arsinoitherium zitteli* Beadnell, CGM-8130, type specimen, juvenile skull, anterior view (1/8th natural size); (*D*) CGM-7805, younger skull, dorsal view (1/8th natural size). (All after Andrews 1906.)

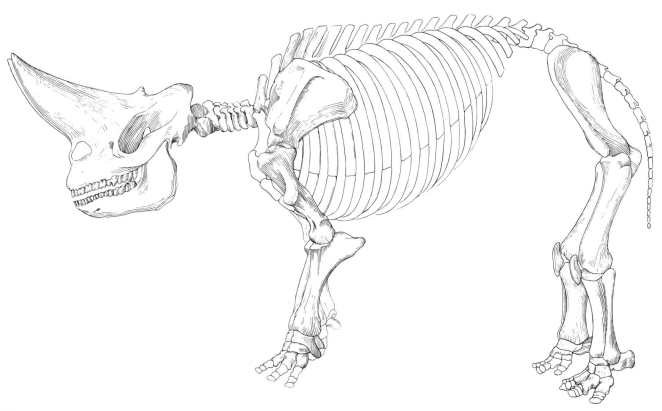

Figure 13.3   Restored skeleton of *Arsinoitherium zitteli*. (After Andrews 1906.)

References

Andrews, C. W. 1906. *A descriptive catalogue of the Tertiary Vertebrata of the Fayûm, Egypt.* Based on the collection of the Egyptian government in the Geological Museum, Cairo, and on the collection in the British Museum (Natural History). London: British Mus. (Nat. Hist.).

Beadnell, H. J. C. 1902. A preliminary note on *Arsinoitherium zitteli* Beadnell, from the Upper Eocene strata of Egypt. *Survey Dept. Public Works Ministry*, Cairo: National Printing Dept, pp. 1–4.

Lankester, E. R. 1903. A new extinct monster. *Sphere*, London, p. 238.

Osborn, H. F. 1908. New fossil mammals from the Fayum Oligocene, Egypt. *Am. Mus. Nat. Hist. Bull.* 24:265–272.

Simons, E. 1967. The earliest apes. *Sci. Amer.* 217:28–35.

Simons, E., and A. E. Wood. 1968. Early Cenozoic mammalian faunas, Fayum Province, Egypt. *Yale Univ. Peabody Mus. Nat. Hist. Bull.* 28:1–105.

14

Hyracoidea

Grant E. Meyer

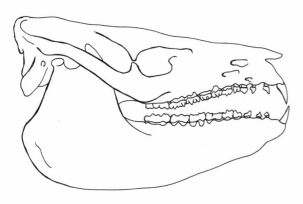

The oldest known hyraxes come from the Lower Fossil Wood Zone of the Jebel el Qatrani Formation in the Fayum Depression of Egypt; they are thought to be of Oligocene age. By that time hyraxes had already evolved into a large and exceedingly diverse assemblage represented by at least six genera including forms ranging in size from a large rabbit to a tapir and varying in tooth morphology from bunodont to selenolophodont. Although the diversity of forms remained the same or was even slightly greater in the fauna of the Upper Fossil Wood Zone, the relative number of hyraxes in the total fauna greatly declined. This decline continued throughout the remainder of geologic time. From Miocene rocks of East Africa two or possibly three genera are known; from Southwest Africa one genus is known. From the Pliocene one genus has been described from Europe, two genera from Asia, and a new form will soon be described from East Africa (M. Pickford, pers. comm.). Two genera are known from Pliocene or Pleistocene deposits in South Africa and one of these extends into recent times where three closely related genera are recognized.

The hyraxes are in many respects quite primitive or generalized subungulates that have changed little since the Oligocene. They have plantigrade feet with an os centrale present in the carpus, features not characteristic of advanced ungulates, and the axis of the limb passes through the third digit. There are five toes in the manus and three in the pes. The inner digit of the pes bears a long curved claw while the other digits terminate in short flattened nails. Although the extant hyraxes can only be described as quick and agile climbers both in trees and on rocky slopes, their feet are not typical of climbing animals. This may be because hyraxes developed their climbing abilities at a relatively recent time in their evolution; their foot structure, especially the naked and rather elastic pads on the soles of the feet and the nails, is similar to that in elephants. Their agility seems to depend on the flexibility of their limb articulations and on a strong grip that has been attributed to abundant skin glands on the surface of the foot pads and a specialized musculature that contracts the pad into a suction cup (Dobson 1876).

Astragalus morphology, unique among mammals, is virtually identical in all fossil and extant forms where it is known. The bone is greatly compressed and possesses a distinctive fossa for the internal malleolus of the tibia; it articulates with the fibula and with the navicular. The femur has a small third trochanter; there is no entepicondylar foramen in the humerus and no clavicle. The vertebral column is convex throughout and the number of vertebrae is exceedingly high for a mammal. In the extant forms the thoracic vertebrae vary between 20 and 23, the lumbar vertebrae between 7 and 9, and the caudals between 4 and 8. The sacrum articulates with the ilium by two vertebrae.

The skull is of rather generalized construction, broad and flattened in the frontal region, nearly vertical in the occipital region, and with wide, rounded zygomatic arches. The postorbital bar is closed (at least with cartilage) in most extant forms and may have been closed in some fossil forms as well. This closing is accomplished by a bony process projecting upwards from the jugal and fusing with the postorbital process of the parietals. The lacrimals form a strong antorbital process projecting distally into the orbits as in elephants and rhinoceros. The jugals form part of the glenoid fossa as a steeply positioned external facet and the glenoid fossa has a squamosal component shaped like a flat horizontal plate anterior to a fairly large postglenoid process. The infraorbital foramen is positioned at the posterior edge of a rather long and deeply engraved sulcus. There is no contact between the premaxillae and the frontals like that in rodents.

The periotic and tympanic bones are ankylosed and form a moderately large bulla; the external auditory meatus is long and narrow. Slightly mesiolateral to the temporomandibular articulation the alisphenoid canal opens, and lateral to this is the foramen ovale and a distinct foramen rotundum. The compressed foramen lacerum medius, which functions as the carotid canal, is only slightly separated from the foramen lacerum posterius. The optic foramen opens in the large orbitosphenoid near its distal margin. The basicranium is very similar in both the fossil and extant forms. Most changes in the extant skull can be attributed to the relatively larger brain size of the living species, the shortening of the snout, and resultant loss of the lateral incisors and canines.

The mandible in hyraxes is quite distinct in its broad ascending ramus, small recurved coronoid process that extends only slightly above the condyle, and the broad, circular and often very deep angular process. Several fossil hyraxes possess a fossa on the internal surface of the mandible, below the dental battery; in some cases a foramen or fenestra opens into a wide and long chamber within the horizontal ramus, and in one species this even extends into the ascending ramus. This feature, although it may be constant within a given species, is extremely variable within the order as a whole. It is not present in *Saghatherium* and *Titanohyrax*, some *Megalohyrax*, *Thyrohyrax*, *Bunohyrax*, and all the extant forms. It is present as a large deep fossa in *Geniohyus*,

Meroëhyrax and *Pliohyrax*. A small and nearly rounded hole under M_3 in *Pachyhyrax* and some *Megalohyrax* leads to a narrow but deep canal that runs forward until it shallows and terminates beneath P_2. In some *Thyrohyrax* the hole is slightly elongate mesiodistally and the canal, which is greatly inflated, not only runs forward in the horizontal ramus but extends dorsodistally into the inflated ascending ramus. In some *Bunohyrax* the hole is extremely large and subtriangular in shape; in others it is missing.

The function of this fossa and fenestra is not fully understood and most of the explanations given previously can be discounted for one reason or another. The most reasonable explanation so far proposed is that these features were used as a resonating chamber for sound production, as first suggested by Andrews (1907). It seems unlikely, however, that a chamber for sound resonation would be located under the tongue.

The dentition in hyraxes is brachyodont and varies from bunodont to selenolophondont. In most fossil forms the complete dental formula is present whereas in Pleistocene and Recent forms reduction by loss of distal incisors and canines has taken place. The teeth form a closed and slightly curved series increasing in size distally from the canine or first premolar to the third molar. The upper molars are subtriangular to trapezoidal in shape with the mesial margin longer than the distal. The parastyle is typically well developed in all known forms and often inflated. The mesostyle is usually well developed, often mesiodistally compressed, and in several early forms is nearly as high as the paracone and metacone. The metastyle is variously developed in the different forms and when present it is best developed in M^3. A hypostyle may be present. The paracone and metacone are buccolingually compressed and attached to the styles by a crista forming a W-shaped ectoloph. The protocone and hypocone are usually well separated, sometimes inflated and bunodont, but more often bunolophodont to bunoselenodont with variously developed preprotocristae (Szalay 1969:202) and pre- and posthypocristae. The protocone and hypocone are nearly as high as the paracone and metacone.

The lower cheek teeth form a continuous series from P_1 distally and range between bunodont and selenodont. The molars increase in size from M_1 to M_3. On the molars the paraconid is small, the protoconid and metaconid are large and opposed, as are the hypoconid and entoconid. The metaconid is the highest cusp. The cristid obliqua is usually well developed and meets the protocristid slightly buccally to the metaconid. The paracristid and protocristid form a crescent that is lingually concave. The cristid obliqua and postcristid form a similar but slightly less acute crescent. In wear the tooth resembles a W or two slightly oblique Vs. A metastylid may be present and a hypoconulid is present on M_3 except in the Procaviidae. The premolars are usually simpler than the molars in early forms but tend to be submolariform in later species.

The upper canines are premolariform and often have three or even four roots. The lower canines are incisiform and usually have a single root, but two-rooted lower canines are found in *Geniohyus, Thyrohyrax,* and *Pliohyrax* where the canines are premolariform and not separated by a diastema from P_1. The lateral incisors, when present, are simple bean-shaped teeth buccolingually compressed and often very small. In males of all known forms the upper central incisor is a large triangular recurved tusk that grows from a persistent pulp. There is a groove on the mesiobuccal surface. Contrary to some reports there is enamel on the lingual side of this tooth, at least at time of eruption, but it is rapidly worn through by contact with the chisel-shaped I_1 and I_2. This wear with the lower incisors maintains sharp points on the triangular upper incisors. In female hyraxes the I^1 is usually not as large and in most forms is only subtriangular in section with a rounded buccal edge. The tooth is not as curved as it is in males and the tip is often more rounded.

History of Research

The earliest known written reference to hyraxes is found in the Bible where they are referred to as conies: "And the rock coney, because it chews the cud but does not part the hoof, is unclean to you" (Lev. 11:5). "Yet of those that chew the cud or have the hoof cloven you shall not eat these: The camel, the hare and the coney, because they chew the cud but do not part the hoof, are unclean for you" (Deut. 14:7). "The high mountains are for the wild goats; the rocks are a refuge for the conies" (Psalms 104:18). "Four things on earth are small but they are exceedingly wise . . . The conies are a people not mighty, yet they make their homes in the rocks" (Prov. 30:24, 26).

The rather rare and obscure living hyraxes were not known by Linnaeus; they were first designated *Procavia* by Storr (1780) who placed them with the rodents near *Cavia.* Cuvier (1798) referred them to his "pachyderms" where they remained for nearly 50 years.

Gaudry, during his 1855 and 1860 excavations in the Pliocene beds at Pikermi, discovered a mandible that was the first fossil hyrax to be found. He named the specimen *Leptodon graecus* (Gaudry 1862) and described it as being closely related to *Palaeotherium* and *Rhinocerus.* In 1868 when the "pachyderms" were separated into several groups, Owen placed the hyraxes in the Perissodactyla and a year later a new order, the Hyracoidea, was created for them (Huxley 1869). Schlosser (1886) argued that *Leptodon* was more closely related to *Titanotherium* than to *Rhinocerus* on the basis of tooth morphology and this was accepted by Zittel (1891–93).

Two notes by Major that were read to the Academy of Science, Paris, in 1891, listed the fauna from Samos but did not mention a hyracoid specimen. In 1894 the British Museum purchased a skull labeled *Rhinocerus pachygnathus* from a dealer and Major at that time assigned it to *Leptodon.* In 1895 Krupp presented a collection of fossils from Samos to the Stuttgart Museum that included the anterior portion of a skull described under the name *Hyrax kruppii* (Rodgers 1896), a name provisionally assigned by Fraas and also mentioned by Smith Woodward (1898). In July 1898 this same skull was given by Fraas to Osborn for formal description. In 1898 before the International Zoological Congress at Cambridge, England, Osborn read what he considered to be the first description of a fossil hyrax, which he named *Pliohyrax kruppii.* He named the new family Pliohyracidae for this genus, separating it from *Procavia* (= *Hyrax*) on the family level; he considered *Pliohyrax* to be an aquatic or subaquatic form. Schlosser (1899) pointed out that the mandible described by Gaudry as *Leptodon graecus* was also a hyrax. Schlosser then described a nearly complete mandible that was included in a large collection of fossils from Samos presented to the Munich Museum by Stützel as belonging to the same species and possibly the same individual as the Stuttgart skull described by Osborn. He thus synonymized *Pliohyrax kruppii* with *Leptodon graecus.* Still later in the same year a notice appeared (Major 1899a) in which it was pointed out that *Leptodon* was preoccupied by Sundevall's genus of Falconidae (1835) and the correct name for these specimens was *Pliohyrax graecus* (Gaudry). The following month Major (1899b) described the London skull and concluded that the London, Paris, Stuttgart, and Munich specimens all belonged to the same species.

Ameghino (1897) had named eight genera of typotheres and toxodonts which he believed to be primitive Hyracoidea. This was refuted by Sinclair (1908); Simpson (1945) regarded the similarities between these South American forms and the true hyraxes to be convergence and not to represent a phylogenetic relationship.

In 1879 Schweinfurth found fossil vertebrates in the Eocene beds on Geziret el Qorn Island in the Fayum Depression of Egypt and in 1886 he discovered fossil mammals in the escarpment behind the ancient Egyptian temple known as Qasr el Sagha Temple. In 1898 Beadnell collected fossil reptiles in the escarpment and during 1901–1904 Beadnell and Andrews returned to these beds, discovered the fossiliferous "Fluvio-Marine Series," and collected many fossil mammals including hyraxes. The first of these hyraxes described (Andrews and Beadnell 1902) was named *Saghatherium antiquum*, a name derived from the Qasr el Sagha Temple. A second specimen, an infant maxilla, was named *Saghatherium minus*. These authors divided the Fayum strata into two units, an upper Fluvio-Marine Series containing the hyraxes, and a lower Qasr el Sagha Series.

During the next several years Andrews described additional forms of Oligocene hyraxes including *Megalohyrax eocaenus* (Andrews 1903), *Geniohyus mirus* (Andrews 1904a), *G. fajumensis* (Andrews 1904a), *G. major* (Andrews 1904b), *Megalohyrax minor* (Andrews 1904b), *Saghatherium magnum* (Andrews 1904b), and *S. majus* (Andrews 1906). He divided these forms into two families; *Saghatherium* and *Megalohyrax* were placed in the Saghatheriidae and *Geniohyus* was mistakenly placed in the Suidae. In 1907 Andrews described specimens he had collected in 1906, particularly a newly found posterior portion of a ramus of *G. mirus*. In noticing the foramen that perforates the base of the ascending ramus behind M_3, he stated, "so far as I am aware, this only occurs in the Hyracoidea and its presence here suggests the possibility that *Geniohyus* may have some relationship with that group" (Andrews 1907:98). Andrews, not Schlosser (1910), first assigned *Geniohyus* to the Hyracoidea.

Sinclair (1908), in discussing the Santa Cruz Typotheria, concluded that they were not closely related to the Fayum Hyracoidea as was commonly proposed at the time. He pointed out that the hyracoid carpus is completely distinctive compared to that of the typothere, the hyracoid astragalus is unique, and that fundamental differences in the dentition, jaw structure, and skull negate any close relationship between the two.

Gregory (1910) discussed many similarities and differences between the Hyracoidea and other orders of mammals: "These resemblances are consistent with the hypothesis that the Hyracoidea and Proboscidea, together with certain other groups, have been derived from unknown basal members of the Condylarth-Amblypod stem."

The same year, Schlosser (1910) named three new genera, *Bunohyrax, Pachyhyrax,* and *Mixohyrax,* based on a large and important fauna collected for the Stuttgart Museum by Markgraf. In the 6 hyracoid genera from the Fayum he distinguished 15 species, naming as new *Pachyhyrax crassidentatus, Mixohyrax andrewsi, M. niloticus, M. suillus, Megalohyrax palaeotheroides,* and *Geniohyus minutus.* He transferred Andrews's *G. fajumensis* and *G. major* to his newly named genus *Bunohyrax*. Although the three new genera are valid names because they are described in the form of a key, Schlosser did not include any definition or description of the specific taxa, nor did he designate any specimen as belonging to these specific taxa. The following year (1911) a revision of the Fayum fauna was published by Schlosser, and he fully described and defined these forms except for *Geniohyus minutus,* which he placed as a senior synonym of *G. micrognathus,* a new species.

Matthew (1910) in a short note reviewing Schlosser's 1910 paper stated: "As the paper is avowedly published in order to secure priority for the results of Dr. Schlosser's studies over the concurrent investigations of other authors, it is unfortunate that other authors, however well disposed, can not accept his species as of this date, without violating the rules of nomenclature."

Any name accepted into nomenclature must meet the publication requirements of the current rules of nomenclature. Any name that does not satisfy the rules at the time is legally unacceptable. Schlosser's specific names (1910) that were published before 1931 must satisfy the conditions of Articles 12 and 16 of the International Code of Zoological Nomenclature. If they do not they should be considered *nomina nuda.*

The species *Pachyhyrax crassidentatus* is a valid taxon because it meets the requirement of being published before 1931 and is accompanied by an indication consisting of a single combined description of a genus and one included species. *Geniohyus micrognathus* is an available name because it first appeared in Schlosser's 1911 paper and was accompanied by a valid description. It is a junior synonym of *G. minutus* but is valid because the latter is a *nomen nudum.* The remainder of the specific names proposed by Schlosser in 1910 are unacceptable because they were published without a description, or a reference to one, or an acceptable indication of its application (*nomina nuda*). These names must therefore be rejected and either replaced by the oldest available synonym that is not a senior homonym or be eliminated.

This matter is further complicated by the fact that nearly all the specimens described by Schlosser (1910, 1911) that belonged to the Stuttgart collection have been lost since they were packed up at the start of World War II. All Schlosser's hyracoid types, as well as his figured specimens, are unavailable for comparison, which makes any revision of the Fayum Hyracoidea extremely difficult. (These specimens may still exist in unopened crates labeled "Fayum Proboscidea" presently stored at the museum at Ludwigsburg, West Germany.)

In Appendix II of a paper on the geology of the Miocene beds of Victoria Nyanza, Andrews (1914) named a new genus and species *Myohyrax oswaldi* and separated it from the Oligocene hyraxes by naming the new family Myohyracidae. The knowledge of the Myohyracidae was enlarged at various times (Stromer 1922, 1926; Hopwood 1929; Whitworth 1954) and a second genus *Protypotheroides* Stromer (1922) was added. Patterson (1965) removed the family Myohyracidae from the Hyracoidea and placed it as the subfamily Myohyracinae in the family Macroscelididae. He demonstrated the similarities between Myohyracinae and the elephant shrews and the complete lack of hyracoid features in the group.

In 1922 Matsumoto pointed out that Schlosser's *Mixohyrax* was a junior synonym of *Megalohyrax* Andrews and that Schlosser's "*Megalohyrax*" was in fact different from the type and needed a new generic name for which he proposed *Titanohyrax*. He named four new species, *Megalohyrax pygmaeus*, *Titanohyrax ultimus*, *T. schlosseri*, and *T. andrewsi*, and made new designations for *Megalohyrax niloticus* (Schlosser), *M. suillus* (Schlosser), and *Titanohyrax palaeotherioides*. When Matsumoto revised the fossil Hyracoidea from the Fayum in 1926, he recognized 6 genera and 24 species. Added to the literature were his new species *Geniohyus gigas*, *G. subgigas*, *G. diphycus*, *Bunohyrax affinis*, *Saghatherium macrodon*, *S. euryodon*, *S. annectens*, and *S. sobrina*. He separated the Hyracoidea into five families that included the existing Procaviidae, Myohyracidae, and Pliohyracidae, to which he added the new Geniohyidae and Titanohyracidae.

A Miocene fauna similar to the East African fauna was reported by Stromer (1926) from Southwest Africa; he described a new genus and species of hyracoid, *Prohyrax tertiarius*, which he included in the family Hyracidae (= Procaviidae). Additional

Miocene fossils were described in 1933 by Arambourg from the area west of Lake Rudolph in Kenya. He named the new species *Pliohyrax championi* based on a mandibular fragment with two teeth. He differentiated this specimen from *Megalohyrax* by the presence of a metastylid on the lower molars.

The following year fossils from Taung associated with *Australopithecus africanus* were described (Broom 1934); many skulls, with teeth more brachyodont than in living forms, were placed in *Procavia antiqua*. In February 1937, Shaw gave the name *Procavia transvaalensis* to a skull and jaws from Sterkfontein Farm that were considerably larger than any extant *Procavia*. The following month, Broom (1937a) described a new form, *Procavia obermeyerae*, from a cave at Uitkomst. Later the same year Broom (1937b) recognized that his *P. obermeyerae* and Shaw's *P. transvaalensis* were conspecific but he continued to use the term *P. obermeyerae* (spelled *P. obermeijerae* in his second paper). Wells (1939) in studying the endocranial casts of two specimens confirmed the placement of *P. transvaalensis* in a species different from extant forms and he described a new and unnamed specimen of Broom's that, on the basis of endocranial morphology, was inseparable from living species. In 1946 Broom figured *P. antiquus* and *P. transvaalensis* as well as a new species, *P. robertsi*. Broom's only statement about this new species was: "A small species of *Procavia* is met with which I am calling *Procavia robertsi*." Because this statement does not fulfill the requirements of the International Code of Zoological Nomenclature, this name must be considered a *nomen nudum*. In 1948 Broom again figured *P. robertsi* and this time a description was given; however, the name was not available.

From Nihowan in China a fauna reported by Teilhard de Chardin and Piveteau (1930) included teeth that they considered belonged to a new and aberrant chalicothere. In 1932 von Koenigswald named these teeth *Postschizotherium chardini*. In 1938 Teilhard de Chardin compared upper molars of this genus with *Palaeotherium* and some lower molars with *Titanotherium* and *Anchitherium* and mentioned a personal communication from Simpson that suggested that these teeth might be referred to the Hyracoidea. Teilhard de Chardin did not accept this referral but did note certain supposed hyracoid similarities.

Viret (1947) published a preliminary notice of an isolated tooth from the Pliocene of Soblay which he considered a chalicothere and closely related to the Chinese form *Postschizotherium*. He designated the name *Neoschizotherium* for this tooth but did not define or describe the tooth, nor did he designate a species until the following year (Viret and Mazenot 1948) when the species *Neoschizotherium rossignoli* was formally proposed and several isolated teeth were figured. The teeth were still believed to belong to the Chalicotheriidae but interestingly they were compared to Brontotheriidae (= Titanotheriidae), as was Gaudry's first *Pliohyrax* jaw. The following year Viret (1949a) transferred this species to *Pliohyrax* and stated that he thought *Postschizotherium* from China was also a hyrax (which further substantiated Simpson's 1939 recommendation). In a later paper (Viret 1949b) the phylogenetic position of *Postschizotherium* was discussed at great length and Viret concluded that the Hyracoidea had in fact lived in China although no forms had until that time been discovered east of Samos. Fossil hyraxes have now been reported in Turkey (Sickenberg and Tobien 1971) and Russia (Gabuniya and Vekua 1966), so this paleogeographical problem no longer exists.

In 1952 Viret and Thenius reported a second species of hyrax that they named *Pliohyrax occidentalis* from the Pliocene of Montpelier. This species is based on a fragment of the ectoloph of an upper molar, which they state is more primitive than *P. graecus* and more hypsodont than *P. rossignoli*.

In 1954 Whitworth reviewed the East African hyraxes and transferred Arambourg's *Pliohyrax championi* to *Megalohyrax*. He recognized two families for this material, the Geniohyidae and the Myohyracidae. His chart (Whitworth 1954:22,23) makes twelve comparisons between *M. championi* and the Oligocene *Megalohyrax* and *Titanohyrax*; *M. championi* more closely agrees with *Titanohyrax* in all cases. He then concluded that the differences outlined in the chart were of limited significance and that "the cheek dentition of the Miocene form may be derived from that of Oligocene *Megalohyraxes*, without radical change." He also named the new genus and species *Meroëhyrax bateae*, based on a new mandible from Rusinga.

In 1956 Churcher discussed the South African hyracoid material that then included over 135 specimens of *P. transvaalensis* and was able without doubt to show that *P. transvaalensis* and *P. obermeyerae* were conspecific. Over 60 specimens of the smaller hyrax were available and Churcher showed that no differences between *P. antiqua* and *P. robertsi* could be considered significant for separating these forms into two species, and *P. robertsi* was therefore considered a junior synonym of *P. antiqua*. *P. antiqua* was believed close to the ancestry of the extant *P. capensis*.

Kitching (1965) reported a new large hyrax from the Limeworks Quarry at Makapansgat that he named *Gigantohyrax maguirei*. This form differs morphologically from *Procavia* and is much larger than *P. transvaalensis*, though contemporaneous.

Melentis (1965, 1966) described new specimens of *Pliohyrax graecus* from Pikermi and Halmyropotamus in Greece. He divided the Hyracoidea into three families; the Procaviidae, the Geniohyidae, and the Myohyracidae. He placed *Pliohyrax* in the Procaviidae and in the subfamily Pliohyracinae.

In 1966 *Postschizotherium* surfaced again in a paper by von Koenigswald in which he named two new species *P. licenti* and *P. intermedia*. This paper gives the most complete description of the genus yet, pointing out that it possesses few morphological traits found in other known hyraxes. From this he concludes that *Postschizotherium* is a unique member of the Hyracoidea. Several other new specimens have recently been discovered in Pliocene rocks of Spain (Crusafont Pairo, pers. comm.), Libya (Savage 1971), Kenya (M. Pickford, pers. comm.), Turkey (Sickenberg and Tobien 1971), and the USSR (Gabuniya and Vekua 1966).

Meyer (1973) described a new genus and species of hyrax from the Upper Fossil Wood Zone of the Fayum, Egypt. The new species named was *Thyrohyrax domorictus*.

Systematic Description

Order Hyracoidea Huxley 1869, p. 101

Family Pliohyracidae Osborn 1899, p. 172, emended.

[including Saghatheriidae Andrews 1906, p. 84; Geniohyidae Matsumoto 1926, p. 259; Titanohyracidae Matsumoto 1926, p. 259; and Pliohyracidae Matsumoto 1926, p. 330.]

TYPE GENUS. *Pliohyrax* Osborn 1899.

INCLUDED GENERA. *Bunohyrax, Geniohyus, Kvabebihyrax, Megalohyrax, Meroëhyrax, Pachyhyrax, Pliohyrax, Postschizotherium, Saghatherium, Thyrohyrax,* and *Titanohyrax*.

DISTRIBUTION. Oligocene of North Africa, Miocene and Pliocene of East Africa, Pliocene of Southern Europe, and Southwestern and Central Asia.

DIAGNOSIS. Hyraxes with the full eutherian dentition; teeth brachydont to slightly hypsodont; upper central incisor forming a triangular, recurved tusk, lower central pair of incisors procumbent, enlarged, spatulate with tripectinate extremity when unworn; lateral incisors simple, reduced; canine simple to premolariform; premolars submolariform to molariform; third lower molars with hypoconulids; skull, when known, with narrow snout, frontal region flattened and broad; palate extending behind posterior molars; sagittal crest often present.

DISCUSSION. Osborn (1899) first mentioned the name Pliohyracidae: "It proves to be not only a new species, but a new genus, which may be termed *Pliohyrax,* and possibly the representative of a new family, *Pliohyracidae,* [his italics] of the Hyracoidea." Most subsequent authors have placed *Pliohyrax* in the family Procaviidae, although most have recognized that *Pliohyrax* was not ancestral to extant forms. Because the sequence *Thyrohyrax— Meroëhyrax—Pliohyrax* may represent a closely related phylogenetic sequence that is quite distinct from the extant forms, *Pliohyrax* is here removed from the Procaviidae as was done by Matsumoto (1926).

Andrews (1906) first placed *Geniohyus* in the Suidae but in 1907 he transferred it to the Hyracoidea when reporting a more complete specimen. In 1906 (p. 84) he had named the family Saghatheriidae to include hyraxes with the full eutherian dentition. In 1926 Matsumoto designated the families Geniohyidae to include *Bunohyrax, Geniohyus,* and *Megalohyrax,* Titanohyracidae to include *Titanohyrax,* and Pliohyracidae, which he claimed as a new family, to include *Pachyhyrax, Saghatherium,* and *Pliohyrax.* This grouping is artificial because *Pachyhyrax* is much more closely related to *Megalohyrax* than to *Saghatherium* although *Pliohyrax* and the latter two genera may be only remotely related. *Geniohyus* exhibits features similar to both *Pliohyrax* and *Bunohyrax* but differs considerably from all other forms in the morphology of the lower teeth. Whitworth (1954) recognized the two families Procaviidae and Geniohyidae but included *Pliohyrax, Saghatherium,* and *Meroëhyrax* in the Procaviidae, thus forming an equally artificial taxon. Of the four familial terms proposed for this group, Pliohyracidae Osborn 1899 has priority and should supplant Geniohyidae in current use.

Subfamily Geniohyinae Andrews 1906 emend.

TYPE GENUS. *Geniohyus* Andrews 1904.

INCLUDED GENERA. *Geniohyus.*

DISTRIBUTION. Oligocene, Jebel el Qatrani Formation, Fayum, Egypt.

DIAGNOSIS. Pliohyracidae with bunodont to only slightly bunoselenodont dentition; dental formula $\frac{3.1.4.3}{3.1.4.3}$; I_2 much larger than I_1; I_3 reduced; lower canine two-rooted; premolars submolariform; lower premolars with hypoconid centrally positioned

on distal border of tooth, connected to protoconid by straight cristid, slightly V-shaped on P_4; upper premolars without mesostyle; small mesoconid present on lower molars; protoconid and metaconid symmetrical, opposed; premetastylid on mesial face of metaconid; paraconid nearly central, poorly developed; hypoconulid on M_3; parastyle and mesostyle of upper molars rounded and blunt; cusps of upper molars in close approximation and centrally located, cusps of lower molars greatly inflated; lower molars without metastylid; ramus extremely deep with large bullate expansion or fossa on the distolingual surface; skull unknown.

DISCUSSION. The separation of *Geniohyus* in a subfamily of its own seems warranted because of its unique cusp pattern on the lower molars, the inflated character of the cusps, its much less molariform premolars, which differ considerably in outline from other hyracoid genera, and its two-rooted lower canine. The two-rooted lower canine and the large fossa on the interior side of the ramus that does not penetrate the jaw are features in common with some members of the Pliohyracinae but the extreme development of the fossa and the exaggerated deepening of the horizontal ramus in *Geniohyus* is unique among mammals. All Pliohyracinae have teeth that are much more selenodont than those of *Geniohyus* and lack additional cuspules. At this time no forms are known that are intermediate between the Geniohyinae and the Pliohyracinae. The earliest known member of the Pliohyracinae, *Meroëhyrax,* was probably derived from a form similar to the more primitive members of the Saghatheriinae and not from members of the Geniohyinae. The Saghatheriinae have shallow to only moderately deep horizontal rami with either no fossa or else a fenestra that penetrates the jaw opening into a chamber or canal in the hollow ramus.

Genus *Geniohyus* Andrews 1904

SYNONYMY.

Geniohyus Andrews 1904, 1906 in part, 1907.

Geniohyus Schlosser 1910, 1911 in part.

Geniohyus Matsumoto 1926 in part.

TYPE SPECIES. *Geniohyus mirus* Andrews 1904.

INCLUDED SPECIES. *Geniohyus mirus, Geniohyus diphycus, Geniohyus magnus.*

DISTRIBUTION. Lower Fossil Wood Zone, Jebel el Qatrani Formation, Fayum, Egypt.

DIAGNOSIS. Same as subfamily.

DISCUSSION. In his original illustration of *Geniohyus mirus,* Andrews (1904, fig. 4) correctly labeled the two-rooted alveolus in front of P_1 as belonging to a canine but he misidentified the alveolae of the incisors, showing no I_3 and mislabeling the alveolae of I_3 and I_2 as I_2 and I_1, respectively. In 1906 (pl. XIX, fig. 1) he further misinterpreted this dentition by labeling the mesial root of the canine as I_3. That the canine of *Geniohyus mirus* is two-rooted is suggested by the extreme closeness of the two alveolae. This is further documented by the jaw of the type specimen of *Geniohyus diphycus* that has a two-rooted alveolus in front of P_1 which could not support two separate teeth in the available space. No known hyrax, whether extant or fossil, has a small isolated I_2 separated from I_1 by a diastema. When there is a difference in size, I_2 is always larger than I_1, not the reverse. The alveolus labeled I_2 by Andrews must belong to I_3.

Schlosser and Matsumoto misinterpreted the generic characteristics of *Geniohyus* and both included in this genus forms that belong in *Bunohyrax*. Schlosser's classification was based to a great extent on the roughness of the enamel of the teeth, a character shown by many subsequent workers to be insignificant for phylogenetic determinations. He placed several specimens into *Geniohyus* on this basis. Matsumoto based his classification on the presence or absence of a fenestra on the interior surface of the ramus. Because by his definition *Bunohyrax* had no fenestra cutting through the mandibular wall, he placed forms possessing a large fenestra but otherwise identical to *Bunohyrax* into *Geniohyus mirus* even though the type does not possess a fenestra. He justified this placement: "The general shape of the mandible of specimen No. 1446 differs considerably from that of Andrew's type mandible, though the size of the teeth and the length of the diastemata of the former are very close to those of the latter" (1926:269). The tooth morphology of this jaw is identical to *Bunohyrax* and not similar to *Geniohyus*. This specimen is associated with a skull that was also described as *Geniohyus mirus* by Matsumoto, and thus the upper dentition of *Geniohyus* was incorrectly diagnosed. Upper dentitions of known species of *Bunohyrax* were consequently assigned to *Geniohyus* and new specific names given for them. That this misinterpretation has in fact been made is illustrated by the fact that Matsumoto's type of *Geniohyus subgigas* (AMNH-3329), a maxilla with P^4 to M^3 on both sides, occludes so well with a mandible of *Bunohyrax fajumensis* (AMNH-13347) from the same locality that these specimens clearly belong not only to the same species but possibly even to the same individual. Because of a probable typographical error in the "Key to genera of Geniohyidae" (Matsumoto 1926:261) there is no mention of the

genus *Geniohyus* and therefore nowhere in Matsumoto's publication are *Geniohyus* and *Bunohyrax* differentiated.

Geniohyus mirus Andrews 1904

SYNONYMY.

Geniohyus mirus Andrews 1904, p. 160, pl. VI, fig. 4.

Geniohyus mirus, Andrews 1906, p. 193, pl. XIX, Fig. 1.

Saghatherium majus Andrews 1906, p. 91 *pars,* pl. VI, fig. 5.

Geniohyus mirus, Andrews 1907, p. 97, fig.1.

Geniohyus mirus, Schlosser 1910, p. 502 *pars.*

Geniohyus mirus, Schlosser 1911, p. 121 *pars, non* pl. XVII, fig. 4, 5.

non Geniohyus mirus, Matsumoto 1926, p. 269, fig. 5-8.

HOLOTYPE. CGM-8634, Cairo Geological Museum, Cairo, Egypt; fragment of a right ramus with root of I_2 and P_1 to M_3.

DISTRIBUTION. Exact stratigraphic level unknown, Oligocene, Jebel el Qatrani Formation, Fayum, Egypt.

DIAGNOSIS. Medium-sized *Geniohyus* with long narrow symphysial region; diastemata between I_2 and I_3, I_3 and canine, and between canine and P_1; I_2 large, procumbent with laterally compressed root giving oval outline; I_3 presumably reduced, represented in type specimen by alveolus only; canine two-rooted; P_1 laterally compressed with high protoconid, with paracristid and postcristid terminating respectively in very small paraconid and hypoconid; P_2 similar with small ridge on distobuccal surface of protoconid creating a small hypoflexid; protoconid just beginning to separate into two cusps; weak cingular ridge developing on mesial face of tooth; mesiobuccal angle of tooth slightly inflated; postcristid higher than in P_1 and hypoconid slightly larger; P_3 and P_4 similar with progressive increase in development of hypoflexid, inflated angle of teeth, and separation of metaconid from protoconid; lower molars bunodont with poorly developed paraconid on mesiolingual face of protoconid and nearly equally developed premetacristid; small cristids on lingual face of protoconid and buccal face of metaconid not quite meeting at centerline of tooth; short, slightly curved postprotocristid and postmetacristid meeting centrally; cristid obliqua with slight thickening, forming mesoconid at midpoint; hypoconid and entoconid possessing cristids similar to those on protoconid and metaconid; symmetrical large hypoconulid on M_3; slight mesial and distal cingula on molars with development of small stylid on latter; horizontal

Figure 14.1 Medial view of *Geniohyus mirus,* showing the large fossa beneath the dental series.

ramus extremely deep bulging out on buccal side with large fossa on lingual side; upper dentition known only from one M², nearly quadrate tooth with strong mesial cingulum, lingual cingulum below protocone ending on mesiolingual face of hypocone, buccal cingula moderately developed with spur distobuccal to parastyle; protocone and hypocone low; preprotocrista and prehypocrista well developed, postprotocrista less so; posthypocrista curving back to form distal cingula; cusps close together, removed from edge of tooth; metastyle moderately developed, parastyle and mesostyle strong, rounded. (See figure 14.1.)

DISCUSSION. This species, known only from three specimens, differs from *Geniohyus diphycus* by its larger size, its longer and narrower symphysial region, and by the diastemata separating the incisors and the canine from each other and from the P₁. It differs from those specimens here placed in *Bunohyrax,* which Schlosser and Matsumoto designated as belonging to this species, by its two-rooted lower canine, its simpler premolars that do not possess V-shaped hypoconids, its more bunodont and inflated cusps, the cusp morphology of the molars, and the greatly deepened horizontal ramus that does not possess a fenestra on its lingual face but rather a large fossa. Both CGM-8634 and AMNH-13349, the type specimen of *G. diphycus,* exhibit cancelous bone on the interior surface of the ramus and neither, therefore, contains a fenestral opening into an inner canal, the latter always lined by smooth inner walls. Measurements are given by Andrews (1906, 1907).

Geniohyus diphycus Matsumoto 1926

SYNONYMY.

Geniohyus diphycus Matsumoto 1926, p. 295, *pars* fig. 9-10, *non* fig. 11.

HOLOTYPE. AMNH-13349, American Museum of Natural History, New York; a left ramus with symphysis and P₁ to M₂.

DISTRIBUTION. Lower Fossil Wood Zone, Jebel el Qatrani Formation, Fayum, Egypt. The type was collected in 1907 west of Quarry A; the paratype, AMNH-14456, was collected in 1908, no precise locality data given.

DIAGNOSIS. Moderately small *Geniohyus* with rather short, narrow and quite deep symphysial region; no diastemata present unless a very short one is present between I₃ and C; tooth morphology of lower dentition similar to *G. mirus* except paraconid better developed on premolars; cingula less developed in molars; ramus deepends rapidly but not as abruptly as in *G. mirus;* upper premolars nearly square, without mesostyle; mesial cingulum extremely weak on P², more developed on P³ and P⁴, not continuous around lingual face of protocone; hypocone present as small spur on distal cingula of P³, larger on P⁴; upper molars with well-developed parastyle and mesostyle; metastyle poorly developed, even on M³; protocone with well-developed preprotocrista and only slightly less developed postprotocrista; hypocone compressed mesiodistally, higher than protocone with strong prehypocrista and with post-hypocrista curving distobuccally to meet metastyle; hypocone close to metacone, protocone close to paracone; M³ nearly triangular; buccal cingula well developed on all teeth with small spur just distal to base of parastyle on all known teeth except P⁴; mesial cingula well developed on upper molars, continuous, though not too strongly, around lingual surface of protocone ending at midline of lingual face of tooth.

DISCUSSION. The tooth morphology of the type is nearly identical except for size to the type of *G. mirus;* it is distinguished from *G. mirus* by its much smaller size and its lack of diastemata. This latter feature precludes the possibility of this specimen being a female *G. mirus;* although sexual size dimorphism is common in nearly all Oligocene hyraxes, the relative size of the diastemata does not vary between the different presumed sexes.

The paratype (AMNH-14456), a maxilla with P² to M³, is here tentatively placed in this species. It differs from *B. fajumensis* by its slightly smaller size, the more triangular shape of M³, the more nearly square premolars, especially P⁴, the slightly more compressed hypocones, and the less developed lingual cingula, especially on the premolars. The mesial dentition is needed to identify this specimen with certainty because *B. fajumensis* has rather large diastemata and *G. diphycus* has none. Measurements are given by Matsumoto (1926).

Geniohyus magnus (Andrews) 1904

SYNONYMY.

Saghatherium antiquum Andrews 1903, p. 340, fig. 2 *non* Andrews and Beadnell 1902.

Saghatherium magnum Andrews 1904, p. 214.

Saghatherium magnum, Andrews 1906, p. 89 *pars* pl. VI, fig. 3 *non* fig. 4.

Saghatherium magnum, Schlosser 1911, p. 110, 113 *pars.*

Saghatherium majus, Schlosser 1911, p. 110, 114 *pars.*

Geniohyus magnus, Matsumoto 1926, p. 208.

HOLOTYPE. BMNH-M-8398, British Museum, Natural History, London; a right maxilla with I^1, C to M^3.

DISTRIBUTION. Lower Fossil Wood Zone, Jebel el Qatrani Formation, Fayum, Egypt. Exact locality of type unknown.

DIAGNOSIS. Small species with closed tooth row from I^3 to M^3, I^1 forming a triangular tusk, I^2 reduced with very small root as seen from alveolus, separated from I^1 by moderate diastema and from I^3 by slightly smaller diastema; I^3 in contact with canine as wear on mesial surface of tooth indicates; canine complex tooth with well-developed parastyle, slight vertical fold on ectoloph, and distinct process on distolingual angle of tooth that wears into fairly broad shelf; p^1 only slightly longer than broad; remainder of cheek teeth similar to *G. diphycus* except lingual cingulum much more pronounced.

DISCUSSION. The mesial portion of the type specimen was first figured by Andrews (1903:340, fig. 2) as belonging to *Saghatherium antiquus.* The distal portion of the same individual was later referred to the new species *Saghatherium magnum* and Andrews (1904:214) subsequently stated that the two pieces belonged to one individual. Matsumoto (1926:299) pointed out that this specimen belongs to *Geniohyus* but that the several mandibles referred to this species by Andrews belonged in *Saghatherium.*

Matsumoto referred an additional specimen (AMNH-13278), a maxilla with P^2 to M^3, to this species: "The general structure of the upper cheek-teeth of the present specimen, as well as Andrews' type, is almost exactly like that of *G. mirus* and *G. pygmaeus;* consequently I refer the present species to *Geniohyus* without any hesitation" (1926:300). This statement highlights the difficulty in sorting upper dentitions that are not associated with lower dentitions. Matsumoto's upper dentition of *G. mirus* appears to be a female *Bunohyrax fajumensis* and the "*G. pygmaeus*" mentioned in the quotation was described by Matsumoto in the same publication as *Megalohyrax pygmaeus;* in this chapter it is referred to *Pachyhyrax.*

The type specimen differs considerably from *Saghatherium* in the shape of the molars, the lack of a well-developed metastyle on M^3, the absence of the spurs on the lingual side of the mesostyle and metastyle that are characteristic of *Saghatherium,* the less developed ridges on the buccal side of the ectoloph opposite the paracone and metacone, and the much stronger lingual cingula. It differs from *B. fajumensis* by its much smaller size, the lack of a diastema, between C and I^3, its more complex canine, its longer more quadrate P^4 with its less developed hypocone, its more compressed hypocones on the molars, and its more complete lingual cingula. It most closely resembles *P. pygmaeus* in size and molar morphology, but it differs in its more strongly developed lingual cingula on the molars, its possession of a buccal cingulum on the molars and premolars, the lack of small simple spurs on the distolingual surface of the paracone of M^1 and M^2, the lack of mesostyles on P^3 and P^4, its much weaker hypocones on the premolars, especially on P^2 and P^3, the much more complex canine, and the lack of a diastema between I^3 and C. The canine of the type specimen is not, in fact, too dissimilar from the P^1 of *P. pygmaeus.* It differs from *G. diphycus* by its slightly smaller size, its better developed lingual cingula, its slightly less compressed hypocones, and its slightly more quadrate M^3.

The specimen referred to this species by Matsumoto differs from the type specimen by its smaller size, its more trenchant parastyle and mesostyle, its less well-developed lingual cingula, and the much less well-developed hypocones on the premolars. The inferred length of the premolar series of this specimen is nearly identical to the length of the premolar series of the type, 31 mm and 32 mm respectively, but the molar series measures 36 mm as opposed to 39 mm for the type and 43 mm for *G. diphycus.* The type of *G. magnus* is therefore intermediate in size between AMNH-13278, *G. magnus,* and AMNH-14456, *G. diphycus.* The difference in size between the two extremes seems too large for one species. Possibly the type specimen is a male and the referred specimen a female because other species of Oligocene hyraxes show a size difference as large as 15% between the two sexes. It is hoped that when more material is collected this problem will be solved.

Subfamily Saghatheriinae Andrews 1906.

TYPE GENUS. *Saghatherium* Andrews and Beadnell 1903.

INCLUDED GENERA. *Bunohyrax, Megalohyrax, Pachyhyrax, Saghatherium, Thyrohyrax, Titanohyrax.*

DISTRIBUTION. Oligocene of North Africa, Miocene of East Africa.

DIAGNOSIS. Pliohyracidae with bunoselenodent to selenodont dentition, dental formula $\frac{3.1.4.3}{3.1.4.3}$; I_2 equal to or slightly larger than I_1, I_3 reduced, lower canine usually with single root; premolars submolariform to molariform with V-shaped hypoconids on at least P_3 and P_4; upper premolars with or without mesostyle; protoconids shorter than metaconids, obliquely placed; no premetastylids; metastylids sometimes developed; no mesoconids present; paraconids central or lingual, often well developed; hypoconulids present on M_3; parastyle and mesostyle on upper molars well developed, often compressed and trenchant; cusps in upper molars well separated, near borders of teeth; cusps of lower molars not very inflated, often compressed; ramus deepens gradually, often rather narrow although quite deep in some forms; fenestra sometimes present in females on lingual side of ramus; skull, when known is long, narrow in snout region, often with frontals pitted with irregular depressions and grooves giving skull roof unusually rough appearance.

DISCUSSION. Included in this subfamily are genera that Matsumoto (1926) separated into three different families. There seems no valid reason to consider these forms diverse enough to warrant such separation since most of the genera are contemporaneous and rather similar morphologically. I have designated three subfamilies within the Pliohyracidae to delineate three possible lineages or radiations. The Saghatheriinae represents the major radiation in the Oligocene, which survived into the middle Miocene. This radiation may well have taken place during the middle or late Eocene because the earliest known fossil hyraxes of this subfamily are represented in the Lower Fossil Wood Zone of the Jebel el Qatrani Formation by five genera. *Pachyhyrax* is represented by one species in the Lower Fossil Wood Zone, one in the upper and a third from the Miocene of East Africa. The remainder of the genera are presently known only from the Fayum and may be considered diverse contemporaneous species if the Lower and Upper Fossil Wood Zones

do not represent much difference in time (Wood 1968:30) or may be sampled from relatively few lineages if a considerable amount of time separates the two fossil wood zones.

Genus *Bunohyrax* Schlosser 1910 *emend.*

SYNONYMY.
Geniohyus, Andrews 1906, *pars.*
Geniohyus, Schlosser 1910, 1911 *pars.*
Bunohyrax Schlosser 1910, 1911.
Geniohyus, Matsumoto 1926 *pars.*
Bunohyrax, Matsumoto 1926.
non Bunohyrax, Whitworth 1954.

TYPE SPECIES. *Bunohyrax fajumensis* (Andrews) 1904, designated by Matsumoto (1926).

INCLUDED SPECIES. *Bunohyrax fajumensis, Bunohyrax major.*

DISTRIBUTION. Lower and Upper Fossil Wood Zones, Jebel el Qatrani Formation, Fayum, Egypt.

DIAGNOSIS. Medium to large Saghatheriinae with long, narrow symphysis and bunoselenodont dentition; dental formula $\frac{3.1.4.3}{3.1.4.3}$; I_2 slightly larger than I_1; I_3 reduced and caniniform, lower canine with one root; large diastemata between P_1 and C, C and I_3, I_3 and I_2, and between canine and all incisors in upper jaw; premolars submolariform; V-shaped hypoconids on all four lower premolars, well developed on P_3 and P_4; metaconid separated from protoconid in P_3 and P_4; upper canine a long, narrow, two-rooted tooth; hypocone developed only slightly on P^1 increasing distally until on P^4 it is a fairly large compressed cone; P^4 nearly same length as P^3 but considerably wider; upper premolars without mesostyle; no mesoconid present on lower molars; protoconid and metaconid only slightly oblique, metaconid higher; no premetacristid present; paraconid only slightly buccal to centerline; hypoconulid on M_3; compressed and slightly oblique parastyle and mesostyle on upper molars rounded and blunt; metastyle present but weak; cusps of upper molars near outer margins, M^3 trapezoidal, distal margin slightly convex; cusps of lower molars not greatly inflated, without metastylid; protocristid and hypocristid interrupted not forming a complete loph; ramus deepens gradually with a rounded fenestra on lingual surface under M_3 in females; skull as described by Matsumoto (1926:276–291). (See figure 14.2.)

DISCUSSION. *Bunohyrax* differs from *Geniohyus* as noted above and from *Megalohyrax* by its more bunodont molars, its incompletely formed protocristid and hypocristid, by the stronger postprotocrista and prehypocrista on the upper molars, and by

Figure 14.2   Comparison of the lower teeth of *Geniohyus mirus,* above, and *Bunohyrax fajumensis,* below. The morphology of P$_4$ is different in the two species. (Not to scale.)

the much less developed metastyle on M^3. It differs from *Pachyhyrax* in lacking a mesostyle on P^3 and P^4, its much simpler premolars, its incomplete protocristid and hypocristid, and the lack of spurs on the lingual side of the ectoloph.

Many specimens placed in *Geniohyus* by Schlosser and by Matsumoto show no difference in tooth morphology from those placed in *Bunohyrax.* Schlosser (1911) stated that some *Bunohyrax* had a fenestra and some did not. Matsumoto (1926) included in *Bunohyrax* only forms lacking fenestra although he stated that *Bunohyrax* had "a very large fenestra-like opening on the inner surface of the ramus" (in his key, p. 261). In the text of his publication he placed those specimens with a large fenestra into *Geniohyus.* The presence or absence of this fenestra does not seem to be a valid phylogenetic character in either *Bunohyrax* or *Megalohyrax.* Among all the known species of these two genera there are forms without a fenestra and also forms, usually 12% to 15% smaller, that have a fenestra but no other morphological differences from the larger specimens. It seems highly unlikely that pairs of species would evolve that consistently show a size difference of 15% or that four different genera could contain forms morphologically identical except for the presence or absence of the fenestra. The few specimens that contain a fenestra and that are complete enough to also contain I$_2$ show a small, less tusklike incisor as found in all living female hyraxes. It is therefore suggested that the females of *Bunohyrax* and *Megalohyrax* possess a fenestra and the males do not.

Bunohyrax fajumensis (Andrews) 1904

SYNONYMY.

Geniohyus fajumensis Andrews 1904, p. 162.
Geniohyus fajumensis, Andrews 1906, p. 195, pl. XIX, fig. 2.
Saghatherium majus Andrews 1906, p. 91 *pars.*
Bunohyrax fajumensis, Schlosser 1910, p. 502.
Geniohyus minutus Schlosser 1910, p. 503, *nomen nudum.*

Bunohyrax fajumensis, Schlosser 1911, p. 119, XI, fig. 8, pl. XII, fig. 2.
Bunohyrax sp., Schlosser 1911, p. 120.
Geniohyus aff. *mirus,* Schlosser 1911, p. 122 *pars.,* pl. XII, fig. 4, 5.
Geniohyus micrognathus Schlosser 1911, p. 123, pl. X, fig. 1, 2.
Geniohyus subgigas Matsumoto 1926, p. 266, fig. 4.
Geniohyus mirus, Matsumoto 1926 (*non* Andrews) p. 269, fig. 5-8.
Bunohyrax fajumensis, Matsumoto 1926, p. 303, figs. 14–17.
Bunohyrax affinis Matsumoto 1926, p. 309, fig. 18, 19.
Megalohyrax suillus, Matsumoto 1926, p. 319 (*non* Schlosser).

HOLOTYPE. BMNH M-8435, British Museum, Natural History, London; fragment of a right ramus with P$_1$ to P$_4$.

DISTRIBUTION. Lower and Upper Fossil Wood Zone, Jebel el Qatrani Formation, Fayum, Egypt. The exact locality of the type specimen is unknown but other specimens are from Quarry A and northwest of Quarry A in the Lower Fossil Wood Zone and Quarries I and M in the Upper Fossil Wood Zone.

DIAGNOSIS. Medium *Bunohyrax* with long, narrow symphysial region and diastemata between incisors and canines; I$_2$ a large, laterally compressed procumbent tooth in males, smaller in females; I$_3$ reduced; lower canine with single root; P$_1$ laterally compressed, with large single protoconid and postcristid very slightly curved to lingual side; prominent ridges on distolingual and distobuccal sides of protoconid; P$_2$ larger, similar except postcristid more V-shaped; P$_3$ with metaconid, indistinct paraconid, and V-shaped postcristid; P$_4$ with well-separated metaconid and protoconid and V-shaped postcristid terminating in small hypoconid; lower molars bunoselenodont with small paraconid on mesial border of teeth, nearly central and larger than the weakly developed premetastylid; small cristids on lingual face of protoconid and buccal face of metaconid; protocristid complete but concave ventrally and not forming a strong loph; cristid obliqua strong, high, without mesostylid; hypoconid and entoconid with similar cristids except hypocristid interrupted and not forming a loph; hypoconulid large, inflated, slightly oblique; well-developed buccal cingula on lower molars; P^4 short, broad with hypocone on distal border of tooth; upper molars with well-developed parastyles and mesostyles, poorly developed metastyle; protocone with well-developed preprotocrista and moderately developed postprotocrista; large spur on mesial cingula runs nearly to top of protocone; hypocone same size as protocone with strong prehypocrista and posthypocrista running distolingually and meeting the distal cingula, which curves buccally to the metastyle; cingula well developed on molars; no fenestra on inner side of ramus in males, present under M_3 in females; fenestra opens into canal in horizontal ramus, which terminates under P_2. Skull described by Matsumoto (1926:269–292). Measurements are given by Matsumoto.

DISCUSSION. As mentioned above, AMNH-13329, the type of "*Geniohyus subgigas*," occludes nearly perfectly with AMNH-13347, *Bunohyrax fajumensis*. Both specimens have both the right and left check teeth preserved, a rare feature for specimens from the Fayum, and both were collected in 1907 northwest of Quarry A. The distinct possibility exists that these two specimens represent one individual; certainly one species. BMM-10187, a nearly complete ramus labeled *Bunohyrax* sp., also belongs to this species, as does BMM-8434, a maxillary fragment with C to P^2, the type of *Saghatherium majus*. Although no large fenestra is known in "males" of this species, a small round hole is found below the unerupted M_3 in AMNH-13336.

In 1910 Schlosser named the species *Geniohyus minutus* but never designated a specimen to this taxon. In his 1911 paper he named and described the species *G. micrognathus* and included under this name *G. minutus nomen nudum*. This specimen, which he figured, is not similar morphologically to *Geniohyus* but rather to *Bunohyrax*. This specimen and Matsumoto's type of *B. affinis* do not possess a fenestra, unlike the specimen that Matsumoto referred to *G. micrognathus* and the specimen he referred to *G. mirus*. There are no significant differences among these four specimens except the fenestra, and they cannot be separated specifically. The fact that two jaws possess a fenestra and that one of these (AMNH-14446) includes a skull that exhibits a small, rounded I^1 typical of living female hyraxes further supports the hypothesis that the fenestra may be a sexual characteristic found only in females.

Recent excavations in the Upper Fossil Wood Zone by Yale University have uncovered two fragments of rami and several nearly complete maxillae that fit well into the size range of the male of *B. fajumensis*. The two rami, however, although fragmentary, show the presence of a large, round fenestra. In YPM-29166 the buccal side of the horizontal ramus is greatly inflated under M_2 and M_3 as in *G. mirus*. A canal within the ramus runs at least into the base of the ascending ramus behind M_3. This feature is found also in *Thyrohyrax*. The morphology of the lower teeth of this specimen differs considerably from that of both *Geniohyus* and *Thyrohyrax* but differs from *B. fajumensis* only in the stronger buccal cingula and the more lingually positioned paraconid. The upper teeth differ from *B. fajumensis* of the Lower Fossil Wood Zone only in the slightly stronger development of the parastyle on P^2 to P^4. These differences do not seem sufficient to separate these specimens from *B. fajumensis*, but the fact that the two rami probably represent females that are the same size as the males in the Lower Fossil Wood Zone suggests that the time difference between these two strata was sufficient for a 10% to 15% size increase in the latter species. Because this is also the approximate size difference between teeth of the two sexes in the Lower Fossil Wood Zone it is not considered enough to warrant designation of a new species. If the specimens from the Upper Fossil Wood Zone were to be considered a different species, one would not be able to separate them from *B. fajumensis* of the Lower Fossil Wood Zone on the basis of the teeth alone, which are the most commonly found elements.

Bunohyrax major (Andrews) 1904

SYNONYMY.
Geniohyus major Andrews 1904, p. 212.
Geniohyus major, Andrews 1906, p. 196, fig. 63.
Bunohyrax major, Schlosser 1910, p. 503.
Bunohyrax major, Schlosser 1911, p. 121.
Geniohyus gigas Matsumoto 1926, p. 264, figs. 1–3.
Bunohyrax major, Matsumoto 1926, p. 300, figs. 12, 13.

HOLOTYPE. CGM-8980, Cairo Geological Museum, Cairo; fragment of a left ramus with P_1 to P_3.

DISTRIBUTION. Lower Fossil Wood Zone, Jebel el Qatrani Formation, Fayum, Egypt. Exact locality of type unknown; referred specimens from Quarries A, B, C and E.

DIAGNOSIS. Very large *Bunohyrax*, females with fairly deep ramus, very strongly bulged outward on buccal side, and with large fenestra on lingual side; males with relatively shallow ramus, no fenestra; P_1 without metaconid, protoconid with prominent ridges on distobuccal and distolingual sides, postcristid moderately V-shaped; P_2 with strongly V-shaped postcristid; metaconid separated from protoconid on P_3 and P_4; cusps on molars inflated; remainder of morphological characters similar to *B. fajumensis*. Measurements are given by Matsumoto.

DISCUSSION. This species differs from *B. fajumensis* by its greater size, its more strongly V-shaped postcristids on the lower premolars and by its more inflated cusps. AMNH-13333, the type of *G. gigas,* exhibits a fenestra under M_3 and the buccal side of the jaw is greatly inflated as in *B. fajumensis* from the Upper Fossil Wood Zone. Its teeth are approximately 12% smaller than those of the type and AMNH-13339 which do not have a fenestra; this size difference suggests that AMNH-13333 is a female and that only females possess a fenestra. The teeth of these specimens are morphologically very similar. The premolars of AMNH-13333 do not exhibit the V-shaped postcristid to the same degree as in the type specimen. The lower premolars of most species of Oligocene hyraxes tend to exhibit considerable variation in detailed morphology; because only four known specimens of *B. major* possess lower premolars, it does not seem wise to separate this specimen from *B. major* on such a variable trait as the degree of curvature of the postcristid in the lower premolars. AMNH-13498, a left M^3, seems referable to this species.

"Bunohyrax sp." Whitworth 1954

Whitworth (1954: 25, pl. 7, fig. 3) described an isolated upper molar from Songhor and referred it to the genus *Bunohyrax.* This tooth is nothing at all like *Bunohyrax* and is, in fact, unlike any known hyracoid molar. The tooth, an upper left second molar, is nearly square; *Bunohyrax* upper molars are trapezoidol. The parastyle is nothing more than a small spur on the mesial cingulum, whereas the parastyle in all known hyracoid molars is large, nearly as high as the paracone, and extends the border of the ectoloph in a mesiobuccal direction. The mesostyle is very short, only slightly taller than the parastyle, and is separate from the ridge between the paracone and metacone. No known hyrax exhibits such a small mesostyle on its molars. There is in fact no real ectoloph, a condition unknown in hyraxes. The paracone and metacone are more bunodont than the protocone and hypocone; the reverse is always true in hyraxes. The protocone shows heavy wear, the hypocone nearly none, an unusual characteristic for a hyrax. The shape of the protocone and hypocone, the development of the preprotocrista, the smaller postprotocrista, the large prehypocrista and the short posthypocrista joining the distal cingula are typical of hyraxes and are similar to the condition in *Bunohyrax.* The lingual half of the tooth is therefore similar to a hyrax, the buccal half is unlike any known hyrax and certainly very different from *Bunohyrax.*

Another specimen in the Kenya National Museum, KNM-RU 343 from Rusinga, is also referred

to *Bunohyrax* sp. This tooth is about half the size of the Songhor molar and contains a protoconule. It appears to be better referred to the Anthracotheridae than to the Hyracoidea.

Genus *Megalohyrax* Andrews 1903

SYNONYMY.
Megalohyrax Andrews 1903.
Megalohyrax, Andrews 1906 *pars.*
Mixohyrax Schlosser 1910, 1911 (*non Megalohyrax* Schlosser).
Megalohyrax, Matsumoto 1922, 1926 *pars.*
TYPE SPECIES. *Megalohyrax eocaenus* Andrews 1903.
INCLUDED SPECIES. *Megalohyrax eocaenus.*
DISTRIBUTION. Lower and Upper Fossil Wood Zones, Jebel el Qatrani Formation, Fayum, Egypt.
DIAGNOSIS. Medium to large Saghatheriinae with long narrow symphysis and brachydont, bunoselenodont dentition; dental formula $\frac{3.1.4.3}{3.1.4.3}$; I_2 larger than I_1; I_3 reduced, caniniform, and separated by large diastemata from I_2 and C; lower canine a single-rooted, compressed tooth separated from P_1 by a large diastema; premolars simpler than molars, laterally compressed; postcristid V-shaped on P_2 to P_4; metaconid only slightly separated from protoconid on P_3 and P_4; P_4 longer than P_3; cusps not inflated; molars long, narrow, with unbroken protocristid and hypocristid forming complete loph; metaconid simple without metastylid; paraconid central; protoconid with strongly developed cristid obliqua and hypocristid; hypoconulid on M_3 large, narrow, not extremely inflated; I^1 a large triangular tusk in males, smaller and more rounded in females; I^2 and I^3 small bullate teeth with large diastemata between each other and C and I^1; C elongate with two outer cusps; upper premolars simpler than molars, without mesostyles; parastyles and metastyles weak; hypocone in P^3 and P^4; upper molars trapezoidal to nearly square; parastyles and mesostyles large, rounded, and stout; metastyle well developed, especially in M^3 where it is stout and rounded; protocone with strong preprotocrista but weak or missing postprotocrista making protocone appear rounded on its distal face, especially when worn; hypocone the same size as protocone with strong prehypocrista and posthypocrista that joins distal cingulum; distal cingulum straight or even concave, not convex as in other forms; round fenestra present in females on lingual side of ramus beneath M_3, not present in males.
DISCUSSION. This genus can be separated from *Bunohyrax* by its more selenodont teeth and by the

complete lophlike nature of the postprotocristids and posthypocristids, the less developed paraconal and metaconal folds in the upper molars, the large rounded metastyle, especially on M³, the concave distal border of the molars, and the weaker postprotocrista. It may be separated from *Titanohyrax* by its more bunodont teeth, its simpler premolars, the lack of mesostyles on the premolars, the relatively larger hypocones on the upper molars, the roundness of the mesostyles, and the diastemata between the anterior teeth that make its snout and symphysis longer. It may be separated from *Pachyhyrax* by its larger metastyles and rounder mesostyles on the molars, its lack of spurs distolingual to the paracone and metacone, its simpler premolars without a hypocone on P² and mesostyles on P³ and P⁴, its less inflated cusps in the lower molars, and by the long, narrow snout and symphysis. As discussed by Matsumoto (1922:841), Schlosser's *Mixohyrax* is really *Megalohyrax* and his *Megalohyrax* differs from Andrews's type. These latter specimens were given the generic name *Titanohyrax* by Matsumoto (1922:844).

Megalohyrax eocaenus Andrews 1903

SYNONYMY.

Megalohyrax eocaenus Andrews 1903, p. 340, fig. 1.

Megalohyrax minor Andrews 1904, p. 213.

Megalohyrax eocaenus, Andrews 1906, p. 92, pl. VI, fig. 1, 2, *non* text fig. 39.

Megalohyrax minor, Andrews 1906, p. 97 *pars,* pl. VII, fig. 1, *non* fig. 2, 3.

Mixohyrax niloticus Schlosser 1910, p. 503, *nomen nudum.*

Mixohyrax suillus Schlosser 1910, p. 503, *nomen nudum.*

Mixohyrax andrewsi Schlosser 1910, p. 503.

Mixohyrax andrewsi, Schlosser 1911, p. 115, pl. X, figs. 9–11.

Mixohyrax niloticus, Schlosser 1911, p. 116, pl. XI, fig. 9, pl. XII, figs. 3, 6, pl. XV, figs. 1, 4, 8.

Mixohyrax suillus, Schlosser 1911, p. 118, pl. X, fig. 6.

Megalohyrax eocaenus, Matsumoto 1922, p. 841.

Megalohyrax minor, Matsumoto 1922, p. 842.

Megalohyrax niloticus, Matsumoto 1922, p. 843.

Megalohyrax suillus, Matsumoto 1922, p. 843.

Bunohyrax major, Matsumoto 1926, p. 300 *pars.*

Megalohyrax eocaenus, Matsumoto 1926, p. 312.

Megalohyrax minor, Matsumoto 1926, p. 313, figs. 20–22.

Megalohyrax niloticus, Matsumoto 1926, p. 316, fig. 23.

HOLOTYPE. BMNH M-8502, British Museum, Natural History, London; right maxilla with C to M³.

DISTRIBUTION. Lower and Upper Fossil Wood Zones, Jebel el Qatrani Formation, Fayum, Egypt. Exact locality of type unknown, but other specimens are from Quarry A, southwest of Quarries A, B, and E in the Lower Fossil Wood Zone, and from Quarries I and R in the Upper Fossil Wood Zone.

DIAGNOSIS. Same as for genus. Measurements are given by Matsumoto (1926).

DISCUSSION. Andrews named the type specimen in 1903 but he incorrectly listed the tooth measurements, transposing the width measurements with the length. In 1904 Andrews named a new species, *M. minor,* based on a left maxilla with P¹ to M³; he stated that although the molars are similar, the premolars of *M. minor* are somewhat simpler and the parastyle is better developed. The first premolar of *M. minor* is definitely simpler than the corresponding tooth in any other *Megalohyrax* specimen and is thought possibly to be anomalous. The remainder of the premolars are quite similar and well within the range of variation expected for a single species. The parastyles of the premolars in the Saghatheriinae tend to be quite variable and the parastyles of *M. minor* and *M. eocaenus* are also within the expected range of variation for a single species. In 1906 Andrews referred a left premaxilla with I¹ and the roots of I² and I³ to *M. eocaenus,* but no other fragment has since been correctly referred to this species. No specimen found since has possessed a P¹ and all have been referred to *M. minor.* The size difference between those specimens referred to *M. minor* and the type of *M. eocaenus* is less than 15% and it is probable that they represent different sexes, not different species. *M. minor* is here considered a junior synonym of *M. eocaenus.* In 1903 Andrews gave the specimen number C-8188 to the type of *M. minor* and in 1906 he used the specimen number C-8818. The specimen was marked 8188 but is entered correctly under 8818 in the catalog of the Cairo Geological Museum. AMNH-13330, referred to *Bunohyrax major* by Matsumoto (1926), is really a left M³ of *M. eocaenus.*

Several specimens of *Megalohyrax eocaenus* have been found in Quarry I of the Upper Fossil Wood Zone. The morphology of the lower teeth is identical to *M. eocaenus* of the Lower Fossil Wood Zone except that the paraconid is placed slightly nearer the lingual border of the tooth with respect to the protocone and P₃ and P₄ are slightly more molariform. The upper molars exhibit a slightly more prominent postprotocrista and the metastyle on M³ is larger and more rounded. These features are not considered enough to warrant designation of a new species for these specimens. Two specimens include the distal portion of both horizontal rami and both lack fenestra; three specimens are approximately 12% to 15% smaller and are otherwise similar except that they possess a fenestra under M_3. Only three specimens from the Lower Fossil Wood Zone are complete enough to show whether a fenestra is present; all possess a fenestra and fall within the size range expected for female *M. eocaenus.*

Once again, as in *Bunohyrax,* we are faced with forms that can hardly be distinguished from each other, except that one possesses and the other lacks a fenestra under M_3. It seems more reasonable to consider these two forms as different sexes than as separate species because a similar situation is also found in *Bunohyrax fajumensis* and *B. major,* and at least a similar size difference is found among *Pachyhyrax crassidentatus, P. championi,* and *Geniohyus magnus.*

The mandibles referred to *Megalohyrax niloticus* by Matsumoto nearly occlude with the type maxilla of *M. minor;* although they are not from the same individual, they certainly cannot be separated from it at the specific level on the basis of size difference as he stated. The measurements given by Schlosser (1911) for *M. niloticus* are only slightly smaller than those of Matsumoto's specimens but they fall within the expected limits of a single species. This skull and mandible almost certainly belong to a female as is shown by the size and shape of I^1 and the lower central incisor pair. Schlosser's measurements of this type specimen differ from those of Matsumoto's specimen only in the length of the molars. Schlosser measured the "mesial part" of the tooth; they would be expected to be less than Matsumoto's, which are measurements of maximum length. The skull and mandible described by Schlosser are unfortunately lost and cannot be directly compared to Matsumoto's specimen.

Schlosser's *M. suillus,* which he differentiates from *M. niloticus* by its slightly smaller size and narrower teeth, does not differ enough to warrant specific separation. AMNH-13344, a left P_1–P_2 that Matsumoto referred to *M. suillus,* seems to be a female *Bunohyrax fajumensis.* Because *M. suillus* and *N. niloticus* cannot be specifically separated by their reported differences and because the *M. niloticus* and *M. minor* size difference is too small to represent two different species, *M. eocaenus* is considered the only valid species of *Megalohyrax.* These four species differ less among themselves in size than do the several hundred specimens of *P. championi* from Rusinga Island.

Andrews (1906:94, fig. 39) referred the cranial portion of a skull to *M. eocaenus.* This skull, which has been included in nearly every review or description of fossil hyraxes since, has caused incorrect definitions and interpretations of Oligocene hyracoid cranial anatomy. Matsumoto (1926) removed this skull from *Megalohyrax* because it was dissimilar to Schlosser's skull of *M. niloticus* and the American Museum skull of "*M. pygmaeus*"; he provisionally placed it in *Titanohyrax* because there was no known skull of this genus. In fact, this skull is not a hyracoid at all but an Anthracothere, here provisionally referred to *Bothreogenys gorringei.*

In all known Oligocene and Recent hyracoid skulls the jugal contributes to the glenoid fossa as a vertical or nearly vertical process that confines lateral movement of the mandible. In the present skull no process of the jugal is involved in the glenoid fossa. The braincase is long and inflated as in *Bothreogenys,* while it is extremely short and not very inflated in *Bunohyrax, Megalohyrax,* and *Pachyhyrax.* The squamosal-parietal suture is shaped like an inverted V and crosses the lambdoidal crest at its midpoint as in *Bothreogenys,* not at its top as in hyraxes. The frontals are broad, unlike those in Oligocene hyraxes. Recent preparation of the skull has shown that the frontal-parietal suture does not extend to the supraoccipital process and the parietals do not form part of the postorbital process as in all hyraxes. Instead the suture follows the temporal ridge for only half its length, then drops well distal to the postorbital process. Two supraorbital foramina also pierce the frontals near the center line and well forward of the junction of the temporal crests with the sagittal crest, foramina that are unknown in hyraxes. The occiput has recently been prepared and it compares with *Bothreogenys* in every detail.

Genus *Pachyhyrax* Schlosser 1910

SYNONYMY.
Pachyhyrax Schlosser 1910, 1911.
Megalohyrax, Matsumoto 1922, 1926, *pars.*
Megalohyrax, Whitworth 1954.
TYPE SPECIES. *Pachyhyrax crassidentatus* Schlosser 1910.

INCLUDED SPECIES. *Pachyhyrax crassidentatus, Pachyhyrax pygmaeus, Pachyhyrax championi.*

DISTRIBUTION. Oligocene, Fayum, Egypt, and Miocene, East Africa.

DIAGNOSIS. Medium-sized Saghatheriinae with relatively short symphysial region and snout in earlier forms, longer in later forms; dental formula

$\dfrac{3.1.4.3}{3.1.4.3}$; diastemata short between mesial teeth; I_1 and I_2 large, subequal, closely approximated; I_3 and C, when present, reduced and separated from adjoining teeth by short diastemata, except in smallest species where C is in contact with P_1; P_1 and sometimes P_2 simpler than molars; postcristid slightly V-shaped in P_1, V-shaped in P_2 to P_4; metaconid separated from protoconid in P_2 to P_4; P_4 and sometimes P_3 molariform; buccal cingula weak on P_2, usually developed on P_3 to M_3; molars bunoselenodont; paraconids small and lingual to center line; protocristid and hypocristid unbroken, forming complete but slightly curved lophs; metaconids high, long, without differentiated metastylids; cristid obliqua without mesostylid; buccal cusps with strong lingual tilt; hypoconulid large, selenodont; ramus deepens gradually with round fenestra under M_3; I^1 triangular in males, more rounded in females; I^2 and I^3 small, simple, separated by short diastemata from each other and from C; C elongate, in contact with P^1; hypocone slightly developed on elongate P^1, well developed on P^2 to P^4; sharp, compressed mesostyle characteristically on P^3 and P^4; molars trapezoidal with large well-developed parastyle, sharp, compressed mesostyle, and poorly to moderately developed metastyle; hypocone and protocone subequal, with well-developed precristae and postcristae; preprotocrista long, cutting mesial cingulum in M^2 and M^3; prehypocrista terminating in valley on lingual side of mesostyle; simple spurs present on distolingual surface of paracone and on mesial surface of distal cingulum; M^3 long with considerable shelf between hypocone and distal cingulum; skull low and long, orbits small, open distally, and frontals broad, rough, and grooved in female *P. pygmaeus* and smooth in male *P. championi*.

DISCUSSION. *Pachyhyrax* was first named by Schlosser (1910) who had only a few isolated teeth for his diagnosis. Because the type specimen comes from the Upper Fossil Wood Zone and the American Museum collection contains no fossils from this level, this genus was not considered by Matsumoto. Thus forms referred to *Megalohyrax* from the Lower Fossil Wood Zone and from the Miocene of East Africa were not recognized as belonging to *Pachyhyrax*. Several new specimens of *Pachyhyrax crassidentatus* from Quarry I of the Upper Fossil Wood Zone were collected by the 1963–1967 Yale expeditions. These include nearly the whole dentition as well as a nearly complete horizontal ramus. It is now possible to transfer *M. pygmaeus* Matsumoto and *M. championi* (Arambourg) to *Pachyhyrax*.

This genus differs from all other hyracoid genera

by the development of the preprotocrista in the molars that cuts the mesial cingulum and rests in a notch at the base of the metastyle of the next mesial tooth, the development of the prehypocristid that terminates in the valley on the lingual side of the mesostyle, and by the small uniquely compressed mesostyles on P^3 and P^4. It further differs from *Megalohyrax* by the molariform third and fourth premolars, the better developed hypocone on P^1 and P^2, the less inflated metastyles on the molars, especially on M^3, the shorter symphysis with much shorter diastemata between the mesial teeth, the inwardly tilted buccal cusps, and the strongly developed buccal cingula on the lower cheek teeth. It differs from *Titanohyrax* by its more bunodont teeth, the absence of a metastylid on the lower cheek teeth, the relatively longer symphysial region and diastemata between the mesial teeth, the subequal hypocones and protocones, the more strongly developed postprotocristae and prehypocristae, the much smaller mesostyles on P^3 and P^4 and lack of mesostyle on P^2, and the presence of a fenestra beneath M_3. It differs from *Saghatherium* by its larger size, shallower mandible with round fenestra under M_3, relatively shorter and broader cheek teeth, narrower, simpler P_1, the less well-developed paraconids on the upper molars, and the absence of complex ridges on the buccal surface of the ectoloph opposite the paracone and metacone.

Pachyhyrax pygmaeus (Matsumoto) 1922

SYNONYMY.
Saghatherium magnum, Andrews 1907, p. 99, fig. 2 (*non* Andrews, 1904).
Megalohyrax pygmaeus Matsumoto 1922, p. 840, fig. 1.
Megalohyrax pygmaeus, Matsumoto 1926, p. 321, figs. 24, 25.

HOLOTYPE. AMNH-14454, American Museum of Natural History, New York; anterior portion of skull and right ramus.

DISTRIBUTION. Lower Fossil Wood Zone, Jebel el Qatrani Formation, Fayum, Egypt. Exact locality of type and referred specimens not known.

DIAGNOSIS. Small species with full eutherian dentition; tooth row closed from canine to M_3^3; small diastemata separating incisors; premolars submolariform; metaconid differentiated on P_1 to P_4; mesostyle present on P^3 and P^4; lower molars slightly more selenodont than those in type species; upper molars with weak metastyle; very weak spur on lingual side of postparacrista of M^1 and M^2; several minute cuspules distolingual to metacone in M^3.

DISCUSSION. This species is here transferred to *Pachyhyrax* because of its lack of large diastemata, the small trenchant mesostyles on P³ and P⁴ that are typical of this genus, the metaconids on the lower premolars, and the lack of a well-developed metastyle on the upper molars that are characteristic of *Megalohyrax*. The much less well-developed spurs on the upper molars of this species are what one would expect in an early and more primitive form and show an intermediate stage between no spurs and the well-developed spurs of *P. crassidentatus*. The slightly more selenodont lower molars suggest that *P. pygmaeus* is not directly ancestral to *P. crassidentatus* but do not remove it from near the possible ancestry of *P. championi*. (See figures 14.3, 14.4.)

Pachyhyrax crassidentatus Schlosser 1910

SYNONYMY.
Pachyhyrax crassidentatus Schlosser 1910, p. 503.
Pachyhyrax crassidentatus, Schlosser 1911, p. 115, pl. XI, figs. 2–6.
Pachyhyrax crassidentatus, Matsumoto 1926, p. 331.
HOLOTYPE. Uncataloged specimen (Schlosser 1911, pl. XI, figs. 2–6), Staatliches Museum für Naturkunde, Ludwigsburg, West Germany; associated left M¹, M², M³, right P³, M², and P₄. Specimen lost.
DISTRIBUTION. Upper Fossil Wood Zone, Jebel el Qatrani Formation, Fayum, Egypt. Exact locality of type unknown; referred specimens from Quarries I and R.
DIAGNOSIS. Medium large *Pachyhyrax*, morphological features as in generic description.
DISCUSSION. This species is now known from specimens including P² to M³ and P₁ to M₃. The buccal cingulum is well developed from the third premolars distally. The upper premolars are not entirely alike as Schlosser (1911) stated, a difficult statement for him to make considering he only possessed one upper premolar. There is a small, sharp mesostyle on P³ and P⁴, not on P². The hypocone is progressively more developed distally on the premolars as is the lingual cingulum.

This species differs from *P. pygmaeus* by its larger size, slightly more bunodont teeth, and the much greater development of the spurs on the upper

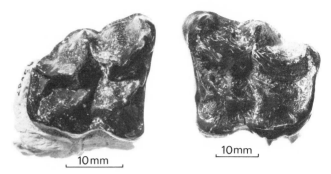

Figure 14.4   Comparison of M³ of *Pachyhyrax crassidentatus,* left, and *Megalohyrax eocaenus,* right, showing spurs on lingual side of ectoloph on *P. crassidentatus* and large rounded metastyle on *M. eocaenus.*

molars. It differs from *P. championi* by its more bunodont teeth, slightly stronger buccal cingula on the lower cheek teeth, and its larger metastyle and thus longer M³.

Among the Yale material from Quarry I is a specimen that includes M² and M³; it might have been considered a separate species on the basis of its smaller size except for the consistent sexual size difference in the Saghatheriinae. Otherwise, this specimen differs from the remainder of *P. crassidentatus* only by slightly more prominent spurs, especially the one on the paracone, and by its slightly more oblique positioning of the cusps (figure 14.5). These features are not considered enough to warrant placement of this specimen in a new species at this time. However, if this specimen may be considered a female, it further strengthens the hypothesis that in the Sagatheriinae females are approximately 15% smaller than males in mean measurements (figure 14.6).

Pachyhyrax championi (Arambourg) 1933

SYNONYMY.
Pliohyrax championi Arambourg 1933, p. 128, fig. 4.
Megalohyrax championi, Whitworth 1954, p. 6.
HOLOTYPE. Fragment of left ramus with P₃–P₄, Museum National d'Histoire Naturelle, Paris.
DISTRIBUTION. Type specimen from Losodok,

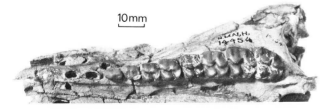

Figure 14.3   Occlusal view of the upper dentition of *Pachyhyrax pygmaeus.*

Figure 14.5   Upper dentition of *Pachyhyrax crassidentatus* showing mesostyles on P³-P⁴ and spurs on molars.

Figure 14.6 Lower dentition of *Pachyhyrax crassidentatus.* Internal mandibular foramen is present on lingual side of ramus beneath M_3.

Kenya; other specimens from Rusinga Island, Maboko Island, Songhor, Moruarot, Karungu, and Mfwanganu Island, all in Kenya; and Bukwa, Uganda.

DIAGNOSIS. Given by Whitworth (1954:6–23), not repeated here.

DISCUSSION. Whitworth (1954:22) adequately showed the difference between this form and *Pliohyrax.* At that time it was not known that *Pachyhyrax* possessed an internal mandibular fenestra nor were any specimens of *Megalohyrax* from the Upper Fossil Wood Zone known. Whitworth recognized that *Geniohyus* possesses a fossa rather than a fenestra and it was logical for him to place this species in *Megalohyrax* even though he recognized differences between this species and the Oligocene *Megalohyrax.* This species is here referred to *Pachyhyrax.* It differs from *Megalohyrax* by the presence of a hypocone on P^1 and P^2, the sharp mesostyle on P^3 and P^4, the small spurs on the lingual side of the paracone and metacone on the upper molars, the greatly reduced metastyle in the upper molars especially in M^3, the resultant trapezoidal rather than quadrangular upper molars, the molariform P_2 to P_4, the longer metaconid in the lower molars, and the more oblique protoconids and hypoconids. In all of these features as well as the presence of an internal mandibular fenestra this species agrees with *Pachyhyrax.* It differs from *P. pygmaeus* by its greater size, its slightly more molariform P_2 and P_3, its slightly stronger lingual cingulum on the lower cheek teeth, and the stronger spur on the distolingual side of the metacone on M^3. The roof of the known skull of *P. pygmaeus,* which is a female, is rough and pitted. The known skull of *P. championi,* which is a male, is smooth. This difference may be either sexual or interspecific.

P. championi and *P. crassidentatus* are nearly equal in size but the former differs by its more selenodont teeth, less prominent cingula, loss of I^3 and the lower C, less well-developed spurs on upper molars, less developed metastyle on M^3, and straighter and stronger lophs on the lower molars.

Whitworth states that the fenestra "is present in all adequately preserved material referred to this species, regardless of sex" (1954:9). Only a very few

of the more than 200 specimens of this species are adequately enough preserved to include that portion of the ramus which shows the fenestra. Of these only one specimen, a female, includes the incisors. Only one ramus of *P. pygmaeus* is known; it has a fenestra and it is associated with a skull with an I^1 of a female. The single ramus of *P. crassidentatus* possesses a fenestra but its sex is unknown. In the three species of *Pachyhyrax,* therefore, only two rami with fenestra can be sexed and both are females. However, no known specimens contain the ramus under M_3 that do not exhibit a fenestra. It is therefore not certain if only the females possess the fenestra, as seems to be the case in *Bunohyrax, Thyrohyrax,* and *Megalohyrax,* of if both sexes possess it. The teeth of *P. championi* range considerably in size, which certainly suggests sexual size difference as found in other Saghatheriinae. Several chalicothere lower deciduous molars in the Nairobi collection were also referred to this species. Although they are nearly the same length, they are narrower than *P. championi* molars and can be further differentiated by their less inflated cusps.

Whitworth (1954:23) describes two mandibular fragments, CM-Hy60 and 5248), and another specimen, CM-Hy25 (p. 25), which is the crown of a right upper incisor, which differ from *P. championi* in their much smaller size. To these specimens may be added KNM-RU 2386, KNM-RU 29, and KNM-RU 35. All these specimen are identical in morphology and in size. Whitworth states that a radiograph of CM Hy60 does not show any unerupted teeth. The bone of all these specimens is striated as only found in juvenile bone and KNM-RU 35 shows a permanent M_1 erupting. The possibility does exist that these specimens are indeed juvenile specimens of *P. championi.*

Genus *Titanohyrax* Matsumoto 1922

SYNONYMY.

Megalohyrax, Andrews 1906, *pars,* (*non* Andrews, 1903).

Megalohyrax, Schlosser 1910, 1911.

Titanohyrax Matsumoto 1922, 1926.

TYPE SPECIES. *Titanohyrax andrewsi* Matsumoto 1922.

INCLUDED SPECIES. *Titanohyrax andrewsi, Titanohyrax ultimus.*

DISTRIBUTION. Oligocene, Jebel el Qatrani Formation, Fayum, Egypt.

DIAGNOSIS. Medium to large Saghatheriinae with short broad symphysis and selenodont dentition; dental formula $\frac{?.?.4.3}{?.?.4.3}$; no diastemata; I_1 and I_2

equal in size; premolars molariform; metaconid present on all lower premolars; metastylid present on P_1 to M_3; entoconid well developed on P_3 to M_3 with slight stylid on mesial face; hypocone present on P^1 to M^3, compressed mesiodistally, smaller than protocones; all cristids well developed on cheek teeth; preprotocrista and posthypocrista strong; postprotocrista very weak on premolars, missing on molars; parastyle and mesostyle present on all premolars and molars, metastyle moderately developed; no ridges on buccal side of paracone and metacone; ramus narrow, without fenestra or fossa in known specimens.

DISCUSSION. This genus was named by Matsumoto in 1922 to include two rami that Andrews had referred to *Megalohyrax* as well as all of the specimens in Schlosser's *Megalohyrax*. *Titanohyrax* differs from all other Hyracoidea by its sharp mesostyles on all the premolars as well as the molars. It differs from *Megalohyrax* by its lack of diastemata in the lower jaw and thus its short snout and symphysis. Contrary to Whitworth's statement (1954:22) there is no reason to suppose that the genus comprises fairly long-snouted forms. The Stuttgart and Cairo specimens are the only specimens that can be identified with certainty that possess the symphyseal regions. Neither of these specimens possess the canines or distal incisors but neither could be considered long-snouted. In fact, there is some question as to whether there is enough room for I_3 and a canine in front of the premolars in the Cairo specimen. The exceedingly elongated premaxilla mentioned by Matsumoto (1922:844; 1926:327, fig. 27) is not elongated along the tooth row but rather along its superior border.

This genus further differs from *Megalohyrax* by its well developed metastylids on the lower premolars and molars, its more molarized premolars, its less developed postprotocristae and prehypocristae, and strikingly compressed hypocones on the upper cheek teeth. The characters of the cheek teeth of *Titanohyrax* can not be reconciled to *Megalohyrax* as stated by Whitworth (1954:23) without considerable changes in tooth morphology, loss of diastemata and mandibular fenestra, and shortening of the snout region. *Titanohyrax* is most certainly a valid genus on morphological grounds.

Titanohyrax andrewsi Matsumoto 1922

SYNONYMY.
Megalohyrax minor, Andrews 1906, p. 97, pl. VII, figs. 2, 3, (*non* Andrews, 1904).
Megalohyrax minor, Schlosser 1910, p. 502 (*non* Andrews 1904).

Megalohyrax eocaenus, Schlosser 1910, p. 502 (*non* Andrews 1903).
Megalohyrax palaeotherioides Schlosser 1910, p. 502, *nomen nudum.*
Megalohyrax minor, Schlosser 1911, p. 105.
Megalohyrax eocaenus, Schlosser 1911, p. 105, pl. XI, fig. 7.
Megalohyrax palaeotherioides, Schlosser 1911, p. 106, pl. XI, fig. 1, pl. XII, fig. 1.
Titanohyrax schlosseri Matsumoto 1922, p. 847.
Titanohyrax palaeotherioides, Matsumoto 1922, p. 847.
Titanohyrax andrewsi Matsumoto 1922, p. 847, fig. 6.
Titanohyrax schlosseri, Matsumoto 1926, p. 325.
Titanohyrax palaeotherioides, Matsumoto 1926, p. 326, figs. 26–28.
Titanohyrax andrewsi, Matsumoto 1926, p. 333.

HOLOTYPE. CGM-8822-3, Cairo Geological Museum, Cairo; partial mandible with left I_1–I_2, P_3–M_2, and right P_3–M_3. The right P_2 was complete at the time of Andrews's publication but was missing in 1963.

DISTRIBUTION. Lower Fossil Wood Zone, Jebel el Qatrani Formation, Fayum, Egypt. Exact locality of the type unknown; AMNH-13328 is from northwest of Quarry A.

DIAGNOSIS. Medium-sized *Titanohyrax;* diagnosis same as for genus.

DISCUSSION. Matsumoto (1922:839) pointed out that Schlosser had misidentified Andrews's genus *Megalohyrax* and proposed the name *Titanohyrax* for those specimens. He recognized four species, accepting Schlosser's *Megalohyrax palaeotherioides* and naming the new species *T. ultimus, T. schlosseri,* and *T. andrewsi. T. ultimus* is a valid species and the remaining three species constitute a second valid species. There is no significant size difference nor any morphological difference among these latter three species and they are here considered subjective synonyms. The name *Megalohyrax palaeotherioides* is a *nomen nudum* because it was not properly described by Schlosser in his 1910 publication. Both *Titanohyrax schlosseri* and *T. andrewsi* were named by Matsumoto in the same publication. Because two names for the same taxon that are published in the same publication are considered as "published simultaneously" (Art. 24a), their priority is determined by the action of the first reviser (Art. 13.4). The specimens on which Matsumoto based his species *T. schlosseri* are lost while the type specimen of *T. andrewsi* is deposited in the Cairo Geological Museum and is well known from figures in Andrews (1906, pl. VII, figs. 2–3) and Matsumoto (1922, fig. 6). It therefore seems best to accept *T. andrewsi* as the valid name.

Titanohyrax ultimus Matsumoto 1922

SYNONYMY.
Titanohyrax ultimus Matsumoto 1922, p. 845, figs. 2–5.
Titanohyrax ultimus, Matsumoto 1926, p. 325.
HOLOTYPE. BMNH M-12057, British Museum, Natural History, London; right M^2.
DISTRIBUTION. Upper Fossil Wood Zone, Jebel el Qatrani Formation, Fayum, Egypt. Exact locality of type specimen unknown; Yale specimens from Quarry I.
DIAGNOSIS. Very large species of *Titanohyrax;* known only from strongly worn teeth; upper molars with characteristic lack of development of postprotocrista and prehypocrista; hypocone mesiodistally compressed; lower molars selenodont with well-developed metastylids; with or without well-developed entostylids.
DISCUSSION. Only two specimens have been found since Matsumoto described the type and paratypes: YPM-31156, a left M_1, and YPM-23889, a maxillary fragment with a left P^4 and M^1, are both from Quarry I in the Upper Fossil Wood Zone. It is doubtful that *T. andrewsi* from the Lower Fossil Wood Zone and *T. ultimus* represent an ancestor-descendant lineage because they are so different in size that an evolutionary rate nearly three times that found in *Megalohyrax* and *Bunohyrax* would be necessary to account for the size increase during the time interval between the deposition of the two fossil wood zones.

Genus *Saghatherium* Andrews and Beadnell 1902

SYNONYMY.
Saghatherium Andrews and Beadnell 1902.
Saghatherium, Andrews 1906.
Saghatherium, Schlosser 1910, 1911.
Saghatherium, Matsumoto 1926.
TYPE SPECIES. *Saghatherium antiquum* Andrews and Beadnell 1902.
INCLUDED SPECIES. *Saghatherium antiquum, Saghatherium sobrina.*
DISTRIBUTION. Lower Fossil Wood Zone, Jebel el Qatrani Formation, Fayum, Egypt.
DIAGNOSIS. Small Saghatheriinae with short, narrow symphysis, bunoselenodont dentition; dental formula $\frac{3.1.4.3}{3.1.4.3}$; I_2 only slightly larger than I_1; I_3 small and reduced; lower canine premolariform, con-

sisting of protoconid, straight postcristid, and slightly V-shaped hypoconid; canine may be separated from I_3 and P_1 by small diastemata; premolars submolariform; P_1 small, narrow; paraconid on lingual border of tooth, connected to protoconid by short straight preprotocristid; hypoconid slightly V-shaped; no metaconid in P_1 or P_2; paraconid on lingual border, metaconid in P_3 and P_4 distinctly distal to protoconid and on lingual border of tooth; P_4 nearly molariform, relatively short and broad; weak to moderate buccal cingula on P_2 to P_4; M_1 and M_2 relatively short, broad, with moderately inflated cusps; protocristid and hypocristid complete, forming lophs; metaconid without metastylid; paraconids on molars weak, near lingual border; cristid obliqua without mesoconid; M_3 long with large hypoconulid; buccal cingula strong on all molars; I^1 a triangular tusk; I^2 and I^3 small, long bladelike teeth with two-pointed buccal wall; upper canine premolariform, three-rooted; upper premolars submolariform; protocone with well-developed preprotocrista on all premolars, moderately well-developed postprotocrista on P^3 and P^4; well-developed hypocone on all premolars; no prehypocrista; very prominent ridges on buccal side of ectoloph opposite paracone and metacone; parastyle well developed on P^1 to P^4; no mesostyle or metastyle on premolars; upper molars trapezoidal, M^3 long; parastyles and mesostyles large, fairly sharp; metastyle very large on M^3; preprotocrista and prehypocrista very strong in all molars, ending in very small conule; postprotocrista and posthypocrista weak or absent; spur on lingual side of mesostyle and metastyle; ramus deepens uniformly to a point behind M_3 under the mandibular foramen; angular process of ramus curves gradually, extending far beyond the tooth row; no fenestra on lingual side of ramus.
DISCUSSION. *Saghatherium* differs from all other fossil hyraxes by the unique spurs on the lingual side of the mesostyle and metastyle. These spurs, which are simple single conules in a few specimens are often complex double or even triple peninsular outgrowths of enamel. These spurs together with the ridges on the buccal side of the paracone and metacone give the *Saghatherium* upper molar the appearance of having five cusps on the buccal side of the ectoloph and four on the lingual side. *Saghatherium* is also unique in having a hypocone on all premolars but a mesostyle on none. The great depth as well as the distal extension of the ramus behind the molar series is unique to this genus. P_1 to M_3 in YPM-18105 is 65 mm long and the distal edge of the ramus at the most dorsal point of the ptery-

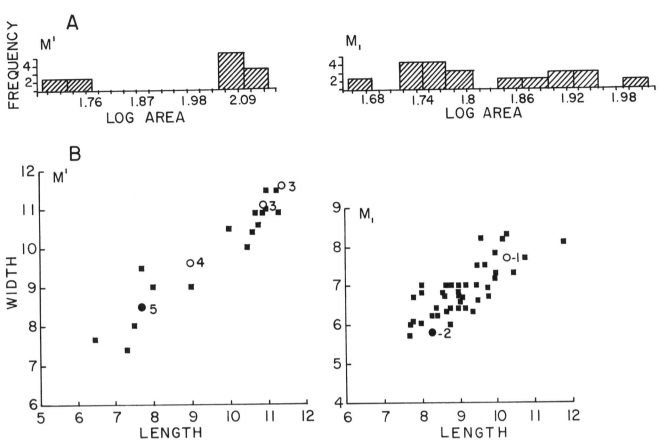

Figure 14.7 (*A*) Histogram of the log of the area of M^1 and M$_1$ of known specimens of *Saghatherium*. (*B*) Scatter diagram of M^1 and M$_1$ of *Saghatherium* specimens. *1, S. euryodon; 2, S. minor; 3, S. antiquum; 4, S. annectens; S. macrodon*, which has a damaged and incomplete M^1, is not shown.

goid crest (in line with the tooth series) is 61 mm behind M$_3$ (figure 14.7*A, B*).

Saghatherium may be differentiated from *Pachyhyrax pygmaeus* by the lack of an internal mandibular fenestra, the more lingually placed paraconids on the premolars, the more distally placed metaconids relative to the protoconids, the smaller, less well-defined entoconid on P$_3$ and P$_4$, the much stronger buccal cingulum on M$_1$ to M$_3$, by the larger, more inflated hypoconulid on M$_3$, the more complex ectoloph in the upper molars, and the lack of mesostyles in P^3 and P^4.

It differs from *Thyrohyrax* by its lack of an internal mandibular foramen, its much more inflated cusps and less well-developed protolophs and hypolophs on the lower molars, lingually placed paraconids, less well-developed entoconids on P$_3$ and P$_4$, lack of a metaconid on P$_1$ and P$_2$, much less inflated horizontal ramus, longer upper molars, and more complex ectoloph with paraconal and metaconal folds and spurs on the inner side.

Saghatherium antiquum Andrews and Beadnell 1902

SYNONYMY.

Saghatherium antiquum Andrews and Beadnell 1902, p. 5, fig. 4.

Saghatherium minus Andrews and Beadnell 1902, p. 7.

Saghatherium antiquum, Andrews 1906, p. 85 *pars*, pl. V4, figs. 4, 5.

Saghatherium magnum, Andrews 1906, p. 89 *pars*, pl. VI, fig. 4 [*non* Andrews, 1904].

Saghatherium minus, Andrews 1906, p. 89.

Saghatherium majus Andrews 1906, p. 91 *pars*, [*non* pl. VI, fig. 5].

Saghatherium antiquum, Osborn 1906, p. 263, fig. 1.

Saghatherium antiquum, Schlosser 1910, p. 503.

Saghatherium magnum, Schlosser 1910, p. 503.

Saghatherium majus, Schlosser 1910, p. 503.

Saghatherium antiquum, Schlosser 1911, p. 112 *pars*, pl. X, fig. 12, P1. XIII, fig. 12.

Saghatherium magnum, Schlosser 1911, p. 113 *pars*.

Saghatherium majus, Schlosser 1911, p. 114, pl. X, fig. 7.

Saghatherium macrodon Matsumoto 1926, p. 333, fig. 29. (*non* p. 334, fig. 30)

Saghatherium euryodon Matsumoto 1926, p. 337, figs. 31, 32.

Saghatherium antiquum, Matsumoto 1926, p. 339, figs. 33–36.

HOLOTYPE. CGM-8635, Cairo Geological Museum, Cairo; right maxilla with P^1 to M^3, left maxilla with M_1 to M_3, and portions of the skull.

DISTRIBUTION. Lower Fossil Wood Zone, Jebel el Qatrani Formation, Fayum, Egypt. Exact locality of type unknown; other specimens from Quarries A, B, C, and E.

DIAGNOSIS. Large *Saghatherium* with moderate to deep horizontal ramus; no fenestra or fossa on ramus beneath M_3. Morphological features are as in diagnosis of genus. Measurements are given by Matsumoto (1926).

DISCUSSION. The separation of *Saghatherium* into different species presents a difficult problem as can be seen by the scatter diagram of M_1^1 tooth measurements in figure 14.7*B*. The known specimens of the genus form a single or slightly hourglass-shaped cluster. The coefficient of variation for this cluster of Oligocene forms is 10.9 while the coefficient of variation for all the living species of hyraxes averages 4.8. Because the coefficient of variation for all the *Saghatherium* teeth is over 10, it seems probable that the Oligocene forms represent more than one species even if a sexual size difference is present. The *Saghatherium* specimens of the Fayum are therefore tentatively separated into two species and the "waist" of the hourglass shape in the cluster diagram is taken as the line of separation for the two species. Because no morphological differences can be found to separate these two species, the possibility does exist that there may be some mixing among males of *Saghatherium sobrina* and females of *Saghatherium antiquum*. That the "waist" of the scatter diagram as a possible separation point is further strengthened by the histogram in figure 14.7*A* where the log of the area of M_1^1 is plotted. Here a more definite separation is evident.

Although *Saghatherium* is more common in the Lower Fossil Wood Zone than all other hyracoid genera together, it has not yet been found in the Upper Fossil Wood Zone. Many fossils of similar size have been found in Quarry I and if *Saghatherium* were present in the fauna it seems probable that it would have been found. The same ecological niche may have been inhabited by *Thyrohyrax*.

Saghatherium sobrina Matsumoto 1926

SYNONYMY.

Saghatherium antiquum, Andrews 1906, p. 85 *pars*, pl. VI, fig. 6.

Saghatherium antiquum, Schlosser 1911, p. 112 *pars*.

Saghatherium minus, Schlosser 1911, p. 112.

Saghatherium annectens Matsumoto 1926, p. 342.

Saghatherium sobrina Matsumoto 1926, p. 347.

HOLOTYPE. AMNH-13282, American Museum of Natural History, New York; partial palate with right P^1 to M^1 and M^3 and left P^2 to P^3.

DISTRIBUTION. Lower Fossil Wood Zone, Jebel el Qatrani Formation, Fayum, Egypt. Type specimen from Quarry A; other specimens from Quarries A and B, south of Quarry A, and 8 km west of Quarry A.

DIAGNOSIS. Small species of *Saghatherium* with moderately deep horizontal ramus. Measurements given by Matsumoto (1926).

Genus *Thyrohyrax* Meyer 1973

TYPE SPECIES. *Thyrohyrax domorictus* Meyer 1973.

INCLUDED SPECIES. *Thyrohyrax domorictus*.

DISTRIBUTION. Upper Fossil Wood Zone, Jebel el Qatrani Formation, Fayum, Egypt.

DIAGNOSIS. Small Saghatheriinae with long narrow symphysis and brachydont, selenolophodont dentition; dental formula $\frac{?.?.4.3}{3.1.4.3}$; I_2 larger and more rounded than I_1; I_3 small, single-rooted, separated from I_2 and C by diastemata; lower C may be two-rooted and is in contact with P_1; P_1 submolariform, with centrally positioned paraconid separated from protoconid by short paracristid, hypoconid not V-shaped, narrow, deep hypoflexid; P_2–P_4 molariform with differentiated metaconid and entoconid, the former with a slight metastylid; paraconid central, with preparacristid on P_2; protocristids and hypocristids complete, forming lophs; lower molars with small paraconid, slightly buccal to centerline, near protoconid; metaconid with metastylid; entoconid projects mesially, partly closing valley between it and metaconid; protocristid and hypocristid complete, forming lophs; M_3 with small, low hypoconulid; all cingula absent or very weak; P^2 with very small mesostyle, much longer than wide, P^3 with well-developed mesostyle, slightly longer than wide, all other upper cheek teeth wider than long; P^2–M^3

molariform, lophodont, with well-developed preprotocristae and prehypocristae which often come in
contact with the ectoloph; parastyles and metastyles
well developed; slight fold on buccal side of paracone, smooth opposite metacone; mesial and lingual
cingulum well developed; small round to subround
fenestra on lingual surface of ramus under M_3 on
some specimens, lacking on others; specimens with
fenestra have greatly inflated ramus with large internal chamber in horizontal ramus that continues
at least into the base of the inflated ascending
ramus. Measurements are given by Meyer (1973).

DISCUSSION. *Thyrohyrax* differs from all other
genera of hyraxes by its greatly inflated ramus in
some specimens that contains a swollen chamber beneath and not connected to the dental series. YPM-
23939 is a fragmentary mandible, but does show
that this chamber extends at least into the base of
the inflated ascending ramus. In this bizarre form,
the internal mandibular fenestra and mandibular
chamber are carried to its extreme; yet there are
also specimens without a fenestra and without an
inflated ramus that differ in no other way. It is
therefore suggested that in this species the internal
mandibular fenestra is also a sexual feature. No
specimens with the fenestra also possess the incisors, so there is no direct evidence as to the sex of
these specimens. Two of the specimens with the fenestra are small and can be considered females, as
in other species. One specimen with a fenestra, however, falls near the top of the size range, at least in
the length and width of M_3 and in the width of M_2.
In all the several dozen hyrax specimens with internal mandibular foramina, only this one specimen is
large for its species. Because some overlap in size between females and males is expected, this specimen
could still be a female; it is suggested that in *Thyrohyrax,* as in other species, the fenestra is found in females (figures 14.8, 14.9).

Thyrohyrax further differs from *Saghatherium* by
its less inflated cusps, more molariform premolars,
more centrally placed paraconids on the lower premolars, more lophodont upper cheek teeth, lack of
spurs on the ectolophs of the upper molars, and relatively shorter and wider upper cheek teeth.

Thyrohyrax domorictus Meyer 1973

Thyrohyrax domorictus Meyer 1973, p. 3, fig. 1–3.
HOLOTYPE. CGM-40001, Cairo Geological Mu-
seum, Cairo; partial right ramus with P_1 to P_4, 1/2
M_2 to M_3.
DISTRIBUTION. Upper Fossil Wood Zone, Jebel
el Qatrani Formation. Type specimen from Quarry

Figure 14.8 Occlusal views of *Thyrohyrax domorictus;*
left maxilla with P^4 to M^3, above, and ramus with symphysis and left P_1 to M_3, below. This ramus does not contain an internal mandibular foramen.

M, referred specimens from Quarries M and I, mid-
dle beds of the Upper Fossil Wood Zone and from
Quarry G, lower beds of the Upper Fossil Wood
Zone.
DIAGNOSIS. Diagnosis same as for genus.

Subfamily Pliohyracinae Osborn 1899

TYPE GENUS. *Pliohyrax* Osborn 1899 [= *Leptodon* Gaudry 1862, *non* Sundevall 1835]
INCLUDED GENERA. *Kvabebihyrax, Meroëhyrax, Pliohyrax, Postschizotherium.*
DISTRIBUTION. Miocene of East Africa, Pliocene
of East Africa (M. Pickford, personal communication), Southern Europe, Russia, and China.
DIAGNOSIS. Pliohyracidae with selendont denti-
tion, dental formula $\frac{3.1.4.3}{3.1.4.3}$; I_2 much larger than I_1,
I_3 reduced, canine with two roots; premolars molari-

Figure 14.9 Occlusal and basal view of a mandible of
Thyrohyrax domorictus with greatly inflated rami, expanded chambers within the rami, and internal mandibular fenestrae.

form with V-shaped hypoconids on all known lower premolars; upper premolars without mesostyles; lower cheek teeth with slightly to well-developed metastylids; no mesoconids; paraconids central or lingual on premolars, lingual on molars, well developed; hypoconulids well developed on M_3; parastyle and mesostyle very strong on upper molars, not compressed; cusps not inflated; ramus deepens gradually; no fenestra present under M_3, although large fossa at least sometimes present on lingual side of ramus under cheek teeth; skull, when known relatively short, wide; diastemata missing or very small.

DISCUSSION. Included in this subfamily is *Meroëhyrax,* which is possibly close to the ancestry of *Pliohyrax,* and those forms from Southern Europe and Asia that are very similar to *Pliohyrax,* if not identical. *Saghatherium* is removed from this subfamily because it is much more similar to *Megalohyrax* and *Pachyhyrax* and is probably not the ancestor of *Meroëhyrax* as suggested by Whitworth (1954).

Meroëhyrax may have been derived from an as yet unknown member of the Saghatheriinae that was dentally similar to *Thyrohyrax. Thyrohyrax* itself was probably not on the direct ancestral line of *Meroëhyrax* because of the possession of an internal mandibular foramen and inflated jaw in the former and a broad fossa in the latter. The family died out in the Pliocene with the large *Pliohyrax*-like forms that ranged from Spain to China.

Genus *Meroëhyrax* Whitworth 1954

TYPE SPECIES. *Meroëhyrax bateae* Whitworth 1954.

INCLUDED SPECIES. *Meroëhyrax bateae.*

DISTRIBUTION. Miocene of Kenya and Uganda.

DIAGNOSIS. Small Pliohyracinae with brachydont, bunoselenodont dentition; P_3 and P_4 molariform, paraconids centrally located; hyperflexid very deep and pronounced; metaconid and entoconid high and very sharp in unworn forms; very slight metastylid in premolars, stronger in molars; molars with more lingually positioned paraconid; valley between entoconid and metaconid large and prominent; large hypoconulid on M_3; mesial lobe of each lower molar wider than distal lobe; cingulum absent or very weak; ramus rather deep with large ovoid fossa on lingual side beneath cheek teeth.

DISCUSSION. *Meroëhyrax* differs from *Saghatherium* by its large fossa on the lingual side of the ramus, its more selenodont dentition with much less inflated cusps, its centrally positioned paraconids on the lower premolars, more molariform premolars, and by the fact that its trigonid is broader than its talonid.

It differs from *Thyrohyrax* by its lack of an internal mandibular foramen and inflated ramus and by its larger size. Dentally the two are very similar. The metaconid and hypoconid are both high and sharp before strong wear has flattened them. The hyperflexid is deeply incised in both genera and the trigonid basin and talonid notch are larger than in most other hyraxes. Both genera have broader trigonids than talonids in M_1 and M_2. In both, M_3 is the longest tooth and M_2 the broadest. M_3 has a relatively large selenodont hypoconulid. The fenestra and internal chamber of *Thyrohyrax* and the fossa of *Meroëhyrax* undoubtedly served the same functional purpose. Something entered both from the dorsolingual direction, as can be seen by the groove at this corner of the fossa in *Meroëhyrax* and by a similar groove and beveled edge of the fenestra in all relevant forms of the Saghatheriinae.

Thyrohyrax is not placed in the same subfamily as *Meroëhyrax* because it possesses an internal mandibular fenestra that is characteristic of the Saghatheriinae. It is possible, however, that a form dentally similar to *Thyrohyrax* may have given rise to *Meroëhyrax* and thus to the Pliohyracinae.

The Pliohyracinae share the possession of an internal mandibular fossa with *Geniohyus* but this form is too large and dentally too different to be ancestral to *Meroëhyrax. Meroëhyrax,* however, may be close to the ancestry of *Pliohyrax. Pliohyrax, Kvabebihyrax,* and possibly even *Postschizotherium* do not seem morphologically diverse enough to separate at the generic level, but a discussion of these forms is beyond the scope of this chapter.

Meroëhyrax bateae Whitworth 1954

Meroëhyrax bateae Whitworth 1954, p. 41, fig. 16, pl. 7, fig. 1.

HOLOTYPE. BMNH M-21338, British Museum, Natural History, London; right ramus with P_3 to M_3.

DISTRIBUTION. Miocene of Rusinga Island, Kenya, and Bukwa, Uganda. Type specimen from locality RI, Rusinga; referred specimens from R3A, Rusinga, and Bukwa, Uganda.

DIAGNOSIS. Same as for genus.

Family Procaviidae Thomas 1892, emended [= Hyracidae Gray 1821]

TYPE GENUS. *Procavia* Storr 1780 [= *Hyrax* Hermann 1783]

INCLUDED GENERA. *Dendrohyrax, Gigantohyrax, Heterohyrax, Procavia, Prohyrax.*

DISTRIBUTION. Miocene of Southwest Africa, Pliocene and Pleistocene of South Africa, and Recent of Africa and Southwest Asia.

DIAGNOSIS. Small to large hyraxes with moderately hypsodont dentitions; permanent dentition reduced to $\frac{1.0.4.3}{2.0.4.3}$, at least in advanced forms; lower incisors subequal in size; premolars molariform; third lower molars without hypoconulids; internal mandibular fenestra and fossa absent.

DISCUSSION. The family Procaviidae as here defined is restricted to the three living genera and to forms possibly close to their ancestry. The three extant genera, *Procavia, Heterohyrax,* and *Dendrohyrax,* are very closely related and must have separated rather recently. It is doubtful if that they would be considered three distinct genera if they were known only from hard parts preserved in the fossil record. *Gigantohyrax* is morphologically very similar to extant *Procavia* except it is nearly three times as large. *Prohyrax* is known only from a few fragmentary specimens and so it is difficult to assess how close it lies to the ancestry of the extant forms, but of all the known genera of hyraxes it is morphologically the most similar to the extant forms and may be close to their ancestry.

Pliohyrax and *Meroëhyrax* are here removed from the Procaviidae because they represent a separate evolutionary sequence that died out with *Pliohyrax* and its closely similar forms from Asia. The very large size of this genus as well as the very large hypoconulid on its M_3 excludes it from the ancestry of *Procavia*. *Saghatherium,* which is probably nearer the ancestry of *Prohyrax* than of *Meroëhyrax,* is included in the Saghatheriinae because it more closely resembles contemporaneous forms from the Fayum and because it cannot be the direct ancestor of *Prohyrax*.

Genus *Prohyrax* Stromer 1926

TYPE SPECIES. *Prohyrax tertiarius* Stromer 1926.

INCLUDED SPECIES. *Prohyrax tertiarius.*

DISTRIBUTION. Miocene, Southwest Africa, Kenya, and Libya.

DIAGNOSIS. Small Procaviidae with moderately hypsodont dentition; P^3 nearly square, protocone bunodont, hypocone small, near center line of tooth, paracone with ridge running down mesiolingual side and curving distally to meet protocone; parastyle well developed, metastyle slight, mesostyle absent, no folds on ectoloph opposite paracone and metacone; P^4 slightly longer, hypocone larger, very near metacone but separated by narrow groove, otherwise similar to P_3; M^1, M^2, and anterior half of M^3 longer than broad, protocone with preprotocrista, hypocone bunodont, very near metacone, strong and

sharp parastyle and mesostyle, metastyle well distal to metacone causing a distal valley behind the metacone and hypocone which is surrounded by ectoloph and a distal cingulum, cingula strong at mesiolingual corner of tooth and along lingual side; valley between protocone and hypocone unusually large and deep in all teeth.

DISCUSSION. *Prohyrax* differs from *Procavia* by its more bunodont hypocone, more prominent mesostyle, and its less molariform premolars. It has in common with *Procavia* the general shape of the upper teeth, the same degree of hypsodonty, the unusual valley behind the hypocone and metacone, the similar-shaped deep valley between the protocone and hypocone, and the absence of a mesostyle in the premolars. The two dentitions are remarkably similar, in fact, for two genera that may have lived 15 to 20 million years apart.

Prohyrax differs from *Thyrohyrax* in that its upper molars are longer than broad, by its shorter preprotocrista, bunodont hypocone, more trenchant parastyle and mesostyle, lack of a mesostyle on the premolars, and a closed valley distal to the hypocone and the larger valley lingual to the mesostyle. It differs from *Saghatherium* by its longer, less broad premolars, bunodont hypocones, much less complex ectoloph without spurs on the lingual side of the mesostyle and metastyle and without folds on the buccal side of the paracone and metacone, and its stronger lingual cingulum (figure 14.10).

Prohyrax could well be close to the ancestry of *Procavia* and *Gigantohyrax* and could have been derived from a form not too different from *Saghatherium*. Because the genus is only known from a few specimens, not much can be said regarding its phylogenetic position, but from what morphological features

Figure 14.10 Maxillary fragment of *Prohyrax* sp. showing M^1 and M^2. Species of this genus are now known from Libya, Kenya, and Southwest Africa.

can be seen in this specimen it seems best to place *Prohyrax* in the Procaviidae.

Prohyrax tertiarius Stromer 1926

SYNONYMY.

Prohyrax tertiarius Stromer 1926, p. 118, fig. 18, pl.41, figs. 33, 34.

Prohyrax tertiarius, Hopwood 1929, p. 2

HOLOTYPE. 1926 X 10, Staatliche Sammlung für Paläontologie und historische Geologie, Munich; left maxilla with I^1, P^3, and 1/2 M^3.

DISTRIBUTION. Miocene, Elisabethfelder, Southwest Africa.

DIAGNOSIS. Same as for genus.

Prohyrax sp.

In 1964 Patterson found a fragment of a right maxilla with M^1 and M^2 that is morphologically very similar to *Prohyrax tertiarius* but approximately 25% smaller and somewhat more primitive in its development of the styles. This specimen (MCZ field no. 27-64) was found in the upper part of the Turkana Grit, in the north fork of the Lamenkwais drainage, approximately 4 mi NNW of Loperot. The specimen was found approximately 20 ft below a basalt dated at 17.5 ± 0.9 m.y. and 16.7 ± 0.8 m.y. This specimen belongs to *Prohyrax* but is probably a species distinct from *P. tertiarius,* which comes from the "early" Miocene of Southwest Africa.

A second specimen, an edentulous right jaw fragment with roots of I_2 and P_2 to P_4 (UCMP-41949) has been assigned to ?*Prohyrax* sp. by Madden (1972). This jaw fragment was found in 1948 by Cooke and Denison and comes from locality V-48100, Bed 14, Moruarot Hill, Kenya. A third specimen, referred to cf *Prohyrax* is reported from the Miocene of Gebel Zelten, Libya (R. Hamilton, pers. comm).

It cannot be determined whether or not these specimens belong to the same taxon, nor is the material complete enough to name a new species at this time. The two identifiable specimens of *Prohyrax* are fragments of maxillae, whereas all the specimens of *Meroëhyrax* are rami or lower teeth. The specimen of *Prohyrax* sp. from Loperot is too small to belong to *Meroëhyrax bateae* and is younger. *Prohyrax tertiarius* is more nearly the same size as *M. bateae* but does not occlude at all with KNM-RU2384 and so no close relationship can be suggested. UCMP-41949 is too fragmentary to tell if it could be assigned to *Meroëhyrax;* although it does not show signs of a fossa, the probable sexual character of the internal mandibular fenestra in the Saghatherinae makes it difficult to exclude it from *Meroëhyrax* because of this lack of a fossa.

Genus *Gigantohyrax* Kitching 1965

TYPE SPECIES. *Gigantohyrax maguirei* Kitching 1965.

INCLUDED SPECIES. *Gigantohyrax maguirei.*

DISTRIBUTION. Plio-Pleistocene, Makapansgat, Republic of South Africa.

DIAGNOSIS. Large Procaviidae with moderately hypsodont, lophoselenodont dentition; dental formula $\frac{1.0.4.3}{?.?.?.?}$; I^2, I^3, and upper canine lost; I^1 massive and short; P^1 long, with very small protocone and paracone, slight parastyle, large hypocone; P^2 to P^4 molariform, without mesostyle except very slight one on P^4, protocone and hypocone consisting of lophs which connect to ectoloph, parastyle moderately developed; upper molars similar with more pronounced parastyles and with blunt mesostyles; M^3 with reduced hypocone and metacone and no talon.

DISCUSSION. *Gigantohyrax* differs from *Procavia* by its longer diastema between I^1 and P^1, by its slightly simpler P^1, by its blunter parastyles and mesostyles on the molars, and by its smaller metastyle on M^3. *Gigantohyrax* is approximately 1.5 times larger than *Procavia transvaalensis* and approximately 3 times larger than *Procavia antiqua* and the extant species.

Gigantohyrax maguirei Kitching 1965

Gigantohyrax maguirei Kitching 1965, p. 91, figs. 39, 40.

HOLOTYPE. M-8230, Bernard Price Institute for Paleontological Research, Johannesburg, Republic of South Africa; anterior two-thirds of a skull with full upper dentition.

DISTRIBUTION. Upper Phase 1 or *Cercopithecoides*-breccia, Limeworks Quarry, Makapansgat, Potgietersrus.

DIAGNOSIS. Same as for genus.

Genus *Procavia* Storr 1780

SYNONYMY.

Procavia Storr 1780.

Hyrax Hermann 1783.

TYPE SPECIES. *Procavia capensis* Storr 1780.

INCLUDED SPECIES. *Procavia antiqua, P. transvaalensis, P. capensis, P. habessinica, P. johnstoni, P. ruficeps.*

DISTRIBUTION. Plio-Pleistocene of South Africa, Recent of Africa and Southwest Asia.

DIAGNOSIS. Medium to small Procaviidae with hypsodont dentition, dental formula $\frac{1.0.4.3}{2.0.4.3}$; dental arcade convergent anteriorly; right and left I^1 close together; P_1 to P_4 shorter than M_1 to M_3; lower incisors subequal to I_2 larger than I_1; premolars molariform; no hypoconulid on M_3.

Procavia transvaalensis Shaw 1937

SYNONYMY.

Procavia transvaalensis Shaw 1937, p. 40, fig. 1.

Procavia obermeyerae Broom 1937, p. 766, fig. 8B.

Procavia obermeyerae, Broom and Schepers 1946, p. 119.

Procavia transvaalensis, Churcher 1956, p. 477.

HOLOTYPE. No. 20, Oral and Dental Hospital, University of the Witwatersrand, Johannesburg, Republic of South Africa; right ramus with complete dentition.

DISTRIBUTION. Plio-Pleistocene of Bolt's Farm, Cooper's Farm, Kromdraai, Sterkfontein, Swartkrans, Makapan, and Taung. Type specimen from Sterkfontein.

DIAGNOSIS. Large *Procavia;* I_1 and I_2 subequal; P_1 retained throughout life; skull 1.5 times the size of extant species; interparietal free from parietals and supraoccipital; postorbital bar open. Morphology of this species is discussed by Churcher (1956).

Procavia antiqua Broom 1934

SYNONYMY.

Procavia antiqua Broom 1934, p. 472, fig. 2, 3.

"*Hyrax*" sp., Wells 1939, p. 365.

Procavia robertsi Broom and Schepers 1946, p. 79.

Procavia antiqua, Broom 1948, p. 32.

Procavia robertsi, Broom 1948, p. 33.

Procavia antiqua, Churcher 1956, p. 488.

HOLOTYPE. The holotype and other specimens on which Broom based his original description have been lost; 65 other specimens in the Transvaal Museum.

DISTRIBUTION. Plio-Pleistocene of Taung, Swartkrans, Sterkfontein, Cooper's Farm, Kromdraai, Makapan, and Bolt's Farm. The type specimen was from Hrdlicka's Cave at Taung.

DIAGNOSIS. Small *Procavia* with brachydont dentition; dental formula $\frac{1.0.4.3}{2.0.4.3}$; skull slightly smaller and lighter than in extant species; molar teeth less hypsodont with smoother ectoloph and smaller mesostyle; protocone and hypocone further from ectoloph than in extant species; well-developed hypocone on M^3.

DISCUSSION. Churcher (1956) states that two species of hyraxes inhabited the late Pliocene—early Pleistocene caves of South Africa. He thought that *Procavia* and *Prohyrax* should be considered

congeneric. In *Procavia* P^2 to P^4 are shaped exactly like M^1 and M^2, while in *Prohyrax* the premolars are more square, less molariform, and with smaller parastyles and mesostyles than the molars. Although these forms are morphologically similar they are more properly generically separated.

Conclusions

Several evolutionary trends may now be seen in the order Hyracoidea. The new material from the Upper Fossil Wood Zone of the Fayum, Egypt, has shown that the internal mandibular fenestra in the Saghatheriinae must be considered a possible sexual characteristic and this has had a bearing on reducing the number of species of hyraxes from the Fayum. The fact that *Saghatherium* is not known from the Upper Fossil Wood Zone, and important morphological differences, tend to substantiate its removal from the ancestry of *Meroëhyrax* Whitworth 1954).

The hyraxes of the Lower Fossil Wood Zone constitute nearly half of all mammal specimens found to date in those deposits; the anthracotheres make up approximately an additional 12% of the fauna. In the Upper Fossil Wood Zone the anthracotheres represent 30% of the fauna, while the hyraxes represent only 16% of the specimens found. Thus, the hyraxes were the most important medium-sized grazing and browsing ungulates during the time of deposition of the Lower Fossil Wood Zone, but their reduced numbers in the Upper Fossil Wood Zone suggests that they were outcompeted by the similar-sized anthracotheres. During the Miocene, at the time of the first radiation of the bovids, the hyraxes were reduced still further to an insignificant part of the fauna. Possibly, "*Pliohyrax*-like" forms managed to reach Europe and spread from Spain to China because they filled a niche that was not in competition with the artiodactyls and with *Hipparion,* which are so characteristic of this fauna.

Nothing important can be added about the origin of the Hyracoidea because no fossiliferous strata have been found in Africa that have yielded hyraxes older than those of the Lower Fossil Wood Zone. At that time their diversity was great; their major radiation probably took place during the middle or late Eocene. Some of the morphological features unique to hyraxes that suggest that this group has maintained a long phylogenetic independence are the extension of the malar into the glenoid fossa, the parietal process of the postorbital process, the V-shaped scapula lacking an acromion, the absence of an entepicondylar foramen, the lack of a clavicle, the ankylosed radius and ulna, the reduced third trochanter of the femur, the steplike articulation between the astragalus and the tibia, and the high vertebral formula. These features far outweigh the convergent similarities between Hyracoidea and Perissodactyla on which Whitworth (1954) advocated phylogenetic relationship. The similarities between the Hyracoidea and the Perissodactyla, Condylarthra and Typotheres may all be attributed to the fact that the hyraxes are a conservative group of subungulates that have retained many primitive characteristics seen in some fossil forms of other ungulate orders. The bulk of the evidence suggests that the closest relative of the Hyracoidea are to be found in the other subungulate orders of mammals indigenous to Africa; the Proboscidea, Embrithopoda, and Sirenia.

The Saghatheriinae represent the basal stock of the Hyracoidea as known, and they flourished during the Oligocene (and probably the Eocene), but died out in the Miocene. The Pliohyracinae, the dominant hyracoid forms of the Pliocene, share many similar features with the Geniohyinae but the morphological differences in their dentitions probably preclude their derivation from *Geniohyus*. The ancestry of the Pliohyracinae should be sought in the Saghatheriinae in a form similar in some respects to *Thyrohyrax*. *Meroëhyrax* may be close to the ancestry of *Pliohyrax*. All the known specimens of *Meroëhyrax bateae* are lower dentitions, whereas the only specimen of *Prohyrax tertiarius* is an upper dentition. These forms do not, at this time, seem conspecific because a difference in size and morphology makes occlusion impossible. If further specimens are found, however, that show a close relationship between them, then *Meroëhyrax* would have to be disregarded as a form close to the ancestry of *Pliohyrax*. The Procaviidae represent a closely related group first known from the Miocene of Kenya, Libya, and Southwest Africa and extending into the Recent.

These relationships are embodied in the following revised classification:

Order Hyracoidea Huxley 1869
  Family Procaviidae Thomas 1892
    Procavia Storr 1780
    Heterohyrax Gray 1868 (Recent only)
    Dendrohyrax Gray 1868 (Recent only)
    Gigantohyrax Kitching 1965
    Prohyrax Stromer 1926
  Family Pliohyracidae Osborn 1899
    Subfamily Geniohyinae Andrews 1906
      Geniohyus Andrews 1904
    Subfamily Saghatheriinae Andrews 1906
      Bunohyrax Schlosser 1910

      Megalohyrax Andrews 1903
      Titanohyrax Matsumoto 1921
      Thyrohyrax Meyer 1973
      Saghatherium Andrews and Beadnell 1902
      Pachyhyrax Schlosser 1911
    Subfamily Pliohyraciinae Osborn 1899
      Meroëhyrax Whitworth 1954
      Pliohyrax Osborn 1899 (Southern Europe)
      Postschizotherium Von Koenigswald 1932 (Asia)
      Kvabebihyrax Gabunia and Vekua 1966 (Southeastern Europe)

I am deeply indebted to the following for allowing me access to material in their collections: the late Dr. Camille Arambourg, Institut de Paléontologie, Paris, France; Dr. Karl Adam, Staatliches Museum für Naturkunde, Ludwigsburg, W. Germany; Mr. Darwish Alfar, Cairo Geological Museum, Cairo, Egypt; Dr. Richard Dehm and Dr. Peter Wellenhafer, Institute für Paläontologie und Historische Geologie, Munich, W. Germany; Dr. James W. Kitching, Bernard Price Institute for Palaeontological Research, University of the Witwatersrand, Johannesburg, South Africa; the late Dr. Louis S. B. Leakey, National Museum of Kenya, Nairobi, Kenya; Dr. Malcolm C. McKenna and Dr. Richard H. Tedford, American Museum of Natural History, New York; Dr. Johann Melentis, then at the University of Athens, Athens, Greece; Professor Bryan Patterson, Museum of Comparative Zoology, Harvard University, Cambridge, Mass.; Dr. Donald E. Savage, University of California, Berkeley, California; and Dr. A. J. Sutcliffe, British Museum (Natural History), London. The manuscript was read by Dr. David Pilbeam and many helpful discussions were held with Dr. David R. Pilbeam and Dr. Elwyn L. Simons of Yale University, Dr. Richard Kay of Duke University, and Dr. Philip Gingerich, the University of Michigan. The manuscript was typed by Mrs. Louise Holtzinger and Mrs. Paula Pope of Yale Peabody Museum. The photographs in figures 14.3 through 14.6 and 14.8 through 14.10 were taken by Al Coleman of Harvard University. Figure 14.7 was drawn by Mrs. Rosanne Rowen, Department of Geology, Yale University. Terminology for tooth morphology is modified from Szalay (1969:202).

Grants that enabled the Yale collections of Fayum fossils to be made included Smithsonian Foreign Currency Program Grants 5, 23, and 1841 and National Science Foundation Grant GP-433 and GP-3547 made to Dr. Elwyn L. Simons. Travel to study the collections at the National Museum of Kenya was made possible by Smithsonian Foreign Currency Grant 1289 to the author.

References

Ameghino, F. 1897. Mammifères crétacés de l'Argentine. *Boll. Inst. Georgr. Argent.* 18:1–117.

Andrews, C. W. 1903. Notes on an expedition to the Fayum, Egypt, with descriptions of some new mammals. *Geol. Mag.* 4(10):337–343.

———. 1904a. Further notes on the mammals of the Eocene of Egypt, II. *Geol. Mag.* 5(1):157–162.

———. 1904b. Further notes on the mammals of the Eocene of Egypt, III. *Geol. Mag.* 5(1):211–215.

———. 1906. *Catalogue of the Tertiary vertebrata of the Fayum, Egypt.* London: Brit. Mus. (Nat. Hist.).

———. 1907. Note on some vertebrate remains collected in the Fayum, Egypt. *Geol. Mag.* 5(4):97–100.

————. 1914. On the lower Miocene vertebrates from British East Africa collected by Dr. Felix Oswald. *Quart. J. Geol. Soc. London* 70:163–186.

Andrews, C. W., and H. J. L. Beadnell. 1902. *A preliminary note on some new mammals from the upper Eocene of Egypt.* Cairo: Survey Dept. Pub. Works Ministry pp. 1–9.

Arambourg, C. 1933. Mammifères miocènes du Turkana (Afrique orientale). *Ann. Paléont.* 22:121–148.

Broom, R. 1934. On the fossil remains associated with *Australopithecus africanus. S. Afr. J. Sci.* 31:471–480.

————. 1937a. On some new Pleistocene mammals from limestone caves of the Transvaal. *S. Afr. J. Sci.* 33:750–768.

————. 1937b. Notices of a few more fossil mammals from the caves of the Transvaal. *Ann. Mag. Nat. Hist.* 10(20):509–514.

————. 1948. Some South African Pliocene and Pleistocene mammals. *Ann. Transv. Mus.* 21:1–38.

Broom, R., and G. W. H. Schepers. 1946. The South African fossil apemen, the Australopithecinae. *Mem. Transv. Mus.* 2:1–272.

Churcher, C. S. 1956. The fossil Hyracoidea of the Transvaal and Taungs deposits. *Ann. Transv. Mus.* 22:477–501.

Cuvier, G. 1798. *Tableau élémentaire de l'histoire naturelle des animaux.* Paris: J. B. Baillière.

Dobson, G. E. 1876. On peculiar structures in the feet of certain species of mammals which enable them to walk on smooth perpendicular surfaces. *Proc. Zool. Soc. London* 1876:526–534.

Gabuniya, L. K., and A. K. Vekua. 1966. [Peculiar representative of hyrax in the upper Pliocene in Eastern Georgia.] *Soobschch. Akad. Nauk. gruz. SSR* 42:643–647. [In Georgian; Russian summary.]

Gaudry, J. A. 1862. *Animaux fossiles et géologie de l'Attique* (1862–1867), Paris, vol. I, pp. 215–218.

Gregory, W. K. 1910. The orders of mammals. *Bull. Am. Mus. Nat. Hist.* 27:1–524.

Hahn, H. 1959. *Von Baum, Busch und Klippschliefern, den kleinen Verwandten der Seekühe und Elefanten.* Wittenberg: Neue Brehm. Bucherei.

Hopwood, A. T. 1929. New and little known mammals from the Miocene of Africa. *Am. Mus. Novit.* 344:1–9.

Huxley, T. H. 1869. *An introduction to the classification of animals.* London: J. Churchill & Sons.

Kitching, J. W. 1965. A new giant hyracoid from the Limeworks Quarry, Makapansgat, Potgietersrus. *Palaeont. Afr.* 9:91–96.

Koenigswald, G. H. R. von. 1932. *Metaschizotherium fraasi,* ein neuer Chalicotheriidae aus dem Obermiocän von Steinheim. *Albuch. Palaeont.* 8:1–23.

————. 1966. Fossil Hyracoidea from China. *Nederl. Akad. Wet.,* ser. B., 9:345–356.

Madden, C. T. 1972. Miocene mammals, stratigraphy and environment of Muruarot Hill, Kenya. *Paleobios* 14:1–12.

Major, C. I. F. 1891a. Considérations novelles sur la faune des vertébrés du miocène supériór dan l'île de Samos. *C. R. Acad. Sci.* (Paris) 113:608–610.

————. 1891b. Sur l'age de la faune de Samos. *C. R. Acad. Sci.* (Paris) 113:708–710.

————. 1889a. Note upon *Pliohyrax graecus* (Gaudry) from Samos. *Geol. Mag.* 4(6):507–508.

————. 1899b. The hyracoid *Pliohyrax graecus* (Gaudry) from the upper Miocene of Samos and Pikermi. *Geol. Mag.* 4(6):547–553.

Matsumoto, H. 1922. *Megalohyrax* Andrews and *Titanohyrax* gen. nov. A revision of the genera of hyracoids from the Fayum, Egypt. *Proc. Zool. Soc. London* 1921:839–850.

————. 1926. Contribution to the knowledge of the fossil Hyracoidea of the Fayum, Egypt, with description of several new species. *Bull. Am. Mus. Nat. Hist.* 56(4):253–350.

Matthew, W. D. 1910. Schlosser on Fayum mammals. A preliminary notice of Dr. Schlosser's studies upon the collections made in the Oligocene of Egypt for the Stuttgart Museum, by Herr Markgraf. *Am. Nat.* 44:700–703.

————. 1912. African Mammals. *Bull. Geol. Soc. Am.* 23:156–162.

Melentis, J. K. 1965. Neue schädel-und Unterkieferfunde aus dem Pont von Pikermi (Attica) und Halmyropotamus (Euböa). *Praklika Akad. Athenon* 40:424–459.

————. 1966. Studien über Fossile Vertebraten Griechenlands. 12. Neue Schädel-und Unterkieferfunde von *Pliohyrax graecus* aus dem Pont von Pikermi (Attica) und Halmyropotamos (Euböa). *Ann. Geol. Pays. Helleniques* 17:182–210.

————. 1966. Studien über Fossile Vertebraten Griechenlands. 15. Fossile "Gehirne" aus dem Pont von Pikermi. *Ann. Geol. Pays. Helleniques* 17:236–246.

Meyer, G. E. 1973. A new Oligocene hyrax from the Jebel el Qatrani formation, Fayum, Egypt. *Postilla* 163:1–11.

Osborn, H. F. 1899. On *Pliohyrax kruppii* Osborn, a fossil hyracoid from Samos, lower Pliocene, in the Stuttgart collection. A new type and the first known Tertiary hyracoid. *Proc. 4th Intern. Congr. Zool., Cambridge* 1898:173–174.

————. 1906. Milk dentition of the hyracoid *Saghatherium* from the upper Eocene of Egypt. *Bull. Am. Mus. Nat. Hist.* 23:263–266.

Patterson, B. 1965. The fossil elephant shrews (family Macroscelididae). *Bull. Mus. Comp. Zool. Harvard* 133(6):295–335.

Savage, R. J. G. 1971. Review of the fossil mammals of Libya. *Symposium on the Geology of Libya.* University of Libya, pp. 216–225.

Schlosser, M. 1886. Beiträge zur Kenntnis der Stammesgeschichte der Hufthiere und Versuch einer Systematik der Paar–und Unpaarhufer. *Morphol. Jahrbuch* 12:1–136.

————. 1899. Über neue Funde von *Leptodon graecus* Gaudry und die systematische Stellung dieses Säugethieres. *Zool. Anz.* 22:378–380, 385–387.

————. 1910. Über einige fossile Säugetiere aus dem Oligocän von Ägypten. *Zool. Anz.* 35(1):500–508.

_____. 1911. Beiträge zur Kenntnis der oligozänen Landsaugetiere aus dem Fayum, Ägypten. *Beitr. Paläont. Geol. Öst.-Ung.* 24:51–167.

Shaw, J. C. M. 1937. Evidence concerning a large fossil hyrax. *J. Dent. Res.* 16(1):37–40.

Sickenberg, O., and H. Tobien. 1971. New Neogene and lower Quaternary vertebrate faunas in Turkey. *Newsl. Stratigr.* 1(3):51–61.

Simpson, G. G. 1945. The principles of classification and a classification of mammals. *Bull. Am. Mus. Nat. Hist.* 85:1–350.

Sinclair, W. J. 1908. The Santa Cruz Typotheria. *Proc. Am. Phil. Soc.* 47:64–78.

Stromer, E. 1922. Erste Mitteilung über tertiäre Wirbeltier-Reste aus Deutch-Sudwestafrika. *S. B. Bayer. Akad. Wiss.* 1921(2):331–340.

_____. 1926. Rest Land- und Süsswasser-Bewohnender Wirbeltiere aus dem Diamantfeldern Deutsch-Sudwestafrikas. In Kaiser, E., *Die Diamantenwüste Sudwestafrikas.* Berlin: D. Reimer, vol. 2:102–153.

Szalay, F. S. 1969. Mixodectedae, Microsyopsidae, and the insectivore-primate transition. *Bull. Am. Mus. Nat. Hist.* 140(4):193–330.

Teilhard de Chardin, P. 1939. New observation on the genus *Postschizotherium* von Koenigswald. *Bull. Geol. Soc. China* 19:257–267.

Teilhard de Chardin, P., and J. Piveteau. 1930. Les mammifères fossiles de Nihowan (Chine). *Ann. Paléont.* 1930:1–134.

Thomas, O. 1892. On the species of Hyracoidea. *Proc. Zool. Soc. London* 1892:50–76.

Vekua, A. K. 1972. [*Kvabeby fauna of the Akchaghylian vertebrates.*] "Nauk" (Science), Moscow. [In Russian; English summary.]

Viret, J. 1947. Découverte d'un nouvel Ancylopode dans le Pontien de Soblay (Ain). *C. R. Acad. Sci.* (Paris) 224: 353–354.

_____. 1949a. Sur le *Pliohyrax rossignoli* du Pontien de Soblay. *C. R. Acad. Sci.* (Paris) 228:1742–1744.

_____. 1949b. Observations complèmentaires sur quelques mammifères fossiles de Soblay. *Eclog. Geol. Helvet.* 42:469–476.

Viret, J., and G. Mazenot. 1948. Nouveaux restes de mammifères dans le gisement de lignite Pontien de Soblay. *Ann. Paléont.* 34:19–58.

Viret, J., and E. Thenius. 1952. Sur la présence d'une nouvelle espèce d'hyracoide dans le Pliocéne de Montpellier. *C. R. Acad. Sci.* (Paris) 235:1678–1680.

Wells, L. H. 1939. The endocranial cast in Recent and fossil hyraces (Procaviidae). *S. Afr. J. Sci.* 36:365–373.

Whitworth, T. 1954. The Miocene hydracoids of East Africa. Brit. Mus. Nat. Hist., *Fossil Mammals of Africa,* 7:1–58.

Wood, A. 1968. Early Cenozoic mammalian faunas Fayum Province, Egypt. Part 2. The African Oligocene Rodentia. *Bull. Peabody Mus. Nat. Hist.* 28:23–105.

Woodward, A. S. 1898. *Outlines of vertebrate paleontology for students of zoology.* Cambridge.

Zittel, K. A. 1891–93. *Handbuch der Palaeontologie.* I. Abth. IV. Bnd. *Mammalia.* Munich and Leipzig.

15

Deinotherioidea and Barytherioidea

John M. Harris

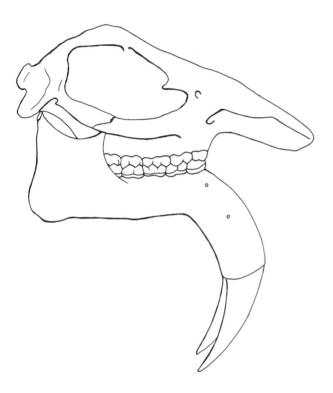

Deinotheres are commonly regarded as undoubted proboscideans that are widely separated from the main evolutionary path of the Elephantoidea (Romer 1966). The overall shape of the deinothere skull, morphology of the occiput and basicranium, position of the orbit, etc., differ widely from those of the elephantoids although in both groups the external nares are situated some distance behind the anterior edge of the premaxilla, thus indicating the presence of a proboscis. The teeth in particular are fundamentally different in the two groups and the lack of upper tusks, orientation of the lower tusks, and bifunctional bilophodont cheek teeth are features restricted to the Deinotheriidae. The postcranial skeleton of the deinotheres is graviportally adapted, as in the elephantoids, but many of the skeletal elements may be distinguished readily from equivalent elephantoid bones.

Deinotheres are restricted geographically and chronologically to Neogene and Quaternary faunas of the Old World and, unlike the elephantoids, failed to penetrate into eastern Asia and North America. It is not obvious whether the limit on deinothere dispersal was imposed by vegetation, climate, terrain, or faunal competition; but deinotheres were manifestly less successful migrants than the elephantoid proboscideans. The most primitive members of the family Deinotheriidae are recorded from early Miocene sites in East Africa, and while this does not preclude a Eurasiatic origin or pre-Miocene distribution, the known Paleogene faunas of Eurasia have yet to yield either deinotheres or elephantoids.

Deinothere remains are not ubiquitously scattered throughout the African continent and would appear to be primarily restricted to East Africa, the Congo, and the North African coast (figures 15.1 and 15.2), though this presumably reflects the location of fossiliferous sites rather than the actual distribution. Deinotheres are relatively abundant in early Miocene faunas, less so in the Plio-Pleistocene, and rather rare in the few sites of middle to late Miocene age in Africa. In Europe deinotheres are not common until the Vindobonian Stage, but isolated occurrences have been recorded in the late Burdigalian of Spain (El Papiol); Hungary (Kotyhaza); and the Loire, Allier, and Aquitaine Basins of France (Savage 1967). The Pontian Stage represents the time of maximum abundance, deinotheres becoming extinct in Europe before the Pleistocene. The earliest records of deinotheres in Asia are from the Bugti Hills and Fatehjang in Pakistan, both localities being equivalent in age to the Burdigalian of Europe. Deinotheres are not abundant in the Siwalik Series of India and Pakistan but are recorded

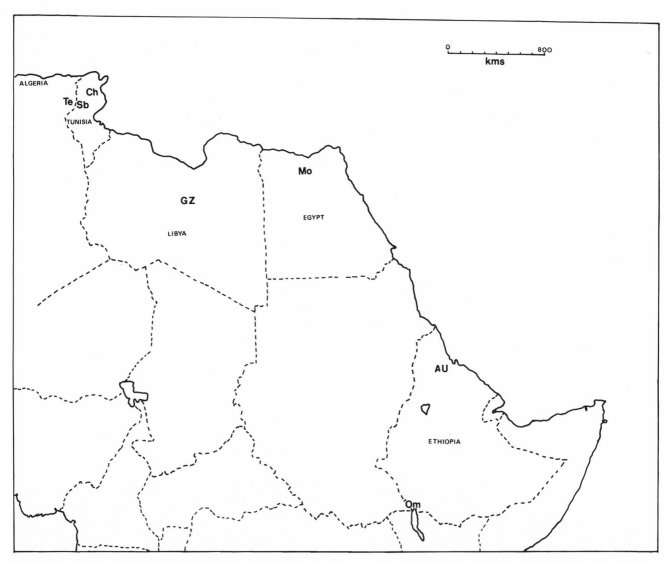

Figure 15.1 Deinothere localities in North and East Africa. Abbreviations: Au: Adi Ugri; Ch: Cherichera; Gz: Gebel Zelten; Mo: Moghara; Om: Omo Basin; Sb: Sbeitla; Te: Tebessa.

from the lower and middle Siwaliks, making their final appearance in the Dhok Pathan Stage. Deinothere teeth found in Miocene accumulations in Israel (Savage and Tchernov 1968) and Turkey (Ozansoy 1957) suggest that the main migratory route from Africa to Eurasia passed through the Middle East.

Family Deinotheriidae

DIAGNOSIS. Large herbivorous graviportal mammals. Dental formula for the deciduous teeth $\frac{0.0.3}{1.0.3}$ and for the permanent dentition $\frac{0.0.2.3}{1.0.2.3}$; dp_2^2 and P_3^3 with well-developed external crest; dP_4^4 and M_1^1 trilophodont, the remainder of the cheek teeth bi-

lophodont. Horizontal tooth replacement not developed so that all permanent teeth may be erupted at the same time (cf. elephantoids). Mandibular symphysis and lower tusks curved downward so that the tusk tips are vertically or near vertically aligned. Skull with deep rostral trough, retracted external nares, low orbit, forwardly inclined occiput, high occipital condyles, elongate paroccipital processes, and diplöe. Numerous but less well characterized distinctions in the postcranial skeleton (see table 15.1).

DISCUSSION. The main evolutionary changes exhibited by the Deinotheriidae, apart from increase in size, are reflected in the cranial and postcranial anatomy. The major cranial changes involve broadening and flattening of the rostrum, retraction of the external nares, shortening and elevation of the cranial vault, excavation of the temporal fossae, and elevation of the occipital condyles. Details of the postcranial changes are listed in table 15.2. The teeth, which provide the most common deinotheriid remains, are highly conservative and there is virtually no change in shape or relative hypsodonty through time despite a general increase in size. The only consistent dental changes occur in the upper dentition. The M^2 of *Prodeinotherium* has a strong ridge extending from the rear of the hypocone to join the lingual edge of the posterior cingulum. This ridge is joined by a posterolabially directed ridge from the metacone and the two ridges thus form a subsidiary and nonfunctional loph behind the metaloph. In M^3 the metacone gives off a strongly developed ridge that does not connect with that from the

hypocone but descends posteriorly and then centrally to the midline of the tooth where it joins the posterior cingulum. The postmetaloph ornamentation of both M^2 and M^3 tends to be reduced in later examples of *Prodeinotherium* and greatly reduced or lost in *Deinotherium*. The postmetaloph ornamentation of M^{2-3} affords some indication of the method of development of the tritoloph of M^1 (Harris 1969) and its reduction in the later deinotheres suggests that it is unnecessary for adequate performance of mastication. The anterior cheek teeth perform a different masticatory function from the posterior cheek teeth (Harris 1975), which presumably explains why deinotheres failed to develop horizontal tooth replacement as in the elephantoids. There is also a tendency in *Deinotherium,* though not *Prodeinotherium,*

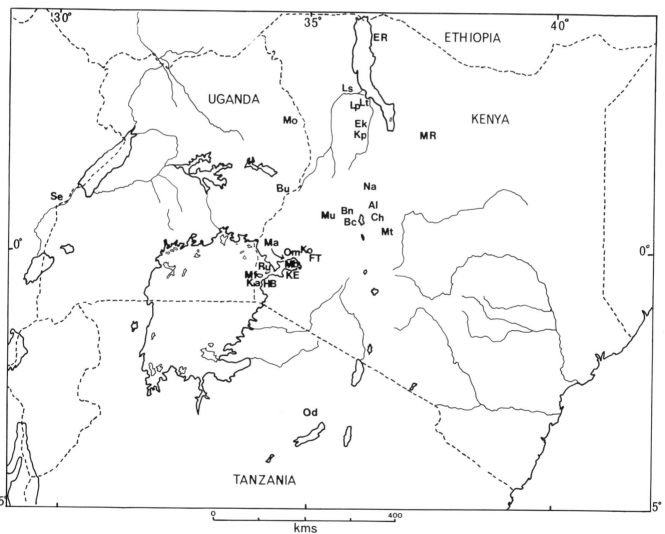

Figure 15.2 Deinothere localities in East Africa. Abbreviations: Al: Alengerr Beds; Bc: Chemeron Beds; Bn: Ngorora Beds; Bu: Bukwa; Ch: Chemoigut; Ek: Ekora; ER: East Rudolf; Ft: Fort Ternan; HB: Homa Bay; Ka: Karungu; Kp: Kanapoi; Ko: Koru; Lp: Loperot; Ls: Losodok; Lt: Lothagam; Ma: Majiwa; Mb: Maboko; Mf: Mfwanganu; Mo: Moroto; MR: Marsabit Road; Mt: Moruorot; Mu: Muruyur; Na: Nakali; Od: Olduvai; Om: Ombo; Ru: Rusinga.

Table 15.1 Major differences between postcranial elements of Deinotheriidae and Elephantoidea.

Bone	Deinotheridae	Elephantoidea
Atlas	condyle facets less concave	—
	neural spine narrow and conical	neural spine forms transverse crest
	rectus muscle scars small and horizontally aligned	rectus muscle scars larger and less horizontal
	odontoid fossa longer and less concave	—
	transverse process not elevated above axis facet, more steeply inclined and stouter	transverse process elevated above axis facet, less stout and more horizontal
Axis	atlas facets wider	—
	neuropophyses elongate	neuropophyses short
	neural spine extremely bifurcate with deep median valley	neural spine less long and less bifurcate, median valley at posterior end only
	postzygapophyses concave	postzygapophyses convex
Cervical Vertebrae	neural arch relatively longer	—
	postzygapophyses concave	postzygapophyses convex
Thoracic Vertebrae	anterior neural spines arched steeply backward	anterior neural spines almost upright (*Loxodonta*) to posteriorly inclined
Scapula	caudal angle higher and less posteriorly elongated	caudal angle lower and posteriorly elongated
	head wider and less concave	—
	metacromion and acromion less protuberant in *Prodeinotherium*, absent in *Deinotherium*	metacromion and acromion always present and stoutly developed
Humerus	distal end more massive, less elongated and less protuberant	distal end slender and elongated but epicondyles relatively more protuberant
Ulna	stout with large olecranon	relatively slender with smaller olecranon
Radius	shaft more curved	shaft straighter
	distal extremity less wide and deeper craniocaudally	distal extremity wider but more slender
Cuneiform	two distinct ulnar facets separated by sagittal ridge	single ulnar facet bisected by transverse groove
Unciform	cuneiform facet triangular and widest anteriorly	cuneiform facet rhomboid and widest posteriorly
Femur	median epicondyle lower, condyles more rounded	median epicondyle high, condyles anteroposteriorly elongated
Tibia	femur facets more concave astragalus facet quadrate	astragalus facet elongate transversely
Astragalus	tibial facet quadrate and more convex sagittally	tibial facet elongate transversely, less convex than in *Prodeinotherium* but more convex than in *Deinotherium*
Magnum	distinct posteromedial projection on lunar facet	posteromedial projection not present

Table 15.2 Comparison of *Prodeinotherium* and *Deinotherium*.

	Prodeinotherium	*Deinotherium*
Skull	rostrum ventrally flexed	rostrum almost straight in *D. giganteum*, flexed in *D. bozasi*
	rostrum narrow and deep	rostrum wide and shallow
	preorbital swelling close to external nares	preorbital swelling more anteriorly placed in *D. giganteum*
	external nares almost as deep as wide	external nares much wider than deep
	external nares surmounted by median projection of nasals	no median projection in *D. giganteum*, slight only in *D. bozasi*
	orbit above P^3	orbit above P^4
	skull roof nearly flat and inclined anteriorly	skull roof shorter and narrower at the temporal fossa
	occiput steeply inclined	occiput gently inclined
	paroccipital processes longer than in elephantoids	paroccipital processes longer than in *Prodeinotherium*
	occipital condyles cut ventrally by Frankfurt plane	occipital condyles elevated above Frankfurt plane
Dentition	P^{3-4} usually lack mesostyles	P^{3-4} often possess mesostyles
	M^{2-3} with well-developed postmetaloph ornamentation	M^{2-3} with reduced postmetaloph ornamentation
	tusks nearly vertical	tusks longer and may be recurved below symphysis
Skeleton	scapula: stout spine, metacromion and acromion; supraspinous fossa well developed	scapula: reduced spine and no metacromion or acromion; supraspinous fossa greatly reduced
	humerus: lateral epicondyle tapers proximally	humerus: lateral epicondyle does not taper proximally
	radius: medial half of head larger than lateral half; distal epiphysis more massive than in *Gomphotherium angustidens*	radius: medial half of head correspondingly larger; distal epiphysis more massive than in *Prodeinotherium*
	lunar: radial facet covers most of proximal surface; magnum facet is concave-convex	lunar: radial facet extends less far posteriorly; magnum facet almost flat
	cuneiform: posterolateral process of similar length to that of *Elephas* but does not articulate with unciform; unciform facet is concave-convex	cuneiform: posterolateral process is relatively longer and more ventrally inclined than in *Prodeinotherium*; unciform facet is biconcave
	unciform: cuneiform facet roughly triangular; largest distal facet is for Mc V	unciform: cuneiform facet extends farther posterolaterally and distally and tapers more abruptly posteriorly; largest distal facet is for Mc IV
	magnum: proximal surface has large posteromedial projection	magnum: posteromedial projection is less pronounced
	metacarpals: laterally compressed but of similar size to *G. augustidens*	metacarpals: more compressed laterally and distinctly more elongate than in *Prodeinotherium*
	manus: more plantigrade	manus: more digitigrade
	femur: of similar length to *Elephas maximus*	femur: proportionately 30% shorter than in *E. maximus*
	astragalus: tibial facet quadrate and convex; prominent posteromedial process	astragalus: tibial facet quadrate but nearly flat; posteromedial process more reduced than in *Prodeinotherium*

for subsidiary median styles to be developed on P[3-4]. Where present these tend to vary greatly in size, shape, and position.

The family Deinotheriidae has a geologic record of about 20 m.y. and is distributed in three continents. Teeth and mandibular fragments are the most common remains. Attempts have been made to define specific taxa on a variety of different features including geographic distribution, size and shape of the tusks, length and degree of flexure of the mandibular symphysis, position of the mental foramina, number of rugose corrugations on the lophs of unerupted cheek teeth, and on the cusp pattern, loph width, and overall size of the cheek teeth. The proliferation of new deinothere taxa during the nineteenth century was such that in Osborn's (1936) monograph on the Proboscidea, 22 species were mentioned and two species, *D. bozasi* (Arambourg 1934) and *D. orlovii* (Sahni and Tripathi 1957), have been proposed subsequently. Most of these species were erected before Weinsheimer (1883) suggested that all deinotheres belonged to a single taxon—*D. giganteum*. Weinsheimer certainly oversimplified deinothere systematics and attracted few disciples, but nevertheless only seven species were proposed after his cautionary note and of these only *D. orlovii* and the African examples appear to be valid.

In view of the preponderance of isolated teeth, a classification based on size and crown morphology of the cheek teeth would seem desirable. However, because of the great variation in minor features of the cusp pattern seemingly present in all deinothere species, this has not yet proved feasible. The teeth of any one species increase in size through time but, with the apparent exception of *D. orlovii*, the length versus breadth ratio remains constant throughout the family regardless of specific taxon. That the cheek teeth fall into two distinct groups on the basis of size has been recognized since Depéret treated them accordingly in 1887. The small size of deinothere teeth from Burdigalian horizons at Kotyhaza and Kirald was one of the factors that influenced Ehik (1930) to create the genus *Prodeinotherium*. The concept of two separate genera within the Deinotheriidae was also recently supported by Harris (1973).

Genus *Prodeinotherium* Ehik 1930

DIAGNOSIS. Small deinotheres. Dental formulae as for the family; M[2-3] with well-defined postmetaloph ornamentation. Skull rostrum turned down parallel to the mandibular symphysis; rostral trough and external nares narrow; preorbital swelling close to orbit; external nares anteriorly placed and nasal bones with median projection; skull roof relatively longer and wider than in *Deinotherium*, occiput more vertically inclined; occipital condyles situated more ventrally than in *Deinotherium* and level with the Frankfurt plane; paroccipital processes short. Postcranial skeleton with graviportal adaptations; scapula with well-developed spine and stout acromion and metacromion; carpals and tarsals narrow but not dolichopodous.

TYPE SPECIES. *Prodeinotherium bavaricum* (von Meyer) 1831.

OTHER RECOGNIZED SPECIES. *P. pentapotamiae* (Falconer) 1868; *P. hobleyi* (Andrews) 1911; *P. (?)orlovii* (Sahni and Tripathi) 1957.

A detailed synonymy of *P. bavaricum* was given by Graf (1957) and of *P. pentapotamiae* by Sahni and Tripathi (1957).

DISCUSSION. Specimens that may be attributed to species of *Prodeinotherium* are observed to increase in size from older to younger horizons. The maximum known measurements for the permanent teeth of *Prodeinotherium* based on personal observation and on data from Graf (1957), MacInnes (1942), and Sahni and Tripathi (1957) are given in table 15.3. Similar figures are given by Bergounioux and Crouzel (1962), who also give known size ranges for the deciduous teeth.

Prodeinotherium hobleyi (Andrews) 1911

SYNONYMY

1911 *Deinotherium hobleyi* Andrews: 943.
1919 *Deinotherium cuvieri* Kaup; Brives: 90.
1957 *Deinotherium bavaricum* v. Meyer; Graf: 152.
1967 *Deinotherium cuvieri* Kaup; Savage: 263.
1972 *Prodeinotherium hobleyi* (Andrews); Harris: 222.

Table 15.3 Maximum measurements for the permanent dentition of *Prodeinotherium* based on all available specimens (in mm).

Tooth	Length	Breadth
P[3]	63	60
P[4]	61	62
M[1]	73	69
M[2]	73	75
M[3]	69	72
P$_3$	51	43
P$_4$	60	52
M$_1$	73	53
M$_2$	75	70
M$_3$	81	68

DIAGNOSIS. A species of *Prodeinotherium* differing from *P. bavaricum* by a more distinct separation of the metaconid from the protoconid in P_3, a P_4 that is shorter in proportion to its width, and by the presence of the outer tubercule of the M_3 talonid being more distinctly and independently developed (after Andrews 1911). This species appears to be confined to Africa.

HOLOTYPE SPECIMEN. Left mandible with P_4 and M_{2-3} from Karungu in the Nyanza District of Kenya. This specimen is in the collections of the British Museum (Natural History), London.

LOCALITIES. Algeria: Tebessa (Brives 1919); Tunisia: Cherichera (pers. comm. C. C. Black), Sbietla (Vialli 1966); Egypt: Moghara (Osborn 1936); Libya: Gebel Zelten (Savage and White 1965; Savage 1967; Harris 1969, 1973); Eritrea: Adi Ugri (Vialli 1966); Congo: Semliki (Hooijer 1963); Uganda: Bukwa (Walker 1969), Moroto (Bishop 1967); Kenya: Alengerr Beds (Bishop et al. 1971), Fort Ternan, Karungu (Andrews 1911, 1914; Bishop 1967), Koru (Bishop 1967), Loperot (Bishop 1967), Losidok (Bishop 1967; Madden 1972b), Maboko (Bishop 1967), Majiwa, Moruorot (Madden 1972a), Mfwanganu (Bishop 1967), Muruyur, Ngorora (Bishop et al. 1971), Ombo (Bishop 1967), Rusinga (MacInnes 1942; Bishop 1967; Harris 1969, 1973), West Stephanie (Harris and Watkins 1974).

DISCUSSION. Andrews (1911) proposed the taxon *D. hobleyi* for an incomplete mandible and isolated teeth from Karungu, Kenya, justifying his new species on the basis of the cusp morphology of the cheek teeth and on the geographical isolation of his specimens from their European relatives. Most subsequent workers have used Andrews's taxon for all small African deinotheres. Brives (1919), however, assigned an isolated tooth from Tebessa, Algeria, to *D. cuvieri* (=*P. bavaricum*) and Savage (1967) followed suit by assigning all early Miocene deinotheres from Africa to this species. Graf (1957) tentatively synonymized *P. hobleyi* with *P. bavaricum* but neither Graf nor Savage attempted justification of their synonymies. If, on the basis of dental similarity, *P. hobleyi* and *P. bavaricum* are considered synonyms for the same species, then *P. pentapotamiae* from Asia must also be considered as synonymous. It is difficult to accept that a single mammalian species existed for some 10 or so m.y. on three continents without undergoing further differentiation. However, diagnoses of species based primarily on geographic distribution must necessarily be both arbitrary and subjective and must also take into account the geologic age and overall faunal composition of the assemblage considered. Other mammals associated with *Prodeinotherium* in Africa include some immigrants from Eurasia, but the overall fauna is basically indigenous to Africa. For this reason, *pro tem,* all small African deinotheres are assigned to *P. hobleyi.*

The majority of known specimens of *P. hobleyi* comprise dentitions or isolated teeth. Relatively complete skulls and mandibles are known from Gebel Zelten, Libya; postcranial elements are known from this site and from several localities in East Africa. Descriptions of the more complete specimens were given by Harris (1973).

It is assumed that in Africa *Prodeinotherium* teeth paralleled those of Eurasian species by showing a tendency to increase in size through time. Certainly the smallest African deinothere teeth come from the East African Miocene sites (table 15.4) from which the earliest radiometric dates have been recorded (see Bishop, Miller, and Fitch 1969). Cranial and postcranial material is rare at present and tooth size plus cusp morphology are therefore the only criteria available for comparison of deinotheres from different sites. Because of the conservative nature of some of the cheek teeth, notably the lower molars, detailed comparisons of tooth morphology are possible only for Gebel Zelten and some of the East African sites. Tooth size forms the sole criterion for comparison between specimens from other sites.

The largest collection of North African deinothere remains comes from Gebel Zelten. Teeth from this site vary in size but the differences do not indicate more than one species. The M^2 from Tebessa, the M_2s from Sbietla and Adi Ugri, and the tooth fragment from Moghara all fall within the size range exhibited by the Gebel Zelten specimens. There are no major differences in crown morphology.

The cusp morphology of the Gebel Zelten *P. hobleyi* teeth differs from that of the East African specimens in several respects. On the Gebel Zelten specimens there is a hypoconulid on P_3, a mesostyle of P_4, and an ectostyle of M^1. None of these features has been recorded on specimens from East Africa. The mesostyle of P^3 is persistent in the Libyan specimens although variable in the East African examples, and the ectoloph of the P^4 of the Libyan specimens is better developed. The relatively larger size of the Gebel Zelten deinothere teeth and their apparently more advanced cusp morphology would suggest that they may be younger than the majority of deinotheres from the early Miocene of East Africa, although Savage (1968) has attributed an early Burdigalian or even late Aquitanian age to the Gebel Zelten specimens.

Table 15.4   Localities yielding *Prodeinotherium hobleyi* in Africa.

Locality	Radiometric Age (m.y.)	Skull	Maxilla	Mandible	Isolated Teeth	Postcranial Elements
Tebessa	—				X	
Cherichera	—				X	
Sbeitla	—				X	
Gebel Zelten[a,b]	—	X	X	X	X	X
Moghara[a]	—				X	
Adi Ugri[c]	—				X	
Semliki	—				X	
Bukwa[a]	22.0				X	
Moroto	12.5–14.3				X	
Alengerr[d]	12.5–14.0				X	
Fort Ternan[d]	14.0–14.7				X	
Karungu[a,d]	22.5			X	X	
Koru	19.5				X	
Loperot[d]	—				X	
Losodok[d,e]	—				X	
Maboko[a,d]	—				X	X
Majiwa	—				X	
Mfwanganu[a,d]	—				X	
Moruorot[e]	—				X	
Muruyur	—				X	
Ngorora	9.0–12.0		X		X	
Ombo	12.3				X	
Rusinga[a,d]	—		X	X	X	X
West Stephanie[d]	—				X	

a. Specimens at British Museum (Natural History), London.
b. Specimens at Department of Geology, University of Bristol.
c. Specimens at Istituto di Geologia e Paleontologia, University of Bologna.
d. Specimens at National Museum of Kenya, Nairobi.
e. Specimens at Department of Paleontology, University of California, Berkeley.

A single middle Miocene deinothere tooth (dP$_4$) is known from Fort Ternan and is slightly larger than fragments of a comparable tooth from Rusinga Island.

Late Miocene deinothere teeth are known only from Ngorora. These teeth are of similar size to those from Gebel Zelten (i.e., larger than the majority of specimens from the early Miocene of East Africa) and approach the upper size range known for *Prodeinotherium*. There is no mesostyle on the P^4 of the Ngorora teeth but the postmetaloph ornamentation of M$^{2-3}$ is simpler than in those of the Gebel Zelten and early Miocene Kavirondo Gulf specimens, and resembles more the stage of development in late Miocene examples of *P. bavaricum*. Although differing slightly in morphology from the early Miocene specimens from East Africa, the Ngorora specimens, like those from Gebel Zelten, merely represent advanced stages of the primitive *P. hobleyi* from East Africa; for the time being, they are placed in the same taxon.

When compared with Asiatic representatives of *Prodeinotherium,* the Gebel Zelten specimens have an ectostyle on M^1 and a mesostyle on P^3 in common with *P. pentapotamiae*. In these two characters *P. pentapotamiae* and the Gebel Zelten specimens are more advanced than the East African forms. But, whereas a tubercule is present on the labial side of the median valley in M^3 of the Gebel Zelten specimens, an outer tubercule is also present in *P. pentapotamiae*. In this respect the Gebel Zelten form may be said to be less advanced than *P. pentapotamiae,* at least morphologically. In terms of size, the early Miocene deinothere teeth from East Africa tend to be smaller than those of *P. pentapotamiae*. The Gebel Zelten examples are larger than those from the Gaj Stage of the Bugti Hills but are similar to those of the Kamlial and Chinji stages of the lower Siwaliks.

If, as suggested by Bergounioux and Crouzel (1962) and Graf (1957), the small deinotheres of Europe are placed in a single species (*P. bavaricum*), then this species exhibits great variability in minor details of cusp morphology of the cheek teeth. The early Miocene teeth from East Africa are generally smaller than known examples of *Prodeinotherium* from Europe. The Gebel Zelten teeth are larger than those from the Burdigalian and Vindobonian sites in Europe but are more primitive in terms of postmetaloph ornamentation reduction than the late Miocene examples of *P. bavaricum*.

Genus *Deinotherium* Kaup 1829

DIAGNOSIS. Large deinotheres. Dental formulae as for the family; tendency for subsidiary styles on P^{3-4}; postmetaloph ornamentation of M^{2-3} greatly reduced or absent. Rostral trough and external nares wide; skull roof short and narrow at the temporal fossae; occipital condyles elevated above the level of the external auditory meatus. Postcranial skeleton with cursorial modifications to graviportal structure; scapular spine with no acromion or metacromion; carpals and tarsals narrow with dolichopodous metapodials exhibiting functional tetradactyly.

TYPE SPECIES. *Deinotherium giganteum* Kaup 1829.

OTHER RECOGNIZED SPECIES. *D. indicum* Falconer 1845; *D. bozasi* Arambourg 1934.

A detailed synonymy of *D. giganteum* was given by Graf (1957) and of *D. indicum* by Sahni and Tripathi (1957).

Specimens that may be attributed to species of *Deinotherium* are observed to increase in size from older to younger horizons. The *minimum* measurements for the permanent teeth of *Deinotherium* based on personal observation and data from Graf (1957), MacInnes (1942), and Sahni and Tripathi (1957) are given in table 15.5.

Deinotherium bozasi Arambourg 1934

SYNONYMY

1928 *Deinotherium giganteum* Kaup; Joleaud: 1001.
1934 *Deinotherium bozasi* Arambourg: 86.
1936 *Deinotherium hopwoodi* Osborn: 117.
1942 *Deinotherium giganteum* var. *bozasi* Dietrich: 90.
1942 *Deinotherium bozasi* Arambourg; MacInnes: 81.

DIAGNOSIS. Species of large deinothere with teeth of similar size to *D. giganteum*. Skull rostrum steeply downturned anteriorly (cf *P. hobleyi*) with narrower external nares and rostral trough than in *D. giganteum;* preorbital swelling reduced and situated just in front of P^3; occiput steeply inclined and nasal bones with slight anterior median projection—both features as in *P. hobleyi* and in contrast to *D. giganteum*. Mandibular symphysis flexed at a right angle.

HOLOTYPE SPECIMEN. Mandible with tusks, M_1, and M_3 from the Omo Basin, Ethiopia. This spec-

Table 15.5 Minimum measurements for permanent dentition of *Deinotherium* based on all available specimens (in mm).

Tooth	Length	Breadth
P^3	63	55
P^4	59	62
M^1	71	64
M^2	70	74
M^3	70	74
P_3	51	43
P_4	61	48
M_1	74	54
M_2	73	63
M_3	75	64

Note: Similar figures are given by Bergounioux and Crouzel (1962), who also define limits for the deciduous teeth.

imen is in the collections of the Muséum National d'Histoire Naturelle, Paris.

LOCALITIES. Ethiopia: Omo Basin (Arambourg 1934, 1947; Arambourg, Chavaillon, and Coppens 1969); Hadar (Taieb et al. 1976); Tanzania: Olduvai Gorge (Osborn 1936; Leakey, L. S. B., 1965; Leakey M. D., 1971; Kenya: Chemeron Formation (Bishop et al. 1971), Chemoigut Beds (Bishop et al. 1971; Carney et al. 1971), East Lake Turkana (Leakey, R. E. F., 1970; Maglio 1971, 1972; Harris 1976), Ekora (Patterson, Behrensmeyer, and Sill 1970), Homa Bay, Kanapoi (Patterson, Behrensmeyer, and Sill 1970), Kanam East (Kent 1942; MacInnes 1942), Kanam (Kendu Bay), Lothagam (Patterson, Behrensmeyer, and Sill 1970), Marsabit Road, Nakali (pers. comm. E. Aguirre). Mozambique: Priai de Morrungusu (Harris in prep.).

DISCUSSION. Large African deinotheres of similar size to *D. giganteum* (Europe) and *D. indicum* (Asia) are known from a number of sites of Plio-Pleistocene age in East Africa (table 15.6) and, with the possible exception of a tooth from Sahabi, Libya (recorded in a provisional list of fauna by Petrocchi [1951] but not mentioned in subsequent pappers on that site), appear to be virtually restricted to that region. The earliest *Deinotherium* specimens would appear to be the few tooth fragments from Nakali—a site that has been tentatively dated as younger than the Ngorora Formation but older than the Mpesida Beds (of Bishop et al. 1971) by E. Aguirre (pers. comm.). African deinotheres survived much longer than their Eurasian relatives, the youngest known specimen being from the Chemoigut Beds (younger than 1.2 m.y.; Carney et al. 1971). All are assigned to *Deinotherium bozasi*. Unlike in Eurasia, *Deinotherium* is not found associated with *Prodeinotherium* in Africa.

Table 15.6 Localities yielding *Deinotherium bozasi* in Africa.

Locality	Radiometric Age	Skull	Maxilla	Mandible	Isolated Teeth	Postcranial Elements
Omo[a]	3.75–1.84		X	X	X	X
Olduvai[b,c]	1.82		X	X	X	X
Chemeron Beds[b]	—			X	X	
Chemoigut Beds[b]	1.2				X	
East Rudolf[b]	2.6–?	X		X	X	X
Ekora[b]	4.0				X	
Homa Bay[b,c]	—				X	
Kanapoi[b]	4.0				X	X
Kanam East[c]	3–3.5				X	
Kanam (Kendu Bay)[c]	3–3.5				X	
Lothagam[b]	—				X	
Marsabit Road[b]	—				X	
Nakali[b]	—				X	
Hadar[d]		X	X	X	X	?

a. Specimens at Muséum National d'Histoire Naturelle, Paris.
b. Specimens at National Museum of Kenya, Nairobi.
c. Specimens at British Museum (Natural History), London.
d. Specimens at National Museum of Ethiopia, Addis Ababa.

The large African deinothere was first discovered in the Omo Basin of Ethiopia in 1903 by the expedition of Vicompte Bourge de Bozas (Haug 1911). Joleaud (1928) attributed the specimens collected to *Deinotherium giganteum*. Osborn (1936) posthumously published a description of similar teeth from Olduvai Gorge, Tanzania, and proposed a new name, *D. hopwoodi,* for them. Before Osborn's descriptions appeared in print Arambourg (1934) described new specimens from Omo as *D. bozasi.* Dietrich (1942) subsequently reduced *D. bozasi* to a variety of *D. giganteum* but later workers have treated *D. bozasi* as a valid species.

Like *P. hobleyi, Deinotherium bozasi* is known mainly from isolated teeth. A skull and mandibles from east of Lake Turkana have been described by Harris (1976). A second skull has recently been discovered at Hadar in the Afar region of Ethiopia (Taieb et al 1976). A badly crushed skull and associated partial skeleton were recovered from Olduvai in 1962, but the skull subsequently disintegrated and the skeletal elements are distorted and as yet incompletely prepared. A partial skeleton has been recovered from the Omo Basin and is currently being investigated by Beden of the Université de Poitiers (pers. comm. Y. Coppens). Odd postcranial elements are known from Kanapoi (pers. comm. V. J. Maglio).

The teeth of *D. bozasi* are much larger than those of *P. hobleyi* and are larger than many specimens of *D. giganteum* and *D. indicum,* though smaller than specimens referred by Ştefănescu and others to *D. gigantissimum* (=*D. giganteum* Graf 1957). The

postmetaloph ornamentation of M^{2-3} is greatly reduced and totally absent in many specimens. P^3 seems to vary greatly in length but too few upper premolars are known to determine if the anterior cheek teeth showed a similar amount of variation in cusp morphology to those of *D. giganteum.*

The mental foramina of the mandibles of both *D. giganteum* and *D. bozasi* are situated farther forward than in specimens of *P. hobleyi,* suggesting a relative increase in the length of the horizontal ramus. The mandibular symphysis of *D. giganteum* is more elongate and more vertically aligned than in species of *Prodeinotherium.* In contrast, *D. bozasi* has a shorter mandibular symphysis than that of *P. hobleyi* and this is flexed more abruptly than in *D. giganteum.* The greater degree of flexure in *D. bozasi* may at least in part relate to the more acutely flexed rostrum of this species as compared to that of *P. hobleyi.* Some deinotheriid mandibles from Europe exhibit a degree of flexure in the symphysis that is strongly reminiscent of the condition in *D. bozasi.* These specimens were originally attributed to *Deinotherium levius* but were later assigned to *D. giganteum* (Bergounioux and Crouzel 1962). Arambourg (1934) and Laskarev (1944) believed the strongly flexed mandibular symphysis of *D. levius* warranted specific separation from *D. giganteum* and Laskarev (1944) concluded that *D. levius* had African rather than European affinities.

Despite the superficial similarity of the mandibles of *D. levius* and *D. bozasi* the latter taxon almost certainly evolved in Africa from *P. hobleyi.* The skulls of *D. bozasi* from East Lake Turkana and Hadar are superficially closer to that of *P. hobleyi* from the Miocene of Gebel Zelten than they are to those of *D. giganteum* from the Pontian of Eppelsheim, Germany, and the Ukraine (Swiston 1974). *D. giganteum* exhibits a number of morphological differences in its postcranial skeleton from that of *Prodeinotherium* that Tobien (1962) and Harris (1973) have interpreted as cursorial modifications. When prepared postcranial material of *D. bozasi* becomes available for study, it will be imperative to determine if similar structural changes are exhibited by the African *Deinotherium*. It is entirely possible that the results of such an investigation will necessitate further revision of deinotheres systematics.

Mode of Life

For most of the present century, deinotheres have been regarded as aberrant elephants that lived in swampy environments, forests, or both (Osborn 1936, p. 113). Early theories regarding the mode of life of deinotheres showed great imagination but were sometimes based on insufficient anatomical information or misidentified material. Buckland's (1835) quaint portrait of an aquatic mammal anchored to a river bank by its tusks while it slept preceded the discovery of deinothere hind limbs. The occurrence of a supposed "epipubic bone" of *Deinotherium* in France pursuaded Solario (1864) that:

Deinotherium was . . . a marsupial. It must have lived upon the branches and trees which the reversed tusks would enable it to bring down within reach. . . Its tusks, being turned downward . . . were especially adapted to strike with heavy blows the animals that were likely to attack it. . . While the neck of the animal was very short, the trunk must have been of great size and . . . its use included, in all probability, the putting of its young into its pouch as well as the feeding of itself.

Deinothere remains have been recovered from fluviatile, lacustrine, and even marine sediments and associated with a variety of terrestrial and aquatic animals. Much of the evidence for mode of life has to be derived from investigation of the anatomy of the animal.

Deinothere skulls differ from those of elephantoids both in overall shape and by different refinements of the facial and cranial region. These differences reflect the phyletic separation of the two groups and also differences in function and mode of life. Adaptations of the facial region of deinothere skulls are concerned with feeding. Differences in the cranial region imply differences in attitude and movement of the deinothere skull from that of elephantoids and are again primarily concerned with feeding. Evolution of feeding mechanisms in the

family Deinotheriidae has been summarized by Harris (1976).

Because of the elongate upper tusks of the elephantoids, development of an elongate and mobile upper lip or proboscis was necessary in order to permit manipulation of the food. In the earliest known deinotheres the superior tusks are lost, the inferior tusks are almost vertically aligned, and the rostrum of the skull and the mandibular symphysis are both flexed ventrally, permitting the mouth to be intimately connected with the food source. Nevertheless, the posterior position of the external nares, foreshortened nasal bones, deep rostral trough, massive preorbital swellings, and gross development of facial nerves V and VII all point to the development of a proboscis. It is impossible to deduce whether the proboscis of deinotheres developed an opposable process at the tip, as in elephantoids, or whether, as in tapirs, it functioned solely to push the food into the mouth, using the tusks and symphysial region of the mandible as inflexible and immobile surfaces against which to gain purchase. Osborn (1910, p. 247) reconstructed the head of *Deinotherium giganteum* with a short proboscis but later (1936, p. 102) preferred a long elephantine proboscis as suggested by Abel (1922, p. 91). It is probable that while *Prodeinotherium* possessed a relatively short proboscis, that of *Deinotherium* was correspondingly more elongate and massive.

The skull in deinotheres (Fig. 15.3) contrasts strongly with that in the elephantoids. The tusks are much smaller, are present only in the mandible, project ventrally rather than anteriorly, and the skulls are low and elongate with high occipital condyles and strongly developed paroccipital processes. The elevated position of the occipital condyles is of major importance. Because the condyles are elevated high above the basicranial region, the lever arms of muscles inserting below the level of the condyles and originating from the underside of the cervical vertebrae are enlarged. Thus the muscles that flex the head ventrally about the atlas-occipital joint are rendered more powerful. Further information about the flexure of the head on the neck and thorax is provided by the paroccipital processes. The degree of enlargement of the paroccipital processes in deinotheres, in particular the increased depth, and the greater distance between these processes and the occipital condyles, contributes to a greater degree of rotation of the head about the condyles than in the elephantoids. Also the depth of the process beneath the condyles prevents any extensor action of the muscles inserted on the paroccipitals, unless such muscles are attached also to portions of the occiput dorsal to the condyles. Because of the elevation of

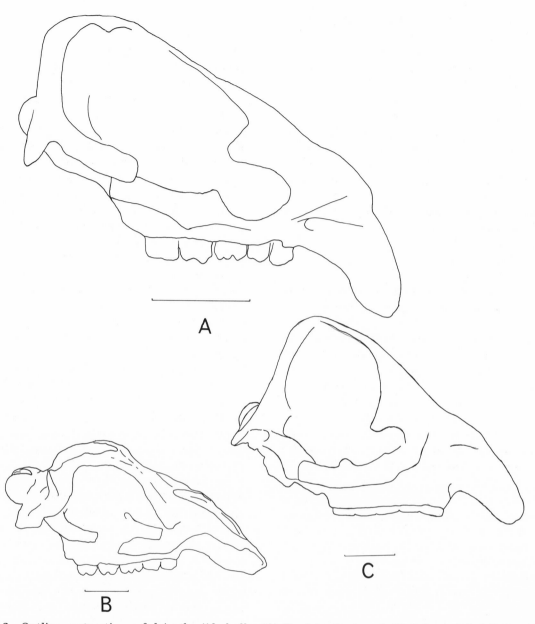

Figure 15.3 Outline restorations of deinotheriid skulls. (*A*) *Prodeinotherium hobleyi* from Gebel Zelten, Libya, (*B*) *Deinotherium giganteum* from Epplesheim, Germany, (*C*) *Deinotherium bozasi* from East Rudolf, Kenya (scale in each figure = 20 cms).

the occipital condyles on deinothere skulls the nuchal fossae are, in compensation, placed considerably higher on the occiput than in the elephantoids. The elevated condyles and deep paroccipital processes of deinotheres reflect specializations for flexor movement of the head on the atlas-occipital joint; the zygapophyses of the anterior cervical vertebrae have reversed curvature when compared to those of elephantoids and also appear to be adapted for a downward movement of the head and neck. These modifications appear to be associated with the function of the tusks and differ significantly from those of the elephantoid proboscideans.

The downturned mandibular tusks of deinotheres are unique in that these are the only mammals to retain the inferior tusks in preference to the superior ones. The shortness of the tusks and their orientation would seem to prevent their use as levers. The presence of enamel on newly erupted permanent tusks but not in adult specimens indicates that the tusks were functional and subjected to wear. Normally, adult deinothere tusks terminate in rounded conical points, although the unworn tip of an immature tusk from Rusinga (specimen RS. 7D in the British Museum [Natural History]) possesses a spatulate, transversely flattened tip with carinate medial and lateral edges. There is apparently no sexual dimorphism of the tusks. Two mandibles of *D. bozasi* from the Omo (Arambourg 1934, 1947) possess tusks of different lengths, but the longer pair belongs to the less mature individual, suggesting differences in wear rather than sex (pers. comm. Arambourg). Wear facets have been observed in a few individuals and are found on the anteromedial tip, suggesting wear by abrasion against an object in front of and between the tusks. The tusks perhaps acted as a source of purchase for the tip of the proboscis and also for stripping bark and vegetation via a downward movement of the head. It is unlikely that the tusks were used for digging (Harris 1975).

Deinothere cheek teeth occlude orthally and may be functionally divided into an anterior crushing battery and a posterior shearing battery. During the occlusal stroke of the mandible some shearing is effected by the premolars during occlusion of the ectolophs and ectolophids. The transverse lophs of the premolars serve to help prevent anterior dislocation of the jaw and also perform a crushing function during the recovery stroke of the mandible. The transverse crests of the deciduous fourth premolars and of the molars have well-developed shearing facets developed on the anterior surface of the lophs and on the rear of the lophids. These perform a shearing function during the occlusal stroke of the mandible. The shearing facets are maintained until the transverse lophs have been almost completely removed by wear. In newly erupted teeth the facets are aligned almost vertically, but these become progressively more horizontal with wear. The change in angle of the shear facets is seen not only in individuals of differing maturity but also within a single toothrow, the facets at the posterior end of the toothrow being more vertically aligned than anterior ones.

The anterior crushing battery and posterior shearing battery are reestablished three times during the life of the individual as follows:

	anterior battery	posterior battery
juvenile	dP2–3	dP4
immature adult	P3–4	M1
mature adult	P2–M1	M2–3

The anterior trilophodont molar is the first permanent tooth to be erupted and performs different functions at different stages of wear. Initially it serves as a shearing tooth in contrast to the crushing function of the deciduous and permanent premolars. By the time the second and third molars are fully erupted and functioning the anterior molar is too worn to provide a similarly efficient shearing action, but

performs as part of the anterior crushing battery. The width of the anterior molar, its attitude in the alveolus, and its alignment in the toothrow all suggest that it functions primarily as a posterior premolar rather than an anterior molar. The shearing action of this tooth is of limited duration but is sufficient to cope during the transition between deciduous and premanent dentition. The bifunctional nature of the toothrow may well explain why deinotheres failed to develop horizontal tooth replacement as in the elephantoids.

Watson (1946, p. 28) observed that the anterior molars of the Oligocene *Phiomia* and *Paleomastodon* were more fully trilophodont than the last molar and inferred that in primitive elephantoids the addition of the extra loph was initiated on the anterior molar. In later elephantoids, of course, horizontal tooth replacement is developed and the posterior teeth become more complex than the anterior. In this respect deinothere teeth are topographically more primitive than the earliest recorded elephantoid proboscideans. Harris (1969) inferred that whereas the tritolophid of M_1 of *Prodeinotherium* probably developed via hypertrophy of the posterior cingulum, the tritoloph of M^1 arose from the postmetaloph ornamentation. The development of three lophs on the anterior molar would seem to have rendered the bifunctional action of the cheek teeth of deinotheres sufficiently successful not to warrant further major modification of the basic crown morphology. The postmetaloph ornamentation of M^{2-3} was subsequently lost. There was a tendency in the upper premolars of *Deinotherium* for development of subsidiary styles, which may have aided the crushing function of these teeth. The presence of similar postmetaloph ornamentation on the upper molars of *Barytherium* supports the relationship of this genus to the deinotheres, although the unilophed upper premolars of the barytheres indicates a different specialization of the masticatory function of these teeth.

The retention of brachydont teeth of similar morphological pattern throughout the known history of the deinotheres suggests a successful adaptation to a diet of clean and relatively soft vegetation.

That deinotheres were terrestrial mammals is shown both by the skeleton, which is fully adapted to bear the large weight of the animal, and by the well-developed lacrimal canals of the orbit (lacrimal canals tending to be lost or reduced in aquatic mammals). One major evolutionary change exhibited by the postcranial skeleton is increase in size. Others apparently involve locomotory adaptations. Dietrich (1916) concluded that the anterior limb of *D. giganteum* was slender and long. Ştefănescu (1907) and

Tobien (1962) pointed out that the carpus was slender and the metapodials elongate. The shorter, relatively broader limb of the early Miocene deinothere from Kotyhaza, Hungary, suggested to Ehik (1930) that the creation of a separate genus *Prodeinotherium* was warranted.

The earliest specimens of *Prodeinotherium* known had already achieved elephantine proportions and graviportal adaptations for supporting the large bulk of the animal. The maneuverability of the head, simple and brachydont cheek teeth, and small proboscis suggest that the early deinotheres may have inhabited jungle and forest habitats and the densely vegetated edges of rivers and swamps. Small and downturned tusks would be more advantageous than large and anteriorly oriented tusks in densely vegetated regions. The traditional semi-amphibious interpretation of the mode of life of deinotheres need not be entirely disregarded. Yet there is no reason to doubt that they also exploited, and perhaps were equally at home in, fully terrestrial habitats.

That *Prodeinotherium* and *Deinotherium* coexisted for some time in Europe suggests that *Deinotherium* enjoyed a different mode of life. The proboscis of *Deinotherium* was probably larger than that of *Prodeinotherium* and may have played a more important, and perhaps different, role in the feeding process. The larger size and apparently more cursorial adaptations of *Deinotherium* in Europe suggest that it may have frequented more open country than did *Prodeinotherium*. Indeed, Tobien (1962) interpreted the foot structure of *D. giganteum* to represent adaptation for swift and enduring migration on savanna and steppelike grasslands. *Deinotherium* and *Prodeinotherium* have not yet been found associated at sites in Africa and available evidence indicates that *D. bozasi* evolved directly and *in situ* from *P. hobleyi*. Once appropriate specimens become available for study it will be interesting to see whether *D. bozasi* underwent similar postcranial changes to those of *D. giganteum*.

Relationships

The discovery of the first few deinothere teeth resulted in a number of diverse theories about the relationship of this group to the remainder of the class mammalia. Cuvier (1799) initially identified *Deinotherium* as a rhinoceros but later (1800) referred to it as a giant tapir. Kennedy (1785) regarded it as a giant mammoth; de Blainville (1837) as a sirenian; and Kaup successively as a hippopotamus (1829), an intermediate between mastodons and sloths (1833), and a herbivorous cetacean (1837). Strauss (1837)

and Dumeril (in de Blainville 1837) considered it between *Rhinoceros* and *Elasmotherium,* whereas Owen (1840) believed it to lie between the tapirs and the proboscideans. However, the elephant-like appearance of the first partial skeleton, discovered in 1853 after the collapse of an embankment on the Prague-Brunn railway (Osborn 1936, p. 91), resulted in the universal acceptance of deinotheres as a distinct family of the order Proboscidea, despite a later report by Solario (1864) of *Deinotherium* "epipubic" bones from the Haute-Garonne in France and assertions by Pictet (1853) and others that large limb bones associated with deinothere teeth in fact belonged to contemporary gomphotheres.

It seems likely that the deinotheres originated in Africa during the Paleogene, but unfortunately African Paleogene faunas are relatively uncommon. The only reasonably close relatives yet known from the African Paleogene would appear to be *Moeritherium, Barytherium,* and the early elephantoids *Phiomia* and *Paleomastodon*. Dental and cranial features seem to suggest that if deinotheres evolved from the Elephantoidea, they did so prior to the appearance of the earliest known African elephantoids, in which event their absence from the Fayum and Dor el Talha faunas is puzzling.

Given the present paucity of African Paleogene mammals, *Moeritherium* seems an ideal structural ancestor for the deinotheres on the basis of dental morphology. There is, of course, a basic difference in the number and development of the pre–cheek teeth, *Moeritherium* retaining all upper incisors and canines plus two pairs of inferior incisors (Andrews 1906). Moreover, *Moeritherium* possesses three permanent premolars. Deciduous second premolars are, however, present in all immature deinotheres and atavistic permanent second premolars have also been recorded (Graf 1957). The P^2 of *Deinotherium* does in fact bear a striking resemblance to that of *Moeritherium*. The remaining premolars and molars of *Moeritherium* are simpler than those of the deinotheres but could conceivably have evolved into the latter. Dentition apart, the skull of *Moeritherium* bears some resemblance to that of *Prodeinotherium*. In both forms the orbits are lower and situated considerably farther forward than in elephantoids and the facial region is proportionately smaller. Selected increase in size of cranial elements plus the development of diplöe could effect the transition relatively easily and the actual change in size from *Moeritherium* to *Prodeinotherium* is not much greater than that from *Phiomia* to *Gomphotherium angustidens*. The cranial and dental morphology of *Moeritherium* and *Prodeinotherium* is probably sufficiently similar to warrant their being placed in the same suborder, although on available evidence direct lineal descent is unlikely.

As with moeritheres, the teeth of *Barytherium* present the most convincing case for close affinity with *Prodeinotherium*. Like *Moeritherium,* the first molar of *Barytherium* lacks a third loph and barytheres also possess two pairs of procumbent lower incisors plus two pairs of upper incisors. The lower premolars, three in number, are superficially similar to those of deinotheres, although simpler in morphology and closer to those of the moeritheres. Except for the lack of the tritolophid on M_1, the inferior molars are virtually identical to those of *Prodeinotherium*. Some specimens of M^{2-3} of *Barytherium* are indistinguishable from those of *Prodeinotherium,* although in general the postmetaloph ornamentation of the former is simpler. Moreover it appears that the orbit of *Barytherium,* like that of *Prodeinotherium,* was placed farther forward than in *Gomphotherium*. However, the premolar toothrow of *Barytherium* is slightly offset in the alveoli from the molar toothrow, whereas in the deinotheres the anterior molar is aligned with the premolars. More important, the upper premolars of *Barytherium* possess single lophs only in contrast to the bilophodont premolars of deinotheres. It is probable, therefore, that the cheek teeth of barytheres performed a simple shearing action in contrast to the separate crushing versus shearing functions indicated by different parts of the deinothere toothrow. It would appear that the specialized shearing action of the cheek teeth of *Barytherium,* among other reasons, precludes direct lineal descent to *Prodeinotherium*. It is likely, however, that barytheres and deinotheres are closely related and derived from the same stem. The absence of deinotheres in the North African Paleogene may be a collecting anomaly but may also indicate that the Deinotheriidae originally evolved in some other part of Africa.

Regardless of any proposed phyletic relationship between moeritheres, barytheres, and deinotheres it is obvious that the teeth, which must currently form the main basis of comparison, exhibit close morphological similarities. All three groups were classified by Simpson (1945) as separate suborders of the Proboscidea. On dental evidence there can be no doubt that these three suborders are closely related and are fundamentally separate from the suborder Elephantoidea. Future discoveries in the Paleogene of Africa may well indicate that the deinotheres, moeritheres, and barytheres should be withdrawn from the order Proboscidea to form a distinct group of separate ordinal rank. The idea that *Barytherium*

and *Moeritherium* belong outside the order is not new. Ameghino (1902) considered that *Moeritherium* and *Barytherium* formed side branches and were not on the main line of descent of the elephantoids. Deraniyagala (1955, p. 15) removed the moeritheres and barytheres from the Proboscidea, considering it advisable to retain in that order only those elephantlike mammals that undoubtedly possessed prosboces. Maglio (1973) extended this exclusion to the deinotheres as well.

Family Barytheriidae

Specimens that may be attributed to the Family Barytheriidae are known from two localities only: the Fayum of Egypt and the Dor el Talha (of Savage 1969) or Jebel Coquin (of Arambourg and Magnier 1961) in Libya. The age of the sediments yielding barytheres has been established at both sites as late Eocene (Simons 1968; Savage 1969).

Genus *Barytherium* Andrews 1901

DIAGNOSIS. Dental formula $\frac{2.0.3.3}{2.0.3.3}$. Upper incisors vertical; outer tooth (I^2) larger than inner tooth and chisel edged. Lower incisors horizontal with inner tooth larger than outer tooth. Diastema between incisors and premolars. P_2 triangular; P_3 with two anterior cusps and posterior loph; P_4 bilophodont; M_{1-3} bilophodont, M_3 with talonid. M^{1-3} rectangular and bilophodont. Skull with retracted external nares. Palate flat with palatonarial border opposite M^3. Mandible massive with elongate symphysis extending backward to M_1. Postcranial skeleton of elephantine proportions.

TYPE SPECIES. *Barytherium grave* Andrews 1901b; late Eocene; Qasr el Sagha Beds of Fayum, Egypt.

Barytherium grave Andrews 1901

REFERENCES

1901a *Barytherium grave* Andrews: 407–408.
1901b *Barytherium grave* Andrews: 577.
1902 *Barytherium grave* Andrews; Andrews: 528.
1906 *Barytherium grave* Andrews; Andrews: 172–177.
1945 *Barytherium grave* Andrews; Simpson: 249–250.
1961 *Barytherium grave* Andrews; Arambourg and Magnier: 1182.
1969 cf *Barytherium grave* Andrews; Savage: 170.

DIAGNOSIS. Large species of *Barytherium,* diagnosis as for genus.

HOLOTYPE SPECIMEN. Cairo Museum no. C. 10012—associated maxilla and mandible with teeth, portions of scapulae, fragment of left humerus, left radius, proximal ulna.

TYPE LOCALITY. Qasr el Sagha Beds, Fayum; late Eocene.

OTHER MATERIAL. The proximal end of a right femur (Cairo Museum no. M.9125; Andrews 1906, p. 177).

Material collected from the Dor el Talha, Libya, and referred to by Arambourg and Magnier (1961) is currently in the collections of the Museum National d'Histoire Naturelle, Paris, and includes a left ascending ramus with roots of M_{1-3}, partial right ramus with M_{2-3}, partial right ramus with M_{1-2} and roots of M_3, partial left maxilla with M^{2-3}, partial right maxilla with roots of P^3-M^2, symphysis fragment with roots of right I_2-P_3 and P_4, anterior symphysis with I_{2-3}, incomplete mandible with roots of I_2, P_2-M_1, and M_2, several isolated teeth, partial scapulae, three ulnae, thoracic and caudal vertebrae, astragali, phalanges, and rib fragments.

Material collected from the Dor el Talha and referred to by Savage (1969) includes a skull, mandibles, isolated teeth, a partial skeleton, and many postcranial elements. This material is currently at the Department of Geology, University of Bristol.

Unpublished material collected by Yale University parties includes a humerus, a pelvis, 19 vertebrae, a sacrum, several ribs, and a few foot bones (pers. comm. E. L. Simons). This material is now at the Peabody Museum, Yale University.

Barytherium sp.

Savage (1969) recorded the occurrence of a second, smaller species of *Barytherium* from the Dor el Talha, Libya. It is apparently less abundant than *B. grave* at that locality but no further information about this taxon has yet appeared.

The only detailed description of specimens of *Barytherium* appears in Andrews's (1906) monograph on the Fayum fauna.

Relationships

Owing mainly to the lack of material, the relationship of *Barytherium* to other mammals has not yet been settled. Andrews (1904) originally placed the barytheres in the Barypoda (subsequently changed to Barytheria) and inferred that this group constituted a subdivision of the Amblypoda of similar rank to the Dinocerata. Simpson (1945) later treated it as a suborder of the Proboscidea. Deraniyagala (1955) thought it advisable to retain in the Proboscidea only those mammals that undoubtedly possessed prosboces, thus removing from Simpson's concept of the Proboscidea the suborders Moeritherioidea and Barytherioidea.

Any attempt to define the affinities of *Barytherium*, though now long overdue, would be premature in light of the relatively complete but as yet unpublished specimens from Libya. However if, as seems likely, the barytheres were entirely African in origin and distribution, then their dentition suggests relationships with both the moeritheres and deinotheres. By virtue of the possession of only two pairs of upper and lower incisors, barytheres are somewhat more specialized than the moeritheres; yet this feature, plus the presence of a second premolar, renders them more primitive than deinotheres. The similarity of *Barytherium* molars to those of deinotheres is remarkable, but the evolution of unilophed upper premolars and the absence of a third loph on the anterior molar suggests different masticatory specializations and probably precludes direct lineal relationship. Detailed investigation of the cranial and postcranial morphology of these enigmatic animals is now possible and it will be interesting to see whether sufficient information is now available to define positively the relationships of this group.

References

Abel, O. 1922. *Lebensbilder aus der Tierwelt der Vorzeit.* Jena.

Ameghino, F. 1902. Linae filogenética de los proboscideos. *Ann. Mus. Nac. Buenos Aires* (3) 1:19–42.

Andrews, C. W. 1901a. Über das Vorkommen von Proboscidiern in untertertiären Ablagerungen Aegyptens. *Tageblatt des V Internationalen Zoologischen-Kongresses, Berlin* No. 6, pp. 4–5.

———. 1901b. A new name for an ungulate. *Nature* 64:577.

———. 1902. Über das Vorkommen von Proboscidiern in untertertiären Ablagerungen Aegyptens. *Verh. V. Int. Zool. Congr., Berlin* p. 528.

———. 1904. Note on the Barypoda, a new Order of ungulate mammals. *Geol. Mag.* (5) 1:481–482.

———. 1906. *A descriptive catalogue of the Tertiary Vertebrata of the Fayûm, Egypt.* London: British Museum (Nat. Hist.).

———. 1911. On a new species of *Dinotherium* (*Dinotherium hobleyi*) from British East Africa. *Proc. Zool. Soc. Lond.* 1911:943–945.

———. 1914. On the Lower Miocene vertebrates from British East Africa collected by Dr Felix Oswald. *Quart. J. Geol. Soc. Lond.* 70:163–186.

Arambourg, C. 1934. Le *Dinotherium* des gisements de l'Omo, *C. R. Soc. Géol. Fr.* 1934:86–87.

———. 1947. *Mission scientifique de l'Omo (1932–1933).* Tome 1, Geologie Anthropologie. Fasc. III, Mus. Natl. Hist. Nat. Paris, pp. 231–562.

Arambourg, C., J. Chavaillon, and Y. Coppens. 1969. Résultats de la nouvelle mission de l'Omo. *C. R. Hebd. Séance. Acad. Sci.* ser. D. 268:759–762.

Arambourg, C., and P. Magnier. 1961. Gisements de vertébrés dans le bassin Tertiaire de Syrte (Libye). *C. R. Hebd. Séance. Acad. Sci.* 252:1181–1183.

Bergounioux, F. M., and F. Crouzel. 1962. Les Deinotherides d'Europe. *Ann. Paléont.*, Paris 48:13–56.

Bishop, W. W. 1967. The later Tertiary in East Africa—volcanics, sediments and faunal inventory. In W. W. Bishop and J. D. Clark, eds. *Background to evolution in Africa.* Chicago: University of Chicago Press, pp. 31–56.

Bishop, W. W., G. R. Chapman, A. Hill, and J. A. Miller 1971. Succession of Cainozoic vertebrate assemblages from the northern Kenya Rift Valley. *Nature* 233:389–394.

Bishop, W. W., J. A. Miller, and F. Fitch 1969. New potassium-argon age determinations relevant to the Miocene fossil mammal sequence in East Africa. *Am. J. Sci.* 267:669–699.

Blainville, H. M. D. de 1837. Note sur la tête de *Deinotherium gigateum* actuellement à Paris. *C. R. Hebd. Séanc. Acad. Sci.* 4(12):421–426.

Brives, A. 1919. Sur la découverte d'une dent de *Deinotherium* dans la sablière du Djebel Kouif prees Tébessa. *Bull. Soc. Hist. Nat. Afr. Nord* 10:90–93.

Buckland, W. 1835. Über den Bau und die mechanische Kraft des Unterkiefers des *Dinotherium. Neues Jb. Min.* 1835:516–518.

Carney, J., A. Hill, A. Miller, and A. Walker 1971. Late australopithicine from Baringo District, Kenya. *Nature* 230:509–514.

Cuvier, G. 1799. Extrait d'un mémoire sur les ossements fossiles de quadrupèdes. *Bull. Soc. Philom., Paris* no. 18, Fructidor 1(2):137–139.

———. 1800. Sur les tapirs fossiles de France. *Bull. Soc. Philom., Paris* no. 34, Nivose 2:73–74.

Depéret, C. 1887. Recherches sur la succession des faunes de vertébrés miocènes de la vallée du Rhône. *Arch. Mus. Hist. Nat. Lyon* 4:45–313.

Deraniyagala, P. E. P. 1955. *Some extinct elephants, their relatives and the two living species.* Colombo: Colombo Mus., Ceylon 1955, pp. 1–161.

Dietrich, W. O. 1916. Ueber der Hand und den Fuss von *Deinotherium. Zeits. Deutsch. Geol.* 68:44–53.

———. 1942. Ältesquartäre Säugetiere aus der südlichen Serengeti, Deutsch-ÖstAfrika. *Paläontogr.* 94(A):43–133.

Ehik, J. 1930. *Prodinotherium hungaricum* n.g., n. sp. *Geol. Hung. ser. Palaeont.* 6:1–24.

Falconer, H. 1845. Description of some fossil remains of *Dinotherium*, giraffe, and other mammalia, from Perim Island, Gulf of Cambay, western coast of India. *Quart. J. Geol. Soc. Lond.* 1:356–372.

———. 1868. Description of some fossil remains of *Dinotherium, Giraffa, Bramatherium*, and other mammalia, from Perim Island, Gulf of Cambay, western coast of India. *Palaeont. Mem.* I:391–411.

Graf, I. E. 1957. Die Prinzipien der Artbestimmung bei *Deinotherium. Paläontogr.* Abt. A 180:131–185.

Harris, J. M. 1969. *Prodeinotherium* from Gebel Zelten, Libya. Unpublished Ph.D. thesis, University of Bristol.

———. 1973. *Prodeinotherium* from Gebel Zelten, Libya. *Bull. Brit. Mus. Nat. Hist.* (Geol.) 23(3):211–274.

———. 1976. Evolution of feeding mechanisms in the family Deinotheriidae (Mammalia; Proboscidea). *Zool. J. Linn. Soc.* 56:331–362.

———. 1976. Cranial and dental remains of *Deinotherium bozasi* from East Rudolf, Kenya. *J. Zool. Lond.* 154:57–75.

Harris, J. M., and R. Watkins. 1974. New early Miocene vertebrate locality near Lake Rudolf, Kenya. *Nature* 252:576–577.

Haug, É. 1911. *Traité de Géologie.* II, Paris. pp. 539–2024.

Hooijer, D. A. 1963. Miocene mammalia of Congo. *Ann. Mus. Roy. de l'Afrique Cent., Tervuren, Belg., Sci. Géol.* 46:1–79.

Jaquemin, E. 1837. Mémoire sur les pachydermes fossiles connus jusqu'a ce jour, et descriptions de nouveau genre *Dinotherium* de M. Kaup. *Mag. Zoologie* (Guerin-Meneville) Sept Année Cl.1:1–30.

Joleaud, L. 1928. Éléphantes et dinothériums pliocènes de l'Éthiopie: contribution à l'étude paléogéographique des proboscidiens africains. *XIV Intern. Geol. Congr. Madrid*, III:1001–1007.

Kaup, J. J., 1829. *Deinotherium giganteum. Isis* (XXII) 4:401–404.

———. 1833. Der Krallen-Phalanx von Eppelsheim, nach welchem Hr. von Cuvier seinen Reisen-Pangolin, *Manis gigantea*, aufstellte, gehört zu *Dinotherium. Neues Jb. Min.* pp. 172–176.

———. 1837. Sur la place que doit occuper le *Dinotherium* dans l'échelle animale. *C. R. Hebd. Séanc. Acad. Sci.* 4:527–529.

Kennedy, I. 1785. Abhandlung von einigen in Baiern gefunden Beinen. Neue Philosophiche. *Abh. Bayer. Akad. Wiss., München* 4:1–48.

Kent, P. E. 1942. The Pleistocene beds of Kanam and Kanjera, Kavirondo, Kenya. *Geol. Mag.* 79:117–132.

Laskarev, V. 1944. Ueber die Dinotherien reste aus Serbien. *Neues Jb. Min. Monatsch* 1944 (2/3) abt. B:67–77.

Leakey, L. S. B. 1965. *Olduvai Gorge, 1951–1961, fauna and background.* Cambridge: Cambridge University Press.

Leakey, M. D. 1971. *Olduvai Gorge—Excavations in Beds I & II, 1960–1963.* Cambridge: Cambridge University Press.

Leakey, R. E. 1970. Fauna and artefacts from a new Plio-Pleistocene locality near Lake Rudolf, Kenya. *Nature* 226:223–224.

MacInnes, D. G. 1942. Miocene and post-Miocene Proboscidea from East Africa. *Trans. Zool. Soc. Lond.* 25(2):33–106.

Madden, C. 1972a. Miocene mammals, stratigraphy and environment of Moruorot Hill, Kenya. *Paleobios.* no. 14:1–12.

———. 1972b. A Miocene mammalian fauna from Lothidok Hill, Kenya. M. A. thesis, University of California, Berkeley.

Maglio, V. J. 1971. Vertebrate faunas from the Kubi Algi, Koobi Fora and Ileret areas, East Rudolf, Kenya. *Nature* 231:248–249.

_____. 1972. Vertebrate faunas and chronology of hominid-bearing sediments east of Lake Rudolf, Kenya. *Nature* 239:379–384.

_____. 1973. Origin and evolution of the Elephantidae. *Trans. Am. Phil. Soc.* 63(3):1–149.

Meyer, H. von 1831. [*Dinotherium bavaricum.*] *Neues Jb. Min. Geol. Pal.* 2:296–297.

Osborn, H. F. 1910. *The age of mammals in Europe, Asia and North America.* New York: MacMillan.

_____. 1936. *Proboscidea*, vol. 1. New York: American Museum of Natural History.

Owen, R. 1840. *Odontography; or a treatise on the comparative anatomy of the teeth; their physical relations, mode of development, and microscopic structure, in the vertebrate animals,* vol. 1. London.

Ozansoy, F. 1957. Faune de mammifères du Tertaire de Turquie et leurs revisions stratigraphiques. *Bull. Min. Res. Inst. Turkey* no. 49:29–48.

Patterson, B., A. K. Behrensmeyer, and W. D. Sill. 1970. Geology and fauna of a new Pliocene locality in northwestern Kenya. *Nature* 226:918–921.

Petrocchi, C. 1951. Paleontologia di Sahabi. *Rendiconti Acad. Naz. dei. XL,* ser. 4, 3:1–115.

Pictet, F. J. 1853. *Traité de Paléontologie,* vol. 1, 2nd ed. Geneva.

Romer, A. S. 1966. *Vertebrate Paleontology,* 3rd ed. Chicago: University of Chicago Press.

Sahni, M. R., and C. Tripathi. 1957. A new classification of the Indian deinotheres and description of *D. orlovii* sp. nov. *Mem. Geol. Surv. India, Paleont. Indica* 33(4):1–33.

Savage, R. J. G. 1967. Early Miocene mammal faunas of the Tethyan region. *Syst. Assoc. Publ.* no. 7:247–282.

_____. 1968. Near shore marine and continental sediments of the Sirte Basin, Libya. *Proc. Geol. Soc. Lond.* no. 1648:81–90.

_____. 1969. Early Tertiary mammal locality in southern Libya. *Proc. Geol. Soc. Lond.* no. 1657:167–171.

Savage, R. J. G., and E. Tchernov, 1968. Miocene mammals of Israel. *Proc. Geol. Soc. Lond.* no. 1648:98–101.

Savage, R. J. G., and M. E. White, 1965. Two mammal faunas from the early Tertiary of central Libya. *Proc. Geol. Soc. Lond.* no. 1623:89–91.

Simons, E. L. 1968. Early Cenozoic mammalian faunas, Fayum Province, Egypt, part 1, African Oligocene mammals: Introduction, history of study and faunal succession. *Bull. Peabody Mus. Nat. Hist. Yale Univ.* 28:1–21.

Simpson, G. G. 1945. Principles of classification and a classification of mammals. *Bull. Am. Mus. Nat. Hist.* 85:1–350.

Solario, P. J. M. S. 1864. Letter in *Les Mondes* of Sept. 29, 1864.

Ştefănescu, G. 1907. Quelques mots sur le *Dinotherium gigantissimum. C. R. Cong. Géol. Intern.,* Mexico, 10:427–430.

Strauss, D. 1837. Considérations sur le genre de vie du *Dinotherium* et sur la place qu'il convient de lui assigner dans une distribution naturelle de mammifères. *C. R. Hebd. Séance. Acad. Sci.* 4:529–532.

Swistun, W. I. 1974. [*Deinotherium* from the Ukraine.] *Bull. Ukraine Acad. Sci.* 1974:1–52. (In Russian.)

Taieb, M., D. C. Johanson, Y. Coppens, and J. L. Aronson. 1976. Geological and palaeontological background of Hadar hominid site, Afar, Ethiopia. *Nature* 260:289–293.

Tobien, H. 1962. Über carpus und tarus von *Deinotherium giganteum* Kaup. *Paläeont. Zeitschr.* Nov. 1962:231–238.

Walker, A. 1969. Lower Miocene fossils from Mount Elgon, Uganda. *Nature* 223:591–593.

Watson, D. M. S. 1946. Evolution of the Proboscidea. *Biol. Rev.* 21:15–29.

Weinsheimer, O. 1883. Ueber *Dinotherium giganteum* Kaup. *Geol. Pal. Abh.* 1:207–282.

Vialli, V. 1966. Sul rinvenimento di Dinoterio nelle ligniti di Adi Ugri (Eritrea). *Giorn. Geol.* ser. 2, 33:447–458.

16

Moeritherioidea

Yves Coppens and Michel Beden

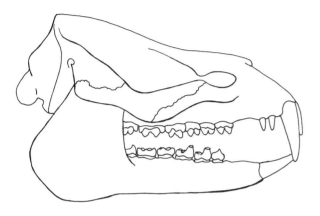

Discovered by Andrews and Beadnell in 1901 in marine sediments of the Fayum (Egypt), the moeritheres were first included among the proboscideans by Andrews in 1906. He distinguished them from all other known members of the order by placing them in a distinct family, the Moeritheriidae. The discovery of new forms and of more complete remains led Osborn (1921) and Matsumoto (1924) to raise them to subordinal rank among the proboscideans, the Suborder Moeritherioidea. Since then their attribution to the Order Proboscidea has been questioned and Deraniyagala (1955) proposed a separate order for them. Other authors, such as Tobien (1971), also exclude them from the Proboscidea and classify them among the sirenians.

More recent studies have suggested that *Moeritherium* shows striking similarities to the Desmostylia, a group of large amphibious mammals known from Miocene and Pliocene deposits of the eastern Tethys and Pacific (D. Domning, pers. comm.). Clearly much remains to be done before the true affinities of this enigmatic group can be elucidated, and for the present we may simply include it as Moeritheriidae: Proboscidea, Sirenia, or Desmostylia *incertae sedis*.

Systematics

Family Moeritheriidae Andrews 1906

The family has only one genus and the distinctive characteristics of the family are the same as for the genus.

Genus *Moeritherium* Andrews 1901

DIAGNOSIS. These animals are of relatively small height, from 0.50 m to 0.70 m to the withers, with a long body, sustained by short rather thin limbs.

They are characterized by a skull with a short facial area (about 1/3 of the skull length) and a reduced facial angle. They have a long sagittal crest, orbits situated above the anterior premolars and not bounded at the back by a postorbital process of the frontal. The cerebral cavity is small, narrow, and low (weak pneumatization of bones). The zygomatic arches are large. The external auditory meatuses open high on the skull. The mandible is massive with a short symphysis and a very broad ascending ramus that slopes forward.

The permanent teeth are almost complete in number, the deciduous teeth remain poorly known. The dental formula is $I_2^3, C_0^1, P_3^3, M_3^3$.

All the permanent teeth are in function simultaneously and there are strong diastemata in the maxillary row between C and P² and in the mandibular row between I₂ and P₂. The first incisors are very reduced, but remain more or less functional; I2 is very well developed and canine-shaped; I3 is preserved but reduced and is present only in the maxillary dentition. Upper C is very reduced, lower C is lost. P2 has two rootlets; P3 and P4 have three tubercles and sometimes a small fourth tubercle. The molars have four tubercles, are bunodont and have a posterior talon which on M₂ and M₃ are well developed into a third transverse crest.

The limbs are of a primitive type and differ in some morphological characters from the true proboscideans (for example, no epicondylar crest on the humerus, rectilinear femoral diaphysis.)

STRATIGRAPHIC RANGE. Only in Africa; all the discoveries of the genus have been made north of the Equator. The distribution is rather large: Egypt —Fayum sites; Libya—Dor el Tahla; Mali—Goa; Senegal—Kaolack. Five species of moritheres have been described from Africa; all occur in the Fayum formations. A sixth species has been described from India but it appears to be a mastodon, *Trilophodon pandionis* (Osborn 1936).

DESCRIBED SPECIES. *M. lyonsi* Andrews 1901; *M. gracile* Andrews 1902; *M. trigodon* Andrews 1904; *M. andrewsi* Schlosser 1911; *M. pandionis* Pilgrim 1912; *M. ancestrale* Petronievics 1923

Of these species, three come from the Qasr el Sagha Formation (late Eocene, probably terminal Bartonian according to Simons 1968:14). These species are *M. lyonsi, M. gracile,* and *M. ancestrale.* Two others occur in the Jebel el Qatrani Formation (early Oligocene).

Recently, Tobien (1971), surveying all the material collected and revising the diagnoses of the different species, has significantly clarified the systematics of *Moeritherium.* Previously, *M. lyonsi* and *M. gracile* respectively were considered small and large Eocene species; likewise for *M. andrewsi* and *M. trigodon* in the Oligocene. The distinction between the species of each pair was without foundation because the evoked criteria did not go beyond the range of individual variation both for morphological and for the biometrical features. The two groups are distinguished from each other essentially by the proportions of premolars and molars; those of the Oligocene form are proportionately narrower than those of the Eocene taxon.

The original diagnosis for *M. ancestrale,* described by Petronievics on a cranial fragment, is based only on few cranial characters that, according to Tobien, lie within the limits of variation observed in *M.*

lyonsi. Consequently the number of species of *Moeritherium* is reduced to two.

Moeritherium lyonsi Andrews 1901

Moeritherium lyonsi Andrews (1901a: 4; 1901b: 528; 1901c: 400–409, 1 fig.; 1906: 120, pl. VIII–X), Matsumoto (1923: 124, fig. 27), Osborn (1936: 72, fig. 18, 43, 44; 1942: 1426), Lehman (1950: 127–140), Vaufrey (1958: 203, fig. 18). Thenius (1959: 255; 1969: 597, fig. 689), Beljaeva (1962: 355), Arambourg (1963: 59), Tobien (1971; 143–152).

Moeritherium gracile Andrews (1902: 291–295; 1906: 127–128, fig. 1–2, pl. XVIII), Matsumoto (1922: 5; 1923: 125), Osborn (1936: 73, fig. 29), Lehman (1950: 127–140), Vaufrey (1958: 203), Thenius (1969: 597).

Moeritherium ancestrale Petronievics (1923: 55–61, fig. 1), Osborn (1936: 76, fig. 47; 1942: 1426), Thenius (1959: 255).

HOLOTYPE. A mandible associated with upper molars and a thoracic vertebra; Geological Museum of Cairo, CGM-10000. Pieces figured in Andrews 1901c, p. 404, fig. 2.

HORIZON. Stratigraphic level of holotype: Fayum Depression, Qasr el Sagha Formation (late Eocene), Egypt. Other specimens from Gao, Mali.

ORIGINAL DIAGNOSIS. (Andrews 1906:120; expanded in Osborn 1936:72); Lower premolars very short; lower molars very long. All the lower cheek teeth very broad. P₂ triangular, its widest part corresponding to the posterior lobe.

EMENDED DIAGNOSIS. (Tobien 1971:144–151); molars with four tubercles without accessory conules, simple, rather broad. Late Eocene species.

Moeritherium trigodon Andrews 1904

Moeritherium trigodon Andrews (1904: 109–115; 1906: 125, fig. 5, pl. IX), Matsumoto (1923: 134), Osborn (1936: 57–74, fig. 30, 46; 1942: 1426), Lehman (1950: 127–140), Vaufrey (1958: 205, fig. 18), Tobien (1971: 143–152).

Moeritherium andrewsi Schlosser (1911: 131, fig. 9), Osborn (1936: 61–74, fig. 36, 37, 46; 1942: 1426), Lehman (1950: 127–140), Vaufrey (1958: 205, fig. 18), Thenius (1959: 257; 1969: 597, fig. 687–688), Arambourg (1963: 59).

Moeritherium lyonsi Andrews, Osborn (1936: 57–74, fig. 30, 46, *partim*).

HOLOTYPE. Part of a right mandible with molars; British Museum (Natural History), M-8499.

HORIZON. Stratigraphic level of holotype: Fayum Depression, Jebel el Qatrani Formation (early Oligocene), Egypt.

ORIGINAL DIAGNOSIS.   (Andrews   1904:118,

1906:128; expanded in Osborn 1936:74); the distinctive criteria concern the skull, on the one hand, and the M₃'s on the other, and thus cannot easily be compared with the diagnostic features for *M. lyonsi.* Lower molars narrow, M₃ with talon of one large tubercle; lower premolars not very short compared to molars; P₂ fusiform in shape, wider in middle.

EMENDED DIAGNOSIS. (Tobien 1971:146, 152); molars with four tubercles with few conules developing, relatively narrower than in *M. lyonsi.* Lower Oligocene species.

Affinities of the Moeritheriidae

For a long time the place of the moeritheres among the mammals has been discussed. They are sometimes placed among the Proboscidea where they are considered to be the most primitive known group and they are sometimes placed among the Sirenia.

One fact appears to be established. These two species seem to have lived in a very restricted habitat. A complete skeleton discovered by Simons in the layers of the Qasr el Sagha Formation (Fayum) shows that the moeritheres were at least partially amphibious (very long body sustained by short thin limbs, skull with aquatic adaptations.) A partially aquatic habit is confirmed by the fact that all the collected remains of moeritheres come from sedimentary marine deltaic deposits. The question arises whether to consider the moeritheres as a side issue of the Sirenia even though there are also true and highly specialized sirenians in the same levels of the Fayum. Or should we view them as a very specialized subungulate group with an amphibious way of life, perhaps of desmostylid affinities; or a group that very early may have diverged from the emerging proboscidean stock. Or, as suggested in chapter 15, should we consider them as a specialized aquatic relative of the barythere-deinothere group?

We do not feel it is possible to draw any firm conclusions at this time, but it is clear that moeritheriid ancestry must be sought among the diverse subungulate groups of the African Eocene. After a brief existence, the group became extinct during the Oligocene, leaving no descendants.

References

Andrews, C. W. 1901a. Fossil vertebrates from Egypt. *Zoologist,* ser. 4, 5:318–319.
_____. 1901b. Tageblatt des V Internationalen Zoologischen-Congresses, Berlin, 6:6.
_____. 1901c. Preliminary note on some recently discovered extinct vertebrates from Egypt, I. *Geol. Mag.,* ser. 4, 8:400–409.
_____. 1902. Preliminary note on some recently discovered extinct vertebrates from Egypt, III. *Geol. Mag.,* ser. 4, 9:291–295.
_____. 1904. Further notes on the mammals of the Eocene of Egypt. *Geol. Mag.,* ser. 5, 1:109–115.
_____. 1906. *A descriptive catalogue of the Tertiary Vertebrata of the Fayum, Egypt.* London: British Museum. 342 pp.
Arambourg, C. 1963. Continental vertebrate faunas of the Tertiary of North Africa. In F. C. Howell and F. Bourlière (eds.), *African ecology and human evolution,* Chicago: Aldine, pp. 55–56.
Beljaeva, E. I. 1962. Moeritherioidea. In A. Orlov (ed.), *Fundamentals of paleontology,* vol. 13. Moskva: Acad. Nauk, S.S.S.R., 355 pp.
Lehman, U. 1950. Uber Mastodontenreste in der Bayerische Staatsammlung in München. *Paläont.* 94A (4/6):121–227.
Matsumoto, H. 1922. Revision of *Palaeomastodon* and *Moeritherium. Palaeomastodon intermedius* and *Phiomia osborni,* n.sp. *Am. Mus. Novit.* 51:1–6.
_____. 1923. A contribution to the knowledge of *Moeritherium. Bull. Am. Mus. Nat. Hist.* 48(4):97–140.
_____. 1924. A revision of *Palaeomastodon* dividing it into two genera, and with descriptions of two new species. *Bull. Am. Mus. Nat. Hist.* 50(1):1–58, 48 figs.
Osborn, H. F. 1921. The evolution, phylogeny and classification of the Proboscidea. *Am. Mus. Novit.* 1:1–15, 4 figs.
_____. 1936. *Proboscidea,* vol. 1. New York: Amer. Mus. Nat. Hist. Press, pp. 1–805, 680 figs., 11 pls.
_____. 1942. *Proboscidea,* vol. 2. New York: Amer. Mus. Nat. Hist. Press, pp. 806–1676, 563 figs., 19 pls.
Petronievics, B. 1923. Remarks upon the skulls of *Moeritherium* and *Palaeomastodon. Ann. Mag. Nat. Hist.* 12(9):55–61, pl. II.
Pilgrim, G. E. 1912. The vertebrate fauna of the Gaj series in the Bugti Hills and the Punjab. *Pal. Indica,* ser. 2, 4(2):1–83.
Schlosser, M. 1911. Beiträge zur Kenntnis des Oligozänen Landsäugetiere aus dem Fayum, Aegypten. *Beitr. Paläont. Geol. Öst.-Ung.* 24:51–117, tabs. 9–16.
Simons, E. L. 1968. Early Cenozoic mammalian faunas in Fayum Province, Egypt. Pt. I, African Oligocene mammals: introduction, history of study and faunal succession. *Bull. Peabody Mus. Nat. Hist.* 28:1–22.
Thenius, E. 1959. Tertiär Wirbeltierfaunen. In *Handbuch der Stratigraphischen Geologie,* vol. 3, pt. 2. Vienna: Fr. Lotzen.
_____. 1969. Stammesgeschichte der Säugetiere (einschliesslich der Hominiden). In *Handbuch der Zoologie,* Bd. 8, Lief 48, 2(1):369–722.
Tobien, H. 1971. *Moeritherium, Palaeomastodon, Phiomia* aus dem Paläogen Nordafrikas und die Abstammung des Mastodonten (Proboscidea, Mammalia). *Mitt. Geol. Inst. Techn. Univ. Hannover* 10:141–163.
Vaufrey, R. 1958. Proboscidea, étude systématique. In J. Piveteau (ed.), *Traité de Paléontologie,* t. 6, vol. 2. Paris: Masson ed.

17

Proboscidea

Yves Coppens, Vincent J. Maglio,
Cary T. Madden, Michel Beden

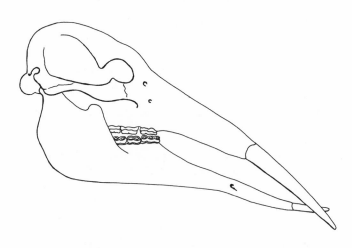

The Order Proboscidea today is a very restricted group, limited to two genera of a single species each. *Loxodonta africana,* the African elephant, is distributed throughout the African continent south of the Sahara. In historic times the species extended northward to the Mediterranean; it appears to have inhabited the forests of ancient Egypt, where it is depicted as a god on standards of the predynastic period (Budge 1969). By dynastic times the elephant seems to have been exterminated in Egypt, for it is not found in Egyptian mythology. Nevertheless, elephants were brought to the North African coast, where they played an important role in the Punic wars between Carthage and Rome.

The Asiatic species, *Elephas maximus,* occurs today throughout southern Asia from India, Ceylon, Burma, and the Malay States south to Sumatra and Borneo. Although the historical data are not clear, it may have been the Asiatic elephant that inhabited Mesopotamia as late as 800 B.C.

The living elephants inhabit a variety of terrain but generally prefer high forest, forest-savanna mosaic, and open woodland. They may penetrate deep into semidesert regions along gallery forests bordering major river systems. With continued deterioration of woodland, as in many parts of Africa, the elephant today is increasingly found in open savanna.

The diet of elephants consists mainly of leaves, fruits, buds, and bark in forests, but in the savannas they will take principally *Acacia* bushes and grasses (Buss 1961; Laws and Parker 1968). The efficiency of their highly developed masticatory apparatus, the hallmark of the family Elephantidae (see Maglio 1972), allows them to adapt to such wide varieties of feeding habits.

The enormous past diversity of the order is evidenced from the fossil record and this group formed a major part of Tertiary and Quaternary faunas of the world. The history of the Proboscidea was punctuated by a number of adaptive shifts that resulted in such varied types as the mastodonts, gomphotheres, stegodonts, and elephants. Several families were extremely successful in dispersing over broad geographic areas so that gomphotheres, mastodonts, and elephants as groups and several genera within each in particular were equally characteristic elements in African as well as in Holarctic Cenozoic faunas.

The semiaquatic moeritheres and the bizarre barytheres and deinotheres, although classically included within this order, are now often excluded and are therefore treated here in separate chapters.

For many years authors have agreed that Africa

was the center of evolution and dispersion for the Proboscidea. It is here that are found early members of groups that appear to be closely related to them, e.g., moeritheres, barytheres, sirenians, hyracoids. The most primitive true proboscideans of the genus *Palaeomastodon* have been recorded in the late Eocene of northern Africa and it is on this continent also that we can trace the origin of all families within the order, except for the stegodonts (figure 17.1). The earliest extra-African record is in the lat-

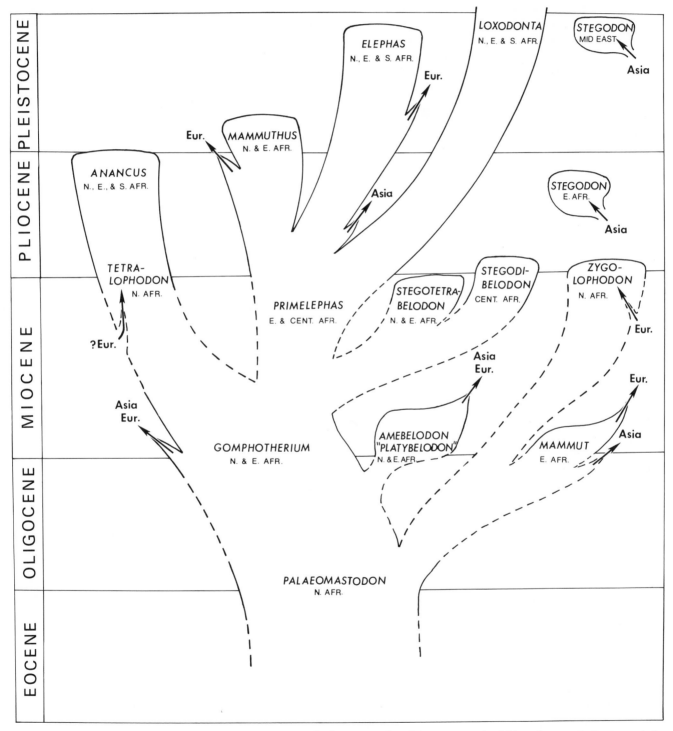

Figure 17.1 Suggested phyletic relationships among the known proboscidean groups in Africa. Arrows indicate periods of major faunal communication with other continents. Broken lines indicate periods where the African record is lacking.

est early Miocene, when gomphotheres and probably also mammutids expanded into Eurasia. There thus seems little reason on present evidence to doubt an African center for the Proboscidea.

Shortly after its origin the order diversified in Africa into two major adaptive groups, the gomphotheres and the mastodonts (= mammutids here). The Oligocene record of both is largely lacking, but they apparently underwent major radiation and dispersion so that by Miocene times both groups were represented by a diverse assemblage extending from Africa through Eurasia and into North America. During the later Cenozoic the Proboscidea experienced a rapid almost "explosive" radiation resulting in new adaptive types that were to dominate most large herbivore faunas of the world. But late in the Pleistocene there occurred in the order a striking and as yet unexplained decline, leaving only the two extant species as relicts of the group's former reign.

In the following pages we present in capsule form current knowledge of the African members of this order. Nomenclature is presently in a state of confusion as new investigations suggest relationships that differ somewhat from older interpretations. Throughout the text such suggestions will be made where appropriate, but on the whole we have chosen to retain a more classical nomenclature until such time as specific data can be published to justify systematic changes. Responsibility for the taxonomic sections below are as follows: *Palaeomastodon* and northern African gomphotheres—Yves Coppens and Michel Beden; *Palaeomastodon,* gomphotheres, and mammutids, plus general evolutionary discussions within these groups—Cary Madden and Vincent Maglio; elephantids, evolutionary discussions, and zoogeography—Vincent Maglio.

There are a number of areas where the data allow several different interpretations on which we did not always agree. These are identified in the text and, where possible, alternative views are presented. The reader should bear in mind that for the most part our approach has been a conservative one subject to change as new data come to light.

Structure

With few exceptions the Order Proboscidea includes some of the largest of land mammals. They have numerous graviportal adaptations, such as laterally expanded ilium with broad crests and downwardly directed acetabula. Limbs are columnar with long proximal and short distal elements. These limb bones lack medullary cavities, the hemopoietic tissues of the narrow occupying spaces between bony

spicules of the cancellous bone filling the interior. Carpals and tarsals are short and broad. Both manus and pes are short, pentadactyl, and semiplantigrade-semidigitigrade (Sikes 1971), the shock of impact being taken up by a complex fibrous connective tissue pad below.

The neck in all proboscideans is short and supports a very large but remarkably light skull with pneumatized diplöe. This latter feature was less developed in more primitive species but is well formed in elephants; it is related to expansion of the outer surface plate during evolution of the masticatory complex and attendant muscular reorientations (Maglio 1973). Even in the most primitive species of the order the nasal orifice is retracted posteriorly to a position between the orbits. Premaxillary bones are expanded along with the maxillae for support of the greatly enlarged tusks. The upper surface of this bony sheath serves as origin for the modified upper lip and nasal musculature, which forms a long, flexible proboscis. Palatal structures are elongated ventrally for housing the high-crowned upper molar teeth; maxillary alveolar borders are long and posteriorly an extended wing of the alisphenoid bone covers the alveolar cavity.

The dentition is reduced to $\frac{1.0.3.3}{0.0.3.3}$ in both living species but was less reduced in some extinct forms. Incisors were reduced to a single pair even in the earliest member of the order. This is believed to be the I^2 of the primitive mammalian dentition. A small milk incisor erupts first and is replaced by the continuously growing permanant tusk, the latter free of enamel except for a tiny cap lost early in wear.

Canines are lacking even in Eocene species. Premolars are present and functional for at least part of an animal's life in most early proboscideans but are lost in the evolution of stegodonts and elephants. In the latter groups the milk or deciduous premolars remain in position until they are pushed out by the erupting permanent molars. These milk teeth are morphologically like the true molars and serve to extend functionally the unique elephantlike masticatory mechanism into earlier ontogenetic stages.

Molar teeth vary within the order from brachyodont and bunodont with accessory columns serving a grinding-crushing function as in gomphotheres, to brachyodont and lophodont with vertical and lateral shearing-crushing functions as in the mastodonts, to hypsodont, but never hypselodont (continuously growing) teeth with parallel enamel bands serving fore and aft horizontal shearing as in elephants and stegodonts. These surfaces consist of flattened, platelike enamel shells filled with dentine and covered externally with cementum. With wear such structures appear in cross-section as alternating intervals of enamel-cement-enamel-dentine, etc. The enamel, being harder than the other substances, stands up as ridges in relief, thus providing the cutting surfaces.

Since teeth are by far the most common elements found in the proboscidean fossil record, taxonomic differentiation has been based primarily on such features as enamel thickness, crown height and width, number of cones or enamel plates, and the latter's relative spacing along the molar length.

The mandible is long in most proboscideans and carries a long pair of incisors in the lengthened symphysis. These incisors may be tusklike, flattened, or variously otherwise modified. Stegodonts and elephants independently lost such mandibular tusks early in their histories, as did several other groups.

Fossil Proboscidea in Africa

The order may be divided into two suborders and four families. For present purposes it is best to avoid the use of superfamilies.

Order Proboscidea
  Suborder Gomphotherioidea
    Family Gomphotheriidae: late Eocene to middle Pleistocene
    Family Elephantidae: late Miocene to Recent
  Suborder Mammutoidea
    Family Mammutidae: early Miocene to sub-Recent
    Family Stegodontidae: middle Miocene to late Pleistocene

The earliest records of the order are in the late Eocene Qasr el Sagha Formation of the Fayum deposits, Egypt, and at Dor el Talha, Libya. Morphologically these are Gomphotheriidae whose proboscidean adaptations are already highly evolved, suggesting a long prior evolutionary history and therefore offer little clue to their ancestry. Presumably they shared some common origin with hyraxes, sirenians, and others in an unknown subungulate group of the African early Eocene or Paleocene. Since deposits of these ages are largely lacking in Africa, such a common ancestry remains speculative at this time.

Suborder Gomphotherioidea

Family Gomphotheriidae Hay 1922

This is a very diverse family known as fossil from every continent except Australia and Antarctica and forms the mainstem stock for the order. In Africa it can be traced from rather primitive species in the late Eocene through the Cenozoic to the middle Pleistocene, when the group became extinct. All members of the family have brachydont cheek teeth with from two to seven pairs of rounded cones and a variable development of enamel columns in the valleys between them. With wear these columns become incorporated into the pretrite cones (lingual on upper molars and buccal on lowers), producing a characteristic three-looped or trifoliate enamel pattern. Premolars erupt forming a functional part of the toothrow, at least during part of the life of an individual. Except for several specialized forms (Anancinae in the Old World and Cuvieroninae in the New World), a single pair of incisors persists in the very elongated lower jaw and a second pair, sometimes with persistent bands of enamel, is present in the upper jaw.

The family includes the subfamilies Gomphotheriinae Hay 1922 in Africa, Asia, Europe, and North America; Cuvieroninae Cabrera 1929 in North and South America; and Anancinae Hay 1922 in Europe, Asia, and Africa. The African representatives of the family may be subdivided as follows:

Subfamily Gomphotheriinae[1]
  Genus *Palaeomastodon* (*Palaeomastodon*), *Palaeomastodon* (*Phiomia*)
  Genus *Gomphotherium*
  Genus *Platybelodon*
  Genus *Tetralophodon*
  Genus aff. *Choerolophodon*
Subfamily Anancinae
  Genus *Anancus*

Subfamily Gomphotheriinae Hay 1922

This subfamily includes the trilophodont[2] gomphotheres with a long straight or downturned mandibular symphysis, bearing rounded to more or less flattened lower incisors. The skull varies from long, narrow, and moderately high to short, broad, and flat with expanded maxillary and premaxillary bones that carry tusks weakly to strongly recurved, with or without remnant enamel bands. The tem-

---

1 It is probable that this diverse assemblage of early proboscideans includes the roots of several different later groups, and Madden has suggested a different classification to reflect this view. According to this interpretation *Palaeomastodon* should be included with *Zygolophodon* as a family of Mammutoidea. Similarly, *Phiomia* would be grouped with *Platybelodon* in the subfamily Amebelodontinae, within the Gomphotheriidae. There is some evidence to support these relationships, but until these data are presented in detail, it seems best to retain the present scheme and to avoid the problems of a vertical classification.

2 The classical terms "trilophodont" or "tetralophodont" refer only to the intermediate cheek teeth, dP4-M2.

poral musculature generally lies behind the occlusal surface and the skull shows little or no basicranial flexure. The jaw is long and the ascending rami generally rather low. The group is generally rather poorly known in Africa, most forms represented only by teeth. They were adapted to moist forest or woodland habitats in which low-level browse was abundant. The rather low-slung body, short neck, and probably short trunk would have restricted these animals to areas with sufficient undergrowth. The dentition, although bearing enormously thick enamel, was relatively unspecialized and suited primarily for dealing with soft browse such as low shrubs and aquatic vegetation.

Genus *Palaeomastodon* Andrews 1901

SYNONYMY. *Palaeomastodon* Andrews 1901a; *Phiomia* Andrews and Beadnell 1902.

TYPE SPECIES. *Palaeomastodon beadnelli* Andrews 1901.

DIAGNOSIS OF THE GENUS. Gomphotheriinae of small to medium height, skull long and narrow; long sagittal crest; nostrils opening just in front of the orbits, bounded behind by a postorbital apophysis of the frontal bone; palate long and narrow. Mandibular symphysis long. Reduced dental formula: $\frac{1.0.3.3}{1.0.3.3}$, premaxillary incisors developed into short recurved tusks, mandibular incisors dorsoventrally flattened; strong diastema between incisors and first premolar; trilophodont molars, the third crest being more (P. [*Phiomia*]) or less (P. [*Palaeomastodon*]) well developed; molars bunodont with some trend toward zygodonty. Limb morphology of proboscidean type.

DISTRIBUTION. Late Eocene, Dor el Talha, Libya and Qasr el Sagha Formation, Fayum, Egypt; early Oligocene, Jebel el Qatrani Formation, Fayum, Egypt, Zella Oasis, Libya; middle Oligocene, Jebel Bou Gobrine, Tunisia.

DISCUSSION. The generic names *Palaeomastodon* and *Phiomia* have undergone many vicissitudes since their establishment. Thus Andrews (1906) followed by Matsumoto (1924) grouped them together in the same family, Palaeomastodontidae. Osborn (1936) put them in two different families, Bunomastodontidae for *Phiomia* and Mastodontidae for *Palaeomastodon*. In 1942 he used the family Palaeomastodontidae for the latter. Continuing the study, Lehman (1950, pp. 150–153) grouped both in the same genus, but later authors (Vaufrey 1958; Thenius 1969) have usually maintained two separate genera. Most recently Tobien (1971, pp. 153–157) has observed that characters usually utilized to

distinguish *Phiomia* from *Palaeomastodon* are essentially typological and do not withstand a study of variability within populations. According to Lehman (1950) and Tobien (1971) features used in the past, such as palatal breadth, length of the mandibular symphysis, and position of the mental foramen, all appear to be extremely variable. On the other hand, other differences between the two genera can be observed; these concern the position of the posterior ridge of the mandibular symphysis, proportional breadth of molars, morphology of the cones, and the cross-sectional shape of the lower tusks.

Lehman (1950, pp. 150–158), after demonstrating the extreme variability of the dental and cranial morphology, deduced that all the described species of *Palaeomastodon* and of *Phiomia* must in fact represent only a single generic group, to which he assigned the former name. He concluded further that there must be only a single species, *Palaeomastodon beadnelli*. Tobien (1971, pp. 153–157), in contrast, accepted the existence of two taxa, which he distinguished at the subgeneric level. This interpretation is followed here.[3]

Thus in many features of skull and tusk morphology *Palaeomastodon* (*Phiomia*) is closer to the *Amebelodon* group of shovel-tusked and flat-tusked gomptheres than to P. (*Palaeomastodon*) or to *Gomphotherium*, and it has been suggested that a true ancestral-descendant relationship exists between the two (Borissiak 1929; Osborn 1936; Tobien 1973). If this is true, it would mean that adaptive divergence among gomphotheres had already begun, albeit not yet extensively progressed, by the late Eocene. Madden proposes to reclassify these early genera to reflect this relationship.

The genus is interesting from the paleogeographic and phyletic points of view since this exclusively African group is the oldest discovered representative of the order. Known species had already acquired diagnostic cranial, dental, and skeletal characters. Their importance is increased further by the fact that some morphological and biometric features of the dentition already show rudimentary trends pointing to differentiation into the two major lineages of middle Cenozoic times.

The localization of the discoveries of *Palaeomastodon* in northeastern Africa is also interesting in re-

---

3 Not all students of the Proboscidea will subscribe to this view, and some still maintain that *Phiomia* can be separated on the generic or even higher level from *Palaeomastodon*. Both views are, of course, only matters of taxonomic preference. In either case, two distinct entities are recognized and to this end most workers are in agreement.

lation to the origin of the Gomphotheriidae. Can it be said that they originated in northeastern Africa as Osborn (1936) thought or do they represent an Asiatic group that migrated into this part of the continent? At this point it is impossible to say since the earlier Eocene record of the order is lacking both in Africa and in Asia and their known Eocene and Oligocene distributions must be considered almost certainly fortuitous. However, for reasons given above, it seems unlikely that earlier members of the Proboscidea will be found anywhere but in Africa.

Palaeomastodon (Palaeomastodon) beadnelli Andrews 1901

SYNONYMY. *Paleomastodon beadnelli* Andrews 1901a (p. 319); *Palaeomastodon minor* Andrews 1904 (p. 115, partim); *Palaeomastodon parvus* Andrews 1905 (p. 562); *Palaeomastodon wintoni* Andrews 1905 (p. 562, partim); *Palaeomastodon minus* Andrews 1905 (p. 562, partim); *Palaeomastodon barroisi* Pontier 1907 (p. 150–154, partim); *Palaeomastodon intermedius* Matsumoto 1922 (p. 1–6, fig. 1).

DIAGNOSIS. (Andrews 1901a; Matsumoto 1924; Tobien 1971); a small species of *Palaeomastodon (Palaeomastodon)* with long mandible, anteriorly positioned symphysis, with short incisors, oval in section. Molars of the "zygodont" type, relatively broad; posterior edge of symphysis far ahead of the anterior-most molars.

DISCUSSION. The holotype of *P. beadnelli* is a left mandible with P_4-M_3. It was destroyed in the Cairo Museum but casts exist in the British Museum of Natural History (BM M-8059) and in the American Museum of Natural History (AMNH-9984). This specimen was figured by Andrews (1901b, fig. 1) and again in 1906 (pls. 12-16). The species has a very limited distribution, the holotype from the Lower Fossil Wood Zone, Jebel el Qatrani Formation (early Oligocene) of the Fayum Depression, Egypt, and more recently discovered material from Zella and Dor el Talha, Libya, of similar age (Savage 1969).

Palaeomastodon (Phiomia) serridens Andrews and Beadnell 1902

SYNONYMY. *Phiomia serridens* Andrews and Beadnell 1902 (pp. 1–9); *Palaeomastodon minor* Andrews 1904 (p. 115, partim); *Palaeomastodon wintoni* Andrews 1905 (pp. 562–563, partim); *Palaeomastodon minus* Andrews 1905 (p. 562, partim); *Palaeomastodon barroisi* Pontier 1907 (pp. 150–154, partim); *Phiomia osborni* Matsumoto 1924 (pp. 40–49).

DIAGNOSIS. (Andrews and Beadnell 1902, pp. 1–9; Matsumoto 1924; Tobien 1971); a small *Palaeomastodon (Phiomia)* with long mandible, anteriorly positioned symphysis with short, broad dorsoventrally flattened incisors. Molars of the bunodont type, relatively narrow; posterior edge of symphysis on the level of the anterior-most premolars.

DISCUSSION. The holotype specimen is an almost complete immature mandible with dI_2 and dP_2-dP_3 in the British Museum of Natural History and was figured by Andrews and Beadnell (1902, pl. 18). The specimen is from the Lower Fossil Wood Zone, Fayum, Egypt, of early Oligocene age. More recently the range has been extended downward into the late Eocene as specimens referable to the taxon have been reported from Dor el Talha, Libya (Savage 1969), and from the Qasr el Sagha Formation of the Fayum (C. T. Madden, University of California at Berkeley collection). Additional material is now also known from the early Oligocene of Zella Oasis, Libya (Arambourg and Magnier 1961), and Djebel Bou Gobrine, Tunisia (Arambourg and Burollet 1962).

An undescribed, incomplete molar in the Yale collection, probably an M_3 (YPM-24847) is the only undoubted specimen from the medial Oligocene, Upper Fossil Wood Zone (Quarries I and J) of the Jebel el Qatrani Formation (E. L. Simons pers. comm.). It is intermediate in size and morphology between *P. (Phiomia)* and Miocene *Gomphotherium* but is too incomplete for generic reference or diagnosis.

Genus *Gomphotherium* Burmeister 1837

SYNONYMY. *Mastodon* Cuvier 1806 (partim); *Gomphotherium* Burmeister 1837; *Trilophodon* Falconer and Cautley 1857; *Bunolophodon* Vacek 1877; *Serridentinus* Osborn 1923.

TYPE SPECIES. *Mastodon angustidens* Cuvier 1806.

DIAGNOSIS OF GENUS. Medium to large proboscideans with large high skull; premaxillary enlarged, bearing a pair of large incisors, oval or triangular in cross-section and with a lateral band of enamel. Dentary elongated, symphysis angulated, proportionately longer than in *Phiomia;* bearing short tusks rounded to oval in section and concave on their dorsal surface; molars bunodont, M_1 and M_2 with three pairs of cones, M^3 may have three to four and one-half, M_3 often with three to five and one-half; never more than three teeth in use simultaneously; well-developed central conules forming with the main cones a trifoliate-shaped wear figure.

DISTRIBUTION. Early to late Miocene, widespread in East and North Africa and on other continents.

DISCUSSION. This genus occupies a central position in the evolution of gomphotheriid proboscideans. It combines both primitive and derived features and apparently stands intermediate between the Oligocene *Palaeomastodon* and later derived groups such as *Tetralophodon,* the Anancinae, and the Elephantidae.

Many of the remains from Africa, especially those from eastern Africa, have been identified only to the generic level; because of the lack of adequate descriptions and figures these determinations cannot be confirmed. This is true (at least in part) for the material collected at Karungu, Ombo, and Songhor. Other specimens that have been described in detail may not all belong in *Gomphotherium,* and it has been suggested that several distinct taxa may actually be involved (Tobien 1973; Maglio 1974; and Madden, in preparation). However, until additional evidence becomes available it is probably best to retain most of the known material in *Gomphotherium sensu lato.*

The earliest records of the genus are in the late early Miocene (19.9 m.y. to 17.8 m.y., Bishop, Miller, and Fitch 1969) of East Africa. This date is some 1 to 3 m.y. earlier than the faunal exchanges between Africa and Europe (approximately 18 to 17 m.y., Berggren and van Couvering 1974), after which the group appeared in Eurasia. It seems likely that the *Gomphotherium*-like forms originated in the African continent in earliest Miocene or late Oligocene times.

Simpson (1945, p. 132) concluded that the genus *Gomphotherium* as described by Burmeister should be the valid name for this group. The name *Trilophodon* was proposed by Falconer and Cautley as a "section" name into which Osborn regrouped a good many early Miocene to Pliocene species formerly referred to *Bunolophon.* It is, however, superceded by Burmeister's nomen. The less commonly used name *Mastodon* of Cuvier is considered a synonym of *Mammut* Kerr, although some confusion still surrounds this designation.

This question of priority of *Gomphotherium* over *Trilophodon* is extremely complex and cannot be settled here. However that may be, most recent authors give preference to the former name. This explains why it is so difficult to give a precise diagnosis of the genus, since past authors have put into it a large number of very diverse species. There have been many, often divergent, emendations to the definition and we quote only the major features above.

cf *Gomphotherium angustidens* (Cuvier) 1806

SYNONYMY. *Mastodon angustidens* Cuvier 1806 (pp. 405–412); *Mastodon angustidens* var. *libyca*

Fourtau 1918 (pp. 84–85); *Trilophodon angustidens kisumuensis* MacInnes 1942 (pp. 51–76, *partim*); *Protanacus macinnesi* Arambourg 1945 (p. 493).

DIAGNOSIS. A large species of *Gomphotherium* with short mandibular incisors with an oval cross-section, the long axis mediolateral, in other features indistinguishable from *G. productum* of North America. Other characters as for the genus.

DISTRIBUTION. Early and middle Miocene of North, Central, and East Africa; known from many localities including Wadi Moghara and ?Siwa in Egypt; Gebel Zelten, Libya; Bled ed Douarah and Chérichira (Beglia Fm.), Tunisia; Maboko, Moruorot Hill, Lothidok Hill, ?Rusinga, ?Ombo, and ?Songhor, Kenya; Napak I, II, IX, and Moroto I, Uganda; and Semliki, Congo.

DISCUSSION. *Gomphotherium angustidens,* type species of the genus, was described by Cuvier from a molar tooth collected at Simore in France. Numerous remains of mastodons were attributed to this taxon in Europe and later in Africa. The observed variability of morphological characters in known samples is very great and has led to the designation of many subspecies, most of which are of questionable validity.

The African material referred to the species differs from that in Europe by the somewhat flatter shape of the mandibular tusks. Tobien (1973, p. 261) believes this feature to indicate that the African species is not only distinct from *G. angustidens,* but that it actually represents an early member of the shovel-tusked group of proboscideans. He thus transferred all this material to the genus *Platybelodon* and utilized MacInnes's name *G. kisumuensis* for it. Madden has recently examined some material from northern Africa with flattened tusks similar to those from Russia and France. He has suggested that the sub-Saharan species should be referred to the European taxon *Platybelodon filholi.* This conclusion is not shared by all the authors of this chapter as regards all the African Miocene material; until such time as evidence is presented in detail, we feel it best to retain the reference of cf *G. angustidens.*

At least two subspecies have been described and the names extensively used in Africa—*G. a. libycus* and *G. a. kisumuensis.* Both, however, are now considered invalid. According to Fourtau (1918), molars of *G. angustidens* from Wadi Moghara are by their morphology and proportions intermediate between those of typical *G. angustidens* from the Burdigalian sands from the Orléans region and those of *G. angustidens pontileviensis* of the Helvetian of France. Since they clearly differed from the teeth of "*G.*" *pygmaeus* of the early Burdigalian of North Africa, especially by their strong development of the talon, Fourtau established the subspecies *libyca* for them.

The East African Miocene gomphotheres from Rusinga and Maboko were described by MacInnes (1942) under the name *G. angustidens kisumensis.* Three years later Arambourg (1945, p. 491) recognized morphological differences within MacInnes's hypodigm and separated the specimens into two groups, one corresponding to the holotype, the other to one of the paratypes. Arambourg considered the second lot as belonging not only to a new species but to a new genus, *Protanancus macinnesi,* which he believed to be tending toward the Pliocene genus *Anancus.* These molars are characterized by weak alternation of lingual and buccal cones and a slight sloping forward of the cones on M3. This determination has been questioned (Maglio 1974) and is not readily substantiated by examination of the original collection.

A number of mandibular tusk fragments from Maboko (MacInnes 1942) and Congo (531) (Hooijer 1963) show a flattened, dorsally grooved structure with high lateral and medial borders, unlike the typical oval tusks usually seen in other specimens referred to *G. angustidens.* These specimens have no special modification of internal dentinal structures. One of the Congo pieces is associated with a mandible containing a partial M_2 indistinguishable from *G. angustidens,* but many early Miocene gomphotheres are difficult to separate on dentition alone. Tobien (1973) considers these flattened tusks to indicate that the African gomphotheres were tending toward the shovel-tusked *Amebelodon* and *Platybelodon.* Madden (manuscript in preparation) has recently reinvestigated the latter group and agrees that at least some of the African material may belong to an early forerunner of this specialized assemblage. However, the majority of the known collection seems to belong to a *Gomphotherium*-like animal in which the upper tusks are typically oval in section. Some of the African gomphothere material may indeed prove to represent a very early stage in shovel-tusker ancestry (as opposed to the highly specialized form described below), but more conclusive evidence beyond the more flattened shape of the lower tusks has not yet been found. For the present we feel it is too early to establish the genus *Amebelodon* in Africa on these present data. We know far too little about the mode of life of these animals to be certain that this is not a parallel development of endemic African Proboscidea to their flat-tusked Asiatic cousins.

Fourtau (1918, p. 89) distinguished a right mandible from the Miocene of Wadi Moghara as *Mastodon spenceri,* separating it from *G. angustidens* by the presence of a mandibular symphysis clearly directed downward and bearing two oval tusk sockets. Such a feature is also seen far better developed in *Rhynchotherium,* leading Osborn (1936, p. 485) to attribute with doubt Fourtau's species to that genus. More recently Tobien (1973) gave good evidence for considering *M. spenceri* as synonymous with *Gomphotherium angustidens* even though atypical, an opinion shared by several other workers.

Arambourg (1945) attributed to the same taxon as Fourtau's holotype the jaw collected by Gaudry in Chérichira (Tunesia), long described as *G. angustidens.* The teeth resemble those of *angustidens* although the symphysis is shorter than expected. Until additional material is available that may prove the distinctness of this taxon, we prefer to follow Tobien in including it within *Gomphotherium,* close to *G. angustidens.*

"*Gomphotherium*" *pygmaeus* (Depéret) 1897

SYNONYMY. *Mastodon angustidens pygmaeus* Depéret 1897 (pp. 518–521); *Phiomia pygmaeus* (Depéret), Osborn 1925 (pp. 17–35); *Trilophodon olisponensis pygmaeus* (Depéret), Bergounioux and Crouzel 1959 (pp. 101–102).

DIAGNOSIS. (Depéret 1897, p. 520, pl. XIX; Osborn 1936, p. 246); a small gomphotheriid with relatively short, narrow, and high-crowned molars, M_2 and M_3 probably with three or four ridges, cement abundant in valleys.

DISCUSSION. The holotype specimen consists of an incomplete right M_3 in the Lyon Museum. It originated from Kabylie, along the road from Chabet el Ameur to Isserville, 4 km from the latter in Algeria, from sandstone of Cartennian (= Burdigalian) age (early Miocene). Other material is now known from Gebel Zelten, Libya, and from Wadi Moghara, Egypt, but the species remains very poorly known.

Osborn (1936), studying a cast of the specimen, believed it to be complete and described it as belonging to an advanced species of *Phiomia, P. pygmaeus.* The probable presence of four well-developed cone-pairs and its larger size, however, suggest that this material belongs among the trilophodont gomphotheres. Bergounioux and Crouzel (1959) reconstructed the holotype and considered it identical with *Trilophodon olisponensis* Zbysewski, to which they transferred it as a subspecies even though the name *pygmaeus* had priority.

The material from North Africa currently referred to the taxon shows great variability in structure and may possibly represent several different forms. This may be particularly true of rather more primitive-looking specimens from Gebel Zelten (Savage and Hamilton 1973) and nearby sites (Hormann 1963; J.

Harris, manuscript in preparation), and perhaps the material from Wadi Moghara also. Whatever the true affinities of this assemblage prove to be, it emphasizes the difficulty in dealing with fragmentary remains.

Tobien (1973) recently suggested that "*G.*" *pygmaeus* may be ancestral to *Choerolophodon*, a widespread Old World middle to late Miocene form. He cites only the thick cement on the M3 as evidence; this feature is notoriously variable in Proboscidea as well as other groups and we must consider this suggestion equivocal at best. However, Madden, who has reexamined the material, believes that this species does not belong in the genus *Gomphotherium*, although its exact affinities remain to be clarified.

"*Gomphotherium*" *ngorora* Maglio 1974

This taxon was based on a single specimen from the late Miocene of Ngorora, Baringo Basin, Kenya, and consists of upper dention, tusk fragments, and skeletal remains. Nevertheless, so distinctive are the elephantlike trends in dental adaptation that there can be little doubt as to the uniqueness of the species among Miocene gomphotheres. In basic morphology "*G.*" *ngorora* lies clearly among the Gomphotheriidae, but the paired cones, three on M2 and four on M3, are closely compressed in the midline and anteroposteriorly compressed to form transverse plates. Vestigial trefoil wear patterns persist, but less so than is typical of *Gomphotherium*.

It is too early to claim that "*G.*" *ngorora* lies on the phyletic line between gomphotheres and later elephants, but it does exhibit features transitional between the two groups. It may represent a collateral evolutionary branch that demonstrates the kind of adaptive changes we would expect to have accompanied the rise of the Elephantidae.

Inclusion of this taxon within the genus *Gomphotherium* is less than satisfactory, but short of proliferating yet another generic name it is best to retain this designation until additional evidence of proper generic affinity becomes available.

Genus ?*Platybelodon* Borissiak 1928

SYNONYMY. *Platybelodon* sp. Maglio 1969 (pp. 1–10); *Torynobelodon* sp., Tobien 1972 (pp. 182–184).

In 1969 Maglio described a mandibular tusk fragment from the early Miocene locality at Loperot, Kenya. We will not repeat here the detailed discussion that led him to attribute this fragment to *Platybelodon*, but the extreme flattening (thickness/width = 0.17) and the complex internal structure of highly organized dentinal rods and laminated dentine are highly diagnostic for this group and seen nowhere else. So unique is this specimen that it stands sharply apart from almost all other gomphothere material in Africa, including the flat (but much less so) tusks discussed above. The latter decidely lack any trace of this unusual internal specialization, as do most lower incisors of *Amebelodon*. Recently, Madden observed true shovel-tusks with specialized internal organization from several North African localities. These specimens remain to be described.

This single tusk is the oldest occurrence of the group and it was suggested that it is indicative of an African origin for the shovel-tusked lineages of Asia and North America. Clearly the evidence is too incomplete to be conclusive, but if an early *Amebelodon* should be confirmed in Africa, then the occurrence of the more specialized members of the same group might be expected.

Reference of the piece to Borissiak's genus *Platybelodon* was based on the close similarities to *P. grangeri* of Central Asia, but reference of this East African specimen to that species would be premature at this time. However, the type species of the genus, *P. danovi* Borissiak, may prove to be congeneric with North American *Amebelodon* (Madden), in which case it might prove desirable to designate a new name for these more evolved forms. We leave this problem to nomenclatural purists.

The presence of this group demonstrates the evolution of yet another adaptive type, this one presumably taking advantage of lowland aquatic or swampy environments. The mode of life of the amebelodonts is far from determined, but their peculiar features are usually interpreted as adaptations for feeding on soft aquatic vegetation that they may have "scooped" up from river banks or lake shores.

Genus *Tetralophodon* Falconer and Cautley 1857

DIAGNOSIS. Intermediate molars (dP4-M2) with four well-developed cone-pairs, third molars with five and one-half pairs or more; teeth bunodont but relatively higher crowned than in other gomphotheres; intravalley columns nearly lacking by incorporation into adjacent cones. Mandibular symphysis may be reduced in length but not as short as that in the Anancinae.

DISCUSSION. Inclusion of this genus within the Gomphotheriinae follows the conservative view that although clearly moving away from the typical plan as seen in *Gomphotherium*, *Tetralophodon* shows some features of progressive adaptation combined with more primitive characters and has not yet achieved any fundamentally new way of life. The genus may be a forerunner of the Pliocene Anancinae (Schlessinger 1917, 1922; Matsumoto 1924) but this alone is not enough to warrant inclusion in that subfamily.

Several localities in Africa have yielded remains referred to *T. longirostris,* all north of the Sahara. These are Zidania and Melka el Ouidane, Morocco; Tozeur, Djebel M'Dilla, Djebel Semene, Sbeitla, Bled ed Douárah, Chérichira, and Foum el Kranga, Tunisia; and Smendou, Algeria. This identification is clearly confirmed at Chérichira (Bergounioux and Crouzel 1956, p. 547) and is probable at Zidania. As for the symphyseal fragment from Djebel Semene described by Bergounioux and Crouzel, its resemblance to *T. longirostris* is based only on the elongation of the bone, its narrowness, and the lack of enamel on the preserved tusk, which cannot be considered as definitive.

In other cases specimens are so incomplete that their determination must be considered questionable, relying more on stratigraphic position of the bed than on morphology of the piece. The material from Smendou and Djebel M'Dilla has never been described.

It is interesting that this species appears for the first time in Africa during the Helvetian equivalent at Chérichira. Bergounioux and Crouzel did not hesitate to suggest an African origin for this group, which, according to Schlessinger (1917, 1922) and Osborn (1936, p. 352), could have derived directly from *Gomphotherium*. However, from its distribution in North Africa it seems more likely to have immigrated from Eurasia, where the genus was widespread.

Genus aff. *Choerolophodon* Schlessinger 1917

Tobien (1973) has reestablished Schlessinger's 1917 genus *Choerolophodon* for material widespread in Eurasia from middle to late Miocene times. It is distinguished from *Gomphotherium* by certain cranial features, a short jaw, small to absent mandibular incisors, and a chevron arrangement of molar cones invested in thick cement. Tobien suggested that a specimen from Henchir Beglia, Tunisia (Robinson pers. comm.), may represent an African member of this group. The material awaits detailed description by Tobien.

Subfamily Anancinae Hay 1922

Members of this Mio-Pliocene to Pleistocene subfamily are characterized by the loss of mandibular tusks and the great shortening of the symphysis, by bunodont molars with a subhypsodont tendency and having four to five pairs of alternating cones on the intermediate molars. The skull is high, domed, and wide, with straight upper incisors lacking enamel bands.

Two genera have been described in Africa, *Anancus* and *Pentalophodon,* but recent authors have included all African materials within the former. Tobien (1973, fig. 1) suggests that *Tetralophodon* be included with this subfamily, but as discussed above it is clearly transitional to *Anancus* and its inclusion in the Gomphotheriinae or Anancinae is a matter of taxonomic preference.

Genus *Anancus* Aymard 1855

SYNONYMY. *Anancus* Aymard 1855 (p. 507); *Pentalophodon* Falconer, MacInnes 1942 (p. 82).

TYPE SPECIES. *Anancus arvernensis* Croizet and Jobert.

DIAGNOSIS. An Anancinae with four pairs of cones, forwardly inclined, and strongly alternating on the lingual and buccal sides.

DISTRIBUTION. The genus is widely distributed in Africa as several species; they have wide stratigraphic range but all are later than middle Miocene.

Anancus osiris Arambourg 1945

SYNONYMY. *Mastodon arvernensis* Croizet and Jobert; Depéret, Lavauden and Solignac 1925 (p. 21); *Anancus arvernensis* (Croizet and Jobert); Dietrich 1943 (pp. 46–48); *Anancus osiris* Arambourg 1945 (pp. 479–489).

DIAGNOSIS. (Arambourg 1945, 1970, pp. 27–34); an *Anancus* with very strong alternation of bunodont cone-pairs strongly sloping forward and mesially converging at their summits, M1 and M2 tetralophodont, M3 with five and one-half to six cone-pairs; molar crowns simple, lacking intravalley accessory columns.

DISCUSSION. The holotype is a right M^3 (Paris Mus. No. 1943-1), figured by Arambourg (1945, fig. 1, pl. I). It was derived from late Pliocene deposits at Gizeh, Egypt, and occurs also at numerous other North African localities from Egypt to Morocco and southward to Chad (see table 17.1).

In our opinion, most of the material from North Africa referred to *Anancus arvernensis* by past authors belongs in *A. osiris*. Some specimens that have neither been described nor figured can only questionably be referred here at this time; these include those from Wadi Natrun, Hamada Damous, and Djebel Mellah. Several skulls were discovered, two in Ferryville (Solignac 1927, p. 397; not collected), a palate from Grombalia, Hamada Damous (molars described by Arambourg 1970, p. 29), and a skull

Table 17.1 Major deposits and localities of occurrence for the fossil Proboscidea in Africa.

Species	Age	West and Central Africa	North Africa	East Africa	South Africa
Palaeomastodon beadnelli	L. Eoc.–E. Olig.		Fayum, Zella, Dor el Talha		
Palaeomastodon serridens	L. Eoc.–E. Olig.	Djebel Bou Gobrine, Fayum, Zella, Dor el Talha			
Gomphotherium angustidens	E.–M. Miocene	Semliki	Wadi Moghara, ?Siwa, Gebel Zelten, Bled ed Douarah, Cherichera	Maboko, Napak, Lothidok, Moruorot, Rusinga, Songhor, Ombo, Moroto I	
"Gomphotherium" pygmaeus	E.–M. Miocene		Isserville, Wadi Moghara, Gebel Zelten		
"Gomphotherium" ngorora	L. Miocene			Ngorora	
Platybelodon sp.	E. Miocene			Loperot	
?Tetralophodon cf *longirostris*	M.–L. Mio.–E. Plio.		Cherichera, Melka el Ouidane, Bled ed Douarah		
Anancus osiris	E. Plio.–M. Pleist.	Bochianga, Atoumanga, Kolinga I, Ouadi Derdemi, Yerki	Gizeh, Si Abd el Azie, Daourat, Oued Akrech, Aïn Boucherit, Lac Ichkeul, Hamada Damous, Djebel Mellah, Ferryville, Wadi Natrun		
Anancus kenyensis	E. Plio.–E. Pleist.			Kanapoi, Kanam, Ekora, Chemeron, Kaiso, Laetolil, Olduvai I, Mursi, Kubi Algi	
Anancus petrocchii	L. Miocene		Sahabi		
Anancus sp.	E. Pliocene				Langebaanweg
Zygolophodon cf *borsoni*	Pliocene		?St. Arnaud, Ferryville (= Menzel Bourguiba)		
Zygolophodon (Turicius) sp.	M.–L. Pliocene		Marceau, Smendou Khenchela		
Stegodon kaisensis	M.–L. Pliocene	Kolinga I		Kaiso, Shungura	
Stegotetrabelodon syrticus	L. Mio.–E. Plio.		Sahabi		

Table 17.1   (*continued*)

Species	Age	West and Central Africa	North Africa	East Africa	South Africa
Stegotetrabelodon orbus	L. Mio.–E. Plio.			Lothagam-1, Mpesida, Kolinga	
Primelephas gomphotheroides	L. Mio.–E. Plio.			Lothagam, Kaiso, Chemeron, Kanam	
Primelephas korotorensis	E. Pliocene	Kolinga, Koulá			
Loxodonta adaurora	M. Plio.–E. Pleist.			Kanapoi, Kubi Algi, Lothagam-3, Chemeron, Kaiso, Kanam, Mursi, Vogel River, Chiwondo beds	
Loxodonta atlantica	M. Pleistocene		Ternifine, Sidi Abder Rahman, Oued Constantine	Shungura	Elandsfontein, Zululand
Loxodonta africana	E. Pleist.–Rec.	Yayo		East Rudolf	
Elephas ekorensis	M. Pliocene			Ekora, Kanapoi, Kubi Algi, Bolt's Farm, Chiwondo beds	
Elephas recki	L. Plio.–M. Pleist.	Toungour, Ouadi Derdemi		Olduvai I–IV, Kaiso, Kikagati, East Rudolf, Homa, Kanjera, Marsabit Rd., Olorgesaillie, Usno fm, Shungura fm, Laetolil, Vogel River	
Elephas iolensis	L. Pleistocene	Zouerate	Port de Mastagnem, Sidi Abder Rahman	Kaiso, Kibish fm, Natodomeri	Vaal River
Mammuthus subplanifrons	E.–M. Pliocene			Kaiso, Vogel River, Kanam, Chemeron, Chiwondo beds	Langebaanweg, Vaal River, Virginia
Mammuthus africanavus	L. Plio.–E. Pleist.	Toungour, Ouadi Derdemi, Koulá	Lac Ichkeul, Kebili, Oued Akrech, Aïn Boucherit, Garet et Tir		
Mammuthus meridionalis	E. Pleistocene		Aïn Hanech, Bel Hacel		

from Atoumanga (figured in Coppens 1965, pl. I); but none of them has been described in detail.

Anancus osiris is the North African equivalent of *A. arvernensis* in Europe and characterizes the middle Villafranchian where both are collected in association with primitive Elephantidae. Both species are recorded as early as the beginning of the Pliocene, the African form differing from the European one only by the greater simplicity of the molar teeth and by their proportions.

Anancus kenyensis (MacInnes) 1942

SYNONYMY. *Pentalophodon sivalensis kenyensis* MacInnes 1942 (p. 42); *Anancus kenyensis* (MacInnes), Arambourg 1948 (p. 305).

DIAGNOSIS. (MacInnes 1942; Arambourg 1945, p. 490); an *Anancus* with four and one-half to five or more cone-pairs on M2 and five and one-half or more on M3; crown complicated by the presence of enamel columns partially fused into the faces of the main cones.

DISCUSSION. The holotype is a second upper molar (K.E. 2) in the British Museum (Nat. Hist.), figured by MacInnes (1942, figs. 5 and 6). It was collected in late Pliocene–early Pleistocene deposits at Kanam in the Kavirondo Gulf region, Kenya; but other specimens are widely distributed in East Africa from Ethiopia to Tanzania (table 17.1) in beds of similar age.

Originally described as a new subspecies of *Pentalophodon* by MacInnes, this taxon was attributed to *Anancus* as early as 1945 by Arambourg. The dental morphology of *A. kenyensis* is scarcely different from *A. osiris* except for the more complicated crown and the development of the M2 talon into a nearly full or full cone-pair. The species is known principally from dentitions and lower jaws.

Several undescribed specimens from earliest Pliocene levels at Lothagam Hill exhibit features considerably more primitive than seen in typical *A. kenyensis*. These are smaller in size and have fewer cone-pairs and simpler crown morphology. They would appear to represent an earlier stage in the evolution of this taxon.

A skull from Atoumanga, Chad, consists of palate and basicranium only. Another skull from the middle Pliocene Kanapoi Formation unfortunately is lacking dentition, but from its structure it cannot be attributed to elephant, stegodont, or deinothere. The general structure is gomphothere, and anancine jaws and teeth are abundant in the same deposits. The skull differs markedly from Eurasiatic members of the subfamily and leads us to suspect that

what we are calling *Anancus* in Africa may in fact represent a distinct African derivative of the Miocene gomphothere stock, paralleling its northern anancine cousins in molar structure.

A. kenyensis was a contemporary of the earliest elephants, representing for East Africa the equivalent of *A. osiris* in North Africa. It apparently became extinct during the beginning of the early Pleistocene, and with it disappeared the entire family Gomphotheriidae in sub-Saharan Africa.

Anancus petrocchii Coppens 1965

SYNONYMY. *Pentalophodon sivalensis* Cautley, Petrocchi 1943 (pp. 137–147); *Anancus (Pentalophodon) petrocchii* Coppens 1965 (pp. 337–338).

DIAGNOSIS. (Coppens 1965, p. 338); an *Anancus* differing from *A. osiris* in possessing five cone-pairs on M2 instead of four (based on Petrocchi's description). Otherwise similar in molar morphology.

DISCUSSION. This species is known only from Sahabi, Libya, of latest Miocene to early Pliocene age. The holotype consists of a mandible with M_3, now in the Libyan Museum of Natural History, Tripoli.

The characteristic second molars were not figured by Petrocchi, but three referred lower third molars were. These have six and one-half pairs of cones, including an undivided talonid. In this last feature they differ from M3 of *A. osiris* (see Arambourg 1970, pl. I, figs. 3 and 3A), which have no individualized talonids. This difference remains very small and it may be asked whether it would be consistant in a larger sample of specimens and consequently whether the presence of five rows of cones on the M2 is significant (Coppens 1965, p. 338). It should be noted that M2 of *A. osiris* shows a large cone on the recurrent anterior fold that with wear may almost form a fifth crest. However, this is a quite different structure than the posterior fifth crest in the Sahabi dentitions. Thus we tentatively maintain this material in the species *A. petrocchii* as a progressive derivation from *A. osiris*.

Anancus sp.

A number of undescribed molar teeth from the Pliocene of Langebaanweg, South Africa (Hendey 1970), represent a line of *Anancus* seemingly divergent from those found farther to the north. The specimens characteristically have somewhat folded enamel and a complex array of accessory tubercles and vertically grooved enamel not seen in other species of the genus. Nevertheless, these specimens bear a typical anancine structure and probably represent a distinct southern group of late gomphothere anancines that evolved in isolation from more northern populations.

Family Elephantidae Gray 1821

This family is known in the fossil record only since the Mio-Pliocene, where it apparently arose in Africa. By the middle Pleistocene it had become nearly worldwide in distribution.

Elephants differ strikingly from most other proboscidean groups in features associated with the masticatory apparatus. The molar teeth may be low to high crowned and are constructed as a series of transversely oriented plates, each consisting of a shell of enamel filled with dentine and separated from adjacent plates by cementum. Upon wear the occlusal surface presents the series of plates in cross-section so that cutting edges are formed that consist of transverse strips of enamel separated by alternating bands of cementum and dentine. Premolars are present only in the earliest members of the family. In later forms these fail to erupt, their position being occupied by persistant molariform milk molars. Because of the great length of permanent molar teeth, eruption proceeds as the tooth is still forming, several plates at a time, in the alveolar crypt behind. Thus the anterior portion of each tooth is erupted and functioning before the posterior portion is completely formed. As tooth formation continues the tooth moves forward in the jaw, pushing more worn molars in front of it until the latter are shed from the mouth. One or two functional teeth are present in each jaw at any one time. This process is a modification of the general mammalian eruption pattern and is in striking contrast to the condition seen in most Gomphotheriidae or Mammutidae, where the tooth enamel shell forms as a unit and where all molars and premolars may remain functional simultaneously. In Anancinae and Stegodontidae the number of teeth in function was already very reduced.

Elephants differ further from the Gomphotheriidae in the loss of mandibular tusks and shortening of the jaw symphysis early in their evolutionary history. The same phenomenon is seen in other groups, such as anancine and cuvierine gomphotheres, later mammutids, and stegodonts. However, such tusks do persist in the earliest known elephants. In the skull the palate is depressed ventrally and the alisphenoid wing is vertically oriented. The basicranium is foreshortened and the glenoid surface is strongly rounded in the anteroposterior direction, bounded in front by the temporal fossa and behind by the newly developed postglenoid depression (Maglio 1972, pp. 644, 653). These changes resulted from a dramatic alteration in masticatory function in this family.

Subfamily Stegotetrabelodontinae Aguirre 1969

This represents the most primitive group of probosocideans now placed in the Elephantidae. The known species retain a number of gomphothere features such as functional premolars, deep median cleft on the molar plates, very thick enamel, V-shaped valleys between plates, and long mandibular symphysis. In contrast to gomphotheres, however, they show several typically elephantlike features. Among these are flattened molar plates (actually two hemiplates), each of which is subdivided by vertical grooves, the obliteration of gomphothere trefoil wear patterns but retention of a single intravalley column behind each plate, a greater number and packing of plates on the crown, and a typically elephant type of tooth formation and eruption.

Genus *Stegotetrabelodon* Petrocchi 1941

SYNONYMY. *Stegotetrabelodon* Petrocchi 1941 (p. 107), *Stegolophodon* Petrocchi 1943 (p. 123).

This genus includes two species, *S. syrticus* (type species) and *S. orbus*. It is distributed in beds of latest Miocene to early Pliocene age in North and East Africa. The major diagnostic characters of the group are the same as those of the subfamily. They are large elephants with low-crowned molars bearing five to seven plates on permanent molars. A deep median cleft is present but does not extend to the base of the crown; thus complete enamel loops form with moderate wear. Single isolated columns are present behind the plates, representing vestigial trefoil folds. The crown height is much less than the width and the enamel thickness is 4 to 7 mm and not folded. Transverse valleys between plates are V-shaped and broad, giving a lamellar frequency of 2.5 to 3.0. The mandible has a long, protruding symphysis and bears a long pair of incisors lacking enamel.

When first proposed by Petrocchi (1941, p. 107), this genus was believed to represent a specialized gomphothere, a not unreasonable interpretation in view of the then known fossil record of elephants. Subsequent studies pointed out the elephantlike features of the group and it soon became clear that in *Stegotetrabelodon* we were dealing with an early elephant of gomphotheriid origin (Aguirre 1969a, p. 1370, 1969b; Maglio 1970a, p. 329). The significance of this conclusion was critical, for it put an end once and for all to the former belief of a *Stegodon* ancestry for the Elephantidae.

The low number of plates is equivalent to the greatest number attained in progressive gomphotheres (i.e., *Anancus*). Although a deep median cleft does persist, it is tightly compressed so that externally a complete platelike structure is formed. Upon moderate wear the two hemiplates fuse, forming a complete enamel loop of the elephant type. Crown height is still not extensive; the hypsodonty index (height × 100/width) is only 60 to 70, no greater than in most gomphotheres.

Stegotetrabelodon syrticus Petrocchi 1941

SYNONYMY. *Stegotetrabelodon syrticus* Petrocchi 1941 (p. 110); 1953–54 (p. 11, pls. 1–4); *Stegotetrabelodon lybicus* Petrocchi 1941 (p. 107); 1953–54 (p. 41); *Stegolophodon sahabianus* Petrocchi 1943 (p. 123, figs. 68, 69); 1953–54 (p. 45).

This was the first described species of the genus and it is known from the latest Miocene Gasr es Sahabi deposits in Libya. It is a large stegotetrabelodont with slender molar plates and columns in every valley, a crown height that is two-thirds or less than the width, and a mandible with very closely spaced slender incisors 2 m long, which form 57% of the total jaw length.

Thus far this species is represented only from a single locality. It provides us with the only cranial evidence for the entire subfamily. Although the frontoparietal region is damaged, a reconstruction (Maglio 1973, fig. 7) suggests that even in this early elephant the premaxilla is already sharply downturned and the temporal region shifted forward to a position more nearly over the dentition than in *Gomphotherium*. These features are particularly characteristic of elephants and are linked with adaptive advances in the dentition.

S. syrticus is the largest species of *Stegotetrabelodon* and in many characters, such as the persistant enamel columns in every molar valley and the very long jaw and lower tusks, it appears to be the least advanced toward the elephantine condition.

Stegotetrabelodon orbus Maglio 1970

SYNONYMY. *Stegotetrabelodon orbus* Maglio 1970b (p. 5, pls. 1 and 2).

Although no skull has yet been found, this form is dentally the best known member of the subfamily. It occurs principally in late Miocene to early Pliocene deposits of Lothagam Hill, Kenya, and other specimens are from the Mpesida and Kaperyon Beds, Baringo, Kenya. Overall molar dimensions are about 12% smaller than in *S. syrticus*. Free columns occur only behind the first two plates on M3, and the relative crown height is 13% greater than in *S. syrticus*. Mandibular incisors are smaller, less than 1 m long, and form only 38% of jaw length. The plate formula is M3 $\frac{6}{7}$, M2 $\frac{5}{5\text{-}6}$, dm4 $\frac{6}{?}$, dm3 $\frac{?}{3}$.

Recent K-Ar dates available for basalts within the Lake Baringo sequence give 7.0 m.y. for the Mpesida Beds and 5.0 m.y. for the Kaperyon Beds. The remainder of the fauna at these localities accords well with that from Lothagam-1 and the species range may be taken as approximately 7 to 5 m.y. On faunal correlations these seem approximately equivalent to the assemblage from Gasr es Sahabi so that *S. orbus* and *S. syrticus* were broadly contemporaneous.

Genus *Stegodibelodon* Coppens 1972

SYNONYMY. *Stegodibelodon* Coppens 1972b (p. 2964).

TYPE SPECIES. *Stegodibelodon schneideri* Coppens 1972.

This genus was recently established by Coppens for material collected near Lake Chad.[4] It is similar to *Stegotetrabelodon* in many features, differing primarily in its shorter mandibular symphysis and lack of lower incisors. This find is particularly important in demonstrating the diversity of the subfamily during the Pliocene and in its experimentation with features characteristic of the later Elephantinae. It seems likely that as the fossil record of the late Miocene and early Pliocene becomes better known this subfamily will prove to have been an even more diverse and widespread group than it is now believed.

Stegodibelodon schneideri Coppens 1972

SYNONYMY. *Stegodibelodon schneideri* Coppens 1972b (p. 2964).

S. schneideri is known from the early Pliocene of Menalla and Kolinga, Chad. It represents a progressive member of the Stegotetrabelodontinae, with a long symphysis, but shorter than in *S. orbus*, and lacking mandibular incisors in adults. It is not certain, however, whether this feature is ontogenetic or sexual. The median sulcus and intravalley

---

4 Maglio (1973) has included this specimen in the genus *Stegotetrabelodon*. With only the holotype specimen available further discussion of the validity of *Stegodibelodon* seems fruitless, and a greater sample is needed before the range of variability within these generic groups can be assessed.

columns are weakly developed in comparison to *S. syrticus* and there are seven plates on M3.

The genus and species were established on a nearly complete mandible that clearly differs from other known forms within the subfamily. However, the molar teeth are remarkably similar to those of *S. orbus* and in spite of the lack of tusks it is certain that *S. schneideri* represents a more highly specialized member of the same general group. The significance of tusk loss is still uncertain and its phylogenetic interpretation remains unsettled.

The deposits at Kolinga seem to be approximately contemporaneous with those at Lothagam, i.e., latest Miocene or early Pliocene in age.

Subfamily Elephantinae Gray 1821

These are elephants with low to high-crowned molars, with from seven to thirty plates per tooth. The crown is more progressive than in gomphotheres or in stegotetrabelodont elephants in that remnants of the old trefoil patterns and crown divisions are now completely lacking. The enamel shell is moderate to thin and folded in all but the earliest forms. A distinguishing feature of the group is the short symphysis of the lower jaw and the very reduced size or absence of the lower incisors. Premolar teeth are generally lacking, although they do occur occasionally in early species.

The subfamily is first recognized in Africa during latest Miocene times and occurs later in Eurasia from the middle Pliocene and in North America from the middle Pleistocene.

Genus *Primelephas* Maglio 1970

SYNONYMY. *Primelephas* Maglio 1970b (p. 10, pls. 3 and 4).

Included here are *Primelephas gomphotheroides* (type species) and *P. korotorensis*. The genus is distributed in the latest Miocene to early Pliocene deposits of East and Central Africa but is too poorly known for this restricted range to be taken too seriously. It is distinguished by having molar teeth that are very low-crowned, being only one-half to three-fourths as high as wide. A median cleft is lacking, but superficial grooves divide the plate faces into vertical pillars. Enamel valleys are V-shaped at the base but are more compressed than in *Stegotetrabelodon* so that the lamellar frequency is 3.0 to 4.0. The enamel is 3 to 6 mm thick and unfolded. Cement is abundant but does not fill the valleys completely. The mandible has a short symphysis and short lower incisors persist, at least in some species.

The genus was originally proposed for the inclusion of species that could not easily be accommodated in other more or less well defined genera of the family. It is possible, however, than when the Mio-Pliocene Proboscidea become better known *Primelephas* may have to be sunk into another taxon. For the present it seems best to retain it so as not to confuse the rather clear-cut features characteristic of other elephant groups.

Primelephas is more progressive than *Stegotetrabelodon* in lacking median clefts on the molar teeth so that the plates wear as complete enamel loops even in early stages. The plates are far less massive in build, more closely spaced, and the enamel is somewhat thinner. The very short mandibular symphysis, with tiny vestigial and widely separated incisors, also differs strikingly from the condition seen in stegotetrabelodonts.

The genus is best known from the Mio-Pliocene of East Africa, but species have also been recovered from Chad. Since *Primelephas* appears to have been largely contemporaneous with *Stegotetrabelodon,* it is not certain at this time whether the former was derived from the latter or whether both groups arose independently from advanced gomphotheres of the Ngorora type.

Primelephas gomphotheroides Maglio 1970

SYNONYMY. *Primelephas gomphotheroides* Maglio 1970b (p. 10, pl. 3, figs. A–D; pl. 4, figs. A and B; pl. 5, fig. A); *Mammuthus subplanifrons,* Cooke and Coryndon 1970 (p. 123).

This is the best known member of the group, even though represented by very limited materials, about 20 specimens in all. The holotype specimen is from the Mio-Pliocene of Lothagam Hill in Kenya, but other specimens are available from beds of similar age elsewhere in Kenya and in Uganda (table 17.1). The diagnostic features of the genus derive mainly from this taxon, and remarks made above apply here as well.

Inclusion of specimens from Kaiso in the present taxon (Cooke and Coryndon 1970, p. 123) may not be adequate when better material is available from those deposits, but the limited evidence seems to differ very little from the Lothagam collection. The best specimen (BM M-25160), the last half of an upper M3, differs only in being broader and slightly lower crowned.

The only specimen from Kanam East (KNM-KE 353) is incomplete but resembles the Kaiso material closely. The degree of difference seen among all the known remains does not justify specific separation at this time, and we may visualize a single species distributed throughout East Africa.

Primelephas korotorensis (Coppens) 1965

SYNONYMY. *Stegodon korotorensis* Coppens 1965b (p. 343, pl. 4, figs. 1–5).

Only two partial molar teeth are presently available for this species from the early Pliocene of Kolinga and upper Pliocene or even lower Pleistocene of Koulá in Chad. It is difficult to diagnose it completely, but the specimens differ from *P. gomphotheroides* in having a relative crown height 11% to 13% greater.

The Chad material was originally described as *Stegodon korotorensis* (Coppens 1965, p. 343) because of the low crown, triangular plate cross-section, and V-shaped transverse valleys. Based on the elephantid material then known, this reference seemed reasonable, especially in view of the known occurrence of *Stegodon* in the Kaiso Formation (MacInnes 1942, p. 84). However, the structure of the plates, their few large apical digitations, and the lack of tightly compressed valley bases suggest that we are dealing with an elephantid of the *Primelephas* type and not a stegodontid.

Since the species differs from *P. gomphotheroides* in having a clearly greater hypsodonty index (75 to 76 versus 61 to 64 for M3) and because of the geographic separation, it would be premature to refer the better East African material to this taxon, and both are provisionally recognized as valid species.

Genus *Loxodonta* F. Cuvier 1825

SYNONYMY. *Loxodonta* F. Cuvier and Geoffroy Saint-Hilaire 1825 (p. 2).

TYPE SPECIES. *Elephas africanus* Blumenbach 1797.

This genus includes only three known species, the living *L. africana* and two extinct forms, *L. adaurora* and *L. atlantica*. Although *Loxodonta* has been claimed to occur in the Pleistocene of Europe (Falconer 1868), the genus now appears to have been exclusively African in distribution. The earliest member of the group is from middle Pliocene beds in East and southern Africa and is already well differentiated in nearly every element of its skeleton.

Loxodonta differs from *Primelephas* in having longer molar teeth with more plates (8 to 15 on M3), a greater hypsodonty index (80 to 110), and generally thinner enamel. The plates are more closely spaced, with a lamellar frequency of four to five. The mandible is long but the symphysis is short and lacks external tusks.

The name of the group derives from the lozenge-shaped enamel wear pattern formed by the plate cross-section on teeth of the living species. Such a pattern is not so well defined in *L. adaurora* because the plates have not yet become expanded in their median portions. However, prominent anterior and posterior enamel loops are present on the partially worn plates (as in other primitive elephants); these result from the partial fusion of intravalley columns into the plate faces.

In the skull the most diagnostic features are the rounded shape, the biconvex frontoparietal surface, the distally flaring tusk sheaths (premaxillae), the lack of a median-parietal depression, and the rounded dorsal borders of the temporal fossae. In these features *Loxodonta* may be distinguished from all other genera of Elephantidae (except *Primelephas,* for which the skull remains unknown).

Throughout the evolutionary history of the *Loxodonta* group of elephants they remained relatively conservative in dental specializations. As other groups either became extinct during the Pliocene or rapidly evolved during the early Pleistocene, *Loxodonta* changed very little from the time of its first appearance until the present (Maglio 1973, pp. 105–106). General molar structure in *L. africana* is no more evolved than in early Pleistocene species of *Elephas* or *Mammuthus.*

Loxodonta adaurora Maglio 1970

SYNONYMY. *Loxodonta adaurora* Maglio 1970b (p. 12, pl. 5, figs. 14–16; pl. 6, figs. 17, 18); *Elephas exoptatus* Dietrich 1942, *partim* (p. 72, figs. 43, 54, 57, 59, 67, 68, 73); *Archidiskodon* cf *meridionalis,* MacInnes 1942 (p. 92, pl. 8, fig. 3); *Archidiskodon planifrons,* MacInnes 1942 (p. 86, pl. 7, fig. 10; pl. 8, figs. 1 and 2); *Archidiskodon africanavus,* Cooke and Coryndon 1970 (pl. 1, figs. D and E).

The earliest record of the genus *Loxodonta adaurora* is known in middle Pliocene to early Pleistocene deposits of eastern and southern Africa. The holotype specimen is a nearly complete skeleton from the Kanapoi Formation in Kenya. An anatomically complete skull has also been recovered from Unit 3 at Lothagam Hill and a complete jaw and partial skull from Kubi Algi, both in the east Lake Turkana region. A second skeleton was excavated from the Chemeron Formation near Lake Baringo.

In most cranial features *L. adaurora* is very close to *L. africana,* differing only in the longer premaxillary and frontal regions, a more posteriorly placed temporal fossa, and in the prominence of frontoparietal ridges lateral to the external naris. The jaw is longer than in the living form and carries a pair of vestigial incisive germ cavities in the symphyseal region. The plate formula for the permanent molars is M3 $\frac{8\text{-}10}{10\text{-}11}$, M2 $\frac{7\text{-}8}{6\text{-}8}$, M1 $\frac{7}{6\text{-}7}$.

Isolated teeth of all elephants passed through similar early stages in their evolution so that it is often difficult to be certain of their identification. Thus the specimens here considered to belong to *Loxodonta adaurora* had been previously assigned to a number of other taxa, including *Archidiskodon meridionalis, Archidiskodon planifrons, Elephas exoptatus,* and *Archidiskodon africanavus.* With large samples of teeth firmer identifications are possible, but only cranial material can sort these early elephants conclusively. It is now reasonably certain that *Elephas planifrons* never occurred in Africa and that *Mammuthus meridionalis* did not penetrate south of the Sahara.

In a review of the Laetolil-Vogel River proboscideans the type collection of *Archidiskodon exoptatus* was shown to consist of two distinct taxa—one referable to *L. adaurora* and the other to *Elephas recki.* A specimen of the latter was chosen as the loctotype specimen (Maglio 1969), thus reducing the name *exoptatus* into the synonymy of *E. recki.*

The elephant collection from the Mursi Formation in Ethiopia has not yet been described but seems certainly to belong to this species. Material from the lower part of the Shungura Formation, plus several specimens from the lower part of the East Lake Turkana section dated at about 2.6 m.y., confirm the persistence of the species from 4.0 m.y. to possibly 2.0 m.y. in Africa. During this period little morphological change is seen, at least in the dentition.

In 1929 Dart described three species of elephant from the Vaal River gravels of South Africa. The holotype of each is based on a single incomplete molar tooth without stratigraphic or faunal association. The taxa to which these specimens belong are morphologically close to *L. adaurora* or *Mammuthus africanavus,* but the type material is inadequate for confident identification. Because of this, these three names, *Archidiskodon vanalpheni, A. milletti,* and *A. loxodontoides,* have been considered *nomina dubia.* A fourth name, *Loxodonta griqua* Haughton 1922, made the type of a new genus *Metarchidiskodon* by Osborn (1934), is similarly indeterminant.

Loxodonta atlantica (Pomel) 1879

SYNONYMY. *Elephas atlanticus* Pomel 1879 (p. 51, pl. 8, figs. 1 and 2); *Elephas (Loxodon) zulu* Scott 1907 (p. 259, pl. 17, fig. 6; pl. 18, fig. 1); *Elephas pomeli* Arambourg 1952, *partim* (p. 413, figs. 7 and 8; pl. 1, fig. 4).

Previously believed to have been exclusively North African in distribution, this species is now recognized in East and southern Africa as well. The holotype specimen is from the middle Pleistocene deposits at Ternifine, Algeria, and other material is from Sidi Abder Rahmane, Morocco, and from the

middle Pleistocene of Elandsfontein in the Cape Province, and in late Pliocene beds at the Omo River, Ethiopia.

Probably derived from *L. adaurora,* the present species diverged rapidly so that by Plio-Pleistocene times it was already very distinct. It is the largest of the loxodont species and also the most progressive in dental features, even more so than the living taxon. Crown height is twice as great as the width and the lamellar frequency is 3.5 to 5.5. Unlike in other members of the genus, however, the enamel is coarsely folded. With wear, the plate cross-sections form enamel figures of the typical lozenge shape with a median loop as in other loxodonts. However, here the loop is generally bifurcated and Y-shaped. Large loops also occur irregularly around the enamel figure. The number of plates on the molar teeth are about as in the living species: $M3 \frac{12\text{-}14}{10\text{-}15}$, $M2 \frac{9\text{-}12}{11\text{-}12}$, $M1 \frac{8\text{-}9}{7\text{-}10}$. An unpublished skull of this form from Ternifine shows typically *Loxodonta* characteristics. It differs from other species in the narrower premaxillary region and in the very large size of the occipital condyles.

Early in this century Scott (1907, p. 259) described as a new species *Elephas zulu* based on two lower molars from South Africa. Later Cooke (1947) recognized the similarity between these teeth and Pomel's (1879) Ternifine species, but he maintained *E. zulu* as distinct because of slight differences in molar crown height and size. In a later revision (Cooke 1960) he followed Arambourg's earlier (1938) suggestion that Scott's species fell completely within the range of variability of the Ternifine collections. As long as Scott's type molars remained the only specimens from southern Africa this issue could not be settled.

The recently collected fauna from Elandsfontein in the Cape Province includes a large sample of elephant remains attributable to Scott's taxon. When this is compared with the large Ternifine assemblage, a number of consistent differences between the two emerge. The North African material tends to have a Y-shaped or bifurcated median loop on worn enamel figures; these loops are almost always single in the southern group. The Ternifine specimens are smaller in size and have a lower lamellar frequency. Finally and most strikingly, the enamel on teeth of the South African material is thinner and nearly completely outside the range of variability of the Ternifine sample (Maglio 1973, tables 9 and 10). These differences do not warrant specific separation of the two populations, however, and they are better given subspecific status as *Loxodonta atlantica atlantica* from North Africa and *L. atlantica zulu* from southern Africa.

Arambourg (1952, p. 413) described *Elephas pomeli,* a new species, based on three molar teeth from Algeria and Morocco. The holotype specimen from Sidi Abder Rahmane belongs to the same taxon as *Elephas iolensis* Pomel (1896) and thus Arambourg's new name falls into the synonymy of the latter species. The remaining specimens in Arambourg's hypodigm of *E. pomeli* from Oued Constantine should be referred to *Loxodonta atlantica atlantica.*

The origin of the species still remains uncertain, and until recently it was seen to appear suddenly in the fossil record in beds of early middle Pleistocene age. The discovery from below Tuff E in the Shungura Formation of two complete teeth attributed to a very early stage of this species (Maglio 1973, p. 29) at least suggests a late Pliocene origin for the species, probably from *L. adaurora.* The rapid exclusion of this form from East Africa with the rise of *Elephas recki* there and its dominance in North and South Africa where *recki* was either lacking or was very rare indeed implies a possible ecological separation of the two genera. This probably explains why the two are almost never found together.

Loxodonta africana (Blumenbach) 1797

SYNONYMY. *Elephas africanus* Blumenbach 1797 (p. 125, fig. C); *Loxodonta prima* Dart 1929 (p. 725, figs. 25 and 26); *Loxodonta atlantica angammensis* Coppens 1965 (p. 360, pls. 11–13).

This is the living African elephant and little need be said about it here. This species first occurs in the fossil record in beds of late early Pleistocene age at Ileret, East Lake Turkana (Maglio 1972), and could easily have derived from *Loxodonta adaurora.* Molar teeth of the extant form are distinguished from those of *L. adaurora* by their higher and narrower crowns and by the development of the typical lozenge shape wear pattern resulting from expansion in the central region of the plates. In the cranium the premaxillaries are greatly reduced, the frontal region is foreshortened, and the lower jaw is shorter and lacks the vestigial incisive chambers of *L. adaurora.*

The subspecies *L. atlantica angammensis* Coppens (1965, p. 364) from deposits at Yayo near the Angamma Escarpment, Chad, was described on a juvenile skull with milk dentition and on several isolated molar teeth. The diagnosis suggests this as an early form of *L. atlantica* but most of its features cannot be distinguished from the living species. In view of the presence of *L. atlantica* as early as Member E of the Shungura Formation, it seems possible

that we are dealing here with an early stage of the Recent species (Coppens 1966, p. 6).

Genus *Elephas* Linnaeus 1758

SYNONYMY. *Elephas* Linnaeus 1758 (p. 11); *Pilgrimia* Osborn 1924 (p. 2); *Palaeoloxodon* Matsumoto 1929 (p. 257); *Omoloxodon* Deraniyagala 1955 (p. 125).

TYPE SPECIES. *Elephas maximus* Linnaeus 1758.

This is the largest and most diverse of the five elephant genera with most of its species being Eurasiatic in distribution. Of the African species, the group includes *Elephas ekorensis, E. recki,* and *E. iolensis;* none of these has yet been recorded outside the continent. Because considerable evolution occurred within the genus during its history, generic diagnosis cannot be precise on teeth alone, but all members of the group can be distinguished on cranial material. The skull is high and the forehead foreshortened. Premaxillary tusk sockets are generally widely separated and the tusks usually gently curved in a single plane. The frontoparietal surface is flat to concave, and the upper borders of the temporal fossae form sharp, raised ridges. In the mandible the corpus is strongly convex laterally and the condyles are transversely elongated and directed upward and inward. The teeth vary from being very primitive to highly evolved. Except in the earliest species, the teeth are distinguished from other African genera by their parallel-sided enamel figures, folded thin enamel (4.0 to 2.0 mm), high lamellar frequency (3.5 to 9.0), and relatively high crowns.

In 1955 Deraniyagala proposed the name *Omoloxodon* for *Elephas recki.* His diagnosis was based on several cranial features, but these do not justify generic separation.

Elephas ekorensis Maglio 1970

SYNONYMY. *Elephas ekorensis* Maglio 1970b (p. 20, pl. 7, figs. 19 and 20).

This is the earliest known species of the genus. The holotype dentition and referred skull are from the middle Pliocene Ekora Formation, Kenya, and other material is from East and southern Africa (table 17.1).

Even at this early stage *Elephas ekorensis* shows features in its dentition that are clearly more progressive than in any contemporaneous species. The crown is 10% to 30% higher than wide (HI = 110 to 130) and the greatest width occurs at the base of the crown. Enamel is 3 to 4 mm thick and smooth except near the base of the tooth, where it is weakly folded. Plates are well separated with a lamellar frequency of four to five, and valleys between plates are U-shaped at the base. The skull is typically *Elephas* in shape, but the forehead is flat. The plate formula is M3 $\frac{11}{12}$, M2 $\frac{?}{?}$, M1 $\frac{7}{8}$.

In all dental and cranial features *Elephas ekorensis* resembles the later *E. recki,* but it is less evolved. There seems little reason to doubt the ancestral position of *E. ekorensis* for the African group of *Elephas* and perhaps for the genus as a whole.

Although referred to the same taxon, the Kubi Algi skull represents a population even more primitive morphologically than the type collection from Ekora. Kubi Algi is somewhat older, with a K-Ar age of 4.55 ± 0.1 m.y. for a tuff immediately underlying the fossiliferous unit, and thus places the origin of the genus well back into the earlier Pliocene. The latest occurrence of the species is from Bolt's Farm, South Africa, which on other faunal evidence appears to be at least 2 m.y. younger. This is significant in that it lessens the value of single faunal elements for correlation purposes. *Elephas recki,* a presumed descendant species in East Africa, does not appear to have ever dispersed south of the Zambesi and it is possible that because of this *E. ekorensis* persisted here with little evolution. Thus its presence is not necessarily indicative of a Pliocene age for the Bolt's Farm deposits, unless of course this proves to contain a mixed fauna. There is, however, no evidence for this.

The earliest extra-African remains of *Elephas* are known from the Bethlehem fauna of Palestine, considered to be of late Pliocene age (Hooijer 1958; Maglio 1973). These fossils have been described as an early stage of *Elephas planifrons,* a species known on excellent material from latest Pliocene deposits of the Pinjor Horizon in the Siwalik Hills Series of India and Pakistan. The Bethlehem dental material is morphologically very close to *Elephas ekorensis* and almost certainly represents an emigrant of this group from Africa. Skulls of *E. planifrons* from the Siwaliks show clear similarities to that of *E. ekorensis* and an ancestral-descendant relationship seems almost certain. We can thus trace the origin of the complex Asiatic *Elephas* group to the Pliocene of Africa, from something close to *E. ekorensis.*

Elephas recki Dietrich 1916

SYNONYMY. *Elephas recki* Dietrich 1916 (p. 22, pls. 1–8); *Elephas zulu,* Hopwood 1926 (p. 32, pl. 3, fig. 1, pl. 4); *Elephas* aff. *meridionalis,* Hopwood 1926 (p. 33, figs. 13 and 14); *Loxodonta (Pilgrimia) antiqua recki,* Osborn 1928 (p. 673); *Archidiskodon griqua,* Hopwood 1939 (p. 311); *Archidiskodon exop-*

tatus Dietrich 1942, *partim* (p. 72, pl. 5, figs. 40, 42, 43, 45, 46, 48, 51–53; pl. 6, fig. 60; pl. 7, figs. 61 and 64; pl. 9, fig. 69; pl. 10, figs. 71 and 72; pl. 11, figs. 76 and 77); *Elephas (Archidiskodon) recki,* Arambourg 1947 (p. 252, pl. 1, fig. 4; pl. 2; pl. 3, figs. 1–4; pl. 4; pl. 5, figs. 1, 2, 5; pl. 6, figs. 1 and 4; pl. 7, figs. 1–4, 6; pl. 8, fig. 5); *Elephas* cf *africanavus,* Leakey 1965 (p. 23, pls. 16, 17).

This now ranks as the best known fossil elephant in Africa, and a definitive study on the abundant materials available is currently underway by Michel Beden.

The species is recorded from late Pliocene to middle Pleistocene age deposits of East and Central Africa. The holotype is from Bed IV, Olduvai Gorge, but the species was very widespread throughout eastern Africa.

The molar teeth of *E. recki* are higher crowned than those of *E. ekorensis,* with a hypsodonty index of 150 to 200, and the plates are more closely spaced, giving a lamellar frequency of four to seven. The enamel is strongly folded except in the earliest stages. The dental formula is M3 $\frac{9\text{-}16}{11\text{-}16}$, M2 $\frac{9\text{-}11}{9\text{-}10}$, M1 $\frac{7\text{-}10}{8\text{-}9}$, dm4 $\frac{7\text{-}10}{8\text{-}10}$, dm3 $\frac{6\text{-}7}{6}$, dm2 $\frac{3}{3}$. In contrast to *E. ekorensis,* the molars attain their greatest width at about one-third up from the crown base. The skull is compressed in the facial plane and parietal and occipital regions are greatly expanded. The forehead is concave and the naris sharply downturned at the sides. As is typical of the genus, strong frontoparietal ridges extend down from the anterior temporal crests to terminate at small, widely spaced orbits. The premaxillary tusk sheaths are very closely spaced and parallel.

Several years ago four evolutionary grades were recognized within this species (Maglio 1970a), designated Stages 1 to 4. These stages grade into one another, making formal taxonomic terminology impossible. The earliest stage (1) is represented only by dentitions and is known from Kikagati, Uganda; it also occurs in the lowest levels of the Shungura Formation (Omo). These teeth show certain advances over the condition of *E. ekorensis* (Cooke and Maglio 1972) in being higher crowned and having more and closer spaced plates and thinner enamel.

Stage 2 is the typical form in the Shungura Formation above Member C and at East Lake Turkana. The plates here are even more closely spaced and the crown is proportionately higher. Enamel is weakly but rather uniformly folded, even in early stages of wear. This is the stage represented in the Vogel River Series of the southern Serengeti, which Dietrich (1942, p. 72) described as *Archidiskodon exoptatus*. Several complete skulls are known and show important advances over *E. ekorensis* but continue the trends initiated there.

A more evolved stage (3) is typical in Olduvai Gorge Beds I and lower II, Ileret (East Lake Turkana) and Kalam (Omo Basin). The enamel is more highly folded and tight loops occur irregularly spaced around the enamel figures. The crown is higher than in Stage 2.

The final stage (4) occurs in middle Pleistocene deposits of Olduvai Bed IV, Olorgesaillie, and Homa Mountain. A great number of plates characterize teeth of this group and these are as high crowned as is ever seen in this species. Median loops, so typical of all earlier stages, occur only rarely here.

This evolutionary story has been complicated by the discovery of several very progressive-looking specimens, one from locality J. M. 90(91) in the Chemeron Formation and one from the lower levels at East Lake Turkana, both in Kenya. In both cases the morphology of the specimen clearly suggests a Stage 3 or 4 *Elephas recki;* but from stratigraphic and associated faunal considerations they would appear to be far too old for such an evolutionary grade. It is too early to satisfactorily resolve this dilemma. It may be that these few specimens represent the extremes of variability within populations of *E. recki* Stage 1 or 2, in which case we must exercise caution in drawing conclusions from any but adequate sized samples. Alternatively, although less likely, these may represent a distinct lineage derived from *E. ekorensis* and paralleling *E. recki* but far more progressive in dental morphology.

Isolated teeth, especially of the earlier stages, can be mistaken for other elephant species and have been identified variously as *Elephas zulu* (Hopwood 1926), *Archidiskodon griqua* (Hopwood 1939), *Palaeoloxodon antiquus* (MacInnes 1942), and *Elephas* cf *africanavus* (Leakey 1965). *Elephas recki* has been referred to five different genera by past authors, but from its cranial adaptations there is little doubt that it is allied to the *Elephas* group. It almost certainly derived from *E. ekorensis* during the middle Pliocene and formed the beginning of the African branch of the genus. From the structure of the skull and dentition, clear relationships seem to tie this species to the Eurasian *Elephas namadicus* (=*antiquus*). The latter makes a rather sudden appearance in northern continents during the earliest middle Pleistocene and may have originated from this common African group (see Maglio 1973, pp. 37, 83).

Elephas iolensis Pomel 1895

SYNONYMY. *Elephas iolensis* Pomel 1895 (p. 32, pl. 5, figs. 3 and 4); *Archidiskodon transvaalensis* Dart 1927 (p. 47, fig. 6 right, 7 left); *Archidiskodon sheppardi* Dart 1927 (p. 48, fig. 6 left, 7 right); *Archidiskodon broomi* Osborn 1928 (p. 672, fig. 2); *Archidiskodon hanekomi* Dart 1929 (p. 713, fig. 24); *Palaeoloxodon kuhni* Dart 1929 (p. 723, figs. 15 and 16); *Pilgrimia yorki* Dart 1929 (p. 717, fig. 19); *Pilgrimia wilmani* Dart 1920 (p. 720); *Pilgrimia archidiskodontoides* Haughton 1932 (p. 4, pls. 1–2); *Pilgrimia subantiqua* Haughton 1932 (p. 8, pl. 4, figs. 1–2, 12, 13); *Elephas pomeli* Arambourg 1952, *partim* (p. 413, fig. 6, pl. 1, fig. 3).

This was the last species of *Elephas* on the African continent and, although known only from dentitions, it appears to have been the terminal member of the *E. ekorensis-E. recki* lineage. It was distributed throughout the continent and is represented in late Pleistocene deposits at Port de Mastaganem, Algeria; Sidi Abder Rahmane, Morocco; Kaiso, Uganda (level unknown); Kibish Formation, Sudan; Zouerate, Mauritania; and the youngest terraces of the Vaal River, South Africa, among other localities. The latest record is from Natodameri, Sudan, provisionally dated by the U-Th method at about 35,000 years B.P. (R. Thurber pers. comm.) and associated with an essentially modern African fauna.

E. iolensis is a large species with relatively broader molar teeth than in *E. recki*. The plates are very thick but the intervals between them are extremely narrow. The crown height is two to three times as great as the width, making these the highest crowned elephant teeth from Africa. The plate number, however, is no greater than in the most progressive *E. recki* nor is the enamel any thinner. Following the trend seen in *E. ekorensis* and *E. recki,* the maximum width of the molars has now moved up to about midheight of the crown. *E. iolensis* can be the descendant of *E. recki* (Coppens et al. 1972, pp. 459, 460).

When Pomel (1895) first described this taxon, he believed the type molar to be an M_3 of a dwarf species, and this concept was long maintained. Later Osborn (1942) allied it with the *E. namadicus* group of Mediterranean pygmy elephants. However, the holotype specimen of *E. iolensis* is a second molar, not a third, and this accounts for its small size and low number of plates.

At the other end of the continent, in South Africa, Dart (1927, 1929) described 11 new species from the Vaal River gravels, most based on teeth of different serial position. Of these, eight appear to belong to a single taxon that cannot be separated from *E. io-lensis* (Maglio 1973, table 15). Thus the names *Archidiskodon transvaalensis, A. sheppardi, A. broomi, A. yorki, A. hanekomi, Palaeoloxodon kuhni, Pilgrimia wilmani,* and *P. yorki* fall into the synonymy of *E. iolensis.*

Another South African species, *Pilgrimia archidiskodontoides* Haughton (1932, p. 4), is based on a very poorly preserved specimen. As Cooke (1960) pointed out, it is not possible to determine with confidence the relationships of this tooth; but from its overall morphology of thick, closely spaced plates it seems best referred to the present taxon.

It was mentioned above that Arambourg's (1952, p. 413) specimens of *E. pomeli* from Sidi Abder Rahmane compare well with undoubted third molars of *E. iolensis,* even though the holotype specimen of that taxon from another site is referred to *Loxodonta atlantica.* Clearly, both *Loxodonta* and *Elephas* were present in northern Africa during the middle Pleistocene, although they are not recorded from the same localities.

Genus *Mammuthus* Burnett 1830

SYNONYMY. *Mammuthus* Burnett 1830 (p. 352); *Archidiskodon* Pohlig 1888 (p. 138).

TYPE SPECIES. *Mammuthus primigenius* (Blumenbach) 1799.

Although this genus is generally thought of in terms of the mammoths of northern continents, like all other elephant groups, this one also had its origin in Africa. The early species of the genus occurring in Africa are *M. subplanifrons, M. africanavus,* and *M. meridionalis.* The group appeared as early as either *Loxodonta* or *Elephas,* that is, in the middle Pliocene. But by early Pleistocene times the mammoths had already become extinct in Africa and were to carry on their spectacular evolution in Eurasia and North America.

The mammoths were medium to large-sized elephants whose dental and cranial specializations were to reach the extreme for the family. Earlier species are difficult to separate from other genera on teeth alone; but when cranial evidence is available, there is little problem in allying them with later, more typical forms. In the teeth the enamel rarely shows the broad enamel folding of *Elephas.* Rather, the enamel is usually very finely wrinkled or wavy and the median loop persists on partially worn plates, but only in the earliest members. The skull is anteroposteriorly foreshortened and the parietals are dorsally expanded. The midsagittal depression so typical of *Elephas* is lacking here and the

forehead is characteristically convex transversely and concave vertically. The temporal constriction is very strong and temporal fossae are bounded dorsally by gently rounded borders, not sharp ridges as in *Elephas.* The premaxillary sheaths are narrow, closely spaced proximally, but divergent distally, and the tusks are spirally twisted.

In his monograph on the Proboscidea, Osborn (1942) separated the species of the present assemblage into three generic groups: *Archidiskodon* (for *subplanifrons, meridionalis,* and *imperator*), *Parelephas* (for *columbi* and *armeniacus* [= *trogontherii*]), and *Mammonteus* (for *primigenius*). Distinctions were based on proportions. However, when the entire group is examined, a continuous transition can be seen from the earliest to the latest of these species, and it is clear that we are dealing with only one or two specific lineages. There seems no reason to separate any of them on the generic level.

Mammuthus subplanifrons (Osborn) 1928

SYNONYMY. *Archidiskodon subplanifrons* Osborn 1928 (p. 672, fig. 1); *Archidiskodon andrewsi* Dart 1929 (p. 711, fig. 14); *Archidiskodon planifrons nyanzae* MacInnes 1942 (p. 86, pl. 7, fig. 9; pl. 8, fig. 1); *Mammuthus (Archidiskodon) scotti* Meiring 1955 (p. 189, pls. 1–4, text figs. 1–8); *Mammuthus subplanifrons,* Maglio and Hendey 1970 (p. 85).

This is one of the least understood species of the family. It is known only from a few dozen specimens, mainly teeth, but the degree of evolution seen here can be matched in (and thus confused with) other early elephants. The best specimens consist of teeth and some skeletal fragments from Langebaanweg, South Africa (originally described as *Stegolophodon* sp. by Singer and Hooijer 1958); and although a jaw shows none of the features typical of *Loxodonta,* it cannot be said for certain that it is not an early *Elephas.* Of the specimens currently referred to *M. subplanifrons,* some suggest *Mammuthus* affinities. For example, the specimen from Virginia, O.F.S., described by Meiring (1955, p. 189) as *M. (Archidiskodon) scotti* includes a tusk of characteristic mammoth shape. For the moment, however, we must view *M. subplanifrons* as an heterogeneous assemblage of convenience in which we are certainly mixing the teeth of other early elephants along with what seems to be the most primitive stage of the mammoth lineage (Maglio and Hendey 1970, p. 87).

Specimens included in this taxon occur in beds of early to middle Pliocene age from the middle(?) terraces of the Vaal River and Langebaanweg, South

Africa, and occur at a number of sites in East Africa. These teeth are all low crowned, the height being only 70% to 90% of the width, and plates are widely separated, giving a lamellar frequency of only 2.5 to 4.5. The enamel is thick (3.5 to 5.5 mm) and unfolded and strong median loops are present on worn enamel figures. The plate formula is M3 $\frac{7\text{-}8}{8\text{-}9}$, M2 $\frac{5\text{-}6}{?}$.

Until additional material of this species becomes available, especially cranial evidence, this will remain a questionable taxon.

Mammuthus africanavus (Arambourg) 1952

SYNONYMY. *Elephas meridionalis,* Pomel 1895 (p. 13, pl. 1, fig. 3); *Elephas planifrons,* Depéret and Mayet 1923 (p. 120, pl. 4, fig. 7); *Elephas africanavus* Arambourg 1952 (p. 413, pl. 1, fig. 2, text fig. 1); *Loxodonta africanava,* Coppens 1965 (p. 348, pls. 5–8; pl. 9, figs. 1–6).

With this form we come to the first clearly definable member of the genus, if the skull described by Arambourg (1970) can be surely related to the teeth described as *E. africanavus* in North and Central Africa. The species occurs in middle to late Pliocene deposits of Algeria, Tunisia, Morocco, and Chad. It is smaller than *M. subplanifrons* and has moderately broad molar teeth that are equal to or up to 20% greater than the width. Plates are still well spaced, with a lamellar frequency of 3.5 to 5.5. The enamel is unfolded but thinner than in *M. subplanifrons* (2.5 to 4.0 mm) and when partially worn, median loops persist on the enamel figures. The sides of the plates taper strongly toward the apex in characteristic fashion. A skull found at Garet et Tir, Algeria (Arambourg 1970, pp. 35–37, pls. 2–3), strongly suggests the *Mammuthus* affinities of this taxon.

The teeth of *M. africanavus* have been mistaken for those of *Loxodonta* and *Elephas,* both of which passed through similar evolutionary stages. When large samples are available, however, such as that from Chad (Coppens 1965, p. 348, pls. 5–9), the sum total of morphological characters can more easily be related to one or another of these groups. As with *M. subplanifrons,* only cranial evidence can adequately identify these early species.

Mammuthus meridionalis (Nesti) 1825

SYNONYMY. *Elephas planifrons,* Doumergue 1928 (p. 114); *Elephas* aff. *meridionalis,* Arambourg 1952 (p. 410, fig. 4; text fig. 5); *Elephas moghrebiensis* Arambourg 1970 (p. 59, pl. 10, fig. 2; pl. 11, fig. 1; text figs. 41–43).

Although this is typically a European species, it or a closely related form did occur in northern Africa during the early Pleistocene, at Aïn Hanech and Bel Hacel, Algeria. These specimens were originally identified as *Elephas planifrons* (Doumergue 1928). Arambourg (1952, p. 410) later recognized their affinities with the European mammoth, *M. meridionalis,* but he subsequently (1970, p. 59) separated it under a new name, *Elephas moghrebiensis,* on several molar characteristics. However, when the degree of variation in large samples of *M. meridionalis* is taken into account (see Maglio 1973, table 30), the North African material cannot adequately be separated from it on the species level.

Although it may reasonably be suggested that European *M. meridionalis* was derived from *M. africanavus* by a cross-Mediterranean expansion during the late Pliocene, it cannot be determined whether the North African early Pleistocene material represents an *in situ* evolved stage from *africanavus* in Africa (and thus a parallel line to the European species) or whether it derived from a back-invasion from Europe after *M. africanavus* became extinct in Africa. Whatever the case may be, *M. meridionalis* did not persist beyond the early Pleistocene here and by middle Pleistocene times had given way to *Loxodonta atlantica* as the dominant North African species.

Suborder Mammutoidea

Family Mammutidae Cabrera 1929

This family includes the nonbunodont mastodonts of North America, Eurasia, and Africa. It is characterized by anteroposterior compression of paired molar cones to form transverse trenchant crests with a closed but not fused median cleft and the loss of intravalley columns and accessory tubercles. The mandibular symphysis is short and lower tusks are reduced or absent. The skull is short and broad.

One genus is currently recognized in Africa, *Zygolophodon.* The group seems to have originated in Africa, possibly from a *Palaeomastodon* stock during the Oligocene. It appears never to have been very abundant on the continent, however, and quickly became extinct here. Later reinvasions of Eurasiatic descendant groups maintained the family's presence in North Africa sporadically during the Pliocene.

This is not the place to enter into the extremely complex arguments concerning the synonymy of the family. Two schools have been followed in recent years, the first keeping the name Mammutidae Cabrera, based on *Mammut* Blumenbach (see Simpson 1945), the other using the name Mastodontidae

Girard, based on Cuvier's genus *Mastodon*. Because the latter is fraught with uncertainty as regards its early usage and because of its use as a common name for the gomphotheres, we follow Simpson in maintaining both Mammutidae and *Mammut* as valid nomia for this assemblage.

Genus *Zygolophodon* Vacek 1877

TYPE SPECIES. *Zygolophodon tapiroides* Cuvier.

DIAGNOSIS. Differs from *Mammut* by its greater number of crests, four to five main pairs forming sharp transverse ridges.

DISCUSSION. Two subgenera are recognized in Europe, *Zygolophodon* and *Turicius,* based on the number of conelets, the tri- or tetralophodonty of the middle molars, and the degree of elongation of the mandibular symphysis. Two species have been reported from Africa, but only one can be confirmed.

Zygolophodon (*Turicius*) sp.

DISCUSSION. Molar teeth have been described as belonging to *Z.* (*Turicius*) from several Algerian localities: Marceau, Khenchela, and Smendou (Arambourg 1945, p. 488; Vaufrey 1958, p. 234; Boné and Singer 1965, p. 293; Thenius 1959, p. 263). From Gervais's (1859) description, quoted in Depéret (1897, p. 518), it appears that the molar attributed by that author to *Mastodon borsoni* should definitely be included in the present taxon. In fact, it shows the several accessory conules and deep valleys between the crests, which are typical. From Khenchela and Marceau only two molars are known and these also resemble those of *Turicius.* Although the material shows that the genus was certainly present in North Africa during the Mio-Pliocene, a specific determination of these teeth would be premature at this time.

?*Zygolophodon* (*Zygolophodon*) *borsoni* (Hays) 1834

DISCUSSION. The European mastodon, *Z.* (*Z.*) *borsoni,* has been reported several times from Plio-Pleistocene deposits of Algeria near Saint-Arnaud (bed of the Sillègue road), in Tunisia at Ferryville, and in Pontian beds of Condé Soreadon. For specimens from the first two sites neither descriptions nor figures are available; they have only been mentioned by Depéret, Lavauden, and Solignac (1925). Arambourg (1970, p. 33) believes the Ferryville (=Menzel Bourguiba) specimen to be referable to *Anancus osiris,* the same may be true of other material.

The tooth described as *Mastodon borsoni* by Pomel (1896) does not exist any longer, but from the de-scriptions published by Pomel, it had its crown entirely within the matrix and the drawing represents it as seen from below. Thus this determination also must remain uncertain, and the presence of the species at all in Africa remains highly questionable.

Mammutidae indet.

A number of definite mammutid specimens (molars and milk dentition) have recently been recovered from various East African sites: Moroto I, Uganda (W. W. Bishop pers. comm.), and Lothidok, Kenya (Madden 1972 and ms.). This material appears surprisingly more like North African *Mammut* than like some European forms, but additional evidence is needed before we would venture an identification. It is significant because of its early Miocene age, which makes it not only the earliest record in Africa but among the earliest representatives of the family known anywhere.

Family Stegodontidae Osborn 1918

The stegodonts were primarily an Asiatic group of mammutid origin that paralleled the elephants in development of a fore-and-aft shearing dentition formed of continuous transverse enamel ridges. Concomitant modifications in cranial architecture show numerous adaptive features similar to those in elephants but with far greater diversity in detail. General characters in the family include low-crowned teeth with thick enamel and apically converging sides completely lacking intravalley columns, division of plate apices into tiny mammillae, wear of plates into highly folded and scalloped enamel figures, close approximation of maxillary tusks, and short jaw lacking incisors.

The family was obviously highly successful in Asia, where it survived until well into the Pleistocene. Only two species appear to have expanded westward, one in the Pliocene and another late in the Pleistocene. The latter has been referred to the Javan species *Stegodon trigonocephalus* and never penetrated south of the Jordan Valley (Hooijer 1961).

The earlier invasion migrated to and differentiated in sub-Saharan Africa to become a distinct African species, *Stegodon kaisensis.* It is reported from the Pliocene Kaiso Formation, Uganda (MacInnes 1942; Cooke and Coryndon 1970), and from Kolinga I in Chad (Coppens 1967a).

The genus *Stegolophodon,* an early member of the *Stegodon* group, has been reported from Langebaanweg (Singer and Hooijer 1958) and from the Congo (Hooijer 1963), but the former specimen is more likely referable to *Mammuthus subplanifrons* and the latter appears to be an early elephant of uncertain identification. The specimen described by Petrocchi (1943) as *Stegolophodon sahabianus* is considered to be conspecific with *Stegotetrabelodon syrticus* (Maglio 1970a, p. 329).

Phylogeny and Evolution

Establishing phyletic sequences is a difficult task at best, with resulting schemes more reflecting authors' personal biases than completely objective criteria. Nevertheless, to students of evolution it is the interrelationship among various fossil groups that is of primary interest. Such schemes for African Proboscidea are possible only on the most generalized level (figure 17.2), except for the family Elephantidae, for which the fossil record is unusually complete.

As alluded to above, *Palaeomastodon* appears to represent the stem group for all later members of the order. In spite of the suggestions that one or another species of this genus lay on the direct line of descent to later groups, it must be emphasized that our present knowledge of the African Eocene and early Oligocene is exceedingly incomplete, and between *Palaeomastodon* and any later genus there remains an enormous time gap (see figure 17.3). At this point it is probably fair to say only that on present evidence *Palaeomastodon* as a genus probably gave rise by collateral branching to both the gomphotheres and mastodonts of the early Miocene.

The African record of Mammutidae is limited to several early Miocene records of uncertain significance. All that can be said here is that the family seems to have had an African origin, from where it gave rise to stegodonts in Asia and mastodonts in Europe during the earlier part of the Miocene. The family subsequently became extinct in Africa. Each of the descendant groups reinvaded from time to time during the Miocene and Pliocene, but none was able to establish itself permanently.

The Gomphotheriidae, after expanding into Eurasia during the early Miocene, must have undergone at least a minor radiation in Africa, of which we have only the vaguest record preserved. It seems almost certain that a conservative lineage, typified by what has been called *Gomphotherium angustidens,* was widespread over large parts of the continent. But as pointed out above, we do not yet have enough evidence to distinguish adequately this taxon in Africa from that in Europe. There are considerable data to suggest that several different species have been masquerading under this name, particularly in East Africa, and it is clear that we have not yet begun really to understand the African history of this group.

In addition to this conservative line we see the early Miocene emergence of the flat-tusked gomphotheres, of which we as yet have only a meager but definite record. Descendants of this stock were to flourish in Asia and North America during the later Tertiary but probably had a very limited history in Africa (figure 17.1).

During the later Miocene we see the reinvasion into North Africa of a more progressive stock from Eurasia, *Tetralophodon,* while resident Gomphotheriidae continued to evolve along two major lineages. One rapidly shortened the mandible, lost the lower tusks, and developed an alternating pattern of cones on the molar teeth. These were the Anancinae and as already suggested, they may have been very different in cranial morphology and thus independent from the group of the same name in Eurasia. It cannot be said for certain whether African anancines derived from resident gomphotheres persistent in the continent or from a *Tetralophodon* that reinvaded from the north. A highly successful line, *Anancus* radiated into at least four species distributed from the Mediterranean to the Cape Province and persisted into the Pleistocene north of the Sahara. With the extinction of *Anancus osiris* came the demise of the family Gomphotheriidae in Africa.

The second major lineage can be seen through scattered later Miocene finds at Fort Ternan, Ngorora, Mpesida, and Lothagam, all in East Africa. Here, as seen in *G. ngorora,* for example, the molar teeth gradually assumed the parallel plate arrangement and loss of intravalley columns typical of Elephantidae even though characteristic gomphothere features such as long symphysis and mandibular tusks and premolars persisted in some species. The record is still grossly inadequate, but it seems likely that this line underwent a minor radiation during latest Miocene time, giving rise to at least three genera, *Stegotetrabelodon, Stegodibelodon,* and *Primelephas.* We venture to guess that these will prove to be the dominant groups of African proboscideans during the later Miocene when this period becomes better known.

By the early Pliocene this radiation was over and only three descendant groups remained, all possibly derived from *Primelephas.* These formed the base of the Elephantinae and marked the emergence of the only three genera of the subfamily: *Loxodonta, Elephas,* and *Mammuthus.*

The most conservative lineage was *Loxodonta,* which retains primitive features in its skeleton and dentition even in the extant form (see figure 17.4).

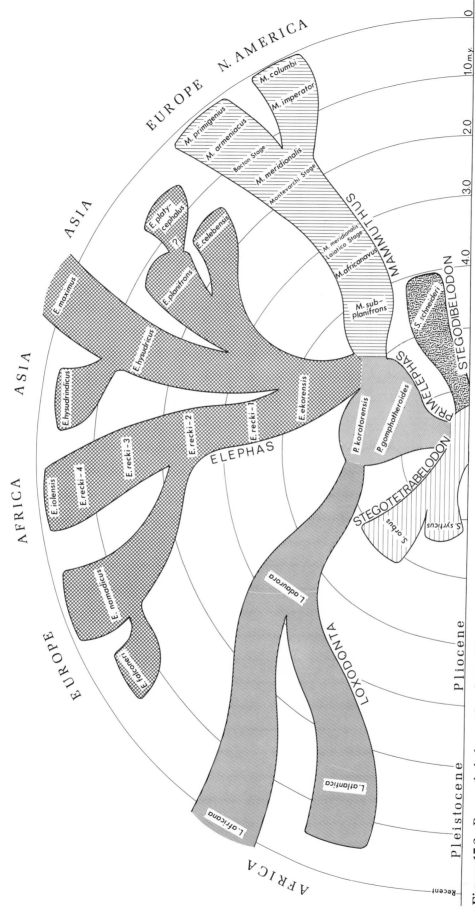

Figure 17.2 Proposed phylogenetic relationships among all species of the family Elephantidae. (From Maglio 1973)

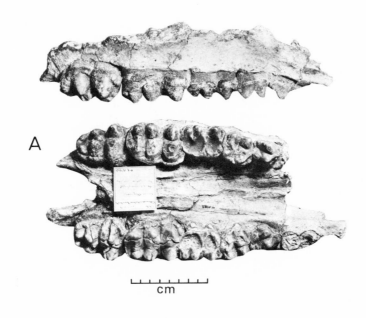

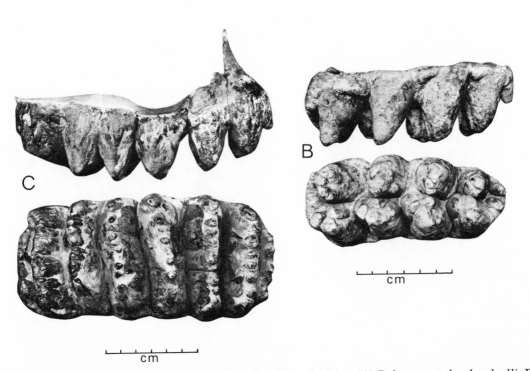

Figure 17.3 Representative dentitions of some fossil Proboscidea of Africa. (*A*) *Palaeomastodon beadnelli,* BM M-8383, palate with P²-M³, early Oligocene, Fayum, Egypt. (*B*) *Gomphotherium angustidens,* BM M-15528, left M³, early Miocene, Maboko, Kenya. (*C*) *Stegodon kaisensis,* BM M-15407, left M², late Pliocene, Kaiso Formation, Uganda.

Three species can be traced, the early Pliocene *L. adaurora,* and derived from it, the middle Pleistocene *L. atlantica,* and the early Pleistocene to Recent *L. africana.*

Elephas can be followed through a fine sequence of species from the early Pliocene *E. ekorensis* through various stages of *E. recki* during the late Pliocene to middle Pleistocene and terminating in *E. iolensis.* The genus gave rise to extra-African lines at least twice, one giving rise to an Asiatic radiation culminating in the living *Elephas maximus* and one leading to the middle Pleistocene southern European elephant, *E. namadicus* (= "*antiquus*").

A third lineage, *Mammuthus,* is less well estab-

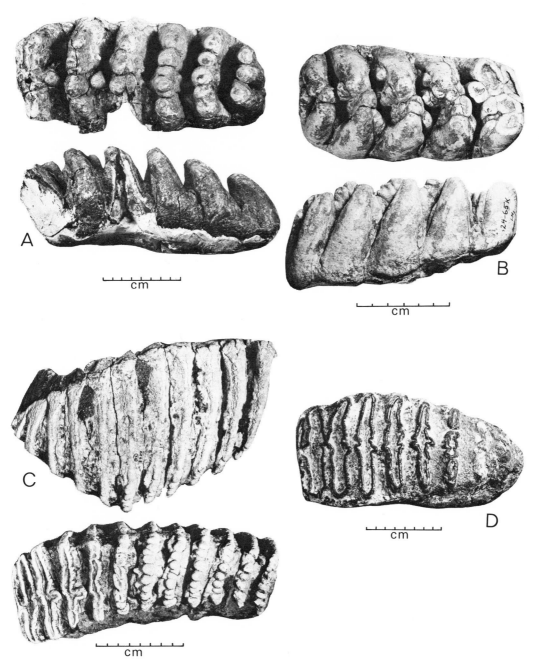

Figure 17.4   Representative dentitions of some fossil Proboscidea of Africa. (*A*) *Stegotetrabelodon orbus,* KNM-LT 359, left M³, early Pliocene, Lothagam-1, Kenya. (*B*) *Anancus kenyensis* KNM-KP 24-65K, left M₂, middle Pliocene, Kanapoi, Kenya. (*C*) *Elephas recki* (2), KNM-ER 342, left M³, early Pleistocene, East Lake Turkana, Kenya. (*D*) *Loxodonta adaurora,* KNM-KP 385, left M³, middle Pliocene, Kanapoi, Kenya.

lished, but if the evidence is interpreted correctly, it appears to include the early Pliocene *M. subplanifrons* and the late Pliocene *M. africanavus.* The latter probably gave rise to European populations of *M. meridionalis* during closing phases of the Pliocene, becoming extinct in Africa except for a progressive population in North Africa.

Thus, by the middle Pleistocene three elephants

persisted in Africa, two species of *Loxodonta* and one of *Elephas.* By the terminal Pleistocene only the living *L. africana* remained.

Zoogeography

Many points concerning the past distribution of the order have already been mentioned and we need only summarize the major events here. Figure 17.1 shows a generalized phyletic scheme for the order and arrows indicate periods of expansion both into and out of the African continent. It is clear from this, as is demonstrated from the Eurasian fossil record, that the Eocene and Oligocene history of the Proboscidea was confined to Africa. By the time the order makes its first appearance on northern continents at least three major lineages had become established. Sometime in the early Miocene *Gomphotherium* expanded into Europe and its more specialized cousins, the shovel tuskers, spread through Africa, northern Asia, and North America. It would seem that at least one and possibly two separate invasions of the north was achieved by the Mammutidae during the early Miocene, once by a generalized zygolophodont ancestor and another by a more specialized stegodont ancestor giving rise to *Stegolophodon* in Eurasia.

In contrast to the early Miocene, the later Miocene and early Pliocene witnessed an influx of several proboscideans into Africa. These represented descendants of groups that emigrated earlier in the epoch. By the late Miocene both the mammutid *Zygolophodon* and the gomphotheriid *Tetralophodon* had entered North Africa. Somewhat later, during the early Pliocene, *Stegodon* penetrated into sub-Saharan Africa but persisted only for about a million years before becoming extinct here.

A second wave of export for African proboscideans occurred during the Pliocene. Early in the epoch, as *Elephas* was emerging as a distinct genus, a close relative of *E. ekorensis* passed via the Middle East, where it is represented at Bethlehem, and then into southern Asia to establish the *E. planifrons* stock and subsequent Asiatic radiation. Later in the epoch an early *Mammuthus* expanded into Europe, probably across the western Mediterranean, to appear in Villafranchian deposits as *M. meridionalis*.

The long early Pleistocene apparently saw no major geographic movements involving African Proboscidea, but by the beginning of the middle Pleistocene a species close to *E. recki* expanded into Europe as *E. namadicus*.

Extinction

An examination of recent literature on the subject demonstrates how complex and confused is the problem of late Pleistocene extinction, and about all that can be said at this point is that no single or simple causative factor can be invoked. This problem is discussed again in the final chapter to this volume, but

as regards Proboscidea, we see a gradual fading out of species over a long period of time, quite unlike the pattern seen for the same order on northern continents (see Maglio 1973 for further discussion of the latter).

The first major wave of extinctions was in the earlier Miocene and involved the specialized descendant groups of shovel-tusked gomphotheres and mammutids. Both seemingly failed to compete with more conservative gomphotheres, deinotheres, and other groups. A second wave is seen in the early Pliocene when anancines (in sub-Saharan Africa), mammutids (in North Africa), and stegotetrabelodontines disappeared due to continued diversification of the Elephantinae. The demise of *Mammuthus* in the early Pleistocene in Africa may relate to competition with *Elephas,* which rapidly moved into all parts of the continent. *Elephas* and *Loxodonta* appear to have survived by being ecologically separated, a suggestion based on the fact that the two are rarely found in the same deposits even though both existed in the same general regions. But by the late Pleistocene, *Elephas* became extinct, leaving only the extant species of *Loxodonta.*

References

Aguirre, E. 1969a. Evolutionary history of the elephant. *Science* 164:1366–1376.

———. 1969b. Revisión sistemática de los Elephantidae por su morfologiá y morfometria dentaria. *Estud. Geol.* 25:123–177; 317–367.

Andrews, C. W. 1901a. Fossil vertebrates from Egypt. *Zoologist* (4) V:318–319.

———. 1901b. Ueber das Vorkommen von Proboscidiern in untertertiären ablagerungen Aegyptens. *V Intern. Zool. Congr.* Berlin, Tagesblatt, no. 6:1–6.

———. 1903. On the evolution of the Proboscidea. *Phil. Trans. Roy. Soc. London* 196(B):99–118.

———. 1904. Further notes on the mammals of the Eocene of Egypt. *Geol. Mag.* n.s., 1:109–115.

———. 1905. Note on the species of *Palaeomastodon. Geol. Mag.* n.s., 2:562–563.

———. 1906. *A descriptive catalogue of the Tertiary Vertebrata of the Fayûm, Egypt.* London: Brit. Mus. (Nat. Hist.), 324 pp.

Andrews, C. W., and H. J. L. Beadnell. 1902. *A preliminary note on some new mammals from the upper Eocene of Egypt.* Cairo: Surv. Dept. Pub. Works Ministry, pp. 1–9.

Arambourg, C. 1938. Mammifères fossiles du Maroc. *Mém. Soc. Sci. Nat. Maroc* 46:1–74.

———. 1945. *Anancus osiris,* un nouveau Mastodonte du Pliocène inférieur d'Égypte. *Bull. Soc. Géol. Fr.* (5) 15:479–495.

————. 1947. Contribution à l'étude géologique et paléontologique du bassin du Lac Rudolphe et de la basse vallée de l'Omo. *Miss. Scient. de l'Omo, 1932–1933* 2:232–562.

————. 1948. Les mammifères pléistocènes d'Afrique. *Bull. Soc. Géol. Fr.* (5) 17:301–310.

————. 1952. Note préliminaire sur quelques éléphants fossiles de Berbérie. *Bull. Mus. Nat. Hist. Nat. Paris* (2) 24:407–418.

————. 1970. Les vertébrés du Pléistocène de 'Afrique du Nord. *Arch. Mus. Nat. Hist. Nat. Paris* (7) 10(1968):1–126.

Arambourg, C., and P. F. Burollet. 1962. Restes de vertébrés Oligocènes en Tunisie centrale. *C. R. Soc. Géol. Fr.* 2:42–43.

Arambourg, C., and P. Magnier. 1961. Gisements de vertébrés dans le bassin tertaire de Syrte (Libie). *C. R. Hebd. Séanc. Acad. Sci.* 252:1181–1183.

Berggren, W., and J. Van Couvering. 1974. The late Neogene. *Paleogeogr., Paleoclimatol., Paleoecol.* Special Issue. 16(1/2):1–216.

Bergounioux, F. M., and F. Crouzel. 1956. Présence de *Tetralophodon longirostris* dans le Vindobonien inférieur de Tunisie. *Bull. Soc. Géol. Fr.* (6)6:547–557.

————. 1959. Nouvelles observations sur un petit Mastodonte du Cartennien de Kabylie. *C. R. Somm. Séanc. Géol. Fr.* 1959:101–102.

Bishop, W. W., J. A. Miller, and F. J. Fitch. 1969. New potassium-argon determinations relevant to the Miocene mammal sequence in East Africa. *Am. J. Sci.* 267:669–699.

Boné, E., and R. Singer. 1965. *Hipparion* from Langebaanweg, Cape Province, and a revision of the genus in Africa. *Ann. S. Afr. Mus.* 48(16):273–397.

Borissiak, A. 1928. On a new mastodon from the Chockrak Beds (middle Miocene) of the Kuban region, *Platybelodon danovi*, n. gen., n. sp. *Ann. Soc. Paleont. Russie* 7:105–120.

————. 1929. On a new direction in the adaptive radiation of mastodonts. *Palaeobiologica* II:19–33.

Budge, E. A. W. 1969. [1904] *The Gods of the Egyptians.* New York: Dover, I: 525 pp., II: 431 pp.

Buss, I. O. 1961. Some observations on food habits and behavior of the African elephant. *J. Wildlife Management* 25(2):131–148.

Cooke, H. B. S. 1947. Variation in the molars of the living African elephant and a critical revision of the fossil Proboscidea of southern Africa. *Am. J. Sci.* 245:434–457; 492–517.

————. 1960. Further revision of the fossil Elephantidae of southern Africa. *Palaeont. Afr.* 7:59–63.

Cooke, H. B. S., and S. Coryndon. 1970. Fossil mammals from the Kaiso Formation and other related deposits in Uganda. In L. S. B. Leakey and R. G. J. Savage (eds.), *Fossil Vertebrates of Africa.* Edinburgh: Academic Press, vol. 2, pp. 107–224.

Cooke, H. B. S., and V. J. Maglio. 1972. Plio-Pleistocene stratigraphy in East Africa in relation to proboscidean and suid evolution. In W. W. Bishop and J. A. Miller (eds.), *Calibration of hominoid evolution,* New York: Academic Press, pp. 303–329.

Coppens, Y. 1965. Les Proboscidiens du Tchad, leur contribution à la chronologie du Quaternaire africain. *Actes du Ve Congr. Panafr. de Préhist. et de l'étude de Quat.* (Santa Cruz de Ternerife), I(5):331–387.

————. 1966. Le *Tchadanthropus.* *L'Anthropol.* 70(1–2):5–16.

————. 1967a. Les faunes de vertébrés quaternaires du Tchad. In W. W. Bishop and J. A. Miller (eds.), *Background to evolution in Africa,* Chicago: Univ. of Chicago Press, pp. 89–97.

————. 1967b. Essai de biostratigraphie du Quaternaire de la région de Koro Toro (Nord Tchad). *Colloques Intern. C.N.R.S.,* 1966, pp. 589–595.

————. 1972. Un nouveau Proboscidien du Pliocène du Tchad, *Stegodibelodon schneideri* nov. gen. nov. sp, et le phylum des Stegotetrabelodontinae. *C. R. Hebd. Acad. Sci.* 274:2962–2965.

Coppens, Y., R. Gouzes, R. Lefloch, and M. Paquet. 1972. Découverte d'un gisement de vertébrés fossiles avec industrie acheulienne prés de Zouérate en Maritanie. *Proc. VI Pan-Afr. Cong. Préhist.,* Dakar, 1967, pp. 457–461.

Dart, R. A. 1927. Mammoths and man in the Transvaal. *Nature Suppl.,* no. 3032:41–48.

————. 1929. Mammoths and other fossil elephants of the Vaal and Limpopo watersheds. *S. Afr. J. Sci.* 26:698–731.

Depéret, C. 1897. Dècouverte du *Mastodon angustidens* dans l'étage Cartennien de Kabylie. *Bull. Soc. Géol. Fr.* 25:518–521.

Depéret, C., L. Lavauden, and M. Solignac. 1925. Sur la découverte de *Mastodon arvernensis* dans le Pliocène de Ferryville (Tunisie). *Bull. Soc. Géol. Fr.* (4) 25:21–22.

Deraniyagala, P. E. P. 1955. *Some extinct elephants, their relatives and the two living species.* Colombo: Ceylon Nat. Mus. Pub., 161 pp.

Dietrich, W. O. 1916. *Elephas antiquus recki* n. f. aus dem Diluvium Deutsch-Ostafrikas. *Arch. Biontol.* 4(1):1–80.

————. 1942. Ältestquartäre Säugetiere aus der südlichen Serengeti, Deutsch-Ostafrika. *Paläontogr.* 94A:43–133.

————. 1943. Über Innerafrikanische Mastodonten. *Zeit. Deut. Geol. Gesell.* 95(1/2):46–48.

Doumergue, F. 1928. Découverte de l'*Elephas planifrons* Falconer à Rachgoun (Dépt. d'Oran). *Bull. Cinquant. Soc. Géog. Archéol. Oran,* pp. 114–132.

Falconer, H. 1868. *Palaeontological memoires and notes of the late Hugh Falconer, with a biographical sketch of the author,* 2 vols. Charles Murchison (ed.), London.

Fourtau, R. 1918. *Contribution à l'étude des vertébrés miocènes de l'Égypte.* Cairo: Egypt Surv. Dept., 121 pp.

Gervais, P. 1859. *Zoologie et Paléontologie Françaises,* 2nd ed. Paris, 544 pp.

Haughton, S. H. 1922. A note on some fossils from the Vaal River gravels. *Trans. Geol. Soc. S. Afr.* 24:11–16.

_____. 1932. On some South African Proboscidea. *Trans. Roy. Soc. S. Afr.* 21:1–18.

Hendey, Q. B. 1970. The age of the fossiliferous deposits at Langebaanweg, Cape Province. *Ann. S. Afr. Mus.* 56(3):119–131.

Hooijer, D. A. 1958. An early Pleistocene mammalian fauna from Bethlehem. *Bull. Brit. Mus. Nat. Hist.* (Geol.) 3(8):267–292.

_____. 1961. Middle Pleistocene mammals from Latemné, Orontes Valley, Syria. *Ann. Archeol. Syrie* 11:117–132.

_____. 1963. Miocene Mammalia of Congo. *Ann. Mus. Roy. Cong. Belge,* ser. 8, Sci. Geol. 46:1–77.

Hopwood, A. T. 1926. The geology and palaeontology of the Kaiso bone-beds. Part II: Palaeontology. *Geol. Surv. Uganda, Occ. Pap.* 2:13–36.

_____. 1939. The mammalian fossils. In T. P. O'Brien, *The prehistory of Uganda Protectorate.* Cambridge: Cambridge Univ. Press, 319 pp.

Hormann, K. 1963. Note on a mastodontid from Libya. *Zeits. f. Säugetierk.* 28:88–93.

Laws, R. M., and J. S. C. Parker. 1968. Recent studies on elephant populations in East Africa. *Symp. Zool. Soc. Lond.* 21:319–359.

Leakey, L. S. B. 1965. *Olduvai Gorge 1951–1961. I: Fauna and background.* Cambridge: Cambridge Univ. Press, 318 pp.

Lehman, U. 1950. Über Mastodontenreste in der Bayerische Staatsammlung in München. *Paleontogr.* 94A(4/6):121–227.

MacInnes, P. G. 1942. Miocene and post-Miocene Proboscidea from East Africa. *Trans. Zool. Soc. Lond.* 25:33–106.

Madden, C. T. 1972. Miocene mammals, stratigraphy and environment of Muruarot Hill, Kenya. *Paleobios* 14:1–12.

Maglio, V. J. 1969. A shovel-tusked gomphothere from the Miocene of Kenya. *Breviora,* no. 310:1–10.

_____. 1970a. Early Elephantidae of Africa and a tentative correlation of African Plio-Pleistocene deposits. *Nature* 225:328–332.

_____. 1970b. Four new species of Elephantidae from the Plio-Pleistocene of northwestern Kenya. *Breviora,* no. 341:1–43.

_____. 1972. Evolution of mastication in the Elephantidae. *Evolution* 26(4):638–658.

_____. 1973. Origin and evolution of the Elephantidae. *Trans. Am. Phil. Soc.,* n.s., 63(3):1–149.

_____. 1974. A new species of Proboscidea from the Miocene of Kenya. *Palaeontology* 17(3):699–705.

Maglio, V. J., and Q. B. Hendey. 1970. New evidence relating to the supposed stegolophodont ancestry of the Elephantidae. *S. Afr. Archaeol. Bull.* 35(3–4):85–87.

Matsumoto, H. 1922. Revision of *Palaeomastodon* and *Moeritherium. Palaeomastodon intermedius* and *Phiomia osborni,* n. sp. *Am. Mus. Novit.,* no. 51:1–6.

_____. 1924. A revision of *Palaeomastodon* dividing it into two genera and with descriptions of two new species. *Bull. Am. Mus. Nat. Hist.* 50:1–58.

Meiring, A. J. D. 1955. Fossil proboscidean teeth and ulna from Virginia, O.F.S. *Res. Nas. Mus. Bloemfontein* 1(8):187–201.

Osborn, H. F. 1923. New subfamily, generic and specific stages in the evolution of the Proboscidea. *Am. Mus. Novit.,* no. 99:1–4.

_____. 1925. Final conclusions on the evolution, phylogeny, and classification of the Proboscidea. *Proc. Am. Phil. Soc.* 64(6):17–35.

_____. 1934. Primitive *Archidiskodon* and *Palaeoloxodon* of South Africa. *Am. Mus. Novit.,* no. 741:1–15.

_____. 1936. *Proboscidea.* I. New York: American Museum Press, pp. 1–802.

_____. 1942. *Proboscidea.* II. New York: American Museum Press, pp. 805–1676.

Petrocchi, C. 1941. I giacimento fossilifero di Sahabi. *Boll. Soc. Geol. Ital.* 60(1):107–114.

_____. 1943. I giacimento fossilifero di Sahabi. *Coll. Scient. Docum. Cura* (Min. A.I., IX) 12.

Pomel, A. 1879. Ossements d'éléphants et d'hippopotames découvertes dans une station préhistorique de la plaine d'Eghis (Province d'Oran). *Bull. Soc. Géol. Fr.* (3)7:44–51.

_____. 1895. Les éléphants Quaternaires. *Carte Géol. l'Algérie, Paléont. Monogr.,* no. 6:1–68.

Pontier, G. 1907. Sur une espèce nouvelle de *Palaeomastodon.* (*Palaeomastodon barroisi*). *Ann. Soc. Géol. Nord* 36:150–154.

Savage, R. J. G. 1969. Early Tertiary mammal locality in southern Libya. *Proc. Geol. Soc. Lond.* 1657:167–171.

Savage, R. J. G., and W. R. Hamilton. 1973. Introduction to the Miocene mammal faunas of Gebel Zelten, Libya. *Bull. Brit. Mus. Nat. Hist.* (Geol.) 22(8):515–527.

Schlessinger, G. 1917. Die Mastodonten des K. K. naturhistorischen Hofmuseums. *Denkschr. Naturhist. Mus. Vienna* I (Geol.-pal. Reihe, 1), 230 pp.

_____. 1922. Die Mastodonten der Budepester Sammlungen. *Geol. Hung.* II:1–248.

Scott, W. B. 1907. A collection of fossil mammals from the coast of Zululand. *Geol. Surv. Natal and Zululand,* 3rd and final rept., pp. 253–262.

Sikes, S. K. 1971. *The natural history of the African elephant.* New York: American Elsevier, 397 pp.

Simpson, G. G. 1945. The principles of classification and a classification of mammals. *Bull. Am. Mus. Nat. Hist.* 85:350.

Singer, R., and D. A. Hooijer. 1958. A *Stegolophodon* from South Africa. *Nature* 182:101–102.

Solignac, M. 1927. *Étude géologique de la Tunisie septentrionale.* Thèse Sciences Paris, 1928. *Mém. Serv. Carte Géol. Tunis,* 756 pp.

Thenius, E. 1959. Tertiär: Wirbeltierfaunen. In *Handbuch der Stratigraphischen Geologie,* herausgegeben von Fr. Lotzen, Bd. III, Teil 2, 528 pp.

_____. 1969. Stammesgeschichte der Säugetiere (einschliesslich der Hominiden). In *Handbuch der Zoologie,* Bd. 8, Lief 48, 2(1), pp. 369–722.

Tobien, H. 1971. *Moeritherium, Palaeomastodon, Phiomia* aus dem Palaeogen Nordafrikas und die Abstammung des Mastodonten (Proboscidea, Mammalia). *Mitt. Geol. Inst. Techn. Univ. Hann.* Heft 10:141–163.

———. 1972. Status of the genus *Serridentinus* Osborn, 1923 (Proboscidea, Mammalia). *Mainzer Geowiss. Mitt.* 1:143–191.

———. 1973. On the evolution of the mastodonts (Proboscidea, Mammalia). Part 1: The bunodont trilophodont groups. *Notizbl. hess. L.-Amt. Bodenforsch.* 101:202–276.

Vaufrey, R. 1958. Proboscidea, étude systématique. In Piveteau, J. (ed.), *Traité de Paléontologie VI,* vol. 2. Paris: Masson.

18

Chalicotheriidae

Percy M. Butler

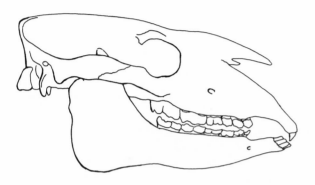

The chalicotheres are a family of perissodactyl ungulates remarkable for their possession of claws instead of hoofs. The claws were embedded in deep clefts in the terminal phalanges in the same way as those of pangolins, and for most of the nineteenth century chalicotherian foot bones were classified as Edentata, while their teeth and skulls were recognized as ungulate. The cheek teeth have an interlocking, chopping type of occlusion like those of *Palaeotherium* and the titanotheres, and they were evidently adapted for eating leaves and other soft vegetation. Chalicotheres are now universally regarded as highly aberrant members of the order Perissodactyla.

The main area of evolution of the family appears to have been Asia north of Tethys. They occur in the Oligocene of Europe and Asia; and their forerunners, the Eomoropidae, are found in the Eocene of Europe, Asia, and North America. The most primitive and earliest known genus is *Schizotherium,* but before the end of the Oligocene there is evidence of more advanced forms in Europe: *Chalicotherium* and *?Phyllotillon.* In the early Miocene these two genera appear in southern Asia (Bugti), and *Chalicotherium* occurs in deposits of similar date in East Africa.

The African Chalicotheres

The following are recorded African species and important references:

Chalicotherium rusingense Butler 1962, 1965. Early Miocene, Songhor, Koru, Rusinga, Mfwanganu, Napak.
Ancylotherium hennigi (Dietrich) 1942; Andrews 1923; Hopwood 1926; George 1950; Thenius 1953; Butler 1965; Webb 1965; Hooijer 1972, 1973; Pickford 1975. Synonyms: *Metaschizotherium hennigi* Dietrich 1942, *Metaschizotherium*(?) *transvaalensis* George 1950. Latest Miocene to early Pleistocene, Chemeron, Omo, Kaiso, Makapansgat, Laetolil, Olduvai.

Chalicotherium rusingense Butler (1965) is known from the dentition, mandible (incomplete posteriorly), facial part of the skull, scaphoid, astragalus, two metacarpals, a metatarsal, and phalanges of manus and pes. It is much smaller than the best-known European species, *C.* (*Macrotherium*) *grande* of the middle Miocene (Vindobonian), agreeing in size with *Schizotherium* and with the Bugti species, *C. pilgrimi.* It resembles *C. grande* rather than *C. pilgrimi* in the upper molar pattern: *C. pilgrimi* is more primitive in that the paracone and the metacone are less displaced from the buccal side of the tooth, giving a pattern very like that of *Schizotherium.* On the lower molars of *C. rusingense* there is nearly always a well-developed metastylid, as in *C. pilgrimi* and *Schizotherium*, whereas in *C. grande* this cusp is usually absent or rudimentary. The premolars are less reduced in comparison with the molars in *C. rusingense* than in *C. grande,* especially P_2. Upper incisors are absent, but there is a well-developed upper canine; whether this was the case in *C. grande* is unknown. There are some differences in the feet, which might be associated with the smaller body size of the African species: thus the trochlea of its astragalus is less depressed and the basal phalanges of its manus are proportionately longer and narrower. The second metatarsal resembles that of *C. pilgrimi* and is more slender than in *C. grande.* There are resemblances in the teeth to the very imperfectly known *C. wetzleri,* from the Aquitanian of Europe, but this is a much larger species and a close relationship is unlikely. *C. rusingense* is more likely to have come from Asia than from Europe, perhaps descended from *C. pilgrimi.*

An incisor and a fragment of molar from the Miocene of Malembe were tentatively referred to *Macrotherium* (i.e., *Chalicotherium*) by Hooijer (1963) following a suggestion by Hopwood. They belong to a much larger animal than *C. rusingense,* but they are so fragmentary that their chalicotherian nature must remain in doubt. No remains of *Chalicotherium* have been found from the late Miocene onward.

As far as is known *Phyllotillon,* a genus associated with *Chalicotherium* at Bugti, did not enter Africa. An upper premolar from Napak has an abnormal pattern similar to a specimen of *P.* (=*Metaschizotherium*) *bavaricum* (von Koenigswald 1932) from Steinheim, and it was, therefore, tentatively referred to that genus by Butler (1962). In the absence of other evidence for the existence of *Phyllotillon* in Africa the tooth is, however, better regarded as an aberrant specimen of *Chalicotherium.*

Nevertheless, the genus *Ancylotherium,* closely related to *Phyllotillon* and perhaps descended from it (Coombs 1974), was present in East Africa in the Pliocene and early Pleistocene. It is known only from very fragmentary remains. The oldest record is a phalanx from the Lukeino Formation (ca 6.5 m.y.) (Pickford 1975), and Hooijer (1972) reported a P^4 from the Chemeron Formation (ca 4 m.y.). A basal phalanx (probably a fused basal and middle phalanx with most of the middle phalanx broken off) from Kaiso was noticed by Andrews (1923) and described, but not named, by Hopwood (1926). From the Laetolil Beds Dietrich (1942) described an upper molar and some phalanges, which he named *Metaschizotherium hennigi.* Because of resemblances to *Ancylotherium pentelicum,* from the Pontian of southeastern

Europe and Iran, Thenius (1953) transferred *M. hennigi* to *Ancylotherium.* A number of associated bones from a manus, found in Olduvai Bed I, were described by Butler (1965), who confirmed the resemblance to *A. pentelicum* but noted that in some ways the African form was less specialized. Hooijer (1973) referred to this species a P^4 and two upper molar fragments from about the 2 m.y. level at Omo.

George (1950) based *Metaschizotherium*(?) *transvaalensis* on three isolated molars and an ungual phalanx from Makapansgat, and subsequently much more material from the same site has been briefly described by Webb (1965). This material includes fragments of the scapula and pelvis, a maxillary fragment with associated but worn teeth and also upper and lower milk molars, permanent molars, an incisor, and other teeth. These specimens indicate an animal of the same size as *Ancylotherium hennigi* and probably conspecific with it, though except for M^2, which is slightly and probably not significantly broader in the Makapansgat form, comparison is difficult because of the fragmentary nature of the Laetolil material. If the incisor is correctly referred, this would indicate a difference from *A. pentelicum* where the incisors and canines are said to be absent (Falconer 1868; Major 1894) and would confirm the primitive nature of the African species in spite of their younger age.

Until the Pliocene and early Pleistocene African chalicotheres are better known we cannot be certain that they are descended directly from the Pontian *A. pentelicum.* As an alternative, they might have been derived from a more primitive species, perhaps a species of *Phyllotillon,* that entered Africa before the Pontian. Too much weight should not be placed on the virtual absence of chalicotheres from the fossil record of the late Miocene and the earlier Pliocene. Not only is knowledge of the African fauna during that span of time limited (see chapter 2, this volume), but even in regions of the world where chalicotheres are known to have existed their remains are localized or rare.

What evidence there is suggests two invasions into Africa, one (a chalicotheriine) probably from Asia at the beginning of the Miocene, the other (a schizotheriine) from Asia or southeastern Europe, perhaps at a later date. Both lines became extinct, the first perhaps before the end of the Miocene, the second lasting into the Pleistocene. We are totally ignorant of the causes of these extinctions. Chalicotheres seem, from their teeth, to have been leafeaters, inhabiting forest or, in the case of *Ancylotherium,* wooded savanna. Their claws may have been used to enable them to cling to the trunks of trees in an upright position like megatheriids and probably *Homalodotherium*. Thus they would have been able to reach a level of vegetation inaccessible to most ungulates except elephants and giraffes; perhaps these were their main competitors.

References

Andrews, C. W. 1923. An African chalicothere. *Nature* 112:696.

Butler, P. M. 1962. Chalicotheriidae. In Bishop, W. W., The mammalian fauna and geomorphological relations of the Napak Volcanics. *Rec. Geol. Surv. Uganda.* 1957–8:11.

———. 1965. East African Miocene and Pleistocene chalicotheres. Fossil Mammals of Africa, no. 18, *Bull. Br. Mus. Nat. Hist.* 10:165–237.

Coombs, M. S. 1974. Ein Vertreter von *Moropus* aus dem europäischen Aquitanien und eine Zusammenfassung der europäischen postoligozänen Schizotheriinae (Mammalia, Perissodactyla, Chalicotheriidae). *Sber. öst. Akad. Wiss.,* Abt. I 182:273–288.

Dietrich, W. O. 1942. Ältestquartäre Säugetiere aus der südlichen Serengeti, Deutsch-Ostafrika. *Paläontogr.* 94A:44–130.

Falconer, H. 1868. On *Chalicotherium sivalense.* In Mur-

chison, C. (ed.), *Palaeontological memoirs and notes,* London, pp. 208–226.

George, M. 1950. A chalicothere from the Limeworks Quarry of the Makapan Valley, Potgietersrust District. *S. Afr. J. Sci.* 46:241–242.

Hooijer, D. A. 1963. Miocene Mammalia of Congo. *Ann. Mus. R. Congo Belge* 46:1–77.

———. 1972. A later Pliocene rhinoceros from Langebaanweg, Cape Province. *Ann. S. Afr. Mus.* 59:151–191.

———. 1973. Additional Miocene to Pleistocene rhinoceroses of Africa. *Zoöl. Med. Leiden* 46:149–178.

Hopwood, A. T. 1926. Fossil Mammalia. In Wayland, E. J., The geology and palaeontology of the Kaiso Bone Beds. *Occ. Pap. Geol. Surv. Uganda* 2:13–36.

Koenigswald, G. H. R. von. 1932. *Metaschizotherium fraasi,* n.g. n.sp. ein neuer Chalicotheriide aus dem Obermiocän von Steinheim a. Albuch. *Paläontogr.* (Suppl.) 8(8):1–24.

Major, C. J. Forsyth. 1894. *Le gisement ossifère de Mityline et catalogue d'ossements fossiles.* Lausanne, 51 pp.

Pickford, M. 1975. Another African chalicothere. *Nature* 253:85.

Thenius, E. 1953. Studien über fossile Vertebraten Griechenlands, III. Das Maxillargebiss von *Ancylotherium pentelicum* Gaudry und Lartet. *Ann. Géol. Pays hell.* 5:97–106.

Webb, G. L. 1965. Notes on some chalicothere remains from Makapansgat. *Palaeont. Afr.* 9:49–73.

19

Rhinocerotidae

Dirk A. Hooijer

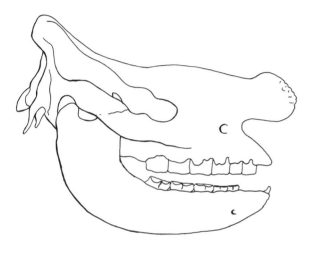

Hitherto there has been no attempt at construction of a phylogeny of African Rhinocerotidae comparable to that of European rhinoceroses as given by Osborn (1900). But then, until some 10 years or so ago very little African rhinocerotid material earlier than Pleistocene was available in the literature, and even this was either not specifically identified or incorrectly named[1].

In the last decade fossil rhinocerotids have turned up in Africa in ever increasing numbers, most of them Miocene in age, such as the Congo species *Aceratherium acutirostratus* (Deraniyagala) and *Brachypotherium heinzelini* Hooijer (1963). Both of these also occur in the East African Miocene as does *Dicerorhinus leakeyi* Hooijer (1966). From the Miocene of Fort Ternan, Kenya, comes *Paradiceros mukirii* Hooijer (1968) and the Miocene rhinocerotid from Loperot, Kenya, originally described on a single tooth as *Chilotherium* spec. (Hooijer 1966, p. 150), proved to be a new genus, *Chilotheridium pattersoni* Hooijer (1971). Then followed two Mio-Pliocene species from Lothagam Hill, Kanapoi and Ekora in Kenya: *Brachypotherium lewisi* Hooijer and Patterson (1972) and *Ceratotherium praecox* Hooijer and Patterson (1972); the latter is also present at Langebaanweg, South Africa (Hooijer 1972). Both *Ceratotherium simum* (Burchell), the modern white rhinoceros, and *Diceros bicornis* (L.), the modern black rhinoceros, proved to date back several million years in well-calibrated sequences of deposits in Kenya, Tanzania, and Ethiopia (Hooijer 1969, 1973). The classical "Pontian" as I use it in this paper is the modern Vallesian and Turolian, around the 10 m.y. level (cf Van Couvering 1972, p. 249).

The Fossil Rhinoceroses

Family Rhinocerotidae Owen 1845

Genus *Brachypotherium* Roger 1904

The ancestral stock from which came the African *Brachypotherium* as well as *Aceratherium* seems to be the Oligocene aceratheres of Europe. It appears now that in the later Oligocene Africa received from

1 *Rhinoceros* spec., Miocene of Moghara, Egypt (Andrews 1900); *Rhinoceros* spec. indet., Miocene of Karungu, Kenya (Andrews 1914); *Teleoceras snowi* Fourtau (1920), Miocene of Moghara, Egypt; Rhinocerine gen. et spec. indet., Miocene of Langental, Southwest Africa (Stromer 1926); *Aceratherium?* spec., Miocene of Losodok, Kenya (Arambourg 1933); *Turkanatherium acutirostratus* Deraniyagala (1951), Miocene of Losodok, Kenya; *Teleoceras* aff. *medicornutum* Osborn, Miocene of Sahabi, Cyrenaica (d'Erasmo 1954).

Eurasia a number of elements in exchange for the export of African stocks to Eurasia (see Cooke 1968 for a conspectus of the evolution of mammals of Africa). Although we have no evidence so far in Africa of any pre-Miocene rhinocerotids (none occur in the rich Fayum deposits), they must have been there already in the (late) Oligocene since the early Miocene African forms are full-fledged species different from other Old World forms of the same genera. The genus *Brachypotherium* is characterized by very brachydont upper molars with ectolophs that are flattened behind the paracone style, antecrochets that are only weakly developed, and very slightly marked protocone constrictions. External cingula and flattened external grooves are usually present on lower molars, and a large pair of central incisors is present in both jaws. In the postcranial skeleton the metapodials are very short, from whence the group gets it name.

When the earliest *Brachypotherium* species of Africa was described (as *Teleoceras snowi* Fourtau [1920] from Moghara in Egypt), the Eurasian brachypotheres were customarily placed in the American genus *Teleoceras* Hatcher of 1894, although already in 1904 Roger had created the genus *Brachypotherium* for them. The then accepted relationship between these two genera was expressed by Osborn (1910, p. 292) thus: *"Teleoceras medicornutus,* discovered in the Pawnee Creek region of Colorado, is a remarkably close successor to the *T. aurelianense* of the lower Miocene of France . . . one of the most brilliant illustrations of the migration theory between the New and Old Worlds." Today we believe that *Teleoceras* is a descendant of *Aphelops* Cope and that the similarity in teeth and brachypody to *Brachypotherium* is due to parallelism. *"Rhinoceros"* *aurelianensis* Nouel (1866), from the early Burdigalian of Europe, was followed in the late Burdigalian and early Vindobonian by *Brachypotherium stehlini* Viret (1961, p. 71), which is generally larger than *Brachypotherium aurelianense* and in which the external groove in the lower cheek teeth between the anterior and posterior lophids may be completely flattened, while the external cingulum is almost invariably present (cf Roman and Viret 1934, p. 33, pl. 10, figs. 7 and 8, as *Brachypotherium* cf *brachypus* [Lartet]). In *Brachypotherium stehlini* the metapodials are more shortened than in *B. aurelianense,* and in the late Vindobonian *Brachypotherium brachypus* proper, and in the "Pontian" *B. goldfussi* (Kaup), metapodial and limb shortening have reached an extreme.

The Moghara *Brachypotherium snowi,* as an early brachypothere, varies in the flattening of the exter-

nal groove and external cingula of the lower molars. A more advanced species is *Brachypotherium heinzelini* Hooijer (1963, p. 45), which occurs in the Sinda Beds of Congo and was afterwards found at Rusinga and Karungu in Kenya, Napak in Uganda (Hooijer 1966, p. 142–150), and identified from Langental in southwestern Africa by Heissig (1971). All these sites in East Africa as well as Langental are considered to be early Miocene in age, tentative correlates of the European Burdigalian. The species was recently found to occur also at Bukwa, another early Miocene Uganda site (Hooijer 1973). From the radiometric dates given in table 19.1 it is evident that this species, *Brachypotherium heinzelini,* existed for at least some 4 m.y., from Bukwa II to Rusinga; we find even more impressive longevities in other Neogene African rhinocerotid species.

Brachypotherium heinzelini, an exclusively Miocene species, resembles the late Vindobonian *B. brachypus* of Europe in its metapodial shortening; only the lateral metacarpals are relatively shorter and wider. It is definitely more advanced than the Moghara *Brachypotherium snowi,* which has only reached the evolutionary stage of the late Burdigalian and early Vindobonian *Brachypotherium stehlini* of Europe.

Its dentition is characterized by variable external cingula and flattened external grooves in the lower cheek teeth, very large upper incisors, the upper cheek teeth by flattened ectolophs behind the paracone style, weak antecrochets, and but slightly constricted protocones. The enamel often displays fine, horizontal striations also seen in European *Brachypotherium.*

There is a gap in the African record of some 6 m.y., for the next *Brachypotherium* is known from the Ngorora Formation in Kenya (Hooijer 1971, p. 364), which is that much younger than the latest *B. heinzelini* at Rusinga (see table 19.1). The Ngorora milk molar is too large for this earlier Miocene species but could belong with *Brachypotherium lewisi* Hooijer and Patterson (1972) from Lothagam-1. There is an upper molar from Sahabi in Libya, originally published as *Teleoceras* aff. *medicornutum* Osborn by d'Erasmo (1954), that apparently represents this large *Brachypotherium* as well (Hooijer and Patterson 1972, p. 17), and we may have it also in the Mpesida Beds in Kenya (Hooijer 1973). *Brachypotherium lewisi* is the terminal form of the *Brachypotherium* lineage in Africa, carrying the genus on into the latest Miocene. (In Europe the genus became extinct already by "Pontian" times, with *B. goldfussi.*) *Brachypotherium lewisi* is very large indeed; the condylobasal length of the type skull is

Table 19.1   The rhinoceroses of Africa, distribution in time and space.

	Million years	*Brachypotherium*	*Aceratherium*	*Dicerorhinus*	*Chilotheridium*	*Diceros*	*Ceratotherium*	*Paradiceros*
Olduvai								
Beds upper II—IV	?1-0.5	.	.	.	.	x	x	.
Beds I—lower II	1.8	.	.	.	.	.	x	.
Chemeron Fm.	4-2	.	.	.	.	.	x	.
Omo								
Shungura Fm.	4-2	.	.	.	.	x	x	.
Mursi Fm.	4	.	.	.	.	x	x	.
Aterir Beds	<4	.	.	.	.	.	x	.
Kanapoi	4	.	.	.	.	.	x	.
Langebaanweg, S.A.	4	.	.	.	.	.	x	.
Lothagam Hill	6	x	.	.	.	.	x	.
Sahabi, Libya	6	x	.	.	.	.	.	.
Mpesida Beds	7	x	.	.	.	.	x	.
Ngorora Fm.	<12	x	x or	x	x	.	.	.
Kirimun	—	.	x or	x	x	.	.	.
Douaria, Tunisia	12	.	.	.	.	x	.	.
Alengerr Beds	14-12	.	x	x	.	.	.	.
Fort Ternan	14	.	.	.	.	.	.	x
Loperot	18	.	.	.	x	.	.	.
Sinda, Congo	—	x	x	?	.	.	.	.
Rusinga	18.5	x	x	x	x	.	.	.
Ombo	—	.	.	x	x	.	.	.
Napak	19	x	x	x	.	.	.	.
Bukwa II	23	x	.	?	x	.	.	.
Moghara, Egypt	—	x	?	.	.	.	.	.

over 70 cm. The upper molars are 90 mm transversely as opposed to 70 mm in *B. snowi* or *B. heinzelini,* from either one of which it may have descended. The nasals are hornless, slender, and not very long; the frontals are flat and hornless; and the inferior squamosal processes unite below the subaural channel.

Genus *Aceratherium* Kaup 1832

The acerathere rhinocerotids (a contradiction in terms) started as Oligocene forms that retained tetradactyl forefeet up into the "Pontian." Metacarpal V (actually the fourth finger, as metacarpal I is but a vestige) remained surprisingly constant as a bone one-half as long as metacarpal IV all through the geological record, which spans some 20 m.y. The earliest representative in Africa may be the Moghara *Aceratherium*(?) spec. recorded by Andrews (1900), which occurs alongside *Brachypotherium snowi;* Dr. Roger Hamilton has indubitable *Aceratherium* material along with *Brachypotherium snowi* from the

early Miocene of Gebel Zelten in Libya. From Moruorot in Kenya, Deraniyagala (1951) described *Turkanatherium acutirostratus,* but this is an *Aceratherium* (Arambourg 1959, p. 74; Hooijer 1963, p. 43). *Aceratherium acutirostratum* (Deraniyagala) is rather widespread in the early Miocene of East Africa—Rusinga, Napak I, Karungu (Hooijer 1966, pp. 136–142)—and ranges up into the Alengerr Beds (Hooijer 1973), some 5 m.y. younger (see table 19.1). Either this species or *Dicerorhinus leakeyi* or both are present at Kirimun and in the Ngorora Formation, which would make for a record span for a single species of 7 m.y. Although the upper teeth and the skull can be told apart easily, it is difficult to distinguish between the lower teeth (except the tusks) and the postcranial material of *Aceratherium acutirostratum* and *Dicerorhinus leakeyi.* This makes identification especially difficult where the two species occur together, as at Napak, Rusinga, and in the Alengerr Beds (Hooijer 1966, 1973).

The skull of *Aceratherium acutirostratum* is

unique in the combination of a shallow nasomaxillary notch (back to above the middle of P³, as in the Aquitanian *Aceratherium lemanense* [Pomel] of Europe) and an elevated occiput (as in the terminal, "Pontian" *Aceratherium incisivum* [Kaup] of Europe). The upper cheek teeth have markedly constricted protocones, prominent antecrochets, and the premolars have strong internal cingula, which distinguish them from those of *Dicerorhinus leakeyi*. The upper incisors are much larger than those in *Dicerorhinus,* as in *Brachypotherium,* but are longer rooted than those in the latter. Both *Aceratherium* and *Dicerorhinus* are non-brachypothere, longer limbed, and longer footed than *Brachypotherium* (Hooijer 1966, 1973).

Genus *Dicerorhinus* Gloger 1841

The Miocene species of this genus, *Dicerorhinus leakeyi* Hooijer (1966, pp. 122–136) from Rusinga, Songhor, and Napak, (afterwards also found at Ombo and in the Alengerr Beds [Hooijer 1973]) combines characters found in different species in Europe. It has the skull shape of *Dicerorhinus sansaniensis* (Lartet) of the Vindobonian, and the teeth of *Dicerorhinus schleiermacheri* (Kaup) of the "Pontian."

The European lineages, from *Dicerorhinus tagicus* (Roman) of the Aquitanian and continuing up into the Pleistocene, are quite distinct from the African forms, which also include *Dicerorhinus primaevus* Arambourg (1959, p. 56) from the "Pontian" of Wadi el Hammam in Algeria and *Dicerorhinus africanus* Arambourg (1970, p. 79) of the North African Villafranchian (Lac Ichkeul). Thus the genus presumably persisted longer in Europe and North Africa than it did in East Africa, where the last occurrence known is in the Alengerr Beds and possibly at Kirimun and Ngorora (see table 19.1). The African forms are two-horned, with upper teeth that have weakly constricted protocones and a basal metacone bulge to M³ giving it a trapezoidal rather than subtriangular outline. The upper incisors, at least in *D. leakeyi,* are quite small and there are small incisors between the lower tusks; the latter differ from those of *Aceratherium* in being less bowed. The fifth metacarpal is reduced to a stump, unlike that in *Aceratherium,* in which it even carried some phalanges, but the articulation facet on metacarpal IV is of the same size in both genera so that even this metapodial cannot be generically identified when it occurs as an isolated bone. The longevity of *D. leakeyi* is similar to that of *Aceratherium acutirostratum,* that is, about 5 m.y.

Genus *Chilotheridium* Hooijer 1971

The genus *Chilotheridium* ranges from the early Miocene of Rusinga, Bukwa II (Hooijer 1971), and Ombo (Hooijer 1973) up into the late Miocene at Ngorora, and thus spans some 10 m.y. The type species, *Chilotheridium pattersoni* Hooijer (1971), comes from Loperot in Kenya. Quite characteristic of the genus are the tiny, horned nasals. The hypsodont teeth have sharply constricted protocones and large, inwardly curving antecrochets as in *Chilotherium* Ringström, with which it was confounded before the skull and postcranial skeleton were available. *Chilotherium* is a Eurasian genus ranging from Burdigalian to "Pontian," and I hold that *Chilotheridium* was its equivalent in Africa and originated from the same stock during the Eurasian Oligocene. But we have as yet no record of this in Africa. In the African *Chilotheridium* metacarpal V is a sizable bone; in *Chilotherium* it is much reduced, a progressive condition in *Chilotherium* as compared to *Chilotheridium*. *Chilotheridium* has a (small) nasal horn and large air sinuses in the frontal and parietal bones, whereas *Chilotherium* is hornless and has very small parietal and frontal air sinuses, which constitute generalized characters compared to *Chilotheridium*. There are no upper but large lower tusks, which are farther apart in *Chilotherium* than in the African *Chilotheridium*. The limb and foot bones of *Chilotheridium* are not as shortened as they are in *Chilotherium*.

Genus *Diceros* Gray 1821

This is the genus of the modern black rhinoceros of Africa, *Diceros bicornis* (L.). The genus as such goes back to the "Pontian" of Europe, southwestern Asia, and northern Africa with forms that are very close morphologically to *Diceros bicornis* and that have been placed in the same genus: *Diceros pachygnathus* (Wagner) from Samos and Maragha (Iran), and *Diceros douariensis* Guérin from Douaria, Tunisia. The modern species appears at the 4 m.y. level at Kanam West, East Africa (Hooijer 1969, pp. 88–89), and has recently been collected at Saragata Deare in the western Afar, Ethiopia, in deposits of about the same age (J. Kalb pers. comm.). The dM⁴ slightly exceeds its recent homologue in size, but the entire upper dentition is within recent size limits. As the teeth are worn we do not know whether they were as high-crowned as those in the living *D. bicornis*. That they were not at this early stage is shown by a slightly worn dM⁴ from the Mursi Formation (Hooijer 1973). Early *D. bicornis* has the dental pattern of the modern form: heavy internal upper premolar cingula, marked paracone styles, produced anterointernal crown angles, proto- and metalophs transverse in position, crista and crochet usually separate, not forming a medifossette, postsinus shallower than medisinus. By 2.5 m.y. ago, the time of deposition of Shungura Formation Member D (cf Cooke and Maglio 1972, p. 306), the molar crowns of *Diceros bicornis* have become as high as they are at present, and the skull from the 3 m.y. level (Shungura Member C) already has the modern characteristics (Hooijer 1973, 1975). We have *Diceros bicornis* from late Olduvai (Bed II, upper part, Bed III, and Bed IV), but not from early Olduvai (Bed I and lower part of Bed II) (Hooijer 1969). *D. bicornis* has not shown up in the Kaiso Formation, not even at later Kaiso, which is just below the 2 m.y. level (Cooke and Coryndon 1970; Cooke and Ewer 1972, p. 230), and neither is it present at Kanapoi (4 m.y. old, Hooijer and Patterson 1972) or in the 4 to 2 m.y. old Chemeron Formation (Hooijer 1969). In the latter, *Ceratotherium* is the only rhinocerotid genus found, which is probably ecological. In the Koobi Fora Formation, from about 3 m.y. ago on up to 1 m.y. ago, *D. bicornis* occurs in all three faunal zones. The molars appear slightly less hypsodont than in recent specimens and the skull shows a number of differences that suggest at least subspecific distinction (Harris 1976, p. 223.)

Genus *Ceratotherium* Gray 1867

Ceratotherium praecox Hooijer and Patterson is the earliest known species of the genus, and it occurs at Kanapoi, Lothagam-1, and Ekora, but also in the Chemeron Formation (locality J.M. 507), the Mursi Formation, the Aterir Beds (Hooijer 1972, pp. 187–189), and the Mpesida Beds (Hooijer 1973). The latter is its oldest occurrence, at the 7 m.y. level (see Cooke and Maglio 1972, p. 306; Bishop 1972, p. 231). It is a species characterized by four none-too-small upper incisors (these teeth have been lost in the modern white rhinoceros); a dorsal skull profile more concave, the posterior portion less extended behind; occiput less posteriorly inclined; and nuchal crest less thickened than in *Ceratotherium simum* (Burchell). The teeth are less high-crowned than those in the modern form, with angular anterointernal crown corners instead of rounded ones and hardly any or no medifossette formation. I believe that *C. praecox* is the immediate ancestor of *C. simum,* and it is most abundantly represented at Langebaanweg, Cape Province (Hooijer 1972), where it is the most common of the large mammals. A single upper molar comes from Swartlintjesfarm,

Hondeklipbaai, Namaqualand, about 160 km north of Langebaanweg.

The modern white rhinoceros of Africa first appears as *C. simum germanoafricanum* (Hilzheimer) in the Shungura Formation Member B (a little more than 3 m.y. old, Hooijer 1973; Guerin 1976); the Chemeron Formation (locality J.M. 91, Hooijer 1969, p. 75); Kanam West (Hooijer 1969, p. 88), which is either at the 4 m.y. level or somewhat younger (Cooke and Maglio 1972, p. 306; Cooke and Ewer 1972, p. 230); and in all levels of the Koobi Fora Formation, from about 3 to 1 m.y. Its cranial characters show several distinct differences from the living white rhinoceros (Harris 1976, p. 223), and the teeth are slightly less hypsodont and the lophs slightly less obliquely placed, although they already have the rounded anterointernal crown corners in which the upper teeth of *C. simum* differ from those of *Diceros bicornis*. As a grazer, *C. simum* has higher crowned teeth than *D. bicornis* the browser, and further the teeth of the former have reduced paracone styles, wavy ectolophs, obliquely placed lophs and lophids, medifossettes formed by the union of crochet and crista, and postsinuses as deep as the medisinuses (Hooijer 1959). The skull of *D. bicornis* has a vertical occiput instead of a posteriorly inclined one as in *C. simum*. From the variations in the skulls of the "Pontian" *Diceros,* Thenius (1955) concluded that *Ceratotherium* split off from the *Diceros* stock probably in the Pliocene, a view that is fully substantiated by finds at Kanapoi, etc., of *Ceratotherium praecox* (for a diagram of the *Diceros* group, see Hooijer and Patterson 1972, fig. 11). The precocious *Ceratotherium* was spread over South Africa in the Mio-Pliocene just as *C. simum* and *D. bicornis* were in the Pleistocene (Makapansgat, Transvaal: Hooijer 1959; Hopefield, Cape Province: Hooijer and Singer 1960).

Genus *Paradiceros* Hooijer 1968

Paradiceros mukirii Hooijer (1968) occurs only at Fort Ternan in Kenya and clearly belongs to the *Diceros* group of genera but is much earlier than either *Diceros* proper or *Ceratotherium*. It is a rather small, two-horned, browsing type in which the molars are lower than those in *D. bicornis* (a primitive character) but with a more abbreviated mandibular symphysis (a progressive feature). The humerus is rather short, suggesting that the Fort Ternan form was a more swift-moving type, more of a generalized, running rhinocerotid than *Diceros* (all species). I hold *Paradiceros* to be a collaterally developed browser from the same ancestral stock as *Diceros;*

the *Diceros* group of genera probably originated in Africa (cf Hooijer and Patterson 1972, fig. 11).

Conclusions

The distribution of the African rhinocerotids is plotted against the chronological sequence of sites in table 19.1. Within most genera there is more than one species, and each of them is different from those in the same genera in Eurasia, showing that evolution in Africa was independent from that in the rest of the Old World. In the early Miocene (Moghara, Bukwa, Napak), some 23 to 19 m.y. ago (van Couvering 1972), there are already in Africa distinct species of *Brachypotherium, Aceratherium, Dicerorhinus,* and *Chilotheridium,* which of course means that their ancestral stocks were there beforehand, in the Oligocene. The rhinocerotid record for this early arrival is a lamentable blank in Africa. *Chilotheridium* is even generically distinct and strictly African in appearance.

Africa further emerges as the continent of origin of the *Diceros* group of genera (*Paradiceros, Diceros,* and *a fortiori Ceratotherium*). *Ceratotherium* (both species) and *Diceros* were spread from southern to northern Africa in the Plio-Pleistocene and so was *Brachypotherium* already in the Miocene, linking up the South African fauna with those of the same period in Central and East Africa.

The species longevities that I have been able to establish (4 m.y. for *Brachypotherium heinzelini,* 7 m.y. for *Brachypotherium lewisi,* 5 m.y. for *Aceratherium acutirostratum* and *Dicerorhinus leakeyi,* and 5 m.y. for *Chilotheridium pattersoni*) are based upon radiometric age determinations, and these corroborate those obtained by means of sophisticated analysis of data in the literature on Neogene faunas of Spain, the Siwaliks, and China (Kurtén 1959). The mean species longevity for the Neogene is 5.2 m.y. This figure contrasts strikingly with the mean species longevity in the Pleistocene, also computed by Kurtén, which is 620,000 years, implying that the evolutionary rate was between eight and nine times faster in the Pleistocene than it was in the Neogene.

In most of the genera that I give in single columns in table 19.1 there is more than one species, but these do not necessarily represent single phyletic lineages even though the species may succeed each other in time (figure 19.1). In these Neogene rhinocerotids evolution was rather slow and even stagnant; two examples, in *Aceratherium* and in *Dicerorhinus,* stand out.

There is none or hardly any evolutionary change in the size of metacarpal V within the genus *Aceratherium* right from the late Oligocene *A. lemanense* up into the "Pontian" *A. incisivum* (Hooijer 1966, p. 153), a time span of 10 m.y. at least. Here evolution seems to have come to a near standstill. The other case in point may be that in the genus *Dicerorhinus,* which is not a single lineage but several (Hooijer 1966, p. 120), the metacone bulge is retained in M^3, which gives this tooth its typical trapezoidal basal outline from the early Miocene East African *D. leakeyi* through the Pleistocene species and even into the Holocene *D. sumatrensis* (Fischer) (Hooijer 1966, pp. 128, 129).

In the genus *Brachypotherium,* which runs up only into the "Pontian" in Europe, there is progressive abbreviation of limbs and feet as well as size increase. *Brachypotherium lewisi* of the late Pliocene, the last survivor of its genus anywhere, is the greatest of them all and may well be the descendant of Miocene *B. heinzelini* or of *B. snowi.* However, there is a gap in the record of 6 m.y. or more (between Rusinga and Ngorora, table 19.1), and the specific advance took place during this interval. It would be of great interest to see if the limb and foot bones of *B. lewisi* are shortened even more than those in the European terminal species *B. goldfussi* of Eppelsheim, but we know next to nothing as yet of the postcranial skeleton of *B. lewisi.*

The closing of the "Pliocene gap" in Africa has given us an insight into the longevity of the living African species. The transformation of *Ceratotherium praecox* into *Ceratotherium simum* took place between 4 and 3 m.y. ago, and *Diceros bicornis* as a species is likewise about 4 m.y. old. There are not very many modern mammalian species that have such a long record; *Hippopotamus amphibius* L. and *Castor fiber* L. date back into the Villafranchian of Europe and are probably between 1 and 3 m.y. old, but some small mammals would seem to have appeared already in the Astian, over 3 m.y. ago (Kurtén 1968, p. 254). In the Plio-Pleistocene of East Africa the extant elephant and suids are not yet in evidence (Cooke and Maglio 1972, pp. 310, 318), which makes the white and the black rhinoceros oldtimers by comparison.

My study of African fossil Rhinocerotidae would have been impossible without the cooperation of the late Louis Leakey and the following colleagues: Peter Andrews (Nairobi), T. H. Barry (Cape Town), W. W. Bishop (London), Frank Brown (Berkeley), P. M. Butler (Englefield Green, Surrey), L. Cahen (Tervuren), H. B. S. Cooke (Halifax, Nova Scotia), Yves Coppens (Paris), John van Couvering (Villa Park, California), P. E. P. Deraniyagala (Ceylon), A. Gautier (Ghent), Roger Hamilton (London), Jean de Heinzelin (Ghent), Brett Hendey (Cape Town), Andrew Hill (London), F. Clark Howell (Berkeley), Richard Leakey (Nairobi), J. Leper-

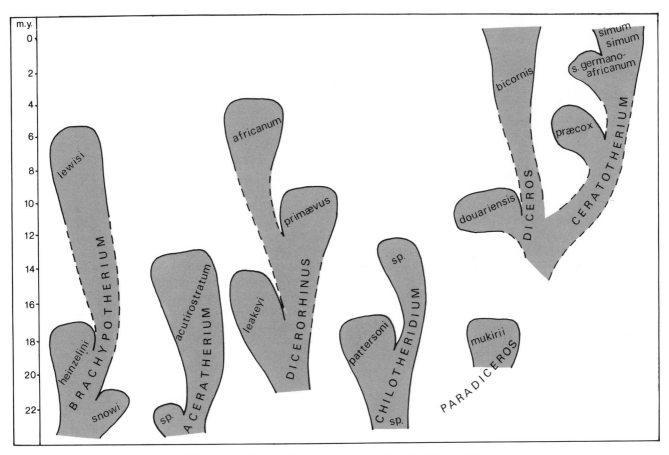

Figure 19.1 Proposed relationships among the fossil and recent species of African rhinoceroses.

sonne (Tervuren), Vincent Maglio (Princeton), Bryan Patterson (Cambridge, Mass.), Martin Pickford (London), Shirley Coryndon and R. J. G. Savage (Bristol), Anthony Sutcliffe (London), and Alan Walker (Nairobi).

Financial support and encouragement by the Wenner-Gren Foundation for Anthropological Research, Inc., New York, is most gratefully acknowledged.

References

Andrews, C. W. 1900. Fossil Mammalia from Egypt. Pt. II. *Geol. Mag.* (4) 7:401–403.

_____. 1914. On the Lower Miocene vertebrates from British East Africa, collected by Dr Felix Oswald. *Quart. J. Geol. Soc. London* 70:163–186.

Arambourg, C. 1933. Mammifères Miocènes du Turkana (Afrique Orientale). *Ann. Paléont.* 22:121–147.

_____. 1959. Vertébrés continentaux du Miocène supérieur de l'Afrique du Nord. *Publ. Serv. Carte Géol. de l'Algérie,* n.s., *Paléontologie, Mém.* 4:1–161.

_____. 1970. Les Vertébrés du Pléistocène de l'Afrique du Nord. Vol. I, Les Faunes Villafranchiennes, fasc. I, historique-stratigraphie-paléontologie (Proboscidiens et Périssodactyles). *Arch. Mus. Nat. Hist. Nat. Paris,* sér. 7, 10:1–126.

Bishop, W. W. 1972. Stratigraphic succession "versus" cal-

ibration in East Africa. In W. W. Bishop and J. A. Miller (eds.), *Calibration of hominoid evolution.* Edinburgh: Scottish Academic Press, pp. 219–246.

Cooke, H. B. S. 1968. Evolution of mammals on southern continents. II. The fossil mammal fauna of Africa. *Quart. Rev. Biol.* 43:234–264.

Cooke, H. B. S., and S. C. Coryndon. 1970. Pleistocene mammals from the Kaiso Formation and other related deposits in Uganda. In L. S. B. Leakey and R. J. G. Savage (eds.), *Fossil Vertebrates of Africa,* London: Academic Press, vol. 2, pp. 107–224.

Cooke, H. B. S., and R. F. Ewer. 1972. Fossil Suidae from Kanapoi and Lothagam, northwestern Kenya. *Bull. Mus. Comp. Zool., Harvard* 143:149–296.

Cooke, H. B. S., and V. J. Maglio. 1972. Plio-Pleistocene stratigraphy in East Africa in relation to proboscidean and suid evolution. In W. W. Bishop and J. A. Miller (eds.), *Calibration of hominoid evolution.* Edinburgh: Scottish Academic Press, pp. 303–329.

Deraniyagala, P. E. P. 1951. A hornless rhinoceros from the Mio-Pliocene deposits of East Africa. *Spolia Zeylan.* 26:133–135.

Erasmo, G. d' 1954. Sopra un molare di *Teleoceras* del giacimento fossilifero di Sahabi in Cirenaica. *Rc. Accad. naz. XL,* ser. 4 (4 and 5):89–102.

Fourtau, R. 1920. Contribution à l'étude des vertébrés

miocènes de l'Égypte. *Min. of Fin., Egypt. Survey Dept. Cairo,* pp. 122.

Guerin, C. 1976. Rhinocerotidae and Chalicotheriidae (Mammalia, Perissodactyla) from the Shungura Formation, lower Omo Basin. In Y. Coppens et al. (eds.), *Earliest man and environments in the Lake Rudolf Basin.* Chicago: Univ. of Chicago Press, pp. 214–221.

Harris, J. M. 1976. Rhinocerotidae from the East Rudolf succession. In Y. Coppens et al. (eds.), *Earliest man and environments in the Lake Rudolf Basin.* Chicago: Univ. of Chicago Press, pp. 222–224.

Heissig, K. 1971. *Brachypotherium* aus dem Miozaen von Südwestafrika. *Mitt. Bayer. Staats. Palaont. Hist. Geol.* 11:125–128.

Hooijer, D. A. 1959. Fossil rhinoceroses from the Limeworks Cave, Makapansgat. *Paleont. Afr.* 6:1–13.

———. 1963. Miocene Mammalia of Congo (with a chapter by A. Gautier and J. Lepersonne). *Ann. Mus. Roy. de l'Afrique Cent., Tervuren, Belg., Sci. Géol.* 46:1–77.

———. 1966. Miocene rhinoceroses of East Africa. *Bull. Brit. Mus. Nat. Hist.* (Geol.) 13:117–190.

———. 1968. A rhinoceros from the late Miocene of Fort Ternan, Kenya. *Zoöl. Med. Leiden* 43:77–92.

———. 1969. Pleistocene East African rhinoceroses. In L. S. B. Leakey (ed.), *Fossil Vertebrates of Africa.* London: Academic Press, vol. 1, pp. 71–98.

———. 1971. A new rhinoceros from the late Miocene of Loperot, Turkana District, Kenya. *Bull. Mus. Comp. Zool., Harvard* 142:339–392.

———. 1972. A late Pliocene rhinoceros from Langebaanweg, Cape Province. *Ann. S. Afr. Mus.* 59:151–191.

———. 1973. Additional Miocene to Pleistocene rhinoceroses of Africa. *Zoöl. Med. Leiden* 46:149–178.

———. 1975. Note on some newly found perissodactyl teeth from the Omo group deposits, Ethiopia. *Koninkl. Nederl. Akad. van Weten.,* ser. B, 78 (3):188–190.

———. 1976. Evolution of the Perissodactyla of the Omo group deposits. In Y. Coppens et al. (eds.), *Earliest man and environments in the Lake Rudolf Basin.* Chicago: Univ. of Chicago Press, pp. 209–213.

Hooijer, D. A., and B. Patterson. 1972. Rhinoceroses from the Pliocene of northwestern Kenya. *Bull. Mus. Comp. Zool., Harvard* 144:1–26.

Hooijer, D. A., and R. Singer. 1960. Fossil rhinoceroses from Hopefield, South Africa. *Zoöl. Med. Leiden* 37:113–128.

Kurtén, B. 1959. On the longevity of mammalian species in the Tertiary. *Soc. Sci. Fennica, Comm. Biol.* 21 (4):1–14.

———. 1968. *Pleistocene mammals of Europe.* London: Weidenfeld and Nicolson.

Maglio, V. J. 1972. Vertebrate faunas and chronology of hominid-bearing sediments east of Lake Rudolf, Kenya. *Nature* 239(5372):379–385.

Nouel, E. 1866. Mémoire sur un nouveau rhinocéros fossile. *Mém. Soc. Agric. Orléans,* sér. 2, 8:241–251.

Osborn, H. F. 1900. Phylogeny of the rhinoceroses of Europe. *Bull. Am. Mus. Nat. Hist.* 13:229–267.

———. 1910. *The age of mammals in Europe, Asia and North America.* New York: MacMillan.

Roger, O. 1904. Wirbeltierreste aus dem Obermiocän der bayerisch-schwäbischen Hochebene. Part 5. *Ber. Naturwiss. Ver. Schwaben Neuburg* 36:1–21.

Roman, F., and J. Viret. 1934. La faune de Mammifères du Burdigalien de la Romieu (Gers). *Mém. Soc. Géol. France,* n.s., 9 (21):1–67.

Stromer, E. 1926. Reste Land- und Süsswasser-bewohnender Wirbeltiere aus den Diamantfeldern Deutsch-Südwestafrikas. In E. Kaiser (ed.), *Die Diamantenwüste Südwest-Afrikas.* Berlin: D. Reimer, vol. 2, pp. 107–153.

Thenius, E. 1955. Zur Kenntniss der unterpliozänen *Diceros*-Arten (Mammalia, Rhinocerotidae). *Ann. Naturhist. Mus. Wien* 60:202–211.

Van Couvering, J. A. 1972. Radiometric calibration of the European Neogene. In W. W. Bishop and J. A. Miller (eds.), *Calibration of hominoid evolution.* Edinburgh: Scottish Academic Press, pp. 247–271.

Viret, J. 1961. Catalogue critique de la faune des mammifères miocènes de la Grive Saint-Alban (Isère) (suite du fascicule III.-1951). *Nouv. Arch. Mus. Hist. Nat. Lyon* 6:53–81.

20

Equidae

C. S. Churcher and M. L. Richardson

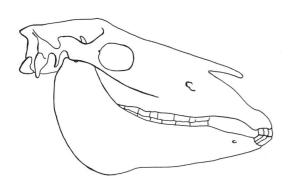

Representatives of the horse family in Africa are known from sites of late middle to late Miocene, Pliocene, Pleistocene, and Holocene ages. The most primitive form is the three-toed *Hipparion* from the later Miocene, descendants of which persisted until the end of the Pleistocene. Primitive members of the genus *Equus* (subgenus *Dolichohippus*) are known from Pliocene deposits in North Africa, and their descendants are known from Pleistocene deposits in eastern, central, and southern Africa. Zebras (subgenus *Hippotigris*) are earliest recorded from latest Pliocene sites in East Africa and appear by the later early Pleistocene both north and south of the Sahara. Asses (subgenus *Asinus*) are first known from northern Africa in the early late Pleistocene and apparently occupied the Mediterranean and Red Sea littorals by the latest Pleistocene.

In all, sixteen equid species are discussed from the late Tertiary and Quaternary of Africa. These are *Hipparion* (eight species); *Equus* (*Dolichohippus*) (four species); *Equus* (*Hippotigris*) (three species); and *Equus* (*Asinus*) (one species). The status of some of these species is insecure.

The compilation of this chapter necessitated the use of data that may be erroneous. Many of the early records are incomplete for locality, age, stratigraphic horizon, identification, recovered element, and final deposition of the material and consequently cannot be checked or assessed adequately. The newer series of dates based on radiometric analyses, mainly from East Africa, have brought about a revision of the time bases for the Miocene, Pliocene, and Pleistocene but are not readily conformable to the older faunal ages, especially for the Maghreb of North Africa. In general, the age determinations for sites that have yielded fossil equids in Africa follow Cooke (Chapter 2).

Identification of a fossil equid from one or a few isolated teeth is unreliable and here again many of the reported occurrences and specific identifications are based on such records. Consequently, this chapter constitutes at best an attempt to review and codify the extant data and to deduce the history of the African Equidae.

As space does not permit comprehensive discussion of all recorded African equid occurrences, these records are summarized in the legends for figures 20.1–20.4 and in table 20.1. The nomenclature used to describe equid teeth in this paper is depicted in figure 20.5.

Family Equidae Gray 1821

DIAGNOSIS. Unguligrade perissodactyls exhibiting progressive skeletal and dental adaptations for

Table 20.1 Selected occurrences of African Equidae.

Column groupings: *Hipparion* (H. primigenium – H. libycum); *Equus (Dolichohippus)* (E. (D.) numidicus – E. (D.) grevyi); *Equus (Hippotigris)* (E. (H.) burchelli – E. (H.) quagga); *Equus* (E. (A.) asinus – Equus sp.).

Site	Age	H. primigenium	H. baardi	H. namaquense	H. sitifense	H. turkanense	H. afarense	H. libycum	E. (D.) numidicus	E. (D.) oldowayensis	E. (D.) capensis	E. (D.) grevyi	E. (H.) burchelli	E. (H.) zebra	E. (H.) quagga	E. (A.) asinus	Equus sp.
Aïn Tit-Melil (3)	L. Pleist.	—	—	—	—	—	—	X	—	—	—	—	—	—	—	—	—
Carrière Marie Feuillet (5)	L. Pleist.	—	—	—	—	—	—	X	—	—	—	—	X	—	—	—	—
Fouarat (6)	E. Pleist.	—	—	—	—	—	—	X	—	—	—	—	—	—	—	—	—
Mugharet el'Aliya (7)	L. Pleist.	—	—	—	—	—	—	—	—	—	—	—	X	—	—	X	—
Grotte de Kiefan Bel Ghomari (8)	L. Pleist.	—	—	—	—	—	—	—	—	—	—	—	X	—	—	X	—
Melka el Ouidane (Gara Zaid: 9)	L. Mio.	X	—	—	—	—	—	—	—	—	—	—	—	—	—	—	—
Guiard (11)	?L. Mio.	X	—	—	—	—	—	—	—	—	—	—	—	—	—	—	—
Lac Karar (12)	M. Pleist.—Holo.	—	—	—	—	—	—	—	X	—	—	—	X	—	—	X	X
Grotte des Troglodytes (14)	Holo.	—	—	—	—	—	—	—	—	—	—	—	X	—	—	X	—
Grotte du Polygone (16)	Holo.	—	—	—	—	—	—	—	—	—	—	—	X	—	—	X	—
Carrière St. Pierre (17)	M. Pleist.	—	—	—	—	—	—	—	X	—	—	—	—	—	—	—	—
Oran Region (21)	?Pleist.	—	—	—	—	—	—	—	X	—	—	—	—	—	—	—	—
Aboukir (24)	?M. Pleist.	—	—	—	—	—	—	—	X	—	—	—	X	—	—	—	—
Bel Hacel (25)	E.—?M. Pleist.	—	—	—	—	—	—	—	X	X	—	—	—	—	—	—	—
Palikao (Ternifine: 26) - Zone 1	E.M. Pleist.	—	—	—	—	—	—	—	X	—	—	—	X	—	—	—	—
Zone 2	?L.L. Pleist.—Holo.	—	—	—	—	—	—	—	—	—	—	—	—	—	—	X	—
Mascara (27)	L. Plio.	—	—	X	—	—	—	—	—	—	—	—	—	—	—	—	—
Oued el Hammam (28)	M./L. Mio.	X	—	—	—	—	—	—	—	—	—	—	—	—	—	—	—
Abd el Kadar (29)	Holo.	—	—	—	—	—	—	—	—	—	—	—	X	—	—	X	—
Carrière des Bains-Romains (30)	L.L. Pleist.	—	—	—	—	—	—	—	—	—	—	—	X	—	—	X	—
Pointe Pescade (32)	L.L. Pleist.	—	—	—	—	—	—	—	—	—	—	—	X	—	—	X	—
Maison-Carrée (33)	?M. Pleist.	—	—	—	—	—	—	—	X	—	—	—	—	—	—	—	—
Grand Rocher (34)	Holo.	—	—	—	—	—	—	—	—	—	—	—	X	—	—	X	—
Djelfa (39)	M.—L. Pleist.	—	—	—	—	—	—	—	—	—	—	—	X	—	—	X	—
Oued Seguin (40)	?M.—L. Pleist.	—	—	—	—	—	—	—	—	—	—	—	X	—	—	X	—
Mansoura (41)	E. Pleist.	—	—	—	—	—	—	—	X	X	—	—	—	—	—	—	—
Grotte Ali Bacha (42)	L. Pleist.—Holo.	—	—	—	—	—	—	—	—	—	—	—	X	—	—	X	—
Setif (46)	E.—M. Pleist.	—	—	—	X	—	—	—	X	X	—	—	—	—	—	—	—
Aïn Jourdel (47)	E. Pleist.	—	—	—	—	—	—	—	X	X	—	—	—	—	—	—	—
St. Arnaud (El Eulma: 48a)	E. Pleist.	—	—	—	X	—	—	—	X	X	—	—	—	—	—	—	—
Aïn Hanech (48d)	L.E. Pleist.	—	—	—	—	—	—	—	X	—	—	—	X	—	—	—	—
Marceau (51)	L. Mio.	X	—	—	—	—	—	—	—	—	—	—	—	—	—	—	—
St. Donat (53)	Plio.	—	—	X	—	—	—	—	—	—	—	—	—	—	—	—	—
Grotte de Bou Zabaouin (54)	Holo.	—	—	—	—	—	—	—	—	—	—	—	X	—	—	X	—
Aïn el Bey (55)	L. Plio.	—	—	X	—	—	—	—	—	—	—	X	—	—	—	—	—
Aïn el Hadj Baba (56)	L. Plio.	—	—	X	—	—	—	—	—	—	—	?	—	—	—	—	—
Garet et Tarf (63)	Holo.	—	—	—	—	—	—	—	—	—	—	—	X	—	—	X	—
Aioun Beriche (65)	Holo.	—	—	—	—	—	—	—	—	—	—	—	X	—	—	X	—
El Mouhaad (66)	L. Pleist.	—	—	—	—	—	—	—	—	—	—	—	X	—	—	X	—
Oued Tamanrasset (67)	?	—	—	—	—	—	—	—	—	—	—	—	—	—	—	X	—
Garet Ichkeul (70)	E. Pleist.	—	—	—	—	—	—	—	X	X	—	—	—	—	—	—	—
Djebel M'Dilla (77)	L. Mio.	X	—	—	—	—	—	—	—	—	—	—	—	—	—	—	—
Bled ed Dourah (78)	L. Mio.	X	—	—	—	—	—	—	—	—	—	—	—	—	—	—	—
Hamada Damous (75)	E. Pleist.	—	—	—	—	—	—	—	—	—	—	—	X	—	—	—	—
Taoudenni (81)	L. Pleist	—	—	—	—	—	—	—	—	—	—	—	X	—	—	—	—
Hagfet et Tera (82)	L. Pleist.	—	—	—	—	—	—	—	—	—	—	—	X	—	—	X	—
Haua Fteah (83)	L.L. Pleist.	—	—	—	—	—	—	—	—	—	—	—	X	—	—	?	—
Koro Toro Region (86)	E. Pleist.	—	—	—	—	—	—	—	X	—	—	—	—	—	—	—	—
Gart el Moluk Hill (87)	L.E. or E.M. Plio.	—	—	—	—	—	—	—	X	—	—	—	—	—	—	—	—
Qâu (88)	Holo.	—	—	—	—	—	—	—	—	—	—	—	—	—	—	X	—
Kom Ombo (91)	L. Pleist.	—	—	—	—	—	—	—	—	—	—	—	—	—	—	X	—
Wadi Halfa - Zone 1 (93a)	?L. Plio. or E. Pleist.	—	—	—	—	—	—	—	—	—	X	—	—	—	—	—	—
Zone 2 (93b)	L. Pleist.	—	—	—	—	—	—	—	—	—	—	—	—	—	—	X	—
Khartoum & Abu Hugar (95)	L. Pleist.	—	—	—	—	—	—	—	—	—	—	—	—	—	—	X	—
Afar (95a)	L. Plio.	—	—	—	—	—	X	—	—	—	—	—	—	—	—	—	—

Table 20.1 (*continued*)

Site	Age	Hipparion							Equus (Dolichohippus)				Equus (Hippotigris)			Equus	
		H. primigenium	*H. baardi*	*H. namaquense*	*H. sitifense*	*H. turkanense*	*H. afarense*	*H. libycum*	*E. (D.) numidicus*	*E. (D.) oldowayensis*	*E. (D.) capensis*	*E. (D.) grevyi*	*E. (H.) burchelli*	*E. (H.) zebra*	*E. (H.) quagga*	*E. (A.) asinus*	*Equus* sp.
Omo - Mursi, Usno & Shungura A-E (96a)	?E.L.—L.L. Plio	X	X	X	X	—	—	—	—	—	—	—	—	—	—	—	X
Shungura F - J (96b)	L. Plio.—Pleist.	—	—	X	X	—	—	X	?	X	—	—	—	—	—	—	X
Kaiso - Early Fauna (97a)	Plio.	X	—	—	X	—	—	—	—	—	—	—	—	—	—	—	—
Later Fauna (Village: 97b)	E. Pleist.	—	—	—	—	—	—	—	X	X	—	—	—	—	—	—	—
Lothagam Hill (100)	L. Mio.	X	—	—	X	X	—	—	—	—	—	—	—	—	—	—	—
Ekora (101)	E. Plio.	X	—	—	X	X	—	—	—	—	—	—	—	—	—	—	—
Kanapoi (102)	E. Plio.	X	—	—	X	X	—	—	—	—	—	—	—	—	—	—	—
Nakali (104)	L. Mio.	X	—	—	—	—	—	—	—	—	—	—	—	—	—	—	—
Baringo Basin - Ngorora Formation (105a)	L.M. Mio.	X	—	—	—	—	—	—	—	—	—	—	—	—	—	—	—
Mpesida Beds (105b)	L. Mio.	—	—	X	—	—	—	—	—	—	—	—	—	—	—	—	—
Lukeino Formation (105c)	L. Mio.	—	—	X	X	—	—	—	—	—	—	—	—	—	—	—	—
Chemeron Formation (105d)	?E. Plio.	X	—	—	X	—	—	—	—	—	—	—	—	—	—	—	—
Aterir Beds (105e)	M. Plio.	X	—	—	—	—	—	—	—	—	—	—	—	—	—	—	—
Kanam (106)	E.L. Pleist.	—	—	—	—	—	—	—	?	—	X	—	—	—	—	—	—
Kanjera (107)	E. Pleist.	—	—	—	—	—	—	—	?	—	X	—	?	—	—	—	—
Gamble's Cave (109)	Holo.	—	—	—	—	—	—	—	—	—	—	?	X	—	—	—	—
Olorgesaillie (110)	M.—L. Pleist.	—	—	—	—	—	—	—	—	X	—	X	?	—	—	—	—
Laetolil (111)	Plio.—Pleist.	—	—	—	—	—	—	—	—	X	—	X	—	—	—	—	—
Olduvai Gorge (112) - Beds I and lower II	E. Pleist.	—	—	—	—	—	—	—	—	X	—	X	—	—	—	—	—
upper Bed II	L.E. or E.M. Pleist.	—	—	—	—	—	—	—	—	X	—	—	X	—	—	—	—
Beds III and IV	M.—E.L. Pleist.	—	—	—	—	—	—	—	—	X	—	—	X	—	—	—	—
Mumba Hills (113)	L. Pleist.	—	—	—	—	—	—	—	—	?	—	—	X	—	—	—	—
Mkujuni (114)	M. Pleist.	—	—	—	—	—	—	—	—	X	—	—	—	—	—	—	X
Chiwondo Beds	L. Plio.	—	—	—	—	—	—	—	—	X	?	—	—	—	—	—	—
Chelmer (119)	L. Pleist.	—	—	—	—	—	—	—	—	—	X	—	X	—	—	—	—
Kalkbank (122)	?L. Pleist.	—	—	—	—	—	—	—	—	—	X	—	X	—	—	—	—
Makapansgat - Limeworks Cave (123a)	E. Pleist.	—	—	—	—	—	—	—	—	—	X	—	X	—	—	—	—
Cave of Hearths (123b)	L. Pleist.—Recent	—	—	—	—	—	—	—	—	—	—	—	?	—	—	—	—
Gladysvale (125)	?M. Pleist.	—	—	—	—	—	—	—	—	—	—	—	?	—	—	—	—
Kromdraai Faunal Site (128)	E./M. Pleist.	—	—	—	—	—	—	—	—	—	X	—	X	—	X	—	—
Sterkfontein Type Site (130)	E./M. Pleist.	—	—	—	—	—	—	—	—	—	—	—	X	—	—	—	—
Sterkfontein Extension (Cave) Site (131)	E./M. Pleist.	—	—	—	—	—	—	—	—	—	—	—	X	—	—	—	—
Swartkrans Australopithecine Site (132)	E. Pleist.	—	—	—	—	—	—	—	—	—	X	—	X	—	?	—	—
Bolt's Farm Dumps (133)	E./M. Pleist.	—	—	—	—	—	—	—	—	—	X	—	X	—	X	—	—
Cornelia (Uitzoek) (134)	M. Pleist.	—	—	—	—	—	—	—	—	—	X	—	X	—	—	—	—
Sheppard Island (136)	M. Pleist.	—	—	—	—	—	—	—	—	—	X	—	X	—	—	—	—
Bankies (138)	L. Pleist.	—	—	—	—	—	—	—	—	—	—	—	X	—	X	—	—
Diamant (141)	M. Pleist.	—	—	—	—	—	—	—	—	—	X	—	—	—	?	—	—
Florisbad (143)	L. Pleist.—Holo.	—	—	—	—	—	—	—	—	—	X	—	X	—	—	—	—
Vlakkraal (144)	L. Pleist.—Holo.	—	—	—	—	—	—	—	—	—	X	—	X	—	—	—	—
Koffiefontein (145)	L. Pleist.	—	—	—	—	—	—	—	—	—	X	—	X	—	—	—	—
Vaal River Gravels (147-167)	M. Pleist.	—	—	—	—	—	—	—	—	—	X	—	X	—	X	—	—
Wonderwerk Cave (168)	L. Pleist.	—	—	—	—	—	—	—	—	—	—	—	X	—	?	—	—
Driefontein Farm (172)	L. Pleist.	—	—	—	—	—	—	—	—	—	—	—	X	—	X	—	—
Glen Craig (173)	?Pleist.	—	—	—	—	—	—	—	—	—	—	—	X	—	X	—	—
Aloes Bone Deposit (174)	L. Pleist.	—	—	—	—	—	—	—	—	—	—	—	X	—	X	—	—
Nelson Bay Cave (175)	L. Pleist.	—	—	—	—	—	—	—	—	—	X	—	X	—	—	—	—
Jakkalsfontein (177)	?Pleist.	—	—	—	—	—	—	—	—	—	—	—	—	—	X	—	—
Cango Caves (179)	?Pleist. or Holo.	—	—	—	—	—	—	—	—	—	—	—	—	—	X	—	—
Springbok Site (Namaqualand: 180)	?Plio. or E. Pleist.	—	X	—	—	—	—	—	—	—	X	—	—	—	—	—	—
Langebaanweg (181)	E. Plio.—?E. Pleist.	—	X	X	—	—	—	—	—	—	X	—	—	—	—	—	—
Yzerplaats (186)	?M. Pleist.	—	—	—	—	—	—	—	—	—	X	—	—	—	—	—	—
Skildegat Cave (187)	L.L. Pleist.	—	—	—	—	—	—	—	—	—	X	—	—	—	?	—	—

Note: E = early, M = middle, and L = late subdivisions of the epochs. Doubtful identifications or occurrences are indicated by "?."

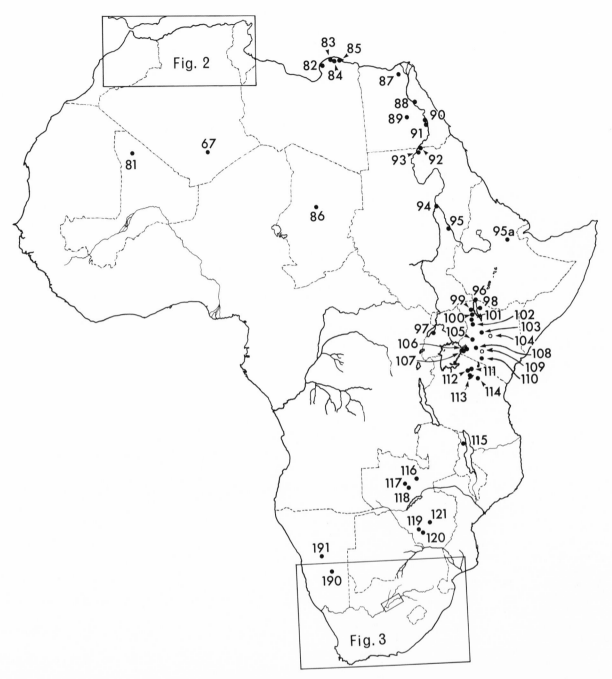

Figures 20.1, 20.2, 20.3, and 20.4 Geographic distribution of African Cenozoic fossil localities from which Equidae have been reported. The sites are listed in numerical order and are keyed to the maps in figures 20.1-20.4. For each site the preferred name is given first, followed by other appellations, its age, and a major reference or the original description. Brief locality data or additional references are given where appropriate. E = early, M = middle, and L = late periods.

1. Aïn Rohr (Aïn Rorh), 15 km east of Oulidia beach; Pleist. Ennouchi 1949a, b.

2. Grotte d'El Kenzira, Mazagan (El Jadida); L. Pleist. Arambourg 1938.

3. Aïn Tit-Melil, Casablanca; L. Pleist. Arambourg 1952.

4. Oued el Akrech, near Argoub el Hafid, Rabat; E. Pleist. Arambourg 1970.

5. Carrière Marie Feuillet, near Rabat; L. Pleist. Arambourg 1952.

6. Fouarat (Kenetra, Port Lyautey, Mina Hassan Tani); E. Pleist. Cooke 1963.

7. Mugharet el'Aliya, Cape Spartel; L. Pleist. Howe and Movius 1947.

8. Grotte de Kiefan Bel Ghomari, near Taza; L. Pleist. Arambourg 1938.

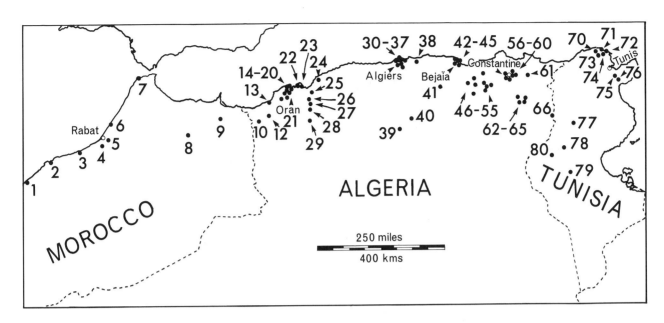

9. Melka el Ouidane (Gara Zaid, Camp Bertaux), near Taourirt; L. Mio. Arambourg 1963.

10. La Mouillah, 5 km north of Maghnia (Marnia); L. Pleist. Pallary 1910.

11. Guiard, south of Koudiat and Tine, on River Tafna; L. Mio. Arambourg 1952.

12. Lac Karâr, near Tlemcen (?Montagnac Remchi, ?Beni Smeil); M. Pleist.–Holo. Estaunie 1941. Arambourg 1952, Cooke 1963.

13. Beni Saf; ?Holo. Romer 1928.

14. Grotte des Troglodytes (Ras el Aïn, Ravin de Noiseaux), near Oran; Holo. Pallary and Tomassini 1892, Romer 1928.

15. Grotte de Ciel Ouvert (Ras el Aïn, Ravin de Noiseaux), near Oran; Holo. Doumergue 1927, Romer 1928.

16. Grotte du Polygone (Ras el Aïn, Ravin de Noiseaux), near Oran; Holo. Doumergue 1927, Romer 1928.

17. Carrière St. Pierre, Charles district, Oran; M. Pleist. Pomel 1897.

18. Grotte du Cuartel, Echmühl district, Oran; Holo. Doumergue 1926.

19. Grotte de la Tranchée, near Oran; Holo. Romer 1928.

20. Aïn el Turk, 18 km northwest of Oran; L. Pleist. Romer 1928.

21. Oran region—unlocated sites; ?Pleist. Pallary 1934, Arambourg 1957, 1970.

22. Puits Kharouby, east of Oran; ?E.–M. Pleist. Arambourg 1970.

23. Ravin Blanc de Gambetta, east of Oran; E. Pleist. Arambourg and Choubert 1965.

24. Aboukir, Mostagnem; ?M. Pleist. Arambourg 1952.

25. Bel Hacel, near Ighil Izane (Relizane); E.–?M. Pleist. Cooke 1963.

26. Palikao (Ternifine), near Mascara; E. M. Pleist.– Holo. Romer 1928, Cooke 1963.

27. Mascara; L. Plio. Forstén 1968.

28. Oued el Hammam, near Bou Hanifa, between Mascara and Mercier Lacombe; M./L. Mio. Hooijer and Maglio 1973.

29. Abd el Kadar (La Grotte de l'Oued Saïda), near Saïda; Holo. Romer 1928.

30. Carrière de Bains-Romains, 7 km west of Algiers; L. L. Pleist. Arambourg 1931.

31. Carrière St. Cloud, near Bains-Romains, Algiers; L. L. Pleist. Arambourg 1931.

32. Grotte de Pointe Pescade (old and new), near Algiers; L. L. Pleist. Pomel 1897, Arambourg 1931.

33. Maison-Carrée (unlocated), near Algiers; ?M. Pleist. Cooke 1963.

34. Grotte de Grand Rocher, near Guyotville; Holo. Romer 1928.

35. Grotte de Carrière Anglade, near Guyotville; L. Pleist. Arambourg 1935.

36. Carrière Sintes, near Guyotville; L. Pleist. Arambourg 1932.

37. Boulevard Bru, Mustapha Supérieur, Algiers; Holo. Romer 1928.

38. Grotte de la Cascade, Bordj Menaiel; Holo. Romer 1928.

39. Djelfa (Oued Djelfa); M.–L. Pleist. Thomas 1884a, c, Romer 1928.

40. Oued Seguin, tributary to River Rummel; ?M.–L. Pleist. Romer 1928.

41. Mansoura; E. Pleist. Romer 1928.

42. Grotte Ali Bacha, near Bejaïa (Bougie); L. Pleist.– Holo. Romer 1928.

43. Pic des Singes, near Ali Bacha and Bejaïa (Bougie); Holo. Romer 1928.

44. Grand Abri, Aiguades, near Bejaïa (Bougie); Holo. Romer 1928.

45. Grotte de Fort Clausel, near Bejaïa (Bougie); Holo. Romer 1928.

46. Setif Region (unlocated); E.–M. Pleist. Arambourg 1948, 1957, Ennouchi and Jeanette 1954.

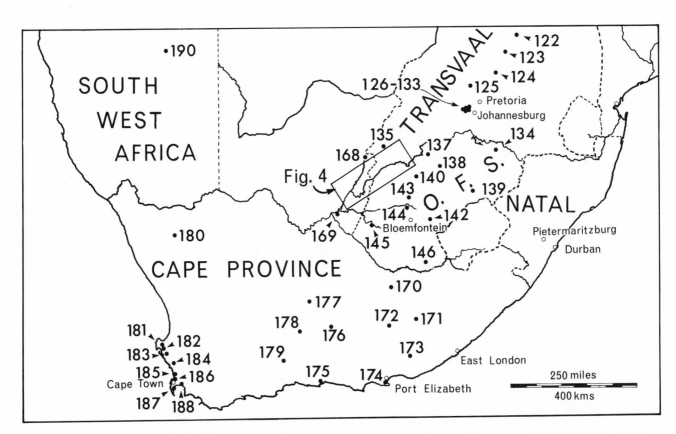

47. Aïn Jourdel, near Setif; E. Pleist. Arambourg 1970.

48a. St. Arnaud (El Eulma); E. Pleist. Dalloni 1940, Ennouchi 1951, Arambourg 1970.

b. St. Arnaud Cemetery; E. Pleist. Arambourg 1970.

c. Beni Foudda (Aïn Boucherit); E. Pleist. Arambourg 1970.

d. Aïn (el) Hanech; L. E. Pleist. Arambourg 1970.

e. Near Aïn Hanech, on road to Sillequé; ?Plio.–Pleist. Arambourg 1949.

49. Grotte des Hyènes, Djebel Roknia, near Setif; Holo. Romer 1928.

50. Djebel Thaya (?Grotte de Djebel Thaya); L. Pleist. Romer 1928.

51. Marceau; L. Mio. Boné and Singer 1965.

52. Mechta el Arbi, Chateaudun du Rhumel; L. Pleist. Romer 1935.

53. St. Donat, near Setif; Plio. Forstén 1968.

54. Grotte de Bou Zabaouin, near Aïn Melila; Holo. Romer 1928.

55. Aïn el Bey, in basin of rivers Smendou and Rummel; L. Plio. Romer 1928, Arambourg 1963.

56. Aïn el Hadj Baba, near Constantine; L. Plio. Forstén 1968, Arambourg 1970.

57. Grotte du Mouflon, near Constantine; L. Pleist. Romer 1928.

58. Grotte des Pigeons, near Constantine; L. Pleist. Romer 1928.

59. Grotte des Ours, near Constantine; L. Pleist. Romer 1928.

60. Constantine Region; ?Pleist. Thomas 1884a, b, c, Joleaud 1918, Ennouchi and Jeanette 1954.

61. Millesimo, Seybouse Valley (?Vallée de l'Oued Skroun), near Guelma; ?L. M. Pleist. Dalloni 1940.

62. Oued Medfoun, Nos. 10 and 14, north of Garet et Tarf; Holo. Romer 1928.

63. Garet et Tarf, north shore (No. 25) and near salt works (No. 26); Holo. Romer 1935.

64. Daoud (Aïn Beida); L. L. Pleist. Romer 1928.

65. Aioun Beriche, Nos. 12 and 51, 17 km north of Aïn Beida; Holo. Romer 1935.

66. El Mouhaad (?Chaachas), 35 km east of Tebessa; L. Pleist. Romer 1935.

67. Oued Tamanrasset, Taunesruft Region; undated. Romer 1935.

68. Junction of rivers Chelif and Mina, near Sidi-Bahim (unlocated); undated. Dalloni 1940.

69. Deux Moulins (unlocated); M. Pleist. Romer 1928.

70. Garet (Lac) Ichkeul, near Ferryville, Bizerte; E. Pleist. Boné and Singer 1965, Coppens 1971.

71. Ferryville, near Bizerte; L. Mio. Ennouchi and Jeanette 1954; ? = Utique of Boné and Singer 1965.

72. Djebel Menzel R'oul, near Bizerte; ?E. Pleist. Solignac 1927.

73. El Ariana, near Bizerte; ?E. Pleist. Solignac 1927.

74. Bizerte; ?E. Pleist. Romer 1928, may refer to 72 and 73 above.

75. Hamada Damous, south of Grombalia; E. Pleist. Coppens 1971.

76. Sidi Bou Koutta, southwest of Grombalia; E. Pleist. Coppens 1971.

77. Djebel M'Dilla, northeast of Sbeitla; L. Mio. Ennouchi and Jeanette 1954.

78. Bled ed Douarah (?Redeyef), near Gafsa; L. Mio. Robinson and Black 1974.

79. Aïn Brimba, near Kebili; E. Pleist. Coque 1957, Coppens 1971.

80. Tozeur; L. Mio. Boné and Singer 1965.

81. Taoudenni, Salines d'Agorgott; L. Pleist. Joleaud 1936.

82. Hagfet et Tera, near Benghazi; L. Pleist. Churcher 1972.

83. Haua Fteah, near Derna; L. L. Pleist. Churcher 1972.

84. Hagfet ed Dabba, near Derna; L. Pleist. Churcher 1972.

85. Wadi Derna; L. Pleist. Churcher 1972.

86. Koro Toro Region; E. Pleist. Arambourg 1970.

87. Gart el Moluk Hill, Wadi Natrun; L. E. or E. M. Plio. Boné and Singer 1965.

88. Qâu (Kau el Kebir); Holo. Churcher 1972.

89. Kharga Oasis, Mound Springs KO 10c and 108a; L. Pleist. Churcher 1972.

90. Edfu (Idfu); Holo. Churcher 1972.

91. Kom Ombo; L. Pleist. Churcher 1974.

92. Debeira East; L. Pleist. Perkins 1965.

93a. Wadi Halfa-Zone 1; ?L. Plio. or E. Pleist.

b. Zone 2; L. Pleist. Joleaud 1933c, Churcher 1972.

94. Khartoum Central Railway Station; L. Pleist. Churcher 1972.

95. Abu Hugar (Abu Higar); L. Pleist. Bate 1951.

a. Afar; L. Plio. Eisenmann 1976.

96. Omo Beds (Lake Turkana, Lake Rudolf); E. L. Plio–?M. Pleist.

a. Mursi, Usno, and Shungura A-E. Cooke 1963, Hooijer and Maglio 1973.

b. Shungura F-J. Cooke 1963.

c. ?Mixed site. Boné and Singer 1965.

97. Kaiso, east shore of Lake Albert; Plio.–E. Pleist. Cooke and Coryndon 1970.

a. Early fauna.

b. Later fauna (village).

98. Lake Turkana (East Rudolf), Koobi Fora and Ileret formations; L. E. Pleist. Maglio 1972, Eisenmann 1976.

99. Moruorot Hill, southwest of Lake Turkana (Lake Rudolf); L. Pleist.–Holo. Madden 1972.

100. Lothagam Hill (Lothagam-1), west of Lake Turkana (Lake Rudolf); L. Mio. Maglio 1974.

101. Ekora, west of Lake Turkana (Lake Rudolf); E. Plio. Hooijer 1975b.

102. Kanapoi, west of Lake Turkana (Lake Rudolf); E. Plio. Hooijer and Maglio 1973.

103. Marsabit Road; ?E. Pleist. Hooijer 1974, pers. comm.

104. Nakali; L. Mio. Aguirre and Leakey 1974.

105. Baringo Basin; L. Mio.–Pleist.

a. Ngorora Formation; L. M. Mio. Hooijer 1974, pers. comm.

b. Mpesida Beds; L. Mio. Maglio 1974.

c. Lukeino (Kaperyon) Formation; L. Mio. Hooijer and Maglio 1973.

d. Chemeron Formation; E. Plio. Hooijer 1974, pers. comm.

e. Aterir Beds; M. Plio. Hooijer 1974, pers. comm.

f. Chemoigut Beds, Chesowanja; Pleist. Bishop et al. 1971.

106. Kanam, 18 km west of Kendu, between Homa Mountain and Lake Victoria; E. Pleist. Boné and Singer 1965.

107. Kanjera, 25 km west of Kendu, between Homa Mountain and Lake Victoria; M. Pleist. Boné and Singer 1965.

108. Naivasha, rivers Morendat and Malewa (?Morendat Station, 8 km northwest of Naivasha); L. Pleist.–Holo. Arambourg 1931.

109. Gamble's Cave, near lakes Elementeita and Nakuru; Holo. Arambourg 1931, Boné and Singer 1965.

110. Olorgesaillie, Lake Magadi; M.–L. Pleist. Cooke 1963, Arambourg 1970.

111. Laetolil (Vogelfluss, Serengeti); Plio.–Pleist. Boné and Singer 1965.

112. Olduvai Gorge; E.–L. Pleist. Boné and Singer 1965, Leakey 1965, Hooijer and Maglio 1973.

113. Mumba Hills, Lake Eyasi; L. Pleist. Boné and Singer 1965.

114. Mkujuni Valley, Lake Manyara; M. Pleist. Kent 1942.

115. Lake Malawi (Lake Nyassa), Chiwondo Beds (Mwanganda, Mwenirondo 3, Mwimbi, and Uruaha Hill sites); L. Plio. Mawby 1970.

116. Broken Hill; L. Pleist. Cooke 1963.

117. Leopard's Hill Cave; L. Pleist. Cooke 1950b.

118. Mumbwa Cave, 145 km northwest of Lusaka; L. Pleist. Cooke 1963.

119. Chelmer, 25 km west-northwest of Bulawayo; L. Pleist. Cooke 1963.

120. Bulawayo Waterworks; L. Pleist. Cooke and Wells 1951.

121. Que Que Quarry; L. Pleist. Cooke and Wells 1951.

122. Kalkbank, near Pietersburg; L. Pleist. Dart and Kitching 1958.

123. Makapansgat, near Potgietersrust.

a. Limeworks Cave; E. Pleist. Churcher 1970.

b. Cave of Hearths; L. Pleist.–Recent. Churcher 1970.

124. Tuinplaas, Springbok Flats; L. Pleist. Haughton 1932.

125. Gladysvale (Uitkomst), near Krugersdorp; ?M. Pleist. Churcher 1970.

126. Clyde Site, near Krugersdorp; ?M. Pleist. Churcher 1970.

127. Minnaar's Cave, near Krugersdorp; E./M. Pleist. Churcher 1970.

128. Kromdraai Faunal (Type) Site, near Krugersdorp; E./M. Pleist. Arambourg 1970, Churcher 1970.

129. Kromdraai II (Coopers), near Krugersdorp; ?E./M. Pleist. Churcher 1970.

130. Sterkfontein Type Site, near Krugersdorp; E./M. Pleist. Churcher 1970.

131. Sterkfontein Extension (Cave) Site, near Krugersdorp; E./M. Pleist. Robinson 1958, Mendrez 1966, Churcher 1970.

132. Swartkrans Australopithecine Site, near Krugersdorp; E. Pleist. Churcher 1970.

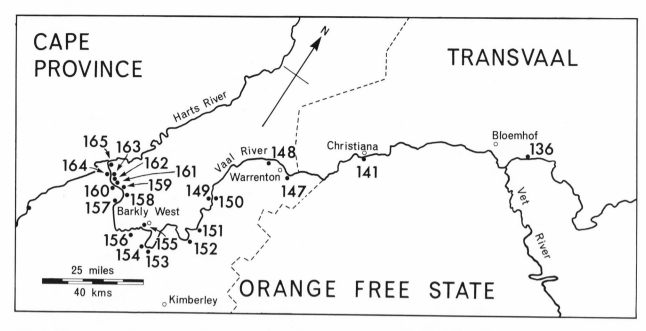

133. Bolt's Farm Dumps, near Krugersdorp; E./M. Pleist. Churcher 1970.

134. Cornelia (Uitzoek), near Cornelia; M. Pleist. Cooke 1974.

135. Wolmaranstad; L. Pleist. Shapiro 1943.

136. Sheppard Island (Bloemhof Site), 18 km upstream from Bloemhof; M. Pleist. Cooke 1955.

137. Bothaville; Pleist. Cooke 1950a.

138. Bankies, near Kroonstad; L. Pleist. Shapiro 1943, Cooke 1950a.

139. Dorpsgrond, Bethlehem; ?Pleist. Haughton 1932.

140. Tierfontein, near Port Allen; M. Pleist. Cooke 1950a.

141. Diamant, near Christina, Vaal River; M. Pleist. Cooke 1949, Boné and Singer 1965.

142. Chubani, near Thaba Nchu; Pleist. Van Hoepen 1930a.

143. Florisbad (Hagenstad Saltpan), 45 km north-northwest of Bloemfontein; L. Pleist.–Holo. Boné and Singer 1965.

144. Vlakkraal, 40 km north-northwest of Bloemfontein; L. Pleist.–Holo. Cooke 1963.

145. Koffiefontein; L. Pleist. Cooke 1948, 1950a.

146. Rouxville; M. Pleist. Van Hoepen 1930a.

147. Schoolplaats No. 1, above Warrenton Weir, Vaal River; M. Pleist. Cooke 1949.

148. Warrenton, Vaal River; M. Pleist. Cooke 1949.

149. Windsorton, 35 ft shaft, lot 197, Vaal River; M. Pleist. Cooke 1949.

150. Riverview Estates, Vaal River; M. Pleist. Cooke 1949.

151. Morris Draai, Vaal River; M. Pleist. Cooke 1949.

152. Riverton, Vaal River; M. Pleist. Cooke 1949.

153. Power's site, near Nooitgedacht, Vaal River; M. Pleist. Cooke 1949.

154. The Bend, Barkly West, Vaal River; M. Pleist. Cooke 1950a.

155. Barkly West, Vaal River; M. Pleist. Cooke 1949.

156. Pniel Estates, Barkly West, Vaal River; M. Pleist. Cooke 1950a, Boné and Singer 1965.

157. Waldeck's Plant, Barkly West, Vaal River; M. Pleist. Cooke 1949.

158. Gong Gong, between Barkly West and Sydney-on-Vaal, Vaal River; M. Pleist. Cooke 1949.

159. Forlorn Hope, Sydney-on-Vaal, Vaal River; M. Pleist. Cooke 1949.

160. Niekerk's Rush, Sydney-on-Vaal, Vaal River; M. Pleist. Cooke 1949.

161. Longlands, Sydney-on-Vaal, Vaal River; M. Pleist. Cooke 1949.

162. Austin's Rush, Sydney-on-Vaal, Vaal River; M. Pleist. Cooke 1949.

163. Winter's Rush, Sydney-on-Vaal, Vaal River; M. Pleist. Cooke 1949.

164. Sydney-on-Vaal, Vaal River; M. Pleist. Cooke 1949, Boné and Singer 1965.

165. Delport's Hope, junction of Vaal and Harts rivers; M. Pleist. Cooke 1949.

166. Schmidtsdrift, Vaal River; M. Pleist. Cooke 1949.

167. Vaal River, unknown localities; M. Pleist. Boné and Singer 1965, Arambourg 1970.

168. Wonderwerk Cave, west of Taung; L. Pleist. Cooke 1941, 1963.

169. Koffiefontein Farm, Douglas; M. Pleist. Dreyer 1931.

170. Middleburg, Karroo; undated. Broom 1913.

171. Hoogstede Farm, Queenstown; ?L. Pleist. or Holo. Cooke 1955.

172. Driefontein Farm, Cradock; L. Pleist. Wells 1970a.

173. Glen Craig, Grahamstown; ?Pleist. Cooke 1950a.

174. Aloes bone deposit, Port Elizabeth; L. Pleist. Wells 1970b.

175. Nelson Bay Cave, Plettenberg Bay, Knysna; L. Pleist. Klein 1972.

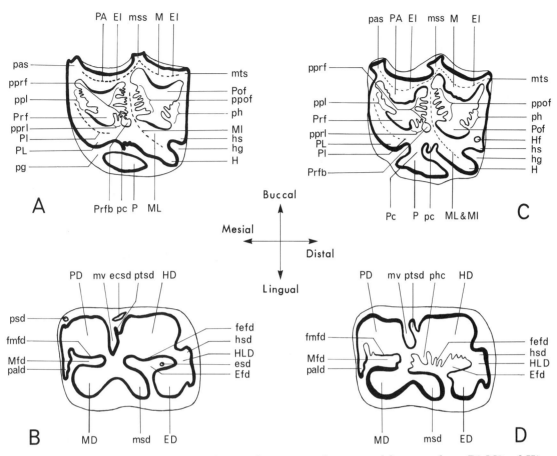

Figure 20.5 Diagrammatic representations of enamel patterns of upper and lower molars, P2-M3, of *Hipparion* and *Equus*. The plis are drawn to illustrate the more complex conditions observable. (*A* and *B*) upper and lower molars of *Hipparion*, (*C* and *D*) upper and lower molars of *Equus*, respectively.

Abbreviations: Upper molars: El-ectoloph, H-hypocone, hg-hypoconal glyph, hypoglyph, hs-hypostyle, M-metacone, ML-metaconule, Ml-metaloph, mts-metastyle, P-protocone, PA-paracone, pas-parastyle, Pc-protoconal commissure, pc-pli caballine, pg-protoconal groove, ph-pli hypostyle, PL-protoconule, Pl-protoloph, Pof-postfossette, ppl-plis protoloph, ppof-plis postfossette, pprf-plis prefossette, pprl-plis protoconule, prf-prefossette, Prfb-prefossette bay. Lower molars: ecsd-ectostylid, ED-entoconid, Efd-entoflexid, fossetula posterior, esd-entostylid, fefd-floor of entoflexid, fmfd-floor of meta-flexid, HD-hypoconid, HLD-hypoconulid, hsd-hypostylid, MD-metaconid, Mfd-metaflexid, mesial valley, fossetula anterior, msd-metastylid, mv-median or buccal valley, vallum medium, outer groove, pald-paralophid, PD-protoconid, phc-pli hypo-conid, psd-protostylid, ptsd-ptychostylid.

Figures 20.1, 20.2, 20.3, and 20.4 (continued)

176. Toorfontein Farm, Murraysburg; ?Pleist. Dreyer 1931.

177. Jakkalsfontein Farm, Victoria West; ?Pleist. Cooke 1950a, 1955.

178. Little England Farm, Beaufort West; ?Pleist. Cooke 1950a.

179. Cango Caves, Oudtshoorn; ?Pleist. or Holo. Cooke 1950a.

180. Springbok site, 75 km east of Springbok, Nama-qualand; ?Plio. or E. Pleist. Boné and Singer 1965.

181. Langebaanweg, 30 km northwest of Hopefield; E. L. Plio.–?E. Pleist. Hendey 1974a, 1974b.

182. Saldanha Bay; ?M.-L. Pleist. Cooke 1955.

183. Elandsfontein Farm, 15 km southwest of Hope-field; E. M. Pleist. Hendey 1970.

184. Bloembosch Farm, Darling; ?M. Pleist.–Holo. Cooke 1955.

185. Melkbos site, north of Melkbosstrand; L. Pleist. Hendey 1968, 1970.

186. Yzerplaats, Table Bay, Maitland; ?M. Pleist. Churcher 1970.

187. Skildegat (Peer's) Cave, Fish Hoek; L. L. Pleist. Cooke 1955.

188. Swartklip, False Bay; L. Pleist. Hendey 1970.

189. Southwestern Cape Province (coastal sands); ?Pleist. Hendey 1974.

190. Kalk Plateau, east of Mariental; ?L. Pleist. Cooke 1955.

191. Usakos; ?M. Pleist. Cooke 1955.

cursorial locomotion and grazing diet respectively. Dentition: $\frac{3.1\text{-}0.4\text{-}3.3}{3.1\text{-}0.4\text{-}3.3}$. Incisors in later forms often possessing infundibula (cups or marks). Canines usually small, sometimes absent or unerupted in females. Premolars 2–4 molarized; P1s simple and peglike, often absent or lost early in life in later forms. Upper molars, except in earliest forms, with three lophs: a mesiodistal ectoloph comprised of two arcuate portions; and two obliquely transverse buccolingual lophs (protoloph and metaloph) becoming J-shaped and elaborated with small folds or plis in later forms. Lower molars with progressively enlarged metaconid, entoconid, and metastylid at junction of both lophids. Body light and slender except in earliest forms. Extremities progressively elongated, especially in manus and pes. In early forms manus and pes tetradactyl and tridactyl respectively; later forms either tridactyl or monodactyl. Lateral digits progressively reduced in size until lost or vestigial; terminal phalanges III short, robust, and hoofed.

AGE. Middle/late Miocene to Recent in Africa; Paleocene/Eocene to Recent elsewhere.

Subfamily Equinae Steinmann et Döderlein 1890

DIAGNOSIS. Skull with postorbital bar. Upper incisors with more or less developed infundibula; in later forms on some or all lower incisors also. Cheek teeth hypsodont with crown height equal to or greater than mesiodistal diameter; cementum in fossettes and in later forms over all of crown; small enamel folds or plis present on lophs, often numerous. Premolars 2 to 4 fully molarized; P1s small or absent. In upper molars protolophs and metalophs join ectolophs even in early stages of wear, but in premolars the transverse lophs may remain free until a late stage of wear; fossettes usually close with wear. Protocone either separate or nearly detached from protoloph; elongate, with major diameter oriented mesiodistally. In lower molars pattern composed of two crescentic lophids joined in middle of tooth; metaconid and metastylid completely fused, mesiodistally elongated, forming a double loop in later forms. Ulna greatly reduced in shaft, sometimes incomplete, fused to radius in nearly all forms. Fibula narrow, reduced, sometimes with only proximal and distal ends remaining; fused distally with tibia in nearly all forms. Metapodial III and phalanges relatively elongated; lateral metapodials and digits narrowed, moved posteriorly to limb axis; distal ends of metapodials bent posteriorly. Lateral hoofed phalanges laterally compressed, not extend-ing to plantar surface of III; in latest forms lateral phalanges lost and lateral metapodials reduced to splint bones.

Genus *Hipparion* de Christol 1832

SYNONYMY. *Eurygnathohippus* van Hoepen 1930, *Stylohipparion* van Hoepen 1932, *Notohipparion* Haughton 1932, *Libyhipparion* Joleaud 1933.

DIAGNOSIS. Three-toed equids of variable size and robustness displaying progressive trends toward increasing hypsodonty of the cheek teeth, development of strong ectostylids in the lower permanent cheek teeth, and strong reduction of all third incisors. Facial region relatively short and deep; preorbital fossae present or absent; sagittal crest low; bony auditory meatus long and directed dorsolaterally; basicranial region usually with prominent longitudinal crest; and occiput with deep fossae above occipital condyles for attachment of nuchal musculature. Coronoid process of dentary high; ascending ramus vertical. Incisors large, cement filled; infundibula typically present on upper and lower I1 and I2; in I3 infundibulum closed or partly closed. Canines usually present in both sexes, possibly dimorphic. Cheek teeth moderately to very hypsodont, slightly bowed, prismatic. Upper cheek teeth with isolated, more or less elongate protocones and almost always with many plis in fossettes. Lower cheek teeth with well-developed double loop, metaflexid with subequal arms and deep median valley; M3 usually with tripartite talonid; well-developed ectostylid in milk lower cheek teeth of all forms where known and in permanent lower cheek teeth of more progressive forms. Limbs with angled joints, ulna complete. Metapodials moderately to very elongate; metapodials III with traces of II and IV contacts along shafts; metapodials II and IV variably developed, apparently without reduction in size through time; vestigial metacarpal V present. Digits II, III, and IV functional, but II and IV shorter and thinner than III. Terminal phalanges (ungules) cut by deep median slit.

DISCUSSION. African hipparion fossils were first discovered in the Maghreb of North Africa during the late nineteenth century and were recorded by Thomas (1884a, b, c) and described by Pomel (1897). During the twentieth century *Hipparion* was found to comprise a significant element of the North African fossil fauna. Further specimens recovered in Ethiopia by du Bourg de Bozas' 1903 expedition to Omo were described by Joleaud (1933a, b). A new species was described from Uganda by Hopwood in 1926, and several new hipparion taxa were described from South Africa during the early 1930s (van Hoepen 1930b, 1932; Haughton 1932). Most recently a series of *Hipparion* fossils from radiometrically dated horizons in Ethiopia, Kenya, and Tanzania have been described (Hooijer and Maglio 1973, 1974; Hooijer 1975a, b; Eisenmann 1976) and others have been recorded from Langebaanweg (Cape Province, South Africa) by Boné and Singer (1965) and Hendey (1974b, 1976).

The now extinct *Hipparion* is classically thought to have evolved in North America from *Merychippus* stock during the Miocene, from whence it spread rapidly across northern Asia to become a prominent member of faunas in southern Asia, Europe, and North Africa during the Pliocene. In Europe the appearance of *Hipparion* was long considered the marker of the Mio-Pliocene boundary, and, although faunas with *Hipparion* had consistently been dated "late Miocene" in North Africa, this was attributed to the practice of French geologists of placing the "Pontian" in the Miocene rather than the Pliocene as originally designated (see Forstén 1968). In North America, *Hipparion* became extinct by the end of the Pliocene and in Europe, the Near East, and India it disappeared from the fossil record in the early Pleistocene ("Villafranchian"). Africa, in contrast, supported somewhat progressive hipparions until the end of the Pleistocene.

More recently (e.g., van Couvering 1972), the advent of the primordial *Hipparion primigenium* in Europe has been radiometrically dated at 12.5 m.y. ago (i.e., late middle Miocene) and this species is known as a fossil in both North Africa (Oued el Hammam or Bou Hanifa) and East Africa (Baringo Basin, Ngorora Formation) by the latest middle or earliest late Miocene. This virtually synchronous appearance of *H. primigenium* in widely separated regions of Africa during the later Miocene seems to indicate extremely rapid proliferation and spreading of the group from a single point of entry (via Spain, southern Italy, or an Afro-Arabian contact prior to complete flooding of the Red Sea trench) or the entry of the species through more than one of these areas (see Dalloni 1915; Pirlot 1956; Forstén 1968; and Azzaroli 1975). In favor of original entry of the species into Africa via a tenacious Arabian contact are recent geological findings that much of the main Red Sea trough is filled with thick (up to 5 km) beds of Miocene evaporite (McKenzie, Davis, and Molnar 1970), indicating at least intermittent desiccation of the Red Sea since its original formation in the middle Miocene (Cooke, Chapter 2).

By the late Miocene two other hipparions, *H. turkanense* and a small hipparion referred by Hooijer and Maglio (1974) to *H.* cf *sitifense*, had appeared in East Africa. The former species, distinguished from *H. primigenium* by the lack of both preorbital fossae in the skull and ectostylids in the lower cheek teeth, appears to bear its closest resemblance to *H. hippidiodum* of China (Hooijer and Maglio 1973, 1974; Hooijer 1975b) and may represent a separate colonization of Africa via the Red Sea region. The ancestry of the "dwarf" hipparion is not clear. However, as no equally small hipparions are known outside Africa during the Miocene and as the same or a similar form continues in Africa during the Pliocene and early Pleistocene, it appears that this "dwarf" form represents an indigenous offshoot of (presumably) *H. primigenium* stock.

All other species of *Hipparion* reported from Africa are autochthonous, indicating a separate evolutionary history of the group in Africa from late Miocene or early Pliocene times.

The taxonomy of African hipparions, once considerably confused, has undergone substantial revision during the past 15 years but remains to be adequately clarified. The generic, subgeneric, specific, and subspecific names as proposed and used by different authors reflect rather diverse viewpoints regarding the history and relationships of African hipparions. For brevity, most of these names will be merely listed as synonyms in consideration of the analyses of Cooke (1950a, 1963), Boné and Singer (1965), Forstén (1968), and Hooijer (1975b). Remaining are eight species to be considered here: *H. primigenium, H. albertense, H. baardi, H. namaquense, H. sitifense, H. turkanense, H. afarense,* and *H. libycum* (including *H. ethiopicum* and *H. steytleri*).

Of these, *H. libycum* from the Pleistocene of North, East, and South Africa is the most advanced form of hipparion known. Its dentition is characterized by highly hypsodont cheek teeth, strong development of ectostylids in the lower cheek teeth, marked reduction of the upper and lower lateral incisors, and finer details. Since teeth described from North Africa as *H. libycum*, from East Africa as *H.* (= *"Libyhipparion"*) *ethiopicum*, and from South Africa as *H.* (= *"Stylohipparion"*) *steytleri* are indistinguishable, these regional species are here regarded as a single species, with the name *H. libycum* taking precedence (Boné and Singer 1965).

Unfortunately, the skull of this advanced species is so far known only from East Africa. Hooijer (1975b) has described advanced *Hipparion* material from Olduvai Gorge, Tanzania, in some detail and has referred it to *H.* cf *ethiopicum*. While he notes (1975b, p. 47) that "the Olduvai *Hipparion* . . . does not differ dentally from either *Hipparion libycum* of the Villafranchian of North Africa or *Hipparion steytleri* (including *Hipparion hipkini*) of southern Africa," he does not place the three forms in a single taxon because it is possible that skulls of the North African and South African advanced hipparions, when known, "may be different from that of the Olduvai form" (1975b, p. 47). This caution being noted, there remains no reason not to consider the three forms a single species, and hence synonymy is accepted here.

On all available evidence *H. libycum* appears to be the direct descendant of *H. afarense,* which is known from the later Pliocene of Ethiopia (Eisenmann 1976). *H. afarense* and *H. libycum* share, among other characters, rather complex enamel patterns in their cheek teeth and the lack of a preorbital fossa. In the finer features of their dentitions they show greatest affinity with the primordial *H. primigenium,* while with *H. turkanense* they share the absence of a preorbital fossa. As both *H. primigenium* and *H. turkanense* are known from the earlier Pliocene of East Africa, each could conceivably have given rise to the progressive forms. We feel, however, that the dental evidence supports more strongly a phylogenetic relationship with *H. primigenium.*

Also seemingly related to *H. primigenium* (or to an *H. primigenium-H. afarense-H. libycum* lineage) are specimens assigned to *H. albertense, H. namaquense,* and *H. baardi. H. albertense* was founded on an inadequate type specimen and is regarded as a *nomen vanum* (see below), while various specimens assigned to this taxon are considered to belong with *H. libycum, H. sitifense, H. baardi, H. primigenium,* or *Equus.* Both *H. namaquense* and *H. baardi* are currently recorded for the Pliocene of southwestern Africa, but it is anticipated that consideration of recently discovered *Hipparion* material of similar age from this region will lead to a revision of these taxa.

The two remaining species, *H. sitifense* and *H. turkanense,* appear to represent distinct lines of African *Hipparion. H. sitifense,* the "dwarf" form, survived in at least North Africa until the early Pleistocene and may even have extended its range to Spain and the Near East. *H. turkanense,* a relatively large form, was sympatric with *H. sitifense* and *H. primigenium* in East Africa during the late Miocene and earlier Pliocene. Its fate since the earlier Pliocene is uncertain.

African hipparions of highly advanced dentitions (i.e., *H. libycum* and its synonyms) have frequently been placed in either the genus or subgenus "*Stylohipparion*" while the remaining species have been grouped under the subgenus "*Hipparion.*" This practice is not followed here for two principle reasons:

first, as Hooijer and Maglio (1973) point out, if *Stylohipparion* van Hoepen 1932 is regarded as a taxon, it must be junior to *Eurygnathohippus* van Hoepen 1930 and second, because the gradual development through time of relative characters (such as hypsodonty of the cheek teeth and strengthening of ectostylids) makes the division between phyletic species of *Hipparion* in certain fossil sequences so arbitrary (see Hooijer 1975b) that generic or subgeneric distinction on the basis of such characters seems unwarranted.

Hipparion primigenium (von Meyer) 1829

SYNONYMY. *Equus primigenius* von Meyer 1829, *Hipparion* aff. *gracile* Arambourg 1951, *Hipparion africanum* Arambourg 1959, *Hipparion primigenius* Forstén 1968, *Hipparion (Hipparion) albertense* Hopwood (partim), Cooke and Coryndon 1970.

AGE. Late middle Miocene to ?later Pliocene in Africa.

LOCALITIES. Morocco (9)[1]; Algeria (11, 28, 51); Tunisia (71, 77, 78, 80); Ethiopia (96a Mursi and Usno formations); Uganda (96a); Kenya (100, 101, 102, 104, 105a, 105d, 105e); Republic of South Africa (?181).

DISCUSSION. *Hipparion primigenium* has been best known in Africa as *Hipparion africanum,* for which Arambourg (1959) described the partially preserved type skull from Oued el Hammam, Algeria. Arambourg considered *H. africanum* more closely related to the advanced African hipparion, *H. libycum,* than to the European *H. primigenium,* but this view was not fully supported by Boné and Singer (1965), who contended that many (but not all) features of the *H. africanum* dentition are found in Eurasian species.

Forstén (1968), drawing largely on the arguments of Boné and Singer, included *H. africanum* in *H. primigenius* together with a large number of Eurasian species, but she noted (e.g., pp. 28–29) that the North African hipparions formed a separate population of this species whose nearest relatives were the temporally equivalent "Vallesian" hipparions of Spain.

Hooijer and Maglio (1974) and Hooijer (1975a and b) accept Forstén's synonymy and assign to *H. primigenium* a number of East African specimens indistinguishable from *H. africanum.* These specimens include a partial juvenile skull from earlier Pliocene deposits of Ekora, Kenya, and isolated

---

1 The numbers in parentheses here and elsewhere in this chapter refer to the particular localities shown in the maps in figures 20.1–20.4 and listed in the legends for those figures.

teeth ranging in age from later middle Miocene (Baringo Basin, Ngorora Formation) to slightly younger Pliocene (Omo Basin, Shunguru Formation).

It seems clear that both North and East African *H. primigenium* are more closely related to each other than to Eurasian populations of this species, from which they may be distinguished on a number of cranial and postcranial features (see Arambourg 1959; Boné and Singer 1965; Forstén 1968). Although it is most probable that the same species originally colonized both Eurasia and Africa, the evolutionary history of the group on the two continents was almost certainly distinct during and since the late Miocene. Hence although we use the name *H. primigenium,* we shall refer only to the African populations of this species.

A fairly complete picture of *H. primigenium* may be compiled from the descriptions of Arambourg (1959), Boné and Singer (1965), Forstén (1968), Hooijer and Maglio (1974), and Hooijer (1975b). It was a large to medium-sized hipparion of rather stocky build possessing a large skull with a moderately elongate face and rostrum and a broad nasal aperture. A pronounced preorbital fossa is present on both the *"H. africanum"* skull from Oued el Hammam and the juvenile skull from Ekora, but this feature is variably developed in European specimens referred to *H. primigenium* (Forstén 1968).

The cheek teeth are only moderately hypsodont. Canines are present and may be sexually dimorphic, with those of males laterally compressed and those of females smaller and round in cross-section. The incisive regions are relatively broad and arcuate and all incisors are moderately large. The mesial incisors (I1 and I2) are anteroposteriorly compressed and slightly grooved both labially and lingually. The upper cheek teeth possess compressed (lenticular to oval) protocones and fossettes characterized by rich plication fore and aft. The pli caballine is variably single or double or with forked folds. Enamel borders are generally smooth. With wear protocones became broader, plication became simpler, and hypocones were indented distally and lingually. Both milk and permanent lower cheek teeth have protostylids and ectostylids, although the latter are never developed to the full height of the crown and may only be detected in well-worn teeth or through the cementum of isolated teeth.

In contrast with the Eurasian *H. primigenium,* which shows the greatest development of ectostylids in earlier populations (Forstén 1968), East African populations show an increase in ectostylid development through time (Hooijer 1975b). Also noteworthy of the East African *H. primigenium* is a progressive increase in crown height, which becomes particularly evident in earlier Pliocene specimens (Hooijer 1975b).

Forstén (1968, p. 29) considered the African *Hipparion* comparatively slender in build and "faintly suggestive of the dolichopodid steppe-hipparions *H. mediterraneus* and *H. matthewi,*" but Arambourg (1959) and Hooijer and Maglio (1974) emphasize the relative robusticity of the limbs of the African form compared with these Eurasian steppe forms. The metapodials are flattened distally and the lateral digits are well developed. Metatarsal III often lacks an ectocuneiform facet, indicating a moderately strong reliance on the lateral digits for weight bearing.

Gromova (1952) interpreted the sturdier limbs of *H. primigenium* as an adaptation to softer ground, presumably of a forest habitat, while she considered its relatively brachydont teeth and moderate development of cingular elements in the cheek teeth indicative of a diet composed at least partially of browse. Forstén (1968) supported her deductions and further showed the distribution of *H. primigenium* in Europe and the Near East to be correlated with regions heavily forested during the Pliocene.

The North African records from Oued el Hammam and Bled ed Douarah are not from faunas that are decidedly forest dwelling, as are those in Europe (Arambourg 1959; Forstén 1972), but may be interpreted as sylvan or savanna woodland rather than closed forest or grass savanna. In East Africa the preponderance of *H. primigenium* in certain locales (e.g., in the Pliocene deposits of Kanapoi and Ekora rather than the terminal Miocene deposits of Lothagam [Hooijer and Maglio 1974]) may likewise indicate the occurrence of favorable wooded habitats only in certain areas and times. Alternatively, East African populations may have been more versatile in their ecological requirements than their North African counterparts, since the development of crown height (hypsodonty) and of strengthening pillars such as ectostylids have been noted, among other characters, to correlate with an adaptation to drier savanna or steppe environments (Gromova 1952; Forstén 1968).

H. primigenium (*sensu lato*) appears to be the most successful of *Hipparion* species, both in terms of its wide geographic distribution and its long temporal range. It was the first hipparion to colonize Eurasia in the late middle Miocene and was among the last to disappear from that continent in the early Pleistocene. It appears in both East and North Africa shortly after it was known in Eurasia and continued to thrive in at least East Africa until well into the Pliocene. In North Africa, however, *H. primigenium* was mainly restricted to late Miocene times; its relatively limited success in this region may have been due to increasing aridity and loss of its natural habitat during the Pliocene.

Some specimens of *H. primigenium* have been tentatively reported from E Quarry at Langebaanweg by Hendey (1976), but Hooijer (1976) assigns them to *H. baardi.* The two Pliocene *Hipparion* species described from southwestern Africa, *H.* (= "*Notohipparion*") *namaquense* and *H. baardi,* both bear strong resemblances to *H. primigenium* and it seems probable that hipparions closely related to, if not indistinguishable from, *H. primigenium* extended their range throughout the grass veld of Africa by the latest Miocene or earliest Pliocene and may have continued to survive in southwestern Africa throughout the Pliocene.

In eastern Africa earlier Pliocene populations of *H. primigenium* appear to have given rise to the advanced Pleistocene hipparion of Africa, *H. libycum,* through an intermediate species, *H. afarense.* Evolutionary progress in this lineage included the reduction and medial migration of the lateral incisors in both jaws in addition to the earlier trends of increased hypsodonty and stronger ectostylid development. By the *H. afarense* stage the preorbital fossa characteristic of *H. primigenium* had almost or entirely disappeared. The highly advanced species spread both north and south during the Pleistocene to occupy the coastal regions of North Africa and the high veld of South Africa.

Hipparion albertense Hopwood 1926

SYNONYMY. *Hipparion albertensis* Hopwood 1926.

AGE. Late Pliocene or early Pleistocene.

LOCALITY. Uganda (97b).

DISCUSSION. Hopwood (1926, p. 17) founded *Hipparion albertense* on the buccal two-thirds of an upper second molar with the diagnosis: "Tooth with very high crown; enamel pattern extremely complicated. Ratio height to diameter 2.8." Accompanying this description of the tooth was a somewhat misleading figure drawn from the undersurface of a sectioned portion of the crown (Cooke and Coryndon 1970).

Because of the inadequacy of the type specimen, Hooijer (1975b, pp. 6–7) correctly states that *H. albertense* "in fact is a *nomen vanum* (in the sense of Simpson 1945, p. 27)" and recommends "that the name *Hipparion albertense* should be either ignored or listed as indeterminate and without known sig-

nificance" or (p. 28) that "it should be kept in the suspense account and not applied to material other than the type." He further elaborates on the confusion that has already arisen through application of the name as the result of many specimens, seemingly of mutually exclusive descriptions, being included in Hopwood's taxon. Hooijer's full documentation of these details will not be repeated here, but several specimens, particularly those described by Cooke and Coryndon (1970) from the type site of Kaiso, Uganda, require further note.

The Kaiso hipparion specimens, all ascribed by Cooke and Coryndon (1970) to *Hipparion (Hipparion) albertense,* were recovered in both Earlier Kaiso and Later (Kaiso Village) assemblages. From the latter fauna came an isolated M³ that Cooke and Coryndon designated the paratype of *H. albertense.* The remaining specimens comprise 13 upper cheek teeth from four individuals and two complete and three fragmentary lower cheek teeth, all found at Earlier Kaiso locales. At least two of the Earlier Kaiso teeth, including the most complete lower molar, are now considered to be *H. sitifense* because of their very small size (Cooke pers. comm. 1975).

Cooke and Coryndon (1970) discuss the Kaiso hipparion teeth in some detail together with teeth from Laetolil and Baard's Quarry, Langebaanweg, which Boné and Singer (1965) had assigned to *Hipparion (Hipparion) albertense* under the subspecific names *serengetense* and *baardi.* Accepting the inclusion of these specimens from Tanzania and South Africa in *H. albertense* (but without affirming separate subspecific designation), Cooke and Coryndon (1970, p. 136) give an emended diagnosis of the species as follows:

A hipparionid with hypsodont cheek teeth; complex to very complex enamel plication in the fossettes of the upper cheek teeth; protocone oval, tending towards sub-triangular or lenticular; antero-internal side of parastyle not marked off by a small flange. Lower cheek teeth constantly lack accessory ectostylid pillar but may develop protostylid to full height of crown.

In this diagnosis the characters of the lower cheek teeth are mainly deduced from Laetolil and Baard's Quarry specimens due to the continued paucity of hipparion lower teeth from the type site. Hooijer (1975b, p. 7) has concluded of these specimens that the Laetolil lower teeth are *Equus,* but the upper teeth belong to "*Stylohipparion*" (i.e., *H. libycum*), while the Baard's Quarry *Hipparion* "appears to represent a species in its own right, and it may henceforth be designated as *Hipparion baardi* Boné and Singer."

Comparing *H. albertense* from Kaiso, Laetolil, and Baard's Quarry with *"H. africanum"* (=*H. primigenium*), Cooke and Coryndon (1970, pp. 144, 146) note "that it is by no means easy to find criteria for distinguishing between them . . . and it may be suspected that there is a genuine and continuous relationship between them." They conclude (p. 146) that "the Kaiso Village hipparionid is a progressive form, very close to the Laetolil variety in both size and in details of enamel pattern" while the Earlier Kaiso form "tends to be even more primitive" than the Baard's Quarry hipparion.

As already noted, Cooke now believes the smallest of the Lower Kaiso teeth to be *H. sitifense* while Hooijer indicates the Laetolil upper cheek teeth belong with *H. libycum*. It is here suggested that the remaining Earlier Kaiso teeth are in fact *H. primigenium* and the Kaiso Village teeth (including the *H. albertense* type specimen) belong in *H. libycum*.

The Kaiso hipparion fauna may thus be seen to resemble the hipparion faunas of other East African sites, with *H. primigenium* and *H. sitifense* both present in the early Pliocene and *H. libycum* appearing in the latest Pliocene or earliest Pleistocene.

Hipparion baardi Boné and Singer 1965

SYNONYMY. *Hipparion (Hipparion) albertense baardi* Boné and Singer 1965, *Hipparion (Hipparion) albertense* Hopwood (partim), Cooke and Coryndon 1970.

AGE. Late Pliocene.

LOCALITY. Republic of South Africa (181 Baard's, C, and ?E Quarries).

DISCUSSION. Boné and Singer (1965) described a sizable collection of isolated hipparion teeth from Baard's Quarry, Langebaanweg, which they assigned to *Hipparion (Hipparion) albertense* as a new subspecies, *baardi*. They noted in their diagnosis (p. 390) that the subspecies was characterized by "rather hypsodont teeth, about 70 mm crown height" and by the "constant absence of the ectostylid and . . . a tendency to form other additional stylids, especially protostylid extending along the total length of the crown."

In evaluating the status of *H. albertense,* Hooijer (1975b) concluded that *H. (H.) albertense baardi* was a species in its own right that he designated *Hipparion baardi* Boné and Singer 1965. He noted that both *H. baardi* and *H. turkanense* apparently lack ectostylids (in at least moderattely worn teeth) but considered other dental characters of the two forms to warrant specific distinction. The recent discovery of a hipparion skull from E Quarry, Langebaanweg, which was taken by Hendey and Hooijer to repre-

sent *H. baardi* (Hooijer 1975b) or *H.* cf *baardi* (Hooijer 1976), lent support to Hooijer's separation of *H. baardi* and *H. turkanense,* as the new skull (unlike that of *H. turkanense*) possesses a pronounced preorbital fossa.

Instead, the Langebaanweg *Hipparion* skull, as described by Hooijer (1975b, 1976) shows affinities with *H. primigenium* in both the presence of a preorbital fossa and a lack of reduction of the upper lateral incisors. A further resemblance between the dentitions of *H. baardi* and *H. primigenium* is the apparent lack of ectostylids on the lower molars found at the same quarry (Hooijer 1976). The proximal phalanx of digit III is elongated more than in *H. primigenium* and approaches the condition seen in advanced *H. "Stylohipparion" libycum,* but the metapodials do not differ significantly from those of known *H. primigenium* (Hooijer 1976).

Boné and Singer (1965) believed that *"H. africanum"* was restricted to the Miocene, that the Langebaanweg *Hipparion* was a Pleistocene species, and that this age differential precluded a close relationship between the two forms. However, the *H.* cf *baardi* skull referred to above is from deposits that are certainly of Pliocene age (Hendey 1974a, b, 1976; Hooijer 1976) and most fossils retrieved from Baard's Quarry are now considered to be late Pliocene (Hendey 1976; Hooijer 1976). As recent discoveries in East Africa have demonstrated that the temporal range of *H. primigenium* extended into the Pliocene (Hooijer and Maglio 1974; Hooijer 1975b), there would seem to be no reason to exclude the possibility of the existence of an advanced *Hipparion* derived from *H. primigenium* in the late Pliocene at Langebaanweg, that is *H. baardi*.

Hipparion namaquense Haughton 1932

SYNONYMY. *Notohipparion namaquense* Haughton 1932.

AGE. Pliocene.

LOCALITIES. Republic of South Africa (180, 181 between E and C quarries).

DISCUSSION. The only specimens that to date have been described for *Hipparion namaquense* are those on which Haughton (1932) founded the species (although he neither designated a holotype nor provided a formal diagnosis). These specimens constitute a fairly complete series of well-worn lower cheek teeth from a single individual. They show a variable appearance of an ectostylid in teeth worn to less than 35 mm crown height; a caballoid enamel configuration; and fossette borders of fairly wavy enamel in the most strongly worn teeth.

Hooijer (1975b) notes a strong resemblance between the dentition described by Haughton (1932) and that of a partial mandible recovered from Shungura Member B 11 of the Omo group. He assigns the Shungura mandible to "*Hipparion* spec.," cautioning (p. 54) that "we badly need skull material to decide what the species are," but he shows the specimen to be clearly "less advanced" than *H. libycum* and suggests it is indistinguishable from *H. primigenium*.

Hooijer (in Hendey 1976) referred an incomplete lower dentition from Langebaanweg to *H. namaquense*. These teeth comprise a left premolar series, P_2–P_4, and derive from a single individual and from a site between E and C quarries, from the uppermost level of Bed 3 (Hendey 1976, p. 240; Hooijer 1976). Hooijer considers them to be practically indistinguishable from those of *H. namaquense* in that they possess ectostylids, unlike the other Langebaanweg teeth, that were not more than 35 mm high and are thus comparable to the low ectostylids in the type of *H. namaquense*, which are 30 mm high or less. Hooijer (1976) assigns these teeth to *H.* cf *namaquense*.

Hipparion sitifense Pomel 1897

SYNONYMY. *Hipparion sitifensis* Pomel 1897, *Hipparion (Hipparion) albertense* Hopwood (partim), Cooke and Coryndon 1970.

AGE. Late Miocene to early Pliocene.

LOCALITIES. Algeria (27, 46, 48a, 48b, 53, 55, 56); Ethiopia (96a Mursi, Usno, and Shungura formations); Uganda (97a); Kenya (100, 101, 102, 105c, 105d).

DISCUSSION. *Hipparion sitifense* is most distinctive for the small size of its cheek teeth and appears (on somewhat slender postcranial evidence) to have been the graceful, diminutive member of the late Tertiary *Hipparion* fauna in northern and eastern Africa. Unfortunately, neither the type specimens of *H. sitifense* from St. Arnaud Cemetery described by Pomel (1897) nor specimens of similar age from other sites in North Africa assigned to this species by Arambourg (1957) provide as clear a diagnosis of *H. sitifense* as could be desired. As a result many specimens of small *Hipparion* from Spain and East Africa referred to or provisionally identified as *H. sitifense* cannot be conclusively demonstrated actually to belong within this taxon.

"True" *H. sitifense* from the later Pliocene and earliest Pleistocene of North Africa, as described and illustrated by Arambourg (1957), is known by upper and lower teeth and by the distal end of a metacarpal III that was originally assigned (along with several teeth) to "*H. gracile*" (=*H. primigenium*) by Thomas (1884a). The most complete specimens come

from Mascara and comprise a palate containing a complete left and partial right molariform series and a symphyseal region with incisors from the same individual. The lower dentition is known from several teeth, still rooted in bone, discovered by Thomas at Aïn el Hadj Baba. A number of isolated cheek teeth from St. Arnaud Cemetery and other North African localities complete the sample. All the teeth are too worn to provide an accurate gauge of hypsodonty for the species, but none of the lower teeth is sufficiently worn or damaged to fully demonstrate, one way or the other, whether weakly developed ectostylids (such as found in approximately 25% of the small hipparion teeth from several East African sites) may have been present (Hooijer and Maglio 1974; Hooijer 1975b).

Arambourg's (1957) description of *H. sitifense* stresses the small size of the teeth, the relatively simple plication found in the upper molars, the oval or flattened shape of the protocone, and an absence of ectostylids in the lower dentition.

Forstén (1968) included two species of *Hipparion* and one subspecies of *H. mediterraneus* (*sic*) from the Pikermian of Spain in *H. sitifense* and on the evidence of this sample provided a diagnosis of the species as follows:

Middle-sized to small. Limbs with proportions as in *H. primigenius*. Preorbital fossa oval. Upper cheek teeth moderately to simply plicated, protocone oval, hypocone defined only posteriorly. Lower cheek teeth with rounded metaconid and metastylid, only seldom is the enamel plicated and then in faintly worn specimens. Hypsodont. Pikermian. Spain and North Africa (p. 33).

In her discussion of *H. sitifense*, Forstén concluded that two subspecies, distinguished on the basis of size, were represented by the Spanish material. These she designated *H. sitifense sitifense* n. ssp. and *H. sitifense gromovae* Villalta and Crusafont 1957. Without providing metrical data to demarcate the size of the Spanish *H. sitifense* material, she diagnosed the nominal species (including, of course, the North African specimens) as "middle-sized" and *H. s. gromovae* as "small."

Alberdi (1974) provides yet another revision of the genus *Hipparion* in Spain. Analyzing a complex series of indices (but again providing no metrical data), she concluded that the three taxa referred by Forstén (1968) to *H. sitifense* are in fact valid species, which she returns to their former taxonomic status.

While there are many indications that a land bridge connected the Iberian Peninsula and North Africa during the Pliocene (Forstén 1968), and hence *H. sitifense*, like *H. primigenium*, may easily have been present in both regions, it is by no means clear that the Spanish *"H. sitifense"* sample represents a single species. Hence it is also unclear that diagnostic characters compiled from a variety of specimens within this sample may be applied to Pomel's taxon.

The postcranial skeleton of the East African "dwarf" hipparion is represented by an almost complete manus, including an entire metacarpal III (Hooijer and Maglio 1974). The lateral digits of this specimen are considerably reduced, while the metacarpal III is both longer and narrower than that of *H. primigenium*. This provides a marked contrast with *H. sitifense* of Spain, in which the limbs "have the proportions seen in *H. primigenium* with short and relatively robust segments" (Forstén 1968, p. 35).

A fair number of small hipparion teeth have been found in East Africa in deposits ranging in age from the late Miocene (Chemeron Formation, Baringo Basin) to latest Pliocene or earliest Pleistocene (Shungura Member G, Omo group). Hooijer and Maglio (1974) and Hooijer (1975b) assign these to *H.* cf *sitifense* or *H.?* aff. *sitifense*, but they take some pains to argue that the teeth are neither metrically nor morphologically distinct from the North African *H. sitifense* insofar as this species is determinable. Hooijer (1975b) notes that these teeth show a trend toward increased hypsodonty and ectostylid development through time.

The most "advanced" East African small hipparions are much the same age as *H. sitifense* populations in North Africa and those considered by Forstén to represent the species in Spain. In the North African specimens, as has already been noted, neither the degree of hypsodonty nor the condition of ectostylids can be determined; but the development of these features in the Spanish specimens appears to be equivalent to an earlier stage in the evolutionary history of the East African group (Forstén 1968; Hooijer 1975b).

Thus faced with the common failings of the fossil record, Hooijer (1975b) refrains from fully including the East African "dwarf" *Hipparion* in *H. sitifense* Pomel and suggests the evidence for this species in Spain is at least equally questionable. Obviously more complete "small *Hipparion*" material (including skulls and postcranial elements) is needed from both North and East Africa before the relationship between the "dwarf" forms known from these regions can be adequately assessed or defined vis-á-vis the small Spanish hipparions.

Nonetheless, the marked similarities in both form and size of teeth between *"H.* cf *sitifense"* from the late Miocene and early Pliocene of East Africa, *"H.?* aff. *sitifense"* from the late Pliocene and earliest Pleistocene of East Africa, and the North African *H. sitifense* known also from the late Pliocene and earliest Pleistocene leaves little doubt that these specimens are very closely related and most probably represent a direct lineage. Even if *"H.?* aff. *sitifense"* and *H. sitifense* Pomel are later demonstrated to be distinct species, a better "ancestral stock" than *"H.* cf *sitifense"* is certainly out of the picture. All these African forms are therefore most conveniently referred to simply as a single species, *H. sitifense*.

H. sitifense, as thus understood, first appears in the late Miocene fossil record of Kenya relatively soon after *H. primigenium* and concurrently with *H. turkanense* (Hooijer 1975a). As no similar forms are known outside Africa from this time, it may be inferred that *H. sitifense* is an indigenous species most probably derived from *H. primigenium*. By the terminal Miocene *H. sitifense* showed cursorial specializations in the elongation of its metapodials and reduction of its lateral digits and hence must have been well adapted to life in a savanna or open woodland environment. Further specialization for grazing in such an environment may be deduced from the progressive increase in hypsodonty and strengthening of the lower molars through development of ectostylids seen to have occurred during the Pliocene.

The same (or a very closely related) species of *Hipparion* had appeared along the North African coast by the later Pliocene and from there may have emigrated to the Iberian Peninsula. Other populations of *H. sitifense*, represented by small teeth from the earliest Pleistocene of Bethlehem (Hooijer and Maglio 1974), may have colonized the Near East after leaving Africa via the Isthmus of Suez.

Hipparion turkanense Hooijer and Maglio 1973

AGE. Late Miocene to earlier Pliocene.

LOCALITIES. Ethiopia (96a); Kenya (100, 101, 102, 105b, 105c).

DISCUSSION. *Hipparion turkanense* is known only from East Africa. Hooijer and Maglio (1973) described as the holotype the skull of an older individual from upper Miocene deposits at Lothagam, Kenya, which was sufficiently well preserved to permit confident restoration. Teeth of *H. turkanense* have been found in slightly older deposits in the Baringo Basin and at earlier Pliocene sites west of Lake Turkana (formerly Lake Rudolf) and in the Omo Basin. *H. turkanense* has consistently been found in association with *H. primigenium* and the "dwarf" hipparion, *H. sitifense,* but only at the type site, Lothagam, is it represented as the dominant *Hipparion* species (Hooijer and Maglio 1974).

H. turkanense is a large hipparion of much the same or slightly greater size than *H. primigenium.* The skull of *H. turkanense* is slightly longer than that of *"H. africanum"* (=*H. primigenium*) from Oued el Hammam (Arambourg 1959; Hooijer and Maglio 1974), but there is no indication that the size ranges of the two species did not overlap. There is also no indication that proportions of the skulls in *"H. africanum"* and *H. turkanense* are essentially different, although comparisons are necessarily limited because the *"H. africanum"* skull lacks the posterior portion of the cranium and has been badly compressed and distorted in and below the nasofrontal region. When measurements provided by Arambourg (1959, p. 76) and Hooijer and Maglio (1974, p. 9) for "prosthion-posterior border of M³," "diastema I³–P²," and "length P²–M³" are compared, it is apparent that the muzzle in both forms was similarly elongated, with the region anterior to P² comprising approximately 40% of the prosthion-posterior border of M³ length.

As in *H. primigenium,* the lateral incisors of *H. turkanense* are approximately the same size as the mesial incisors. Unfortunately, most of the incisors in the *H. turkanense* type skull are too damaged to provide useful measurement, but they appear to have been smaller and more rounded in cross-section than those of *H. primigenium* (Hooijer and Maglio 1974; Hooijer 1975b).

Common to the skulls of both *H. turkanense* and *H. primigenium* are a relatively broad external nasal opening (of which the anterior margin is arcuate, rather than ogive, when viewed from above) and the position and development of the nasal notch, which lies above the anterior border of P² and is formed by the premaxillary bone. Arambourg (1959, p. 77) considered these features highly distinctive in the skull of *"H. africanum"* when comparing the Oued el Hammam specimen with Eurasian hipparionid skulls.

In the *H. turkanense* skull, however, the nasal notch lies relatively "high on the side of the snout" (Hooijer and Maglio 1974, p. 8), while in *"H. africanum"* the well-preserved premaxillary region shows the nasal notch to have been relatively low (Arambourg 1959, pl. X). This suggests that the maxillary region was deeper in *H. turkanense* and may reflect a greater development of hypsodonty in this species than in the North African *H. primigenium* specimen.

Hooijer and Maglio (1973, 1974) and Hooijer

(1975b) have defined other morphological distinctions between these two forms. The skull in *H. turkanense* is most noteworthy for its lack of a preorbital fossa, while in *"H. africanum"* from Oued el Hammam and the juvenile *H. primigenium* specimen from Ekora a preorbital fossa is present and distinct. In the upper cheek teeth the protocone in *H. turkanense* is more pronounced than in *H. primigenium* and is D-shaped (with the internal or lingual enamel border flattened) rather than lenticular or oval. Fossette plis are fewer and of lower amplitude in *H. turkanense,* while the anteroposteriorly oriented enamel, especially of the inner and outer fossette borders, is usually wrinkled in *H. turkanense* but smooth or only slightly wavy in *H. primigenium.* Lower cheek teeth assigned to *H. turkanense* also show marked wrinkling of the buccal enamel borders (making the ptychostylids inconspicuous) and heavy wrinkling of the borders of the flexids. No ectostylids have been found on lower cheek teeth clearly belonging to *H. turkanense.*

Certainly some "overlapping" in dental characters between *H. turkanense* and *H. primigenium* occurs, at least during latest Miocene–earliest Pliocene times. Of this, Hooijer and Maglio (1974, p. 18) write,

Not all of the isolated teeth, however, show clear characters one way or the other, and assigning them to a particular species comes close to sophisticated guesswork, which does not so much increase our knowledge of the two large species in question as merely showing that there is individual variation in dental characters in either one of them.

The postcranial skeleton is hardly, or not, known. The distal end of a metatarsal III and an astragalus were found at Lothagam and Kanapoi respectively. These are slightly larger than the range for these elements known for *"H. africanum"* from North Africa and *H. primigenium* from Europe and may have represented *H. turkanense* (Hooijer and Maglio 1974).

Despite the similarities between *H. turkanense* and *H. primigenium* elaborated upon above, it is reasonably clear that the two forms are specifically distinct when *H. turkanense* makes its debut, along with the small form, *H. sitifense,* in the African fossil record. Hooijer and Maglio (1973, 1974), setting considerable store by the presence or absence of a preorbital fossa, believe *H. turkanense* more closely related to *H. hippidiodum* of the "Pontian" of China (which also lacks the fossa) than to *H. primigenium* and hence suggest that *H. turkanense* represents a second colonization by hipparions from Asia. They note, however, that internal flattening of the protocone is more pronounced in *H. turkanense* than in the otherwise similar Chinese species.

In originally describing *H. turkanense*, Hooijer and Maglio (1973, p. 313) also remarked on a resemblance (again notably in the common lack of a preorbital fossa) between their new species and the advanced African *"Stylohipparion"* (=*H. libycum*) and hypothesized an ancestor-descendant relationship between these two forms. As *H. libycum* is characterized by strongly developed ectostylids, while *H. turkanense* apparently lacked ectostylids in even weakly developed form, they deduced a progressive development of these elements through time.

With the discovery of a fair sample of *H. primigenium* teeth from East Africa and closer examination of these teeth and those of *"Stylohipparion"* Hooijer and Maglio (1974) later changed their opinion and suggested that *"Stylohipparion"* more probably evolved from *H. primigenium*, losing the preorbital fossa in the process. Again, Hooijer (1975b, p. 21) leaves both possibilities open but then (e.g., pp. 51–52) weights the odds in favor of *H. primigenium* as the ancestral species.

Recently discovered hipparionid skulls and mandibles from the later Pliocene of Ethiopia described by Eisenmann (1976) provide new evidence that may help to elucidate the fates of these earlier Pliocene species. Eisenmann indicates that two forms are represented in the Ethiopian sample. One, which she names a new species, *H. afarense*, is more clearly and closely ancestral to the advanced *"Stylohipparion"* than is either *H. turkanense* or *H. primigenium* but seems to bear closer affinities with the latter species as it is known from East Africa (see below); the other, for which several cranial and mandibular specimens are available, she has referred to *"Hipparion* spec." pending discovery of better preserved material as she feels "it is not yet possible to decide whether these (specimens) represent individual variation of the species *H. afarense*, an evolutionary stage, or an independent species" (translated from Eisenmann 1976, p. 590).

The most complete *"Hipparion* spec." specimen (AL 340-8) is the skull and mandible of a senile individual, probably female. Unfortunately, lateral compression of the skull makes it impossible to tell if preorbital fossae existed. However, several features of the dentition in this specimen, as described and illustrated by Eisenmann (1976, pp. 580–582, pls. 1 and 4), suggest an affinity with *H. turkanense*. These include the smaller number of low-amplitude fossette plis and the presence of very prominent, markedly D-shaped protocones in the upper cheek teeth

and the presence of relatively small, ungrooved incisors of subequal size in both jaws. Ectostylids are present on all lower cheek teeth in the partial mandible, but because these teeth are in an advanced state of wear, it is impossible to determine the degree to which the ectostylids had been developed.

Closer examination of this *"Hipparion* spec." material, relative to the possibility that at least some of the hipparionid specimens from the later Pliocene of Ethiopia represent an evolved form of *H. turkanense*, is warranted.

Hipparion afarense Eisenmann 1976

AGE. Late Pliocene.

LOCALITY. Ethiopia (95a).

DISCUSSION. During the past few years international expeditions to the Afar region of Ethiopia have provided a sizable collection of new hipparionid material from later Pliocene deposits of the Hadar Formation. *Hipparion* fossils were collected from upper members of the formation that lie above a tuff dated at ca 3 m.y. but are probably no younger than 2.5 m.y. (Eisenmann 1976). Hence the Hadar hipparions are the only known representatives of their genus in East Africa between *H. primigenium*, *H. turkanense*, and *H. sitifense* of the earlier Pliocene and the advanced form, *H. libycum*, of the Pleistocene.

Eisenmann (1976) has recently described from this collection five skulls and four mandibles in various states and degrees of preservation. Considering primarily cranial—and secondarily dental—characteristics, she has concluded that at least one and potentially two new species are represented in the Hadar Formation. One of these she has named *Hipparion afarense*, designating as the holotype a partially preserved skull. One virtually complete mandible is referred to this species; the other cranial and mandibular specimens are temporarily assigned to *"Hipparion* spec." Numerous isolated teeth and postcranial elements await description and analysis.

The *H. afarense* type skull lacks much of the right side and is broken in front of the P^2s, but the anterior portion of the rostrum containing the three left incisors is preserved. The left side is cracked but otherwise undeformed. Wear of the teeth indicates an adult of middle age. The complete mandible assigned to this species is from a nearly adult individual with M_3s erupted but unworn and I_3s erupting. Small canines are present.

Eisenmann (1976:590) diagnoses *H. afarense* thus:

Large size: the length basion-P^2 is 377 mm. Occipital fossa large. Glenoid processes (=postglenoid processes) large and flat. Vomerine notch acute, not arcuate as in

Equus, vomerine index high (140). No preorbital fossa. Incisors of large size, upper cheek teeth relatively small: the ratio I¹/M¹ is approximately 90. No reduction of I³. Lower teeth caballine with ectostylids present but little developed. Lower incisors grooved, little incurved, of large size: the ratio I_2/M_2 is approximately 85. I_3 not atrophied. Symphysis elongated and flattened.

The teeth of *H. afarense* are in most morphological features close to those of *H. primigenium* on the one hand and to the advanced Pleistocene hipparion, *H. libycum,* on the other. The incisors are grooved and relatively large; the upper cheek teeth show an elaboration of fossette plication and possess relatively unpronounced protocones of more or less elliptical section (Eisenmann 1976, pp. 583–584, pl. 4A). In the mandible, ectostylids are visible on all cheek teeth except P_2 and P_4. Hypsodonty (gauged primarily from the depth of the mandibular ramus) appears well advanced but less extreme than in the middle Pleistocene *Hipparion* specimens (Eisenmann 1976, p. 590). The most primitive character of the *H. afarense* dentition is the lack of reduction of the third incisors.

The size of the *H. afarense* type skull is somewhat larger than that of *H. turkanense* from Lothagam, in which basion-P² length is 348 mm (calculated from Hooijer and Maglio 1974, table I). As the rostral region of the Lothagam skull is slightly longer than that of the *"H. africanum"* skull from Oued el Hammam, *H. afarense* may have been significantly larger that the North African *H. primigenium.*

In the size of the occipital fossa and the flatness (fore and aft) and large size of the postglenoid processes, the *H. afarense* skull contrasts with the most complete of the *"Hipparion* spec." specimens from Afar but agrees with the advanced Pleistocene *Hipparion* discussed below. Unfortunately, these regions are lacking in the *"H. africanum"* type skull (Arambourg 1959) and are neither described nor clearly illustrated for either *H. turkanense* or the *H. primigenium* juvenile skull from Ekora (Hooijer and Maglio 1974). The postglenoid processes in a cast of the Lothagam skull we have examined are not greatly enlarged or flattened posteriorly, but unfortunately, the occipital region of this cast was damaged in shipment and provides no useful information.

Eisenmann (1976, pp. 583, 587) considers the morphology of the vomerine region of considerable systematic interest and notes that in both *H. afarense* and the advanced *Hipparion* the vomerine notch is acute rather than arcuate, as in horses, and the vomer is short, giving a high vomerine index (notch to basion over notch to palatum durum). In horses,

which also have a relatively short vomer, the index is over 100, while in asses the vomer is relatively long and the index is below 100. In the *H. turkanense* skull from Lothagam only the right portion of the vomerine "notch" is preserved (Hooijer and Maglio 1974, pl. 1), but it shows the notch to have been almost as acute as in *H. afarense* and the advanced *Hipparion* (Eisenmann 1976, pls. 2 and 3). The vomerine index in *H. turkanense* is, however, notably lower, approximately 105 compared with 140 in *H. afarense* and 150 in the advanced *Hipparion.* Neither the shape of the vomer nor the vomerine index is known for African specimens of *H. primigenium* (Arambourg 1959, pl. XI-4; Hooijer and Maglio 1974, pl. 4).

It is reasonably clear, as Eisenmann (1976, p. 590) suggests, that *H. afarense* of the later Pliocene represents a stage *préethiopicum* and is, on present evidence, the most likely immediate ancestor of the Pleistocene *Hipparion* of Africa. *H. afarense* differs from the advanced form only in the "nonreduction of the third incisors, the slighter development of ectostylids, and the inferior height of the mandibular ramus, which denotes a lesser degree of hypsodonty" (translated, p. 590).

The origin of *H. afarense* is, however, less well defined. Both *H. turkanense* and *H. primigenium* are present in earlier Pliocene deposits of East Africa and both have been considered possible ancestral forms to the advanced Pleistocene hipparions of Africa (Hooijer and Maglio 1973, 1974; Hooijer 1975b). With *H. turkanense* the major line of evidence lies in the shared absence of preorbital fossae; in *H. primigenium* the finer features of the dentition imply a closer phylogenetic relationship to the advanced species. The condition of the preorbital fossa in *H. afarense* is diagnosed as absent above, but Eisenmann (1976, p. 583) writes in her description of the type skull that "la fosse pré-orbitaire est faible." More convincing in support of a closer relationship with *H. primigenium,* the dentition of *H. afarense* possesses the finer morphological features of both *H. primigenium* and the advanced Pleistocene *Hipparion* while in development and proportions it provides an almost ideal transitional form. Unfortunately, few of the cranial features considered distinctive of the later hipparions are known in African specimens of *H. primigenium.*

We have already considered the possibility that at least some of the *"Hipparion* spec." specimens that appear together with *H. afarense* in the Hadar Formation represent an advanced form of *H. turkanense* (see above). It seems to us possible that the hipparion fauna of the later Pliocene in Ethiopia represents the products of parallel evolutionary advance by both *H. turkanense* and *H. primigenium* populations, with the latter lineage culminating in the advanced Pleistocene *Hipparion*.

Hipparion libycum Pomel 1897

SYNONYMY. *Hipparion massoesylium* Pomel 1897, *Hipparion ambiguum* Pomel 1897, *Hipparion albertensis* Hopwood 1926, *Hipparion crassum* Gervais (partim), Solignac 1927, *Hipparion steytleri* van Hoepen 1930, *Eurygnathohippus cornelianus* van Hoepen 1930, *Stylohipparion hipkini* van Hoepen 1932, *Stylohipparion steytleri* (van Hoepen), van Hoepen 1932, *Libyhipparion ethiopicum* Joleaud 1933, Non *Equus (Hippotigris)* sp. Joleaud 1933, *Libyhipparion steytleri* (van Hoepen), Joleaud 1933, *Stylohipparion* cf *albertense* (Hopwood), Hopwood 1937, *Stylohipparion albertense* (Hopwood), Arambourg 1947, *Hipparion (Stylohipparion) libycum* Pomel (partim), Boné and Singer 1965, *Hipparion (Hipparion) albertense serengetense* (partim) Boné and Singer 1965, *Hipparion (Hipparion) albertense* Hopwood (partim), Cooke and Coryndon 1970, *Hipparion* cf *ethiopicum* Hooijer 1975, *Hipparion ethiopicum* Hooijer 1975.

AGE. ?Latest Pliocene to late Pleistocene.

LOCALITIES. Morocco (3, 4, 5, 6); Algeria (12 Zone 1, 17, 21, 22, 23, 24, 25, 26 Zone 1, 33, 41, 46, 47, 48a, 48c, 48d, 48e, 61); Tunisia (72, 73, 74, 79); Chad (86); Egypt (87); Ethiopia (96b Shungura Members F–J); Uganda (97b); Kenya (98 Koobi Fora Formation, ?106, ?107, 110); Tanzania (111, 112 Beds I–IV, ?113, 114); Malawi (115); Republic of South Africa (123a, 128, 132, 133, 134, 141, 156, 164, 167).

DISCUSSION. A highly advanced hipparion of a form probably restricted to Africa has long been recognized on teeth from Pleistocene deposits in northern, eastern, and southern regions of the continent. These teeth are uniformly characterized by extreme hypsodonty (over 70 mm unworn height), the strong development of broad, spindle-shaped ectostylids, and, where known, the marked reduction and medial migration of the lateral incisors in both jaws. Skulls of this advanced *Hipparion* are known only from Olduvai Gorge, Tanzania (L. S. B. Leakey 1965; M. D. Leakey 1971; Hooijer 1975b), and the Koobi Fora Formation of East Lake Turkana, Kenya (Maglio 1971, 1972; Eisenmann 1976). The lower incisive region of the Pleistocene *Hipparion* is known from a specimen originally described as *Eurygnathohippus cornelianus* from Cornelia (= Uitzoek), South Africa (van Hoepen 1930b; Cooke 1950a) and from a hemimandible discovered at East

Turkana (Eisenmann 1976). Various postcranial elements of this species are recorded from sites throughout its known African range and include two lateral metapodial fragments, formerly described as the acromial extremities of hominid clavicles, from Makapansgat. Save for the metapodials, the postcranial skeleton of the Pleistocene *Hipparion* has not been described.

The rather complex synonymy of the advanced hipparions of Africa, which we here refer to the single species *Hipparion libycum* Pomel, interestingly reflects more fundamental agreement than disagreement among various authors. Virtually every author who has carefully considered Pleistocene hipparionid specimens in Africa has agreed that the teeth of this advanced form are indistinguishable throughout the continent. The only significant divergence of opinion has stemmed from the recognition of this form as a genus or subgenus distinct from *Hipparion* and the recognition of regional species of advanced hipparions from North, East, and South Africa.

We follow Hooijer (1975b) in referring all the Pleistocene hipparionid specimens to the genus *Hipparion* without subgeneric distinction. In describing the gradual development of advanced characters in hipparion teeth from sequential members of the Shungura Formation, Ethiopia, Hooijer has demonstrated that generic distinction of the advanced form is unwarranted. We also follow Hooijer (1975b) in regarding the species *Hipparion albertense* Hopwood as a *nomen vanum* and in recognizing some specimens formerly attributed to this species as advanced hipparions. However, we do not agree with Hooijer (1975b, pp. 10 and 66–67) that the lack of cranial material from North and South Africa that, when discovered, may permit geographic distinction of forms now indistinguishable on dental evidence from those of East Africa necessitates the inference of three distinct species at this time. We prefer, therefore, to use the senior synonym *Hipparion libycum* Pomel for the advanced Pleistocene hipparion and allow for potential geographic variations under subspecific distinction. Thus we suggest the names *Hipparion libycum libycum, Hipparion libycum ethiopicum,* and *Hipparion libycum steytleri* be used to designate regional populations among advanced Pleistocene hipparions in North, East, and South Africa respectively.

As noted above, cranial evidence for *H. libycum* is limited to specimens from East Africa. These have been referred to "*H.* cf *ethiopicum*" by Hooijer (1975b) and Eisenmann (1976) but we here recognize them as belonging to *H. l. ethiopicum*. The available cranial specimens comprise a broken and much distorted adult skull and a partial subadult skull from Olduvai described by Hooijer (1975b) and a nearly complete juvenile skull from East Turkana described by Eisenmann (1976). The Olduvai specimens both lack canines and probably represent females (Petrocchi 1952), while the East Turkana skull appears to possess erupting canines (Eisenmann 1976, pl. 3B) and may be considered male. The latter specimen was found just below the KBS Tuff of the Koobi Fora Formation at East Turkana and hence is of earliest Pleistocene age according to the latest determinations (Curtis et al. 1975). The Olduvai specimens were recovered from upper Bed II and are of significantly younger (late middle Pleistocene) age (Hay in Leakey 1971).

In describing the two Olduvai skulls Hooijer (1975b) emphasizes that they both lack a preorbital fossa, although he observes that in the subadult skull there is a slight depression along the nasomaxillary suture in front of the lacrimal. On the East Turkana juvenile skull, however, Eisenmann (1976, p. 586) describes a very feebly marked preorbital fossa. It is not clear from the photographs of this skull (Eisenmann 1976, pl. 3) whether the fossa in question represents a depression similar to that in the Olduvai subadult skull or a true preorbital fossa.

The poor preservation of the Olduvai specimens permits few useful comparisons. One exception is found in the morphology of the external nasal opening, which seems to have agreed very closely with that of both "*H. africanum*" and *H. turkanense* and may be distinctive of African hipparions (see Arambourg 1959, pp. 77–78 and 95). Hooijer (1975b, pp. 29 and 33) notes that such measurements as can be obtained from the adult skull are very close to those of *H. turkanense* in both size and proportions and that "the dentition . . . provides the means of distinguishing between the two."

A more complete description is available for the East Turkana juvenile skull, of which Eisenmann (1976) writes "the most striking characters are the shortness of the face, the length of I¹ and I² and the vomerine region. Various characters recall the skull (of *H. afarense*) described before" (translated, p. 585).

The first of these characters is judged on the index of length prosthion to posterior orbital margin over length posterior orbital margin to external occipital protuberance (Eisenmann and de Giuli 1974; Eisenmann 1976). In describing this character, Eisenmann notes that in *Equus* the index obtained is not influenced by age and hence "one may suppose a face equally short in an adult of the same species" (trans-

lated, p. 586). Even if this analogy is accepted, it is not clear by which standard Eisenmann makes her comparison. The only other (African) *Hipparion* skull complete enough to potentially provide the requisite index is that of *H. turkanense* from Lothagam, and in this skull the preserved (left) orbit is distorted with the lower margin pushed in and the posterior margin damaged (Hooijer and Maglio 1974, pp. 8–9). Indices obtained from photographs of this skull (Hooijer and Maglio 1974, pl. 1) are 160 for the reconstructed side and 180 for the distorted side, while it seems the "true" index must be approximately 170. In the East Turkana juvenile skull the index obtained from measurements provided by Eisenmann (1976, table I) is 162. While no direct index can be obtained for *H. primigenium*, it may be remembered that proportions of the muzzle in this form are virtually identical to those of *H. turkanense* and hence it is probable that the face in these two forms was similarly elongated or "short" and not significantly different in proportion than that of *H. l. ethiopicum*. We therefore find it difficult to accept this character as distinctive of the advanced *Hipparion*.

The mesial incisors are indeed large, with the ratio of occlusal lengths of I¹ and M¹ near 100 (Eisenmann 1976, p. 586). However, it may be noted that these incisors are not entirely out of the alveolus and hence are near their maximum occlusal dimensions (Hooijer 1975b); in the more worn teeth of older individuals the mesial incisors would appear relatively smaller.

The vomer in the East Turkana juvenile skull has an acute rather than arcuate notch, as in *H. afarense* and *H. turkanense*, but is less elongate posteriorly, providing a slightly higher vomerine index than in *H. afarense* (150 versus 140) and a considerably higher index than in *H. turkanense* (ca 105). The condition of the vomer in *H. primigenium* from either North or East Africa cannot be determined (see above).

The other cranial features shared with *H. afarense* include a large occipital fossa and a large, flat postglenoid process. Eisenmann (1976, p. 590) notes that these features distinguish *H. afarense* from "*Hipparion* spec." of the Hadar Formation and our examination of a cast of the *H. turkanense* skull from Lothagam (see above) leads us to believe that at least the condition of the postglenoid processes may also have differed in *H. turkanense*. In *H. primigenium* the relevant portions of North and East African skulls are again missing.

Of the various cranial characteristics attributed to the advanced African *Hipparion* by Eisenmann

(1976), only the relative shortness of the vomer and possibly the size and shape of the occipital fossa and postglenoid processes appear to be clearly distinctive, and these features are shared with *H. afarense*. *H. turkanense* is apparently much the same size as at least females of the advanced form and shares with the Pleistocene *Hipparion* a "shortness" of the face, the absence of a pronounced preorbital fossa, and an acute vomerine notch. It differs from *H. libycum* (*ethiopicum*) in possessing relatively small incisors (see above) and in a less marked posterior elongation of the vomer. In *H. primigenium,* for which only the partially preserved skulls from Oued el Hammam and Ekora are available for comparison, only the condition of the preorbital fossa and the size of the incisors are actually known, although a "shortness" of the face similar to that of other African hipparions may be inferred. The *H. primigenium* skulls differ from those of *H. l. ethiopicum, H. afarense,* and *H. turkanense* in possessing pronounced preorbital fossae; but the "*H. africanum*" skull shares with the two former species the relatively larger, if less pronounced, size of the mesial incisors.

In contrast to the cranial features of *H. libycum* (*ethiopicum*), which for the most part seem essentially primitive and possibly distinctly "African" in nature, the teeth of this species are evolutionarily advanced and provide to date the best available means of discerning the phylogenetic status of this group.

Aside from the relative characteristics of extreme hypsodonty and stronger ectostylid development, which distinguish the cheek teeth of the advanced *Hipparion,* a considerable array of other features in these teeth appear to indicate a closer affinity with *H. afarense* and *H. primigenium* than with *H. turkanense*. In the upper series the protocone is a less conspicuous element than in *H. turkanense* and is more elongate and elliptical than D-shaped. There is an elaboration of high amplitude fossette plication while the enamel borders are not, or are scarcely, wrinkled. In the upper premolars the anterior horn of the postfossette usually extends outward beyond the posterior horn of the prefossette, a condition scarcely observed in *H. turkanense* but readily seen in *H. primigenium* (Arambourg 1959; Hooijer and Maglio 1974; Hooijer 1975b). In the lower cheek teeth the presence of an ectostylid, even if only weakly developed, appears to distinguish these forms from *H. turkanense,* for which no ectostylids are known.

The incisor regions of *H. libycum* as known from both East and South Africa are most distinctive in the reduction and medial migration of the lateral incisors, which come to lie behind the I2s in both jaws. The mesial incisors (I1 and I2) in the advanced form are, however, morphologically very similar to those of *H. afarense* and *H. primigenium* in showing anteroposterior compression and grooving on both labial and lingual surfaces. The mesial incisors of all three forms are large relative to the cheek teeth and become progressively larger through time. In contrast, the incisors of *H. turkanense,* though severely damaged in the type skull, appear to have been rounded in occlusal view and were almost certainly smaller relative to the cheek teeth than in the primitive *H. primigenium*.

The postcranial skeleton of *H. libycum* has received relatively little attention in the literature, although both Arambourg (1970) and Hooijer (1975b) provide measurements for third metapodials and Hooijer discusses other distal limb elements. Hooijer (1975b, pp. 49–52) considers the limbs of *H. libycum* to have remained essentially primitive, with no apparent reduction of the lateral metapodials through time, but he notes (p. 52) that the proximal phalanx of the median digit is more elongated in the Pleistocene species than in *H. primigenium*. Elongation of only this phalanx would be indicative of a more cursorial habit with minimal interference of the lateral digits during running.

In recognizing but a single species of advanced *Hipparion* ranging from northern to southern Africa, we discount the possibility that hipparion populations in North, East, and South Africa independently acquired the distinguishing features of the Pleistocene form. While the relative characters of extreme hypsodonty and stronger ectostylid development could well have developed in separate populations responding to widespread ecological conditions, it seems much less likely that general selective pressure would have uniformly favored the reduction and medial migration of the lateral incisors, which is seen in advanced hipparionid specimens from both East and South Africa. In North Africa the condition of the incisors in advanced hipparions is not known, but here *H. primigenium* is restricted to the Miocene and the only hipparion found in Pliocene deposits is the "dwarf" form, *H. sitifense*. Isolated teeth of *H. libycum* first appear in the North African fossil record during the Pleistocene and seem to represent a new invasion of this region rather than an indigenously evolved form.

Pliocene hipparions are known from southern Africa and have been considered by various authors to be ancestral to or early representatives of the

Pleistocene *"Stylohipparion steytleri"* (Cooke 1950a; Boné and Singer 1965; Hooijer 1975b). However, the Pliocene specimens so far described from southern Africa as either *H. baardi* or *H. namaquense* are morphologically closer to *H. primigenium* from East Africa than to the advanced Pleistocene *Hipparion* (see above) and on present evidence seem rather to represent a separate line or vestigial populations of the earlier form.

In contrast, the East African fossil record provides not only the earliest known appearance of the advanced *Hipparion* (Omo Basin, Shungura Member F and East Turkana, Koobi Fora Formation), but also the clearly transitional species, *H. afarense,* and, in sequences where only isolated teeth are known, evidence for the gradual and progressive development of the special dental features.

The advanced characteristics of the *H. libycum* dentition suggest an increasingly strong reliance on savanna environments and a grazing habit. Ecological conditions in the nonforested regions of Africa near the Plio-Pleistocene boundary may have been particularly favorable for the dissemination of savanna animals and a number of grazing ungulate species rapidly spread throughout the continent. In addition to the advanced *Hipparion,* these included *Equus* and members of the Rhinocerotidae and Bovidae. Explosive radiation of the latter family in Africa during the Pleistocene, which led to a proliferation of antelope species, probably provoked the final extinction of the three-toed equids toward the end of this epoch.

Genus *Equus* Linnaeus 1758

DIAGNOSIS. Facial region usually shallower than in *Hipparion,* with lachrymal fossae absent or rudimentary. Sagittal crest absent. Fossae on occiput for nuchal attachment small, and those directly above occipital condyles poorly developed. Bony auditory meatus variable in length and orientation. Basicranial region with or without low longitudinal crest. Coronoid process of dentary lower than in *Hipparion* and ascending ramus with obliquely posterior orientation. Marks on lower incisors variably developed. Canines usually absent in females. Cheek teeth very hypsodont. Upper cheek teeth with protocone connected to protolophs by a narrow isthmus and few to no enamel plis in fossettes. Lower cheek teeth with well-developed double loops, almost always lacking parastylids; mesial valley (metaflexid) with unequal arms, distal long and mesial short and transverse; buccal valley variably developed; M_3 with bipartite talonid; ptychostylid lacking on lower milk molars. Limbs straighter than in *Hipparion,* ulna often discontinuous in shaft; extremities always monodactyl, lateral digits lacking, and lateral metapodials reduced to shortened splints that show contact for no more than two-thirds of the length of metapodial III. No rudiment of metacarpal V. Hoofed phalanges lack developed median slits on anterior margin.

AGE. Late Pliocene to Recent.

DISCUSSION. Identification of the various species of horses, asses, onagers, zebras, etc., is usually founded on their coat colors and general conformation. Identification of these animals from fossilized elements is less certain. The proportions of the long bones will often indicate whether the animal was slim as in onagers and half-asses, small as in donkeys and asses, large as in horses, or heavily built or stocky as in zebras. The dental elements all conform to a uniform pattern in which variation is found in the development of folds in the enamel. Such variation in the characteristic pattern of the teeth is as great between teeth in the same individual, between individuals, and dependent on the state of wear of the teeth, as between species (Petit 1939). Because of this variation, the identification of a fossil equine to species is best founded upon a series of teeth, some of which are known to derive from a single individual and some of which come from other individuals, so that variation between and within individuals and variation due to wear can be allowed for. Wells (1959) considered all equine species founded on single teeth or on small series of teeth to be indeterminable *nomina vana* (*sensu* Simpson 1945) and preferred either dental series within a jaw or assemblages of teeth as determinable types.

Horses (*Equus* [*Equus*] spp.) typically have the pli caballine evident at the base of the protocone at the head of the postprotoconal valley (Hopwood 1936, p. 898) but this pli may be lost in well-worn teeth and may occasionally be absent in some individuals (Petit 1939). The lingual sinus between the metastylid and metaconid in the lower cheek teeth of horses is usually wide open and may be U-shaped. A small pli on the buccal base of the hypoconid, the ptychostylid (or pli caballinid), is present.

Zebras and asses (*Equus* [*Dolichohippus*] spp., *E.* [*Hippotigris*] spp., and *E.* [*Asinus*] spp.) typically lack the pli caballine, although it may be present in reduced form in some dentitions of zebras (Cooke 1950a). The lingual sinus between the metastylid and metaconid is more or less a pointed V and usually lacks any small plis. The conformation of the buccal enamel walls of the upper cheek teeth can be used to separate zebras and asses (Arambourg 1947, pp. 310–311). The valleys between the protostyle, mesostyle, and metastyle lie at the same level but with the distal valley tilted posteriorly in zebras and are parallel but with the distal valley situated more buccally in asses. As with the foldings in the enamel, the conformation of these valleys can become distorted or obscured in well-worn specimens and identification may then only be to "*Equus* sp." or possibly "*Equus* sp., not a horse."

Remains identified as deriving from true horses (*Equus* [*Equus*] spp.) in Africa are recorded from deposits of late Pliocene age in North Africa and from deposits of early to late Pleistocene ages in East and South Africa. Such horses probably became extinct in North Africa at the latest by the end of the early Pleistocene and in sub-Saharan Africa by the end of the late Pleistocene or early Holocene. Introduction of true horses is known to have taken place in North Africa following the invasion of Egypt by the Hyksos kings ca 1580 B.C. (Zeuner 1963). This horse, *E. caballus,* was spread along the north shore of Africa and across the Sahara to the soudan and sahel regions by man. A second introduction of *E. caballus* took place at the Cape of Good Hope by Dutch settlers in the late seventeenth century, whence they were spread by European explorers and settlers throughout the savannas of southern, central, and eastern Africa.

Many fossils are assigned to one Pliocene species (*E. numidicus*) and to at least two Pleistocene species (*E. oldowayensis* in East Africa and *E. capensis* in South Africa; see below for discussion of other specific names available). It appears likely that these equids were the antecedents of the extant Grevy's zebra, *Equus* (*Dolichohippus*) *grevyi,* which is now restricted to the arid and semiarid regions of the Horn of Africa.

E. grevyi Oustalet 1882 was placed in the subgenus *Dolichohippus* by Heller (1912). Skinner (1972) regards the similarities between skulls and dentitions of the extinct North American *Equus* (*Dolichohippus*) *simplicidens* (= *Plesippus simplicidens*) and the living *E.* (*D.*) *grevyi* so marked that he:

consider(s) *Plesippus* Matthew 1924 a junior synonym of *E.* (*Dolichohippus*). Such differences as exist are mainly those of the skull, but when temporal and geographic separation are considered, these differences seem slight, barely of specific value. In superficial characteristics *E.* (*D.*) *grevyi* resembles a mule with a very long head and ears, but divested of the soft parts and hide, the skull, teeth, and postcranial elements indicate that it was a direct descendant of the extinct equids found in the North American Pleistocene (p. 118).

However, the occurrence of *Dolichohippus* equids in Africa by the middle to late Pliocene would appear to preclude the latter assertion.

True zebras (including quaggas and bontequaggas) comprise three living or historically known species grouped by modern taxonomists within the subgenus *Hippotigris*. These are Burchell's zebra or bontequagga, *Equus* (*Hippotigris*) *burchelli;* the mountain zebra, *E.* (*H.*) *zebra;* and the quagga, *E.* (*H.*) *quagga.* The several fossil taxa erected for hippotigrine zebras and some asses have now been synonymized with the extant species *E. burchelli.* All true zebras are now restricted to sub-Saharan Africa.

E. burchelli, the most common zebra, occupies the savannas north of the Orange River to the Congo rain forest and the Somali Arid Zone. The northeastern portion of its range overlaps parts of the *E. grevyi* range. The fossil record for *E. burchelli* is continuous throughout the Quaternary in Africa, with known records for the early and middle Pleistocene from northern, eastern, and southern Africa and for the late Pleistocene for eastern, central, and southern Africa, where its range appears to coincide generally with the present range discounting human interference.

E. zebra was found in the mountainous country from the Serra da Neve in Angola through coastal South West Africa, to the southern Cape Province as far as Cathcart District. It is still numerous in Angola and South West Africa, but only survives in Cape Province in and near the Cradock Mountain National Park, where a protected herd of 98 was present in 1970 (Millar 1970).

E. quagga became extinct around 1880 when the last surviving individuals died in the Orange Free State. Formerly the species inhabited the southern and eastern parts of the Great Karroo and possibly Great Namaqualand.

The fossil records of *E. zebra* and *E. quagga* are all from southern Africa and fall for the most part within the known ranges for these species. The interrelationships of the hippotigrine zebras are not clear. One is tempted to hypothesize that *E. zebra* and *E. quagga* result from an early speciation of southern populations of *E. burchelli,* with *E. zebra* representing an adaptation to conditions of broken montane country while *E. quagga* presumably arose in response to life in the semiarid Karroo.

The historic and present distribution of the African asses, *Equus* (*Asinus*) *asinus* (= *E. africanus*) included the North African coastal areas and mountains, the Nile Valley, and the Red Sea Mountains south to the Somali Arid Zone. Unconfirmed reports of asses from northern Chad may represent feral donkeys (Ansell 1967) and the asses of Socotra may be descended from introduced Nubian stock (de Winton, Forbes, and Ogilvie-Grant 1903). Several North African fossil species originally described as asses are now thought to represent *E. burchelli.* It seems likely that the zebra was replaced in North Africa by the ass during the late Palaeolithic, with the ass owing much of its success here to human activity.

Subgenus *Dolichohippus* Heller 1912

SYNONYMY. *Plesippus* Matthew 1924.

DIAGNOSIS. The original diagnosis is based on the distinctive pelage of *E. grevyi* and is inapplicable for a fossil species.

Skinner (1972) considered the subgenus to be characterized by a larger head; a wider, heavier, and more posteriorly extended occipital region; a longer and more slender rostral region, with a nasal notch formed more completely within the nasal bones; and more strongly sexually dimorphic canines than in equids of the subgenera *Equus, Hippotigris, Pseudoquagga, Hemionus, Asinus, Quagga, Amerhippus, Hesperohippus,* and *Tomolabis.* However, he noted that the presence or absence of marks (cups or infundibula) on the incisors is not diagnostic at the subgeneric level in equids.

In addition, the postcranial skeleton is distinctive in that the ulna and radius in the living *Dolichohippus* are entirely free, while an entire fibula of *E. (D.) simplicidens* (AMNH-82037) recorded by Skinner (1972) suggests that other postcranial features may be as characteristic and distinctive for the subgenus. Metapodials III of *Dolichohippus* are heavier and have longer and stronger splint metapodials II and IV than in the other subgenera of *Equus.*

Equus (Dolichohippus) numidicus Pomel 1897

AGE. Middle/late Pliocene to early Pleistocene.

LOCALITIES. Algeria (25, 41, 46, 47, 48a, 48b, 48c, 48d, 55, ?56, 60); Tunisia (70); Sudan (93a); Ethiopia (?96b Shungura Member G); Uganda (97a).

DISCUSSION. This is a medium-sized to small horse, about the size of a large zebra. In the dentition the protocone is mediodistally short, especially distal to its junction with the protoloph; the plis are small and simple and the pli caballine is absent; the hypoglyph is shallow, simple, and V-shaped and the styles are rounded, slightly set off at angles from ectolophs, which are shallowly concave buccally. The cementum is not thick. No postcranial materials are reported or described, nor complete skulls, so comparisons with either *E. (D.) simplicidens* of the Pleistocene of North America or *E. (D.) grevyi* from the Horn of Africa are impossible or inadequate.

Romer (1929, p. 120) refers all the North African Plio-Pleistocene horse remains to "*Equus stenonis*" saying "I have grouped here citations to this or related species of primitive members of the genus *Equus.* They appear to have been contemporaneous with *Hipparion,* the latter genus in fact, apparently persisting longer."

Cooke (1963, p. 72) refers to the North African Villafranchian horse as *Equus numidicus* and indicates that it was extinct by the middle of the early Pleistocene in North Africa. Boné and Singer (1965) refer to *Equus numidicus* from Aïn Boucherit and Aïn Jourdel and date these deposits as middle Pliocene (p. 286) or early Pleistocene (p. 322), while Cooke (Chapter 2) dates these sites as late Pliocene. Boné and Singer (1965, p. 293), citing Arambourg (1947, 1949), list only "*Equus* sp." and "*Equus*" from Aïn Hanech and Garet Ichkeul respectively. These occurrences may also represent *E. numidicus,* although those from Aïn Hanech are here assigned to *E. burchelli.* Arambourg (1970) records *Equus numidicus* from Aïn Boucherit (Beni Foudda), Aïn Jourdel, and possibly Aïn el Bey and places it in the subgenus *Dolichohippus,* which accords with the views of more recent works.

Lydekker (1887) describes a newly worn upper molar of a generalized early species of *Equus* from the late Pliocene or lowest Villafranchian deposits at Wadi Halfa in Sudanese Nubia, which he compared to "*E. stenonis*" from Algeria, Val d'Arno, and the Norfolk Forest Bed, and to *E. sivalensis* from India, with the latter of which he considered it "absolutely indistinguishable from the first true molar of the maxilla" (p. 162).

Hopwood (1926) identified *Equus zebra* on a large, well-worn right M^2 from Kaiso, Uganda. He remarked that although "the grinding surface is obscured in many of the finer points by deposits of iron ore, enough is visible to make identification certain" (p. 18). Cooke and Coryndon (1970) consider this tooth to match closely some fossil teeth from South Africa tentatively referred to *E. burchelli* and some of comparable size from Olduvai referred to *E. oldowayensis* by Hopwood (1937). Since the Kaiso deposits are late Pliocene in age, it is unlikely this specimen represents *E. zebra, E. grevyi,* or *E. burchelli;* and it is here assigned to *Equus* sp., as is Lydekker's Wadi Halfa specimen, although both teeth may well represent *E. numidicus.*

Equus (Dolichohippus) oldowayensis Hopwood 1937

AGE. Early/middle Pliocene to late middle Pleistocene.

LOCALITIES. Ethiopia (96b Shungura members G–J); Kenya (103, 106, 107, 110); Tanzania (112 Beds I–IV); Malawi (?115 Mwanganda).

DISCUSSION. This form was originally described

by Hopwood (1937, p. 117, translated) as "a zebra, whose molar teeth correspond well with those of *Equus grevyi,* in which the muzzle is also very broad and is comparable to that of draught horses rather than to that of *Equus grevyi.*" Little has been reported on *E. oldowayensis* since Hopwood's (1937) original description, although many specimens were sent to the late R. A. Stirton at Berkeley, California, U.S.A., prior to his death for description and identification (Leakey 1958).

Hopwood's original paper illustrates side and occlusal aspects of the type specimen, a complete mandible, and provides measurements of teeth, mandible, and a metatarsus compared with those of other equids, along with descriptions of specimens. Unfortunately, these data serve only to emphasize the massiveness of the teeth and the size of the animal and do not distinguish it specifically. However, Hopwood considered one metatarsal to be closely similar to that of *E. stenonis* and suggested that *E. oldowayensis* may be descended from that species. Measurements of two metatarsals show that *E. oldowayensis* was as large, or nearly as large, as *E. numidicus* and longer-legged than *E. stenonis;* but all three species seem to have an equivalent bulk as the metatarsals are similar in breadth.

Hooijer (1973, p. 2) considered the larger teeth found in Shungura Members G through J of the Omo group deposits and states that "no evolution is observable or may be expected in such a short time span" (about 0.5 to 0.7 m.y.) and "the large teeth are not quite the same as those in either *Equus zebra* L. or *Equus grevyi* Oustalet of the modern African Fauna." Later, Hooijer (1974, pers. comm.), commenting on the evolution of the Perissodactyla of the Omo group deposits, argues that:

In the absence of skulls from the Omo group deposits the species in question cannot be determined: even a cf. determination may impart a false sense of precision and has been omitted. The protocones are short, the metaconid-metastylid valley V-shaped, but occasionally the protocone is long (14 mm in an upper third or fourth premolar, half the median length of the crown), and the lower internal valley U-shaped. It is not even certain that there is only one species of large *Equus* in the Omo Beds, but characters may vary individually within the same species. *Equus stenonis* Cocchi, one of the two Villafranchian horses of the Upper Valdarno (Azzaroli, 1965: pp. 5–9, pl. V, figs. 1–2), is very close to the Upper Omo *Equus* in the characters of both the upper and lower teeth, but in the teeth of this evidence I am not prepared to accept conspecificity.

Hooijer assigns these teeth to "*Equus* spec. indet." and considers that the two very large *Equus* teeth

from Marsabit Road may be similarly assigned. The Omo horse teeth have also been assigned to *Equus oldowayensis* by Cooke (1963).

The relations of *E. oldowayensis* have thus been suggested to be a zebra near to *E. grevyi,* derived from *E. stenonis* or a similar form or possibly *E. numidicus* (=*E. stenonis* of North Africa). The teeth of *E. oldowayensis* as illustrated by Leakey (1958) and other specimens in the National Museums of Kenya, Nairobi, show the same ranges of variation and characteristics as do those of *E. capensis* from Makapansgat, Transvaal, or elsewhere in South Africa. It is likely, therefore, that *E. oldowayensis* is a synonym of *E. capensis* and that both represent intermediates on the evolutionary lineage from *E. numidicus* to *E. grevyi.*

Equus (Dolichohippus) capensis Broom 1909

SYNONYMY. Between 1909 and 1950 twenty species of *Equus* were described from South Africa. Of these, five are zebras rather than large equids, probably *E. burchelli* (Cooke 1963), and are considered later. The remainder are here considered to represent a single species, *E. capensis,* for reasons discussed below.

Because of the complexity of the synonymy for this species, the binomials and references are given more fully than for other species:

Equus capensis Broom 1909. *Ann. S. Afr. Mus.,* 7: 281–282.

E. harrisi Broom 1928. *Ann. S. Afr. Mus.,* 22: 441, fig. 2, B[1–3].

E. cawoodi Broom 1928. *Ann. S. Afr. Mus.,* 22: 443–444, fig. 3A.

E. kuhni Broom 1928. *Ann. S. Afr. Mus.,* 22: 444, fig. 3B.

E. gigas van Hoepen 1930a. *Paleont. Nav. Nas. Mus., Bloemfontein,* 2: 2–3, fig. 1.

Sterrohippus robustus van Hoepen 1930. *Paleont. Nav. Nas. Mus., Bloemfontein,* 2: 6–7, fig. 8.

Kolpohippus plicatus van Hoepen 1930. *Paleont. Nav. Nas. Mus., Bloemfontein,* 2: 3–10, fig. 10.

E. louwi van Hoepen 1930. *Paleont. Nav. Nas. Mus., Bloemfontein,* 2: 19–20, fig. 6–11.

E. westphali Dreyer 1931. Dreyer and Lyle. New fossil mammals and man from South Africa. Dept. Zool., Grey Univ. Coll., Bloemfontein, p. 36.

E. helmei Dreyer 1931. Dreyer and Lyle. New fossil mammals and man from South Africa. Dept. Zool., Grey Univ. Coll., Bloemfontein, p. 36.

E. sandwithi Haughton 1932. *Ann. S. Afr. Mus.,* 27: 419–421, fig. 4.

E. poweri Cooke 1939. *S. Afr. J. Sci.,* 36: 412–414, fig. 1.

E. fowleri Wells 1941. *Trans. Roy. Soc. S. Afr.,* 28: 301–306, fig. 1, pl. IV.

E. zietsmani Broom 1948. *Ann. Trans. Mus.,* 21: 23–25, fig. 16.

E. broomi Cooke 1950. *Ann. S. Afr. Mus.,* 31: 469–471, fig. 30.

AGE. Late Pliocene to late Pleistocene.

LOCALITIES. Zambia (118); Rhodesia (119); Republic of South Africa (122, 123a, 123b, 124, ?125, 128, 129, 130, 132, 133, 134, 135, 136, 137, 138, 140, 142, 143, 144, 145, 147–167 selected, 168, 170, 171, 172, 173, 174, 176, 177, 181 Baard's Quarry, 183, 184, 185, 186, 187, 188); South West Africa (191).

DISCUSSION. Broom (1909, 1913) and Broom and le Riche (1937) give no formal diagnosis of *E. capensis,* although Broom's remarks may be summarized to constitute an original diagnostic description as a species of horse with a skull considerably larger (10 to 15%) than *E. caballus,* with the pattern of the lower molars similar to that of modern horses but lacking any traces of the protostylid. The hypolophid is very large, the entostylid (= ?distolingual stylid) is small, and the parastylid scarcely extends lingually anterior to the metaconid (Broom 1909). *E. capensis* was "more powerfully built but did not stand so high" as *E. caballus* (Broom 1913, p. 438).

Cooke (1950a, pp. 448, 450) gives an effective diagnosis:

The form of the upper dentition of *E. capensis* (as now understood) is very similar to *E. burchellii,* of which it is virtually an enlarged version. The halves of the ectoloph are concave inwards and curve easily into the styles, except for a distinct tendency for an anterior overhand of the mesostyle. The parastyle is obliquely flattened anteriorly in the premolars but less noticeably so in the true molars. The metastyle is small. The hypocone is moderately small but the hypoglyph may be rather deep. The caballine fold is sometimes present but may often be absent. The pli-protoconule and pli-postfossette are well marked, but the pli-protoloph and pli-hypostyle are very small and may disappear completely with wear. The pli-postfossette and secondary small postfossette are distinct in very early wear but vanish rapidly with abrasion. The protocone is elongate oval in form, with rather more than one-third of its total length lying anterior to the junction with the protoconule. The range in size is, as far as observation goes, that deduced from the dimensions of the lowers.

In 1932 Haughton attempted the first review of South African fossil Equidae in which he listed 17 described species and added a new horse and a new hipparion. He noted resemblances between some of the *E. capensis* teeth from Saldanha Bay and Bloembosch, Cape Province, in which the plications were small and the pli caballine absent, with "members of the *Equus quagga* group, in which the pli caballine fold has not been developed" (Haughton 1932, p. 412).

Cooke (1950a) originally recognized as separate species *Equus capensis, E. kuhni, E. harrisi, E. plicatus, E. fowleri, E. sandwithi, E. poweri,* and *E. broomi* and considered them to be horses rather than quaggas. Cooke's (1950a) appendix, added some five years after the paper was originally written and just prior to publication, notes Broom's (1948) *E. zietsmani* and concludes that it and *E. plicatus, E. harrisi,* and *E. fowleri* are synonymous with *E. kuhni.*

Wells (1959, p. 64) discussed the determinability of species founded on a single tooth or a few isolated teeth and considered it "justifiable to argue that species cannot legitimately be named from such specimens." He therefore considered that only the species "*plicatus* Van Hoepen, *helmei* Dreyer, *lylei* Dreyer, *sandwithi* Haughton, *fowleri* Wells, and *zietsmani* Broom are determinable. Of these, *lylei* is probably a synonym or at most a subspecies of either *quagga* or *burchelli,* while *fowleri* and *zietsmani* are almost certainly synonyms of *plicatus*" (p. 65).

Cooke (1963) accepts only *E. plicatus* and *E. helmei.* Since the type specimen of *E. capensis* comprises a well-preserved mandibular molar series, Churcher (1970) rejects Wells's (1959) opinion that it is indeterminable and concludes that *E. capensis* should include *E. helmei, E. cawoodi, E. kuhni, E. zietsmani,* and some of the teeth referred to *E. harrisi* and *E. plicatus* and records both *E. capensis* and *E. plicatus* in the Transvaal cave deposits.

E. capensis was founded on a left lower P_3–M_2 series and *E. plicatus* on a newly worn right lower P_2–M_1 series. The major difference between these two rows is in the complex folding of the wall of the entoflexid in the type of *E. plicatus* and its slightly sinuous conformation in the type of *E. capensis.* Other specimens of lower cheek teeth series of both *E. plicatus* and *E. capensis* illustrated by Cooke (1950a, pp. 447, 461) show folding similar to that in the type of *E. capensis.* Petit (1939, pp. 176–221) showed that in *E. caballus* the ranges of variation in the morphology of the enamel patterns for the lower cheek teeth encompass both more and less complex patterns than those considered characteristic of the two South African species. Similarly, in *E. caballus* the ranges of size for length and breadth of the tooth vary sufficiently between individuals or as the tooth changes in conformation because of wear for the variations observed in the South African species to be encompassed within a single species.

The upper cheek teeth assigned to *E. capensis, E. plicatus,* etc., by Cooke (1950a) show variation similar to that in the lower dentitions, and measurements for both upper and lower teeth of these species are included within the ranges described by Petit (1939) for *E. caballus.* There is thus no reason to consider that there was more than a single large equid extant in South Africa during the Pleistocene and all the described variants may be attributed to variation within a single species because of age, wear, or temporal or geographic variation. This species should be referred to as *Equus capensis* Broom 1909.

As noted above, the horse from Olduvai, *E. oldowayensis,* may be considered also to belong within *E. capensis,* as the enamel patterns of the lower cheek teeth illustrated by Leakey (1958, pl. 19 bottom) are similar to those illustrated by Cooke (1950a, p. 456) for *E. harrisi.* Examination of both upper and lower molar rows of *E. oldowayensis* from various levels and sites in East Africa that are now housed in the National Museums of Kenya confirms that these dentitions vary in both size and pattern in a similar manner to those assigned to *E. capensis* from southern Africa.

Equus (Dolichohippus) grevyi Oustalet 1882

AGE. Early Pleistocene to Recent.

LOCALITIES. Ethiopia (?96c); Kenya (?107, 108, ?109, ?110).

DISCUSSION. This extant species possesses large upper cheek teeth that are larger than those in *E. zebra, E. quagga,* or *E. burchelli* of the subgenus *Hippotigris.* The protocone is large, varying in form from subtriangular to bilobate or simple but is seldom elongate. The parastyle and the mesostyle are flatly truncate. The parastyle flows into the buccal surface of the ectoloph at a right angle while the mesostyle joins the ectoloph at a sharp mesial angle and is smoothly rounded distally. The forms of the hypocone and hypoglyph are variable. The pli caballine is usually present but may be vestigial, the plis within the fossettes are variably present and are often better developed on premolars than molars, and the pli protoconule may be twinned. The lower cheek teeth usually have convex buccal walls and a ptychostylid is developed. The valley between the metaconid and metastylid is open, with a rounded end to the V; the metastylid is rounded or squared; and the metaflexid and entoflexid are smooth or have slightly wavy walls.

A possible relationship of *E. capensis* and *E. grevyi* has been considered by some authors. For example, Cooke (1950a) remarked on the resemblance between *E. kuhni* and *E. grevyi* and noted that individual teeth of *E. grevyi* also resembled those of *E. fow-*

leri, E. plicatus, and *E. harrisi.* He said, "There can be no doubt that the fossil *E. kuhni* (and its presumed synonyms) is closely related to *E. grevyi . . .* for the present, however, it is felt that it will be better to retain the designation *E. kuhni* for the fossil material" (p. 478).

E. grevyi is considered a zebra because of its striped pelage and, among other dental characters, because of the variability in the development of the pli caballine. In *E. numidicus* the pli caballine is reported to be typically absent; in *E. oldowayensis* it is reportedly variably present to vestigial; in *E. capensis* it is absent, small, or variably developed; and in *E. grevyi* it is usually present or vestigial. In *Equus (Equus)* spp. the pli caballine is typically present but may occasionally be absent, while in the zebras *Equus (Hippotigris)* spp. it is typically lacking. Such a variable characteristic is not absolute for distinguishing between horses and living zebras and does not preclude *E. grevyi* from being the lineal descendant of *numidicus-oldowayensis/capensis* stock. Such a conclusion was anticipated by Hopwood (1937, p. 117), Cooke (1950a, p. 478; 1950b, p. 138), and Arambourg (1937) and was endorsed by Skinner (1972).

E. grevyi has long been considered the most distinct of the zebras, mainly on the basis of a finer pattern of striping and on its proportions. It is known as a large-headed species with fairly short legs, but unfortunately very little is recorded in the literature on the postcranial proportions or dental characters of *E. grevyi* and few specimens are available for comparison. Also, little is known or recorded of the fossils other than dental material, although Broom (1913) reports a metatarsus and Churcher (1970) phalanges of *E. capensis,* and Hopwood (1937) reports metatarsals of *E. oldowayensis.* These elements support Broom's (1909) and Hopwood's (1937) original contentions that these horses were large and massive but do not allow comparison in detail with other equids. However, the large heads suggested for *E. capensis* and *E. oldowayensis* and the more complete skeletal material described for *E. simplicidens* (Skinner 1972) conform to expectations for precursors of the modern *E. grevyi.*

Although consideration of the dental material available seems to indicate an evolutionary lineage from *E. numidicus* to *E. oldowayensis* and *E. capensis* with *E. grevyi* as the surviving form, this lineage cannot be confirmed without complete fossil skulls and more postcranial material for the subgenus *Dolichohippus* from Africa. It is hoped such material would also determine the status of the African *Dolichohippus* species vis-à-vis the North American *E. simplicidens.*

Subgenus *Hippotigris* H. Smith 1841

DIAGNOSIS. Size intermediate between *Equus (Equus)* and *E. (Dolichohippus)* on the one hand and *E. (Asinus)* on the other. Marks on lower incisors variably developed, but generally less well developed than in *E. (E.) caballus.* Ectoloph variable: in *E. (H.) burchelli* generally convex medially as in *E. caballus;* in *E. (H.) zebra* and *E. (H.) quagga* generally straight or slightly convex laterally as in *E. (A.) asinus.* Parastyle and metastyle generally angled to ectoloph, usually furrowed; wearing to be broadly flattened, angular, or pointed. Ptychostylid more prominent than in *E. asinus.* Protocone generally smaller or narrower and more elongate than in *E. caballus* and *E. (D.) grevyi.* Lower cheek teeth with median (buccal) valley extending past midline to level of metaconid and metastylid. Metastylid pointed to bluntly pointed. Metaflexid and entoflexid walls possessing few folds.

DISCUSSION. The present and past distribution of hippotigrine equids has already been discussed. Few characters other than pelage and body proportions have been elucidated that reliably distinguish members of the subgenus *Hippotigris* from those of the subgenera *Equus, Dolichohippus,* and *Asinus.* In sub-Saharan Africa the fossil *Equus* record is limited to the subgenera *Hippotigris* and *Dolichohippus,* which are represented primarily by dental material, often in the form of isolated teeth. *E. (Hippotigris)* has generally been distinguished from *E. (Dolichohippus)* by its significantly smaller size. Haughton (1932) and Cooke (1943, 1950a) present dental criteria for distinguishing hippotigrine species (summarized in table 20.2). Eisenmann and de Giuli (1974) were able to separate *E. zebra* and *E. burchelli antiquorum* on cranial characteristics but considered only the presence or absence of infundibula (marks) in the lower incisors to be dental characters useful in separating these species. They found that in *E. zebra* infundibula are always closed in I_1 and I_2, and often in I_3, while in *E. burchelli antiquorum* infundibula are never present in I_3 and are rare in I_1 and I_2. Specific identification of postcranial elements without associated dental material is considered completely unreliable.

In North Africa the picture is further complicated by the presence of *E. (Asinus)* in the middle Pleistocene, while the relatively small *E. (Dolichohippus) numidicus* persisted until the late Pleistocene. Fossils of the subgenus *Hippotigris* are common but apparently represent only a single species, *E. burchelli*

(*mauritanicus*). Arambourg (1938) considers several dental characters useful in separating zebras and asses: (1) the condition of the ectoloph wall (convex medially in zebras, straight or convex laterally in asses); (2) the shape of the metastylid (broadly flattened or angular in zebras, rounded in asses); (3) the development of the ptychostylid (less well developed and less prominent in asses); and (4) the lingual extent of the median (buccal) valley (past the midline in zebras, before the midline in asses).

Equus (Hippotigris) burchelli (Gray) 1824

SYNONYMY. *Asinus burchelli* Gray 1824, *Equus mauritanicus* Pomel 1888, *Equus quagga wahlbergi* auctorum van Hoepen 1930, *Equus platyconus* van Hoepen 1930, *Equus simplex* van Hoepen 1930, *Equus simplicissimus* van Hoepen 1930, *Kraterohippus elongatus* van Hoepen 1930, *Equus lylei* Dreyer 1939, *Equus quagga quagga* Dreyer 1931, *Equus quagga* var. Haughton 1932, *Equus burchellii* Cooke 1932, *Equus (Asinus) tabeti* Arambourg 1970.

AGE. ?Late Pliocene to Recent.

LOCALITIES. Morocco (1, 2, 3, 5, 7, 8); Algeria (10, 12, 13, 14, 15, 16, 20, 21, 24, 26, 29, 30, 31, 32, 33, 34, 35, 36, 37, 38, 39, 40, 42, 43, 44, 45, 48d, 49, 50, 52, 54, 62, 63, 64, 65, 66); Tunisia (75, 76); Mali (81); Libya (82, 83, 84, 85); Kenya (109); Tanzania (111, 112 Beds II–IV, 113); Zambia (116); Rhodesia (119); Republic of South Africa (122, 123b, 125, 128, 131, 133, 134, 136, 138, 139, 142, 143, 144, 145, 146, 148, 150a, 150b, 150e–h, 152, 153, 155, 156, 157, 159, 160, 161, 164, 166, 168, 169, 171, 177, 178); South West Africa (190).

DISCUSSION. On cranial characters *Equus (Hippotigris) burchelli* is distinguished from *Equus (Hippotigris) zebra* in possessing a convex frontal region, a strong infraorbital rim, a V-shaped nasofrontal suture, a relatively weak and posterodorsally oriented external auditory meatus, and a horizontally oriented nasopremaxillary suture (Eisenmann and de Giuli 1974).

Fossil material has most frequently been identified on dental evidence. A few relative (and often variable) dental characteristics distinguish *E. burchelli* from other hippotigrine zebras (see table 20.2). Marks (infundibula) in the lower incisors are absent or less well developed and when present are rapidly lost with wear. The upper first premolar is more frequently lost relatively early in life. The protocone is larger and forms an elongate oval with the mesial portion contributing a third of the total length. The hypocone is small and may bulge slightly into the medial valley. The parastyle is obliquely flattened and, with the mesostyle, may be buccally grooved.

Both the parastyle and the mesostyle are usually smoothly confluent with the ectoloph. The pli caballine is frequently present in premolars. Within the fossettes, the plis hypostyle and plis postfossette are usually present and well developed. In the lower series the ptychostylid fold in the hypoconid is small and squared, the metaconid is oval, and the metastylid is small, pear-shaped, and bluntly pointed. Some folding of the buccal walls of the entoflexid may be found in unworn teeth.

Fossil remains of *E. burchelli* are the commonest in Africa, with localities known from the Maghreb to the Cape of Good Hope. The species was originally reported from North Africa as a horse, *Equus mauritanicus,* by Pomel (1888a,b, 1897) but was shown by Boule (1899 [1900], 1900) to be more similar to the bontequagga, *E. burchelli,* than to a caballine horse. In South Africa the specimens described by van Hoepen (1930a) as *E. platyconus* and *Kraterohippus elongatus* were considered probable synonyms of *E. burchelli* by Haughton (1932), who also regarded *E. lylei* Dreyer 1931 as indistinguishable from *E. burchelli.* Cooke (1950a) included van Hoepen's (1930a) *E. simplicissimus* and *E. simplex* in *E. burchelli* and noted that most fossil specimens described as quaggas by other authors in fact represent bontequaggas. Wells (1964) considered that all of Cooke's (1949) reports of *E. burchelli* from the Vaal River Younger Gravels should be referred to as *E.* cf *burchelli* because few of them were established on more than isolated teeth and thus were not diagnostic (Wells 1959).

Arambourg's (1970) description of *Equus (Asinus) tabeti* from Aïn Hanech, Algeria, is founded on a dorsoventrally compressed cranium, a palate, a mandible, some isolated teeth, and many postcranial elements. Arambourg assigned these specimens to an ass because the styles on the buccal face of the upper molars project strongly and overhang the ectoloph wall at right angles while the base of the ectoloph is straight or slightly convex. Such characteristics, although present in asses (Arambourg 1938), are also typical of some zebras (see table 20.1), while the *E. tabeti* teeth illustrated by Arambourg do not preclude their assignment to a zebra and in fact suggest *E. burchelli* as described by Cooke (1950a). Arambourg notes that fossil asses are not rare in middle and late Pleistocene deposits in North Africa and that their remains are generally attributed to "*E. mauritanicus*" (i.e., *E. burchelli*); but he provides no comparison of *E. tabeti* with this species.

The measurements of the elements given by Arambourg for *E. tabeti* are seldom extensive (e.g., for 9 humeral specimens, 2 sets of measurements; for 56 metatarsal specimens, 5 sets of measurements) and no indications of the observable ranges or deviations accompany them. The measurements as presented indicate an animal at least as large as any of the zebras and greater than any of the asses.

Arambourg (1970) and Coppens (1971) date *E. tabeti* as Upper Villafranchian (late early Pleistocene) and thus as the earliest record of ass in Africa. Because of its large size, its early date, and the lack of conclusive published evidence to the contrary, *E. tabeti* as Upper Villafranchian (late early Pleistocene) *icus*), which is known from the early Pleistocene both elsewhere in North Africa and in other parts of the continent.

Equus (Hippotigris) zebra Linnaeus 1758

AGE. ?Middle Pleistocene to Recent.

LOCALITIES. Republic of South Africa (173, 179, 184, ?187, 189).

DISCUSSION. Eisenmann and de Giuli (1974) consider the skull of *E. zebra* distinguishable from that of *E. burchelli antiquorum* by a flat frontal region, a strong lateral projection of the posterior orbital margins (i.e., greater bizygomatic breadth), a straight nasofrontal suture, a strong development and lateral projection of the external auditory meatus, a vertical orientation of the nasopremaxillary suture, and the more consistent occurrence and greater development of infundibula (marks) on the lower incisors.

The upper cheek teeth are slightly broader than in *E. burchelli.* The mesial portion of the protocone comprises less than one-third its length. The hypocone is only slightly smaller than the protocone. The floor of the ectoloph wall is straight or slightly convex laterally and the parastyle is commonly flattened mesially and not directed obliquely as in *E. burchelli.* The pli caballine is often, but not consistently, present. Plis within the fossettes are small and reduced and the pli protoloph is typically lacking or very small. In the lower cheek teeth the ectostylid is lacking or small, the entoconid and metaconid are rounded, and the metastylid is pear-shaped and somewhat pointed.

E. zebra is rather sparsely recognized as a fossil, possibly because of the lack of clearly diagnostic dental characteristics, because the preservation of montane species, except in caves, is not common, or because much of its range has not been thoroughly investigated for fossils. An exception is seen in the southwestern Cape Province where Hendey (1974a, 1974b) recorded the occurrence of *E. zebra* together with *E. capensis* in late Pleistocene and Holocene assemblages. Both *E. burchelli* and *E. quagga* appear

Table 20.2 Distinguishing characteristics of the teeth of zebras, including the subgenera *Hippotigris* and *Dolichohippus*, after Haughton (1932), Arambourg (1938), and Cooke (1943, 1950a) and from observations on specimens in the American Museum of Natural History, New York, and the National Museum of Kenya, Nairobi.

Characteristic	*E. zebra* Mountain zebra	*E. quagga* Quagga	*E. burchelli* Bontequagga	*E. grevyi* Grevy's zebra
Incisors:				
All upper incisors have marks	marks in I_1 and I_2 at least	marks in I_1 and I_2 at least; in I_3 may be reduced or absent	marks usually absent from all lower incisors	all lower incisors usually have marks; I_3's mark reduced or open distally
Upper cheek teeth:				
Ratio of diastema between I^3 and P^2 to length of premolar series[a]	considerably greater than	generally less than	approximately subequal or slightly greater than	approximately 2/3 to 3/4 of
Condition of P^1	usually absent	usually absent	frequently absent; lost late in life	present only in young individuals
Ectolophs[b]	straight or slightly convex buccally	straight or slightly convex buccally	convex lingually	floor flat to slightly convex lingually, with rounded junctions with styles
Styles (parastyle, mesostyle)[b]	set off at right angles from ectolophs; flat or convex on crests	markedly overhang ectolophs; usually furrowed on crests	generally flow smoothly into ectolophs; furrowed on crests	set off at right angles mesially or flow into floors distally; styles squarely or obliquely truncated
Protocone	small and reduced; subtriangular on mesial portion	larger, bilobate; mesial portion larger than distal	larger, more elongate; not markedly bilobate	larger, sub-triangular and bilobate, or with mesial portion small to absent or rounded
Pli-caballine	often present	usually absent	usually absent	usually present or vestigial in older individuals

to be absent from this region, and Hendey hypothesizes that *E. zebra* may have formed a "species-barrier" between *E. quagga* in the Karroo and *E. capensis* in the southwestern Cape, permitting the latter species to survive in this region longer than elsewhere.

Equus (Hippotigris) quagga Gmelin 1788

AGE. ?Early/middle Pleistocene to Recent (ca. 1880 A.D.).

LOCALITIES. Republic of South Africa (?132, 138, ?141, 143, 145, ?150a, ?153, ?157, ?167, 168, ?172, 174, 175).

DISCUSSION. Cooke (1950a) suggests that the dentition of *E. quagga* is on average slightly larger than that of *E. burchelli* but exhibits a greater size range than does the more common species. The upper cheek teeth show slightly more lateral compression than in *E. burchelli* and *E. zebra*. The protocone is bilobate, with the mesial portion comprising one-third to one-half the mesiodistal length. The parastyle is usually flattened distally with its buccal face parallel to the axis of the molar row. The mesostyle is rounded and overhangs mesially and distally. The hypocone is smaller than the protocone and the hypoglyph varies in size and shape. The pli caballine is usually, but not always, absent. Plis within the fossettes are often simple, small, or few, though all are usually present except the pli prefossette, which is typically absent or reduced. The lower cheek teeth resemble those of *E. burchelli* rather than those of *E. zebra*. The metaconid is rounded or oval and the metastylid is pear-shaped and separated from the metaconid by a rounded

Table 20.2   (*continued*)

Characteristic	*E. zebra* Mountain zebra	*E. quagga* Quagga	*E. burchelli* Bontequagga	*E. grevyi* Grevy's zebra
Pli-protoloph	absent	usually present	present	usually present on premolars; variably developed on molars
Pli-prefossette (unreliable)	weak or absent	weak or absent	variable	variably developed
Pli-protoconule	small or absent	small or absent	normally strong	well-developed; may be double or bifurcated
Pli-hypostyle	absent	usually present	present	variably present
Length P²-M³ (mm)[c]	155–165	136–171	142–162	170–190
Lower cheek teeth[d]:				
Protoconid walls	well-rounded, convex	rounded or slightly flattened	flattened or concave	convex to slightly convex
Hypoconid walls	well-rounded, convex	flattened	flattened or concave	convex to slightly convex
Valley between metaconid and metastylid[e]	acutely pointed	rounded at apex, open	fairly acute apex	open, narrow V with rounded apex
Metastylid	pointed	bluntly pointed or rounded	pear-shaped or bluntly pointed	rounded to subquadrate
Metaflexid and entoflexid walls	walls possess few folds	walls possess few folds	walls possess few folds	smooth or with occasional to developed folds
Transverse diameters across the cheekteeth	In all but *E. grevyi*, which is larger, the diameters lie within the same ranges; the upper dentition measuring 21 to 26 mm across the enamel from the mesostyle to the protocone and the lower dentition 11 to 15 mm across from the protoconid or hypoconid to the metastylid.			

[a]Mandibular diastema is slightly shorter than palatine diastema.
[b]More marked in premolars than in molars, and distinction may be impossible on M¹'s.
[c]Sample size in *E. zebra* and *E. quagga* is too small for reliability.
[d]Characters in the lower dentitions are more variable and less reliable than in the upper dentitions.
[e]Valley becomes broader as wear takes place, e.g., in older individuals.

groove. The entoconid is rounded to quadrate and is larger in premolars than in molars. A ptychostylid is often present.

The extinction of natural populations of *E. quagga* in the 1870s (Willoughby 1966; Rau 1974) took place before any collections of specimens had been made. Cooke (1950a) records that a skull in the Stuttgart Museum, a cast of which is in the Transvaal Museum, Pretoria, unfortunately represents a young individual. Owen (1869) gives natural size figures of the dentition of an adult specimen then in the Royal College of Surgeons, London, and Robert Broom examined and made drawings of this specimen. Some old jaws and skulls in the McGregor Museum, Kimberley, and in the Kingwilliamstown Museum agree closely with Owen's published and Broom's manuscript figures.

Cooke (1950a, p. 441) remarks of fossils attributed to *E. quagga* that:

probably largely in consequence of the confused ideas which have hitherto existed regarding the characters of the teeth of *Equus quagga,* none of the teeth referred to by various authors as "*E. quagga*" or "*E. quagga quagga*" can be regarded as actually belonging to Gmelin's species. These so-called "quagga" teeth are mainly those of *E. burchellii* (Gray) but, remarkably enough, none of the fossil teeth described under other names can be regarded as belonging to *E. quagga* either, so that it would appear that this recently extinct species has not hitherto been recognised to any notable degree in our fossil collection.

Cooke (1950a) recognized *E. quagga* from only three sites: Koffiefontein, Bankies, and Wonderwerk Cave. In 1949 he referred a number of isolated and/or fragmentary teeth from the Vaal River gravels to *E.* cf *quagga* but thought the occurrence of the species doubtful. Churcher (1970) identified as *E. quagga* a somewhat distorted palate with seven teeth *in situ* and portions of a mandible retaining a deciduous molar dentition from the Swartkrans Australopithecine site on the criteria described by Cooke (1943, 1950a). If these Swartkrans specimens are correctly assigned, they seemingly represent an early Pleistocene occurrence of *E. quagga,* making it coeval with the earliest *E. burchelli* reported from southern Africa. However, in the context of viewing the African hippotigrine fossils as a whole, it seems best to regard the presence of *E. quagga* at so early a date and so great a distance from its known range with skepticism. As Hendey (1974a, p. 46) remarks:

Yet another curious feature of the record of the South African Pleistocene zebras is the reported presence of both *E. burchelli* and *E. quagga* in the early Pleistocene of the Transvaal . . . even though the latter is often regarded as being no more than the most southerly variety of the plains zebra group. Even if it is accepted that *E. quagga* and *E. burchelli* are specifically distinct, it is difficult to accept that they were recognizably different and could have co-existed as far back as the early Pleistocene . . . It is worth noting that if *E. quagga* did once extend its range as far north as the Transvaal, its absence from the fossil record of the southwestern Cape is more unexpected than ever.

An even less likely identification of *E. quagga* is that of Dietrich (1942) from Laetolil. This material is considered to derive from *E. burchelli.*

Subgenus *Asinus* Frisch 1775

DIAGNOSIS. Equines of small size and stocky build. Upper permanent molars square. Ectoloph of relatively "low relief," with styles lower and valleys shallower than in the subgenera *Equus* and *Hippotigris.* Ectoloph valleys flat or slightly convex laterally; parastyle and mesostyle squared and tilted mesially. Pli caballine typically lacking or reduced. Fossettes small, with small, simple plis. Protocone with mesial arm comprising approximately 40% of length. Lower molars with pli caballinid. Median (buccal) valley ending before flexid reentrants except in earliest wear; ptychostylid small. Metaflexid with sharp buccal angle. Metaconid elongate, lingually flattened, directed mesiolingually; metastylid rounded, directed distolingually; entoconid rounded and constricted at base; hypoconulid and hypostylid prominent. Reentrant between metastylid and metaconid open and V-shaped. (After Arambourg 1938; Quinn 1955, 1957).

Equus (Asinus) asinus Linnaeus 1758

SYNONYMY. *Equus africanus* Fitzinger 1857, *Equus (Asinus) hydruntinus* Regàlia of Blanc (1956).

AGE. ?Middle Pleistocene to Recent.

LOCALITIES. Morocco (?7, 8); Algeria (12, 14, 16, 18, 19, 26, 29, 30, 32, 34, 39, 40, 42, 54, 63, 65, 66, 67, 69); Libya (82, ?83); Egypt (88, 89, 90, 91); Sudan (92, 93b, 94, 95).

DISCUSSION. There is some disagreement on the origin of the ass. Zeuner (1963) and Ansell (1967) consider the ass indigenous to Africa, while Churcher (1974) provides evidence supporting the view that the African ass derived from an equine inhabiting the eastern Mediterranean area of Arabia and southward east of the Red Sea. This equine later spread across North Africa and south along the Nile Valley and western Red Sea coast to occupy its present distribution.

The North African record of asses suggests that they replaced Burchell's zebra (*E. burchelli mauritanicus*) in that region during the late Pleistocene. Bate (1955) records *E. burchelli* in the middle and upper Palaeolithic and *E. asinus* in the upper and latest Palaeolithic at Hagfet el Tera, Libya. Higgs (1967) records *E. burchelli* only from the Levalloiso-Mousterian, upper Palaeolithic (45,000–41,500 years B.P.) and *Equus* sp. only from later levels at the cave of Haua Fteah, Libya. Some of the latter specimens may represent asses but none is recognized with certainty. Romer (1928) recorded *E. burchelli* from Mousterian deposits and ass from upper Palaeolithic deposits in Algeria, and in 1935 he identified *E. burchelli* and *E. asinus* from the Mousterian and Capsian (= Aurignacian) levels at Ali Bacha and other sites in the Aïn Beida region, Algeria. Romer (1928) concluded that the zebra persisted in the Maghreb (= Mauritania) at least until the older Neolithic, but notes (p. 122) that "citations (of *E. asinus*) before the Neolithic are rare, and it is probable that the Neolithic finds are in great measure those of domestic animals."

In the Nile Valley Churcher and Smith (1972) and Churcher (1972) record *E. asinus* cf *africanus* from the late Palaeolithic of Kom Ombo, Upper Egypt. The history of the African asses may therefore be one of entrance into Africa during late middle or early upper Palaeolithic times, their gradual spread westward across North Africa during upper Palaeolithic and Neolithic times, replacing *E. burchelli* in Libya during the middle upper Palaeolithic and in the Maghreb during the Neolithic. Romer's (1928) suggestion that the later Neolithic replacement of *E. burchelli* by *E. asinus* may have been due to man

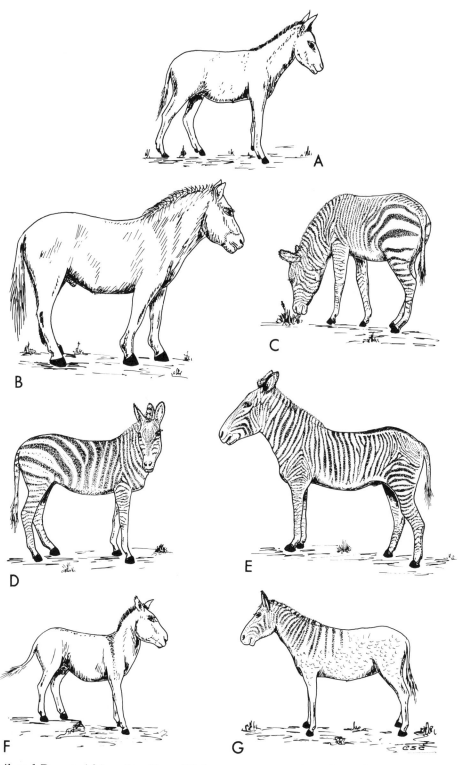

Figure 20.6   Fossil and Recent African Equidae. All figures are approximately to the same scale; numbers in parentheses give heights at the shoulders. (*A*) *Hipparion* sp., hipparion or three-toed horse (?135 cm), (*B*) *Equus* (*Dolichohippus*) *capensis*, extinct Cape horse (?190 cm), the pelage may have been striped similarly to those of Grevy's zebra and its build may have been lighter, (*C*) *Equus* (*Hippotigris*) *zebra*, zebra or mountain zebra (118-128 cm), (*D*) *Equus* (*Hippotigris*) *burchelli chapmani,* Chapman's variant of Burchell's zebra or bontequagga (125-137 cm), (*E*) *Equus* (*Dolichohippus*) *grevyi*, Grevy's zebra (170 cm), (*F*) *Equus* (*Asinus*) *africanus*, Nubian ass (90-150 cm), (*G*) *Equus* (*Hippotigris*) *quagga*, recently extinct quagga (?150 cm).

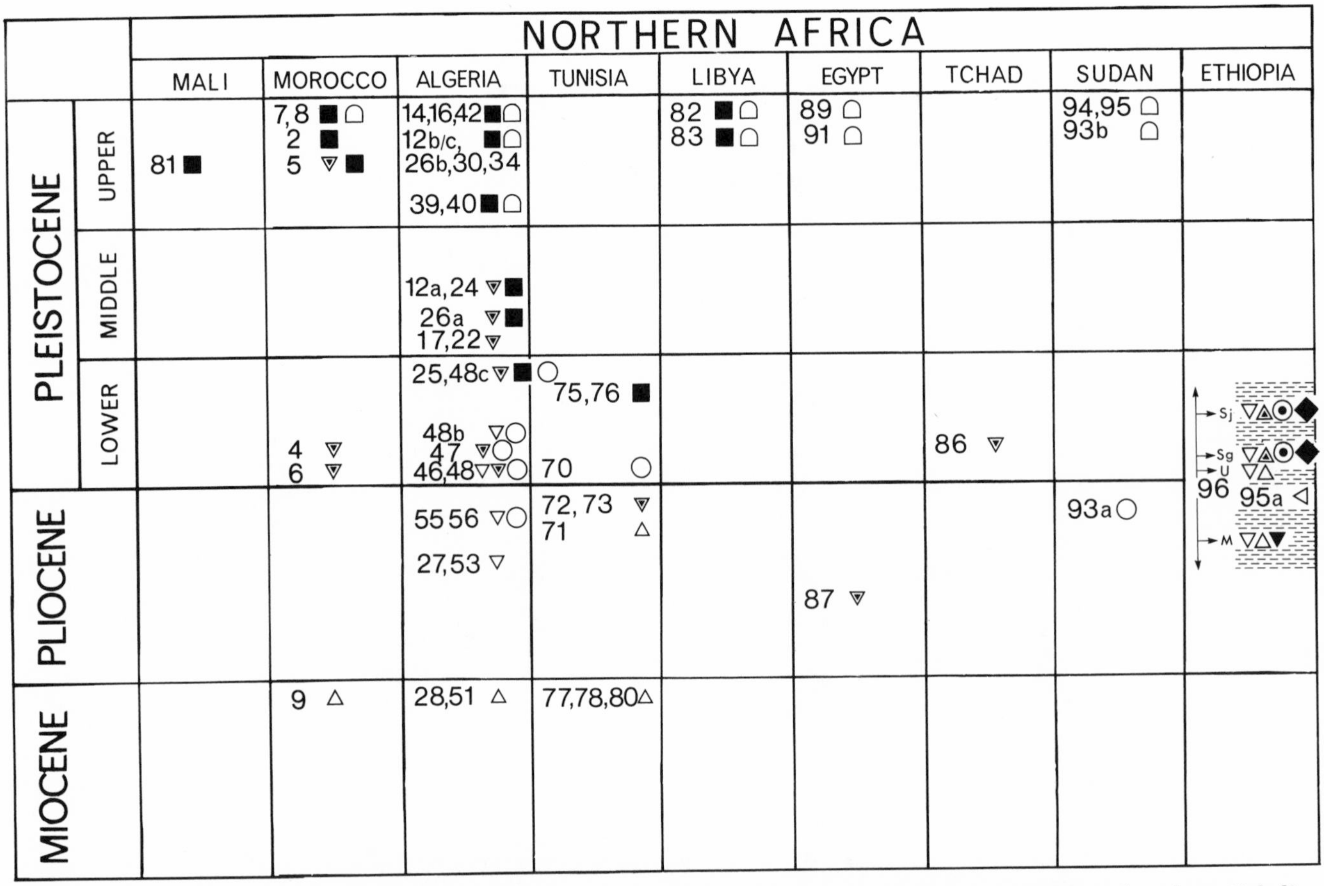

		NORTHERN AFRICA							
		MALI	MOROCCO	ALGERIA	TUNISIA	LIBYA	EGYPT	TCHAD	SUDAN	ETHIOPIA
PLEISTOCENE — UPPER		81 ■	7,8 ■□ / 2 ■ / 5 ▽■	14,16,42 ■□ / 12b/c, ■□ / 26b,30,34 / 39,40 ■□		82 ■□ / 83 ■□	89 □ / 91 □		94,95 □ / 93b □	
PLEISTOCENE — MIDDLE				12a,24 ▽■ / 26a ▽■ / 17,22 ▽						
PLEISTOCENE — LOWER			4 ▽ / 6 ▽	25,48c ▽■○ / 48b ▽○ / 47 ▽○ / 46,48 ▽▽○	75,76 ■ / 70 ○			86 ▽		96 95a ⊲ (Sj ▽▲◉◆ / Sg ▽▲◉◆ / U ▽▲ / M ▽▲▼)
PLIOCENE				55 56 ▽○ / 27,53 ▽	72,73 ▽ / 71 △		87 ▽		93a ○	
MIOCENE			9 △	28,51 △	77,78,80 △					

Figure 20.7 Stratigraphic distribution of African fossil Equidae. Selected sites show both temporal and spatial distributions but correlations are at best tentative because of unclear stratigraphic relationships. The numbers represent the sites shown in figures 20.1-20.4. Sites in which there is a major sequence of continuous deposition are shown with horizontal hatching. Site 93, Omo Beds, Lake Turkana, includes four major horizons: Mursi Formation (M), Usno Formation (U), and Shungura Members G and J (Sg and Sj); Site 99, Lake Baringo Beds, has five of its six horizons shown: Ngorora

raises the question whether the spread of asses southward up the Nile River and into the central Sahara may not also have been assisted, at least in part, by man. The differentiation of the wild (or ?feral) ass populations into the modern races or subspecies would have taken place during the late Pleistocene and perhaps early Holocene.

The report of *Equus* (*Asinus*) sp. from Broken Hill, Zambia (Leakey 1959), is founded on a tooth and an astragalus, which are rather larger than those of living *E.* (*A.*) *asinus africanus*. The identification is unreliable as single teeth are not diagnostic (Wells 1959) and, because of the larger size of the astragalus, these specimens are assigned to *E. burchelli*, which is also present in the deposit.

Summary and Discussion

The African fossil Equidae have been derived from descendants of immigrants that crossed into northeastern Africa during late middle Miocene

times or later. The earliest known representatives of these immigrants are three-toed hipparions (*Hipparion primigenium*) known from late middle or earliest late Miocene beds in East Africa and the Maghreb (figures 20.6 and 20.7). This species is the same as the primordial Eurasian hipparion but seems to have had a separate evolutionary history in Africa during and since the late Miocene. Two other lines of *Hipparion* appeared shortly after *H. primigenium* in the African fossil record.

Evolution of all hipparions in Africa is characterized by increased hypsodonty of the cheek teeth, by a mesiodistal elongation of the protocone to a lenticular or lozenge-shaped cross-section, and by the addition of large ectostylids and other additional pillars in the lower molars. Forstén (1968, 1972) suggests that the original hipparions were forest dwellers (i.e., *H. primigenium* in Africa) and that from these there evolved geographically separate lineages that became more adapted to steppe or savanna conditions by becoming extremely long and slim in the

		EAST AFRICA			SOUTHERN AFRICA					
		KENYA	UGANDA	TANZANIA	ZAMBIA	RHODESIA	TRANSVAAL	ORANGE FREE STATE	CAPE OF GOOD HOPE	SOUTH-WEST AFRICA
PLEISTOCENE	UPPER	108 ◆ — 109 ■◆ — 106 ⚠ ☉		114 ⚠ ☉? — 113 ⚠ ■	116 ■ — 118 ●	119 ●■	122 ●■ — 123b ●■■ — 135 ● — 124 ●	144 ●■ — 145 ●◈■ — 138 ◈●■ — 143 ●◈■	174 ●◇ — 188 ●□ — 173 ●◈ — 185 ●	190 ■
PLEISTOCENE	MIDDLE	110 ⚠☉◆		IV ⚠☉■ — III ⚠☉■ — II ⚠☉■			125 ●■ — 136 ●■–	146 ■ — 140 ●■ — 134 ▽●■ — 141 ▽◇■	169 ■ — 183 ● — 150 ●◇■ — 156 ▽●■	191 ●
PLEISTOCENE	LOWER			112 — I ⚠☉	MALAWI — 115 ⚠☉		128 ▽●◇ ■ — 133 ▽●■ — 132 ▽●◇ — 123a ▽●			
PLIOCENE			97a △▽○	111 ⚠■					181 ⚠▷● — 180 ▷	
PLIOCENE		102 ▽⚠▼ — f ☉ — 107 ⚠☉ — e ⚠ — 105 d ▽⚠	97b ⚠▽							
MIOCENE		100 ▽⚠▼ — b ▼ — a ⚠								

Legend

△ H. primigenium	⚠ H. baardi	■ E. burchelli
▼ H. turkanense	◁ H. afarense	◈ E. zebra
▽ H. sitifense	▷ H. namaquense	□ E. (Hippotigris) sp.
▽ H. libycum	○ E. numidicus	◇ E. quagga
⚠ H. ethiopicum	☉ E. oldowayensis	∩ E. asinus
▽ H. steytleri	● E. capensis	☉? Equus sp. s.l.
▲ H. albertense	◆ E. grevyi	

Formation (*a*), Mpesida Beds (*b*), Chemeron Formation (*d*), Aterir Beds (*e*), and Chemoigut Beds (*f*); and Site 106, Olduvai Gorge, is divided into Beds I, II, III and IV (indicated by Roman numerals).

Species are indicated by symbols shown in the legend. *Hipparion ethiopicum, H. libycum,* and *H. steytleri* are used *sensu* Hooijer (1975b).

leg in some forms. Gromova (1952) and Forstén (1968) discuss the variability and adaptations of mainly the Holarctic species of *Hipparion* and Boné and Singer (1965) the species of *Hipparion* that might have occupied the African Miocene, Pliocene, and Pleistocene biotypes. Even the largest of the African forms were relatively small when compared to modern horses and, since the smaller modern equids (e.g., asses and zebras) usually inhabit relatively dry grassy environments, hipparions may also have preferred such habitats.

When larger and smaller species of hipparion have been found in paleontological association (e.g., *H. sitifense* and *H. primigenium*), they may represent two species that were ecologically separate in life. Parallel development of hypsodonty and the addition of stylids in the three lines of African hipparions may be correlated to synchronous adaptations to harder, drier food, such as is found in savanna or wooded grassland. In response to tougher food, the gross size and volumes of the cheek

teeth will increase, with the abrasion of both upper and lower teeth removing equal volumes of tooth if both upper and lower teeth are equally resistant to wear. The mass of the teeth relative to the body will increase if size is increased so that the volumes of the teeth will correspond to the body volumes if lifespans are comparable, since it is the available grinding material of the tooth that determines the amount of food that may be processed and that corresponds to the life of the individual and its body volume. The smaller hipparions therefore had proportionately and absolutely smaller teeth than did the larger hipparions. This differential will be increased if the smaller forms are browsers and the larger grazers, as leaves of herbs and trees are softer than those of grasses, with their contained silica cells, and browsing introduces less mineral particles from the soil than does grazing.

The isolated protocone typical of hipparions probably weakened the upper cheek teeth (Stirton 1941) and this was compensated for by the increased cementum on the lingual surface. The protocones of the African hipparions are consistently more mesio-distally elongate and buccolingually narrower than those of most non-African species (Boné and Singer 1965, figs. 12 and 13). As the protocone wears, it becomes more oval and less elongate, suggestive of a lanceolate three-dimensional conformation. The buccal and lingual enamel crests may provide additional triturating surfaces that function when the jaw moved laterad-mediad. The changes noted in the protocones and stylids of hipparion teeth from Africa would agree with the interpretation that these hipparions were adapting to an increasingly tougher diet and that this could have been the result of either or both increasingly arid conditions or adaptation to more extreme ecological situations. The latter could be interpreted as a response to either increased population density or to competition with other species.

The development of the enamel plications within the fossettes of the upper cheek teeth correlates similarly with the suggestion of increasingly harsh diets. These plis are also oriented mesiodistally, as is the long diameter of the cross-section of the protocone, and would present the best triturating surface during laterad-mediad motions of the jaw. The metaconid-metastylid double knot develops toward a caballoid conformation in the African hipparions, with the metastylid tending to elongate mesiodistally to correspond with the elongation of the protocone against which it occludes.

The development of elongated metapodials in the "dwarf" form, *H. sitifense,* and elongation of the proximal median phalanges in the *H. primigenium–H. afarense–H. libycum* lineage by late Pliocene or early Pleistocene times are cursorial adaptations that would increase the chances of the individual's survival in a more open grassy habitat.

The later African hipparions (*H. libycum libycum–ethiopicum–steytleri*) were probably restricted to dry semiarid savanna or bush veld much like that inhabited by gazelles and zebras in East Africa today. The extinction of the African hipparions by the end of the Pleistocene cannot be explained with confidence. It may have come about from competition with the differentiating lineages of zebras, the increasingly successful radiation of the small and middle-sized antelopes, and changing ecological conditions. Since we have no evidence for zebras from levels older than late Pliocene (e.g., *E. burchelli mauritanicus* from Aïn el Hadj Baba, Algeria), but much more evidence for increasingly varied antelope taxa, the replacement of the hipparions appears to have been the result of increasing competition for the available food and the relatively less efficient use of that food by the hipparions in comparison with that of the evolving ruminating antelopes.

The history of the large African equids (subgenus *Dolichohippus*) is simple in that there appear to be only two main forms, *Equus numidicus* from the late Pliocene of northern Africa and the *Equus capensis–Equus oldowayensis–Equus grevyi* complex present at least during the late early Pleistocene to late late Pleistocene or Recent times in eastern and southern Africa (figure 20.8). The origin of this lineage is North American as far as may be known at present, with resemblances between *E. numidicus* and the Asiatic *E. sivalensis,* as noted by some authors (e.g., Lydekker 1887) and between *E. grevyi* and *E. simplicidens* (Skinner 1972).

None of the authors who have written on the phylogeny of these African equids doubt that *E. numidicus* was ancestral to the later forms, and many have suspected a relationship with Grevy's zebra (*E. grevyi;* Cooke 1950a). The main problem appears to have been that *E. numidicus* and the *E. capensis* complex was placed with the horses in the subgenus *Equus* while *E. grevyi* was placed in its own separate subgenus *Dolichohippus* and is now related to some extinct North American horses originally placed in the genus *Plesippus*. The occurrence of the supposedly diagnostic characteristic of the subgenus *Equus,* the pli caballine on the upper cheek teeth, is variable in zebras of the subgenus *Hippotigris,* within the subgenus *Dolichohippus,* and even in true caballine horses of the subgenus *Equus*.

The diagnostic striped coat of the zebra is not always fully developed in zebras (e.g., *E. quagga*) and may occur in other subgenera (e.g., *Asinus* as in *E. (A.) asinus africanus*) including *Equus,* where stripes on the shoulders or hind quarters are occasionally reported and even common in some isolated populations (e.g., Sable Island horses, Welsh 1973; Arabian horses from the Caspian coast of Iran, Forbes 1970; Firouz 1971). There is therefore no strong argument for not relating the larger African fossil equids and Grevy's zebra as a single lineage more closely related to the origin of the true horses than to the living zebra subgenus *Hippotigris*.

Grevy's zebra, *E. grevyi,* has no wide or shadow stripes over its hindquarters and is striped equally finely over most of its body. It does have a heavier head and body and is more caballine in its body conformation, characters that indicate a relationship closer to the horses than to the zebras. Its pattern would then represent a parallel development of a disruptive "zebra" camouflage pattern from a stock that might have been partially striped over the

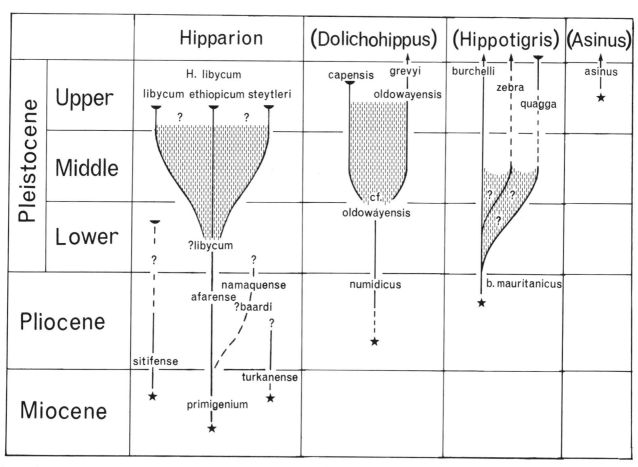

Figure 20.8   Relationships of African Equidae. The genus *Equus* is shown by the subgenera *Dolichohippus, Hippotigris,* and *Asinus. Hipparion libycum* is subdivided into the northern *libycum,* the eastern *ethiopicum,* and the southern *steytleri* populations; these groups are classified by some authors within the subgenus *Stylohipparion.* Living lineages terminate in arrows, extinct lineages in caps. Doubtful relationships between lineages or species are indicated by hatched zones. Question marks indicate lack of specimens, specimens doubtfully identified, or species of doubtful status. Asterisks denote first appearances of lineages in the African fossil rocord.

hindquarters but was self-colored over the head and forequarters. This stock could be a late Pliocene North African population of *E. numidicus* or even an earlier North American population of "*Plesippus.*"

The zebras of the subgenus *Hippotigris* are well known from the *Equus "mauritanicus"* populations of the Maghreb and middle and later Pleistocene records of Burchell's zebra, *E. burchelli,* from eastern and southern Africa. The North African *Equus burchelli mauritanicus* represents the original population descended from the original Asian immigrants and the ancestors of the later and modern *E. burchelli* subspecies of sub-Saharan Africa. The origins of the zebra, *E. zebra,* and the quagga, *E. quagga,* are obscure. Little fossil evidence is available that is reliably ascribed to either of these species and it comprises mainly isolated teeth, which do not constitute reliable material for taxonomic diagnosis. The pattern of stripes suggests that *E. quagga,* with stripes only on the head and variably over the neck, forequarters, and withers (Rau 1974), is less specialized than *E. zebra,* in which the stripes cover the whole body. However, in *E. zebra* the stripes over the rump and hindquarters are broader and more widely separated than over the forequarters, as in *E. burchelli,* but there are also shadow stripes intercalated between the main stripes on the hindquarters of the latter species. This suggests a cline of variation from *E. quagga* through *E. zebra* to *E. burchelli,* of which *E. quagga*'s pattern is nearer that of a horse and may represent the more or less ancestral condition in *Equus burchelli mauritanicus.*

The asses, subgenus *Asinus,* are known only from the late Pleistocene and Holocene from northern Africa and the Red Sea and Somali coastal zones. Asses appear to have replaced Burchell's zebra during the latest Pleistocene in North Africa, between middle Paleolithic and latest late Paleolithic times in Libya (e.g., Bate 1955; Higgs 1967). Romer (1928) showed that *E. burchelli* lasted into the early Neolithic in the Maghreb and was replaced by *E. (Asinus) asinus africanus* in later Neolithic times. There is no evidence for such a succession along the Red Sea or Somali littoral and the presence of asses on the island of Socotra east of the Horn of Africa attests to man's aid in extending their distribution along the coast (Churcher 1974).

All the known African Equidae originated from immigrants from Asia, probably over the region of the Red Sea and Arabia. It is possible that some hipparions entered Iberia from North Africa or that there was transmigration of populations across the Mediterranean (Azzaroli 1975) or that the entry of horses, zebras, or asses could have been from Arabia Felix over the Afar Triangle and Danakil Depression but there is little evidence to suggest any of these routes in preference to the obvious northeastern entry route.

V. J. Maglio provided the impetus and many useful suggestions for this chapter and D. A. Hooijer of the Rijksmuseum van Natuurlijke Historie, Leiden, made many constructive suggestions on the taxonomy and importance of the East African equids. R. van Gelder of the American Museum of Natural History, New York, and J. M. Harris of the National Museums of Kenya, Nairobi, allowed Churcher to examine specimens of *Equus grevyi* and *E. burchelli* in their care. Mary Jackes of the Department of Anthropology, University of Toronto, took a series of comparative measurements on the Nairobi *E. grevyi* skulls. Ann Forstén of the Department of Zoology, University of Helsinki, Q. B. Hendey of the Department of Vertebrate Palaeontology, South African Museum, Cape Town, and R. Singer of the Department of Anatomy, University of Chicago, read earlier versions of this chapter and offered critical comments in the light of their experiences with the African Equidae. W. J. White and T. S. Woolf, Churcher's assistants during the period of original preparation for the chapter, reviewed the known localities and the relevant references for occurrences of Equidae in Africa. H. B. S. Cooke of the Department of Geology, Dalhousie University, Halifax, Nova Scotia, gave us extensive advice and, as co-editor, made suggestions and gave us time to rework the whole chapter. We record our sincere appreciation to all these individuals for their varied assistance.

References

Aguirre, E., and P. Leakey. 1974. Nakali: nueva fauna de *Hipparion* del Rift Valley de Kenya. *Estud. Geol.* 30:219–227.

Alberdi, M. T. 1974. *El genero* Hipparion *en España; nuevas formas de Castilla y Andalucia, revisión e historia evolutiva. Trabajos sobre Neogeno-Cuaternario 1.* Madrid: Sección de Paleontologia de Vertebrados y Humana, Instituto Lucas Mallado, Consejo Superior de Investigaciónes Científicas, pp. 1–146.

Ansell, W. F. H. 1967. Order Perissodactyla, Part 14. In J. Meester and H. W. Setzer (eds.), *The Mammals of Africa; an identification manual.* Washington, D.C.: Smithsonian Institution Press, pp. 1–14.

Arambourg, C. 1931. Observations sur une grotte à ossements des environs d'Alger. *Bull. Soc. Hist. Nat. Afr. Nord* 22:169–176.

———. 1932. Note préliminaire sur une nouvelle grotte à ossements des environs d'Alger. *Bull. Soc. Hist. Nat. Afr. Nord* 23:154–162.

———. 1935. La grotte de la Carrière Anglade à Guyotville (Dt. d'Alger). *Bull. Soc. Hist. Nat. Afr. Nord* 26:15–22.

———. 1938. Mammifères fossiles du Maroc. *Mém. Soc. Sci. Nat. Maroc* 46:1–74.

———. 1947. Contribution à l'étude géologique et paléontologique de bassin du Lac Rudolphe et de la basse vallée de l'Omo. Deuxième partie, Paléontologie. *Miss. Scient. Omo, 1932–1933,* I. Géol. Anthrop.: 232–562.

———. 1948. Les mammifères pléistocènes d'Afrique. *Bull. Soc. Géol. Fr.,* Sér. 5, 17:301–310.

———. 1949. Les gisements de vertébrés villafranchiens de l'Afrique du Nord. *Bull. Soc. Géol. Fr.,* Sér. 5, 19:195–203.

———. 1951. Observations sur les couches à *Hipparion* de la vallée de l'Oued el Hammam (Algérie) et sur l'époque d'apparition de la faune de vertébrés dite "pontienne." *C. R. hebd. Acad. Sci.* 252:2464–2466.

———. 1952. La paléontologie de vertébrés en Afrique du Nord Française. *19e Congr. Géol. Internat.,* Monog. Région., s.i. Algérie; hors série: 1–63.

———. 1957 (1956). Sur des restes d'*Hipparion sitifense* Pomel, des calcaires lacustres de Mascara (Oran). *Bull. Soc. Géol. Fr.,* Sér. 6, 6:817–827.

———. 1959. Vertébrés continentaux du Miocène supérieur de l'Afrique du Nord. *Mém. Publ. Serv. Carte Géol. Algérie* (Alger), (n.s.), *Paléont.* no. 4:1–161.

———. 1963. Continental vertebrate faunas of the tertiary of North Africa. In F. C. Howell and F. Bourlière (eds.), *African ecology and human evolution.* Chicago: Aldine Publishing Company, pp. 55–60.

———. 1970. Les vertébrés du Pléistocène de l'Afrique du Nord. *Arch. Mus. Hist. Nat. Paris* 10:1–126.

Arambourg, C., and G. Choubert. 1965. Les faunes de mammifères de l'étage moghrebien du Maroc Occidental. *Notes Mém. Serv. Géol. Maroc* 185:29–33.

Azzaroli, A. 1975. Late Miocene interchange of terrestrial faunas across the Mediterranean. Trabajos sobre Neogeno-Cuaternario 4. *Actas Colloquiales Internaciónales sobre Biostratigraphia Continentale, Neogeno Superior y Cuaternario Inferior.* Montpelier 25-IX-Madrid 11-X, 1974, M. T. Alberdi and E. Aguirre (eds.): 67–72.

Bate, D. M. A. 1951. The mammals from Singa and Abu Hugar, pp. 1–28. In D. M. A. Bate, The Pleistocene fauna of two Blue Nile sites. *Brit. Mus. Nat. Hist.; Fossil Mammals of Africa,* no. 2:1–50.

———. 1955. Vertebrate faunas of Quaternary deposits in Cyrenaica, App. A. In C. B. M. McBurney and R. W. Hey (eds.), *Prehistory and Pleistocene geology in Cyrenaican Libya.* Cambridge: Cambridge University Press, pp. 274–291.

Bishop, W. W., G. R. Chapman, A. Hill, and J. A. Miller. 1971. Succession of Cainozoic vertebrate assemblages from the northern Kenya Rift Valley. *Nature* 233:389–394.

Blanc, G. A. 1956. Sulla esistenza di *Equus (Asinus) hydruntinus* Regàlia nel Pleistocene del Nord Africa. *Boll. Soc. Geol. Ital.* 75:176–187.

Boné, E. L., and R. Singer. 1965. *Hipparion* from Langebaanweg, Cape Province, and a revision of the genus in Africa. *Ann. S. Afr. Mus.* 48(16):273–397.

Boule, M. 1899 (1900). Observations sur quelques équidés fossiles. *Bull. Soc. Géol. Fr.* 27(3):531–542.

_____. 1900. Étude paléontologique et archéologique sur la station paléolithique du Lac Karâr (Algérie). *Anthropologie* 11:1–21.

Bourg de Bozas, see du Bourg de Bozas, Vicomte R. 1903.

Broom, R. 1909. On evidence of a large horse recently extinct in South Africa. *Ann. S. Afr. Mus.* 7:281–282.

_____. 1913. Note on *Equus capensis* Broom. *Bull. Am. Mus. Nat. Hist.* 32:437–439.

_____. 1928. On some new mammals from the diamond gravels of the Kimberley District. *Ann. S. Afr. Mus.* 22:439–444.

_____. 1948. Some South African Pliocene and Pleistocene mammals. *Ann. Transv. Mus.* 21(1):1–38.

Broom, R., and H. le Riche. 1937. On the dentition of *Equus capensis* Broom. *S. Afr. J. Sci.* 33:769–770.

Churcher, C. S. 1970. The fossil Equidae from the Krugersdorp Caves. *Ann. Transv. Mus.* 26 (6):145–168.

_____. 1972. Late Pleistocene vertebrates from archaeological sites in the Plain of Kom Ombo, Upper Egypt. *Life Sci. Contrib., Roy. Ont. Mus.,* No. 82:1–172.

_____. 1974. Relationships of the late Pleistocene vertebrate fauna from Kom Ombo, Upper Egypt. In Rushdi Said and B. H. Slaughter (eds.), Contributions to the paleontology of Africa. Proc. 75th. Anniv. Geol. Surv. Egypt, *Ann. Geol. Surv. Egypt* 4:363–384.

Churcher, C. S., and P. E. L. Smith. 1972. Kom Ombo: preliminary report on the fauna of the late Paleolithic sites in Upper Egypt. *Science* 177:259–261.

Cooke, H. B. S. 1939. On a collection of fossil mammalian remains from the Vaal River gravels at Pniel. *S. Afr. J. Sci.* 36:412–416.

_____. 1941. A preliminary account of the Wonderwerk Cave, Kuruman District. Section II. The fossil remains. *S. Afr. J. Sci.* 37:303–312.

_____. 1943. Cranial and dental characters of the recent South African Equidae. *S. Afr. J. Sci.* 49:243–262.

_____. 1948. The Fowler collection of fossils from Koffiefontein, O.F.S. *S. Afr. J. Sci.* 2 (4):96–98.

_____. 1949. Fossil mammals of the Vaal River deposits. *Mem. Geol. Surv. S. Afr.* 35 (3):1–117.

_____. 1950a. A critical revision of Quaternary Perissodactyla of southern Africa. *Ann. S. Afr. Mus.* 32 (4):393–479.

_____. 1950b. Quaternary fossils from Northern Rhodesia, Appendix B. In J. D. Clark (ed.), *The stone age cultures of Northern Rhodesia.* Claremont, Cape Province: South African Archaeological Society, pp. 137–142.

_____. 1955. Some fossil mammals in the South African Museum collections. *Ann. S. Afr. Mus.* 42 (3):161–168.

_____. 1963. Pleistocene mammal faunas of Africa with particular reference to southern Africa. In F. C. Howell and F. Bourlière (eds.), *African ecology and human evolution.* Chicago: Aldine Publishing Company, pp. 65–116.

_____. 1974. The fossil mammals of Cornelia, O. F. S., South Africa. In K. W. Butzer, J. D. Clark, and H. B. S. Cooke, The geology, archaeology and fossil mammals of the Cornelia Beds, O. F. S. *Mem. Nas. Mus., Bloemfontien* 9:1–88.

Cooke, H. B. S., and S. C. Coryndon. 1970. Pleistocene mammals from the Kaiso Formation and other related deposits in Uganda. In L. S. B. Leakey and R. J. G. Savage (eds.), *Fossil Vertebrates of Africa.* London: Academic Press, vol. 2, pp. 107–224.

Cooke, H. B. S., and L. H. Wells. 1951. Fossil remains from Chelmer, near Bulawayo, Southern Rhodesia. *S. Afr. J. Sci.* 47:205–209.

Coppens, Y. 1971. Les vertébrés villafranchiens de Tunisie: gisements nouveaux, signification. *C. R. hebd. Acad. Sci.,* Sér. D, 273:51–54.

Coque, R. 1957. Découverte d'un gisement de mammifères Villafranchiens dans le Sud-Tunisien. *C. R. hebd. Acad. Sci.,* Sér. D, 245:1069–1071.

Curtis, G. M., T. Drake, T. Cerling, and J. H. Hampel. 1975. Age of KBS Tuff in Koobi Fora Formation, East Rudolf, Kenya. *Nature* 258:395–398.

Dalloni, M. 1915. Le Miocène supérieur dans l'ouest de l'Algérie; couches à *Hipparion* de la Tafna. *C. R. hebd. Acad. Sci.,* Sér D, 161:639–641.

_____. 1940. Notes sur la classification du Pliocène supérieur et du Quaternaire. *Bull. Soc. Géog. Archéol. Oran.* 61:8–43.

Dart, R. A., and J. W. Kitching. 1958. Bone tools at the Kalkbank middle stone age site and the Makapansgat australopithecine locality, Central Transvaal. Pt. II, The osteodontokeratic contribution. *S. Afr. Archaeol. Bull.* 13:94–119.

de Winton, W. E., H. O. Forbes, and W. R. Ogilvie-Grant. 1903. Mammals. In H. O. Forbes (ed.), *The natural history of Sokotra and the Abd-el-Kuri.* Liverpool: Young and Sons, pp. 5–116.

Dietrich, W. O. 1942. Ältestquartäre Säugetiere aus der südlichen Serengeti, Deutsch-Ostafrika. *Palaeontogr.* 94(Abt. A):43–133.

Doumergue, F. 1926. Grotte du Cuartel (Oran). *Bull. Soc. Géog. Archéol. Oran* 46:185–204.

_____. 1927. La grotte du Polygone (Oran). *Bull. Soc. Géog. Archéol. Oran* 48:205–254.

Dreyer, T. F. 1931. Equidae. In T. F. Dreyer and A. Lyle, *New fossil mammals and man from South Africa.* Bloemfontein: Department of Zoology, Grey University College, pp. 1–60.

du Bourg de Bozas, Vicomte R. 1903. D'Addis Abbaba au

Nile par le Lac Rudolphe. *La Géographie, Bull. Soc. Géog. Oran* 7:91–112.

Eisenmann, V. 1976. Nouveaux crânes d'hipparions (Mammalia, Perissodactyla) Plio-Pléistocènes d'Afrique Orientale (Ethiopie et Kenya): *Hipparion* sp., *Hipparion* cf *ethiopicum*, et *Hipparion afarense* nov. sp. *Geobios* 9(5):577–605.

Eisenmann, V., and C. de Giuli. 1974. Caractères distinctifs entre vrais zèbres (*Equus zebra*) et zèbres de Chapman (*Equus burchelli antiquorum*) d'après l'étude de 60 têtes osseuses. *Mammalia* 38:509–543.

Ennouchi, E. 1949a. Le gisement de vertébrés pléistocènes d'Aïn Rohr. *C. R. Soc. Géol. Fr.* 1949:111–112.

———. 1949b. Une deuxième faune pléistocène à Aïn Rohr (Maroc). *C. R. Soc. Géol. Fr.* 1949:237–238.

———. 1951. Découverte d'un *Hipparion* dans les environs de Rabat. *Notes Mém. Serv. Géol. Maroc* 5(85):139–144.

Ennouchi, E., and A. Jeannette. 1954. L'*Hipparion* de Camp-Bertaux, près de Taourirt (Maroc oriental). *Notes Mém. Serv. Géol. Maroc* 10(122):65–75.

Estaunie, M. D. 1941. Nouvelles stations préhistorique du départment d'Oran. *Bull. Soc. Géog. Archéol. Oran* 62:177–184.

Firouz, L. 1971. Osteological and historical implications of the Caspian Sea Pony to early domestication in Iran. *3rd Internat. Agric. Mus., Budapest,* pp. 309–315.

Forbes, J. F. 1970. *Hoofbeats along the Tigris.* London: Allen and Company, pp. 1–129.

Forstén, A. M. 1968. Revision of the Palearctic *Hipparion. Acta Zool. Fenn.* 119:1–134.

———. 1972. *Hipparion primigenium* from southern Tunisia. *Notes Serv. Géol. Tunis* 5(1971:1):7–28.

———. 1973. New systematics and classification of Old World *Hipparion. Zeits. Säugetierk.* 38(5):289–294.

Gromova, V. I. 1952. Le genre *Hipparion.* Inst. Paleont. Acad. Sci. U. S. S. R. 36. Translated by St. Aubin Bur. Recherch. Min. Géol., *Ann. C. E. D. P.* 12:1–288.

———. 1962. Family Equidae Gray 1821. Subfamily Equinae Steinmann et Döderlein 1890. In Yu. A. Orlov (chief ed.), *Osnovy paleontologii (Fundamentals of Paleontology)*; V. I. Gromova (ed.), Akad. Nauk. S.S.S.R., Moskva, pp. 300–303 (410–414). Translated by Trans. Israel Program Sci., Jerusalem, 1968.

Haughton, S. H. 1932. The fossil Equidae of South Africa. *Ann. S. Afr. Mus.* 28:407–427.

Hendey, Q. B. 1968. The Melkbos site: an upper Pleistocene fossil occurrence in the southwestern Cape Province. *Ann. S. Afr. Mus.* 52(4):89–119.

———. 1970. A review of the geology and palaeontology of the Plio-Pleistocene deposits at Langebaanweg, Cape Province. *Ann. S. Afr. Mus.* 56(2):75–117.

———. 1974a. The late Cenozoic Carnivora of the southwestern Cape Province. *Ann. S. Afr. Mus.* 63(1):1–369.

———. 1974b. Faunal dating of the late Cenozoic of southern Africa, with special reference to the Carnivora. *Quatern. Res.* 4(2):149–161.

———. 1976. The Pliocene fossil occurrences in "E" Quarry, Langebaanweg, South Africa. *Ann. S. Afr. Mus.* 69(9):215–247.

Higgs, E. S. 1967. Environment and chronology—the evidence from mammalian fauna. In C. B. M. McBurney, *The Haua Fteah (Cyrenaica) and the stone age of the southeast Mediterranean.* London: Cambridge University Press, pp. 16–44.

Hooijer, D. A. 1973. Stratigraphy, paleoecology and evolution in the Lake Rudolf Basin. Mimeographed paper presented at Wenner-Gren Foundation for Anthropological Research, Sept. 8–20, New York, 1973, pp. 1–5.

———. 1975a. The hipparions of the Baringo Basin sequence. *Nature* 254(5497):211–212.

———. 1975b. Miocene to Pleistocene hipparions of Kenya, Tanzania and Ethiopia. *Zoöl. Verhandel.,* No. 142:1–80.

———. 1976. The late Pliocene Equidae of Langebaanweg, Cape Province, South Africa. *Zoöl. Verhandel.,* No. 148:1–39.

Hooijer, D. A., and V. J. Maglio. 1973. The earliest *Hipparion* south of the Sahara, in the late Miocene of Kenya. *Proc. Kon. Ned. Akad. Wetensch.,* Ser. B., 76(4):311–315.

———. 1974. Hipparions from the late Miocene and Pliocene of northwestern Kenya. *Zoöl. Verhandel.,* No. 142:1–80.

Hopwood, A. T. 1926. Fossil Mammalia. In The geology and palaeontology of the Kaiso Bone Beds, Pt. II. Palaeontology. *Geol. Surv. Dept., Uganda Protect., Occ. Pap.* No. 2:13–36.

———. 1936. The former distribution of caballine and zebrine horses in Europe and Asia. *Proc. Zool. Soc. London* 1936:897–912.

———. 1937. Die fossilien pferde von Oldoway. *Wiss. Ergebn. Oldoway-Exped. 1913* (N.F.) 4:111–136.

Howe, B., and H. L. Movius. 1947. A stone age cave site in Tangier. *Pap. Peabody Mus. Amer. Archeol. Ethnol.* 28(1):21–23.

Joleaud, L. 1918 (1917/1916). Notice géologique et paléontologique sur la grotte des Pigeons (Constantine). *Rec. Not. Mém. Soc. Archéol. Constantine* 50(Sér. 5;7): 25–35.

———. 1933a. Un nouveau genre d'équidé quaternaire de l'Omo (Abyssinie): *Libyhipparion ethiopicum. C. R. Soc. Géol. Fr.* 1933:13–14.

———. 1933b. Un nouveau genre d'équidé quaternaire de l'Omo (Abyssinie): *Libyhipparion ethiopicum. Bull. Soc. Géol. Fr.* 3(5):7–28.

———. 1933c. Progrès récents de nos connaissances sur la géologie du quaternaire et sur la préhistorie de l'Égypte. *Rev. Gen. Sci. Pur. Appl. Paris* 44:601–616.

———. 1936. Gisements de vertébrés quaternaires du Sahara. *Bull. Soc. Hist. Nat. Afr. Nord* 26:23–29.

Kent, P. E. 1942. A note on Pleistocene deposits near Lake Manyara, Tanganyika. *Geol. Mag.* 79:72–77.

Klein, R. G. 1972. The late Quaternary mammalian fauna of Nelson Bay Cave (Cape Province, South Africa): its implications for megafaunal extinctions and environmental and cultural change. *Quatern. Res.* 2(2):135–142.

Leakey, L. S. B. 1958. Some East African Pleistocene Suidae. *Brit. Mus. Nat. Hist., Fossil Mammals of Africa,* No. 14:1–69.

———. 1959. A preliminary reassessment of the fossil fauna from Broken Hill, N. Rhodesia, Appendix. In J. D. Clark, Further excavations at Broken Hill, Northern Rhodesia. *J. Roy. Anthrop. Inst. Gr. Brit.* 89:225–231.

———. 1965. *Olduvai Gorge 1951–1961. Fauna and background.* Cambridge: Cambridge Univ. Press, pp. 1–118.

Leakey, M. D. 1971. *Olduvai Gorge 1971,* vol. 3, *Excavations in Beds I and II, 1960–1963.* Cambridge: Cambridge Univ. Press, pp. 1–299.

Lydekker, R. 1887. On a molar of a Pliocene type of *Equus* from Nubia. *Quart. J. Geol. Soc. London* 43:161–164.

Madden, C. T. 1972. Miocene mammals, stratigraphy and environment of Muruarot Hill, Kenya. *Paleobios* 14:1–12.

Maglio, V. J. 1971. Vertebrate faunas from the Kubi Algi, Koobi Fora and Ileret areas, East Rudolf, Kenya. *Nature* 231:248–249.

———. 1972. Vertebrate faunas and chronology of the hominid-bearing sediments east of Lake Rudolf, Kenya. *Nature* 239:379–385.

———. 1974. Late Tertiary fossil vertebrate successions in the northern Gregory Rift, East Africa. In R. Said and B. H. Slaughter (eds.), Contributions to the paleontology of Africa. Proc. 75th Anniv. Geol. Surv. Egypt. *Ann. Geol. Surv. Egypt* 4:269–286.

Mawby, J. E. 1970. Fossil vertebrates from northern Malawi: preliminary report. *Quaternaria* 13:319–323.

McKenzie, D. P., D. Davies, and P. Molnar. 1970. Plate tectonics of the Red Sea and East Africa. *Nature* 226:243–248.

Mendrez, C. 1966. On *Equus* (*Hippotigris*) *burchelli* from "Sterkfontein Extension," Transvaal, South Africa. *Ann. Transv. Mus.* 25(5):93–97.

Millar, J. C. G. 1970. Census of Cape Mountain zebras. *Afr. Wildlife* 24(1):17–25 and (2):105–114.

Owen, R. 1869. Description of the cavern of Bruniquell and its organic contents. Pt. II. Equine remains. *Phil. Trans. Roy. Soc. London* 159:535–557.

Pallary, P. 1910 (1907). Note sur un gisement paléolithique de la province d'Oran. *Anthropologie* 22:317.

———. 1934. L'abris Alain près d'Oran (Algérie). *Arch. Inst. Paléont. Hum.* 12:1–50.

Pallary, P., and P. Tommasini. 1892 (1891). La grotte des troglodytes (Oran). *C. R. Assoc. Franç. Avanc. Sci.* 20(2):633–649.

Perkins, D. 1965. Three faunal assemblages from Sudanese Nubia. *Kush* 13:56–61.

Petit, M. 1939. Anatomie des molaires des équidés. Thèse de Docteur és Sciences Naturelles, Université de Paris. Toulouse: Toulousaine Lion, pp. 1–328.

Petrocchi, C. 1952. Les canines chez *Hipparion* et l'apparition d'une caractère sexuel secondaire des mammifères. *Bull. Mus. Nat. Hist. Nat. Paris* 24(2):419–422.

Pirlot, P. L. 1956. Les formes européennes du genre *Hipparion. Mem. Comm. Inst. Geol. Barcelona* 14:1–129.

Pomel, N. A. 1888a. Excursion special de la 11ᵉ section. Visite faite à la station préhistorique de Ternifine (Palikao), par le groupe excursioniste D(I). *C. R. Assoc. Franç. Avanc. Sci.* 17(I):208–212.

———. 1888b. La station quaternaire de Palikao (départment d'Oran; sous-préfecture de Mascara). II. Note géologique et paléontologique. *Matér. Hist. Prim. Nat. Homme* 22:224–232.

———. 1897. Les équidés. *Carte Géol. Algérie, Paléont. Monog.,* No. 12:1–44.

Posnansky, M. 1959. A Hope Fountain site at Olorgesailie, Kenya Colony. *S. Afr. Archaeol. Bull.* 14(55):83–89.

Quinn, J. H. 1955. Miocene Equidae of the Texas Gulf coastal plain. *Bur. Econ. Geol., Univ. Texas Publ.,* No. 5516:1–102.

———. 1957. Pleistocene Equidae of Texas. *Bur. Econ. Geol., Univ. Texas,* Rept. Investig. No. 33:1–51.

Rau, R. E. 1974. Revised list of the preserved material of the extinct Cape Colony quagga, *Equus quagga quagga* (Gmelin). *Ann. S. Afr. Mus.* 65(2):41–87.

Robinson, J. T. 1958. The Sterkfontein tool-maker. *The Leech* 28:94–100.

Robinson, P., and C. C. Black. 1974. Vertebrate faunas from the Neogene of Tunisia. In R. Said and B. H. Slaughter (eds.), Contributions to the paleontology of Africa. Proc. 75th. Anniv. Geol. Surv. Egypt, *Ann. Geol. Surv. Egypt* 4:319–332.

Romer, A. S. 1928. Pleistocene mammals of Africa. Fauna of the Paleolithic station of Mechta-el-Arbi. *Bull. Logan Mus.* 1:80–163.

———. 1935. Mammalian remains from some Paleolithic stations in Algeria, part II. In A. W. Pond, L. Capius, A. S. Romer, and F. C. Baker, Prehistoric habitations in the Sahara and North Africa. *Bull. Logan Mus.* 5:165–184.

Shapiro, M. M. J. 1943. Fossil mammalian remains from Bankies, Kroonstad District, O.F.S. *S. Afr. J. Sci.* 39:176–181.

Simpson, G. G. 1945. The principles of classification and a classification of mammals. *Bull. Am. Mus. Nat. Hist.* 85:1–350.

Skinner, M. F. 1972. Order Perissodactyla. In M. F. Skinner and C. W. Hibbard, et al., Early Pleistocene preglacial and glacial rocks and faunas of North-Central Nebraska. *Bull. Am. Mus. Nat. Hist.* 148(1):117–130.

Solignac, M. 1927. Étude géologique de la Tunisie septentrionale. *Div. Gén. Trav. Publ., Serv. Mines, Carte Géol.:* 1–756.

Stirton, R. A. 1941. Development of characters in horse teeth and the dental nomenclature. *J. Mamm.* 22(4):434–446.

Thomas, P. 1884a. Recherches stratigraphiques et paléontologiques sur quelques formations d'eau douce de l'Algérie. *Mém. Soc. Géol. France* 3:1–51.

———. 1884b. Sur quelques formations d'eau douce tertiaires d'Algérie. *C. R. hebd. Acad. Sci.* 1884:311–314.

———. 1884c. Sur quelques formations d'eau douce quaternaires d'Algérie. *C. R. hebd. Acad. Sci.* 1884:381–383.

Van Couvering, J. A. 1972. Radiometric calibration of the European Neogene. In W. W. Bishop and J. A. Miller (eds.), *Calibration of hominoid evolution.* Edinburgh: Scottish Academic Press, pp. 247–271.

van Hoepen, E. C. N. 1930a. Vrystaatse fossiele perde. *Paleont. Navors. Nas. Mus. Bloemfontein* 2:1–11.

_____. 1930b. Fossiele perde van Cornelia, O. V. S. *Paleont. Navors. Nas. Mus. Bloemfontein* 2:13–24.

_____. 1932. Die stamlyn van die sebras. *Paleont. Navors. Nas. Mus. Bloemfontein* 2:25–37.

Wells, L. H. 1941. A fossil horse from Koffiefontein, O.F.S. *Trans. Roy. Soc. S. Afr.* 28:301–306.

_____. 1959. The nomenclature of South African fossil equids. *S. Afr. J. Sci.* 55:64–66.

_____. 1964. The Vaal River "Younger Gravels" faunal assemblage: a revised list. *S. Afr. J. Sci.* 60:91.

_____. 1970a. A late Pleistocene faunal assemblage from Driefontein, Cradock District, C. P. *S. Afr. J. Sci.* 66:59–61.

_____. 1970b. The fauna of the Aloes Bone Deposit: a preliminary note. *S. Afr. Archaeol. Bull.* 25:22–23.

Welsh, D. A. 1973. The life of Sable Island's wild horses. *Nature* (Canada) 2(2):7–14.

Willoughby, D. P. 1966. The vanished quagga. *Nat. Hist.* 75:60–63.

Zeuner, F. E. 1963. *A history of domesticated animals.* London: Hutchinson and Company, pp. 1–560.

21

Anthracotheriidae

Craig C. Black

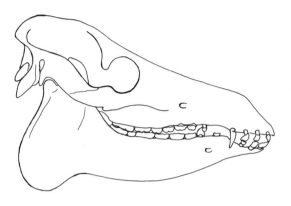

Anthracotheres are a group of Tertiary suiform ungulates that achieved a Holarctic and African distribution during the Oligocene and Miocene. The family became extinct in the Pleistocene and showed a general decline in diversity as hippopotamids evolved. Members of the Anthracotheriidae ranged from small, terrier-sized animals to beasts approaching the hippopotamus in bulk and at least some types may have had a hippopotamid mode of existence.

An extensive record of the family occurs in Africa spanning the time from the early Oligocene to the late Pliocene or early Pleistocene, a period of some 35 m.y. The group first appeared in Europe in the middle Eocene but became extinct there in early Miocene times. The anthracothere record in Asia is long but broken, with a diversity of forms first appearing in the late Eocene Pondaung fauna of Burma. No Oligocene representatives are yet known but there is a good Miocene and Pliocene record from Dera Bugti in Baluchistan through the Indian Siwalik Series. In North America the anthracothere record is quite restricted; they first appear in the early Oligocene and disappear during the early Miocene.

This family has never been comprehensively reviewed. Hence, the interrelationships of the many described taxa are only poorly understood. There is undoubtedly an overabundance of generic names, particularly as regards the Asian forms from Dera Bugti, Baluchistan, described by Foster-Cooper (1924). In the absence of a general study of the family, Simpson (1945) refrained from using subfamilial or tribal designations in his classifications of the group.

As this review is the first to consider all of the African members of the family, certain taxonomic changes have been necessary; however, no new names are proposed. Much of the material considered while preparing this review is undescribed and some of it represents new species, although there do not appear to be any new genera of African anthracotheres presently awaiting recognition. Therefore, the broader outlines of anthracothere migration to, and evolution in, Africa seem reasonably clear.

General Characteristics

Anthracotheres retain a relatively primitive body structure with few of the more advanced artiodactyl specializations. Elements of the limbs are unmodified, retaining in many cases five digits on the front foot and four behind. The metacarpals and metatarsals are short as in many suiforms and there is no fusion of these elements. The astragalus in at least some anthracotheres, i.e., *Merycopotamus,* resembles that of the modern hippopotamus to a remarkable degree. The limb bones are heavy and relatively short. While anthracotheres do not show any cursorial specializations, they were undoubtedly capable of periods of rapid locomotion, just as hippos are today.

There is a general tendency to elongation of the snout in many members of the family. As this occurs rather long diastemata develop between the incisors and canines, between the canine and the first premolar, and occasionally between the anterior premolars. Some anthracotheres display a considerable degree of sexual dimorphism quite similar to that seen in hippos (Laws 1968), with large recurved canines in the males and much smaller straighter canines in the females. In some later species there is a reduction in incisor number and the development of an elongate heavy pair of lower incisors much as in the modern hippopotamus.

The dentition of older animals attests to a tremendous amount of wear and this suggests a diet of extremely abrasive vegetation. Either the plant material itself contained large amounts of silica or it was covered with large quantities of fine sand particles adhering to its outer surface. It is of interest that individuals of *Hippopotamus amphibius* show the same pattern of tooth wear (Laws 1968) as seen in *Merycopotamus* and other anthracotheres.

Occurrence

The distribution of species of anthracotheres through time in Africa is presented in table 21.1. The earliest record is found in the Qasr el Sagha Formation of the Fayum (Simons 1968), where one species of *Bothriogenys* has been reported (as *Brachyodus*). Several species of *Bothriogenys* are present in the slightly younger and overlying Jebel el Qatrani Formation. By Miocene time anthracotheres had spread into East and Central Africa, although they could well have been there earlier. The genera *Hyoboops* and *Masritherium,* the latter first described as *Brachyodus,* are both known from the early Miocene of Libya, Egypt, and Kenya with *Masritherium* also reported from the Congo (Hooijer 1963). Anthracotheres have not been reported from the Miocene of southwestern Africa.

All of the East and Central African occurrences are essentially contemporaneous and date at about 18 to 20 m.y. (van Couvering 1972, p. 261). Although a younger Miocene fauna is well known from Fort Ternan and several Pliocene faunas are now known (Maglio 1970), no anthracotheres have been reported in East Africa later than the Rusinga-Ombo-

Napak-Congo early Miocene. Those that are present during the early Miocene are species congeneric with those known from North Africa at about the same time.

In North Africa species of *Hyoboops* and *Masritherium* are known from Djebel Zelten in Libya and from Wadi Moghara in Egypt, two faunas that are probably approximately contemporaneous. These faunas may also be temporal equivalents of those in East Africa, although this is less certain. One species of *Hyoboops, H. moneyi,* may be common to all of these early Miocene faunas. *Masritherium depereti* present at Wadi Moghara may also be present in the Djebel Zelten fauna but the latter specimens are too fragmentary to be certain of the species assignment.

Later in the Miocene anthracotheres are known only from localities in Tunisia, which are some 10 to 12 m.y. old (Robinson and Black 1974). Only one species is known but it is present at several localities, being best represented by large quarry samples recovered near Gafsa (Black 1972). Although faunas of similar age are well known in Algeria (Arambourg 1959) and Morocco (Lavocat 1961), no anthracotheres have been reported as members of these assemblages.

The Tunisian anthracothere, *Merycopotamus anisae,* was quite probably ancestral to *Merycopotamus petrocchii,* the latest African anthracothere presently recognized. *M. petrocchii* is known from Sahabi, Libya, together with *Stegotetrabelodon syrticus* (Maglio 1970) and is considered to be of Pliocene age between 5 and 6 m.y. old. Coppens (1972) has reported an anthracothere from Chad. This specimen comes from Agrange and is believed to be of similar age to the Sahabi occurrence and may represent a population of *M. petrocchii.*

Systematic Review

Class: Mammalia
  Order: Artiodactyla
    Suborder: Suiformes

Family Anthracotheriidae Gill 1872

Genus *Bothriogenys* Schmidt 1913

TYPE SPECIES. *Brachyodus* (*Bothriogenys*) *fraasi* Schmidt 1913, p. 158.

GENERIC DIAGNOSIS. Small to medium-sized anthracotheres; skull and jaws moderately elongate; diastema between P–C present, generally short, between C–I_3 absent; no descending flange of the angle; no reduction in dental formula; no incisor enlargement; premolars simple, elongate; molars bunodont to bunoselenodont; enamel somewhat rugose; proto-

Table 21.1 Representative observed ranges of measurements (in mm) of molars of African anthracotheres.

| | Length | Width | | |
		Anterior loph (id)	Midline	Posterior loph (id)
Bothriogenys from Schmidt 1913				
B. fraasi				
M^1 (3)	22.0–24.5	21.9–22.5	22.3–23.0	15.4–17.5
M^2 (2)	27.6–29.3	28.6–29.0	27.8–28.0	19.0–20.8
M^3 (2)	30.7–31.0	33.0–34.1	30.0–32.2	24.8–26.5
B. gorringei				
M^1 (2)	16.0–19.3	18.9–19.0	19.4–21.9	—
M^2 (2)	22.0	22.0–24.8	22.4–25.1	—
M^3 (2)	25.0–26.4	25.0–28.1	26.0–22.0	—
B. rugulosus				
M^1 (1)	17.0	16.7	16.0	14.1
M^2 (1)	19.1	21.3	19.3	14.6
M^3 (1)	22.0	22.7	20.3	18.5
B. andrewsi				
M^2 (1)	30.0	31.4	30.3	19.0
B. parvus				
M^1 (1)	18.1	17.5	15.2	10.5
M^2 (1)	15.8	16.0	15.0	10.8
M^3 (1)	18.5	19.6	18.6	12.3
B. fraasi				
M_1 (4)	18.9–19.6	(3) 10.0–11.7	11.5–12.5	—
M_2 (4)	20.0–24.5	(3) 13.6–15.4	13.8–15.1	—
M_3 (2)	37.0–38.0	(3) 15.3–15.9	15.4–15.8	12.3–13.7
B. gorringei				
M_1 (4)	15.7–18.9	(3) 9.2–11.9	—	—
M_2 (4)	18.4–22.5	(3) 11.6–15.5	—	—
M_3 (4)	29.1–39.4	(3) 13.9–18.4	—	—
B. rugulosus				
M_1 (3)	11.0–15.0	(1) 8.7	9.6	—
M_2 (3)	16.0–17.8	9.8–11.2	10.5–11.6	—
M_3 (3)	25.8–26.8	11.8–12.9	7.9–8.6	—
B. andrewsi				
M_3 (1)	44.0	18.8	18.7	11.9
B. parvus				
M_1 (3)	13.2–17.0	6.5–9.8	7.4–11.0	—
M_2 (1)	15.1	8.0	8.8	—
M_3 (1)	27.4	12.1	11.6	8.2

Table 21.1 (*continued*)

	Length	Width		
		Anterior loph (id)	Midline	Posterior loph (id)
Bothriogenys africanus from Fourtau 1920				
M_1-M_3 (2)	112.0–114.0			
M_1 (3)	25.0–30.0	17.0–21.0	—	—
M_2 (7)	29.0–41.0	20.0–28.0	—	—
M_3 (6)	42.0–56.0	22.0–28.0	20.0–27.0	13.0–18.0
Masritherium depereti from Fourtau 1920				
M_1-M_3 (1)	127			
M_1 (4)	30.0–33.0	24.0–44.0	—	—
M_2 (6)	33.0–42.0	25.0–46.0	—	—
M_3 (5)	53.0–64.0	(4) 29.0–35.0	20.0–31.0	14.0–22.0
M^3 (1)	53.3	55.6	57.0	45.7
M. aequitorialis				
M^1	30.0	29.5–30.0	—	—
M^2 type skull	30.7	—	—	—
M^3	34.0	—	34.5	—
M_1 (3)	27.3–32.2	18.9–22.0	18.2–22.3	—
M_2 (3)	31.0–37.4	21.4–25.7	21.4–25.2	—
M_3 (2)	44.2–48.7	23.8–26.3	22.1–24.1	14.0–14.3
Hyoboops moneyi from Fourtau 1920				
M_1-M_3 (2)	59.0			
M_1 (3)	13.0–15.0	9.0–10.0	—	—
M_2 (3)	18.0–20.0	11.0–12.5	—	—
M_3 (2)	28.0	13.0–14.0	12.0–13.0	7.5–10.0
H. africanus				
M_3 (3)	32.1–33.5	15.0–15.8	—	—
M^3 (2)	22.5–24.1	21.4–24.8	—	—
Merycopotamus anisae from Black 1972				
M^1 (18)	21.4–27.3	(20) 23.0–29.2	24.1–30.6	—
M^2 (22)	28.4–35.5	(26) 31.3–39.9	30.3–38.9	—
M^3 (27)	30.5–39.1	(16) 34.6–43.3	(27) 32.5–43.7	(27) 29.0–38.0
M_1 (8)	16.6–26.5	(7) 15.8–18.0	(8) 16.0–18.8	—
M_2 (25)	26.3–36.4	(25) 19.0–24.1	(24) 18.8–24.7	—
M_3 (38)	30.0–50.5	(36) 20.0–27.0	(38) 18.5–26.9	(38) 15.1–21.8
Merycopotamus petrocchii referred specimen				
M^1 (1)	31.1	33.3	35.4	—
M^2 (1)	39.5	42.0	42.0	—
M^3 (1)	44.4	48.8	44.9	—

Note: Unless otherwise noted the measurements were made by the author. Numbers in parentheses refer to the sample size.

conule well developed; parastyle and mesostyle flattened; lower molars with short anterior protoconid and hypoconid arms (figures 21.1, 21.2, 21.3).

INCLUDED SPECIES. *Bothriogenys fraasi, B. gorringei, B. rugulosus, B. andrewsi,* and *B. parvus,* all from the Jebel el Qatrani Formation of the Fayum, and *B. africanus,* from the Moghara.

RANGE. Early Oligocene to early Miocene of North Africa.

Bothriogenys fraasi Schmidt 1913, p. 158

TYPE. BM M-10186, skull, figured by Schmidt (1913, pl. II, figs. 1–3).

Bothriogenys andrewsi Schmidt 1913, p. 168

TYPE. Stuttgart S. 16(168) LM², figured by Schmidt (1913, pl. I, fig. 12). Type not seen.

Bothriogenys parvus (Andrews) 1906, p. 189

TYPE. CGM-8821, right mandible with M_1-M_2 figured by Andrews (1906, fig. 62). Type not seen.

Bothriogenys gorringei (Andrews and Beadnell) 1902, p. 7

TYPE. CGM-8629, right and left mandibles figured by Andrews (1906, pl. XVII, figs. 1 and 2). Type not seen.

Bothriogenys rugulosus Schmidt 1913, p. 165

TYPE. Stuttgart S. 14(166), LM²–M³, figured by Schmidt (1913, pl. III, figs. 11–12). Type not seen.

Bothriogenys africanus (Andrews) 1899, p. 481

TYPE. CGM-2849, right mandible with P_4-M_3 (BM M-7343 is a cast of the type). Type not seen.

DISCUSSION. It is presently impossible for me to give adequate specific diagnoses for the species here referred to *Bothriogenys* as I have not examined any of the type specimens with the exception of the skull of *B. fraasi.* In addition there is a wealth of new material from the Fayum in the Yale University collection and this must be considered in any revision of the Fayum anthracotheres.

On the basis of published figures and descriptions there appear to be at least three valid species of *Bothriogenys* within the Fayum assemblages. However, I have refrained from offering any synonymies until a complete revision can be undertaken. *Bothriogenys africanus* evidently represents a Miocene carryover from the Oligocene complex.

Genus *Rhagatherium* Pictet 1855

TYPE SPECIES. *Rhagatherium aegypticum* Andrews 1906, p. 192.

TYPE. BM M-8449, LM³.

RANGE. Early Oligocene, Jebel el Qatrani Formation, Fayum, Egypt.

DISCUSSION. This is the only African specimen referred to the European genus *Rhagatherium.* One upper molar is entirely insufficient for such a determination. I seriously doubt that this tooth is, in fact, outside the range of variation for one of the smaller *Bothriogenys* species but again I am refraining from such synonymy until all the Fayum anthracotheres have been restudied.

Genus *Masritherium* Fourtau 1918

TYPE SPECIES. *Masritherium depereti* Fourtau 1918, p. 65.

GENERIC DIAGNOSIS. Large-sized anthracotheres; skull and jaws elongate with long diastema between P–C, shorter between C–I; shallow flange of the angle; some reduction in incisor number; P_1 occasionally lost; premolars simple, elongate, not transversely expanded; no accessory premolar cusps, molars bunoselenodont; large, separate protoconule; parastyle and mesostyle somewhat flattened to bulbous, but generally distinct, not formed by wings from the paracone and metacone; lower molars with short anterior hypoconid arm not reaching lingual border; anterior protoconid arm not joining base of metaconid.

Figure 21.1 Occlusal view of LP³–M³ of *Bothriogenys* sp., YPM-30128. (Approx. ×1.)

Figure 21.2 Occlusal view of right mandible with P_2, P_4–M_3 of *Bothriogenys* sp., YPM-18097. (Approx. ×1/2.)

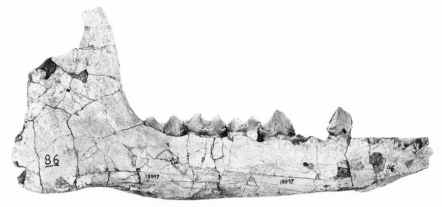

Figure 21.3 Lateral view of right mandible of *Bothriogenys* sp., YPM-18097. (Approx. ×1/2.)

INCLUDED SPECIES. *Masritherium depereti* from Moghara, *M. aequitorialis* from Rusinga, and *M.* sp. from Loperot.

RANGE. Early Miocene of Egypt, Libya, and Kenya.

Masritherium depereti (Fourtau) 1918

TYPE. Geological Museum Cairo, complete right and left mandibles (figured by Fourtau 1920, fig. 96; BM M-29714 is a cast of the type).

REVISED DIAGNOSIS. Largest African anthracothere; molars selenodont, premolars simple, massive; protoconule large, distinctly separate from protocone and paracone for half its crown height; I_2 enlarged, C possibly absent in females; long P_1–Canine diastema; mandible relatively shallow; small flange of angle (figures 21.4, 21.5).

Masritherium aequitorialis (MacInnes) 1951

SYNONYMY. *Brachyodus aequitorialis* MacInnes 1951, p. 3.

TYPE. KNM-RU 1009, skull.

REVISED DIAGNOSIS. Smaller than *Masritherium depereti;* skull elongate; long diastemata between P^1–C and C–I^3; snout tubular; presence of P_1 variable; long P_1–C diastema, short C–I_3; molars bunoselenodont; mandible shallow (figures 21.6, 21.7, 21.8).

DISCUSSION. This genus appears to have been confined to Africa and is known only from the early Miocene. It probably evolved from within the *Bothriogenys* species complex. A species of *Masritherium,* possibly new, is also present in the Loperot fauna.

Genus *Hyoboops* Trouessart 1904

TYPE SPECIES. *Hyoboops palaeindicus* (Lydekker) 1883, p. 158.

GENERIC DIAGNOSIS. Small to medium-sized anthracotheres; skull and jaws generally short; diastema between P–C short, between C–I_3 absent; sexually dimorphic; no flange of the angle; premolars developing accessory cusps and somewhat expanded transversely; molars selenodont; protoconule small to absent; parastyle formed by enlargement of anterior cingulum and anterior wing of paracone; mesostyle broadly convex, formed by crests from paracone and metacone; lower molars with long anterior hypoconid arm reaching lingual border; anterior protoconid arm long, fused into base of metaconid.

INCLUDED AFRICAN SPECIES. *Hyoboops africanus* from Rusinga, *H. moneyi* from Moghara, and *H.* sp. from Djebel Zelten, Libya.

Figure 21.4 Occlusal view of RM^3 of *Masritherium depereti,* CGM-7210. (Approx. ×1.)

Figure 21.5 Occlusal view of RP₂–M₃ of *Masritherium depereti*, CGM-7210. (Approx. ×1/3.)

Hyoboops moneyi (Fourtau) 1918

SYNONYMY. *Brachyodus moneyi* Fourtau 1918.

TYPE. Geological Museum Cairo, left mandible with P_3–M_3 (figured by Fourtau 1920, pl. 1, fig. 1; BM M-29379 is a cast of the type).

DIAGNOSIS. Smaller than *Hyoboops africanus;* C–P diastema short; molars selenodont; anterior hypoconid arm shorter than in *H. africanus,* occasionally fails to reach lingual border behind metaconid on M_1–M_3 (figure 21.9).

Hyoboops africanus (Andrews) 1914

SYNONYMY. *Merycops africanus* Andrews 1914, p. 172; *Hyoboops africanus* (Andrews), MacInnes 1951, p. 19.

TYPE. BM M-10613, a fragment of left mandible with M_3.

DIAGNOSIS. Larger than *H. moneyi;* anterior hypoconid arm reaches lingual border behind metaconid.

DISCUSSION. There are specimens from Rusinga and Ombo in the Kenya National Museum collections that are probably referable to *Hyoboops moneyi.* If this assignment is correct, these specimens demonstrate that there is no protoconule in the upper molars of this species and also that the mesostyle is quite bulbous. However, as no upper dentitions of this species are known in the hypodigm, this

reference of the East African material must remain tentative. In addition, there are other specimens from Libya in the Bristol University paleontological collections that may pertain to *H. moneyi* and these also have no protoconule.

A third species of *Hyoboops* is also present in the Libyan Miocene but has not yet been described. This species has a small protoconule in the upper molars and is close to the genotypic species *H. palaeindicus* in this regard. The mesostylar loop is broader in the Djebel Zelten species than in *H. moneyi* and *H. africanus* and the premolars are more complex in the undescribed species.

The African species are close to those from Dera Bugti. The direction of migration within this genus is uncertain.

Genus *Merycopotamus* Falconer and Cautley 1847

TYPE SPECIES. *Merycopotamus dissimilis* Falconer and Cautley 1847, pl. LXVII.

GENERIC DIAGNOSIS. Medium to large size; skull and jaws not elongate; diastema between C-I absent, between C–P short; sexually dimorphic; deep flange of the angle; premolars with serrate crests from protocones and protoconids; premolars expanded transversely; molars selenodont; no protoconule; mesostyle closed and bulbous, larger than parastyle; lower molars with long anterior hypoconid arm to lingual border (figure 21.10, 21.11).

INCLUDED AFRICAN SPECIES. *Merycopotamus anisae* from Tunisia and *M. petrocchii* from Libya.

RANGE. Late Miocene to Pliocene, Tunisia, Libya, Chad, India, and Pakistan.

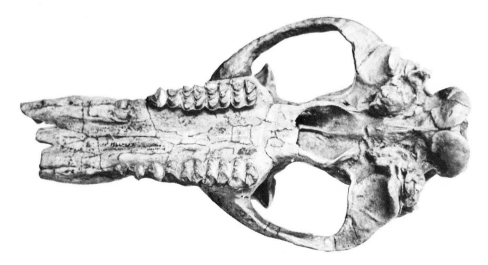

Figure 21.6 Palatal view of the skull of *Masritherium aequitorialis*, KNM-RU 1009. (Approx. ×1/4.)

Figure 21.7   Occlusal view of partial mandible of *Masritherium aequitorialis*, KNM-RU 1014. (Approx. ×1/3.)

Figure 21.8   Lateral view of right mandible of *Masritherium aequitorialis*, KNM-RU 1014. (Approx. ×1/3.)

Merycopotamus anisae **Black 1972**

SYNONYMY. *Merycopotamus anisae* Black 1972, p. 7.

TYPE. Tunisia Geological Survey T-3566, right mandible with P_1-M_2.

DIAGNOSIS. Black (1972, p. 7): "Medium-sized anthracothere larger than *Merycopotamus dissimilis* of the Siwaliks; upper and lower molar cusps selenodont; ribs on upper molars reduced; no protoconules; mesostyle of M^1-M^3 fused, transverse valley closed buccally; diastema C1/1–P1/1 short; upper and lower premolars with serrate crests from protocones and protoconids; lower premolars short; anterior hypoconid crests of M_1-M_3 reach lingual borders; great sexual dimorphism with variation in canine and incisor size; angle descends steeply below M_3."

Merycopotamus petrocchii **(Bonarelli) 1947**

SYNONYMY. *Libycosaurus petrocchii* Bonarelli 1947, p. 26.

TYPE. Museo Coloniale Roma, anterior portion of a skull with P^1-M^3 (cast of type in the British Museum of Natural History).

DIAGNOSIS. Larger than *M. anisae;* upper premolars more complexly serrate; mesostyle less prominent; central transverse valley more open buccally than in *M. anisae* (figure 21.12).

DISCUSSION. Only two African species of *Merycopotamus* are presently recognized and these are restricted to North Africa. They are clearly very closely related to the Asian member of the genus and most probably represent immigrants from Asia into North Africa.

Relationships

Little recent work has been done on the numerous anthracotheres reported from the Eocene of Europe and Southeast Asia. A variety of forms is known from England, France, and Burma. It is from within one of these groups that the Fayum species of *Bothriogenys* descended. A very brief and preliminary survey of specimens from the Isle of Wight in the Sedgwick Museum at Cambridge suggests that the anthracotheres of the genus *Bothriodon* (Foster-Cooper 1925) are close to the Fayum species but warrant generic separation. That these specimens, as well as the North African material, have little in common with the Burdigalian *Brachyodus* is also clear.

As the specimens from the Hempstead Beds, from Ronzon (Filhol 1881), and from the Fayum are approximately contemporaneous, a common ancestry for *Bothriodon* and *Bothriogenys* must be sought in earlier horizons. Such an ancestry may be present in the species of *Anthracotherium* known from the Eocene of France. However, descent of *Bothriogenys* and *Bothriodon* from some member of the Asian anthracothere assemblage as known from the Pondung cannot be ruled out. Colbert (1938, p. 392) has suggested that *Anthracokeryx* might have been ancestral to *Bothriogenys* primarily on the basis of the elongate facial region in both genera. This material needs to be restudied.

Between the early Oligocene anthracotheres of the Fayum and the Burdigalian forms known from Djebel Zelten, Libya, and Moghara, Egypt, there is a gap of some 15 m.y. (van Couvering 1972, p. 266). Whether the assemblages from East Africa

Figure 21.9   Occlusal view of LP$_3$–M$_3$ of *Hyoboops moneyi*, CGM. (Approx. ×1.)

("Rusinga-like fauna" of van Couvering) are contemporaneous with or slightly older than the North African faunas is still open to some question. Regardless, there is probably no more than 2 to 3 m.y. separating all of the early Miocene African faunas, approximately 18-20 million years old (van Couvering 1972; van Couvering and Miller 1969). All of these faunas show a remarkable "Africanness," with greater Eurasian affinities not appearing until later in the Miocene as shown in the assemblages from Fort Ternan (14 m.y. old) and those from Tunisia, Morocco, and Algeria, some 10 to 12 m.y. old.

Among the early Miocene anthracotheres, *Bothriogenys africanus* is a little-modified carryover from the Fayum *Bothriogenys* group (figure 21.13). Also descended from *Bothriogenys* is the genus *Masritherium* with two species, one in North Africa, *M. depereti,* and one in East Africa, *M. aequitorialis.* MacInnes (1951) in describing the Rusinga species referred it to the European genus *Brachyodus,* but with the statement that it bore no direct resemblance to the European form but was descended from an African Oligocene ancestor. He suggested that additional material, including a complete lower jaw, might show that the East African species was related to *Masritherium.* In this he was correct. The species is close to, but smaller than, *M. depereti.*

A second group of early Miocene anthracotheres is more difficult to attribute to the endemic Africa fauna. These are species here assigned to the Asian genus *Hyoboops.* The genotypic species *H. palaeindicus* (Lyddeker 1883, as *Hyopotamus palaeindicus*) is from the lower Siwaliks of Sind. Pilgrim (1912) recognized species of *Hyoboops* in the Dera Bugti fauna generally considered to be Burdigalian in age. This would suggest that the North African faunas at Zelten and Moghara are of approximately the same age as the Asian Dera Bugti fauna while the Rusinga *Hyoboops* may be slightly older. There are no known Oligocene anthracotheres in Africa that seem likely to have been ancestral to *H. moneyi* and *H. africanus.* However there are a number of quite closely related forms in the Asian Dera Bugti and Siwalik faunas, indicating a probable Asian origin for this group. If, indeed, *Hyoboops* first appeared in Asia, then it represents one of the earliest immigrants into Africa, crossing onto the Arabian Peninsula and then into Africa before the Afro-Arabian crustal plate had moved completely against Eurasia. This event is dated as occurring at about 18 m.y. ago (van Couvering 1972, p. 262). Perhaps some individuals were able to cross from Asia into Africa before a complete dry land union had been accomplished. The other alternative, that *Hyoboops* evolved from an unknown African ancestor and then moved into Asia at about 18 m.y. to then give rise to the diverse

Figure 21.10   Lateral view of right mandible of *Merycopotamus dissimilis,* BM M-18442. (Approx. ×1/3.)

Figure 21.11   Occlusal view of right mandible of *Merycopotamus dissimilis,* BM M-18442. (Approx. ×1/3.)

Figure 21.12 Palatal view of a partial skull of *Merycopotamus petrocchii,* referred specimen in the Florence Museum. (Approx. ×1/3.)

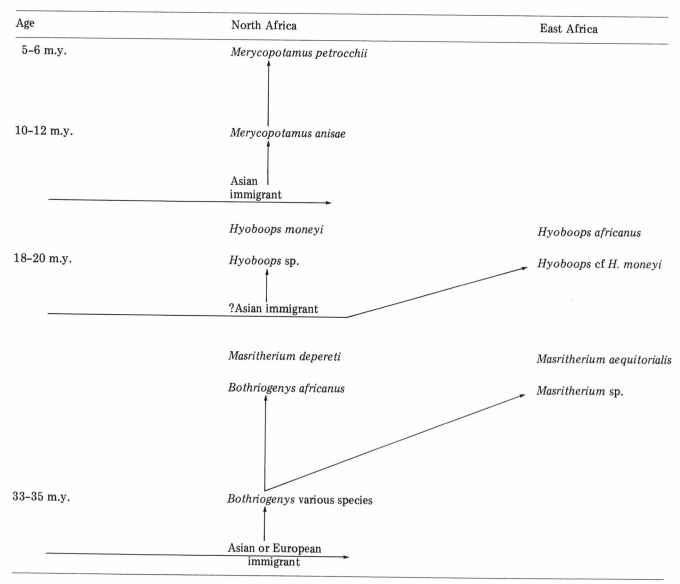

Age	North Africa	East Africa
5–6 m.y.	*Merycopotamus petrocchii*	
10–12 m.y.	*Merycopotamus anisae*	
	Asian immigrant	
18–20 m.y.	*Hyoboops moneyi* *Hyoboops* sp.	*Hyoboops africanus* *Hyoboops* cf *H. moneyi*
	?Asian immigrant	
	Masritherium depereti *Bothriogenys africanus*	*Masritherium aequitorialis* *Masritherium* sp.
33–35 m.y.	*Bothriogenys* various species	
	Asian or European immigrant	

Figure 21.13 Temporal and geographic distribution of African Anthracotheres.

anthracothere assemblage known at Dera Bugti, appears to me to be unlikely. This is particularly difficult to visualize in light of the early and diverse record of anthracotheres in the late Eocene of Burma and the much greater Miocene diversity in Asia than in Africa. This would seem to indicate a continuous history of successful evolution in Asia.

The final group of anthracotheres to appear on the scene in Africa are species of the genus *Merycopotamus*. The earliest of these, *M. anisae* from Tunisia, is definitely an immigrant from Asia, being quite close to the Siwalik *Merycopotamus dissimilis* (Black 1972) and undoubtedly arrived over dry land once the Afro-Arabian–Eurasia contact had been established. Species of *Merycopotamus* are known from Tunisia, Libya, and Chad.

The anthracothere story in Africa comes to a close at about the time that hippopotamids begin to flourish, some 5 to 6 m.y. ago. The last large sample of anthracotheres is found in the Bled Douara in Tunisia in a fauna that is probably some 10 to 12 m.y. old. Two specimens of *M. petrocchii* are known from Sahabi, Libya, and Coppens (1972) reports an anthracothere from Chad of about the same age, approximately 5 to 6 m.y.

As Shirley Coryndon points out in the chapter on hippopotamids, they were probably in existence some 10 to 12 m.y. ago in East Africa. As they evolved and diversified, anthracotheres evidently declined. In general body proportions and in dental morphology the two groups are similar in many ways. Perhaps they had similar habits.

It is probable, although as yet unproven, that hippopotamids evolved from an anthracothere, perhaps from some *Merycopotamus*-like form although this is pure speculation at present. The short facial region in *Merycopotamus* with a relatively short diastema between canines and premolars is similar to the condition in hippopotamids. The two also show similarity in reduction in incisor number, enlargement of one pair of lower incisors, and great sexual dimorphism. The enlarged angular flange of the lower jaw is also found in both *Merycopotamus* and *Hippopotamus*. All of these features, plus a great similarity in bones of the hind foot, are found in *Merycopotamus* but not in other anthracotheres. In any case, it appears most reasonable that the two groups shared a similar mode of life and that hippos replaced anthracotheres in later Neogene African faunas.

There are three paramount problems yet to be addressed concerning the evolution of African anthracotheres.

1. The origin of the early Oligocene Fayum *Bothriogenys* and its relationship to the European and Asian Eocene forms.

2. The time and direction of movement of *Hyoboops* into, or out of, Africa.

3. The relationship of anthracotheres and hippopotamids.

Colleagues in many museums were most helpful in supplying access to collections and information on specimens. I wish particularly to thank R. E. F. Leakey and J. M. Harris of the Kenya National Museum (KNM), M. Azzouz and Mlle. Memmi of the Tunisian Geological Survey (TGS), H. W. Ball and W. R. Hamilton of the British Museum of Natural History (BM M), R. J. G. Savage of Bristol University, and E. L. Simons and G. E. Meyer of the Peabody Museum, Yale University (YPM). S. C. Coryndon and W. R. Hamilton read the manuscript and offered valuable comments for which I am grateful.

References

Andrews, C. W. 1899. Fossil Mammalia from Egypt. *Geol. Mag.* (4)6:481–484.

———. 1906. *A descriptive catalogue of the Tertiary Vertebrata of the Fayûm, Egypt.* London: British Museum, xxxviii+342 pp.

———. 1914. On the Lower Miocene vertebrates from British East Africa, collected by Dr. Felix Oswald. *Quart. J. Geol. Soc. London* 70:163–186.

Andrews, C. W., and H. J. L. Beadnell. 1902. *A preliminary note on some new mammals from the Upper Eocene of Egypt.* Egypt Surv. Dept., Pub. Works Min., 9 pp.

Arambourg, C. 1959. Vertébrés continentaux du Miocène supérieur de l'Afrique du Nord. *Publ. Serv. Carte Géol. Algérie,* (n.s.) *Paléont. Mém.* 4, 161 pp.

Black, C. C. 1972. A new species of *Merycopotamus* (Artiodactyla: Anthracotheriidae) from the late Miocene of Tunisia. *Notes Serv. Géol. Tunisie* 37(2):5–39.

Bonarelli, G. 1947. Dinosauro fossile del Sahara Cirenaico. *Riv. Biol. Colon. Roma* 8:23–33.

Colbert, E. H. 1938. Fossil Mammals from Burma in the American Museum of Natural History. *Bull. Am. Mus. Nat. Hist.* 74(6):255–436.

Coppens, Y. 1972. Tentative de zonation du Pliocène et du Pléistocène d'Afrique par les grands mammifères. *C. R. Hebd. Séanc. Acad. Sci.* 274:181–184.

Falconer, H., and P. T. Cautley. 1847. *Fauna antiqua sivalensis.* Atlas. London: Smith Elder and Co.

Filhol, H. 1881. Mammifères de Ronzon. *Bibl. École Hautes Études* 24(4):1–87.

Forster-Cooper, C. 1924. The Anthracotheriidae of the Dera Bugti deposits in Baluchistan. *Pal. Indica,* (n.s.) 8(2):1–60.

———. 1925. Notes on the species of *Ancodon* from the Hempstead Beds. *Ann. Mag. Nat. Hist. Lond.,* Ser. 9, 16:113–138.

Fourtau, R. 1920. *Contribution à l'étude des Vertébrés Miocènes de l'Égypte.* Survey Dept. Cairo, vii+122 p.

Hooijer, D. A. 1963. Miocene Mammalia of Congo. *Annls. Mus. R. Congo belge,* Ser. 8, 46:1–77.

Lavocat, R. 1961. Le gisement de vertébrés Miocènes de Beni Mellal (Maroc). Étude systématique de la faune de mammifères et conclusions. *Notes et Mém. Ser. Géol. Maroc* 155:29–144.

Laws, R. M. 1968. Dentition and aging of the hippopotamus. *E. Afr. Wildlife* 6:19–52.

Lydekker, R. 1883. Indian Tertiary and post-Tertiary Vertebrata. Siwalik selenodont Suina, etc. *Pal. Indica,* Ser. x, II(V):142–176.

MacInnes, D. G. 1951. Miocene Anthracotheriidae from East Africa. *Brit. Mus. Nat. Hist., Fossil Mammals of Africa,* No. 4:1–24.

Maglio, V. 1970. Early Elephantidae of Africa and a tentative correlation of African Plio-Pleistocene deposits. *Nature* 225(5230):328–332.

Pilgrim, G. E. 1912. The vertebrate fauna of the Gaj Series in the Bugti Hills and the Punjab. *Pal. Indica* 8:1–65.

Robinson, P., and C. C. Black. 1974. Vertebrate faunas from the Neogene of Tunisia. *Ann. Geol. Surv. Egypt.* 4: 319–332.

Schmidt, M. 1913. Ueber Paarhufer des fluviomarinen Schichten des Fajum, odontographisches und osteologisches Material. *Geol. Pal. Abh.* 11:153–264.

Simons, E. L. 1968. Early Cenozoic mammalian faunas Fayum Province, Egypt. Part I. African Oligocene Mammals: Introduction, History of Study, and faunal succession. *Bull. Peabody Mus. Nat. Hist. (Yale Univ.)* 28:1–21.

Simpson, G. G. 1945. The principles of classification and a classification of mammals. *Bull. Am. Mus. Nat. Hist.* 85:xvi+350.

Van Couvering, J. A. 1972. Radiometric calibrations of the European Neogene. In *Calibration of Hominoid Evolution.* W. W. Bishop and J. A. Miller, eds. Edinburgh: Scottish Academic Press, pp. 247–271.

Van Couvering, J. A., and J. A. Miller. 1969. Miocene stratigraphy and age determinations, Rusinga Island, Kenya. *Nature* 221(5181):628–632.

22

Suidae and Tayassuidae

H. B. S. Cooke and A. F. Wilkinson

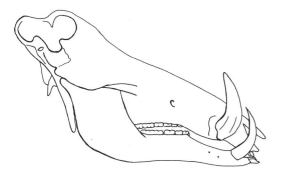

The Living Suidae[1]

General Characteristics

The living Suidae are artiodactyls of medium size with a stout body clad in thick, tough skin that is covered, sparsely or extensively, with bristly hair. The tail is short and thin and carries a tuft of bristles at its tip. The lower parts of the limbs are relatively slender, with narrow feet, each of which possesses four toes. The central pairs of toes are longer than the laterals, have flattened hooves, and are used most of the time in locomotion. The lateral toes have smaller nails and do not function in normal walking, although they may assist on soft ground. The stomach is simple, except for a cardiac pouch. There is no dorsal gland and there are at least four mammary glands.[2] The head is long and pointed, terminating in a mobile snout at the end of which is an oval cartilaginous disc that is penetrated by the nostrils. This disc, which functions in digging, is supported by a peculiar isolated "prenasal" or "rostral" bone. The upper canines curve more or less outward and upward, usually abrading against the lowers to maintain a sharp edge. The full dentition has 44 teeth, but this may be reduced by suppression of one or more pairs. In primitive forms the molars are brachydont and bunodont, with a small talon or talonid on the third molar, but in the fossil forms there may be intermediate, elongate, and hypsodont last molars.

The fossil ancestors of the pigs and peccaries first appear in the Oligocene. Some peccarylike fossils occur in Eurasia but in general the Neogene forms belong to the Suidae, and the Tayassuidae are restricted to the Americas. In Africa, pig remains are not yet known from the Oligocene but are found in a number of lower to middle Miocene deposits. Suidae are fairly plentiful in Pleistocene deposits and also occur in the Pliocene. There is at present a gap in our knowledge of the later Miocene and it is difficult to see any close relationships between the earlier and the later groups. It is thus convenient here to consider the African Suidae in two separate sections, one primarily concerned with the early Miocene forms and the other with those of the Pliocene-Pleistocene.

1 Sections on the living Suidae and upper Miocene-Pleistocene Suidae are by H. B. S. Cooke, and the section on lower-middle Miocene Suidae is by A. F. Wilkinson.

2 The peccaries (Family Tayassuidae or Dicotylidae) have only three functional toes, a complex stomach, a dorsal gland, two or four mammae, and the canines are in the more primitive vertical position.

The Four African Genera

The living pigs of Africa belong to four genera, each represented only by a single species, although it is possible that the western warthog belonged to a separate species that was exterminated in historic times.

The wild boar *Sus scrofa* ranged along the Mediterranean coastal region and some way up the Nile Valley. Although reported to have occurred in the central Sudan, there is no firm evidence to suggest that this was really part of its prehistoric range and it seems that specimens found probably represent escaped domestic forms. In North Africa, the wild boar is found in the fossil state at a number of localities, the oldest of which are of mid-Pleistocene age, but it has been suggested that the species in the oldest beds is different from *scrofa*. Curiously, it does not seem to be an element of the Neolithic sites, although the warthog does occur there, and the domestic variety was probably introduced in historic times. In sub-Saharan Africa the escape of domestic pigs has led to the existence of feral herds in a number of areas, although few survive today. *Sus* is a "primitive" type of pig with a long Asiatic fossil record and it is generally supposed to have invaded North Africa during the Pleistocene. It has not been recorded as a fossil in the sub-Saharan region.

The bush pig, *Potamochoerus porcus,* is the least specialized of the African suids, retaining a simple brachydont dentition and a long, fairly narrow skull. It is widespread in Africa south of the Sahara but has not been reported in North Africa. Although plentiful as a fossil in later Pleistocene and Holocene sites, it is not firmly recognized in earlier deposits. In Madagascar there is also a wild bush pig, sometimes regarded as a separate species, *P. larvatus,* but probably only subspecifically distinct. Bush pigs are sometimes called river hogs because of their liking for damp and shady places, and they range from light forest and open woodland to the bushy margins of the forest-savanna mosaic. They are also found in the upland savanna if there are adequate water and riverine bush.

The forest hog, *Hylochoerus meinertzhageni,* is the largest of the wild pigs and is limited in its range to the forested belt of equatorial Africa, including areas of more or less isolated montane forest. Specialized features of the skull and of the dentition make it difficult to determine its ancestral relationships, a situation not assisted by its virtual absence from the fossil record.

The warthog, *Phacochoerus africanus,* is very widespread in Africa, occurring practically everywhere except in very arid regions or in the tropical forest habitat favored by *Hylocheorus. Phacochoerus* certainly occurred in North Africa during the Pleistocene, as also did some of its extinct relatives. The skull is highly specialized, with elevated orbits and very hypsodont third molars, and it has been suggested (E. C. N. and H. E. van Hoepen 1932) that it should be separated from the other pigs at the subfamily level as the *Phacochoerinae*. This suggestion has been followed by Arambourg (1947) and Viret (1961) but not by Simpson (1945) and is not even mentioned by Ansell (1968) in his recent review of the Artiodactyla. At present it does not seem feasible to divide the African fossil forms on this basis.

Some of the main characteristics of the skull and dentition are useful in considering the fossil material.

Both *Sus scrofa* and *Potamochoerus porcus* have skulls that are typically elongate and pointed, or wedge-shaped, with an almost straight profile from the tips of the nasals to the parietals. The braincase is gently domed behind the orbits and the parietal constriction is narrow. In *S. scrofa* the snout is long and the nasals are more or less parallel sided, whereas in *P. porcus* the nasals widen above the canine flanges and in the males there is an expanded rugose area on the nasals and the adjoining part of the maxillae. In the wild boar there are only small maxillary flanges around the roots of the upper canines, even in the male, whereas the male bush pig has an elevated lateral crest on the canine flanges, the upper part of the crest being rugose. The rugose areas are covered with cartilaginous tissue and there is a pair of hard calloused swellings on either side of the face, but no warts. The female bush pig has small canine flanges and lacks the facial callosities. In *S. scrofa* the zygomatic arches are narrow and not expanded laterally, whereas in *P. porcus* the maxillary root of the zygoma juts out sharply and may be quite inflated in male animals. In the wild boar, the area of origin of the levator rostri muscles is high and deeply scooped, with a strong ridge separating it from a weakly scooped area of origin for the depressor rostri muscles, whereas in the bush pig the depressor rostri muscle areas are deeply scooped while the levator rostri attachment area is not very high and only moderately scooped, and the ridge below it is weak and blunt. In both species the canines are relatively small. A full complement of incisors is retained in both species and the cheek teeth are bunodont and brachydont. The mandibles are very similar, with a spoon-shaped front to the symphysis and a moderate constriction in the diastema between the canine and P_2. In the bush pig there is a distinct small shelf at the back of the symphysis for the attachment of the genioglossus and geniohyoideus muscles, which is not present in *S. scrofa*.

Hylochoerus meinertzhageni and *Phacochoerus africanus* have distinctive specializations in the skull and in the dentition. In both the premolars are reduced and only one pair of upper incisors remains, even these being commonly shed in the adult forest hog. The striking features of the forest hog skull, apart from its larger size, are the wide parietal area, concave between the orbits instead of convex, the wide occiput, the robust and expanded zygoma with a bulging inflated area of hollow bone, and the large outflaring canines; all these characters are strongest in male animals. The skull of *P. africanus* is quite highly specialized, notably through extreme elevation of the orbits, accompanied by some elevation also of the occipital condyles above the palatal plane. The maxilla is heightened to accommodate the very hypsodont third molars and the root of the zygoma is raised, but also merges very gently with the maxilla. The orbits are set back as well as raised, resulting in the development of a broad platy area below them. The jugal flares laterally and its outer and lower margins are thickened into bosses, which may be large in male animals; in females the morphology is similar but the jugals are not as enlarged. In life the face is adorned with large warty thickenings of the skin below the eyes, between the eyes and the upper tusks, and on the borders of the

mandible. The frontal area is very broad and flattened between the orbits. The canines are large and curved and free of enamel except at the tips in juvenile animals. The mandible in both forest hog and warthog has a broad anterior border that is only gently curved and the incisors are much smaller than in the wild boar or the bush pig. In the warthog the incisors lie in a shelflike projection below which the symphysis is quite strongly curved, unlike the rather flat profile of the symphysis in the forest hog.

The occipital aspects of the skull of these four living suids are somewhat different and reflect differences in the carriage of the head. Despite the differences in gross size of the skulls, the actual height of the nuchal crest above the lower borders of the foramen magnum is very much the same (about 110-120 mm) but the crest breadth varies: in *Sus scrofa* it is about 60-70% of the height, in *Potamochoerus porcus* 80-100%, in *Hylochoerus meinertzhageni* 100-130%, and in *Phacochoerus africanus* 80-100%. In *Sus* and *Potamochoerus* the temporal condyles are about on a level with the top of the foramen magnum, but in *Phacochoerus* they are placed relatively higher, and in *Hylochoerus* somewhat lower, about on the level of the lower border of the foramen magnum. The external ear openings are also in relatively different positions (figure 22.1) and this affects the angle at which the auditory canal emerges (see table 22.1). In both *Sus* and *Potamochoerus* the

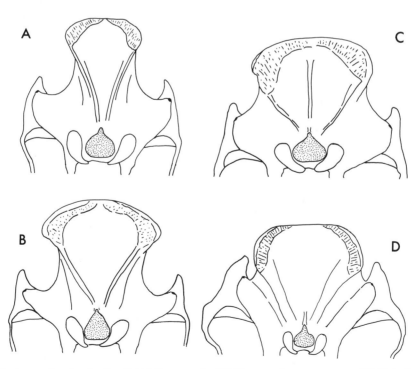

Figure 22.1   Occipital views of male skulls of: (*A*) *Sus scrofa;* (*B*) *Potamochoerus porcus;* (*C*) *Hylochoerus meinertzhageni;* (*D*) *Phacochoerus africanus.* (1/3 natural size.)

Table 22.1 Summary of characters in living suid skulls.

Character	*Sus*	*Potamochoerus*	*Hylochoerus*	*Phacochoerus*
Lateral profile	Almost straight, slight change of slope at nasal/frontal suture	As in *Sus*	Parietals elevated above nasal plane; slope change at nasal/frontal suture	Profile concave near nasal/frontal suture; convex at parietals
Orbits	Well below fronto-parietal surface	As in *Sus*	Below fronto-parietal surface	Elevated above fronto-parietal surface
Frontal and parietal area	Gently arched	As in *Sus*	Parietal area flat or concave	Gently arched; frontals flat and wide
Parietal constriction	Narrow but much wider than snout	Narrow, about equals snout width	Very broad	Narrow, about equals nasal breadth
Braincase below parietal crest	Sides almost vertical	Bulges outward	Parietal crest overhangs braincase	Bulges slightly outward
Occiput	Narrow; no marked median ridge	Broader than *Sus* but similar	Broad; strong median ridge	Much like *Potamochoerus*
Angle of auditory canal	Steeper than 45° from horizontal	Less than 45° (c. 35° – 40°)	Low, close to 20°	Close to 45°
Nasals and muzzle shape	Nasals parallel-sided or narrowed in middle; gently rounded, curving into near-vertical side walls	Nasals widen posteriorly, expanded in middle in males; flat and project over vertical or hollowed side walls	Nasals wide in middle, taper slightly towards back; well rounded; muzzle broad, widely arched above canine area	Nasals expanded above canine, narrowed in middle; well rounded; muzzle slightly wider above canine area
Premaxilla/nasal opening	Premaxilla upper border almost straight; nasals not downcurved at tips	As in *Sus*	Premaxilla border strongly concave; tips of nasals strongly downcurved	Premaxilla border moderately concave; nasals droop towards tips
Back of premaxilla	Almost to position above P^4	Ends above $C – P^2$ diastema	Ends above C	Ends above $C – P^3$ diastema
Position of infraorbital foramen	Above P^4	Above M^1	Above M^1	Above front of M^2

central squamous area is deeply scooped and there is a sharp ridge on either side of this triangular area separating it from the lateral part of the occipital and squamous temporal areas, which are set forward. In *Phacochoerus* there is some anteroposterior compression so that the ridges are less prominent, but the general morphology is similar and there is no median ridge. In *Hylochoerus* the occipital area is not only broadened but also flattened and the left and right sides of the rectus capitis dorsalis muscles are distinct and are separated by a median ridge (Ewer 1970). Some of the more important features of the skulls are summarized in table 22.1.

The incisors in all forms of the living species under discussion are essentially similar in structure and, although there are minor differences, they have not proved very diagnostic. Some features of the

varieties and distinctions are discussed by Bouet and Neuville (1930). The main differences in the upper canines are summarized in table 22.1 and will be discussed below. The lower canines are all of scrofic type.

In the four genera of living Suidae presently under discussion, only *Sus* has a complete complement of premolars and the lower P_1 may be lost in adults. In *Potamochoerus porcus,* the lower P_1 is always absent and the upper P^1 commonly fails to appear or is shed during life. Although the teeth are generally similar, the premolars of *Sus* are more bladelike or sectorial in character than those of the bush pig, which are more dominated by a robust main cusp (figure 22.2 *A–D*). The P^1 is very alike in both and P^2 differs mainly in the presence of a median internal fovea in *Sus* which is largely obliter-

Table 22.1   (continued)

Character	Sus	Potamocheorus	Hylochoerus	Phacochoerus
Canine flanges	Small	Small in female; high crest in male	Broad with low lateral crest	Large and tubular
Canine form	Small, curve laterally	Small, curve laterally	Large, curve laterally	Large, curve laterally and upwards
Zygoma structure	Root curves at 45° to axis of skull; then smooth sweep to lateral border, parallel to skull axis; lateral walls nearly vertical	Root curves sharply so that anterior border approximately perpendicular to skull axis; lateral walls parallel to skull axis, flare slightly (10°) downwards	Root curves at 45° to skull axis; smooth sweep in female but curve sharply to perpendicular in males; lateral walls flared and thickened in older males	Root sweeps gently back from just behind infraorbital foramen, continuing at 30° in female, 45° in male; lateral walls broad below orbits, flared outwards
Lower border of zygoma	Parallel to occlusal plane	Slopes down towards back	Slopes markedly down	Parallel to palate but elevated
Origin of levator rostri	High, deeply scooped	Not very high, moderate scooping	Small and shallow	Elongate, triangular, shallow
Ridge below levator rostri	Strong, sharp	Moderate, blunt	Small, weak	Strong
Origin of depressor rostri	Weak, shallow	Deeply scooped	Weak, shallow	Moderate
Incisors	6 upper incisors	6 upper incisors	2 upper incisors in adult, often shed	2 upper incisors, retained in adult
Premolars	P^1 retained, but sometimes shed	P^1 absent	P^1 absent, P^2 shed early	P^1 absent, P^2 and P^3 shed early
Upper canine structure	Flattened oval cross-section; strong median dorsal groove, weak ventral groove; strong ventral enamel band, ribbed; inset dorso-median and dorso-lateral enamel strips	Rounded quadrilateral cross-section; weak dorsal median groove; strong median groove, weaker lateral groove, both high up sides; enamel bands as in *Sus*	Sub-rectangular cross-section, higher than wide; strong dorsal and median grooves, weaker lateral groove; enamel bands as in *Sus*	Sub-rectangular cross-section, long axis diagonally upward; strong, broad, dorsal and ventral grooves, weaker anterior grooves; no enamel except at tips in juveniles

ated in *Potamochoerus porcus*. P^3 in the two forms is decidedly different, *Sus* developing a wide parallel-sided external enamel island with wear whereas in the bush pig an expanded central cone is the obvious feature. The anterior and posterior cingulum cusps are more isolated and lower crowned in *Potamochoerus porcus* so that they wear somewhat like T-shaped attachments to the central cusp (a tendency also shown in P^2). The enamel is thicker in *Potamochoerus* than in *Sus,* so less dentine is exposed with wear. In P^4 the structure can be fairly similar and in particular cases distinction can be difficult. In *Sus,* the two main cusps on the external side (paracone and metacone) are well separated and even after the enamel unites into a single island, the dentine can be seen to be divided between the primary cones. In the bush pig these two elements are divided only near their tips and they soon fuse into a figure 8 that develops into a more or less continuous oval. As the enamel is thicker in the bush pig tooth, dentine is seldom exposed in the posterior cingulum cusps, as it is in *Sus*. In *Sus,* also, the protocone is of modest size and usually remains isolated from the paracone until wear is very advanced, whereas in the bush pig it tends to unite with the paracone even before any dentine shows within it. In the lowers, the anterior and posterior cingula are greatly elevated in *Sus* so that they unite with the two central cones to form an elongate narrow ridge, whereas they are lower in the bush pig and thus emphasize the development of the strong central main cone. In the P_4 in *Sus* and to a lesser extent in P_3, the main cone is divided into anterior and posterior tubercles that are not aligned fore and aft but the anterior element is displaced in-

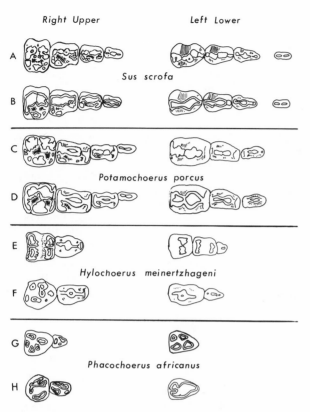

Figure 22.2 Right upper and left lower premolar series in four living suids. For *Sus scrofa* and *Potamochoerus porcus* the upper set is in early wear and the lower in normal wear. Two variations are shown for *Hylochoerus meinertzhageni* and *Phacochoerus africanus*. (2/3 natural size.)

wards (lingually), producing a kink in the elongate enamel island that results from wear. In *Potamochoerus porcus,* the main cone is undivided and strongly dominant, forming an oval enamel island that fuses with the posterior cingulum cusp only at a late stage in the exposure of the dentine (figure 22.2*A–D*). The thicker enamel assists in showing the differences but with advanced wear the distinction becomes difficult to make.

Both *Hylochoerus meinertzhageni* and *Phacochoerus africanus* have suffered reduction of the premolars and modification in structure. In both species the premolars are normally reduced to P^3 and P^4 only in the upper jaw, although P^2 sometimes appears in *Hylochoerus*. P_3 (and occasionally also P_2) is retained in the lower jaw in *Hylochoerus* but in the warthog only P_4 usually survives. In the forest hog P^3 is an inflated tooth generally resembling that of *Potamochoerus porcus* and the normal P_3 and P_4 are also essentially like the corresponding teeth of the bush pig; a distant relationship seems possible. The usual P^4 comprises a number of separate small pillars whose mutual relationships vary and thus

make it very difficult to determine analogues or relationships (figure 22.2*F*). In a small proportion of individuals (whose geographic or racial variation has not been investigated), the P^4 is strongly "molarized" and consists of two pairs of columnar cusps very like those of the true molars (figure 22.2*E*) and a similar structure affects the lower premolars; this obscures completely any possible indication of relationships.

In the warthog, the surviving upper P^3 and P^4 and lower P_4 consist of a number of slender rounded pillars, like those of the true molars, set in a substantial body of cementum. The teeth are often rotated out of their normal orientation, so it is difficult to determine possible analogues. However, P^4 commonly exhibits five enamel islands, of which the two outer (buccal) ones are probably the paracone and metacone, while one of the inner (lingual) ones is presumably the protocone (figure 22.2*G*). With wear two islands are produced, one incorporating the supposed paracone and metacone and a small posterior cone, while the two antero-internal cones form into a separate island, leaving a valley between them more in *Sus* fashion than that of *Potamochoerus porcus* (figure 22.2*H*). No clear inferences on relationships can be made. In the lower P_4 there are commonly four small enamel islands that unite into an irregular pattern and there is no particular resemblance to the other living suids (figure 22.2*G–H*). The deciduous teeth do not provide additional clues.

The molars of *Sus scrofa* and *Potamochoerus porcus* are essentially similar and are relatively unspecialized bunodont and brachydont teeth with the talon and talonid a good deal shorter than the body of the crown. In *P. porcus* the talon and talonid tend to be relatively less developed than in *S. scrofa,* often consisting of one robust terminal pillar and a few small angular pillars, whereas in *S. scrofa* there is one moderately developed third lateral pillar, or even a third pair of laterals. In *S. scrofa* the enamel is significantly thinner than in *P. porcus* and thus in the earlier stages of wear the main pillars have a more folded stellate appearance, with relatively more complex dentine islands (figure 22.3). The main pillars in *S. scrofa* taper upwards more than is the case in *P. porcus,* in which the crown appears a little more columnar and less compact than in *S. scrofa.* Cement is virtually absent.

The molars of *Hylochoerus* are peculiar both in structure and in wear. The talon and talonid are about as long as the body of the crown and there is a fully developed third pair of lateral pillars and sometimes an incipient fourth pair (usually terminal). Each of the lateral and main median pillars consists of a comparatively slender column of dentine and relatively thin enamel around which there is a heavy coating of cement, as if it were building normal-sized pillars from cement. The main pillars are thus widely spaced and they bite into the transverse valleys of the opposing tooth to wear deep transverse furrows (figure 22.3) that are very characteristic and are associated with the unusual chewing action (Ewer 1970).

In the living *Phacochoerus africanus* the molars are highly specialized and the second and third molars very hypsodont. The upper first molar is also high-crowned but is worn to a stump before the third molar is fully in wear and is shed so as to bring M² against P⁴; later M² is shed and the M³ comes next to P⁴, which may also be shed. In the lower jaw the same sequence is followed but lags a little behind the uppers. In old animals only the third molars remain in the rami. The talon and talonid are twice as long as the body of the crown and carry about five additional pairs of laterals and the appropriate me-

dians; exceptionally there may be a total of only six (2 + 4) or as many as ten (2 + 8) laterals, but the extremes are very rare. A little less uncommon is the occurrence of an "extra" column on only one side of the crown. Additional medians may also be found and rarely there is a double row of medians in upper third molars, or in the back part of the upper third molar. Both the lateral and the median column forms are simple, varying from subtriangular to oval as a rule but sometimes rounded; the laterals are occasionally almost Y-shaped.

The crown height is about equal to the crown length (about 50 mm ± 10 mm) but the full height is seldom seen as root formation is commonly delayed until the columns are already in wear (see figure 22.12). Roots form first on the anterior pair and eventually become closed; these roots remain independent of the more or less continuous open root that forms progressively at the base of the remaining columns. As temporary growing roots exist and are resorbed, it is sometimes difficult to determine if a particular crown is really fully formed or not. Once the anterior root has formed, the molar tends to rotate by rising only slowly at the front and faster at the back, where it is still growing. In consequence, the plane of wear becomes more and more oblique with respect to the axis of the columns. The enamel islands progressively become more elongate and the medians eventually form into a lengthening central complex.

Lower-middle Miocene Suidae[3]

Although the mid-Tertiary suids from Europe have received considerable attention their contemporaries in Africa have been relatively neglected. Stromer (1926) described the first occurrence of a new genus, *Diamantohyus* from Southwest Africa, and reported the probable presence of *Propalaeochoerus* Stehlin. Arambourg (1933) described *Kubanochoerus jeanneli* from Turkana and, more recently, *K. massai* (1961) from Gebel Zelten, while Hooijer (1963) described a solitary tooth of *Hyotherium dartevellei* from Malembe, Congo. The supposed presence of the Asian form *Sivachoerus* (Pilgrim 1926) was recorded from the Pliocene of Gasr-es-Sahabi, Libya, by Leonardi (1954) and Wadi Natrun, Egypt, by Tobien (1936).

Suids are a common element in the extensive fossil collections made in East Africa, for the most part by L. S. B. Leakey and his colleagues. To this must be added the material collected by R. J. G. Savage

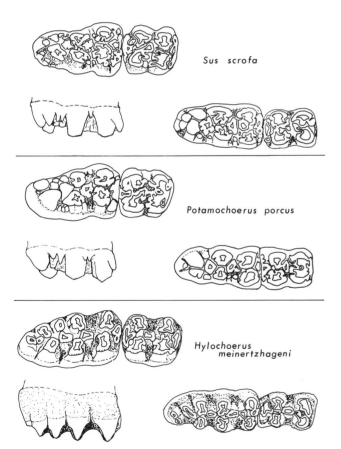

Figure 22.3 Occlusal views of right upper and left lower second and third molars and outer lateral views of upper third molars in three living suids, *Sus scrofa, Potamochoerus porcus,* and *Hylochoerus meinertzhageni.* (2/3 natural size.)

Sus scrofa

Potamochoerus porcus

Hylochoerus meinertzhageni

---

3 By A. F. Wilkinson.

from Gebel Zelten in Libya (Savage 1967, 1971; Savage and White 1965).

No suids are known from pre-Miocene strata in Africa. The East African and Libyan finds have recently been described in detail (Wilkinson 1976).

Localities

Suids have been discovered at most of the well-documented East African Miocene sites (see Bishop 1967, 1968; Bishop et al. 1969). Table 22.2 tabulates the broad distribution of middle Tertiary suids in Africa. However, the majority have come from Rusinga Island (see figure 2.5, chapter 2).

Major Features of Groups and Subgroups

An examination of the Burdigalian faunas of Africa, India, and Europe reveals the suids as an important and diverse element. Several clearly distinguishable genera can be determined. Before the early Miocene, back through the Oligocene, and into the Eocene, the picture becomes increasingly obscure. The ancestors of the Miocene forms lose their individuality and become difficult to separate. Indeed, confusion with Old World Tayassuidae is possible as certain species of *Doliochoerus* persist into the middle Miocene. Retention of a basically similar bunodont dentition in a large number of suid genera adds to the difficulty of delineating valid groups.

The proliferation of highly variable minor cusp ornaments in many genera has led to a welter of specific names, most of which must be considered invalid. Simpson (1945), following the work of Colbert (1935), erected five subfamilies within the Suidae. Although some reorganization of these subfamilies is now necessary the basic concept and division is essentially sound. Each subfamily is characterized as follows:

1. Subfamily Hyotheriinae—small size, simple bunodont molars, canines nearly vertical, and "scrofic," tympanic bullae ovoid and directed horizontally, no elongation of the snout.

2. Subfamily Listriodontinae—primitively retain a molar structure closely analogous to Hyotheriinae, but show an evolutionary trend toward great size in one direction and development of marked lophodonty in another.

3. Subfamily Tetraconodontinae—great enlargement of posterior premolars relative to molars.

4. Subfamily Sanitheriinae—originally contained the single genus *Sanitherium*, characterized by a labial cingulum and development of crescentic lophs. It seems probable that the primitive *Xenochoerus* should also be included here and also the *Hyosus-Hippohyus* group characterized by hypsodont and complexly folded molars.

5. Subfamily Suinae—contains a number of genera showing a wide range of variation but typified by elongation of the skull, complication of cheek teeth, and various developments of canine tusks.

A checklist of the African middle Tertiary suid species known to date is given below:

Hyotherium dartevellei (Hooijer) 1963
Hyotherium kijivium Wilkinson 1976
Kubanochoerus jeanneli (Arambourg) 1933
Kubanochoerus massai (Arambourg) 1961
Kubanochoerus khinzikebirus Wilkinson 1976
Listriodon akatikubas Wilkinson 1976
Listriodon akatidogus Wilkinson 1976
Lopholistriodon kidogosana Pickford and Wilkinson 1975
Lopholistriodon moruoroti (Wilkinson) 1976
Xenochoerus africanus (Stromer) 1926
Sanitherium nadirum Wilkinson 1976

Subfamily Hyotheriinae Cope 1888

The subfamily Hyotheriinae in Africa is represented by one genus, *Hyotherium,* containing two species, a large and small.

Table 22.2 Distribution of African Miocene Suidae (numbers of individuals given for each locality).

Species	Rusinga	Koru	Mbwagathi	Maboko	Napak	Bukwa	Karungu	Moruorot	Mfwanganu	Songhor	Namib	Ombo	Gebel Zelten
Hyotherium dartevellei	67[a]	—	—	3	—	3	—	3	4	—	—	—	—
Hyotherium kijivium	10	3	—	—	4	—	—	—	3	6	—	—	—
Kubanochoerus jeanneli	21	—	—	1	1	—	—	—	1	—	—	—	—
Kubanochoerus massai	—	—	—	—	—	—	—	—	—	—	—	—	16
Kubanochoerus khinzikebirus	—	—	—	2	—	—	—	—	—	—	—	—	12
Listriodon akatikubas	—	—	7	1	—	—	—	—	—	—	—	—	—
Listriodon akatidogus	1	—	2	—	—	—	—	—	—	—	—	—	1[b]
Xenochoerus africanus	9	—	—	—	—	1[c]	6	—	2	1	1[d]	—	2
Sanitherium nadirum	—	—	—	—	—	—	—	—	—	—	—	1	—
Lopholistriodon moruoroti	—	—	—	—	—	—	6	—	—	—	—	—	—

[a] Postcranial material rated as three individuals.
[b] Doubtfully assigned mandible.
[c] Walker 1968.
[d] Stromer 1926.

Genus *Hyotherium* H. von Meyer 1834

SYNONYMY. *Palaeochoerus* Pomel, *Choerotherium* Lartet, *Choeromorus* Gervais, *Amphichoerus* Bravard.

TYPE SPECIES. *Hyotherium soemmeringi* H. von Meyer 1834.

DIAGNOSIS. Slender limbs and vertebral column; preorbital skull length equals postorbital length; large orbits; short, backwardly directed paroccipital processes; prominent parietal crests which join posteriorly to form a sagittal crest (new diagnosis).

REMARKS. *Hyotherium* is a well-documented genus from European and Asian Miocene faunas. The status of the genus in Europe is confused by use of the well-established name *Palaeochoerus* Pomel 1846, which appears to be synonymous with *Hyotherium*.

Gervais (1859) concluded that *Palaeochoerus* was synonymous with *Hyotherium*. Peters (1869) corroborated this view. However, Gaudry (1878) and later Stehlin (1899) preferred to retain two separate genera although admitting that there was no clear distinction between them. This view has been widely accepted by subsequent authors (Pilgrim 1926, Colbert 1935a, Viret 1929, Hünermann 1968, Schmidt-Kittler 1971). The reasons advocated to support such a division may be summarized as: (1) greater size in *Hyotherium;* (2) greater complexity in the molars of *Hyotherium;* (3) more advanced canine structure in *Hyotherium;* and (4) different stratigraphical range.

Certainly end forms such as *H. typum* Pomel from the early Miocene and *H. palaeochoerus* Kaup from the lower Pliocene show great disparity in size and the elaboration of the molars. However, when intermediate species—for example, *H. waterhousi* or *H. dartevellei*—are inserted between them the increase in size and molar complexity is seen to be gradational; where is the boundary between *Palaeochoerus* and *Hyotherium* to be placed?

Hünermann (1968) realized the futility of using molar teeth for diagnosis and based his separation into two genera on the structure of the canines. *Hyotherium* sensu stricto shows marked sexual dimorphism: the male lower canine is hypsodont and openrooted, whereas the male upper canine and both upper and lower canines in the female have closed roots. The female canines tend to be broad and premolariform with only slightly elevated crowns. In cross-section the male lower canine is always clearly "scrofic." The concept of "scrofic" and "verrucosic" types of canine was first propounded by Stehlin (1899, pl. VII). The division was based on a study of modern pigs that possess one of two cross-sectional

patterns: "scrofic" (after *Sus scrofa*), where the shape is more nearly equilateral, the posterolabial face being broadest after the lingual face; and "verrucosic" (after *Sus verrucosus*), where the shape approximates to a scalene triangle, the labial face broadest after the lingual face. Stehlin was of the opinion that the scrofic canine was peculiar to the *Hyotherium-Sus* group. All the Oligocene (*Propalaeochoerus*) and early Miocene (*Palaeochoerus/ Hyotherium*) pigs possess a pronounced scrofic canine, as does *Hyotherium palaeochoerus* from the European Pontian. From this Stehlin concluded that *Propalaeochoerus* was the lineal ancestor of *Sus scrofa* through the intermediate stages outlined above. Forsyth Major (1897) and Pilgrim (1926) recognized that the verrucosic type is clearly more primitive and that the transformation to the advanced scrofic pattern takes place not only at different rates in different lines of suid evolution, but also within the ontogenetic development of individual animals. Nevertheless the very early attainment of the scrofic pattern in the *Propalaeochoerus-Hyotherium* group is of familial diagnostic value. *Palaeochoerus* is not known to possess an openrooted canine and this remains as its only distinctive feature and one both difficult to determine and rarely preserved. The acquisition of hypsodont canines can be considered as a transitional feature along with the attainment of increased complexity and size in the molars.

With regard to geographical and stratigraphical separation, within Europe acknowledged *Palaeochoerus* species range from the Aquitanian to the Helvetian (Stehlin 1899), whereas *Hyotherium* ranges from the Burdigalian (Thenius 1956) to Pannonian (Hünermann 1968). In Asia *Palaeochoerus* has been reported from deposits of Tortonian to Pontian age (Pilgrim 1926).

There thus appears to be no case for maintaining two genera. *Hyotherium,* by virtue of its priority, should be retained and *Palaeochoerus,* despite being well established in literature, should be abandoned.

The problem of synonymy does not finish at generic level. Intraspecific variation has resulted in a confusing number of specific names. Stehlin (1899) recommended retention of three arbitrary species based on size; *Hyotherium typum, H. waterhousi,* and an intermediate *H. meissneri.* However, it is probable that only two distinct species are present at any one horizon.

Hyotherium dartevellei (Hooijer) 1963

H. dartevellei is closely analogous to *H. waterhousi* in molar structure but is slightly larger. It is more comparable in size to *H. soemmeringi,* but differs from the latter in its greater simplicity. The holotype, a lower M3 (R.G. 6412, Hooijer 1963:50, pl. IX, fig. 3) from Malembe, Congo, is worn. However, excellent material abounds from East African localities, including an almost perfect skull and skeleton from Rusinga (figure 22.4). The skull shows a persistent temporal canal with postglenoid foramina and the presence of a mastoid foramen (Wilkinson 1976). It may prove convenient to declare Hooijer's specimen a nomen nudum and to erect a nomen novum to include his and the much superior East African material.

Hyotherium kijivium Wilkinson 1976

A new, small African *Hyotherium* species, equivalent to the European *H. aurelianensis* and Asian *H. pascoei*, is now also recognized at Gebel Zelten. Apart from size, it differs from *H. dartevellei* in the nature of M_3, which is simple with closely applied cusps lying in transverse pairs and with a symmetrical posterior taper. The talon on M^3 is rudimentary or absent.

Pearson (1928) recognized another small hyotherine, *Chleuastochoerus*, from the lower Pliocene of China. Apart from a few quite exceptional features of the skull and dentition, *Chleuastochoerus* is primitive and clearly closely related to *Hyotherium*.

Genus *Propalaeochoerus* Stehlin 1899

TYPE SPECIES. *Propalaeochoerus leptodon* Stehlin 1899.

DIAGNOSIS. Small primitive suid; simple quadricuspidate upper molars (sometimes reduced to three cusps on M^3) with incipient to well-developed anterolabially directed lophs on the lingual cusps; M^2 larger than M^3; P_4 tall and robust with a deeply cleft summit; anterior premolars reduced, P_1 probably absent; the symphysis of the mandible short and steeply sloping; pentadactyl foot.

REMARKS. The genus *Propalaeochoerus* from the upper Oligocene of Europe is not definitely recorded from Africa, although Stromer (1926) doubtfully assigned to it a lower third molar from the Namib Desert. Despite the proposed demise of *Palaeochoerus*, *Propalaeochoerus* remains a valid genus. It is our view that it forms the basal stock for two major and divergent lines of suid evolution. This proposal is further considered below. Regrettably the status of the genus is so confused that it is difficult to provide a satisfactory diagnosis for *Propalaeochoerus*. Stehlin originally proposed the following: "The same characters as *Palaeochoerus* but with the internal roots of the upper molars fused." Viret (1929) added to the diagnosis the fused nature of the lower molars. However, these features are sometimes found in small *Hyotherium* and *Xenochoerus* specimens. A revised diagnosis is suggested above but may require modification as the group becomes better known.

Subfamily Listriodontinae Simpson 1931

The subfamily forms a major branch of suid evolution that split off very early from the *Hyotherium* stock. It contains three genera: *Listriodon, Kubanochoerus,* and *Lopholistriodon.*

Genus *Listriodon* H. von Meyer 1846

SYNONYMY. *Bunolistriodon* Arambourg 1963 (*partim;* holotype *B. lockharti*).

TYPE SPECIES. *Listriodon splendens* H. von Meyer 1846.

REMARKS. *Listriodon* s.s. is a characteristic form from the Helvetian of Europe. It has not been definitely described from Africa. However, there are rare teeth from East Africa and Gebel Zelten with intermediate lophobunodont crowns. These are included tentatively in the genus *Listriodon* as two new species, a large and small, with essentially similar dentitions. The large species differs from the Lophodont listriodonts (*L. splendens* H. von Meyer, *L. pentapotamiae* Lydekker 1876, in Pilgrim 1926, *L. theobaldi* Lydekker 1878, 1884, *L. mongoliensis* Colbert 1934) in several ways that testify to its intermediate status: (1) the transverse lophs are imperfectly formed, the individual cusps are still discernible at a late stage of wear; (2) the oblique connecting ridges are comparatively larger and more bunodont; (3) the talonid of M_3 is generally more symmetrical; (4)

Figure 22.4 *Hyotherium dartevellei* Hooijer; lateral aspect of skull from Rusinga Island (Kenya National Museum KNM-RU 2701 A and B). (Slightly less than 1/2 natural size.)

slightly larger size; and (5) extreme simplicity with smooth enamel, no linear corrugations, and no development of extra accessory tubercles.

Two other *Listriodon* species have been advocated as intermediate types. Pilgrim (1917) at one time considered *L. guptai* from the Kamlial horizon of India (Vindobonian) to be a likely transitional type, but subsequently (1926) changed his mind. However, Viret (1961), on reexamination of the Listriodontiinae, found that although *L. guptai* closely resembled *L. lockharti* and *L. latidens* it had several features which indicated an intermediate position. An Asian Vindobonian listriodont of intermediate type would fit in with what is known of *Listriodon* phylogeny.

Liu Tung-sen and Lee Yu-ching (1963) report another intermediate form, *L. intermedius* from Kou Chia Tsun, Lantien Hsien, of middle Miocene age. This determination is based on one apparently broken tooth and must be considered as suspect for the time being.

Crusafont and Lavocat (1954) reported a new genus, *Schizochoerus* from the Pontian of Vallés Penedés, Spain. *S. vallesensis* is comparable in size to the smaller East African intermediate listriodont. Furthermore it resembles the latter in several morphological features: (1) the presence of rounded cusps joined transversely by ridges; (2) a knife-edge ridge connecting the hypoconid to the posterior face of the anterior loph; (3) lack of cingulum on the labial and lingual sides; and (4) short, simple talonid.

Crusafont and Lavocat recognized the similarity of *Schizochoerus* with *Listriodon*, but considered its features distinctive enough to warrant generic rank. It seems likely that *Schizochoerus* represents the European equivalent (or survivor) of intermediate types found in Africa and Asia. It may be convenient at a later date to refer all those listriodonts with neither true bunodont nor lophodont dentitions to the genus *Schizochoerus*.

Listriodon akatikubas Wilkinson 1976

This is the larger of the two East African species of *Listriodon*, being slightly bigger than *L. splendens* and *L. pentapotamiae*. The upper and lower molars show a structure intermediate between lophodont and bunodont forms. They are exceptionally simple in structure, the enamel of the cusps and ridges being completely smooth, with little sign of accessory tubercles and no trace of linear corrugations. M_3 has four cusps arranged in pairs and the talonid is small and asymmetrical. Cingula are weakly developed or absent. It is known by a limited number of specimens, most of them from Mbagathi.

Listriodon akatidogus Wilkinson 1976

This species is considerably smaller than *L. akatikubas*, which it resembles in general morphology. P_3 has a broad posterolingual talon and the cingulum is continuous around the whole tooth. It occurs at Mbagathi and Rusinga in Kenya and is tentatively also recognized at Gebel Zelten in Libya.

Genus *Kubanochoerus* Gabunia 1955

SYNONYMY. *Libycochoerus* Arambourg 1961; *Bunolistriodon* Arambourg 1963 (*partim*).

TYPE SPECIES. *Kubanochoerus robustus* Gabunia 1955.

DIAGNOSIS. Dentition complete, more or less continuous; massive stocky canines; broad spatulate incisors; brachydont, bunodont, and simple molars; talon on M^3 reduced to a single lingual cusp, talonid on M_3 short with a single accessory tubercle and wide crenulated cingula; simple premolars with the principle cusp of P^4 split into two. Cingulum on all sides except lingually. Tendency toward massive size. Cranium with preorbital region very elongated, narrow, and with parallel tooth rows; reduced postorbital region, the parietals typically not overhanging the occiput; strongly developed zygomatic arches. (Mostly from Arambourg 1963.)

REMARKS. In 1933 Arambourg described a crushed skull from Losodok, Turkana, which he named *Listriodon jeanneli*. In this paper he suggested that the bunodont listriodonts might be separated under a generic name such as *Bunolistriodon*, but he did not formally establish the name at that time. In 1961 he created the genus *Libycochoerus* for a large suid, *L. massai*, and two years later applied his name, *Bunolistriodon*, to it. As Leinders (1975) has pointed out, this does not validate that name with effect from 1933, but only from 1963, so *Libycochoerus* has priority. However, it now seems likely that the latter is congeneric with *Kubanochoerus* Gabunia 1955 (and Gabunia 1960) and the latter name thus applies also to the African material.

Kubanochoerus jeanneli (Arambourg) 1933

This is the smallest of the three African species and has a general resemblance to *Listriodon lockharti* of Europe, although appearing more primitive. Arambourg (1963) included *L. lockharti* in his *Bunolistriodon* (indeed, he made it the type species) but Leinders (1975) has shown that it is closer to *Listriodon splendens* and is to be retained in the latter genus. Several specimens of *Kubanochoerus jeanneli* have been found at Rusinga, including a fine mandible (figure 22.5). Both the upper and the lower denti-

Figure 22.5 *Kubanochoerus jeanneli,* occlusal aspect of mandible (Kenya National Museum KNM-RU 2785). (3/10 natural size.)

tions are somewhat like those of *Hyotherium.* The talon of M³ is small and there is a well-developed cingulum, even on the labial face. P³ is elongated, with its main cone aligned at 45° to the dental row. The upper canine is fairly stout. This species has not so far been recorded from Gebel Zelten.

Kubanochoerus massai (Arambourg 1961)

In 1961 Arambourg described a species from Gebel Zelten that was considerably larger than *K. jeanneli.* This species, *K. massai,* provided valuable knowledge of the skull, which displays conical protuberances and, in the males, an additional frontal horn. This is very similar to the skull of *K. robustus* from the Caucasus described by Gabunia in 1960 and is one of the definitive characters of the genus. The molars are simple but show some advances over *K. jeanneli:* increases in the number and depth of the linear furrows, a few additional accessory elements, and a relatively larger talon(id). Further isolated teeth have been found at Gebel Zelten.

Kubanochoerus khinzikebirus (Wilkinson 1976)

Gebel Zelten has also yielded a remarkable new species of gigantic proportions, from 10–20% larger than the biggest listriodont previously known—*K. gigas* Pearson 1928 from China. Other than in size, this species resembles *K. gigas:* the molars are simple, the talon of P³ and P² are weak and flat, P¹ is comparatively small, and there is a short diastema between P¹ and P². This species also occurs at Maboko in East Africa.

Genus *Lopholistriodon* Pickford and Wilkinson 1975

TYPE SPECIES. *Lopholistriodon kidogosana* Pickford and Wilkinson 1975.

DIAGNOSIS. Small size, extreme development of the transverse crests in the molars and fourth premolar and suppression of accessory cusps. Premolars possess enlarged cingular platforms. Nasal ridge narrow.

REMARKS. The skull is primitive, being close to *Hyotherium* but the dentition is extremely special-

ized. Features distinguishing this genus from *Listriodon* include: (1) small size; (2) relatively well-developed talonid on M₃; (3) no marked diastema between I₃ and canine; (4) narrow lower incisors; (5) small lower canine; (6) broad ridge connecting the transverse lophs. The lower canines are received in niches in the upper jaw.

Lopholistriodon kidogosana Pickford and Wilkinson 1975

The skull is about two-thirds the size of that in *Listriodon splendens* and is light and slender. The muscles for manipulating the rhinarium were weak, so the snout was not used for digging and rooting. The laevator lateralis muscle was remarkably well-developed, suggesting that the lip was raised frequently to allow food to be gathered from the side of the mouth. The high glenoid allows occlusion of the dental rows along their entire length, encouraging vertical chopping with the sharp transverse ridges of the teeth. This indicates a diet of foods such as soft grass, leaves and marsh plants that need a minimum of grinding.

Lopholistriodon moruoroti (Wilkinson) 1976

This species is slightly smaller and more primitive than *L. kidogosana* and is very probably its direct ancestor.

It is possible that *Lophochoerus* Pilgrim 1926, a genus of hitherto unknown affinities from the Sarmatian-Pannonian (Chinji to Nagri horizons) of the Siwaliks, is related to *Lopholistriodon.* Comparison of the type specimen of *Lophochoerus himalayensis,* a right mandibular ramus, with the ramus of *Lopholistriodon moruoroti* reveals some striking similarities. The M₂ of both are approximately the same size, simple with smooth enamel, and no significant accessory tubercles. Both have cusps in pairs that, when worn, coalesce to expose the dentine in dumbbell-shaped lobes. The lobes are connected by a low ridge running from hypoconid to metaconid. The P₄ of both specimens are also very similar, robust with a main central cusp, wide talonid, and a broad posterior rear facet.

Subfamily Sanitheriinae Simpson 1931

In Simpson's classification (1945) the s.f. Sanitheriinae contains a single genus, *Sanitherium* H. von Meyer 1866. However, it can be demonstrated that the subfamily occupies a crucial position in the phylogeny of the Suidae and should be expanded by including the genus *Xenochoerus* (=*Diamantohyus*), hitherto placed in the Suidae incertae sedis, and of the genera *Hyosus, Sivahyus,* and *Hippohyus,* formerly placed in the s.f. Suinae.

Genus *Xenochoerus* Zdarsky 1909

SYNONYMY. *Diamantohyus* Stromer 1926.

TYPE SPECIES. *Xenochoerus leobensis* Zdarsky 1909.

DIAGNOSIS. Small; dental formula $\frac{3.1.?3.3}{3.1.?3.3}$; upper molars with high isolated bunodont labial cusps and low crescentic lingual cusps, opposite in the lower molars; strongly developed labial cingulum; upper premolars show incipient molarization.

REMARKS. Zdarsky (1909) described and named a small maxilla with suid affinities from the early Miocene of Leoben, Austria, as *Xenochoerus leobensis.* In 1913, Forster Cooper placed a maxilla (BMNH M-11989), provisionally in the genus *Hyotherium,* under the new name *H. jeffreysi.* This specimen came from the Gaj horizon of the Bugti Hills, recognized as lower Miocene in age. Stehlin, in a personal communication to Pilgrim, was the first to recognize the resemblance between these two specimens. Furthermore, the presence of a double inner cusp of P^4 clearly distinguishes the Gaj specimen from *Hyotherium.* Subsequently, Pilgrim (1926) tentatively included Cooper's Bugti specimen in the genus *Xenochoerus.* Because this specimen includes part of a skull, the only one known of *Xenochoerus,* it has considerable value.

Xenochoerus africanus (Stromer) 1926

In 1926 Stromer described a small unusual suid from the diamond fields of the Namib Desert, Southwest Africa. He gave it the new generic designation *Diamantohyus* and specific name *africanus.* The general agreement in shape and size of the molars and their highly distinctive lingual cusps complete with transverse lophs clearly relate the Southwest African with the European and Asian specimens of *Xenochoerus.* As Zdarsky's genus has undisputed priority, it is proposed that Stromer's generic name *Diamantohyus* be relegated to the synonymy of *Xenochoerus.* The Southwest African specimen should be known as *Xenochoerus africanus.*

X. africanus is now a relatively well-known species; several specimens have been recovered from East Africa and Gebel Zelten (figure 22.6). It is possible that the left mandible from Southwest Africa described by Stromer (1926) as *Propalaeochoerus* sp. also belongs to *Xenochoerus.*

Figure 22.6 *Xenochoerus africanus,* occlusal aspect of upper dentition: (*a*) right M^2 (Kenya National Museum KNM-KA-46); (*b*) right M^2 and M^1 (Bristol University BU-6407 2). (1.9 × natural size.)

Genus *Sanitherium* H. von Meyer 1866

TYPE SPECIES. *Sanitherium schlagintweitii* H. von Meyer 1866.

DIAGNOSIS. A very small suid, distinguished by its narrow teeth, folded enamel, the strong beaded cingulum of the molar teeth, and caniniform P_1 (mainly from Colbert 1935a).

REMARKS. In 1866 von Meyer published a description of *Sanitherium schlagintweitii,* a new genus and species from the Chinji Zone of Kushalgar, Punjab. This little known genus was again reported from India by Pilgrim (1920) and enlarged upon by him in 1926. He recognized a second species, *S. cingulatum,* very much like *S. schlagintweitii* but with more compressed teeth and a more slender mandible. Colbert (1935a) described a crushed but otherwise complete mandible with a milk dentition from the Chinji Zone of the Lower Siwaliks. He noted a rather peculiar dental formula—one deciduous incisor and one deciduous molar appearing to be absent. The lower true canine was clearly verrucosic. This specimen confirmed the generic rank of *Sanitherium* and its close relationship to *Xenochoerus.*

Sanitherium nadirum Wilkinson 1976

A single tooth (M_2) of the recently described *Sanitherium nadirum* has been recovered from Ombo, Kenya, a site of middle Miocene age. It is slightly larger than *S. cingulatum* with less complex enamel folding. This specimen adds little to our knowledge of the morphology of the genus but extends its geographical range into Africa.

The similarity of *Sanitherium* with *Xenochoerus* can be summarized as follows: (1) the presence of a labial cingulum; (2) labial cusps with anterior and posterior directed ridges wearing to a crescentic pattern; (3) relatively insular bunodont lingual cusps; (4) accessory tubercles developed on the posterolingual face of the metaconid; and (5) oblique wear planes on the molars.

Phylogeny and Classification

Many authors have contributed to the elucidation of suid phylogeny. Pilgrim (1926), in an effort to bring order to a chaotic group, looked hopefully for a factor that showed gradual and significant evolutionary changes. He finally hit upon the last lower premolar and proceeded to erect a complex and ramifying polyphyletic "tree" for the Suidae. Although his work produced a good basis for further studies it contained glaring anachronisms, such as a wide and early divergence of the *Potamochoerus* and *Sus* lines. Matthew (1929) recognized the unduly complex nature of Pilgrim's arrangements. It is obviously unrealistic to place such importance on a single detail. From Colbert's revision (1935a) of Pilgrim's work a more realistic approach to this group has followed.

Colbert's classification (1935a) was essentially monophyletic, returning to the broad concepts of Stehlin (1899). He considered the Suidae as a single line through the early part of the Tertiary, not beginning their adaptive radiation until some time in the Miocene.

A combination of both Pilgrim's and Colbert's ideas is probably closer to the truth. Although applauding the greater simplicity and realism of Colbert's scheme, we think it likely that the Suidae split very early in the Tertiary, shortly after the divergence of the Tayassuidae, and from here the family pursued a polyphyletic evolution.

Status of *Propalaeochoerus* Stehlin.

Stehlin (1899) recognized several migrant "prosuid" groups in the Eocene of Europe and considered them, probably correctly, as terminal species. Stehlin thought that *Propalaeochoerus* of the Oligocene of Europe was also a migrant group but he considered it to be the stem of all the Miocene and later pigs. He advocated a direct evolutionary sequence from *Propalaeochoerus* to modern *Sus* through the intermediate stages of *Palaeochoerus* and *Hyotherium*. Strongly divergent forms were separated under different generic names—e.g., *Listriodon*, *Tetraconodon*, *Phacochoerus*—but other than the derivation of *Tetraconodon* from *Conohyus* few suggestions were made on the affinities of these to the hypothetical *Propalaeochoerus-Sus* main stem. Many recent students dispute the validity of this single evolutionary line and consider *Hyotherium* to be an end form terminating in the late Miocene or early Pliocene.

It has been established that species of *Palaeochoerus* and *Hyotherium* certainly do lie on a continuous sequence, justifying retaining only one generic name, *Hyotherium*. It is also beyond reasonable doubt that *Propalaeochoerus* forms the direct Oligocene ancestor of that line. It differs from the earlier *Hyotherium* species mainly in its primitiveness and generally smaller size. Why then retain *Propalaeochoerus* as a distinct genus? Why not include it as the earliest representative of *Hyotherium*?

Pilgrim (1941), when advocating an African origin for the Suidae, perceptively suggested that *Xenochoerus* (=*Diamantohyus*) *africanus* was a remnant of the ancestral pig stock. At that time *Xenochoerus* was a rare and little known genus restricted to Southwest Africa. It is clear now that, although still a rare element, *Xenochoerus* has a widespread distribution in Burdigalian times: central Europe, Asia, and the whole of the African continent. It is also clear that Pilgrim was correct in proposing *X. africanus* as a primitive pig, but even so, in lower Miocene times *Xenochoerus* had evolved far along its own particular line. It could not be the ancestor of the contemporaneous *Hyotherium*.

Africa is impoverished with respect to fossiliferous deposits older than early Miocene. The Oligocene and Eocene deposits of the Fayum, Dor-el-Tahla, and odd, for the most part one-specimen, sites in North Africa of similar age yield a highly specialized fauna obviously environmentally controlled. Until a more broadly representative early Tertiary site is found, the early history of the Suidae in this continent will remain conjectural.

Turning to Europe, where Oligocene and Eocene faunas are fairly abundant, the only true pig is *Propalaeochoerus*. A close examination of *Propalaeochoerus* reveals a marked similarity to *Xenochoerus*, so much so that isolated teeth would be hard to distinguish. Thus the molars have a structure amenable to interpretation as the lineal ancestor of either *Hyotherium* or *Xenochoerus*. *Propalaeochoerus* may be the common ancestor of both these divergent genera, rather than a single line of descent leading through *Hyotherium* to the higher pigs with *Xenochoerus* a remnant of a pro-*Propalaeochoerus* stock. It is proposed that from the Oligocene *Propalaeochoerus* the Suidae split into two branches: one branch, *Hyotherium*, itself relatively short-lived, contained the progenitor of modern pigs; the other branch, *Xenochoerus*, led ultimately to the highly evolved *Hyosus-Hippohyus* group in the Pliocene and Pleistocene of Asia. By early Miocene times the two genera had traveled sufficiently far along their separate evolutionary paths to be clearly distinguishable. The genus *Propalaeochoerus* is retained to include the antecedents of *Hyotherium* and *Xenochoerus*.

The Divergence of Two Basic Suid Stocks

By the early Miocene the *Hyotherium* and *Xenochoerus* lines were fully differentiated. The subfamily Hyotheriinae is characterized by a trend toward increased size; development from the transverse ridges of *Propalaeochoerus* to rounded, isolated, and tall tubercles of subequal height; a multiplication of these accessory elements, especially on the talonid; reduction or disappearance of the labial cingulum on the lower molars; the retention of relatively simple, trenchant, bladed premolars in the upper and lower jaws; precocious development of "scrofic" hypsodont lower canines. The subfamily Sanitheriinae, represented at this stage by *Xenochoerus* and *Sanitherium,* shows a retention of a primitive selenodont molar structure and subsequently its elaboration and perfection; persistent small size (until later Tertiary times); development of a prominent labial cingulum on upper and lower molars with a marked inclination to preferential wear on the lingual cusps of the upper and labial cusps of the lower molars; molarization of the premolars and, in later forms, achievement of marked hypsodonty in the molars.

Stehlin (1899) was of the opinion that the bunodont condition found in *Hyotherium* and many later pigs was secondary, derived from a primitive selenodont condition. He proposed the term "neobunodonty." Gregory (1910) arrived at the same conclusion. This is proved to be true by the primitive nature of the *Propalaeochoerus* molars which have no distinct accessory tubercles, but instead have oblique, anteriorly directed ridges varying in stature from incipient to strongly developed. The logical progression of this structure is seen in the *Xenochoerus-Sanitherium-Hippohyus* line. This proposed evolutionary sequence will be expounded upon below. The delimitation of the primitive ridges into insular cusps and their subsequent elaboration—"neobunodonty"—is a departure from the main line of evolution, albeit a very successful one.

Simons (1964) noted that the Old World primate superfamilies, the cercopithecoids and hominoids, could be reasonably distinguished from each other by broad differences in the lower molar crown pattern. The Suidae can be split in a similar manner and with increasing confidence from Burdigalian times onwards. Hünermann (1968), recognizing the basic similarity in tooth morphology between the bunodont genera *Hyotherium, Kubanochoerus, Potamochoerus, Dicoryphochoerus,* and *Sus,* provided a *furchenmuster* ("blueprint") furrow pattern for all these genera illustrating the essential constituents of the tooth and their elaborations. It is not possible to fit this pattern to the sanitherine genera, nor

to the ancestral *Propalaeochoerus.* An analogous and complementary scheme is required to include those forms which retain the basic ancestral pattern, although elaborations upon it are so complex in later periods that it becomes almost unrecognizable. Such a scheme is provided in figure 22.7, which illustrates the evolution of the furrow pattern in the subfamily Sanitheriinae. The layout and notation of Hünermann is closely followed so that comparison can be readily made. The insertion of secondary, intermediate furrows is irregular in the Sanitheriinae and these tend to be crowded along the anterior and posterior ridges.

To summarize: in the neobunodont pigs there are four subequal cusps that are elaborated by the regular insertion of furrows and secondary furrows. These tend to cut off secondary tubercles, which become rounded and isolated. In the sanitherine pigs the cusps are unequal, with those on one side forming fore and aft ridges. Elaboration of these teeth also takes place by the insertion of furrows and secondary furrows, but in this case the ridges are retained and elongated. The furrows become deeply incised but do not cut off isolated tubercles; rather the lobes are produced, become elongated lengthwise and remain contiguous for the greater part of their length. The secondary furrows are very irregular and produce a variety of complications on the basic pattern.

Figure 22.8 shows the proposed evolution of the neobunodont and sanitherine conditions from the *Propalaeochoerus* stock in the upper and lower second molar.

The Position of *Sanitherium* and the Origin of *Hyosus-Hippohyus*

The close relationship of *Sanitherium* to *Xenochoerus* has already been discussed above. The African *Sanitherium nadirum* is too far advanced to be the successor of *X. africanus* but is so close to that species that it must have arisen along a side branch late in the Burdigalian.

The Siwalik genera *Hyosus* Pilgrim 1926, *Sivahyus* Pilgrim 1926, and *Hippohyus* Falconer and Cautley 1847 are essential to the understanding of suid evolution and will be considered briefly. *Sivahyus* is probably synonymous with *Hyosus* (Colbert 1935a). Pilgrim (1926) advocated *Sivahyus* as the direct lineal ancestor of *Hippohyus*. Lydekker (1883) first realized the resemblance in the folding of the enamel between *Sanitherium* and *Hippohyus*, but Pilgrim (1926) pointed out that the resemblance between *Sanitherium* and *Hyosus* is even more striking. Pilgrim mentions two advances in structural

Upper Right 2nd Molar   Lower Left 2nd Molar

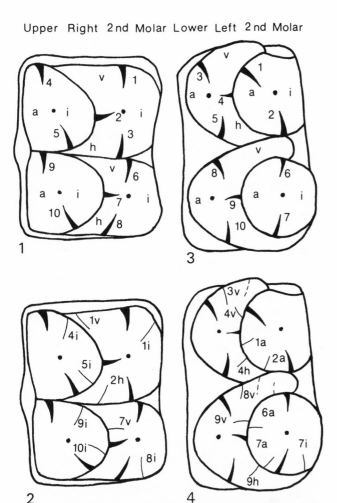

Figure 22.7   Furrow pattern of sanitherine suid molars. Key: *1* and *3*, *notation of the major furrows:* (*a*) Aussenfeld (outer area); (*i*) Innenfeld (inner area); (*v*) Vorderfeld (front area); (*h*) Hinterfeld (rear area). *2* and *4*, *notation of the minor furrows:* each index refers to the nearest narrow line on the figure. Dotted lines represent supernumerary, unnotated furrows. (From a similar figure in Hünermann 1968 for bunodont suid molars.)

complexity of *Sivahyus* (=*Hyosus*) *punjabicus* over *Sanitherium:* a greater lateral compression of the cusps and the greater number and depth of the enamel folds.

The talon of *Sanitherium* has two equal cusps, a more complex arrangement than is found in *Sivahyus,* but the talon is notoriously variable and even in *Hippohyus,* although well developed, it does not achieve the complexity and importance of the *Sus-Dicoryphochoerus* line. A reduced talon in *Sivahyus* would not be incompatible with this scheme of evolution. The only serious difference between *Sanitherium* and *Sivahyus* is the absence in the latter of the

labial cingulum. *Sanitherium* appears to be a highly evolved side branch of the *Xenochoerus-Hippohyus* line that paralleled it and reflected many of its features. The labial cingulum in *Xenochoerus* is diagnostic but not as prominent as that of *Sanitherium* where it reaches its acme of development.

Thus the supposition that *Xenochoerus-Sanitherium-Hippohyus* form a distinct evolutionary branch of the Suidae can be convincingly demonstrated. The evolution of minor details of crown morphology can be traced with some confidence through this sequence.

There is ample and good quality skull material of *Hyppohyus*, both of the earlier *H. sivalensis* and the advanced terminal form *H. grandis.* Lydekker (1884), Pilgrim (1926), and Colbert (1935b) noted the apparent clear distinction between the skull of *Hippohyus* and *Sus.* They assumed that this was due in part to the retention of primitive features. They retained *Hippohyus* in the subfamily Suinae. *Hippohyus* does retain some primitive features in its skull —e.g., long sagittal crest and broad occiput. However, this probably does not indicate a relationship to *Hyotherium,* but rather a retention of some of these features found in ancestral pigs.

During the late Tertiary period climatic changes accompanied by general continental uplift produced a change from lush lowland forests to vast grassy plains. Therefore, it is not unreasonable to assume that a branch of the Suidae would adopt a grazing habit, thus taking advantage of this new niche. In the dentition of *Hippohyus,* the acquisition of complexly folded enamel, together with hypsodonty, suggests such an adaptation. The salient features of the skull may also be interpreted in this light. Grazing requires a grinding action in the jaws: possibly rotary, a combination of propalinal and orthal movements, but often a predominantly orthal motion. The *Hippohyus* skull may be interpreted as adapted to the latter. *Phacochoerus* is also an orthal grinder and in many ways resembles *Hippohyus.* The zygomas are widely separated from the cranium and in addition a prezygomatic shelf extends as far forward as P^4. Both these features indicate a powerful masseter muscle. The snout is relatively short compared to the postorbital length and the nasal bones are wide. This indicates that the snout was not used as a digging organ as in *Sus* and *Potamochoerus.* The dental row has marked convexity in profile, which guides the mandible over the upper jaw in sideway movements and finds its analogue in the "stop" of M_1 in *Phacochoerus. Hippohyus* is also characterized by elevated orbits and flat frontal bones, again also

Upper Right 2nd Molar

Lower Left 2nd Molar

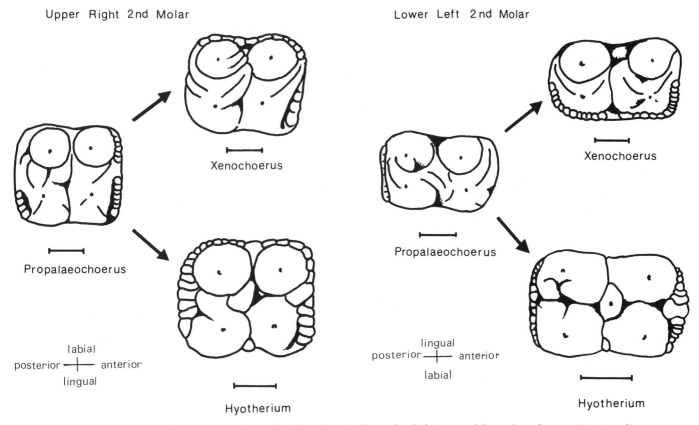

Xenochoerus

Propalaeochoerus

Hyotherium

Xenochoerus

Propalaeochoerus

Hyotherium

labial
posterior —┼— anterior
lingual

lingual
posterior —┼— anterior
labial

Figure 22.8   Diagrammatic representation of suid molars to show the dichotomy of *Propalaeochoerus* into two divergent lines of evolution: the "neobunodont" condition (*Hyotherium*) and the "sanitherine" condition (*Xenochoerus*).

found in *Phacochoerus*. The elevated orbits follow as a consequence of the high position of the glenoid fossa which gives increased power in grinding. The significance of the flat frontal bones is not fully understood, but may reflect a method of fighting similar to that of *Phacochoerus*.

The degree of difference in skull and tooth structure between *Hippohyus* and *Sus* is greater than between genera hitherto considered as widely divergent—e.g., *Sus* and *Listriodon*, *Sus* and *Tetraconodon*. There thus seems little justification for the inclusion of *Hippohyus* in the s.f. Suinae.

The Origin of the Subfamily Tetraconodontinae

Simpson (1931) erected the subfamily Tetraconodontiinae to contain suid genera with very much enlarged premolars. The Asian genus *Conohyus* Pilgrim 1926 differs little from *Hyotherium* except for its relatively enlarged P_3 and P_4 and can be considered the ancestral form. *Tetraconodon* Falconer 1868, from the Pleistocene of India, is the culmination of this specialization, with relatively enormously enlarged premolars and great body size.

Pilgrim (1926) erected a third genus, *Sivachoerus,* within the subfamily Tetraconodontinae. *Sivachoerus* and *Tetraconodon* were both derived from *Conohyus* but were divergent, the latter evolving more rapidly so that it became much larger and more specialized than the rather conservative *Sivachoerus* (Colbert 1935b).

Sivachoerus has also been described from the Pliocene of North Africa (Tobien 1936 and Leonardi 1954), although Cooke and Ewer (1972) consider this material to belong to a different genus, *Nyanzachoerus* (see the next section, Upper Miocene-Pleistocene Suidae).

The skulls of the more advanced tetraconodonts show similarities with the Hyaena. The latter is a specialized carnivore capable of crushing bone. The tetraconodonts were conceivably specialized to cope with analogous hard food—perhaps cracking nuts, cones, or bamboo.

The Origin of the Subfamily Suinae

The subfamily Suinae is clearly on the "neobunodont" line of descent. Having removed the hypsodont group, *Hyosus-Hippohyus,* the genera *Sus, Dicory-*

phochoerus, Microstonyx, and the Pliocene-Pleistocene–Recent genera remain.

Pilgrim (1926) erected the genus *Dicoryphochoerus* to contain a large number and variety of Siwalik Suidae characterized by a P$_4$ with a cleft main cusp, which could not be placed in *Hyotherium.* Because the nature of the P$_4$ and canine are unreliable as diagnostic criteria, Pilgrim's genera must be considered suspect. Primitive *Dicoryphochoerus* species are clearly ancestral to *Sus* and to later suine genera (Colbert 1935a, Hünermann 1968). A form such as *D. haydeni* may be considered as the stem for all these genera. The advanced dicoryphochoerids—such as *D. titan* and *D. robustus*—seem to merit generic distinction. However, it is not unlikely that the early members of the genus are synonymous with, or at least derived from, a small bunolistriodont. Both show, for example: (1) simple bunodont molars with few accessory elements; (2) the P$_4$ with a bifid central conical cusp and a large tubercle on the talonid; (3) the remaining lower premolars with pointed blades and short talonids; (4) the P^4 with two cusps, a labial and lingual, both compressed laterally and tuberculated; and (5) the remaining upper premolars with a single flattened bladelike cusp and broad posterolingual talons.

Of the extant suine genera, *Sus, Phacochoerus,* and *Babirussa* can be traced, via *Dicoryphochoerus,* to the subfamily Listriodontinae rather than having an independent origin from the Hyotheriinae. *Potamochoerus* and *Hylochoerus* appear to be a more ancient lineage with its origin closer to the ancestral bunolistriodont stock (although Cooke, in the next section, presents a different phylogeny).

Microstonyx Pilgrim 1926 is an upper Miocene to lower Pliocene form so far only reported from Europe and Asia. It is closely related to *Dicorphochoerus* (Schmidt-Kittler 1971, Hünermann 1968).

Center of Origin and Migration of the Early Suidae

It is impossible to state unequivocally which continent witnessed the emergence of the true suids. Certainly during the Oligocene, the probable ancestral stock, the Dichobunids, are most abundant in Europe. It is only here too that *Propalaeochoerus,* the first true pig, can be called common. It seems likely therefore that the suids made their debut here. By the early Miocene, *Hyotherium* and *Xenochoerus,* the basal members of the two succeeding lines of evolution, were widespread in Europe, Asia, and Africa. Africa, in addition, could boast of a

highly evolved and differentiated stock of bunolistriodonts and intermediate listriodonts although Europe and Asia also had equivalents of the most conservative of these animals. It is clear that the center for this assemblage had shifted from Europe to Africa. Pilgrim (1941) came to this conclusion but he used as evidence the then unique presence of *Xenochoerus* (=*Diamantohyus*) in Africa, which he considered a remnant of the prosuid stock.

The intermediate listriodonts in Africa are most convincing for this transitory stage. However, the fully evolved *Listriodon* is a typically Eurasian middle Miocene animal not represented in Africa. It seems probable that in the middle and later Miocene, movement of suids between Eurasia and Africa occurred in both directions. Different lines were evolving in separate localities. The probable successor of *Lopholistriodon, Lophochoerus* Pilgrim 1926, is an Indian genus whereas *Sanitherium nadirum,* a very typical Siwalik form, is also found in Africa. Nevertheless it is certain that by the later Miocene the center of radiation had moved decisively to Asia. It is here, from Tortonian onward, that a second great burst of adaptive radiation took place. This concerned, for the most part, the subfamilies Tetraconodontinae and Sanitheriinae. Elements of these stocks found their way to Europe and occasionally to North Africa (*Nyanzachoerus,* Cooke and Ewer 1972), but communication with Africa as a whole was very much restricted. Africa remained relatively impoverished until the late Pliocene–early Pleistocene.

It is probable that the subfamily Suinae also arose in Asia from a dicoryphochoerid stock, although via a bunolistriodont ultimately of African origin. By middle Pliocene-Pleistocene times the suids, represented mainly by the subfamily Suinae, had repenetrated Africa. There followed a third burst of adaptive radiation that gave rise to an enormous variety of types and contained the immediate ancestors of the modern African suid genera.

Palaeoecology

Although *Sus* and *Potamochoerus,* perhaps the most ubiquitous of the extant suids, show a wide tolerance of habitus, many suid genera show a marked specific environmental preference.

Hyotherium is a widespread genus, especially abundant in East African fossil faunas. However, it is not found at Gebel Zelten. The palaeoenvironment of Gebel Zelten is thought to have been a series of estuaries, the rivers possibly lined with riverine forest and backed by floodplains. Beyond this would be savanna (Selley 1966, Doust 1968). Most of these ecological niches are also found at Rusinga; extensive arid floodplains with streams bordered with forest (Andrews and van Couvering, pers. comm.). In addition there were the volcanic cones of Rangwa and Kisingiri with steep slopes that probably supported forests. Evidence derived from a functional analysis of the postcranial skeleton of *Hyotherium* from Rusinga (Wilkinson 1972) indicates a slender, fleet-footed animal. The structure of the foot supports the supposition that *Hyotherium* was adapted to life on steep, rocky slopes, although Schmidt-Kittler (1971) proposes that *H. soemmeringi,* from the upper Miocene locality of Sandelzhausen, was a creature of marshy woodlands. This conclusion is based on the presence of lignite deposits (Falbush and Gall 1970), and it is difficult to reconcile with the situation in East Africa.

The extensive broad estuaries at Gebel Zelten were also instrumental in diversification of the bunolistriodonts. The smallest form, *Kubanochoerus jeanneli,* remained conservative with very similar relatives in Europe and Asia. The simplicity, bunodonty, and brachydonty of the dentition, all primitive features, suggest an unspecialized diet. It seems that these early offshoots from the *Hyotherium* line probably lived close to rivers and fed on soft vegetation. *K. jeanneli* and its close relatives perhaps occupied the niche of today's African river hog (*Potamochoerus porcus*), feeding on, among other foodstuffs, soft bankside vegetation (Phillips 1926). This habitus would allow the small bunolistriodonts a wide geographical range, especially in the early Miocene, a time of widespread tropical conditions. The attainment of large and ultimately enormous dimensions suggests a greater commitment to an aquatic existence. One can imagine the giant Gebel Zelten bunolistriodont in a similar niche to the present-day hippopotamus. This diet of water vegetation would require no great dental adaptions; simple crushers would suffice and could lead to great size. Such a mode of life might explain the rarity and localized nature of the fossilized giant pigs. The forests shrank during the harsher, arid Pliocene, which perhaps contributed to the demise of the bunolistriodonts.

The marked divergence of *Listriodon* coincident with drier climatic conditions has been equated with a transition to a grazing habit. However, lophodont teeth are not typically associated with grazers (e.g., the tapir and kangaroo), and it may be better to consider *Listriodon* as a specialized browser. Neverthe-

less it would be surprising if, within such a varied group, the grazing habit had not been adopted at some time. The modern *Phacochoerus* is a rather specialized grazer but its similarity in skull morphology with *Hippohyus* is marked. In addition, the dentition of *Hippohyus* is hypsodont and complexly folded. Although *Xenochoerus* must be considered a nonspecialized browser, the more evolved members of the subfamily Sanitheriinae indicate a marked adaption to a grazing habit.

Upper Miocene-Pleistocene Suidae[4]

Subfamily uncertain (?Tetraconodontinae)

From late Miocene and Pliocene deposits in East Africa, North Africa, and South Africa, large extinct suids have been referred to the two genera *Nyanzachoerus* and *Notochoerus,* which share certain features that separate them from the other Pliocene and Pleistocene fossils. *Nyanzachoerus* seems to be descended from a form like *Kubanochoerus massai,* which Wilkinson (1976) placed in the subfamily Listriodontinae but which may equally well be regarded as a hyotherine. Both in *Nyanzachoerus* and in *Notochoerus,* the tympanic bullae are relatively slender and forwardly directed, recalling the condition in the Hyotheriinae. The enlarged premolars may be taken to suggest an affinity with the Tetraconodontinae, which in turn is closely related to the Hyotheriinae.

Genus *Nyanzachoerus* Leakey 1958

SYNONYMY. *Sivachoerus giganteus* Tobien 1936, *Sivachoerus* cf *giganteus* Leonardi 1952, *Sivachoerus syrticus* Leonardi 1952 (non *Sivachoerus* Pilgrim 1926).

TYPE SPECIES. *Nyanzachoerus kanamensis* Leakey 1958.

DIAGNOSIS. A genus of Suidae with cheek teeth similar to *Potamochoerus* in basic structure but tending to be more hypsodont and with main cusps of molars more distinctly columnar; third and fourth premolars relatively enlarged and more hypsodont than in *Potamochoerus;* lower fourth premolar with three roots. Incisors similar to those of *Potamochoerus.* Upper canines oval to flattened oval in cross-section, bearing little or no enamel, and with closed or nearly closed roots. Lower canines U-shaped or heart-shaped in cross-section near base,

---

4 By H. B. S. Cooke.

becoming more verrucose or triangular towards the tips; thin, weakly grooved enamel on two lateral faces only; roots partially closed or open. Strong sexual dimorphism exhibited in size of canines and massiveness of skull. Hollow bony protuberances or bosses on zygomatic arches in male, weak or absent in female. Corpus of mandible heavy and contrasting markedly with unusually thin bone forming angle.

REMARKS. This genus seems to be a characteristic Pliocene one and it is doubtful if it extends into the Pleistocene, assuming the base of the latter to lie at about 2 million years ago. The morphology of the skull and dentition was discussed fully in a recent paper by Cooke and Ewer (1972). The time relationships are shown in figure 22.9.

Nyanzachoerus kanamensis Leakey 1958

The genotype species, *Nyanzachoerus kanamensis*, is still known certainly only by the type right mandibular ramus from Kanam, Kenya, together with the few associated specimens. It differs from *N. pattersoni* (below) mainly in the greater elongation of the third premolar and narrowing of P_4, while the molars are a little narrower but morphologically very similar. The diastema is longer than in any of the known material of *N. pattersoni* and it seems that the animal had a longer and narrower snout than in the Kanapoi species. In the Kaiso Formation, Uganda, teeth from the Upper Kaiso are inseparable from *N. kanamensis* but those from the earlier Kaiso have a closer resemblance to *N. pattersoni*, perhaps suggesting that *N. kanamensis* represents a late stage rather than a lateral offshoot. More material with the characteristics of the type are needed to resolve the problem and it is possible that *pattersoni* may be a junior synonym of *kanamensis*.

Nyanzachoerus devauxi (Arambourg) 1968

From the upper Miocene of Bou Hanifia (Wadi el Hammam) in Algeria, Arambourg (1968) described

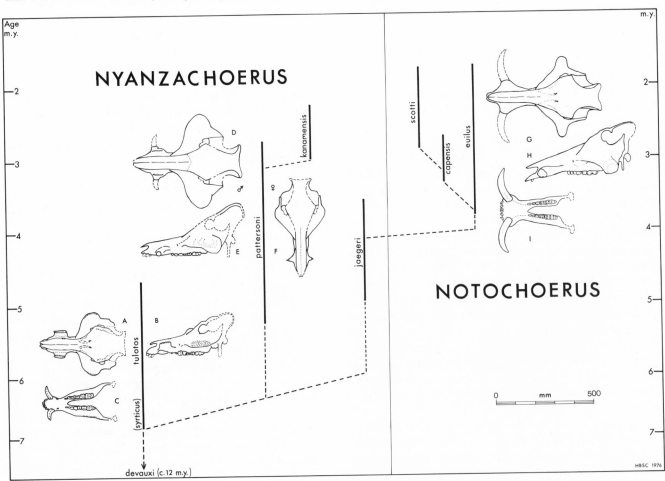

Figure 22.9 Time relationships of species of *Nyanzachoerus* and *Notochoerus*. *Nyanzachoerus tulotos:* male skull (*A–B*); male mandible (*C*). *Nyanzachoerus pattersoni:* male skull (*D–E*); female skull (*F*). *Notochoerus euilus:* male skull (*G–H*); male mandible (*I*). (Approximately 1/19 natural size; drawn scale equals 500 mm.)

the parts of a disrupted mandible under the name *Propotamochoerus devauxi.* The detached symphysis has the two central incisors and one canine socket, as well as a small but damaged P_1. The incisors are morphologically similar to those of *Potamochoerus porcus,* but an isolated lateral incisor is proportionally larger. An isolated canine is gently curved, having a long closed root and a short triangular tip, bearing smooth enamel on two sides; it may well represent a female. The cheek teeth of both sides are well preserved, with a reduced P_2 followed by an enlarged P_3 that is as wide as M_1, and a P_4 that is slightly wider. Both the premolars and the molars are extremely like those of *Nyanzachoerus pattersoni* and there can be little doubt that they belong to the same genus. The overall lengths of the individual teeth are slightly less than in *N. tulotos,* but they are not as broad, and the talonid of M_3 is less developed than in *N. pattersoni.* It would, in fact, provide an almost ideal ancestor for all three of the later *Nyanzachoerus* species, developing into *tulotos* by broadening the crowns, into *N. pattersoni* by some growth in the talonid, and into *N. jaegeri* by substantial enlargement of the third molars, accompanied by slight reduction in the premolars.

Although Arambourg (1968) placed this species in *Propotamochoerus,* his reasons for doing so are not clear as the Indian material does not show the enlargement of the P_3 and P_4 exhibited in the Algerian specimen, although there are some very general resemblances. If Arambourg is correct, *Nyanzachoerus* would fall away as a junior synonym of *Propotamochoerus,* but it seems more probable that it is an African genus derived from a *Kubanochoerus* stock.

Nyanzachoerus syrticus (Leonardi) 1952

In 1952, Leonardi described a skull and a lower jaw from Sahabi, in Libya, referring the skull to *Sivachoerus* cf *giganteus* and naming the mandible *Sivachoerus syrticus.* He separated the two for reasons that appear valid, including: the massiveness of the mandible, which seems inappropriate for the skull; the presence of a well developed P_1 in the mandible, whereas P^1 is clearly not present in the skull; differences in the state of fossilization. The skull is in the Istituto di Geologia e Paleontologia of the Universita degli Studi in Rome; reexamination (Cooke and Ewer 1972) suggests that it may be an early ancestor of *Nyanzachoerus jaegeri.* However, the holotype mandible cannot be traced and apparently was destroyed during the war. Illustrations and published measurements show that it is very close to *Nyanzachoerus tulotos* in dental characters, although the symphysis is decidedly longer in the Sahabi jaw, and

it almost certainly represents an early stage in the same lineage. Under normal circumstances it would be reasonable to regard *N. tulotos* as a synonym of *N. syrticus* but as the only known specimen of the latter is lost, it seems best to regard it essentially as a nomen vanum and to apply that specific name only to the missing holotype.

Nyanzachoerus tulotos Cooke and Ewer 1972

Nyanzachoerus tulotos is based on a good collection from the Lothagam sediments in northwestern Kenya, with an estimated age of about 5-6 m.y. (basal Pliocene or uppermost Miocene). Fragmentary remains, still undescribed, occur in the Lukeino Beds of the Lake Baringo area and seem to represent an earlier form of *N. tulotos* or an even more primitive species. The lower jaw described by Leonardi (1952) from Sahabi as *Sivachoerus syrticus* is close to *Nyanzachoerus tulotos.*

The skull of the male *N. tulotos* is much the same size as that of *Hylochoerus meinertzhageni,* but the zygomatic arches are more robust and inflated, with rugose areas on the inflated jugal area, also seen in the male forest hog (figure 22.9*A–B* and figure 22.11*P–Q*). The nasals are expanded above the canines, as in the forest hog, but the canine flanges have large vertical extensions reminiscent of those in the bush pig, but rising vertically instead of diagonally backward. The whole posterior region of the skull is very broad, with the orbits, as in the forest hog, just lower than the frontoparietal surface. In front of the postorbital process the upper margin of the orbit is inflated and very slightly overhangs the orbital plane. Behind this ridge lies another marginal elevation of the parietal, giving the skull the "lumpy" appearance from which the specific name is derived. The full complement of incisors is retained. The male canine is short but not strongly curved, and carries a ventral band of ribbed rugose enamel, as well as some enamel on the dorsal surface, but not the two inset narrow enamel bands found in the bush pig.[5] The lower jaw is *Sus*-like in general morphology (figure 22.9*C*), as are the incisors. The lower canines do not have the usual triangular cross-section but are U-shaped or ovate with a flattened posterior face. The first and second premolars are retained and are small elongate teeth with a low main cusp. The third and fourth premolars are greatly enlarged, particularly in breadth, being broader than

---

5 The isolated right upper canine KNM-LT 317 described as "Suid indet" by Cooke and Ewer (1972) is now believed to belong to the holotype skull, the alveoli of which are now known to have been distorted by slight crushing.

the first molar and about as wide as the second molar. Their morphology is characteristic, P^4 having a strong anteroposterior valley between the closely fused outer elements and the protocone. The lower P_3 is essentially just a stout cone with no anterior cingulum but possesses a small posterior one: P_4 is *Potamochoerus*-like but the posterior cone is more isolated. The main pillars in the molars are more columnar than in *Sus* and the talonid is better developed, but the general morphology is fairly similar, although with some distinctive characteristics.

A number of features of the skull and dentition resemble those which characterize the lower Miocene *Kubanochoerus massai* (Arambourg 1961) from Gebel Zelten, Libya. The molars are very similar both in size and structure and the premolars are large and also similar in general morphology, although not relatively as enlarged as in *Nyanzachoerus* and not showing the strong disparity of size between the second and the third premolars. P^4 is lower crowned but the metacone and paracone are similarly fused and there is a similar wide valley separating it from the protocone; as in *Nyanzachoerus*, this valley is not blocked at the ends by elevations of the cingulum. In P_3 the main cone is weaker in *Kubanochoerus* and the posterior element well separated. The male upper canine is more strongly curved, and the enamel is not grooved or rugose; the lower canines are reportedly triangular. *K. massai* is notable for its hornlike cranial bosses above the eyes and the "lumpiness" of the *Nyanzachoerus tulotos* cranium may well be analogues. The differences are such as to require separation of this material from *Kubanochoerus,* but the resemblances suggest strongly that *Nyanzachoerus tulotos* is derived from a *Kubanochoerus* stock.

Comparison of the various species of *Nyanzachoerus* with *Sivachoerus* of the Siwaliks is rendered difficult by the inadequate and sometimes confused accounts of the material. The third and fourth premolars of *Sivachoerus* are enlarged and the general structure is similar to those of *Nyanzachoerus*, although in *Sivachoerus* the P^4 has a relatively smaller and more divided paracone-metacone. Apart from the absolute size, the teeth of *Propotamochoerus* are about as similar. In the lower jaw of *Sivachoerus giganteus* there is little or no diastema behind the canines, the corpus is less robust, and the lower canine is of the normal triangular type, quite different from that of *Nyanzachoerus*. As far as can be determined, the holotype skull of *Sivachoerus giganteus* must have been short-snouted to fit what Pilgrim (1926) regards as the mandible most typical

of the genus. This is very different from the elongate muzzle of *Nyanzachoerus*. In the *Sivachoerus* skull, the rounded frontal area is of modest width, compared with the wide flat frontal area in *Nyanzachoerus,* and the parietals are narrow and domed, whereas in *Nyanzachoerus* the parietal constriction is wide and flat. The zygomatic arches, although robust, lack any sign of lateral inflation and are more like those of *Sus verrucosus*. Accordingly, although it is possible that *Sivachoerus* and *Nyanzachoerus* are related in common descent from a *Kubanochoerus*-like stock, they do seem to represent distinct lineages.

Nyanzachoerus pattersoni Cooke and Ewer 1972

Nyanzachoerus pattersoni is the typical suid of the Kanapoi sediments in northwestern Kenya and is found in the upper part of the Lothagam sequence. In Ethiopia it occurs in the Hadar Formation in the Afar area, and in the Mursi Formation and lower part of the Shungura Formation of the Omo area. It is found in the East Rudolf region of Kenya in the Kubi Algi Formation, in the Lake Baringo basin in the Chemeron Formation and in some other deposits, and in Uganda in the Kaiso Formation and in the Kazinga Channel. Undescribed material from Langebaanweg in South Africa is close to *Nyanzachoerus pattersoni* and probably represents a slightly earlier southern race (Q. B. Hendey, pers. comm.).

By comparison with *N. tulotos, N. pattersoni* has a larger and more "normal" skull, lacking the peculiar lumps and elevated cranial flanges, but remarkable for the extreme sexual dimorphism displayed in the exaggerated lateral inflation of the zygomatic arch in the male (figure 22.9*D*). The female canines are small rooted teeth but in the male they are larger and like those of *N. tulotos*. The canine flanges in the male are like those of the female forest hog in general morphology. The third and fourth premolars are not as expanded as those of *N. tulotos*, but are similar in form (figure 22.10*A*). In the third molars the talon is better developed and in the lower molars there is a strong third pair of laterals. The teeth are also more columnar and a little higher crowned. The lower canines are U-shaped in their early part but become more triangular towards the base, but still with a rounded vertex, and the cross-section is best described as heart-shaped. Postcranial material from Kanapoi is generally similar to the corresponding elements in *Hylochoerus* but the fossil form did not have the broad feet of the forest hog nor the cursorial adaptations characteristic of *Phacochoerus*.

Nyanzachoerus jaegeri Coppens 1971

Nyanzachoerus jaegeri was named while the account of Cooke and Ewer (1972) was in the press, and it thus has priority over the latter's species *N. plicatus,* with which it seems to be synonymous. The type consists of much of a mandible and associated M² and M³ of the same individual and came from Hamada Damous, Tunisia, in deposits assigned to the lower Villafranchian. The material described by Cooke and Ewer is mainly from Kanapoi and includes some skull material, from which a tentative reconstruction was made, as well as canines and partial mandibles. The skull was about the same size as that of *N. pattersoni* but the root of the zygoma originated a little higher and the jugal arch seems to have flared laterally. The snout was elongate and narrow, with a transverse section involving laterally curved nasals that meet the maxilla in a curved "shoulder" that overhangs the ventral part of the maxilla. The canines are surrounded by a small flange and were directed outward at about 45° to the axis of the skull and somewhat downward. The canines are peculiar, being flattened oval in cross-section and shaped rather like a banana; the roots are closed and there is no enamel preserved, although there may originally have been a thin covering. The lower canines resemble those of *N. pattersoni* but with a stronger tendency towards becoming triangular. The third and fourth premolars are morphologically like those of *N. pattersoni* but relatively less enlarged and show a *Sus*-like kink in the main cone of the P₄. The third molars are larger, more elongated, and somewhat higher crowned than those of *N. pattersoni* and both uppers and lowers have a third pair of lateral pillars. The enamel of the molars is more obviously folded than in *N. pattersoni,* giving a more complex pattern of the enamel islands produced by wear, particularly in M³. This species also occurs in the upper part of the Lothagam sequence, perhaps at Ekora, near Kanapoi, and has been found in the Kazinga Chennel in Uganda, and in the Chiwondo Beds of Malawi. The suid skull from Sahabi, regarded by Leonardi (1952) as *Sivachoerus* cf *giganteus* and the material from Wadi Natrun that appears essentially similar (Tobien 1936) may represent an early form of *N. jaegeri,* somewhat closer to *N. pattersoni.* From Lukeino in the Baringo Basin of Kenya there is a palatal fragment with most of the cheek teeth that also resembles the Sahabi and Wadi Natrun material. One specimen from Langebaanweg, from the pelletal phosphate member, may belong to this species (Q. B. Hendey, pers. comm.). *N. jaegeri* also occurs in the Chiwondo Beds of Malawi, represented by a single RM³ from Uraha Hill that was described by Hopwood (1931) as a lower third molar of *Hylochoerus* and by Leakey (1958) as an upper molar of *"Mesochoerus paiceae."*

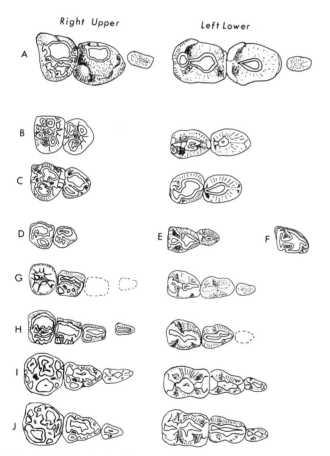

Figure 22.10 Occlusal views of premolar series in some extinct African suids. (A) *Nyanzachoerus pattersoni;* (B–C) *Notochoerus euilus;* (D–F) *Metridiochoerus jacksoni;* (G–H) *Potamochoeroides shawi;* (I–J) *Kolpochoerus limnetes.* (2/3 natural size.)

Genus *Notochoerus* Broom 1925

SYNONYMY. *Gerontochoerus* Leakey 1943.

TYPE SPECIES. *Notochoerus capensis* Broom 1925.

DIAGNOSIS. A genus of Suidae of large size, possessing hypsodont third molars in which the main lateral pillars are strongly folded, tending to produce dumbbell or H-shaped enamel islands, particularly in the lowers. Premolars reduced, with only the third and fourth premolars normally retained in the adult. P⁴ with small, closely apressed, paracone and metacone well separated from small rounded protocone. Zygomatic arches robust and possessing thimble-shaped lateral projections in the male. Upper canines dorsoventrally flattened, carrying a ventral enamel band, and flaring outward in a strong flat curve not much above the palatal plane. Mandible robust with long, wide symphysis; anterior border almost straight and incisors small and well separated. Lower canines heart-shaped to U-shaped, at least in early stages of growth, and flaring laterally parallel to the uppers.

REMARKS. *Notochoerus* occurs as a contemporary of *Nyanzachoerus pattersoni* at Kanapoi and has also been found in the Kubi Algi Formation in the East Rudolf area in Kenya and in the Mursi Formation of the Omo area in Ethiopia, all of which are about 4 m.y. or a little older. It ranges up through the Omo sequence into the lower Pleistocene. Paucity of skulls, or even good mandibles, from the younger strata, hampers the evaluation of differences observed chiefly in the commonest element found as a fossil, namely isolated third molars.

Notochoerus capensis Broom 1925

Notochoerus capensis was founded by Broom (1925) on an isolated upper right third molar from the Vaal River gravels, South Africa. The tooth lacks an unknown amount from the anterior and was restored by Broom with an additional pair of laterals, making a total of five pairs (the last rather small) and six medians, but Shaw (1939) reconstructed it with the addition only of an anterior complex. It is not possible to be sure which is correct as both counts of laterals can be matched in *Notochoerus* teeth from other areas, but on balance Broom's interpretation is favored because of the robust nature of the talon and large size of the medians. The tooth is moderately hypsodont, with the maximum crown height at least 1.5 times the maximum basal breadth. A feature of the type that distinguishes it from most other material assigned to *Notochoerus* is the anteroposterior "length" of the lateral wall of the lingual enamel islands. This feature is also shown by a pair of lower third molars from the Makapansgat Limeworks, South Africa, in which there are five well-developed pairs of laterals, followed by two small pairs of columns. Ewer (1958) ascribed these teeth to *"Gerontochoerus" euilus*, but it seems more probable that they represent *Notochoerus capensis*. An undescribed incomplete RM₃ from Makapansgat matches the other teeth and has a crown height that is almost exactly twice the anterior basal breadth. Two fragments of upper molars also occur and match quite well with Broom's type.

Two partial skulls from Sibilot in the Kubi Algi Formation of the East Rudolf area must be assigned to *Notochoerus*, but they differ both in absolute size and in the lengths of the third molars. The larger specimen consists of the front two-thirds of a skull carrying third molars with five pairs of laterals and, although a little less hypsodont, the teeth are similar enough to Broom's reconstructed type for the specimen to be assigned tentatively to *N. capensis*. Two premolars are present, but there is no sign of P² or of a scar where it might have been. Enough of the zygomatic arches is preserved to show that they carried projections or "knobs" like those found in skulls of *N. euilus*. The animal was half as big again as the giant forest hog, or nearly twice the size of a warthog. The maxillary part of the zygoma is a little like that of the warthog, as also is the broad short braincase, but the orbits are not as elevated. The canine sheaths extend smoothly back so that their bases lie above the premolars. The tusks sweep outwards in an even curve, but not as sharply sideways as in the warthog, and they rise only a little above the palatal plane. The canines themselves are dorsoventrally flattened and have a cross-section best described as resembling a squashed bell. The nasals taper toward the tips and the premaxilla apparently projected beyond them, which is unusual.

Coppens (1967) has referred some teeth from the Lake Chad area to *N. capensis*. The third molars show the elongate lateral wall of the lingual enamel islands, and also in some of the specimens, the rather poor "pairing" of the main lateral columns exhibited in the holotype (the enamel islands on the lingual side being stouter and more complexly folded, while those in the buccal side are smaller and are also displaced a little toward the front of the tooth). However, the Chad teeth are distinctly lower crowned than the type. The upper third molars have four fully developed pairs of laterals, with an incipient fifth pair in some. The lower third molars have four or five pairs of laterals, with an incipient sixth pair in one specimen, and the laterals on the two sides are more symmetrically arranged than in the uppers. The Chad teeth also show that with increasing wear the minor complexities of the enamel folding decrease, although the islands remain rather irregular in the uppers; in the lowers the main lateral pillars acquire a crude approximation to the dumbbell or H-shape that seems typical of the genus. The Chad third molars are larger than any of the material ascribed to *N. euilus* and may be left provisionally in *N. capensis*. The material from the Chiwondo Beds which Mawby (1970) placed in *N. capensis* is more probably *N. euilus*, and the specimens from the Omo area that Arambourg (1947) regarded as *N. capensis* are here referred to *N. scotti*.

Notochoerus euilus (Hopwood) 1926

The species *euilus* was established by Hopwood (1926b) as *Hylochoerus euilus* on the basis of incomplete lower third molars from the Kaiso Formation, Uganda. Cooke and Coryndon (1970) restudied Hopwood's type material and concluded that *N. euilus* differed from *N. capensis* in possessing somewhat lower crowned and more symmetrical teeth. Since that time much new material belonging to *Notochoerus* has been found at several localities, including the East Rudolf area in Kenya, the lower Omo basin in Ethiopia, and a magnificent collection from Hadar in the Afar region. The latter includes several skulls and mandibles (figure 22.9*G–H*) and may provide a good concept of the species (Cooke 1978).

The skull is considerably larger than that of *Hylochoerus meinertzhageni*, with a vertex length almost half as great again. The skull table is broad and flat, with the orbits very slightly elevated above the top of the braincase. The temporal crest sweeps in sharply behind the eyes, although not as abruptly as in the warthog, and the occipital crest is wide. The nasals are straight and parallel-sided but taper sharply to the tips. The muzzle is very similar to that in *Nyanzachoerus*. The canines are rooted in long sheaths that sweep smoothly back to the level of the infraorbital foramen. The dorsal surface of the flange has a long, low crest like that of the forest hog. Posteriorly, the flange curves smoothly into the root of the zygoma, much as it does in the warthog, and the zygoma curves out rather like that of the female forest hog. Large thimble-shaped lateral projections (or "knobs") project abruptly from the jugal area, perpendicular to the axis of the skull but drooping downward so that their rounded tips are close to the palatal plane. These knobs are made of hollow bone and when they are broken away (as is often the case) it is difficult to determine whether they were present or not. Consequently, it is not certain that they are confined to males. The area below the orbits is broad and platelike; a curious feature is that the area where there would normally be a hollow for the levator rostri muscles is, instead, inflated in a manner resembling the so-called lacrimal bulla of the ox. The depressor rostri origin is very weakly scooped and it is clear that *N. euilus* did not use its snout for rooting. The palate is broadly similar to that of the forest hog.

The canines are large and fairly strongly curved, as in the warthog and forest hog, but their orientation is different as they extend horizontally almost in the plane of the palate, with only a small rise at the tips. They are unusual in being dorsoventrally compressed, with the customary inset ventral enamel band displaced toward the inner (back) part of the curve, so that the cross-section resembles a diagonally squashed bell. Only one pair of upper incisors is present, P^1 is absent and P^2 is lacking in all but one of the specimens. P^3 is a subtriangular tooth with the paracone and metacone separated from the well-developed protocone by a wide and deep fovea, open anteriorly. P^4 is rounded in outline and the paracone and metacone are barely separated at the tips; the protocone is moderately small and separated from the paracone-metacone by a narrow but well developed transverse valley. The three main cones tend to become columnar in some specimens and additional conelets may be present. The resemblance to the corresponding teeth of *Nyanzachoerus* is clear, as also is the case with the lower premolars (figure 22.10*B–C*).

The mandible has a long, wide symphysis, much like that of the forest hog except that the anterior border is even straighter. Although six incisors are retained, they are very much reduced and almost peglike in character, heavily reinforced with cement. The canines are robust, emerging at a low angle, not much above the occlusal plane, and are not systematically in contact with the upper canines. The cross-section is a flattened oval with a slightly flattened posterior surface that does not carry enamel; the enamel on the rest of the tooth is very thin. P_1 is not present and P_2 has been noted only in one specimen. The upper third molars have three or four pairs of lateral pillars, lower third molars normally four pairs of laterals, plus a small terminal pillar or complex. The laterals produce stellate enamel islands in early wear, becoming more H-shaped with advancing wear, especially in the lowers. The maximum crown height is less than 1.5 times the anterior basal breadth.

In 1972, Cooke and Ewer described a large incomplete mandible and several fragments from Kanapoi as *Notochoerus* cf *capensis*. The mandible has the typical morphology of this genus and is unlike the known mandibles of *Nyanzachoerus*. There is also a typical upper canine and a detached zygomatic "knob" of large size, so the presence of a *Notochoerus* seems highly probable. The teeth are mostly lost and the only two third molars are very worn or damaged. However, they contain only three well-developed pairs of lateral pillars and a posterior complex of several smaller pillars, which would place them more appropriately with *N. euilus*, despite the massive size of the mandible and of the "knob." It seems clear that *Notochoerus* is derived from *Nyanzachoerus*, almost certainly from the *N. jaegeri* stock as the premolars in this species are relatively less enlarged than in the other species of that genus. Although the skull and mandible of *N. jaegeri* are not adequately known, the premolars in the *jaegeri* material at Kanapoi are decidedly larger than those fortunately preserved in the "cf *capensis*" mandibles. Accordingly, this material cannot be ascribed to *N. jaegeri* and is here regarded as the earliest stage of *Notochoerus euilus*.

N. euilus is well represented in the Usno Formation in the lower Omo basin, Ethiopia, including the back of a skull illustrated by Cooke and Ewer (1972) as "cf *capensis*." Comparable material occurs at the base of the Shungura Formation and occasional specimens are found up to Member G, but there is a tendency for some simplification and reduction in size, for which the varietal name *"minor"* is proposed (Cooke and Coppens forthcoming). In Kenya, *N. euilus* is also represented in the Kubi Algi Formation in the East Rudolf area and in the Chemeron Formation in the Baringo basin. The type material comes from Kaiso in Uganda. Dietrich's *Hylochoerus euilus* (1942) material from Laetolil has some unusual features but is retained in this species.[6] The material from the Chiwondo Beds of Malawi described by Mawby as *Notochoerus capensis* is more likely *N. euilus*.

Notochoerus scotti (Leakey) 1943

This species was established by Leakey (1943) as the genotype species of *Gerontochoerus*, subsequently removed to synonymy with *Notochoerus* (Leakey 1958), and was based on a pair of third molars from Shungura, Omo. The type upper right third molar has five pairs of laterals, with an extra one on the lingual side, and there are also five pairs of good laterals on the lower third molar, followed by a small sixth pair and two terminal pillars. Very similar material from the Shungura Formation had already been described by Arambourg (1947) as belonging to *Notochoerus capensis*, but it seems better to separate it under Leakey's specific designation. The lateral pillars in *N. scotti* tend to be in close mutual contact whereas they are well separated in typical *N. euilus*. These teeth are all higher crowned than in third molars of *N. euilus*, and in the middle levels of the Shungura Formation (members D to G) other specimens are found that are still higher crowned and add further pairs of lateral pillars.

Seven pairs is common but the later ones may carry as many as eleven pairs of lateral pillars and come to resemble *Phacochoerus* molars in hypsodonty, although they are much larger and retain the H-shaped enamel islands in the lateral pillars.

Notochoerus scotti has its closest affinities with *N. capensis*, from which it was most probably derived. It is not common even in the Shungura Formation and is only known otherwise as a rare element in the Koobi Fora Formation in the East Rudolf area.

Subfamily Suinae

Genus *Sus* Linnaeus 1758

TYPE SPECIES. *Sus scrofa* Linnaeus 1758.

DIAGNOSIS. Suidae of moderate to large size with elongate skull, zygomatic arches not inflated; upper canines small to moderate in size and surrounded by small canine flanges, tips truncated by wear against lower canines; premolars, except P[4], elongate and relatively narrow with cutting edges and high cingulum in lowers; molars brachydont and bunodont; main pillars with radial plications giving a scalloped appearance to occlusal pattern; enamel of moderate thickness; cement normally absent.

Sus scrofa Linnaeus 1758

The wild boar is much the same size as the bush pig and shares many characteristics, including rather small upper tusks against which the lower canines make full contact so as to maintain a sharp cutting edge or point. *Sus scrofa* lacks the pointed ears, tuberosities on the snout, light gray facial pattern and whitish dorsal crest, and also the vertical flange of bone found above the tusks in male *Potamochoerus porcus*. The upper and lower P[1] are normally retained and the dental formula is $\frac{3.1.4.3}{3.1.4.3}$. General features of the skull have been described in the first section of this chapter and summarized in table 22.1.

Remains of *Sus scrofa* are found at many later Pleistocene sites in North Africa. It also occurs at the mid-Pleistocene locality of Ternifine (Palikao), where it was named *Sus barbarus* by Pomel (1896); that name is preoccupied but in any case the material is not apparently separable from *S. scrofa*. Pomel's *Sus algeriensis* (1896), from the cave of Pointe-Pescade, is also not distinguishable from *S. scrofa*. The oldest record of this species in Africa is in the Villafranchian of Mansoura, Constantine (Joleaud 1933).

---

6 *"Notochoerus" serengetensis*, Dietrich 1942, from Laetolil, and also two supposed *Notochoerus* species from Olduvai, *N. compactus* Leakey 1958, and *N. hopwoodi* Leakey 1958, are now placed in the *Metridiochoerus* complex (below).

Genus *Potamochoerus* Gray 1854

TYPE SPECIES. *Potamochoerus porcus* Linnaeus 1758.

DIAGNOSIS. Suidae with skull generally resembling that of *Sus* but distinguished by strong canine flanges in male, abrupt lateral sweep of zygomatic arches, which may be inflated; premolars stout, posterior cingulum not as high as in *Sus;* molars more columnar, slightly higher crowned, enamel relatively thin and folding more apparent.

Potamochoerus porcus (Linnaeus) 1758

The characters of the soft parts that distinguish the bush pig from the wild boar have been given above. In the skull there is a general resemblance to that of *Sus scrofa*, expecially in the female of *Potamochoerus porcus*. The most obvious difference lies in the sideways projection of the zygoma, especially marked in the adult male bush pig (figure 22.11*A–D*). Another important distinction from *Sus scrofa* is the backward widening of the nasals with, in males, an expanded rugose area on the nasals and adjoining maxillary. This lies opposite the top of a very characteristic elevated lateral crest on the canine flanges, the upper part of the crest being likewise rugose; in rare aged animals, the canine and maxillary exostoses may even fuse. In life, these rugose areas are covered by cartilaginous tissue and there is a pair of hard, calloused swellings on either side of the face, but no warts; the female lacks these callosities. The occiput is wider than in the wild boar but morphologically similar; however, the ear openings are relatively lower. The canines are very similar in size and morphology to those of *Sus scrofa*, but differ slightly in cross-section. The upper P^1 is almost in-

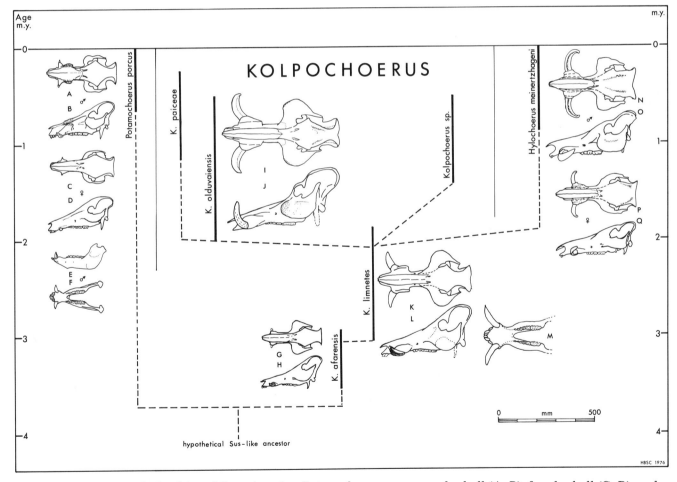

Figure 22.11 Time relationships of the suine pigs. *Potamochoerus porcus:* male skull (*A–B*); female skull (*C–D*); male mandible (*E–F*). *Kolpochoerus afarensis:* male skull (*G–H*). *Kolpochoerus olduvaiensis:* male skull (*I–J*). *Kolpochoerus limnetes:* male skull (*K–L*); male mandible (*M*). *Hylochoerus meinertzhageni:* male skull (*N–O*); female skull (*P–Q*). (Approximately 1/19 natural size; drawn scale equals 500 mm.)

variably, and the lower P_1 always, absent in *Potamochoerus porcus* so that the dental formula becomes $\frac{3.1.3\text{-}4.3}{3.1.2\text{-}3.3}$.

The mandible of *Potamochoerus porcus* is very like that of *Sus scrofa* in general shape, as well as in the rounded spatulate anterior border and long thick symphysis (figure 22.11*E–F*); however, at the back of the symphysis there is a distinct small "shelf" below the attachments for the genioglossus and geniohyoideus muscles. There is a long diastema between the canine and P_2 (or P_3 if P_2 is lost) and this area is more "pinched in" or constricted than in *S. scrofa*. Compared with the rather slender mandibular rami of the wild boar, those of *Potamochoerus porcus* are somewhat inflated behind the mental foramen. The posterior angle is not expanded and there is only a slight outward flare.

Fossil remains of the bush pig are surprisingly rare in sub-Saharan Africa, even in later Pleistocene deposits. This is probably due at least in part to its particular ecology and, as Ansell (1968) put it, "its secretive habits." As *P. porcus* is, in many ways a "primitive" suine, it must have had a fairly long period of isolated evolution in Africa. It is usually assumed that *P. porcus* was descended from an immigrant Asiatic *Propotamochoerus*. However, in Ethiopia, and perhaps elsewhere, the earliest small suine, *Kolpochoerus afarensis*, is in some respects *Potamochoerus*-like, but also has features of the dentition that resemble *Sus* and could not reasonably be derived from a *Propotamochoerus* ancestor in which they were not present. It seems more probable that the ancestor was more *Sus*-like or came from a stock closer to *Dicoryphochoerus*.

Genus *Kolpochoerus* E. C. N. and H. E. van Hoepen 1932

SYNONYMY. *Mesochoerus* Shaw and Cooke 1941; *Omochoerus* Arambourg 1942; *Promesochoerus* Leakey 1965; *Ectopotamochoerus* Leakey 1965.

TYPE SPECIES. *Kolpochoerus paiceae* (Broom 1931) (syn. *K. sinuosus* E. C. N. and H. E. van Hoepen 1932).

DIAGNOSIS. Suidae of moderate to large size with skull architecture generally resembling that of *Potamochoerus* in early forms but zygoma expanded laterally and drooping, especially in the male. Male canines resemble those of *Hylochoerus* in structure and cross-section, but relatively shorter and stouter; female canines much smaller than in the male and primitively rooted in some species. Cheek teeth resemble those of *Sus* or *Potamochoerus* in general

structure but molars higher crowned and have lateral columns that are distinct and well separated. Talon of third molar tends to become more developed than in *Sus scrofa* or *Potamochoerus porcus,* exceeding the length of the main body of the crown in advanced species. Molar brachydont or moderately hypsodont, always strongly rooted. Little cement in more brachydont forms, abundant in hypsodont molars. Premolars rather more *Sus*-like than *Potamochoerus*-like; P^2 and P^3 triangular with well-developed protocone; P^4 equidimensional with a strong protocone and tendency toward the development of multituberculate and complex islands; P_4 has elevated anterior and posterior cingulum cusps and a double central cusp with the two elements displaced laterally relative to one another.

REMARKS. *Kolpochoerus* is widely distributed in the later Pliocene and Pleistocene, ranging from South Africa to North Africa. This generic name has priority over *Mesochoerus,* which has been used in most of the literature. The type species is one of the most advanced in the genus.

Kolpochoerus paiceae (Broom) 1931

The type was a fragment of a lower jaw with a well-worn third molar from the Vaal River gravels, South Africa, and was placed by Broom tentatively in his genus *Notochoerus*. On the basis of part of a right mandible, also from the Vaal River gravels, with an unworn third molar, second molar, damaged fourth premolar, stump of M_1 and roots of P_2 and P_3, Shaw and Cooke (1941) established the genus *Mesochoerus,* to which so much other material has since been referred. The distinctive features of the Vaal River specimens are also shown in material from Elandsfontein (Hopefield) in the southwestern Cape, which were named *Mesochoerus lategani* by Singer and Keen (1955), but they subsequently (Keen and Singer 1956) identified one specimen as *M. paiceae,* and it now seems desirable to include all the specimens under the latter species. The range of variation found in these collections now makes it clear that four teeth from Cornelia, O.F.S., which were designated as types for *Kolpochoerus sinuosus* (E. C. N. and H. E. van Hoepen 1932, Cooke 1974) must be included in *paiceae,* which has specific priority. However, as this does not belong to the genus *Notochoerus,* it must be transferred to *Kolpochoerus,* which has priority over *Mesochoerus.*

The symphysis resembles that of the bush pig but is relatively wider and is also shorter than in the East African species, with the back of the symphysis well in front of the premolars. The anterior border is less arcuate than in *Potamochoerus porcus* but the incisors themselves are very similar in size and structure. The premolars are poorly known but conform to the pattern seen in the East African material; however, the premolar series is a little reduced as compared with *Kolpochoerus olduvaiensis,* which is similar in size. The third molars usually have four, sometimes five, well-developed pairs of lateral pillars that are well separated down to the base of the crown, where there is a good cingulum. The enamel islands are folded in early wear but become rather simple trilobate islands as wear advances, with the lateral wall thickened and becoming rounded with advancing wear. The crown is slightly hypsodont, with the maximum height about 1.6 times the maximum basal breadth in the lowers and 1.3 in the uppers. There is abundant cement.

The canines were not described by Singer and Keen. The uppers are short and rather strongly curved (almost listriodont in aspect) with heavy enamel bands in *Potamochoerus* fashion. The roots are narrowed, but not closed. These canines are so different from the strongly flared hylochoerine ones of *K. olduvaiensis* that they form one of the reasons for maintaining the specific distinction. The lower canines are robust and fairly strongly arched, with a cross-section that tends to be more U-shaped than V-shaped; the pulp cavity is small and the roots may have closed eventually. The female canines, if different, are not known.

This species is known only from South Africa, where it appears to have evolved independently (figure 22.11).

Kolpochoerus limnetes (Hopwood) 1926

A few teeth from Kaiso Village, Uganda, were described by Hopwood (1926b) as *Sus limnetes* and the species was transferred to *Mesochoerus* by Leakey in 1958. Cooke and Coryndon (1970) reconsidered the Kaiso material and showed that it could not be separated from *Mesochoerus heseloni* Leakey 1943, from Shungura, Omo. Arambourg (1943) had almost simultaneously named the same Omo form *Omochoerus pachygnathus,* but both *heseloni* and *limnetes* have priority. Arambourg based his genus on the absence of cement but in fact cement does occur in the Omo material, becoming more important in the later horizons; *Omochoerus* is thus redundant. Arambourg (1947) has given an excellent account of the dentition but it is only during the past few years that good skulls have been found at Omo and in the Koobi Fora Formation in the East Rudolf area (all undescribed).

The skull is about the size of the living *Hylochoerus,* or a little larger, and has resemblances

both to it and to the bush pig. The braincase is long and is domed in the normal fashion, not depressed as in *Hylochoerus;* the parietal constriction is of moderate width. The orbits are not raised significantly more than in the male forest hog, but the cranium rises sharply and results in a very characteristic "dished" profile (figure 22.11*L*).

The snout is generally rather hylochoerine in appearance, as also are the canine sheaths except that there is a lateral projection of the back of the flange, a little in front of the infraorbital foramen, a character not shown by *Hylochoerus* but seen in a weaker form both in *Sus scrofa* and *Potamochoerus porcus.* In the female the canine sheaths are extremely small. In some specimens there are closer resemblances to the bush pig snout, with the nasals tapering forward and expanded above the infraorbital foramen. The expanded area is rugose and in one specimen there is a moderately high flange on the outer side of the canine sheaths. The nasals tend to overhang the maxilla. The sweep of the zygoma is hylochoerine but the jugal area is still more inflated in the males and "droops" even lower than in the forest hog. The lower jaw (figure 22.11*M*) has a long symphysis, as in the bush pig, but it is wider at the canines and not as constricted behind them, being thus about midway between forest hog and bush pig in the morphology of the symphysial region. The hylochoerine expansion of the corpus lateral to the anterior molars, followed by narrowing level with the middle of the third molar, is shown quite clearly and is a useful diagnostic feature of *Kolpochoerus.*

The premolars are of normal size in relation to the molars and are more like those of *Sus* than *Potamochoerus.* The P⁴ is peculiar for two reasons; the first is that the paracone and metacone are even more distinct than in *Sus* and the metacone tends to be reduced in size so that the crown is dominated by the paracone and protocone; the second oddity is the tendency for the development of a number of separate cingulum cusps, giving the whole crown a multituberculate appearance, especially in early wear (figure 22.10*I–J*). The lower P₁ is normally absent, at least in the adult. The remaining premolars are *Sus*-like but rather less sectorial and broader towards the base of the crown so that with advanced wear they are more like those of bush pig. The main cone in P₄ is divided into lingual and buccal elements and their expansion tends to form lateral bulges united to the posterior cingulum island by a narrow "bridge," and the bridge is flanked by a marked fovea on either side (figure 22.10*I–J*). The molars are like those of *Sus* and *Potamochoerus* in basic structure but with the main pillars taller and more columnar, and there is less tendency toward infilling of the spaces between the main pillars by accessory basal columns. In the main pillars the lateral enamel wall is smooth and very thick, that of the median portion rather thinner and weakly folded. The lateral enamel islands tend to be trilobate or mushroom-shaped, with the "stem" splayed out in early wear stages, but the whole column may become rounded in advanced wear. In the early forms the talon is short and at the end of the time range is about as long as the body of the crown. Normally, in a "typical" *Kolpochoerus limnetes,* M³ has a well-developed third lateral on the lingual side but on the buccal side there may be a pair of smaller pillars. The maximum height of the crown is approximately equal to the anterior basal breadth. In the lower M₃ there are normally three pairs of laterals, the third pair weak in the early horizons whereas in the younger levels there is an incipient or smaller fourth pair developed.

The male canines are very similar to those of *Hylochoerus* but in the female the tusks are small. The female upper canines are rooted and project diagonally downward and sideways, the front being truncated by a small vertical wear facet. The enamel is smooth, extending well back on the ventral side and indented by a V-shaped notch on the upper surface, coinciding with a dorsal groove on the root. The lowers are also small in the female, triangular at the tip in early wear but becoming U-shaped toward the roots, which close in the adult. External wear facets develop as a result of digging.

In 1965 Leakey described three new genera of suids from Bed I, Olduvai, which he named *Promesochoerus mukiri, Ectopotamochoerus dubius,* and *Potamochoerus intermedius.* More material has been found since then and it is clear that only a single species is represented. *Promesochoerus mukiri* is an old female, *Ectopotamochoerus dubius* a young female, and *Potamochoerus intermedius* a young male. The Bed I form is less robust than the Omo and East Rudolf material but it seems best to refer all of it to *Kolpochoerus limnetes.* An isolated lower left third molar from the "Earlier Kaiso" assemblage at North Nyabrogo was named *"Sus" waylandi* by Cooke and Coryndon. Although there are some morphological differences, it is regarded as an early stage of *Kolpochoerus limnetes.* The specimen described by Ennouchi (1954) as *Omochoerus maroccanus* from Guyot, near Rabat, does not seem to be separable as the main distinguishing character (the terminal pillar in the lower third molar) is variable and easily matched in Omo specimens. *Kolpochoerus limnetes*

is also known from the Chemeron Formation in the Baringo Basin.

Kolpochoerus phacochoeroides (Thomas) 1884

Part of a right lower jaw with the three molars preserved was described by Thomas (1884) as *Sus phacochoerides* and Arambourg (1947) placed it in his genus *Omochoerus.* It came from lacustrine limestone at Aïn-el-Bey, south of Constantine, Algeria, and there was also a fragment of symphysis with the left canine and alveolus for P₂. The dimensions given by Arambourg for the M₃ would permit it to fall within the known range for *Kolpochoerus limnetes* but Arambourg regarded the jaw as having a more *Sus*-like character and considered it distinct. More material from North Africa is required to determine the issue, which is important as the inappropriate specific name would have priority.

Kolpochoerus olduvaiensis (Leakey) 1942

The holotype of this species is the broken left side of a mandible from Bed II, Olduvai, with the second and third molars complete and the first damaged; the paratype is a similar fragment of the right side stated to be from Bed I. Although a characteristic species in Bed II, it is not particularly abundant and is known mainly by fragments and isolated teeth, apart from a good mandible from the junction of Beds III and IV described and figured by Leakey (1958, pl.6) and the incomplete mandible from Bed II placed by Leakey in *Potamochoerus majus* (1958, pl.2). The species differs from *K. limnetes* primarily in the elongation of the talon, so that typically there are four pairs of laterals in the upper third molars and five in the lowers, although in some lowers there are only four well-developed pairs. The increased length of the talon is accompanied by a slight increase in crown height and also, in some of the material from the higher horizons in the Shungura Formation at Omo and the Koobi Fora Formation at East Rudolf, by thinning of the enamel and an increase in the complexity of folding in the upper part of the lateral pillars.

The skull of *K. olduvaiensis* is larger than that of *K. limnetes* and shows several morphological changes, the most important of which is elevation of the orbits so that their rims now lie flush with or above the frontal area. In the males the parietal area is flattened, or even depressed to the extent seen in *Hylochoerus* females. The zygomatic arches are also elevated and no longer droop to the level of the occlusal plane. The expansion of the jugals is greater and there is a rugose area on the anterolateral part of the boss. The canine flanges and snout are very hylochoerine. A fine skull from East Rudolf is shown in figure 22.11*I–J*. The female, illustrated by a beautiful undescribed specimen from Peninj, Lake Natron, demonstrates the strong sexual dimorphism characteristic of the genus. The profile is typical, with the orbits well elevated, but the braincase is generally rounded and not flattened like that in the male. The zygomatic arches are expanded but the expansion is lateral and does not produce the flaring bosses of the male. The difference in the snout is very marked; the small-rooted canines of the female do not require supporting flanges, which are much like those of a female *Sus* or *Potamochoerus*. The lower jaw is unchanged in morphology.

Kolpochoerus olduvaiensis is undoubtedly descended from *K. limnetes* but the retention of Leakey's specific name for the more evolved form seems desirable as the difference between the early *K. limnetes* and the *Kolpochoerus* of Bed II is very substantial. In the material from Bed I there appears to be a bimodal distribution, as if the two forms occur synchronously but with different ecological preferences. In the Shungura Formation in the Omo area, the slow enlargement of the third molars that is apparent through Members B to F is followed in Member G by a marked increase in size range, suggesting a dichotomy rather than a continuous change. Although somewhat arbitrary, the changes in the number of pairs of pillars can be used to define practical limits for the respective species. *K. olduvaiensis* is now known from Bed I (rare) and Beds II–IV at Olduvai, and from Peninj in Tanzania, from Olorgesailie and East Rudolf in Kenya, and Omo in Ethiopia.

A lower right third molar, broken anteriorly, was described by Hopwood (1929) as *Hylochoerus grabhami,* but was first figured by Leakey (1958), who placed it in *Mesochoerus.* It came from alluvial deposits near the White Nile at Kosti and is of unknown date. Although the specimen is narrower than most teeth of *K. olduvaiensis,* it is possible that the Nile specimen belongs to the same species. However, it is a very inadequate type and is best regarded as a nomen vanum.

In 1934 Hopwood described a species which he named *Koiropotamus majus,* using as the holotype a partial left mandibular ramus from Olduvai Bed IV, containing P_3–M_1 and the alveolus for P_2. Other specimens were assigned to it subsequently by Leakey (1942, 1958) but some of these were later withdrawn (Leakey 1965). If the holotype were found today, it would be referred with little hesitation to

Kolpochoerus olduvaiensis, but in fact it possesses those teeth that are least diagnostic, except at the generic level. It seems best to consider it as an inadequate holotype and hence to regard it as a nomen vanum. This is unfortunate, as there is a distinctive small unnamed species of *Kolpochoerus* in the top of Bed II and in Bed IV, and also in the Baringo basin, to which the *K. majus* holotype does not belong.

Kolpochoerus afarensis Cooke 1978

In the lower part of the Hadar Formation, in the Afar region of Ethiopia, there is a small suid that is in marked contrast to the accompanying large *Nyanzachoerus pattersoni* and *Notochoerus euilus.* The skull size is very similar to the living bush pig and wild boar and the teeth are also similar in general size and complexity. The zygomatic arches jut out abruptly as in *Potamochoerus porcus,* but the form of the root of the zygoma and the lateral expansion of the jugals is more like that in *Kolpochoerus limnetes* (figure 22.11*G–H*). The incisors are like the bush pig's, and the canines are small, with flanges like those of *Sus scrofa.* The first premolars are usually present, but are sometimes missing in the stratigraphically younger horizons. The premolars are intermediate in form between those of bush pig and wild boar, with the lowers not as expanded as in the bush pig. P_4 has a strong main cone consisting of two elements that are laterally displaced, producing the typical *Sus* "kink," but not the lateral buttress developed in *Kolpochoerus limnetes.* The third molars are relatively a little larger than in the bush pig but match closely with those of the wild boar in dimensions. However, the enamel is thicker, as in the bush pig, but the main pillars are smoother, less plicate and better separated than in either of these living suids, as might be expected in an early stage of *Kolpochoerus.* However, the talonid is not expanded and the third pair of laterals, typical of *K. limnetes* is not developed, although specimens from the higher horizons may have the terminal pillar divided.

Although the affinities of this material place it closest to *Kolpochoerus* limnetes as the probable ancestor of that lineage, it is also basically similar to the bush pig in some aspects of skull structure and to the wild boar in some dental features. It may thus provide a possible clue to the ancestry of the living bush pig and suggests that both *Potamochoerus* and *Kolpochoerus* were derived from a common ancestor that possessed basic *Sus*-like features.

The only described material that can be regarded as probably representing this species are the two third molars from Laetolil, Tanzania, described and figured by Dietrich (1942, figures 150, 157) as *Potamochoerus* sp cf *major*. An undescribed maxillary fragment with LM$^{2-3}$ in the same collection in East Berlin is clearly the same as the unnamed late *Kolpochoerus* from Beds II-IV at Olduvai, already mentioned above.

Genus *Hylochoerus* Thomas 1904

TYPE SPECIES.   *Hylochoerus meinertzhageni* Thomas 1904.

DIAGNOSIS. Suidae of moderate to large size with skull differing from *Sus* and *Potamochoerus* in the possession of a wide dorsally flattened parietal area, zygomatic arches more expanded, especially in the male, canines robust and outflaring, surrounded by long, low flanges. Lower canines in contact only with the lower part of the upper canines, leaving the upper tips free from mutual attrition. Upper incisors and premolars reduced; molars with strongly columnar structure, columns widely separated, slender and slightly folded, strengthened with abundant cement. Lower jaw with wide symphysis and lateral expansion of the corpus level with the premolars and first and second molars.

REMARKS. It is curious that a number of fossil specimens of suid had been found in the north African region well before the turn of the century, whereas one of the living African pigs, *Hylochoerus,* was not described until 1904, and was not recorded in West Africa until two decades later. It is a large animal and is, as Ansell (1968) says, "more unobtrusive than really scarce." It has an extremely poor fossil record, doubtless because of its forest habitat.

Hylochoerus meinertzhageni (Thomas) 1904

The skull of the forest hog differs markedly in general appearance from that of the bush pig, particularly in the broad occiput, wide dished parietal area—so unlike the usual domed braincase—and the large outflaring canines (figure 22.11*N–O*). The occiput has a strong vertical ridge (figure 22.1). The profile is not straight but kinked at the nasal/frontal boundary and again a little behind the orbits, where the raised rim of the parietals begins. Other features of difference have been summarized in table 22.1. The robust zygomatic arches originate much like those of *Potamochoerus porcus* but are broader and deeper below the orbits with, in the male in particular, a lateral bulging inflated area of hollow bone that is very characteristic. The muzzle has a smoothly arched cross-section, broad across the canine area where the canines are surrounded by wide long flanges extending back to the infraorbital fora-

men. The canines are structurally like those of *Sus* and *Potamochoerus* but much larger, even those of the female being considerably bigger than in the male bush pig. Whereas both in *Sus* and *Potamochoerus* the lower and upper canines abrade at the tips, maintaining both points by attrition, the lower canines of the forest hog merely produce a large wear facet on the uppers but usually leave the points of the upper tusk more or less intact. The dentition is specialized, with loss of incisors and early shedding of premolars so that the dental formula is

$$\frac{1.1.3.3}{2\text{-}3.1.2.3}.$$

Hylochoerus meinerzhageni has a mandible that is much wider and longer than in *Sus* or *Potamochoerus,* tending toward a rectangular aspect in the occlusal view rather than the V-shape of the other two genera. The symphysis is not significantly longer than in *Potamochoerus* and, in view of the relative sizes, is thus proportionally shorter; it is not as thick. There is no shelf at the base of the symphysis but the muscle insertions are strongly marked. The great width across the canines (about equal to the length of the symphysis) is accompanied by a gently curved anterior border, much straighter than in *Sus* or *Potamochoerus*. There is only a very slight constriction behind the canines. The six incisors are retained but they are often irregularly arranged and one or more may be lost. P_2 is present in young individuals, lying midway along the diastema between the canines and P_3, but it is usually shed fairly early. A notable feature of the lower jaw is the inflation of the mandibular ramus lateral to the anterior molars, with a fairly abrupt narrowing opposite the middle of M_3 where the insertion for the large superficial masseter muscle begins. The posterior angle flares slightly outward. The height of the jaw is not very great, considering its length, and the ascending ramus is short because of the low position of the temporal condyles already referred to above. This provides a shortened lever arm and suggests the importance of a shearing action in mastication rather than crushing or grinding. Almost the only published record of *Hylochoerus meinertzhageni* is by Hopwood (1931) from the late Pleistocene of Gamble's Cave, Naivasha, Kenya, but it does also occur in other cave deposits in Kenya. A fragment of left mandible with the second and third molar preserved was found on the surface at Kanjera, Kenya, and was described by Leakey (1958) as *Hylochoerus antiquus*. The teeth are larger and more robust than in the living species and the enamel-clad pillars are stouter, but they are still slender by comparison with other genera, widely spaced, and abraded in the typical fashion. It could certainly be an early representative of the living forest hog and it is unfortunate that it was a surface find. However, it is mineralized in the same way as other specimens from the Kanjera Beds and probably belongs to the late middle Pleistocene. The holotype is the only specimen so far recorded from Kanjera.

The rather unusual specialization of the molars involves reduction in the diameter of the main enamel-dentine pillars, which are then greatly strengthened by abundant cement. The third molars normally have three pairs of laterals and the *Hylochoerus* structure can be derived very easily from that of *Kolpochoerus limnetes,* which is regarded as the probably ancestral stock. The male canines are similar in structure, but those of *Hylochoerus* are greatly enlarged. The tusks of female *Hylochoerus* are also large, unlike the small-rooted tusks of females in *Kolpochoerus,* and substantial genetic change is demanded if this origin is correct.

Genus *Phacochoerus* Cuvier 1817

SYNONYMY. *Tapinochoerus* E. C. N. and H. E. van Hoepen 1932.

TYPE SPECIES. *Phacochoerus aethiopicus* (Pallas 1767).

DIAGNOSIS. Medium-sized Suidae with the skull differing from *Sus* and *Potamochoerus* mainly by the elevated orbits, broad frontal area, more anteroposteriorly compressed cranial region, broad and thickened zygomatic arches, and enlarged bony bosses around the sockets of the upper canines. Upper canines large, outflaring, and free of enamel except at the tips. Premolars reduced and commonly shed in adult. Molars, especially third molars, hypsodont and formed of closely packed little-folded oval to subtriangular columnar elements, well cemented. Talon of third molar considerably elongated. Incisors reduced, lying in shelflike projection in mandible. Lower canines verrucose, more compressed laterally than in *Sus* or *Potamochoerus,* wearing against anterior face of uppers and not normally affecting the tips of the upper tusks.

REMARKS. E. C. N. and H. E. van Hoepen (1932) demonstrated that in the third molars of the living warthog, which is widespread in the savanna areas of Africa, anterior root formation occurs at about the time that the last column of the third molar reaches the grinding surface, or a little earlier (figure 22.12*A*). However, in remains believed to belong to the true Cape warthog (now extinct) and also in fossil and subfossil teeth from localities within its former range, root formation is delayed until well after the last columns come into wear (figure 22.12*B*).

This latter phenomenon has not been noted in any of the living races and because the Cape warthog was exterminated more than a century ago, there is strong presumptive evidence that it differed from the living races in this respect. Shaw (1939) confirmed the reality of this difference, which also included further reduction in the incisors. Both Allen (1939) and Ansell (1968) refer all the living warthogs to a single species, which they call *Phacochoerus aethiopicus* (Pallas). Because the Cape Warthog is the type of *Aper aethiopicus* (Pallas 1767) for which Cuvier created the genus *Phacochoerus* in 1817, there seems to be no doubt about the correct specific designation for the extinct form with the delayed root formation. Gmelin's *Sus africanus* dates from 1788 and the type came from Senegal, where the warthog still exists. It would seem best to apply Gmelin's specific name to the living races and to restrict the use of *aethiopicus* to the exterminated Cape warthog and to fossils with similar characteristics. From a palaeontological viewpoint the difference warrants specific distinction and it is particularly useful to be able to refer to the teeth of the living form as of the "*africanus* type" as has been

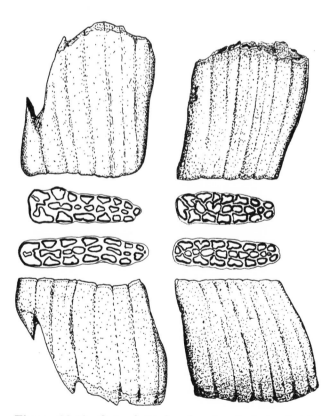

Figure 22.12 Outer lateral and occlusal views of right upper and left lower third molars of *Phacochoerus africanus* (*left*) and *P. aethiopicus* (*right*) to show differences in wear at time of root formation. (2/3 natural size.)

done by the van Hoepens, Shaw, Cooke (1949), Ewer (1958), and others (see Hopwood and Hollyfield 1954), and to imply by the *"aethiopicus* type" the features of the extinct form (figure 22.12).

Phacochoerus africanus (Gmelin) 1788

The skull of the living warthog is very characteristic, with its orbits set back and greatly elevated, while the braincase is anteroposteriorly compressed, the zygoma deep and platelike, and the large tusks flaring outwards (figure 22.13*A–B*). The occiput is more or less intermediate between that of the bush pig and forest hog, but has no median ridge, as the latter does. The profile is "dished" and the front of the snout droops below the occlusal plane. The nasals are flat, slightly expanded above the canines and narrowed above the infraorbital foramen. The muzzle is gently arched in cross-section and the flanges around the canines are almost tubular and

directed sideways. The temporal condyles lie well above the foramen magnum, resulting in an unusual lengthening of the ascending ramus of the lower jaw (figure 22.13*C*).

The mandible is roughly V-shaped on the occlusal aspect but the canines are widely spaced and there is a marked constriction toward the back of the diastema (figure 22.13*D*). The anterior border is only slightly curved and the incisors lie in a shelflike projection below which the symphysial contour is quite strongly curved, unlike the rather flat profile in the *Hylochoerus* symphysis. The length of the symphysis is slightly greater than the width across the canines and it is thick; there is no basal posterior shelf at the insertions of the genioglossus and geniohyoideus muscles. The mandibular rami are not inflated but they flare outward toward the base of the corpus and the posterior angles flare outward quite markedly. The posterior angle rises toward the back

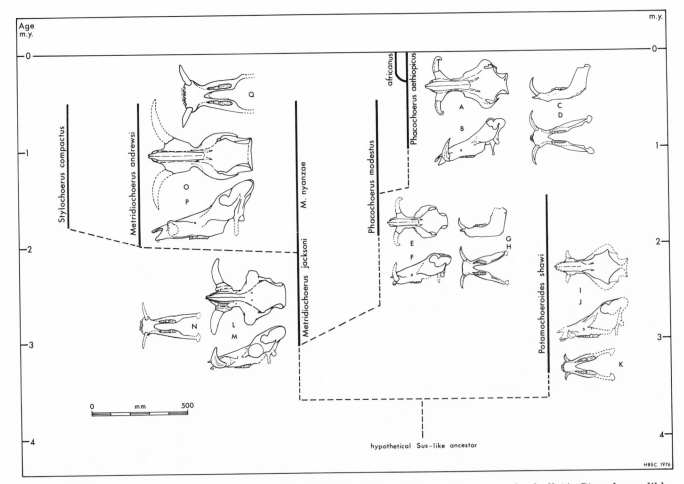

Figure 22.13 Time relationships of the "phacochoerine" pigs. *Phacochoerus africanus:* male skull (*A–B*) and mandible (*C–D*). *Phacochoerus modestus:* skull (*E–F*) and mandible (*G–H*), probably male. *Potamochoeroides shawi:* male skull (*I–J*) and mandible (*K*). *Metriodiochoerus jacksoni:* male skull (*L–M*) and mandible (*N*). *Metridiochoerus andrewsi:* female skull (*O–P*) and mandible (*Q*). (Approximately 1/19 natural size; drawn scale equals 500 mm.)

and the very long ascending ramus has a backward rake, providing a long lever arm as compared with *Hylochoerus,* which, as Ewer (1958) has pointed out, is advantageous for a grinding action of the jaws.

The dentition is specialized for the large third molars and is $\frac{1.1.3.3}{2\text{-}3.1.2.3}$ but the premolars and anterior molars are shed progressively during the aging of the animal. The single pair of upper incisors is retained in the adult and roots form on the anterior pillars of the third molars well before the hindmost columns reach the grinding surface. Although fossilized skulls are very rare, isolated third molars are common in upper Pleistocene sites throughout the continent.

The phacochoere bones and teeth from the North African littoral are mainly in early Neolithic contexts but *Phacochoerus* has also been found at a site called Aïn Tit Mellil near Casablanca in association with hand axes (Arambourg 1938). However, it cannot be separated from *P. africanus* and both the species described by Pomel in 1897, *P. mauritanicus* and *P. barbarus,* are referred by Joleaud (1933) to *P. africanus.* The Southwest African *Phacochoerus stenobunus* of Pia (1930) also belongs here, although it is smaller than the average. Still smaller, but capable of being matched by living specimens is *P. congolensis* Van Straelen 1924, from the Katanga region of the Congo. The third molars from Florisbad, Vlakkraal, and a few other sites in the Orange Free State, South Africa, that were referred to *P. helmei* Dreyer and Lyle 1931 are regarded by Ewer (1957) as different enough to be subspecifically distinct from *P. africanus.* The syntypes of *P. laticolumnatus* E. C. N. & H. E. van Hoepen 1932 are also to be placed in *P. africanus helmei* (a referred tooth from Cornelia, however, belongs to *Stylochoerus compactus*).

Phacochoerus aethiopicus (Pallas) 1767

The extinct Cape warthog differs from the typical *P. africanus* in the loss of the last pair of upper incisors, possibly some reduction in the lower incisors, and delay in formation of roots on the anterior columns of the third molars until well after the posterior columns had reached the grinding surface (E. C. N. and H. E. van Hoepen 1932, Shaw 1939, Cooke 1949). It is fairly common as a fossil on open sites in southern Africa, particularly in the Orange Free State and adjoining areas, and has also been found in stone age caves with "Middle Stone Age" or later assemblages. *P. venteri* Dreyer and Lyle 1931, from Florisbad is not separable even at the subspecific level (Ewer 1957) and *P. dreyeri* and *P.*

meiringi (Dreyer and Lyle 1931), from other O.F.S. sites are also regarded as synonyms of *P. aethiopicus.*

Third molars inseparable from *P. aethiopicus* or *P. africanus* are found at a number of sites of considerable antiquity, including the later Kaiso, the Shungura Formation (Members G and above), and near the top of the upper members of the Koobi Fora Formation. None has been found in situ and they have been dismissed as being derived from overlying younger horizons. There is at least a suspicion that this explanation is an evasion and that a form close to the living warthog may have existed for much of the Pleistocene. The two partial third molars from Beds II and III at Olduvai referred by Leakey (1958) to *P. africanus* belong to *Stylochoerus compactus.*

Phacochoerus kabuae Whitworth 1965

From old lake beds at Kabua, west of Lake Rudolf, Whitworth (1965) described two fragments with the upper and lower second molars preserved that represent a phacochoere with teeth about three-quarters of the size of those at the lower end of the size range for *P. africanus.* The teeth are in rather early wear and the pattern shown in Whitworth's drawing is probably more irregular than it would be later on. Nevertheless the pattern is a little unusual and the size is such that they would be difficult to absorb into *P. africanus.* They could be derived from *P. modestus,* which also has the laterals rather well separated, but the crowns are narrower. The status of this species is uncertain at the present time.

Phacochoerus modestus (E. C. N. & H. E. van Hoepen) 1932

This species was named by the van Hoepens (1932) for a pair of upper third molars from Cornelia, Orange Free State, and this was the type of a new genus *Tapinochoerus.* Both teeth have five pairs of lateral pillars, a row of six stout median pillars, and a single terminal pillar. The arrangement and shape of the pillars is like that of *P. africanus,* but there are fewer columns, which are better separated and the crown height is proportionally lower, although still hypsodont. Strong roots are formed on the two anterior pillars on the buccal side and are beginning to develop on the lingual side. The front two pairs of laterals are in a fairly early stage of wear, but the posterior pillars are well below the grinding surface and are incompletely formed, with open roots. Although not definitely associated with them are two canines very similar in size and morphology to those of a female warthog and probably belonging with the molars (Cooke 1974).

In 1948, Broom described an immature skull, with

lower jaws, from the Kromdraai faunal site in the Sterkfontein area, South Africa, and named it *Phacochoerus antiquus*. This short account was amplified by Ewer (1956) and she established the subgenus *Potamochoerops* for this species; a few isolated teeth and one anterior mandibular fragment from Swartkrans were also included. There are two adult skulls now known, but undescribed, one from Bolt's Farm, near Swartkrans and Kromdraai, and the other from near the top of Olduvai Bed I; the Olduvai skull is used for the views in figure 22.13E–G. The skull is 15 to 20% smaller than that of *P. africanus* and differs mainly in the lesser elevation of the back of the skull (although the orbits are raised), the more normal width of the frontoparietal area, the little expanded and less platelike zygomatic arches, and the short thick snout, which does not droop much below the occlusal plane. The symphysis is shorter with a straighter anterior border, and the posterior angle of the mandible is like that of *Potamochoerus*. There are only four incisors even in the juvenile. The canines are very like those of *P. africanus* but a little smaller and are directed a little more forward. The dentition shows strong premolar reduction, as in the living warthog, but the lower P$_4$ is a simple conical tooth with a small posterior cingulum cusp. The third molars have four or five pairs of laterals and the islands are relatively longer than in the living warthog, the medians almost as big as the posterior laterals. The laterals are more folded or irregular than in *P. africanus,* but become more oval in later wear. The roots form on the anterior pillars before the posterior columns are erupted, but root formation proceeds quite rapidly and the roots are strongly developed, as in the *Metridiochoerus* group.

It is clear that this fossil already has most of the definitive features of the genus *Phacochoerus*. However, it appears unlikely that it is actually ancestral to the living warthogs as it seems to be a fairly stable species with a range from Bed I to Bed IV at Olduvai. The Bolt's Farm skull has a slightly longer snout and may be less specialized than the Olduvai Bed I fossil and is possibly the earliest member of the species so far known.

In addition to the syntypes from Cornelia and the material from the Transvaal cave breccias, an upper third molar from the Vaal River gravels that Shaw and Cooke (1941) named *Notochoerus broomi* must be regarded as representing *Phacochoerus modestus*. Also placed here are the two third molars from Olduvai Bed IV that Leakey (1958) named *Tapinochoerus minutus*.

Genus *Potamochoeroides* Dale 1948

TYPE SPECIES. *Potamochoeroides shawi* (Dale 1948) (= *P. hypsodon* Dale 1948).

DIAGNOSIS. Suidae that in general structure of the teeth and skull resemble *Potamochoerus* but differ in the following points: the third molars are longer and more hypsodont, the enamel of the lateral pillars is more deeply infolded, and a cement covering may be present; the mandibular symphysis is shorter and there is [little or][7] no constriction of the mandible and palate in the region of the canines (Ewer 1958a). [Three premolars retained in the adult. Orbits and braincase elevated, zygoma deep below orbits and somewhat resembling that of *Phacochoerus*. Parietal constriction narrow.]

REMARKS. Two suid species from the Limeworks Quarry, Makapansgat, South Africa were described by Dale (1948), who referred most of the material to Leakey's genus *Pronotochoerus* as a new species *P. shawi,* but she also created a new genus *Potamochoeroides* for a maxillary fragment with teeth smaller than those of *"Pronotochoerus"* shawi. Only one specimen—a mandibular ramus with P$_3$–M$_3$—has since been referred to *Potamochoeroides hypsodon* (Ewer 1958a). Ewer showed that *shawi* did not belong to *Pronotochoerus* and very properly assigned it to the same genus as *hypsodon*. It now seems clear that only a single species is represented and *shawi* has page priority over *hypsodon,* so the latter becomes a junior synonym of the former. This does not invalidate the generic status of the material, which is distinctive.

Potamochoeroides shawi (Dale) 1948

The type is a right mandibular ramus with P$_3$–M$_3$ and came from the "grey breccia" at Makapansgat. Much additional material has been described by Ewer (1958). The lower M$_3$ has a distinct resemblance to that of *"Pronotochoerus"* (=*Metridiochoerus*) *jacksoni;* because only the type of that species was known in 1948, Dale had good reason to place the specimen in Leakey's genus. However, as Ewer has pointed out, there is substantial premolar reduction in *M. jacksoni,* whereas in the Makapansgat material the premolar series is normally developed. It may be added that whereas the upper canines of *Metridiochoerus* are more or less phacochoerine, those of *shawi* are stout, short tusks, strongly curved and carrying a wide ventral band of ribbed enamel, as well as the two thin inset upper

---

7 Bracketed material added by Cooke.

enamel bands typical of *Sus, Potamochoerus* and *Hylochoerus*. The strong curvature results in a few of the upper tusks having a wear facet on the anterior face and not at the tips, which is the more usual situation, but even in the latter case the wear facet is much more oblique than in *Sus* or *Potamochoerus*. The canine flanges are small and *Sus*-like.

From a good part of a skull, several fragments, and partial mandible, it is possible to make a good reconstruction (figure 22.13*I–K*). The orbits are more elevated than in the bush pig, giving a warthoglike profile, but the snout is short and does not droop much below the palatal plane, and the back of the braincase is not abnormally elevated. The frontals are broad and the parietal constriction narrow, unlike that of *Metridiochoerus,* and the braincase is well rounded and not as anteroposteriorly compressed as in the warthog. The zygoma is broad below the orbits and its maxillary root is much like that of *Sus verrucosus*. There is no diastema behind the upper canine and no gap between P[1] and P[2]. In the mandible P[1] lies in the middle of a short gap between the canine and P[2]. The lower canines are verrucose and rise much as in the bush pig. The symphysis is relatively short and not much constricted behind the canines. The mandibular rami are thick but not very deep.

P[1] is not known but was a small two-rooted tooth. P[2] and P[3] are both rather more like the corresponding teeth of *Sus* and *Potamochoerus* (figure 22.5*G–H*). P[4] consists initially of a multicusped external complex separated from an internal complex by a deep valley. With wear these unite to produce a more or less *Sus*-like pattern, but not really very close either to *Sus* or to *Potamochoerus*. P[1] is unknown but P[2] is very similar to that in *Sus scrofa*. In the third and fourth lower premolars the anterior cingulum is rather weak and the posterior one strong. The crest is long and narrow with the sectorial character found in *Sus,* but the main cone is stout and broadens rapidly, maintaining a strong fovea on each side at the back, reminiscent of that seen in *Metridiochoerus jacksoni*. The lower premolars are thus not really like *Potamochoerus* or *Sus* but in some respects are intermediate in character. The upper third molar is moderately hypsodont, with only two well-developed pairs of lateral pillars. A marked feature is the wide separation of the first and second main lateral pillars on the buccal side. The talon is very variable and commonly consists of a small median pillar flanked by four to six oval pillars arranged in a semicircle around it, but sometimes there is a well-developed third lateral on the buccal side, or, more rarely, a moderately well-developed third pair. The anterointernal main pillar has a well-developed X shape, as also may some of the other laterals, and the islands are usually four-lobed, but there is considerable variation. In the lower third molars there are two well-defined pairs of lateral pillars and a third pair that is rather variable in development. On the lingual side the first and second pairs of laterals are well separated but on the buccal side the gap is filled by a small accessory pillar that may later unite with one of the laterals. A variable number of small pillars form the back of the tooth. The enamel islands are mushroom-shaped, but the "stem" typically has two prongs, so the overall shape is four-lobed. The anteroexternal main lateral pillar is more nearly X-shaped.

Potamochoeroides shawi is known only in the australopithecine cave breccias of South Africa and is rare except at Makapansgat. There are decided resemblances to *Metridiochoerus jacksoni,* which suggest some kind of relationship, but the retention of the premolars, the very suine canines, narrow parietal constriction, and short snout are considered to warrant generic separation. The South African occurrences are probably too late for *shawi* to be directly ancestral to *jacksoni*.

Genus *Metridiochoerus* Hopwood 1926

TYPE SPECIES. *Metridiochoerus andrewsi* Hopwood 1926.

DIAGNOSIS. Suidae of moderate or large size with broad parietal and occipital area; orbits elevated, but less than in the warthog, accompanied by elevation of the maxillary root of the zygoma. Lower jaw resembling forest hog in general form but symphysis and corpus without marked inflation. Premolars greatly reduced, only P[4] normally retained in adult. Third molars moderately hypsodont to hypsodont with complex talon or talonid, strongly rooted. Pillars tend to unite to produce complex enamel islands and areas of dentine with scattered lakes[8] of enamel, especially in first and second molars and in the anterior part of the third molars. Upper canines with thin or no enamel. Lower canines verrucose.

REMARKS. There are a number of fossil suids that share such common features as reduction of the premolars and a tendency toward a characteristic

---

8 The term "lake" is employed to designate a ring of enamel surrounded by dentine (i.e., the opposite of an enamel island). The center of the lake is filled with cement and is analagous to the fossette in equine molars.

pattern in well-worn molars, and it is felt that they share a common heritage from an as yet unknown ancestor. It is convenient to consider them all as parts of a group, here termed the "*Metridiochoerus* complex" and including the genera *Metridiochoerus* and *Stylochoerus* and possibly also *Potamochoeroides*.

Metridiochoerus andrewsi Hopwood 1926

Metridiochoerus andrewsi was established by Hopwood (1926a) on the basis of an isolated upper left third molar found in 1911 by Felix Oswald "near Homa Mountain" and presented to the British Museum; there are also two talons of similar teeth. His original diagnosis emphasized that the enamel pattern was complicated, "the degree of folding being such that very little more would be necessary to cause the figures to split up into separate pillars. Talon, and probably the whole tooth, with an enamel pattern of numerous small rings in the early stages of wear." Unfortunately most suids have a tendency toward occasional splitting of the columns, especially toward the tips and this is not in itself a reliable guide, even if it is usual in *Metridiochoerus;* the two other talons do not have the columns subdivided as much as in the type and are probably more normal. The same is true for the fine maxilla fragment with LM$^{1-3}$ referred to this species by Leakey (1958); it came from Kagua, near Homa Mountain and supposedly belonged to the middle Pleistocene, but this is uncertain.

Undescribed material from the upper member of the Koobi Fora Formation in the East Rudolf area, and from Member G in the Shungura Formation, Omo, is considered to represent *Metridiochoerus andrewsi* and provides information on the character of the skull and lower jaw, as well as of the dentition. A typical skull and mandible are shown in figure 22.13*O–Q*. The lateral aspect is reminiscent of that of the warthog because of elevation of the orbits and braincase, as well as of the occipital condyles. The root of the zygoma is raised and the maxilla is high to accommodate the hypsodont third molar. However, the snout does not droop, although it is long, nor is the elevation of the orbits as extreme. The zygomatic arches are morphologically like those of a female forest hog, but are wider below the orbits; one specimen from East Rudolf shows lateral knobs developed on the zygoma. The braincase is very different from that of the warthog which, though broad across the frontals, has a narrow parietal constriction and is anteroposteriorly compressed. In *M. andrewsi* the braincase is long and broad in the parietal and occipital areas, although not as broad as in *Hylochoerus,* nor does it have the depressed fronto-

parietal area of the forest hog. The canine flanges are more like those of warthog than anything else but the canines do not flare out as sharply and are robust so that the roots extend back to the infraorbital foramen and cause a substantial bulge. The canine is strongly curved and in its lower part has a distinctive trefoil shape, like an ace of clubs upside down. There seems to have been some enamel on all three surfaces. The canines flare outward and sideways, with a slight spiral twist that leads to the tips pointing a little upward and backward. The anterior surface is worn over a considerable area by contact with the lowers but it is not a facet of continuous contact. The lower canines are verrucose and are placed much as in *Hylochoerus* but are very little curved and do not extend back into the ramus, being rather "implanted" in the symphysis. They are also worn anteriorly, presumably by being used for digging. The symphysis is very broad across the canines and appears constricted behind them but is, in fact, as wide as at the premolars. The front edge of the symphysis is very straight and forms a ledge that carries six incisors, which are unusually small and morphologically almost like milk teeth. The morphology of the symphysis is otherwise very like that of the warthog in profile. Only one pair of incisors seems to have been retained in the upper jaw.

Although the upper fourth premolar was apparently present, it was squeezed out at an early stage and its morphology is not known. There are a few examples of the lower P$_4$, which was a short, wide, oval tooth with its main body consisting of a stout cone divided into a number of tubercles, usually four, that wear to form a cross-shaped island separated by a rather deep groove from a posterior cingulum cusp; however, it is variable in detail. The first and second molars are very worn in most specimens and consist essentially of a rim of enamel surrounding an area of dentine within which are three rows of enamel lakes. As the same pattern is shown on the body of the third molar, it may be assumed that the original structure was similar. The first and second molars clearly expanded upwards in the same way as in the warthog, whereas in *Notochoerus* there is a strongly bulging posterior cingulum.

In the upper third molar, the laterals are well developed on the lingual side, but on the buccal side the laterals on the talon tend to be subdivided for some distance from the tips, occasionally nearly to the base. The lateral pillars are in close contact for almost their full height. The anteroexternal main pillar is initially H-shaped, with the sides curved in at the top and bottom of the H, while the anterointernal main pillar is set a little behind it and is shaped more like an italic *x*; the posteromedial arm of the X is a separate small pillar in early wear. The laterals tend to form enamel islands with the shape of an old-fashioned collar stud or a mushroom but as wear advances this becomes distorted into a Y shape in the third or subsequent lateral pillars. As wear proceeds, the median portions of the laterals unite with the rounded medians to produce complex patterns. The medians also fuse and eventually so do the laterals. This general morphology is common to all the members of the *Metridiochoerus* complex. The lower teeth are essentially similar but the dominant main laterals are on the buccal side and the "pairing" or matching on the two sides is better. There are four well-developed laterals. As the crown is narrower, the pillars appear more elongate and more laterally compressed than in the uppers and the laterals tend more toward the adoption of the Y shape. Fusion is slower than in the uppers. There is a good deal of variation both in uppers and in lowers with respect to rates of fusion as many teeth maintain the separateness of most of the lateral and median islands until fairly advanced wear, whereas others may produce quite complex patterns in moderately early wear.

The third molars erupt in the same way as in the warthog and rotate as they wear, so the islands appear to become progressively more elongate. The occlusal length also increases with wear but, as the length of the first and second molars diminishes, the effective length of the grinding surface is maintained throughout the adult life of the animal. Anterior roots form quite early and strong roots develop on the second pair of pillars. In *M. andrewsi* the upper third molars have a maximum crown height equal to, or greater than, the basal length; the lowers are about three-quarters as high as their basal length. The production of complex patterns in the third molars usually begins anteriorly as soon as the posterior columns reach the grinding surface.

In 1928, Broom described an isolated lower left third molar from the Vaal River gravels as *Notochoerus meadowsi,* although commenting that it more probably belonged to a different genus. The species has been put into *Phacochoerus* and into *Kolpochoerus* by various authors and I assigned it (Cooke 1949) to the van Hoepens's genus *Tapinochoerus,* thus inadvertently creating continuing confusion so that it has come to be known widely as "*Tapinochoerus meadowsi*" in subsequent literature. It now seems clear that this is a late stage of *Metridiochoerus andrewsi* and it is questionable whether a separate specific name should be maintained. The holotype of *Notochoerus dietrichi* (Hopwood 1934) and its former paratype, which Leakey (1958) made the type of a new species, *Notochoerus hopwoodi,* also belong here.

No complete skull is known but there is a fine mandible and accompanying palate from Olorgesailie, Kenya, which may be regarded as typical; it has been well described and figured by Leakey (1958, pls. 15, 16). Compared with the "typical" *M. andrewsi* of the upper Koobi Fora Formation, the mandible is not as wide across the canines and is morphologically somewhere between that of the forest hog and the warthog, but it does show some outward flaring of the angle. The lower canines have a rounded triangular cross-section and are moderately curved, although they do not pass back into the corpus. The upper canine sheaths are damaged but they extend back to the infraorbital foramina and bulge out as in the typical East Rudolf *M. andrewsi* skulls; the alveoli show that the tusks had a trefoil cross-section, as in the East Rudolf material.

The teeth of this stage differ from those of the typical East Rudolf material in being relatively longer, in having the lateral columns separated for much of their height, and in maintaining the individuality of the pillars to a fairly advanced stage of wear. In the upper third molars the two anterior main pillars are more equidimensional and the walls of the enamel islands less curved, although maintaining the H and X shapes described above. In the lower third molars, the anterointernal main pillar is H- or X-shaped but the anteroexternal pillar is smaller and tends to be shaped like a squashed collar stud. The two main anterior pillars unite to form a complex island in moderate wear and the lateral pillars tend to become increasingly Y-shaped.

The type of *M. andrewsi* is from Homa Mountain, Kavirondo Gulf, Kenya, and Leakey (1958) described another specimen from Kagua in the same area. The "typical" stage is also found in the *Metridiochoerus andrewsi* zone at East Rudolf, in Member G of the Shungura Formation, Omo, in Olduvai Bed I, and at Bolt's Farm in the Sterkfontein area of South Africa. The more advanced variety occurs at Swartkrans in the latter area, in the Vaal River gravels and at Hopefield (Elandsfontein), Cape Province. In East Africa it occurs at Rawe, Kanjera, and Olorgesaile, as well as in Beds II to IV at Olduvai, in the *Loxodonta africana* zone of the Koobi Fora Formation, and in the upper members of the Shungura Formation at Omo.

Metridiochoerus jacksoni (Leakey) 1943

Metridiochoerus jacksoni is a species somewhat smaller than *M. andrewsi* and was named by Leakey in 1943 on material from Shungura, Omo; he created for it the genus *Pronotochoerus.* Arambourg

(1947) correctly recognized the resemblances to Hopwood's genus and referred his own Omo specimens to *M. andrewsi;* however, they differ from the type and from the East Rudolf material and are here referred to *M. jacksoni.* At Omo, skulls of both *M. jacksoni* and *M. andrewsi* occur in Member G and are clearly distinct. *M. jacksoni* is considered ancestral to *M. nyanzae,* which occurs along with *M. andrewsi.*

Arambourg's collection included three good mandibles, which are very similar to those of *M. andrewsi* but smaller (figure 22.13*H*). The canines are verrucose and laterally compressed. They are not strongly curved and do not extend far back within the horizontal portion of the ramus. The lower border of the symphysis is curved, as in the warthog, with the incisors set in a forwardly directed flange, but the incisors themselves are not known. The lower premolars are known from undescribed Omo material. P_3 is smaller than in the bush pig and relatively broader, consisting of a short main cone, a small anterior cingulum, and a fairly well-developed posterior cingulum. P_4 is rectangular, much smaller than in the bush pig, with a robust cruciform main cone; there is a weak anterior cingulum and a narrow posterior one with a central cusp flanked by two persistent foveae that become lakes in advanced wear (figure 22.10*E–F*).

The skull is smaller than that of *M. andrewsi* but shares the broad braincase and occiput. The snout is shorter even in the early forms and considerably shorter in the later ones (figure 22.13*L–M*). There are strongly developed zygomatic knobs in the males and the zygoma juts out more abruptly than in *M. andrewsi.* The canine flanges are very different from those of *M. andrewsi* and are intermediate in form between those of the warthog and the forest hog, bearing elevated ridges like those of the latter but with the canine alveoli directed diagonally outward and the tusks curve laterally. The canines are dorsoventrally flattened but have thin ribbed enamel on the ventral surface. The upper P^3 and P^4 are small and a little like those of *Notochoerus* (figure 22.10*D*).

A feature of the upper third molars of *Metridiochoerus jacksoni* is the presence of an unusually marked gap between the anterior and the second lateral pillars on the buccal side. The enamel island formed by the anteroexternal main pillar has a characteristic italic *x* shape but the anterointernal pillar varies from an H to a collar stud shape and is somewhat compressed, as in *M. "meadowsi".* The second pair of laterals is rather variable, usually with a good X or H shape on the lingual side but less regular on the buccal side. A third pair of laterals is usually present, but often it is well developed only

on the lingual side and the rest of the talon is a complex of rounded pillars. The talon is normally no longer than the body of the crown, usually a good deal shorter. The lowers are far more regular and typically have three pairs of well-developed laterals, but sometimes a weaker fourth pair. The collar stud shape is well-developed, even on the anterior lateral columns. Although the lateral columns do fuse with the medians, especially in the uppers, this does not lead to the development of complex patterns like those of *M. andrewsi* except in advanced wear. In the undescribed collection from Omo there is a tendency toward progressive elongation of the crown, but the height remains less than the basal length in the upper molars and less than three-quarters of the basal length in the lowers; in Member G the basal length in upper third molars is about 55 ± 5 mm and in the lowers is 60 ± 5 mm.

The status of *M. pygmaeus* Leakey 1958, from Kanam, is uncertain but it is probably an aberrant early *M. jacksoni.* Typical *M. jacksoni* material is known from Omo and from the *"Mesochoerus limnetes"* zone of the Koobi Fora Formation at East Rudolf.

Metridiochoerus nyanzae (Leakey) 1958

In 1958, Leakey separated from *"Pronotochoerus" jacksoni* what he regarded as a new and somewhat larger species, *P. nyanzae.* The type is the left side of a mandible from Rawi, in the Kavirondo Gulf area of Lake Victoria, Kenya. In the same monograph he named another new species *Notochoerus compactus,* with the first syntype a damaged mandible from Bed II at Olduvai. Although there are minor differences between them, there seems to be no reason to consider that they represent more than one species, which is very close to *Metridiochoerus jacksoni.* Indeed *M. nyanzae* could be regarded as a synonym of the latter but it is useful to retain the name for this late stage in the lineage as there are some features of difference. In the Rawi specimen and in occasional undescribed teeth from Olduvai, there are only four well-developed pairs of laterals, with an incipient fifth pair, there are usually five pairs of laterals, and the teeth are slightly narrower than those attributed to *M. jacksoni* from earlier horizons. There is also a stronger tendency for the posterior lateral pillars to form Y-shaped enamel islands in the lowers, and the lower premolar series is reduced to the retention of P_4 only. The latter is even smaller than in the typical *jacksoni* material, though generally similar in structure but with the main cone initially formed from several small tubercles; the lateral foveae are not marked. The upper third molars show more clearly the pattern seen in *M. jacksoni*, but they are longer and relatively narrower.

In addition to the occurrences at Rawi and in Beds II and IV at Olduvai, this form occurs at Chesowanja in the Baringo Basin, Kenya, and in the upper part of the Shungura Formation, Omo.

Genus *Stylochoerus* E. C. N. & H. E. van Hoepen 1932

SYNONYMY. *Afrochoerus* Leakey 1942, *Orthostonyx* Leakey 1958.

TYPE SPECIES. *Stylochoerus compactus* E. C. N. & H. E. van Hoepen 1932.

DIAGNOSIS. Large Suidae with very specialized canines, having an oval cross-section, both in uppers and lowers, and possessing a core of cancellous osteodentine in the interior of the tusk, extending for at least half its length. Upper canines rise at about 45° from the palatal plane and perpendicular to the axis of the skull. Upper and lower canines do not wear against each other. Third molars very hypsodont and composed of closely packed series of cusps or pillars arranged much as in the living warthog. The enamel pattern of the lateral pillars, both buccal and lingual, is not subcylindrical, as in *Phacochoerus*, but is roughly Y-shaped, the fork of the Y being formed by a fold in the enamel anteriorly and is not due to fusion of pillars. Fusion of pillars into more complex patterns takes place with advancing wear. Teeth thickly coated with cement.[9]

REMARKS. The third molars of *Stylochoerus* resemble those of the advanced *Metridiochoerus andrewsi* and may be difficult to separate in some instances, if the abnormal hypsodonty of the *Stylochoerus* teeth is not apparent. The resemblances in the molars might suggest that *S. compactus* should be included in *Metridiochoerus* but separate generic status seems justified by the unique structure of the canines, as well as by their attitude relative to the skull and jaw. Furthermore *"Notochoerus"* compactus is a true *Metridiochoerus* and the use of this specific name would cause confusion.

Stylochoerus compactus E. C. N. & H. E. van Hoepen 1932

The type is an RM³ from Cornelia, Orange Free State, South Africa, but there are other specimens of similar character from the same locality, including some that were originally described as *Synaptochoerus hieroglyphicus* (E. C. N. & H. E. van Hoepen 1932). Shaw (1939) regarded *Stylochoerus compactus* as a probable synonym of *Phacochoerus aethopicus* and Wells and Cooke (1942) considered it a distinct species of *Phacochoerus*. This assessment fails to take account of the abnormal hypsodonty and has been revised recently by Cooke (1974). In 1942, Leakey founded the genus *Afrochoerus*, with the type species *A. nicoli*, on the basis of three large lower molars from Beds II and IV at Olduvai, and in 1958 he recognized the large, almost elephantlike canines of this peculiar suid. At that time he also established *Orthostonyx brachyops*, in which the unusual orientation of the canines in the maxilla was detected for the first time; the general structure of the canines was like that of *"Afrochoerus"* but the canines themselves were smaller. I have found both types of canine in the Cornelia collections and there can be no doubt that they belong with *Stylochoerus compactus*, the molars of which are very close to those of *Orthostonyx*, and there are also two fragments of third molars of *Afrochoerus* type. It would thus appear that only a single species is represented, probably with strong sexual dimorphism so that *"Orthostonyx"* represents the female and *"Afrochoerus"* the male variations.

The skull is not known but was probably fairly like that of *Metridiochoerus jacksoni*. The type maxilla fragment of *"Orthostonyx"* showed that the upper canine was set in a flange not unlike that of the warthog, but projected outward at right angles to the axis of the skull and at an angle of about 45° above the palatal plane, almost like *Babirussa*. The female canine is fairly short, oval in cross-section, dorsoventrally flattened, and curves very slightly up towards the tip. There is an external area of wear, clearly through use and not by abrasion against the lower canines. The latter are subtriangular in cross-section, with a rounded posterior face, and carry thin enamel on the lateral faces. They are gently curved but their exact orientation in the mandible is not known. They show signs of wear along the outer edge as if used for digging. The core is of cellular osteodentine and the surrounding dentine is concentrically banded. The male tusks are very much larger, almost egg-shaped in cross-section, with the wider end inward and the narrower one outward; there is a fairly strong groove on the lower side and a weaker groove toward the roots on the upper side. The tusk has a slightly spiral twist from the roots to the middle and then curves upwards and slightly backwards. Although normally 40-50 cm long, some are as long as 75 cm, making this a formidable ani-

---

9. The description of the molars is taken almost verbatim from Leakey's 1942 diagnosis of *Afrochoerus*, the canines of which were only described for the first time in 1958 and are here regarded as of prime significance for the recognition of the genus.

mal. One undescribed maxilla fragment from East Rudolf shows that they were carried at the same angle as in the female.

The female mandible is not known, although there is a symphysis with the lower canines, which are oval in cross-section, gently curved and directed diagonally outward and upward. The male mandible is well represented, including a fine specimen from Kanjera illustrated by Leakey (1958), by a good undescribed mandible from near the top of the upper member of the Koobi Fora Formation at East Rudolf, and by an assemblage from Ternifine in Algeria, still undescribed. The animal must have been almost twice the size of the living warthog. The morphology of the lower jaw is strongly phacochoerine and the ascending ramus is long, suggesting elevation of the cranium. In the mandible, the large tusks are implanted at the sides of the symphysis, rising at about 45° relative both to the axis of the skull and to the horizontal plane. The lower canines are very nearly straight and have a flattened oval cross-section, lacking the shallow grooves found in the upper canines. There are traces of thin enamel, and the teeth are worn at the tips through use in digging but not by attrition against the uppers. There is a core of cancellous material.

The premolars are not known and were probably not retained in life. The first and second molars are hypsodont, with a strong posterior "bulge" more marked even than in living warthogs. The enamel is thinner than usual and produces a rather distinctive enamel pattern. The lateral pillars form mushroom-shaped enamel islands that may become H-shaped, with curved uprights to the H. In early wear there is a complex of small pillars at the back but as the tooth expands to the "bulge" mentioned above, this becomes a third pair of laterals—unusual in first or second molars. In the third molars the main feature is the unusual hypsodonty, with the anterior columns reaching quite an advanced stage of wear before the posterior columns are fully erupted (see Cooke 1974). In normal wear the enamel islands seem to be very crowded and in some instances the fork of the Y-shaped laterals projects a little to form a ridge down the lateral pillar. In advanced wear the enamel islands unite to form complex patterns, rather like those found in *Metridiochoerus andrewsi*. In the uppers there is commonly, but not always, a double row of median columns in the talon, a feature not noted in *M. andrewsi*. The extreme hypsodonty is more obvious in the upper than in the lower third molars, perhaps because of the relatively greater anteroposterior compression.

The type of *Phacochoerus altidens* from the Vaal River gravels (Shaw and Cooke 1941) is clearly *Stylochoerus,* as also is *Phacochoerus complectidens* (Leakey 1942) from Olduvai, which Leakey (1958) recognized as a synonym of *P. altidens;* his two subspecies *altidens* and *robustus* also fall away. In addition to its occurrence at Cornelia and in the Vaal River gravels, South Africa, *Stylochoerus compactus* is well represented at Olduvai, in the upper part of the Koobi Fora Formation at East Rudolf, and at Kanjera in Kenya, and in the upper part of the Shungura Formation, Ethiopia. From Laetolil, Tanzania there are a few rather hypsodont third molars, one of which was made the type of *Notochoerus serengetensis* by Dietrich in 1942; these clearly belong to the *Metridiochoerus* complex and most probably to *Stylochoerus,* but they are unlikely to come from the Laetolil Beds proper and must belong to deposits of about the same age as Olduvai Bed II.

Family Tayassuidae

The living peccaries are confined to North and South America, where they occupy an ecological niche resembling that of the Suidae in other parts of the world. Like the Suidae, they have an elongate, mobile, cartilaginous snout with a flattened terminal area around the nostrils. There is a line of erectile hairs and bristles in the neck and back and a musk gland some distance in front of the tail. The stomach is two-chambered and nonruminating, but more complex than in the Suidae. Females have four or only two mammae and nurse their young from the rear instead of laterally. The legs are long and slender with four toes on the front feet and three on the hind feet, terminated by rather small hooves. The third and fourth metatarsals are united, as in the ruminants. The skull has an elevated, backward-sloping, occipital crest, a long narrow palate, and small postorbital processes. The canines are small and lie almost vertically, with the gently curved lowers wearing against the front of the uppers to maintain sharp points; there is a notch in the maxilla to allow for the reception of the lower canine. There is normally a short gap between the canine and the anterior premolar, but the premolars are in full mutual contact. The molars have simple bunodont crowns with four cusps and the last premolar tends to be molariform. The upper incisors are short and stout, with curved roots, while the lowers are long and slender with straight roots.

Fossil peccaries occur in North America from the Oligocene to the present, in Europe in the Oligocene and Miocene, in the Pliocene of Asia, and in the Pleistocene and Recent of South America. One species has been reported recently from South Africa in the Pliocene Varswater Formation at Langebaan-weg, Cape Province (Hendey 1976), but there are some undescribed specimens from Fort Ternan and Ngorora that may also belong to the Tayassuidae (Pickford, pers. comm.).

Pecarichoerus africanus Hendey 1976

The Pliocene Varswater Formation at Langebaan-weg, South Africa, contains part of a skull and most of the mandible of a very small peccary in which the entire series of five cheek teeth is only about 45 mm long. The lower incisors are directed forward from a fairly long shallow symphysis. The first three premolars are single-cusped, rather high-crowned teeth with double roots. P^3 has a strong posterointernal cingulum, while P^4 is three-rooted and has two main cusps lateral to each other, and both a weak anterior and a strong posterior cingulum. P_4 is two-rooted and also has two laterally arranged cusps and a posterior cusp. The molars are basically four-cusped, with an additional terminal cusp in M_3 and smaller terminal accessory cusps in M_2 and even in M_1. Other fragments confirm these characters. A damaged metatarsal shows complete fusion proximally in III/IV and apparently almost complete fusion distally, a condition more extreme than in other fossil and living specimens. The generic status of the specimen is uncertain but seems closest to the Miocene (Chinji) *Pecarichoerus orientalis* of the Siwaliks of India, but the Langebaanweg species is smaller.

Discussion

Geologic Range and Phylogeny. The stratigraphic framework in East Africa has been worked out in some detail during the past few years and is well controlled by radiometric dates (Cooke and Maglio 1972, Maglio 1973). This has made possible a reasonable estimate of the known and inferred time ranges for the main suid genera and species shown in figures 22.9, 22.11, and 22.13. The distribution among the more important localities is summarized in table 22.3. In the figures, some attempt has been made to indicate possible relationships between the genera and species within each diagram.

In considering the possible relationships between the various genera, it seems that characteristics of the premolars and of the canines may be especially useful. The molars are perhaps more subject to development of convergent features through adaptation to similar diets but are most useful for discrimination at the species level. The establishment of a meaningful phylogeny at the present time is hampered by serious gaps in the Pliocene record and by the scrappy nature of some of the earlier material.

However, it seems clear that *Nyanzachoerus* and *Notochoerus* (figure 22.9) stand apart from the other fossil suids and are probably descended from a middle Miocene form such as *Kubanochoerus*. The undoubted suines *Potamochoerus, Kolpochoerus,* and *Hylochoerus* share upper canines that are essentially *Sus*-like in the possession of a strong ventral band of ribbed enamel, as well as two thin inset enamel strips on the lateral surfaces. The premolars are also essentially *Sus*-like, although the P_4 of *Kolpochoerus* is somewhat like that of *Potamochoerus;* the P_4 of *Hylochoerus* demands rather imaginative interpretation. It thus seems very probable that this group is descended from a common *Sus*-like ancestor, perhaps with closer affinities to *S. verrucosus* than to *S. scrofa. Potamochoerus porcus* could as easily be derived from such a stock as from a hypothetical *Propotamochoerus,* which would not be suitable as an ancestor for *Kolpochoerus. Potamochoeroides* also has some *Sus*-like features and the dentition is relatively conservative and not far removed from a simple bunodont ancestral stock. It could be that the same stock provided the ancestors for both the *Kolpochoerus* line and for the members of the *Metridiochoerus* complex, the former remaining fairly conservative while the latter differentiated more dynamically. However, this is still highly speculative and it is to be hoped that more discoveries in beds of 4 to 5 m.y. age may resolve the problem.

Environment and Habitat. Although in the present revision the number of suid genera and species has been reduced drastically, it is probable that continued study may result in further simplification; however, it is also likely that there will be new discoveries, particularly in the Pliocene and later Miocene, requiring the recognition of additional species and perhaps genera also. There remains an astonishing picture of variety among the fossil suids with many contemporaneous genera and species through most of the Pliocene and Pleistocene. In any particular stratigraphic unit, it is common to find at least three different species, usually representing three different genera. This does not, of course, mean that all three were competitors in a common habitat, for seasonal or other climatic factors, or even the mere presence of water, may bring together in a geologically short time the products of somewhat different local environments.

What these environments were we can only guess, although we can perhaps gain some idea from a study of the habitats of the three living suids of the same region. Ewer (1958b, 1970) has made a notable contribution in trying to relate anatomical and functional characters to feeding habits and in con-

Table 22.3 Occurrence of suid taxa at principal African sites.

Site	*Nyanzachoerus* devauxi	syrticus	tulotos	pattersoni	kanamensis	jaegeri	*Notochoerus* capensis	euilus	scotti	*Sus scrofa*	*Potamochoerus porcus*	*Kolpochoerus* afarensis	limnetes	phacochoeroides	olduvaiensis	paiceae	sp.	*Hylochoerus meinertzhageni*	*Phacochoerus* modestus	kabuae	aethiopicus	africanus	*Potamochoeroides shawi*	*Metridiochoerus* jacksoni	nyanzae	andrewsi	*Stylochoerus compactus*	*Percarichoerus africanus*
Sahabi	—	X	—	—	—	X	—	—	—	—	—	—	—	—	—	—	—	—	—	—	—	—	—	—	—	—	—	—
Wadi Natrun	—	—	—	—	—	X	—	—	—	—	—	—	—	—	—	—	—	—	—	—	—	—	—	—	—	—	—	—
Chad	—	—	—	X	—	—	X	—	—	—	—	—	—	—	—	—	—	—	—	—	—	—	—	—	—	—	—	—
Ternifine	—	—	—	—	—	—	—	—	—	X	—	—	—	—	—	—	—	—	—	—	—	—	—	—	—	?	X	—
Lukeino	—	—	cf	—	—	cf	—	—	—	—	—	—	—	—	—	—	—	—	—	—	—	—	—	—	—	—	—	—
Lothagam 1	—	X	cf	—	—	—	—	—	—	—	—	—	—	—	—	—	—	—	—	—	—	—	—	—	—	—	—	—
3	—	—	—	—	—	X	—	cf	—	—	—	—	—	—	—	—	—	—	—	—	—	—	—	—	—	—	—	—
Kanapoi	—	—	—	X	X	—	X	—	—	—	—	—	—	—	—	—	—	—	—	—	—	—	—	—	—	—	—	—
Ekora	—	—	cf	—	—	—	—	—	—	—	—	—	—	—	—	—	—	—	—	—	—	—	—	—	—	—	—	—
Hadar	—	—	—	X	—	—	X	—	—	—	—	X	—	—	—	—	—	—	—	—	—	—	—	—	—	—	—	—
Kanam	—	—	—	—	X	—	—	—	—	—	—	—	—	—	—	—	—	—	—	—	—	—	—	cf	—	—	—	—
Kaiso early	—	—	—	X	—	—	cf	—	—	—	—	—	X	—	—	—	—	—	—	—	—	—	—	—	—	—	—	—
later	—	—	—	X	—	—	X	—	—	—	—	—	X	—	—	—	—	—	—	—	—	—	—	—	—	—	—	—
Chemeron	—	—	—	X	—	cf	X	—	—	—	—	—	X	—	—	—	—	—	—	—	—	—	—	—	—	—	—	—
Mursi formation	—	—	—	X	—	cf	—	—	—	—	—	—	—	—	—	—	—	—	—	—	—	—	—	—	—	—	—	—
Usno formation	—	—	—	X	—	—	X	—	—	—	—	—	X	—	—	—	—	—	—	—	—	—	—	—	—	—	—	—
Shungura formation A – C	—	—	—	X	—	—	X	—	—	—	—	?	X	—	—	—	—	—	—	—	—	—	—	X	—	—	—	—
D – G	—	—	—	—	—	—	X	X	—	—	—	—	X	—	—	—	—	—	—	?	—	—	—	X	X	—	—	—
H – L	—	—	—	—	—	—	—	—	X	—	—	—	—	X	X	—	—	—	—	?	—	—	—	—	X	X	X	—
Kubi Algi formation	—	—	—	X	—	—	X	X	—	—	—	—	X	—	—	—	—	—	—	—	—	—	—	X	—	—	—	—
Koobi Fora fm. Zone 3	—	—	—	—	—	—	X	—	—	—	—	—	X	—	—	—	—	—	—	—	—	—	—	X	—	—	—	—
Zone 2	—	—	—	—	—	—	X	X	—	—	—	—	X	—	X	—	—	—	—	—	—	—	—	—	—	X	?	—
Zone 1	—	—	—	—	—	—	—	—	—	—	—	—	X	—	X	—	—	—	X	?	—	—	—	—	—	X	X	—
Laetolil	—	—	—	—	—	—	—	X	—	—	—	cf	—	—	—	—	—	—	—	—	—	—	—	cf	—	—	—	—
Olduvai Bed I, Lr. II	—	—	—	—	—	—	—	—	—	—	—	—	X	—	X	—	—	—	X	—	—	—	—	—	—	X	X	—
Upper II	—	—	—	—	—	—	—	—	—	—	—	—	—	—	X	—	—	—	X	—	—	—	—	—	—	X	X	—
Beds III–IV	—	—	—	—	—	—	—	—	—	—	—	—	—	—	X	—	—	—	—	—	—	—	—	—	—	X	X	—
Kanjera	—	—	—	—	—	—	—	—	—	—	—	—	—	—	—	—	—	X	—	—	—	—	—	—	—	X	X	—
Chiwondo beds	—	—	—	—	X	—	—	X	—	—	—	—	—	—	—	—	—	—	—	—	—	—	—	—	—	X	X	—
Langebaanweg	—	—	—	cf	—	cf	—	—	—	—	—	—	—	—	—	—	—	—	—	—	—	—	—	—	—	—	—	X
Makapansgat	—	—	—	—	—	—	X	—	—	—	—	—	—	—	—	—	—	—	—	—	—	—	X	—	—	—	—	—
Taung	—	—	—	—	—	—	—	—	—	—	—	—	—	—	—	—	—	—	—	—	—	—	X	—	—	—	—	—
Sterkfontein	—	—	—	—	—	—	—	—	—	—	—	—	—	—	—	—	—	—	—	—	—	—	cf	—	—	—	—	—
Swartkrans	—	—	—	—	—	—	—	—	—	—	—	—	—	—	—	—	—	—	X	—	—	—	—	—	X	—	—	—
Kromdraai	—	—	—	—	—	—	—	—	—	—	—	—	—	—	—	—	—	—	X	—	—	—	—	—	—	—	—	—
Bolt's Farm	—	—	—	—	—	—	—	—	—	—	—	—	—	—	—	—	—	—	X	—	—	—	X	—	X	—	—	—
Vaal River (mixed)	—	—	—	—	—	—	X	—	—	—	—	—	—	—	X	—	—	—	—	—	—	—	—	—	X	—	X	—
Cornelia	—	—	—	—	—	—	—	—	—	—	—	—	—	—	X	—	—	—	X	—	—	—	—	—	X	—	—	—
Elandsfontein (Hopefield)	—	—	—	—	—	—	—	—	—	—	—	—	—	—	X	—	—	—	—	—	—	—	—	—	X	—	—	—
Florisbad	—	—	—	—	—	—	—	—	—	—	—	—	—	—	—	—	—	—	—	—	X	X	—	—	—	—	—	—
Miscellaneous sites	X	—	—	—	—	X	—	—	—	X	X	—	X	X	—	—	—	X	—	X	X	X	X	—	X	X	—	—

sidering some rather striking features as related primarily to display rather than to dietary needs. One of the prominent features of the fossil suids is the tendency for exaggerated development of the zygomatic arches, reaching absurd limits in *Nyanzachoerus pattersoni* and *Notochoerus euilus*. This is probably related to recognition and to display and is most marked in the males. Strong sexual dimorphism is another feature of the Pliocene-Pleistocene suids, particularly apparent in the canines of *Kolpochoerus* but also reflected in the zygomatic inflation, or lack of it. *Stylochoerus,* with its enormous tusks in the male, is another example.

The teeth may perhaps serve as a fairly simple guide to diet and hence to habitat. Small brachydont, bunodont teeth probably reflect a life-style like that of modern *Sus* or *Potamochoerus*. The more elongate, but still brachydont third molars may suggest either a greater tendency towards browsing, as in *Hylochoerus,* or may mark the beginning of a trend towards increasing hypsodonty as an adaptation to the consumption of increasing amounts of harsh grass. The very hypsodont molars, by analogy with the warthog, must surely be taken to indicate a corresponding diet and habitat. It certainly appears that phacochoerelike teeth have evolved convergently, but independently, in *Notochoerus* and in the *Metridiochoerus* complex (including *Stylochoerus*). Elevation of the orbits, useful in a grassland environment, appears in the *Metridiochoerus* complex in a moderate degree, but also in *Potamochoeroides shawi* and in the advanced *Mesochoerus olduvaiensis,* not necessarily indicating any real affinity with *Phacochoerus.*

Many individuals and institutions have granted access to material, much of it as yet unpublished, and many colleagues have participated in stimulating discussions that have helped to shape the views expressed here. It would be invidious to name only a few and tedious to name all, but the writer's indebtedness is none the less real and his thanks are sincere. Particular mention must be made of the Wenner-Gren Foundation for Anthropological Research, New York, without whose aid the work would not have been accomplished. Support has also been provided by Dalhousie University and by the National Research Council of Canada.

References

Allen, G. M. 1939. A checklist of African mammals. *Bull. Mus. Comp. Zool. Harvard* 83:1–763.

Ansell, W. F. H. 1968. Artiodactyla. In *Smithsonian Institution preliminary identification manual for African mammals,* ed. J. Meester. Washington: U. S. National Mus.

Arambourg, C. 1933. Mammifères Miocènes du Turkana (Afrique orientale). *Ann. Paléont.* 22:123–146.

———. 1938. Mammifères fossiles du Maroc. *Mém. Soc. Sci. Nat. Maroc* 46:1–74.

———. 1943. Observations sur les Suidés fossiles du Pléistocène d'Afrique. *Bull. Mus. Hist. Nat. Paris* (2) 15:471–476.

———. 1947. Contribution à l'étude géologique et Paléontologique du bassin du lac Rodolphe et de la basse vallée de l'Omo. Deuxième partie. Paléontologie. *Mission scient. Omo, 1932–1933, 1, Géol. Anthrop.*: 232–562.

———. 1961. Note préliminaire sur quelques vertébrés nouveaux du Burdigalien de Libye. *C. R. Somm. Soc. Géol. Fr.* 4:107–109.

———. 1963. Le genre *Bunolistriodon* Arambourg 1933. *Bull. Soc. Géol. Fr.* (7) 5:903–911.

Bishop, W. W. 1967. The Later Tertiary in East Africa—volcanics, sediments and faunal inventory. In *Background to Evolution in Africa,* ed. W. W. Bishop and J. D. Clark. Chicago: University of Chicago Press, pp. 31–56.

———. 1968. The evolution of fossil environments in East Africa. *Trans. Leicester Lit. Phil. Soc.* 62:22–44.

Bishop, W. W., J. A. Miller, and F. J. Fitch. 1969. New potassium-argon age determinations relevant to the Miocene fossil mammal sequence of East Africa. *Am. J. Sci.* 267:669–699.

Bouet, G., and H. Neuville. 1930. Recherches sur le genre *Hylochoerus. Arch. Mus. Nat. Hist. Nat. Paris* (6) 5: 215–301.

Broom, R. 1925. On evidence of a giant pig from the late Tertiaries of South Africa. *Rec. Albany Mus.* (Grahamstown) 3:307–308.

———. 1928. On some new mammals from the diamond gravels of the Kimberley district. *Ann. S. Afr. Mus.* 22:439–444.

———. 1931. A new extinct giant pig from the diamond gravels of Windsorton, South Africa. *Rec. Albany Mus.* (Grahamstown) 4:167–168.

———. 1948. Some South African Pliocene and Pleistocene mammals. *Ann. Transv. Mus.* 21:1–38.

Colbert, E. H. 1934. An Upper Miocene suid from the Gobi Desert. *Am. Mus. Novit.* 690:1–7.

———. 1935a. Distributional and phylogenetic studies of Indian fossil mammals, IV: the phylogeny of Indian Suidae and the origin of the Hippopotamidae. *Am. Mus. Novit.* 799:1–24.

———. 1935b. Siwalik mammals in the American Museum of Natural History. *Trans. Am. Phil. Soc.* n.s. 26: 1–401.

Cooke, H. B. S. 1949. The fossil Suina of South Africa. *Trans. Roy. Soc. S. Afr.* 32:1–44.

———. 1974. The fossil mammals of Cornelia, O.F.S., South Africa. *Nat. Mus. Bloemfontein, Mem.* 9:63–84.

———. 1978. Pliocene-Pleistocene Suidae from Hadar, Ethiopia. *Kirtlandia* (in press).

Cooke, H. B. S., and Y. Coppens. (forthcoming). Fossil Suidae of Omo, Ethiopia.

Cooke, H. B. S., and S. C. Coryndon. 1970. Pleistocene mammals from the Kaiso Formation and other related deposits in Uganda. In *Fossil Vertebrates of Africa,* vol.

2, ed. L. S. B. Leakey and R. J. G. Savage. 107–224. London and New York: Academic Press.

Cooke, H. B. S., and R. F. Ewer. 1972. Fossil Suidae from Kanapoi and Lothagam, northwestern Kenya. *Bull. Mus. Comp. Zool. Harvard* 143(3):149–295.

Cooke, H. B. S., and V. J. Maglio. 1972. Plio-Pleistocene stratigraphy in East Africa in relation to proboscidean and suid evolution. In *Calibration of Hominoid Evolution,* ed. W. W. Bishop and J. A. Miller. Edinburgh: Scottish Academic Press, pp. 303–329.

Coppens, Y. 1967. Les faunes de vertébrés quaternaires du Tchad. In *Background to Evolution in Africa,* ed. W. W. Bishop and J. D. Clark. Chicago: University of Chicago Press, pp. 89–96.

———. 1971. Une nouvelle espèce de Suidé du Villafranchien du Tunisie, *Nyanzachoerus jaegeri* nov. sp. *C. R. Hebd. Acad. Sci.* (Paris) 272:3264–67.

Crusafont Pairo, M., and R. Lavocat. 1954. *Schizochoerus* un nuevo género de Súidos del Pontiense inferior (Vallesiense) del Vallés Pendés. *Notas Comun. Inst. Geol. Min. Esp.* 36:79–90.

Cuvier, G. 1817. *Le Regne Animal—Les Mammifères.* Paris: Fortin-Masson.

Dale, M. M. 1948. New fossil Suidae from the Limeworks quarry, Makapansgat, Potgietersrust. *S. Afr. Sci.* 2: 114–116.

Dietrich, W. O. 1937. Pleistozäne Suiden-Reste aus Oldoway, Deutsch-Ostafrika. *Wiss. Ergebn. Oldoway-Exped.* 1913 (N.F.) 4:91–104.

———. 1942. Ältestquartäre Säugetiere aus der südlichen Serengeti, Deutsch-Ostafrika. *Paläontogr.* 94(A):43–133.

Doust, H. 1968. Palaeoenvironmental studies in the Miocene of Libya, vol. 1. Ph.D. thesis, Imperial College, London.

Dreyer, T. F., and A. Lyle. 1931. *New fossil mammals and man from South Africa.* Bloemfontein: Dept. Zool., Grey Univ. College, pp. 1–60.

Ennouchi, E. 1954. *Omochoerus maroccanus* nov. sp. nouveau Suidé marocain. *Bull. Soc. Géol. Fr.* (6) 3:649–656.

Ewer, R. F. 1956. The fossil suids of the Transvaal caves. *Proc. Zool. Soc. Lond.* 127:527–544.

———. 1957. The fossil pigs of Florisbad. *Researches Nat. Mus. Bloemfontein* 1 (10):239–257.

———. 1958a. The fossil Suidae of Makapansgat. *Proc. Zool. Soc. Lond.* 130:329–372.

———. 1958b. Adaptive features in the skulls of African Suidae. *Proc. Zool. Soc. Lond.* 131:135–155.

———. 1970. The head of the Forest Hog *Hylochoerus meinertzhageni.* *E. Afr. Wildlife J.* 8:43–52.

Fahlbusch, V., and H. Gall. 1970. Die obermiozäne Fossillagerstätte Sandelzhausen 1. Entdeckung, Geologie, Faunenübersicht und Grabungsbericht fur 1969. *Mitt. Bayer. Staatssamml. Paläont. Hist. Geol.* 10:365–396.

Falconer, H. 1868. *Palaeontological memoirs,* vol. 1. *Fauna Antiqua Sivalensis.* London: Smith, Elder and Co.

Falconer, H., and P. T. Cautley. 1847. *Fauna Antiqua Sivalensis,* Pt. VIII, Pls. 69–71.

Forster Cooper, C. 1913. New anthracotheres and allied forms from Baluchistan—preliminary notice. *Ann. Mag. Nat. Hist. Lond.* 12:514–522.

Forsyth Major, C. 1897. On *Sus verrucosus* and allies from the Eastern Archipelago. *Ann. Mag. Nat. Hist. Lond.* 19:521–542.

Gabunia, L. K. 1955. Novyi pryedstsvityel' Suidae iz sryendnyego miotsyena Byelmyechyetskoy (Syevyernyy Kavkaz). *Doklady Akad. Nauk* 102 (6).

———. 1960. Kubanochoerinae, nouvelle sous-famille de porcs du Miocène moyen du Caucase. *Vertebr. Palasiat.* 4:87–97.

Gaudry, A. 1878. *Les Enchainements du Monde Animal dans Les Temps Géologiques. Mammifères Tertiaires.* Paris.

Gervais, P. 1859. *Zoologie et Paléontologie Françaises.* 2nd ed. Paris.

Gmelin, H. 1788. *Systema Naturae de Linne.* Paris, 13 Ed. 1(1):220.

Gray, J. E. 1854. On the painted pig of the Cameroons (*Potamochoerus penicillatus*). *Proc. Zool. Soc. Lond.* 20 (1852):129–132.

Gregory, W. K. 1910. The orders of mammals. *Bull. Am. Mus. Nat. Hist.* 27:1–524.

Hendey, Q. B. 1976. Fossil peccary from the Pliocene of South Africa. *Science* 192:787–789.

Hoepen, E. C. N. van, and H. E. van Hoepen. 1932. Vrystaatse wilde varke. *Paleont. Navors. Nas. Mus. Bloemfontein* 2 (4):39–62.

Hooijer, D. A. 1963. Miocene Mammalia of Congo. *Ann. Mus. R. Afr. Cent. Sci. Geol.* 46:1–77.

Hopwood, A. T. 1926a. Some Mammalia from the Pliocene of Homa Mountain, Victoria Nyanza. *Ann. Mag. Nat. Hist.* (9) 18:266–272.

———. 1926b. Fossil Mammalia. In The geology and palaeontology of the Kaiso Bone Beds, ed. F. J. Wayland. *Geol. Surv. Uganda Occ. Papers* 2:13–36.

———. 1929. *Hylochoerus grabhami,* a new species of fossil pig from the White Nile. *Ann. Mag. Nat. Hist.* (10) 4:289–290.

———. 1931. Pleistocene mammalia from Nyasaland and Tanganyika Territory. *Geol. Mag.* 68:133–135.

———. 1934. New fossil mammals from Olduvai, Tanganyika Territory. *Ann. Mag. Nat. Hist.* (10) 14:546–550.

Hopwood, A. T., and J. P. Hollyfield. 1954. An annotated bibliography of the fossil mammals of Africa (1742–1950). British Mus. (Nat. Hist.), *Fossil Mammals of Africa* 8:1–194.

Hünermann, K. A. 1968. Die Suidae (Mammalia, Artiodactyla) aus den Dinotheriensanden (Unterpliozän = Pont.) Rheinhessens (Südwestdeutschland). *Schweiz. Palaeont. Abh.* 86:1–96.

Joleaud, L. 1933. Études de géographie zoologique sur le Berbérie. Les Pachydermes—1. Les sangliers et les phacochères. *Rev. Géogr. Maroc* 17:177–192.

Keen, E. N., and R. Singer. 1956. Further fossil Suidae from Hopefield. *Ann. S. Afr. Mus.* 42 (4):350–360.

Leakey, L. S. B. 1942. Fossil Suidae from Olduvai. *J. E. Afr. Uganda Nat. Hist. Soc.* 16:178–196.

_____. 1943. New fossil Suidae from Shungura, Omo. *J. E. Afr. Uganda Nat. Hist. Soc.* 17:45–61.

_____. 1958. Some East African Pleistocene Suidae. Brit. Mus. (Nat. Hist.), *Fossil Mammals of Africa* 14:1–133.

_____. 1965. *Olduvai Gorge 1951–1961:* Vol. 1: *Fauna and background.* Cambridge University Press: 1–109.

Leinders, J. 1975. Sur les affinités des Listriodontinae bunodontes de l'Europe et de l'Afrique. *Bull. Mus. Natn. Hist. Nat., Paris,* 3ᵉ ser. 341, *Sciences de la Terre* 46:197–204.

Leonardi, P. 1952. I suidi di Sahabi nella Sirtica (Africa Settentrionale). *Rc. Accad. Naz.* 40 (4) 4:75–88.

_____. 1954. Resti fossili di *Sivachoerus* del giaccimento di Sahabi in Cirenaica (Africa Settentrionale), Notizie preliminari. *Atti. Acad. Naz. Lincei Rend.* (8) 13:166–169.

Linnaeus, C. 1758. *Systema Naturae.* 10 Ed., 1:49.

Liu, T. S., and Y. C. Lee. 1963. New species of *Listriodon* from Miocene of Lantien, Shensi, China. *Vertebr. Palasiat.* 7:291–309.

Lydekker, R. 1878. Notice of Siwalik mammals. *Rec. Geol. Surv. India* 11:64–104.

_____. 1884. Indian Tertiary and post-Tertiary Vertebrata. Siwalik and Narbada Bunodont Suina. *Mem. Geol. Surv. India, Palaeont. Indica* (10), 3(2):35–104.

Maglio, V. J. 1973. Origin and evolution of the Elephantidae. *Trans. Am. Phil. Soc.* n.s. 63(3):1–149.

Matthew, W. D. 1934. Reclassification of the Artiodactyl families. *Bull. Geol. Soc. Am.* 40:403–408.

Mawby, J. 1970. Fossil vertebrates from northern Malawi: preliminary report. *Quaternaria* 13:319–323.

Meyer, H. von. 1834. Die fossilen Zähne und Knochen und ihre Ablagerung in der Gegend von Georgensmünd in Bayern. Mus. senckenb. 1.

_____. 1846. [Letter on various fossils.] *Neues Jahrb. Min. Geol. Pal.* 1846:462–476.

_____. 1866. Ueber die fossilen Reste von Wirbelthieren, welche die Herren von Schlagenweit von ihren Reisen in Indien und Hochasien mitgebracht haben. *Paläontogr.* 15:1–40.

Pallas, P. S. 1767. Spicalegia Zool. 2:2.

Pearson, H. S. 1927. On the skulls of Early Tertiary Suidae, together with an account of the otic region in some other primitive Artiodactyla. *Phil. Trans. Roy. Soc.* B215:389–460.

_____. 1928. Chinese fossil Suidae. *Palaeont. Sin.* (C) 5 (5):1–75.

Peters, K. F. 1869. Zur Kenntniss der Wirbelthiere aus den Miozänschichten von Eibiswald in Steiermark, II: Amphicyon, Viverra, Hyotherium. *Denkschr. Akad. Wiss. Wien* 29:189–214.

Phillips, J. F. V. 1926. Wild Pig (*Potamochoerus choiropotamus*) at the Knysna: notes by a naturalist. *S. Afr. J. Sci.* 23:655–660.

Pia, J. 1930. Eine neue quartäre Warzenschwein-Art aus Südwestafrika. *Cbl. Min. Geol. Palaont.* 1930B:76–83, 205–206.

Pickford, M., and A. F. Wilkinson. 1975. Stratigraphic and phylogenetic implications of new Listriodontinae from Kenya. *Neth. J. Zool.* 25 (1):128–137.

Pilgrim, G. E. 1917. Preliminary note on some recent mammal collections from the basal beds of the Siwaliks. *Rec. Geol. Surv. India* 48:98–101.

_____. 1926. The fossil Suidae of India. *Mem. Geol. Surv. India, Palaeont. Indica* 8(4):1–68.

_____. 1941. The dispersal of the Artiodactyla. *Biol. Rev.* 16:134–163.

Pomel, A. 1846. Note sur des animaux fossiles découverts dans le département de l'Allier. *Bull Soc. Géol. Fr.* 4:378–385.

_____. 1896. Les Suilliens: Porciens. *Carte géol. Algérie, Paléont. Monogr. Alger.,* pp. 1–39.

Savage, R. J. G. 1969. Early Tertiary mammal localities in southern Libya. *Proc. Geol. Soc. Lond.* 1657:167–171.

Savage, R. J. G., and M. White. 1965. Two mammal faunas from the Early Tertiary of Central Libya. *Proc. Geol. Soc. Lond.* 1623:89–91.

Schmidt-Kittler, N. 1971. Die obermiozäne Fossilagerstätte Sandelzhausen 3. Suidae (Artiodactyla, Mammalia). *Mitt. Bayer. Staatssamml. Paläont. Hist. Geol.* 11:129–170.

Selley, R. C. 1966. The Miocene Rocks of the Marada and Gebel Zelten Area, Central Libya. Tripoli, Libya: The Petroleum Exploration Society of Libya, pp. 1–30.

Shaw, J. C. M. 1939. Growth changes and variations in warthog third molars and their palaeontological importance. *Trans. Roy. Soc. S. Afr.* 27:51–94.

Shaw, J. C. M., and H. B. S. Cooke. 1941. New fossil pig remains from the Vaal River gravels. *Trans. Roy. Soc. S. Afr.* 28:293–299.

Simons, E. L. 1964. The early relatives of man. *Scient. Am.* 622:2–14.

Simpson, G. G. 1931. A new classification of mammals. *Bull. Am. Mus. Nat. Hist.* 59:259–293.

_____. 1945. The principles of classification and a classification of mammals. *Bull. Am. Mus. Nat. Hist.* 85:xvi, 1–350.

_____. 1967. The Tertiary Lorisiform primates of Africa. *Bull. Mus. Comp. Zool. Harvard* 136:39–62.

Singer, R., and E. N. Keen. 1955. Fossil Suiformes from Hopefield. *Ann. S. Afr. Mus.* 42 (3):169–179.

Stehlin, H. G. 1899–1900. Über die Geschichte des Suiden-Gebisses. *Abh. Schweiz. Paläont. Gessel.* 26/27:1–527.

Straelen, V. van. 1924. Sur les premièrs restes de Phacochères fossiles recueilles au Congo Belge. *Bull. Acad. Roy. Belg.* (Cl. Sci.) (5) 10:360–365.

Stromer, E. 1926. Reste Land- und Süsswasser-Bewohnender Wirbeltiere aus den Diamentenfeldern Deutsch-Südwestafrikas. In *Die Diamantenwüste Südwestafrikas,* vol. 2., ed. E. Kaiser. Berlin.

Thenius, E. 1956. Die Suiden und Tayassuiden der steirischen Tertiärs. Beiträge zur Kentnis der Säugetierreste des steirischen Tertiärs VIII. *Sber. Akad. Wiss. Wien* 1:165, 337–382.

Thomas, O. 1904. On *Hylochoerus,* the forest-pig of Central Africa. *Proc. Zool. Soc. Lond.* 1904 (2):193–199.

Thomas, Ph. 1884. Sur quelques formations d'eau douce tertiares d'Algérie. *C. R. Acad. Sci.* (Paris) 98:311–314.

Tobien, H. 1936. Mitteilungen über Wirbeltierreste aus dem Mittelpliocän des Natrontales (Aegyptien). 7. Artiodactyla: A. Bunodontia: Suidae. *Z. dtsch. geol. Ges.* Berlin 88:42–53.

Viret, J. 1929. Les Faunes de Mammifères de l'Oligocène supérieur de la Limagne Bourbonnaise. *Ann. Univ. Lyon.* 47:13–325.

_____. 1961. Artiodactyla. In *Traité de Paleontologie,* ed. J. Piveteau. Paris: Masson & Cie, vol. 6 (1).

Walker, A. 1968. The Lower Miocene fossil site of Bukwa, Sebei. *Uganda J.* 32(2):149–156.

Wells, L. H., and H. B. S. Cooke. 1942. The associated fauna and culture of Vlakkraal thermal springs, O.F.S.: III—The faunal remains. *Trans. Roy. Soc. S. Afr.* 29:214–232.

Whitworth, T. 1965. The Pleistocene lake beds of Kabua, northern Kenya. *Durham Univ. J.* 57 (2):88–100.

Wilkinson, A. F. 1972. The Lower Miocene Suidae of Africa. Ph.D. thesis, University of Bristol.

_____. 1976. The Lower Miocene Suidae of Africa. In *Fossil vertebrates of Africa,* ed. R. J. G. Savage and S. C. Coryndon. vol. 4, pp. 173–282. London and New York: Academic Press.

Zdarsky, A. 1909. Die miozäne Säugetierfauna von Leoben. *Jb. K. K. Geol. Reichsanst. Wien* 59:245–288.

23

Hippopotamidae

Shirley C. Coryndon[1]

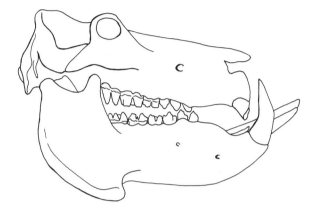

The family Hippopotamidae contains two living genera, both monospecific and confined to sub-Saharan Africa. *Hexaprotodon liberiensis,* the West African pygmy hippopotamus, is found only in the forests and coastal plains from Guinea to the Ivory Coast; the larger and more common *Hippopotamus amphibius* is found in many regions south of the Sahara where there are suitable aquatic environments. Fossil hippopotamids are known from Pleistocene deposits throughout a greater part of the Old World, and although two records of isolated teeth are known from China it seems that, like the Rhinocerotidae but unlike the Proboscidea, the family did not reach the New World.

Only two genera are recognized in the family Hippopotamidae: *Hippopotamus* and *Hexaprotodon.* The West African pygmy hippopotamus is usually referred to *Choeropsis* Leidy 1853, but Coryndon (1977) has shown that it belongs to the genus *Hexaprotodon* Falconer and Cautley 1836. Deraniyagala (1949) described a specimen dredged from the Nile as the mandible of a new genus *Prechoeropsis;* it is considered here, however, that this specimen is most likely a juvenile and possibly pathological individual of *Hippopotamus amphibius,* having no bearing on the ancestry of *Hexaprotodon.*

General Features of Hippopotamidae

As with so many mammalian groups, remains of earliest Hippopotamidae as well as being rare consist of fragmentary and isolated teeth with very few cranial or postcranial elements; thus comparison between one fauna and another has to be determined mainly on dental characteristics. In *Hippopotamus,* the dentition is remarkably conservative throughout the known evolutionary history of the family, particularly as regards the molar teeth, which are quite distinct from the pattern seen in either the Suidae or the Anthracotheriidae, the two most closely allied families. The incisors and canines of Hippopotamidae show rather more change than the molars, and the premolars probably show most change of all throughout the known history.

Early forms of hippopotamid, unlike many mammals, have a complete dentition, that is $\frac{3.1.4.3}{3.1.4.3}$. In more advanced forms the incisors tend to be reduced and the first premolar lost, though this tooth is retained frequently as a persistent milk element even

1 Shirley C. Coryndon died in October 1976 and the draft manuscript, written in 1973, has been updated by her husband, R. J. G. Savage.

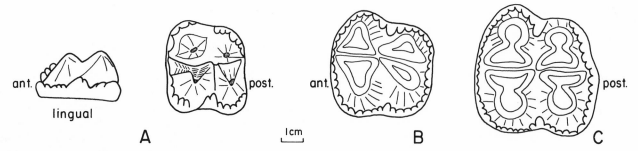

Figure 23.1   Left upper molars in hippopotamids: (*A*) unworn M² of Rusinga hippopotamid; (*B* and *C*) patially worn M³ demonstrating differences in enamel figures between (*B*) *Hexaprotodon* and (*C*) *Hippopotamus*.

in the living species. In both fossil and living forms the first molars are worn at an early stage, little enamel remaining when the second molar comes into wear. Because fragmentary and isolated dentitions form the major part of the known early fossil specimens, tooth size and morphology are used as the main basis for any comparisons. An isolated tooth is obviously easier to measure than one that is part of a dentition where the overlap of one tooth with another may make accurate measurement difficult. On the other hand it is not always possible to orientate an isolated tooth or even to find the correct position in the jaw—second and third premolars and first and second molars being particularly difficult to distinguish when isolated. The third upper molar has a high posterior cingulum and the third lower a distinct talonid. When possible, the following criteria are used in measuring the teeth of Hippopotamidae:

1. *Incisors.* At alveolar margin: (a) superior-inferior diameter in lowers, anterior-posterior diameter in uppers; (b) lateral diameter.

2. *Canines.* At alveolar margin: (a) anterior-posterior diameter; (b) lateral diameter.

3. *Premolars.* At level of and including cingulum: (a) greatest length; (b) greatest width at right angles to (a); (c) height, taken from tip of main cusp to base of cingulum.

4. *Molars.* At level of cingulum: (a) greatest length; (b) width of anterior half of tooth at right angles to length; (c) width of posterior half of tooth; (d) height of anterior lingual cusp (metaconid) of lower molars, anterior buccal cusp (paracone) of uppers; (e) length of talonid in lower third molar. (See figure 23.1).

The teeth of both genera of living hippopotamus are basically similar in that the morphology of the cusp pattern is unique to the family. The dental formula differs slightly between *Hippopotamus* and *Hexaprotodon,* the former having $\dfrac{2.1.3 \text{ or } 4.3}{2.1.3 \text{ or } 4.3}$ and the latter $\dfrac{2.1.3.3}{1.1.3.3}$ (figure 23.2). There are important differences in the bones of the cranium, those of *Hexaprotodon* following the basic pattern seen in early members of the family and those of the extant *Hip-*

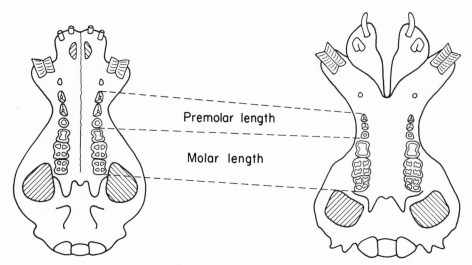

Figure 23.2   Diagrammatic comparison of the cranium of the two living hippopotamus species: (*left*) *Hexaprotodon liberiensis;* (*right*) *Hippopotamus amphibius.* Ventral view.

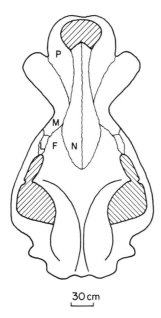

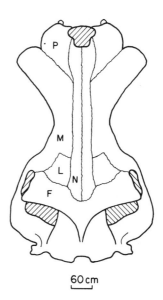

Figure 23.3 Diagrammatic comparison of the cranium of the two living hippopotamus species: (*left*) *Hexaprotodon liberiensis*; (*right*) *Hippopotamus amphibius*. Dorsal view.

popotamus being rather more specialized (figure 23.3); differences in postcranial elements reflect the differences in size and habitat between the two. The position of the upper incisors in *Hexaprotodon* is rather different from that in *Hippopotamus*, the former being set in a gentle arc in the premaxilla, with direct tip-to-tip occlusion with the single lower incisor, whereas in *Hippopotamus* the incisors are positioned one behind the other, and occlusion with the

two lower incisors is scissorlike, with the lateral borders of the teeth in contact. The occlusion of the incisors in *Hexaprotodon* is like that seen in most early fossil hippopotami, which usually have a full incisor complement and can be considered an unspecialized arrangement. *Hexaprotodon liberiensis* is unspecialized in many features and does not appear to have any close affinities with the earlier fossil hippopotamids (figure 23.4).

The main differences in the cranium are tabulated below:

Hippopotamus amphibius	*Hexaprotodon liberiensis*
Orbits high and in posterior half of cranium	Orbits on side of head and more or less midway along length
Lacrimal bone large and adjoining nasal	Lacrimal bone small and separated from nasal by anterior extension of frontal
Frontal bone separated from contact with maxillary by lacrimal	Frontal bone in contact with nasal and maxillary
(In neither genus is there a lacrimal duct).	

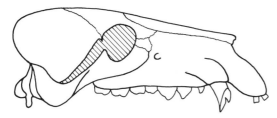

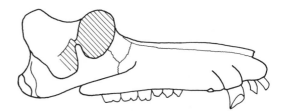

Figure 23.4 Diagrammatic comparison of the cranium of the two living hippopotamus species: (*left*) *Hexaprotodon liberiensis*; (*right*) *Hippopotamus amphibius*. Lateral view.

Fossil Hippopotamidae

The origin of the Hippopotamidae is not clear, although it is most likely that the family arose from an anthracotherelike ancestor in the late Paleogene of Africa. The supposed presence of an anthracotherelike animal in the fauna of Fort Ternan (14 m.y. K-Ar date) is entirely unfounded (Leakey 1961; Hooijer 1966; Bishop 1967; Cooke 1968). The report of this animal was based on some large *Entelodon*-like canines, later referred to the giant creodont *Megistotherium* Savage, and some molar teeth now shown to be proboscidean milk molars. In fact, no anthracothere or hippolike animal has been recognized at Fort Ternan, and in view of the supposed preference of hippopotami and anthracotheres to a habitat near water and the evidence of a plains and savanna fauna from Fort Ternan, it is unlikely that members of either the Hippopotamidae or the Anthracotheriidae would be recovered from these deposits. A member of either family at this point in time and space would be very interesting.

In a paper describing a new rhinoceros from the middle Miocene of Loperot, Kenya, Hooijer (1971) mentions some phalanges that he concludes are hardly distinguishable from those of the modern *Hippopotamus amphibius*. He goes on to say that the Loperot hippopotamus is the earliest known specimen in the world. I have seen these phalanges and cannot agree with Hooijer. The bones are morphologically more similar to those of a large anthracothere, although it is true that early hippos and large anthracotheres are difficult to distinguish on phalanges alone. However, as there are a number of specimens representing the dentitions of at least one species of large anthracothere from these deposits and no hippolike teeth or any other parts have been recovered, it seems more cautious at this stage to refer the phalanges to the Anthracotheriidae, and together with the specimens from Fort Ternan to take them away from proposed hippopotamus ancestry.

The earliest member of the Hippopotamidae so far recognized was found in the lower Miocene Kulu Waregu deposits of Rusinga Island, Kenya. Only one specimen is known, although some isolated teeth from Maboko Island, Kenya, may also belong to the same genus. The molar teeth are low-crowned, bunodont with a distinct pustulate cingulum and an enamel pattern that in wear demonstrates a triangular trefoil pattern characteristic of the molar teeth in all species of hippopotamus, fossil or living. In the Rusinga specimen a primitive feature is recognized in molars that retain a weak hypocone as in many anthracotheres, although this seems to be the

only point of similarity (figure 23.1A). It is unusual to find anthracotheres and hippopotami in the same deposits, and in the Rusinga deposits the former are relatively common, whereas this new and basic hippopotamus is the only specimen so far known. Anthracotheres in East Africa are not known in any deposits after the lower part of the middle Miocene (Loperot), although in North African deposits the family continues into the late Tertiary. In the Pliocene deposits of Sahabi, Libya, a large anthracothere and a small hippopotamus are found, although it is not clear if they come from the same level. Generally speaking, the two families appear to be mutually exclusive.

In other parts of the Old World where fossil hippopotamus has been found, none is known earlier than the Pikermian (Turolean) deposits of Spain. In Asia the earliest known hippopotamus is not found until the Burmese equivalent of the Dhok Pathan levels of the Siwaliks from which only two or three specimens have been recovered, although hippopotamus becomes common in the later levels of the Tatrot and Pinjor, alongside the more rare anthracothere, *Merycopotamus* (Hooijer 1950). Thus it seems that anthracotheres survived in North Africa and Asia some time after their disappearance in East Africa, whereas in East Africa the beginnings of the hippopotamus line in the lower Miocene seems to have been a dominant event terminating any further anthracothere evolution in this area.

Although in East Africa many fossil localities contain radiometrically dated deposits, in Europe this form of absolute dating can be applied to only a small proportion of the Neogene mammal localities (van Couvering 1972). No radiometric dates have been obtained from the Siwalik deposits and thus in any interpretation of intercontinental contemporaniety faunal criteria must be used in comparative dating of localities.

In East Africa there is growing evidence, in sediments that have been radiometrically dated, that a hippopotamus fauna can be traced back almost to Fort Ternan times. In the Baringo area of Kenya's Rift Valley fragmentary teeth that clearly represent a hippopotamid have been recovered from the Ngorora Formation (Coryndon 1978). Lower incisors, two broken premolars, and an upper canine are represented. The slender evidence so far seen is of a medium-sized hippopotamid with very large, pustulate premolars, cylindrical and rather small lower incisors, and an upper canine with a pronounced posterior groove. Rather late beds, also in the Baringo area at Mpesida (Bishop et al. 1971), have been dated at about 7 m.y. and have produced a slightly more complete collection of hippopotamids, but still only isolated teeth. It is clear from these new remains that even after a gap of some 4 m.y. the hippopotamid of Mpesida is still medium-sized and may be only slightly advanced from the evolutionary stage seen at Ngorora. The premolars are large, robust, and very cuspidate, and the fourth upper premolar has a large accessory cusp that forms a greater part of the occlusal surface than the main cusp—a primitive character seen also in anthracotheres. The lower canines have very fine grained enamel with no ridges, and the upper canine is large with a deep posterior groove. The molar teeth are low-crowned with thick enamel and an enamel pattern showing a poorly formed trefoil shape. Evidence from a juvenile specimen suggests that these animals had a hexaprotodont anterior dentition, that is, six incisors in each jaw.

The specimens from Mpesida are interesting from the fact that it is in deposits of roughly the same age that the earliest hippopotamus is found outside Africa. These are in the Turolean sediments of Teruel and Granada in Spain, from which two species have been described: *Hippopotamus crusafonti* Aguirre and *Hippopotamus primaevus* Crusafont. Like the specimens from the Baringo deposits, these two species are represented by only isolated teeth; and though in the descriptions of both species they are said to be hexaprotodont, only one incisor is known from one of the species and from the published accounts there is no evidence for the above assumption.

The original description of *H. crusafonti* is illustrated, but that of *H. primaevus* has little morphological description and no illustration, but I am fortunate in having been able to study casts of the latter species. These Spanish species are probably conspecific and are smaller than the Baringo specimens, with higher-crowned molars and proportionally smaller premolars. Evidence from African specimens suggests that the most primitive hippos have large premolars, low-crowned molars, deeply grooved upper canines, and lower canines with smooth enamel. The evidence also tends toward the probability of a hexaprotodont anterior dentition. Thus the hippopotamus from Spain would appear to be morphologically more advanced than those from East Africa at that level. No other pre-Pleistocene specimens are known from intermediate geographical sites, the nearest being those from Sahabi and Wadi Natrun in North Africa, where the age of the fossiliferous deposits may be in the region of 5 to 6 m.y. or younger and thus later than the Spanish or the Baringo specimens so far mentioned. The hippo-

potamus from Wadi Natrun represents a small species that Arambourg (1947) described as a subspecies of the tetraprotodont hippopotamus from the Shungura Formation of the Omo Basin, Ethiopia, *Hippopotamus protamphibius andrewsi*. The hippopotamus from Sahabi, found together with a very large anthracothere though not necessarily from the same stratigraphic levels, is about the same size as *H. p. andrewsi* and might also be tetraprotodont. On morphological evidence of isolated teeth it appears that the hippopotami from North Africa in the late Tertiary, that is, Sahabi and Wadi Natrun, are closer to the Spanish species than to those of East Africa. If this is true, then there seems to have been a geographical barrier in the region of the Sahara even at this time and a real possibility of a link from northwestern Africa to the Iberian Penninsula, as suggested by other mammalian groups such as the Giraffidae. The Spanish hippopotamus appears to be older than the two hexaprotodont species from the late Pliocene of Casino in Italy and of Sicily, *H. pantanelli* and *H. siculus,* which may be derived from African forms either via the Middle East or Asian deposits of Dhok Pathan age.

It is not until the upper Villafranchian that the hippopotamus becomes common in European deposits, and by this time they are all of tetraprotodont form allied to *H. amphibius* and probably derived from the late Pliocene tetraprotodont forms of Africa such as *H. kaisensis* (figure 23.5).

The most complete collection of early hippopotamids comes from Lothagam Hill, west of Lake Turkana in Kenya, where sediments have been faunally dated at about 6 m.y., and the hippopotamid fauna includes far more complete material than in any other earlier fauna anywhere. The Lothagam mammal-bearing sediments are divided into three main units (Patterson, Behrensmeyer, and Sill 1970), the lowest being by far the richest in mammals. These sediments indicate fluvio-deltaic conditions and contain remains of a large hexaprotodont hippopotamus that give the first indication of the morphology of skull and postcranial elements in early hippopotamids. Although the remains from the earlier sediments of the Baringo area are so fragmentary, there appears to be a strong similarity to the more complete Lothagam form, and the Baringo hippopotamus could well be directly ancestral.

The Lothagam hippopotamus, *Hexaprotodon harvardi,* is named in Coryndon (1977). It comes from the lowest Unit 1 and is of an animal only slightly smaller than the living *H. amphibius* but morphologically very different (figure 23.5). Specimen number KNM-LT 57-67K (Patterson field number) is the only

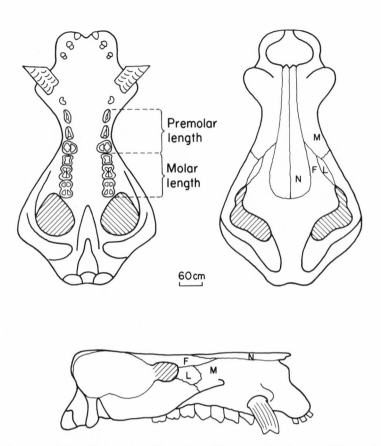

Premolar length

Molar length

60 cm

Figure 23.5   The earliest known African hippopotamid cranium, *Hexaprotodon harvardi* Coryndon, from the early Pliocene of Lothagam Hill, Kenya.

known almost complete hippopotamus cranium from deposits of this age anywhere in the world. It is damaged on the left side but is otherwise whole. The cranial vault is flat and the orbits are not elevated as in the modern species but are positioned to the side of the skull, though not to the extent seen in *Hexaprotodon liberiensis.* The postorbital portion of the cranium is proportionally longer than in *H. amphibius,* being about one-third of the total length from premaxilla to occiput, whereas in *H. amphibius* this portion is about one-sixth the total length. The parietals are long and narrow and the whole appearance is of an unspecialized, rather lightly built creature. The arrangement of the facial bones is as found in all unspecialized suiformes, that is, with the frontal bone extending forward to separate the lacrimal from the nasal bone. The lacrimal bone is small and as in all Hippopotamidae there is no sign of a lacrimal duct. The rostrum is not markedly constricted, but the premaxilla tends to bend down anteriorly. The zygoma is long and slender.

There is a complete dentition, $\frac{3.1.4.3}{3.1.4.3}$. The upper incisors are small, peglike, and positioned in a shallow arc in the premaxilla. The upper canine is of medium size and has a deep posterior groove. The pre-

molars are large, robust, rugose, and pustulate, very similar to the premolars from Ngorora and Mpesida in Baringo. The first upper premolar is double-rooted and the upper fourth premolar has a large secondary cusp as well as a stellate main cusp, both surrounded by a strong and pustulate cingulum. The molar teeth are low-crowned and sharply tapered from cingulum to occlusal surface both in antero-posterior and lateral aspect, with an enamel figure roughly triangular in shape. The length of the molars (M1-M3) is roughly equal to the premolar length (P2-P4) giving an index $\frac{P}{M} \times 100 = 100$.

The mandible is also lightly built with a shallow horizontal ramus, long and shallow symphysis, and an angle that is not turned forward. The six incisors are subequal in size and cylindrical. The lower canine has finely crenulated enamel, similar to that seen some suidae, and quite unlike the roughened and ridged enamel of lower canines in *H. amphibius.*

Postcranial elements from Lothagam indicate that *H. harvardi* had slender limbs with the extremities longer and more gracile, indicating an animal that was faster moving, more agile, and probably less amphibious than the living *H. amphibius.*

This collection from Lothagam comprises the most

completely known hippopotamid remains from the later Tertiary and must be taken as the general pattern for the early form of hippopotamid. This probably holds as far as East Africa is concerned, but it is not so clear where the hippopotamids from contemporaneous levels in North Africa and Spain fit into the general pattern. It is probable that the Lothagam deposits are slightly later in time than those of the Spanish Turolean, although the hippopotamid from East Africa at this level appears to be more primitive than those farther north. It can be postulated that the Hippopotamidae arose in sub-Saharan Africa in the early Miocene and evolved in isolation there, to the exclusion of the anthracotheres, until about 8 m.y. ago. At this time the hippopotamus may have entered North Africa and then Europe, possibly via Arabia and the Middle East rather than the Nile Valley, and across North Africa to Spain (figure 23.6). The reason for the scarcity of hippopotamus outside East Africa at this time might be the continued presence of large anthracotheres, which constituted a formidable barrier to hippopotamid expansion except in the East African region.

By far the greatest accumulation of evidence on the evolution of Hippopotamidae comes from East Africa, where a diversity of species becomes apparent at about 4 m.y. ago. Deposits at Kanapoi (Patterson 1966), south of Lothagam, have yielded hippopotamid remains that show a slightly more specialized form than those from Lothagam but are clearly on the same evolutionary line. The advanced characters seen in the skull of the Kanapoi hippopotamus, which is of similar size to that from Lothagam, are: (1) shortening of premolar row in comparison with length of the molar row; this is shown by the index of premolar/molar length, which in the Lothagam hippopotamus is near 100, indicating a premolar row virtually the same length as the molars, whereas in the Kanapoi specimens the index is about 78, demonstrating the molar length is the greater; (2) the first premolar probably has only one root, although there are indications that this single root is formed from the fusion of two roots; (3) the upper fourth premolar is smaller, although of similar morphology; (4) the central incisor is larger than the other two; (5) the orbits are very slightly elevated. These differences are not considered to be more than of subspecific standing.

It is probably from a hippopotamid similar to those of Lothagam and Kanapoi that the species found in the Irrawaddy Beds of Burma and the Siwalik Hills of India are derived. These Asiatic forms have been described as *Hexaprotodon iravaticus* from deposits of Dhok Pathan age in Burma and *Hexaprotodon sivalensis* from the Tatrot and later

levels of the Siwalik Hills. There are several morphological features that link the Asian specimens to those of the late Tertiary of East Africa. The premolars are pustulate but not as large as in the Lothagam specimens, but the first premolar has a double root, the premolar/molar index in *H. sivalensis* from the Tatrot is 87. Other unspecialized features, such as the arrangement of the facial bones, the hexaprotodont anterior dentition, and the gracile limb structure, are also retained.

The Siwalik hippopotamus evolved in Asia with no further invasion from Africa, the animal getting progressively larger and the second incisor becoming gradually reduced so that in the highest late Pleistocene levels this tooth is very insignificant, although it never entirely disappears. In the early Pleistocene, *Hexaprotodon* migrated to the Far East where fossil forms similar to those from the Siwaliks are found in Java. There was no diversification of species in Asia; all the known forms except for the scanty remains from Burma can be accommodated in the species *H. sivalensis* (figure 23.6).

Taking the *Hexaprotodon harvardi* from Lothagam and Kanapoi as the generalized basic pattern for East African hippopotamids, the later hexaprotodonts from East Africa can be seen as rather more advanced forms. In the late Tertiary and early Quaternary of Africa the hexaprotodont hippopotamus was dominant over all other forms and found throughout deposits of the Baringo Basin, the Lake Turkana Basin in the west and north, in the Kaiso and Kazinga Channel deposits of Uganda, and from Rawe and Kanam in the Kisumu Gulf region of Kenya. Many of these small hexaprotodonts have been referred to the species *Hippopotamus imagunculus* Hopwood, but there is unfortunately no positive evidence that *H. imagunculus* is truly hexaprotodont, although it is most likely to be so.

Outside East Africa the small hexaprotodont *Hippopotamus hipponensis* from Bône in Algeria is probably closely allied to *H. imagunculus,* and this may also be true of the late Tertiary specimens from Casino in Italy and Gravitelli in Sicily, as well as the specimens recovered from early Pleistocene deposits in the Lake Chad area. I have not seen any of these specimens.

The extensive hippopotamus fauna from the Omo Basin in southern Ethiopia probably derives from a generalized hexaprotodont similar to that seen at Kanapoi and exemplified by a cranium from the Mursi Formation (dated at 4 m.y.) in the Omo Series (Bonnefille et al. 1973). The later Usno Formation and the lower levels (Members A and B) of the Shungura Formation also produce a medium-sized hexaprotodont (Howell, Fichter, and Eck 1969). However

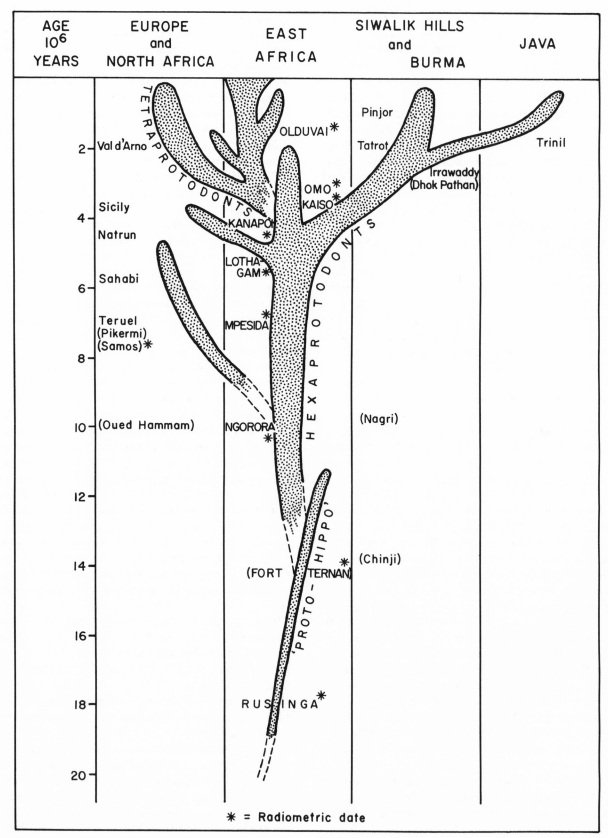

Figure 23.6 Tentative correlation of some Eurasian and African fossil mammal localities relative to the history of the Hippopotamidae. * = Radiometric date. ( ) = no hippopotamid recorded.

in the higher levels of Shungura Formation (Members C–G), an evolutionary trend can be seen in which the characteristic hippopotamid of the main part of the Shungura has become tetraprotodont while still retaining many of the unspecialized features of the earlier forms such as the facial bone arrangement, only slightly elevated orbits, and gracile limb bones. This hippopotamus was described in detail by Arambourg (1947) and named *Hippopotamus protamphibius*. A diversity of hippopotamus species unknown in earlier deposits appears in the Omo deposits and in addition to *H. protamphibius* there are at least two and possibly three other species represented, though all are rare. One species (species "A," Coryndon and Coppens 1973) is hexaprotodont and very similar to *H. imagunculus*. A second species, *Hippopotamus aethiopicus* (Coryndon and Coppens 1975) is a very small tetraprotodont that occurs as a rarity in the Shungura Formation from Member C upward into I. In the upper levels alone, above Tuff G where *H. protamphibius* disappears, a large tetraprotodont very like *Hippopotamus amphibius* is recorded.

In the lower Pleistocene levels of the Kaiso Formation in Uganda, as well as the small and possibly hexaprotodont *H. imagunculus* there appears for the first time in hippopotamus history an undoubted large tetraprotodont, *H. kaisensis,* which is the most likely ancestor for the common hippopotamus of the present day, *H. amphibius* (Cooke and Coryndon 1970). With *H. kaisensis* forming the base of the tetraprotodont stock this species gave rise not only to *H. amphibius* but also to the specialized *H. gorgops,* which is found commonly throughout the deposits of Olduvai Gorge in Tanzania, with a few isolated specimens known from other East African Pleistocene localities and one specimen from Cornelia in South Africa. The earliest (Bed I) of the Olduvai hippos can scarcely be distinguished from the basic tetraprotodont hippopotamus as exemplified by *H. kaisensis*. Differences from the hexaprotodonts, seen as specialized characters, are not only the number of incisors but the way in which they are engaged in the premaxilla and the mandible. The upper incisors are set with the second rather posterior as well as lateral to the first incisor and are more curved than in the hexaprotodonts and occlude in a different way. The facial bones are as in the modern *H. amphibius* with the frontal bones entirely excluded from contact with the nasals by the enlarged lacrimal bone. The orbits are elevated, to an extreme extent in the later forms of *H. gorgops,* the premolar row is shortened, and the upper fourth premolar tends to lose the large accessory cusp. The premolar/molar index

is around 71 in *H. amphibius* and down to 56 in advanced *H. gorgops*. The limb bones are shortened and much more robust than in the hexaprotodonts, the whole animal being more adapted for an amphibious habitat. In *H. gorgops* an evolutionary trend is seen in which throughout the deposits of Olduvai from Bed I through to Bed IV the animal gets progressively larger, the preorbital region of the cranium elongates, and the postorbital length diminishes. The orbits are elevated to a marked degree, and in fact the profile of the skull in the most advanced form is very similar to that of a crocodile (Coryndon 1970, fig. 5).

No fossil hippopotamus has been recovered from the extensive deposits of South Africa earlier than the middle Pleistocene, by which time the species is clearly *H. amphibius*. But in Central Africa the early Pleistocene sediments of Uraha in Malawi have yielded two species of hippopotamus very similar to *H. kaisensis* and *H. imagunculus* of the Kaiso deposits. It is possible that representatives of the two main hippopotamus lines, hexaprotodont and tetraprotodont, were widespread in Africa in the late Tertiary and early Pleistocene, and although the hexaprotodonts do not seem to have reached South Africa, the tetraprotodonts as typified by *H. kaisensis* and its descendants succeeded in the south.

The oldest sediments in the Omo Basin to yield hippopotami belong to the Mursi Formation, roughly equivalent to the Kanapoi sediments. They are known from only a few specimens but can be referred to an early form of the Kanapoi species, *Hexaprotodon protamphibius;* this species is also found in the Usno Formation and throughout beds of the Shungura Formation (Coryndon and Coppens 1973). From Tuff C upward this hippopotamus is tetraprotodont in its anterior dentition, with the lower central incisor about twice the size of the lateral. The upper incisors are subequal in size and arranged in the maxilla in a shallow arc as in the hexaprotodonts, not one behind the other as in the tetraprotodonts proper. The facial bones of the Shungura hippopotamus have the unspecialized arrangement seen in the hexaprotodonts, and although the orbits are slightly elevated, they do not reach the specialization seen in the extant species. The limb bones are more slender than in *H. amphibius,* although more robust than in the Lothagam and Kanapoi species. Thus a single evolutionary line can be postulated to link the Lothagam species through Kanapoi, Mursi, and Usno, culminating in typical *H. protamphibius* of Shungura, Member G.

In Members D–H of the Shungura Formation, the Kalem Beds, and in the upper levels of East Turkana is found a pygmy tetraprotodont, *Hippopotamus aethiopicus* (Coryndon and Coppens 1975). The remains include a fairly complete mandible and recently a cranium has been discovered. Another hippopotamus from the Omo Basin is a small hexaprotodont, *H.* sp. "A," which though poorly preserved appears to differ from both *H. imagunculus* and *H. protamphibius* (Coryndon and Coppens 1973).

Recently hippopotami have been discovered among the Plio-Pleistocene faunas of the Afar in Ethiopia. Three species are present: a large hexaprotodont found only at Hadar, which shows an advance on the Kanapoi species; a second smaller species from Amado, Hadar, and Geraru that is very similar to the Omo sp. "A"; and a true tetraprotodont, also known from the Shungura Formation.

Another very specialized hippopotamus whose ancestry is not altogether clear is found in the Plio-Pleistocene deposits of East Turkana in Kenya (Coryndon 1976, 1977). Here, even in the earliest levels of Kubi Algi in the south, which are roughly contemporaneous with Kanapoi on the other side of the lake (Vondra et al. 1971), it is surprising to find a tetraprotodont hippopotamus that although superficially like *H. protamphibius* of the Shungura Formation of the Omo (Maglio 1971) is in fact the forerunner of a quite distinct species that attains its most specialized form in the Koobi Fora deposits of early Pleistocene age. The Koobi Fora hippopotamus is superseded in the upper levels of middle Pleistocene age at East Turkana by *H. gorgops* in much the same way and at the same level at which *H. protamphibius* is replaced by *H.* cf *amphibius* in the upper levels of the Shungura Formation of Omo.

The East Turkana hippopotamus *Hexaprotodon karumensis* is a large hippopotamus about the size of the living *H. amphibius*. There are only two incisors in the lower jaw, which are very large, widely spaced, and are separated by a bony protuberance in the midline of the anterior face of the symphysis. The lower canines are small, narrow, and set in a long slender alveolus very laterally projected. There are four upper incisors that are very small, peglike, and occlude in pairs with a single lower incisor. In spite of the very specialized anterior dentition, the facial bones are as in the hexaprotodonts, with the lacrimal bone small and separated from the nasal. The molar teeth are very low-crowned with a poorly formed enamel trefoil pattern. In its most specialized form the East Turkana hippopotamus has elevated orbits approaching the size seen in early *H. gorgops,* an interesting case of parallelism, particularly as it was the latter species that became dominant in the upper levels of East Turkana. This

change of dominance is probably due to changing conditions of Lake Turkana. The East Turkana hippopotamus with its low-crowned molar teeth and specialized incisors was probably a browsing animal and its remains are found principally in fluviatile deposits. It may not have ventured into the open lake where *H. gorgops* would have been more at home; the East Turkana diprotodont with its very long and slender limbs, in particular the extremities, would have had no difficulty in moving around on land with fair speed and agility, but its feet would have been unsuitable for a muddy lake shore. As well as the large browsing diprotodont of the rivers and the later grazing *H. gorgops* of the lake, there is also found in the East Turkana sediments of Pleistocene age a very small tetraprodont hippopotamus, smaller than the extant pygmy hippopotamus of West Africa and most probably the same species as the small *Hippopotamus aethiopicus* from Omo (Coryndon and Coppens 1975). It has been suggested that as there were very small hippopotami in East Africa during the Pleistocene, these could be considered as ancestors of the pygmy hippos of the Mediterranean islands rather than the large *H. amphibius antiquus* of the European Pleistocene, but these ideas need much more investigation (figure 23.7).

The unspecialized hexaprotodont hippopotamus fauna became extinct in the middle Pleistocene and has no modern equivalent. However the specialized tetraprotodont line, with *H. kaisensis* as the basic form, flourished and by the late Pleistocene *H. amphibius* was widespread not only in Africa, including North Africa and parts of the Sahara, but also spread via Israel to Europe, where it reached northeast England. With the coming of the last glaciation the hippopotamus again retreated to the warmer climes of Africa and never returned to more northerly latitudes.

The small subfossil *Hippopotamus lemerlei* of Malagasy has clear affinities with the basic tetraprotodont stock and could well be derived from a late Miocene tetraprotodont immigrant from the mainland. The lack of Tertiary fossil deposits in Malagasy makes interpretation of the ancestry of *H. lemerlei* extremely conjectural.

Evolutionary trends in African Hippopotamidae are not always clear; with such fragmentary material in the pre-Pleistocene deposits and no evidence yet of the transitional stages between the hexaprotodont and tetraprotodont lines, there are major gaps in the sequence that can only be filled by further discoveries and research. However, the terms "hexaprotodont" and "tetraprotodont" refer to character states and do not necessarily reflect phylogeny.

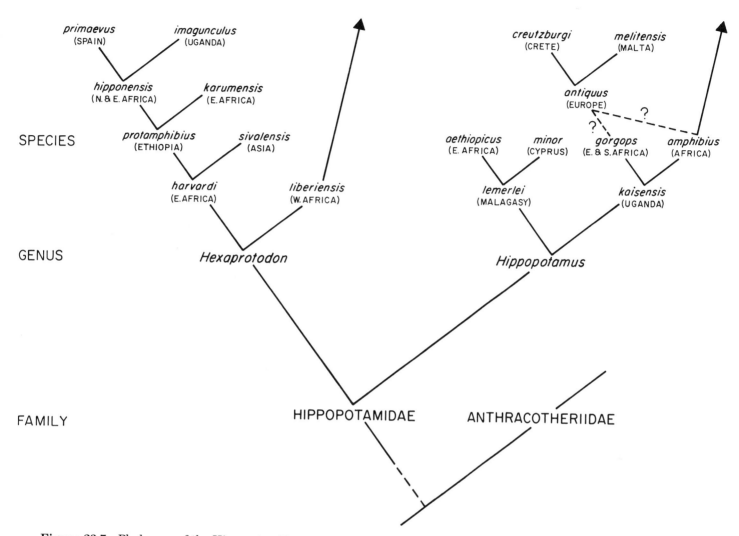

Figure 23.7 Phylogeny of the Hippopotamidae.

No species of *Hippopotamus* is known that possesses more than two pairs of incisors (tetraprotodont) in either upper or lower jaw; in several species the dental formula is unknown and in *H. gorgops* there is only one pair of lower incisors. Within the genus *Hexaprotodon* primitive species possess a full complement of three pairs of incisors (hexaprotodont) in both upper and lower jaws; these are variously reduced through tetraprotodont to diprotodont states. The main evolutionary trends are summarized in table 23.1 and figure 23.7.

The reason for the change in dominance from the early hexaprotodonts to the later tetraprotodonts may well be linked with the rise to power of the Bovidae in the late Tertiary and Pleistocene. The land-living generalized hexaprotodonts found their old habitats and food taken over by the Bovidae and they could not compete; the more specialized tetraprotodonts, however, were able to evolve their am-

phibious tendencies and thus use to clear advantage an ecological environment untouched by the Bovidae.

During Neolithic times, when the climate of the Sahara was more clement, the hippopotamus was known across North Africa along the length of the Nile and even in areas where now complete desert prevails. Although never a popular subject for prehistoric artists, the hippopotamus is portrayed in rock etchings in the Tibesti Mountains and elsewhere, many hundred miles beyond their range today. In ancient Egypt the hippopotamus was venerated as a semblance of the god Tuares, and even as recently as Roman times mosaics from North Africa show clearly that the hippo was well known and frequently captured. Today, however, the common *Hippopotamus amphibius* is known only south of the Sahara but is fairly extensive wherever water is suitably abundant. The pygmy hippopotamus *Hexa-*

Table 23.1 Summary of some evolutionary trends in African Hippopotamidae.

	EARLY HIPPOPOTAMID	"HEXAPROTODONTS" *H. imagunculus* Lothagam Kanapoi Omo East Turkana *H. liberiensis*	TETRAPROTODONTS *H. kaisensis* *H. gorgops* *H. amphibius* Teruel Sahabi Natrun
Miocene	low-crowned molars large premolars		
Pliocene		low-crowned molars	molars slightly hypsodont
		enamel figure triangular	enamel figure weakly grooved
		large pustulate premolars	medium-sized premolars not pustulate
		molar row length = premolar row	molar row length > premolar row
		upper canine with deep posterior groove	upper canine with shallow posterior groove
		lower canine with smooth enamel	lower canine with ridged enamel
		usually 6 incisors	4 incisors
		lacrimal small, excluded from nasal	lacrimal large, in contact with nasal
		frontal with anterior extension	frontal with no anterior extension
		preorbital length < post-orbital	preorbital length > postorbital
		orbits low	orbits slightly raised
		limbs gracile	limbs short and robust
Pleistocene		molar row > premolar row	molar row much > premolar row
		enamel figure weakly grooved	enamel figure strongly grooved
		lacrimal just touching nasals	lacrimal large and in full contact with nasals
		orbits slightly raised	orbits elevated
		limbs gracile	limbs short, very robust

protodon liberiensis is not known outside West Africa nor is it represented in historical art form.

Both species of living hippopotamus are considered great gastronomic delicacies. *H. liberiensis* is probably reasonably safe from extinction due to its remote habitat, and while it seems always to have been a rare element in the West African fauna, the numbers do not appear to be dwindling. As regards the more familiar, more common, and widespread *Hippopotamus amphibius*, this beast, too, is very acceptable as tasty food. However, it seems not only to be holding its own wherever suitable habitats are found but also to increase alarmingly if the numbers are not kept in check in areas where it is protected. With the chief item of food being grass, the erosion caused by the overgrazing of a herd of hippopotami can be considerable and in certain areas the animals have to be carefully controlled by regular culling, the meat of the slaughtered animals providing a much needed local source of protein. With man as the only predator of the hippopotamus, *Homo sapiens* has a clear responsibility to ensure that *Hippo-*

potamus amphibius does not become extinct like so many of its ancestors.

The background research for this chapter has been greatly helped over several years by scientists in Kenya, U.S.A., Canada, and England, who have generously offered material for study and given much practical advice for which I am very grateful. The National Science Foundation (through F. Clark Howell), the National Geographic Society (through L. S. B. Leakey and R. E. F. Leakey), the Royal Society (through W. W. Bishop), the Wenner-Gren Foundation, and the Appleyard Foundation of the Linnean Society have given financial help toward expenses on various occasions. To all those bodies grateful thanks are given.

References

Arambourg, C. 1947. Contribution à l'étude géologique et paléontologique du bassin du lac Rodolphe et de la basse vallée de l'Omo. Deuxième partie, Paléontologie. *Mission scientifique de l'Omo, Géol.-Anthrop. I 1932–1933.* pp. 231–562.

Bishop, W. W. 1967. The later Tertiary in East Africa—volcanics, sediments and faunal inventory. In W. W. Bishop and J. D. Clark, eds., *Background to Evolution in Africa.* Chicago: Univ. Chicago Press, pp. 31–56.

Bishop, W. W., G. R. Chapman, A. Hill, and J. A. Miller. 1971. Succession of Cainozoic vertebrate assemblages from the northern Kenya Rift Valley. *Nature* 233:389–394.

Bonnefille, R., F. H. Brown, J. Chavaillon, Y. Coppens, P. Haesarts, J. de Heinzelin, and F. C. Howell. 1973. Situation stratigraphique des localités à hominidés gisements plio-pléistocènes de l'Omo en Ethiopie. *C. R. Hebd. Séanc. Acad. Sci.,* Sér. D, 276:2781–2784, 2879–2882.

Cooke, H. B. S. 1968. Evolution of mammals on southern continents. II. The fossil mammal fauna of Africa. *Quart. Rev. Biol.* 43 (3):234–264.

Cooke, H. B. S., and S. C. Coryndon. 1970. Pleistocene mammals from the Kaiso Formation and other related deposits in Uganda. In L. S. B. Leakey and R. J. G. Savage, eds., *Fossil Vertebrates of Africa.* London: Academic Press, 2:107–224.

Coryndon, S. C. 1970. The extent of variation in fossil *Hippopotamus* from Africa. *Symp. Zool. Soc. London,* no. 26:135–147.

———. 1976. Fossil Hippopotamidae from Pliocene/Pleistocene successions of the Rudolf Basin. In Y. Coppens et al., eds., *Earliest Man and Environments in the Lake Rudolf Basin.* Chicago: Univ. Chicago Press, pp. 238–250.

———. 1977. The taxonomy and nomenclature of the Hippopotamidae (Mammalia, Artiodactyla) and a description of two new fossil species. *Proc. Kon. Ned. Akad. Wetensch.,* Ser. B, 80(2):61–88.

———. 1978. Fossil Hippopotamidae from the Baringo Basin and relationships within the Gregory Rift, Kenya. In W. W. Bishop, ed., *Geological background to fossil man.* London: Geol. Soc. Lond. (in press).

Coryndon, S. C., and Y. Coppens. 1973. Preliminary report on Hippopotamidae (Mammalia, Artiodactyla) from the Plio/Pleistocene of the lower Omo Basin, Ethiopia. In L. S. B. Leakey, R. J. G. Savage, and S. C. Coryndon, eds., *Fossil Vertebrates of Africa,* London: Academic Press 3:139–157.

———. 1975. Une espèce nouvelle d'Hippopotame nain du Plio-Pléistocène du bassin du lac Rodolphe (Ethiopie, Kenya). *C. R. Hebd. Acad. Sci.,* Sér. D, 280:1777–80.

Deraniyagala, P. E. P. 1949. Some scientific results of two visits to Africa. *Spol. Zeyl.* 25 (2):1–42.

Hooijer, D. A. 1950. The fossil Hippopotamidae of Asia, with notes on the recent species. *Zoöl. Verhandel.* 8:1–124.

———. 1966. Miocene rhinoceroses of East Africa. *Fossil Mammals of Africa,* no. 21, *Bull. Brit. Mus. Nat. Hist. (Geol.)* 13 (2):117–190.

———. 1971. A new rhinoceros from the late Miocene of Loperot, Turkana District, Kenya. *Bull. Mus. Comp. Zool. Harvard* 142 (3):339–392.

Howell, F. C., L. S. Fichter, and G. Eck. 1969. Vertebrate assemblages from the Usno Formation, White Sands and Brown Sands localities, lower Omo Basin, Ethiopia. *Quaternaria* 11:65–87.

Leakey, L. S. B. 1961. A new lower Pliocene fossil primate from Kenya. *Ann. Mag. Nat. Hist. Lond.* 4 (47):689–694.

Maglio, V. J. 1971. Vertebrate faunas from the Kubi Algi, Koobi Fora and Ileret areas, East Rudolf, Kenya. *Nature* 231:248–249.

Patterson, B. 1966. A new locality for early Pleistocene fossils in northwestern Kenya. *Nature* 212:577–578.

Patterson, B., A. K. Behrensmeyer, and W. D. Sill. 1970. Geology and fauna of a new Pliocene locality in northwestern Kenya. *Nature* 266:918–921.

Van Couvering, J. A. 1972. Radiometric calibration of the European Neogene. In W. W. Bishop and J. A. Miller, eds., *Calibration of Hominoid Evolution,* Edinburgh: Scottish Academic Press, pp. 247–271.

Vondra, C. F., G. D. Johnson, B. E. Bowen, and A. K. Behrensmeyer. 1971. Preliminary stratigraphical studies of the East Rudolf Basin, Kenya. *Nature* 231:245–248.

24

Cervidae and Palaeomerycidae

W. R. Hamilton

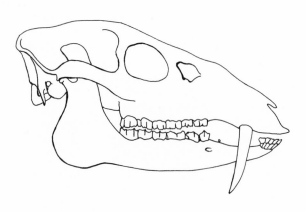

Family Cervidae

The Family Cervidae includes those ruminants in which the paired frontal appendages each consist of a permanent, bony, skin-covered "pedicle" and a deciduous distal region or "antler." The antler usually has two or more branches or tines, and near the base of the antler is a rugose region known as the "burr" or "rosette." The cervids originated in late Oligocene or early Miocene times and by middle Miocene times *Dicrocerus* was common in European faunas. This genus retains large upper canines and has relatively small, two-pointed antlers that are carried on long pedicles as in the living muntjac (*Muntiacus*), a form that is closely related to *Dicrocerus*. In more advanced cervids the upper canines were lost and the pedicles became very short as the antlers became dominant in combat.

The deer have enjoyed their greatest success in the Northern Temperate Zone and in Eurasia they are at least as important as the bovids. However, they are mainly animals of woodland and forest habitats, which may in part explain their relatively minor role in the African mammalian fauna. The family probably entered North Africa in late Pliocene or early Pleistocene times when much of the area was wooded; their failure to penetrate into sub-Saharan Africa may be explained by their restriction to woodland conditions.

North African Deer

Cervus elaphus Linnaeus 1758

This is the only surviving North African deer and is now restricted to northwestern Tunisia and the adjoining region of Algeria (Dupuy 1967) where it is represented by a relatively small population with a probable maximum of 400 individuals in 1953 (Salez 1959). The North African form is assigned to the subspecies *C. elaphus barbarus* and is distinguished from the European "red deer" by its smaller body size; differences of the coat, which retains juvenile spotting through adult life (Joleaud 1926); and by the antlers, which are relatively smaller than in the European red deer and have smaller brow tines (Meyer 1972).

C. elaphus probably reached North Africa during late Pliocene times but the point of entry is uncertain. A route across the Straits of Gibraltar seems possible, but Joleaud (1935) and Salez (1959) regard this as unlikely because remains of *C. elaphus* are unknown west of Tlemcen. However, Arambourg (1938a) records *C. elaphus* from Khenzira, Morocco. Joleaud (1926) and Salez favor a route to North Africa via Sicily and this is supported by similarities among the Italian, Sicilian, and North African Pleistocene faunas and also by the close relationship of *C. elaphus barbarus* to the Corsican deer. The earliest record of *C. elaphus* in Africa is from the Palaeolithic–Acheulian–Lake Karâr and there are several Mousterian records (quoted by Romer 1928, p. 108). Later records of the species indicate a relatively restricted distribution extending from the Moroccan-Algerian border eastward to northern Algeria and Tunisia. *C. elaphus* was abundant during the Neolithic but declined during historical times.

Dama dama Linnaeus 1758

Dama[1] probably entered Africa from the Middle East, via Suez, in late Pliocene times (Salez 1959); the earliest record for the species is supposedly of late Pliocene age (Lydekker 1887), though the evidence for this dating is not very strong. During the wet periods *Dama* spread across Egypt and into Cyrenaica and southward into Ethiopia, but with increasing dryness in North Africa its range became progressively restricted to the Nile Valley (Joleaud 1935). *Dama* became extinct in Africa in late Pleistocene times but was reintroduced in historical times (Joleaud 1926). A detailed review of the history of *Dama* in North Africa is given by Joleaud (1935).

The North African form of *Dama* was assigned to a separate species, *D. schaeferi,* by Hilzheimer (1926), but Joleaud (1935) regarded this as a subspecies of *Dama dama.*

Megaloceros algericus (Lydekker) 1890

Joleaud (1914) recognized that the holotype maxilla of *Cervus algericus* (Lydekker 1890, p. 602, pl. 1, fig. 11) and the holotype mandible of *Cervus pachygenys* (Pomel 1892, p. 157) belonged to the same species. He established the genus *Megaceroides* but this is listed as a synonym of *Megaloceros* by Romer (1966). *M. algericus* is characterized by its short face with extremely thickened mandibles (Joleaud 1930) and by large antlers with flattened tines. It first occurs in the Mousterian (Romer 1928), but according to Joleaud (1935) it probably entered Africa from Sicily at the same time as *C. elaphus* or slightly earlier. It is represented throughout a restricted part of North Africa, mainly in Algeria around Constantine. Arambourg (1938a) records it from Morocco at Casablanca and Taza and from Tit (1938b). The species became extinct at the end of the Pleistocene or the beginning of the Holocene (Arambourg 1962).

Capreolus matheronis (Gervais) 1850

This species is recorded from the Pontian of South Tunisia by Roman (1931) on the basis of an isolated upper molar that exhibits medium wear. Joleaud (1935) mentions this record, but Dietrich (1950, p. 52, fn. 8) suggests that the molar could well be a bovid deciduous fourth premolar and Gentry (pers. comm.) agrees with this. Therefore the record of *Capreolus* from North Africa is probably not valid.

Family Palaeomerycidae

The Family Palaeomerycidae was established by Lydekker (1883) as part of his description of *Propalaeomeryx sivalensis*. In this description Lydekker included comparisons with *Palaeomeryx bojani* (Von Meyer 1834) and *Palaeomeryx eminens* (Von Meyer 1834) but he did not list the genera to be included in his family. In his later description of *Palaeomeryx* Lydekker (1885) included *Dremotherium* (Geoffroy 1833), *Dicroceros* (Lartet 1851), *Micromeryx* (Lartet 1851), and *Propalaeomeryx* (Lydekker 1883) as synonyms of *Palaeomeryx* and later in the same description *Prox* (Hensel 1859) and *Dremotherium* (Geoffroy 1833; Pomel 1853; Geoffroy 1835) were added to this synonymy.

Stirton (1944) revised the family and attempted to assess its relationships. He included in the Palaeomerycidae the genera *Lagomeryx*[2] (Roger 1904), *Procervulus* (Gaudry 1878), *Climacoceras* (MacInnes 1936), *Amphitragulus* (Pomel 1853), and *Dremotherium,* as well as fourteen North American genera.

Stirton's (1944, p. 645, fig. 2) phylogeny of the Palaeomerycidae demonstrates his understanding of relationships within the family and its relationships with other groups. Stirton clearly regarded the family as an evolutionary grade with the Giraffidae, Moschidae and Cervidae each derived independently from different parts of the group. Simpson's (1945) arrangement of the group differed markedly from that given by Stirton. He placed the Palaeomerycinae as a subfamily of the Cervidae and included within it *Eumeryx* (Matthew and Granger 1923), *Blastomeryx* (Cope 1877), *Machaeromeryx* (Matthew 1926), *Longirostromeryx* (Frick 1937) as the tribe Blastomerycini, and *Amphitragulus, Dremotherium,* and *Palaeomeryx* in the Palaeomerycini. Other genera included in the group by Stirton were

---

1 The generic name *Dama* is used here in preference to *Cervus* after Morrison-Scott (1949, p. 94).

2 *Lagomeryx* was established by Roger (1904) to accommodate small species of *Palaeomeryx*. Roman and Viret (1934) suggested that the genus was a synonym of *Palaeomeryx* and this was accepted by Stirton and later by Whitworth (1958).

placed by Simpson in the Lagomerycidae (Pilgrim 1941), which he identified as a family of the Giraffoidea. Simpson placed *Procervulus, Lagomeryx,* and *Climacoceras* in this family. This system agrees closely with that proposed by Pilgrim (1941). Palaeomerycids were recorded from Africa by Whitworth (1958), who established the species *Palaeomeryx africanus* to accommodate a small ruminant from Songhor, Koru, Moruorot, and Rusinga. Ginsburg and Heintz (1966) suggested that this species should be removed from the genus *Palaeomeryx*. They based their suggestion on interpretation of features of the premolars, particularly the presence in *"Palaeomeryx" africanus* of a P_1 and the primitive condition of the other anterior premolars. Ginsburg and Heintz suggested that this species should be placed in a new genus, *Kenyameryx*. I have argued (Hamilton 1973a) that *Palaeomeryx africanus* and *Walangania gracilis* (Whitworth 1958) are synonymous and that the species resulting from this synonymy, *Walangania africanus,* is probably a bovid. *Walangania africanus* is described and discussed by Gentry in this volume. Whitworth also described several isolated cheek teeth, which he identified as *"?Palaeomeryx sp."* In my description of the ruminants from Gebel Zelten (Hamilton 1973a) I identified the Palaeomerycidae as a family of the Giraffoidea and described a new genus, *Canthumeryx,* which I placed in the family. I also suggested that the genus *Palaeomeryx* was represented in the Gebel Zelten fauna by two molar fragments (BM M-26691 and BU-20112). A pair of ossicones (BM M-26690) was identified as Palaeomerycidae indet. In my discussion of the Palaeomerycidae I suggested that the "Oligocene genera which lack ossicones" should be removed from the Palaeomerycidae and that the African genus *Propalaeoryx* and the Iberian genus *Triceromeryx* should be included in the family. This left the Palaeomerycidae with the genera listed below:

Palaeomerycidae

Climacoceras MacInnes 1936, middle–upper Miocene; Africa
Canthumeryx Hamilton 1973, lower Miocene; Africa
Heterocemas[3] Young 1937, upper Miocene; Asia
Palaeomeryx Von Meyer 1834, middle–upper Miocene; Europe. ?lower Miocene; Africa

---

3 *Heterocemas* was synonymized with *Lagomeryx* by Chardin (1939), but Young (1965) in a revision of the "lagomerycids" maintained *Heterocemas* as a valid genus. Hamilton (1973a) synonymized all the Asian "lagomerycids" with *Heterocemas,* but it seems that Chardin's specimens, which include mandibles, may more correctly be placed with *Lagomeryx* and that *Heterocemas* should be restricted in its use as suggested by Young.

Procervulus Gaudry 1878, lower–middle Miocene; Europe
Propalaeoryx Stromer 1926, lower Miocene; Africa
Triceromeryx Villalta, Crusafont, and Lavocat 1946, lower Miocene; Europe

Incertae Sedis

Progiraffa Pilgrim 1908, lower Miocene; Asia

This account is concerned primarily with the genera *Canthumeryx, Climacoceras, Propalaeoryx,* and *Palaeomeryx*. Even in 1973 I expressed some doubts about the group, stating (1973a, p. 136) that "the family is best regarded as a level of ruminant evolution." At this stage it must be pointed out that a change in my understanding of phylogeny and systematics coupled with study of more complete material of *Canthumeryx* and *Climacoceras* has resulted in rejection of the Palaeomerycidae as a valid monophyletic group of ruminants and has also led to a rethinking of the relationships of the genera *Canthumeryx* and *Climacoceras*. Reasons for these changes will become clear in the descriptions; my new understanding of relationships will be explained toward the end of this chapter.

Terminology

The cranial appendages of deer, giraffes, pronghorns, and bovids are referred to as antlers, ossicones, horncores, and horns respectively. Nomenclature of cusps on the cheek teeth generally follows that used in Hamilton (1973a) but cusps on the lower premolars are named as in figure 24.1. In the discussion of relationships the terms plesiomorphic, apomorphic, and sister-group are used as in Hennig (1966).

Specimens in collections of museums and other institutions are referred to by prefixes as listed below followed by specimen numbers. Where field numbers of specimens have been used in publications by other authors these are given with their equivalent collec-

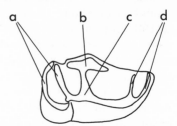

Figure 24.1 Cusp and crest names on the ruminant P_4. (*a*) posterior paired transverse crests, (*b*) central lingual cuspid, (*c*) central labial cuspid, (*d*) anterior paired transverse crests.

tion number so that cross-reference to these works will be possible.

BM M	British Museum of Natural History, Department of Palaeontology, Fossil Mammal Collection.
BU	Department of Geology, University of Bristol
IPP	Institut de Paléontologie, Paris
KNM	Kenya National Museum, Nairobi
MM	Bayerische Stratssammlung für Paläontologie und Historische Geologie, Munich
SAM-PQ-N	South African Museum, Department of Palaeontology, Southwest African Collection
UCB-V	University of California, Berkeley, Fossil Vertebrate Collection.

Description

Climacoceras africanus MacInnes 1936

MacInnes (1936) described *Climacoceras africanus* from Maboko (Kiboko), Kavirondo Gulf, Lake Victoria. He based his description almost entirely on ossicone fragments and the holotype of the species BM M-15301) is an almost complete left ossicone (figure 24.2). The beam of each ossicone shows slight lateral compression at its base so that the cross-section is oval. The beam tapers distally and the short tines are irregularly positioned and of different sizes. They project almost at right angles from the beam either anteriorly or posteriorly but not laterally. MacInnes sectioned several of these ossicones and was able to demonstrate that their structure was continuous from the frontal bone to the ossicone tip, in contrast to the structure of the cervid antler, which shows a junction between the pedicle region and the antler proper.

MacInnes also described and figured upper and lower cheek teeth, which he identified with *Climacoceras africanus*. The lower molars (BM M-15311, 15312, 15313) are subhypsodont with finely striated enamel. The metaconid and entoconid are strongly compressed transversely and the metaconid has well-developed anterior and posterior crests while the entoconid has a strong anterior crest and a weaker posterior crest that does not fully close the posterior fossettid. The labial cuspids are crescentic with the anterior region slightly lingually displaced relative to the posterior region. The accessory column of the M_3 is simple, as in *Prolibytherium* (Hamilton 1973a, pl. 10, figs. 3 and 4), lacking the crescentic form found in more advanced ruminants. Whitworth (1958, p. 26) describes and figures a P_4 (KB-781:52 = BM M-21367). He identified this tooth as an "Indeterminate bovid" but it agrees very closely

Figure 24.2 *Climacoceras africanus* MacInnes. Ossicone fragment BM M-15301, holotype. Maboko, Kenya.

with the P_4 of *Climacoceras* sp. (KNM-FT 2946) and is here identified with *Climacoceras africanus*. The crest pattern of this tooth consists of paired anterior and posterior crests that are joined by a labial anteroposterior crest to the high central labial cuspid. A posterolingual crest is produced from this central cuspid and this crest is slightly swollen at its lingual end but lacks any development of an independent central lingual cuspid.

Whitworth (1958, p. 25) figures an upper molar (KB-783:52 = BM M-15314a) and mentions another similar molar (KB-784:52 = BM M-15314b). These molars have high crowns on which the parastyle, paracone rib, and mesostyle are strong. The protocone and metaconule are crescentic and the enamel in the median valley shows slight complication. A maxillary fragment with P^{3-4} is present in the collections of the Kenya National Museum (KNM-MB 552). The P^4 has a fully developed lingual crescent that lacks accessory crests. The styles and central rib are strong on the labial face. The P^3 shows some anteroposterior compression. Its lingual crescent is strong and the labial ribs and styles are well developed.

Climacoceras sp. nov.

Gentry (1970) mentions an ossicone and mandible from Fort Ternan and suggests that they might belong with a new species of *Climacoceras*. After a detailed study of the Fort Ternan material (Hamilton, in press) I have confirmed Gentry's suggestion. I identified 38 specimens in the Fort Ternan collections with this species and to this were added 15 specimens in the Baringo collection.

The left ossicone (KNM-FT 3146 = 64.463.4) is complete (figure 24.3) but was heavily weathered prior to fossilization. This resulted in very heavy ridging of much of the surface. Part of the original surface is, however, preserved between the first and second tines and this shows that the surface was finely ridged with grooves running along the beam. The ossicone carries long slender tines and it is very similar to the antler of *Cervus elaphus*. However, there is no sign of a basal burr and the ossicone was not shed.

An almost complete mandible (KNM-FT 2946 = 64.55.5) has a very shallow horizontal ramus with a relatively long diastema (figure 24.4). The I$_3$ and C̄ as well as all the cheek teeth are preserved. The I$_3$ and C̄ have been glued back into the mandible and are more vertical than they would have been in life. The orientation of the undisturbed roots suggests that these teeth were procumbent. The canine is bilobed. Its anterior region is spatulate and similar to the cervid or bovid type of lower canine but the posterolabial region of the tooth carries a very small accessory lobe, which agrees closely in size with a similar lobe on the canine of *Canthumeryx* and is thus far smaller than the accessory lobes on the lower canines of *Palaeotragus, Samotherium, Sivatherium, Giraffa*, or *Okapia*. The I$_3$ is spatulate with a simple crown.

Figure 24.4 *Climacoceras* sp. nov. Left mandible KNM-FT 2946. Fort Ternan, Kenya.

Upper molars (KNM-FT 2952) and premolars (KNM-FT 2953) of *Climacoceras* sp. nov. were identified from Fort Ternan (Hamilton, in press) and they are easily distinguished from those of *Palaeotragus primaevus* on the basis of crown height and size. There is, however, the chance of confusion with *Oioceros* and *Protragocerus*, from which they differ only in the greater strength of the labial ribs and style and details of the median valley. On the molars the paracone rib, parastyle, meostyle, and metastyle are all strong in *Climacoceras* sp. nov. while the metacone rib is weak but extends to the base of the crown. The P^4 is compressed anteroposteriorly and its labial and lingual faces are more nearly vertical than those of *P. primaevus*, reflecting the greater crown height in *Climacoceras*. The P^3 has strong labial styles and ribs and it agrees in general shape with the P^3 of *Palaeotragus primaevus*, being more elongate than in *Protragocerus* and *Oioceros*. An isolated second upper premolar (KNM-FT 3221) is also identified with *Climacoceras* sp. nov. This tooth is similar in shape to the P^2 of *Prolibytherium* (Hamilton 1973a, pl. 9) and has a relatively compressed anterior region. This feature probably reflects the shortening of the premolar row in *Climacoceras* relative to that of *Palaeotragus primaevus*.

Postcranial elements from Fort Ternan that are identified with *Climacoceras* sp. nov. indicate that the animal was smaller than *P. primaevus* but larger than the Fort Ternan bovids described by Gentry (1970). Cervical vertebrae suggest that there was some elongation of the neck but not enough is known of the limb bones to give indications of corresponding increases in length of the limbs.

Canthumeryx sirtensis Hamilton 1973

I established this genus and species (Hamilton 1973a) on the basis of two immature mandibles from the lower Miocene deposits of Gebel Zelten, Libya.

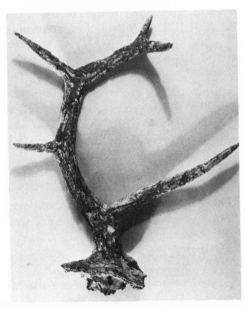

Figure 24.3 *Climacoceras* sp. nov. Ossicone KNM-FT 3146. Fort Ternan, Kenya.

In this same work I also established *Zarafa zelteni* (Hamilton 1973a), basing this genus and species on an almost complete but edentulous skull, two sets of upper cheek teeth, and isolated lower cheek teeth. Between 1973 and 1975 I studied material from Moruorot, East Africa, which is conspecific with the material of *Canthumeryx sirtensis*. The Moruorot collection includes a partial skeleton that consists of an almost complete mandible with P_2 to M_3, the lower canine and the roots of the incisors, an upper molar, and upper premolars and postcranial elements. The lower cheek teeth agree closely with those from Libya while the upper cheek teeth agree closely with those identified as *Zarafa zelteni*. I have therefore (Hamilton, in press) suggested that *Zarafa zelteni* is a synonym of *Canthumeryx sirtensis* and the name *Canthumeryx sirtensis* is retained as it has page precedence.

The skull of *Canthumeryx sirtensis* (*Zarafa zelteni*) is described in detail in Hamilton (1973a) and I do not redescribe it here. The lower and upper cheek teeth are also described in detail elsewhere so here I will give only a brief description with mention of those features that are relevant to assessment of relationships of this species.

The left lower canine (figure 24.5) is preserved on the specimen from Moruorot (UCB-V4899/42058). The anterior part of the crown is spatulate and relatively narrow and a small accessory lobe is present on the posterolabial edge of this tooth. As a result this canine resembles that of *Climacoceras*. The lower premolars have the same basic crest pattern as *Climacoceras* but the P_4 differs considerably in detail from that of *Climacoceras*. The anterior part of the tooth carries paired transverse crests that meet at their labial ends and join the labial anteroposterior crest. The posterior region also consists of paired transverse crests but the back of the tooth is separated from the central region by a deep vertical notch on the posterior labial face of the tooth and a marked depression of the anteroposterior crest. A

similar vertical notch is found in *Palaeotragus primaevus, Giraffokeryx,* and the other advanced giraffids in which the posterior region is clearly separated from the rest of the tooth. The lower molars have low crowns on which the lingual rib of the metaconid is strong and the metastylid is very strong in contrast to the flatter lingual faces of the lower molars in *Palaeotragus primaevus* and other giraffids. The posterior crest of the entoconid is very poorly developed and the posterior fossettid has a wide lingual opening. The labial cusps are of the usual crescentic form but the accessory column of the M_3 is more advanced than that of *Climacoceras* with a high hypoconulid that produces a strong labial crest and a lower lingual crest. These two crests both curve anteriorly and meet the back of the hypoconid, thus enclosing a shallow fossettid. The upper molars are low crowned and have the usual four-cusped pattern with strongly crescentic lingual cusps and transversely compressed labial cusps. The labial face of each molar carries a strong parastyle, paracone rib, and mesostyle and the labial face of the metacone is flattened with only a weak swelling indicating the metacone rib. The P^4 is subtriangular and its labial face carries strong anterior and posterior styles and a strong central rib. The P^3 is elongate with a strong anterolabial style and a strong rib on the face of the cusp. The central lingual region is slightly swollen and is joined by narrow crests to the anterior and posterior ends of the tooth. There are several small accessory crests in the fossette. The poor development of the lingual region on the P^3 is a primitive feature and the P^3 of *P. primaevus* carries a much stronger lingual region. There are two upper third premolars from Rusinga (BM M-30222 = 825.50 and BM M-30223 = 1099.51), which are described by Whitworth (1958). Both of these teeth have relatively poorly developed lingual regions and I have suggested (Hamilton, in press) that these teeth may be identified with *Canthumeryx*.

The P^2 is similar in general appearance to the P^3 but the anterior and posterior lingual crests are not fully developed. The anterior crest is formed from a series of small tubercles while the posterior crest is very weak and low. In contrast the P^2 of *P. primaevus* has strongly expanded anterolingual and posterolingual regions and is thus much more advanced toward the condition found in such advanced giraffids as *Palaeotragus rouenii*.

The postcranial skeleton suggests that *Canthumeryx* was a slenderly built ruminant exhibiting some lengthening of its neck and limbs but only comparable to *Okapia* in the development of these features. Arambourg (1933) described postcranial

Figure 24.5 *Canthumeryx sirtensis.* Anterior region of left mandible with canine, UCB-V 4899/42058. Moruorot, Kenya.

elements of a medium-sized ruminant from Turkana (= Moruorot). These agree closely with corresponding elements in the Berkeley collection and it is possible that all the postcranial elements could be from the same individual. Using the postcranial material from Moruorot it was possible to reidentify much of the material from Gebel Zelten. There are, however, a few specimens from the Gebel Zelten collections that cannot easily be identified with *Canthumeryx* and these suggest that a larger giraffoid may also have been present in the North African fauna.

Whitworth (1958) identified five isolated cheek teeth as "?*Palaeomeryx* sp." and Walker (1968, 1969) includes *Palaeomeryx* in the Mount Elgon fauna. Madden (1972) mentions *Palaeomeryx* in the Moruorot fauna and I (Hamilton 1973a, 1973b; Savage and Hamilton (1973) identified the genus from Gebel Zelten.

Two of the specimens identified by Whitworth (BM M-825.50 and BM M-1099.51) have been described above and are identified with *Canthumeryx*. The mandibular fragment and P_4 (BM M-776.52) is also identified with *Canthumeryx*. It has a crest pattern very similar to that of UCB-V 4899/42058 and has an expanded central lingual cuspid; BM M-442.51 is an M_3 that is more brachydont than the M_3 of *Canthumeryx*. I do not think that this tooth can be identified as *Canthumeryx sirtensis* but equally I do not think that it can be identified with *Palaeomeryx*. The upper molar (BM M-644.49) may belong with *Canthumeryx* and agrees with upper molars from Moruorot and Gebel Zelten. Walker's records of *Palaeomeryx* from Mount Elgon were based on a ruminant tibia that agrees very closely with tibias from Moruorot in the Berkeley and Institut de Paléontologie collections and is therefore identified with *Canthumeryx*. Records of *Palaeomeryx* from Gebel Zelten are based on two isolated molars. As I have indicated (Hamilton 1973a), there are difficulties in identifying these two lower third molars with either *Prolibytherium* or *Canthumeryx*. These same difficulties, however, would exclude the two specimens from *Palaeomeryx*.

Propalaeoryx austroafricanus Stromer 1926

Propalaeoryx austroafricanus was described by Stromer (1926) from the diamond fields of southwestern Africa. The species was based on a heavily weathered mandible (MM-1926 × 507) from the Elisabethfeld locality. This mandible contains the series of cheek teeth from P_2 to M_3 and an alveolus for a P_1. Stromer reconstructed the diastema region and also tentatively identified the I_3 and C. The cheek teeth of *Propalaeoryx* appear relatively high

crowned but this is probably a factor of the relatively small size of the animal and in degree of hypsodonty they do not approach the molars of *Climacoceras*. On each lower molar the metastylid and metaconid rib are very strong and the entoconid has only a weakly developed posterior crest so that the posterior end of the back fossettid has a wide lingual opening. The accessory column of the M_3 is crescentic with an enclosed fossettid as in *Canthumeryx* and in contrast to that of *Climacoceras*.

The P_4 carries paired anterior and posterior crests and its central lingual cuspid is well developed, resembling that of *Canthumeryx* and thus being stronger than that of *Climacoceras*. The vertical groove in the posterior labial region of the P_4 is strong. The P_3 has well-developed paired transverse crests at the anterior and posterior ends and a well-developed lingual wing from the central labial cuspid. This wing does not expand at its lingual end. The P_2 has a single anterior crest but the posterior region carries paired transverse crests. The presence of a P_1 in the holotype is suggested by a well-developed alveolus in front of the P_2, however a P_1 was not developed in the other specimens that are identified with this species.

Stromer (1926) described two anterior teeth as having shovel-shaped crowns that were narrow, particularly that of the canine. Had an accessory lobe been present on this tooth it is unlikely that as careful a worker as Stromer would have missed or failed to mention it. That Stromer does not mention such a lobe raises a problem. The tooth may in fact be a canine, in which case the absence of an accessory lobe is important and contrasts with the condition in *Climacoceras* and *Canthumeryx*. Alternatively the tooth may have been incorrectly identified as a canine and Stomer certainly had strong reservations over his identification of this tooth as a canine. In a recent visit to Munich I was unable to find these teeth and it is likely that they were destroyed together with much of the collection from the Miocene of southwestern Africa.

In his description of the southwestern African fauna Stromer also described cf *Strogulognathus sansaniensis* (Stromer 1922, 1924, 1926). This description was based on a large number of specimens from the Langental locality. On a recent expedition to southwestern Africa (Hamilton and Van Couvering, in preparation) we collected many specimens from the Langental locality, including upper and lower jaws (SAM-PQ-N 51, 57, 58, 59). We also recovered a single mandible from the Elisabethfeld locality (SAM-PQ-N 50). A single mandible from Langental that was already in the collections of the

South African Museum (SAM-PQ-G 8356) completes the list of important additions to our knowledge of the dentition of this ruminant. Comparison of this new material with Stromer's holotype of *Propalaeoryx austroafricanus* and also comparison of his figured specimen of cf *Strogulognathus sansaniensis* (MM-1926 × 509) suggests that they all belong to the same species. I can see no reason why this material should be identified with the genus *Strogulognathus* and I therefore identify all the southwestern African medium-sized ruminant material as *Propalaeoryx austroafricanus*. A detailed description of the new material from the diamond fields is in preparation.

Propalaeoryx nyanzae Whitworth 1958

Propalaeoryx nyanzae was described by Whitworth (1958) on the basis of several lower cheek teeth with an M_{1-2} (BM M-21368 = 324.47) as the holotype. This specimen is smaller than the holotype of *Propalaeoryx austroafricanus* and the metaconid rib and metastylid are slightly stronger. Other than this there are no marked differences. Other specimens of *Propalaeoryx* from East Africa appear to bridge the size difference between the two species and features of the lingual faces of the molars are variable. It is therefore possible that *Propalaeoryx nyanzae* is a synonym of *Propalaeoryx austroafricanus;* however, the discovery of much new material from southwestern Africa means that careful study of the material will be required before the synonymy or otherwise can be established.

Upper cheek teeth of *Propalaeoryx nyanzae* were described by me (Hamilton 1973a, p. 271). These teeth carry strong cingula and the paracone rib and mesostyle of the molars are also strong. The median valley and fossettes carry many small accessory crests and tubercles. This is unusual but could be an individual variation.

Palaeomerycidae indet.

A pair of ossicones (BM M-26690) from Gebel Zelten was described (Hamilton 1973a) as Palaeomerycidae indet. These ossicones are long and slender and diverge gently from near the midline of the skull. Each ossicone has an almost circular cross-section and a lightly grooved outer surface. The attachment of the ossicones to the skull roof suggests that they do not belong with the genus *Climacoceras* and identification of the skull BM M-26670 as *Canthumeryx* excludes these ossicones from this genus. They could represent the juvenile or female form of the ossicones of *Prolibytherium*. The attachment, angle of divergence, and posterior slope sug-

gest this interpretation but the absence of any ridge or lateral swelling is against it.

In my recent revision of the Giraffoidea I have suggested that there is no basis for the identification of *Prolibytherium* as a member of the Sivatheriinae. Indeed, I now think that there is no case for placing *Prolibytherium* in the Giraffoidea and I did not include it in my consideration of the group (Hamilton, in press). The discovery of the anterior dentition of *Prolibytherium* would provide important evidence on the relationships of this genus.

Conclusions

Whitworth (1958) gave a long diagnosis of the family Palaeomerycidae and I (Hamilton 1973a) gave a shorter version of this. Whitworth's diagnosis is as good as any other yet presented for the family, but it does not include a single apomorphic feature that can be used to indicate that the family is monophyletic. The family is thus based on symplesiomorphies. Even without further investigation this suggests that the group is probably polyphyletic, a suggestion that is supported by interpretation of Stirton's (1944) figure of the phylogeny of the group. Nelson (1970) has explained why groups based on symplesiomorphies are usually invalid. Even a cursory inspection of the genera included in this group by Stirton (1944) and myself (Hamilton 1973a) indicates that the group is at least paraphyletic in the sense of Nelson (1971) because *Canthumeryx* and *Climacoceras*, which are clearly giraffoids, are included whereas the other giraffoids are excluded. Similarly, Simpson's Lagomerycidae is paraphyletic and there is nothing to suggest that his interpretation of the Palaeomerycidae is valid. For these reasons I have rejected the Palaeomerycidae as a valid grouping. Having done this, however, I am faced with the problem of assessing the relationship of the genera and species that were once included in the family. Assessment of the European and Asian genera is outside the scope of this work but the relationships of the African genera must be assessed.

Relationships in the Pecora have not yet been assessed using Hennigian techniques and this leads to problems, as the sister-group of the Giraffoidea has not been identified. Features of the frontal appendages offer the possibility of discovering the sister-group of the Giraffoidea. Frontal appendages—horns, antlers, horncores, and ossicones—are found in the bovids, cervids, pronghorns, and giraffes respectively. Of these appendage types the ossicones appear to be the most plesiomorphic as they have a skin covering and it is likely that the precursors of both antlers and horns must have been skin covered (Coope 1968). I have previously (Hamilton 1973a) suggested that this is the case but I do not now think that it stands up to further investigation. For the Pecora as a group the absence of any frontal appendage is the plesiomorphic condition and the presence of any kind of frontal appendage may be interpreted as a synapomorphy indicating that the cervids, bovids, pronghorns, and giraffes are more closely related to each other than any is to, for example, *Dremotherium* or *Blastomeryx.* However this suggestion immediately runs into difficulties because two extant pecorans, *Moschus* and *Hydropotes,* lack frontal appendages. This situation may be explained by either of two hypotheses. Either frontal appendages evolved only once or frontal appendages evolved independently at least twice. The first hypothesis carries the implications either that *Moschus* and *Hydropotes* are not advanced pecorans and certainly not cervids or that *Moschus* and *Hydropotes* are advanced pecorans but have secondarily lost their frontal appendages. The number of possible interpretations is increased by the exclusion of either *Moschus* or *Hydropotes* from these considerations. The suggestion that *Moschus* is not an advanced pecoran and is not a cervid has had many advocates including Stirton (1944) and indeed there is little to support grouping it with the cervids. It seems likely that the presence of large saberlike upper canines in this genus led to its grouping with the Cervoidea but the presence of such canines is plesiomorphic. Removal of *Hydropotes* from the Cervoidea, however, would be more difficult to defend as *Hydropotes* resembles the cervids in details of the formation of the lingual crests of the upper premolars, in other features of its dentition, and in some anatomical features. It will probably never be possible to show whether *Moschus* and *Hydropotes* have secondarily lost their frontal appendages, but fossil genera such as *Dremotherium* lack frontal appendages and yet resemble *Hydropotes* closely in many features.

The early development of antlers differs markedly from that of ossicones and horns. Bubenik (1966) gives a detailed description of antler development and summarizes current knowledge on the development of horns and ossicones. Antlers develop directly from the frontal bones; horns and ossicones begin their development as independent ossifications in the skin on the top of the head, these ossifications later fuse to the frontal, frontal and parietal, or parietal bones. There is no indication of such an independent ossification during the development of antlers.

From the above I suggest that the plesiomorphic pecoran condition is absence of any frontal appendage and that this condition persisted until after the split had occurred between the ancestors of the cervoids on the one hand and of the giraffes and bovoids on the other hand. I also suggest that *Hydropotes* is a cervoid but became separated from the rest of the cervoids before frontal appendages had evolved in this group. This interpretation indicates that the Giraffoidea and Bovoidea are sister-groups; the Cervoidea is possibly the sister-group of the bovoid-giraffoid stem. This leads to a relationship diagram as in figure 24.6. There are two main problems in assessing these relationships. One is the relationship of *Moschus* and the other is the relationship of the Antilocapridae. Simpson (1945) placed the Antilocapridae in the Bovoidea, which implies sister-group relationship to the bovids, and this suggestion is supported by the interpretation of the horncores of *Antilocapra* given by O'Gara and Matson (1975). However Gaastra (1975) and Beintema (pers. comm.) present evidence suggesting that *Antilocapra* is the sister-group of the giraffes with these two as the sister-group of the bovids and the cervoids as the sister-group of these three. In general the interpretation by Gaastra (1975) supports my interpretation of the frontal appendages in the cervoids, bovoids, and giraffoids.

For this work the Bovoidea will be used as the sister-group of the Giraffoidea, and features of the bovoids and cervoids can thus be used to assess features of the giraffoids. The lower canine has a simple narrow crown and is incisiform in all cervoids and

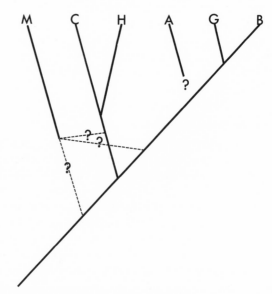

Figure 24.6 Relationships of the major pecoran groups. (*M*) *Moschus,* (*C*) Cervoidea, (*H*) *Hydropotes,* (*A*) Antilocapridae, (*G*) Giraffoidea, (*B*) Bovidae.

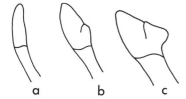

Figure 24.7 Transformation series of the lower canine. (*a*) Plesiomorphic condition found in all pecorans except giraffoids, (*b*) lower canine condition found in *Climacoceras* and *Canthumeryx*, (*c*) apomorphic giraffid condition found in all advanced giraffids.

bovids. In contrast, the lower canines of the extant *Giraffa* and *Okapia* are bilobed with a large posterolabial accessory lobe. This feature is identified as apomorphic and the presence of such a lobe is used to indicate relationships in the Giraffoidea. The lower canines of *Climacoceras* and *Canthumeryx* are very similar, as each has a small accessory lobe. Canines like this are not found in any other ruminant and a small accessory lobe is identified as part of a transformation series (figure 24.7) from the incisiform canine to the canine like that found in *Giraffa*, in which the accessory lobe forms about one-third of the tooth crown. The presence of a small accessory lobe in *Canthumeryx* and *Climacoceras* therefore indicates that they are giraffoids, but within the giraffoids this feature is plesiomorphic relative to the large accessory lobe found on the canine of *Okapia*, *Giraffa*, *Palaeotragus*, *Sivatherium*, *Giraffokeryx*, and *Samotherium*. Therefore the lower canine cannot be used to assess relationships between *Climacoceras* and *Canthumeryx* and any of the three possible relationships shown in figure 24.8 is possible.

To assess further the relationships of these two genera an independent feature must be used. For this, details of the crest patterns of the lower premolars and particularly the P_4 are used. The plesiomor-

phic crest pattern of the P_4 is identified as in figure 24.5, in which there are paired transverse crests in the anterior and posterior regions of the tooth and a short posterolingual to lingual crest produced from the central labial crest. In the Giraffoidea the most apomorphic P_4 crest pattern is shown by *Palaeotragus rouenii,* in which a fully molariform condition exists with the anterior and posterior lobes fully separated and each carrying crescentic labial and lingual cuspids. An intermediate condition is shown by the P_4 pattern found in *Palaeotragus tungurensis, Palaeotragus primaevus,*[4] and *Giraffokeryx punjabiensis* (figure 24.9), in which the central lingual cuspid is fully developed and independent but small accessory crests and features of the posterior region still indicate the plesiomorphic crest pattern. If this is accepted, then interpretation of the P_4 crest patterns in *Climacoceras* and *Canthumeryx* is possible. The crest pattern of *Climacoceras* shows very little change from the plesiomorphic condition. A central lingual cuspid is not developed and there is very little indication of the separation of the posterior region as the posterior labial vertical notch is only weakly developed. In *Canthumeryx* the vertical notch in the posterior region is strong and is reflected by depression of the anteroposterior labial crest at the top of the notch.

The lingual end of the central lingual crest is expanded, showing the development of a small but clearly defined central lingual cuspid. Thus the P_4 of *Canthumeryx* is clearly more apomorphic than that of *Climacoceras* and relationships as in figure 24.4 are indicated, with *Climacoceras* as the sister-group of the giraffids plus *Canthumeryx*, while *Canthu-*

4 I have recently suggested (Hamilton, in press) that *P. tungurensis* (Colbert 1936) and *P. primaevus* (Churcher 1970) are synonymous.

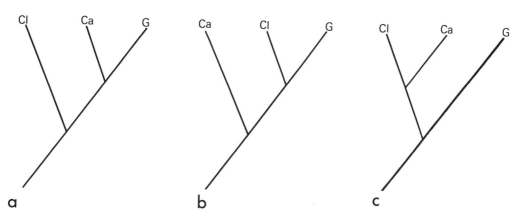

Figure 24.8 Possible relationships of *Climacoceras* (*Cl*), *Canthumeryx* (*Ca*), and the advanced giraffids (*G*), as indicated by interpretation of the lower canines only.

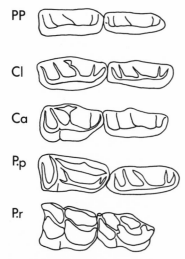

Figure 24.9 Transformation series demonstrated by crest and cuspid patterns of the lower third and fourth premolars. *(PP)* Plesiomorphic pecoran condition, *(Cl) Climacoceras, (Ca) Canthumeryx, (P.p) Palaeotragus primaevus, (P.r) Palaeotragus rouenii.*

meryx is the sister-group of the giraffids. Details of relationships in the giraffids are given in my recent publication on giraffoid phylogeny (Hamilton, in press). Use of the P_4 may be open to criticism, as the pecorans show considerable convergence in the evolution of the crest pattern on this tooth; for example, the P_4 of *Rangifer* is very similar to that of *Palaeotragus rouenii* even though review of all other apomorphies in these two ruminants indicates that *Rangifer* is a cervoid while *Palaeotragus rouenii* is a giraffoid. The convergences of crest pattern in the P_4 cause problems when an attempt is made to assess the relationships of *Propalaeoryx* and *Palaeomeryx*. Using the P_4 *Propalaeoryx* would be identified as very closely related to *Canthumeryx*, but without knowing the form of the $\bar{C}$, I am forced to retain *Propalaeoryx* as Pecora *incertae sedis*. The same applies to *Palaeomeryx* and ultimately to all European and Asian genera that have been grouped in the Palaeomerycidae, although a detailed revision of these genera may reveal other apomorphic features that could be used.

Figure 24.10 summarizes my understanding of the relationships of African species *Climacoceras africanus, Climacoceras* sp. nov., and *Canthumeryx sirtensis. Propalaeoryx austroafricanus, Propalaeoryx nyanzae,* and *Prolibytherium magnieri*[5] are placed as Pecora *incertae sedis* until either their anterior dentitions are discovered or interpretation of their features allows other assessments to be made.

5 In Chapter 25, Churcher considers *Prolibytherium magnieri* a giraffid of the subfamily Sivatheriinae—Ed.

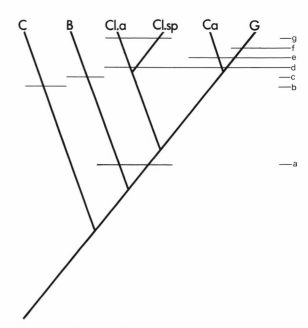

Figure 24.10 Summary diagram of relationships indicated in this paper. Horizontal lines indicate autapomorphic and synapomorphic features that are used to define groups or indicate relationships. *(C)* Cervoidea, *(B)* Bovidae, *(Cla) Climacoceras africanus, (Cl.sp.) Climacoceras* sp. nov, *(Ca) Canthumeryx, (G)* advanced giraffids. *(a)* Ossicones present during development of frontal appendages, *(b)* antlers present, *(c)* true horns present, *(d)* lower canines bilobed, *(e)* central lingual cuspid of P_4 well developed, *(f)* accessory lobe of lower canine forming about one-third of crown, *(g)* ossicones carrying tines.

References

Arambourg, C. 1933. Mammifères miocènes du Turkana. *Annl. Paléont.* 22:1–26.

———. 1938a. Mammifères fossiles du Maroc. *Mém. Soc. Sci. Nat. Maroc* 46:1–74.

———. 1938b. La faune fossile de l'Aïn Tit Mellil (Maroc). *Bull. Soc. Préhist. Maroc* 12(1–4):1–6.

———. 1962. Les faunes mammalogiques du Pléistocène d'Afrique. *Coll. Int. Cent. Nat. Rech. Scient.* 104:369–376.

Bubenik, A. B. 1966. *Das Geweih.* Hamburg, Berlin: P. Parey Verlag, pp. 1–214.

Chardin, T. de. 1939. The Miocene cervids from Shantung. *Bull. Geol. Soc. China* 19(3):269–278.

Churcher, C. S. 1970. Two new upper Miocene giraffids from Fort Ternan, Kenya, East Africa: *Palaeotragus primaevus* sp. nov. and *Samotherium africanum* sp. nov. In L. S. B. Leakey and R. J. G. Savage, eds., *Fossil Vertebrates of Africa* 2:1–105.

Colbert, E. H. 1936. *Palaeotragus* in the Tung Gur Formation of Mongolia. *Am. Mus. Novit.* 874:1–17.

Coope, G. R. 1968. The evolutionary origin of antlers. *Deer* 1:215–217.

Cope, E. D. 1877. Descriptions of new Vertebrata from the upper Tertiary Formations of the West. *Proc. Am. Phil. Soc.* 17:219–231.

Dietrich, W. O. 1950. Fossile Antilopen und Rinder äquatorialafrikas. (Material der Kohl-Larsen'schen Expeditionen). *Palaeontogr. Abt. A,* 99:1–61.

Dupuy, A. 1967. Repartition actuelle des espèces menacés de l'Algérie. *Bull. Soc. Sci. Nat. Phys. Maroc* 47:355–385.

Frick, C. 1937. The horned ruminants of North America. *Bull. Am. Mus. Nat. Hist.* 69:1–669.

Gaastra, W. 1975. Giraffe pancreatic ribonuclease. Groningen: Veenstra-Visser Offset, pp. 1–94.

Gaudry, A. 1878. *Les echaînement du monde animal dans les temps géologiques; Mammifères tertiares.* Paris, pp. 1–293.

Gentry, A. W. 1970. The Bovidae (Mammalia) of the Fort Ternan fossil fauna. In L. S. B. Leakey and R. J. G. Savage, eds., *Fossil Vertebrates of Africa* 2:243–323.

Geoffroy St. Hilaire, Étienne-F. 1833. *Palaeontographie.* Rev. encycl. vol. 59. Paris, pp. 76–95.

———. 1835. Études progressives d'un naturaliste. Paris, pp. 1–189.

Ginsburg, L., and Heintz, E. 1966. Sur les affinités du genre *Palaeomeryx* (Ruminant du Miocène européen). *C. R. Hebd. Séanc. Acad. Sci.,* Ser. D, 262:979–982.

Hamilton, W. R. 1973a. The lower Miocene ruminants of Gebel Zelten, Libya. *Bull. Brit. Mus. Nat. Hist. (Geol.)* 21(3):75–150.

———. 1973b. North African lower Miocene rhinoceroses. *Bull. Brit. Mus. Nat. Hist. (Geol.)* 24(6):351–396.

———. In press. Fossil Giraffes from the Miocene of Africa and a revision of the phylogeny of the Giraffoidea. *Bull. Brit. Mus. Nat. Hist. (Geol.).*

Hennig, W. 1966. *Phylogenetic systematics.* Translated by D. D. Davis and L. Zangler. Urbana: Univ. of Illinois Press, pp. 1–263.

Hensel, R. F. 1859. Ueber einen fossilen Muntjac aus Schlesien. *Zeits. Deutsch. Geol. Ges.* 11:251–279.

Hilzheimer, M. 1926. Säugetierkunde und Archäologie. *Zeits. Säugetierk.* 1:140–169.

Joleaud, L. 1914. Sur le *Cervus (Megaceroides) algericus* Lydekker, 1890. *C. R. Soc. Biol. Paris* 76:737–739.

———. 1926. Études de géographie zoologique sur la Berbérie. *Glasn. hrv. pirodosl. Dr.* 37–38:263–322.

———. 1930. On the "pachygenes" or "pachygnathes" (thick-jawed Quaternary deer from Africa and Asia). *Bull. Soc. Geol. China* 9(3):195–203.

———. 1935. Les ruminants cervicornes d'Afrique. *Mém. Inst. Égypte* 27:1–85.

Lartet, E. 1851. Notice sur la colline de Sansan. *Auch* 1851, pp. 1–42.

Lydekker, R. 1883. Siwalik selenodont Suina. *Pal. Indica* (10):142–177.

———. 1885. Catalogue of fossil Mammalia in the British Museum (Natural History). Part II. London: British Museum Nat. Hist., pp. 1–530.

———. 1887. On a molar of a Pliocene type of *Equus* from Nubia. *Quart. J. Geol. Soc. London* 43:161–164.

———. 1890. On a cervine jaw from Algeria. *Proc. Zool. Soc. Lond.* 1890:602–604.

MacInnes, D. 1936. A new genus of fossil deer from the Miocene of Africa. *J. Linn. Soc. London* 39:521–530.

Madden, C. T. 1972. Miocene mammals, stratigraphy and environment of Muruorot Hill, Kenya. *Paleobios* 14:1–12.

Matthew, W. D. 1926. On a new primitive deer and two traguloid genera from the lower Miocene of Nebraska. *Am. Mus. Novit.* 215:1–8.

Matthew, W. D., and Granger, W. 1923. The fauna of the Ardyn Obo Formation. *Am. Mus. Novit.* 98:1–5.

Meyer, H. von. 1834. *Die fossile Zähne und Knochen von Georgensgmünd.* Frankfurt: Museum Senckenberg, pp. 1–126.

Meyer, P. 1972. Zur Biologie und Okologie des Althirsches *Cervus elaphus barabarus,* 1883. *Zeits. Säugetierk.* 37(2):101–116.

Morrison-Scott, T. C. S. 1949. Technical names for the fallow deer and the Virginian deer. *J. Mammal.* 30(1):94.

Nelson, G. J. 1970. Outline of a theory of comparative biology. *Syst. Zool.* 19(4):373–384.

———. 1971. Paraphyly and Polyphyly: Redefinitions. *Syst. Zool.* 20(4):471–472.

O'Gara, B. W., and Matson, G. 1975. Growth and casting of horns by pronghorns and exfoliation of horns by bovids. *J. Mammal.* 56(4):829–846.

Pilgrim, G. E. 1908. The Tertiary and post-Tertiary fresh-water deposits of Baluchistan and Sind with notices of new vertebrates. *Rec. Geol. Surv. Ind.* 37(2):139–166.

———. 1941. The relationship of certain variant fossil types of "horn" to those of living Pecora. *Ann. Mag. Nat. Hist. Lond.* (2) 7:172–184.

Pomel, A. 1853. Catalogue méthodique et descriptif des vertébrés fossiles. Paris, pp. 1–193.

———. 1892. Sur deux ruminants de l'époque néolithique de l'Algérie. *C. R. Acad. Sci. Paris* 115:213–216.

Roger, O. 1904. Wirbeltierreste aus dem Obermiocän der bayerisch-schwabischen Hochebene. *Ber. naturw. Var. Schwaben.* 36:1–22.

Roman, F. 1931. Description de la faune pontique du Djerid (El Hamma et Nefta). *Ann. Univ. Lyon* n.s., 1, 48:30–42.

Roman, F., and Viret, J. 1934. La faune de Mammifères du Burdigalien de la Romieu (Gers). *Mèm. Soc. Géol. Fr.* n.s., 9, 21:1–67.

Romer, A. S. 1928. Pleistocene mammals of Algeria. Fauna of the Paleolithic station of Mechta-el-Arbi. *Bull. Logan Mus.* 1:80–163.

———. 1966. *Vertebrate paleontology,* 3rd. ed. Chicago: Univ. of Chicago Press.

Salez, M. 1959. Notes sur la distribution et la biologie du cerf de Barbarie (*Cervus elephus barbarus*). *Mammalia* 23:134–138.

Savage, R. J. G., and Hamilton, W. R. 1973. Introduction to the Miocene mammal faunas of Gebel Zelten, Libya. *Bull. Brit. Mus. Nat. Hist. (Geol.)* 22(7):483–511.

Simpson, G. G. 1945. The principles of classification and a

classification of mammals. *Bull. Am. Mus. Nat. Hist.* 85:1–350.

Stirton, R. 1944. Comments on the relationships of the cervoid family Palaeomerycidae. *Am. J. Sci.* 242:633–655.

Stromer, E. 1922. Erste Mitteilung über Tertiäre Wirbeltier-Reste aus Deutsch-Südwestafrika. *S. B. Bayer. Akad. Wiss.* 1921:331–340.

———. 1924. Ergebnisse der Bearbeitung mitteltertiärer Wirbeltier-Reste aus Deutsch-Südwestafrika. *S. B. Bayer. Akad. Wiss.* 1923:253–270.

———. 1926. Reste Land- und Süsswasserbewohnender Wirbeltiere aus den Diamantenfeldern Deutsch-Südwestafrikas. In E. Kaiser, ed., *Die Diamantenwüste Südwest-Afrikas,* vol. 2, pp. 107–153.

Villalta Comella, J. de F., Crusafont Pairo, M., and Lavo-cat, R. 1946. *Primer hallazgo en Europa de Ruminantes fosiles tricornois.* Sabadell: Publ. Mus. Sabadell, Communicaciones Cient., Paleont., 4 pp.

Walker, A. 1968. The Lower Miocene fossil site of Bukwa, Sebei. *Uganda J.* 32(2):149–156.

———. 1969. Fossil mammal locality on Mount Elgon, Eastern Uganda. *Nature* 223:591–593.

Whitworth, T. 1958. Miocene ruminants of East Africa. *Bull. Brit. Mus. Nat. Hist., Fossil Mammals of Africa* 15:1–50.

Young, C. C. 1937. On a Miocene fauna from Shantung. *Bull. Geol. Soc. China* 17:209–238.

———. 1965. On a new *Lagomeryx* from Lantian, Shensi. *Vert. Palasiat.* 8:329–340.

25

Giraffidae

C. S. Churcher

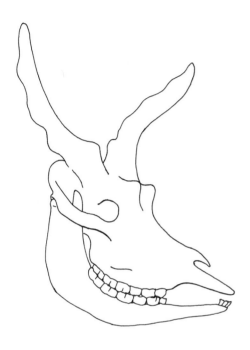

The earliest records of African giraffids are from the early Miocene at Gebel Zelten, Libya, from which two forms are recognized. (See figures 25.1 and 25.8 and table 25.1). One is a smaller and more lightly built animal, *Zarafa,* with simple supraorbital ossicones resembling *Palaeotragus,* and the other, *Prolibytherium,* is a larger and more robust form with palmate ossicones. These two forms persist in Africa as two distinct lines: the first including *Palaeotragus* from the middle Miocene of Kenya and Algeria and the extant *Okapia* of the Ituri Forests of Zaire; the second including the massively built *Sivatherium* of the Pleistocene of North, East, and South Africa, but which was not a direct descendant of *Prolibytherium.* The earliest specimens of the more widely known *Giraffa* are known from the late Miocene and early Pliocene of Kenya and the early Pliocene of South Africa. *Giraffa* continues to be reported throughout the Pleistocene in East and South Africa and by the middle Pleistocene was known in North Africa. It became extinct north of the Sahara in Neolithic times and now inhabits only the sub-Saharan African bush veld. *Samotherium* has also been reported on scant evidence from Miocene deposits in Algeria, Egypt, and Kenya, and *Giraffokeryx,* previously known only from the Siwaliks of India, is recognized from Kenya.

The condition of the ossicones in the extinct giraffids is not known for certain but is assumed to be similar to that in *Okapia* and *Giraffa,* i.e., skin-covered and vascularized throughout the life of the individual. The pointed tips of the ossicones may be bare of skin in *Okapia* and there are some indications that similar conditions may have been present in *Palaeotragus* or the similar *Zarafa,* but the blunt-ended or palmated ossicones of *Prolibytherium, Sivatherium,* and *Giraffa* show no signs that any of the surface was bare of skin.

In all, nine genera of the family Giraffidae are reported from the fossil and extant record of Africa (*Zarafa, Palaeotragus, Giraffokeryx, Samotherium, Okapia, Giraffa, Prolibytherium, Helladotherium,* and *Sivatherium*). All these genera are represented by single species except *Palaeotragus,* for which *P. primaevus* and *P. germaini* are described; *Sivatherium,* for which *S. maurusium* and *S. hendeyi* are recorded; and *Giraffa,* with the extinct *G. stillei, G. jumae, G. gracilis,* and *G. pygmaea,* and both fossil and extant *G. camelopardalis.* (See Hopwood and Hollyfield 1954; Cross and Maglio 1975; and Harris 1976b for many references.)

Family Giraffidae Gray 1821

DIAGNOSIS. Medium-sized to large, ruminating artiodactyls, attaining 2 to 3 m or more in shoulder

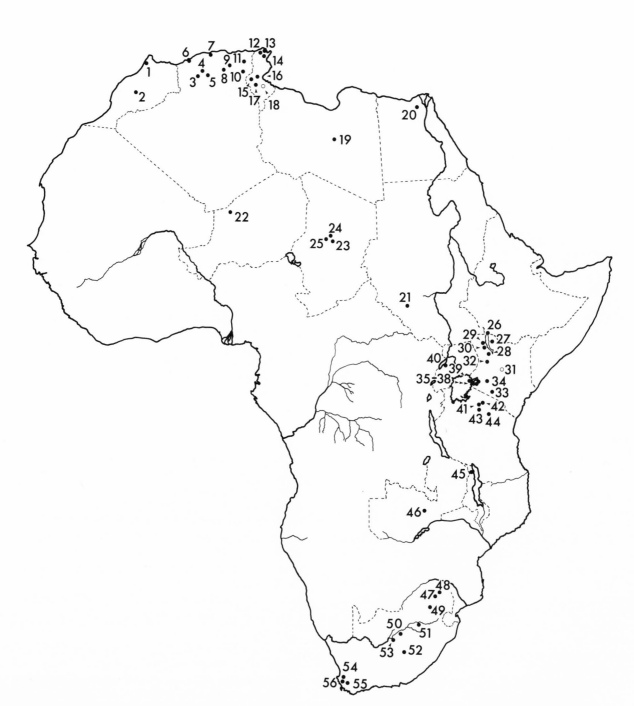

Figure 25.1 Geographic distribution of African Cenozoic fossil localities from which Giraffidae have been reported, with major references.

1. Mugharet el'Aliya, Cape Spartel, Tangier. Howe and Movius 1947.

2. Beni Mellal. Lavocat 1961.

3. Oued el Hammam, near Mascara, Oran. Arambourg 1959.

4. Bou Hanifa, opposite Oued el Hammam, near Mascara, Oran. Arambourg 1959.

5. Palikao (=Ternifine), Lake Karâr, near Mascara, Oran. Arambourg 1952.

6. St. Charles, Oran. Singer and Boné 1960.

7. Boulevard Bru, Mustapha Superieur, Algiers. Romer 1928.

8. Setif, near Constantine. Arambourg 1948a; Singer and Boné 1960.

9. Aïn (el) Hanech, near St. Arnaud, Constantine. Singer and Boné 1960.

10. Chaachas, Tebessa, Constantine. Arambourg 1960 (invalid record).

11. Smendou, Constantine. Joleaud 1937.

12. Garet (Lac) Ichkeul, near Ferryville and Bizerté. Singer and Boné 1960.

13. Douaria, near Bizerté. Roman and Solignac 1934.

height. Skull usually with ossicones; if lacking, then lacking in females or both sexes; if present, then usually simple, small, conical to subconical, but may be palmate, complex, and more than a single pair; covered by skin in living forms (except for tip in adult males of *Okapia*). Cranial vault bones are pneumatic, with sinuses passing into bases of ossicones. Sagittal crest and preorbital fossae lacking, except in *Prolibytherium,* which has fossae; lachrymal foramina single or double. Dentition $\frac{0.0.3.3}{3.1.3.3}$.

Outer incisors slightly overlap inner ones; crowns elongated, palmate; lower canines bilobed. Cheek teeth brachydont to moderately hypsodont, enamel heavy, wrinkled, or rugose, cement lacking. Premolars resemble those of Bovidae. *Palaeomeryx* fold usually absent. Limbs and neck usually elongated; feet mesaxonic, didactylous, metapodials III and IV fused, II and V absent, rudimentary or fused to III and IV respectively; digits I, II, and V lacking.

AGE. Early Miocene to present.

Subfamily Palaeotraginae Pilgrim 1911— Okapis

DIAGNOSIS. Primitive, medium-sized giraffids, usually with a single pair of supraorbital ossicones borne on the frontals. A second pair may be present

---

Figure 25.1 (continued)

14. Hamada Damous, west of Bou Arkoub, south of Grombalia. Coppens 1971.

15. Bled ed Douarah, near Gafsa. Robinson and Black 1974.

16. Djebel Sehib, south of Gafsa and of Metlaoun ridge. Burollet 1956.

17. Aïn Brimba, north of Mansoura. Coppens 1971.

18. "Sud tunisien," possibly near Djerid. Arambourg 1948a.

19. Gebel Zelten. Hamilton 1973.

20. Garet el Moluk Hill, Wadi Natrun. Joleaud 1937.

21. Bahr el Ghazel. Arambourg 1960.

22. In-Azaoua (Azaoua). Joleaud 1936.

23. Goz-Kerki, near Mortcha. Coppens 1960.

24. Koro-Toro Region. Coppens 1960.

25. Ouadi Derdemy and Koualà, near Koro-Toro. Coppens 1967.

26. Bourillé and Todenyang, Lake Turkana (= Lake Rudolf), Omo Basin. Arambourg 1948c; Leakey 1965; Harris 1976a.

27. East Turkana (= East Lake Rudolf), Ileret, Koobi Fora and Koobi Algi Formations. Maglio 1974; Harris 1976a.

28. Kanapoi, southwest of Lake Turkana. M. L. Richardson, pers. comm.

29. Moruorot Hill (Losodok), near Lodwar. Churcher 1970.

30. Lothagam, west of Lake Turkana. Churcher, this paper.

31. Marsabit Road. Singer and Boné 1960.

32. Baringo Basin, Ngorora and Chemeron Formations, Karmosit and Chemoigut Beds. Maglio 1974; Bishop et al. 1971; Bishop and Pickford 1975.

33. Olorgesaillie, near Lake Magadi. Vaufrey 1947; Cooke 1963.

34. Fort Ternan. Maglio 1974.

35. Kanjera, Homa Mountain, Lake Victoria. Cooke 1963.

36. Kanam, Homa Mountain, Lake Victoria. J. M. Harris, pers. comm.

37. Rawe (Rawi), Homa Mountain, Lake Victoria. Kent 1942b; Leakey 1970.

38. Kagua, Homa Mountain, Lake Victoria. Kent 1942b.

39. Kiahera Hill, Rusinga Island, Lake Victoria. Churcher 1970.

40. Kaiso, near Lake Albert. Cooke and Coryndon 1970.

41. Laetolil (Vogelfluss, Serengeti). Dietrich 1942; Leakey 1965.

42. Olduvai Gorge. Leakey 1965, 1970; Hendey 1970a; Harris 1976b.

43. Mumba Hills. Cooke 1963.

44. Mkujuni (Makujuni) Valley, Lake Manyara. Kent 1942a.

45. Mwimbi (45a) and Mwenirondo 3 (45b) Chiwondo Beds, Karonga District. Mawby 1970.

46. Broken Hill. Cooke 1963.

47. Makapansgat (Limeworks), Potgietersrus, Transvaal. Cooke 1963.

48. Kalkbank, near Pietersburg, Transvaal. Dart and Kitching 1958.

49. Swartkrans Australopithecine Site, Krugersdorp, Transvaal. Churcher 1974.

50. Tierfontein, near Port Allen, Vet River, Orange Free State. Cooke 1974.

51. Cornelia (Uitzoek), near Frankfort, Skoonspruit, Orange Free State. Cooke 1974.

52. Florisbad, Orange Free State. Singer and Boné 1960; Cooke 1963.

53. Barkly West (?Waldeck's Plant or Gong Gong), Cape Province. Cooke 1963.

54. Elandsfontein (Hopefield), Cape Province. Hendey 1968, 1969.

55. Bloembosch, Darling District, Cape Province. Cooke 1955.

56. Langebaanweg ("C," "E," and Baard's quarries), 30 km northwest of Hopefield, Cape Province. Hendy 1970a, 1974a, 1976; Harris 1976b.

For stratigraphic ages and species recovered from these sites, see figure 25.8 and table 25.1.

Table 25.1   Distribution of African giraffid taxa by stratigraphic horizon and locality.

Site	Age	Zarafa zelteni	Palaeotragus primaevus	Palaeotragus germaini	Samotherium africanum	Giraffokeryx cf punjabiensis	Giraffa jumae	Giraffa stillei	Giraffa gracilis	Giraffa pygmaea	Giraffa camelopardalis	Prolibytherium magnieri	Helladotherium duvernoyi	Sivatherium maurusium
Mugharet el 'Aliya (1)	L. Pleist.	—	—	—	—	—	—	—	—	—	X	—	—	—
Beni Mellal (2)	L. Mio.	—	—	X	—	—	—	—	—	—	—	—	—	—
Oued el Hammam (3)	L. Mio.	—	—	X	X	—	—	—	—	—	—	—	—	—
Bou Hanifa (4)	L. Mio.	—	—	X	—	—	—	—	—	—	—	—	—	—
Palikao (= Ternifine: 5)	M. Pleist.	—	—	—	—	—	—	—	—	—	X	—	—	—
St. Charles (6)	L. Plio.	—	—	—	—	—	—	—	—	—	—	—	—	X
Boulevard Bru, Algiers (7)	L. Pleist.–Holo.	—	—	—	—	—	—	—	—	—	X	—	—	—
Setif (8)	M. Pleist.	—	—	—	—	—	—	—	—	—	X	—	—	X
Aïn (el) Hanech (9)	E. Pleist.	—	—	—	—	—	—	—	—	—	X	—	—	X
Chaachas (10)	L. Pleist.	—	—	—	—	—	—	—	—	—	—	—	—	X
Smendou (11)	L. Mio.–E. Plio.	—	—	—	—	—	—	—	—	—	—	—	?cf.	—
Garet (Lac) Ichkeul (12)	L. Mio.–E. Plio.	—	—	—	—	—	—	—	—	—	—	—	—	X
Douaria (13)	L. Mio.–E. Plio.	—	—	—	—	—	—	—	—	—	—	—	?cf.	—
Hamada Damous (14)	L. Plio.	—	—	—	—	—	—	—	—	—	—	—	—	X
Bled ed Douarah: Lower fauna (15)	M. Mio.	—	—	X	—	—	—	—	—	—	—	—	—	—
Upper fauna (15)	L. Mio.	—	—	X	X	—	—	—	—	—	—	—	—	—
Djebel Sehib (16)	L. Mio.–E. Plio.	—	—	—	—	—	—	—	—	—	—	—	—	X
Aïn Brimba (17)	L. Plio.	—	—	—	—	—	—	—	—	—	—	—	—	X
"Sud tunisien" (18)	Pleist.	—	—	—	—	—	—	—	—	—	—	—	—	X
Gebel Zelten (19)	E. Mio.	X	—	—	—	—	—	—	—	—	—	X	—	—
Wadi Natrun (20)	L. Mio.	—	—	X	—	—	—	—	—	—	—	—	—	—
Bahr el Ghazel (21)	E. Pleist.	—	—	—	—	—	—	—	—	—	—	—	—	X
In-Azaoua (22)	Holo.	—	—	—	—	—	—	—	—	—	X	—	—	—
Goz-Kerki (23)	L. Plio.	—	—	—	—	—	—	—	—	—	X	—	—	—
Koro-Toro Region (24)	E. Pleist.	—	—	—	—	—	—	—	—	—	X	—	—	X
Ouadi Derdemy and Kouala (25)	M. Pleist.	—	—	—	—	—	—	—	—	—	X	—	—	X
Bourillé and Todenyang (26)	L. Plio.–E. Pleist.	—	—	—	—	—	X	—	X	X	—	—	—	X
East Turkana (= East Lake Rudolf): Zone 1 (27b)	E. Pleist.	—	—	—	—	—	X	—	X	X	—	—	—	X
Zone 2	L. Plio.	—	—	—	—	—	X	—	X	—	—	—	—	X
Zone 3 (27a)	M. Plio.	—	—	—	—	—	—	—	X	X	—	—	—	—
Kanapoi (28)	E. Plio.	—	—	—	—	—	cf.	?cf.	—	—	—	—	—	—
Moruorot Hill (29)	?E. Mio.	—	X	—	—	—	—	—	—	—	—	—	—	—

on the anterior extremities of the frontals in *Giraffokeryx* and *Palaeotragus quadricornis*. Ossicones in the form of simple, conical spikes, oriented laterally, dorsally, or posteriorly or a combination; sexually dimorphic, longer in males and frequently absent in females. Skull dolichocephalic, facial region elongated; cranial vault with small sinuses, cheek teeth brachydont to slightly hypsodont, enamel with moderately coarse rugosity. Neck and limbs slightly elongated, fore and hind limbs subequal or hind limbs slightly longer (after Singer and Boné 1960; Godina 1962).

Genus *Zarafa* Hamilton 1973

DIAGNOSIS. A very primitive palaeotragine with flattened, laterally expanded frontals and pneumatic frontal sinuses in the supraorbital region. Ossicones simple, pointed, placed supraorbitally, projecting dorsolaterally. Lachrymal foramina paired, located on anterior margin of orbit. Basicranial and palatal planes almost parallel. Cheek teeth strongly brachydont, enamel finely rugose, parastyles generally well developed; strong accessory crest present on the posterior region of the metaconule. Scapula with acromial process on spine (after Hamilton 1973).

Table 25.1 (*continued*)

Site	Age	*Zarafa zelteni*	*Palaeotragus primaevus*	*Palaeotragus germaini*	*Samotherium africanum*	*Giraffokeryx cf punjabiensis*	*Giraffa jumae*	*Giraffa stillei*	*Giraffa gracilis*	*Giraffa pygmaea*	*Giraffa camelopardalis*	*Prolibytherium magnieri*	*Helladotherium duvernoyi*	*Sivatherium maurusium*
Lothagam (30)	L. Mio.	—	—	X	—	—	—	—	—	—	—	—	—	—
Marsabit Road (31)	E. Pleist.	—	—	—	—	—	—	—	—	—	—	—	—	X
Baringo Basin: Lukeino Fm. (32)	L. Mio.	—	—	—	X	X	cf.	—	—	—	—	—	—	—
Ngorora Fm. (32)	L. Mio.	—	—	X	?X	—	—	—	—	—	—	—	—	—
Olorgesaillie (33)	M. Pleist.	—	—	—	—	—	—	—	—	—	X	—	—	X
Fort Ternan (34)	M. Mio.	—	X	—	X	—	—	—	—	—	—	—	—	—
Kanjera (35)	M. Pleist.	—	—	—	—	—	—	—	X	—	—	—	—	—
Kanam (36)	E. Pleist.	—	—	—	—	—	—	—	X	—	—	—	—	—
Rawe (37)	E./M. Pleist.	—	—	—	—	—	—	—	X	—	—	—	—	X
Kagua (38)	?E. Pleist.	—	—	—	—	—	?X	—	—	—	—	—	—	X
Kiahera Hill (39)	E./M. Pleist.	—	X	—	—	—	—	—	—	—	—	—	—	—
Kaiso (40)	M. Plio.	—	—	—	—	—	—	?sp.	—	—	—	—	—	—
Laetolil (41)	L. Plio.–E. Pleist.	—	—	—	—	—	X	X	—	X	—	—	—	X
Olduvai Gorge: Beds I and lower II (42b)	E. Pleist.	—	—	—	—	—	—	?X	X	X	—	—	—	X
Upper II (42a)	M. Pleist.	—	—	—	—	—	X	—	X	—	—	—	—	—
Mumba Hills (43)	L. Pleist.	—	—	—	—	—	—	—	—	—	X	—	—	—
Mkujuni (44)	M. Pleist.	—	—	—	—	—	—	—	—	—	sp.	—	—	—
Mwimbi and Mwenirondo 3 (45)	?Pleist.	—	—	—	—	—	—	—	—	—	sp.	—	—	X
Broken Hill (46)	L. Pleist.	—	—	—	—	—	—	—	—	—	X	—	—	X
Makapansgat (Limeworks: 47)	L. Plio.	—	—	—	—	—	—	X	—	X	—	—	—	X
Kalkbank (48)	L. Pleist.	—	—	—	—	—	—	—	—	—	X	—	—	—
Swartkrans Australopithecine Site (49)	E. Pleist.	—	—	—	—	—	—	—	—	—	—	—	—	X
Tierfontein (50)	?M. Pleist.	—	—	—	—	—	—	—	—	—	—	—	—	X
Cornelia (= Uitzoek: 51)	L. Pleist.	—	—	—	—	—	—	—	—	—	X	—	—	X
Florisbad (52)	M./L. Pleist.	—	—	—	—	—	—	—	—	—	X	—	—	X
Barkly West (53)	M. Pleist.	—	—	—	—	—	—	—	—	—	—	—	—	X
Elandsfontein (= Hopefield: 54)	M. Pleist.	—	—	—	—	—	—	—	—	—	X	—	—	X
Bloembosch (55)	?M. Pleist.	—	—	—	—	—	—	—	—	—	—	X	—	—
Langebaanweg (E Quarry: 56)	E. Plio.	—	—	—	—	X	—	—	—	—	—	—	—	X

Zarafa zelteni Hamilton 1973

AGE. Early Miocene.

LOCALITY. Libya (19)[1].

DISCUSSION. *Zarafa zelteni* is the earliest known palaeotragine[2] as well as the earliest known from Africa. It is founded on an almost complete but edentulous skull that also lacks the premaxillary and diastematic regions of the rostrum and on which the generic and specific diagnoses are based. The frontals are laterally expanded and flattened dorsally and bear simple, spikelike ossicones that project dorsolaterally. The projecting frontals are buttressed anteriorly by concave lachrymals, producing an arrangement of ossicones, frontals, and lachrymals that is characteristic of the species.

The cheek teeth have finely plicated or rugose enamel, are brachydont as in *Palaeomeryx* and more so than in *Palaeotragus* (figure 25.2). The molars are

1 The numbers in parentheses here and elsewhere refer to the particular localities shown in the map in figure 25.1 and listed in its legend.

2 In chapter 24, Hamilton considers *Zarafa zelteni* as a synonym of *Canthumeryx sirtensis,* thereby transferring it to the Family Palaeomerycidae—Ed.

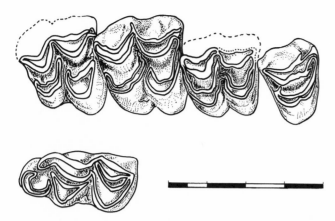

Figure 25.2   Check teeth of *Zarafa zelteni:* (*above*) right upper P⁴–M³, damaged series, (*below*) isolated right lower M₃. Redrawn from Hamilton's (1973) Gebel Zelten specimens. (Scale = 4 cm.)

four-rooted with fused lingual roots in the uppers. The upper molars possess strong parastyles, and there are strong buccal ribs to the paracones and parastyles, as in *Palaeomeryx* and stronger than in *Palaeotragus*. P⁴ is three-rooted, also with a strong buccal rib to the parastyle as in *Palaeomeryx* and stronger than in *Palaeotragus,* and a paracone similar to that of *Palaeomeryx*. The metacone of P⁴ is slightly swollen distal to the paracone rib, whereas in *Palaeomeryx* the two are fused and there is no swelling in *Palaeotragus*. The lower molars have weaker mesostylids than in *Palaeotragus* and weaker metaconid ribs than in *Palaeomeryx* but stronger than in *Palaeotragus*. M₂ lacks an ectostylid (interlophar external accessory column) and M₃ has strong mesial and weaker distal ectostylids and a well-developed talonid.

The characteristics of the cheek teeth of *Zarafa zelteni* suggest that it is intermediate between *Palaeomeryx* and *Palaeotragus*, with ribs and styles intermediately developed between the conditions observed in the two latter genera. The conditions of the mesostyles, metacones, and cingula and the degree of brachydonty-hypsodonty confirm this conclusion.

The postcranial skeleton is incompletely known but resembles that of *Palaeotragus* while possessing a few distinctive characteristics. There is an acromial process on the scapular spine, as in *Palaeotragus* but not *Okapia*. The proportions of the scapula are otherwise characteristically palaeotragine. There is no radial stop-facet on the anterior surface of the distal end of the humerus as in *P. primaevus,* but the coronid fossa is shallower in *Zarafa*. Vestigial metacarpal V and metatarsals II and V may have been present and free or fused to the metapodials III and IV. The long bones are similar to those of *Pa-*

laeotragus but apparently are somewhat more slender. The astragalus is more slender than in *Okapia* or *Palaeotragus,* and the calcaneum resembles that of *Palaeomeryx* or *Capreolus* with a very long tuber calcis, whereas in *Palaeotragus* and *Samotherium* it is shorter and in *Okapia* and *Giraffa* much shortened. The malleolar facet resembles that in *Palaeomeryx,* with a well-developed proximomedial angle, which is reduced in *Palaeotragus* and *Okapia*. The neck of *Zarafa* may have been proportionately longer than in *Palaeotragus,* as the axis is smaller and more elongate than in either *Palaeotragus* or *Okapia*.

The evidence of the postcranial skeleton suggests that the body of *Zarafa zelteni* resembles that of a lightly built deer or antelope, with generally slender proportions and a light build to all parts. The number of primitive and *Palaeomeryx*-like features suggests that the postcranial skeleton was conservative and less intermediate between *Palaeomeryx* and *Palaeotragus* than was the dentition.

Genus *Palaeotragus* Gaudry 1861

SYNONYMY. *Achtiaria* Borissiak 1914.

DIAGNOSIS. Small to large giraffids with long facial regions, but less so than in *Samotherium*. Ossicones paired, simple, pointed, and widely separated, borne on frontals in supraorbital position or possibly lacking, occasionally two pairs. Frontals dished, nasals straight or slightly convex; superior margin of orbit at or above nasal-frontal plane, anterior margin above M² or M³. Cheek teeth moderately brachydont, P⁴ molarized, P₄ partly molarized. Neck and limbs more elongate than in *Samotherium;* scapular spine terminates low, acromion present; radius and ulna fused by diaphysis only; metapodials thin and elongate, posterior faces concave, usually less broad than deep in midshaft.

Palaeotragus primaevus Churcher 1970

AGE. Early to middle Miocene.

LOCALITIES. Kenya (29, 34, 39).

DISCUSSION. This is a small giraffid that probably lacked ossicones. The cheek teeth have rugose enamel and are laterally compressed and moderately brachydont. The premolars are only slightly molarized except for P⁴. There is an acromial process on the scapular spine, as in *Zarafa,* but in neither *Okapia* nor *Giraffa*. Splint metapodials II and V are present, at least on the metacarpals, and there are marked tendinous attachments on the carpals and especially the tarsals.

Palaeotragus primaevus is known from some 243 specimens, including 25 dental rows, 83 isolated teeth, and 60 postcranial elements from the Fort Ternan volcanic beds. There is thus a comparatively numerous sample of bones of this animal on which to base a description. Unfortunately the skull is not known and the absence of ossicones can only be inferred, since the only possible ossicones preserved in the deposits are larger than recorded for *Palaeotragus* and match best those given for *Samotherium* (Bohlin 1926).

The cheek teeth are brachydont, of about the same size as those in *Zarafa,* especially the molars, as the degree of molarization of the premolars affects their mesiodistal dimensions (figure 25.3). In comparison with *Okapia* the teeth are similarly proportioned but are generally smaller. The enamel is roughened or plicated in wavy lines that disappear from projecting surfaces (e.g., cingula and styles) with advancing wear. Buccal and lingual styles may be present but are less strongly marked than in *Zarafa.* The molars are four-rooted with the lingual pair fused. Endostyles occur in about 50% of M^1s, 30% of M^2s, and are small and rare in M^3s. The lower molars have weak stylids and ribs, and ectostylids

are always present on M_1s, present in 75% of M_2s, and in M_3s are present in about 75% of the mesial and 33% of the distal positions. The dentition is primitive when compared with other *Palaeotragus* spp. or *Okapia* but progressive in comparison with that of *Palaeomeryx* or *Zarafa zelteni.*

The postcranial skeleton of *Palaeotragus primaevus* is remarkable for its elongated cubiti, tibiae, and metapodials. These are slender and elegant and suggest that the animal was a fast and agile runner. The conformation of the astragalus, navicular, calcaneum, and proximal end of the metatarsals III and IV show insertions for strong tendons of the Achilles system and, together with the slim but strong tuber calcis, suggest a powerful running or leaping capability.

P. primaevus differs from the more numerous species of *Palaeotragus* described from Asia and Europe in having buccolingually narrower toothrows, and in this reflects a more primitive condition of the palaeotragines such as that known to exist in *Zarafa zelteni* on the available evidence. The cranial skeleton of the former is not known but if it were hornless

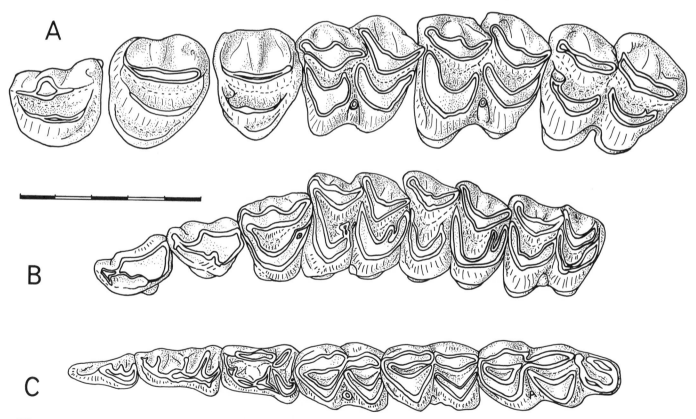

Figure 25.3   Cheek teeth of *Palaeotragus* species: (*A*) left upper cheek teeth of *Palaeotragus germaini,* redrawn from Arambourg (1959), (*B*) left upper cheek teeth of *Palaeotragus primaevus,* (*C*) left lower cheek teeth of *Palaeotragus primaevus.* (*B*) and (*C*) redrawn from Churcher (1970). (Scale = 5 cm.)

as is supposed, then the presence of ossicones in the latter would suggest separate lineages. The minor characters of the postcranial skeleton and dentitions also suggest this, although the known sample assigned to *Zarafa zelteni* comprises only 12 teeth and 26 postcranial elements (including some six long bones, three astragali, and a calcaneum) in addition to the type skull and two other cranial specimens. A better founded comparison will only be possible when a skull of *P. primaevus* and better preserved specimens of the postcranial skeleton of *Z. zelteni* are available. However, it may be suggested that *Z. zelteni* represents an intermediate stage in the transition from a dremotheriid such as *Dremotherium* or *Eumeryx* and that it shared an ancestor with *P. primaevus*.

Palaeotragus germaini Arambourg 1959

AGE. Late Miocene.

LOCALITIES. Morocco (2); Algeria (3, 4, ?11); Tunisia (?13, 15); Kenya (30).

DISCUSSION. *P. germaini* was a giraffid of considerable size with elongate neck and legs. Its limbs were subequal with the forelimb slightly longer than the hind limb. It possessed long, slender supraorbital ossicones of subtriangular section and a strongly brachydont dentition.

The permanent upper cheek teeth have strong buccal walls and lingual cingula with lingual interlophar endostyles on the molars. Metacones are present and separated from the paracones at the junction of the lophs on the buccal surfaces. The cement lakes or fossettes are deep and usually open lingually. Premolars 3 and 4 are molarized, as in *P. rouenii*, while P2s exhibit a degree of molarization intermediate between those of *P. primaevus* and *P. rouenii*. The upper milk cheek teeth are laterally compressed and mesiodistally elongate, with a crown pattern close to that of *P. primaevus*.

The limb bones are characteristically giraffid, with the radius longer than the tibia and the femur short. The metacarpus and metatarsus are subequal and are long, slender, and laterally compressed with subrectangular section. The ectocuneiform is free and the meso- and endocuneiforms are free of the navicular-cuboid. The navicular-cuboid has a lateral groove on its external face for passage of the peroneus longus tendon.

P. germaini is known from a palate (holotype), anterior and posterior limbs with radius, carpus, metacarpus, tibia, tarsus, and metatarsus (paratypes), and some 57 other specimens. It is larger than any of the other described species of *Palaeotragus* and is as large or larger than *Samotherium* spp. The single ossicone known from this species is similar in size to those of other *Palaeotragus* and therefore proportionately smaller than those of *Samotherium*. The dentition is advanced in the degree of molarization of the premolars but the fossettes generally lack any plis on the protolophs or hypolophs, possibly suggesting a divergent and slightly different evolutionary history (figure 25.3). Elements of the vertebral column are similar to those of *Samotherium sinense* (Bohlin 1926). The appendicular skeleton is large and generally intermediate in size between those of *Samotherium* and *Giraffa*.

Arambourg (1959) considered *P. germaini* to resemble recent *Giraffa*, but it exhibited a parallel development separate from the true giraffes and was related to *Palaeotragus* or *Samotherium*, in which the long supraorbital ossicones, the moderately elongated limbs and neck, and the nearly equal lengths of the forelimbs and hind limbs were characteristic. This species was assigned to *Palaeotragus* because of its limb proportions, especially those of the hind limb.

The stratigraphic levels from which *P. germaini* has been obtained in North Africa (late Miocene) are younger than those at Ft. Ternan, Kenya (middle Miocene) from which *P. primaevus* is best known. It is therefore possible that *P. germaini* is a direct descendant of *P. primaevus*.

An undescribed left upper molar (probably M^2) from the late Miocene of Kenya, recovered by B. Patterson at Lothagam in 1967, strongly resembles those illustrated by Arambourg (1959) for *P. germaini* and is of the right size for both *P. germaini* and *Samotherium*. It differs from *Giraffa* in the presence of a rib on the buccal surface of the metacone and the fusion of the paracone and metacone into a straight ectoloph, as well as in other characters.

Genus *Giraffokeryx* Pilgrim 1910

DIAGNOSIS. A medium-sized giraffid with four ossicones, two being at the anterior extremities of the frontal bones and two on the fronto-parietal region. Posterior ossicones overhanging the temporal fossae; anterior ossicones in front of orbits. Basicranium as in other Palaeotraginae. Teeth brachydont with rugose enamel as in the other Giraffidae. Limbs and feet presumably of medium length. (After Colbert 1933, 1935.)

Giraffokeryx cf *punjabiensis* Pilgrim 1910

AGE. Late middle Miocene.

LOCALITY. Kenya (32).

DISCUSSION. A fragment of the right mandible with P_3 and P_4, associated right P^2–P^4, and an isolated magnum of a giraffid assigned to this genus were recovered from the Karbarsero Beds of the

Ngorora Formation at Chepkesin in the Baringo Basin, Kenya, in 1969, by Aguirre and Leakey (1974). They are numbered NGO '69 43 to 46 and 148. These specimens are from a relatively large palaeotragine in which P⁴s and P₄s are in the advanced conditions depicted by Churcher (1970). Casts of the specimens resemble Colbert's (1935) illustrations of *G. punjabiensis* in patterns and proportions and have similar dimensions.

The three upper premolars from the Baringo Basin have small cingular shelves; the P² bears a small shelf or rudimentary style on the buccal re-entrant of the metacone, while the P³ and P⁴ bear narrow, thin-edged shelves on the mesiolingual surfaces of the protocones. These are variable in *G. punjabiensis*. The two lower premolars are also very similar to those of the Siwalik form and, although the hypoconid on P₄ is more selenodont than that illustrated by Colbert, he noted its width to be variable.

The teeth are more advanced than those of *Palaeotragus,* being robustly brachydont and transversely broader. No differences can be observed between those from the Baringo Basin and those from the Siwaliks that allow certain specific separation of the two forms. Therefore, although Aguirre and Leakey refer the Baringo Basin specimens to *Giraffokeryx* sp. (nova), they are here conserved within *G.* cf *punjabiensis* until further material is available.

Genus *Samotherium* Forsyth Major 1888

SYNONYMY. *Alcicephalus* Rodler and Weithofer 1890, *Shansitherium* Killgus 1922, and *Chersenotherium* Alexajew 1916.

DIAGNOSIS. Medium-sized giraffids with long facial regions, especially between P² and orbit. Ossicones paired, simple, pointed, and widely separated; borne on frontals above supraorbital bar; directed laterally, dorsally, or posterodorsally; ossicones sexually dimorphic, better developed in males and poorly developed or lacking in females. Rudimentary or small paired ossicones may lie anterior to main ossicones. Forehead dished, nasals straight, orbit with superior margin at or above nasal-frontal plane, and with anterior margin above or posterior to M³. Cheek teeth moderately hypsodont, with mediumly rugose enamel. Neck and limbs elongate, usually broader than deep in middle.

Samotherium africanum Churcher 1970

AGE. Middle Miocene to middle Pliocene.

LOCALITIES. Algeria (3); Tunisia (15); Egypt (20); Kenya (32, 34).

DISCUSSION. The presence of *Samotherium* in Africa cannot be considered fully established on the reported evidence. Stromer (1907) reported a form near *Samotherium* and *Libytherium* from the middle Pliocene deposits of Wadi Natrun and described the buccal surface of a left M₃ and some postcranial elements. The tooth fragment and the proximal end of a right femur are illustrated as "?*Libytherium* Pomel." Arambourg's (1959) identification of *Samotherium* sp. in Algeria is founded on a worn maxillary toothrow with P²–M¹, two fragments of a metacarpus, and the distal end of a femur. Arambourg considered his *Samotherium* to be distinct from *P. germaini* because of its larger size, the symmetry of the premolars and their conformation, and the lack of cingula. He assigned it to *Samotherium* because it lacked the deep nutrient foramen on the posterior face of the metacarpal, which defines the genus *sensu* Bohlin (1926). Churcher's (1970) description is founded on ossicones and cervical vertebrae that are giraffid and too large for *P. primaevus*. The ossicones are larger than those reported by Arambourg (1959) for *P. germaini*. Churcher therefore considered that the Fort Ternan ossicones and atlas best agreed with those of *Samotherium* species described by Bohlin (1926) and assigned them to *Samotherium* on these resemblances.

A partial M¹ or M², an incomplete calcaneum, a malleolar and an ulnare were collected in 1969 from the Nakali Tuffs of the Ngorora Formation, Baringo Basin, Kenya, by Aguirre and Leakey (1974). These are tentatively identified as cf *Samotherium* sp. because the tooth resembles other specimens of *Samotherium* in its strong parastyle and mesostyle, strong paracone rib, reduction of the cingulum and metacone rib, and free margins between the mesostyle and distal margin of the paracone loph.

Genus *Okapia* Lankester 1901

DIAGNOSIS. A medium-sized palaeotragine with simple, pointed, paired ossicones placed supraorbitally on frontals; present in males only. Ossicones separate in young and fused to frontals in adult individuals; covered except at tips by hairy integument. Nasal bones bearing a median tumescence but no ossicone. Basicranial and basipalatal planes approximately parallel. Dental arch of lower incisors and canines small and narrow; individual incisors and canines relatively small. External ears large. Limbs and neck moderately elongate; limbs subequal (after Lankester 1901).

"*Okapia stillei*" Dietrich 1941

See *Giraffa stillei* and *G. pygmaea* below.

Okapia sp.

AGE. Early Pliocene.

LOCALITY. Uganda (40).

DISCUSSION. Fossil *Okapia* is almost unknown. *O. stillei* has been reported from East Africa but is considered by Harris (1976b) to be misidentified *Giraffa* (see *G. stillei* and *G. pygmaea* below). The remaining possible fossil *Okapia* is represented by a right ossicone from the Kaiso Formation, Uganda (Cooke and Coryndon 1970, p. 199), from which little diagnostic or descriptive information can be gleaned. The extant *Okapia johnstoni* is restricted to the Ituri Forest, Zaire (figure 25.4).

Subfamily Giraffinae Zittel 1893—Giraffes

DIAGNOSIS. Medium-sized to large giraffids with moderately brachycephalic skulls. Ossicones variously developed, paired over frontoparietal sutures, median on nasals; ossicones rounded or flattened on ends, covered with hair. Roof with highly developed sinuses. Cheek teeth brachydont or moderately so; enamel heavily rugose. Limbs and neck greatly elongated, forelimbs longer than hind limbs (after Singer and Boné 1960; Godina 1962).

Genus *Giraffa* Brisson 1756

SYNONYMY. *Camelopardalis* Schreber 1784 and *Orasius* Oken 1816.

DIAGNOSIS. Ossicones paired, short, on frontoparietal suture; sometimes with single frontal ossicone between or just behind level of orbits; all ossicones variable in form, but usually straight and bluntly ended if paired and rounded or tumescent if single. Exostotic occipital "horns" may be developed from nuchal crest and azygous "horns" from the orbital boss. Lower incisors and canines robust; buccal enamel rugose; lower canine occasionally trifid.

Cheek teeth variable in size and moderately brachydont; premolars progressively molariform and complex. Basicranial and basipalatal planes not parallel.

Giraffa jumae Leakey 1965

AGE. ?Late Miocene to middle Pleistocene.

LOCALITIES. Ethiopia (26); Kenya (27, ?28, ?32, 35, 36, 37, ?38); Tanzania (41, 42a upper Bed II and Bed IV); Republic of South Africa (47, 56).

DISCUSSION. *Giraffa jumae* is generally more massive than the largest recorded specimens of *G. camelopardalis,* while its teeth approximate those of the largest known specimens in the living species (figure 25.5). The width of the skull roof between the orbits is wider than in *G. camelopardalis.* The ossicones extend posteriorly from the posterodorsal rims of the orbits, possess terminal knobs, and project more posteriad and straighter than in *G. camelopardalis.* In males the "third horn," set on the midline near the upper limit of the nasal bones, is either not at all developed or much less strongly developed than in the adult male giraffe of the living species and the frontal bone is therefore nearly flat. There is no secondary bone apposition of the ossicones. The basilar process of the occipital is narrow and not inflated. The length of the premolar-molar series is greater than the transverse diameter of the palate over the M^2s. The ascending ramus of the mandible is wide and stout, and the corpus is deep and long, with the diastematic region inclining upward to the symphysis and then downward to the incisor region (after Leakey 1965, 1970; Harris 1976b).

This species is founded on a nearly complete skull and mandible, together with a large part of the post-

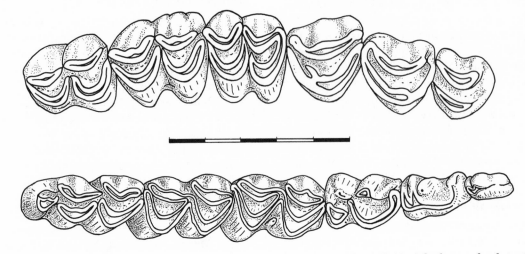

Figure 25.4 Check teeth of *Okapia johnstoni:* (*above*) right upper cheek teeth, (*below*) right lower cheek teeth. Redrawn from Lankester (1910). (Scale = 5 cm.)

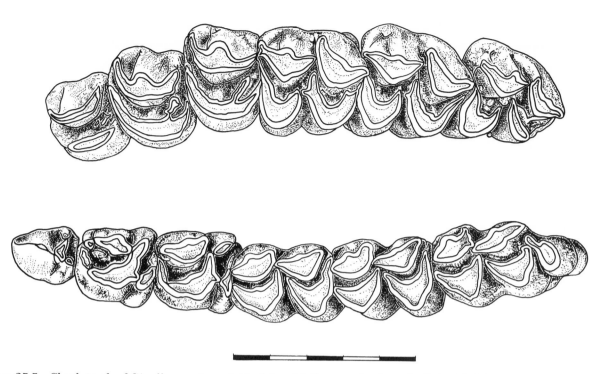

Figure 25.5   Cheek teeth of *Giraffa camelopardalis:* (*above*) left upper cheek teeth, (*below*) left lower cheek teeth. Drawn from a Recent specimen in the Royal Ontario Museum. (Scale = 5 cm.)

cranial skeleton, from Rawe, Kenya (BM M-21466), which is confirmed by the discovery of a very well preserved skull from upper Bed II, Olduvai. Leakey (1965) notes that Hopwood's (1936) reported occurrence of *Giraffa* cf *capensis* in Bed I, Olduvai, was founded on a specimen from the surface of Bed I and restricts the occurrence to Beds II and IV. He considers that this material is referrable to *G. jumae.* Similarly, Leakey considers Arambourg's 1947 report of *G. camelopardalis* from Omo to be based on its obvious differences from *G. gracilis.* He considered that the larger Omo specimens, as well as Dietrich's (1942) fragmentary material from Laetolil all belong to *G. jumae.* Harris (1976b) describes a frontlet of *G. jumae* from East Turkana in which the left ossicone is triangular in section with keels on the lateral and posterior edges, the right ossicone is convex anteriorly but the proximal third of the posterior surface is flattened, and both ossicones narrow distally to terminate in knobs, but without any areas of deposition of secondary bone. He also describes other ossicones of similar conformation. The ossicones of *G. jumae* are appreciably more parallel to the plane of the skull roof than are those of *G. camelopardalis* and arise from a somewhat more anterior point. The presence or absence of a median protuberance or ossicone may be a regional or sexually dimorphic character.

The record of *G.* cf *jumae* (Harris 1976a) from Langebaanweg is founded mainly on isolated teeth and ossicones that are more robust than those of extant giraffes. Harris (1976b) discusses Arambourg's (1947) tentative identifications of *G. camelopardalis* from Todenyang on the west side of Lake Turkana (Lake Rudolf) and from the Omo Basin collected by Bourg de Bozas' 1902 expedition and concludes that perhaps both should be more correctly referred to *G. jumae,* although he indicates that he did not examine the materials (Harris 1976b, p. 285).

Harris (1976b) has described a tibia, metatarsal, and astragalus of *G. jumae* from East Turkana and listed measurements taken for these bones and those of a scapula, metacarpals, and femora of the Rawe type specimen. The lengths of these bones, when available, fall within or above the range of *G. camelopardalis,* while the dimensions of the proximal and distal epiphyses appear to be proportionately smaller. Some minor differences are noted between the articular surfaces in the fossil and modern giraffe bones.

Undescribed limb bones referable to *G. jumae* were recovered at Kanapoi, Kenya in 1966 by B. Patterson. These include portions of the major elements of a left forelimb, an almost complete right tibia, and the proximal third of a left radioulna. Where dimensional comparisons can be made, these limb bones are as large, if not larger, than those of male *G. camelopardalis* and the tibial morphology compares well with Harris's description of the *G. jumae* tibia from East Turkana (M. L. Richardson, pers. comm.). Along with the material assigned to *G.* cf *jumae* from Langebaanweg, and undescribed *G.* cf *jumae* from late Miocene sediments in the Baringo Basin, Kenya (Pickford 1975), the Kanapoi postcranial specimens confirm the very early occurrence of undoubted *Giraffa* in Africa.

Giraffa stillei (Dietrich) 1941

SYNONYMY. *Okapia stillei* Dietrich 1941 partim.
AGE. ?Early Pliocene to early middle Pleistocene.
LOCALITIES. Kenya (?28); Uganda (40); Tanzania (41, ?42b).
DISCUSSION. *Giraffa stillei* was first described as *Okapia stillei* on milk and permanent teeth, a left astragalus, a proximal phalanx, and the distal end of a metacarpal by Dietrich (1941, 1942), who diagnosed it as "An *Okapia* species with palaeotragine characteristics" (Dietrich 1942, p. 115, translated). The original specimens were further described as having a dentition more brachydont than in *Samotherium* or other palaeotragines; P^2 primitive in the absence of a mesostylar cingulum; P^4 possessing buccal styles and midribs on the paracone and the metacone more strongly developed than in *O. johnstoni;* a very shallow groove between the mesostyle and metacone not present in *O. johnstoni;* mesial fossettes in P_3 and P_4 closed as in *O. johnstoni* and advanced palaeotragines; and lower milk premolars with ectostylids probably less strongly developed than those of *O. johnstoni* (after Dietrich 1942, pp. 113–115).

Additional specimens of this species were collected by Leakey at Laetolil in 1935 and specimens provisionally assigned to *O. stillei* were reported from Bed I Olduvai (Leakey 1965), as were an ossicone and P^3 from Kaiso, Uganda (Cooke and Coryndon 1970).

However, Harris (1976b, p. 307) considers the specimens originally described as *O. stillei* to in fact represent *Giraffa* on the basis of their premolar morphology, which appears to differ from *Okapia* and resemble *Giraffa* in the following characters: the paraconid and metaconid on P_3 nearly meet, while they are separated in *Okapia;* the well-developed P_3 parastylid of *Okapia* is absent; in P_2 and P_3 the anterior arm of the protocone meets the paracone; the talonid of P_4 is similar in width to that of the trigonid, while it is narrower in *Okapia;* and there are no plis in the prefossettes. Harris (1976b, p. 307) has thus rediagnosed Dietrich's Laetolil specimens as belonging to

"A species of *Giraffa* with teeth of similar morphology but slightly smaller size to those of *G. gracilis*."

Of the other "*O.*" *stillei* specimens reported, Harris considers Leakey's Laetolil specimens and the Kaiso P^3 to be *G. stillei* but the Olduvai specimens to belong to *G. pygmaea*.

Possibly referrable to Dietrich's taxon are three small, isolated giraffid teeth recovered from early Pliocene deposits at Kanapoi, Kenya, by B. Patterson in 1966. Of these, a right M^3 is virtually indistinguishable from that illustrated by Dietrich (1942, pl. XXI, no. 170). It shows clear palaeotragine affinities, differing from the same tooth in *P. primaevus* only in the very strong development of a paracone rib, while greater contrasts are found in comparisons with *O. johnstoni* and particularly *G. camelopardalis*. Although the other teeth from this site, an associated M_2 and M_3, are smaller than those of *G. stillei* listed by Harris (1976b, p. 308), they are not morphologically distinct from this species and a peculiar ϵ-shaped talonid on the M_3 closely matches that on one of Dietrich's specimens (1942, pl. XXI, no. 169). (M. L. Richardson, pers. comm.)

Examinations of the *Giraffa* and *Okapia* materials in the United States National Museum and the Kenya National Museum and the *Okapia* specimens illustrated in the atlas for Lankester's (1910) monograph of the *Okapia*, suggest that the characters Harris is using to distinguish *Giraffa* and *Okapia* are highly variable. Moreover, as Dietrich (1942) originally noted, the species appears to be very close to both fossil *Palaeotragus* and extant *O. johnstoni*, differing only from the latter in characters capable of wide variation on the evidence of other palaeotragines. Without further diagnostic evidence of the cranial and postcranial skeleton of this animal, the validity of removing "*Okapia*" *stillei* from the Palaeotraginae is questionable.

Giraffa gracilis Arambourg 1947

AGE. Late Pliocene to late Pleistocene.
LOCALITIES. Ethiopia (26); Kenya (27); Tanzania (42 Beds I–IV); Republic of South Africa (?47, ?54).
DISCUSSION. Arambourg originally described *G. gracilis* as a "species comparable in length of the limb-bones and neck to *G. camelopardalis,* but with a more lightly constructed skeleton and finer proportions in all parts. The dentition is smaller and the mandible more slender than in *G. camelopardalis.* P_3 possessing a metaconid and a continuous lingual crest. Metapodials remarkable for the narrowness of their articular extremities" (translated from Arambourg 1947, p. 364).

Harris (1976b) subsequently described the area between the orbits as convex, with or without a median ossicone, which, when present, is separated from the orbital margin by a wide, shallow groove. The lateral ossicones are shorter and smaller than in *G. jumae* or *G. camelopardalis,* are oval in basal section, and appear to lack terminal bosses. They arise from above the posterior orbital margin and are oriented as in *G. camelopardalis* (i.e., at ± 75° to the basicranial plane). The ossicones exhibit secondary apposition of bone, as do those of the modern species. Small nuchal protuberances may be present on the laterodorsal margins of the nuchal plate as in *G. camelopardalis.* Entostyles may be present on upper molars, again as in *G. camelopardalis.*

G. *gracilis* was considered to be ancestral to *G. jumae* by Leakey (1965), who reported it from the lower part of Bed II at Olduvai and *G. jumae* from the upper part of Bed II. However, recent discoveries in the Omo Basin and at East Turkana suggest that the two forms were coeval, at least from late Pliocene to middle Pleistocene times, in eastern and possibly southern Africa (Harris 1976b). So this phylogenetic relationship must be disallowed.

Photographs by Leakey (1965, pls. 33–35) of a humerus, radioulna, metacarpal, femur, and tibia of *G. gracilis* placed aside the same bones of recent *G. camelopardalis* show the fossil bones to be shorter than but proportionately similar to the modern bones. *G. gracilis* is generally said to be smaller in all its elements than both *G. jumae* and *G. camelopardalis,* but measurements given by Harris (1976b, tables VII–IX, pp. 302–305) do not support this. Cranial measurement of the skull KNM-ER 779 for dimensions of the braincase suggest that this skull, assigned to *G. gracilis,* is about 10% smaller in general than that of the *G. camelopardalis* specimen with which it is compared when only dimensions that are not sexually dimorphic (i.e., those not associated with the ossicones) are considered. The upper dental elements are similarly inconclusively divergent, with the mean given for *G. camelopardalis* (sample size and range not given) falling within, above, or below the range of variation reported for *G. gracilis* (from 2 to 8 measurements). The postcranial elements are composed of 14 to 17 male, and 5 to 7 female specimens, for each of which measurements of two selected individuals are given (Harris 1976b, table II, p. 287). These comparisons also suggest that the ranges of variation of both species overlap, since the means for both sexes of *G. camelopardalis* do not fall regularly between the measurements given for the two selected specimens and often lie close to or within the ranges reported for the few (2 to 6) ele-

ments assigned to *G. gracilis.* When Harris's measurements for the type of *G. jumae* are compared, they are usually, but not always, larger than any of those for *G. gracilis,* being sometimes smaller than either those for *G. camelopardalis* or *G. gracilis.*

On the basis of Harris's data it appears that measurements alone do not distinguish *G. gracilis* from *G. camelopardalis* or even *G. jumae.* Although *G. gracilis* may be diagnosed from its ossicone morphology, it appears that the majority of assignations to this species have been intuitive.

Giraffa pygmaea Harris 1976

SYNONYMY. *Giraffa pygmaeus* Harris 1976 and *Okapia stillei* Dietrich 1941 partim.

AGE. Early Pleistocene.

LOCALITIES. Ethiopia (26); Kenya (26, 27); Tanzania (?41, 42b Beds I and lower II)

DISCUSSION. Harris (1976b) describes this new species as having a dental morphology similar to that of *G. camelopardalis* with ectostylids on the lower molars but appreciably smaller in size than *G. gracilis.* The lateral ossicones bear terminal knobs as in *G. jumae* and *G. camelopardalis,* are flattened at their posterolateral surfaces, and are shorter and slighter than in *G. gracilis.* Secondary bone apposition on the lateral ossicones appears poorly developed and confined to the presumed males, as in *G. camelopardalis* and *G. gracilis* but not in *G. jumae.*

Some specimens from Olduvai that were assigned by Leakey (1965) to *Okapia stillei* are considered to be *G. pygmaea* by Harris because of their small size.

The major diagnostic characteristics of this species are its small size, smaller dentition, and slender ossicones with reduced or absent secondary bone apposition but with terminal knobs. Two ossicone specimens are known. Harris (1976b) considers them to represent male and female individuals and that this sexual dimorphism justifies their taxonomic separation from *G. gracilis.* The mandibular specimen that is the holotype contains teeth that may be separated from *Okapia* because of their typical giraffine morphology but differ only in size from other species of *Giraffa.*

Additional material is known from the Shungura Formation at Omo (three ossicones, dental and fragmentary postcranial material) and the Hadar Formation in Ethiopia (four ossicones, dental and postcranial material) that support Harris's description of *G. pygmaeus* as a separate species (Harris pers. comm.).

This species is insufficiently described in the literature at present for certain identification. The measurements available (Harris 1976b, tables XI and

XII, pp. 312–313) show that the specimens assigned to *G. pygmaea* are generally smaller than those of *G. gracilis.*

Giraffa camelopardalis Brisson 1756

AGE. ?Early Pleistocene to Recent.

LOCALITIES. Morocco (1); Algeria (5, 7, 8, 9); Tunisia (17); Niger (22); Chad (23, 25); Kenya (33); Tanzania (43, 44); Malawi (45); Zambia (46); Republic of South Africa (48, 51, 52, 55).

DISCUSSION. This is the large modern giraffe possessing paired supraorbital ossicones on the frontoparietal sutures of both sexes; these are usually smaller in females than males. Paired nuchal exostotic "horns" or eminences may also be present and single frontal "horns" with low rounded to crested profiles that are covered with skin may be present in males. The neck and legs are greatly elongated and robust.

Many specimens of giraffes were first assigned to *G. camelopardalis* but have subsequently been considered as belonging to the extinct species *G. jumae* or *G. gracilis.* Most of these reassignments concern the eastern or southern African specimens, so that *G. camelopardalis* is now not recorded from levels earlier than early middle Pleistocene in the Transvaal and Orange Free State of South Africa, the late middle Pleistocene in Kenya, and the early late Pleistocene in Tanzania and Zambia (table 25.1). Other early occurrences of *G. camelopardalis,* such as those reported from the upper Pliocene of Chad and late lower or middle Pleistocene deposits of Algeria, may also represent *G. jumae* or *G. gracilis.*

The taxonomic status of the species of *Giraffa* requires comment. Singer and Boné (1960) remark on the wide variability in ossicone morphology and conformation in modern *G. camelopardalis.* Size is also widely variable within the modern populations, both as an expression of sexual dimorphism and as individual variation, either within or between populations, as is also the development of the bony structure beneath the ossicones (Walker 1969; Dagg and Foster 1976). Since no recognizable qualitative characteristics have been identified by which teeth of these species may be distinguished, and since their sizes appear to intergrade into one another, the taxonomic separation finally becomes based on the shape, conformation, cross-sectional outline, and presence or absence of secondary bone on the ossicones. Further material may substantiate the identity of the described species of *Giraffa* or may provide such a continuum of characters and sizes that all *Giraffa* may best be considered but one species. For the present the status of *G. jumae* appears most

secure because of the apparently constant angle and morphology of the ossicones, while the status of the other fossil specimens may be alternatively considered but examples of clinal variation that in its extreme, might include both *G. jumae* and *G. camelopardalis.* One grouping of *G. camelopardalis* and *G. jumae* and another of *G. pygmaea* and *G. stillei* appear plausible and are founded mainly on the variation in the dentitions, as well as being possible over the time span from late Pliocene to the Holocene. *G. gracilis* may belong within the second grouping if it is synonymous with *G. stillei,* as suggested by Harris (1976b).

Subfamily Sivatheriinae Zittel 1893— Sivatheres

DIAGNOSIS. Heavy giraffids with highly developed, often palmate, ossicones with tubercules on margins. Ossicones paired, borne on frontals and parietals, and on males only in some genera. Skull large, dolichocephalic to brachycephalic, with roof having highly developed sinus cavities in advanced forms. Cheek teeth moderately hypsodont to hypsodont, enamel heavily rugose. Lachrymal fossa in primitive forms (after Matthew 1929; Singer and Boné 1960; Godina 1962).

Genus *Prolibytherium* Arambourg 1961

DIAGNOSIS. A primitive sivathere of small size. Cranium narrow with very little facial flexion. Frontals support large, aliform ossicones that extend anteriorly in the supraorbital region and posteriorly over the parietal and occipital regions. Occipital condyles large and massive. Lachrymal fossae and paired lachrymal foramina present. Dentition relatively brachydont, cheek teeth smaller and more hypsodont than in *Zarafa,* enamel finely rugose (after Arambourg 1961; Hamilton 1973).

Prolibytherium magnieri Arambourg 1961

AGE. Early Miocene.

LOCALITY. Libya (19).

DISCUSSION. *Prolibytherium magnieri* represents the earliest known sivathere from any continent or horizon, as well as being the earliest known in Africa.[3] It is characterized by each ossicone being flattened, alate, or winglike, the smaller anterior and much larger posterior areas, together forming a "butterflylike" conformation when viewed in plan view. These ossicones are constructed with thicker antero- and posteromedial margins and a thinner

---

3 In chapter 24, Hamilton regards this species as Pecora, *incertae sedis*—Ed.

web between them. The surface of the ossicones is covered with fine radiating grooves with larger grooves near the base and along the thickened margins, suggestive of the vascularization for the dermal covering of the ossicones. *P. magnieri* is unique as a giraffid in that it possesses an antorbital lachrymal fossa, a feature that is absent in all other known giraffids but is present in cervids and some bovids and may be considered a primitive characteristic in this context.

The characters of the cheek teeth include stout cusps and ribs with weak metastyles and stout parastyles and suggest that the dentition is more advanced toward the *Palaeotragus* level than in *Zarafa*. Endostyles are present on each upper molar and an anterior ectostylid is present on M_3 (figure 25.6).

There was a strong acromial process on the scapular spine, and the bones of the limbs indicate that *Prolibytherium* was built for strength rather than speed or agility. The presence of large alate ossicones would necessitate a large head and neck and thus massive forelimbs to support these structures. If the ossicones were used in intraspecific contests, then the strength requirement would be further enhanced. Details of the skull and atlas indicate that considerable forces were encountered by the head of *Prolibytherium,* supporting the earlier deductions. The neural spines of the seventh cervical vertebra were vertical during life and suggest that the head and neck extended almost horizontally from the shoulders, with perhaps a slight rise toward the head. The stout vertebral centra and short axis indicate that the neck was short and heavy.

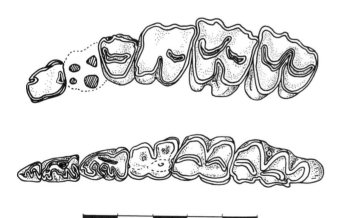

Figure 25.6 Cheek teeth of *Prolibytherium magnieri:* (*above*) left upper cheek teeth, very well worn and with P³ missing, (*below*) right lower cheek teeth, well worn, especially M₁. Both series redrawn from Hamilton's (1973) Gebel Zelten specimens. (Scale = 5 cm.)

Genus cf *Helladotherium* Gaudry 1860

cf *Helladotherium duvernoyi* Gaudry 1860

SYNONYMY. ?syn. of *Palaeotragus germaini* Arambourg 1959 or *Samotherium africanum* Churcher 1970.

AGE. Late Miocene to ?Pliocene.

LOCALITIES. Algeria (11); Tunisia (13).

DISCUSSION. Two localities only have yielded remains assigned to *Helladotherium duvernoyi:* Smendou in Algeria (Joleaud 1937) and Douaria in Tunisia (Roman and Solignac 1934; Joleaud 1937) and the materials from these sites appears to have never been described. Most other authors who have considered African Giraffidae have ignored the reports of this taxon (e.g., Arambourg 1948a, 1949; Singer and Boné 1960) although Arambourg (1952) mentions the occurrences and Romer (1966) notes a questionable Pliocene occurrence in Africa. Roman and Solignac state "*Helladotherium duvernoyi* Gaudry is represented by a set of three upper molars and part of P₄, in place in a cranial fragment, and by an incomplete mandible, with M₃ and M₂ with preserved buccal walls and fragmentary M₁ and P₄" (1934, p. 1650, translated).

Joleaud (1926) lists *Helladotherium* sp. for "Barbary" and mentions "The existence in this land of remains of *Helladotherium duvernoyi* Gaudry from Smendou, near Constantine, and from the west of Bizerte" (1937, p. 90, translated).

Since the descriptions in the literature of these materials are inadequate for any judgment on their taxonomic status and since no additional materials have been reported subsequently, the report of this taxon from Africa must be considered doubtful, especially in the light of the diagnostic materials described as *Palaeotragus germaini* by Arambourg (1959) from similar horizons. It may be that the *Helladotherium duvernoyi* specimens in reality belong within either *P. germaini* or even within *Samotherium,* both of which are reported from sites of late Miocene or Pliocene age in Africa. Since *Helladotherium* is not known to have possessed ossicones, it is even possible that African specimens ascribed to this genus represent females of one of these species.

Genus *Sivatherium* Falconer et Cautley 1832

SYNONYMY. *Indratherium* Pilgrim 1910, *Libytherium* Pomel 1892, *Griquatherium* Haughton 1922, and *Orangiatherium* van Hoepen 1932.

DIAGNOSIS. Gigantic giraffid; skull brachycephalic, face short, nasals retracted. Ossicones in two pairs in males, absent in females; anterior pair arising from frontals and massive palmate posterior pair arising from parietals. Males—frontals broad, flat, or slightly dished; nasals short, convex; sinuses extending throughout main stems of ossicones. Females—skull longer and lower, not markedly broadened, frontals convex. Deep muscular pits in temporal and supraoccipital areas. Facial region relatively short, anterior margin of orbit above M². Cranial region deeper than facial, especially in males; basicranial and palatal planes not parallel. Teeth large; enamel coarsely rugose. Body, neck, and limbs heavy, neck and limbs not elongated (after Colbert 1935; Singer and Boné 1960; Godina 1962).

DISCUSSION. The Sivatherinae of Africa are considered by Arambourg (1948b), Dietrich (1942), and Singer and Boné (1960) to comprise a single species referred to as *Sivatherium olduvaiense,* with the possible exception of *Libytherium maurusium,* which is represented by a single mandibular fragment in which only two premolars are present in the molar row. Arambourg (1960), at the suggestion of Singer and Boné (1960, p. 521) shows this to be an incorrect reconstruction and considers that the type of *L. maurusium* does not differ from other African *Sivatherium* in having only P₃–P₄ (figure 25.7). He then refers all the African sivatheres to the species *Libytherium maurusium,* considering that *Libytherium* is distinct from *Sivatherium* because of its less complex ossicone morphology and more massive distal limb elements.

Leakey (1965) refers all the African Pleistocene sivatheres to *Libytherium* but does not justify the separation of *Libytherium* from *Sivatherium* at the generic level. He refers those from Olduvai to *L. olduvaiense* and thus differs from Arambourg in not considering *L. maurusium* as the correct name. Godina (1962) lists two species of *Sivatherium*: *S. giganteum* from the upper Pliocene of the Siwaliks, India, and *S. olduvaiense* from the Pleistocene of Africa. Since *Sivatherium* Falconer et Cautley 1835 antedates *Libytherium* Pomel 1892, the former generic name is applied here; and since *L. maurusium* Pomel 1892 antedates *Helladotherium olduvaiense* Hopwood 1934 (= *Sivatherium olduvaiense* [Hopwood] 1934), the African Pleistocene sivatheres are referred to *S. maurusium* (Pomel) 1892.

Arambourg (1960) considers that *Libytherium* is not known from outside the Villafranchian, but this does not take into account Vaufrey's (1947) record from Olorgesaillie, Kenya, the middle Pleistocene occurrence at Ileret, East Turkana (Harris 1974), the Broken Hill, Zambia, occurrence (Leakey 1959; Cooke 1963), or the Florisbad, South Africa, specimens originally referred to *Orangiatherium* (Singer and Boné 1960). It is therefore likely that giraffid material from the early Pleistocene in Africa may represent *Sivatherium* or *Giraffa* (e.g., *jumae, gracilis, stillei,* or *pygmaea*).

Sivatherium maurusium (Pomel) 1892

SYNONYMY. *Libytherium maurusium* Pomel 1892, *Helladotherium olduvaiense* Hopwood 1934, *Sivatherium olduvaiense* (Hopwood) 1934, *Orangia-*

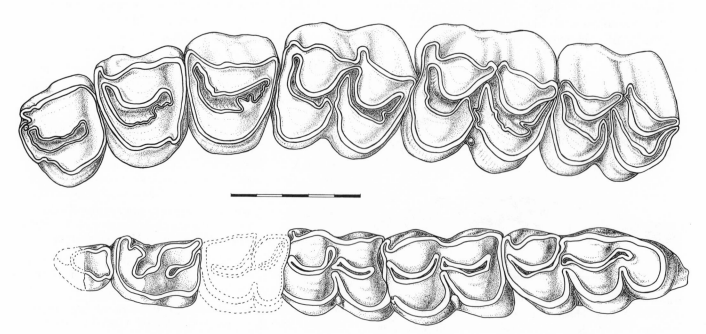

Figure 25.7 Cheek teeth of *Sivatherium maurusium:* (*above*) left upper cheek teeth, moderately well worn (drawn from the incomplete skull from Ileret, East Turkana [KNM-ER 797A], courtesy of J. M. Harris), (*below*) left lower cheek teeth, well worn (redrawn after Pomel [1893] and modified after Arambourg [1960].) (Scale = 5 cm.]

therium vanrhyni van Hoepen 1932, *Griquatherium cingulatum* Haughton 1922, *G. haughtoni* Cooke 1949, *Sivatherium hendeyi* Harris 1976.

AGE. Early Pliocene to middle Pleistocene.

LOCALITIES. Algeria (6, 8, 9, 10); Tunisia (12, ?13, 14, 16, 17, 18); Egypt (?20); Sudan (21); Chad (24, 25); Ethiopia (26); Kenya (27, 31, 33, 37, 38); Tanzania (41, 42 Beds I–VI); Malawi (45b); Zambia (46); Republic of South Africa (47, 49, 50, 51, 52, 53, 54, 56).

DISCUSSION. *Sivatherium maurusium* is the largest and most massive giraffid known from Africa. Its skeleton is not particularly variable, but the ossicones are highly variable in conformation. Singer and Boné (1960, p. 492–494) describe the ossicones and some of the observable variation. The anterior ossicones are short, with sinuses projecting slightly into the bases, laterally compressed and saddle-shaped, with a prominent superior margin to the anterior border and the posterior border concave; with one surface grooved, especially anteriorly. The posterior ossicones are large (560 to 640 mm on the posterior border) and stout (330 to 410 mm circumference at the base), with the base circular or pear-shaped in section. The anterior border is concave and twisted helically so that the anterior border becomes superior near the tip and the posterior becomes inferior. There is usually a flangelike projection, which may be variably developed, on the anterior border above the base of the ossicone, and there are usually three rounded projections along the anterior border above the flange. Harris (1976b) considers the ossicones of *S. maurusium* to be less palmate and more variably oriented than in *S. giganteum,* the anterior ossicones more ridgelike, and the face more prognathous.

The ossicones assigned to *S. maurusium* from sites such as Garet Ichkeul, Olduvai, Makapansgat (Limeworks), Tierfontein, and Elandsfontein show as wide a range of variation as that observable in the modern *Giraffa camelopardalis.* The degree of torsion varies from 0° to 90°, is usually mainly at the base, and the curvature is widely variable. The specimens from Olduvai cover most of this range, although most specimens show the anteroposterior curvature and only slight or medium helical torsion. A few are so twisted that the planes of parts of the palmation lie in markedly different orientations and the knobs come to resemble tines. The number and distribution of the knobs are also variable, with some individuals having smooth beams and a single protuberance only at the base of the ossicone, others having the knobs evenly distributed along the anterior border, and in those that are highly twisted, one or more lower knobs coming to lie posteriorly.

Harris (1976b) describes three major formations of the ossicones: type A, those that extend posterolaterally from the base, are only slightly twisted and bear anterior and lateral flanges and knobs; type B, those that extend laterally, then dorsally, and finally medially and anteriorly to a position anterodorsal of the base and bear flanges and knobs on the anterior and lateral margins and with a posteroventral knob; and type C, those that extend directly posterior but bear more numerous knobs on both medial and lateral margins.

It has been suggested that sexual dimorphism is present in the ossicones (Singer and Boné 1960), with males possessing the most twisted and varied ossicone conformation, and that the ossicones were used for display and possibly intraspecific sparring. However, ossicones are generally believed to have been absent in the females (see diagnosis of genus), although the presence of gracile ossicones suggests that some females may have borne less impressive cranial ornamentation (see Harris 1976b, pls. 12–18). However, another possibility is that Harris's three types represent young (type C), intermediate (type A), and mature (type B) individuals and are stages in the maturation of the ossicones that are complicated and obscured by both sexual variation and individual variation in form and distribution of the ornamentation.

The East Turkana deposits have yielded sivathere remains from all levels, including a complete skull from the upper member (*Metridiochoerus andrewsi* Zone) of the Koobi Fora Formation, of middle Pleistocene age (Harris 1974, 1976b). Its dental morphology and size are equivalent to those of *S. maurusium* and fall within the range of size variation given by Singer and Boné (1960). The skull is badly cracked but is still the only extensively preserved specimen of a skull of a late sivathere known from Africa. The anterofrontal ossicones arise from behind the orbit at the level of the nasal-frontal suture, are saddle-shaped in anteroposterior orientation, and extend posteriorly to the base of the posterior parietal ossicones. The latter are 880 mm (left) and 780 mm (right) long, and extend posteriorly and laterally, then dorsally, and finally medially and anteriorly, to give a bowed or arcuate conformation, and the left side is more curved than the right. The left posterior ossicone is closer in shape to those previously reported for *S. maurusium* (Singer and Boné 1960; Leakey 1965) and resembles Harris's (1976b) type A, while the right resembles type C in lateral view.

?Libytherium is reported from the middle Pliocene of Garet el Moluk Hill, Wadi Natrun, Egypt, by Stromer (1907). Since this would represent the earliest record of *Sivatherium* if the tentative identifi-

cation was confirmed and since Stromer remarks that the animal resembles both *Samotherium* and *Libytherium,* this record is considered probably to represent the former genus which is recorded from the late Miocene of Africa and early Pliocene of Europe and Asia Minor.

Harris (1976a) describes *S. hendeyi* from the early Pliocene of Quarry E at Langebaanweg, South Africa. This species is founded on a virtually complete left ossicone and supported by other cranial, dental, and postcranial specimens. It is differentiated from *S. maurusium* by separate conical anterior ossicones, as in *S. giganteum,* short posterior ossicones that extend laterally backward from the cranium and are unornamented by knobs and flanges or palmate digitations, and its longer metacarpals. While Harris admits that sivathere ossicones are quite variable in morphology, that the degree of or-

namentation may well be a function of the age of the individual, and that it is therefore possible that the Langebaanweg sivathere ossicones are all from immature individuals (Harris 1976a, p. 328), he considers that a new species is warranted. Since the postcranial and dental materials will appear undiagnostic for species, *S. hendeyi* is here included within *S. maurusium* until its establishment as more than a probable variant of *S. maurusium* can be certainly attested. The Langebaanweg sivathere appears to be somewhat divergent from typical *S. maurusium* in its ossicone and postcranial morphology and may represent an ancestral African sivathere population.

The status of *Griquatherium cingulatum, G. haughtoni,* and *Orangiatherium vanrhyni* require comment.

Griquatherium cingulatum was described by

		NORTHERN AFRICA							
		MOROCCO	ALGERIA	TUNISIA	LIBYA	EGYPT	SUDAN	NIGER	CHAD
HOLOCENE			7 ●					22 ●	
PLEISTOCENE	LATE	1 ●	10 □						
	MIDDLE		8 ● □						
			5 ●						25 ● □
	EARLY		9 ● □	16 □ ◉					24 □
				12 □			21 □		
PLIOCENE			6 □	14 □					23 ●
				17 □		20 ▽			
			11 ▣	13 ▣					
MIOCENE		2 △	3 △▽	15 △▽					
			4 △						
				19 ◆ ◇					

Figure 25.8 Stratigraphic distribution of African fossil Giraffidae. This chart is at best tentative since correlation of many sites depends on their correct stratigraphic determination. The numbers 1 through 57 represent the sites shown in figure 25.1 and listed in its legend. The taxa are indicated by symbols identified in the legend included in the second part of the figure.

Haughton (1922) on an isolated left M² or M³ (Cooke 1949) or M³ (Singer and Boné 1960) from Makapansgat (Limeworks), Transvaal. *G. haughtoni* was described by Cooke (1949) on an isolated tooth fragment from Cornelia, near Port Allen, Orange Free State, which he considered to represent a lower molar but which Singer and Boné (1960) considered was probably P³. Harris (pers. comm.) considers the fragment to represent an M₃, probably a left M₃. The teeth of neither of these species differ in size or general characters from others assigned to *Sivatherium maurusium* and may be included in this species (Singer and Boné 1960). Isolated teeth of ungulate herbivores probably provide a very unsound foundation on which to base specific and sometimes generic descriptions and even if individually unique may represent anomalous elements uncharacteristic of the normal individual. Such descriptions may be conveniently considered as *nomina vana* as they are inherently indeterminable (Wells 1959).

Orangiatherium vanrhyni was described by van Hoepen (1932) on ossicones from Tierfontein, near Port Allen, Orange Free State. Four molars (right M², M³, left M², and left M₁) were obtained from the same site and probably derive from the same individual. These teeth are extremely worn but are larger than other Olduvai or Elandsfontein specimens (Singer and Boné 1960). These specimens are assigned to *S. maurusium* because they differ only in size from known material.

Discussions and Conclusions

The history of the Giraffidae in Africa begins with *Zarafa zelteni* and *Prolibytherium magnieri* in North Africa (figures 25.8 and 25.9). Hamilton (1973) remarks on the resemblances between *Z. zelteni* and *Palaeomeryx* on the one hand and *Palaeotragus* on the other and suggests that *Z. zelteni* represents an early stage in the evolution of the Giraffidae from an Oligocene member of the family Dremotheriidae. Churcher (1970) follows Colbert (1935, 1936) and derives the less specialized *Palaeotragus* species (*P. primaevus* of Africa, *P. expectans* and *quadricornis*

		ETHIOPIA	KENYA	UGANDA	TANZANIA	MALAWI	ZAMBIA	TRANSVAAL	ORANGE FREE STATE	CAPE OF GOOD HOPE
EASTERN AND SOUTHERN AFRICA										
HOLOCENE										
PLEISTOCENE	LATE				43 ●		46 ●□	48 ●	52 ●□	
	MIDDLE		33 ●□ 28 ⊙⊙		42 ⊙○□ 44 ◐			49 □	51 ●□ 50 □	55 ● 53 □ 54 ○□
	EARLY		37 ⊙□ 27a ○⊙⊙●□ 31 □ 36 ⊙ 38 ⊙□ 27b ○⊕		42 ○⊙⊕	45b □		47 ○⊙□		
PLIOCENE		26 ○⊙⊙□		40 ▼	41 ⊙⊙⊕□	45a ◐□				56 ⊙□
MIOCENE			30 △ 34 ▲▽ 32 △▽⊙ 39 ▲ 29 ▲							

Legend

◆	*Zarafa zelteni*	▣	cf. *Helladotherium*
△	*Palaeotragus germaini*	□	*Sivatherium maurusium*
▲	*P. primaevus*	○	*Giraffa gracilis*
▼	*Okapia* sp.	⊙	*G. jumae*
◬	*Giraffokeryx*	⊚	*G. stillei*
▽	*Samotherium africanum* or sp.	⊕	*G. pygmaea*
◇	*Prolibytherium magnieri*	●	*G. camelopardalis*
		◐	*Giraffa* spp.

of Europe, and *P. tungurensis* of Asia) from Oligocene *Eumeryx* via early Miocene *Palaeomeryx*.

The discovery of *Z. zelteni* in African lower Miocene deposits eliminates *Palaeomeryx* and the Palaeomerycidae as the likely direct ancestors from the evolutionary line leading to the Giraffidae and suggests that the lineage may have originated in the Dremotheriidae and be close to *Eumeryx-Dremotherium-Zarafa-Palaeotragus*, although *Zarafa* is probably not on the direct line (see figure 25.9).

Heintz (1975) concludes that Africa was the original center for the evolution of the Palaeotraginae, which gave rise to the giraffe, sivathere, and okapi lineages. Thus the palaeotragine "stock" probably gave rise to varied Eurasian forms as well as providing the origins in Africa for the advanced giraffe and sivathere populations that came to be common in Africa and migrated through Arabia into southern Asia. The subsequent evolution of the Palaeotraginae in Africa led to the living and very similar *Okapia johnstoni* of the Ituri (Congo) forests and perhaps to the Miocene and Pliocene *Giraffa* spp. through a form such as *Samotherium*.

Arambourg's (1959) description of *Palaeotragus germaini* from the late Miocene of North Africa is of a palaeotragine that was larger than any other

known species ascribed to the genus and rivaled *Samotherium* in size. He considered *P. germaini* to be a type related to fossil forms such as *Palaeotragus* or *Samotherium,* which show analogous characters and proportions and which because of its height and elongation of the neck and limbs recalled the living *Giraffa*. However, *P. germaini* retains the strong metacone rib on the upper molars that *Samotherium* and *Giraffa* lack and thus is likely a convergent rather than ancestral giraffelike form.

Prolibytherium magnieri, from the same deposits as *Z. zelteni,* is a much more specialized giraffid and bears little resemblance to the palaeotragine conformation of *Palaeotragus* or *Okapia*. Hamilton (1973) considers that *Prolibytherium* was an ancestral sivathere, although not on the main lineage, and that it represents a separate line of evolution that originated in a *Dremotherium*-like form of middle to late Oligocene age from Asia. *Prolibytherium* itself might represent the degree of morphological complexity from which the Asiatic sivatheres subsequently evolved. The African late Pliocene *Sivatherium* possibly originated in Africa from an early sivathere in which there may have been as many as four ossicones, one pair on the frontals and another on the parietals, but evidence for this linkage is

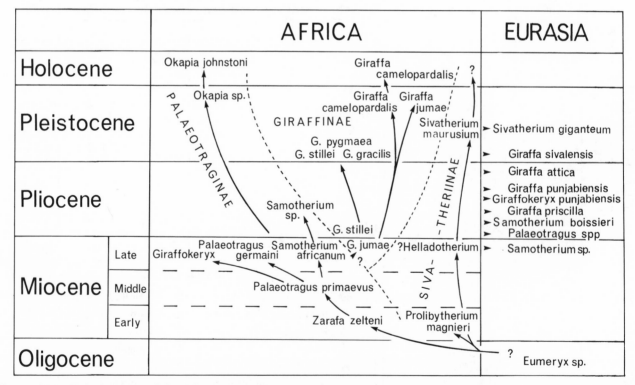

Figure 25.9 Relationships of the African Giraffidae, revised on the basis of an original Miocene radiation from which Europe and Asia were subsequently colonized. Arrows suggest possible lineages but not necessarily direct descent; dotted lines separate subfamilies; carets indicate possible migrations from northern Africa into Eurasia; and question marks indicate putative origins or occurrences.

lacking. Because the Asian *Helladotherium* (early to ?middle Pliocene) lacked or possessed only small ossicones, *Bramatherium* (late Pliocene) had paired upwardly directed frontoparietal ossicones and paired laterally directed parietal ossicones, and *Hydaspitherium* (late Pliocene) had two ossicones that were fused at their bases into a unit, while in *Vishnutherium* (late Pliocene) the condition is not recorded, it is unlikely that we can identify in an Asiatic genus the ancestral form from which *Sivatherium* may have derived. The African sivatheres therefore appear to belong to two indirectly related genera for which ancestral genera are hard to identify.

The four-ossiconed *Giraffokeryx* is now known from Africa from late Miocene deposits at Lothagam, Kenya. There are now no known major palaeotragine forms unique to non-African areas and there are three primitive forms (*Zarafa*, *P. primaevus*, and *Prolibytherium*) known only from Africa. The giraffid radiation thus appears to have begun in Africa.

The genus *Giraffa* is first known in the fossil record from the late Miocene of Africa. So little is known of these early giraffes that their direct origin and characteristics are still uncertain, although it is probable that they derived from a *Samotherium*-like ancestor as far as size and the characteristics of the dentition indicate. The *Giraffa* cf *jumae* from Baringo Basin, Kanapoi, and Langebaanweg represent the late Miocene to early Pliocene African population. The earliest Asiatic record of *Giraffa* is *G. priscilla* (Pilgrim 1910) from the early Pliocene Chinji Zones of the lower Siwaliks of India. Early Pliocene *G. attica* from the Pontian of Pikermi and Salonika in Greece and in Asiatic Turkey represent the earliest Eurasian Mediterranean populations. The wide geographic distribution of *Giraffa* by the end of the early Pliocene suggests that the genus had been in existence for some considerable time and had become well adapted to a widespread niche.

G. priscilla is founded on a left M³, a fragment of mandible, and a left M² and a right M³. Matthew concluded that it may represent a primitive giraffine, although he inclined to consider it more probably a palaeomerycine. Colbert (1935) recorded *Giraffa punjabiensis* from the Dhok Pathan Zone of the Middle Siwaliks, of late middle Pliocene age, and confirmed Pilgrim's (1910) original establishment of *G. priscilla*. This was a small animal, smaller than *G. camelopardalis* or *G. sivalensis*, and possibly similar in size to *G. gracilis*. Its premolars are narrower and its lower molars bear well-developed cingula and buccal ectostylids. *Giraffa sivalensis* (syn. *Camelopardalis affinis*) is known from the Upper Siwaliks of early Pleistocene age. It is slightly smaller

than *G. camelopardalis* and larger than *G. punjabiensis,* and both Pilgrim (1911) and Matthew (1929) considered it to be more advanced than the modern *G. camelopardalis.*

Giraffa attica from the early Pliocene of Pikermi, Salonika, and Turkey, is smaller than *G. gracilis*, although it somewhat resembles that species, and perhaps these two species and *G. sivalensis* are closely related.

Many of the early species, whether Asiatic (*G. priscilla* or *G. punjabiensis*), Eurasian (*G. attica*), or African (*G. gracilis*, *G. stillei*, or *G. pygmaea*), were more lightly built than the Pleistocene to Recent *G. camelopardalis* or *G. sivalensis*, although the extinct *G. jumae* apparently coexisted from late Miocene to late Pleistocene times with these smaller species. The modern African and now extinct late Pleistocene Asiatic giraffes derived from a savanna-dwelling ancestor, apparently resident in the Saharan-Arabian region during the late Miocene. *G. priscilla* would represent the Asian stock from which *G. punjabiensis, G. sivalensis,* and possibly the Chinese *Honanotherium* may have derived. *G. attica* would represent the Eurasian population corresponding to *G. sivalensis* in morphology. *G. camelopardalis* may derive from the *G. jumae* or *G. gracilis* stocks.

Unfortunately the variation in size and morphological characters of modern *G. camelopardalis* is such as to render any conclusions on the limits of variability of the extinct *Giraffa* populations inconclusive. It is not inconceivable that the *G. gracilis* and *G. jumae* specimens represent the lesser and greater limits of size and morphological variations of a single population, the modern descendants of which we call *G. camelopardalis.*

It is impossible on the present state of knowledge to do more than suggest stages in the giraffine lineage. A derivation from a condition of size and morphological complexity as seen in *Palaeotragus primaevus* through that seen in *Samotherium* to that of *G. gracilis* and ending in *G. camelopardalis* appears possible. *Palaeotragus germaini* would then represent a parallel or convergent evolution similar to *Samotherium* and *G. stillei*, a lineage of smaller giraffids that were probably distinct from the *G. gracilis*–*G. camelopardalis* lineage, while *G. jumae* may have represented the large individuals of that lineage. Excavations presently being carried out in the Siwalik deposits by many workers should throw further light on the chronology and relationships of the Asiatic Giraffidae with each other and with the African species.

The three types of giraffid known from Africa probably represent adaptations to three different modes of life. The smaller and more lightly built *Pa-*

laeotragus seems to have been a cursorial browser adapted to an open woodland environment, while the modern palaeotragine, *Okapia johnstoni,* had become progressively adapted to a closed forest environment. Large size might be a hindrance in the latter environment, and this factor may be responsible for *Okapia* being more or less similar in size to *Palaeotragus.*

The larger *Giraffa* is adapted to an open environment where neither trees nor broken ground prevents the animal from escaping any potential predator. The long neck and legs allow the animal to browse on foliage that is out of reach of most of the other large herbivores and ensure the animal an adequate supply of food even in times of scarcity.

The heavier and also larger sivatheres probably lived in habitats similar to that of giraffes. Meladze (1964) considered that sivatheres were savanna forms while Hamilton (1973) concludes that they were woodland or forest forms that fed on low vegetation or grasses on the woodland floor. The presence of *Sivatherium* and *Giraffa* spp. (including "*Okapia*" *stillei*) in the Olduvai and Laetolil deposits suggests that they may have shared the same habitat. The short neck of *Sivatherium* suggests a grazing or low browsing habit of foraging, similar to that of cattle or some of the larger African antelope or buffalo, and an adaptation to open woodland or bush veld.

It is possible, therefore, to consider that the palaeotragines were specializing toward life in a thick bush or forest environment, that sivatheres lived in more open woodland or forest in which they could move without obstruction because of their size, and that giraffes occupied the open orchard savanna or bush veld and that all of these forms browsed but that the palaeotragines and sivatheres both ate forbs, herbs, low woody plants, and grasses, while giraffes mainly ate the foliage located at higher levels of the taller woody plants and trees.

The variation in shape and the presence or absence of ossicones may reflect different habits or methods for intraspecific combat (figure 25.10). Sparring and fighting in *Giraffa* (Innis 1958; Dagg and Foster 1976) and *Okapia* (Walther 1960, 1962) involves lateral display leading to heavy blows of the head and ossicones to the sides of the neck and body of the opponent. Lateral display is observed in tylopods and ruminants to be the early stage in combat and may be considered primitive (Geist 1965). The change from lateral display to a swinging lateral buffet requires little behavioral modification and, in fact, would be enhanced by one opponent retreating. A long neck would help in delivering strong blows with the head and blunt ossicones would ensure that the opponent received bruises but was not badly damaged. The elongation of the neck in giraffes could then reflect selective advantages in both feeding and sparring habits and would have become a characteristic of the family by the late Miocene.

The flattened and expanded ossicones of the sivatheres may indicate that the display of these ornaments, either laterally or frontally, was more important than their use as bludgeons and that the increased size was a major factor in intimidating the opponent. Frontal presentation may have led to head-to-head contact that developed into a shoving match. A short massive neck supported by a similarly short and massive body would provide the purchase and mass necessary to compete in such a trial of strength, as they would resist compression, be less likely to bend, and require great force to dislodge. In the early Miocene *Prolibytherium magnieri* the morphology of the cervical vertebrae and head suggests that already the sivathere line had adopted this mode of contest, and the continued success of the line and the increased complexity of the ossicones reached in the Plio-Pleistocene Indian and African sivatheres indicates that the shoving form of intraspecific combat was not abandoned. Some of the sivatheres were without ossicones (e.g., *Helladotherium*) and in *Sivatherium* the female may lack ossicones. The genus *Indratherium* was originally described as separate from *Sivatherium* but is now generally considered to represent the sexually dimorphic female of *Sivatherium* (Bohlin 1926; Matthew 1929; Colbert 1935). Both these conditions suggest that the size of the creatures was sufficient to reduce predation and to free the ossicones from any interspecific function.

The pointed ossicones of modern *Okapia* (and probably of the extinct palaeotragines) are used in lateral display and the shafts for buffeting a rival opponent, but the points are brought into use only in interspecific combat with predators (Walther 1960, 1962). Dependence mainly on lateral display for intraspecific confrontations reduces the chance of damage to the opponent by the pointed ossicones and, when confronted with interspecific combat, maximizes the chance of seriously damaging the enemy. As the various lineages evolved, the larger giraffes and sivatheres reduced their susceptibility to predation and could depend on their large size, massiveness, and hooves to protect them. They could then specialize their intraspecific defenses. Because the okapis remained in a denser, forested habitat, their smaller size was an advantage that outweighed the advantages of increased size, and there-

Figure 25.10   Fossil and living giraffids: (*A*) *Zarafa zelteni* from Gebel Zelten, Libya; early Miocene. (*B*) *Samotherium* sp., after *S. boissieri* from Samos, Greece; Pliocene. (*C*) *Giraffa camelopardalis rothschildi,* Baringo or Rothschild's giraffe, from Uasin Gishu Plateau, Kenya, showing the three-horned condition; Recent. (*D*) *Prolibytherium magnieri,* from Gebel Zelten, Libya; early Miocene. (*E*) *Sivatherium maurusium,* from North, East, and South Africa; Pleistocene. (*F*) *Palaeotragus primaevus,* from Fort Ternan, Kenya; middle Miocene. (*G*) *Okapia johnstoni,* okapi from the Semliki Forest, Congo Basin, Zaire; Recent. (*H*) *Giraffa jumae,* extinct giraffe, from Rawe, Kenya, and Olduvai Bed II, Tanzania; Pliocene and Pleistocene. (*I*) *Giraffa camelopardalis camelopardalis,* Nubian giraffe, from the Sudan; Recent. Heads not drawn to same scale; (*E*), (*G*), and (*I*) drawn to scale.

fore they retained the pointed ossicones suitable for this habitat. Some of the palaeotragines (e.g., *Giraf-fokeryx punjabiensis, Palaeotragus quadricornis*) "experimented" with four ossicones, as did some of the sivatheres, but without increasing their size to any great extent. Presumably these forms still used their ossicones for display and for buffeting, since the shafts were still more vertical than lateral and the tips were still pointed.

I am most appreciative for all the assistance that I have received from W. R. Hamilton of the British Museum of Natural History, London, and from J. M. Harris of the National Museum of Kenya, Nairobi, both of whom allowed me to study their reports on African Giraffidae when they were still unpublished.

J. M. Harris also spent many hours discussing the taxonomy of the East African species of *Giraffa,* the morphology of *Giraffa* and *Sivatherium,* and the South African giraffid records. I am indebted to H. B. S. Cooke of the Department of Geology, Dalhousie University, Halifax, Nova Scotia, for assisting me with the names of the geological stages and for editorial comments, many of which he offered before he became an editor for this volume. I thank R. Singer of the Department of Anatomy, University of Chicago, Illinois, and P. Robinson of the University of Colorado Museum, Boulder, Colorado, for reading drafts of the paper and commenting on various aspects of the geology and palaeontology of Africa and the Giraffidae. P. Robinson allowed me to examine specimens from Tunisia that have assisted me in my comprehension of the evolution of the African Giraffidae.

I wish to thank W. J. White and T. Woolf, my assistants during the period when the paper was first prepared, for their help with geographic records and the bibliography, and M. L. Richardson, my assistant during the final preparation and revision, for her help and for allowing me to examine and cite materials in her care and for her critical reading of the paper.

The chapter was originally completed in June 1974, and finally revised in November 1976.

References

Abadie, J., J. Barbeau, and Y. Coppens. 1959. Une faune de vertébrés villafranchiens au Tchad. *C. R. hebd. Séances Acad. Sci.,* Sér. D, 248:3328–3330

Aguirre, E., and P. Leakey. 1974. Nakali, nueva fauna de *Hipparion* del Rift Valley de Kenya. *Estud. Geol.* 30: 219–227.

Arambourg, C. 1947. Contribution à l'étude géologique et paléontologique du bassin du Lac Rudolphe et de la basse vallée de l'Omo. Deuxième partie, Paléontologie. *Miss. Scient. Omo, 1932–1933, I, Géol.-Anthrop.:* 232–562.

———. 1948a. Les vertébrés fossiles des formations des plateaux Constantinois (Note préliminaire). *Bull. Soc. Hist. Nat. Afr. Nord* 38:45–48.

———. 1948b. Un sivatheriné nord africain: *Libytherium maurusium* Pomel. *C. R. Soc. Géol. France* 1948:177–179.

———. 1948c. Les mammifères pléistocène d'Afrique. *Bull. Soc. Géol. France,* Sér. 5, 17:301–310.

———. 1949. Les gisements de vertébrés villafranchiens de l'Afrique du Nord. *Bull. Soc. Géol. France,* Sér. 5, 19:195–203.

———. 1951. Observations sur les couches à *Hipparion* de la vallée de l'Oued el Hammam (Algérie) sur l'époque d'apparition de la faune de vertébrés dite "pontienne." *C. R. hebd. Séances Acad. Sci.,* Sér. D, 232:2464–2466.

———. 1952. La paléontologie des vertébrés en Afrique du Nord Française. *Congr. Géol. Internat., 19: Monog. Région., s.i., Algérie;* hors série, 1–63.

———. 1954. La faune à *Hipparion* de l'Oued el Hammam (Algérie). Congr. Géol. Internat., 19 Sess., Alger 1952. *Assoc. Serv. Géol. Afr.,* Pt. 2, fasc. 21:295–302.

———. 1958. La faune de vertébrés Miocènes de l'Oued el Hammam (Oran, Algérie). *C. R. Soc. Géol. France* 6: 116–119.

———. 1959. Vertébrés continentaux du Miocène supérieur de l'Afrique du Nord. *Publ. Serv. Carte. Géol. Algérie* (Alger), N.S., *Paléont. Mém.,* No. 4:1–161.

———. 1960. Précisions nouvelles sur *Libytherium maurusium* Pomel, Giraffidé du Villafranchien d'Afrique. *Bull. Soc. Géol. France,* Sér. 7, 2:888–894.

———. 1961. *Prolibytherium magnieri,* un Velléricorne nouveau du Burdigalien de Libye (Note préliminaire). *C. R. Somm. Géol. France* 1961(3):61–63.

Arambourg, C., and M. Arnauld. 1949. Note sur les fouilles paléontologiques exécutées en 1947–48 et 1949 dans les gisements villafranchiens de la Garet Ichkeul. *Bull. Soc. Sci. Nat. Tunisie* 2:149–157.

Arambourg, C., and R. Coque. 1958. Le gisement villa-franchien de l'Aïn Brimba (Sud-Tunisien) et sa faune. *Bull. Soc. Géol. France* 8:607–614.

Bishop, W. W., G. R. Chapman, A. Hill, and J. A. Miller. 1971. Succession of Cainozoic vertebrate assemblages from the northern Kenya Rift Valley. *Nature* 233: 389–394.

Bishop, W. W., and M. H. L. Pickford. 1975. Geology, fauna and palaeoenvironment of the Ngorora Formation, Kenya Rift Valley. *Nature* 254:185–192.

Bohlin, B. 1926. Die Familie Giraffidae. *Palaeont. Sinica,* Ser. C, 4:1–179.

Boné, E. L., and R. Singer. 1965. *Hipparion* from Lange-baanweg, Cape Province, and a revision of the genus in Africa. *Ann. S. Afr. Mus.* 48:273–397.

Burollet, P. F. 1956. Contribution à l'étude stratigraphique de la Tunisie Centrale. *Ann. Mines Géol. Tunisie* 18:1–350.

Churcher, C. S. 1970. Two new upper Miocene giraffids from Fort Ternan, Kenya, East Africa: *Palaeotragus primaevus* n. sp. and *Samotherium africanum* n. sp. In L. S. B. Leakey and R. J. G. Savage, eds., *Fossil Vertebrates of Africa.* London: Academic Press, Vol. 2, pp. 1–109.

———. 1974. *Sivatherium maurusium* (Pomel) from the Swartkrans Australopithecine site, Transvaal (Mammalia, Giraffidae). *Ann. Transv. Mus.* 29:65–70.

Colbert, E. H. 1933. A skull and mandible of *Giraffokeryx punjabiensis* Pilgrim. *Am. Mus. Novit.* No. 632:1–14.

———. 1935. Siwalik mammals in the American Museum of Natural History. *Trans. Am. Phil. Soc.,* N.S., 26: 1–401.

———. 1936. *Palaeotragus* in the Tung Gur Formation of Mongolia. *Am. Mus. Novit.* No. 894:1–17.

Cooke, H. B. S. 1949. Fossil mammals of the Vaal River deposits. *Mem. Geol. Surv. S. Afr.* 35:1–117.

———. 1955. Some fossil mammals in the South African Museum collections. *Ann. S. Afr. Mus.* 42:161–168.

———. 1963. Pleistocene mammal faunas of Africa, with particular reference to southern Africa. In F. C. Howell and F. Bourlière, eds., *African ecology and human evolution.* Chicago: Aldine, pp. 65–116.

———. 1974. The fossil mammals of Cornelia, O.F.S., South Africa. In K. W. Butzer, J. D. Clark, and H. B. S. Cooke, The geology, archaeology and fossil mammals of the Cornelia Beds, O.F.S. *Mem. Nas. Mus., Bloemfontein* 9:63–88.

Cooke, H. B. S., and S. C. Coryndon. 1970. Pleistocene mammals from the Kaiso Formation and other related deposits in Uganda. In L. B. S. Leakey and R. J. G. Savage, eds., *Fossil Vertebrates of Africa.* London: Academic Press, vol. 2, pp. 107–224.

Cooke, H. B. S., and L. H. Wells. 1947. Fossil mammals from the Makapan Valley, Potgietersrust. III. Giraffidae. *S. Afr. J. Sci.* 43:232–235.

Coppens, Y. 1960. Le quaternaire fossilifère de Koro-Toro (Tchad). Résultats d'une première mission. *C. R. hebd. Séances Acad. Sci.,* Sér. D, 251:2385–2386.

———. 1962. Deux gisements de vertébrés villafranchiens

du Tchad. *Actes IV Pan-Afr. Congr. Préhist.,* Leopold-ville, 1959:229–315.

_____. 1967. Les faunes de vertébrés quaternaires du Tchad. In W. W. Bishop and J. D. Clark, eds. *Background to evolution in Africa.* Chicago: University of Chicago Press, pp. 89–97.

_____. 1971. Les vertébrés villafranchiens de Tunisie: gisements nouveaux, signification. *C. R. hebd. Séances Acad. Sci.,* Sér. D, 273:51–54.

Coque, R. 1957. Découverte d'un gisement de mammifères villafranchiens dans le Sud-Tunisien. *C. R. hebd. Séances Acad. Sci.,* Sér. D, 245:1069–1071.

Coryndon, S. C. 1966. Preliminary report on some fossils from the Chiwondo Beds of the Karonga District, Malawi. *Am. Anthrop.* 68:59–66.

Cross, M. W., and V. J. Maglio. 1975. A bibliography of the fossil mammals of Africa, 1950–1972. Princeton: Princeton Univ., 291 pp.

Crusafont, M. 1952. Los jiráfidos fósiles de España. *Mem. Comm. Inst. Geol. Barcelona* 8:1–239.

Dagg, A. I., and J. B. Foster. 1976. The giraffe: its biology, behavior, and ecology. New York: Van Nostrand Reinhold, 210 pp.

Dietrich, W. O. 1941. Die säugetierpaläontologischen Ergebnisse der Kohl-Larsen'schen Expedition, 1937–1939 in nördlichen Deutsch-Ostafrika. *Zentbl. Min. Geol. Paläont.* 1941B:217–223.

_____. 1942. Ältestquartäre Säugetiere aus der südlichen Serengeti, Deutsch-Ostafrika. *Palaeontographica,* Abt. A, 94:1–133.

Dreyer, T. F., and A. Lyle. 1931. *New fossil mammals and man from South Africa.* Bloemfontein: Dept. Zool., Grey Univ. Coll., 60 pp.

Flamand, G. B. M. 1902 (1901). Sur l'utilisation, comme instruments néolithiques, de coquilles fossiles à taille intentionale (Littoral du Nord-Africain). *C. R. Assoc. Franç. Avanc. Sci.* 30:729–734.

Geist, V. 1965. The evolution of horn-like organs. *Behaviour* 27:175–214.

_____. 1971. The relation of social evolution and dispersal in ungulates during the Pleistocene, with emphasis on the Old World deer and the genus *Bison. Quartern. Res.* 1:285–315.

Godina, A. Y. 1962. Family Giraffidae Gray, 1821. Giraffes. In *Osnovy paleontologii (Fundamentals of paleontology),* Orlov, Yu A., chief ed.; Vol. 13, Gromova, V. I., Ed. Moskva: Akademiya Nauk SSSR, pp. 379–383. Trans. Israel Program Sci. Translat., Jerusalem, 1968, pp. 523–530.

Hamilton, W. R. 1973. The lower Miocene ruminants of Gebel Zelten, Libya. *Bull. Brit. Mus. Nat. Hist.* (Geol.) 21:73–150.

Harris, J. M. 1974. Orientation and variability in the ossicones of African Sivatheriinae (Mammalia; Giraffidae). *Ann. S. Afr. Mus.* 65(6):189–198.

_____. 1976a. Pliocene Giraffoidea (Mammalia, Artiodactyla) from the Cape Province. *Ann. S. Afr. Mus.* 69(12):325–353.

_____. 1976b. Pleistocene Giraffidae (Mammalia, Artiodactyla) from East Rudolf, Kenya. In R. J. G. Savage, ed., *Fossil Vertebrates of Africa.* London: Academic Press, vol. 4, pp. 283–332.

Haughton, S. H. 1922. A note on some fossils from the Vaal River gravels. *Trans. Geol. Soc. S. Afr.* 24:11–16.

Heintz, E. 1975. Origine, migration et paléobiogéographie des Palaeotraginae (Giraffidae, Artiodactyla) antévallésiens de l'ancien monde. *Coll. Internat., C.N.R.S.,* No. 218:723–730.

Hendey, Q. B. 1968. New Quaternary fossil sites near Swartklip, Cape Province. *Ann. S. Afr. Mus.* 52(2):43–73.

_____. 1969. Quaternary vertebrate fossil sites in the southwestern Cape Province. *S. Afr. Archaeol. Bull.* 24(3 & 4):96–105.

_____. 1970a. A review of the geology and palaeontology of the Plio/Pleistocene deposits at Langebaanweg, Cape Province. *Ann. S. Afr. Mus.* 56(2):75–117.

_____. 1970b. The age of the fossiliferous deposits at Langebaanweg, Cape Province. *Ann. S. Afr. Mus.* 56(3):119–131.

_____. 1972. Further observations on the age of the mammalian fauna from Langebaanweg, Cape Province. In E. M. van Zinderen Bakker, Sr., ed., *Palaeoecology of Africa, the surrounding islands and Antarctica.* Cape Town: Balkema, vol. 6, pp. 172–176.

_____. 1974a. The late Cenozoic Carnivora of the southwestern Cape Province. *Ann. S. Afr. Mus.* 63:1–369.

_____. 1974b. Faunal dating of the late Cenozoic of Southern Africa, with special reference to the Carnivora. *Quartern. Res.* 4(2):49–161.

_____. 1976. The Pliocene fossil occurrences in "E" Quarry, Langebaanweg, South Africa. *Ann. S. Afr. Mus.* 69(9):215–217.

Hopwood, A. T. 1928. Mammalia. In W. P. Pycraft et al., eds., *Rhodesian man and other associated remains.* London: Trustees of the British Museum, pp. 70–73.

_____. 1934. New fossil mammals from Olduvai, Tanganyika Territory. *Ann. Mag. Nat. Hist. Lond.,* Ser. 10, 14:546–550.

_____. 1936. New and little-known fossil mammals from the Pleistocene of Kenya Colony and Tanganyika Territory-I. *Ann. Mag. Nat. Hist. Lond.,* Ser. 10, 17:636–641.

Hopwood, A. T., and J. P. Hollyfield. 1954. An annotated bibliography of the fossil mammals of Africa (1742–1950). *Brit. Mus. Nat. Hist., Fossil Mammals of Africa,* No. 8:1–194.

Howe, B., and H. L. Movius. 1947. A stone age cave in Tangier. *Pap. Peabody Mus. Amer. Archaeol. Ethnol. Harvard* 28(1):1–23.

Innis, A. 1958. The behaviour of the giraffe, *Giraffa camelopardalis,* in the eastern Transvaal. *Proc. Zool. Soc. London* 131(2):245–278.

Jaeger, J. G., J. Michaux, and B. David. 1973. Biochronologie du Miocène moyen et supérieur continental du Maghreb. *C. R. hebd. Séances Acad. Sci.,* Sér. D, 277:2477–2480.

Joleaud, L. 1926. Études de géographie zoologique sur la Berbérie (Les ruminants cervicornes). *Glasn. Hrv. Pirodasl. Dr.,* Zagreb, 37/38:263–322.

———. 1934. Vertébrés subfossiles de L'Azaoua (Colonie du Niger). *C. R. hebd. Séances Acad. Sci.,* Sér. D, 198:599–601.

———. 1936. Gisements de vertébrés quaternaires du Sahara. *Bull. Soc. Hist. Nat. Afr. Nord* 26:23–29.

———. 1937. Remarques sur les giraffidés fossiles d'Afrique. *Mammalia* 1:85–96.

Kent, P. E. 1942a. A note on Pleistocene deposits near Lake Manyara, Tanganyika. *Geol. Mag.* 79:72–77.

———. 1942b. The Pleistocene beds of Kanam and Kanjera, Kavirondo, Kenya. *Geol. Mag.* 79:117–132.

Kohl-Larsen, L. 1939. Vorläufiger Bericht über meine Afrika-Expedition 1937–1939. *Forsch. Fortschr.* 15:339–340.

Lankester, E. R. 1901. On *Okapia,* a new genus of Giraffidae, from central Africa. *Trans. Zool. Soc. London* 16(1):279–314.

———. 1910. *Monograph of the okapi.* Atlas. London: Trustees of the British Museum, pp. i–viii.

Lavocat, R. 1961. Le gisement de vertébrés de Beni Mellal (Maroc). Étude systématique de la faune de mammifères et conclusions générales. *Notes Mém. Serv. Géol. Maroc* 155:29–94 & 109–145.

Leakey, L. S. B. 1959. A preliminary reassessment of the fossil fauna from Broken Hill, Northern Rhodesia, pp. 225–231. In J. D. Clark, ed., Further excavations at Broken Hill, Northern Rhodesia. *J. Roy. Anthrop. Inst. Gr. Brit.* 89:201–232.

———. 1965. *Olduvai Gorge 1951–1961, Vol. 1. A preliminary report on the geology and fauna.* Cambridge: Cambridge University Press, 109 pp.

———. 1970. Additional information on the status of *Giraffa jumae* from East Africa. In L. B. S. Leakey and R. J. G. Savage, eds., *Fossil Vertebrates of Africa.* London: Academic Press, vol. 2, pp. 325–330.

Madden, C. T. 1972. Miocene mammals, stratigraphy and environment of Muruorot Hill, Kenya. *Paleobios* 14:1–12.

Maglio, V. J. 1971. Vertebrate faunas from the Kubi Algi, Koobi Fora and Ileret Areas, East Rudolf, Kenya. *Nature* 231:248–249.

———. 1972. Vertebrate faunas and chronology of the hominid-bearing sediments east of Lake Rudolf, Kenya. *Nature* 239:379–385.

———. 1974. Late Tertiary fossil vertebrate successions in the northern Gregory Rift, East Africa. *Ann. Geol. Surv. Egypt* 4:269–286.

Matthew, W. D. 1929. Critical observations upon Siwalik mammals. *Bull. Am. Mus. Nat. Hist.* 56:437–560.

Mawby, J. E. 1970. Fossil vertebrates from northern Malawi: preliminary report. *Quaternaria* 13:319–323.

Meladze, G. K. 1964. [About the paleobiological study of Sivatheriinae.] *Soob. Akad. Nauk Gruz. SSR* 33(3):567–600.

Oakley, K. P. 1954. Study tour of early hominid sites in southern Africa, 1953. *S. Afr. Archaeol. Bull.* 9(35):75–87.

Pallary, P. 1900. Note sur le girafe et le chameau du quaternaire algérien. *Bull. Soc. Géol. France,* Sér. 3, 28:908–909.

Pickford, M. 1975. Late Miocene sediments and fossils from the northern Kenya Rift Valley. *Nature* 256:279–284.

Pilgrim, G. E. 1910. Notices of new mammalian genera and species from the Tertiaries of India. *Rec. Geol. Surv. India* 40(1):63–71.

———. 1911. The fossil Giraffidae of India. *Palaeont. Indica,* N.S. 4(1):1–29.

Pomel, A. 1892. Sur le *Libytherium maurusium,* grand ruminant du terrain pliocène de l'Algérie. *C. R. hebd. Séances Acad. Sci.,* Sér. D, 115:100–102.

———. 1893. Caméliens et Cervidés. *Carte Géol. Algérie, Paléont., Monogr.,* pp. 1–52.

Reck, H., and L. Kohl-Larsen. 1936. Erster Uberblick über die jungdiluvialen Tier- und Menschenfunde Dr. Kohl-Larsens im Nordostlichen Teil des Njarasa-Grabens (Östafrika) und die Geologischen Verhältnisse des Fundgebietes. *Geol. Rundschau* 27:401–441.

Robinson, P., and C. C. Black. 1969. Note préliminaire sur les vertébrés fossiles du Vindobonien (formation Béglia), du Bled Dourah, Gouvernorat de Gafsa, Tunisie. *Notes Serv. Géol. Tunisie* 31:67–70.

———. 1974. Vertebrate faunas from the Neogene of Tunisia. *Ann. Geol. Surv. Egypt* 4:319–332.

Roman, F. 1931. Description de la faune pontique de Djerid (El Hamma et Nefta). *Ann. Univ. Lyon,* N.S., Sci. Med., 1(48):30–42.

Roman, F., and M. Solignac. 1934. Découverte d'un gisement de mammifères pontiens à Douaria (Tunisie septentrionale). *C. R. hebd. Séances Acad. Sci.,* Sér D, 199:1649–1650.

Romer, A. S. 1928. Pleistocene mammals of Algeria. Fauna of the paleolithic station of Mechta-el-Arbi. *Bull. Logan Mus.* 1:80–163.

———. 1966. *Vertebrate paleontology.* Edit. III. Chicago: University of Chicago Press, 468 pp.

Singer, R. 1957. Recent discoveries in southern Africa. *Mitteil. Anthrop. Gesellsch. Wien* 88/89:133–136.

Singer, R., and E. L. Boné. 1960. Modern giraffes and the fossil giraffids of Africa. *Ann. S. Afr. Mus.* 45(4):375–603.

Singer, R., and J. R. Crawford. 1958. The significance of the archaeological discoveries at Hopefield. *J. Roy. Anthrop. Inst. Gr. Brit.* 88(1):11–19.

Stromer, E. 1907. Fossile Wirbeltier-Reste aus dem Uadi Fâregh und Uadi Natrûn in Ägypten. *Abh. Senkenb. Naturf. Ges.* 29(2):97–132.

Suess, M. 1932. Sur la présence de gastropodes et de vertébrés dans le grès de Bou Hanifa, feuille de Mascara (départment d'Oran). *C. R. hebd. Séances Acad. Sci.,* Sér. D, 194:1970–1972.

Thenius, E. 1967. Die Giraffen und ihre Vorfahren. *Kosmos* 63(4):160–164.

Trevor, J. C. 1953. A new fossil man from South Africa. *Nature* 172:652.

van Hoepen, E. C. N. 1932. Voorlopige beskryving van Vrystaatse soogdiere. *Paleont. Navors. Nas. Mus., Bloemfontein* 2:63–65.

Vaufrey, R. 1947. Olorgesaillie. Un site acheuléen d'une exceptionelle richesse. *Anthropologie* 51:367.

Walker, E. P. 1969. Mammals of the world, vol. II. Baltimore: Johns Hopkins Press, pp. 647–1500.

Walther, F. 1960. "Antilopenhafte" Verhalfensweisen im Paarungszeremoniell des Okapi (*Okapia johnstoni* Sclater, 1901). *Zeits. Tierpsychol.* 17:188–210.

_____. 1962. Uber ein Spiel bei *Okapia johnstoni*. *Zeits. Säugetierk.* 27:245–251.

Wells, L. H. 1959. The nomenclature of the South African fossil equids. *S. Afr. J. Sci.* 55(3)64–66.

26

Tragulidae and Camelidae

Alan W. Gentry

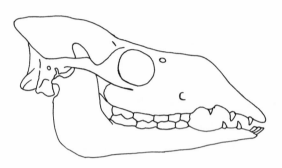

Families Tragulidae and Gelocidae

The main living subgroup of the Ruminantia, the infraorder Pecora, comprises the families Giraffidae (giraffe and okapi), Cervidae (deer), Antilocapridae (north American pronghorn), and Bovidae (cattle, sheep, goats, and antelopes). Also surviving today is the less advanced family Tragulidae (Asian chevrotains *Tragulus* and African water chevrotain *Hyemoschus*) belonging to the infraorder Tragulina. Apart from some poorly known Eocene-Oligocene European fossils referred to the family Amphimerycidae, fossil tragulines are referable either to the surviving and more primitive family Tragulidae or to a more advanced extinct family Gelocidae. Both families occur in the African fossil record, where nearly all occurrences are in East Africa.

The Tragulidae may be defined as ruminants lacking frontal appendages, with a bunodont or primitive selenodont dentition, large upper canines in males, with fusion of the two central metapodials to form a cannon bone in the hind leg and sometimes in the front leg as well, complete side metapodials, the ectocuneiform joined with the naviculo-cuboid, and often a bony carapace above the pelvic girdle in males.

The mainly Asiatic Gelocidae differ from tragulids in having more advanced selenodonty, strong cingula on the cheek teeth, P_1 separate from the other premolars by a short diastema, no fusion of ectocuneiform with naviculo-cuboid, and the side metatapodials with entirely atrophied shafts. *Gelocus whitworthi* Hamilton (1973, p. 140), known by mandibular pieces from Songhor and Rusinga, is the first member of this family to be recognized in Africa. Hopwood (1929, p. 5) had referred a much worn dentition from the Miocene of Namibia to "tragulid indet. cf *Bachitherium,*" but the family status of this piece cannot be regarded as established.

Fossil Tragulids of Africa

All African fossil tragulids are referred to *Dorcatherium,* a Miocene form only doubtfully distinguishable from *Hyemoschus* by the probable presence of a preorbital fossa (now lost), the premaxilla contacting the nasal, more compressed bladelike lower premolars, and the occasional persistence of P_1. It differs from *Tragulus* only by the probable preorbital fossa, the occasional P_1, and the lack of a cannon bone in the forelimb. There seems every likelihood that *Tragulus* and *Hyemoschus* will eventually be sunk in *Dorcatherium*. No described African *Dorcatherium* is as large as the European Vindobonian species *D. peneckei* Hofmann (including *D.*

rogeri?) with M_1-M_3 up to 50 mm long, or the Siwaliks Chinji species *D. majus* Lydekker. On the other hand, African *Dorcatherium* includes species smaller than those fossilized in the Miocene of other continents and as small as the living *Tragulus* of tropical Asia.

Dorcatherium chappuisi Arambourg (1933) was described on a mandible from Losodok (= Moruorot), and Whitworth (1958) referred to it much material from Rusinga (mainly Hiwegi Formation as understood by van Couvering and Miller [1969] but also from the Kiahera Formation), Mfwanganu, and possibly Maboko. Gentry (1970, p. 301) thought it was present at Fort Ternan, but perhaps this material is from a more advanced species. *D. chappuisi* has M_1-M_3 with an occlusal length of about 43 mm, about the same size as in *D. naui* Kaup, the type species of the genus from Eppelsheim, Germany (which includes *D. crassum* from Sansan, France). However, *D. chappuisi* differs from the latter by its more bunodont molars, less prominent mesostyle on the upper molars, and a shallower mandibular ramus, all of which are more primitive characters. It also differs by having no basal pillars on the lower molars and no cingulum on the upper or lower molars. *D. chappuisi* differs by the same characters, insofar as they are known, from the similarly sized Chinji to Dhok Pathan *D. minus* Lydekker. It is interesting that this African form is so very distinctive; its apparently advanced condition in not having basal pillars or cingula suggests that it could be generically separated from the main *Tragulus-Dorcatherium-Hyemoschus* stock. It is also interesting that it retained an independent ectocuneiform in its tarsus, as noted by Whitworth (1958, p. 41), unlike other tragulids. In many of its characters it is like the Chinji and Nagri *Dorcabune* Pilgrim, which is not represented in Africa and which has been taken as a tragulid (Colbert 1935, pp. 301–306). However, the alternative possibility that *Dorcabune* is an anthracothere has not yet been eliminated.

The extent to which P_1 occurs in Miocene tragulids is not known. It is supposed to be absent in most of the Sansan specimens but occurs in some and also in the Eppelsheim holotype of *Dorcatherium naui*. Probably this tooth was tending to become smaller and to fall out earlier in life, and this process may have gone on at different evolutionary rates in different stocks. Without large samples for study it is not possible to determine the significance of this feature. The holotype of *D. chappuisi* has a larger P_1 than the holotype of *D. naui*.

A large tragulid other than *D. chappuisi* has been found at Bukwa, Uganda (Walker 1969, p. 592).

Hopwood (1929, p. 5) recorded a large tragulid from Namibia.

D. libiensis Hamilton (1973) from Gebel Zelten, Libya, has teeth almost as large as in *D. chappuisi* but without the latter's distinctive characters.

D. pigotti Whitworth (1958) is a smaller species than *D. chappuisi* with M_1-M_3 having an occlusal length of about 30 mm. It is recorded from the Hiwegi and Kiahera Formations of Rusinga, Mfwanganu, Moruorot (Madden 1972), Karungu, and Ombo, all in Kenya, and from Bukwa, Uganda. It is slightly smaller than the European fossils named *D. puyhauberti* Arambourg and Piveteau 1929 from Salonica.

D. parvum Whitworth (1958) is a still smaller species with M_1-M_3 having an occlusal length of about 20 mm. Some individuals are known to have lacked P_1. It is known from the Hiwegi and Kiahera Formations of Rusinga, Karungu, Maboko, and Moruorot in Kenya and from Bukwa in Uganda.

D. songhorensis Whitworth (1958) is the Songhor species in which the M_1-M_3 is about 24 mm long and thus intermediate in size between *D. pigotti* and *D. parvum*.

A second species of *Dorcatherium* at Fort Ternan, smaller than *D. chappuisi,* was recorded by Gentry (1970, p. 302). Bishop and Pickford (1975) recorded two species of tragulids at Ngorora, Kenya.

Nobody has yet investigated the possible relationship of these small African tragulids to the living tropical Asian *Tragulus*. At the present time the only surviving African tragulid, *Hyemoschus aquaticus,* has become restricted to parts of West Africa. The tragulids disappeared from the Palaearctic region in the Miocene, and their place was taken, according to Flerov (1971), by the cold-adapted musk deer, *Moschus,* an isolated form with bovidlike teeth.

Family Camelidae

Camels and llamas are placed in a distinct suborder, Tylopoda, of the Artiodactyla. There are two living camels, the two-humped Bactrian camel, *Camelus bactrianus,* and the one-humped dromedary, which is not known in the wild state. The dromedary is found in North Africa and parts of southwestern Asia and is distinguished from the Bactrian camel mainly by its orbits being directed more downward (at least in many skulls), a smaller or completely absent ethmoidal fissure, less relative displacement of anterior and posterior parts of the medial wall of the lower molars (Grattard, Howell, and Coppens 1976,

p. 269), and more slender limb bones with relatively longer distal elements.

Until a decade ago camels were believed to have been introduced into sub-Saharan Africa relatively recently. Fossil evidence for their earlier presence is only just coming to light.

Fossil Camels of Africa

An extinct wild camel is known in Africa as early as the Pliocene. Howell, Fichter, and Wolff (1969) recorded a damaged lower molar from Member B and a distal metatarsal from Member F in the Shungura Formation, lower Omo Valley, Ethiopia. Subsequently Grattard, Howell, and Coppens (1976, p. 268, table 1) established its presence in Members D and G as well. Gentry and Gentry (1969) recorded two camel molars from Marsabit Road, Kenya, and from site BK, upper Bed II, Olduvai Gorge, Tanzania, the latter being the southernmost record in Africa.

A camel is also known in North Africa, from Pliocene beds equivalent in age to the lower Villafranchian (Arambourg 1962, p. 104; Coppens 1971). North African material is usually referred to *Camelus thomasi* Pomel (1893), a species described and well illustrated on rather scrappy material principally from the lower Pleistocene of Ternifine, Algeria. It is a large camel, held to have some similarities to the Bactrian camel. Gautier (1966, table 1) showed that the proportions of a Ternifine metatarsal agreed with the Bactrian camel but acknowledged that the specimen was not complete. *C. thomasi* is known until the end of the Pleistocene in North Africa (Arambourg 1962), and Gautier (1966) has identified as this species a camel skeleton from deposits in the northern Sudan dated to 22,000 years B.P. This skeleton is large for a camel and the proportions of its calcaneum agree with that of the Bactrian camel. Gautier also referred to a fragmentary large camel metapodial found by Coppens at Bochianga, near Koro Toro, in Chad.

It seems safe to admit that the fossil camel of Africa was large, but other evidence for resemblance to the Bactrian camel does not appear to be conclusive. It amounts to the proportions of a calcaneum and the characters of the lower molars discovered by Grattard, Howell, and Coppens (1976). However, the Bactrian camel is the only surviving wild species, so that the resemblances may testify less to a *C. thomasi-bactrianus* relationship than to the distinctiveness of the characters of domestication in the dromedary. Further, the restricted distributional area of *C. bactrianus* as a relict species weakens its value

for zoogeographical analysis. Finally, samples of camels in any one museum collection are small and sometimes the identifications and provenances are suspect. We have to be sure that even combined samples from several museums will be adequate for the differentiation of wild Bactrian camels, domestic Bactrian camels, and dromedary. We will probably have to turn to fossil evidence to answer the vital question of whether there was a separate wild species from which the dromedary is descended. When better African fossils are available, it will be profitable to compare them with the Pliocene *C. sivalensis* Falconer and Cautley from the Pinjor Formation of the Siwaliks. Complete metapodials of this species are large but appear to have proportions closer to the dromedary than to the Bactrian camel (Colbert 1935, figs. 133 and 134). Fossil camels from elsewhere in the Old World (see Howell, Fichter, and Wolff 1969 for references) have less reduced premolar rows than do living camels, and it is likely that any Pliocene or Pleistocene African camel will be similar.

References

Arambourg, C. 1933. Mammifères miocènes du Turkana (Afrique orientale). *Ann. Paléont.* 22:121–148.

———. 1962. Les faunes mammalogiques du Pléistocène circumméditerranéen. *Quaternaria* 6:97–109.

Arambourg, C., and J. Piveteau. 1929. Les vertébrés du Pontien de Salonique. *Ann. Paléont.* 18:59–138.

Bishop, W. W., and M. H. L. Pickford. 1975. Geology, fauna and palaeoenvironments of the Ngorora Formation, Kenya Rift Valley. *Nature* 254:185–192.

Colbert, E. H. 1935. Siwalik mammals in the American Museum of Natural History. *Trans. Am. Phil. Soc.* n.s. 26:1–401.

Coppens, Y. 1971. Les vertébrés Villafranchiens de Tunisie: gisements nouveaux, signification. *C. R. Hebd. Séanc. Acad. Sci.,* Sér. D 273:51–54.

Flerov, C. C. 1971. Evolution of certain mammals in the Cenozoic. In K. T. Turekian (ed.), *The late Cenozoic glacial ages.* New Haven: Yale University Press, pp. 479–491.

Gautier, A. 1966. *Camelus thomasi* from the northern Sudan and its bearing on the relationship *C. thomasi–C. bactrianus. J. Paleont.* 40:1368–1372.

Gentry, A. W. 1970. The Bovidae (Mammalia) of the Fort Ternan fossil fauna. In L. S. B. Leakey and R. J. G. Savage, eds., *Fossil Vertebrates of Africa.* London: Academic Press, vol. 2, pp. 243–323.

Gentry, A. W., and A. Gentry. 1969. Fossil camels in Kenya and Tanzania. *Nature* 222:898.

Grattard, J. L., F. C. Howell, and Y. Coppens. 1976. Remains of *Camelus* from the Shungura Formation, lower Omo Valley. In Y. Coppens, F. C. Howell, G. Ll. Isaac, and R. E. F. Leakey, eds., *Earliest man and environ-*

ments in the Lake Rudolf Basin. Chicago: Univ. Chicago Press, pp. 268–274.

Hamilton, W. R. 1973. The lower Miocene ruminants of Gebel Zelten, Libya. *Bull. Brit. Mus. Nat. Hist. (Geol.)* 21 (3):73–150.

Hopwood, A. T. 1929. New and little-known mammals from the Miocene of Africa. *Am. Mus. Novit.,* no. 344: 1–9.

Howell, F. C., L. S. Fichter, and R. Wolff. 1969. Fossil camels in the Omo Beds, southern Ethiopia. *Nature* 223:150–152.

Madden, C. T. 1972. Miocene mammals, stratigraphy and environment of Muruarot Hill, Kenya. *Paleobios* 14: 1–12.

Pomel, A. 1893. Caméliens et cervidés. *Carte Géol. Algér. Paléont. Monogr.* pp. 1–52.

Van Couvering, J. A., and J. A. Miller. 1969. Miocene stratigraphy and age determinations, Rusinga Island, Kenya. *Nature* 221:628–632.

Walker, A. C. 1969. Lower Miocene fossils from Mount Elgon, Uganda. *Nature* 223:591–596.

Whitworth, T. 1958. Miocene Ruminants of East Africa. Brit. Mus. Nat. Hist., *Fossil Mammals of Africa,* no. 15, pp. 1–50.

27

Bovidae

Alan W. Gentry

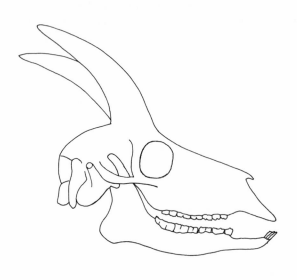

Many bovids are found in open country and even in arid regions, and it is this adaptive facility that undoubtedly explains the great success of the family. Linked with life in open areas is the tendency to live in herds. The only continents without native bovids are South America, Australia, and Antarctica.

Bovids have horns consisting of a hollow keratinized sheath fitting over a bony core that may or may not have internal sinuses; neither sheath nor core is branched or seasonally shed. Horns may be present in both sexes or only in males. The horns of different species are distinctive and permit easy identification; in fact with fossil cores it is often more difficult to group than to separate the specimens. There are no upper incisors or canines, the upper and lower first premolars are missing, and the cheek teeth are selenodont (with crescentic cusps) and frequently hypsodont (high-crowned). Originally in artiodactyls the upper and lower incisors were used together for biting off foliage, the premolars for cutting it into smaller pieces, and the molars for crushing and grinding the small pieces. In bovids the lower incisors and tongue seize and tear off the food in different ways, and the cutting function of the premolars has declined. The premolars come to occupy a shorter length of mandible to allow more space for the relatively larger molars. However the premolars are not so reduced as in camels, nor do they become so molarized as in perissodactyls. The third and fourth metapodials have fused to form cannon bones in the front and back legs, and the lateral and medial metapodials have become more reduced than in deer. Cursorial adaptations of the limb bones, a feature of all artiodactyls, are most pronounced in some bovids.

Bovid Classification

A modified version of Simpson's (1945) classification will be followed, with some improvements, mainly from Ansell (1971), and some of my innovations.

Family Bovidae

Subfamily Bovinae
    Tribe Tragelaphini   bushbuck and allies
    Tribe Boselaphini   Indian nilgai and four-horned antelope
    Tribe Bovini   cattle and buffaloes
Subfamily Cephalophinae
    Tribe Cephalophini   duikers
Subfamily Hippotraginae
    Tribe Reduncini   reedbuck group
    Tribe Hippotragini   roan and allies

Subfamily Alcelaphinae
  Tribe Alcelaphini   wildebeest and hartebeest group, impala

Subfamily Antilopinae
  Tribe Neotragini   dik-dik group
  Tribe Antilopini   [incl. Saigini] gazelles, Springbok and others

Subfamily Caprinae
  Tribe "Rupicaprini"   goral, serow group. *Rupicapra* itself might be better placed in the Caprini.
  Tribe Ovibovini   muskox, takin, extinct allies
  Tribe Caprini   sheep and goats

In the account of each tribe the fossils will be considered in reverse time order. A high proportion of the known bovid species are alive today and they give a lot of information about way of life that would not be available from fossils alone. It is therefore appropriate to use the living forms as a starting point for the fossil history of each tribe.

Fossil Localities

The Pleistocene will here be divided into lower, middle, and upper parts as suggested in Butzer and Isaac (1975) and taken as having started around 1.8 m.y. (Berggren and van Couvering 1974). As a consequence the expression "lower Pleistocene" now denotes a period of time from 1.8 to 0.7 m.y., whereas in many recent studies of fossil mammals it has denoted the Villafranchian and equivalent periods from about 3.5 to 1.8 m.y.

African sites of middle and upper Pleistocene age are mostly represented in Morocco, Algeria, and Tunisia in the north, and in South Africa. Important South African sites with bovids are Florisbad and Cornelia in the Orange Free State and Elandsfontein, Melkbos, and Swartklip on the coast of Cape Province. The later assemblages from the Sterkfontein, Swartkrans, and Kromdraai cave sites in the

Transvaal are also of middle and upper Pleistocene age (Vrba 1975). Faunal spans have been set up for South African sites (see chapter 1), to which the mammal ages of Hendey (1974) are equivalent. The latest of these, the Florisbad faunal span, or Florisian mammal age, is likely to run from 125,000 years B.P. or more until the Pleistocene-Holocene boundary, and is therefore coincidental with the upper Pleistocene. The preceding Cornelia faunal span dates from 700,000 years B.P. or earlier and is therefore nearly entirely of middle Pleistocene age, but it may have begun late in the lower Pleistocene.

Some major East African sites are of Pliocene to Pleistocene age going back to about 3.0 m.y. They embrace the Shungura Formation in Ethiopia, East Turkana in Kenya, and Olduvai Gorge Beds I–IV in Tanzania. Some deposits in this group have slight temporal overlap at their upper levels with those of the South African Cornelian faunal span. South Africa also has sites that by faunal correlation are thought to be of Pliocene or lower Pleistocene age, for example the earlier assemblages from Sterkfontein, Swartkrans, and Kromdraai, as well as the other important site at Makapansgat Limeworks. There is considerable North African material of Pliocene to lower Pleistocene age, but the definitive account of the bovids has yet to be published as a posthumous continuation of Arambourg (1970).

Among the sites older than 3.0 m.y. are Langebaanweg on the coast of Cape Province, Laetolil in Tanzania, Afar in Ethiopia, and most of the localities of the Baringo sequence in Kenya. A well-preserved early bovid fauna is known from Fort Ternan, Kenya, and has been dated to 14 m.y. (Bishop, Miller, and Fitch 1969, p. 685), but it comprises only four species. Nonbovid ruminants were still present in some numbers at this period, and only a few bovids are known from sites older than Fort Ternan.

The following localities are mentioned in this chapter:

Aboukir	Algeria	(?middle) Pleistocene
Abu Hugar	Sudan	upper Pleistocene
Afar	(see Hadar Formation)	
Aïn Boucherit	Algeria	Pliocene (LV)
Aïn Brimba	Tunisia	Pliocene (LV)
Aïn Hanech	Algeria	Pliocene (UV)
Aïn Jourdel, near Constantine	Algeria	Pliocene (LV)
Bel Hacel	Algeria	Pliocene (UV)
Beni Mellal	Morocco	middle Miocene
Bled Douarah	Tunisia	middle Miocene
Bloembos	South Africa: CP	(?upper) Pleistocene
Bolt's Farm	South Africa: TR	lower Pleistocene
Border Cave	South Africa: Natal	upper Pleistocene
Broken Hill (now Kabwe)	Zambia	middle or upper Pleistocene

Buffalo Cave	South Africa: TR	Pliocene or Pleistocene
Chelmer	Rhodesia	upper Pleistocene
Cornelia	South Africa: OFS	middle Pleistocene
Djerid	Tunisia	middle Miocene
East Turkana (= East Rudolf)	Kenya	Pliocene to Pleistocene
Elandsfontein (= Hopefield)	South Africa: CP	middle Pleistocene, also a little upper Pleistocene material
Florisbad	South Africa: OFS	upper Pleistocene
Fort Ternan	Kenya	middle Miocene
Garaet Ichkeul	Tunisia	Pliocene (LV)
Gebel Zelten	Libya	early Miocene
Hadar Formation, Afar	Ethiopia	Pliocene
Isimila	Tanzania	lower Pleistocene
Kaiso Formation	Uganda	Pliocene, with two faunal levels
Kanam	Kenya	Pliocene to Pleistocene
Kanjera	Kenya	lower Pleistocene
Karmosit Beds, Baringo	Kenya	Pliocene
Kibish Formation, Omo	Ethiopia	upper Pleistocene to Recent
Kom Ombo	Egypt	upper Pleistocene
Koobi Fora Formation, East Turkana	Kenya	Pliocene to lower Pleistocene
Kranskraal	South Africa: OFS	upper Pleistocene
Kromdraai faunal site (KA)	South Africa: TR	lower (?middle) Pleistocene
Laetolil	Tanzania	Pliocene
Langebaanweg	South Africa: CP	Pliocene
Losodok (= Moruorot)	Kenya	early Miocene
Lukeino Formation, Baringo	Kenya	late Miocene
Maboko	Kenya	(?middle) Miocene
Mahemspan	South Africa: OFS	upper Pleistocene to Recent
Makapansgat Limeworks	South Africa: TR	Pliocene or lower Pleistocene
Mansoura, near Constantine	Algeria	Pliocene (UV)
Marceau	Algeria	late Miocene
Marsabit Road	Kenya	Pliocene or lower Pleistocene
Melkbos	South Africa: CP	upper Pleistocene
Mockesdam	South Africa: OFS	upper Pleistocene
Modder River	South Africa: OFS	middle or upper Pleistocene
Mpesida Beds, Baringo	Kenya	late Miocene
Mursi Formation, Omo	Ethiopia	Pliocene
Namib Desert	Namibia	early Miocene
Nelson Bay Cave	South Africa: CP	upper Pleistocene to Recent
Ngorora Formation, Baringo	Kenya	middle to late Miocene
Olduvai Gorge, Beds I–IV	Tanzania	lower Pleistocene
Oued Bou Sellam, near Setif	Algeria	middle or upper Pleistocene
Oued el Atteuch	Algeria	Pliocene
Oued el Hammam (= Bou Hanifia)	Algeria	middle Miocene
Peninj	Tanzania	lower Pleistocene
Power's site, Vaal River younger gravels	South Africa: CP	middle Pleistocene
Rusinga Island	Kenya	early Miocene
Sahabi	Libya	late Miocene
Shungura Formation, Omo	Ethiopia	Pliocene to lower Pleistocene
Sidi Bou Kouffa	Tunisia	Pliocene (UV)
Singa	Sudan	upper Pleistocene
Songhor	Kenya	early Miocene
Sterkfontein type site (STS)	South Africa: TR	Pliocene or lower Pleistocene
Sterkfontein "upper quarry"	South Africa: TR	upper Pleistocene or Recent
Steynspruit	South Africa: OFS	upper Pleistocene
Swartklip	South Africa: CP	upper Pleistocene
Swartkrans main assemblage (SKa)	South Africa: TR	lower Pleistocene
Taung	South Africa: CP	Pliocene or lower Pleistocene

Temara	Morocco	upper Pleistocene
Ternifine	Algeria	lower Pleistocene
Vlakkraal	South Africa: OFS	upper Pleistocene
Wadi Natrun	Egypt	late Miocene
Wonderwerk Cave	South Africa: CP	upper Pleistocene

The abbreviations used in the list are CP = Cape Province, LV = lower "Villafranchian," OFS = Orange Free State, UV = upper "Villafranchian," TR = Transvaal, and KA, SKa, STS = site designations used in Vrba (1975).

Systematics

Subfamily Bovinae

Tribe Tragelaphini

The Tragelaphini are medium to large-sized browsing antelopes almost confined to Africa and tending to live where there is cover or even in forests. Living ones have spiraled horn cores in which the torsion is anticlockwise on the right side and in which there are two or three keels (anterior, posterolateral, and posteromedial). Other major characters are absence of internal sinuses in the frontals or horn pedicels, braincase not very angled on the face axis, teeth rather brachydont, basal pillars on molars small or absent, central cavities without a complicated outline, upper molars without prominent lateral ribs between the styles, lower molars with narrowly pointed lateral lobes and without goat folds, long premolar rows and large anterior premolars, P_4's often with paraconid-metaconid fusion that closes the anterior part of the medial wall, and mandibles with shallow horizontal rami (see figure 27.1).

There are two living genera, *Tragelaphus* (including *Strepsiceros* and *Boocercus*) and *Taurotragus*. The living species are *Tragelaphus scriptus*, the widespread bushbuck of small to medium size with anteroposteriorly compressed horn cores in which the posterolateral keel is the most prominent; *T. spekei*, the sitatunga that is the most aquatic of antelopes, larger than the bushbuck but with similar horns; *T. angasi*, the nyala with a restricted range in southeastern Africa and very similar cranially to the sitatunga; *T. imberbis*, the lesser kudu of parts of East Africa and perhaps Arabia (Harrison 1972, p. 629), with horn cores unlike the sitatunga by being longer, more spiraled, less divergent, and with a less strong posterolateral keel; *T. strepsiceros*, the greater kudu with strongly spiraled horn cores with the anterior keel more pronounced than the others and no anteroposterior compression; *T. buxtoni*, the mountain nyala of Ethiopia with some features of its horn cores like the greater kudu; *T. eurycerus*, the bongo in which both sexes have horns; *Taurotragus oryx*, the eland, the largest known antelope, with horns in both sexes, a strong anterior keel, without anteroposterior compression, quite tightly twisted but not openly spiraled. Of these species, probable fossil ancestors are known for all except *Tragelaphus buxtoni* and *T. eurycerus*.

In the middle and upper Pleistocene of South Africa there is an eland that can be safely referred to the living species. In the abundant material from Elandsfontein the insertion of the horn cores may not be at quite such a low angle as in the extant form of the species. An eland, apparently the living species, is known from the later Pleistocene of North Africa but disappeared at the start of the Holocene (Arambourg 1962).

Taurotragus arkelli L. S. B. Leakey (1965) is known as a single cranium from Olduvai Bed IV, Tanzania. It is less advanced than the living eland and has horn cores slightly more uprightly inserted and a braincase less drastically shortened. It must almost certainly predate the Elandsfontein *T. oryx*. Arambourg (1962, p. 106) refers to an eland at Ternifine, Algeria, but details have yet to be published. Remains of eland from other sites are only fragmentary.

Another tragelaphine in the later Pleistocene of the Cape Province is an enigmatic kudu known only from horn cores and teeth (Hendey 1968, p. 108). The horn cores agree quite well with the greater kudu but are much more tightly spiraled, and the teeth are of a size between greater and lesser kudu. It may be a subspecies of *Tragelaphus strepsiceros* or a separate species.

Tragelaphus strepsiceros is the most frequently found tragelaphine at Olduvai but is sufficiently different from living examples for subspecific names to be applicable. *T. s. grandis* L. S. B. Leakey (1965, p. 38) is the large form from middle and upper Bed II, represented by the holotype cranium, horn cores, and dental remains. The horn cores are more divergent at their bases and have less mediolateral compression than in living kudus. There is a posterolateral keel for a short distance at the base of some horn cores. Linear measurements of the holotype indicate a size about 10 to 20% greater than at the present day. Appropriately sized teeth for this form

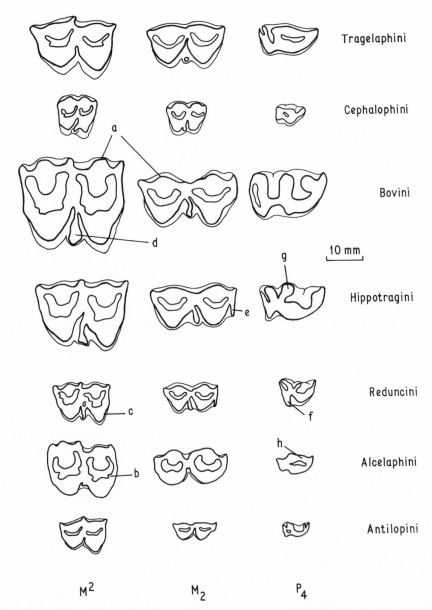

Figure 27.1 Occlusal views of teeth of the right side in African tribes of Bovidae. The bovine P_4 is shown with the close approach of paraconid and metaconid seen in African bovines, but not in those of other regions. The molars of *Aepyceros* are very like those of Antilopini, but there is fusion of paraconid and metaconid on P_4. Neotragine teeth differ from genus to genus and have not been illustrated.

Apart from overall size, degree of hypsodonty, and relative lengths of premolar and molar rows, the following characters help in identifying bovid teeth: (*a*) development of ribs on lateral side of upper molars and medial side of lowers, (*b*) a complicated or simple outline of central cavities, (*c*) medial lobes of upper molars constricted or not, (*d*) presence or absence of basal pillars, (*e*) a transverse "goat fold" at the front of the lower molars, (*f*) projecting hypoconid on P_4, (*g*) size and shape of metaconid on P_4, and (*h*) paraconid-metaconid fusion on P_4.

occur also at Peninj, Tanzania, and Makapansgat Limeworks, South Africa, but subspecific assignment for teeth is perhaps unwise.

Bed I and the lower part of Bed II contain *T. s. maryanus* L. S. B. Leakey (1965, p. 40), about the same size or slightly smaller than present day kudus. It has strong mediolateral compression of its horn cores and a braincase roof more bent down on the face axis than in later kudus. It is curious that out of seven known P_4 specimens of this subspecies paraconid-metaconid fusion occurs in five (including two in late wear) and is just starting on a further two in middle wear; the same fusion was seen in only 13 of 24 examples of the living species. The only complete lower cheek toothrow of *T. s. maryanus* has a premolar row shorter than in most living greater kudus.

T. strepsiceros is well represented in the Koobi Fora Formation of East Turkana, both below and above the KBS Tuff (Harris 1976, p. 295).

Only a single horn core fragment from Member G represents *T. strepsiceros* in the Shungura Formation, Ethiopia. The common kudu in Members E to G of this formation is a smaller form. Compared with later kudus it has horn cores spiraled more closely to their longitudinal axis but not tightly spiraled. The horn cores are less mediolaterally compressed and inserted at a lower angle. The braincase is relatively narrower, presumably an allometric effect of smaller overall size. This kudu is unlike the lesser kudu in its strong anterior keel and in its tendencies for its horn cores to become more mediolaterally compressed from Member E up to Member G and to reduce the posterolateral keel. Nonetheless it could be ancestral to *T. imberbis* if it showed a degree of evolutionary reversal in response to competition from the closely related *T. strepsiceros*. The holotype horn core of *T. gaudryi* (Thomas 1884, p. 15) came from Aïn Jourdel, and a frontlet came from Mansoura (Gervais 1867–69, pl. 19, fig. 4), both sites being of Villafranchian-equivalent age. *T. gaudryi* can be used provisionally to include the Omo kudu, but this may need changing if the Algerian kudu should turn out to be an early stage of the *T. strepsiceros* lineage. So far kudus have not been reported from later deposits in North Africa. Some horn core pieces from the Mursi Formation, Ethiopia, appear to belong to a kudu.

Horn cores from various sites have been referred to *Tragelaphus* cf *spekei* or *T.* cf *angasi*, but the remains are too scrappy to give much information about their evolution. At one time there was an unfigured skull from Olduvai described as *T. spekei stromeri* Schwarz (1932, 1937), but this was destroyed during the Second World War. The poorly preserved Laetolil cranium and horn cores of *T.* cf *buxtoni* Dietrich (1942, p. 118) is of this stock and is not related to the mountain nyala, which has horn cores like those of greater kudu in their open spiraling, upright insertions, and slight anteroposterior compression, but has its main keel posterolateral instead of anterior. It does not belong to the early fauna of the Laetolil Beds (Leakey et al. 1976). Pieces of horn cores of *T.* cf *spekei* or *T.* cf *angasi* are known from the early fauna of the Kaiso Formation

near North Nyabrogo in Uganda (Cooke and Coryndon 1970, p. 200), Kanam East, Kanam East Hot Springs, and Kanjera in Kenya, and Makapansgat Limeworks. The Makapansgat Limeworks specimen shows less anteroposterior compression than in living sitatunga and nyala. There are also some appropriately sized teeth at Makapansgat Limeworks, already referred to *T.* cf *angasi* by Wells and Cooke (1956, p. 10).

The only definite evidence for the ancestry of *T. scriptus* is a pair of horn cores from Member C of the Shungura Formation, Omo, and some tooth remains from Makapansgat Limeworks. The horn cores differ from those of the living species by being less anteroposteriorly compressed and more uprightly inserted. However, they agree with the living species in the strong posterolateral keel and the slightly less strong anterior keel. The dentitions at Makapansgat Limeworks were referred by Wells and Cooke (1956, p. 12, fig. 5) to *Cephalophus pricei* but are clearly tragelaphine by their less massive premolars, paraconid-metaconid fusion on P_4, P_2 insufficiently large relative to P_3 and P_4, and less rounded outbowings on the medial walls of the lower molars. The holotype for the name *C. pricei* was one of the mandibles, but the paratype was a horn core of a fossil species of *Raphicerus*. It remains to be seen if and when horn cores conspecific with the mandibles are discovered at Makapansgat Limeworks whether the name *pricei* can be applied to the Shungura species.

Tragelaphus nakuae Arambourg (1941, 1947, p. 418) comes from Members B to H of the Shungura Formation and also from the Koobi Fora Formation as high as "the lower part of the *M. andrewsi* zone" (Harris 1976, p. 295), which would be just above the KBS Tuff. It is not a kudu and differs from the contemporaneous *T. gaudryi* by its larger size, a transverse ridge across the cranial roof above the occipital, and horn core characters. The material in Members C to G differs from that in Member B by increased size and horn cores shorter and more curved with less twisting of the keels, a weaker anterior keel arising from a more anterolateral insertion, greater anteroposterior compression, and wider basal divergence. *T. nakuae* is very interesting in possessing skull characters in common with certain boselaphines and unlike other tragelaphines. The braincase is set high and not at all inclined to the axis of the face, there are strong temporal ridges on the cranial roof, the dorsal parts of the orbital rim project quite strongly, the supraorbital pits are not narrowed and drawn out anteroposteriorly, and one specimen shows a large preorbital fossa. There are strong similarities to *Selenoportax vexillarius* Pilgrim (1937) from the Nagri Formation of the Siwaliks,[1] and the question arises of whether *T. nakuae* could be better classified as a boselaphine. It seems best to leave it as a tragelaphine until more is known about tragelaphines and boselaphines in Africa prior to the Pliocene. *S. vexillarius* could as easily be related to Bovini as to Tragelaphini, and its similarities to *T. nakuae* lie more with the Members C to G representatives than with that from Member B. Moreover, the teeth of *T. nakuae* are adequately known and are satisfactorily tragelaphine in morphology. What appears to be an ancestral species to *Tragelaphus nakuae* has recently been discovered earlier in the Pliocene of the Afar area of Ethiopia (see figure 27.2).

A number of tragelaphine horn cores at Langebaanweg, South Africa, have less anteroposterior compression and more upright insertions than in living *T. angasi* or *T. spekei*.

Tragelaphine teeth are known from the Mpesida Beds and apparently from Ngorora, Kenya (KNM-MP 071, 072; KNM-BN 1235), making them the oldest record of the tribe. One or more boselaphine

___

1 The holotype, a skull roof with horn cores, cannot be later than the Nagri Formation according to Pilgrim. I incorrectly wrote that it was from the Dhok Pathan Formation (Gentry 1970a, pp. 260, 315), but this formation contains only teeth attributed to *Selenoportax*.

lineages must have developed tragelaphine tooth characters quite early, and this will lead to difficulties in tribal classification.

Tribe Boselaphini

The Boselaphini are a mainly fossil group with only two living species, the large *Boselaphus tragocamelus,* the Indian nilgai, and *Tetracerus quadricornis,* the small Indian four-horned antelope. However, the tribe has occurred in the past in Africa. The main characters of the tribe are horn cores with keels but no transverse ridges, braincases little angled on the face axis, strong temporal ridges on the cranial roof, brachydont cheek teeth with enamel that is often rugose, lower molars generally without goat folds, and premolar rows long.

The most interesting boselaphine in Africa is *Mesembriportax acrae* Gentry (1974) from Langebaanweg. It is large with short divergent horn cores, very mediolaterally compressed, with a prominent anterior keel in their lower part but with a circular reduced cross-section terminally. The horn sheaths were probably bifurcated, unlike in any living bovid. There is an advanced system of internal sinuses in the horn pedicels and frontals, which raise the frontals so much that the inclination of the dorsal part of the orbital rims becomes almost vertical. The skull is low and wide and the temporal ridges are strong. The premolar rows are long with large upper and

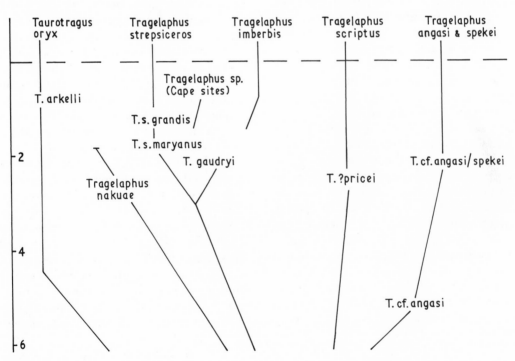

Figure 27.2 Tentative phylogeny for Tragelaphini. Living species are entered above the dashed line and age is shown on the left in millions of years. No fossils are known related to *Tragelaphus buxtoni* and *T. eurycerus*.

lower P2s and no paraconid-metaconid fusion on P$_4$, and the first incisors are not greatly enlarged. The species is not very close to other boselaphines but may be descended from *Protragocerus labidotus.*

An M$_3$ from Wadi Natrun, Egypt, identified by Andrews (1902, p. 437, pl. 21, figs. 7 and 8) as *Hippotragus ?cordieri* appears to be a large boselaphine or tragelaphine. It chances to resemble living *Oryx* except in being lower crowned, and the anterior goat fold is not usual in boselaphines or tragelaphines. *Parabos cordieri* is a French Pliocene bovine or boselaphine, but there is no reason to link the Wadi Natrun tooth with this species in particular.

Protragocerus labidotus Gentry (1970a, p. 247) from the Fort Ternan Miocene is a small species and has mediolaterally compressed horn cores with the anterior keel extended downward as a ridge onto the pedicel, no internal hollowing of the frontals, upper canine alveoli less reduced than in later bovids, and enlarged first incisors in the mandible. The limb bones are not advanced in cursorial specialization. It is congeneric with a European and Siwaliks species and close to the origins of *Tragoportax, Selenoportax,* and *Pachyportax.* A slightly later *Pro-*

tragocerus comes from the Ngorora Formation, Kenya (Bishop et al. 1971), and an earlier member of the genus has been recorded from Libya (see "The Earliest Bovids of Africa" below).

A frontlet, apparently of the well-known later Miocene Eurasian boselaphine *Miotragocerus,* has been found at Sahabi, Libya, and is now in Rome. This genus was formerly called *Tragocerus* until Kretzoi discovered that the name was preoccupied by a beetle (see Gentry 1971, p. 234). *Miotragocerus* is not otherwise known from Africa (see figure 27.3). Roman's (1931, p. 34, pl. 4, figs. 1–4) record from the Djerid area of Tunisia is based on teeth that could easily belong to the caprine *Pachytragus solignaci.* The same applies to a horn core from Djerid that Boule (1910, p. 50) thought was similar to *Miotragocerus* or *Hemitragus.* The *Miotragocerus* from Marceau, Algeria (Arambourg 1959, pl. 17, figs. 5–7), is based on some cheek teeth that are not exclusively identifiable as *Miotragocerus.*

Tribe Bovini

The Bovini comprise large-sized descendants of boselaphines known from the later Tertiary onward.

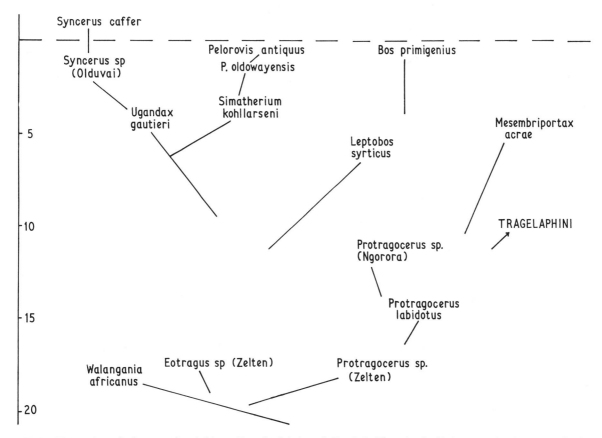

Figure 27.3 Tentative phylogeny for African Boselaphini and Bovini. The single living species is entered above the dashed line and age is shown on the left in millions of years.

The living ones have low and wide skulls (allometrically linked with their large size), horn cores in both sexes emerging transversely from the skulls, frontals and horn cores with internal sinuses, a short braincase and triangular basioccipital, molars with basal pillars and complicated central cavities, upper molars with prominent outbowed ribs between the styles, and lower molars without large goat folds.

There are three living genera, *Bos* for the cattle and bison of Eurasia with one species in North America, *Bubalus* for the Asiatic water buffaloes (one mainland species and two allopatric island forms), and *Syncerus* for the African buffalo, *S. caffer*. The latter is shortfaced, has short horn cores inserted just behind the orbits, and the paraconid of P_4 has usually fused with the metaconid to close the anterior part of the medial wall. It is a complexly variable species (see Grubb 1972).

Even in the more recent deposits there are no signs of buffaloes with the uparched frontals, large basal horn bosses, and downturned horns found in living *S. caffer caffer* south of Ethiopia. A nearly complete *S. caffer* skeleton from the Kibish Formation, Omo (Leakey 1969, p. 1,132), resembles the less advanced West African and Ethiopian savanna buffaloes yet it is as large as a large male *S. c. caffer*. Another large *S. caffer* is represented by a cranium, horn core pieces, and tooth remains at Melkbos, South Africa (Hendey 1968, p. 104). It has internally hollowed horn cores and surface rugosity of the frontals, but again the horn cores pass outward at their bases and do not turn immediately downward. A large bovine lower jaw from the Zululand coast, *Bubalus andersoni* Scott (1907), is likely to be yet another large *S. caffer*. *Syncerus* has not been definitely recorded north of the Sahara Desert.

In Olduvai middle and upper Bed II are a number of crania and one or two other finds of a buffalo, presumably ancestral to *Syncerus caffer*, in which the horn cores are internally hollowed only near their bases, have a triangular cross-section formed by upper, lower, and front surfaces, P_4 with paraconid and metaconid growing closer to one another but not fused, and the basioccipital perhaps not so narrowed anteriorly. The species has horn cores emerging transversely and without large basal bosses; in this it resembles the less advanced forms of *S. caffer*. Skull parts from Members B to G of the Shungura Formation have short horn cores and show a substantially more primitive morphology than the living species by their less shortened braincases and strong temporal ridges.

Ugandax gautieri Cooke and Coryndon (1970, p. 206) from deposits of unknown age in the Kaiso For-

mation was published as a hippotragine but appears to be an early member of the *Syncerus* lineage (see figure 27.3). It has a braincase even less shortened and less low and wide than in the Shungura specimens and differs from the living buffalo by its smaller, less dorsoventrally flattened horn cores that pass more upward and backward than outward. It is like *Proamphibos* Pilgrim from the Tatrot Formation of the Siwaliks, ancestor of *Hemibos* and *Bubalus*, but differs in its shorter horn cores with less marked keels, less triangular basioccipital, and upper molars with smaller basal pillars and less pronounced styles and ribs on the lateral walls. An interesting bovine, apparently a more advanced *Ugandax* than *U. gautieri*, has been discovered at Afar.

In the middle and upper Pleistocene of northern, eastern, and southern Africa are remains of an extinct lineage of larger, long-horned buffaloes. With the possible exception of the abundant Elandsfontein remains, they may be regarded as belonging to one species, *Pelorovis antiquus* (Duvernoy 1851), which includes *Bubalus baini* Seeley and *B. nilssoni* Lönnberg. Besides the longer horn cores, they also differ from *S. caffer* by their horn cores being less dorsoventrally compressed and without basal bosses and by longer metapodials. The holotype came from Oued Bou Sellam near Setif, Algeria, and is of middle or upper Pleistocene age. Other North African specimens have been illustrated by Pomel (1893). A horn core, WK-East 2305, from Bed IV and possibly some teeth from Beds III and IV provide evidence of *P. antiquus* at Olduvai. *Homoioceras singae* Bate (1949, 1951) from Singa and Abu Hugar, Sudan, was based on a skull with horn bases thought to be of one of these long-horned buffaloes. The new generic name was intended to apply to all of them and its founding emphasized that they were not related to the south Asian water buffaloes of the genus *Bubalus*. It is unfortunate that the holotype of *H. singae*, the type species, probably belongs to a large member of the *Syncerus* stock (an idea suggested to me by J. W. Simons) as shown by the great dorsoventral compression of the horn bases and the closeness of their dorsal edges on the top of the skull. However, the older generic name *Pelorovis* Reck (1928) is available, and Bate is correct about the dissimilarity to *Bubalus*. *Pelorovis* is more like *Syncerus* in its shorter face, irregular or absent keels on the horn cores, nasals without lateral flanges anteriorly, premaxillae without a long nasal contact, vomer not fused to the back of the palate, lower and wider occipital surface, and tendency to fuse paraconid and metaconid on P_4.

Pelorovis oldowayensis Reck 1928 (including *Bu-*

larchus arok Hopwood according to Gentry 1967) is known from Olduvai middle and upper Bed II, Kanjera, and high in the Koobi Fora Formation, East Turkana (Harris 1976, p. 295). The holotype is said to come from Bed IV at Olduvai. It has horn cores inserted close together and very posteriorly near to or above the occipital surface. With the toothrow in a horizontal plane, the horn cores curve backward on leaving the skull; this makes their support difficult, especially in an animal that had attained the size of *P. oldowayensis*. Probably this species went on to give rise to *P. antiquus* with its generally downsweeping horn cores. The change in their course was linked with the insertions moving forward again to a more primitive position just behind the orbits. The occlusal pattern of the cheek teeth of *P. oldowayensis* is simpler than in *P. antiquus,* except for the Elandsfontein examples. There are few other differences, and we can regard *P. antiquus* and *P. oldowayensis* as congeneric. The priority of the generic name *Pelorovis* circumvents the problem of the type species of *Homoioceras* being probably a *Syncerus*, and *Pelorovis* need not be abandoned even if *P. antiquus* were not the descendant of *P. oldowayensis*. Teeth belonging to *P. oldowayensis* in Olduvai Bed II are well differentiated from those of *Syncerus* by their greater size and less pronounced occlusal complexity. This distinction between the two lineages is less apparent in teeth from the Shungura Formation, where a few horn core fragments show the presence *P. oldowayensis* or its ancestor.

Simatherium kohllarseni Dietrich (1941, p. 222, 1942, p. 119) is based on a badly preserved Laetolil cranium that could be ancestral to *Pelorovis*. There are three differences from *P. oldowayensis*: the horn cores are inserted closer to the orbits, are wider apart, and diverge less toward their tips. The first and third are to be expected in an ancestor, and the second is not incompatible with that form being ancestral. Similar horn core fragments are known from Langebaanweg and were mistakenly attributed by Gentry (in Hendey 1970, p. 114) to a kudu. In fact, the keel is much too prominent for a kudu but is not unlike the condition on the Shungura Formation horn core fragments or even the irregular keels or ridges that occur occasionally on the Elandsfontein *Pelorovis*.

Bos primigenius is known from the late Pleistocene of North Africa, where it survived into Neolithic and perhaps Roman times. Pomel (1894a) illustrated some remains. *Bubalus vignardi* Gaillard (1934, pl. 5, fig. 1) from Kom Ombo, upper Egypt, also looks like a *Bos primigenius,* and Churcher and

Smith (1972) and Churcher (1972) confirm *B. primigenius* at this site. The species is known in North Africa back to early in the Amirian (= middle Pleistocene), but it is not found at Ternifine (Arambourg 1962, p. 106). In earlier Villafranchian-equivalent horizons is a large bovine related to *Leptobos* Rütimeyer (Arambourg 1962, p. 104); the *Bos* sp. teeth from Sidi Bou Kouffa, Tunisia, mentioned by Coppens (1971) could be the same species.

Petrocchi (1956) described *Leptobos syrticus* from Sahabi, Libya. Several characters of the three complete or partial crania of this species suggest that it is rather primitive compared with other *Leptobos*: the closeness of the horn core insertions to the orbits in the anteroposterior plane, the absence of any lengthening of the horn pedicels, the retention of an anterior keel and a posterolateral keel in parts of the horn core's course, and the closeness of the supraorbital pits to the longitudinal midline of the skull. These are all resemblances to *Parabos boodon* (Gervais) from Perpignan, France, particularly to the cranium in Paris figured by Depéret (1890, pl. 7, fig. 4), but *L. syrticus* is a separate lineage from *Parabos*. *L. syrticus* is unlikely to be related to *Syncerus* because of the smoothness of the frontal bone surfaces between the horn core bases, insufficient inclination of the braincase roof, the excessively developed temporal ridges, and perhaps the weakness of the posteromedial (= posterodorsal) keel. Pilgrim (1937, p. 817) has pointed out the lingering of faint keels in a specimen of *L. falconeri*.

Subfamily Cephalophinae

Tribe Cephalophini

Cephalophines or duikers are mostly forest-living small to moderate-sized antelopes, rather heavily built and with short horn cores. All but one of the species are placed in the genus *Cephalophus*. The exception is *Sylvicapra grimmia*, which has more upright horn cores and longer legs and lives in areas with less dense cover than those inhabited by *Cephalophus*. Speciation has been extensive in cephalophines, but they are extremely uncommon as fossils.

They are seen in some late archaeological contexts, e.g., at Nelson Bay, Cape Province (Klein 1972). *Cephalophus parvus* Broom (1934, p. 477, fig. 7) was described on a maxilla with P^3 and P^4 from Taung, South Africa. Wells (1967, p. 101) believed it to be conspecific with the living blue duiker, *C. "caerulus"* (= *C. caeruleus*, a junior synonym of *C. monticola*). From Makapansgat Limeworks there is a piece of a right mandible, BPI M-22, with part of

P₃ to M₂ described by Wells and Cooke (1956, p. 15) as cf *Cephalophus caeruleus*. Cooke (1949, p. 38) referred to a mandible and a broken lower M3 of *Sylvicapra grimmia* from the Vaal River younger gravels at Power's site, but Vrba (1973, p. 288) was satisfied that the mandible belongs to *Antidorcas bondi*. Schwarz (1937, p. 25) referred a mandible and vertebrae from Olduvai to *Philantomba* (= *Cephalophus*) *monticola,* but without illustrating them. The material was destroyed during the Second World War.

A fragmentary maxilla from the Ngorora Formation, KNM-BN 96, may represent a duiker. The two surviving molars are brachydont with rugose enamel and have a pronounced rib between parastyle and mesostyle. The fossil is unlikely to belong to a neotragine or gazelle. The tooth of *Cephalophus* sp. Arambourg (1959, p. 127, pl. 17, figs. 8, 8a, and 8b) from Marceau is smaller than the contemporaneous gazelle species but otherwise indeterminate.

Subfamily Hippotraginae

Tribe Reduncini

Reduncines are moderate to large-sized antelopes commonly found near water and now confined to Africa. Living ones have horn cores only in males, without keels or spiraling but with transverse ridges, and mostly inserted at a low angle in side view. They also have little internal hollowing within the frontals, temporal ridges on the braincase roof often approaching closely to one another, a large maxillary tuberosity prominent in ventral view, palatal ridges close together on the maxillae in front of the toothrow, cheek teeth rather small in relation to skull and mandible size, and quite hypsodont, basal pillars on upper and lower molars, medial lobes of upper molars and lateral lobes of lowers constricted, upper molars with small but well-protruding ribs between styles, lower molars with goat folds, upper and lower P2s small, lower premolars with the appearance of being anteroposteriorly compressed, P₄ with a strongly projecting hypoconid, and rarely with paraconid-metaconid fusion.

There are two living genera, *Redunca* and *Kobus*. *Redunca redunca* the bohor reedbuck has short horns over most of its range, inserted at a very low angle; *R. arundinum* has longer horns, some other distinguishing skull characters, and a more southerly distribution; the small *R. fulvorufula* has short and little divergent horn cores. Species of *Kobus* are more gregarious. *K. kob,* the kob and puku, has mediolaterally compressed horn cores inserted rather uprightly in side view and close together; *K. leche,* the Central African lechwe, has mediolaterally com-

pressed horn cores inserted widely apart and at a low angle and with an initial backward curvature; *K. ellipsipyrmnus,* the waterbuck, is the largest living reduncine and has horn cores with little mediolateral compression inserted widely apart and without any basal backward curvature; *K. megaceros,* the Nile lechwe, has horn cores with little mediolateral compression inserted farther behind the orbits than in other reduncines, wide nasals, and very strong longitudinal ridges on the basioccipital.

Fossils are known that could be ancestral or related to two reedbucks and to all living species of *Kobus* except the Nile lechwe, and there are two or more lineages without living descendants.

Redunca arundinum is common as a fossil at Elandsfontein and other sites (Hendey 1968, p. 110; Hendey and Hendey 1968, p. 51). The Elandsfontein form has horn cores more anteroposteriorly compressed, less divergent, and possibly shorter than in living representatives of the species. The last two characters, and possibly the first, give it some resemblance to *R. redunca,* suggesting a recent common ancestry.

Redunca redunca is present in the late Pleistocene and early Holocene of North Africa, far to the north of its present limits (Arambourg 1938, p. 44). *Antilope (Oegoceros) selenocera* Pomel (1895, pl. 6, figs. 1–3), *A. (Dorcas) triquetricornis* Pomel (1895, p. 28, pl. 11, figs. 1 and 2) and the teeth called *A. (Nagor) maupasii* Pomel (1895, p. 38, pl. 10, figs. 1–11) are all synonyms of *R. redunca.*

Redunca darti Wells and Cooke (1956, p. 17) is an earlier species from Makapansgat Limeworks. It differs from living *R. redunca* and *R. arundinum,* to which it may be ancestral, by its horn cores being more uprightly inserted and the posteromedial basal flattened surface lying more medial than posterior. It is larger than *R. fulvorufula* and has larger, more divergent, and less upright horn cores. Two pieces, BPI M-690 and M-2798, suggest that it may have had a small preorbital fossa unlike any living *Redunca.* A horn core from Kanam East Hot Springs, BM M-15928, is similar to *R. darti.* A few tooth remains likely to be of *Redunca* are known from Olduvai. Horn cores from Kanam and a horn core base from Kanjera, BM M-26930, M-26931, M-26932, M-26934, resemble *R. redunca* rather than the more primitive *R. darti.* Coppens (1971) refers to *Redunca* from Villafranchian-equivalent sites in Tunisia, of which further details are to be given by Arambourg (to be published posthumously).

Other fossil reduncines form a more heterogeneous group and are related or assignable to *Kobus.* Potentially very interesting is a late Pleistocene horn core base from Abu Hugar, Sudan (Bate 1951, fig. 3), which Wells (1963) discussed and identified as "cf *Kobus* sp." The fossil looks reduncine, but I cannot identify it with any known extant species. It is somewhat like a very large kob but shows pronounced mediolateral compression.

A second reduncine in the upper Pleistocene of South Africa is *Kobus venterae* Broom (1913, p. 13), of which the holotype cranium and some horn cores come from Florisbad and other horn cores from Mahemspan, Mockesdam, and Vlakkraal. It appears not to be distinguishable from the living *K. leche,* and it presumably indicates the former presence of seasonally inundated floodplains.

Reduncines are common in the Shungura Formation but rather rare at Olduvai Gorge. *Kobus sigmoidalis* Arambourg (1941) has long, mediolaterally compressed, divergent horn cores with an initial curve backward at the base. Its teeth are not as advanced as in present-day reduncines in that upper molars have central cavities that are often not very complicated and ribs between the styles that are not very localized or accentuated. The lower molars frequently have less constricted lateral lobes than in the living waterbuck. Horn cores and teeth of *K. sigmoidalis* are scarce in Members C and D, abundant in Members E to G of the Shungura Formation, and occur in Olduvai Beds I and lower II. A few horn cores of the same lineage but with mediolaterally thicker bases and scarcely any backward curvature are known from Bed III and perhaps Bed I Olduvai and from the upper levels of the Shungura Formation. Thus *K. sigmoidalis* was probably evolving into *K. ellipsiprymnus.* Remains from the Koobi Fora Formation are comparable with those in Member G and above in the Shungura Formation (Harris 1976, pp. 296–297). *K. leche* may also descend from *K. sigmoidalis.*

In middle and later Bed II at Olduvai are found sparse remains of *K. kob.* It was smaller than *K. sigmoidalis* and had horn cores inserted at a low angle with little backward basal curvature. Metapodials were a little shorter than at the present day. The kob from Bed III was larger and had larger, more mediolaterally compressed horn cores curving backward at the base from a slightly more upright insertion. Compared with the living kob, it is larger and the horn cores less upright at their bases and their insertions less close together. The evolutionary change in compression and insertion plane of the horn cores is the reverse of what occurred in the waterbuck lineage.

Menelikia lyrocera Arambourg (1941, 1947, p. 392) is a medium-sized extinct reduncine from Members E to J of the Shungura Formation, the Koobi Fora Formation (Harris 1976), and Marsabit Road, Kenya. Its most surprising character is the extensive internal hollowing of the frontals, almost as much as in alcelaphines. Its horn cores rise close together, then diverge outward, and finally curve upward at the tips. The braincase is a little angled on the facial axis, and a small preorbital fossa is present (absent in all living reduncines). The anterior tuberosities of the basioccipital are large and close together. Arambourg founded a new subfamily for this form, but it is satisfactory as a reduncine by its horn cores with transverse ridges, temporal ridges approaching closely on the cranial roof, prominent maxillary tuberosity in ventral view, the extreme closeness of the palatal ridges on the maxilla in front of the toothrow, large anterior tuberosities of the basioccipital, basal pillars on the molars, constricted medial lobes of the upper molars, lower molars with goat folds, lower premolars with the appearance of being anteroposteriorly compressed, P_4 with a strongly projecting hypoconid and often a deep and narrow lateral valley in front of it, and no paraconid-metaconid fusion on the P_4. An earlier species of *Menelikia* with longer horns less distinct from those of *K. sigmoidalis* occurs in Member C of the Shungura Formation. A short-horned variety of *M. lyrocera* is known from Members H and J.

The Kaiso Village locality of the Kaiso Formation has yielded horn cores of cf *Menelikia lyrocera* (Cooke and Coryndon 1970, p. 214). Also, a horn core collected by Bishop from Nyawiega lower in the same formation (referred by Cooke and Coryndon to *Pultiphagonides* cf *africanus*) appears to belong to *Kobus sigmoidalis.*

Kobus ancystrocera (Arambourg 1947, p. 416) occurs in Members B to J of the Shungura Formation, Omo, and above the KBS Tuff at East Turkana (Harris 1976, p. 296). It has long horn cores inserted at a very low angle and curving strongly upward and even slightly forward at the tips. They are very divergent and have some mediolateral compression with a flattened lateral surface. Arambourg placed the species in *Redunca,* but it seems rather remote from the neat group formed by the three living species and *R. darti.* Its larger size and long horn cores with their low insertion angle, great divergence, curvature at the tips, and flattened lateral surface are all unlike *R. darti.*

Thaleroceros radiciformis Reck (1935, 1937) is known only from the holotype frontlet with right horn core from Olduvai Bed IV. It is the only Olduvai antelope to have survived the Second World War in Munich and is perhaps the oddest antelope to have come from Olduvai. It is moderate to large-sized with massive horn cores passing backward then upward from the base and parallel to one another. There is a sudden diminution of cross-sectional area near the tips. The massive united horn core pedicel is without internal sinuses and has paired anterior protuberances just below the horn core bases. It is possible that this Bed IV frontlet represents a descendant of *K. ancystrocera,* specialized in its massive horn cores and united pedicel. A pair of unidentified horn cores, BM M-15925, from Kanam "Museum Cliff" are structurally intermediate. *"T. radiciformis (?)"* of Leakey (1965, p. 65) is an alcelaphine and is unrelated to the present species.

Another reduncine, common in Member B of the Shungura Formation, shows a short braincase hardly at all angled on the face axis, long horn cores with some anteroposterior compression, their insertions close together, and small supraorbital pits. This form has not been found elsewhere, but a related or ancestral species occurs at Afar. It is abundant in the middle part of the Hadar Formation.

A short-horned *Kobus,* perhaps related to *K. sigmoidalis,* is known from Langebaanweg. A similar form, but with slightly longer horn cores, comes from Sahabi, and a cast of a right horn core from Wadi Natrun, BM M-8200 labeled *"? Gazella,"* is also very similar.

Reduncines with koblike characters are known from the Dhok Pathan, Tatrot, and Pinjor Formations of the Siwaliks. They are difficult to group into specieslike taxa, a problem that Pilgrim (1939, pp. 99–129) solved by giving almost every specimen a different name. The Pinjor forms seem well fitted to be ancestral to both kobs and *K. sigmoidalis,* but without knowing their stratigraphic levels of origin or having radiometric dates, one cannot be sure if they are ancestral or merely primitive Indian survivals already isolated from African forms. They have strong temporal ridges on the braincase roofs, which suggests the likelihood of a boselaphine origin, and early tooth remains are also not unlike Boselaphini. It is interesting that the male skulls BM M-487 and M-2402, but not the females M-36673 and M-39569, retain small preorbital fossae.

The Tatrot and Dhok Pathan remains are of smaller animals and have been described under the names *Dorcadoxa porrecticornis* (Lydekker), *? Gazella superba* Pilgrim, cf *Indoredunca theobaldi* Pilgrim and cf *Hydaspicobus auritus* Pilgrim. Horn cores of similar reduncines have been found at Baard's Quarry, Langebaanweg, and a similar frontlet, KNM-LU 011, in the Lukeino Formation of the Lake Baringo sequence, Kenya. The horn cores are somewhat mediolaterally compressed, inserted above the back of the orbits, fairly divergent, with a slight and regular backward curvature. They can be distinguished from gazelles only by their larger size, greater divergence, deep postcornual fossa, and larger supraorbital pit. These data suggest the near contemporaneous occurrence of similar primitive reduncine forms in Asia and Africa during the earliest Pliocene (see figure 27.4).

Tribe Hippotragini

The Hippotragini are a tribe of medium to large, rather stocky antelopes. The few species cover a wide range of habitats. Both sexes have long, not very divergent horn cores without keels or transverse ridges, hollowed pedicels to the horn cores, teeth quite hypsodont without much premolar reduction, molars with basal pillars, lower molars with a tendency to have anterior goat folds, and P_4s without fusion of paraconid and metaconid.

There are three living genera, *Hippotragus, Oryx,* and *Addax.* The species of *Hippotragus* comprise *H. leucophaeus,* the small South African blaauwbok exterminated in about 1799; *H. equinus,* the roan antelope with uprightly inserted and backwardly curved horn cores; *H. niger,* the sable antelope with longer, more mediolaterally compressed horn cores and higher frontals between the horn bases than in the roan; *Oryx gazella,* the gemsbok and beisa, with a lower and wider skull and straighter, more anteroposteriorly compressed horn cores inserted very obliquely; *O. dammah,* the scimitar oryx with more curved horn cores; *O. leucoryx,* the Arabian oryx; and *Addax nasomaculatus,* the addax of the Sahara with spiraled horns.

Fossils are known related to living *Oryx* and *Hippotragus,* and there is also an extinct lineage of *Hippotragus.*

In the later Pleistocene of South Africa three species of *Hippotragus* are known. A small one having teeth with an advanced occlusal pattern and long premolar rows occurs at Swartklip (Hendey and Hendey 1968, p. 54), Elandsfontein, and more recent sites, and Cooke (1947) referred similar mandibles from Bloembos to *H. problematicus.* All these remains are most likely to belong to the extinct blaauwbok. Mohr's (1967, p. 64) belief that *H. problematicus* had different teeth from *H. leucophaeus* arose from a comparison of the fossil with a skull in Glasgow, identified as *H. leucophaeus.* However, Mohr's illustrations suggest that the Glasgow skull could well be a sable, in which case the difference of *problematicus* from *leucophaeus* remains to be demonstrated.

Hippotragoides broomi Cooke (1947) from the

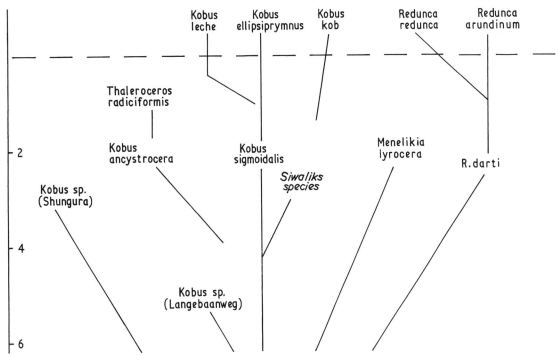

Figure 27.4 Tentative phylogeny for Reduncini. Living species are entered above the dashed line and age is shown on the left in millions of years. Related non-African reduncines are shown in italics. No fossils are known related to *Kobus megaceros* or *Redunca fulvorufula.*

Sterkfontein "upper quarry" (not the australopithecine site) is likely to be an example of the living roan antelope. In a most interesting paper, Klein (1974) has demonstrated both that *H. equinus* occurs in the late Pleistocene and Holocene of the southern Cape Province and that at a number of sites it existed contemporaneously with *H. leucophaeus.*

A third species of *Hippotragus* is the extinct *H. gigas* L. S. B. Leakey (1965), which occurs abundantly at Elandsfontein and as a few teeth at Florisbad but appears not to have survived into the Holocene. It differs from sable or roan by less mediolaterally compressed horn cores, a larger oryxlike dentition with relatively shorter premolar rows, P_2 as small as in *Oryx,* a simpler occlusal pattern of the molars, strong temporal ridges on the braincase roof, a low and wide occipital surface, and a short basioccipital with anterior tuberosities localized more as in *Oryx* than as in other *Hippotragus.*

Further back in the Pleistocene there is no record of *H. leucophaeus,* but *H. gigas* is found in Beds I through III at Olduvai, its type locality. This material has some differences from the Elandsfontein remains, so that the latter should certainly have subspecific status. At Olduvai the temporal ridges are less strong, the occipital less low and wide, and the horn cores larger. The dimensions of the horn cores probably exceed all but those of the largest

sable. The holotype (Leakey 1965, p. 49, pls. 56 and 58) is from Bed II, probably above the Lemuta Member. The adult females and the young of both sexes probably had more inclined horn core insertions than the males, and females probably had less backward curvature. It is difficult to distinguish hippotragine teeth from those of bovines in Olduvai Beds I and II. They seem to be smaller and the premolar rows shorter; the lower molars may have less narrowed lateral lobes, less pronounced outbowings on the medial walls, and goat folds extending down to the neck of the teeth. When Bovini have goat folds, these are smaller and disappear before the tooth is completely worn down. The hippotragine lower P_4 has a large bulbous metaconid (less apparent in living *Hippotragus*) but little tendency for paraconid and metaconid to fuse. In hippotragine upper molars the mediolateral width is smaller nearer the base and the central cavities are less curved.

Horn cores of *H. gigas* are also known from East Turkana (Harris 1976), Kanjera, and Peninj and of *H. gigas* or an ancestral species from Member G of the Shungura Formation, Omo. It is interesting that some large *Hippotragus* upper molars from Kanjera, BM M-25711, M-25695, and M-25702 are more advanced than *H. gigas* teeth from other sites. Perhaps *H. equinus* and *H. niger* descended from some population of *H. gigas,* while other populations elsewhere evolved into the later subspecies of *H. gigas* as known from Elandsfontein. *H. niger* is more specialized than the other two species in its narrow braincase, compressed horn cores of the males, narrower dorsal orbital rims, and frontals raised between the horn core bases. Presumably it evolved more recently than *H. equinus*.

Oryx sp. indet. is known from Olduvai Bed I by one or possibly two horn cores and from above the KBS Tuff at East Turkana by two partial crania (Harris 1976, p. 297). There is a frontlet from Member G of the Shungura Formation, but the upper molar that Arambourg (1947, p. 410) referred to *Oryx* is bovine. An oryx horn core was recorded from Mansoura near Constantine, Algeria (Joleaud 1918, p. 90, fig. 1), in deposits of Villafranchian-equivalent age. Coppens has referred to hippotragines of the same age at Bel Hacel and Aïn Hanech, Algeria, and at Garaet Ichkeul, Tunisia, but further details have yet to be published in the continuation of Arambourg (1970). Balout (1942) attributed a palate from Algiers to *Addax nasomaculatus*, but the identity cannot be taken as established.

The horn core of *Praedamalis deturi* Dietrich (1950, pl. 2, fig. 23) from Laetolil is likely to be hippotragine. It is inserted rather uprightly and has almost no curvature, hence it is difficult to know whether to relate it to *Oryx* or *Hippotragus*. Teeth of *Aeotragus garussi* Dietrich (1950, pl. 3 figs. 37–40, 42) may be conspecific, although the horn core assigned to this name is indeterminate. A slender, backwardly curved horn core from the earlier North African site of Sahabi is possibly on the *Hippotragus* lineage. Less certainly hippotragine are the Laetolil horn cores *Gazella kohllarseni* and Aepycerotinae gen. et sp. indet. (Dietrich 1950, pl. 1, fig. 7; pl. 4, fig. 45).

A Wadi Natrun tooth assigned by Andrews (1902) to *Hippotragus ?cordieri* has already been mentioned. Studer (1898) thought a Wadi Natrun horn core was of the oryx group, and Andrews tentatively associated it with the above tooth, but it is from an unidentified alcelaphine. "*Hippotragus* sp." from Langebaanweg (Gentry in Hendey 1970, p. 115) now appears to be an alcelaphine.

The holotype cranium of *Sivatragus bohlini* Pilgrim (1939) from the Pinjor Formation of the Siwaliks would be suitable as an ancestor of *H. gigas*. It is smaller but has a short braincase and a short, small basioccipital with localized anterior tuberosities. Its stronger temporal ridges and a braincase less inclined than in *H. gigas* are compatible with descent from the Boselaphini. *S. brevicornis* Pilgrim (1939) is a form of uncertain tribal affinity. Since *S. bohlini*

is the type species and appears to be congeneric with *H. gigas,* it should be possible to sink *Sivatragus* in *Hippotragus* (see figure 27.5).

A fossil *Oryx* is also known from the Pinjor Formation, represented by the two conspecific skulls, *Antilope sivalensis* Lydekker (1878) and the immature *Sivoryx cautleyi* Pilgrim (1939). *Oryx sivalensis,* as it should probably be known, has a more angled braincase roof without temporal ridges and horn core insertions more inclined than in *Hippotragus bohlini,* but a large shallow preorbital fossa would fit boselaphine ancestry.

Subfamily Alcelaphinae

Tribe Alcelaphini

The Alcelaphini are medium to large antelopes of open country and are now confined to Africa. Their main osteological features are a long skull, horns in both sexes, horn cores often of irregular course and often with transverse ridges, generally no keels, extensive internal sinuses of frontal and horn pedicels, short braincase strongly angled on the face axis, zygomatic bars usually deepened anteriorly under the orbits, a central longitudinal groove on the basioccipital, cheek teeth strongly hypsodont, without basal pillars, complicated central cavities, rounded medial lobes of upper molars, and lateral lobes of lowers, widely outbowed ribs of upper molars, lower molars without goat folds, short premolar rows with P2s reduced or absent, P_4 with fused paraconid and metaconid, mandible with deep horizontal rami, limb bones with markedly cursorial adaptations.

There are four living genera, *Alcelaphus, Damaliscus, Connochaetes,* and *Beatragus,* and I would include *Aepyceros,* which is normally put in the Antilopini. *Alcelaphus buselaphus* is the widespread hartebeest with a united horn pedicel and immense local variation in horn core morphology; *A. lichtensteini* is the Central African Lichtenstein's hartebeest with horn cores set widely apart; *Damaliscus lunatus* includes the topi, tiang, and tsessebe with horn cores inserted more forward, nearer to the orbits, and without the abrupt alterations in course seen in hartebeests; *D. dorcas* is the smaller South African bontebok and blesbok with skull morphology very similar to *D. lunatus; Connochaetes gnou,* the black wildebeest, has a lower and wider skull than any of the preceding species and horn cores with broad bases passing forward to end in sharply recurved tips; *C. taurinus,* the larger blue wildebeest, also has a wide skull but a longer face, and horn cores without such large bases and not passing forward; *Beatragus hunteri,* the herola or Tana

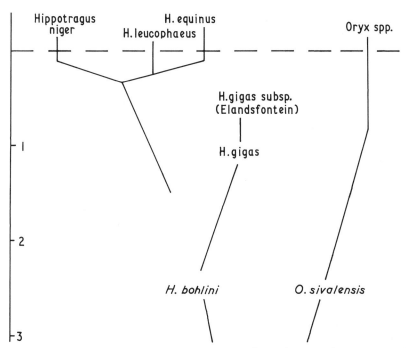

Figure 27.5 Tentative phylogeny for Hippotragini. Living or recently extinct species are entered above the dashed line and the age is shown on the left in millions of years. Non-African hippotragines are shown in italics. No fossils are known related to *Addax nasomaculatus*.

River hartebeest, with long horns diverging outward then having long parallel or subparallel distal parts. *Aepyceros melampus,* the impala, is a smaller antelope with horns somewhat like those of the preceding species but only found in males.

Alcelaphine fossils assignable to all the living and several extinct lineages are abundant.

Megalotragus priscus (Broom 1909) is the terminal species of an extinct lineage of very large alcelaphines. It is known from late Pleistocene sites of South Africa but not from northern Africa, and in my opinion it includes *Pelorocerus helmei* (Lyle), *P. elegans* Van Hoepen, *Megalotragus eucornutus* Van Hoepen, *Lunatoceras mirum* (Van Hoepen), and probably *Pelorocerus broomi* Cooke (1949). It had a relatively narrow skull for its great size, being larger than any living alcelaphine. It had molar teeth without a complicated occlusal pattern, very short premolar rows with reduction of anterior premolars, and long legs. The long curved horn cores are quite like those of the bovine *Pelorovis oldowayensis,* but do not reach the great size of the Olduvai Bed II representatives of that species and have a stronger upward component to their curvature. No complete skull has been found. The degree of basal divergence in the horn cores is weak in specimens from Cornelia, Elandsfontein, Steynspruit, and Mahemspan but strong in those from Florisbad, Mockesdam, and Kranskraal and in the holotype

cranial piece from the Modder River (all localities except Elandsfontein being in the Orange Free State). It remains to be seen whether this distribution will stand up to an interpretation of the two varieties as temporal subspecies or whether those horn cores with greater divergence, being also the longer ones, are from older individuals.

In the earlier Shungura-Olduvai time span *Megalotragus* is represented by *M. kattwinkeli* (Schwarz 1932, p. 4), which includes as synonyms *Alcelaphus howardi* Leakey (1965) and *Xenocephalus robustus* Leakey (1965). It has short horn cores inserted not quite as far back as in *M. priscus. M. kattwinkeli* is known from Olduvai middle Bed II to Bed IV, and the same species or its ancestor is known from the Koobi Fora Formation (Harris 1976), earlier Olduvai levels, and the later levels of the Shungura Formation.

Wildebeest survived in North Africa until the late Pleistocene (Ficheur and Brives 1900, reporting toothrows figured by Pomel 1894b, pl. 2, figs. 1–4), and perhaps into the Neolithic (e.g., Débruge 1906, an identification based on a metacarpal and teeth). Arambourg (1938, p. 42) is doubtful about Neolithic records. Wildebeests that can best be classified as belonging to extinct subspecies of *Connochaetes gnou* are known from later South African sites. *C. gnou antiquus* Broom (1913, p. 14) from Florisbad has horn cores passing less markedly forward from the base and tips perhaps less recurved. *C. g. laticornutus* (Van Hoepen 1932, p. 65) is an earlier stage from Cornelia and Elandsfontein, known from crania and horn cores but not from facial parts. The crania resemble *C. taurinus* at about the time level of Olduvai Beds II through IV, but the base of the horn core has begun to be expanded in dorsal view and a rugose surface texture has begun to spread across the frontals. These features are scarcely visible in the holotype cranium from Cornelia (Van Hoepen 1932, fig. 3), which I attribute to its being a female.

The wildebeest from Olduvai middle Bed II to Bed IV is *Connochaetes taurinus olduvaiensis,* differing from the living subspecies by its horn cores being inserted at a slightly less posterior level and passing less downward as they emerge from the skull. A cranium from the upper Pleistocene of Temara in Morocco, assigned by Arambourg (1938, p. 38) to the Ternifine (lower Pleistocene) form *C. taurinus prognu* Pomel, is very similar to that from Olduvai. However, Pomel (1894b) did not illustrate any of the Ternifine (= Palikao) horn cores, so that more information is required before we can adequately assess the validity of these two subspecific names. *Rhynotragus semiticus* Reck (1935, p. 218, fig. 1) is an odd skull fragment from Olduvai that Schwarz (1937, pp. 60, 85) took as a subspecies of *C. taurinus* and to which he assigned much Olduvai material. The absence of a sharply outlined temporal fossa between horn core base and orbit and the sloping braincase roof of Reck's fossil make it unlikely to have been a wildebeest, but its true affinities remain uncertain.

In the Koobi Fora Formation above the KBS Tuff (Harris 1976, p. 298) and the earlier sediments of Olduvai, up to and including the lowest levels of middle Bed II in the HWK area, the wildebeest present has horn cores inserted less far back on the skull and their distal parts not so markedly turned upward and inward. It is presumably a precursor of *C. taurinus.* A still earlier wildebeest is *Oreonagor tournoueri* (Thomas 1884, pl. 7, fig. 1) from Aïn Jourdel, Algeria, in which the horn cores are less extremely divergent. A problematical specimen is the Olduvai Bed II holotype skull of *Connochaetes africanus* (Hopwood 1934, p. 549), originally the type species of the genus *Pultiphagonides.* The face has closer resemblances to *C. gnou* than to *C. taurinus,* and the rather small horn cores are inserted extremely widely apart—almost as if space were being left for descendants to develop the basal bosses of living *C. gnou.* Thus the *C. gnou* lineage formerly may have occurred north of South Africa, as did *Antidorcas,* and the identification of fossil wildebeest,

especially less complete specimens, may be more uncertain than has been implied here.

Alcelaphus buselaphus is known only from very late sites in southern and northern Africa; like *Syncerus caffer caffer,* it can have appeared only in the very recent past. I discovered *Alcelaphus lichtensteini* horn cores among material from Broken Hill, Zambia, hitherto placed with *Connochaetes;* the horn cores differ from those of the extant members of the species by being less curved upward and passing more forward as they rise.

Rabaticeras arambourgi Ennouchi (1953) is known from Morocco, Elandsfontein, and Olduvai Bed III. It has horn cores arising close together and above the orbits, inserted uprightly and curving forward with a torsion that is clockwise from the base upward on the right side, and with a short braincase much angled on the facial axis. *R. porrocornutus* Vrba (1971) from Swartkrans, South Africa, is similar. I believe that both the living species of hartebeest descended from *Rabaticeras,* in which case the united horn pedicel of *A. buselaphus* must have appeared very quickly in geological time. It is noteworthy that the characters whereby the Broken Hill fossils of *A. lichtensteini* differ from extant specimens make the fossils less removed from *Rabaticeras.* Also, a frontlet from Aboukir, Algeria, of *Boselaphus probubalis* Pomel (1894b, pl. 4, figs. 14 and 15) is certainly more primitive than the living hartebeest and may be a subspecies of *R. arambourgi,* although Arambourg (1938, p. 37) had made Pomel's name a synonym of *A. buselaphus.* An earlier *Rabaticeras*-like cranium from the Lemuta Member at the top of lower Bed II, Olduvai, has horn cores inserted more widely apart and at a lower angle and has a less inclined braincase. A possible interpretation of this fossil is that it represents the ancestry of *Alcelaphus lichtensteini,* already separated from that of *A. buselaphus.* If this were true, *Alcelaphus* would become a diphyletic genus and nomenclature would need to be changed.

Horn cores of *Damaliscus dorcas,* or a probable ancestor of that species, are known from Florisbad and Vlakkraal; they are larger and less mediolaterally compressed, with the lyration nearer to the base. A new and unnamed species of *Damaliscus* is represented in Olduvai Beds II through IV, and possibly I. It is about the size of *D. dorcas* but differs by some details of skull morphology and, surprisingly, by shorter premolar rows without P_2. Except for its premolar characters, it seems suitable as an ancestor of both living *Damaliscus* species.

Damaliscus niro (Hopwood 1936, p. 640) belongs

to an extinct lineage known from Olduvai lower Bed II to Bed IV, Peninj, and later South African sites. It has horn cores curved backward, very compressed with flattened medial and lateral surfaces, and generally with well-spaced strong transverse ridges on the front surface. Some horn core variations are encountered at Olduvai in which there is not an even backward curvature but a sharper alteration in course near their midpoints. Some males reached a large size, but average size seems to have become smaller over the time span from Cornelia to Florisbad. At Florisbad and Vlakkraal its horn cores are smaller than those of the other *Damaliscus,* the medial surface less rounded, and the whole horn core more mediolaterally compressed, the widest part of the transverse section more anteriorly placed, and without the slight lyration. A horn core from Wonderwerk Cave, Cape Province ("cf *Capra walie,*" Wells 1943, p. 268; also see Cooke 1941), still had part of the horn sheath attached. It is unfortunate that there is not a single toothrow definitely associated exclusively with *D. niro* horn cores. The species is not known in northern Africa.

Damalops palaeindicus (Falconer) from the Pinjor Formation of the Siwaliks seems most likely to be related to the *Alcelaphus–Rabaticeras–Damaliscus* group of alcelaphines.

Parmularius Hopwood is an extinct stock of medium-sized alcelaphines with horn cores tending to have basal swellings, a tendency to small bosses on the parietal roofs, small preorbital fossae, short premolar rows, and deep mandibles. They are known principally from Olduvai. *P. angusticornis* (Schwarz 1937, p. 55), formerly placed in *Damaliscus* and including *D. antiquus* Leakey (1965), is a well-known species from middle and upper Bed II with heavy, straight horn cores and a much shortened braincase. It is also known from Kanjera and Isimila. It is probably descended from the slightly smaller type species *P. altidens* Hopwood (1934) of Olduvai Bed I and lower Bed II and the highest part of the Shungura Formation. This latter form has less massive and more backwardly turned horn cores. It is possibly a different lineage that produced *P. rugosus* Leakey (1965) of Beds III and IV and possibly earlier and that appears to have outlived *P. angusticornis* (see figure 27.6).

A frontlet, SAM-16561, and some horn cores from

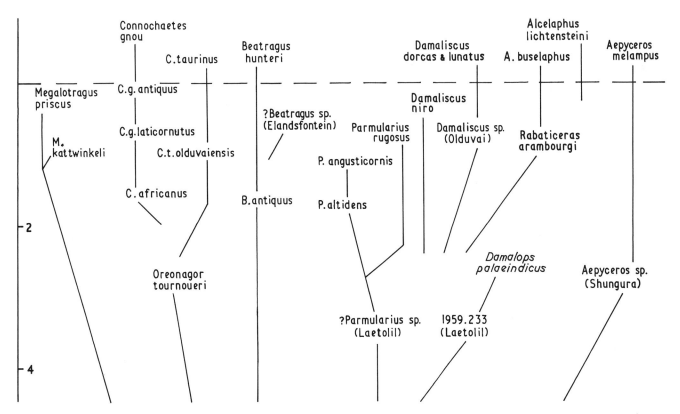

Figure 27.6   Tentative phylogeny for Alcelaphini. Living species are entered above the dashed line and age is shown on the left in millions of years. The single non-African alcelaphine is shown in italics.

Elandsfontein seem most likely to be a species of *Beatragus* with short and very divergent horn cores, however other interpretations are possible.

Beatragus antiquus Leakey (1965) was evidently more widespread than the living *B. hunteri,* being known from Olduvai Beds I to II, and as an immature cranium from just below Tuff H of the Shungura Formation. It differed from *B. hunteri* by its larger size, wider skull, and more uprightly inserted horn cores and may have been its ancestor.

Aepyceros, apparently *A. melampus,* is known by a few small horn core pieces from Olduvai Gorge and Peninj. An impala is present in the Koobi Fora Formation (Harris 1976) and abundant in the Shungura Formation, both localities being north of the present-day range of *Aepyceros.* The material in the earlier part of these sequences is sufficiently distinct to justify specific separation from *A. melampus.* In Shungura Member B it is smaller, has a less lengthened face, supraorbital pits placed less widely apart, the vestiges of a preorbital fossa, back of the nasals more narrowly pointed, and a longer premolar row. Horn cores of an even more primitive species occur in the Mursi and Hadar Formations, Ethiopia, and in the Karmosit Beds, Kenya. It is also possible that the supposed gazelle at Fort Ternan could be connected with *Aepyceros* instead (cf Gentry 1970a, p. 292).

Alcelaphines other than *Aepyceros* are known from African sites predating the Shungura Formation. A Laetolil cranium, 1959.277, could be primitive in its rather long, little-angled braincase, and other characters. It has a low parietal boss and could be an ancestor of *Parmularius altidens* and *Damaliscus niro;* for the present it is best classified as *?Parmularius* sp. The Laetolil horn core called "Reduncini gen. et sp. indet." by Dietrich (1950, pl. 2, fig. 21) is probably conspecific with the latter taxon. The alcelaphine species that Dietrich named *Parestigorgon gadjingeri* will remain of uncertain identity until better finds become available.

An alcelaphine found at Afar may be conspecific with a partial cranium and horn cores, 1959.233, from Laetolil, and it may become possible to link this stock with *Damalops* mentioned above.

Two species of primitive alcelaphines occur at Langebaanweg. Their frontals between the horn core bases are higher than the orbital rims, and the horn pedicels are internally hollowed as in later alcelaphines. One species has a braincase strongly angled on the facial axis, the distal parts of the horn cores tending to pass outward rather than backward and the long axis of the horn core cross-section set at a considerable angle to the long axis of the skull.

The other species has a less angled braincase, horn cores with a backward curvature, and normally orientated pedicels. The alcelaphine teeth at Langebaanweg are markedly primitive in the upper molars with a more pronounced mesostyle and less rounded medial lobes than in later forms, basal pillars on M1 and occasionally on the upper molars, less constriction across the central cavities, and less rounded medial and lateral lobes of the lower molars. They are more primitive than alcelaphine teeth at Laetolil.

A cast of an M_3 from Wadi Natrun, BM M-12361, which Andrews (1902, p. 439) thought was from a large gazellelike form, could perhaps represent an alcelaphine at the Langebaanweg level of development. It is less hypsodont than Pleistocene or living alcelaphines. A badly preserved horn core base, BM M-8199 (cast), labeled *Hippotragus ?cordieri,* also appears to be alcelaphine by the large, smooth-sided internal cavity in its horn pedicel.

Subfamily Antilopinae

Tribe Neotragini

The Neotragini includes 14 living small-sized antelopes, possibly not very closely related to one another. They have simple, short, and small horn cores that are straight or curved slightly forward. They have large preorbital fossae. There are generally no basal pillars on the molars and no outwardly bowed ribs between the styles on the lateral walls of the upper molars; the enamel outer walls of the molars tend to be straight with pointed rather than rounded corners. The living genera are *Neotragus* (including *Nesotragus*) with three species, *Madoqua* (including *Rhynchotragus*) with five living species, *Oreotragus,* *Dorcatragus,* and *Ourebia* with one species each, and *Raphicerus* with three species. Fossils related to most of these genera are known but are not common.

The Vaal rhebbok, *Pelea capreolus,* now recognized as not a reduncine (Roberts 1937, p. 86; Wells 1967, p. 100), might be related to Caprinae (Gentry 1970b, p. 65) or it might be a large neotragine. Fossil *Pelea* occurs in the Transvaal caves (Vrba 1975) as far back as the Kromdraai faunal site (KA) and the Swartkrans main assemblage (SKa), i.e., probably back to the lower Pleistocene.

A *Raphicerus* at Melkbos and Swartklip is not distinguishable from living grysbok or steinbok, but the Elandsfontein *Raphicerus* was larger and had more inclined horn cores with a tendency toward posterolateral keels. The paratype horn core of *"Cephalophus" pricei* from Makapansgat Limeworks appears to be a *Raphicerus* with a keeled horn core like some of those from Elandsfontein. Some horn cores of *Raphicerus* at Langebaanweg are larger than at Elandsfontein and have a still more irregular cross-section. A few mandibular pieces at Laetolil are about the right size for *Raphicerus*.

At Olduvai the only neotragine remains are two horn cores from FLKN in Bed I, perhaps of *Raphicerus*.

Oreotragus major Wells (1951, p. 167) is based on a skull from a red breccia of unknown age near Makapansgat. It is probably a late form, judging by its closeness to the living klipspringer. The Makapansgat Limeworks frontlet assigned to *Oreotragus major* by Wells and Cooke (1956, p. 35) has very short horn cores with some degree of anteroposterior compression and is not very like the holotype; it would be better referred to *Oreotragus* sp. The associated dentitions at Makapansgat Limeworks have premolar rows that are probably slightly longer relative to the molar rows than in living *O. oreotragus*.

Palaeotragiscus longiceps Broom (1934, p. 477, fig. 6) is a maxilla from Taung, thought to be an odd neotragine. A right upper molar, BM M-25708 from Kanjera, is probably of *Ourebia*.

Madoqua avifluminis (Dietrich (1950, p. 34) is known by many horn cores and dentitions from Laetolil. Dietrich referred them to *Praemadoqua* gen. nov., but the horn cores and proportionate lengths of the premolar and molar rows agree with living *Madoqua*. There is a small back lobe on the M_3 agreeing more with subgenus *Rhynchotragus* than with *Madoqua*. A complete metatarsal in the Nairobi collection is shorter and thicker than in living dik-diks.

A maxilla fragment from the Ngorora Formation, KNM-BN 92, is small enough to fit Neotragini and appears not to belong to a duiker.

Tribe Antilopini

The Antilopini are small- to medium-sized antelopes that generally live in arid areas. The main features of the living species are braincase not strongly bent down on the face axis, complicated midfrontal and parietofrontal sutures, teeth hypsodont and without basal pillars, central cavities with a simple outline, and P_4 normally without paraconid-metaconid fusion. The horn cores of the constituent species differ so much from one another that it is impossible to generalize about them. The living genera are *Antilope*, *Gazella*, *Antidorcas*, *Litocranius*, and *Ammodorcas*, but only *Gazella*, the gazelles, has more than one living species. *Antilope cervicapra* is the spiral-horned Indian blackbuck and *Antidorcas marsupialis*, the South African springbok. An African fossil belonging to *Antilope* is known. Other

African fossils are of *Antidorcas* and *Gazella*, the former having sinuses in the frontals that raise their level between the horn bases, male horn cores usually bent backward quite strongly, higher-crowned teeth, shorter premolar rows, P_2 generally absent, and deeper mandibular rami.

Antidorcas australis Hendey and Hendey (1968, p. 56) from Swartklip, Elandsfontein, and other Cape Province coastal sites, was originally described as a subspecies of *A. marsupialis* but is now thought to be a separate species (Hendey 1974, p. 52). It differs from *A. marsupialis* by its smaller and more mediolaterally compressed horn cores without any sharp bending backward and outward and by its smaller dentitions. It could be another form peculiar to the southern Cape, like *Hippotragus leucophaeus* and the unnamed kudu.

Some horn cores from Elandsfontein and Bolt's Farm are indistinguishable from the Olduvai species *A. recki* (see below). They are less common at Elandsfontein than is *A. australis*, and Hendey (1974, p. 52) believes that the *A. recki* may have been more a species of the inland plateau and ancestral to the living springbok.

Almost certainly to be included in *Antidorcas* is *Gazella bondi* Cooke and Wells (1951), originally described from Chelmer, Rhodesia, but also at Vlakkraal and Florisbad. It has small teeth, but they are so extremely hypsodont that the lower edge of the mandibular ramus in subadults is seriously distorted. Vrba (1973) has described the first known horn cores and crania from Swartkrans and concludes that *A. bondi* is not ancestral to *A. marsupialis*. Klein (pers. comm.) has identified it from Late Stone Age levels of Border Cave, South Africa, dated to about 36,000 B.P. (Beaumont 1973, p. 45). The holotype of *Gazella wellsi* Cooke (1949, p. 38, fig. 11), a left adult mandible from Power's Site on the Vaal River, is possibly conspecific with *A. bondi* or *A. recki* and is discussed by Vrba (1973, p. 311).

The common antilopine at Olduvai is *Antidorcas recki* (Schwarz 1932, 1937, p. 53), originally placed in a new genus, *Phenacotragus*. It was smaller than *A. marsupialis* and had horn cores more mediolaterally compressed and often more sharply bent backward in their distal parts, premolar rows less reduced, and shorter radii, tibiae, and metapodials. The Olduvai horn cores show considerable variability, part of which may be due to a temporal succession of different varieties. Gentry (1966, p. 56) had referred most of the Bed I examples to *Gazella wellsi*, but it now seems wisest not to complicate the East African story with the introduction of this name. It is possible that extensive horn core variability in time and space has characterized other *Antidorcas* species. *A. recki* is also at East Turkana (Harris 1976), Peninj, and Kanjera. It, or its immediate ancestor, is known from the Shungura Formation, Omo, by a few horn cores with much compression and a slight but localized bending back of their distal parts. It seems that an *Antidorcas* occurred in Villafranchian-equivalent deposits of Aïn Brimba, Tunisia (Arambourg and Coque 1958); Coppens (1971) indicates that it was also present at Oued el Atteuch and Aïn Boucherit in Algeria, and Bayle (1854) had earlier referred to what may be this species from Mansoura. *Antidorcas* sp. Arambourg (1947, p. 390) from the Omo is the mandible of a small alcelaphine. The description of *Gazella gazella praecursor* Schwarz (1937) from Olduvai suggests that it could have been *A. recki*, but the material was lost in Germany during the Second World War.

Leakey (in Clark 1959, p. 230) identified *Litocranius* on some limb bones from Broken Hill, Zambia, but *Gazella* or *Antidorcas* seem a more likely identification.

Fossil gazelles are well known in North Africa. *G. atlantica* Bourguignat (1870, p. 84), abundant in the middle and late Pleistocene, seems to have been about the size of a springbok and had rather short, thick horn cores with strong backward curvature and a flattened lateral surface. It is difficult to know which living gazelle to link it with, but it ought to be compared with the early Holocene *G. decora* and *G. arista* Bate (1940, pp. 419, 429) of Palestine.

G. tingitana Arambourg (1957) is another late Pleistocene North African species, having very long, backwardly curved, and slender horn cores. It might be ancestral to the living *G. leptoceros,* which has long and slender but straight horn cores. Other late Pleistocene fossils belong to gazelles that survived in the region until recent centuries (*G. dorcas, G. rufina, G. cuvieri*). Many names proposed by Pomel (1895) for North African gazelles have been amended and synonymized by Arambourg (1957).

G. thomasi (Pomel 1895, p. 18), the corrected name for the junior homonym *G. atlantica* Thomas 1884, comes from Villafranchian-equivalent deposits of North Africa and is a horn core with strong mediolateral compression. The horn cores of *G. praethomsoni* Arambourg (1947, p. 387) from the Shungura Formation are very similar but with more backward curvature; the mandibular piece assigned to *G. praethomsoni* by Arambourg (1947, p. 388, pl. 27, fig. 1) is antilopine, but probably *Antidorcas.*

G. setifensis (Pomel 1895, p. 15, pl. 10, figs. 14, 15) is another North African gazelle of similar age to *G.*

thomasi, somewhat larger and with less mediolateral compression and more curvature of its horn cores.

The fossil gazelles of eastern and southern Africa are somewhat easier to arrange in tentative lineages. An extinct *Gazella* is known from several horn cores and mandibles at Elandsfontein far to the south of the present limits for the genus in Tanzania. The same species is found in Olduvai Bed II, Peninj, and Kanam West, and it is less common than *Antidorcas* at Olduvai, as at Elandsfontein. It has only slightly compressed horn cores, inserted at a low angle and little curved backward. The level of the greatest mediolateral diameter lies more anteriorly in the cross-section than in living *G. dorcas;* this and the low inclination might align it with the living African *G. thomsoni* and *G. rufifrons.* A horn core from Olduvai Bed I and another from the base of middle Bed II are more mediolaterally compressed. The mandibles assigned to this species have a shallower ramus and longer premolar row than *Antidorcas.*

G. janenschi Dietrich (1950, p. 25, pl. 2, fig. 22) is known by horn cores and dentitions from Laetolil and could be ancestral to the Olduvai and Elandsfontein gazelle (see figure 27.7). It differs by its slightly smaller horn cores, divergence lessening a little toward the tips, no flattening of the lateral surface, and stronger backward curvature. Teeth of *Gazella hennigi* Dietrich (1950, pl. 5, fig. 47) are probably conspecific, but the *G. hennigi* horn cores are perhaps of an *Antidorcas.*

A second gazelle lineage is represented by *G. vanhoepeni* (Wells and Cooke 1956, p. 43) from Makapansgat Limeworks, originally placed in *Phenacotragus* but later removed to *Gazella* (Wells 1969). It is a large gazelle with strongly compressed horn cores, and Wells was disposed to link it with the three large living African gazelles, *G. granti, G. soemmerringi,* and *G. dama.* I think he was right, and I think that the Makapansgat Limeworks *G. gracilior* Wells and Cooke (1956, p. 37) is likely to be the female of *G. vanhoepeni.* The species would have had rather less sexual dimorphism of its horn cores than living gazelles.

A Langebaanweg gazelle has very compressed horn cores that are curved backward. These features cause it to resemble *G. vanhoepeni,* to which it may be ancestral. It is possible that the North African *G. thomasi* and the Omo *G. praethomsoni,* already mentioned above, are related to the *G. vanhoepeni* lineage, but the available material of them is insufficient for decision. A large Laetolil antilopine is represented by the teeth of *Gazella kohllarseni* Die-

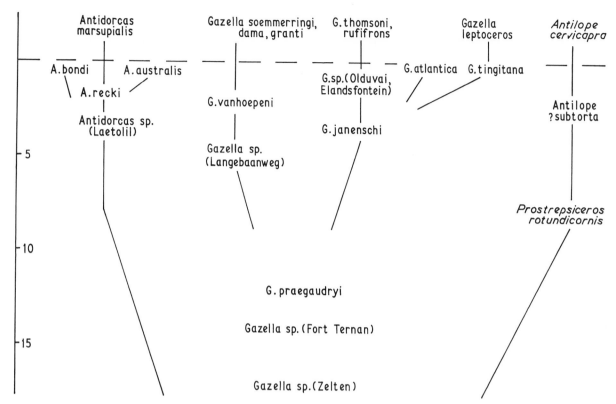

Figure 27.7 Tentative phylogeny for African Antilopini. Living species are entered above the dashed line and age is shown on the left in millions of years. Related non-African antilopines are shown in italics. The North African *Gazella dorcas, G. cuvieri,* and *G. rufina* are known from the late Pleistocene of North Africa but are of unknown ancestry; they have not been shown. *G. praethomsoni* and some other enigmatic, but named, horn cores have not been entered.

trich (1950, pl. 2, fig. 16, pl. 5, fig. 49), although the horn core assigned to this name is probably alcelaphine.

A gazelle mandible, KNM-MP 129, comes from the Mpesida Beds, Kenya, and *G. praegaudryi* Arambourg (1959, p. 123, pl. 17, figs. 9–11) was described from Oued el Hammam. Early gazelles have relatively small and little-compressed horn cores and rather brachydont teeth with long premolar rows and lower M3 without a very noticeably enlarged back lobe. The gazelle at Fort Ternan (Gentry 1970a, pp. 292, 299) has several distinctive features and has already been mentioned.

Gazella helmoedi Van Hoepen (1932, p. 65, fig. 2) from Cornelia is a long horn core, very narrow for its length and slightly curved in two planes. It is not a gazelle and I can only suggest that it may be an oryx horn core, distorted in life or after fossilization. Wells (1967, p. 102) has suggested that it might belong to an alcelaphine genus (*? Parmularius*).

Two *Antilope* horn cores from Member C of the Shungura Formation, Omo, can be rather doubtfully likened to *A. subtorta* Pilgrim (1937), a Siwaliks species from the Pinjor Formation. *A. subtorta* is an-cestral to the Indian blackbuck, *A. cervicapra* (Linnaeus), from which it differs by less twisted horn cores with a trace of a posterolateral keel.

Subfamily Caprinae

Tribe Ovibovini

The Ovibovini are a very diverse group of mostly moderate to large-sized bovids. They have short and often divergent horn cores, which are strongly specialized in some genera. Other characters are a high infraorbital foramen, large occipital condyles and basioccipital, hypsodont cheek teeth without basal pillars, with a short premolar row and paraconid-metaconid fusion on P_4. The only two living species are *Ovibos moschatus* and *Budorcas taxicolor,* the muskox of the North American Arctic and the takin of Tibet. There are many Eurasian fossils (see Gentry 1971, p. 289) and some in North America.

There is growing evidence of the presence of Ovibovini in the African fossil record, although with one exception they are only poorly known. The exception is *Makapania broomi* Wells and Cooke (1956) from Makapansgat Limeworks. This is a moderate to large-sized antelope with horn cores emerging almost transversely from the skull just behind the orbits. Gentry (1970b) believed that it was an ovibovine related to the European Villafranchian *Megalovis latifrons* Schaub, mainly on details of basioccipital anatomy. Its teeth are fairly hypsodont, without basal pillars, with poor to moderate styles and ribs on the lateral walls of the upper molars, and having a fused paraconid and metaconid on P_4. Vrba (1975) records *Makapania* from the Sterkfontein type site.

Bos makapaani Broom (1937, p. 510, figured) from Buffalo Cave in the Makapan Valley appears to be a second ovibovine species, but of uncertain generic identity. The position of sutures on the top of the holotype frontlet suggest that the convex edge of the short curved horn cores is anterior or anterodorsal, whereas Broom evidently thought that the convex edge was posterior. A horn core from Bed I Olduvai, BM M-14531, could be the same species.

A fine ovibovine skull, not of *Makapania broomi*, has been found in the Hadar Formation, Ethiopia.

A damaged Langebaanweg cranium, SAM-L 13105, resembles the cranium of the living takin.

Damalavus boroccoi Arambourg (1959, p. 120) from Oued el Hammam was described as an alcelaphine but is quite possibly ancestral to *Palaeoryx* Gaudry, a genus best known from the Samos and Pi-

kermi *Hipparion* faunas (see figure 27.8). I consider it to be an early ovibovine (Gentry 1970a, p. 132, 1971, p. 284).

Tribe Caprini

Caprini are moderate-sized bovids, tending to have rather high and narrow skulls, both horn cores and frontals internally hollowed in living species, wide anterior tuberosities of the basioccipital, and a small angle on the mandible. They have hypsodont cheek teeth without basal pillars, rather flat lateral walls between the styles of the upper molars, lower molars often with goat folds, short premolar rows, P_4 with paraconid and metaconid fused, and little enlargement of the central incisors. Living species, which are mostly Eurasian, are usually found in hilly habitats or in rocky open areas.

Capra, or *Ammotragus, lervia* is known from North African sites in the late Pleistocene, normally in association with Mousterian stone implements. Churcher (1972) found it in the late Pleistocene of southern Egypt. Romer's (1928, p. 119) reference to it in the Villafranchian appears to be a misinterpretation of a possible *Antidorcas* reported by Bayle (1854). Fossils related to either of the living African ibexes have not been discovered.

Numidocapra crassicornis Arambourg (1949),

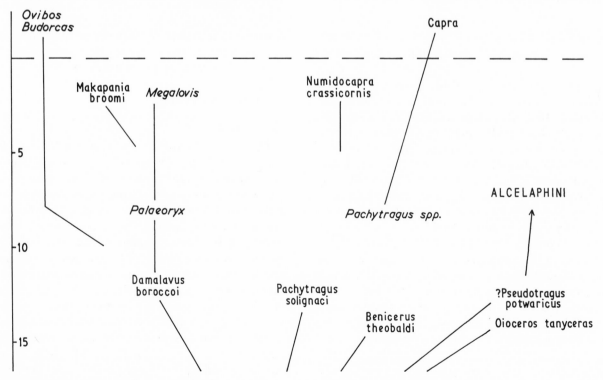

Figure 27.8 Tentative phylogeny for African Ovibovini and Caprini. Living genera are entered above the dashed line and age is shown on the left in millions of years. Non-African forms are shown in italics.

based on a frontlet from Aïn Hanech, Algeria, is large with long, thick horn cores curving upward and forward from their close insertions and with very little divergence. The top of the orbital rim is well below the level of the frontals between the two pedicels, and the short braincase is strongly inclined. Its affinities are unknown, but it is not totally unlike the smaller *Procamptoceras brivatense* Schaub from the Villafranchian of Senèze, France, which could be related to the chamois, *Rupicapra rupicapra,* which I think is likely to belong to the Caprini. Therefore I tentatively include *Numidocapra* in the Caprini.

Coppens (1971) refers to a *Capra* at Aïn Brimba, for which Arambourg will give further details.

The lower of two faunas in the Beglia Formation at Bled Douarah, Tunisia, contains predominantly savanna forms (Robinson and Black 1969), including an abundant caprine *Pachytragus solignaci* Robinson (1972). Its horn cores are like those of the Samos *P. crassicornis* Schlosser but have stronger mediolateral compression and more marked anterior keels. The teeth are more brachydont with more rugose enamel, basal pillars on the lower molars, the medial lobes of the upper molars not joined to one another or to the lateral side of the teeth until late in wear, and a smaller hypoconid on P4. Probable references to this species by Boule (1910) and Roman (1931) have already been mentioned. *P. solignaci* is a problematical species. That it has been correctly assigned to the Caprini is shown by the downward slope of the braincase in profile, the upright insertion of the horn cores, the fairly deep ramus of the mandible, and the absence of boselaphinelike specializations of the horn cores. However, the combination of primitive teeth and precociously compressed horn cores is striking, and the animal must be a different lineage from the Samos *Pachytragus,* which is likely to be at least 3 m.y. younger. One can see analogies in the differences of *Hemitragus* (tahrs) from *Capra* (goats) at the present day.

Heintz (1973) has named a horn core from Beni Mellal, Morocco, as *Benicerus theobaldi.* He did not give an opinion about its tribal or subfamily affiliation but pointed out that its torsion was in the same direction as in *Oioceros* and that it had similarities to the Chinese *Prosinotragus.*

Oioceros tanyceras Gentry (1970a) is well represented at Fort Ternan and is very like *O. grangeri* Pilgrim (1934) and *O. noverca* Pilgrim from Tung Gur, Mongolia. It has horn cores curving upward and outward with a slight clockwise torsion on the right side, a short premolar row, and limb bones with cursorial adaptations. It is not very like the

Oioceros species from the later *Hipparion* faunas of Pikermi, Maragha, and Samos.

?Pseudotragus potwaricus (Pilgrim 1939, p. 86) from the Nagri Formation of the Siwaliks also occurs at Fort Ternan (Gentry 1970a, p. 284). It has backwardly curved horn cores but may be fairly closely related phyletically to *Oioceros tanyceras.* A slightly larger descendant species is represented by a skull from the Ngorora Formation, KNM-BN 100. Both *?P. potwaricus* and *O. tanyceras* may be ancestral to Alcelaphini, and their transfer to that tribe may become necessary to preserve monophyly.

The Earliest Bovids in Africa

Cooke (1968, p. 248) had expected that a basic tragulid-pecoran stock would be first known in Africa in the later Oligocene, and it is now certain that bovids occur in the African Miocene before Fort Ternan.

Hamilton (1973, pp. 126–128) has described from Gebel Zelten, Libya, three horn cores of *Protragocerus,* two of *Eotragus,* and a mandible and mandibular fragment of *Gazella.* The gazelle has a notably deep mandibular ramus, deeper than in the gazelles from Oued el Hammam or Fort Ternan, so it may represent a lineage that later gave rise to *Antidorcas.* Of course more information is needed to be sure of this. *Protragocerus* is not known in Europe before the Vindobonian, and *Eotragus* occurs from the Burdigalian to the Vindobonian. Savage (in Selley 1969, p. 458) and Savage and Hamilton (1973, p. 525) correlated Gebel Zelten with the early Burdigalian of Europe or even the Aquitanian but did not suppose that it was of an earlier age than the East African Miocene sites that predate Fort Ternan.

Well known as a possible early bovid is *Walangania africanus* (Whitworth 1958, p. 19), a species in which Hamilton (1973, p. 146) has amalgamated "*Palaeomeryx*" *africanus* and *W. gracilis* Whitworth (1958, p. 30). The holotype is a mandible from Songhor. Many other tooth remains are known from Songhor, Rusinga, and other sites, and there are also an incomplete dentition and limb bones of a single juvenile individual from Rusinga. The femur and tibia are about the size of an adult of the Fort Ternan species *Protragocerus labidotus* so would have a size appropriate for the young of a slightly larger species. The morphology of the cheek teeth does not suggest much difference at a generic level from *Eotragus* Pilgrim (see Thenius 1952). In fact the morphology of P3 and P4 shows more of a boselaphine than a caprine pattern (see Gentry 1970a, p. 265, fig. 2). On the other hand the transversely narrow rear part of the top articular surface of the metatarsal does not suggest an affinity with the Boselaphini (see Gentry 1970a, pp. 248, 280, fig. 12). Possibly *Walangania* was a boselaphine ancestor of some caprine lineage. Alternatively it may not be a bovid but an early pecoran close to *Dremotherium* or *Amphitragulus* as known from the Oligocene to Miocene of Europe.

Other possible remains of early African bovids include the mandibular fragment of cf *Strogulognathus* (=*Eotragus*) *sansaniensis* from the Namib Desert of Namibia (Stromer 1926, p. 115), if this is not identical with *Propalaeoryx austroafricanus* (see below). There are two partial bovid crania from Maboko, BM M-15543 and M-15544, which must belong to different species. One of them could be a late occurrence of *Walangania*. Bovid teeth from Maboko have already been referred to by Whitworth (1958, p. 25) and Gentry (1970a, p. 303).

Stromer (1926, p. 117) described *Propalaeoryx austroafricanus* from the Namib Desert as a bovid, but Whitworth (1958, p. 26) assigned it and an East African species, *P. nyanzae*, only to the Pecora, and Hamilton (1973, p. 142) has now placed *Propalaeoryx* in the giraffoid family Palaeomerycidae (see chapter 25). The shallow ramus of the first species does not easily suggest bovid affinities.

An indeterminate bovid from Losodok (Arambourg 1933, p. 142, 1947, pl. 22, figs. 5, 6) is more likely to be giraffoid (Gentry 1970a, p. 302).

In Europe the earliest bovid is *Eotragus artenensis* Ginsburg and Heintz (1968) from the Burdigalian of Artenay, France. In Mongolia supposed bovids have been described from the lower Miocene of the western Gobi (*Gobiocerus mongolicus* Sokolov 1952) and from the Hsanda Gol Formation of Oligocene age at Tatal Gol (*Palaeohypsodontus asiaticus* Trofimov 1958). The identity at family level of the Asiatic species, which are based only on teeth, cannot be regarded as certain, for middle Tertiary nonsuiform artiodactyls on several continents were showing trends such as widening of the premolars, increasing selenodonty of the molars, development of hornlike appendages on the top of the skull, fusion of carpal bones, and appearance of cannon bones. It is not yet clear what characters can be taken as diagnostic of what groups of what time levels, to what extent the Bovidae are monophyletic, or whether they evolved in one place. Can early bovidlike teeth, especially in Asia, be presumed not to belong to ancestors or relatives of the musk-deer? The final words on the origin of the Bovidae can be taken from Hamilton (1973, p. 134):

The Pecora probably originated from the Traguloidea during the Upper Eocene or Lower Oligocene and of the two traguloid families the Gelocidae are the most likely to have given rise to the Pecora. In the gelocids true selenodonty is developed from more bunodont forms; thus *Lophiomeryx* has very bunoid lower molars showing few signs of true selenodonty while *Bachitherium* and *Prodremotherium* have molars which are very similar to those of *Dremotherium*. A detailed study of this group is needed and it is here that the divergence of the Bovoidea and other higher ruminants probably occurred.

Discussion

Classification and Zoogeography

Schlosser (in Zittel 1925, p. 208), Pilgrim (1939, p. 10), and Simpson (1945, p. 270) have discussed the extent to which the bovids may be divided into two groups called Böodontia and Aegodontia. This grouping, although not used in the formal classification of this chapter, does reflect important features of bovid evolution. The limits of the two groups have not been agreed upon, but böodonts could be conceived as comprising the Bovinae, Cephalophinae, and Hippotraginae, and aegodonts the Alcelaphinae, Antilopinae, and Caprinae. Böodonts tend to have lower and wider skulls, braincases little angled on their faces, horn cores more frequently keeled, internal sinuses of the frontals less frequent, less hypsodont teeth, rugose enamel, basal pillars persisting and declining later in evolution if at all, a slower rate of fusion between the lobes of the molar teeth in ontogeny, longer premolar rows, shallower mandibles, and less cursorial limb bones. (Gentry 1970a, pp. 277–282, figs. 12–15 has listed cursorial characters of bovid limb bones.) Aegodonts have the converse characters. The differentiating characters of böodonts and aegodonts reach their clearest manifestation in the late Miocene to early Pliocene. Böodonts have not exploited such ecologically extreme environments as aegodonts, and it seems likely that aegodonts evolved from early bovids with predominantly böodont characters. *Walangania* could be an incipient aegodont. Figures 27.9 and 27.10 show representative horn cores of various bovid groups.

As far as Africa is concerned, the basic split between browsing böodont Boselaphini and more cursorial aegodont Caprini was clear by 14 m.y. ago at Fort Ternan. At this stage there was still much similarity among the bovid faunas everywhere in the Old World. Only the species, not the genera, differed from continent to continent. Regional variation began to be more marked by the time of the famous

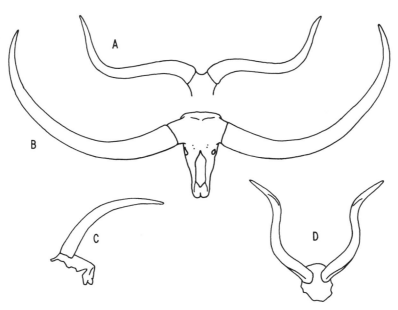

Figure 27.9 Some African fossil bovids. (*A*) *Megalotragus priscus* (Alcelaphini), (*B*) *Pelorovis antiquus* (Bovini), (*C*) *Hippotragus gigas* (Hippotragini), (*D*) *Tragelaphus strepsiceros grandis* (Tragelaphini). (×1/20.)

Eurasian *Hipparion* faunas of Pikermi, Samos (8 to 9 m.y., van Couvering and Miller 1971), Maragha, and northern China. In Europe and China *Miotragocerus* had become the only boselaphine, but there had appeared various ovibovines such as *Palaeoryx* and *Criotherium,* the earliest Caprini, and spiral-horned Antilopini. The equivalents in Africa of these *Hipparion* faunas are not at all well known, but later at Langebaanweg and Laetolil tragelaphines, reduncines, and alcelaphines had appeared, the first two probably descended from boselaphines and the third from caprines. However boselaphines and ovibovines did not die out in Africa (cf Gentry 1970a, p. 313).

In connection with this faunal divergence between Africa and Palaearctic Eurasia, it is interesting that Siwaliks bovids have more resemblances to African than to other Eurasian forms. This is evidenced by the similarity of *Ugandax* to *Proamphibos* (but the difference of either from the *Bos* ancestry is unknown), at least one reduncine lineage from the Dhok Pathan to the Pinjor Formation, the presence of two hippotragines in the Pinjor Formation, the Pinjor alcelaphine *Damalops,* and the presence of *Antilope* in the Shungura Formation. The presence of *Oryx* in Arabia, the former presence of an alcelaphine, probably hartebeest, in Palestine and Jordan (Garrod and Bate 1937, p. 215; Clutton-Brock 1970, p. 26), and the possible lesser kudu record for Arabia also suggest a faunal link from Africa across Arabia. Kurtén (1957, p. 223) believed that the Indian fauna

has only acquired its present resemblance to the Eurasian one since the Villafranchian. The decline of "African" bovids in that region may be connected with the burgeoning of deer.

Bearing in mind the böodont-aegodont split and the affinity between Siwalik and African bovids, we may consider the tribes of antelopes from a zoogeographical standpoint.[2]

The Tragelaphini are now almost confined to Africa. All fossil records from other continents are incorrect or unsubstantiated. The most long-standing misidentifications are of Eurasian Tertiary spiral-horned Antilopini (see Gentry 1971). Pilgrim (1939, p. 131) claimed that the Chinji frontlet *Sivoreas eremita* was tragelaphine, but it now seems more likely (cf Gentry 1970a, p. 257) that it is boselaphine and perhaps even conspecific with *Protragocerus gluten,* also from the Chinji. Tragelaphines probably descended from early boselaphines, and basal pillars on the teeth became less prominent in tragelaphines than in other böodont tribes.

The Boselaphini is the tribe that probably includes the oldest bovid fossils. They seem to have disappeared from most of the Palaearctic soon after the Samos time level but were abundant in the Dhok Pathan Formation (*Miotragocerus, Tragoportax, Pachyportax, Selenoportax*). At least one lineage lasted well into the Tertiary in Africa and had resem-

2 This discussion will not take into account the fact that some bovids have immigrated into North America.

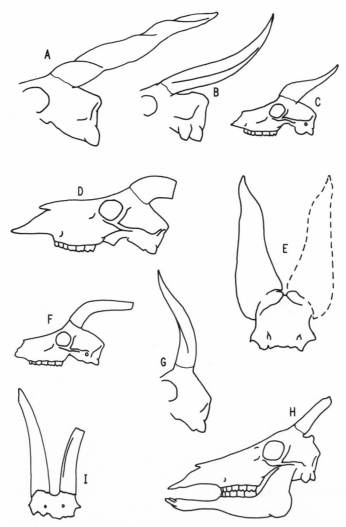

Figure 27.10 Some African fossil bovids. (*A*) *Taurotragus arkelli* (Tragelaphini), (*B*) *Tragelaphus nakuae* (Tragelaphini), (*C*) *Protragocerus labidotus* (Boselaphini), (*D*) *Menelikia lyrocera* (Reduncini), (*E*) *Thaleroceros radiciformis* (Reduncini), (*F*) *Antidorcas recki* (Antilopini), (*G*) *Rabaticeras arambourgi* (Alcelaphini), (*H*) *Parmularius altidens* (Alcelaphini), (*I*) *Pachytragus solignaci* (Caprini). (× 1/8.)

blances to Siwaliks forms. It is noteworthy that the only two present-day boselaphines are Indian (but see Clutton-Brock 1970, p. 25, for an upper Pleistocene record in Jordan) and that they therefore now constitute an element of faunal difference from Africa. The extensive internal sinuses of the frontals in the Langebaanweg boselaphine are unusual for a böodont.

The Bovini are mostly large-sized descendants of Boselaphini that appeared in the later Tertiary. Two closely related lineages, *Simatherium → Pelorovis* and *Ugandax → Syncerus,* are confined to Africa, although *Ugandax* is quite like the Tatrot *Proamphibos,* which probably gave rise through the Pinjor *Hemibos* to *Bubalus. Hemibos* has been recorded from Palestine (Pilgrim 1941). The other bovine stock is the one that gave rise to *Bos,* a Palaearctic

and Indian genus not found in Africa south of the Sahara Desert. Its ancestry is unknown (Siwalik *Pachyportax,* European Pliocene *Parabos?*), and it is puzzling that the Villafranchian and Pinjor "ox," *Leptobos,* has hornless females, whereas in *Bos* the females are horned. With the Bovini, as with Tragelaphini, it is difficult to differentiate early members from the Boselaphini.

Cephalophini are unknown outside Africa and, being mainly forest dwellers, are almost unknown in the fossil state. They are quite likely to have a boselaphine ancestry.

Reduncines are found in Africa and the Siwaliks, and it is interesting that as a group they are dependent on habitats close to water, yet the intervening area of Arabia is today so arid. The earliest reduncines come from the Dhok Pathan Formation, Wadi

Natrun, Sahabi, the Lukeino Formation, and Lange-baanweg. Khomenko (1913) claimed reduncines from the Russian Tertiary, but I agree with Pilgrim's (1939, p. 95) reservations about their tribal assignations; at least one of them may be a *Palaeoryx*.

Hippotragini are likewise confined to the Siwaliks and Africa. Gentry (1971) has argued that alleged Tertiary hippotragines from Samos and other Palaearctic *Hipparion* faunas are misidentified ovibovines and caprines. The earliest likely hippotragine comes from Sahabi. The Pinjor ones give hints of a boselaphine ancestry, but there are no earlier boselaphines that appear to be the actual ancestors.

Alcelaphini, the first of the aegodont tribes in the list, are also limited to the Siwaliks, Africa, and the intervening areas. The only Siwalik representative, *Damalops palaeindicus,* could be related to the hartebeests, so this is the only stock for which an extra-African representation is likely. The earliest adequately preserved alcelaphines are the interesting undescribed species from Langebaanweg, but they are also known from Wadi Natrun, and earlier African and Siwalik caprines may include alcelaphine ancestors.

The small Neotragini are only known from Africa.

The Antilopini is a difficult tribe zoogeographically. *Gazella* is a very early bovid and has survived in arid areas of the Palaearctic, Indian, and Ethiopian realms. *G. rufifrons* and *thomsoni* on the one hand and *G. dama, soemmerringi,* and *granti* on the other constitute two African groups of gazelles. *G. dorcas* and related species constitute one or more Palaearctic groups. *Procapra* (which I would include in *Gazella*), *Saiga,* and *Pantholops* are Asiatic genera probably descended from *Gazella,* while *Antidorcas, Ammodorcas,* and *Litocranius* are African genera. Gentry (1970a, pp. 295–300) hypothesized that Eurasian fossil gazelles may be divisible into two groups, one ancestral to *Procapra* and the other to the *G. dorcas* group. In Africa, fossils are known that appear to be ancestral to *Antidorcas* and probably to both groups of African gazelles. Spiral-horned Antilopini are known from the Samos and other *Hipparion* faunas of the Palaearctic. By the Pleistocene they seem to have resolved themselves into two stocks, the more northerly *Spirocerus* (Asia) and *Gazellospira* (Europe) and the more southerly *Antilope,* which was now present in India (did it and the *Leptobos*/*Bos* bovines immigrate from elsewhere in Eurasia?) and northeastern Africa.

Ovibovini are more abundant in the fossil record than at present. One archaic group with peculiar horn cores comprises the Miocene genera *Criother-* *ium, Palaeoreas, Urmiatherium* (including *Parurmiatherium* and *Plesiaddax*), and *Tsaidamotherium,* none of which has been found in Africa. A second and longer-lasting group has been present in Africa. The only adequately known African species is *Makapania broomi;* together with *Megalovis* and *Budorcas* it may have formed a more southerly ranging stock than *Ovibos.* No ovibovines have been recorded from the Siwaliks.

Caprini are basically a Palaearctic group that have scarcely penetrated Africa south of the Sahara. However, early African bovids from which Alcelaphini probably descended, have been "horizontally" classified with their contemporaneous Eurasian relatives as Caprini (*Oioceros* and ?*Pseudotragus* at Fort Ternan). North Africa contained Caprini at least at certain periods.

The tribe called Rupicaprini is an essentially Eurasian stock and has not been recorded from Africa.

In summary, and still ignoring Nearctic forms, we see that the Siwaliks and Africa contain rather more Tertiary boselaphines than does the Palaearctic. From such boselaphines there probably evolved the Tragelaphini, one southeastern Asian and two African lineages of Bovini, the Cephalophini, the Reduncini, and the Hippotragini. Cattle, bison, and their ancestors comprise the only major Palaearctic böodont group. Among the aegodonts the Antilopini can be broken down into Palaearctic and African-Siwaliks forms at generic level, except in *Gazella,* where the zoogeographical separation is at species or species-group level. The Alcelaphini and Neotragini are totally African-Siwalik, and the former tribe descended from early Caprini, which may be regarded as the aegodont parallel to early Boselaphini. Of the Ovibovini, one early group is Palaearctic, but another is found in Africa and the Palaearctic. The tribe called Rupicaprini is totally Palaearctic and southeast Asian, and the Caprini became almost totally Palaearctic after its beginning in the Miocene. The tribal changes in the Siwalik-Africa area from early Boselaphini to Tragelaphini and their other descendants, and from early Caprini to Alcelaphini, are an artificial result of the "horizontal" classification of the earliest bovids.

Faunal Evolution in Bovids

Although nearly all tribes and genera of bovids are characteristic of particular faunal realms, it is difficult to generalize about regional bovid faunas in units smaller than realms. Admittedly the southern part of Cape Province appears to have had a number of endemic species in the recent past, as witnessed by the extinct kudu, the blaauwbok, the grysbok, and the extinct springbok, *Antidorcas australis*. This possible exception apart, it appears that *regional* groupings of species are not clear cut within the faunal realms, although species differ from habitat to habitat. Arising from such considerations, together with the inadequate knowledge of former species' ranges allowed by the fossil record, it is impossible to give precise information about where particular bovids evolved. It is worth repeating that the involvement of the Siwaliks in the history of African bovids does not necessitate the opinion that African bovids originated in India. It is better to say no more than that the two areas have shared many of their antelopes, at least at the generic level.

Faunas in local areas must certainly have evolved as integrated units with a degree of interdependence between the species. Theoretically one may expect that a morphological change in one lineage, which enables or reflects a modification in its ecology, might well produce smaller compensating changes in a number of contemporaneous, sympatric, but phylogenetically unrelated lineages. Such a phenomenon would be difficult to identify in fossil faunas, and it would produce gradual evolution rather than dramatic changes. Any marked faunal changes that do occur in the fossil record are likely to reflect only local changes of ecology and habitats.

North Africa is faunally interesting not for incipient endemism like South Africa but as the area where the Palaearctic and Ethiopian realms meet. The oldest known bovids in the region are those of Gebel Zelten, which are not zoogeographically distinctive, so far as they are known at present, and could correlate with other Old World sites down to Fort Ternan at 14 m.y. Later in North Africa the *Benicerus* of Beni Mellal, the *Pachytragus* of the Beglia Formation, and the *Damalavus* of Oued el Hammam are Caprinae with more or less convincing affinities with Eurasian forms. Later still at Sahabi and Wadi Natrun the affinities change. Sahabi has *Miotragocerus, Leptobos,* a reduncine, and a possible hippotragine, and Wadi Natrun has a reduncine and alcelaphine. One could align the *Miotragocerus* and *Leptobos* with European faunas and the rest with African faunas, or one could align all the bovids with the Siwaliks. It will be interesting to see if future research can substantiate a faunal change of some kind preceding the Sahabi and Wadi Natrun faunas or clarify the zoogeographical affinities of those faunas.

Character Evolution

It looks as if changes in particular characters in bovid lineages have not always been from the more

primitive to the more advanced. Three examples are: (1) *Pelorovis oldowayensis* has a more posterior insertion of its horn cores than the later *P. antiquus;* (2) paraconid-metaconid fusion of kudu P_4 occurs in a higher percentage of Olduvai than extant specimens; and (3) the premolar row of *Kobus sigmoidalis* may have been shorter than in the living waterbuck. A partial explanation of the tooth characters may be that some living antelopes have less narrow adaptive ranges than their ancestors. A greater number of species in the past would have allowed narrower specializations of individual species. Also, living species may sometimes have evolved from atypical populations of their predecessors, or antelope lineages may have consisted of series of sequentially replacing species (and not chronospecies transitional to one another), as Martin (1972, p. 316) has postulated for lemurs. This latter possibility would allow slightly different adaptive characters in successive species (but, if correct, would necessitate splitting among some species names as used here).

Extinctions

Tragelaphus nakuae, Menelikia, Makapania, and *Numidocapra* appear to have become extinct sometime in the early Pleistocene and *Mesembriportax* probably earlier. Apart from these, there may have been few extinctions until the end of the Pleistocene. Klein (1972, pers. comm.) has produced definite evidence of *Pelorovis* and *Megalotragus* in the Cape Province until about 12,000 to 11,000 B.P. and 15,000 to 14,000 B.P. respectively, and *Antidorcas australis, A. bondi,* and *Damaliscus niro* may have been other late surviving species. In East Africa there is no evidence for the extinction of *Thaleroceros, Hippotragus gigas,* or *Parmularius* within the sequence of Olduvai Beds I to IV. *Beatragus hunteri* could be a species naturally on the verge of extinction at the present time.

Correlations

The contribution of bovids to faunal correlations has been implicit in many of the statements in this chapter. Only a summary of major points can be given.

At Elandsfontein the eland is definitely more advanced than the Olduvai Bed IV *Taurotragus arkelli,* the *Rabaticeras arambourgi* corresponds closely with that in Olduvai Bed III, and the gazelle appears identical with that of middle and upper Bed II. Many Elandsfontein bovids are not greatly different from living South African species. A likely conclusion from all this is that much of the Elandsfontein time span corresponds to Olduvai Bed IV or later.

Makapansgat Limeworks has a *Tragelaphus* with rather a primitive horn core cross-section, *Redunca darti* that is substantially less advanced than living *Redunca*, *Gazella vanhoepeni* that is perhaps descended from the Langebaanweg gazelle, and *Makapania broomi* whose closest relative is in the European Villafranchian. None of this is very conclusive, but an age about the same or a little earlier than Olduvai Bed I would probably best fit the bovid evidence.

The Shungura Formation is known by potassium-argon dating to immediately predate and overlap Olduvai. Its bovids comprise mainly tragelaphines, reduncines, and impala and so cannot be compared easily with those of Olduvai, which are mainly alcelaphines (not impala) and antilopines.

At Laetolil the *Simatherium kohllarseni*, *?Parmularius* sp., *Beatragus* sp., and *Gazella janenschi* and the supposed hippotragine teeth are all more primitive than Olduvai forms and suggest that some horizons substantially predate Olduvai. No reduncines are known from Laetolil, a fact that is ecologically interesting.

The chief interest of the Langebaanweg bovids is that the alcelaphine teeth are more primitive than any at Laetolil, which would suggest an earlier date. The *Tragelaphus* has a primitive horn core cross-section, and the gazelle predates that at Makapansgat Limeworks.

In the Kaiso Formation the *Menelikia lyrocera* of the later fauna would best fit a time level around Member F of the Shungura Formation. The tragelaphine horn core of the early fauna seems insufficiently primitive to match the supposed age of the fauna and the *Kobus sigmoidalis* is notably early.

The Contribution of Bovids to the Definition of Land Mammal Ages

The most recent faunal division presenting itself in the history of African bovids is that marked by the extinctions of a number of lineages near the end of the Pleistocene. So far reasonably precise dates have been secured for only two lineages, but should other extinctions be found to have occurred within a few millenia of them, then the boundary of two land mammal ages could reasonably be fixed here.

In South Africa the immediately preceding part of the later Pleistocene, in which are present the extinct forms and slightly different subspecies of some living species, has been called the Florisbad faunal span or Florisian mammal age. It seems doubtful that the bovids of the still earlier Cornelia faunal span (Elandsfontein and Cornelia sites) are sufficiently distinct from Florisbad-Florisian ones to help in the definition of another land mammal age.

Passing still further back into the lower Pleistocene, one notices increasing differences among the bovids. Bed IV at Olduvai contains *Taurotragus arkelli* instead of the living *T. oryx*, and upper Bed II contains *Pelorovis oldowayensis* instead of the later *P. antiquus*. At a level somewhere between the top of Bed I and the later parts of Bed II the differences become sufficiently marked to justify having a boundary between two land mammal ages. There is a faunal change at this level at Olduvai, and it may be coeval with the ending of the Sterkfontian faunal span in South Africa.

In the preceding period back as far as, say, Member C of the Shungura Formation the bovids are definitely different and often more archaic than living ones. One has *Tragelaphus pricei* instead of *T. scriptus*, *Kobus sigmoidalis* and *Redunca darti* instead of *K. ellipsiprymnus* and the living *Redunca* species, a *Connochaetes* ancestral to *C. taurinus*, *Parmularius altidens* instead of the later *P. angusticornis* or *P. rugosus*, as well as lineages that may scarcely have survived into the Pleistocene at all: *Tragelaphus nakuae*, *Menelikia lyrocera*, and *Makapania broomi*.

As studies proceed it is becoming clear that another distinct stratum of bovids exists at such earlier sites as Laetolil, Afar, and certainly Langebaanweg. It may become convenient in the future to visualize this faunal level as extending upward to the top of Member B of the Shungura Formation. Member B contains a distinctive variety of *Tragelaphus nakuae* and a *Kobus* species, which both have antecedents at Afar.

Thus the three boundary lines one can draw across the flow of African bovid evolution lie at the end of the Pleistocene, around the Olduvai lower–middle Bed II junction corresponding to the end of the Sterkfontian in South Africa, and low in the Shungura Formation. It would be premature to extend this process any further back into the Tertiary for the present.

I am grateful to G. B. Corbet, J. Clutton-Brock, W. R. Hamilton, and my wife for reading the first draft of this paper.

References

Andrews, C. W. 1902. Note on a Pliocene vertebrate fauna from the Wadi-Natrun, Egypt. *Geol. Mag.* (4) 9:433–439.

Ansell, W. F. H. 1971. Artiodactyla. In J. Meester and H. W. Setzer, eds., *The mammals of Africa: an identification manual*, part 15. Washington: Smithsonian Institution Press, pp. 1–93.

Arambourg, C. 1933. Mammifères miocènes du Turkana (Afrique orientale). *Ann. Paléont.* 22:121–148.

———. 1938. Mammifères fossiles du Maroc. *Mém. Soc. Sci. Nat. Phys. Maroc* 46:1–74.

———. 1941. Antilopes nouvelles du Pléistocène ancien de l'Omo (Abyssinie). *Bull. Mus. Natn. Hist. Nat. Paris* (2) 13:339–347.

———. 1947. Contribution à l'étude géologique et paléontologique du bassin du lac Rudolphe et de la basse vallée de l'Omo. *Mission Scient. Omo 1932–1933 I, Géol. Anthrop.*, pp. 232–562.

———. 1949. *Numidocapra crassicornis*, nov. gen., nov. sp., un Ovicapriné nouveau du Villafranchien constantinois. *C. R. Soc. Géol. Fr.* no. 13:290–291.

———. 1957. Observations sur les gazelles fossiles du Pléistocène supérieur de l'Afrique du Nord. *Bull. Soc. Hist. Nat. Afr. Nord* 48:49–81.

———. 1959. Vertébrés continentaux du Miocène supérieur de l'Afrique du Nord. *Mém. Carte Géol. Algérie,* n.s., *Paléont.* 4:1–159.

———. 1962. Les faunes mammalogiques du Pléistocène circumméditerranéen. *Quaternaria* 6:97–109.

———. 1970. Les vertébrés du Pléistocène de l'Afrique du Nord. *Archs. Mus. Natn. Hist. Nat. Paris* (7) 10:1–126.

Arambourg, C., and R. Coque. 1958. Le gisement villafranchien de l'Aïn Brimba (sud Tunisien) et sa faune. *Bull. Soc. Géol. Fr.* (6) 8:607–614.

Arambourg, C., and J. Piveteau. 1929. Note préliminaire sur un ruminant du Pliocène inférieur du Roussillon. *C. R. Soc. Géol. Fr.* no. 10:144–146.

Balout, L. 1942. Note sur la présence des restes fossiles d'une *Addax nasomaculata* Blainv. parmi des ossements découverts au Parc d'Hydra, commune de Birmandreis (Alger). *Bull. Soc. Hist. Nat. Afr. Nord.* 33:138–140.

Bate, D. M. A. 1940. The fossil antelopes of Palestine in Natufian (Mesolithic) times, with descriptions of new species. *Geol. Mag.* 77:418–443.

———. 1949. A new African fossil long-horned buffalo. *Ann. Mag. Nat. Hist. Lond.* (12) 2:396–398.

———. 1951. The mammals from Singa and Abu Hugar. Brit. Mus. Nat. Hist., *Fossil Mammals of Africa* no. 2:1–28.

Bayle, E. 1854. Sur une collection d'ossements fossiles trouvés près de Constantine. *Bull. Soc. Géol. Fr.* (2) 11:343–345.

Beaumont, P. B. 1973. Border Cave—a progress report. *S. Afr. J. Sci.* 69:41–46.

Berggren, W. A., and J. A. van Couvering. 1974. The late Neogene. *Palaeogeogr. Palaeoclimat. Palaeoecol.* 16:1–216.

Bishop, W. W., G. R. Chapman, A. Hill, and J. A. Miller. 1971. Succession of Cainozoic vertebrate assemblages from the northern Kenya Rift Valley. *Nature* 233:389–394.

Bishop, W. W., J. A. Miller, and F. J. Fitch. 1969. New potassium-argon age determinations relevant to the Miocene fossil mammal sequence in East Africa. *Am. J. Sci.* 267:669–699.

Boule, M. 1910. Sur quelques vertébrés fossiles du Sud de la Tunisie. *C. R. Soc. Géol. Fr.*, pp. 50–51.

Bourguignat, J. R. 1870. *Histoire du Djebel Thaya et des ossements fossiles recueillis dans la grande Caverne de la Mosquée.* Paris, pp. 1–108.

Broom, R. 1909. On a large extinct species of *Bubalis. Ann. S. Afr. Mus.* 7:279–280.

———. 1913. Man contemporaneous with extinct animals in South Africa. *Ann. S. Afr. Mus.* 12:13–16.

———. 1934. On the fossil remains associated with *Australopithecus africanus. S. Afr. J. Sci.* 31:471–480.

———. 1937. Notices of a few more new fossil mammals from the caves of the Transvaal. *Ann. Mag. Nat. Hist. Lond.* (10) 20:509–514.

Butzer, K. W., and G. Ll. Isaac, eds. 1975. *After the australopithecines: stratigraphy, ecology and culture change in the middle Pleistocene.* The Hague: Mouton.

Churcher, C. S. 1972. Late Pleistocene vertebrates from archaeological sites in the plain of Kom Ombo, upper Egypt. *Contr. Life Sci. Div. Roy. Ont. Mus.* 82:1–172.

Churcher, C. S., and P. E. L. Smith. 1972. Kom Ombo: preliminary report on the fauna of late Paleolithic sites in upper Egypt. *Science* 177:259–261.

Clark, J. D. 1959. Further excavations at Broken Hill, northern Rhodesia. *J. Roy. Anthrop. Inst. Gr. Brit.* 89:201–232.

Clutton-Brock, J. 1970. The fossil fauna from an upper Pleistocene site in Jordan. *J. Zool. Lond.* 162:19–29.

Cooke, H. B. S. 1941. A preliminary account of the Wonderwerk Cave, Kuruman District. Section 2. The fossil remains. *S. Afr. J. Sci.* 37:303–312.

———. 1947. Some fossil hippotragine antelopes from South Africa. *S. Afr. J. Sci.* 43:226–231.

———. 1949. Fossil mammals of the Vaal River deposits. *Mem. Geol. Surv. S. Afr.* 35, part 3:1–109.

———. 1968. Evolution of mammals on southern continents. II. The fossil mammal fauna of Africa. *Quart. Rev. Biol.* 43:234–264.

Cooke, H. B. S., and S. C. Coryndon. 1970. Pleistocene mammals from the Kaiso Formation and other related deposits in Uganda. In L. S. B. Leakey and R. J. G. Savage, eds., *Fossil Vertebrates of Africa.* London: Academic Press, vol. 2, pp. 107–224.

Cooke, H. B. S., and L. H. Wells. 1951. Fossil remains from Chelmer, near Bulawayo, Southern Rhodesia. *S. Afr. J. Sci.* 47:205–209.

Coppens, Y. 1971. Les vertébrés Villafranchiens de Tunisie: gisements nouveaux, signification. *C. R. Hebd. Séanc. Acad. Sci.,* Sér D, 273:51–54.

Débruge, A. 1906. La grotte du Fort Clauzel. *C. R. Assoc. Franç. Avanc. Sci.* (1905) 34:624–632.

Depéret, C. 1890. Les animaux pliocènes du Roussillon. *Mém. Soc. Géol. Fr., Paléont.* no. 3:1–194.

Dietrich, W. O. 1941. Die säugetierpaläontologischen Ergebnisse der Kohl-Larsen'schen Expedition 1937–1939 im nördlichen Deutsch-Ostafrika. *Zentbl. Miner. Geol. Paläont.* Ser. B, no. 8:217–223.

———. 1942. Ältestquartäre Säugetiere aus der südlichen Serengeti, Deutsch Ost-Afrika. *Palaeontogr.* Abt. A 94:43–133.

———. 1950. Fossile Antilopen und Rinder Aquatorialafrikas. *Palaeontogr.* Abt. A 99:1–62.

Duvernoy, G. L. 1851. Note sur une espèce de buffle fossile, *Bubalus (Arni) antiquus,* découverte en Algérie. *C. R. Hebd. Séanc. Acad. Sci.* 33:595–597.

Ennouchi, E. 1953. Un nouveau genre d'Ovicapriné dans un gisement Pléistocène de Rabat. *C. R. Soc. Géol. Fr.* no. 8:126–128.

Ficheur, E., and A. Brives. 1900. Sur la découverte d'une caverne à ossements, à la carrière des Bains-Romains, à l'ouest d'Alger. *C. R. Hebd. Séanc. Acad. Sci.* 130: 1485–1487.

Gaillard, C. 1934. Contribution à l'étude de la faune préhistorique de l'Égypte. *Arch. Mus. Hist. Nat. Lyon* 14, 3:1–126.

Garrod, D. A. E., and D. M. A. Bate. 1937. *The Stone Age of Mount Carmel I.* Oxford: Clarendon Press.

Gentry, A. W. 1966. Fossil Antilopini of East Africa. *Bull. Brit. Mus. Nat. Hist. (Geol.)* 12:45–106.

————. 1967. *Pelorovis oldowayensis* Reck, an extinct Bovid from East Africa. *Bull. Brit. Mus. Nat. Hist. (Geol.)* 14:243–299.

————. 1970a. The Bovidae (Mammalia) of the Fort Ternan fossil fauna. In L. S. B. Leakey and R. J. G. Savage, eds., *Fossil Vertebrates of Africa.* London: Academic Press, vol. 2, pp. 243–323.

————. 1970b. Revised classification for *Makapania broomi* Wells and Cooke (Bovidae, Mammalia). *Palaeont. Afr.* 13:63–67.

————. 1971. The earliest goats and other antelopes from the Samos *Hipparion* fauna. *Bull. Brit. Mus. Nat. Hist. (Geol.)* 20:229–296.

————. 1974. A new genus and species of Pliocene boselaphine (Bovidae, Mammalia) from South Africa. *Ann. S. Afr. Mus.* 65:145–188.

Gervais, P. 1867–1869. *Zoologie et paléontologie générales.* 1st series, Paris, 263 pp.

Ginsburg, L., and E. Heintz. 1968. La plus ancienne antilope d'Europe, *Eotragus artenensis* du Burdigalien d'Artenay. *Bull. Mus. Natn. Hist. Nat. Paris* (2) 40: 837–842.

Grubb, P. 1972. Variation and incipient speciation in the African buffalo. *Zeits. Saügetierk.* 37:121–144.

Hamilton, W. R. 1973. The lower Miocene ruminants of Gebel Zelten, Libya. *Bull. Br. Mus. Nat. Hist. (Geol.)* 21:73–150.

Harris, J. M. 1976. Bovidae from the East Rudolf succession. In Y. Coppens, F. C. Howell, G. Ll. Isaac, and R. E. F. Leakey, eds., *Earliest man and environments in the Lake Rudolf Basin.* Chicago: University of Chicago Press, pp. 293–301.

Harrison, D. L. 1972. *The mammals of Arabia,* vol. 3. London: Benn.

Heintz, E. 1973. Un nouveau bovidé du Miocène de Beni Mellal, Maroc: *Benicerus theobaldi* n. g., n. sp. (Bovidae, Artiodactyla, Mammalia). *Ann. Scient. Univ. Besancon* (3) 18:245–248.

Hendey, Q. B. 1968. The Melkbos site: an upper Pleistocene fossil occurrence in the southwestern Cape Province. *Ann. S. Afr. Mus.* 52:89–119.

————. 1970. A review of the geology and palaeon-

tology of the Plio-Pleistocene deposits at Langebaanweg, Cape Province. *Ann. S. Afr. Mus.* 56:75–117.

————. 1974. The Late Cenozoic Carnivora of the southwestern Cape Province. *Ann. S. Afr. Mus.* 63: 1–369.

Hendey, Q. B., and H. Hendey. 1968. New Quaternary fossil sites near Swartklip, Cape Province. *Ann. S. Afr. Mus.* 52:43–73.

Hopwood, A. T. 1934. New fossil mammals from Olduvai, Tanganyika Territory. *Ann. Mag. Nat. Hist. Lond.* (10) 14:546–550.

————. 1936. New and little-known fossil mammals from the Pleistocene of Kenya Colony and Tanganyika Territory. *Ann. Mag. Nat. Hist. Lond.* (10) 17:636–641.

Joleaud, L. 1918. Études de géographie zoologique sur la Berbérie. III: Les hippotraginés. *Bull. Soc. Géogr. Oran* 38:89–118.

Khomenko, J. P. 1913. La faune Méotique du village Taraklia du District de Bendery. *Ezheg. Geol. Mineral. Rossii* 15:107–143.

Klein, R. G. 1972. The late Quaternary mammalian fauna of Nelson Bay Cave (Cape Province, South Africa). *Quatern. Res.* 2:135–142.

————. 1974. On the taxonomic status, distribution, and ecology of the blue antelope, *Hippotragus leucophaeus* (Pallas, 1766). *Ann. S. Afr. Mus.* 65:99–143.

Kurtén, B. 1957. Mammal migrations, Cenozoic stratigraphy, and the age of Peking man and the Australopithecines. *J. Paleont.* 31:215–227.

Leakey, L. S. B. 1965. *Olduvai Gorge 1951–61. I. Fauna and Background.* Cambridge: Cambridge University Press.

Leakey, M. D., R. L. Hay, G. H. Curtis, R. E. Drake, M. K. Jackes and T. D. White. 1976. Fossil hominids from the Laetolil Beds. *Nature* 262:460–466.

Leakey, R. E. F. 1969. Early *Homo sapiens* remains from the Omo River region of southwest Ethiopia. *Nature* 222:1132–1133.

Lydekker, R. 1878. Crania of ruminants from the Indian Tertiaries, and supplement. *Pal. Indica* (10) 1:88–181.

Martin, R. D. 1972. Adaptive radiation and behaviour of the Malagasy lemurs. *Phil. Trans. Roy. Soc. London,* Series B, 264:295–352.

Mohr, E. 1967. *Mammalia depicta, der Blaubock* Hippotragus leucophaeus *(Pallas, 1766), eine Dokumentation.* Hamburg: Verlag Paul Parey.

Petrocchi, C. 1956. I *Leptobos* di Sahabi. *Boll. Soc. Geol. Ital.* 75(1):206–238.

Pilgrim, G. E. 1934. Two new species of sheep-like antelope from the Miocene of Mongolia. *Am. Mus. Novit.* no. 716:1–29.

————. 1937. Siwalik antelopes and oxen in the American Museum of Natural History. *Bull. Am. Mus. Nat. Hist.* 72:729–874.

————. 1939. The fossil Bovidae of India. *Pal. Indica* (N.S.) 26, 1:1–356.

————. 1941. A fossil skull of *Hemibos* from Palestine. *Ann. Mag. Nat. Hist. Lond.* (11) 7:347–360.

Pomel, A. 1893. *Bubalus antiquus. Carte Géol. Algér., Paléont. Monogr.* pp. 1–94.

_____. 1894a. Les Boeufs Taureaux. *Carte Géol. Algér., Paléont. Monogr.* pp. 1–108.

_____. 1894b. Les Bosélaphes Ray. *Carte Géol. Algér., Paléont. Monogr.* pp. 1–61.

_____. 1895. Les Antilopes Pallas. *Carte Géol. Algér., Paléont. Monogr.* pp. 1–56.

Reck, H. 1928. *Pelorovis oldowayensis* n.g., n.sp. *Wiss. Ergebn. Oldoway-Exped. 1913* (N.F.) 3:57–67.

_____. 1935. Neue Genera aus der Oldoway-Fauna. *Zentbl. Miner. Geol. Paläont.*, Ser. B, 1935, pp. 215–218.

_____. 1937. *Thaleroceros radiciformis* n.g., n.sp. *Wiss. Ergebn. Oldoway-Exped. 1913* (N.F.) 4:137–142.

Roberts, A. 1937. The old surviving types of mammals found in the Union. *S. Afr. J. Sci.* 34:73–88.

Robinson, P. 1972. *Pachytragus solignaci,* a new species of caprine bovid from the late Miocene Beglia Formation of Tunisia. *Notes Serv. Géol. Tunisie* 37:73–94.

Robinson, P., and C. C. Black. 1969. Note préliminaire sur les vertébrés fossiles du Vindobonien. *Notes Serv. Géol. Tunisie* 31:67–70.

Roman, F. 1931. Description de la faune pontique du Djerid (El Hamma et Nefta). In M. Solignac, Le Pontien dans le sud Tunisien. *Ann. Univ. Lyon* (N.S.) I 48:1–42.

Romer, A. S. 1928. Pleistocene mammals of Algeria. Fauna of the paleolithic station of Mechta-el-Arbi. *Bull. Logan Mus.* 1:80–163.

Savage, R. J. G., and W. R. Hamilton. 1973. Introduction to the Miocene mammal faunas of Gebel Zelten, Libya. *Bull. Brit. Mus. Nat. Hist. (Geol.)* 22:513–527.

Schwarz, E. 1932. Neue diluviale Antilopen aus Ostafrika. *Zentbl. Miner. Geol. Paläont.* ser. B, 1932, pp. 1–4.

_____. 1937. Die fossilen Antilopen von Oldoway. *Wiss. Ergebn. Oldoway-Exped. 1913* (N.F.) 4:8–104.

Scott, W. B. 1907. A collection of fossil mammals from the coast of Zululand. *Rep. Geol. Surv. Natal Zululand* 3:253–262.

Selley, R. C. 1969. Near-shore marine and continental sediments of the Sirte basin, Libya. *Quart. J. Geol. Soc. Lond.* 124:419–460.

Simpson, G. G. 1945. The principles of classification and a classification of mammals. *Bull. Am. Mus. Nat. Hist.* 85:1–350.

Sokolov, I. I. 1952. Remains of Bovidae, Mammalia from lower Miocene deposits of western Gobi. *Trudý paléont. Inst. Akad. Nauk. SSSR* (Moscow) 41:155–158.

Stromer, E. 1926. Reste Land- und Süsswasser-Bewohnender Wirbeltiere aus den Diamantfeldern Deutsch-Südwestafrikas. In E. Kaiser, *Die Diamantenwüste Südwest-afrikas,* Berlin, vol. 2, pp. 107–153.

Studer, T. 1898. Ueber fossile Knochen vom Wadi-Natrun Unteregypten. *Mitt. Naturf. Ges. Bern* 1460:72–77.

Thenius, E. 1952. Die Boviden des steirischen Tertiärs. *Sber. Öst. Akad. Wiss.* Abt. I, 161:409–439.

Thomas, P. 1884. Recherches stratigraphiques et paléontologiques sur quelques formations d'eau douce de l'Algérie. *Mém. Soc. Géol. Fr.,* sér. 3, 3(2):1–51.

Trofimov, B. A. 1958. New Bovidae from the Oligocene of Central Asia. *Vert. Palasiat.* 2:244–247.

Van Couvering, J. A., and J. A. Miller. 1971. Late Miocene marine and nonmarine time scale in Europe. *Nature* 230:559–563.

van Hoepen, E. C. N. 1932. Voorlopige beskrywing van Vrystaatse soogdiere. *Paleont. Navors. Nas. Mus. Bloemfontein* 2(5):63–65.

Vrba, E. S. 1971. A new fossil alcelaphine (Artiodactyla: Bovidae) from Swartkrans. *Ann. Transv. Mus.* 27:59–82.

_____. 1973. Two species of *Antidorcas* Sundevall at Swartkrans (Mammalia, Bovidae). *Ann. Transv. Mus.* 28:287–352.

_____. 1975. Some evidence of chronology and palaeoecology of Sterkfontein, Swartkrans and Kromdraai from the fossil Bovidae. *Nature* 254:301–304.

Wells, L. H. 1943. A further report on the Wonderwerk Cave, Kuruman. Section 2, Fauna. *S. Afr. J. Sci.* 40:263–270.

_____. 1951. A large fossil klipspringer from Potgietersrust. *S. Afr. J. Sci.* 47:167–168.

_____. 1963. Note on a bovid fossil from the Pleistocene of Abu Hugar, Sudan. *Ann. Mag. Nat. Hist. Lond.* (13) 6:303–304.

_____. 1967. Antelopes in the Pleistocene of southern Africa. In W. W. Bishop and J. D. Clark, eds., *Background to evolution in Africa.* Chicago: Univ. of Chicago Press, pp. 99–107.

_____. 1969. Generic position of "*Phenacotragus*" *vanhoepeni. S. Afr. J. Sci.* 65:162–163.

Wells, L. H., and H. B. S. Cooke. 1956. Fossil Bovidae from the Limeworks Quarry, Makapansgat, Potgietersrust. *Palaeont. Afr.* 4:1–55.

Whitworth, T. 1958. Miocene ruminants of East Africa. Brit. Mus. Nat. Hist., *Fossil Mammals of Africa,* no. 15, pp. 1–50.

Zittel, K. A. 1925. *Textbook of Palaeontology,* vol. 3, *Mammalia.* Translated and revised by Arthur Smith Woodward, London.

28

Sirenia

Daryl P. Domning

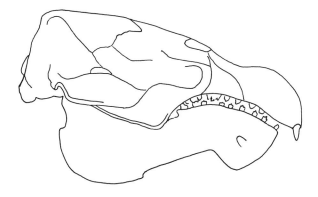

The order Sirenia is the only extant group of mammals adapted to feed exclusively on aquatic plants. In view of the worldwide abundance of aquatic macrophytes and the few other large herbivores competing for this resource, it is noteworthy that Recent sirenians comprise only three genera and five species. One of these, Steller's sea cow (*Hydrodamalis gigas*) of the North Pacific, was exterminated by man in the eighteenth century. Uniquely among sirenians, it was adapted to cold climates and a diet of kelp and other algae (Domning 1977). All the living sirenians are tropical forms that feed preferentially on angiosperms, and this appears to have been the primitive condition for the order.

The Indian Ocean and West Pacific tropics are today inhabited by a single species, *Dugong dugon*, distributed in nearshore marine waters from East Africa and the Red Sea to Japan, Micronesia, and Australia. The three species of manatees (*Trichechus*) occur on both sides of the tropical Atlantic: *T. manatus* in fresh and salt water from the southeastern United States through the Caribbean to beyond the eastern tip of Brazil; *T. senegalensis* in rivers and coastal waters of West Africa; and *T. inunguis* in the Amazon Basin of South America (Bertram and Bertram 1973).

The fossil record of sirenians is extensive but uneven in both geographic and taxonomic distribution, and many of the described genera, subfamilies, and families are monotypic. The order seems to have reached its peak diversity in the Miocene with about a dozen known genera. Sirenians first appear in the Eocene, and the abundance of sirenian remains in the middle and upper Eocene of North Africa early led to the labeling of that area as the center of origin of the group. This was corroborated by recognition of anatomical similarities between sirenians and other characteristically African groups, such as moeritheres, proboscideans, and hyracoids. However, the earliest known sirenian fossils are from the lower Eocene of Hungary (Kretzoi 1953), and the presently known distribution of Eocene sirenians from the Caribbean to the East Indies (Reinhart 1976) points to the shores of the Tethyan Seaway, probably in the Old World, as the specification of choice for a center of origin. Significantly, the marine angiosperms also show signs of an originally Tethyan distribution, although they entered the water during the Cretaceous, well before the appearance of sirenians (den Hartog 1970).

The last appearing (late Oligocene) group of "subungulates," the desmostylians, were formerly included within the Sirenia but are now accorded ordinal status in close alliance with moeritheres, proboscideans, and sirenians. These extinct marine herbivores, of possible eastern Tethyan origin, were apparently restricted to the North Pacific except for a limited foray into the Caribbean (Reinhart 1976). All of the so-called subungulates presumably share a Paleocene condylarth ancestry.

I provisionally follow Sickenberg (1934) and Simpson (1945) in recognizing four families of sirenians: Prorastomidae, Protosirenidae, Trichechidae, and Dugongidae. The former two are known only from middle Eocene deposits; the manatees have a scanty Miocene to Recent record; while the dugongids, comprising the majority of known forms, are fairly well documented from the middle Eocene to the present and are divided into several subfamilies.

Prorastomus from Jamaica shows resemblances to condylarths in its ear region and atlas; however, the inflated rostrum, pachyostotic skull, and bilophodont molars are shared with later sirenians (Savage 1977). *Sirenavus* from the middle Eocene of Hungary has also been questionably referred to the Prorastomidae (Kretzoi 1941). *Protosiren* from Egypt is more advanced and in most ways a good structural ancestor for the other two families, although too late in time to be the actual ancestor. The trichechids are a small, conservative group apparently confined to South America from the Eocene until their dispersal to North America and Africa in the Quaternary. The dugongids, the most successful family, maintained a pantropical distribution throughout the Tertiary. Their main stem is generally considered to be the subfamily Halitheriinae (middle Eocene to Pliocene, pantropical), which gave rise to at least two specialized side branches, the Hydrodamalinae (middle Miocene to Recent, North Pacific) and the Dugonginae (Recent, Indopacific; fossil record lacking). I regard *Rytiodus* (early Miocene, Europe and Africa) as referable to the Halitheriinae rather than to a subfamily of its own, and I divide Reinhart's (1959) "Halianassinae" between the Halitheriinae and Hydrodamalinae. *Miosiren* (middle Miocene, Europe) is an aberrant, possibly molluscivorous form placed in its own subfamily (Miosireninae) within the Dugongidae, though separate descent from protosirenids may be an alternate possibility.

Structure

The Sirenia, together with the Cetacea, are the only obligatorily aquatic mammals, sharing fusiform bodies, finlike forelimbs, horizontally expanded tail fins, and loss of hind limbs. The latter

were reduced and may have been nonfunctional even in *Protosiren*. Unlike many cetaceans, however, sirenians lack dorsal fins and show no aptitude or adaptations for echolocation. As slow-swimming herbivores, they have refined the art of hydrostasis to a unique degree in order to minimize energy expenditure in locomotion, evolving massive pachyostotic skeletons for ballast and horizontal diaphragms, elongate lungs, and monopodially-branching bronchial trees (Pick 1907) for maintaining fore-and-aft trim. Muscular contraction of the thorax evidently can adjust specific gravity at will and permit silent, vertical submergence. In conjunction with these apparent adaptations for statically keeping the body axis close to a horizontal position, the degree of downward flexion of the rostrum and anterior mandible is adjusted to the feeding habits of the species: *Dugong,* with a strongly downturned snout, is adapted to habitats (sea grass meadows) where the dominant growth habit of plants is short and bottom hugging, whereas sirenians with slight snout deflections are found in habitats with abundant floating or other near-surface vegetation (Domning 1977).

The earliest sirenians, including *Prorastomus* (Savage 1977), *Protosiren,* and *Eotheroides* (Sickenberg 1934), had a dentition consisting of $\frac{3.1.5.3}{3.1.5.3}$. This soon became reduced, first by loss of P5 and retention of DP5 in the adult dentition, then by progressive loss of the anterior permanent and deciduous premolars and the canines and incisors. One pair of upper incisors, apparently the first, is retained as short tusks in many dugongids; their degree of development is at least sometimes (as in *Dugong*) sexually dimorphic; but the sexes never seem to differ to the extent of presence versus complete absence of tusks. In all sirenians the fronts of the upper and lower jaws bear tough pads that serve with the prehensile upper lip to pull food into the mouth, with or without the help of the flippers. While halitheriines always retain an adult cheek dentition of at least M2-3 (more commonly DP5-M3 in Neogene forms) and the teeth retain the primitive bunobilophodont-brachydont condition, the other dugongid subfamilies show more specialization: M3 of *Miosiren* is reduced to a simple, stout peg; adult *Dugong* retain only M2-3, lacking roots, enamel, and (after initial wear) cusps; and *Hydrodamalis* was completely toothless. Trichechids, in contrast, after reaching a halitheriine-like degree of reduction by the late Miocene, evolved a unique adaptation to increased tooth wear by continuing to produce supernumerary brachydont, bilophodont molars at the rear of each toothrow. The entire row migrates forward, worn teeth falling out at the front and new ones replacing them at the rear, throughout the animal's life. This process has often been misleadingly compared to tooth replacement in elephants; but while forward tooth movement or "mesial drift" occurs to some degree in many mammals, the combination of this process with unlimited production of supernumerary teeth is unique to the advanced trichechids (*Ribodon* and *Trichechus*).

Trichechus also differs from dugongids in having a rounded rather than whalelike tail fin (probably a primitive feature), in having only six cervical vertebrae, and in the structure of the shoulder joint and pelvis, features suggesting a long independent history.

The most common sirenian fossils are fragments of the swollen, pachyostotic ribs (usually indeterminable), vertebrae, and skullcaps (fused parietals and supraoccipital). The latter have, *faute de mieux*, been much used in sirenian systematics, but I consider their diagnostic value very limited. Isolated teeth are also difficult to interpret. Indeed, so morphologically similar are most sirenians that precise identification may be impossible without relatively large portions of skulls or mandibles. Considerable numbers of high-quality specimens will be necessary to elucidate the relations between the many monotypic forms known and to sort out the apparently frequent instances of parallel evolution in sirenian history.

Fossil Sirenia in Africa

The earliest geological record of African sirenians is that of *Eotheroides aegyptiacum* (Owen 1875) from the middle Eocene Nummulitic Beds of the Mokattam Hills near Cairo. Later discoveries were predominantly from the late Eocene marine beds of the Fayum. With the exception of three Miocene occurrences of "*Halitherium* sp."—from the Isthmus of Suez (Gervais 1872), Madagascar (Collignon and Cottreau 1927), and the Congo (Dartevelle 1935)—the known distribution of fossil sirenians in Africa remained restricted to the Egyptian Eocene until the reports of Pliocene material from Morocco (Ennouchi 1954), Eocene, Oligocene, and Miocene material from Libya (Savage and White 1965; Savage 1967, 1969, 1971), and Eocene and Oligocene remains from Somalia (Savage 1969). Fragmentary remains are now also known from the Oligocene and Miocene of Tunisia. Although the Egyptian material has been thoroughly described, the sirenian history of the rest of the continent has barely begun to be investigated. A synopsis of the published African material follows.

Order Sirenia Illiger 1811

Family Protosirenidae Sickenberg 1934

Genus *Protosiren* Abel 1904

This family appears to contain but a single species, *P. fraasi* Abel 1904, from the middle Eocene (basal Mokattam Formation) of Egypt. *P. dolloi* Abel 1904, from the upper Eocene of Italy, later referred to *Mesosiren* Abel 1906, has been most recently (Simpson 1945) considered synonymous with the dugongid *Prototherium veronense*. Sickenberg (1934) suggested the name *?Protosiren dubia* for teeth from the middle Eocene of France, but their affinities are indeed dubious.

Protosiren fraasi, after *Prorastomus* the most primitive known sirenian, is described by Sickenberg (1934) as having five (or possibly six) single-rooted premolars anterior to the three molars. Unlike later forms, it possesses an alisphenoid canal but shows a clear advance over *Prorastomus* in that the periotic is no longer fused to the rest of the skull (Savage 1977). A descending process of the frontal (the lamina orbitalis) forms part of the wall of the orbit, a feature seen also in trichechids and in *Miosiren*. The inside of the braincase roof, however, lacks a bony falx cerebri, and this probably derived condition together with *Protosiren*'s near-contemporaneity with the first dugongids seems to rule it out of direct ancestry to the later families. Since, to my knowledge, *Protosiren* shares no derived characters with either modern family, I favor retaining the Protosirenidae as a primitive stem group from which both could have been derived.

Family Dugongidae Gray 1821

Subfamily Halitheriinae Abel 1913

Genus *Eotheroides* Palmer 1899

SYNONYMY. *Eotherium* Owen 1875, *nec* Leidy 1853, *Eosiren* Andrews 1902, *Archaeosiren* Abel 1913, *Masrisiren* Kretzoi 1941.

TYPE SPECIES. *Eotherium aegyptiacum* Owen 1875.

Eotheroides contains five nominal species from the Eocene of Egypt: *E. aegyptiacum* (Owen 1875) (including "*Manatus*" *coulombi* Filhol 1878), *E. abeli* Sickenberg 1934, and "*Eotherium*" *majus* Zdansky 1938 from the middle Eocene Mokattam Formation (above the *Protosiren* horizon; Kellogg 1936, p. 235), and *E. libyca* (Andrews 1902) and *E. stromeri* (Abel 1913) from the late Eocene Qasr el Sagha Formation of the Fayum. (*"Eotherium markgrafi"* Abel 1913, p. 337 is a *nomen nudum.*) The genus may also occur in southern Europe (Sickenberg 1934; Richard 1946; Fuchs 1973) along with its close late Eocene relative, *Prototherium* de Zigno 1887; the latter, however, has not been reported from Africa. All these names are in need of reexamination and some should probably be placed in synonymy. The specimen of *E. majus* in particular, consisting only of an isolated M², is probably referable to one of the other species, as is the *"Eotheroides* sp. indet." skull from the Qasr el Sagha Formation described by Reinhart (1959).

Eotheroides is the oldest and most primitive dugongid known and is probably generalized enough to have given rise to all the later forms. Its rostrum is moderately deflected and bears a pair of small tusks. Five premolar teeth are present, but DP5 does not appear to have still had a permanent replacement (Sickenberg 1934). The tooth cusp morphology is typical of Paleogene dugongids. The permanent premolars are single-rooted with one major cusp surrounded by a low multicuspate cingulum. Upper molariform teeth have a relatively straight protoloph composed of protocone, protoconule, and paracone; an unobstructed transverse valley; a straight or convex-forward metaloph formed by hypocone, hypoconule, and metacone; and small, variably cuspate anterior and posterior cingula. Lower molariform teeth similarly have two lophs slightly oblique to the tooth's axis, a distinct crista obliqua intruding on the transverse valley, a well-developed hypoconulid lophule (largest on M_3), and no anterior cingulum. The skull is principally marked by characters primitive for sirenians in general: expanded supraorbital processes; well-developed nasals and lacrimals; jugals with greatest ventral expansion posterior to orbit; prominent sigmoid ridge on posterior flank of squamosal; unexpanded dorsolateral rim of exoccipital; slender horizontal mandibular ramus; and several accessory foramina posterior to the large mental foramen. External hind limbs were completely lacking (Siegfried 1967).

A new genus of primitive sirenian, possibly allied to *Eotheroides,* has been discovered in middle Eocene beds at Bu el Haderait, Libya, and is being described by G. Heal (Savage 1977).

Genus *Halitherium* Kaup 1838

SYNONYMY. *Pugmeodon* Kaup 1834 (*nomen oblitum*), *Manatherium* Hartlaub 1886.

TYPE SPECIES. *Hippopotamus dubius* G. Cuvier 1824.

Halitherium is the common Oligocene sirenian of Europe (Lepsius 1882; Spillmann 1959) and also occurs in North America (Reinhart 1976). As the senior available generic name proposed for a fossil sirenian, *Halitherium* has also served in its time as a wastebasket name for a variety of Eocene to Miocene specimens, many fragmentary and indeterminable. At least seven nominal species are currently in the literature, and many more names have been erected and consigned to various degrees of oblivion; thorough revision is badly needed.

Halitherium can be easily derived from dugongids of the *Eotheroides* type; apart from a greater snout deflection, larger tusks, loss of the most anterior premolars, and some reduction of the nasals and lacrimals, there is little to distinguish the two genera. *H. christoli* from the Austrian upper Oligocene (Spillmann 1959), with a somewhat deeper mandible, could be transitional to Miocene forms of the *Metaxytherium* grade.

Only three specimens referred to *Halitherium* sp. have been reported from the African region. None of these, however, is sufficiently complete to justify generic assignment, and their post-Oligocene ages make assignment to *Halitherium* additionally questionable.

Gervais (1872:341) records ribs, cf *Halitherium,* from the *Carcharodon megalodon* beds at Chalouf (= El Shallûfa?), Isthmus of Suez, received by the Paris Museum. He also (p. 352) alludes to other ribs from Lower Egypt cited by de Blainville. In fact, de Blainville (1840:43, 51) was originally inclined to refer the fragmentary vertebrae and ribs in question to a pinniped, but later (1844:119–120) concluded that they were sirenian. These remains, from the "left bank of the Nile valley" and of uncertain age, appear to be the earliest recorded sirenian remains from Africa.

The second *Halitherium* sp. deserves mention for its zoogeographic interest. A fragmentary skull and skeleton were recorded from the island of Makamby on the northwest coast of Madagascar by Collignon and Cottreau (1927), who considered the deposits Miocene (Burdigalian to Helvetian). They compared the specimen with European *Halitherium,* pointing out that the latest known of the latter was *H. bellunense* from the basal Burdigalian of Italy. The Madagascar skullcap is elongated, with temporal crests meeting in the midline, suggesting the stage of evolution represented by *Halitherium.* This record at least establishes the presence in the Indopacific region of a *Halitherium*-like form from which *Dugong* might have been derived, but connecting links are lacking.

Dartevelle (1935) recorded *Halitherium*(?) sp. from beds of probable Burdigalian age at Malembe, Congo, but as the material consisted only of ribs the only tenable identification is Sirenia indet.

Genus *Rytiodus* Lartet 1866

TYPE SPECIES. *Rytiodus capgrandi* Lartet 1866.

This poorly known dugongid was described on the basis of tusks and other fragments from the Aquitanian of France. A fragmentary skeleton was later reported by Delfortrie (1880), but no further material has come to light in Europe. The genus is chiefly notable for its flat, bladelike tusks, relatively large size (estimated at nearly 5 m), and (as restored by Delfortrie) straight, undeflected rostrum. That the latter interpretation is erroneous, however, was apparent from Delfortrie's own illustration (1880, pl. 5), which clearly shows a pronounced deflection of the anterior end of the maxilla, indicating a strongly downturned snout.

This was confirmed by the discovery of an intact skull and other remains in the lower Miocene at Gebel Zelten, Libya, which are being described as a new species of *Rytiodus* by G. Heal (pers. comm.). The rostrum is indeed very sharply deflected, adding one more genus to the roster of dugongids apparently specialized for bottom feeding. *Rytiodus* was evidently an offshoot of Oligocene *Halitherium*, but its specializations do not impress me as warranting the subfamilial distinction usually accorded to it.

Genus *Metaxytherium* de Christol 1840

SYNONYMY. *Cheirotherium* Bruno 1839, *nec* Kaup 1835, *Fucotherium* Kaup 1840, *Pontotherium* Kaup 1840, *Felsinotherium* Capellini 1871, *Halysiren* Kretzoi 1941.

TYPE SPECIES. *Hippopotamus medius* Desmarest 1822.

As *Halitherium* has often served as a wastebasket term, usually connoting "Oligocene dugongid," *Metaxytherium* and *Felsinotherium* have served the same function for the Miocene and Pliocene respectively. The virtual impossibility of separating the latter two nominal taxa has long been acknowledged, and even Capellini himself regarded them as synonymous; but only recently has *Felsinotherium* been dropped from usage (Fondi and Pacini 1974; Reinhart 1976). The synonymy of *Metaxytherium* and *Halianassa* Studer 1887, to which many nominal species of *Metaxytherium* have at times been assigned, is a more complex problem, summarized by Kellogg (1966); the holotype of *Halianassa* may in fact pertain to yet another form, *Thalattosiren* (see

Thenius 1952). Clearly the Neogene halitheriines are as badly in need of revision as the Paleogene ones; at least a dozen nominal species of *Metaxytherium* are still recognized.

As noted above, *Eotheroides*, *Halitherium*, and *Metaxytherium* seem to form a straightforward sequence of structural stages, although the details on the specific level remain to be worked out. The European *Metaxytherium* continued the tradition of tusked dugongids with moderately deflected snouts, but with derived characters such as reduced supraorbital processes, nasals, and lacrimals, more anterior suborbital expansion of the jugals, distinct and slightly inturned processi retroversi of the squamosals, deeper mandibles, and absence of accessory mental foramina. The adult cheek dentition is typically reduced to DP5-M3, and the cusp patterns are somewhat more complex than in *Halitherium*. The upper teeth generally feature partial obstruction of the transverse valley by the metaconule, while the lophs of the lowers tend to be more crescentic. North American *Metaxytherium*, which may well deserve separate generic status, generally lacked tusks and, at least in the Pliocene, developed quite sharply downturned snouts.

This genus has so far been encountered twice in Africa. A partial skull of *Metaxytherium* sp. indet. is known from the lower Miocene of Gebel Zelten, Libya (Savage pers. comm.). A lower molar and a skullcap from the Pliocene of Dar bel Hamri, Morocco, were referred to *"Felsinotherium"* cf *serresi* by Ennouchi (1954). *M. serresi* (Gervais 1859) is a contemporary form from France (Depéret and Roman 1920), and the identification is reasonable. This record serves principally to underline the not surprising fact that both sides of the Mediterranean shared a common sirenian fauna.

Indeterminable Sirenian Remains

At this point may be listed several African occurrences of Tertiary sirenians that, unfortunately, do not yet include diagnostic material.

Eocene: Isolated cheek teeth and other remains have been collected at Mogadishu and at Callis and Carcar, Somalia (Savage 1977), and further fragments in the middle Eocene Nautilus Beds at Daban, Somali Republic (MacFayden in Haas and Miller 1952). Upper Eocene beds at Dor el Talha, Libya, have also yielded rib fragments (Savage 1971).

Oligocene: A tusk, possibly sirenian, has been found at Bedeil, Somali Republic (Savage 1969). A skeleton has been found in the Fortuna Sandstone at

Djebel ech Cherichira, Tunisia (P. Robinson, pers. comm. to G. Heal).

Miocene: Fragments have been collected from both the *Hipparion* and the pre-*Hipparion* levels of the Vindobonian Beglia Formation, Bled ed Douarah, Tunisia, by Robinson (pers. comm. to G. Heal; Robinson and Black 1969).

Subfamily Dugonginae Simpson 1932

Genus *Dugong* Lacépède 1799

SYNONYMY. *Platystomus* Fischer v. Waldheim 1803, *nec Platystoma* Meigen 1803, *Amblychilus* Fischer v. Waldheim 1814, *Dugungus* Tiedemann 1808, *Halicore* Illiger 1811.

TYPE SPECIES. *Trichechus dugon* Müller 1776.

Dugong dugon (Müller) 1776 has, as noted above, no known fossil record, and the few fragmentary fossil sirenians known from the Indopacific region have shed no real light on its history. The dugong, however, is reasonably viewed as a descendant of halitheriines that became isolated in the eastern remnant of Tethys during the Miocene and specialized for feeding on the extensive sea grass meadows of the Indopacific. It is anatomically quite conservative, apart from minor derived characters such as narrow supraorbital processes and strongly inflected processi retroversi of the squamosals and the obvious dental specializations (adult cheek teeth reduced to M2-3, lacking roots and enamel). Populations in different parts of the Indopacific, once distinguished by specific names, are now all regarded as but a single species. A convenient summary of the literature on *Dugong* is provided by Husar (1975). Dugongs and manatees (see below) have been traditionally hunted by man (see, for example, Petit 1927), and archaeologists working at coastal sites in tropical Africa should expect to encounter sirenian bones (for illustrations see Kaiser 1974).

Family Trichechidae Gill 1872

Genus *Trichechus* Linnaeus 1758, *nec* 1766

SYNONYMY. *Manatus* Brünnich 1772, *Oxystomus* Fischer v. Waldheim 1803, *Neodermus* Rafinesque 1815, *Halipaedisca* Gistel 1848.

TYPE SPECIES. *Trichechus manatus* Linnaeus 1758.

What little Tertiary record there is of manatees is restricted to South America. They appear in North America only in the Pleistocene and lack any fossil record at all in Africa. The close similarities among the three living species, however, and the fact that anatomically the most distinct of the three is not the West African but the South American, suggest that the African form is a very recent immigrant from the New World. Hatt (1934) provides a useful summary of the anatomy and distribution of the African species *T. senegalensis* Link 1795.

Phylogeny and Evolution

Much has already been said about the phylogeny of the sirenians discussed in this chapter; the general relationships are summarized in figure 28.1. It must be noted that sirenian paleontology to date has been purely descriptive, and actual documentation of phylogeny—as opposed to mere considerations of "stage of evolution"—is only beginning to become possible. As the record becomes more completely known, the ecology as well as the cladistic relationships of fossil sirenians can be clarified, and the sirenian record can begin to contribute ideas and data to other areas of paleontological interpretation. But much basic taxonomic work remains to be done; the many described species of genera such as *Eotheroides*, *Halitherium*, and *Metaxytherium* doubtless require some lumping, and the genera themselves may need to be split in order to better reflect their kinship to some of the monotypic forms.

Sirenian diversity seems to have been constrained by the lack of diversity in suitable marine food plants and habitats, which made niche partitioning difficult, and by the vagility of sirenians themselves, which minimized opportunities for geographic isolation of populations. The pattern that seems to be emerging is one of repeated parallel evolution of adaptations for a very small number of possible niches, e.g., bottom feeding in forms with strongly deflected snouts such as in *Rytiodus*, *Thalattosiren*, *Dugong*, and some *Metaxytherium*. Such parallelism, of course, makes the record all the more difficult to unravel.

Africa's role in sirenian evolution has been very limited due to that continent's relative emergence throughout the Cenozoic. Apart from whatever importance the south shore of Tethys may have had as the, or a, scene of original entry of the water by sirenian ancestors, the main events of sirenian history seem to have been played out in areas peripheral to Africa—notably Europe and the Caribbean, as the Indopacific record is still inadequate. The newly discovered Eocene to Miocene sirenian-bearing deposits in Africa, however, could fill some of the intriguing gaps in the record preserved in the more thoroughly explored parts of the Tethyan realm.

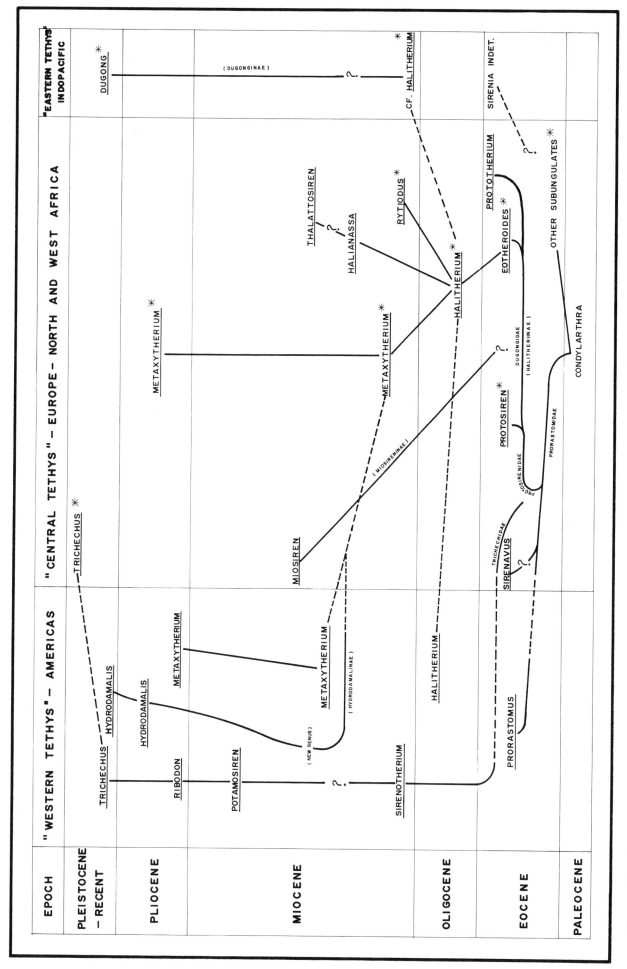

Figure 28.1 Simplified phylogeny of sirenians. Solid lines indicate inferred phyletic relationships. Broken lines indicate inferred major dispersals. Asterisks indicate forms reported from Africa (including Madagascar).

References

Abel, O. 1913. Die eozänen Sirenen der Mittelmeerregion. Erster Teil: Der Schädel von *Eotherium aegyptiacum*. *Paläontogr.* 59:289–360.

Andrews, C. W. 1902. Preliminary note on some recently discovered vertebrates from Egypt (part III). *Geol. Mag.* (4) 9:291–295.

Bertram, G. C. L., and Bertram, C. K. R. 1973. The modern Sirenia: Their distribution and status. *Biol. J. Linn. Soc. Lond.* 5(4):297–338.

Blainville, H. M. D. de. 1840. *Ostéographie, Livr. 7, Des Phoques (G. Phoca, L).* Paris.

———. 1844. *Ostéographie, Livr. 15, Des Lamantins 4(Buffon), (Manatus, Scopoli), ou gravigrades aquatiques.* Paris.

Collignon, M., and Cottreau, J. 1927. Paléontologie de Madagascar. XIV. Fossiles du Miocène marin. *Annls. Paléont.* 16(4):135–171.

Dartevelle, E. 1935. Les premiers restes de mammifères du tertiaire du Congo: la faune Miocène de Malembe. *C. R. Congr. Sci. Bruxelles* 1:715–720.

Delfortrie, E. 1880. Découverte d'un squelette entier de *Rytiodus* dans le falun Aquitanien. *Actes Soc. Linn. Bordeaux* 34:131–144.

Depéret, C., and Roman, F. 1920. Le *Felsinotherium serresi* des sables pliocènes de Montpellier et les rameaux phylétiques des siréniens fossiles de l'Ancien Monde. *Arch. Mus. Hist. Nat. Lyon* 12(4):1–56.

Domning, D. P. 1977. An ecological model for late Tertiary sirenian evolution in the North Pacific Ocean. *Syst. Zool.* 25(4):352–362.

Ennouchi, E. 1954. Un sirénien, *Felsinotherium* cf *serresi*, à Dar bel Hamri. *Notes Serv. Géol. Maroc* 9 (Notes et Mém. no. 121):77–82.

Filhol, H. 1878. Note sur la découverte d'un nouveau mammifère marin (*Manatus coulombi*) en Afrique, dans les carrières de Mokattam près du Caire. *Bull. Soc. Phil. Paris* (7)2:124–125.

Fondi, R., and Pacini, P. 1974. Nuovi resti di Sirenide dal Pliocene antico della Provincia di Siena. *Palaeontogr. Ital.* 67(n.s. 37):37–53.

Fuchs, H. 1973. Contributions à l'étude des siréniens fossiles du bassin de la Transylvanie. (IV) Sur un fragment d'humérus de Cheia Baciului (Cluj). *Studia Univ. Babes-Bolyai, Ser. Geol.-Mineral.* 18(2):71–77.

Gervais, P. 1859. *Zoologie et paléontologie françaises.* 2nd ed. Paris.

———. 1872. Travaux récentes sur les sirénides vivants et fossiles. *Jour. de Zool.* 1:332–353.

Haas, O., and Miller, A. K. 1952. Eocene nautiloids of British Somaliland. *Bull. Am. Mus. Nat. Hist.* 99:313–354.

Hartog, C. den. 1970. The sea-grasses of the world. *Verh. kon. Nederl. Akad. Wet., Afd. Natuurk.* (2)59(1):1–275.

Hatt, R. T. 1934. A manatee collected by the American Museum Congo expedition, with observations on the Recent manatees. *Bull. Am. Mus. Nat. Hist.* 66(4):533–566.

Husar, S. L. 1975. A review of the literature of the dugong (*Dugong dugon*). *U.S. Fish & Wildl. Serv., Wildl. Res. Rept.* 4:1–30.

Kaiser, H. E. 1974. *Morphology of the Sirenia: a macroscopic and x-ray atlas of the osteology of Recent species.* New York: S. Karger.

Kellogg, R. 1936. A review of the Archaeoceti. *Carnegie Inst. Wash. Publ.* 482:1–366.

———. 1966. Fossil marine mammals from the Miocene Calvert Formation of Maryland and Virginia. 3. New species of extinct Miocene Sirenia. *Bull. U.S. Nat. Mus.* 247(3):65–98.

Kretzoi, M. 1941. *Sirenavus hungaricus* n.g., n.sp., ein neuer Prorastomide aus dem Mitteleozän (Lutetium) von Felsögalla in Ungarn. *Ann. Hist.-Nat. Mus. Nat. Hungarici (Min. Geol. Pal.)* 34:146–156.

———. 1953. Le plus ancien vestige fossile de mammifère en Hongrie. *Földt. Közlöny* 83(7–9):273–277.

Lartet, E. 1866. Note sur deux nouveaux Siréniens fossiles des terrains tertiaires du bassin de la Garonne. *Bull. Soc. Géol. Fr.* (2)23:673–686.

Lepsius, G. R. 1882. *Halitherium schinzi*, die fossile Sirene des Mainzer Beckens. *Abh. Mittelrhein. Geol. Ver.* 1:vi + 200.

Owen, R. 1875. On fossil evidences of a sirenian mammal (*Eotherium aegyptiacum*, Owen) from the Nummulitic Eocene of the Mokattam Cliffs, near Cairo. *Quart. J. Geol. Soc. London* 31(1):100–105.

Petit, G. 1927. Nouvelles observations sur la pêche rituelle du dugong à Madagascar. *Bull. Mém. Soc. d'Anthrop. Paris* (7)8:246–250.

Pick, F. K. 1907. Zur feineren Anatomie der Lunge von *Halicore dugong*. *Arch. f. Naturgesch.* 73(1)(2):245–272.

Reinhart, R. H. 1959. A review of the Sirenia and Desmostylia. *Univ. Calif. Publ. Geol. Sci.* 36(1):1–146.

———. 1976. Fossil sirenians and desmostylids from Florida and elsewhere. *Bull. Fla. St. Mus. Biol. Sci.* 20(4):187–300.

Richard, M. 1946. Les gisements de mammifères tertiaires d'Aquitaine. *Mém. Soc. Géol. Fr.* 52:xxiv + 380.

Robinson, P., and Black, C. C. 1969. Note préliminaire sur les vertébrés fossiles du Vindobonien (formation Béglia), du Bled Douarah, Gouvernorat de Gafsa, Tunisie. *Notes Serv. Géol. Tunisie* 31:67–70.

Savage, R. J. G. 1967. Early Miocene mammal faunas of the Tethyan region. *Syst. Assoc. Publ.* 7:247–282.

———. 1969. Early Tertiary mammal locality in southern Libya. *Proc. Geol. Soc. Lond.* 1657:167–171.

———. 1971. Review of the fossil mammals of Libya. In "Symposium on the geology of Libya," University of Libya, pp. 215–225.

———. 1975. *Prorastomus* and new early Tertiary sirenians from North Africa. *Amer. Zool.* 15(3):824.

———. 1976 (1977). Review of early Sirenia. *Syst. Zool.* 25(4):344–351.

Savage, R. J. G., and White, M. E. 1965. Exhibit: Two mammal faunas from the early Tertiary of central Libya. *Proc. Geol. Soc. Lond.* 1623:89–91.

Sickenberg, O. 1934. Beiträge zur Kenntnis tertiärer Sirenen. *Mém. Mus. Roy. Hist. Nat. Belgique* 63:1–352.

Siegfried, P. 1967. Das Femur von *Eotheroides libyca* (Owen) (Sirenia). *Paläont. Zeits.* 41(3/4):165–172.

Simpson, G. G. 1945. The principles of classification and a classification of mammals. *Bull. Am. Mus. Nat. Hist.* 85:xvi + 350.

Spillmann, F. 1959. Die Sirenen aus dem Oligozän des Linzer Beckens (Oberösterreich), mit Ausführungen über "Osteosklerose" und "Pachyostose." *Oesterr. Akad. Wiss., math.-nat. Kl., Denkschr.* 110(3):1–68.

Thenius, E. 1952. Die Säugetierfauna aus dem Torton von Neudorf an der March (ČSR). *Neues Jb. Geol. Pal., Abh.* 96(1):27–136.

Zdansky, O. 1938. *Eotherium majus* sp.n., eine neue Sirene aus dem Mitteleozän von Aegypten. *Palaeobiologica* 6(2):429–434.

29

Cetacea

Lawrence G. Barnes and
Edward Mitchell

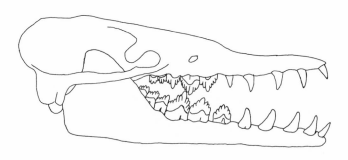

The majority of fossil whale species reported from Africa were collected in Eocene rocks of Egypt and belong in the primitive cetacean suborder Archaeoceti. Most were collected or described in the scientific literature before or around the first part of the Twentieth Century, and some of the primary literature is out of date. Recent publications dealing with specimens from India (Sahni and Mishra 1972, 1975; Satsangi and Mukhopadhyay 1975) have probably not referred extensively enough to the literature on north African Tethys fossils. The studies by Dames, Fraas, Stromer, and Andrews have described most of the archaeocetes of the world, the majority of which were found in Egypt. These species have figured prominently in theories on the origin and evolution of the order Cetacea. In marked contrast, the post-Eocene cetacean record of Africa is scanty and undoubtedly biased by a lack of field work in appropriate rock units.

A discussion of the fossil whales, dolphins, and porpoises of Africa must consider that in life these animals, the fossil remains of which are now found in rocks on the African continent, were inhabitants of the marine waters surrounding it and the fresh water on it. Whether these cetaceans were fluvial, near-shore, or pelagic, their distribution was not necessarily influenced by the same dietary, climatic, or geographic constraints as was the dispersal of the terrestrial animals of Africa. Cetacean systematics is sometimes biased by the land mass on which fossils are found, without full consideration of the ancient ocean basin in which they lived. Therefore, distributions of fossil marine mammals (listed in summary works like Simpson 1945; Romer 1966) should more correctly be recorded as Eocene of Tethys instead of Eocene of Egypt, or Miocene of North Atlantic instead of Miocene of Europe. The fossil record of cetaceans is now too incomplete to recognize and document precise, limited distributions. The safest initial assumption is that a species was probably distributed throughout an ocean basin or water mass rather than having a restricted distribution near a continental margin.

The orientation of this essay must be termed "gradistic," as the limited fossil evidence is insufficiently documented and interpreted at present to be quantified or subjected to cladistic analyses.

Some Morphological Specializations of Cetacea

The following characterization is based on many general and summary works on the history and morphology of the Cetacea, especially those of Miller

(1923), Kellogg (1928a,b, 1936, 1938), Slijper (1936, 1962), Simpson (1945), Tomilin (1957, 1967), Dechaseaux (1961), Rice (1967), and Trofimov and Gromova (1962, 1968).

Living cetaceans are obligative aquatic mammals, having lost efficient means of locomotion on land. Their anatomy and physiology are modified for moving, feeding, and reproducing in water and breathing at the surface. They are one of the most highly modified and specialized of the mammalian orders. The neck is generally short, sometimes with fused vertebrae, and the body is fusiform. Hair is present in modern Cetacea in the form of isolated hairs in patterns on the head and jaws and chin of many species. The presence of hair on early archaeocetes is debatable but entirely possible. Horizontal caudal flukes supported by fibro-cartilage are present on all Recent Cetacea and presumably on extinct species. Fossil skeletons show the change in proportion and shape of the terminal caudal vertebrae that in modern species is associated with the base of the caudal fluke. A single, median dorsal fin is present in most but not all Recent species and may have been secondarily lost in various lineages. The forelimb is flattened into a flipper with the bony digits enclosed within it. Blubber is usually well developed in modern species. Evolutionary trends have been toward the development of a "telescoped" skull, in which the bones of the rostrum extend over and under the cranium. This is likely the result of changes in cranial kinetics associated with specialized feeding habits. The narial opening moved from the typical, anterior position of most mammals to a dorsoposterior position, apparently associated with the need to breathe quickly and efficiently at the air-water interface. In many species the middle ear air sinus extends into various parts of the ventral surface of the skull.

Three suborders have long been recognized. The primitive, extinct suborder Archaeoceti has skulls with the bones not telescoped, but retaining the usual mammalian relationships to each other. The external narial opening is at midpoint on the skull or further forward. The intertemporal region is long and narrow, and the lambdoidal crests are large. The full eutherian dental formula is present, with loss of M^3 in some species. The posterior cheek teeth are multiple rooted with pronounced cusps.

The living suborder Mysticeti, the baleen whales, have vestigial teeth, generally lost *in utero,* and baleen in the adult. Baleen is a neomorph, comprised of fused fibrous tubules of epidermal origin rooted in and suspended from the palate. The fringe of the baleen becomes frayed and interwoven, forming a sieve which living species use to capture food, mainly fish and small crustaceans. The rostral portions of the maxillae are extended laterally in a thin plate to support the baleen, and the cranial portions do not extend far posteriorly onto the cranium. The occipital shield extends dorsoanteriorly over the cranium and in some species shortens the length of the primitive interorbital constriction, thereby changing the size of the temporal fossae. The external bony naris is anterior to the orbit and the external fleshy nostrils are paired in all mysticetes.

The living suborder Odontoceti, the toothed whales, have multiple to single-rooted teeth with complex to simple conical crowns. Some species have greatly reduced functional dentition in the adult. The rostral portions of the maxillae are thick and not greatly extended laterally, and in most families they are extended posteriorly over the front of the cranium nearly to, or touching, the occipital shield. The bony external nares migrate to a dorsal position at the level of the orbits. The occipital shield does not extend far dorsoanteriorly over the cranium. The primitive intertemporal constriction is obliterated by the rostral and frontal bones. The external fleshy nostril is a single opening in living odontocetes.

Some of the characters above are not reflected in bones, and their existence in fossil species can only be implied. There are, however, sufficient osteological characters given to separate the three suborders.

The Fossil Cetacea of Africa

The fossil cetaceans that have been reported from the continent of Africa are neither numerous nor taxonomically diverse. They do however include specimens and taxa that have been extremely important in studies of cetacean evolution. Most African cetacean fossils are Eocene Archaeoceti. Post-Eocene fossils are more scanty and not well studied. The potential exists for further exploration to obtain new fossils, and for much interpretation.

Part of the reason for this incomplete record of African cetacean evolution results from the nearly continuous history of uplift of the African continent. Marine transgressions have been relatively small in area, shallow, and short-lived. Most African cetacean fossils have been found in near-shore and even estuarine and freshwater facies. This precludes the systematic sampling of truly pelagic cetacean assemblages but does not obviate the finding of pelagic species in near-shore facies. On the other hand, it allows unusually precise correlations with chronologies based upon terrestrial vertebrates.

Fossils of African cetaceans occur in Eocene, Oligocene, Miocene, and Pliocene rocks. Eocene marine transgressions left bands of marine deposits on the African continental margin. There were large embayments from the Atlantic Ocean in Nigeria and from Tethys in northern Africa. Archaeocete fossils have been discovered in the former Atlantic embayment represented by the middle Eocene Ameki Formation (Andrews 1920; Van Valen 1968). Élouard (1966) has reported archaeocete fossils from near Kaolack in Senegal. In Libya and Egypt, the Tethys Sea left a regressive marine sequence of middle through late Eocene rocks containing archaeocete fossils. These fossils record the major evolutionary features of the group. They are associated with other marine organisms and in some cases with terrestrial or freshwater animals, allowing extensive interpretations of paleoecology and chronology. The geology of Eocene rocks in Egypt has been described by Beadnell (1905), Andrews (1906), Kellogg (1936), Said (1962), and Simons (1968). These fossils have been collected and studied periodically since the turn of the twentieth century. Kellogg, Andrews, and Simons summarize the research on earlier collections. Specimens were collected by expeditions from Germany (including Richard Markgraf), the Egyptian Geological Survey (under Beadnell), the British Museum (Natural History) (Andrews 1901a,b, 1907b; Andrews and Beadnell 1902; Anonymous 1901), the American Museum of Natural History (Granger 1908, 1910), the University of California (Deraniyagala 1948; Phillips 1948), and Yale University (Simons 1961, 1962, 1964; Simons and Ostrom 1963, 1967; Moustafa 1974).

Two additional Eocene localities in Africa have produced cetacean fossils that have not been identified below the level of order. Van Valen (1968:37) quoted a statement by Reyment that in 1965 the latter had seen "many bones of a whale *in situ*" in the middle Eocene Ameki Formation in Nigeria. The Ameki Formation produced the mandibles and vertebra of the primitive archaeocete *Pappocetus lugardi* Andrews 1920 (see Kellogg 1936:243–244), and it is imperative that more specimens (such as reported by Halstead and Middleton 1974) be collected to properly place the species in evolutionary context. The fossils reported by Van Valen, Reyment, and Halstead and Middleton demonstrate the necessity for further field work in Nigeria. The second locality is in Libya. Savage (1971:219–220) has reported cetaceans from a late Eocene deposit at Dor el Talha (= Gebel Coquin). The material, which was found with sirenians and late Eocene terrestrial mammals, has not been further identified.

During Oligocene time, Tethys retreated northward and the Atlas Mountains were formed. Rocks representing this age in northern Africa are rare and have often been covered by more recent deposits. No fossil cetaceans of this age are reported in the formally published literature from Africa, nor indeed from any southern continent except Australia and New Zealand. Oligocene fossil cetaceans are reported from marine rocks of North America, Europe, and Asia, where their scarcity has been discussed by several writers (Mead 1975; Orr and Faulhaber 1975; Lipps and Mitchell 1976; Whitmore and Sanders 1976; Barnes 1976).

In Miocene time, the Mediterranean Sea in the Tethys area covered parts of northern Africa, but this transgression was not as extensive as the previous Eocene transgression. At the end of the Miocene Epoch, the Mediterranean Sea's connection with the Atlantic Ocean was blocked and evaporation began. Cetaceans have been reported from the early Miocene Marada and Moghara formations, representing the Mediterranean transgression in Libya and Egypt, but these fossils are few. One occurrence of a Miocene ziphiid or beaked whale has been recorded from rocks deposited in a freshwater river system far inland in Kenya (Mead 1975).

Pliocene records are few. Cetacean fossils have been reported but not further identified from Pliocene rocks at Gasr es Sahabi in Libya by Savage (1971:221–222). There are several species of cetaceans in collections from Pliocene marine deposits at Langebaanweg in South Africa (see Hendey 1970:103, 1973; 1974:39; 1976:237), but these have not been studied in detail.

Hendey has commented on the late Cenozoic record of the southwestern Cape Province as follows:

Apart from the Carnivora, the only other mammals which are recorded locally as fossils are Cetacea. Holocene cetacean remains are not uncommon in hominid occupation sites and other deposits adjacent to the present coast and have also been recovered during building operations on Cape Town's reclaimed foreshore area. Heavily mineralized cetacean remains are frequently washed ashore on the beach at Milnerton near Cape Town, in association with other marine fossils and the remains of terrestrial mammals. The latter have included the gomphothere tooth fragment referred to earlier. Most of this material is in private collections and is unstudied. The cetacean remains from Langebaanweg (Hendey 1970a:103), all of which are from Bed I of the Varswater Formation are also unstudied (Hendey 1974:53).

While Pleistocene records of cetaceans in Africa exist in collections and likely occur in various faunal listings and other compilations, we have been unable to find any certainly identified records readily indexed in the formal world literature.

Bones and teeth of uncertain geologic age have

been dredged off the west coast of Cape Peninsula, northward as far as Cape Columbine, Saldanha Bay, South Africa, and along the southwestern slope of Agulhas Bank, at a depth of 150 to 200 fathoms. These remains have apparently not yet been studied and published, but a listing of those on exhibit at the South African Museum (Barnard 1954:33) included "vertebrae of moderate sized whales, ear-bones of Right Whales and Fin Whales, portion of lower jaw resembling that of a Killer Whale, portions of the skulls of Beaked Whales (*Mesoplodon*), portion of a skull closely resembling that of Cuvier's Beaked Whale (*Ziphius*)."

Systematic Account

Order Cetacea Brisson 1762

Suborder Archaeoceti Flower 1883

Family Protocetidae Stromer 1908

Genus *Protocetus* Fraas 1904

SYNONYMY. *Protocetus* Fraas 1904a:201.
TYPE SPECIES. *Protocetus atavus* Fraas 1904a.

Protocetus atavus Fraas 1904

SYNONYMY. *Protocetus atavus* Fraas 1904a:201, pls. 10–12.

The geologically oldest and the most primitive fossil cetacean known by a skull from Africa is *Protocetus atavus*. The type species is also the only species of the genus known from Africa. The holotype (no. 11084, Württembergische Naturaliensammlung, Stuttgart) is a well-preserved skull (see Fraas 1904a, pl. 10, figs. 1–2, pl. 11, fig. 1; Kellogg 1936, pl. 34) from near Cairo, Egypt. It was collected from the same stratum as *Protosiren fraasi*, in the basal portion of the lower Mokattam Formation (= Mokattam Series of Beadnell 1905; Kellogg 1936), the age of which is early middle Eocene (= early Lutetian age). Vertebrae, ribs, a tooth, and part of a second skull have been referred to the species (Fraas 1904a; Stromer 1908b; Pompeckji 1922; Kellogg 1936; Slijper 1936, figs. 64, 199). The skull fragment referred to *Protocetus atavus* by Stromer (1908b:108–109) is slightly larger than the holotype (Kellogg 1936:237). The total body length can be estimated at approximately 2.5 meters (see also Kellogg 1936:240, 276) based on the vertebral centra lengths and the schematic skeletal reconstruction given by Slijper (1936, fig. 217a). This is small for whales, but about the size of many fossil and Recent porpoises.

The primitive characters shown by *P. atavus* in-

clude: the anterior position of the external narial opening, trenchant cheek teeth lacking serrations such as are present on teeth of the Dorudontinae and Basilosaurinae, lack of an enlarged air sinus in the basicranium surrounding the periotic and auditory bulla, short vertebral centra with broad and erect neural spines and large zygapophyses, and apparently an articulation between the innominate bones and sacral vertebrae. The primitive nature of *P. atavus*, particularly in the dentition, has been used as evidence for its derivation from creodonts (Fraas 1904a; Andrews 1906; Kellogg 1936). Kellogg (1936:276, 279) believed that *Protocetus* was ancestral to the later Dorudontidae (*Dorudon* and *Zygorhiza*, Dorudontinae of this paper), citing as evidence the small body size and relatively unmodified vertebrae with short centra common to all three genera.

Genus *Pappocetus* Andrews 1920

SYNONYMY. *Pappocetus* Andrews 1920:309.
TYPE SPECIES. *Pappocetus lugardi* Andrews 1920.

Pappocetus lugardi Andrews 1920

SYNONYMY. *Pappocetus lugardi* Andrews 1920: 309, text-fig. 1, pl. 1.

This poorly known species was based upon an incomplete mandible (holotype, BM(NH) M-11414, our figure 29.1), and a referred mandible, tooth, and axis vertebra (see Andrews 1920, text fig. 1, pl. 1; Halstead and Middleton 1974, figs. 4, 5), all collected from the Ameki Formation (Reyment 1965; Van Valen 1968:37) in the Ombialla District of Nigeria (Andrews 1920:309). These specimens, from an animal larger than *Protocetus atavus*, are of middle Eocene (= Lutetian) age (see Kellogg 1936:243; Van Valen 1968:37). They are probably younger geologically than *Protocetus atavus*, which is of early Lutetian age. Andrews (1920) realized that because of disparate parts, the type mandible of *Pappocetus lugardi* could not be objectively compared with *Protocetus atavus*. Kellogg (1936:279–280) regarded the two genera as distinct, but Van Valen (1966:90; 1968:37) questioned the validity of *Pappocetus*. Subsequently, Sahni and Mishra (1972) described a new species of middle Eocene *Protocetus* from India, and they considered the two genera "independent but related" (1972:494). They qualified this distinction by pointing out that the mandible fragments referred to their new species, while differing from those of *Pappocetus*, were only tentatively associated with the type skull of their new species.

The type jaw of *P. lugardi* definitely represents a primitive morphology of protocetid whale. Andrews (1920) cited the "carnivore"-like teeth, which bear

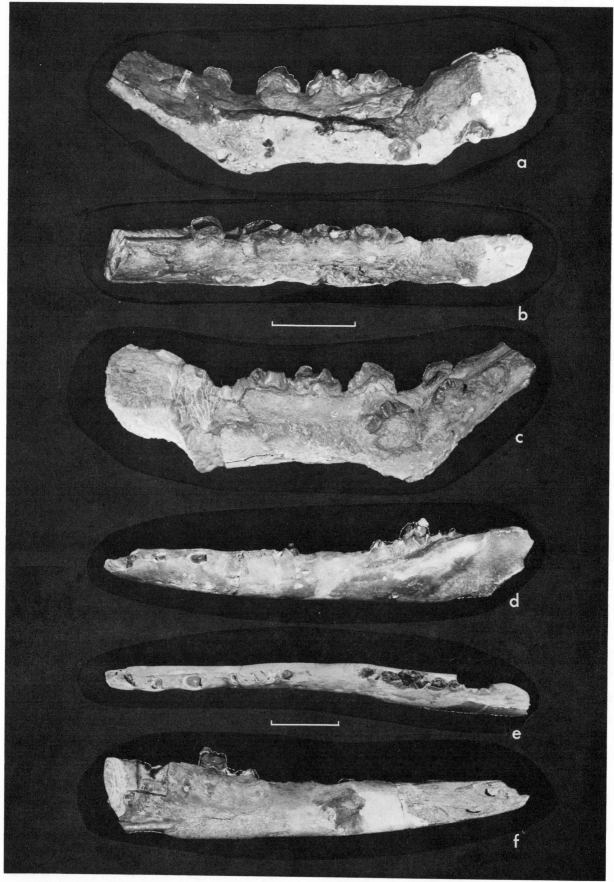

Figure 29.1 *Pappocetus lugardi* Andrews 1920. Holotype, BM(NH) M-11414, incomplete mandible: (*a*) left lateral, (*b*) occlusal, and (*c*) right lateral views. Referred left dentary, BM(NH) M-11086: (*d*) lateral, (*e*) occlusal, and (*f*) medial views. Scale lines equal 10 cm.

cingulae and are present in the full eutherian complement (I_{1-3}, C_1, P_{1-4}, M_{1-3}). More fossils are needed to resolve its uncertain relationships. Van Valen (1968:37) mentioned that in 1965 R. A. Reyment saw bones of a cetacean at the type locality of the Ameki Formation. Dorsal vertebrae and a rib collected near the type locality of *Pappocetus lugardi* indicated to Halstead and Middleton (1974) that the species had a short body.

Genus *Eocetus* Fraas 1904

SYNONYMY. *Mesocetus* Fraas 1904a:217, *nec* van Beneden, 1880, *nec* Moreno 1892; *Eocetus* Fraas 1904b:374.

TYPE SPECIES. *Mesocetus schweinfurthi* Fraas 1904a.

Eocetus schweinfurthi (Fraas) 1904

SYNONYMY. *Mesocetus schweinfurthi* Fraas 1904a:217, pl. 1, fig. 3; *Eocetus schweinfurthi*, Fraas 1904b:374.

Eocetus schweinfurthi is known only by the holotype, a skull (Württembergische Naturaliensammlung, Stuttgart, No. 10986). Two lumbar vertebrae described by Stromer (1903b:83–85) were referred to the species by Kellogg (1936:232). All three fossils were collected from the same horizon, in the upper part of the lower Mokattam Formation (= Mokattam Series of Beadnell 1905; Kellogg 1936), which is late middle Eocene in age (Kellogg 1936:232, 280), and they are therefore stratigraphically higher and geologically younger than the holotype of *Protocetus atavus*. The locality is near Cairo, Egypt, and geographically near the type locality of *Protocetus atavus*.

The skull of *Eocetus schweinfurthi* (see Fraas 1904a, pl. 10, fig. 3; Kellogg 1936, pl. 33) is much larger than that of *Protocetus atavus* and is intermediate in size between skulls of the dorudontine *Dorudon osiris* and the basilosaurine *Prozeuglodon isis* (Kellogg 1936:280). Like *Protocetus atavus*, the skull of *Eocetus schweinfurthi* has the full eutherian dental formula (I^{1-3}, C^1, P^{1-4}, M^{1-3}). The cheek teeth resemble those of *Protocetus*, but their crowns bear small crenulations on the cutting edges (see Fraas 1904a, pl. 11, figs. 10, 11). These serrations have been interpreted as precursors of the enlarged accessory cusps on the cheek teeth of the Basilosauridae (including Dorudontidae in this chapter, Kellogg 1936:231, 233, 280). Additional advanced characters differentiating *E. schweinfurthi* from *Protocetus* include the large size of the skull, its elongate rostrum, more posteriorly positioned external nares, long and narrow intertemporal region,

and high sagittal crest (Andrews 1906:xxiii; Kellogg 1936:232–280). Fraas (1904a:219) regarded *Eocetus* as an intermediate between *Protocetus* and "*Zeuglodon*" (=*Prozeuglodon* and *Dorudon*), but Kellogg (1936:276, 280), citing the near contemporaneity of *Eocetus* with *Protocetus*, suggested that at least two different lines of descent existed, with the protocetid *E. schweinfurthi* being the oldest member of a morphological series leading to the increasingly derived basilosaurine genera *Prozeuglodon*, *Basilosaurus*, and *Platyosphys*. Van Valen (1968:37) believed that *Eocetus* is "clearly related to the Basilosauridae and could equally well be referred to the . . . family."

The statements by Kellogg and Van Valen, however, rely mainly on the fact that the vertebrae Kellogg referred to *Eocetus schweinfurthi* have anteroposteriorly elongate centra like those of *Basilosaurus* and *Prozeuglodon*. We believe that the referral of these vertebrae to the species is tenuous, and that the many small serrations on the cheek teeth of *Eocetus schweinfurthi* are not necessarily homologous with the three to four accessory cusps on the teeth of basilosaurids. We retain *Eocetus* in the Protocetidae, and we caution that the species still is represented with certainty only by the holotype skull. An associated skeleton is needed to provide additional data on the vertebrae of this species.

Family Basilosauridae Cope 1868

SYNONYMY. Zeuglodontidae Bonaparte 1849: 618; Hydrarchidae Bonaparte 1850:1; Basilosauridae Cope 1868:144; Stegorhinidae Brandt 1873:334; Dorudontidae Miller 1923:13; Prozeuglodontidae Moustafa 1954:87.

Species placed in the genera *Dorudon*, *Zygorhiza*, *Prozeuglodon*, and *Basilosaurus* have very similar dentition and cranial anatomy. This was recognized by Kellogg (1936) and Moustafa (1954). Most of these species were originally placed in the genus "*Zeuglodon*" and in the family Zeuglodontidae, both now considered invalid. Miller (1923:13, 40) separated out the species of "*Zeuglodon*" with short or "normal" vertebral centra and placed them in a new family, Dorudontidae, in contrast to the Basilosauridae, which had vertebral centra "greatly elongated." Kellogg (1936) made diagnoses and generic allocations following Miller's (1923) classification, and he has been followed by most subsequent writers (Simpson 1945; Thenius and Hofer 1960; Dechaseaux 1961; Trofimov and Gromova 1962, 1968; Romer 1966; Simons 1968). Moustafa (1954) reunited the two families, but put them in a new family, the Prozeuglodontidae, "without any implication that the genus [*Prozeuglodon*] is in any sense the type of the family" (1954:87, brackets ours). We agree that it is proper to recognize one family, because of the similarities in skulls and teeth, but believe that the elongated vertebral centra in the larger archaeocetes is not simply a function of large size, as stated by Moustafa (1954:87). For example, vertebral structure, number, and centrum length in species of modern Delphinidae vary widely and are probably a reflection both of phylogenetic relationships and of locomotor function

The new family name Prozeuglodontidae, proposed by Moustafa (1954:87), is unnecessary. Article 23, d, 1 of the International Code of Zoological Nomenclature (Stoll et al. 1964) specifies that when a family group is formed by the union of two or more taxa, the oldest valid family group name of any included taxa shall be used. Therefore, Basilosauridae Cope 1868 should be the family name. We do believe that the genera in the previously recognized families Dorudontidae and Basilosauridae form two groups, and we place them in the subfamilies Dorudontinae and Basilosaurinae within the family Basilosauridae. This arrangement is essentially that used by Slijper (1936:540), who reduced Dorudontidae to a subfamily that he included with the subfamily Zeuglodontinae in the family Zeuglodontidae.

Subfamily Dorudontinae Miller 1923

SYNONYMY. Dorudontidae Miller 1923:13; Dorudontinae, Slijper 1936:540, as a subfamily of Zeuglodontidae.

Genus *Dorudon* Gibbes 1845

SYNONYMY. *Dorudon* Gibbes 1845:254–256; *Basilosaurus,* Gibbes 1847:5–14; *Doryodon* Cope 1868:144, 155; *Durodon* Gill 1872:93 (typog. error); *Zeuglodon,* Dames 1894:204 (part, and see Kellogg 1936:184); *Zeuglodon,* Smith 1903:322; *Zeuglodon,* Stromer 1903a:(part); *Protocetus,* Fraas 1904a:(part); *Zeuglodon (Dorudon),* Stromer 1908a:81–88; *Zeuglodon,* Abel 1914:204; *Zeuglodon,* Dart 1923:616–618, 627; *Prozeuglodon,* Kellogg 1928:40.

TYPE SPECIES. *Dorudon serratus* Gibbes 1845.

Dorudon intermedius (Dart) 1923

SYNONYMY. *?Zeuglodon osiris* Dames 1894:35, 36 (part); *Zeuglodon osiris,* Stromer 1908b:110, 114–118 (part); *Zeuglodon intermedius* Dart 1923:617, 629–632; *Dorudon intermedius* (Dart 1923), Kellogg 1936:222.

Dorudon intermedius is the geologically oldest African species in the genus *Dorudon.* It was originally based upon characters of an endocast from a skull. The holotype skull (Kellogg 1936, pls. 30–31;

BM[NH] M-10173; our figure 29.2a) and a referred skull figured by Stromer (1908b, as *Zeuglodon osiris*) are both from the Birket el Qurun Formation of the Fayum, Egypt, although Moustafa (1974:72) listed the species from the Qasr el Sagha Formation. They are of early late Eocene age, and therefore roughly contemporaneous with the basilosaurine *Prozeuglodon isis.* The skull of *D. intermedius,* a relatively small archaeocete, is approximately 795 mm (31 inches) long, and its skeleton was probably no more than 4.9 m (16 feet) long. According to Kellogg (1936:287), it differs from the geologically younger *Dorudon osiris* by having a more highly vaulted cranium and lambdoidal crests that do not project so far posteriorly and are less constricted medially in occipital view.

The vertebrae that Kellogg (1936:223) referred to the species had been previously referred by Dames (1894:35–36) to *Zeuglodon* (=*Dorudon*) *osiris.* Kellogg did not discuss the reassignment of them to *D. intermedius.* We consider this referral tenuous in the absence of associated skeletons of *D. intermedius.*

Dorudon osiris (Dames) 1894

SYNONYMY. *Zeuglodon osiris* Dames 1894:204; *Zeuglodon (Dorudon) osiris* (Dames 1894), Stromer 1908b:81–88 (part); *Dorudon osiris* (Dames 1894), Kellogg 1936:184.

Dorudon osiris is the largest species of *Dorudon* from Egypt. It is represented by the lectotype mandible and premaxillae (Geologisch-Paläeontologisches Inst. und Museum der Universität, Berlin, no. M 16) and 21 referred specimens as listed by Kellogg (1936:185–186). All are from the Qasr el Sagha Formation (see Kellogg 1936:184–186; Simons 1968:13–14) of late Eocene age (late Bartonian correlative), and therefore stratigraphically above and geologically younger than *D. intermedius* from the Birket el Qurun Formation of early late Eocene age. *D. osiris* is one of three species of *Dorudon* recognized by Kellogg (1936) from the Qasr el Sagha Formation. It differs from the contemporaneous *D. stromeri* and the older *D. intermedius* in three respects: it has a less highly vaulted cranium; the lateral parts of the lambdoidal crests on the cranium project strongly posteriorly and are constricted medially in posterior view; and it is larger in size.

Most of the bones in the postcranial skeleton are included in the referred specimens listed by Kellogg (1936), but few if any were found associated with the skulls that comprise the majority of the referred specimens. The vertebrae that Dames (1894:197–201) assigned to *D. osiris* were reassigned to other

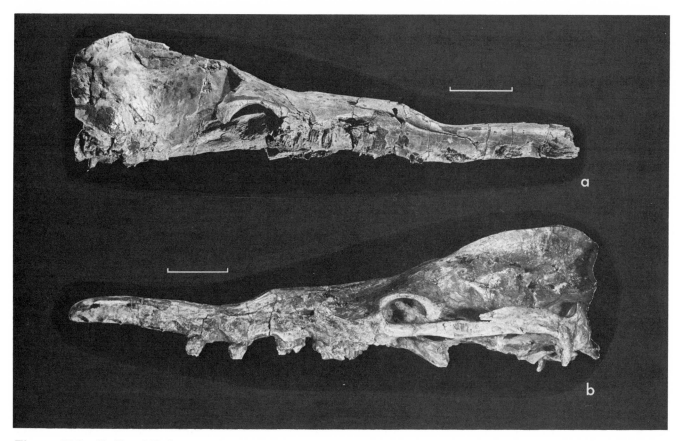

Figure 29.2 Skulls of Tethys Basilosauridae: (*a*) *Dorudon intermedius* (Dart 1923), holotype, BM(NH) M-10173; (*b*) *Dorudon osiris* (Dames 1894), referred specimen, BM(NH) M-10228. Scale lines equal 10 cm.

species by Kellogg (1936:184). Skeletons associated with skulls are therefore needed before the postcranial osteology of *D. osiris* can be objectively studied and compared.

Mitchell and El-Khashab could not locate some specimens of *D. osiris* specifically looked for in the Cairo Geological Museum (CGM), 20–21 July 1977. These included the partial skull CGM-10018 and the endocast therefrom (Kellogg 1936:186), and the portion of the ramus of the left mandible anterior to the crack containing canine and incisor alveoli, illustrated by Andrews (1906, fig. 77), specimen CGM-10207. (See also discussion under *D. elliotsmithii,* below.) Élouard (1966) described vertebrae and a cheek tooth from near Kaolack in Senegal which he identified as *Zeuglodon* cf *osiris*. The fossils may indeed represent *Dorudon osiris* but more study is necessary for positive identification. The tooth (Élouard 1966, fig. 2) is not a premolar, but a posterior molar (see Stromer 1903a:346; Kellogg 1936, pls. 19, 21, 22).

Dorudon zitteli (Stromer) 1903

SYNONYMY. *Zeuglodon zitteli* Stromer 1903b:

70; *Protocetus zitteli* (Stromer 1903), Fraas 1904a: 216, 217, 219; *Dorudon zitteli* (Stromer 1903), Kellogg 1936:212.

Dorudon zitteli is known from the late Eocene Qasr el Sagha Formation. It therefore was contemporaneous with *D. osiris* and *D. stromeri* and geologically younger than *D. intermedius* from the early late Eocene Birket el Qurun Formation. It is about the size of *D. stromeri* and *D. intermedius* and has a skull similar to theirs in general conformation. Like them, it differs from *D. osiris* by being smaller. Kellogg (1936:214) summarized the unique characters of *D. zitteli* as: a different curvature of the posterior margin of the supraorbital process of the frontal, a bilobed root on P^1, unusually large auditory bullae, a high coronoid process of the mandible, and a scapula with a narrow prescapular fossa.

The species is represented by the holotype, a weathered skull with vertebrae and ribs (Paläontologische Sammlung, Alte Akademie, Munich, no. 1902.XI.60.), and four confidently or tentatively referred specimens. Kellogg's (1936:214) distinctions separating *D. zitteli* from *D. osiris* and *D. stromeri* were not strongly made, and more specimens may be needed to reevaluate the supposed diagnostic characters.

Dorudon stromeri Kellogg 1936

SYNONYMY. *Zeuglodon osiris* (Dames 1894), Stromer 1908:110, 111–125 (part); *Prozeuglodon stromeri* Kellogg 1928:40 (*nomen nudum); Dorudon stromeri* Kellogg 1936:203.

The holotype of *D. stromeri* (Paläontologische Sammlung, Alte Akademie, Munich, no. 1904. XII.134e.) is a front half of a skeleton nearly complete back to the mid-thoracic region. A second specimen was only tentatively referred to the species by Kellogg (1936:204).

It is our opinion that Kellogg's (1928:40) mention of the name *Prozeuglodon stromeri* and his reference to Stromer's (1908) publication do not meet the criteria for a species-group name published prior to 1931 (cf Article 16, Stoll et al. 1964). Stromer's (1908) report includes more than one specimen, and Kellogg (1928:40) did not cite page, plate, or specimen numbers, although he did mention "a fairly complete skeleton . . . measuring about ten feet in length including the skull." We attribute the name *Dorudon stromeri* to Kellogg (1936:203).

This *Dorudon* species is one of the three recognized by Kellogg (1936) from the late Eocene Qasr el Sagha Formation (see Simons 1968:13–15), and, like *D. osiris* and *D. zitteli,* it is younger geologically than *D. intermedius. D. stromeri* is very similar to *D. intermedius,* but is smaller, and does not have its lambdoidal crests constricted at all toward the midline in occipital view. *D. osiris,* which is contemporaneous with *D. stromeri,* differs from it by being substantially larger with a less highly vaulted cranium and with lambdoidal crests projecting more strongly posteriorly and much more constricted medially in occipital view.

Dorudon (?) *sensitivus* (Dart) 1923

SYNONYMY. *Zeuglodon sensitivus* Dart 1923: 616; *Dorudon sensitivus* (Dart 1923), Kellogg 1936: 221.

The species was named solely on the basis of an isolated endocranial matrix cast (BM(NH) M-12123), which had weathered out of a skull. The specimen was collected in the Fayum, Egypt, but the precise geologic horizon is unknown (Kellogg 1936:221). Simons (1968:14) and Moustafa (1974:72) included it in lists of animals from the Qasr el Sagha Formation, but there appear to be no published data to support this. It is distinct from endocranial casts known from skulls of *Dorudon osiris, D. intermedius,* and *Prozeuglodon isis* and from an isolated endocranial cast of *D.*(?) *elliotsmithii.* Endocranial

casts are not known, or have not been prepared and published, for *Protocetus atavus, Pappocetus lugardi, Eocetus schweinfurthi, Dorudon zitteli,* or *D. stromeri,* and synonymy of *D. sensitivus* with one of these is possible. Kellogg (1936:222) suggested in passing that *D.*(?) *sensitivus* resembled his new species, *D. stromeri.*

Because of the disparate nature of the fossils, precluding comparison of the holotype of *D.*(?) *sensitivus* with several other taxa of archaeocetes, we query its generic assignment and call for the preparation of suitable endocranial casts from identified skulls to help resolve this problem of disparate fossils.

Dorudon (?) *elliotsmithii* (Dart) 1923

SYNONYMY. Archaeoceti undet., Smith 1903: 322, fig. 1; *Zeuglodon elliotsmithii* Dart 1923:618; *Dorudon elliotsmithii* (Dart 1923), Kellogg 1936:220.

Like *Dorudon*(?) *sensitivus,* this species was based solely upon a natural endocranial matrix cast (CGM-C.10018) from Fayum, Egypt. Its original geologic horizon is unknown, but it may be from the late Eocene Qasr el Sagha Formation (see Kellogg 1936:220; Simons 1968:14; Moustafa 1974:72). Kellogg (1936:220) reported a high degree of similarity between this holotype endocast (Smith 1903, fig. 1; Dart 1923, figs. 5–7) and one from a juvenile skull of *Dorudon zitteli.*

The lack of a referred skull of *D. elliotsmithii,* and the possibility that it may be conspecific with one of the other species of archaeocetes from the Qasr el Sagha Formation or older rocks, places its validity in question. We query its placement in the genus *Dorudon.*

Subfamily Basilosaurinae Cope 1868, new rank

SYNONYMY. Basilosauridae Cope 1868:144; Zeuglodontinae, Slijper 1936:540, as a subfamily of Zeuglodontidae.

Genus *Prozeuglodon* Andrews 1906

SYNONYMY. *Prozeuglodon* Andrews 1906:243. TYPE SPECIES. *Prozeuglodon atrox* Andrews 1906.

Prozeuglodon isis (Andrews) 1904

SYNONYMY. *Zeuglodon isis* Andrews 1904:214–215; *Prozeuglodon atrox* Andrews 1906:243–257; *Prozeuglodon isis* (Andrews 1904), Kellogg 1936:75; ?*Protocetus isis* (Beadnell), Trofimov and Gromova 1968, fig. 174 on p. 229.

As Kellogg (1936) revised the genus *Prozeuglodon* it contains but one species, *P. isis* (Andrews) 1904, represented by the holotype mandible (CGM-10208) and at least 19 known referred specimens, mostly from the Birket el Qurun Formation in the Fayum, Egypt, of early late Eocene age, although Moustafa (1974:72) listed it from the Qasr el Sagha Formation. The taxon is geologically younger than *Protocetus atavus* and *Eocetus schweinfurthi,* and differs from both of these protocetids by having cheek teeth bearing large accessory cusps on the anterior and posterior cutting edges of the central cusp. The centra of the posterior thoracic, all the lumbar, and the sacral vertebrae are anteroposteriorly elongate, approaching the proportions found in *Basilosaurus.* It is contemporaneous with *Dorudon intermedius* and older than *D. osiris, D. zitteli,* and *D. stromeri.* Kellogg (1936) placed *P. isis* in the Basilosauridae. It is therefore the most primitive member of the Basilosaurinae (as used here) and its only representative in the Tethys region.

Mitchell and El-Khashab could locate only the atlas and (?) third cervical (CGM-9329, -9332) of a referred specimen in Cairo on 20–21 July 1977; but the casts of all three cervicals of the referred specimen were examined in the British Museum, BM(NH) M-9372a–c. Mitchell also noted some damage to the skull of the type on display in Cairo (CGM-9319). (See also Andrews 1908a,b, for data on casts.)

Kellogg (1936:281–282) summarized the derived characters that separate *Prozeuglodon* from the Protocetidae. In addition to the modified teeth and elongate vertebral centra, *P. isis* differs from Protocetidae in having: an enlarged air sinus in the bone surrounding the middle ear, elongate mandibular symphysis, a single-rooted rather than bilobe-rooted P^1, a more typically cetacean-like posterior end of the mandible and scapula, and by the loss of M^3. The species may have reached nearly 12.2 m (40 feet) in length (Kellogg 1936:281).

Archaeoceti, genus undetermined

"Zeuglodon" cf *"Z."* *brachyspondylus* Müller 1849

SYNONYMY. *Zeuglodon* cfr *brachyspondylus,* Stromer 1908b:36; *"Zeuglodon"* cfr *brachyspondylus,* Kellogg 1936:262.

The three vertebrae discussed by Stromer (1908b) and Kellogg (1936) under the name *Zeuglodon* cfr *brachyspondylus* are from the early late Eocene Birket el Qurun Formation of Fayum, Egypt. They are thus contemporaneous with *Dorudon intermedius* and *Prozeuglodon isis.* Kellogg (1936:262) found that the dorsal vertebra resembles those of *Zygorhiza kochii* from North America and does not resemble those of the basilosaurines *Prozeuglodon* or *Basilosaurus.* The vertebrae are possibly, therefore, referable to the Dorudontinae.

The species *Pontogeneus brachyspondylus* (Müller 1849) is known from late Eocene rocks in North America, and Kellogg (1936:248) placed it in the Archaeoceti, *incertae sedis.*

Suborder Odontoceti Flower 1867

?Family Acrodelphidae Abel 1905

Genus *Schizodelphis* Gervais 1861

SYNONYMY. *Schizodelphis* Gervais 1861; *Cyrtodelphis* Abel 1900.

TYPE SPECIES. *Delphinus sulcatus* Gervais 1853.

Schizodelphis aff. *S. sulcatus* (Gervais) 1853

SYNONYMY. *Schizodelphis* aff. *sulcatus,* Stromer 1907:100–102; *Cyrtodelphis* aff. *sulcatus,* Fourtau 1920:36; *Schizodelphis* aff. *sulcatus,* Hamilton 1973: 276.

Mandible fragments of this long-snouted, dolphinlike cetacean have been reported from three sites in Egypt: Wadi Faregh (Stromer 1907), Moghara (Fourtau 1920), and Siwa Oasis (Hamilton 1973). Hamilton (1973:276) compared measurements of them and found the three specimens to be similar. Fourtau (1920) had given the age of the fossils from Moghara as late Miocene, but Hamilton (1973) stated that the terrestrial mammals from there were similar to those from Siwa Oasis and from Gebel Zelten in Libya (see also Savage 1969), and all were early Miocene in age. Hamilton interpreted the three Egyptian sites as being in the Moghara Formation, which may be an eastern continuation of the Marada Formation in Libya. Hamilton (1973:280) declared *Schizodelphis* aff *S. sulcatus* characteristic of early Miocene deposits in North Africa.

Schizodelphis sulcatus has been reported from many localities in Europe. Fossils from early Miocene rocks in Italy (Dal Piaz 1901, 1903, 1905) may represent populations in contact with those represented by the sample from the early Miocene shoreline in Egypt. The possibility exists of using this and other species of Miocene cetaceans that lived in the Tethyan region to help correlate between European and African nearshore and terrestrial vertebrate-bearing deposits.

Schizodelphis has also been reported in Miocene rocks of North America (Allen 1921; Case 1934), but restudy of these specimens is needed. Different investigators have allied the genus with the platanistids and with *Zarhachis* and *Pomatodelphis* in North America. It may be related more nearly to

Eurhinodelphis, but the systematic position of *Schizodelphis* needs much further study.

Family Ziphiidae Gray 1865

Genus and species undetermined

An incomplete skull of a beaked whale, family Ziphiidae, (Mus. Comp. Zool., Harvard Univ., field no. 14-64K), was reported by Mead (1975) from near Loperot, 50 miles from Lodwar, Turkana District, Kenya. It was collected from the Turkana Grit below basalt with a Potassium-Argon date of 16.7 ± 1.0 million years. The family has been essentially cosmopolitan since Miocene time (Simpson 1945:101). An interesting aspect of the Loperot ziphiid is its occurrence in a freshwater deposit, in association with freshwater and terrestrial organisms including mammals (Maglio 1969). Since all Recent and fossil ziphiids inhabit marine waters, Mead (1975:750–751) concluded the fossil represents an animal that strayed far up a large river.

Odontoceti, genus undetermined

"*Delphinus*" *vanzelleri* Fourtau 1918

SYNONYMY. *Delphinus van Zelleri* Fourtau 1920:36.

The holotype of this small odontocete cetacean from the Moghara Formation in Egypt was originally regarded as late Miocene in age. Hamilton (1973; see also discussion, above, under *Schizodelphis* aff. *S. sulcatus*) has shown the Moghara Formation to be early Miocene in age. The species was based upon a fragment of a mandible with nine teeth either in place or represented by alveoli (Fourtau 1920, fig. 25). This specimen is probably generically, if not specifically, unidentifiable, and cannot be compared critically with holotypes of many other valid, nominal species of small odontocetes. The true identity of "*Delphinus*" *vanzelleri* will remain unknown until more, and more complete, topotypic specimens are discovered in the Moghara Formation.

Cetacean Origins and the Phylogenetic Position of African Archaeocetes

Cetaceans were early regarded as fish, then as descendants of marine reptiles. Recent investigators have regarded cetaceans as descendants of terrestrial mammals whose precise ancestry has been obscured by greatly modified morphology and by lack of a fossil record of transitional forms. The discovery of primitive archaeocetes, particularly the Protocetidae, has led to interpretations that they were

aquatic creodonts (Fraas 1904a) or were descended from such creodonts (Andrews 1906; Kellogg 1936:236) or from insectivores (Gregory 1910). Kellogg (1936:278) postulated an origin from much earlier (i.e., Mesozoic) mammals, because of the advanced stage of aquatic adaptation of middle Eocene archaeocetes and the time supposedly necessary for their evolution.

The origination of Cetacea from "hoofed mammals" has been proposed by Flower (1883), Anthony (1926), and Mossman (1937). Neontological studies based upon tooth enamel microstructure (Carter 1948), chromosomes (Makino 1948), insulin (Ishihara et al. 1958), serology (Boyden and Gemeroy 1950), and fetal blood fructose (Goodwin 1952) consistently show closest relationships between Cetacea and ungulates, particularly Artiodactyla and in many cases suids (Zhemkova 1965).

Recent studies of primitive mammals have led to reinterpretations of the Order Creodonta. Van Valen (1966:90–93) presented strong paleontological evidence that protocetid cetaceans were derived from mesonychid condylarths, the probable ancestors of modern ungulates (hoofed mammals). The Paleocene mesonychid condylarth *Dissacus navajovius,* from North America, figured strongly in Van Valen's discussion, and he showed Cetacea as derived from Mesonychidae on a phylogeny (1966, fig. 16). Szalay (1969a, b) elaborated upon Van Valen's (1966) interpretation and argued that a good mesonychid ancestor for archaeocetes is the small and rare hapalodectine *Halpalodectes,* which may have had its ancestry near *Dissacus. Hapalodectes* is known from early Eocene rocks in North America and late Eocene rocks in Asia. Szalay used as evidence the conformation of the zygomatic arch and the presence of embrasure pits, to receive lower tooth cusps, between cheek teeth on the palate. Szalay showed archaeocetes on a phylogeny (1969a, fig. 19) derived from the Mesonychidae near Hapalodectinae in middle Paleocene time. A major morphological jump from mesonychids to archaeocetes is the loss of the protocone on upper cheek teeth; the protocone was not known on intermediate archaeocete fossils at the time Szalay made his proposal.

Protocetus atavus has until recently been recognized as the most primitive fossil cetacean for which a skull is known. Sahni and Mishra (1972, 1975) recently named two additional species of *Protocetus* and the most primitive putative cetacean yet described, *Indocetus ramani* Sahni and Mishra 1975, from middle Eocene (Lutetian correlative) rocks in India. These are broadly contemporaneous with the occurrences of *Protocetus* and *Pappocetus* in Africa.

Indocetus ramani is geographically close to the sites of discovery of several Eocene mesonychids. Sahni and Mishra (1975:18–19) interpreted its skull as intermediate between that of *Protocetus atavus* and the Eocene mesonychine mesonychid *Harpagolestes orientalis* Szalay and Gould 1966. Sahni and Mishra (1975) accepted an origin for archaeocetes from Mesonychidae in the broad sense; however, they did not say that this was proposed by Szalay (1969a, b) and Van Valen (1966), nor did they credit Szalay's suggestion that archaeocetes are derived from hapalodectine mesonychids near *Dissacus*. We think that to derive *Indocetus* from *Harpagolestes* major differences in structural details and proportions must be overcome. Among these are the width of the supraorbital process of the frontals and the fact that in all archaeocete cetaceans the upper cheek tooth row diverges and extends onto the base of the zygomatic arch, whereas in *Harpagolestes,* as in essentially all generalized mammals, the cheek tooth rows do not extend onto the zygomatic arch. Like all mesonychids, and unlike all other archaeocetes, teeth of *Indocetus* have three roots and have a protocone present on crowns in the row P³–M². Because of the incomplete nature of the holotype skull, it is not known whether M³ was present. M³ is lost in *Harpagolestes orientalis* (Szalay and Gould 1966, pl. 17–18), but is present in protocetids. This alone precludes *Harpagolestes orientalis* as an ancestor of protocetids. The illustrations presented by Sahni and Mishra (1972, 1975) are of insufficient quality to substantiate their descriptions or to allow detailed comparisons with other specimens.

Sahni and Mishra (1975) also compared *I. ramani* with the large mesonychid *Andrewsarchus*. Examining the illustrations presented by Osborn (1924) and Sahni and Mishra (1975, pl. 4, figs. 1a, 1b, 2c), we find that in posterior view there is very little similarity between *Andrewsarchus* and *I. ramani*. The occipital condyles of *I. ramani* are large and relatively much closer together than in *Andrewsarchus*, and the degrees of divergence of the squamosal and zygomatic regions in posterior view are quite different. In dorsal view the bulge in the intertemporal region of *Andrewsarchus* is not to be found in the illustration of *I. ramani*. The tooth rows are widely divergent posteriorly in *Indocetus* and nearly parallel in *Andrewsarchus*. In *Andrewsarchus* the internal narial opening is just posterior to the last cheek tooth, whereas in *Indocetus* there is a posteriorly prolonged, tube-like, internal narial opening floored by the pterygoid bones.

The type species of *Protocetus, P. atavus,* is only known by a skull without mandibles. This is the most primitive skull of any known African protocetid, having the narial opening farther forward and having the primitive eutherian dental formula, with no serrations on the teeth. It cannot be directly compared with *Pappocetus lugardi,* which is based only on mandibles. Andrews (1920) recognized this fact. Sahni and Mishra (1975) have found mandible fragments that they have tentatively associated with skulls of two new species of *Protocetus* from India. The length of the mandibular symphysis and the spacing of teeth indicated that the mandibles they called *Protocetus* were different from the holotype of *Pappocetus lugardi.* Therefore, they reasserted the distinction of both genera, contradicting Van Valen's (1968) suggestion that they might be synonymous. The named Indian species of *Protocetus* have an uncertain relation to the African *Protocetus atavus.* Kellogg (1936:276) suggested that *Protocetus* is ancestral to *Dorudon* and *Zygorhiza* (both Dorudontinae).

The single known skull of *Eocetus schweinfurthi* is different from that of *Protocetus,* being larger with the external narial opening farther posterior on the skull. It must be regarded as a more derived taxon than *Protocetus.* Kellogg (1936) assigned separate vertebrae with elongate centra to *Eocetus schweinfurthi* and used these as evidence that the species is ancestral to the Basilosauridae (= Basilosaurinae).

Dorudontine basilosaurids in Africa are represented only by the genus *Dorudon,* in which there are a total of six recognized species (as synonymized by Kellogg 1936). One of these, *D. intermedius,* is from the lower middle Eocene Birket el Qurun Formation. Three are from the overlying Qasr el Sagha Formation, and two others are from unknown stratigraphic levels. *Dorudon intermedius,* being from the lower formation, is the oldest species of *Dorudon* in the world. *Dorudon serratus,* the type species of the genus, is from the late Eocene of North America. *Dorudon intermedius* is probably characteristic of the primitive morphology; and the more conservative dorudontines in the overlying Qasr el Sagha Formation appear to be *D. zitteli* or *D. stromeri,* both of which are small and neither of which have lambdoidal crests greatly constricted medially. *Dorudon osiris* is more modified, being larger and having a higher sagittal crest, a higher occipital shield, and more medially constricted lambdoidal crests. *D. sensitivus* and *D. elliotsmithii* are based only upon endocranial casts. Dart (1923) and Andrews (1923) have shown that these two species are not the same as *D. osiris,* an endocranial cast of which was made from a skull. Kellogg (1936) has followed their taxonomy; however, he has cautioned that there is no way to

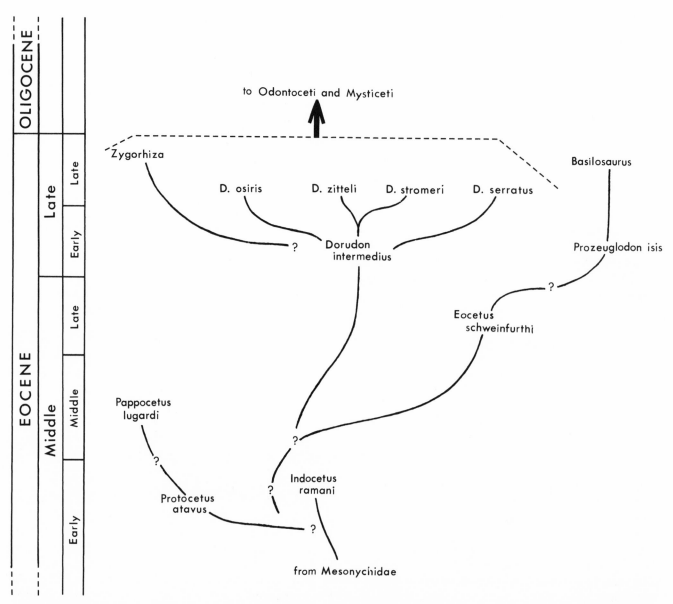

Figure 29.3 Tentative phylogeny of some archaeocetes, based on stated or inferred evolutionary relationships and age or stratigraphic positions given by Andrews (1906, 1920), Kellogg (1936), and others. The age and stratigraphic designations of Moustafa (1974) and Simons (1968) would alter some aspects of this phylogeny. The origin of the modern suborders, Mysticeti and Odontoceti, is believed to be within the dorudontine basilosaurids, but is not more precisely known.

compare *D. elliotsmithii* and *D. sensitivus* with *D. zitteli* and *D. stromeri,* and he has said that they are rather similar in size. Kellogg, in naming *Dorudon stromeri,* has said (1936) that there are slight differences between it and contemporaneous *D. zitteli,* citing a difference in the curvature of the posterior margin of the supraorbital process. Thus, there were apparently the very specialized *Dorudon osiris* and as many as four other contemporaneous species of *Dorudon,* which may be synonymized as one species or which may conceivably represent only sexual differences from typical *D. osiris.* A third possibility is

that there were sympatric species, exhibiting in life character displacement of some kind that does not show up in the fossil materials so far described, similar to the modern diversity of *Stenella* or *Balaenoptera* species. This is at present a moot point.

Kellogg (1936:276) proposed the phylogenetic lineage *Eocetus* to *Prozeuglodon* to *Platyosphys,* implying that the latter would be the most derived basilosaurine and that the subfamily evolved from Protocetidae. If we reject the uncertain evidence that the referred vertebrae of *Eocetus* indicate it is an ancestor of the Tethys *Prozeuglodon,* the latter is still the oldest basilosaurine. *Basilosaurus cetoides* is almost certainly derived from *Prozeuglodon* (see Kellogg 1936).

Debate has continued without resolution as to whether the Archaeoceti were ancestral to the Odontoceti, the Mysticeti, or both. Weber (1904) cogently argued for a common origin from Archaeoceti. Beddard (1900), True (1908), Thompson (1889), Gidley (1913), Miller (1923), Simpson (1945), Kleinenberg (1958), Kleinenberg and Yablokov (1958), and Yablokov (1964, 1965) have argued that archaeocetes are not ancestral whales, and that the order Cetacea is therefore polyphyletic. Taxonomic treatment has varied from Archaeoceti being retained as a suborder of Cetacea (e.g., Simpson 1945) to their being placed in a separate order (Cabrera 1922). Van Valen (1968) has argued against polyphyly, and the recent discoveries of very primitive mysticetes (*Aetiocetus* Emlong 1966; as reallocated by Van Valen 1968) and of very primitive odontocetes (Whitmore and Sanders 1976) make their origin from generalized Dorudontinae plausible. Mchelidze (1976) also has pointed to relationships between Archaeoceti and Mysticeti, but the genera he calls "archaeocetes" (*Aetiocetus, Ferecetotherium,* and *Mirocetus*) have crania (Mchedlidze 1976, pl. 27) more like those of the mysticetes *Cetotheriopsis* and *"Mauicetus"* than like those of Archaeoceti. The dorudontine basilosaurids *Dorudon* and *Zygorhiza,* with serrate cheek teeth, expanded pterygoid middle ear air sinuses, large mandibular foramina, and short vertebral centra, are good structural ancestors for Agorophiidae or Aetiocetidae, but taxa showing intermediate stages of cranial architecture await discovery or interpretation.

Most authors who have conceived an archaeocete origin for odontocetes and mysticetes have postulated an origin from *Protocetus* (e.g., Raven and Gregory 1933:3; Van Valen 1968). In considering the earliest Archaeoceti, we are dealing with very primitive taxa that lack serrate cheek teeth and the air sinus around the ear region, and that have the nostril opening farther anterior. However, if we compare Agorophiidae (*Agorophius* and *Xenorophus,* for example) and Aetiocetidae with *Zygorhiza* and *Dorudon,* we find similarities in elongate mandibles, narial opening far back on the rostrum, serrate cheek teeth, enlarged air sinus around the ear, and a conservative postcranial skeleton with short vertebral centra. Thus, Dorudontinae may comprise suitable ancestors for Odontoceti and Mysticeti (fig. 29.3).

Modern cetacean suborders have shared characters that indicate common origin, including the horizontal caudal fluke; single median dorsal fin; formation of the flipper from the distal end of the pectoral limb; absence of rotational movement in the elbow joint, with a transverse keel dividing the facets on the distal end of the humerus; the extension of the middle ear air sinus system into the areas around the petrotympanic bones; enlargement of the mandibular foramen posteriorly; the reduction of external hair; polydonty; presence of blubber; absence of gall bladder; and similar placental structure (Zhemkova 1965).

Zoogeographic Significance of African Fossil Cetaceans

With the discovery of several primitive taxa, Sahni and Mishra (1972, 1975) shifted attention regarding sites of origin of archaeocete whales in the Tethys from Africa to India. The taxa included *Indocetus ramani* and two species of *Protocetus* in middle Eocene (Lutetian correlative) rocks that are broadly contemporaneous with occurrences of *Protocetus* and *Pappocetus* in Africa. A vertebra from Texas, identified as *Protocetus* sp. by Kellogg (1936:242–243, pl. 15, fig. 4–6), is also of middle Eocene age (Gimbrede 1962). A supposedly older cetacean, *Anglocetus beatsoni* Tarlo 1964, from England has been reidentified as a turtle (Halstead and Middleton 1976:133).[1] The probable place of origin for cetaceans and of diversification for the Protocetidae is still Tethys, but not limited to northern Africa (*fide* Halstead and Middleton 1976:132).

Later Dorudontinae are most diverse in North African Tethys, but they are also known from England (Seeley 1876, 1881; Andrews 1907a; Kellogg 1936; Halstead and Middleton 1972), the southeastern United States (see Kellogg 1936 for summary), Germany, apparently (Kuhn 1935; as *"Zeuglodon"* cf *"Z." osiris*), and Senegal (Élouard 1966). A conservative interpretation of dorudontine history could include an early late Eocene origin in North Africa Tethys and a dispersal into Atlantic waters in late Eocene time. *Kekenodon* from middle Oligocene rocks (Keyes 1973:381) in New Zealand has been called a dorudontid (Kellogg 1936, = our Dorudontinae) but we are not convinced that its unusual morphology warrants that allocation.

The early late Eocene *Prozeuglodon isis* in North African Tethys is the oldest basilosaurine. Kellogg

---

1 We here formally transfer the taxon *Anglocetus beatsoni* Tarlo 1964 from Mammalia: Archaeoceti to Reptilia; and we suggest that future workers base new cetacean names on cranial material.

has suggested an origin for the Basilosaurinae from the protocetid *Eocetus*. Therefore, western Tethys may have been the center of origin or dispersal for the Basilosaurinae. *"Zeuglodon"* cf *"Z." isis* has been reported from late Eocene rocks in Germany (Kuhn 1935; the same specimen is called *Pachycetus robustus* by Van Beneden 1883 and Geinitz 1884), and late Eocene *Basilosaurus* is known from the southeastern United States (Kellogg 1936) and England (Halstead and Middleton 1972).

Other records of archaeocetes from British Columbia, the southeastern United States, Europe, and Antarctica (Kellogg 1936) are yet too poorly known to relate to Tethys species.

The zoogeographic information provided by the post-Eocene African cetaceans is not as significant. The Miocene *Schizodelphis sulcatus* is widely reported from Europe and the genus has been recorded from North America. It probably had at least a Mediterranean-North Atlantic distribution, and Abel's (1905) referral of *Cyrtodelphis* (=*Schizodelphis*) to the family Acrodelphidae did little to clarify its relationships. It is definitely an advanced, long-snouted, dolphin-like odontocete and may have relationships with Eurinodelphidae or with Platanistidae in the broad sense.

The ziphiid from Kenya cannot be placed into generic context, but it records the presence of the family in the Indian Ocean in the Miocene. Fossil Miocene Ziphiidae are known from both the North Atlantic and the North Pacific.

Morphology and Habitus of Archaeocetes

The Archaeoceti includes small species, the Protocetidae and Dorudontinae, whose body proportions are not unlike those of Squalodontidae and Ziphiidae. The Basilosaurinae are among the largest of cetaceans, approximating lengths of the largest living odontocetes. It has been assumed that they were long, thin, and snake-like, but we believe that in life they were slender animals, with body proportions very much like the larger balaenopterids. The head of basilosaurines, however, is proportionally smaller than in balaenopterids (see Kellogg 1936, pl. 1; and our fig. 29.4).

There has been much speculation regarding the structure of the archaeocete (particularly basilosaurine) tail. Kellogg (1936) discussed the ventrolateral orientation of the transverse process on lumbar and anterior caudal vertebrae of basilosaurines. This orientation, however, is not unique and occurs also in *Eurhinodelphis* (Kellogg 1925, pl. 8), *Zarha-*

chis (Kellogg 1924, pl. 15), and *Zygorhiza*. There is no evidence for lateral fin folds, nor for vertical fins in the caudal region, since these structures do not leave traces as fossils. In most modern mammals with tails modified for aquatic propulsion, the tails are expanded in a transverse plane. There is clear cut evidence for horizontal caudal flukes on the tail, based on analogy with all modern cetaceans. This analogy depends on the abrupt change in length versus diameter and the change from flat-faced to round-faced vertebral centra within the tail. The change occurs at the anterior edge of the caudal flukes; apparently this marks one of the two centers of rotation in the tail (see Parry 1949). The other is in the anal region between the lumbars and caudals. Our restoration (fig. 29.4) shows that while the osteological evidence only shows the length of the base of the caudal flukes, if a wide caudal fluke of about the same proportions as that on a large balaenopterine whale is restored on a basilosaurine it makes a reasonable appearing propulsive organ even though the base of the fluke is much shorter.

The restoration of an archaeocete presented here shows an external hind limb. A femur and an innominate bone with an acetabulum have been described for *Basilosaurus cetoides* (Lucas 1900; Gidley 1913; Kellogg 1936), one of the most derived species of archaeocetes. The sacral vertebrae, however, have only rugosities on their extremities. As the sacral vertebra of the more primitive *Protocetus atavus* has a definite facet for articulation with the innominate bone, it is probable that the species had a much less reduced pelvic limb, which may have included more distal elements. Recent Cetacea possess vestigial pelves and femora (Struthers 1893), and anomalous projecting rudimentary pelvic limbs have been reported for example in dolphins (Ohsumi 1965; Slijper 1958), sperm whales (Ogawa and Kamiya 1957; Nemoto 1963), and a humpback whale (Andrews 1921). Struthers (1881) reported a tibia and fibula in *Balaena mysticetus,* and Andrews (1921) reported tarsal and metatarsal bones in the humpback whale. The pelvic limb has obviously undergone considerable reduction during cetacean history, and we believe it protruded externally in archaeocetes. In the restoration of a basilosaurine it is based upon the limb described by Andrews (1921, fig. 1) for a humpback whale.

The elbow joint was capable of movement in an anteroposterior plane because the capitular and trochlear surfaces on the humerus formed a continuous curved surface without a transverse ridge, which appears in Odontoceti and Mysticeti. Presence or absence of a median dorsal fin in Archaeoceti

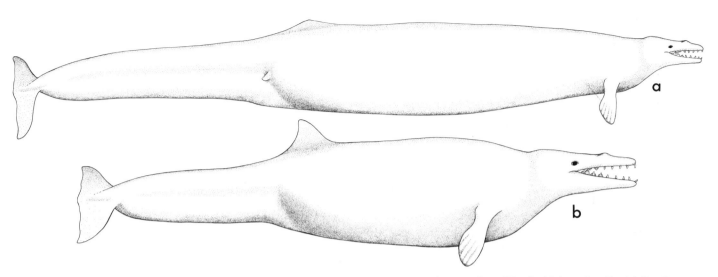

Figure 29.4 Reconstructions of Eocene Basilosauridae similar to species known from North African fossils: (*a*) Basilosaurinae; (*b*) Dorudontinae. Based upon *Basilosaurus* and *Zygorhiza,* from Kellogg 1936, pl. 1, drawn to different scales.

is a moot point. The structure is absent in some specialized cetaceans (Mitchell 1970), and small or absent in various primitive species (i.e., Platanistidae, Iniidae). A large fin is a specialized character.

The ear of advanced archaeocetes is at least partly surrounded by a peribullary sinus. The mandible of archaeocetes is like that of odontocetes, the posterior part of which has been interpreted by Norris (1968) as having a function in acoustic reception by acting as a window to guide incoming sound to the ear. This may indicate that Archaeoceti had at least the capability of passive reception of signals, if not the ability to echolocate actively.

Invasion of the aquatic zone was likely undertaken by protocetids and their ancestors to exploit rich food resources. Lipps and Mitchell (1976) have set out oceanographic and ecologic parameters for a trophic model predicting bursts of diversity. The biological attributes of whale populations that allow them to diversify are numerous. In the relatively short period of time from middle to late Eocene, archaeocetes diversified into several lineages and reached the southern oceans and both sides of the Atlantic Ocean. Much diversity existed, particularly among the Dorudontinae, whose lack of extreme morphological distinctions have complicated their taxonomy. Similar species diversity with little morphological differentiation may be seen in Recent species of the baleen whale genus *Balaenoptera,* in which some species, like the blue whale, feed deeply on euphausiids, while some, like the sei whale, feed more shallowly on copepods, and some, like the fin whale, feed on fish at any depth. All are closely related and exhibit character displacement, but osteo-

logically they are hard to tell apart. Differences of size, pigmentation, behavior, and reproductive time effectively separate the species. The Basilosauridae, for example, may have diversified as the *Balaenoptera* have; it may be merely that the small number of fossils have not yet provided sufficient evidence for us to distinguish species, much less recognize character displacement and evolutionary diversity.

Archaeocetes probably fed in surface waters and perhaps on their sides. When catching fish, or any active prey near the surface, it is more efficient for a narrow-snouted but wide-gaped animal to turn on its side as do modern *Platanista,* crocodiles, fin whales, and other organisms. When feeding on more than one food organism, particularly those which school laterally near the surface, an efficient feeding method would be to open the jaws 30 to 40 degrees, then close them over all items in that area.

Some Problems in African Cetacean Paleontology

As the pace of research on fossil cetaceans increases worldwide, it will become increasingly important to refer to the fossil cetaceans of Africa, particularly the Eocene archaeocetes, in order to place various taxa in phylogenetic context. The Neogene cetaceans of the northern hemisphere have been studied intensively, and they can be organized into faunal and evolutionary sequences (Mitchell 1966; Barnes 1976). Much less is known of Neogene fossil cetaceans of the southern hemisphere.

The taxonomy of the Tethys (northern African) archaeocetes was hindered for years by the almost uniform practice of putting all species in the now invalid genus *"Zeuglodon."* That situation continued until Kellogg's monograph of 1936 appeared, and the concepts of multiple genera and lineages were proposed. To this day, however, some enigmatic species remain in the genus *"Zeuglodon"* for want of a better allocation. Despite Kellogg's extensive review, 'critical' questions remain regarding the Tethys archaeocetes. These are subjects that require future field work and research.

The nomenclature of rock units in the Fayum region is not yet stabilized and the most recent summaries (Simons 1968; Moustafa 1974), do not place the described archaeocete species in the same units as did Kellogg (1936). The Birket el Qurun and overlying Qasr el Sagha Series of Beadnell (1905) were called stages by Kellogg (1936). Simons (1968: 13) and Moustafa (1974:46) call these the "Birket Quarun" and Qasr el Sagha Formations. Moustafa (1974:72) places all described African species of *Dorudon* and *Prozeuglodon isis* in the Qasr el Sagha Formation and does not list vertebrates from the Birket el Qurun Formation. Simons (1968:14) omits *Dorudon intermedius* and *Prozeuglodon isis* (species Kellogg [1936] reported from the Birket el Qurun

Formation) from his list of taxa from the Qasr el Sagha Formation, but does include *Dorudon(?) sensitivus* and *Dorudon(?) elliotsmithii*, species based on specimens whose stratigraphic origins are respectively unknown and uncertain. It is beyond the scope of the present study to rectify these inconsistencies, and we have accepted the stratigraphic data given by Kellogg (1936).

A complete skeleton, including the mandible and forelimb, is not known for *Protocetus atavus*. Variation in the species and its relationship to *Protocetus* from India need investigation. The skull and more bones of the skeleton of *Pappocetus lugardi* are needed to compare the species objectively with the nearly contemporary *P. atavus* and with *Indocetus ramani*, known only by skulls. A postcranial skeleton of *Eocetus schweinfurthi* must be discovered to learn whether Kellogg (1936) was correct in assigning some vertebrae to the species, and whether he and Van Valen (1968) were correct in calling the species an ancestor of the Basilosaurinae. The evolutionary positions of these species within the Protocetidae need detailed study, as do their relationships to suggested terrestrial ancestors.

Among the Dorudontinae, more specimens and

Table 29.1 Nature of sample of African fossil Cetacea.

Species	Endocranial cast	Cranium	Mandible	Teeth	Vertebrae	Ribs	Sternebrae	Scapula	Humerus	Radius	Ulna	Manus	Innominate
Protocetidae													
Protocetus atavus	0	3	0	3	2	2	0	0	0	0	0	0	0
Pappocetus lugardi	0	0	3	3	2	2	0	0	0	0	0	0	0
Eocetus schweinfurthi	0	3	0	3	1	0	0	0	0	0	0	0	0
Basilosauridae													
Dorudon intermedius	3	3	0	3	1	0	0	0	0	0	0	0	0
Dorudon osiris	2	3	3	3	2	2	0	0	2	2	2	0	0
Dorudon zitteli	0	3	3	0	3	3	0	2	1	1	0	0	0
Dorudon stromeri	0	3	3	3	3	3	3	3	3	1	1	0	0
Dorudon(?) sensitivus	3	0	0	0	0	0	0	0	0	0	0	0	0
Dorudon(?) elliotsmithii	3	0	0	0	0	0	0	0	0	0	0	0	0
Prozeuglodon isis	2	2	3	3	2	2	2	2	2	2	2	0	0
Archaeoceti undet.													
"Zeuglodon" cf *"Z."* *brachyspondylus*	0	0	0	0	1	0	0	0	0	0	0	0	0
?Acrodelphidae													
Schizodelphis aff. *S. sulcatus*	0	0	1	0	0	0	0	0	0	0	0	0	0
Ziphiidae													
Gen. et sp. undet.	0	2	0	0	0	0	0	0	0	0	0	0	0
Odontoceti undet.													
"Delphinus" vanzelleri	0	0	3	3	0	0	0	0	0	0	0	0	0

Note: Disparate fossil specimens create problems in comparing taxa and some species may be synonyms. Symbols are: 0, unknown; 1, questionably referred; 2, confidently referred; 3, holotype or lectotype.

endocasts are needed for the majority of the species in order to determine their interrelationships and even their validity. This is particularly true for *Dorudon zitteli, D. stromeri, D. sensitivus,* and *D. elliotsmithii.* The identity of *"Zeuglodon" brachyspondylus,* known only from vertebrae, must be learned. Kellogg pointed out the similarity between these vertebrae and those of *Zygorhiza kochii* from North America. Numerous skulls and skeletons of *"Prozeuglodon"* and *Dorudon* were seen but not collected in the Birket el Qurun Formation by the University of California expedition in 1947 (Phillips 1948:667; Deraniyagala 1948:2, 16, pl. 1).

The total known cetacean record in Oligocene through Pleistocene marine rocks in Africa is pitifully meager. It has been shown herein that although suitable sedimentary deposits are rare, they do exist and cetaceans have been reported from some of them. It is an understatement to say that research on Neogene fossil cetaceans in Africa is a wide open field.

We thank the following persons for providing Mitchell with access to and assistance with collections: R. Said, R. A. Eissa, and B. El-Khashab of the Geological Museum, Cairo; and A. J. Sutcliffe, A. Gentry, A. Currant, and J. Hooker of the British Museum (Natural History). We thank D. E. Savage and J. H. Hutchison for assisting Barnes with collections at the University of California, Berkeley. R. E. Fordyce helped with literature searches, and V. M. Kozicki took the photographs from which L. Reynolds made the prints. Ms. P. Zeadow and Ms. Daphene Cowan aided in manuscript preparation, and Ms. M. Butler prepared the archaeocete restorations.

References

Abel, O. 1900. Untersuchungen über die fossilen Platanistiden des Wiener Beckens. *Denkschr. Acad. Wiss. Vienna* 68:839–874, pls. 1–4 (for 1899).

Abel, O. 1905. Les odontocètes du Boldérien (Miocène supérieur) d'Anvers. *Mém. Mus. Roy. d'Hist. Nat. de Belgique* 3(2):1–155.

Abel, O. 1914. Die Vorfahren der Bartenwale. *Denkschr. Akad. Wiss. Vienna* 90:155–224, pls. 1–12 (for 1913).

Allen, G. M. 1921. Fossil cetaceans from the Florida phosphate beds. *Jour. Mamm.* 2(3):144–159, pls. 9–12.

Andrews, C. W. 1901a. Preliminary note on some recently discovered extinct vertebrates from Egypt. Part I. *Geol. Mag.,* n.s., decade 4, 8:400–409.

Andrews, C. W. 1901b. Preliminary note on some recently discovered extinct vertebrates from Egypt. Part II. *Geol. Mag.,* n.s., decade 4, 8:436–444.

Andrews, C. W. 1904. Further notes on the mammals of the Eocene of Egypt. Part III. *Geol. Mag.,* n.s., decade 5, 1(5):211–215.

Andrews, C. W. 1906. *A descriptive catalogue of the Tertiary Vertebrata of the Fayûm, Egypt.* London: British Museum (Natural History), xxxvii + 324 p., 26 pls.

Andrews, C. W. 1907a. Note on the cervical vertebra of a *Zeuglodon* from the Barton Clay of Barton Cliff (Hampshire). *Quart. Jour. Geol. Soc. London* 63:124–127.

Andrews, C. W. 1907b. The recently discovered Tertiary Vertebrata of Egypt. *Smithsonian Report* 1906:295–307.

Andrews, C. W. 1908a. Note on a model of the skull and mandible of *Prozeuglodon atrox* Andrews. *Geol. Mag.,* ser. 5, 5:209–212, pl. 9.

Andrews, C. W. 1908b. Model of the skull and mandible of *Prozeuglodon atrox* And. *Proc. Zool. Soc. London* 1908:203.

Andrews, C. W. 1920. A description of new species of zeuglodont and of leathery turtle from the Eocene of southern Nigeria. *Proc. Zool. Soc. London* 22:309–319, pls. 1–2 (for 1919).

Andrews, C. W. 1923. Note on the skulls from which the endocranial casts described by Dr. Dart were taken. *Proc. Zool. Soc. London* 1923:648–654.

Andrews, C. W., and H. J. L. Beadnell. 1902. *A preliminary note on some new mammals from the upper Eocene of Egypt.* Cairo: Egypt Surv. Dept. Pub. Works Min., 9 pp., 4 figs.

Andrews, R. C. 1921. A remarkable case of external hind limbs in a humpback whale. *Amer. Mus. Novit.* 9:1–6.

Anonymous. 1901. Geological discoveries in Egypt. *London Times* (4 Oct.).

Anthony, R. 1926. Les affinités des cétacés. *Ann. Inst. Océanogr. Monaco, Paris,* ser. 2, 3:93–135, pl. 1.

Barnard, K. H. 1954. *A guide book to South African whales and dolphins.* South African Museum, Guide no. 4, 33 pp.

Barnes, L. G. 1976. Outline of eastern North Pacific fossil cetacean assemblages. *Syst. Zool.* 25(4):321–343.

Beadnell, H. J. L. 1905. *The topography and geology of the Fayûm province of Egypt.* Cairo: Egypt Survey Dept., Pub. Works Min., 101 pp.

Beddard, F. E. 1900. *A book of whales.* New York: G. P. Putnam's Sons, xv + 320 p.

Boyden, A., and D. Gemeroy. 1950. The relative position of the Cetacea among the orders of Mammalia as indicated by precipitin tests. *Zoologica* 35:145–151.

Cabrera, Á. 1922. *Manual de Mastozoologia.* Madrid and Barcelona: Manuales—Gallach, cxx; 440 + 12 pp. pl.

Carter, J. T. 1948. Comparison of the microscopic structure of the enamel in the teeth of *Zeuglodon osiris* Dames, and of *Prosqualodon davidi* Flynn. In T. Flynn, Description of *Prosqualodon davidi* Flynn, a fossil cetacean from Tasmania. *Trans. Zool. Soc. London* 26:192–193 + pls. 5–6.

Case, E. C. 1934. A specimen of a long-nosed dolphin from the Bone Valley Gravels of Polk County, Florida. *Univ. Michigan Contrib. Mus. Paleontol.* 4(6):105–113, pls. 1–2.

Cope, E. D. 1868. An addition to the vertebrate fauna of the Miocene period, with a synopsis of the extinct Cetacea of the United States. *Proc. Acad. Nat. Sci. Philadelphia* 19(4):138–156.

Dal Piaz, G. 1901. Di alcuni resti di *Cyrtodelphis sulcatus* dell'arenaria di Belluno. Pt. 1. *Palaeont. Italica* 9:187–7:287–292, pl. 34.

Dal Piaz, G. 1903. Sugli avanzi di *Cyrtodelphis sulcatus* dell'arenaria di Belluno. Pt. 1. *Palaeont. Italica* 9:187–219, pls. 28–31.

Dal Piaz, G. 1905. Sugli avanzi di *Cyrtodelphis sulcatus* dell'arenaria di Belluno. Pt. 2. *Palaeont. Italica* 11:253–279, pls. 18–21.

Dames, W. 1894. Über Zeuglodonten aus Äegypten und die Beziehungen der Archaeoceten zu den übrigen Cetaceen. *Palaeontol. Abhandl., Jena,* neue folge, 1(5):189–222 (reprint numbered 1–36), pls. 30–36.

Dart, R. A. 1923. The brain of the Zeuglodontidae (Cetacea). *Proc. Zool. Soc. London* 42:615–648.

Dechaseaux, C. 1961. Cetacea, p. 831–886. In J. Piveteau, ed., *Traité de Paléontologie,* vol. 6, pt. 1. Paris: Masson et Cie.

Deraniyagala, P. E. P. 1948. Some scientific results of two visits to Africa. *Spolia Zeylanica* 25(2):1–42, pls. 1–14.

Élouard, P. 1966. Découverte d'un archéocète dans les environs de Kaolack. *Notes Africaines* 109:8–10.

Emlong, D. 1966. A new archaic cetacean from the Oligocene of northwest Oregon. *Univ. Oregon Mus. Nat. Hist. Bull.* 3:1–51.

Flower, W. H. 1883. Cetacea. In *Encyclopaedia Britannica,* 9th ed., vol. 15, pp. 391–400. Edinburgh.

Fourtau, R. 1918. Études sur les vertébrés fossiles du Wadi Natroun. *Bull. Inst.* Égypte 6(1):195–196.

Fourtau, R. 1920. *Contribution à l'étude des vertébrés miocènes de l'Égypte.* Cairo: Egypt. Surv. Dept., 121 pp. + 3 pls.

Fraas, E. 1904a. Neue Zeuglodonten aus dem unteren Mitteleocän vom Mokattam bei Cairo. *Geol. Palaeont. Abhandl.,* neue folge, 6(3):199–220, pls. 10–12.

Fraas, E. 1904b. (Abstract No. 1048). *Geol. Centralbl., Leipzig* 5(8):374.

Geinitz, H. B. 1884. Nachträge zu den Funden in den phosphatlagern von Helmstedt, Büddenstedt u.a. *Sitz.-Ber. Abhandl. naturwiss. Ges. Isis* 1883:105–111.

Gervais, P. 1853. Sur quelques ossements fossiles de phoques et de cétacés du Languedoc et de l'Aquitaine. *Bull. Soc. Géol. France,* ser. 2, 10:311–313.

Gervais, P. 1861. Sur différentes espèces de vertébrés fossiles observées pour la plupart dans le midi de la France. *Mém. Acad. Sci. Lett. Montpellier* 5:117–132, pl. 4.

Gibbes, R. W. 1845. Description of the teeth of a new fossil animal found in the Green-sand of South Carolina. *Proc. Acad. Nat. Sci. Philadelphia* 2:254–256, pl. 1.

Gibbes, R. W. 1847. On the fossil genus *Basilosaurus,* Harlan, (*Zeuglodon,* Owen,) with a notice of specimens from the Eocene Green Sand of South Carolina. *J. Acad. Nat. Sci. Philadelphia,* n.s., 1(2):5–15, pls. 1–5 (1848?).

Gidley, J. W. 1913. A recently mounted *Zeuglodon* skeleton in the United States National Museum. *Proc. U.S. Nat. Mus.* 44:649–654, pls. 81–82.

Gill, T. 1872. Arrangement of the families of mammals. *Smithsonian Misc. Coll.,* No. 230, 11(1):i–vi, 1–98.

Gimbrede, L. de A. 1962. The Hurricane lentil: source of many Claiborne fossils. *Jour. Paleontol.* 36(5):1116–1120.

Goodwin, R. F. 1952. Foetal fructose in various mammals. *Nature* 170:750.

Granger, W. 1908. A preliminary notice of the Fayûm collection. *Nat. Hist.* 8:8–14.

Granger, W. 1910. [Report on the exploration of the Fayûm.] *Trans. New York Acad. Sci.* 19:303–304.

Gregory, W. K. 1910. The orders of mammals. *Bull. Amer. Mus. Nat. Hist.* 27:1–524.

Halstead, L. B., and J. A. Middleton. 1972. Notes on fossil whales from the Upper Eocene of Barton, Hampshire. *Proc. Geol. Assoc. (London)* 83(2):185–190.

Halstead, L. B., and J. A. Middleton. 1974. New material of the archaeocete whale, *Pappocetus lugardi* Andrews, from the middle Eocene of Nigeria. *Jour. Mining and Geol.* 8(1 + 2):81–85.

Halstead, L. B., and J. A. Middleton. 1976. Fossil Vertebrates of Nigeria. Part II. 3. Eocene Vertebrates. 3.4. Archaeocete whale: *Pappocetus lugardi* Andrews, 1920. *The Nigerian Field* 41(3):131–133, figs. 4–6.

Hamilton, W. R. 1973. A lower Miocene mammalian fauna from Siwa, Egypt. *Palaeontology* 16(2):275–281.

Hendey, Q. B. 1970. A review of the geology and palaeontology of the Plio/Pleistocene deposits at Langebaanweg, Cape Province. *Ann. S. African Mus.* 56:75–117.

Hendey, Q. B. 1973. Fossil occurrences at Langebaanweg, Cape Province. *Nature* 244:13–14.

Hendey, Q. B. 1974. The late Cenozoic Carnivora of the south-western Cape Province. *Ann. S. African Mus.* 63:i–ii, 1–369.

Hendey, Q. B. 1976. The Pliocene fossil occurrences in "E" Quarry, Langebaanweg, South Africa. *Ann. S. African Mus.* 69:215–247.

Ishihara, Y., T. Saito, Y. Itô, and M. Fujino. 1958. Structure of sperm and sei whale insulins and their breakdown by whale pepsin. *Nature* 181:1468–1469.

Kellogg, A. R. 1924. A fossil porpoise from the Calvert Formation of Maryland. *Proc. U.S. Nat. Mus.* 63(14):1–39, pls. 1–18.

Kellogg, A. R. 1925. On the occurrence of remains of fossil porpoises of the genus *Eurhinodelphis* in North America. *Proc. U.S. Nat. Mus.* 66(26):1–40, pls. 1–17.

Kellogg, A. R. 1928a. The history of whales: their adaptation to life in the water. *Q. Rev. Biol.* 3(1):29–76.

Kellogg, A. R. 1928b. The history of whales: their adaptation to life in the water (concluded). *Q. Rev. Biol.* 3(2):174–208.

Kellogg, A. R. 1936. A review of the Archaeoceti. *Carnegie Inst. Washington Publ.* 482:i–xv + 1–366, pls. 1–37.

Kellogg, A. R. 1938. Adaptation of structure to function in whales. *Carnegie Inst. Washington Publ.* 501:649–682.

Keyes, I. W. 1973. Early Oligocene squalodont cetacean from Oamaru, New Zealand. *New Zealand Jour. Marine and Freshwater Res.* 7(4):381–390.

Kleinenberg, S. E. 1958. K voprosu o proiskhozhdenii kitoobraznykh [The origin of the Cetacea]. *Doklady Akad. Nauk S.S.S.R.* 122(5):950–952.

Kleinenberg, S. E., and A. V. Yablokov. 1958. O morfologii verkhnikh dykhatel'nykn putei kitoobraznykh [The morphology of the upper respiratory passages in the cetaceans]. *Zool. Zhurnal* 37, edit. 7:1091–1099.

Kuhn, O. 1935. Archäoceten aus dem norddeutschen Alttertiär. *Zentralblatt für Mineralogie, Geologie und Paläontologie, Abteilung B: Geologie und Paläontologie:* 219–226.

Lipps, J. H., and E. D. Mitchell. 1976. Trophic model for the adaptive radiations and extinctions of pelagic marine mammals. *Paleobiology* 2(2):147–155.

Lucas, F. A. 1900. The pelvic girdle of *Zeuglodon, Basilosaurus cetoides* (Owen), with notes on other portions of the skeleton. *Proc. U.S. Nat. Mus.* 23:327–331, pls. 5–7.

Maglio, V. J. 1969. A shovel-tusked gomphothere from the Miocene of Kenya. *Breviora, Mus. Comp. Zool., Harvard Coll.* 310:1–10.

Makino, S. 1948. The chromosomes of Dall's Porpoise, *Phocoenoides dallii* (True), with remarks on the phylogenetic relation of the Cetacea. *Chromosoma* 3(3):220–231.

Mchedlidze, G. A. 1976. *Osnovnyye cherty paleobiologicheskoi istorii Kitoobraznykh* [Basic features of the paleobiological history of the Cetacea]. Akademia Nauk Gruzinskoi S.S.R., Inst. Paleobiol., "Metsniereba" Press, 136 pp., 32 pls. [In Russian, English summary.]

Mead, J. G. 1975. A fossil beaked whale (Cetacea: Ziphiidae) from the Miocene of Kenya. *Jour. Paleont.* 49(4):745–751.

Miller, G. S., Jr. 1923. The telescoping of the cetacean skull. *Smithsonian Misc. Coll.* 76(5):1–70.

Mitchell, E. D. 1966. Faunal succession of extinct North Pacific marine mammals. *Norsk Hvalfangst-Tidende* 1966 (3):47–60.

Mitchell, E. D. 1970. Pigmentation pattern evolution in delphinid cetaceans: an essay in adaptive coloration. *Canadian J. Zool.* 48(4):717–740.

Mossman, H. W. 1937. Comparative morphogenesis of the fetal membranes and accessory uterine structures. *Publ. Carnegie Inst. Washington* 482:130–246.

Moustafa, Y. S. 1954. Additional information on the skull of *Prozeuglodon isis* and the morphological history of the Archaeoceti. *Proc. Egyptian Acad. Sci.* 9:80–88, pl. 1 (for 1953).

Moustafa, Y. S. 1974. Critical observations on the occurrence of Fayum fossil vertebrates. *Ann. Geol. Surv. Egypt* 4:41–78.

Müller, J. 1849. *Über die fossilen Reste der Zeuglodonten von Nordamerica, mit Rücksicht auf die europäischen Reste aus dieser Familie.* Berlin: G. Reimer, iv + 38 pp., 27 pls. (folio).

Nemoto, T. 1963. New records of sperm whales with protruded rudimentary hind limbs. *Sci. Repts. Whales Res. Inst.* 17:79–81.

Norris, K. S. 1968. The evolution of acoustic mechanisms in odontocete cetaceans, p. 297–324. In E. T. Drake, ed., *Evolution and Environment.* New Haven: Yale Univ. Press, 470 pp.

Ogawa, T., and T. Kamiya. 1957. A case of the cachalot with protruded rudimentary hind limbs. *Sci. Repts. Whales Res. Inst.* 12:197–208.

Ohsumi, S. 1965. A dolphin (*Stenella caeruleoalba*) with protruded rudimentary hind limbs. *Sci. Repts. Whales Res. Inst.* 19:135–136, pls. 1–2.

Orr, W. N., and J. Faulhaber. 1975. A middle Tertiary cetacean from Oregon. *Northwest Sci.* 49:174–181.

Osborn, H. F. 1924. *Andrewsarchus,* giant mesonychid of Mongolia. *Amer. Mus. Novit.* 146:1–5.

Parry, D. A. 1949. The anatomical basis of swimming in whales. *Proc. Zool. Soc. London* 119:49–60.

Phillips, W. 1948. Recent discoveries in the Egyptian Faiyum and Sinai. *Science* 107(2791):666–670.

Pompeckj, J. F. 1922. Das Ohrskelett von *Zeuglodon. Senckenbergiana* 4(3–4):43–100, pl. 2.

Raven, H. C., and W. K. Gregory. 1933. The spermaceti organ and nasal passages of the sperm whale (*Physeter catodon*) and other odontocetes. *Amer. Mus. Novit.* 677:1–18.

Reyment, R. A. 1965. *Aspects of the geology of Nigeria.* Ibadan: Ibadan Univ. Press, 145 pp.

Rice, D. W. 1967. Cetaceans. In S. Anderson and J. K. Jones, Jr., eds., *Recent mammals of the world: a synopsis of families,* pp. 291–324. New York: Ronald Press Co.

Romer, A. S. 1966. *Vertebrate paleontology,* 3rd ed. Chicago and London: Univ. Chicago Press, viii + 468 pp.

Sahni, A., and V. P. Mishra. 1972. A new species of *Protocetus* (Cetacea) from the middle Eocene of Kutch, western India. *Palaeontology* 15(3):490–495.

Sahni, A., and V. P. Mishra. 1975. Lower Tertiary vertebrates from western India. *Monogr. Palaeontol. Soc. India* 3:1–48 + 6 pls.

Said, R. 1962. *The geology of Egypt.* New York: American Elsevier, 377 pp.

Satsangi, P. P., and P. K. Mukhopadhyay. 1975. New marine Eocene vertebrates from Kutch. *J. Geol. Soc. India* 16(1):84–86.

Savage, R. J. G. 1969. Early Tertiary mammal locality in southern Libya. *Proc. Geol. Soc. London* 1657:167–171.

Savage, R. J. G. 1971. Review of the fossil mammals of Libya. In C. Gray, ed., *Symposium on the Geology of Libya,* pp. 215–225. Faculty of Science, Univ. of Libya.

Seeley, H. G. 1876. Notice on the occurrence of remains of a British fossil zeuglodon (*Z. Wanklyni,* Seeley) in the Barton Clay of the Hampshire Coast. *Quart. Jour. Geol. Soc. London* 32, pt. 4 (128):428–432.

Seeley, H. G. 1881. Note on the caudal vertebra of a cetacean discovered by Prof. Judd in the Brockenhurst beds, indicative of a new type allied to *Balaenoptera* (*Balaenoptera juddi*). *Quart. Jour. Geol. Soc. London* 37(148):709–712.

Simons, E. L. 1961. [Untitled.] *Soc. Vert. Paleo. News Bull.* 61:13–14.

Simons, E. L. 1962. An expedition to the Egyptian desert. *Yale Sci. Mag.*

Simons, E. L. 1964. [Untitled.] *Soc. Vert. Paleo. News Bull.* 70:14–15.

Simons, E. L. 1968. Early Cenozoic mammalian faunas, Fayum Province, Egypt. Part 1. African Oligocene mammals: introduction, history of study, and faunal succession. *Peabody Mus. Nat. Hist. Yale Univ. Bull.* 28:1–21.

Simons, E. L., and J. H. Ostrom. 1963. [Untitled.] *Soc. Vert. Paleo. News Bull.* 67:18–19.

Simons, E. L., and J. H. Ostrom. 1967. [Untitled.] *Soc. Vert. Paleo. News Bull.* 80:8–10.

Simpson, G. G. 1945. The principles of classification and a classification of mammals. *Bull. Am. Mus. Nat. Hist.* 85:i–xvi, 1–350.

Slijper, E. J. 1936. Die Cetaceen. Vergleichend—anatomisch und systematisch. *Capita Zool.* 6–7:i–xvi, 1–590.

Slijper, E. J. 1958. *Walvissen.* Amsterdam: Centen's Uitge-versmaatschappij, 524 pp.

Slijper, E. J. 1962. *Whales.* London: Hutchinson & Co., 475 pp.

Smith, G. E. 1903. The brain of the Archaeoceti. *Proc. Roy. Soc. London* 71:322–331.

Stoll, N. R., et al., eds. 1964. *International Code of Zoological Nomenclature adopted by the XV International Congress of Zoology.* London: International Trust for Zoological Nomenclature. 176 pp.

Stromer, E. 1903a. Bericht über eine von den Privatdozenten Dr. Max Blanckenhorn und Dr. Ernst Stromer von Reichenbach ausgeführte Reise nach Aegypten. Einleitung und ein Schädel und Unterkiefer von *Zeuglodon osiris* Dames. *Sitz. mathemtisch-physikalische Classe der kgl. Bayerischen Akad. Wissensch.* 32(3):341–352.

Stromer, E. 1903b. *Zeuglodon*-Reste aus dem oberen Mitteleocän des Fajûm. *Beitr. Paläontol. Geol. Osterr. Ung.* 15(2 + 3):59–100, pls. 8–11.

Stromer, E. 1907. Fossile Wirbeltier-Reste aus dem Uadi Fâregh und Uadi Natrûn in Ägypten. *Abh. Senkenberg. Naturf. Gesell.* 29:97–132, pl. 20.

Stromer, E. 1908a. Die Urwale (Archaeoceti). *Anat. Anz.* 33(4–5):81–88, pl. 1.

Stromer, E. 1908b. Die Archaeoceti des ägyptischen Eozäns. *Beitr. Paläontol. Geol. Osterr.-Ung. Orients.* 21:106–177, pls. 4–7.

Struthers, J. 1881. On the bones, articulations, and muscles of the rudimentary hind-limb of the Greenland right-whale (*Balaena mysticetus*). *Jour. Anat. Physiol., London* 15:141–176, 301–321, pls. 14–17.

Struthers, J. 1893. On the rudimentary hind-limb of a great fin-whale (*Balaenoptera musculus*) in comparison with those of the Humpback whale and the Greenland right whale. *Jour. Anat. Physiol.* 27:291–335, pls. 17–20.

Szalay, F. S. 1969a. The Hapalodectinae and a phylogeny of the Mesonychidae (Mammalia, Condylarthra). *Amer. Mus. Novit.* 2361:1–26.

Szalay, F. S. 1969b. Origin and evolution of function of the mesonychid condylarth feeding mechanism. *Evolution* 23:703–720.

Szalay, F. S., and S. J. Gould. 1966. Asiatic Mesonychidae (Mammalia, Condylarthra). *Bull. Am. Mus. Nat. Hist.* 132(2):129–173, pls. 9–12.

Tarlo, L. B. H. 1964. A primitive whale from the London Clay of the Isle of Sheppey. *Proc. Geol. Assoc.* 74:319–323 (for 1963).

Thenius, E., and H. Hofer. 1960. *Stammesgeschichte der Säugetiere. Eine Übersicht über Tatsachen und Probleme der Evolution der Säugetiere.* Berlin: Springer-Verlag, 322 pp.

Thompson, D'A. W. 1889. On the systematic position of *Zeuglodon. Stud. Mus. Zool. Univ. College Dundee* 1(9):1–8.

Tomilin, A. G. 1957. *Kitoobraznye* [Cetacea]. In *Zveri SSSR i prilezhashchikh stran* [Mammals of the U.S.S.R. and adjacent countries], vol. 9, Moscow: Izdatel'stvo Akademi Nauk S.S.S.R.

Tomlin, A. G. 1967. *Cetacea.* In V. G. Heptner, ed., *Mammals of the U.S.S.R. and adjacent countries,* vol. 9. Jerusalem: Israel Program for Scientific Translations, 717 pp.

Trofimov, B. A., and V. I. Gromova. 1962. Otryad Cetacea (Kitoobraznye). In Y. A. Orlov, ed., *Osnovi Paleontologii* vol. 13, pp. 171–182. Moscow: Gosudarstvennoe Nauchno-Tekhnicheskoe Izdatel'stvo Literatury po Geologii i Okhrane Nedr.

Trofimov, B. A., and V. I. Gromova. 1968. Order Cetacea. In V. I. Gromova, ed., *Fundamentals of paleontology,* vol. 13, *Mammals,* pp. 225–241. Jerusalem: Israel Program for Scientific Translations.

True, F. W. 1908. On the classification of the Cetacea. *Proc. Amer. Philos. Soc. Philadelphia* 47(189):385–391.

Van Beneden, P. J. 1883. Sur quelques ossements de cétacés fossiles, recueillis dans les couches phosphatées entre l'Elbe et le Weser. *Bull. Acad. Roy., Belgique,* sér. 3, 6:27–33.

Van Valen, L. 1966. Deltatheridia, a new order of mammals. *Bull. Am. Mus. Nat. Hist.* 132(1):1–126, pls. 1–8.

Van Valen, L. 1968. Monophyly or diphyly in the origin of whales. *Evolution* 22(1):37–41.

Weber, M. C. W. 1904. *Die Säugetiere. Einführung in die Anatomie und Systematik der Recenten und Fossilen Mammalia.* Jena: Gustav Fischer, 866 pp.

Whitmore, F. C., and A. E. Sanders. 1976. Review of the Oligocene Cetacea. *Syst. Zool.* 25(4):304–320.

Yablokov, A. V. 1964. Konvergentsiya ili parallelizm v razvitii Kitoobraznykh. [Convergence or parallelism in the evolution of cetaceans.] *Paleont. Zhur.* 1964 (1):97–106.

Yablokov, A. V. 1965. Convergence or parallelism in the evolution of cetaceans. *Int. Geol. Rev.* 7(8):1461–1468.

Zhemkova, Z. P. 1965. O proiskhozhdenii Kitoobraznykh (*Cetacea*) [On the origin of whales (*Cetacea*)]. *Zool. Zh.* 44(10):1546–1552. [In Russian, English summary.]

30

Patterns of Faunal Evolution

Vincent J. Maglio

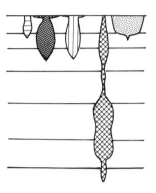

From the preceding chapters it is clear that the so-called "African fauna" is in many ways no more African than Eurasian. Its development was closely linked with mammalian evolution in northern continents and its history was characterized by slow accretion punctuated by quantum shifts in composition. Yet throughout the later Cenozoic Africa gave as much to the north as it received in return. The exact sequence of events that led to the various modern Old World faunas is not entirely clear, and perhaps never will be, but from continued studies of the kind discussed in this volume, at least the broad outlines of these events are emerging.

In spite of tremendous strides over the last several decades, the record still remains limited and certainly grossly underrepresented in some cases. Cooke (1972) has pointed out that for the entire African Tertiary we know only about 150 fossil genera and fewer than 250 species, compared to 256 genera and 740 species living in the continent today. The record improves drastically in the Pleistocene, where another 150 fossil genera are seen. Clearly we must approach any discussion of Cenozoic African faunas with due skepticism, keeping in mind that probably upwards of 80% or more of the continent's former diversity remains undiscovered.

The inadequacy of the record is seen principally on the species level. But many genera also are unknown, and even on the family or ordinal levels we must certainly be lacking knowledge, especially of earlier Tertiary groups that were either endemic to the continent or that, after entering from Eurasia, failed to survive. We need only recall the order Pholidota for a dramatic example. This group probably existed in Africa since the late Oligocene or early Miocene (Patterson, in press) and yet remains unrepresented as fossil except in upper Pleistocene deposits (Klein 1972). The Embrithropoda, a uniquely African order, would have remained completely unknown were it not for a single Oligocene locality in Egypt.

In the midst of this problem we may still find it instructive to analyze major faunal events that shaped the biological profile of Africa. Beginning with an original stocking of primitive placental mammals, there followed a sequence of new immigrations from Eurasia, plus *in situ* evolution of endemic groups, each subsequent step further complicating the faunal array.

Several authors have examined various aspects of this history. Cooke (1968, 1972) discusses major fossil localities and faunas through the Cenozoic within the framework of paleogeographic and paleoecologic studies. Coryndon and Savage (1973) analyze the extent of faunal communication between African and Eurasiatic plates across the closing Tethys Seaway that separated them and give data on major periods of mammalian dispersion into and out of Africa. It would be pointless to repeat here what is said by these authors, and this chapter will attempt only to pinpoint the most important episodes in this history and to summarize overall patterns in the evolution of modern African mammals.

Most of the data used here were drawn directly from preceding chapters in this volume and from sources cited therein. The reader is referred to these references so that specific bibliographic citations need not be repeated and can be kept to a minimum here.

Faunal Successions

In table 30.1 are listed "typical" faunas of Cenozoic epochs in Africa. These are typical only in the sense that the genera represented have their primary temporal distributions as shown, and a fauna of any particular age will include many of the genera shown for that epoch. Actual assemblages will depend largely on geographic region, ecologic setting, and the presence or absence of competing forms. (For tables, see pp. 613–619.)

It can be seen that the entire Cenozoic record is punctuated by "sudden" appearance and disappearance of taxa on all levels from order down to genus. Some of these represent true evolutionary or dispersal events, whereas others are artifacts of the record. Without discussing details of fauna or geology, both discussed admirably by Cooke (1972), I will briefly review the major features in the origin of the modern African assemblage.

Mesozoic

The mammalian record in Africa properly begins in the Triassic, where forms transitional between therapsid reptiles and true mammals can be found. Two undoubted mammals have been described from southern Africa. *Erythrotherium parringtoni* (Crompton 1964) from beds of late Triassic age in Lesotho appears to be closely related to *Morganucodon* of similar age in England, and a common ancestor has been suggested for both. Similarly, *Megazostrodon rudnerae* from upper Triassic deposits of Lesotho is very similar to European contemporaries (Crompton and Jenkins 1968). The only other described Mesozoic mammal from Africa is an edentulous jaw from the late Jurassic of Tendaguru, Tanzania, described by Dietrich (1928) as *Brancatherulum tendagurense,* of uncertain affinities.

It is unlikely that any of these early mammalian records had anything to do directly with the origin of later mammals in the continent, but they do demonstrate a close faunal tie between Africa and Europe during this time interval, a tie that was to persist probably at least until the end of the Paleocene.

Paleocene-Eocene

Cretaceous and Paleocene records of mammals in Africa are totally lacking and we can only speculate as to the course of events that was to shape the structure of later faunas so characteristic of the Eocene and Oligocene. It is likely that the Paleocene and earlier Eocene witnessed immigration into Africa of at least three groups from the north—one or more primitive condylarthran stocks that were to give rise to later African subungulates, a prosimian primate stock, and creodont carnivores.

During the middle and late Eocene we catch glimpses of a record, still grossly incomplete and confined entirely to northern Africa from Egypt to Senegal (see Savage 1969). These faunas are limited to only nine families, of which five are marine mammals (Cetacea and Sirenia). These are the earliest records of the orders and consist of already highly specialized forms, suggesting a long prior history, perhaps from late Paleocene times. Of the remaining four families, one is the hyaenodontid carnivore *Apterodon* and the others are members of endemic orders, probably with a common origin in the middle Eocene. These are gomphotheriid proboscideans and, only distantly related to them, *Moeritherium* and *Barytherium.*

Without an earlier Paleocene record we can only guess that the whales derived from some creodont stock of which no trace remains and that sirenians, moeritheres, barytheres, and proboscideans arose in the early to middle Eocene from earlier subungulate invaders. The hyaenodont undoubtedly represents a late Eocene migrant from Europe.

Just what was going on in the interior of the continent is not known, and it is intriguing to wonder what new groups wait to be discovered there.

Oligocene

By the early Oligocene a great change had occurred in African faunas. Here, too, the record is underrepresented and confined to North Africa. Only two families persist from the Eocene, gomphotheres and hyaenodonts, and to these 14 new families are added. Eight of the latter are rodents, insectivores, or bats, most of endemic origin, derived from somewhat earlier invasions from Eurasia. Of the larger mammals, creodonts are abundant, with four genera and nine species known from the Fayum alone. Primates now appear in the record, represented by three endemic families, Parapithecidae, Pongidae, and Hylobatidae, all apparently derived from earlier prosimian stock. The only artiodactyls are anthracotheres, represented by two genera with close relatives in Eurasia.

Two new endemic orders make their appearance here, the Embrithropoda and Hyracoidea, each with a single family. The former filled the large herbivore niche but never amounted to more than a single genus. Hyracoids, filling a medium-size browser niche, appear in full radiation showing the greatest diversity they will ever achieve, with seven genera known only from northeastern Africa. Both of these orders clearly evolved in the continent and their origins must have occurred much earlier, in the later half of the Eocene.

Of earlier groups that do not continue into the Oligocene record, cetaceans and sirenians almost cer-

tainly persisted in some form, although not known as fossil, and two groups, Moeritherioidea and Barytherioidea, became extinct without issue.

Earlier Miocene

By far the most dramatic faunal upheaval on the continent occurred in the late Oligocene–early Miocene (figures 30.1 to 30.4). Fossil localities are concentrated in East Africa, with several in North Africa and one in the Namib. Only 14 families present earlier are also recorded here. Twenty-nine new families and 79 new genera make their appearance, and of these, 12 families and 31 genera are micromammals. In terms of overall generic resemblances for this fauna, Savage (1967, p. 277) has shown a greater link between Africa and Asia during the Burdigalian than between Africa and Europe.

One order, Embrithropoda, failed to survive from the Oligocene, and two new ones are recorded. One, Tubulidentata, is probably autochthonous and of African subungulate origin; the other, Perissodac-

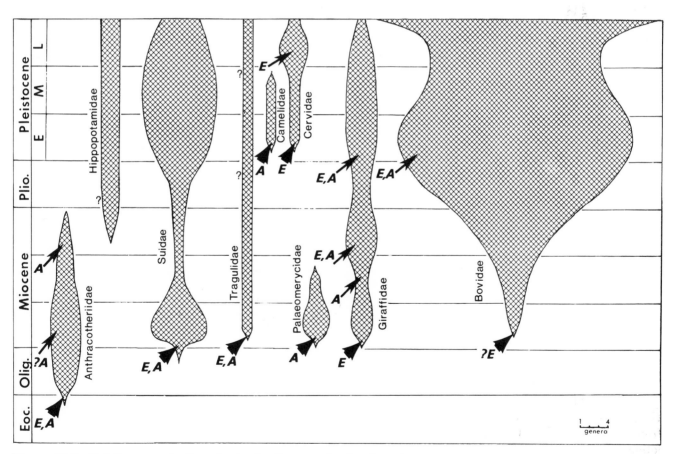

Figure 30.1 Relative generic diversity in the Cenozoic for families of artiodactyls in Africa. Large arrowheads indicate times when families first entered the continent. Small arrowheads indicate periods of later immigration. Families lacking arrowheads are autochthonous. (*E*) Europe, (*A*) Asia.

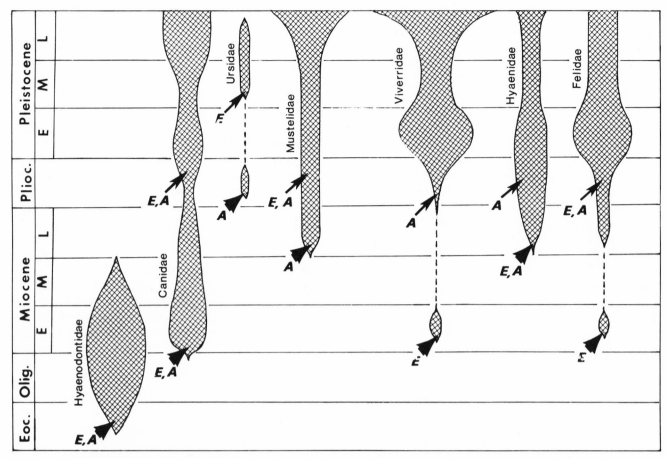

Figure 30.2 Relative generic diversity in the Cenozoic for families of carnivores in Africa. Symbols as in figure 30.1.

tyla, certainly entered from Eurasia and is represented by two families, Chalicotheriidae and Rhinocerotidae. The former was a chance invasion and the family subsequently became extinct here. The latter was far more successful, with four genera penetrating the continent independently.

Pliohyracidae persist but are drastically reduced, only three genera being represented; the extant family Procaviidae is now also present, derived from some earlier hyracoid stock, perhaps in late Oligocene times. Proboscidea have by this time begun a radiation into several basic groups, with Mammutidae certainly derived from gomphotheres late in the Oligocene.

Creodont carnivores remain diverse with four new genera added to two that carry through from earlier times. The first fissiped groups enter at this time from Eurasia and include canids, viverrids, and felids. But together these families barely equal contemporary creodonts in generic diversity.

Primates remain about as abundant as earlier. Hylobatids are still rather rare and pongids comprise two genera, of which *Dryopithecus* is the more

diverse. Monkeys (of uncertain family reference) are recorded for the first time, as are Lorisidae, although the latter may have existed earlier.

Among artiodactyls, anthracotheres remain with three genera, one probably representing a new immigrant from Asia. Archaic suids entered at about this time with four Eurasiatic genera recorded and two additional genera probably representing *in situ* evolution from primitive Eurasiatic stocks. Tragulids and palaeomerycids also expanded into Africa at this time, with similar genera occurring in Africa and Eurasia. At least two endemic giraffes are seen in lower Miocene deposits, representing the Palaeotraginae and Sivatheriinae; both appear to have derived from African ancestors. Of the Bovidae only *Walangania* is recorded in East Africa and slightly later *Protragocerus* and *Eotragus* appear in North Africa, the latter two also known in Europe at about the same time.

Later Miocene

By the late Miocene another 18 families have appeared in Africa. Eleven of these are micromammals, of which five are bats known in Europe much earlier. They were perhaps present but unrecorded in the African early Miocene or even Oligocene.

Of the primates, Hominidae are now present although rare, and hylobatids appear not to have survived in Africa. Mustelid and hyaenid carnivores have now entered the continent, and among the hyraxes the family Procaviidae has completely replaced the earlier pliohyracids.

Elephants of a primitive type had only recently emerged from earlier gomphotheres, but the latter also persist in the form of short-jawed Anancinae. Rhinocerotids are still a diverse group, with a new form related to Recent genera emerging from some unknown earlier stock. From Eurasia *Hipparion* populated the continent during middle Miocene times as the first African equid.

Artiodactyls for the first time are beginning to dominate the African landscape. Hexoprotodont Hippopotamidae, recently evolved possibly in Africa, are now seen both here and in Asia, and more advanced suids of the subfamily Suinae are characteristic. Nine distinct tribes of Bovidae are now recognized, mostly with Asiatic ties. At least nine genera are recorded by latest Miocene times, compared to only one in the early Miocene. Of these, five are unknown outside the continent, suggesting some *in situ* evolution of recently arrived stocks. Four new giraffid genera are seen here for the first time, at least two probably evolved in Africa from earlier members of the family. Anthracotheres and palaeomerycids are now rare and fail to survive the epoch.

Pliocene

By Pliocene time faunal changes on the family level had tapered off, with only three new families appearing. But a major revolution was occurring on the generic level as 81 new genera are recorded for the first time. Thus 76% of the land-mammal Pliocene fauna is new, and this represents the greatest single faunal change since the revolution of the Oligocene-Miocene transition. About 53% of these genera are endemic to Africa.

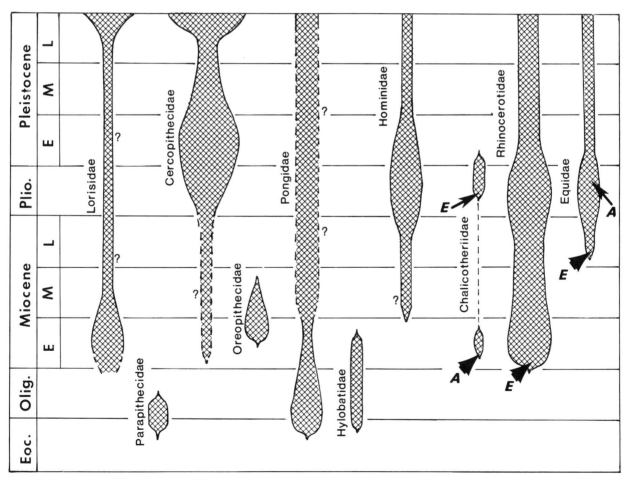

Figure 30.3 Relative generic diversity in the Cenozoic for families of primates and perissodactyls in Africa. Symbols as in figure 30.1.

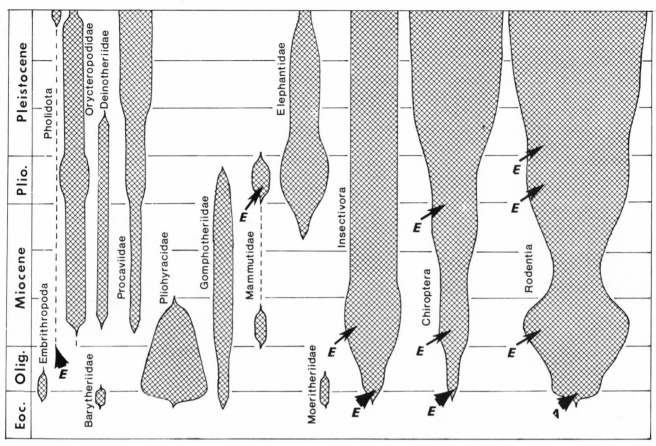

Figure 30.4 Relative generic diversity in the Cenozoic for families of varied subungulate orders and for micromammals. Symbols as in figure 30.1.

Of groups evolved in Africa, monkeys definitely assignable to the Cercopithecidae are now widespread, as are at least two types of hominids. Although pongids are lacking from the record, they certainly were present. Two new endemic genera of orycteropodids have evolved and a large deinothere has replaced the smaller Miocene form. Two new genera of procaviid hyraxes are seen for the first time, and three new genera of elephant have emerged from a Mio-Pliocene radiation of the family. Suine genera number about six, all endemic but of uncertain relationship. Fifteen bovid genera have been added to previous ones. Of these all but five are unknown outside the continent although most have close relatives in Asia, suggesting immigration during the late Miocene.

New groups recently arrived in Africa include the camel, clearly of Asiatic origin, the modern *Giraffa,* a chalicothere from Asia, the genus *Equus,* a Eurasiatic saber-toothed carnivore, and probably an agriotherine bear.

Earlier Pleistocene

The early Quaternary as a whole saw relatively minor changes in the African fauna on the family level, only four new ones appearing at this time. Three of the latter, Leporidae, Rhizopodidae, and Cervidae, entered from Europe, and the fourth, the bat family Myzopodidae, is endemic and now confined to Madagascar. Deinotheres, chalicotheres, and gomphotheres finally vanish from the African scene at the end of this period.

The major changes seen in African faunas are on the generic level. About 53 new genera appear here, but since 89 Pliocene genera persisted into the Quaternary, the total character of the fauna is less drastically altered than in the Mio-Pliocene transition. About half of these new appearances were immigrants from the north and half were products of *in situ* evolution. Of the 16 carnivore genera first recorded here, at least nine seemingly entered from Eurasia, where they are also present. Two new suid genera, eleven bovids, a number of monkeys, bats, and rodents also penetrated the continent at this time.

Later Pleistocene

During the last half of the Pleistocene very little occurred to alter the complexion of African mammalian faunas. Only one family new to the continent appears in the record. This group, the Pholidota, although not recorded earlier, must certainly have been present since earlier Tertiary times.

On the generic level, again only minor changes took place, with the disappearance of older forms being as important as the appearance of new ones. Most of these latter evolved from earlier groups in Africa and therefore do not occur outside the continent.

Endemism

On the basis of overall faunal comparisons with Eurasia many authors have recently concluded that Africa was relatively isolated during Oligocene and earliest Miocene time (van Couvering 1972, p. 264; Coryndon and Savage 1973, p. 123). Such isolation must have followed a period of limited faunal exchange with continents to the north during the Paleocene and early Eocene, for by late Eocene and early Oligocene times a number of specialized and uniquely African terrestrial groups had appeared, descended from primitive Eurasiatic stocks. Even though present along the southern shores of Tethys, most of these groups remained confined to the continent, giving Africa its greatest endemism of the Cenozoic.

In figures 30.5 and 30.6 the degree of endemism for terrestrial mammals is shown. Family endemism reached 81% by early Oligocene times and for genera the figure was 84%; these values may be higher still when a fossil record for deeper continental regions becomes available. Of the three nonendemic families, one is a bat, leaving only two, Anthracotheriidae and Hyaenodontidae, for which a sweepstakes dispersal across the Tethys Seaway must be postulated.

A sudden influx of mammals from northern continents during the Oligocene-Miocene transition dramatically marked the end of Africa's isolation. Family endemism dropped to 33% and generic endemism to 62%. More than half of the new families are micromammals that evolved rapidly to new family status, accounting for much of the observed endemism. If only larger mammals are considered, however, endemism drops to about 25%, nearly one-fourth that of the Oligocene.

Subsequent patterns of familial evolution involved minor reduction in endemism, reaching a low of 20% by the Pliocene. During the later Pleistocene and early Holocene family endemism rose slightly again to 23%. But rather than resulting from evolution or immigration of new groups, this rise was due principally to extinction in Eurasia of families such as hippopotamids and giraffids, giving the impression of an increase in African endemics. This has also been the reason why modern Ethiopian mammals appear to represent a relict Pliocene fauna by European standards.

Generic-level endemism reached its lowest value of 46% in the middle Miocene, probably as a result of continued relatively unhampered immigration from the north. Between the Pliocene and Recent, generic endemism rose again, reflecting progressive *in situ* evolution of numerous stocks that entered earlier. Toward the end of the Pleistocene and in the Holocene endemism reached its highest level since Oligocene times, and presumably followed development of the Sahara Desert, which served as a major barrier against north-south dispersal (Coryndon and Savage 1973, p. 134). This idea is supported by the fact that North Africa, partially isolated from the remainder of the continent, has continued to exchange faunal elements with Eurasia to the present day (Cooke 1972, p. 122).

Faunal Turnover

In figures 30.5 and 30.6 the entire Quaternary is plotted as a single point for turnover and extinction curves (shaded areas) in order to make time intervals for Cenozoic subdivisions more nearly equal. Even so, later periods tend to be shorter than earlier ones, but their better record compensates somewhat for these inequities. Broken lines give values plotted for subunits within the Pleistocene, but it should be noted that the short time intervals involved are mainly responsible for the lower values here.

Viewing these data in terms of faunal replacement, we see on the generic level extremely high rates of turnover in which 80 to 90% of the fauna was renewed between Eocene and Oligocene times and again between the Oligocene and early Miocene. As discussed above, the essential difference between these two transitions was a turnover caused by local evolution and involving mainly autochthonous and endemic groups in the former event but encompassing mostly immigrants and their slightly modified endemic descendants in the latter.

By middle and late Miocene times less than one-half of the African fauna was renewed in each suc-

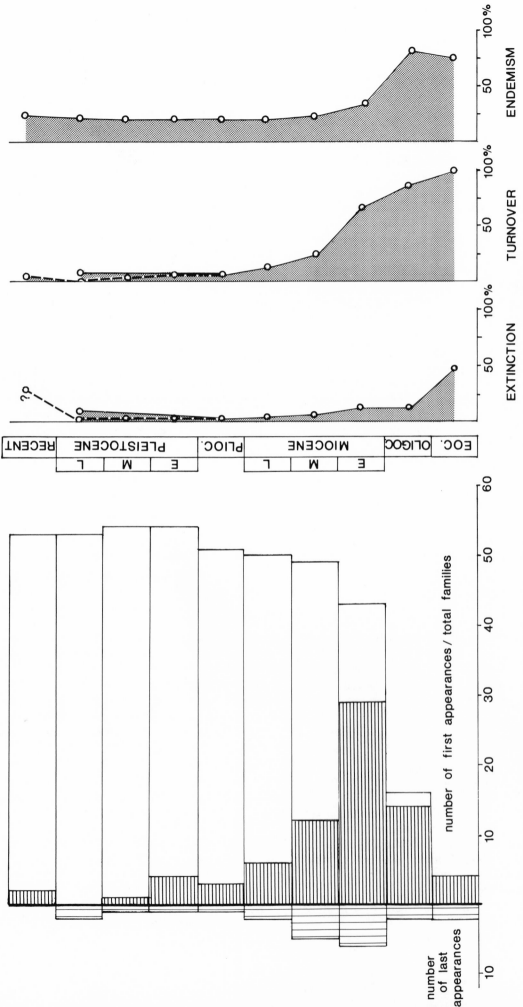

Figure 30.5 Distribution of terrestrial mammalian families in the African Cenozoic. To the left total numbers of families are shown with first and last appearances in each epoch. To the right the proportion of families in each epoch is shown for extinction (percent families permanently disappearing from the record), turnover (percent families in each epoch that are new to the record), and endemism (percent families occurring only in Africa). Taxa are assumed to have been present between first and last appearances, even if not actually recorded in deposits of intermediate age.

ime period, although shorter time scales
somewhat distort true values. Neverthe-
e Pliocene, in spite of its limited duration,
observe nearly 80% turnover. This un-
reflected massive influx of new genera
asia. It was a principal cause of an in-
end in endemism that was initiated at
time as newly arrived forms evolved into
African types. During the subsequent
y, faunal turnover rates fell once again as
fewer immigrants were able to penetrate
l the growing Sahara. At this stage turn-
ependent principally on *in situ* evolution.
igh apparent turnover seen for the Recent
ct. It is caused by the presence of numer-
for which there is no fossil record.
rnover rates followed a similar pattern
er Cenozoic but, contrary to generic rec-
ued to fall through the Pliocene and
Such a phenomenon is more nearly un-
terms of a larger world pool from which
its immigrants. Such was the Old World
ugh a total of 75 terrestrial families exi-
a at one time or another since the late
about 20 of these were both autoch-
endemic. Thus more than two-thirds of
mal families originated elsewhere, en-
ntinent via southern Asia or Europe. By
en, 34 African autochthons (plus sev-
a fossil record), were potentially avail-
rt to the north. Of these, fourteen were

lists numbers of land mammal families
endemic to Africa and shared with
an be seen in columns 4 and 5, Africa
pported fewer families than were pres-
l World pool. Of particular interest is
of Old World families resident in
each epoch (column 6). In Eocene
of total Old World families occurred
an region. By early Oligocene times
tripled, due almost entirely to African
the early Miocene suddenly 79% of
families for that period are repre-
ca, mostly as a result of extensive
al dispersion. Subsequent faunal ex-
the figure only very slightly, to 86%
ne. It is clear that once four-fifths of
ool had already penetrated into the
ss (that is, by the late Miocene), the
al pool for new families was so signi-
that only minor turnover could be
is resulting as much from autoch-
n as from further immigration. This

is precisely the pattern seen in figure 30.5. The
slight increase in family diversity during the Quaternary is artifactual, due to a "sudden" appearance
in latest Pleistocene and Holocene time of several
endemic families lacking a fossil record. All of these
were certainly present much earlier.

Extinction

Much has been written in recent years on causes
of extinction during the Pleistocene in northern
continents and in Africa. No conclusions can be
seriously entertained at this point, but several facts
emerge from present data that contribute to and further complicate this interesting problem. Figure
30.5 plots family extinctions based on last appearances and figure 30.6 does the same for genera. Similar data are shown in different form in tables 30.3
and 30.4. In both graphs the highest rates of extinction are seen in Eocene to early Miocene times and
correspond to periods of major faunal turnover. Presumably as new groups entered or evolved in the
continent they replaced older ones with which they
came into competition. Family extinctions fell to
very low levels by middle Miocene times and remained low to the present day. A slight rise in the
curve during the Quaternary is due to disappearance from Africa of several long-resident groups
(e.g., gomphotheres and deinotheres), restriction of
myzopodid bats to Madagascar, and withdrawal to
Asia of two families that had only recently entered
Africa, Camelidae, and Ursidae (the former reintroduced by man).

Generic extinctions for larger mammals also
show sharp decline to a basal level in the Pliocene
when only 17% of known genera in that epoch disappeared. Quaternary extinctions rose to 33% overall,
a figure similar to the 30% quoted by Martin (1966).
If we divide the Pleistocene into smaller units, it can
be seen that most of this extinction, 33 out of 56 genera, occurred during the early Pleistocene (figure
30.6, broken line). In the subsequent middle Pleistocene 12 more genera disappeared, and in the late
Pleistocene another 11 genera. These data are consistent with Martin's (1967) observation that Quaternary extinctions occurred in Africa significantly
earlier than elsewhere, although on a smaller scale.

These 56 extinct genera represent the greatest
number to disappear in any one epoch of the entire
Cenozoic, except for the early Miocene when 57 genera apparently died out. Nevertheless we cannot
view this simply as an unusually high "extinction
rate" but rather as an unusually high "replacement
rate," for during the same Quaternary period 77 new

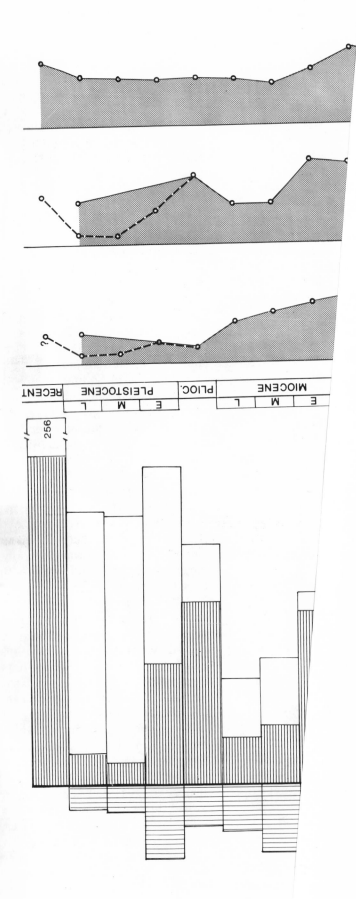

RECENT | PLEISTOCENE | PLIOC. | MIOCENE

E | M | L | | E | M | L

256

ceeding
here may
less, by th
we again
doubtedly
from Eur
creased t
about this
distinctive
Quaternar
fewer and
Arabia an
over was d
The very h
is an artifa
ous genera
Family t
in the earli
ords, conti
Pleistocene
derstood in
Africa drew
fauna. Alth
sted in Afri
Eocene, onl
thonous and
Africa's man
tering the co
the same to
eral lacking
able for expo
successful.
Table 30.2
through time
Eurasia. As
has always su
ent in the Ol
the proportio
Africa during
times only 9%
in the Ethiop
the figure had
autochthons.
all Old World
sented in Afr
southward fau
changes raised
in the Pleistoc
the Old World
African landma
potential residu
ficantly reduce
expected, and t
thonous evoluti

genera made their first appearance in the record. Of these 77, 53 appear in the early and the remaining 24 since the middle Pleistocene. Thus in many ways Pleistocene faunas are as "depauperate" compared to Recent assemblages as the latter are with respect to earlier Pleistocene ones. This epoch gained more new taxa than it lost.

As to the causes of these extinctions, a number of hypotheses have been put forward. Leakey (1965) suggested climatic alteration and drought as a major factor in extinctions seen in the Olduvai Gorge record. Martin (1967) and Edwards (1967) have argued for human cultural expansion as a principal determinant. Whatever the answer, we presently know far too little about detailed climatic changes and consequent ecologic shifts in the African Pleistocene or of early hominid hunting behavior to attempt any firm conclusions. To do so would be a disservice to the science.

Conclusions

From this brief outline of faunal evolution in Africa several features worthy of reiteration emerge. First, Africa's mammalian assemblage can be divided into two basic groups, an autochthonous "archaic fauna" that dominated during the Eocene and Oligocene and a "neofauna" arising more or less suddenly during the early Miocene, mainly by immigration from the north. Some degree of overlap existed between the two, but by middle Miocene times the transition was essentially complete.

The neofauna can be further subdivided into a "primitive fauna" consisting of early Miocene members of modern families, a "middle fauna", beginning with Fort Ternan and lasting with some modernization through the Pliocene and encompassing many precursors of extant genera, and a "modern fauna" in the Pleistocene and Holocene during which modern genera passed from early to extant species.

During a history of more than 50 m.y. Africa gained much of its distinctive faunal character from Eurasia while giving equally to the north. It thus emerges not as a refuge for dying Asiatic Cenozoic groups nor as an isolated faunal province unto itself, but as an integral constituent of Old World mammalian history. What we consider to be uniquely African today is only partly so. For devastating extinctions of the last 2 m.y. in Eurasia have relegated to Africa numerous genera that were formerly also at home on the steppes of Asia or the forests of southern Europe. Had it not been for the Sahara as a formidable late Pleistocene barrier to north-south movements, many more northern species, unfortunately now extinct, might have found refuge in an even richer modern African fauna.

My sincere thanks to Rosanne Leidy for her help in compiling the data for this chapter and for discussions of interpretation.

Table 30.1 Approximate distribution of mammalian genera recorded as fossil in the African Cenozoic.

	Eocene	Oligocene	Early Miocene	Middle Miocene	Late Miocene	Pliocene	Early Pleistocene	Middle Pleistocene	Late Pleistocene
INSECTIVORA									
Ptolemaiidae									
Qarunavus	—	X	—	—	—	—	—	—	—
Ptolemaia	—	X	—	—	—	—	—	—	—
?Kelba	—	—	X	—	—	—	—	—	—
Macroscelididae									
Metoldobotes	—	X	—	—	—	—	—	—	—
Rhynchocyon	—	—	X	—	—	—	—	—	—
Myohyrax	—	—	X	—	—	—	—	—	—
Protypotheroides	—	—	X	—	—	—	—	—	—
Palaeothentoides	—	—	—	—	—	X	—	—	—
Elephantulus	—	—	—	—	—	X	X	—	—
Macroscelides	—	—	—	—	—	X	X	—	—
Mylomygale	—	—	—	—	—	X	X	—	—
Erinaceidae									

Table 30.1 (*continued*)

	Eocene	Oligocene	Early Miocene	Middle Miocene	Late Miocene	Pliocene	Early Pleistocene	Middle Pleistocene	Late Pleistocene
Galerix	—	—	X	—	—	—	—	—	—
Lanthanotherium	—	—	X	—	—	—	—	—	—
Amphechinus	—	—	X	—	—	—	—	—	—
Gymnurechinus	—	—	X	—	—	—	—	—	—
Protechinus	—	—	—	X	X	—	—	—	—
Erinaceus	—	—	—	—	—	—	X	—	—
Soricidae									
Crocidura	—	—	X	—	—	X	X	—	—
"Sorex"	—	—	—	X	X	—	—	—	—
Myosorex	—	—	—	—	—	—	X	X	—
Sylvisorex	—	—	—	—	—	—	X	—	—
Suncus	—	—	—	—	—	—	X	X	—
Diplomesodon	—	—	—	—	—	—	X	—	—
Tenrecidae									
Protenrec	—	—	X	—	—	—	—	—	—

Table 30.1 (*continued*)

	Eocene	Oligocene	Early Miocene	Middle Miocene	Late Miocene	Pliocene	Early Pleistocene	Middle Pleistocene	Late Pleistocene
Erythrozootes	—	—	X	—	—	—	—	—	—
Geogale	—	—	X	—	—	—	—	—	—
Chrysochloridae									
Prochrysochloris	—	—	X	—	—	—	—	—	—
Proamblysomus	—	—	—	—	—	X	X	—	—
Chlorotalpa	—	—	—	—	—	X	X	—	—
Amblysomus	—	—	—	—	—	—	X	—	—
CHIROPTERA									
Pteropodidae									
Propotto	—	—	X	—	—	—	—	—	—
Microchiroptera									
Vampyravus	—	X	—	X	—	—	—	—	—
Emballonuridae									
Taphozous	—	—	X	—	—	—	—	—	—
Megadermatidae									
Afropterus	—	—	—	X	—	—	—	—	—
Cardioderma	—	—	—	—	—	—	X	—	—
Rhinolophidae									
Rhinolophus	—	—	—	X	X	X	—	—	—
Hipposideridae									
Hipposideros	—	—	—	X	—	—	—	—	—
Asellia	—	—	—	X	—	—	—	—	—
Myzopodidae									
Myzopoda	—	—	—	—	—	—	X	—	—
Vespertilionidae									
gen. indet.	—	—	—	X	—	—	—	—	—
Myotis	—	—	—	—	—	X	X	—	—
cf *Nycticeis*	—	—	—	—	—	—	X	—	—
cf *Pipistrellus*	—	—	—	—	—	—	X	—	—
Eptesicus	—	—	—	—	—	—	X	—	—
Miniopterus	—	—	—	—	—	—	X	—	—
Molossidae									
gen. indet.	—	—	—	X	—	—	X	—	—
RODENTIA									
Phiomyidae									
Phiomys	—	X	—	—	—	—	—	—	—
Andrewsimys	—	—	X	—	—	—	—	—	—
Gaudeamus	—	X	—	—	—	—	—	—	—
Thryonomyidae									
Paraphiomys	—	X	X	—	X	—	—	—	—
Epiphiomys	—	—	X	—	—	—	—	—	—
Thryonomys	—	—	—	—	—	—	X	—	—
Diamantomyidae									
Metaphiomys	—	X	—	—	—	—	—	—	—
Diamantomys	—	—	X	—	—	—	—	—	—
Pomonomys	—	—	X	—	—	—	—	—	—
Kenyamyidae									
Kenyamys	—	—	X	—	—	—	—	—	—
Simonimys	—	—	X	—	—	—	—	—	—

Table 30.1 (*continued*)

	Eocene	Oligocene	Early Miocene	Middle Miocene	Late Miocene	Pliocene	Early Pleistocene	Middle Pleistocene	Late Pleistocene
Myophiomyidae									
Phiocricetomys	—	X	—	—	—	—	—	—	—
Miophiomys	—	—	X	—	—	—	—	—	—
Elmerimys	—	—	X	—	—	—	—	—	—
Bathyergoididae									
Bathyergoides	—	—	X	—	—	—	—	—	—
Bathyergidae									
gen. indet.	—	—	—	X	—	—	—	—	—
Proheliophobius	—	—	—	?	—	X	X	—	—
Gypsorhychus	—	—	—	—	—	X	X	—	—
Heterocephalus	—	—	—	—	—	X	X	—	—
Hystricidae									
gen. indet.	—	—	—	—	X	—	—	—	—
Hystrix	—	—	—	—	—	X	X	—	—
Xenohystrix	—	—	—	—	—	X	X	—	—
Pedetidae									
Parapedetes	—	—	X	—	—	—	—	—	—
Megapedetes	—	—	X	—	—	—	—	—	—
Pedetes	—	—	—	—	—	X	—	—	—
Cricetodontidae									
Paratarsomys	—	—	X	—	—	—	—	—	—
Afrocricetodon	—	—	X	?	—	—	—	—	—
Notocricetodon	—	—	X	—	—	—	—	—	—
Mellalomys	—	—	—	X	—	—	—	—	—
Dakkamys	—	—	—	X	—	—	—	—	—
Cricetodon	—	—	—	X	—	—	—	—	—
Myocricetodon	—	—	—	X	—	—	—	—	—
Nesomyidae									
Leakeymys	—	—	—	X	—	—	—	—	—
Mystromys	—	—	—	—	—	X	X	—	—
Otomys	—	—	—	—	—	—	X	X	—
Cricetidae									
Kanisamys	—	—	—	?	—	—	—	—	—
Meriones	—	—	—	—	X	—	X	—	—
Protatera	—	—	—	—	X	—	—	—	—
Tatera	—	—	—	—	—	X	X	—	—
Gerbillus	—	—	—	—	—	—	X	—	—
Dipodidae									
Protalactaga	—	—	X	X	—	—	—	—	—
Jaculus	—	—	—	—	—	—	X	—	—
Rhizomyidae									
gen. indet.	—	—	—	—	—	—	X	—	—
Muridae									
gen. indet.	—	—	—	X	—	—	—	—	—
Arvicanthis	—	—	—	—	X	—	X	—	—
Apodemus	—	—	—	—	—	—	—	—	X
Lemniscomys	—	—	—	—	—	—	—	—	X
Ellobius	—	—	—	—	—	—	X	—	—
Dendromus	—	—	—	—	—	—	X	—	—
Malacothrix	—	—	—	—	X	X	—	—	—
Steatomys	—	—	—	—	—	—	X	—	—
Progonomys	—	—	—	—	X	X	—	—	—

Table 30.1 (*continued*)

	Eocene	Oligocene	Early Miocene	Middle Miocene	Late Miocene	Pliocene	Early Pleistocene	Middle Pleistocene	Late Pleistocene
Paraethomys	—	—	—	—	—	X	X	—	X
Mus	—	—	—	—	—	X	X	—	—
Praomys	—	—	—	—	—	—	X	X	—
Pelomys	—	—	—	—	—	X	—	—	—
Gliridae									
Microdyromys	—	—	—	X	—	—	—	—	—
Eliomys	—	—	—	—	—	—	—	X	—
Sciuridae									
Vulcanisciurus	—	—	X	—	—	—	—	—	—
Heteroxerus	—	—	—	—	X	—	—	—	—
Atlantoxerus	—	—	X	—	—	—	—	—	—
Xerus	—	—	—	—	—	X	—	—	—
Ctenodactylidae									
Africanomys	—	—	—	X	X	—	—	—	—
Metasayimys	—	—	—	—	X	—	—	—	—
Irhoudia	—	—	—	—	—	X	X	—	—
Testouromys	—	—	—	X	—	—	—	—	—
Anomaluridae									
gen. indet.	—	X	X	X	—	X	—	—	—
LAGOMORPHA									
Ochotonidae									
Australolagomys	—	—	X	—	—	—	—	—	—
Kenyalagomys	—	—	X	—	—	—	—	—	—
Prolagus	—	—	—	—	—	X	—	—	—
Leporidae									
Serengetilagus	—	—	—	—	—	—	X	—	—
Lepus	—	—	—	—	—	—	X	X	X
Pronolagus	—	—	—	—	—	—	—	—	X
PRIMATES									
Lorisidae									
Progalago	—	—	X	—	—	—	—	—	—
Komba	—	—	X	—	—	—	—	—	—
Mioeuoticus	—	—	X	—	—	—	—	—	—
gen. nov.	—	—	—	X	—	—	—	—	—
Galago	—	—	—	—	—	X	X	—	—
Parapithecidae									
Apidium	—	X	—	—	—	—	—	—	—
Parapithecus	—	X	—	—	—	—	—	—	—
Cercopithecidae									
Cercopithecus	—	—	—	—	—	—	X	X	—
Macaca	—	—	—	—	X	—	—	—	—
Theropithecus	—	—	—	—	—	—	X	X	X
Libypithecus	—	—	—	—	X	—	—	—	—
Parapapio	—	—	—	—	—	—	X	X	—
Dinopithecus	—	—	—	—	—	—	—	X	—
Gorgopithecus	—	—	—	—	—	—	—	X	—
Cercocebus	—	—	—	—	—	—	X	X	—
Cercopithecoides	—	—	—	—	—	—	—	X	—
Colobus	—	—	—	—	—	?	X	X	—
Paracolobus	—	—	—	—	—	—	—	X	—

Table 30.1 (*continued*)

	Eocene	Oligocene	Early Miocene	Middle Miocene	Late Miocene	Pliocene	Early Pleistocene	Middle Pleistocene	Late Pleistocene
Family indet.									
Victoriapithecus	—	—	?	X	—	—	—	—	—
Prohylobates	—	—	X	—	—	—	—	—	—
Pongidae									
Oligopithecus	—	X	—	—	—	—	—	—	—
Propliopithecus	—	X	—	—	—	—	—	—	—
Aegyptopithecus	—	X	—	—	—	—	—	—	—
Proconsul	—	—	X	—	—	—	—	—	—
Limnopithecus	—	—	X	—	—	—	—	—	—
Hylobatidae									
Aeolopithecus	—	X	—	—	—	—	—	—	—
Dendropithecus	—	—	X	—	—	—	—	—	—
Hominidae									
Ramapithecus	—	—	—	X	—	—	—	—	—
Australopithecus	—	—	—	—	X	X	X	—	—
Homo	—	—	—	—	—	X	X	X	X
CARNIVORA									
Hyaenodontidae									
Metasinopa	—	X	—	—	—	—	—	—	—
Masrasector	—	X	—	—	—	—	—	—	—
Anasinopa	—	—	X	—	—	—	—	—	—
Dissopsalis	—	—	—	X	—	—	—	—	—
Tetratodon	—	—	X	—	—	—	—	—	—
Apterodon	X	X	—	—	—	—	—	—	—
Leakitherium	—	—	X	—	—	—	—	—	—
Pterodon	—	X	X	—	—	—	—	—	—
Hyainailouros	—	—	X	X	—	—	—	—	—
Megistotherium	—	—	—	X	—	—	—	—	—
Canidae									
Hecubides	—	—	X	—	—	—	—	—	—
Afrocyon	—	—	X	—	—	—	—	—	—
Canis	—	—	—	—	—	—	X	X	X
Vulpes	—	—	—	—	—	—	X	—	X
Fennecus	—	—	—	—	—	—	—	—	X
Lycaon	—	—	—	—	—	—	—	X	X
Otocyon	—	—	—	—	—	—	X	—	X
Ursidae									
Agriotherium	—	—	—	—	—	X	—	—	—
Ursus	—	—	—	—	—	—	—	X	X
Mustelidae									
Mustela	—	—	—	—	—	—	—	—	X
Mellivora	—	—	—	—	X	—	—	X	X
Lutra	—	—	—	—	—	X	X	X	X
Aonyx	—	—	—	—	—	—	X	X	X
Enhydriodon	—	—	—	—	—	X	—	—	—
Viverridae									
Genetta	—	—	—	—	—	—	—	—	X
Viverra	—	—	—	—	—	X	X	—	—
Pseudocivetta	—	—	—	—	—	—	—	X	—
Kichechia	—	—	X	—	—	—	—	—	—
Suricata	—	—	—	—	—	—	X	—	X

Table 30.1 (*continued*)

	Eocene	Oligocene	Early Miocene	Middle Miocene	Late Miocene	Pliocene	Early Pleistocene	Middle Pleistocene	Late Pleistocene
Herpestes	—	—	—	—	—	—	X	—	X
Atilax	—	—	—	—	—	—	—	—	X
Mungos	—	—	—	—	—	—	X	—	—
Crossarchus	—	—	—	—	—	—	X	—	—
Ichneumia	—	—	—	—	—	—	X	—	—
Cynictis	—	—	—	—	—	—	—	—	X
Hyaenidae									
Ictitherium	—	—	—	—	X	—	—	—	—
Leecyaena	—	—	—	—	—	X	—	—	—
Euryboas	—	—	—	—	—	X	X	—	—
Hyaena	—	—	—	—	X	X	X	X	X
Crocuta	—	—	—	—	—	—	X	X	X
Felidae									
Afrosmilus	—	—	X	—	—	—	—	—	—
Dinofelis	—	—	—	—	—	X	X	—	—
Megantereon	—	—	—	—	—	—	X	—	—
Machairodus	—	—	—	—	X	X	X	—	—
Homotherium	—	—	—	—	—	—	X	—	—
Acinonyx	—	—	—	—	—	—	?	—	X
Felis	—	—	—	—	—	—	X	X	X
Panthera	—	—	—	—	—	—	X	X	X
Phocidae									
Pristiphoca	—	—	—	—	—	X	—	—	—
Prionodelphis	—	—	—	—	—	X	—	—	—
HYRACOIDEA									
Pliohyracidae									
Geniohyrax	—	X	—	—	—	—	—	—	—
Bunohyrax	—	X	X	—	—	—	—	—	—
Megalohyrax	—	X	—	—	—	—	—	—	—
Pachyhyrax	—	—	X	—	—	—	—	—	—
Titanohyrax	—	X	—	—	—	—	—	—	—
Saghatherium	—	X	—	—	—	—	—	—	—
Thyrohyrax	—	X	—	—	—	—	—	—	—
Meroehyrax	—	—	X	—	—	—	—	—	—
Procaviidae									
Prohyrax	—	—	X	—	—	—	—	—	—
Gigantohyrax	—	—	—	—	—	X	X	—	—
Procavia	—	—	—	—	—	X	X	—	—
PROBOSCIDEA									
Gomphotheriidae									
Palaeomastodon	X	X	—	—	—	—	—	—	—
Gomphotherium	—	—	X	X	X	—	—	—	—
Platybelodon	—	—	X	—	—	—	—	—	—
Tetralophodon	—	—	—	X	—	—	—	—	—
Anancus	—	—	—	—	X	X	X	—	—
Mammutidae									
Zygolophodon	—	—	—	—	—	X	—	—	—
Mammut	—	—	X	—	—	—	—	—	—
Stegodon	—	—	—	—	—	X	—	—	—
Elephantidae									
Stegotetrabelodon	—	—	—	—	X	—	—	—	—

Table 30.1 (*continued*)

	Eocene	Oligocene	Early Miocene	Middle Miocene	Late Miocene	Pliocene	Early Pleistocene	Middle Pleistocene	Late Pleistocene
Stegodibelodon	—	—	—	—	X	—	—	—	—
Primelephas	—	—	—	X	X	—	—	—	—
Loxodonta	—	—	—	—	—	X	X	X	X
Elephas	—	—	—	—	—	X	X	X	X
Mammuthus	—	—	—	—	—	X	X	X	—
MOERITHERIOIDEA									
Moeritherium	X	—	—	—	—	—	—	—	—
BARYTHERIOIDEA									
Barytherium	X	—	—	—	—	—	—	—	—
DEINOTHERIOIDEA									
Prodeinotherium	—	—	X	X	X	—	—	—	—
Deinotherium	—	—	—	—	X	X	—	—	—
TUBULIDENTATA									
Leptorycteropus	—	—	—	X	—	—	—	—	—
Myorycteropus	—	—	X	—	—	—	—	—	—
Orycteropus	—	—	?	?	—	—	X	X	X
Plesiorycteropus	—	—	—	—	—	—	—	—	X
PHOLIDOTA									
Phataginus	—	—	—	—	—	—	—	X	X
EMBRITHROPODA									
Arsinoitherium	—	X	—	—	—	—	—	—	—
PERISSODACTYLA									
Chalicotheriidae									
Chalicotherium	—	—	X	—	—	—	—	—	—
Ancylotherium	—	—	—	—	X	X	—	—	—
Rhinocerotidae									
Brachypotherium	—	—	X	X	X	X	—	—	—
Aceratherium	—	—	X	X	—	—	—	—	—
Dicerorhinus	—	—	X	—	—	—	—	—	—
Chilotheridium	—	—	X	—	—	—	—	—	—
Paradiceros	—	—	—	X	—	—	—	—	—
Diceros	—	—	—	—	—	X	X	X	X
Ceratotherium	—	—	—	—	X	X	X	X	X
Equidae									
Hipparion	—	—	—	X	X	X	X	X	X
Equus	—	—	—	—	—	—	X	X	X
ARTIODACTYLA									
Anthracotheriidae									
Bothriogenys	—	X	X	—	—	—	—	—	—
Rhagatherium	—	X	—	—	—	—	—	—	—
Masritherium	—	—	X	—	—	—	—	—	—
Hyoboops	—	—	X	—	—	—	—	—	—
Merycopotamus	—	—	—	—	X	—	—	—	—
Suidae									
Hyotherium	—	—	X	X	—	—	—	—	—

Table 30.1   (*continued*)

	Eocene	Oligocene	Early Miocene	Middle Miocene	Late Miocene	Pliocene	Early Pleistocene	Middle Pleistocene	Late Pleistocene
Propalaeochoerus	—	—	?	—	—	—	—	—	—
Listriodon	—	—	?	—	—	—	—	—	—
Kubanochoerus	—	—	X	X	—	—	—	—	—
Lopholistriodon	—	—	X	—	—	—	—	—	—
Xenochoerus	—	—	X	—	—	—	—	—	—
Sanitherium	—	—	—	X	—	—	—	—	—
Nyanzachoerus	—	—	—	—	X	X	—	—	—
Notochoerus	—	—	—	—	—	X	X	—	—
Metridiochoerus	—	—	—	—	—	X	X	X	—
Kolpochoerus	—	—	—	—	—	X	X	X	—
Stylochoerus	—	—	—	—	—	—	—	X	—
Potamochoeroides	—	—	—	—	—	X	X	—	—
Phacochoerus	—	—	—	—	—	X	X	X	X
Sus	—	—	—	—	—	—	X	X	X
Potamochoerus	—	—	—	—	—	—	—	X	X
Hylochoerus	—	—	—	—	—	—	X	X	X
Tayassuidae									
Percarichoerus	—	—	—	—	—	X	—	—	—
Hippopotamidae									
Hexaprotodon	—	—	—	—	X	X	X	X	—
Hippopotamus	—	—	—	—	—	X	X	X	X
Tragulidae									
Dorcatherium	—	—	X	X	—	—	—	—	—
Camelidae									
Camelus	—	—	—	—	—	X	X	X	—
Cervidae									
Cervus	—	—	—	—	—	—	—	—	X
Dama	—	—	—	—	—	—	X	X	X
Megaloceros	—	—	—	—	—	—	—	—	X
Palaeomerycidae									
Climacoceras	—	—	—	X	—	—	—	—	—
Propalaeoryx	—	—	X	—	—	—	—	—	—
Canthumeryx	—	—	X	—	—	—	—	—	—
Giraffidae									
Zarafa	—	—	X	—	—	—	—	—	—
Palaeotragus	—	—	X	X	X	—	—	—	—
Giraffokeryx	—	—	—	X	—	—	—	—	—
Samotherium	—	—	—	X	X	X	—	—	—
Okapia	—	—	—	—	—	—	X	X	—
Giraffa	—	—	—	—	?	X	X	X	X
Prolibytherium	—	—	X	—	—	—	—	—	—
cf *Helladotherium*	—	—	—	—	X	?	—	—	—
Sivatherium	—	—	—	—	—	X	X	X	—
Bovidae									
Taurotragus	—	—	—	—	—	—	—	X	X
Tragelaphus	—	—	—	—	—	—	X	X	X
Eotragus	—	—	—	X	—	X	X	—	—
Protragocerus	—	—	—	X	X	—	—	—	—
Miotragus	—	—	—	—	—	X	—	—	—
Syncerus	—	—	—	—	—	—	—	—	X
Ugandax	—	—	—	—	—	X	X	—	—
Pelorovis	—	—	—	—	—	—	X	X	X
Simatherium	—	—	—	—	—	X	X	—	—

Table 30.1   (*continued*)

	Eocene	Oligocene	Early Miocene	Middle Miocene	Late Miocene	Pliocene	Early Pleistocene	Middle Pleistocene	Late Pleistocene
Bos	—	—	—	—	—	—	—	—	X
"*Leptobos*"	—	—	—	X	—	—	—	—	—
"*Hemibos*"	—	—	—	—	—	—	X	—	—
Sylvicapra	—	—	—	—	—	—	—	X	—
Cephalophus	—	—	—	—	—	X	X	X	—
Redunca	—	—	—	—	—	X	X	X	X
Kobus	—	—	—	—	—	X	X	X	X
Menelikia	—	—	—	—	—	X	X	X	—
Thalerocerus	—	—	—	X	X	—	—	—	—
Hippotragus	—	—	—	—	—	—	—	X	X
Oryx	—	—	—	—	—	X	X	X	X
Megalotragus	—	—	—	—	—	—	X	X	X
Connochaetes	—	—	—	—	—	—	X	X	X
Oreonagor	—	—	—	—	—	X	—	—	—
Alcelaphus	—	—	—	—	—	—	—	—	X
Rabaticeras	—	—	—	—	—	—	X	X	—
Damaliscus	—	—	—	—	—	—	X	X	X
Parmularius	—	—	—	—	—	—	X	X	X
Beatragus	—	—	—	—	—	X	—	—	—
Aepycerus	—	—	—	—	—	X	X	X	—
Raphicerus	—	—	—	—	—	—	X	X	—
Oreotragus	—	—	—	—	—	—	X	—	X
Ourebia	—	—	—	—	—	—	—	X	—
Madoqua	—	—	—	—	—	X	X	—	—
Antidorcas	—	—	—	—	—	—	X	X	X
Gazella	—	—	—	—	X	X	X	X	X
"*Antilope*"	—	—	—	—	—	X	—	—	—
Tossunnoria	—	—	—	—	—	X	—	—	—
Numidocapra	—	—	—	—	—	—	—	X	—
cf *Capra*	—	—	—	—	—	—	X	X	—
Pachytragus	—	—	—	—	X	X	—	—	—
Benicerus	—	—	—	—	—	X	—	—	—
Oioceros	—	—	—	—	—	X	—	—	—
?*Pseudotragus*	—	—	—	—	X	X	—	—	—
Makapania	—	—	—	—	—	—	X	—	—
Damalavus	—	—	—	—	—	X	—	—	—
Walangania	—	—	X	—	—	—	—	—	—

SIRENIA

Protosirenidae

	Eocene	Oligocene	Early Miocene	Middle Miocene	Late Miocene	Pliocene	Early Pleistocene	Middle Pleistocene	Late Pleistocene
Protosiren	X	—	—	—	—	—	—	—	—

Dugongidae

	Eocene	Oligocene	Early Miocene	Middle Miocene	Late Miocene	Pliocene	Early Pleistocene	Middle Pleistocene	Late Pleistocene
Eotheroides	X	—	—	—	—	—	—	—	—
Halitherium	—	—	X	?	—	—	—	—	—
Rytiodus	—	—	X	—	—	—	—	—	—
Metaxytherium	—	—	X	—	—	X	—	—	—

CETACEA

Protocetidae

	Eocene	Oligocene	Early Miocene	Middle Miocene	Late Miocene	Pliocene	Early Pleistocene	Middle Pleistocene	Late Pleistocene
Protocetus	X	—	—	—	—	—	—	—	—
Eocetus	X	—	—	—	—	—	—	—	—
Pappocetus	X	—	—	—	—	—	—	—	—

Table 30.1 *(continued)*

	Eocene	Oligocene	Early Miocene	Middle Miocene	Late Miocene	Pliocene	Early Pleistocene	Middle Pleistocene	Late Pleistocene
Basilosauridae									
Dorudon	X	–	–	–	–	–	–	–	–
Prozeuglodon	X	–	–	–	–	–	–	–	–
Archaeoceti indet.									
"Zeuglodon"	X	–	–	–	–	–	–	–	–
Zaphiidae									
gen. indet.	–	–	X	–	–	–	–	–	–

Table 30.1 *(continued)*

	Eocene	Oligocene	Early Miocene	Middle Miocene	Late Miocene	Pliocene	Early Pleistocene	Middle Pleistocene	Late Pleistocene
Acrodelphidae									
Schizodelphis	–	–	X	–	–	–	–	–	–
Odontoceti indet.									
"Delphinus"	–	–	X	–	–	–	–	–	–
Balaenopteridae									
Balaenoptera	–	–	X	–	–	–	–	–	–

Table 30.2 Distribution of endemic land mammal families in the Old World. Data are modified from Simpson 1945 and the present volume.

	African endemics	Eurasiatic endemics	Present in both	Total Old World families	Total families in Africa	Percent Old World families in Africa	Percent Old World families endemic to Africa
Pleistocene	10	9	44	63	54	86	15
Pliocene	10	14	42	66	51	79	15
Miocene	22	16	39	77	61	79	29
Oligocene	12	43	4	59	16	27	20
Eocene	3	40	1	44	4	9	7

Table 30.3 Cenozoic distribution of land mammal families in Africa.

	Eocene	Oligocene	Early Miocene	Middle Miocene	Late Miocene	Pliocene	Early Pleistocene	Middle Pleistocene	Late Pleistocene
Total number	4	16	43	49	50	51	54	54	53
First appearances	4	14	29	12	6	3	4	1	0
Last appearances	2	2	6	5	2	1	1	1	2
Endemics	3	13	14	11	10	10	11	11	11

Table 30.4 Cenozoic distribution of land mammal genera in Africa.

	Eocene	Oligocene	Early Miocene	Middle Miocene	Late Miocene	Pliocene	Early Pleistocene	Middle Pleistocene	Late Pleistocene	Recent
Total number	4	31	87	56	47	107	142	119	121	256
First appearances	4	29	79	26	21	81	53	10	14	146
Last appearances	2	23	57	30	21	18	33	12	11	?
Endemics	3	26	54	26	24	57	72	63	65	184

References

Cooke, H. B. S. 1968. Evolution of mammals on southern continents. II. The fossil mammal fauna of Africa. *Quart. Rev. Biol.* 43(3):234–264.

———. 1972. The fossil mammal fauna of Africa. In Keast, A., F. C. Erk, and B. Glass, eds., *Evolution, mammals and southern continents*, Albany: State Univ. of N.Y. Press, pp. 89–139.

Coryndon, S. C. and R. J. G. Savage. 1973. The origin and affinities of African mammal faunas. In *Organisms and continents through time*. London: Syst. Assoc. Publ. no. 9, Special Papers in Palaeontology 12:121–135.

Crompton, A. W. 1964. A preliminary description of a new mammal from the upper Triassic of South Africa. *Proc. Zool. Soc. Lond.* 142:441–452.

Crompton, A. W. and F. A. Jenkins, Jr. 1968. Molar occlusion in late Triassic mammals. *Biol. Rev.* 43:427–458.

Dietrich, W. O. 1928. *Brancatherulum* n.g.—ein Proplacentalier aus dem obersten Jura des Tendaguru in Deutsch-Östafrika. *Zentr. Mineral. Geol. Palaeontol.* 1927B:423–426.

Edwards, W. E. 1967. The late-Pleistocene extinction and diminution in size of many mammalian species. In Martin, P. S. and H. E. Wright, Jr., eds., *Pleistocene extinctions. The search for a cause*, New Haven: Yale Univ. Press, pp. 141–154.

Klein, R. G. 1972. The late Quaternary mammalian fauna of Nelson Bay (Cape Province, South Africa): its importance for megafaunal extinctions and environmental and cultural changes. *Quatern. Res.* 2:134–142.

Leakey, L. S. B. 1965. *Olduvai Gorge 1951–1961*. Cambridge: Cambridge Univ. Press.

Martin, P. S. 1966. Africa and Pleistocene overkill. *Nature* 212:339–342.

———. 1967. Prehistoric overkill. In Martin, P. S. and H. E. Wright, Jr., eds., *Pleistocene extinctions. The search for a cause*, New Haven: Yale Univ. Press, pp. 75–120.

Patterson, B. 1975. The fossil aardvarks (Mammalia: Tubulidentata). *Bull. Mus. Comp. Zool., Harvard* 147:185–237.

Savage, R. J. G. 1967. Early Miocene mammal faunas of the Tethyan region. In Adams, C. G. and D. V. Ager, eds., *Aspects of Tethyan biogeography*. London: Syst. Assoc. Publ. no. 7, pp. 247–282.

———. 1969. Early Tertiary mammal localities in southern Libya. *Proc. Geol. Soc. Lond.* 1657:167–171.

Simpson, G. G. 1945. Principles of classification and a classification of mammals. *Bull. Am. Mus. Nat. Hist.* 85:1–360.

———. 1965. *The geography of evolution*. Philadelphia: Chilton Books, 249 pp.

Van Couvering, J. A. 1972. Radiometric calibration of the European Neogene. In Bishop, W. W. and J. A. Miller, eds., *Calibration of hominoid evolution*, Edinburgh: Scottish Academic Press, pp. 247–271.

Index

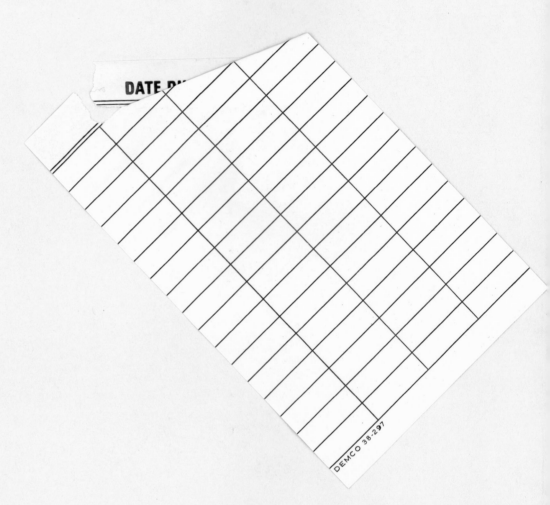

DATE DUE

DEMCO 38-297